Metric-English Conversions

Length

ENGLISH (USA)	=	METRIC
inch	=	2.54 cm, 25.4 mm
foot	=	0.30 m, 30.48 cm
yard	=	0.91 m, 91.4 cm
mile (statute) (5280 ft)	=	1.61 km, 1609 m
mile (nautical) (6077 ft, 1.15 statute mi)	=	1.85 km, 1850 m

METRIC	=	ENGLISH (USA)
millimeter	=	0.039 in
centimeter	=	0.39 in
meter	=	3.28 ft, 39.37 in
kilometer	=	0.62 mi, 1091 yd, 3273 ft

Weight

ENGLISH (USA)	=	METRIC
grain	=	64.80 mg
ounce	=	28.35 g
pound	=	453.60 g, 0.45 kg
ton (short—2000 lb)	=	0.91 metric ton (907 kg)

METRIC	=	ENGLISH (USA)
milligram	=	0.02 grain (0.000035 oz)
gram	=	0.04 oz
kilogram	=	35.27 oz, 2.20 lb
metric ton (1000 kg)	=	1.10 tons

Volume

ENGLISH (USA)	=	METRIC
cubic inch	=	16.39 cc
cubic foot	=	0.03 m³
cubic yard	=	0.765 m³
ounce	=	0.03 liter (3 ml)*
pint	=	0.47 liter
quart	=	0.95 liter
gallon	=	3.79 liters

METRIC	=	ENGLISH (USA)
milliliter	=	0.03 oz
liter	=	2.12 pt
liter	=	1.06 qt
liter	=	0.27 gal

1 liter ÷ 1000 = milliliter or cubic centimeter (10⁻³ liter)
1 liter ÷ 1,000,000 = microliter (10⁻⁶ liter)
*Note: 1 ml = 1 cc

Fahrenheit–Celsius Conversion

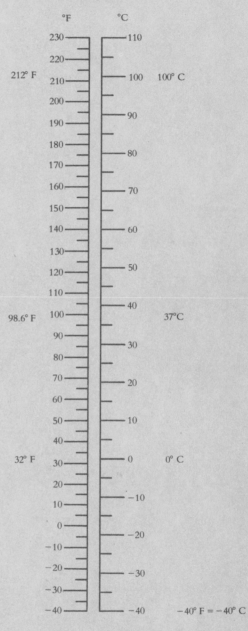

To convert temperature scales:

Fahrenheit to Celsius $°C = \frac{5}{9} (°F - 32)$

Celsius to Fahrenheit $°F = \frac{9}{5} (°C) + 32$

UNDERSTANDING BIOLOGY

This forest, at seven thousand feet (2120 m) at Cibodja in Java, is interrupted by an opening through which steams a hot 176° F (80° C) spring, seen flowing along the bottom of the photograph. The colorful hue of the stream reflects a flourishing population of cyanobacteria (blue-green algae). These bacteria, known from the earliest fossil records, are similar to those which first introduced oxygen to the earth's atmosphere more than two billion years ago.

UNDERSTANDING
BIOLOGY

PETER H. RAVEN

Director, Missouri Botanical Garden;
Engelmann Professor of Botany,
Washington University, St. Louis, Missouri

GEORGE B. JOHNSON

Professor of Biology,
Washington University, St. Louis, Missouri;
Director, Education and Discovery Center,
St. Louis Zoological Park

Illustration program developed by
WILLIAM C. OBER,
Crozet, Virginia

Times Mirror/Mosby College Publishing
ST. LOUIS • TORONTO • SANTA CLARA • 1988

Editor: David Kendric Brake
Editorial Assistant: Mary Huggins
Project Manager: Patricia Gayle May
Production Editor: John P. Rowan
Art Director: Kay Michael Kramer
Design Coordinator: Susan E. Lane
Production: Gail Morey Hudson, Susan Trail
Illustrators: William C. Ober and Ronald J. Ervin
with Molly K. Ryan, George J. Venable, and Claire Garrison
Cover art: "Giant Panda" © 1985 Robert Bateman. Courtesy of the artist and Mill Pond Press, Inc., Venice, FL 33533
Back cover art: "Pair of Skimmers" © 1980 Robert Bateman. Courtesy of the artist and Mill Pond Press, Inc., Venice, FL 33533

Copyright © 1988 by Times Mirror/Mosby College Publishing

A division of The C.V. Mosby Company
11830 Westline Industrial Drive
St. Louis, Missouri 63146

Printed in the United States of America

Library of Congress Cataloging-in-Publication Data

Raven, Peter H.
    Understanding biology/Peter H. Raven, George B. Johnson; illustration program developed by William C. Ober.
        p.   cm.
    Rev. ed. of: Biology.   1986.
    Includes index.
    ISBN 8-8016-2518-1
    1. Biology.   I. Johnson, George B. (George Brooks).
II. Raven, Peter H.   Biology.   III. Title.
QH308.2.R38   1988
574—dc19

C/VH/VH   9   8   7   6   5   4   3   2   01/A/078

PREFACE

Most teachers who lecture to introductory biology classes have been tempted at one time or another to begin by reading to the class the "Science" section of the daily newspaper. Today's paper, for example, reports the identification of the gene that causes cystic fibrosis, a fatal genetic disorder that affects more than 30,000 people in the United States today. For years scientists have been searching for the gene, which is carried silently by an estimated 1 in every 20 Caucasians; studying it may provide clues that will eventually lead to a cure. This discovery is of great interest to every student in a freshman biology class; to premeds and English majors and students majoring in dance. They all want to know about cystic fibrosis, and also about AIDS and cancer and genetic engineering, because these issues affect their lives. They want to learn enough biology to understand the newspapers and magazines they read.

This is not how introductory biology is usually taught to prospective biology majors. What about worms and the parts of a flower and dissecting frogs? Prospective biology majors are traditionally introduced to the subject by exposure to a large body of information, much of it process-oriented, about animals and plants. This information, the nuts and bolts of the subject of biology, is not composed of attention-grabbing stories such as one reads in the newspapers, but rather of a set of principles that explain why biological processes work the way they do—why brothers and sisters look alike, why you become dehydrated when you drink too much alcohol, why people grow old. Biology as a science is not so much a body of information as it is a set of working rules, principles that we understand only imperfectly. In training our next generation of doctors, veterinarians, and biologists, we try to impart a clear view of this framework as we now understand it, both so that they can do their jobs and also so that some of them can in turn further improve the clarity of our understanding.

Both kinds of biology need to be taught to freshmen, many of whom will take only one biology course. The "trendy" biology gets into the newspapers for a reason that we should not ignore: because it affects many people in important ways. In 1987 voters faced ballot initiatives concerning the barring of students afflicted with AIDS from classrooms, the imposition of mandatory testing for the AIDS virus, and alternative proposals concerning the protection of the confidentiality of those who test AIDS-positive. Of the many thousands of voters casting ballots, few have had the biology necessary to understand the issues. We need to teach all of our students the sort of biology that will give them the tools to be informed citizens and to live in a world where biological issues are of increasing importance. Few instructors of potential biology majors would disagree. The problem is one of limited time in a freshman class. To teach the more traditional "classical" material to potential majors usually requires two full semesters. Why not trim the traditional material a little, to squeeze in a little of the new? Because it takes time for students to assimilate the basic principles that constitute the core of biology. Students learn Mendelian genetics only by tracing in detail the results of genetic crosses, by puzzling through problems and considering unusual results. It is not possible for even the best of students to learn all of this at one sitting; the cleverest of words cannot substitute for student involvement.

So it is a case of apples and oranges: two different approaches, both valid and necessary. Many schools respond to this problem by offering two sorts of introductory biology courses: a "trendy" course for non-majors and a "principles" course for majors. In our opinion, this is not the best solution to the problem. A "trendy" course without a foundation of basic principles gives a student no tools with which to make decisions five years later, when the stories in the newspapers are different; the course simply replaces the newspaper as a source of current information. Nor are the potential majors well served by two courses. Current issues add both excitement and perspective, and students are not being properly introduced to biology unless they are exposed to these issues.

The different approaches employed in teaching introductory biology to prospective majors and to non-majors comes into sharp focus in courses intended for both majors and nonmajors—the so-called "mixed-majors" courses that are becoming increasingly popular around the country. How ought one to teach such a course? Over the last five years we have been involved in preparing introductory biology texts. We were originally induced to leave our labs for this task by what we perceived as a need for a new sort of introductory biology text, one in which the central biological principle of evolution is integrated into every chapter, rather than being relegated to a few brief pages. In 1986 we published *Biology,* a comprehensive text organized around an evolutionary theme, intended for use by prospective biology majors. Since then, we have addressed ourselves to the preparation of a short text for a one-semester course—the text you now hold in your hands. Writing this text, we have had to address directly the issue of balancing the need to teach principles with the need to be current in a rapidly changing science. Here is what we chose to do:

Rearrange traditional material. Much of the beginning of the text is devoted to general principles, with detailed consideration of the anatomy and physiology of animals and plants delayed to the end of the course. The student is introduced to cell biology, metabolism, genetics, evolution, and ecology, the principles of which apply to all organisms, before focusing on the biology of any particular organism. This both ensures that the key material gets taught, and, interestingly, provides a better framework for consideration of recent developments. Teaching why smoking leads to lung cancer uses a knowledge of genetics—of how mutations alter a cell's DNA—rather than a knowledge of descriptive information about human lungs.

Limit paradigms. To make room for teaching new developments, while at the same time preserving a detailed treatment of basic principles, we limit ourselves to examining only those principles that are absolutely essential. The student is given a detailed treatment of Darwinian evolution, Mendelian genetics, and the role of DNA as the genetic material, but is not treated to a similarly detailed presentation of all areas of biology that are well understood. This is a key distinction between this book's approach and that of our more comprehensive text for majors. Choosing the proper material to present in depth is the essence of a successful one-term introductory biology course.

Stress evolution. It has often been said that biology only makes sense in the context of evolution, and it is certainly true that learning the principles of biology is much easier if the material is presented to the student within an evolutionary context. With this in mind, we have maintained the evolutionary emphasis of our majors text. The student is provided with a detailed treatment of Darwin in the introductory chapters, and even the chemistry is presented within an evolutionary framework.

The section of the text devoted specifically to evolution begins with a chapter explicitly presenting the evidence that Darwin was right—if the principle of evolution is central to biology, then we ought to be able to convince our students of its validity. They will be voting in a society where many voters support "scientific creationism" (even the Supreme Court was not unanimous in rejecting it), and we have attempted to give them the training to evaluate those views.

Focus on current material. In the necessary trade-off between classical material and recent developments, we have leaned far toward the new. We incorporate detailed discussions of the mechanisms leading to cancer and AIDS, progress in genetic engineering, accounts of current ecological issues such as destruction of the tropical rain forests and acid rain, and other topics of current interest.

Integrate new developments with basic principles. Each new development is introduced as an extension of basic principles, rather than nakedly, so that students can truly understand what is going on, rather than simply garnering more information. Such integration of recent developments with general principles is critical to successful teaching because, in most cases, it is not possible to simply tack on new developments as a linear extension of what we already teach. Why not? Because fast-moving areas in biology tend to coalesce. As we learn more, we often come to view areas traditionally considered distinct as different facets of the same phenomenon. Only by coupling advances to basic principles can instructors of introductory biology convey how advances relate to one another, and so convey some of the excitement and potential gripping biological scientists today.

Present biology in an engaging way. The most important quality of any text is that it be written clearly. In teaching a broad spectrum of students,

Interesting stuff

a second important quality of a successful text is that the writing style be relaxed and enjoyable to read, as students' motivation and background will vary widely. Our goal in preparing this text has been to write each chapter so that a history or art major, as well as a potential biology major, will be interested enough to want to keep on reading. While an attractive illustration program can help win student interest, there is no substitute for writing at the proper level.

Increase pedagogy. As students are exposed to more and more modern information, they find it progressively harder to see the structure of the course. Almost all of what they are learning is new to them, and they have no perspective to appreciate what is most important, what less so. Learning aids are thus much more important than they used to be. Spot summaries spaced throughout the text of each chapter, for example, can help guide a student's learning. These sorts of aids have not been used often in introductory nonmajors texts in the past, because space in an introductory text is at a premium. Yet, such aids are essential. The text is a teaching *tool,* and should be crafted as such. Believing this, we have made extensive use of pedagogical aids in this text.

We thus attempt to balance the need to teach principles to all students with the desire to be current in a rapidly changing science by first limiting the number of principles we attempt to teach, and then by carefully developing the connection of new developments to these principles, so that what is going on now becomes for the student a natural extension, a further development. Our goal is both to promote a fuller understanding by the student of what is happening now, and to present the student with a dynamic picture of how science works, with ideas constantly in a state of flux—a hazy picture becoming clearer.

DEVELOPMENT OF THIS TEXT

No successful text is written in isolation. In writing this one, we have had the benefit of literally hundreds of reviews and comments, both by reviewers commissioned by the publisher and by scientific acquaintances. This input not only ensures the accuracy of what is said, but also helps to mold the text's presentation to better suit its teaching purpose.

It is unlikely that we could have agreed to undertake this project—we have learned from preparing our majors text that writing a text takes an enormous chunk out of our lives—except for one significant advantage: the majors text that we had just finished provided us with an invaluable resource from which to garner material. Used by more than 70,000 students in its first year, *Biology* has been gone over more carefully than any manuscript can be, and input by hundreds of instructors all over the country from Harvard and Stanford to Forest Park Community College has cleaned our majors text of all the glitches, errors, and misstatements that any first edition is prone to contain. We had an extraordinarily clean base from which to prepare this text.

Because *Understanding Biology* was not to be simply a cut-down version of *Biology,* but rather a differently organized book aimed at a far broader audience, the project was launched with a qualifying review of our detailed book plan and proposed table of contents by several instructors chosen to help us refine the proposed contents and organization. In a focus group session, eight of the reviewers met with us and the editors for two days to advise us in detail on our plans and approach. The first draft was reviewed by 15 instructors, representing a broad array of institutions. From their extensive comments we developed a second draft, which was in turn reviewed by a second panel of 15 reviewers. The comments from these reviews were used to hone and refine the final draft. To ensure that no errors had been inadvertently introduced, the final draft was reviewed yet one more time by 10 additional technical reviewers who read the manuscript with but one mission: catch errors and suggest corrections for any misstatements or ambiguous passages. It has been our goal that *Understanding Biology* be as free of errors, typos, misprints, and other glitches as a *second* edition would be.

TEACHING AND LEARNING AIDS

Getting the science correct and up to date is only one phase of developing a biology text. To be an effective teaching tool, the text must present the science in a way that can be clearly taught and easily learned. With this in mind, we have included a variety of different teaching and learning aids.

General Approach

Readability. A text is only as easy to use as it is easy to read. This text has a "discovery" approach and a lively writing style intended to engage the interest of students. Simple analogies are used extensively as an aid to understanding. The level of difficulty has been carefully controlled to correspond to the actual abilities of today's undergraduates.

Vocabulary. The large number of terms encountered in studying biology is a real stumbling block for many students. This book lessens the problem by addressing and reinforcing a student's vocabulary development at three levels. First, no term is used without first being clearly defined. A student does not have to be familiar with the terminology of biology to use this book. Second, terms that are particularly important are printed in **boldface** at the point they are first used and defined. Finally, a complete and up-to-date glossary of terms is included at the end of the book.

Illustration. Because illustrations are so critical to a student's learning and enjoyment of biology, we have used full-color photographs throughout the text. All line art in *Understanding Biology* was done under the direct supervision of the authors, and planned to illustrate and reinforce specific points in the text. Consequently, every illustration is directly related to the text narrative and specifically cited in the text when the illustrated point is discussed. Great care has been taken in page make-up to place illustrations as close as possible to the text discussion. All illustrations and captions have been independently and carefully reviewed for accuracy and clarity.

Individual Chapter Elements

Chapter Overviews. Each chapter is introduced by a brief paragraph that tells the student what the chapter is about and why the subject is important. Not a summary, the overview allows students to place the chapter in perspective as they begin it.

For Review. As students progress through the text, they acquire an increasing vocabulary of terms and concepts, many of which play an important role in later chapters. To aid students in reviewing particularly important concepts, we have included a unique pedagogical device: each chapter is preceded by a short list of terms and concepts from earlier chapters that will be critical to understanding the present chapter, with an indication of where students should look to review any terms or concepts that are unfamiliar.

Concept Summaries. Another unique learning element that we have employed in each chapter is brief concept summaries. Throughout the text, important discussions and key points are succinctly summarized, with the summaries set off from the text and in boldface for emphasis. These concept summaries recap concisely the essential points that a student should learn.

Boxed Essays. To engage the student's interest and make the learning more exciting and fun, most chapters contain boxed essays on topics of special interest. Only a few paragraphs long, these essays provide an opportunity to briefly examine specialized topics for which there would otherwise be no space in the text.

Chapter Summaries. At the end of each chapter, a numbered summary provides a quick review of key concepts in the chapter.

Review and Self-Quiz Questions. A list of objective questions follows each chapter, intended to spot-check key information in the chapter. These questions can be used by students to test the degree to which they have read the chapter with sufficient attention and retention. The answers are in an appendix at the back of the book, so that students may check their performance.

Thought Questions. A brief list of thought-provoking questions and problems follows each chapter. These questions are not simply review, but rather are intended to challenge the student to think about the lessons of the chapter. Where chapters treat material such as genetics, in which solving problems plays an important role in learning, more problems are presented. The answers to these questions appear in the Instructor's Manual.

For Further Reading. Each chapter is followed by a short annotated list of books and articles, to which interested students can refer for additional information. Many of the articles are from *Scientific American* or other sources at a level appropriate to students with a limited science background.

SUPPLEMENTS

Few introductory courses attempt to cover the enormous amount of material that biology texts typically address. In the range of topics covered, introductory biology is almost unique. For this reason, we have provided a complete package of supplements for use with this text, to aid both student and instructor in dealing with what must sometimes seem an immense amount of material.

For the Student

Understanding Biology Study Guide, written by Susan M. Feldkamp and David Whitenberg, both of Southwest Texas State University. A direct companion to the text, the study guide provides students with significant additional study aids, including expanded chapter overviews, learning objectives, capsule summaries of key points, vocabulary reviews, additional self-study review questions, and other features intended to support motivated self-study of the textual materials.

Biology Laboratory Manual, written by Darrell Vodopich and Randall Moore of Baylor University to accompany Raven/Johnson, *Biology,* and equally useful as an accompaniment to *Understanding Biology.* The 35 laboratory exercises in the manual illustrate basic concepts and focus on experiments that will actually work when attempted. The emphasis in the laboratory manual is on manageability of equipment and content, with extensive visual support through numerous illustrations.

For the Instructor

Instructor's Manual, prepared by Ronald S. Daniel of California Polytechnic University at Pomona and Sharon Callaway Daniel of Orange Coast Community College, provides text adopters with substantial support in preparing for and teaching introductory biology with this text. The manual contains suggested course outlines, extensive sources of supplementary materials and additional resources such as film and computer software, 108 overhead transparency masters to supplement the acetates available with the text; suggested learning objectives for each chapter; and chapter-by-chapter notes.

Overhead Transparency Acetates of 100 of the text's most important four-color illustrations are available to instructors from the publisher for use as teaching aids. These transparencies were selected with the assistance of a number of teachers of introductory biology, to provide the maximum utility to the instructor. The instructor's manual contains suggestions for the use of these transparencies.

A printed *Test Bank,* prepared by Richard Van Norman, University of Utah, provides an extensive battery of more than 1800 objective test items that may be used by instructors as a powerful instructional tool. Each chapter has between 40 and 50 questions, including multiple choice, short answer, and classification formats. For each question, in addition to the answer, we have identified the subject tested, given an approximate difficulty rating, and indicated the type of question (factual or conceptual) and the text page on which the question's information appears.

Microtest II, a computerized version of the complete test bank, is available to instructors from the publisher on disks compatible with the IBM-PC or the Apple IIc and IIe. It also provides instructors with the opportunity to add questions of their own as well as to modify or delete questions already on file.

A FEW WORDS OF THANKS

This is the second book that we have written together, and we have been fortunate to have had continued excellent support within our publishing company, with editors who put in hours as long as ours and production staff who care deeply how the book looks. The art was again in the sensitive and gifted hands of Bill Ober. Again, our wives and families have tolerated neglect so that the book could be finished in a timely manner. At every stage, excellent and dedicated reviewers suggested countless improvements. We are especially grateful to those individuals who participated in our Focus Group; the names of the reviewers and the Focus Group members follow. Without the help of all these people, we could not have written this book, and we thank them sincerely.

Peter Raven
George Johnson

REVIEWERS

Ann Antelfinger
University of Nebraska at Omaha

Mary Berenbaum
University of Illinois at Urbana

Brenda C. Blackwelder
Central Piedmont Community College

Richard K. Boohar
University of Nebraska at Lincoln

John S. Boyle
Cerritos College

Donald Collins
Orange Coast College

Joyce Corban
Wright State University

Michael Corn
College of Lake County

John Crane
Washington State University

Ronald S. Daniel
California State Polytechnic University

Rose Davis
St. Louis Community College

Katherine Denniston
Towson State University

William Dickison
University of North Carolina

Fred Drewes
Suffolk County Community College

Paul Elliott
Florida State University

Larry Friedman
University of Missouri at St. Louis

Judy Goodenhough
University of Massachusetts

Gene Goselin
Diablo Valley College

Lane Graham
University of Manitoba

Thomas Gray
University of Kentucky

John P. Harley
Eastern Kentucky University

Holt Harner
Broward Community College

Terry Harrison
Central State University

Fred Hinson
Western Carolina University

Wilfred Iltis
San Jose State University

Alan Journet
Southeast Missouri State University

Peter Kareiva
University of Washington

Ann Lumsden
Florida State University

Constance Murray
Tulsa Junior College–Metro Campus

Steve Murray
California State University at Fullerton

Jim Peck
University of Arkansas at Little Rock

Gary Peterson
South Dakota State University

Ronald Rak
Morraine Valley Community College

Jonathan Reiskind
University of Florida

Martin Rochford
Fullerton College

Samuel Rushforth
Brigham Young University

Larry St. Clair
Brigham Young University

Ed Samuels
Los Angeles Valley College

Donald Scoby
North Dakota State University

Erik Scully
Towson State University

Russell Skavaril
Ohio State University

Gerald Summers
University of Missouri at Columbia

Jay Templin
Widener University

Richard R. Tolman
Brigham Young University

Richard Van Norman
University of Utah

David Calvin Whitenberg
Southwest Texas State University

Dana L. Wrensch
Ohio State University

FOCUS GROUP PARTICIPANTS

Ann Antelfinger
University of Nebraska at Omaha

Brenda Blackwelder
Central Piedmont Community College

Fred Drewes
Suffolk County Community College

Holt Harner
Broward Community College

Alan Journet
Southeast Missouri State University

Steve Murray
California State University at Fullerton

Richard Tolman
Brigham Young University

Richard Van Norman
University of Utah

Contents in Brief

Overview

Part One — INTRODUCTION

1 Your Study of Biology   3
2 Biology As a Science   17

General Principles

Part Two — CELL BIOLOGY

3 The Chemistry of Life   37
4 The Origin of Life   65
5 Cells   77
6 How Cells Interact With the Environment   105

Part Three — ENERGY

7 Energy and Metabolism   129
8 How Cells Make ATP   145
9 Photosynthesis   165

Part Four — GENETICS

10 How Cells Reproduce   185
11 Mendelian Genetics   205
12 Human Genetics   223
13 The Mechanism of Heredity   243
14 Genes and How They Work   263
15 How Genes Change   281

Part Five — EVOLUTION

16 The Evidence for Evolution   305
17 How Species Form   329
18 The Evolution of Life on Earth   345
19 How We Evolved: Vertebrate Evolution   361

Part Six — ECOLOGY

20 How Species Interact With Their Environment   387
21 How Species Interact With One Another   405
22 Communities and Ecosystems   419
23 The Future of the Biosphere   445

The Biological World

Part Seven — BIOLOGICAL DIVERSITY

24 The Five Kingdoms of Life   467
25 The Invisible World: Bacteria and Viruses   483
26 The Origins of Multicellularity: Protists and Fungi   497
27 Plants   513
28 Animals   531

Part Eight — PLANT BIOLOGY

29 The Structure of Plants   561
30 Flowering Plant Reproduction   577
31 How Plants Function   591

Part Nine — ANIMAL BIOLOGY

32 The Vertebrate Body   607
33 How Animals Move   623
34 How Animals Digest Food   637
35 How Animals Capture Oxygen   653
36 Circulation   667
37 The Vertebrate Immune System   685
38 How Animals Transmit Information   701
39 The Nervous System   713
40 Hormones   735
41 The Control of Water Balance   749
42 Sex and Reproduction   761
43 Behavior   785

Appendices

A Classification of Organisms   A-1
B Genetics Problems   B-1
C Answers   C-1

Glossary   G-1
Illustration Credits   IC-1
Index   I-1

CONTENTS

Overview

Part One
INTRODUCTION

1   YOUR STUDY OF BIOLOGY   3
   Biology is the study of life   3
   How is biology important to you?   7
      An overcrowded world   7
      The biology of cancer   10
      The battle against AIDS   12
   Your study of biology   13
   Summary   14

2   BIOLOGY AS A SCIENCE   17
   The nature of science   17
      Testing hypotheses   18
      Theories   19
      The scientific method   20
   History of a biological theory: Darwin's theory of
      evolution   20
   Darwin's evidence   22
      What Darwin saw   22
      Darwin and Malthus   23
      Natural selection   24
   Publication of Darwin's theory   24
   Evolution after Darwin: testing the theory   24
      The fossil record   25
      The age of the earth   25
      The mechanism of heredity   27
      Comparative anatomy   27
      Molecular biology   27
   The limits of science: scientific creationism   27
   How this text is organized to teach you biology   29
      Levels of organization   29
   Summary   32

Part Two
CELL BIOLOGY

3 THE CHEMISTRY OF LIFE 37
  Atoms 37
  Electron orbitals 38
  Molecules 39
  Nature of the chemical bond 40
  Ionic bonds 40
  Covalent bonds 41
  The cradle of life: water 43
    Water is a powerful solvent 45
    Water organizes nonpolar molecules 46
    Water ionizes 47
  The chemical building blocks of life 48
  Carbohydrates 49
    Sugars 49
    Starches 51
    Cellulose 51
  Fats 53
    Other lipids 54
  Proteins 55
    Amino acids 55
    Polypeptides 57
  Nucleic acids 59
  Summary 61

4 THE ORIGIN OF LIFE 65
  The origin of organic molecules: carbon
    polymers 66
  Origin of the first cells 69
  The earliest cells 69
    A living fossil 70
    Methane-producing bacteria 71
    Photosynthetic bacteria 71
    The origin of modern bacteria 71
  The appearance of eukaryotic cells 72
  Is there life on other worlds? 73
  Summary 74

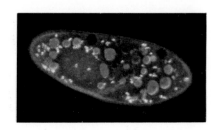

5 CELLS 77

General characteristics of cells 77

Cell size 78

    The cell theory 78

    Why aren't cells larger? 79

    The structure of simple cells: bacteria 82

    Cell walls 83

    Simple interior organization 83

    A comparison of bacteria and eukaryotic cells 84

    The interior of eukaryotic cells: an overview 86

    Boundary of the eukaryotic cell: the plasma
        membrane 86

The endoplasmic reticulum 86

    Rough ER 87

    Smooth ER 87

The nucleus 88

    The nuclear envelope 88

    Chromosomes 89

    The nucleolus 89

The Golgi complex 90

Microbodies 91

Lysosomes 91

Energy-producing organelles 92

    Mitochondria 93

    Chloroplasts 94

The cytoskeleton 95

Flagella 98

An overview of cell structure 101

Summary 102

Boxed Essay: Microscopes 80

**6 HOW CELLS INTERACT WITH THE
ENVIRONMENT 105**

The lipid foundation of membranes 106

    Phospholipids 106

    Phospholipids form bilayer sheets 106

    The lipid bilayer is a fluid 107

Architecture of the plasma membrane 108

How a cell's membranes regulate interactions
    with its environment 111

The passage of water into and out of cells 112

Bulk passage into the cell 114

Selective transport of molecules 114

    The importance of selective permeability 116

    Facilitated diffusion 116

    Active transport 117

    Transfer between cell compartments 119

Reception of information 120

Expression of cell identity 121

Physical connections between cells 121

How a cell communicates with the outside
    world 124

Summary 124

Boxed Essay: Chloride channels and cystic fibrosis 118

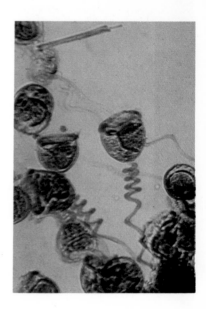

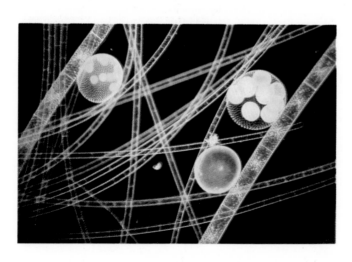

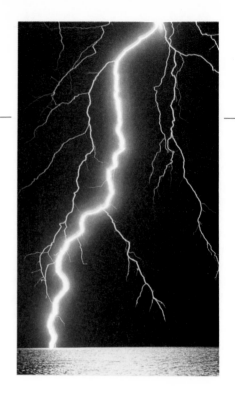

Part Three
ENERGY

7 ENERGY AND METABOLISM 129
   What is energy? 130
   The laws of thermodynamics 131
   Oxidation-reduction 133
   Activation energy 134
   Enzymes 135
      How enzymes work 137
      Factors affecting enzyme activity 137
      The regulation of enzyme activity 138
      Coenzymes 139
   The energy currency of organisms: ATP 140
   Summary 142

8 HOW CELLS MAKE ATP 145
   Using chemical energy to drive metabolism 146
   How cells make ATP: an overview 146
   The fate of a candy bar 148
   Glycolysis 149
      An overview of glycolysis 150
      The universality of the glycolytic sequence 150
   The need to close the metabolic circle 152
   Oxidative respiration 152
      The oxidation of pyruvate 152
      The oxidation of acetyl-CoA 155
      The reactions of the citric acid cycle 155
      The products of the citric acid cycle 155
   Using the electrons generated by the citric acid cycle
      to make ATP 156
   An overview of glucose catabolism: the balance
      sheet 158
   Fermentation 160
   Summary 161

9 PHOTOSYNTHESIS 165

The biophysics of light 166
Capturing light energy in chemical bonds 167
An overview of photosynthesis 170
    Absorbing light energy 170
    Fixing carbon 170
    Replenishing the pigment 171
How light drives chemistry: the light reactions 171
    Evolution of the photocenter 171
    Where does the electron go? 172
Light reactions of plants 172
    The advent of photosystem II 173
    How the two photosystems work together 173
    The formation of oxygen gas 174
    Comparing plant and bacterial light
       reactions 175
    Accessory pigments 176
How the products of photosynthesis are used to build
    organic molecules from CO_2 176
The chloroplast as a photosynthetic machine 178
A look back 180
Summary 181

Part Four
GENETICS

10 HOW CELLS REPRODUCE 185
Cell division in bacteria 185
Cell division among eukaryotes 186
 The structure of eukaryotic chromosomes 188
 The cell cycle 189
Mitosis 189
 Preparing the scene: interphase 190
 Formation of the Mitotic Apparatus:
   prophase 191
 Separation of sister chromatids: metaphase 192
 Separation of the chromatids: anaphase 193
 Re-formation of nuclei: telophase 193
Cytokinesis 194
Meiosis 196
The stages of meiosis 197
 The first meiotic division 197
 The second meiotic division 200
The importance of meiotic recombination 200
Summary 201

11 MENDELIAN GENETICS 205
Early ideas about heredity: the road to Mendel 206
Mendel and the garden pea 206
Mendel's experimental design 207
What Mendel found 208
How Mendel interpreted his results 208
 The F_1 generation 211
 The F_2 generation 211
 Further generations 212
The test cross 212
Independent assortment 213
Some key terms in genetics 215
Chromosomes: the vehicles of Mendelian
  inheritance 215
Crossing-over 217
 Genetic maps 218
Genetics 219
Summary 220

12 HUMAN GENETICS 223
Human chromosomes 224
 Down syndrome 225
 Sex chromosomes 226
Patterns of inheritance 227
 A pedigree: albinism 227
Multiple alleles 229
 ABO blood groups 229
Genetic disorders 231
 Cystic fibrosis 232
 Sickle cell anemia 233
 Tay-Sachs disease 234
 Phenylketonuria 234
 Hemophilia 234
 Huntington's disease 236
Genetic counseling 236
 Genetic therapy 239
Summary 240
Boxed Essay: Human races 230

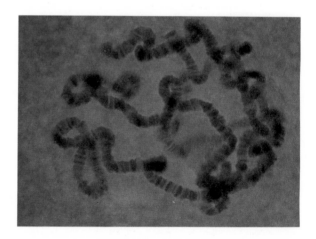

13 THE MECHANISM OF HEREDITY 243
An experimental journey 243
Where do cells store hereditary information? 244
What component of the chromosomes contains the
    hereditary information? 246
    The Griffith-Avery experiments 246
    The Hershey-Chase experiment 247
    The Fraenkel-Conrat experiment 249
How is the information in DNA reproduced so
    accurately? 250
What is the unit of hereditary information? 252
How do genes interact to produce a phenotype? 256
The gene concept 258
Summary 260

14 GENES AND HOW THEY WORK 263
Cells use RNA to make protein 264
An overview of gene expression 265
    Transcription 265
    Translation 266
The genetic code 266
The mechanism of protein synthesis 270
Protein synthesis in eukaryotes 273
Regulating gene expression 273
The architecture of a gene 275
How DNA replicates 276
Summary 278
Boxed Essay: The discovery of introns 272

15 HOW GENES CHANGE 281
Mutation 282
    DNA damage 282
Mutation and cancer 284
Transposition 288
    The impact of transposition 289
Mutations altering chromosome organization 290
    Deletions 290
    Chromosomal rearrangement 291
Gene transfer 291
    Plasmids 292
    Gene transfer among bacteria 293
Genetic engineering 294
    Restriction enzymes 295
Putting genes into plasmids 296
Recent progress in genetic engineering 297
The importance of genetic change 299
Summary 300

Part Five
EVOLUTION

16   THE EVIDENCE FOR EVOLUTION   305
Evolution and adaptation   306
Variation in nature   307
  The Hardy-Weinberg principle   308
Why do allele frequencies change?   309
  Mutation   309
  Migration   310
  Genetic drift   310
  Nonrandom mating   311
  Selection   312
  Which force is most important in evolution?   313
The evidence that natural selection produces
  microevolutionary change   313
  Sickle cell anemia   314
  Peppered moths and industrial melanism   314
  Multiple drug resistance   315
  Lead tolerance   316
  An overview of adaptation   316
The evidence of macroevolution   317
  The fossil record   319
  The molecular record   320
  Homology   320
  Development   321
  Vestigial structures   322
  Parallel adaptation   322
  Patterns of distribution   324
Summary   325
Boxed Essay: Homology and the evolution of the eye   318

17   HOW SPECIES FORM   329
The nature of species   330
The divergence of populations   331
Reproductive isolating mechanisms   333
  Prezygotic mechanisms   334
  Postzygotic mechanisms   337
The evidence that divergence leads to species
  formation   338
  Ecological races   338
  Clusters of species   340
Does evolution occur in spurts?   341
The evidence for evolution: an overview   341
Summary   342
Boxed Essay: Extinction   338

18   THE EVOLUTION OF LIFE ON EARTH   345
Fossils   347
The early history of life on earth   347
The Paleozoic Era   348
  Origins of major groups of organisms   348
  The invasion of the land   350
  Mass extinctions   351
The Mesozoic Era   352
  The history of plants   353
  The movement of continents   354
  The extinction of the dinosaurs   355
The Cenozoic Era: the world we know   356
Summary   358
*Boxed Essay: The discovery of a living
                coelacanth*   346

19 HOW WE EVOLVED: VERTEBRATE
EVOLUTION   361
General characteristics of the vertebrates   362
Jawless fishes   363
The appearance of jawed fishes   363
    The evolution of jaws   363
    Sharks and rays   363
Bony fishes   364
The invasion of the land   365
Amphibians   366
Reptiles   367
Temperature control in land animals   370
Birds   370
Mammals   372
    Characteristics of mammals   373
    Monotremes   373
    Marsupials   373
    Placental mammals   374
The evolution of primates   374
    Anthropoids   375
The appearance of hominids   379
    The Australopithecines   379
    The use of tools: *Homo habilis*   379
    Homo erectus   379
    Modern humans: *Homo sapiens*   381
Summary   382

Part Six
ECOLOGY

20 HOW SPECIES INTERACT WITH THEIR
ENVIRONMENT   387
Selection in action   388
  Forms of selection   389
  The limits to selection   391
The concept of population   391
How populations grow   392
  Biotic potential   393
  Carrying capacity   394
  Density-dependent and density-independent
    effects   395
  r Strategists and K strategists   396
  Human populations   397
Mortality and survivorship   397
Demography   398
Factors limiting the distribution of species   399
  Climatic factors   400
  Edaphic factors   400
  Biotic factors   401
Summary   401

21 HOW SPECIES INTERACT WITH
ONE ANOTHER   405
Competition between species   406
  Competition in nature   406
Coevolution   407
Predator-prey interactions   407
  Predation   407
  Plant defenses against herbivores   409
  The evolution of herbivores   409
  Chemical defenses in animals   410
  Warning and protective coloration   411
  Mimicry   412
Symbiosis   412
  Commensalism   413
  Mutualism   414
  Parasitism   415
Summary   416

22 COMMUNITIES AND ECOSYSTEMS 419

Ecosystems 420
The cycling of nutrients in ecosystems 420
    The water cycle 420
    The carbon cycle 421
    The nitrogen cycle 422
    The phosphorus cycle 423
The flow of energy in ecosystems 424
    Trophic levels 424
Ecological succession 427
The impact of climate on ecosystems 429
    Major circulation patterns 430
    Patterns of circulation in the ocean 431
The oceans 432
    The neritic zone 432
    The surface zone 433
    The abyssal zone 433
Fresh water 434
Biomes 434
    Tropical rain forests 435
    Savannas 437
    Desert 437
    Grasslands 438
    Temperate deciduous forests 438
    Taiga 439
    Tundra 441
The fate of the earth 441
Summary 442

23 THE FUTURE OF THE BIOSPHERE 445

The population explosion 445
    The present situation 447
Food and population 449
The future of agriculture 451
The prospects for more food 452
The tropics 454
Pollution 459
What biologists have to contribute 462
Summary 462

Part Seven
BIOLOGICAL DIVERSITY

24 THE FIVE KINGDOMS OF LIFE   467
The classification of organisms   468
  The polynomial system   468
  The binomial system   468
What is a species?   469
How many species are there?   471
The taxonomic hierarchy   471
The five kingdoms of organisms   473
The kingdoms of eukaryotic organisms   474
Features of eukaryotic evolution   474
Symbiosis and the origin of the eukaryotic
  phyla   477
The evolution of multicellularity and sexuality   478
  Multicellularity   478
  Sexuality   478
Organisms and evolution   480
Summary   480

25 THE INVISIBLE WORLD: BACTERIA
AND VIRUSES   483
Bacterial structure   484
Bacterial ecology and metabolic diversity   485
  Photosynthetic bacteria   485
  Chemoautotrophic bacteria   485
  Heterotrophic bacteria   485
  Nitrogen-fixing bacteria   486
Bacteria as pathogens   486
Viruses   488
Are viruses alive?   488
The diversity of viruses   490
Viruses and disease   491
The infection cycle of an AIDS virus   491
  Attachment   492
  Entry   492
  Replication   492
Simple but versatile organisms   493
Summary   494
Boxed Essay: The eradication of smallpox   489

26 THE ORIGINS OF MULTICELLULARITY:
PROTISTS AND FUNGI   497
The protists   498
  Symbiosis and the origin of eukaryotes   499
Major groups of protists   500
  Multicellular protists   500
  Unicellular protists   500
The most diverse kingdom of eukaryotes   504
The fungi   504
  Fungal ecology   506
  Fungal structure   506
Major groups of fungi   507
Lichens   508
Mycorrhizae   509
Summary   510
*Boxed Essay: Cyclosporine: a modern "wonder
          drug"   502*

27 PLANTS 513

The green invasion of the land 515
The plant life cycle 516
    Alternation of generations 516
    The specialization of gametophytes 517
Mosses and other bryophytes 519
Vascular plants 520
    Growth of the vascular plants 521
    Conducting systems of the vascular plants 521
    What is a seed? 521
    Seedless vascular plants 522
    Gymnosperms 523
The flowering plants 524
    Monocots and dicots 526
    A very successful group 527
Summary 528

28 ANIMALS 531

Some general features of animals 532
The sponges—animals without tissues 532
Cnidarians: the radially symmetrical animals 534
    Nematocysts 535
    Extracellular digestion 536
The evolution of bilateral symmetry 536
Solid worms: the acoelomate phyla 537
The evolution of a body cavity 538
    Nematodes 539
Advent of the coelomates 539
    The advantage of the coelom 539
    An embryonic revolution 540
Mollusks 541
The rise of segmentation 543
Annelids 543
Arthropods 544
Diversity of the arthropods 545
    Chelicerates 546
    Mandibulates: crustaceans 547
    Mandibulates: insects and their relatives 548
Deuterostomes 550
Echinoderms 551
    Basic features of echinoderms 553
Chordates 554
    Tunicates 555
    Lancelets 555
    Vertebrates 556
Summary 557

Part Eight
PLANT BIOLOGY

29 THE STRUCTURE OF PLANTS   561
The organization of a plant   561
Tissue types in plants   563
Types of meristems   564
Plant cell types   564
   Ground tissue   564
   Conducting cells   566
Shoots   567
   Leaves   567
   Stems   569
   Roots   572
Summary   573

30 FLOWERING PLANT REPRODUCTION   577
Flowers and pollination   578
   Pollination by animals   578
   Pollination by wind   579
   Self-pollination   580
Seed formation   580
Fruits   581
Germination   583
Growth and differentiation   583
Reproductive strategies of plants   585
   Factors promoting outcrossing   585
   Self-pollination   587
Summary   587
Boxed Essay: The flowering of bamboos and the starvation
of pandas   586

31 HOW PLANTS FUNCTION 591

  Water movement 592
    Transpiration 593
    The absorption of water by roots 593
    The regulation of transpiration rate 595
  Carbohydrate transport 595
  Plant nutrients 596
    Essential nutrients 596
  Regulating plant growth: plant hormones 597
    The discovery of the first plant hormone 597
    How the shoot apex uses auxin to control
      growth 597
  The major plant hormones 598
    Auxin 598
    Synthetic auxins 598
    Cytokinins 599
    Gibberellins 599
    Ethylene 600
    Abscisic acid 600
  Tropisms 600
    Phototropism 600
    Gravitropism 601
    Thigmotropism 601
  Turgor movements 601
  Photoperiodism 602
  Dormancy 602
  Summary 603
  Boxed Essay: Plant responses to flooding 594

Part Nine
ANIMAL BIOLOGY

32   THE VERTEBRATE BODY   607
The human animal   608
   Levels of organization   608
Tissues   608
   Epithelium   608
   Connective tissue   611
   Muscle   613
   Nerve   613
Summary   620

33   HOW ANIMALS MOVE   623
The mechanical problems posed by movement   624
Skin and bones   626
   Skin   626
   Bone   626
Muscle   627
   Smooth muscle   627
   Skeletal muscle   628
   Cardiac muscle   629
How cells move   629
   The structure of microfilaments   630
   How myofilaments contract   630
   How striated muscle contracts   632
   How nerves signal muscles to contract   633
Looking ahead   633
Summary   634

34   HOW ANIMALS DIGEST FOOD   637
The nature of digestion   638
Where it all begins: the mouth   639
The journey of food to the stomach   640
Preliminary digestion: the stomach   640
Terminal digestion and absorption: the small
   intestine   642
Concentration of solids: the large intestine   644
Nutrition   645
Summary   649
Boxed Essay: Dangerous eating habits   646

35   HOW ANIMALS CAPTURE OXYGEN   653
The evolution of respiration   654
   Creating a water current   654
   Increasing the diffusion surface   655
The gill as an aqueous respiratory machine   655
From aquatic to atmospheric breathing:
   the lung   657
   Evolution of the lung   658
The mechanics of human breathing   660
How respiration works: gas transport
   and exchange   662
   Hemoglobin and gas transport   662
Summary   663

36 CIRCULATION 667
The evolution of circulatory systems 667
The cardiovascular system 668
   Arteries 669
   Arterioles 669
   Capillaries 669
   Veins and venules 671
   The lymphatic system 671
The contents of vertebrate circulatory systems 673
   Blood plasma 673
   Types of blood cells 673
The evolution of the vertebrate heart 674
The human heart 675
   Circulation through the heart 676
   How the heart contracts 678
   Monitoring the heart's performance 679
The central importance of circulation 682
Summary 682
*Boxed Essay: Diseases of the heart and
         blood vessels 680*

37 THE VERTEBRATE IMMUNE SYSTEM 685
Discovery of the immune response 685
The cells of the immune system 687
The architecture of the immune defense 689
   Sounding the alarm 690
   The cell-mediated immune response 691
   The humoral immune response 691
How do immune receptors recognize antigens? 691
   Antibody structure 692
How can the immune system respond to so many
   different foreign antigens? 693
Defeat of the immune system 695
Monoclonal antibodies 696
Summary 697
Boxed Essay: Allergy 694

**38 HOW ANIMALS TRANSMIT
INFORMATION 701**
The neuron 702
The nerve impulse 703
   The resting potential 704
   Initiating a nerve impulse 704
   Why transmission occurs 705
   The action potential 705
   Saltatory conduction 706
Transferring information from nerve to tissue 706
   Nerve-to-muscle connections 708
   Nerve-to-nerve connections 709
Summary 710

39 THE NERVOUS SYSTEM 713
   Organization of the vertebrate nervous system   713
   Evolution of the vertebrate brain   716
      The advent of a dominant forebrain   716
      The recent expansion of the cerebrum   717
   Anatomy and function of the human brain   718
      The cerebral cortex   718
      Associative organization of the
         cerebral cortex   718
      Memory and learning   720
   Sensory information   720
   Sensing internal information   721
      Temperature change   721
      Blood chemistry   721
      Pain   721
      Muscle contraction   722
      Blood pressure   722
      Touch   722
      Balance   722
      Motion   722
   Sensing the external environment   724
      Taste and smell   724
      Hearing   724
      Vision   725
   Other environmental senses in vertebrates   728
   The peripheral nervous system   728
   Neuromuscular control   729
      Reflex arcs   729
   Neurovisceral control   729
      The autonomic nervous system   730
      Antagonistic control of the autonomic
         nervous system   731
   Summary   731

40 HORMONES 735
   Neuroendocrine control   735
   Hormones released by the posterior pituitary   738
   Hormones released by the anterior pituitary   739
      Control over hormone production   740
   Nonpituitary hormones   741
   How neuroendocrine control works   741
   Regulation of glucose levels in the blood   742
   The regulation of physiological functions:
      water balance   745
   Summary   746

41 THE CONTROL OF WATER BALANCE 749
   Osmoregulation   750
      The problems faced by osmoregulators   750
      How osmoregulation is achieved   751
   The organization of the vertebrate kidney   751
      Filtration   752
      Reabsorption   752
      Excretion   752
   Evolution of the vertebrate kidney   752
      Freshwater fishes   752
      Marine fishes   753
      Sharks   753
      Amphibians and reptiles   754
      Mammals and birds   754
   How the mammalian kidney works   755
   Integration and homeostasis   757
   Summary   758

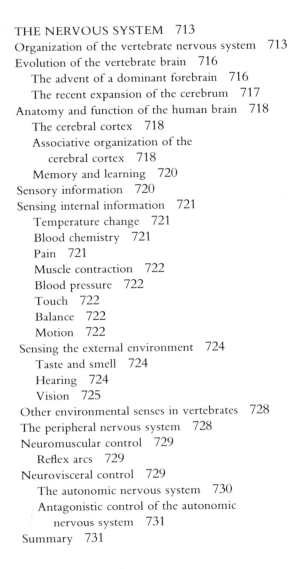

43 BEHAVIOR 785
　Is behavior learned, or are we ruled by
　　our genes? 786
　　Ethology 787
　　Behavioral psychologists 787
　　How learning and instinct interact to shape
　　　behavior 789
　Sociobiology: the biological basis of
　　social behavior 789
　　Aggression 791
　　Altruistic behavior 794
　The genetic basis of behavior 794
　Many vertebrate behaviors are adaptive 795
　　Territoriality 795
　　Migratory memory 796
　　Dominance hierarchies 796
　　Reproductive behaviors 797
　The biological basis of human behavior 797
　Summary 798
　*Boxed Essay: Naked mole rats—a rigidly organized
　　　　　vertebrate society* 790

42 SEX AND REPRODUCTION 761
　Sex evolved in the sea 762
　Vertebrate sex and reproduction: four strategies 762
　　Fishes 762
　　Amphibians 762
　　Reptiles and birds 763
　　Mammals 764
　Sex hormones 764
　The reproductive cycle of mammals 765
　The human reproductive system 767
　　Males 767
　　Females 769
　The physiology of human sexuality 771
　　Excitement 771
　　Plateau 772
　　Orgasm 772
　　Resolution 772
　Contraception and birth control 772
　Abortion 773
　Paths of embryonic development 777
　The course of human development 777
　　First trimester 777
　　Second trimester 779
　　Third trimester 779
　Summary 781
　Boxed Essay: AIDS on the college campus 774

Brent Antinori as
a child

_____*Appendices*_____

A   Classifications of Organisms   A-1
B   Genetics Problems   B-1
C   Answers   C-1

Glossary   G-1
Illustration Credits   IC-1
Index   I-1

INTRODUCTION

Part One

FIGURE 1-1   A female green turtle leaves Ascension Island in the South Atlantic Ocean after laying her eggs in the sand.

YOUR STUDY OF BIOLOGY Chapter 1

The green turtle on the opposite page has laid eggs during the night on the upper beach; as dawn breaks, she is returning to the sea, laboriously dragging her 400-pound body across the sand (Figure 1-1). She is on Ascension Island, in the middle of the Atlantic Ocean, far from home. Every year great numbers of green turtles swim to Ascension Island to lay their eggs—from the coast of Brazil, 1400 miles across the ocean. Their journey lasts about 2 months; 2 months of churning through strong equatorial currents with no food. There are seven small sandy beaches on this tiny rocky island, and the turtles seek them out as if drawn by a beacon. Once there, they lay their eggs and then turn their backs on them to begin the long swim home.

In Brazil, home of the green turtles, there are countless miles of warm, sandy beaches. What is it that so draws the turtles to these isolated, cold strips of sand? How do they find Ascension Island, plowing head down through the waves for long days and weeks? The island is a mere speck in the ocean, hard for an airplane or ship to find even with modern navigational equipment. How can they know the island is out there, over the horizon, more than a thousand miles away? How do the newborn turtles then find their way back over open sea to a home in Brazil that they have never seen?

At first glance, the turtle on the opposite page is just a turtle on a beach. It is only by observation and study that we learn enough to appreciate the mystery that is so much a part of its life. Similar mysteries are common in human life, and many of them have a biological basis: the more we know about the biology of other animals, the better we can understand ourselves. Biology is a science devoted to this larger vision. It provides knowledge about the living world of which we are a part, and, ceaselessly, it bombards us with questions. Slowly, a little at a time, it sometimes gives us answers. As we proceed through this text we will weave a tapestry of questions and answers, a picture of life on our planet. When we have completed our journey, we will ask again about the long voyage of the green turtles.

BIOLOGY IS THE STUDY OF LIFE

Biology is a science that attempts to understand the teeming diversity of life on earth, of which we are a part. It is important that mankind learn how to live in harmony with earth's other residents. The science of biology has much to contribute to this effort. A good way to start your study of biology is to focus for a moment on biology's subject, life. What is life? What do we mean when we use the term? This is not as simple a question as it appears, largely because life itself is not a simple concept. Pause for a moment and try to write a simple definition of "life." You will find that it is not an easy task. The

FIGURE 1-2

Movement. A graceful gull gliding through the air, a killer whale exploding from the sea—animals have evolved mechanisms that allow them to move about. Whether it is the awkward galumphing of a giraffe, the laborious crawling of a weevil, or a human's first faltering steps, we have grown to expect some kind of movement from all land-dwelling animals. Although not all kinds of organisms move from place to place, movement is one of the common properties of living things.

problem is not your ignorance, but rather the loose manner in which the concept "life" is used. For example, imagine a situation in which two astronauts encounter a large, formless blob on the surface of some other planet. One might say to the other, "Is it alive?" We can try to answer the question of what life is by observing what the astronauts do to find out whether the blob has life. Probably they would first observe the blob to see whether it moves.

1. *Movement.* Most animals move about (Figure 1-2). A horse winning the Kentucky Derby, a dog chasing a car, you rolling over in bed—movement seems an integral part of living. Movement from one place to another is not in itself a sure sign of life, however. Many animals, and most plants, do *not* move about, and many nonliving objects such as clouds can be observed to move. The criterion of movement is thus neither *necessary*—possessed by all life forms—nor *sufficient*—possessed only by life forms, even though it is a common attribute of many kinds of organisms.

The astronauts might prod the blob to see whether it responds, and thus test for another criterion, sensitivity.

2. *Sensitivity.* Almost all living things respond to stimuli (Figure 1-3). Plants grow toward light, and animals retreat from fire. Not all stimuli produce responses, however. Imagine kicking a redwood tree or singing to a mushroom. Different kinds of organisms often react to the same stimuli in different ways. This criterion, although superior to the first one, is still inadequate to define life.

The astronauts might watch the blob to see whether it changes, and thus test yet another criterion, development.

3. *Development.* Most multicellular organisms exhibit development, an orderly progressive change in form and degree of specialization. You, for example, started life as a single cell. As you grew to be an embryo and then a baby, different cell types developed, giving rise to brain and lung and bone, and finally to your adult form. Without development, you would be simply a large blob of similar cells. Not all living things exhibit development, however. A single-celled bacterium, for example, does not develop from a simpler form; its parent cell simply divides into two identical daughter cells. Nor are all things alive that undergo progressive, orderly change. The progressive, orderly series of rocks that can be seen on the walls of the Grand Canyon does not indicate that the canyon was ever alive.

The astronauts might think that the motionless blob had once been alive, but is now dead.

4. *Death.* All living things die, whereas no inanimate objects do. Death is not the same as disorder, however. A car that breaks down does not die; we may say, "The car died on me," or "I killed the engine," but the now-broken car was never alive. Death is simply the termination of life. Unless one can detect life, death is a meaningless concept. Death is a terribly inadequate criterion.

Finally, the astronauts might attempt to pick up the blob and examine it more carefully, to see how complex it is.

5. *Complexity.* All living things are complex. Even the simplest bacterium contains a bewildering array of molecules, organized into many complex structures. Complexity is not diagnostic of life, however. A computer is also complex, but it is not alive. Complexity is a necessary condition of life, but not sufficient in itself to identify living things, since many complex things are not alive.

Movement, sensitivity, development, death, and complexity are all characteristics of some organisms, but they do not define life.

FIGURE 1-3

Sensitivity. The father is responding to a stimulus—he has just been bitten on the rump by his cub. As far as we know, all organisms respond to stimuli, although not always to the same ones. Had the cub bitten a tree instead of his father, the response would not have been as dramatic, but there would have been one nonetheless.

FIGURE 1-4

Cellular organization. The *Paramecium*, a complex organism that consists of a single cell, has just ingested several yeast cells, stained red in this micrograph. Yeast is a single-celled fungus. Within the *Paramecium,* the yeast cells are enclosed inside specialized compartments. Several other internal structures of the *Paramecium* are also visible.

To determine whether the blob is alive, the astronauts must learn much more about it. The best thing they could do would be to examine it more carefully and determine the ways in which it resembles living organisms. All organisms that we know about share certain general properties, ones that we think must ultimately have been derived from the first organisms that evolved on earth. It is by these properties that we recognize other living things, and to a large degree these properties define what we mean by the process of life. Four fundamental properties shared by all organisms on earth are:

1. *Cellular organization.* All organisms are composed of one or more cells, complex organized assemblages of **molecules**—the smallest units of a chemical compound that still have the properties of that compound—within membranes (Figure 1-4). The simplest organisms possess only a single cell; your body contains about 100 trillion.

FIGURE 1-5

Metabolism. The energy that almost all organisms use to grow is obtained through the process of photosynthesis, which is carried out by plants, algae, and some bacteria. In this process the energy of light is captured by chlorophyll and other pigments and used to build carbon-containing molecules from carbon dioxide gas, which is a relatively minor component of the earth's atmosphere.

2. *Growth and metabolism.* All living things assimilate energy and use it to grow in a process called **metabolism** (Figure 1-5). Plants, algae, and some bacteria utilize the energy of sunlight to create the more complicated molecules that make up living organisms from carbon dioxide (CO_2) and water (H_2O). The process by which they do this is known as **photosynthesis**. Nearly all other organisms obtain their energy by consuming these photosynthetic organisms, or one another, in a constant ebb and flow of energy. All of the organisms you see about you drive the processes of life within themselves by using chemical energy first captured by photosynthesis. In all living things, this energy is transferred from one place to another by means of special, small, energy-carrying molecules called **ATP molecules.**

3. *Reproduction.* Some organisms live for a very long time. Some of the bristlecone pines *(Pinus longaeva)* growing near timberline in the western Great Basin of the United States have been alive for nearly 5000 years. But no organisms live forever, as far as we know. Because all individual organisms ultimately die, life as an ongoing process is impossible without reproduction (Figure 1-6).

4. *Homeostasis.* All living things maintain an internal environment quite different from their surroundings, with more of certain chemicals and less of others. Maintaining relatively stable internal conditions in an organism is called **homeostasis.** If someone were to grind you up into a soup, you would not be alive, even though all the same molecules would be present. The relationship between the molecules, which forms the stable internal environment necessary for you to live, would have been destroyed.

Are these properties adequate to define life? Is a membrane-enclosed entity that grows and reproduces alive? Not necessarily. Soap bubbles in water solution spontaneously form hollow spheres, membranes that enclose a small volume of air. These spheres may grow and subdivide, and maintain an internal environment quite different from the water surrounding them. Despite these features, the soap bubbles are certainly not alive (Figure 1-7). Therefore the four criteria just listed are necessary for life but are not sufficient to define life. One ingredient is missing: heredity, a mechanism for preserving features that determines what an organism is like.

5. *Heredity.* All organisms on earth possess a "genetic" system that is based on the replication (duplication) of a complex linear molecule called **DNA.** The order of the subunits making up the DNA contains, in code, the information that determines what an individual organism will be like, just as the order of letters on this page determines the sense of what you are reading. Blocks of coded information in the DNA contain the directions for creating the molecules that determine what organisms are like. These subunits of DNA are called **genes.**

To understand the role of heredity in our definition of life, let us return for a moment to soap bubbles. When examining an individual bubble, we see it at that precise moment in time, but we learn nothing of its predecessors. It is likewise impossible to

FIGURE 1-6

Reproduction. The reptile crawling out of this egg represents successful reproduction. All organisms reproduce, although not all hatch from eggs. Some organisms reproduce several generations each hour; others, only once in a thousand years.

guess what future bubbles will be like. The bubbles are the passive prisoners of a changing environment, and it is in this sense that they are not alive. The essence of being alive is the ability to reproduce permanently the results of change. **Heredity,** the transmission of characteristics from parent to offspring, therefore provides the basis for the great division between the living and the nonliving. A genetic system that enables this transmission to occur is the sufficient condition of life.

A natural consequence of heredity is adaptation. An **adaptation** is any peculiarity of structure, physiology (life processes), or behavior that promotes the likelihood of an organism's survival and reproduction in a particular environment. Organisms seem remarkably well suited to the environments in which they live. In the course of the progressive adaptation of organisms to the conditions of life on earth—a process known as **evolution**—those organisms that were less suited to particular places have not persisted. The ones that are found there today are the "winners," for the moment. When we look at any living organism, therefore, we see in its features a record of its history. Not only does life evolve, evolution is the very essence of life.

> **All living things on earth are characterized by cellular organization, growth, reproduction, homeostasis, and heredity. These characteristics define the term "life."**

HOW IS BIOLOGY IMPORTANT TO YOU?

Biologists do more than simply prod blobs and ask whether they are alive. They live with gorillas and collect fossils and listen to whales. They isolate viruses and grow mushrooms and grind up insects. They read the message encoded in the long molecules of heredity, and count how many times a hummingbird's wings beat each second. Perhaps most importantly, biologists attempt to describe the way in which human beings fit into the whole picture of life on earth. Only by understanding this can we deal appropriately with our own destiny. Life exists on earth in incredible diversity, and we are part of that diversity.

Biology is one of the most interesting of subjects because of its great variety. But not only is it fun, it also is an important subject for you and for everyone, simply because biology will affect your future life in many ways. The knowledge that biologists are gaining is of fundamental importance to our ability to manage the world's resources in a suitable manner, to prevent or cure diseases, and to improve the quality of our lives and those of our children and grandchildren. We will illustrate these relationships by discussing three examples of the kinds of problems that concern biologists today. Biologists are working on many problems that critically affect our lives, and these three are intended simply as illustrations of the importance and relevance of the work that is going on. Because the activities of biologists alter our lives in so many ways, an understanding of biology is becoming increasingly necessary for any educated person.

An Overcrowded World

When the authors of this text took freshman biology, the world's population was less than 3 billion people. As you take freshman biology, it stands at more than 5 billion people. When your children study the subject, the population will be in excess of 8 billion people. In this short span of some 70 years, the population of the world will have nearly tripled. Such enormous numbers of people are changing the face of the earth in ways that could not even have been imagined a few hundred years ago. The skill with which we manage the fate of the earth and guard its ability to produce the food and other substances that make our lives possible will determine the kind of lives that our children and grandchildren are able to enjoy. One of the great tasks of biology is to learn the basic information that we need to develop this skill.

Part of the problem is that more than half of the world's people live in countries that are completely or partly tropical, and their numbers are increasing much more rapidly than those of people living in other parts of the world. For each person living in an industrialized country like the United States in 1950, there were two people living

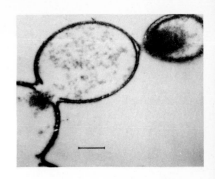

FIGURE 1-7

These *coacervates,* microscopic spheres formed from protein molecules, possess many of the characteristics of living cells. Each is bounded by a membrane, which actually has the double-layered (bilayer) structure characteristic of the plasma membranes that bound all cells. The first cells are thought to have evolved from coacervates.

FIGURE 1-8

Traditional methods of cultivating areas occupied by tropical forests depend on growing crops in the cleared areas for no more than several years and then moving on to clear a new area. When too many people are engaged in such practices, the forests do not have time to recover.

elsewhere; by 2020, there will be five. The result of these changing relationships is a very rapidly increasing pressure on the resources of tropical regions. Many tropical soils are infertile or difficult to manage by conventional agricultural methods. For this reason, the widespread cutting of the forests that is occurring in the tropics and subtropics (Figure 1-8) often leads not to the development of new areas for agriculture and forestry, but rather to wastelands that do not contribute significantly to the welfare of the rapidly growing populations of these regions.

Attaining stable population levels is of fundamental importance, but many people are hungry even with the current numbers of people. We need to develop new technologies in agriculture and forestry that will help to overcome these problems. Such efforts obviously need to be coupled with a decrease in the rate of population growth throughout the world. Even assuming that population growth can be brought under control, about a third of the people living in the tropics exist in a state of absolute poverty now. For many of them, malnutrition is also a problem; for these people, solutions will not wait. If you were they, you would appreciate why it is important that we soon find a way to farm these areas successfully. To do so will require a major investment in biological research. This problem poses one of the greatest challenges facing biologists today. If the problem is not solved, a large and growing proportion of the world's population is going to be increasingly hungry, and the wealthy nations will certainly be increasingly affected by the consequences.

One promising avenue of biological research that may help to feed an increasingly hungry world is the creation of new crop plants. By collecting many different varieties of the principal crops of the tropics and then breeding them to select for desirable traits, biologists often have been able to improve the plants' yields. For example, corn, wheat, and rice together provide about 60% of the world's food, so that improvements in their yields are of particular importance. All three grains now produce higher yields than were thought possible at the end of World War II. This improvement has been referred to as the *Green Revolution* (Figure 1-9). The development of these new strains and their distribution to many parts of the world have significantly improved the ability of many countries to feed themselves.

Efforts continue in the laboratory as well as on the farm. Molecular biologists, using techniques for modifying genes **(genetic engineering)** to be described in Chapter 15, have been learning to transfer genes from one plant to another. The hope is to assemble within a crop plant a battery of additional genes that will provide it with a desirable combination of new traits that cannot easily be obtained by selection of plants within the crop itself. Among these traits are improved pest resistance, increased yield, reduced requirements for certain key nutrients, and faster growth rate. Crops that are

FIGURE 1-9

Some of the most dramatic products of the Green Revolution have been varieties of rice that are easier to harvest and more resistant to disease, thus producing a higher yield of grain. Rice is the most important grain of Asia and certain other tropical and subtropical regions. The introduction of more productive strains has had an important effect on feeding large numbers of people there. Much of tropical Asia is hilly, and rice is grown there in terraces cut into hillsides.

naturally resistant to their pests could be grown without the addition of huge amounts of chemical herbicides and pesticides—dangerous substances that often damage the environment and may lower the long-term productivity of areas where they are used. These genetic engineering methods, combined with conventional methods of improving crops and with the utilization of new crops for food and other products, will make valuable contributions to the solution of the serious problems identified above.

Attaining stable population levels, alleviating widespread poverty and malnutrition, and managing the earth's resources better, partly through the production of improved crops, are all important elements in stabilizing the effects of humans on the world that supports us.

The problems of coping with the effects of a record human population that is growing at an explosive rate will be discussed again in Chapter 23. One of the important aspects of the problem is coping with the loss of biological diversity through extinction. It has been calculated that we may lose more than a million different kinds of organisms **(species),** representing as much as a quarter of the diversity of life on earth, during the next 20 years or so. More importantly, what we have is the only base for what our children can ever have. The loss of diversity cripples the future. Biologists have a major role to play in promoting the preservation of tropical forests and taking other steps to preserve biological diversity, which is so filled with interest, often beauty, and usefulness for human beings. The young field of **conservation biology** is addressing the theoretical and practical aspects of such preservation effectively. Even when the techniques of genetic engineering are applied fully to the improvement of crops and production of other useful organisms, genes will be needed for transfer from

FIGURE 1-10

Cigarette smoker. The smoke this man is drawing into his lungs may someday kill him.

one kind of organism to another. The preservation of as many different kinds of organisms as possible, and thus of their genes, will be critical to the future development of such useful organisms.

The Biology of Cancer

Cancer is perhaps the most dreaded human disease. Of the children born in the late 1980s, as this book is being written, one third will contract cancer, and more than one quarter will someday die of cancer. Each of us has had family or friends affected by the disease. Many of you using this text will die of cancer. Not surprisingly, a great deal of effort has been expended to learn the cause of this disease. With the new techniques of molecular biology, a great deal of progress has recently been made.

The picture of cancer that emerges from recent work is one of a disease that results from a loss of control over the body's hereditary information. All of the instructions that dictate how your body works are present within each cell in coded form as genes in long DNA molecules. Each segment of information is a separate gene. As you grow older, your body's cells shut off certain types of genes that are no longer useful, such as genes that direct cells to divide and grow rapidly. If these genes were not shut off in the normal course of human development, you would never stop growing! The way the cell does this is with special "regulatory genes," whose only function is to keep the undesired gene segments shut off. If such a regulatory gene is damaged or changed, however, the controls that normally restrict cell growth and division do not operate. Change in genes is called **mutation,** and it now appears that the critical factor in inducing cancer is the creation of mutations, because some mutations inevitably damage regulatory genes. Mutations and their effects will be considered in more detail in Chapter 15. At any event, the kind of overcrowded world we discussed in the preceding section is one in which the kinds of damaging substances that cause mutations are more abundant than they ever have been in the past.

How can we prevent cancer? The most obvious strategy is to minimize the occurrence of mutations. We can best do this by avoiding those things that cause mutations. Anything that we do to increase our exposure to radiation or to chemicals—called **mutagens**—that cause mutations will result in an increased incidence of cancer, for the unavoidable reason that such exposure increases the probability of mutating a potential cancer-causing gene.

Of all the environmental mutagens to which we are exposed, perhaps the most tragic are the chemicals we encounter as a result of cigarettes, cigars, pipes, and other products that we smoke. They are tragic because the cancers they cause are preventable. About one third of all cancers in the United States can be attributed directly to smoking. The association is particularly striking for lung cancer. More than 160,000 cases of lung cancer will be diagnosed in 1988, and 90% of these individuals will die within 3 years. Of those who die, 96% will be smokers.

Smoking is a popular pastime in the United States (Figure 1-10). One third of the U.S. population smokes. American smokers consume more than 450 billion cigarettes a year. Cigarettes emit in their tobacco smoke some 3000 chemical components, a number of which are potent mutagens. Smoking introduces these mutagens to the tissues of your lungs. Figure 1-11, a photograph of lung cancer in an adult human, illustrates the result. As you might imagine, a lung in this condition doesn't function well, but lack of breathing is rarely the cause of death from lung cancer. As the cancer grows within the lung, its cells invade the surrounding tissues and eventually break through into the lymph and blood vessels. Once the cancer cells have done this and enter the circulatory system, they spread rapidly through the body, lodging and growing at many locations, particularly in the brain. Death soon follows.

The relationship between smoking and lung cancer in male smokers (Figure 1-12) shows a highly positive correlation: the risk of lung cancer increases with increasing amounts of smoking. For those people smoking two or more packs of cigarettes a day, the risk of contracting lung cancer is 40 times greater than it is for nonsmokers. Note that lung cancer is essentially absent among those people who do not smoke at all. Life insurance companies have computed that, on a statistical basis, smoking a single ciga-

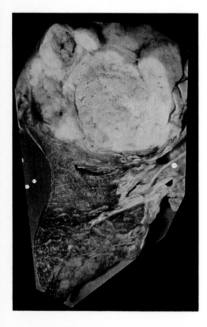

FIGURE 1-11

This is the lung of a cigarette smoker. The solid whitish mass of tissue is the cancer that killed him. The tissue below is blackened by the tars of years of smoking.

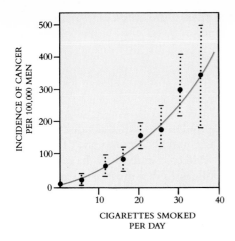

FIGURE 1-12

The annual incidence of lung cancer per 100,000 men is clearly related to the number of cigarettes and other tobacco products smoked per day. In addition, smoking is a major contributor to heart disease. Of every 100 American college students who smoke cigarettes regularly, 1 will be murdered, about 2 will be killed in automobile accidents, and about 25 will be killed by lung cancer and other diseases related to smoking.

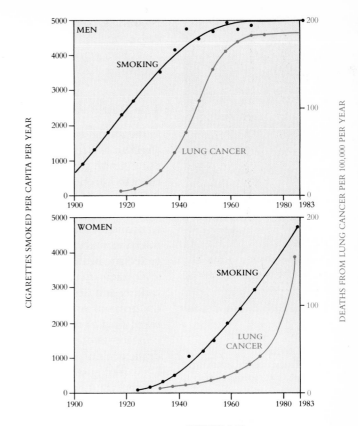

FIGURE 1-13

Incidences of smoking and of lung cancer in the United States from the turn of the century until 1983.

rette lowers your life expectancy 10.7 minutes. (That's more than the time it takes to smoke the cigarette!) Every pack of 20 cigarettes thus bears an unwritten label:

"The price of smoking this pack of cigarettes is three and a half hours of your life."

Particularly among manufacturers of cigarettes and other tobacco products, it has been popular to argue that the causal connection between smoking and cancer has not been proved, that somehow the relationship is coincidental. Look carefully at the data presented in Figure 1-13 and see if you agree. The upper of the two graphs presents data collected for American males. It presents the incidence of smoking from the turn of the century until now, and the incidence of lung cancer over the same period. Note that as late as 1920 lung cancer was a rare disease. With a lag of some 20 years behind the increase in smoking, it became progressively more common.

Now look at the lower graph, which presents data on American women. Because of social mores, significant numbers of American women did not smoke until after World War II, when many social conventions changed. As late as 1963, when lung cancer in males was already near its current levels, this disease was still rare in women. In the United States that year, only 6588 women died of lung cancer. But as their smoking increased, so did their incidence of lung cancer, with the same inexorable lag of about 20 years. Female Americans today have achieved equality with their male counterparts in the numbers of cigarettes that they smoke—and their lung cancer death rates are now no different from men's. In 1986, more than 40,000 women died of lung cancer in the United States.

Smoking is very much like going into a totally dark room, standing still, and then calling in someone with a gun and closing the door. The person with the gun can't see you, doesn't know where you are, and so just shoots once in a random direction and leaves the room. Every time an individual smokes a cigarette, he or she is shooting mutagens at their genes. Just as in the dark room, a hit is unlikely, and most shots will miss potential cancer-causing genes. As one keeps shooting, however, the odds of eventually scoring a hit get larger and larger. Nor do statistics protect any one individual.

Nothing says the first shot won't hit. Old people are not the only ones to die of lung cancer.

Except for eating radioactive chemicals, there is probably no more certain way to get cancer than to smoke.

The Battle Against AIDS

In the spring of 1981 a new and vicious disease, acquired immune deficiency syndrome **(AIDS),** was first reported in the United States. It is rapidly becoming one of the most acute health problems in the history of humanity. Affected individuals have no resistance to infection, and all of them eventually die of illnesses that most of us easily ward off. No one who has contracted AIDS has ever survived more than a few years. Because AIDS is infectious, it is a very dangerous disease. There is essentially no risk of transmission of AIDS from an infected individual to a healthy one in the course of day-to-day contact, but the transfer of bodily fluids, such as blood or semen, between infected and healthy individuals poses a severe risk.

In normal individuals, an army of specialized cells patrols the bloodstream, attacking and destroying any invading bacteria or viruses. In AIDS patients, however, this army of defenders stands helpless. One special kind of cell, called a "helper T cell," is required to rouse the defending cells to action, and in AIDS patients these helper T cells are silent. Like a football team without a quarterback, defender cells mill about doing nothing to prevent the spread of infection, as if unaware of the problem.

Biologists all over the world began working to determine the cause of AIDS. It was not long before the infectious agent was identified by laboratories in France and in the United States. It proved to be a virus (Figure 1-14). Study of the virus revealed it to be closely related to an African vervet or green monkey virus, perhaps introduced to humans in central Africa from a monkey bite. The virus homes in on helper T cells, infecting and killing them until none are left. The virus is transmitted from one person to another with the transfer of body fluids—sexually in semen and vaginal fluid, or as a result of small wounds, in blood transfusions, or as a result of the use of nonsterile needles. As mentioned above, there is virtually no risk of transmission of AIDS between individuals in the course of routine, day-to-day contact. Nonetheless, the incidence of AIDS is growing very rapidly in the United States, and it is already very high in many countries of Africa. The implications of the AIDS epidemic for college students are discussed in Chapter 42.

In the United States, an estimated 2 million people had been exposed to the AIDS virus by the year 1987, and most of them are thought to harbor the virus. Many—perhaps all—of them will eventually come down with the disease. There is also growing evidence that the virus may also attack brain cells, leading to progressive dementia. There is a long latency period after infection before clinical symptoms develop, typically from 2 to 5 years. During this long interval, carriers of the AIDS virus have no clinical symptoms but are thought to be fully infectious, which makes the spread of the disease very difficult to control (Figure 1-15).

Many different kinds of efforts are being made to combat AIDS. For example, sexual abstinence by infected individuals, the use of condoms, and strictly hygienic procedures in the use of needles and in certain medical practices are all important. In addition, a major attempt is being made to produce a vaccine against AIDS. A **vaccine** is a substance that prevents a disease when injected into the body of an individual who does not already have the disease. The action of vaccines will be discussed further in Chapter 37. In the case of AIDS, biologists are inserting fragments of the AIDS virus into otherwise harmless viruses; such fragments of the AIDS virus may be sufficient to prevent infection, but not to cause the disease. Ideally, the engineered virus would fool the human body into developing a defense against AIDS without actually experiencing the disease. When humans are inoculated with the engineered virus, their bodies produce proteins called **antibodies,** which bind to the exterior of the virus, marking it for destruction by the body's defender cells. In this case, the infected cells display on their surfaces proteins specified by genes on the inserted fragment of AIDS virus. There they are recognized by the body's immune defenses, and antibodies are made against them.

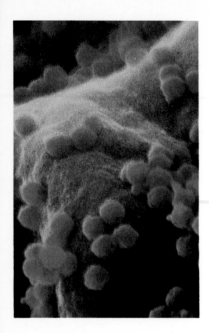

FIGURE 1-14

AIDS viruses released from infected helper T cells, blue in this micrograph, soon spread over neighboring helper T cells, infecting them in turn. The individual AIDS particles are very small; over 200 million would fit on the period at the end of this sentence.

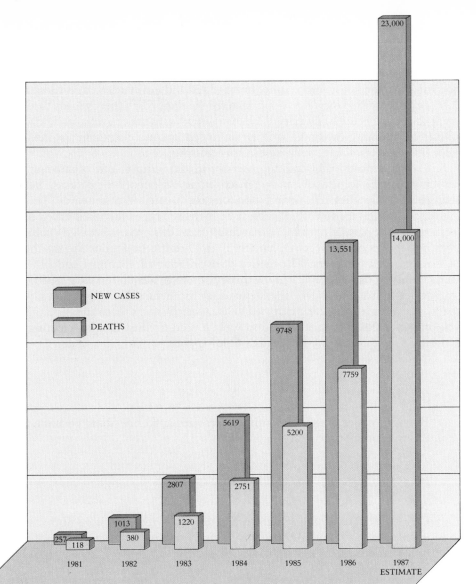

FIGURE 1-15

The incidence of new AIDS cases is rising dramatically each year— followed inescapably by death. The average survival time after diagnosis is less than 2 years. In 1987, nearly 20 people died each day of AIDS in the United States. The U.S. Public Health Service estimates that by 1992, 10 years after the outbreak started, there will have been 179,000 deaths in the United States caused by AIDS.

After inoculation, the bloodstream builds up high levels of antibody directed against the AIDS virus surface. If any real AIDS viruses come along, they will be immediately bound by these antibodies and attacked. Engineered viruses such as these are not yet available for use, and it is not clear that they will be effective. There may be many strains of AIDS virus, each requiring a different vaccine for immunity. Work continues at a high pitch.

YOUR STUDY OF BIOLOGY

In its broadest sense, biology is the study of living things. Life, however, does not take the form of a uniform green slime covering the surface of the earth; rather, it consists of a diverse array of living forms. A biologist tries to understand the sources of this diversity, and in many cases to harness particular life forms to perform useful tasks. Even the narrowest study of a seemingly unimportant life form is a study in biological diversity.

As a science, biology is devoted to understanding biological diversity and its consequences. Like all other scientists, biologists achieve understanding by observing nature and drawing deductions from these observations. In doing so, biologists also attempt to explain the unity of structure and function that underlies the diversity, and the consequences of that unity. It is the special relationship between unity and diversity that underlies and distinguishes the science of biology.

A scientist attempts to account for the observations with possible explanations—**hypotheses**—that explain the observations in terms of what we already know. A hypothesis that stands the test of time, often tested and never rejected, is called a **theory.** A theory is a verified hypothesis. Long-term validation of a theory leads to a **law:** thus we speak of the "Law of Gravity." The way in which science grows and develops, the so-called "scientific method," will be discussed at some length in the next chapter, since this matter lies at the very root of scientific progress.

From centuries of biological observation and inquiry, one organizing principle has emerged: biological diversity reflects *history,* a record of success, failure, and change extending back to a period soon after the formation of the earth. The weeding out of failures and the reward of success by increased reproduction is called **natural selection,** and the pattern of changes that result from this process is called **evolution.** Evolution is the special subject of Chapters 16 to 19 of this book, but also the theme of all of the remaining chapters. The theory of evolution will form the backbone of your study of biological science, just as the theory of the covalent bond is the backbone of the study of chemistry, or the theory of quantum mechanics that of physics. It is a thread that runs through everything that you will learn in this text. In the next chapter, the development of the theory of evolution will be used to illustrate the scientific method. Evolution is the essence of the science of biology.

SUMMARY

1. Biology is a science that attempts to describe and understand both the unity and the diversity of life on earth.

2. Movement, sensitivity, development, death, and complexity are all characteristics of many living things, but they do not define life, either singly or in combination.

3. Cellular organization, growth and metabolism, reproduction, homeostasis, and heredity characterize all living things, and together they define life. Of these, heredity is perhaps the key characteristic.

4. A large and rapidly growing human population threatens the stability of life on earth. The problem is most severe in the tropical areas of the world, where most people live. Attaining stable population levels, alleviating poverty and malnutrition, and managing natural resources in a sustainable way are the keys to the human future. Biology has important contributions to make in each of these areas.

5. Cancer results from mutations, or changes in genes that cause cells to grow in an uncontrolled fashion. Avoiding substances (mutagens) that cause mutations, such as those associated with smoking, is therefore the most important factor in avoiding cancer.

6. AIDS, or acquired immune deficiency syndrome, is a disease of the immune system that is transferred between individuals through bodily fluids such as blood or semen. AIDS is caused by a virus. It can be controlled by changes in human behavior and by the development of appropriate vaccines.

7. A scientist attempts to account for observations by constructing possible explanations, or hypotheses. A hypothesis that stands the test of time, often tested and never rejected, is called a theory.

8. Evolutionary change, achieved through the process of natural selection, has progressively shaped the character of life on earth.

REVIEW

1. List five fundamental properties shared by all organisms on earth.

2. _____ is the complex linear molecule responsible for heredity.

3. As you read this book, approximately how many people are alive in the world?

4. Currently, about what percentage of the world's population lives in tropical or subtropical regions?

5. According to current estimates, what percentage of the total world population will be living in non-industrial regions in the year 2020?

6. As of 1987, how many people in the United States are thought to have been exposed to the AIDS virus?

SELF-QUIZ

1. It is easy to think of ways that animals respond to stimuli, but plants also respond to stimuli. Which of the following are examples of plants responding to a stimulus? (Give three.)
   (a) Plants grow toward light.
   (b) The roots of plants grow generally down.
   (c) Cultivated plants sometimes escape from cultivation.
   (d) The trap of a Venus flytrap closes on a fly or other insect that has landed on it.
   (e) Plants die if they get no water.

2. A natural consequence of heredity is:
   (a) Development
   (b) Growth
   (c) Adaptation
   (d) Death
   (e) Movement

3. What percentage of the U.S. population expose themselves to powerful mutagens by smoking cigarettes?
   (a) 10%
   (b) 25%
   (c) 33%
   (d) 50%
   (e) 70%

4. How is the virus responsible for AIDS transmitted? (Choose two.)
   (a) sexually
   (b) insect bites
   (c) toilet seats
   (d) blood
   (e) coughs or sneezes

FOR FURTHER READING

ATTENBOROUGH, D.: *Life on Earth*. Boston: Little, Brown, & Co., 1979. The companion volume to a television series, a history of nature from the emergence of the first tiny, one-celled organisms to the emergence of upright human beings. Full of interesting evolutionary stories and exceptional photographs.

FIGURE 2-1 Charles Darwin, famous English naturalist, set forth on H.M.S. *Beagle* in 1831, at the age of 22. During the 5 years of this voyage that went around the world, but which mainly explored the coasts and coastal islands of South America, Darwin formulated and began to test his hypothesis of evolution by means of natural selection. This is a replica of the *Beagle* off the southern coast of South America.

BIOLOGY AS A SCIENCE — Chapter 2

Overview

One of the most exciting aspects of studying biology is that biology is an active science, dynamic and ever-changing. Like all sciences, biology does not give us absolute truths, but rather "best guesses," which we call hypotheses. As we learn more about nature we sometimes revise our thinking, leading to the construction of new hypotheses. When hypotheses have been tested over a long period of time and still have not been shown to be false, they are called theories. Perhaps the most important theory in biology is the theory of evolution by natural selection. First proposed by Charles Darwin more than a hundred years ago, this theory has been widely examined and is now almost universally accepted.

Biology, like literature, examines life in many different ways. In this text, for example, we will learn how a poet thinks, and why a butterfly's wings shine. We will learn that good biology depends on careful observation, just as good writing does. The difference between biology and literature is not so much in *what* they look at as in *how* the looking is done. Unlike literature, biology is a science and looks at nature in a special way, one very different from the gaze of a poet or novelist. In this chapter we will briefly consider what is unique about the way in which science views the world. We will ask what science is and examine the ways in which scientific inquiry has shaped the process of human thought itself (Figure 2-1).

THE NATURE OF SCIENCE

When we study geometry in high school, we formulate specific relationships about angles and lines by applying general principles called **theorems.** This sort of analysis of specific cases by the use of general principles is called **deductive reasoning**. It is the reasoning of mathematics and of philosophy, and the way in which the validity of general ideas is tested in all branches of knowledge. General principles are constructed and then used as the basis for examining specific cases to which they apply.

For the construction of scientific principles, a different logical process is used. Webster's Dictionary defines "science" as systematized knowledge derived from observation and experiment carried on in order to determine the principles underlying what is being studied. Said briefly, a scientist determines principles from observations. This mode of discovering general principles by careful examination of specific cases is called **inductive reasoning.** It first became popular in the 1600s in Europe, when Francis Bacon, Isaac Newton, and others began to use the results of particular experiments that they carried out to infer general principles about how the world operates. If you release an apple from your hand, what happens? The apple falls to the ground. This result may not stun you, but it is the sort of observation on which science is built. From a host of particular observations, each no more difficult to observe than the falling of an apple, Newton inferred a general principle: that all objects fall toward the center of the earth. What Newton did was to try to construct a mental model of how the world works, a family of working rules, general principles consistent with what he could see and learn. And, like Newton, that is what scientists do today—they are makers of models, and observations are the materials from which they build these models.

Testing Hypotheses

How do scientists learn which general principles are actually true, from among the many that might be true? They do this by attempting systematically to demonstrate

FIGURE 2-2

This diagram illustrates the way in which scientific investigations proceed. A number of potential explanations (hypotheses) are suggested in answer to a question; experiments are carried out in an attempt to eliminate one or more of these hypotheses; predictions are made based on these hypotheses; and further experiments are carried out to test these predictions. As a result of this process, the most likely hypothesis is selected. If it is validated by numerous experiments and stands the test of time, the hypothesis may eventually be considered a theory.

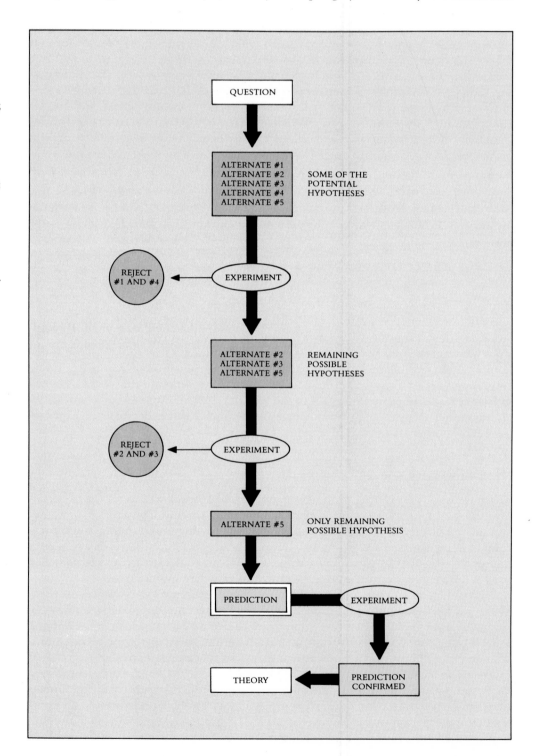

that certain proposals are *not* valid—not consistent with what they learn from experimental observation. A great deal of careful and creative thinking is necessary for the construction of hypotheses that account for the facts, observations, and experiments that are available concerning a particular area of science. Those hypotheses that have not yet been disproved are retained. They are useful because they fit the known facts, but they are always subject to future rejection if, in the light of new information, they are found to be incorrect.

A hypothesis is a proposition that might be true. We call the test of a hypothesis an **experiment** (Figure 2-2). An experiment evaluates alternative hypotheses. "There is no light in this dark room because the light switch is turned off" is a hypothesis; an alternative hypothesis is "There is no light in the room because the bulb is blown"; a third alternative might be "You are going blind."*An experimental test works by eliminating one of the hypotheses.* For example, we might test these alternatives by reversing the position of the light switch. Let us say that when we do this the light does not come on. The result of our experiment is thus to disprove the first of the hypotheses: something other than the setting of the light switch must be at fault. Note that a test such as this does not prove that any one alternative is true, but rather demonstrates that one of them is not. In this instance, the fact that the light does not come on does not establish that the switch was in fact in the "on" position (it might have been either "on" or "off" if the bulb was burnt out), but rather that the switch setting alone is not the sole reason for the darkness. A successful experiment is one in which one or more of the alternative hypotheses is demonstrated to be inconsistent with experimental or observational results and thus rejected. Scientific progress is made the same way a marble statue is, by chipping away unwanted bits.

> **The process of science involves the rejection of hypotheses that are not consistent with experimental results or observations. Hypotheses consistent with available data are provisionally accepted.**

As you proceed through this text, you will encounter a great deal of information, often coupled with explanations. These explanations are hypotheses that have stood the test of experiment. Many will continue to do so; others will be revised. Biology, like all healthy science, is in a constant state of ferment, with new ideas bubbling up and replacing old ones.

Theories

Hypotheses that stand the test of time, often tested and never rejected, are called **theories.** Thus one speaks of the general principle first noted by Newton as the theory of gravity. Theories are the solid ground of science, that of which we are most certain. There is no absolute truth in science, however, only varying degrees of uncertainty. The possibility always remains that future evidence will cause a theory to be revised. A scientist's acceptance of a theory is always provisional.

The word "theory" is thus used very differently by scientists than by the general public. To a scientist, a theory represents that of which he or she is most certain, to the general public, the word *theory* implies the *lack* of knowledge, or a guess. As you can imagine, confusion often results. In this text, the word "theory" will always be used in its scientific sense, in reference to a generally accepted scientific principle.

> **A theory is a hypothesis that is supported by a great deal of evidence.**

Some theories are so strongly supported that the likelihood of their being rejected in the future is very small. The theory of evolution by natural selection, for example, is so broadly supported by different lines of inquiry that most biologists accept it with as much certainty as they do the theory of gravity. We will examine this particular theory later in this chapter, as an example of how science is carried out. It is a particularly important theory to biologists, since the theory of evolution provides the conceptual framework that unifies biology as a science.

The Scientific Method

It used to be fashionable to speak of the "scientific method" as consisting of an orderly sequence of logical "either/or" steps, each step rejecting one of two mutually incompatible alternatives, as if trial-and-error testing would inevitably lead one through the maze of uncertainty that always slows scientific progress. If this were indeed true, a computer would make a good scientist—but science is not done this way. As British philosopher Karl Popper has pointed out, if you ask successful scientists how they do their work, you will discover that without exception they design their experiments with a pretty fair idea of how the experiments are going to come out—they use what Popper calls an "imaginative preconception" of what the truth might be. A hypothesis that a successful scientist tests is not just any hypothesis, but rather a "hunch" or educated guess in which the scientist integrates all that he or she knows and also allows his or her imagination full play, in an attempt to get a sense of what *might* be. It is because insight and imagination play such a large role in scientific progress that some scientists are so much better at science than others—for precisely the same reason that Beethoven and Mozart stand out above most other composers.

> **The scientific method is the experimental testing of a hypothesis formulated after the systematic, objective collection of data. Hypotheses are not usually formulated simply by rejecting a series of alternative possibilities, but rather often involve creative insight.**

HISTORY OF A BIOLOGICAL THEORY: DARWIN'S THEORY OF EVOLUTION

The idea of evolution—the notion that kinds of living things on earth change gradually from one form into another over the course of time—provides a good example of how an idea, an educated guess, is developed into a hypothesis, tested, and eventually accepted as a theory.

Charles Robert Darwin (1809–1882; Figure 2-3) was an English naturalist who, at the age of 50, after 30 years of study and observation, wrote one of the most famous and influential books of all time. The full title of this book, *On the Origin of Species by Means of Natural Selection, or the Preservation of Favoured Races in the Struggle for Life,* expressed both the nature of its subject and the way in which Darwin treated it. The book created a sensation when it was published in 1859, and the ideas expressed in it have played a central role in the development of human thought ever since.

In Darwin's time, it was traditional to believe that the various kinds of organisms and their individual structures resulted from direct actions of the Creator. Species were held to be specially created and unchangeable, or immutable, over the course of time. However, a number of philosophers before and in Darwin's time had presented the view that living things must have changed during the course of the history of life on earth. Darwin, though, was the first to present a coherent, logical explanation for this process—natural selection—and was the first to bring these ideas to wide public attention. His book, as you can see from its title, presented a conclusion that differed sharply from conventional wisdom. Although his theory did not challenge the existence of a Divine Creator, Darwin argued that this Creator did not simply create things and then leave them forever unchanged. Darwin's God expressed Himself through the operation of natural laws that produced continual change and improvement. These views put Darwin at odds with most people of his time, who believed in a literal interpretation of the Bible and accepted the idea of a fixed and constant world. His theory was a revolutionary one that troubled many of his contemporaries, and Darwin himself, deeply.

The story of Darwin and his theory begins in 1831, when he was 22 years old. On the recommendation of one of his professors at Cambridge University, he was selected to serve as naturalist on a 5-year voyage around the coasts of South America, the voyage of H.M.S. *Beagle* (1831 to 1836; Figure 2-4). During his long journey, Darwin had the chance to study plants and animals widely on continents and islands, and in far-flung seas. He was able to experience first-hand the biological richness of the tropical

FIGURE 2-3

Charles Darwin at the age of 29, 2 years after his return from the voyage of the *Beagle*. Darwin had just been married to his cousin Emma Wedgewood, and was hard at work studying the materials he had gathered on the voyage.

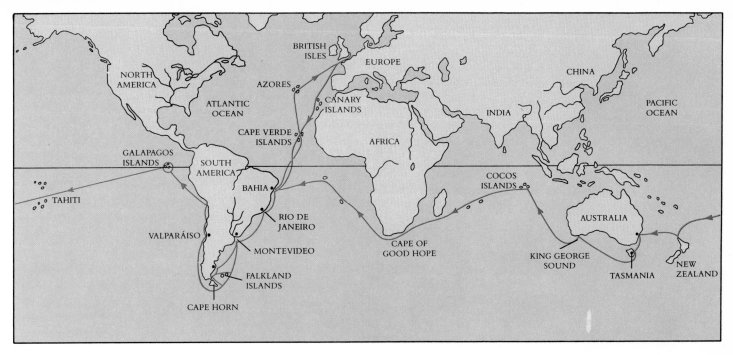

FIGURE 2-4

The 5-year voyage of H.M.S. *Beagle*. Most of the time was spent exploring the coasts and coastal islands of South America, such as the Galapagos Islands. Darwin's studies of the animals of the Galapagos Islands played a key role in his eventual development of the theory of evolution by means of natural selection.

forests, the extraordinary fossils of huge extinct mammals in Patagonia (Figure 2-5), and the remarkable series of related but distinct forms of life on the Galapagos Islands, off the west coast of South America. Such an opportunity clearly played an important role in the development of his thoughts about the nature of life on earth.

When Darwin returned from the voyage, at the age of 27, he began a long life of study and contemplation. During the next 10 years, he published important books on several different subjects, including the formation of oceanic islands from coral reefs and the geology of South America. He also devoted 8 years of study to barnacles, a group of marine animals, writing a four-volume work on their classification and natural history. In 1842, Darwin and his family moved a short distance out of London to a country home at Downe, in county of Kent. In these pleasant surroundings he lived, studied, and wrote for the next 40 years. During this period, he formulated for the first time a consistent theory of the process of evolution, an explanation that provided a mechanism for evolution. He presented his ideas in such convincing detail that they could logically be accepted as explaining the diversity of life on earth, the intricate adaptations of living things, and the ways in which they are related to one another.

FIGURE 2-5

Reconstruction from a fossil of a glyptodont, a 2-ton South American armadillo (a member of a distinctive group of mammals), compared with a modern armadillo, which averages 10 pounds. Finding the fossils of forms such as the glyptodonts and noting their similarity to living animals found in the same regions, Charles Darwin concluded that evolution had indeed taken place.

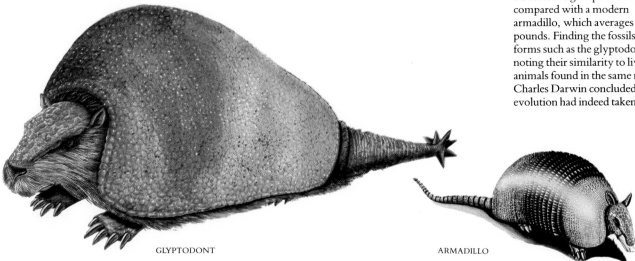

GLYPTODONT   ARMADILLO

B

FIGURE 2-6

A One of Darwin's finches, the medium ground finch.
B The blue-black grassquit, which is found in grasslands along the Pacific Coast from Mexico to Chile. This bird may be the ancestor of Darwin's finches.

TABLE 2-1   DARWIN'S EVIDENCE THAT EVOLUTION OCCURS

FOSSILS

1. Extinct species, such as the fossil armadillos of Figure 2-5, most closely resemble living ones in the same area, suggesting one had given rise to the other.
2. In rock strata (layers), progressive changes in characteristics can be seen in fossils from progressively older layers.

GEOGRAPHICAL DISTRIBUTION

3. Lands that have similar climates, such as Australia, South Africa, California, and Chile, have unrelated plants and animals, indicating that differences in environment are not creating the diversity directly.
4. The plants and animals of each continent are distinctive, although there is no reason why special creation should create this association: all South American rodents belong to a single group, structurally similar to the guinea pigs, for example, whereas most of the rodents found elsewhere belong to other groups.

OCEANIC ISLANDS

5. Although oceanic islands have few species, those they have are very often unique ("endemic") and show relatedness to one another, such as the tortoises of the Galapagos (see Figure 2-7). This suggests that the tortoises and other groups of endemic species formed after their ancestors reached the islands, and are therefore directly related to one another.
6. Species on oceanic islands show strong affinities to those on the nearest mainland. Thus the finches of the Galapagos (such as the one in Figure 2-6, *A*) closely resemble a finch seen on the western coast of South America (Figure 2-6, *B*). The Galapagos finches do *not* resemble birds of the Cape Verde Islands, islands in the Atlantic Ocean off Africa that are very similar to the Galapagos. Darwin visited the Cape Verde Islands and many other island groups personally and was able to make such comparisons on the basis of his own observations.

DARWIN'S EVIDENCE

So much information had accumulated by 1859 that the acceptance of the theory of evolution now seems, in retrospect, to have been inevitable. Darwin was able to arrive at a successful theory, where many others had failed, because he rejected supernatural explanations for the phenomena that he was studying. One of the obstacles that blocked the acceptance of any theory of evolution was the incorrect notion, still widely believed at that time, that the earth was only a few thousand years old. The discoveries of thick layers of rocks, evidences of extensive and prolonged erosion, and the increasing numbers of diverse and unfamiliar fossils found during Darwin's time were making this assertion seem less and less likely, however. For example, the great geologist Charles Lyell (1797-1895), whose works Darwin read eagerly while he was sailing on the *Beagle,* outlined for the first time the story of an ancient world of plants and animals in flux. In this world, species were constantly becoming extinct, while others were emerging. It was this world that Darwin sought to explain.

What Darwin Saw

When the *Beagle* set sail, Darwin was fully convinced that species were unchanging and immutable. Indeed, he later wrote that it was not until 2 or 3 years after his return that he began to consider seriously the possibility that they could change. Nevertheless, during his 5 years on the ship, Darwin observed a number of phenomena that were of central importance to him in reaching his ultimate conclusion (Table 2-1). For example, in rich beds of fossils in southern South America, he observed extinct armadillos that were related directly to the armadillos that still lived in the same area (see Figure 2-5). Why would there be living and fossil organisms, directly related to one another, in the same area, unless one had given rise to the other?

Repeatedly, Darwin saw that the characteristics of different species varied from place to place. These patterns suggested to him that organisms change gradually as

A

B

they migrate from one area into another. On the Galapagos Islands, off the coast of Ecuador, Darwin encountered giant land tortoises. Surprisingly, these tortoises were not all identical. Indeed, local residents and the sailors who captured the tortoises for food could tell which island a particular animal had come from just by looking at it (Figure 2-7). This pattern of variation suggested that all of the tortoises were related, but had changed slightly in appearance after they had become isolated on the different islands.

In a more general sense, Darwin was struck by the fact that on these relatively young volcanic islands there was a profusion of living things, but that these plants and animals resembled those of the nearby coast of South America. If each one of these plants and animals had been created independently and simply placed on the Galapagos Islands, why did they not resemble the plants and animals of faraway Africa, for example? Why did they resemble those of the adjacent South American coast instead?

The patterns of distribution and relationship of organisms that Darwin observed on the voyage of the *Beagle* ultimately made him certain that a process of evolution had been responsible for these patterns.

Darwin and Malthus

It is one thing to observe evolution, quite another to understand how it happens. Darwin's great achievement lies in his perception that evolution occurs because of natural selection. Of key importance to the development of Darwin's insight was his study of Thomas Malthus' *Essay on the Principles of Population.* In this book Malthus pointed out that populations of plants and animals, including human beings, tend to increase geometrically, while, in the case of people, our food supply increases only arithmetically. A **geometric progression** is one in which the elements progress by a constant factor, as 2, 6, 18, 54, and so forth; in this example, each number is three times the preceding one. An **arithmetic progression,** in contrast, is one in which the elements increase by a constant difference, as 2, 6, 10, 14, and so forth. In this progression, each number is 4 larger than the preceding one.

Virtually any kind of animal or plant, if it could reproduce unchecked, would cover the entire surface of the world within a surprisingly short period of time. In fact, this does not occur; instead, populations of species remain more or less constant year after year, because death intervenes and limits population numbers. Malthus' conclusion was very important in the development of Darwin's ideas. It provided the key ingredient that was necessary for Darwin in developing the hypothesis that evolution occurs by natural selection.

A key contribution to Darwin's thinking was Malthus' concept of geometric population growth. The fact that real populations do not expand at this rate implies that nature acts to limit population numbers.

FIGURE 2-7

Galapagos tortoises. Tortoises with large, domed shells **(A)** are found in relatively moist habitats, where the competition for food may not be as intense as it is in drier areas. The lower, saddleback-type shells **(B)** in which the front of the shell is bent up, exposing the head and part of the neck, are found among the tortoises of dry habitats. Taken together, differences of these kinds make it possible to identify the races of tortoises that inhabit the different islands of the Galapagos.

Natural Selection

Sparked by Malthus' ideas, Darwin saw that although every organism has the potential to produce more offspring than are able to survive, only a limited number of offspring actually survive and produce their own offspring. Combining this observation with what he had seen on the voyage of the *Beagle,* as well as with his own experiences in breeding domestic animals, Darwin made the key association: *those individuals that possess superior physical, behavioral, or other attributes are more likely to survive that those that are not so well endowed.* In surviving, these individuals have the opportunity to pass on their favorable characteristics to their offspring. Since these characteristics will increase in the population, the nature of the population as a whole will gradually change. Darwin called this process **natural selection,** and he referred to the driving force that he had identified as the *survival of the fittest*.

> **Natural selection is the increase in succeeding generations of the traits of those organisms that leave more offspring. Its operation depends on the traits being inherited. The nature of the population gradually changes as more and more individuals with those traits appear.**

Darwin was thoroughly familiar with variation in domesticated animals and began his *Origin of Species* with a detailed discussion of pigeon breeding. He knew that varieties of pigeons and other animals, such as dogs, could be selected to exhibit certain characteristics. Once this had been done, the animals would breed true for the characteristics that had been concentrated in them. Darwin had also observed that the differences that could be developed between domesticated races or breeds in this way were often greater than those that separated wild species. The breeds of domestic pigeon are much more different from one another in various ways than are all of the hundreds of wild species of pigeons found throughout the world. Such relationships suggested to Darwin that evolutionary change could occur very rapidly under the right circumstances.

PUBLICATION OF DARWIN'S THEORY

Darwin drafted the overall argument for evolution by natural selection in 1842 and continued to enlarge and refine it for many years. The stimulus that finally brought it into print was an essay that he received in 1858. A young English naturalist named Alfred Russel Wallace (1823-1913; Figure 2-8) sent the essay to Darwin from Malaysia: it concisely set forth the theory of evolution by means of natural selection! Like Darwin, Wallace had been greatly influenced in his development of this theory by reading Malthus' 1798 essay. After receiving Wallace's essay, Darwin arranged for a joint presentation of their ideas at a seminar in London, and proceeded to complete his own book, on which he had been working for so long, for publication in what he considered an abbreviated version.

Darwin's book appeared in November, 1859, and caused an immediate sensation (Figure 2-9). Many people were deeply disturbed by the idea that human beings were descended from the apes, for example. This idea was not discussed by Darwin in his book, but it followed directly from the principles that he did outline. It had long been accepted that humans closely resembled the apes in all of their characteristics, but the possibility that there might be a direct evolutionary relationship between them was unacceptable to many people. Darwin's arguments for the theory of evolution by natural selection were so compelling, however, that his views were almost completely accepted within the intellectual community of Britain after the 1860s.

EVOLUTION AFTER DARWIN: TESTING THE THEORY

Darwin did more than propose a mechanism that accounts for how evolution has generated the diversity of life on earth. He also assembled masses of facts, otherwise seemingly without logic, that began to make sense when they were viewed in the light of his

FIGURE 2-8

Alfred Russel Wallace in 1902.

FIGURE 2-9

In his time, Darwin was often portrayed unsympathetically, as in this drawing from an 1874 publication.

FIGURE 2-10

Fossil of an early bird, *Archaeopteryx*. A well-preserved fossil of this bird, about 150 million years old, was discovered within 2 years of the publication of *The Origin of Species*. *Archaeopteryx* provides an indication of the evolutionary relationship that exists between birds and reptiles.

theory. After publication of his book, other biologists continued this process, and it soon became evident that the theory of evolution was supported by a wide variety of biological information gathered by many investigators. These observations included the following three points: 1) the fact that members of different biological groups often share common features; 2) the ways in which embryos (organisms developing from fertilized eggs) are more similar at earlier embryo stages and diverge through later stages; and 3) the increasing complexity that is observed to develop in the fossil record through time. In the century since Darwin, evolution has become the main unifying theme of the biological sciences. It provides one of the most important insights that human beings have achieved into their own nature and that of the earth on which they have evolved.

More than a century has now elapsed since Charles Darwin's death in 1882. During this period, the evidence supporting his theory has grown progressively stronger. There have also been many significant advances in our understanding of how evolution works. These advances have not altered the basic structure of Darwin's theory, but they have taught us a great deal more about the mechanisms by which evolution occurs.

The Fossil Record

Darwin explicitly postulated that the fossil record should yield intermediate links (Figure 2-10) between the great groups of organisms—for example, between fishes and the amphibians thought to have arisen from them, and between reptiles and birds. The fossil record is now known to a degree that would have been unthinkable in the nineteenth century. Recent discoveries of microscopic fossils have extended the known history of life on earth back to more than 3.5 billion years. The discovery of other fossils has shed light on the ways in which organisms have evolved from the simple to the complex over the course of this enormous time span. For vertebrate animals—those with backbones—especially, the fossil record is rich and exhibits a graded series of changes in form, with the evolutionary parade visible for all to see.

The Age of the Earth

In Darwin's day, some physicists argued that the earth was only a few thousand years old. This bothered Darwin, as the evolution of all living things from some single orig-

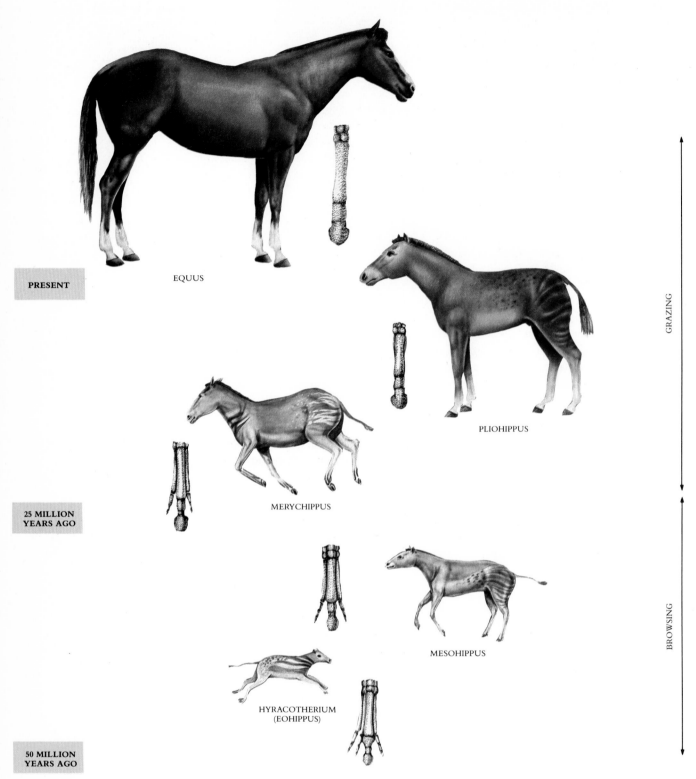

PRESENT

EQUUS

PLIOHIPPUS

MERYCHIPPUS

25 MILLION
YEARS AGO

MESOHIPPUS

HYRACOTHERIUM
(EOHIPPUS)

50 MILLION
YEARS AGO

GRAZING

BROWSING

FIGURE 2-11

Reconstructions of fossils, illustrating the evolution of horses. Animals called **hyracotheres,** which included the earliest member of the evolutionary line that is illustrated, *Hyracotherium,* gave rise to several groups of mammals, including tapirs and rhinoceroses, in addition to the horses. Horses provide an excellent example of the way in which abundant fossil evidence has allowed the reconstruction of the evolution of a particular vertebrate group.

inal ancestor would have required a great deal more time. Using evidence obtained by studying rates of radioactive decay, we now know the earth to have been formed some 4.5 billion years ago.

The Mechanism of Heredity

It was in the area of heredity that Darwin received some of his sharpest criticism. Since at that time no one had any concept of genes, of how heredity works, it was not possible for Darwin to explain completely how evolution occurs. Theories of heredity current in Darwin's day seemed to rule out the possibility of genetic variation in nature, a critical requirement of Darwin's theory. Genetics was established as a science only at the start of the twentieth century, 40 years after the publication of Darwin's *Origin*. When the laws of inheritance became understood, the problem with Darwin's theory vanished, since the laws of inheritance (discussed in Chapter 11) account in a neat and orderly way for the production of new variations in nature required by Darwin's theory.

Comparative Anatomy

Comparative studies of animals have provided strong evidence for Darwin's theory. As vertebrates have evolved, for example, the same bones sometimes get put to different uses—and yet they can still be seen, betraying their evolutionary past (Figure 2-11). Thus the forelimbs in Figure 2-12 are all constructed from the same basic array of bones, modified in one way in the wings of birds, in another way in the fins of whales, and in yet another way in the legs of horses. The bones are said to be **homologous** or to exhibit **homology** in the different vertebrates—that is, having the same evolutionary origin, although now differing in structure and function.

Molecular Biology

Biochemical tools have become of major importance in our efforts to reach a better understanding of how evolution has occurred. Within the last few years, for example, evolutionists have begun to "read" genes, much as you read this page, by recognizing in order the "letters" of the long DNA molecules that store genetic information. When the DNA sequences of different groups of animals or plants are compared, the degree of relationship between the groups can be specified more precisely than by any other means. In many cases, detailed **phylogenies** (family trees) can be constructed. When study of the DNA that encodes the structure of two different molecules leads to the same phylogeny, this provides strong evidence that the phylogeny is telling the evolutionary history of the group. Indeed, it is often possible to measure the rates at which evolution is occurring in the different groups.

> **In the century since Darwin, a large body of evidence contributed by many branches of science has supported his view that evolution has occurred and that it has taken place by means of the mechanism of natural selection that he proposed.**

THE LIMITS OF SCIENCE: SCIENTIFIC CREATIONISM

In the century since he proposed it, Darwin's theory of evolution has become nearly universally accepted by biologists as the best available explanation of biological diversity. Its predictions have been supported by the experiments and observations of generations of scientists. The theory of evolution is as well accepted as the theory of gravity, and its operation is no more doubted than predictions that a dropped apple will fall or that the sun will rise tomorrow morning.

Darwin's theory is not, however, the only possible opinion. The argument advanced in Darwin's day, that the Bible provides the correct explanation of biological diversity, is still widely accepted by many people today, people who prefer a religious to a scientific explanation. In our time, as in Darwin's time, the scientific perspective is only one of many world views. It is important in this regard to understand the limits of

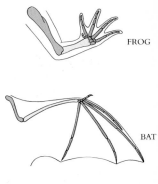

MAN

FROG

BAT

PORPOISE

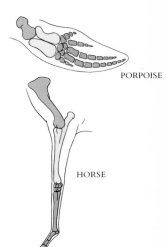

HORSE

FIGURE 2-12

Homologies between the forelimbs of four mammals and a frog, showing the ways in which the proportions of the bones have changed in relation to the particular way of life of the organism.

science. What science provides is a coherent means of organizing observations, of making predictions about how the world is going to behave. It is not a substitute for religion, which addresses a different arena of human concerns: questions of ethics and of ultimate causes. Religion and science do not preclude one another, but rather are regarded by many as complementary ways of viewing the world.

The clear distinction between science and religion sometimes gets muddled. Thus a number of individuals, mainly in the United States and starting largely in the 1970s, have put forward a view they title *Scientific Creationism,* which holds that the biblical account of the origin of the earth is literally true, that the earth is much younger than most scientists believe, and that all species of organisms were individually created just as they exist today. Scientific creationists are arguing in the courts that their views should be taught alongside evolution in classrooms. If both evolution and scientific creationism provide scientific explanations of biological diversity, they argue, then teachers have an obligation to present *both* views, taught side-by-side, so that students can choose knowledgeably between them.

This does not seem to be a bad argument, if you accept the premise, which is that the view of the "scientific creationists" is indeed scientific. The confusion is not in the beliefs of the scientific creationists, which are religious beliefs that many people hold, but rather in their labeling of these beliefs as "scientific." There is no scientific evidence to support the hypothesis that the earth is much younger, and none that indicates that every species of organism was created separately. These conclusions can be reached only on the basis of arbitrary faith; they are untestable, and, as such, they lie outside of the realm of science.

Science, as represented by the observations of scientists, has come to different conclusions. There are virtually no differences of opinion among modern biologists, and indeed nearly all scientists, on the following major points: (1) the earth has had a history of approximately 4.5 billion years; (2) organisms have inhabited it for the greater part of that time; and (3) all living things, including human beings, have arisen from earlier, simpler living things. The antiquity of the earth and the role that the process of evolution played in the production of all organisms, living and extinct, are accepted by the overwhelming majority of scientists today.

Besides the fact that it contradicts the considered judgment of almost all scientists, there is an even more fundamental problem associated with labeling "scientific creationism" as science: scientific creationism implicitly denies the intellectual basis of science itself, the reasoning on which the operation of science depends. Science insists on acceptance of the most predictive explanation of biological diversity, which is evolution; scientific creationism says "Yes, but God just made it look that way." Perhaps so, but this is simply substituting a religious argument for a scientific one. Science consists of inferring principle from observation, and when faith is substituted for observation, the conclusion is not science. It is in just this sense that scientific creationism is not science. On the basis that scientific creationism is in fact religion thinly disguised, the Supreme Court in June 1987 ruled that states could not require their schools to teach creationism alongside evolution, as if they were two valid alternatives.

> **Scientific creationism should not be labeled science for two reasons:**
> 1. **It is not supported by any scientific observations.**
> 2. **It does not infer its principles from observation, as does all science.**

"Scientific creationism" implicitly denies the whole intellectual basis for the set of facts that human beings have assembled over the centuries about the nature of life on earth. Certainly there is ample controversy among serious students as to the details of how evolution has occurred, just as there is controversy in every active scientific field. There is not controversy, however, about the facts presented in this section or about Darwin's basic finding that natural selection has played, and is continuing to play, the central role in the process of evolution.

The future of the human race depends largely on our collective ability to deal with the science of biology and all the phenomena that it includes. We need the information that we have gained to deal with the problems, challenges, and uncertainties of the world in an appropriate way. We cannot afford to discard the advantages that this knowledge gives us because some individuals wish to do so as an act of what they construe as religious faith. Instead, we must use all of the knowledge that we are able to gain for our common benefit. With the help of this knowledge, we can come to understand ourselves and our potentialities better. In no way should such rational behavior be taken as denial of the existence of a supreme being; it should rather be considered by those who do have religious faith as a sign that they are using their God-given gifts to reason and to understand.

HOW THIS TEXT IS ORGANIZED TO TEACH YOU BIOLOGY

In the century since Darwin's publication of the *Origin of Species,* biology has exploded as a science, presenting today's student with a wealth of information and theory. There are all sorts of ways in which a beginner can be introduced to biology and can see the things that are available to learn. In an introductory biology course you encounter a wealth of experiment and observation that you *could* learn, and from this you must select a small body of information that you *will* learn. Your target is the basic body of principles that unite biology as a science.

Levels of Organization

A very good way to begin your examination of basic biological principles is to focus on complexity, one of the essential features of life discussed in Chapter 1. Arranging the wealth of information about biology in terms of the level of complexity leads to what has been called a "levels of organization" approach to the field.

At the most basic, or **molecular,** level, atoms—the smallest particles into which a chemical element can be divided and retain the properties characteristic of that element—are organized into molecules, and the molecules are grouped together in complex ways.

At the **subcellular** level, molecules are organized into the membranes and small compartments or structures called organelles that are the building blocks of cells. Cell membranes are composed of soaplike molecules in which protein molecules (the main constituent of your muscles) float like icebergs.

At the **cellular** level, spaces enclosed by these membranes called *cells* exist both independently and as components of cell groupings or aggregations known as tissues. Each of the individual cells in tissues includes numerous subcellular organelles.

At the **organismal** level, independent cells and many-celled organisms reproduce and evolve. A cow and the grass on which it grazes, are individual organisms. Each cow's body contains trillions of cells acting together as an organized unit.

At the **population** level, individuals of the same type (a species) interact, forming an evolutionary unit. The large herds of wildebeests that migrate across the African veldt are part of a single, enormous population.

At the **community** level, different populations of organisms interact to form communities. For example, a kelp community consists of photosynthetic organisms, the kelp (a kind of large alga), together with fishes and many other kinds of animals that feed on them. The kelp beds provide shelter, nesting sites, and places for these animals to hide from their predators, in addition to food.

At the **global** level, there are different groups of organisms in different climates; these groups have a characteristic appearance and are distributed over a wide area on land. They are known as **biomes.** The dunes of the Namib desert of South-West Africa are part of the desert biome, dry and harsh but often surprisingly rich with life.

TABLE 2-2   LEVELS OF BIOLOGICAL ORGANIZATION

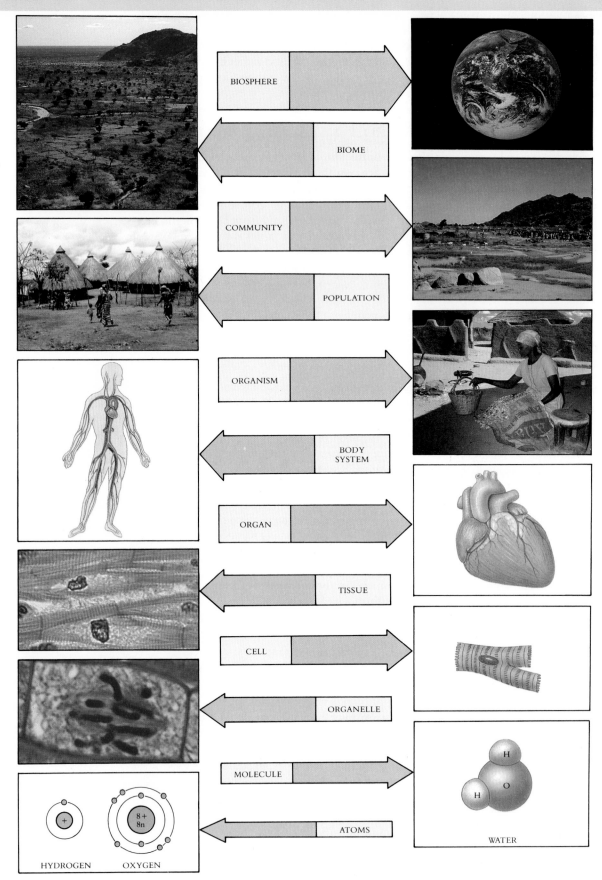

BIOSPHERE

BIOME

COMMUNITY

POPULATION

ORGANISM

BODY SYSTEM

ORGAN

TISSUE

CELL

ORGANELLE

MOLECULE

ATOMS

WATER

HYDROGEN   OXYGEN

Part I—Basic Principles

The first half of this text is devoted to a description of the basic principles of biology. It uses a levels-of-organization framework (Table 2-2) to introduce different principles, each at the level it is most easily understood. At the molecular, subcellular, and cellular levels of organization, you will be introduced to the principles of **cell biology,** learning how cells are constructed and how they grow, divide, and communicate. At the organismal level, you will learn the principles of **genetics,** which deals with the way in which the traits of an individual are transmitted from one generation to the next. At the population level you will study **evolution,** a field that is concerned with the nature of population changes from one generation to the next as a result of selection, and the way in which this has led to the biological diversity we see around us. Finally, at the community and global levels you will study **ecology,** which deals with how organisms interact with their environments and with one another to produce the complex communities characteristic of life on earth.

Part II—Biological Diversity

The second half of this book is devoted to an examination of organisms, the products of evolution. The diversity of living organisms is incredible. It is estimated that at least 5 million different kinds of plants, animals, and microorganisms exist. This teeming diversity is divided into **five kingdoms.**

Among these five kingdoms, the **bacteria** (kingdom Monera) are simplest in structure; their cells, which are termed **prokaryotic,** have no membrane-bound compartments within them. For billions of years, bacteria were the only living things on earth; **eukaryotes**—organisms with more complicated cells—evolved from them about 1.5 billion years ago. The first eukaryotes were organisms that consisted of a single cell, but many different groups evolved in which the individuals consisted of more than a single cell. Such organisms are called **multicellular** organisms.

Most single-celled eukaryotes are placed in the kingdom Protista—the **protists**—along with some of their multicellular descendants. Among these are the various lines of protozoa and algae. Three evolutionary lines that consist of almost exclusively multicellular organisms, however, are so large and distinctive in their structure and in their way of life that they are considered separate kingdoms: Animalia, the **animals;** Plantae, the **plants;** and Fungi, the **fungi.** Although they each have many unique structures and other features, these kingdoms can perhaps be characterized most easily by their ways of obtaining food: (1) animals **ingest** their food, taking it inside their bodies and digesting it there; (2) plants **manufacture** their food (as do certain protists called algae, and a few groups of bacteria), capturing the sun's energy by using a special kind of molecule called chlorophyll; and (3) fungi **absorb** their food, secreting special chemical substances (enzymes) from their bodies into their environment, digesting food there, and then taking it back into their bodies. There are about 250,000 kinds of plants, perhaps a similar number of fungi, and at least 3 million kinds of animals.

In this book, we shall concentrate primarily on animals and plants, which are the two kingdoms most familiar to us. We shall take a particularly detailed look at the vertebrate body and how it functions, stressing the human body, since this is information that is of interest and importance to all students. Because of their importance in feeding the world and providing the energy to power all biological processes, we shall also consider plants in some detail.

As you proceed through this course, what you learn at one stage will give you tools to tackle the next. In the following chapter we shall examine some simple chemistry. You are not subjected to chemistry first to torture you, but rather to make what comes later easier to comprehend. To understand lions and tigers and bears, you first need to know the basic chemistry that makes them tick, for they are chemical machines, as are you.

SUMMARY

1. Science is the determination of general principles from observation and experiment.

2. From among all possible hypothetical principles, or hypotheses, scientists select the best ones by attempting to reject some of the alternatives as inconsistent with observation.

3. Hypotheses that are supported by a large body of evidence are called theories. Unlike the everyday use of the word, the term "theory" in science refers to what we are most sure about.

4. Because even a theory is accepted only provisionally, there are no sure truths in science, no propositions that are not subject to change.

5. One of the most central theories of biology is Darwin's theory that evolution occurs by natural selection. Proposed over a hundred years ago, this theory has stood up well to a century of testing and questioning.

6. On an extended voyage, particularly while studying the animals and plants of oceanic islands, Darwin accumulated a wealth of evidence that evolution has occurred.

7. Sparked by Malthus' ideas, Darwin proposed that evolution occurs as a result of natural selection. Some individuals have heritable traits that let them produce more offspring in a given kind of environment than other individuals that lack these traits. As a result of this process, the traits will increase in frequency through time.

8. A wealth of evidence since Darwin's time has supported his twin proposals: that evolution occurs and that its agent is natural selection. Together these two hypotheses are now usually referred to as Darwin's theory of evolution.

9. It is important to distinguish scientific theories such as Darwin's from nonscientific ones. A nonscientific theory is one whose principles are not derived from observation or supported by observation. Scientific creationism is an example of a nonscientific theory.

10. Biology may be considered at many levels of organization. This text treats basic principles first and then considers biological diversity.

11. Organisms are classified into five kingdoms. Members of Monera—the bacteria—are prokaryotic; organisms in the other four kingdoms are eukaryotic. Protista consists mainly of groups that are generally unicellular and have given rise to three primarily multicellular kingdoms, the plants (kingdom Plantae), the animals (kingdom Animalia), and the fungi (kingdom Fungi).

REVIEW

1. A scientific _____ is a hypothesis that is supported by a great quantity of data.

2. Because _____ is not supported by any scientific observations and does not infer its principles from observation, as does all science, it cannot be called science.

3. Current evidence indicates that the earth is approximately _____ years old.

4. The arm of a man and the fin of a porpoise are considered to be _____ structures, made of the same bones, which have been modified in size and shape for different uses.

5. How many of the five kingdoms of living things contain multicellular organisms?

SELF-QUIZ

1. Hypotheses that are consistent with the available data are _____ accepted by scientists.
   (a) faithfully
   (b) provisionally
   (c) never
   (d) always
   (e) sometimes

2. Your dog suddenly drops to the ground in front of you. Five hypotheses suggest themselves to you: (1) your dog wants to play; (2) your dog has gone to sleep; (3) your dog is playing a trick on you; (4) your dog is chasing a bug; or (5) your dog just died. To find out what is going on, you try to eliminate one or more of these hypotheses by conducting an experiment. Which of the following experiments would eliminate at least one hypothesis, whatever the outcome?
   (a) Ask the dog why he dropped to the ground.
   (b) Push the the dog and see if it responds.
   (c) Look and see if you can detect a bug.
   (d) Wait and see if the dog moves.
   (e) Feed him.

3. Which of the following is *not* one of Darwin's pieces of evidence that evolution occurs?
   (a) The earth was created in 4004 BC.
   (b) Extinct species most closely resemble living species in the same area.
   (c) Species on oceanic islands show strong similarities to species on the nearest mainland.
   (d) The plants and animals of each continent are distinctive.
   (e) Progressive changes in characteristics of plants and animals can be seen in successive rock layers.

4. Darwin was greatly influenced by an essay written by _____, which pointed out that population growth is geometric while increase in food is arithmetic.
   (a) Emma Wedgewood
   (b) Alfred R. Wallace
   (c) James Usher
   (d) Thomas Malthus
   (e) Charles Lyell

5. Darwin believed that the major driving force in evolution was
   (a) natural selection
   (b) scientific creation
   (c) homologies
   (d) uniformitarianism
   (e) molecular biology

PROBLEM

1. Malthus' *Essay on the Principles of Population* pointed out that populations of both plants and animals tend to increase geometrically. What is the next number in this geometric progression: 2, 8, 32, 128, _____?

THOUGHT QUESTIONS

1. It is sometimes argued that Darwin's reasoning is circular, that he first defined the "fittest" individuals as those that leave the most offspring and then turned around and said that the fittest survive preferentially (that is, leave the most offspring). Do you think this is a fair criticism of Darwin's theory as described in this chapter?

2. On the Galapagos Islands Darwin saw a variety of different kinds of finch, but few other small birds. Imagine that you are visiting another group of islands about as far away from the South American mainland as the Galapagos, but upwind, so that no birds travel between the two island groups. Do you expect that on your visit to this second island group you will find a variety of finches? Comment on how your knowledge of the birds of the Galapagos, as discussed in this text, aids you in predicting what you will find on the second island group, if indeed it does.

3. Imagine that you sat on the Supreme Court in the fall of 1986, hearing a case in which it was argued that creation science should be taught in public schools alongside evolution as a legitimate alternative scientific explanation of biological diversity. What is the best case that lawyers might have made for and against this proposition? A decision of the Supreme Court was announced in June, 1987. How would you have voted, and why?

FOR FURTHER READING

DARWIN, C.R.: *The Origin of Species by Means of Natural Selection, or the Preservation of Favoured Races in the Struggle for Life,* Cambridge University Press, New York, 1975 reprint. One of the most important scientific books of all time, Darwin's long essay is still comprehensible and interesting to modern readers.

DARWIN, C.R.: *The Voyage of the Beagle,* Natural History Press, Garden City, N.Y., 1962 reprint. Darwin's own account of his observations and adventures during the famous 5-year voyage he took in his twenties.

FUTUYMA, D.: *Science on Trial: The Case for Evolution,* Pantheon Books, New York, 1983. An excellent exposition of the basic reasons that the creationist argument is flawed by serious errors.

GILKEY, L.: *Creationism on Trial: Evolution and God at Little Rock,* Winston Press, Minneapolis, 1985. Excellent book, written by a theologian, outlining the case for creationism as argued in the courts at Little Rock.

GOULD, S.: "Darwinism Defined: The Difference Between Fact and Theory," *Discover,* January 1987, pages 64–70. A clear account of what biologists do and do not mean when they refer to the theory of evolution.

IRVINE, W.: *Apes, Angels, and Victorians,* McGraw-Hill Book Co., New York, 1954. The story of Darwin and the early years of the theory of evolution; beautifully written.

MOORE, J.A.: "Science as a Way of Knowing—Evolutionary Biology," *American Zoologist,* vol. 23, pages 1–68, 1983. An outstanding exposition of the whole field of evolution.

CELL BIOLOGY

Part Two

FIGURE 3-1   The universe is thought to have been formed about 14 billion years ago. Every dot in this galaxy is a star like our sun—and the universe contains over a billion galaxies.

THE CHEMISTRY OF LIFE Chapter 3

Overview

Organisms are made up of molecules, which are collections of atoms bound to one another. The molecule most important to the evolution of life is water. Living organisms are built by assembling large molecules, much as a house is assembled from prefabricated building blocks. Some of these building blocks are long polymers, chains of similar units joined in a row. Among the polymers that make up the bodies of organisms are starches (used to store chemical energy), proteins (molecules that speed up specific chemical reactions), and nucleic acids (molecules in which hereditary information is stored).

Biology is the study of life, of ants and polar bears and roses. To start our study of biology, we begin with a brief look into chemistry—not a lot, just a taste. Organisms are chemical machines; to understand them, we must start by considering chemistry. Organisms are composed of molecules, which are collections of smaller units called atoms, bound to one another. All atoms now in the universe are thought to have been formed long ago, as the universe itself evolved (Figure 3-1). Every carbon atom in your body was created in a star.

ATOMS

All matter is composed of small particles called **atoms.** Atoms are very small and hard to study, and for a long time it was difficult for scientists to figure out their structure. It was only early in this century that experiments were carried out that suggested the first vague outlines of what an atom is like. We now know a great deal about the complexities of atomic structure, but the simple view proposed in 1913 by the Danish physicist Niels Bohr provides a good starting point. Bohr proposed that every atom possesses an orbiting cloud of tiny subatomic particles called **electrons** whizzing around the core like planets of a miniature solar system. At the center of each atom is a small, very dense nucleus formed of two other kinds of subatomic particles, **protons** and **neutrons.**

Within the nucleus, the cluster of protons and neutrons is held together by subatomic forces that work only over very short distances. Each proton carries a positive (+) charge. The number of charged protons (atomic number) determines the chemical character of the atom, because it dictates the number of electrons orbiting the nucleus and available for chemical activity: there is one electron for each proton. Neutrons are similar to protons in mass, but as their name implies, they are neutral and possess no charge. The atomic mass of an atom consists of the combined weight of all of its protons and neutrons. Atoms that occur naturally on earth contain from 1 to 92 protons

FIGURE 3-2

A   The smallest atom is hydrogen (atomic mass, 1), whose nucleus consists of a single proton.

B   Hydrogen also has two naturally occurring isotopic forms, which possess neutrons as well as the single proton in the nucleus: deuterium (one neutron) and tritium (two neutrons).

C   A nucleus that contains two protons is not an isotope of hydrogen, but rather the next element in the periodic table, helium. The largest naturally occurring atom is uranium (atomic mass, 238), whose nucleus contains 92 protons and 146 neutrons.

HELIUM (He)

DEUTERIUM (^{2}H)

HYDROGEN (H)

TRITIUM (^{3}H)

NUCLEUS OF URANIUM 238 (^{238}U)

A NUCLEUS OF HYDROGEN   **B** ISOTOPES OF HYDROGEN   **C** SIZE RANGE OF ATOMS

and up to 146 neutrons (Figure 3-2). Atoms that have the same number of protons but different numbers of neutrons are called **isotopes.** Isotopes of an atom differ in atomic mass but have similar chemical properties.

The positive charges in the nucleus of an atom are counterbalanced by negatively ($-$) charged electrons orbiting the atomic nucleus at various distances. The negative charge of one electron exactly balances the positive charge of one proton. Thus atoms with the same number of protons and electrons have no net charge. An atom in which the number of protons in the nucleus is the same as the number of orbiting electrons is known as a neutral atom.

Electrons have very little mass (only $\frac{1}{1840}$ of the mass of a proton). Of all the mass contributing to your weight, the portion contributed by electrons is less than the mass of your eyelashes. Electrons stay in their orbits because they are attracted to the positive charge of the nucleus. This attraction is sometimes overcome by other forces, and one or more electrons fly off, lost to the atom. Sometimes atoms gain additional electrons. Atoms in which the number of electrons does not equal the number of protons are known as **ions,** and ions do carry an electrical charge. For example, an atom of sodium (Na) that has lost an electron becomes a positively charged sodium ion (Na^+), because the positive charge of one of the protons is not balanced by the negative charge of an electron.

> **An atom is a core (nucleus) of protons and neutrons surrounded by a cloud of electrons. The number and distribution of its electrons largely determine the chemical properties of an atom.**

ELECTRON ORBITALS

The key to the chemical behavior of atoms lies in the arrangement of the electrons that spin around them. It is convenient to visualize individual electrons as spinning in well-defined circular orbits around a central nucleus, the way Earth, Mars, and Venus circle the sun. Such a simple picture is not realistic, however. Physicists have learned that it is impossible to locate precisely the position of any individual electron at any given time. In fact, theories indicate that at a given instant a particular electron can be located anywhere from close to the nucleus to infinitely far away from it.

A particular electron is not equally likely to be located at all positions, however. Some locations are much more probable than others. For this reason, it is possible to say where an electron is *most likely* to be. The volume of space around a nucleus where

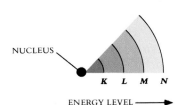

NUCLEUS

K L M N

ENERGY LEVEL ——▶

2+
2n

K

HELIUM

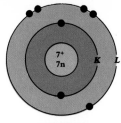

7+
7n

K L

NITROGEN

ELEMENT	ATOMIC NUMBER	NUMBER OF ELECTRONS IN EACH ENERGY LEVEL		
		FIRST (K)	SECOND (L)	THIRD (M)
Hydrogen (H)	1	1	—	—
Helium (He)	2	2	—	—
Carbon (C)	6	2	4	—
Nitrogen (N)	7	2	5	—
Oxygen (O)	8	2	6	—
Neon (Ne)	10	2	8	—
Sodium (Na)	11	2	8	1
Phosphorus (P)	15	2	8	5
Sulfur (S)	16	2	8	6
Chlorine (Cl)	17	2	8	7

an electron is most likely to be found is called the **orbital** of that electron. Atoms can have many electron orbitals. Some are simple spheres enclosing the nucleus like a wrapper. Others resemble dumbbells and other complex shapes.

In a schematic drawing of an atom (Figure 3-3), the nucleus is represented as a small circle with the number of protons and neutrons indicated. The orbitals are depicted as concentric rings. In helium, for example, there is a single orbital containing two electrons. A given orbital may contain no more than two electrons. Such schematics are only diagrams; the actual orbitals have complex, three-dimensional shapes.

Most atoms have more than one orbital. When electrons from two different orbitals are the same distance from the nucleus, they are placed on the same ring by convention. When electrons in two orbitals are different distances from the nucleus, they are placed in separate concentric rings. These rings are called **energy levels,** or **shells.** The farther an electron is from the nucleus, the more energy it has, as if it were spinning faster.

MOLECULES

Much of the earth's core is thought to consist of atoms of iron, nickel, and other heavy elements. An **element** is defined as a substance that cannot be separated into different substances by ordinary chemical methods. The composition of the earth's crust is quite different from the core, the crust being made up primarily of lighter elements (see Table 3-1). Thus, by weight, 74.3% of the earth's crust (its land, oceans, and atmosphere) consists of oxygen and silicon. Most of these elements are combined stable associations of atoms called **molecules.**

The combining activity of atoms that form molecules reflects three chemical forces: (1) the tendency of electrons to occur in pairs, (2) the tendency of atoms to balance positive and negative charges, and (3) the tendency of the outer shell, or energy level, of electrons to be full. This third chemical force is often called the **octet rule.** "Octet" refers to a group of eight objects, and the outer electron shell of most atoms contains a maximum of eight electrons. While the octet rule does not apply to all atoms, it does apply to all biologically important ones.

Only those elements that have equal numbers of protons and electrons (and hence no net charge), no unpaired electrons, and full outer-electron energy levels can exist as free atoms. This is because only elements that satisfy these requirements do not experience chemical forces that lead them to combine. Atoms with this set of characteristics are called **noble elements** because they do not associate with other elements. The noble elements are all gases, and most are rare in the earth's crust. All other elements are usually found on earth in molecular combinations that satisfy the three fundamental chemical tendencies just listed.

NATURE OF THE CHEMICAL BOND

An atom that does not carry an electrical charge and that is not a noble element can satisfy the octet rule in one of three ways:

1. It can gain one or two electrons from another atom.
2. It can lose one or two electrons to another atom.
3. It can share one or more electron pairs with another atom.

A **chemical bond** is a force holding two atoms together. The force can result from the attraction of opposite charges, called an ionic bond, or from the sharing of one or more pairs of electrons, a covalent bond. Other, weaker kinds of bonds also occur.

IONIC BONDS

Ionic bonds form when atoms are attracted to one another by opposite electrical charges. Common table salt, sodium chloride (NaCl), is a lattice of ions in which atoms are held together by ionic bonds. Sodium atoms (Na) have 11 electrons (Figure 3-4). Two of these are in the inner energy level, eight are at the next level, and one is at the outer energy level. Because of this distribution of electrons, two powerful chemical tendencies result:

1. The outer electron is unpaired ("free") and has a strong tendency to form a pair; and
2. The octet rule is not satisfied.

FIGURE 3-4

A The formation of ionic bonds in sodium chloride. When a sodium atom donates an electron to a chlorine atom, the sodium atom, lacking that electron, becomes a positively charged sodium ion; the chlorine atom, having gained an extra electron, becomes a negatively charged chloride ion. These positive and negative ions cluster so that each charge is surrounded by ions of opposite charge.
B When water evaporates from solutions containing high concentrations of sodium chloride, a highly regular lattice of alternating Na^+ and Cl^- ions forms—a crystal. You are familiar with these crystals as table salt.
C Viewed as they form, such sodium chloride crystals can be very beautiful.

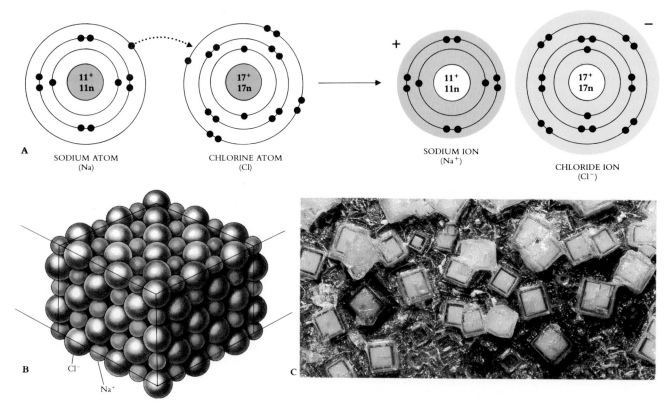

| SODIUM ATOM (Na) | CHLORINE ATOM (Cl) | SODIUM ION (Na^+) | CHLORIDE ION (Cl^-) |

B Cl^-
    Na^+

A stable configuration is achieved, however, if the outer electron is lost. The loss of this electron results in the formation of a positively charged sodium ion (Na^+).

The chlorine atom faces a similar dilemma. It has 17 electrons: 2 at the inner energy level, 8 at the next energy level, and 7 at the outer energy level. The outer energy level of the chlorine atom has an unpaired electron, and the octet rule is not satisfied. The addition of an electron to the outer level, however, satisfies both requirements and causes the formation of a negatively charged chloride ion (Cl^-).

When placed together, metallic sodium and gaseous chlorine react swiftly and explosively, with the sodium atoms donating electrons to chlorine atoms. The result is the production of Na^+ and Cl^- ions. Because opposite charges attract, electrostatic neutrality (the second chemical tendency) can be achieved by the association of these ions with each other. The ions aggregate or come together to form a crystal matrix, which has a precise geometry. Such aggregations are what we know as crystals of salt. If a salt such as Na^+Cl^- is placed in water, the electrical attraction of the water molecules (for reasons we shall point out later in this chapter) disrupts the forces holding the salt ions in their crystal matrix, causing the salt to dissolve into a roughly equal mixture of free Na^+ and free Cl^- ions. Approximately 0.06% of the atoms in your body are free Na^+ or Cl^- ions—about 40 grams, the weight of your fingernails.

An ionic bond is an attraction between ions of opposite charge.

COVALENT BONDS

Covalent bonds form when two atoms share electrons. Consider hydrogen (H) as an example. Each hydrogen atom has an unpaired electron and an unfilled outer electron shell. For these reasons the hydrogen atom is chemically unstable. When two hydrogen atoms are close enough to one another, however, each electron can orbit both nuclei. In effect, nuclei in close proximity are able to share their electrons. The result is a diatomic molecule (one with two atoms) of hydrogen gas (H_2) (Figure 3-5).

The diatomic hydrogen gas molecule that is formed as a result of this sharing of electrons is not charged, however. It still contains two protons and two electrons. Each hydrogen atom can be considered to have two orbiting electrons in the outer electron shell. This relationship satisfies the octet rule because each shared outer-shell electron orbits both nuclei and therefore is included in the outer shell of *both* atoms. The relationship also results in the pairing of the two free electrons. The two hydrogen atoms thus form a stable molecule. Note, however, that this stability is conferred by the electrons that orbit *both* nuclei; stability of this kind occurs only when the nuclei are very close. For this reason the strong chemical forces tending to pair electrons and satisfy the octet rule will act to keep the two hydrogen nuclei near one another. The bond between the two hydrogen atoms of diatomic hydrogen gas is an example of a covalent bond.

A covalent bond is a chemical bond formed by the sharing of one or more pairs of electrons.

Covalent bonds can be very strong, that is, difficult to break. Covalent bonds that share *two* pairs of electrons, called **double bonds,** are stronger than covalent bonds sharing only one electron pair, called **single bonds.** Energy is required to form larger aggregations of atoms from smaller ones because of the need to establish new orbitals for the electrons. This energy is released when the bonds are broken. Covalent bonds are represented in chemical formulations as lines connecting atomic symbols. Each line between two bonded atoms represents the sharing of one pair of electrons. Hydrogen gas is thus symbolized H—H and oxygen gas, O=O.

Molecules are often made up of more than two atoms. One reason larger molecules may form is that a given atom is able to share electrons with more than one other atom. An atom that requires two, three, or four additional electrons to fill its outer energy level completely may acquire them by sharing its electrons with two or more other atoms. For example, carbon (C) atoms (atomic number 6) contain six electrons, two of them at the inner level and the other four in the outer shell. To satisfy the octet

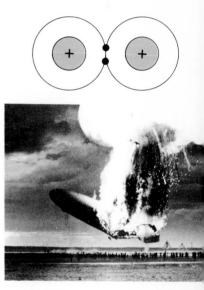

FIGURE 3-5

Hydrogen gas in a diatomic molecule composed of two hydrogen atoms, each sharing its electron with the other. Even more stable molecules are possible when the two hydrogen atoms share their electrons with an oxygen atom, forming water. The flash of fire that consumed the *Hindenburg* occurred when the hydrogen gas used to inflate the airship combined explosively with oxygen gas in the air to form water.

rule a carbon atom must gain access to four additional electrons. It must form the equivalent of four covalent bonds. Because there are many ways that four covalent bonds may form, carbon atoms are able to participate in many different kinds of molecules.

Of the 92 kinds of atoms (elements) that form the crust of the earth, only 11 are common in living organisms. Table 3-1 lists the frequency with which various elements occur in the earth's crust and in the human body. Unlike the elements that occur

TABLE 3-1   THE MOST COMMON ELEMENTS ON EARTH AND THEIR DISTRIBUTION IN THE HUMAN BODY

ELEMENT	SYMBOL	ATOMIC NUMBER	APPROXIMATE PERCENT OF EARTH'S CRUST BY WEIGHT	PERCENT OF HUMAN BODY BY WEIGHT	IMPORTANCE OR FUNCTION
Oxygen	O	8	46.6	65.0	Required for cellular respiration; component of water
Silicon	Si	14	27.7	Trace	—
Aluminum	Al	13	6.5	Trace	—
Iron	Fe	26	5.0	Trace	Critical component of hemoglobin in the blood
Calcium	Ca	20	3.6	1.5	Component of bones and teeth; triggers muscle contraction
Sodium	Na	11	2.8	0.2	Principal positive ion bathing cells; important in nerve function
Potassium	K	19	2.6	0.4	Principal postive ion in cells; important in nerve function
Magnesium	Mg	12	2.1	0.1	Critical component of many energy-transferring enzymes
Hydrogen	H	1	0.14	9.5	Electron carrier; component of water and most organic molecules
Manganese	Mn	25	0.1	Trace	—
Fluorine	F	9	0.07	Trace	—
Phosphorus	P	15	0.07	1.0	Backbone of nucleic acids; important in energy transfer
Carbon	C	6	0.03	18.5	Backbone of organic molecules
Sulfur	S	16	0.03	0.3	Component of most proteins
Chlorine	Cl	17	0.01	0.2	Principal negative ion bathing cells
Vanadium	V	23	0.01	Trace	—
Chromium	Cr	24	0.01	Trace	—
Copper	Cu	29	0.01	Trace	Key component of many enzymes
Nitrogen	N	7	Trace	3.3	Component of all proteins and nucleic acids
Boron	B	5	Trace	Trace	—
Cobalt	Co	27	Trace	Trace	—
Zinc	Zn	30	Trace	Trace	Key component of some enzymes
Selenium	Se	34	Trace	Trace	—
Molybdenum	Mo	42	Trace	Trace	Key component of many enzymes
Tin	Sn	50	Trace	Trace	—
Iodine	I	53	Trace	Trace	Component of thyroid hormone

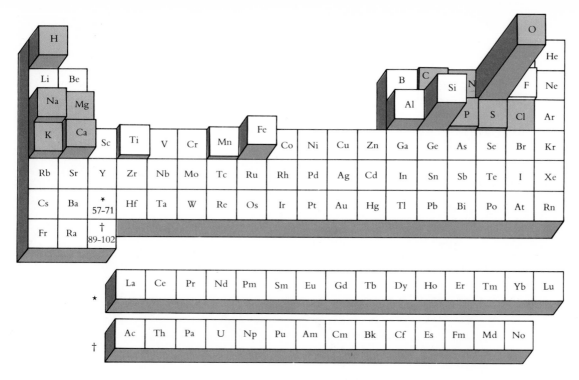

most abundantly in the earth's crust, all of the elements common in living organisms are light. Each has an atomic number less than 21 and thus a low mass. The great majority of atoms in living things, 99.4% of the atoms in the human body, for example, are either nitrogen, oxygen, carbon, or hydrogen (Figure 3-6). You can remember these elements by their first initials, NOCH.

THE CRADLE OF LIFE: WATER

If you were to count the atoms in your body, and then ask in what molecules those atoms were present, you would find that most of your atoms are parts of water molecules. The most common atoms in living things are oxygen and hydrogen atoms; the great majority of them are combined together in water molecules. Water has the chemical formula H_2O. As you will see, this seemingly simple molecule has many surprising properties. For example, of all the common molecules on earth, only water exists as a liquid at the relatively cool temperatures prevailing on the earth's surface (Figure 3-7). When life on earth was beginning, water, because it is a liquid at such temperatures, provided a medium in which other molecules could move around and interact without being bound by strong covalent or ionic bonds. Life evolved as a result of these interactions.

Life as it evolved on earth is inextricably tied to water (Figure 3-8). Three fourths of the earth's surface is covered by liquid water. You yourself are about two-thirds water, and you cannot exist long without it. All other organisms also require water. It is no accident that tropical rain forests are bursting with life, whereas deserts are almost lifeless except when water becomes temporarily plentiful, such as after a rainstorm. Farming is possible only in areas of the earth where rain is plentiful or water can be supplied by irrigation. No plant or animal can grow and reproduce in any but a water-rich environment.

The chemistry of life, then, is water chemistry. The way that life evolved was determined largely by the chemical properties of the liquid water in which its evolution occurred. The single most outstanding chemical property of water is its ability to form weak chemical associations with only 5% to 10% of the strength of covalent bonds. This one property of water, which derives directly from its structure, is responsible for much of the organization of living chemistry.

FIGURE 3-6

Chemists order the elements in an arrangement known as the periodic table. In this representation, the frequency of elements that occur in the earth's crust in more than trace amounts is indicated in the vertical dimension. Thus the most frequent element, oxygen, rises the farthest above the plane of the page, whereas iron, not as common, does not rise as far, and chlorine, relatively rare, does not rise at all. Elements found in significant amounts in living organisms are shaded in color. Many elements that are common on the surface of the earth, such as silicon and iron, are not present in living organisms in significant concentrations.

FIGURE 3-7

Water takes many forms. As a liquid, it fills our rivers and runs down over the land to the sea, sometimes falling in great cascades, such as at Victoria Falls in Zimbabwe, Africa. The iceberg on which the penguins are holding their meeting was formed in Antarctica from huge blocks of ice breaking away into the ocean water. When water cools below 0° C, it forms beautiful crystals, familiar to us as snow and ice. However, water is not always plentiful. At Badwater, Death Valley, there is no hint of water except for the broken patterns of dry mud.

FIGURE 3-8

Life originated in the warm waters of the ancient earth, and many kinds of organisms, like these small frogs, seen through the transparent walls of their eggs, begin life in water.

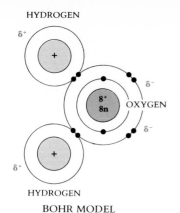

HYDROGEN

OXYGEN

HYDROGEN

BOHR MODEL

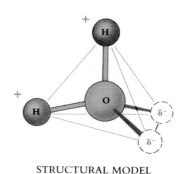

+

STRUCTURAL MODEL

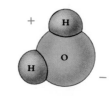

+

MOLECULAR MODEL

Water has a simple atomic structure, one oxygen atom bound by covalent single bonds to two hydrogen atoms. The resulting molecule is stable: it satisfies the octet rule, has no unpaired electrons, and does not carry a net electrostatic charge.

The remarkable story of the properties of water does not end there. The electron-attracting power of the oxygen atom (referred to by chemists as its electronegativity) is much greater than that of the hydrogen atom. As a result, the electron pair shared in each of the two single oxygen-hydrogen covalent bonds of a water molecule is more strongly attracted to the oxygen nucleus than to either of the hydrogen nuclei. Although electron orbitals encompass both the oxygen and hydrogen nuclei, the negatively charged electrons are far more likely, at a given moment, to be found near the oxygen nucleus than near one of the hydrogen nuclei. This relationship has a profoundly important result: the oxygen atom acquires a partial negative charge. It is as if the electron cloud were more dense in the neighborhood of the oxygen atom and less dense around the hydrogen atoms. This charge separation within the water molecule creates negative and positive electrical charges on the ends of the molecule (Figure 3-9). These partial charges are much less than the unit charges of ions.

The water molecule thus has distinct "ends," each with a partial charge, like the two poles of a magnet. Molecules such as water that exhibit charge separation are called **polar molecules** because of these magnetlike poles. Water is one of the most polar molecules known. The polarity of water underlies its chemistry and thus the chemistry of life.

> **Much of the biologically important behavior of water results because the oxygen atom attracts electrons more strongly than the hydrogen atoms do, with the result that the water molecule has electron-rich (−) and electron-poor (+) regions, giving it magnetlike positive and negative poles.**

Polar molecules interact with one another. The partial negative charge at one end of a polar molecule is attracted to the partial positive charge of another polar molecule. This weak attraction is called a **hydrogen bond.** Water forms a lattice of such hydrogen bonds. Each hydrogen bond is individually very weak and transient. A given bond lasts only 1/100,000,000,000th of a second. Although each bond is transient, a very large number of such hydrogen bonds can form, and the cumulative effects of very large numbers of these bonds can be enormous. The cumulative effect of very large numbers of hydrogen bonds is responsible for many of the important physical properties of water (Figure 3-10).

Water Is a Powerful Solvent

Water molecules gather closely around any molecule that exhibits an electrical charge, whether the molecule carries a full charge (ion) or a charge separation (polar molecule). For example, sucrose (table sugar) is composed of molecules that contain slightly polar hydroxyl (OH^-) groups. A sugar crystal dissolves rapidly when it is placed in water because water molecules can form hydrogen bonds with the polar hydroxyl groups of the sucrose molecules. Every time a sugar molecule dissociates or breaks away from

FIGURE 3-9

Water has a simple molecular structure. Each molecule is composed of one oxygen atom and two hydrogen atoms. The oxygen atom shares a pair of electrons with each participating hydrogen atom, contributing one electron of each pair. The δ^+ and δ^- represent the partial charges at the corners of the water molecule.

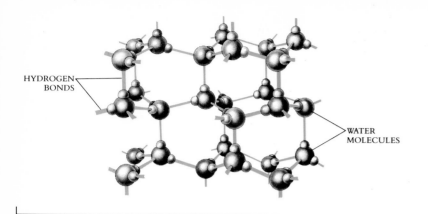

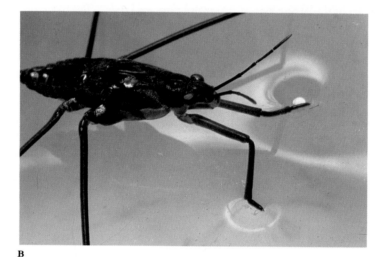

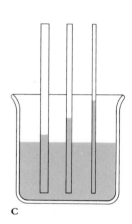

FIGURE 3-10

Many of the physical properties of water depend on hydrogen bonding.

A Ice formation. When water cools below 0° C, it forms a regular crystal structure in which the partial charges of each atom in the water molecule interact with opposite charges of atoms in other water molecules to form H bonds.

B Surface tension. Some insects, such as the water strider, literally walk on water. In this photograph you can see the dimpling its feet make on the water as its weight bears down on the surface. Because the surface tension of the water is greater than the force that one foot brings to bear, the water strider does not sink, but rather glides along.

C Adhesion. Capillary action will cause the water within a narrow tube to rise above the surrounding fluid; the adhesion of the water to the glass surface, drawing it upward, is stronger than the force of gravity, drawing it down. The narrower the tube, the greater the surface/volume ratio and the more the adhesion counteracts the force of gravity.

the crystal, water molecules orient around it in a cloud. Such a **hydration shell,** formed by the water molecules, prevents every sucrose molecule from associating with other sucrose molecules. Similarly, hydration shells form around all polar molecules. Polar molecules that dissolve in water in this way are said to be **soluble** in water. Nonpolar molecules are not water-soluble. Oil is an example of a nonpolar molecule. Life originated in water not only because it is a liquid, but also because so many molecules are polar or ionized and thus are water-soluble.

Water Organizes Nonpolar Molecules

Water molecules in solution always tend to form the maximum number of hydrogen bonds possible. When nonpolar molecules, which do not form hydrogen bonds, are placed in water, the water molecules act to exclude them. The water molecules preferentially form hydrogen bonds with other water molecules. The nonpolar molecules are forced to associate with one another, minimizing their disruption of the hydrogen bonding of water. The result is a sort of molecular ghetto where all the nonpolar molecules are crowded together. This is why oil and water do not mix but always separate. It seems almost as if the nonpolar compounds shrink from contact with the water, and for this reason they are called **hydrophobic** (Greek *hydros,* water + *phobos,* hating; "water-hated" might be a more apt description). The tendency for nonpolar molecules to band together in water solution is called **hydrophobic bonding.** Hydrophobic forces determine the three-dimensional shapes of many biological molecules, which are often surrounded by water within organisms.

Water Ionizes

The covalent bonds of water sometimes break spontaneously. When this happens, one of the protons (hydrogen atom nuclei) dissociates from the molecule. Because the dissociated proton lacks the negatively charged electron that it had shared in the covalent bond with oxygen, its own positive charge is not counterbalanced; it is a positively charged hydrogen ion, H^+. The remaining bit of the water molecule retains the shared electron from the covalent bond and has one less proton to counterbalance it; it is a negatively charged hydroxyl ion (OH^-). This process of spontaneous ion formation is called **ionization.**

$$H_2O \rightarrow OH^- + H^+$$

To measure numerically the concentration of H^+ ions in solution, a scale based on the slight degree of spontaneous ionization of water has been constructed. Roughly one water molecule in every 550 million is ionized at any instant in time. In one liter of water this corresponds to one ten-millionth of a mole of H^+ ions. (A **mole** is defined as the weight in grams that corresponds to the summed atomic weights of all the atoms of a molecule, its molecular mass. In the case of H^+ the molecular mass equals 1, and a mole of H^+ ions would weigh 1 gram.) The molar concentration of hydrogen ions in pure water, 1 in 10,000,000, can be written more easily by using exponential notation. This is done by counting the number of zeros after the digit "1" in the denominator:

$$[H^+] = \frac{1}{10,000,000}$$

Since there are seven zeros, the molar concentration is 10^{-7} moles per liter.

Any substance that dissociates to form H^+ ions when dissolved in water is called an **acid.** The exponents in the exponential notation of H^+ ion concentrations are used as a convenient indication of acid strength, called the **pH scale.** pH is normally expressed as a positive number and is determined by taking the negative value of the exponent of the molar hydrogen concentration. Thus pure water has a molar H^+ concentration of 10^{-7} and a pH of 7. The stronger an acid is, the more H^+ ions it produces and the *lower* (smaller) its pH. Hydrochloric acid (HCl), which is abundant in your stomach, ionizes completely, so the molar concentration of H^+ in water containing $\frac{1}{10}$

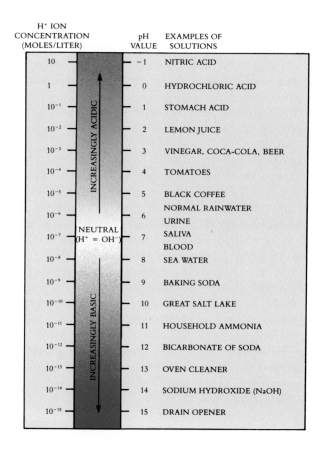

H⁺ ION CONCENTRATION (MOLES/LITER)		pH VALUE	EXAMPLES OF SOLUTIONS
10		−1	NITRIC ACID
1		0	HYDROCHLORIC ACID
10^{-1}	INCREASINGLY ACIDIC	1	STOMACH ACID
10^{-2}		2	LEMON JUICE
10^{-3}		3	VINEGAR, COCA-COLA, BEER
10^{-4}		4	TOMATOES
10^{-5}		5	BLACK COFFEE
10^{-6}		6	NORMAL RAINWATER / URINE
10^{-7}	NEUTRAL ($H^+ = OH^-$)	7	SALIVA / BLOOD
10^{-8}		8	SEA WATER
10^{-9}		9	BAKING SODA
10^{-10}	INCREASINGLY BASIC	10	GREAT SALT LAKE
10^{-11}		11	HOUSEHOLD AMMONIA
10^{-12}		12	BICARBONATE OF SODA
10^{-13}		13	OVEN CLEANER
10^{-14}		14	SODIUM HYDROXIDE (NaOH)
10^{-15}		15	DRAIN OPENER

FIGURE 3-11

The pH scale, in which a fluid is assigned a value according to the number of hydrogen ions present in a liter of that fluid. The scale is logarithmic, so that a change of only 1 means a tenfold change in the concentration of hydrogen ions; thus lemon juice is 100 times more acidic than tomato juice, and seawater is 10 times more basic than pure water.

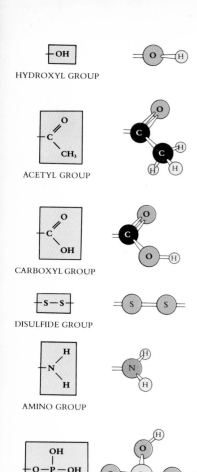

HYDROXYL GROUP

ACETYL GROUP

CARBOXYL GROUP

DISULFIDE GROUP

AMINO GROUP

PHOSPHATE GROUP

FIGURE 3-12

The principal functional chemical groups. These groups tend to act as units during chemical reactions and to confer specific chemical properties on the molecules that possess them. Hydroxyl groups, for example, make a molecule more basic, whereas carboxyl groups make a molecule more acidic.

of a mole of hydrochloric acid per liter is 10^{-1} moles per liter, corresponding to a pH of 1. Some acids such as nitric acid are even stronger, although such very strong acids are rarely found in living systems (Figure 3-11). The pH of champagne, which bubbles because of the carbonic acid dissolved in it, is about 2.

pH refers to the relative concentration of H$^+$ ions in a solution. The numerical value of the pH is the negative of the exponent of the molar concentration. Low pH values indicate high concentrations of H$^+$ ions (acids), and high pH values indicate low concentrations.

Recall that H$^+$ ions are not the only type of ion produced when water ionizes. Negatively charged OH$^-$ ions are also produced in equal concentration. Any substance that combines with H$^+$ ions, as OH$^-$ ions do, is said to be a **base.** In pure water the concentrations of H$^+$ and OH$^-$ ions are both 10^{-7} mole per liter, reflecting the spontaneous rate of dissociation of water. Any increase in base concentration has the effect of lowering the H$^+$ ion concentration, because base and H$^+$ ions join spontaneously. Bases, therefore, have pH values above 7. Strong bases such as sodium hydroxide (NaOH) have pH values of 12 or more.

THE CHEMICAL BUILDING BLOCKS OF LIFE

The basic chemical building blocks of organisms, the mortar and bricks used to assemble a cell, are made of molecules, just as a house is built of bricks. The molecules formed by living organisms, which contain carbon, are called **organic molecules.** Although many organic molecules are used in the construction of a cell, they need not confuse you. A good way to see through the tangle of different molecules is to focus on those bits of the molecules that are important, like focusing on who has the ball in a football game. Much of the complex structure of an organic molecule may have little to do with the biological process you are studying. It is often helpful to think of an organic molecule as a carbon-based core with special bits attached, groups of atoms with definite chemical properties. We refer to these groups of atoms as **functional groups.** For example, a hydrogen atom bonded to an oxygen atom, —OH, is a hydroxyl group. The most important functional groups are illustrated in Figure 3-12. Most chemical reactions that occur within organisms involve the transfer of a functional group from one molecule to another, or the breaking of a carbon-carbon bond. Proteins called kinases, for example, transfer phosphate groups from one kind of molecule to another.

Some molecules that occur in organisms are simple organic molecules, often with a single reactive functional group protruding from a carbon chain. Other molecules are far larger and are called **macromolecules.** This is particularly true of molecules that play a structural role in organisms or that store information. Most of these macromolecules are themselves composed of simpler components, just as a wall is composed of individual bricks.

A **polymer** is a macromolecule built by forming covalent bonds between a long chain of similar components, like coupling cars to form a railroad train. Many of the most important building blocks of organisms are polymers. **Complex carbohydrates,** for example, are polymers of simple ring molecules called sugars. **Proteins** are polymers of molecules called amino acids. **DNA,** the molecule that stores hereditary information, is one version of a long chain molecule called a nucleic acid, a polymer composed of a long series of molecules called **nucleotides.**

In contrast to polymers, **composite molecules** are made up of several distinct elements that are joined to form an operational whole. **Lipids** are composite molecules, as are the nucleotides discussed in Chapters 7 and 8. Four of the major classes of macromolecules that occur in organisms are presented in Table 3-2.

In discussing the polymers that make up the bodies of organisms, we will start with carbohydrates. Some carbohydrates are simple, small molecules. Others are long polymers. Carbohydrates are important both as structural elements and because of their role in energy storage. After we have discussed carbohydrates, we shall address lipids, amino acids and proteins, and nucleic acids.

TABLE 3-2 MACROMOLECULES

MACROMOLECULE	SUBUNIT	FUNCTION	EXAMPLE
POLYSACCHARIDES			
Starch, glycogen	Glucose	Energy storage	Potatoes
Cellulose	Glucose	Cell walls	Paper
Chitin	Modified glucose	Cell walls	Crab shells
LIPIDS			
Fats	Glycerol + 3 fatty acids	Energy storage	Butter
Phospholipids	Glycerol + 2 fatty acids + phosphate	Cell membranes	Soap
Steroids	4 carbon rings	Membranes; hormones	Cholesterol; estrogen
Terpenes	Long carbon chains	Pigments; structural	Chlorophyll; rubber
PROTEINS			
Globular	Amino acids	Enzyme catalysts	Hemoglobin
Structural	Amino acids	Support	Hair; silk
NUCLEIC ACIDS			
DNA	Nucleotides	Encodes genes	Chromosomes
RNA	Nucleotides	Operating blueprint of genes	Flu virus

CARBOHYDRATES
Sugars

Carbohydrates are a loosely defined group of molecules that contain carbon, hydrogen, and oxygen. Because they contain many carbon-hydrogen (C—H) bonds, carbohydrates are well suited for energy storage. Such C—H bonds are the ones most often broken by organisms to obtain energy. Most carbohydrates contain the three elements carbon, hydrogen, and oxygen in the molar ratio 1:2:1. A chemist would say that the empirical formula (a list of the atoms in a molecule with a subscript to indicate how many of each) is $(CH_2O)_n$, where n is the number of carbon atoms. Among the simplest of the carbohydrates are the simple sugars or **monosaccharides** (Greek *monos,* single + *saccharon,* sweet). As their name implies, monosaccharides taste sweet. Simple sugars may have as few as three carbon atoms, but the molecules that play the central role in energy storage have six. They have the empirical formula

$$C_6H_{12}O_6, \text{ or } (CH_2O)_6$$

Sugars can exist in a straight-chain form, but in water solution they almost always form rings. The primary energy-storage molecule is glucose (Figure 3-13), a six-carbon sugar with seven energy-storing C—H bonds.

> **Among the most important energy-storage molecules in organisms are sugars. Many simple sugars contain six carbon atoms and seven energy-storing C—H bonds.**

Glucose is not the only sugar with the formula $C_6H_{12}O_6$. Other sugars having this same formula are fructose and galactose (Figure 3-14). Because these molecules have the same molecular formula as glucose but different structural formulas, they are called **isomers,** or alternative forms, of glucose. Glucose and fructose are **structural isomers.** In fructose the double-bonded oxygen is attached to a carbon located within the chain when the sugar exists in a straight-chain condition. In glucose the double-bonded oxygen is attached to a terminal carbon. Your taste buds can tell the difference: fructose tastes much sweeter than glucose (Figure 3-15).

FIGURE 3-13

Structure of the glucose molecule. Glucose is a linear six-carbon molecule that forms a ring in solution. The structure of the ring can be represented in many ways, of which these are some of the most common.

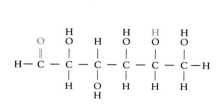

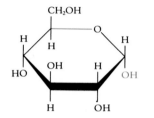

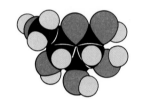

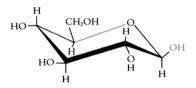

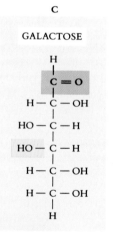

FIGURE 3-14

A structural isomer of glucose **(A)**, such as fructose **(B)**, has identical chemical groups bonded to different carbon atoms, whereas a stereoisomer of glucose, such as galactose **(C)**, has identical chemical groups bonded to the same carbon atoms in a different orientation.

A	**B**	**C**
GLUCOSE	FRUCTOSE	GALACTOSE

FIGURE 3-15

This hungry butterfly has its uncoiled mouthparts extended down into a sugar solution, which it is sucking from the flowers of this wild relative of a sunflower in California.

Unlike fructose, galactose has the same bond structure as glucose. The only difference between them is the orientation of one hydroxyl (—OH) group. Because the two orientations involve the same groups in different positions, glucose and galactose are called **stereoisomers** (see Figure 3-14, *C*). Again, this seemingly slight difference has important consequences. This hydroxyl group is one that is often involved when links are created to form long polymers called polysaccharides.

Many organisms transport sugars within their bodies. In human beings, glucose circulates in the blood. In many other organisms glucose is converted to a **transport form** before it is moved from place to place within the organism. In transport form, glucose is less readily consumed (metabolized) while it is being moved. Transport forms of sugars are commonly formed by linking two monosaccharide molecules to form a **disaccharide** (Greek *di,* two). Sucrose (table sugar) is a disaccharide formed by linking a molecule of glucose to a molecule of fructose. It is the common transport form of sugar in plants. If a glucose molecule is linked to its stereoisomer galactose, the resulting disaccharide is lactose. Lactose is the molecule many mammals use to feed their babies.

Starches

Organisms store the metabolic energy contained in glucose by converting it to an insoluble form and depositing it in specific storage areas. Sugars are made insoluble by joining them together into long polymers called **polysaccharides** composed of monosaccharide sugar subunits. If the polymers are branched, that is, if they have side chains coming off of a main chain, the molecules are even less soluble. **Starches** are polysaccharides formed from glucose.

The starch with the simplest structure is amylose. Amylose is made up of many hundreds of glucose molecules linked together in long unbranched chains. Potato starch is about 20% amylose. When amylose is digested by a sprouting potato plant (or by you), proteins called enzymes first break it into fragments of random length. These shorter fragments are soluble. Baking or boiling potatoes has the same effect.

Most plant starch, including 80% of potato starch, is a more complicated variant of amylose called amylopectin. **Pectins** are branched polysaccharides. Amylopectin is a form of amylose with short, linear amylose branches consisting of 20 to 30 glucose subunits.

Animals also store glucose in branched amylose chains. However, the average chain length is much longer in animals, and there are more branches. This results in a highly branched structure called **glycogen** (Figure 3-16).

> **Starches are polysaccharides formed from glucose. Because starches form long chains, they are relatively insoluble. If starches are branched, which is often the case, they are still less soluble.**

Cellulose

Imagine that you could draw a line down the central axis of a starch molecule, like threading a rope through a pipe. Because all the glucose subunits of the starch chain are joined in the same orientation, all the CH_2OH groups would fall on the same side of the line (see Figure 3-16). There is another way to build a chain of glucose molecules, however, in which the glucose subunit orientations switch back and forth (the CH_2OH groups alternate on opposite sides of the line). The resulting polysaccharide is **cellulose,** the chief component of plant cell walls. Cellulose is chemically similar to amylose, with one important difference (Figure 3-17): the starch–degrading proteins that occur in most organisms cannot break the bond between two sugars in opposite orientation. It is not that the bond is stronger, but rather that its cleavage requires the aid of a different protein, one not usually present. Because cellulose cannot readily be broken down, it works well as a biological structural material and occurs widely in this role in plants. For those few animals able to break down cellulose, it provides a rich source of energy.

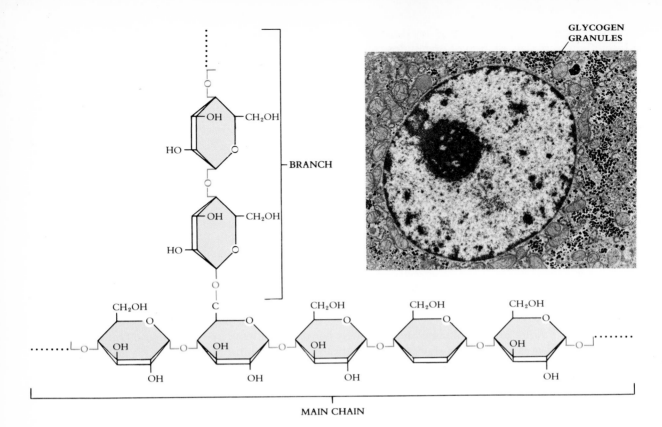

BRANCH

MAIN CHAIN

FIGURE 3-16

Storage polymers of glucose are called starches. The simplest starches
are long chains of glucose called amylose. Most plants contain more
complex starches called amylopectins, which contain branches. The
nucleus of the liver cell in the micrograph insert is surrounded by
dense granules of animal starch called glycogen, which is even more
highly branched.

FIGURE 3-17

The jumble of cellulose fibers *(right)* is from a ponderosa pine. Each
fiber is composed of microfibrils *(left)*, which are bundles of cellulose
chains. Cellulose fibers can be very strong; they are quite resistant to
metabolic breakdown, which is one reason why wood is such a good
building material.

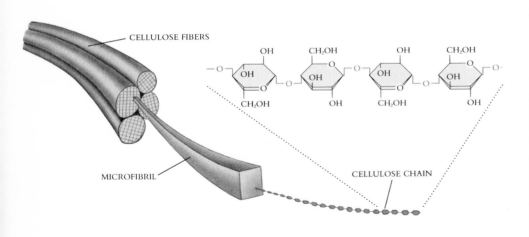

CELLULOSE FIBERS

MICROFIBRIL

CELLULOSE CHAIN

The structural material in insects, many fungi, and certain other organisms is **chitin** (Figure 3-18). Chitin is a modified form of cellulose in which a nitrogen group has been added to the glucose units. Chitin is a tough, resistant surface material. Few organisms are able to digest chitin.

LIPIDS
Fats

When organisms store glucose molecules for long periods of time, they usually convert glucose into another kind of insoluble molecule that contains more C—H bonds than carbohydrates do. These storage molecules are called **fats**. The ratio of H to O in carbohydrates is 2:1, but in fat molecules it is much higher. Like starches, fats are insoluble and can therefore be deposited at specific storage locations within the organism. The insolubility of starches results from the fact that they are long polymers. The insolubility of fats, in contrast, arises from the fact that they are nonpolar. Unlike the H—O bonds of water, the C—H bonds of carbohydrates and fats are nonpolar and cannot form hydrogen bonds. Because fat molecules contain a large number of C—H bonds, they are hydrophobically excluded by water, because water molecules tend to form hydrogen bonds with other water molecules. The result is that the fat molecules cluster together, insoluble in water.

Fats are one kind of **lipid,** a loosely defined group of molecules that are insoluble in water but soluble in oil. Oils and waxes are also classified as lipids. Fats are composite molecules; each molecule is built from two different kinds of subunits:

1. *Glycerol:* a three-carbon alcohol with each carbon bearing a hydroxyl (—OH) group. The three carbons form the backbone of the fat molecule, to which three fatty acids are attached.
2. *Fatty acids:* long **hydrocarbon** chains (chains consisting only of carbon and hydrogen atoms) ending in a carboxyl (—COOH) group. Three fatty acids are attached to each glycerol backbone.

The structure of an individual fat molecule, like the one diagrammed in Figure 3-19, consists simply of a glycerol molecule with a fatty acid joined to each of its the three carbon atoms:

$$
\begin{array}{c}
\text{H} \\
| \\
\text{H—C—fatty acid} \\
| \\
\text{H—C—fatty acid} \\
| \\
\text{H—C—fatty acid} \\
| \\
\text{H}
\end{array}
$$

Because there are three fatty acids, the resulting fat molecule is called a triglyceride.

FIGURE 3-18

Chitin, which may be considered a modified form of cellulose with nitrogen groups added to the sugar subunits, is the principal structural element in the external skeletons of many invertebrates, such as this crab, and in the cell walls of fungi.

FIGURE 3-19

Triglycerides are composite molecules, made up of three fatty acid molecules coupled to a single glycerol backbone.

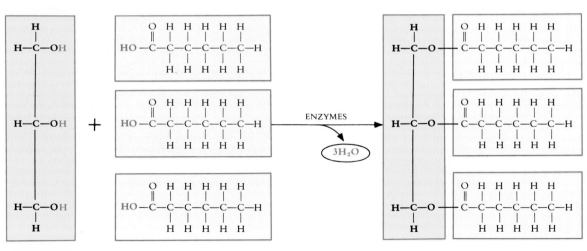

GLYCEROL   FATTY ACIDS   TRIGLYCERIDE MOLECULE

A

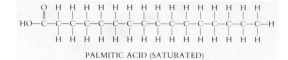

PALMITIC ACID (SATURATED)

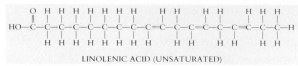

LINOLENIC ACID (UNSATURATED)

B

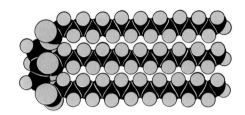

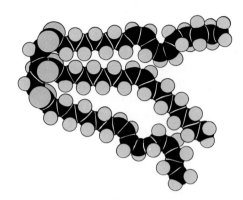

FIGURE 3-20

A Palmitic acid, with no double bonds and thus a maximum number of hydrogen atoms bonded to the carbon chain, is a saturated fatty acid. Linoleic acid, with three double bonds and thus fewer than the maximum number of hydrogen atoms bonded to the carbon chain, is an unsaturated fatty acid.

B Many animal triglyceride fats are saturated. Because their fatty acid chains can fit closely together, these triglycerides form immobile arrays called *hard fat*. Plant fats, in contrast, are typically unsaturated, and the many kinks that the double bonds introduce into the fatty acid chains prevent close association of the triglycerides and produce oils such as linseed oil, which is obtained from flax seed.

Fatty acids vary in length. The most common are even-numbered chains of 14 to 20 carbons. Fatty acids with all internal carbon atoms having two hydrogen side groups are called **saturated,** because they contain the maximum number of hydrogen atoms possible. Some fatty acids have double bonds between one or more pairs of successive carbon atoms (Figure 3-20). Fats composed of fatty acids with double bonds are said to be **unsaturated** because the double bonds replace some of the hydrogen atoms. The fatty acids therefore contain fewer than the maximum number of hydrogen atoms. If a given fat has more than one double bond, it is said to be **polyunsaturated.**

Polyunsaturated fats have low melting points because their chains bend at the double bonds and the fat molecules cannot be aligned closely with one another. Consequently the fat may be fluid. A liquid fat is called an **oil.** Many plant fatty acids, such as oleic acid (a vegetable oil) and linolenic acid (a linseed oil) are unsaturated. Animal fats, in contrast, are often saturated and occur as hard fats.

It is possible to convert an oil into a fat by adding hydrogen. The peanut butter that you buy in the store has usually been hydrogenated to convert the peanut fatty acids to hard fat. This prevents the fatty acids from separating out as oils while the jar sits on the store shelf.

Fats are very efficient energy-storage molecules because of their high concentration of C—H bonds. Most fats contain more than 40 carbon atoms. The ratio of energy-storing C—H bonds to carbon atoms is more than twice that of carbohydrates, making fats much more efficient vehicles for storing chemical energy. Fats usually yield about twice the amount of chemical energy per gram that carbohydrates do. As you might expect, the more highly saturated fats are richer in energy than less saturated ones. Animal fats contain more calories than vegetable fats. Human diets with large amounts of saturated fats appear to upset the normal balance of fatty acids in the body, a situation that can lead to heart disease.

Other Lipids

Fats are just one example of the oily or waxy class of molecules called lipids. Your body contains many different kinds of lipids. The membranes of your cells are composed of a kind of modified fat called a phospholipid, a molecule that will play a key role in Chapter 6 when we discuss membrane structure. Membranes often also contain a different kind of lipid called **steroids,** composed of four carbon rings; most of your cell membranes contain the steroid cholesterol. Male and female sex hormones (**hormones** are molecules produced in one part of an organism that trigger a specific reaction in target tissues and organs some distance away) are also steroids. Hormones will be discussed in Chapter 40. A very different kind of lipid forms many of the biologically important pigments, such as the photosynthetic pigment chlorophyll in plants and the pigment retinal, which absorbs light in your eyes. These pigments are examples of long-chain lipids called **terpenes.** Rubber is also a terpene.

PROTEINS

Proteins are the third major group of macromolecules that make up the bodies of organisms. Perhaps the most important proteins are **enzymes,** proteins capable of speeding up specific chemical reactions. Enzymes lower the energy required to activate or start the reaction, but are unaltered themselves in the process. Enzymes are biological **catalysts,** a more general term for substances that affect chemical reactions in this way. Other kinds of proteins also have important functions. Cartilage, bones, and tendons are made up of a protein called collagen. Keratin, another protein, forms both the horns of a rhinoceros and the feathers of a bird. The fluid within your eyeballs contains still other proteins. Short proteins called peptides are used as chemical messengers within your brain and throughout your body. Despite their diverse functions, all proteins have the same basic structure: a long polymer chain of amino acid subunits linked end to end.

Amino Acids

Amino acids are small molecules with a simple basic structure. An **amino acid** can be defined as a molecule containing an amino group ($-NH_2$), a carboxyl group ($-COOH$), a hydrogen atom, and a functional group designated R, all bonded to a central carbon atom:

$$
\begin{array}{c}
R \\
| \\
H_2N-C-COOH \\
| \\
H
\end{array}
$$

The identity and unique chemical properties of each amino acid are determined by the nature of the R group linked to the central carbon atom. An amino acid can potentially have any of a variety of different R groups, often called "side groups." Although many different amino acids occur in nature, only 20 are used in proteins. These 20 "common" amino acids and their side groups are illustrated in Figure 3-21. The different functional groups present on the side groups of the 20 amino acids give each amino acid distinctive chemical properties. For example, when the side group is $-H$, the amino acid (glycine) is polar, while when the side group is $-CH_3$, the amino acid (alanine) is nonpolar. The 20 amino acids that occur in proteins are commonly grouped into five chemical classes based on the chemical nature of their side groups.

 The way that each amino acid affects the shape of a protein depends on the chemical nature of the amino acid's side group. Portions of a protein chain with many nonpolar amino acids tend to be shoved into the interior of the protein by hydrophobic interactions because polar water molecules tend to exclude nonpolar amino acid side groups.

> **Proteins can contain up to 20 different kinds of amino acids. These amino acids fall into five chemical classes, which have properties quite different from one another. These differences determine what the proteins are like.**

 Note that in addition to its R group, each amino acid, when ionized, has a positive (amino, or NH_3^+) group at one end and a negative (carboxyl, or COO^-) group at the other end. These two groups can undergo a chemical reaction, losing a molecule of water and forming a covalent bond between two amino acids. A covalent bond linking two amino acids is called a **peptide bond.**

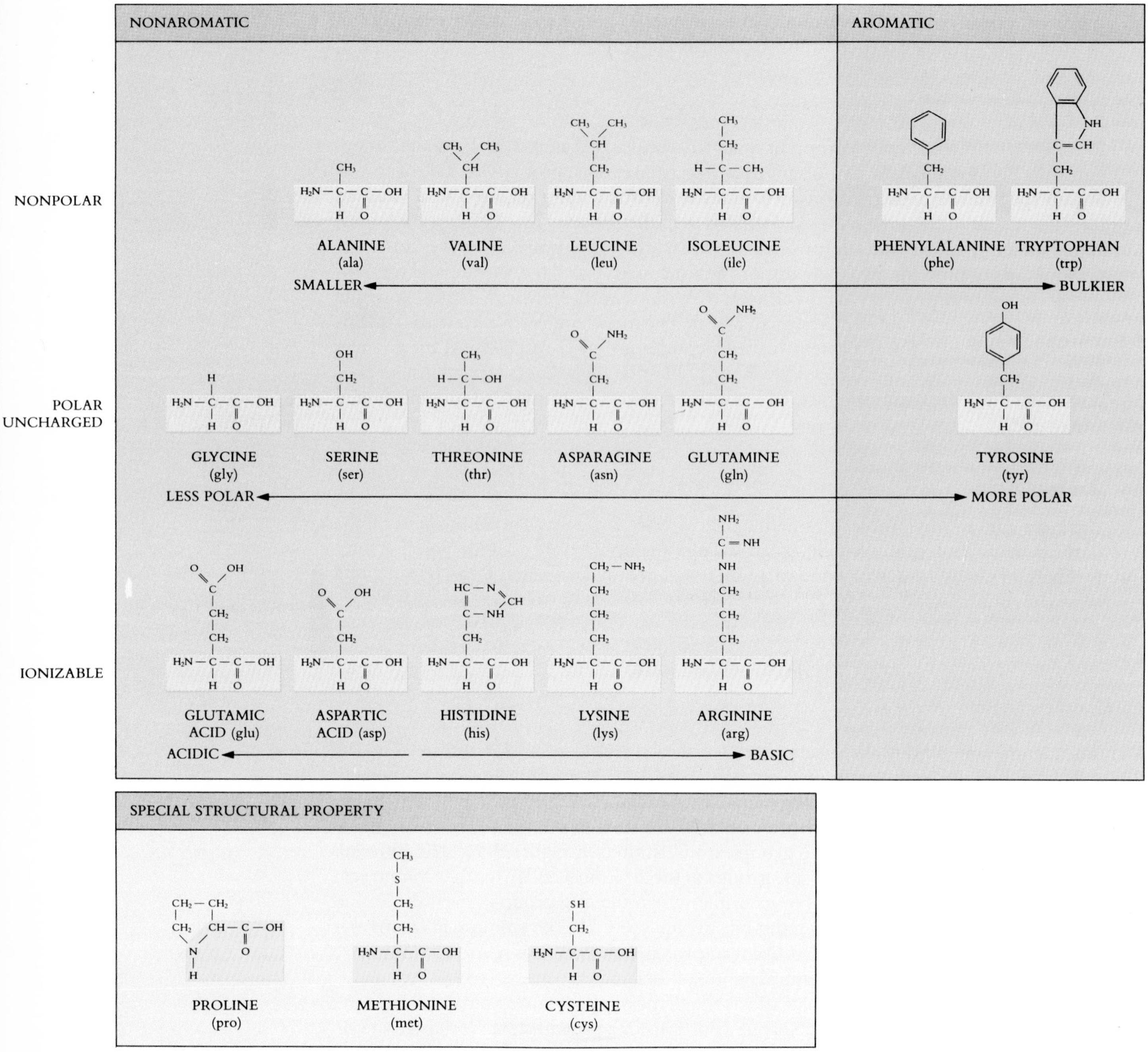

FIGURE 3-21

The 20 common amino acids. Each amino acid has the same chemical backbone, but differs from the others in the side or R group that it possesses. Six of the amino acid R groups are nonpolar, some more bulky than others (particularly the ones containing ring structures, which are called the aromatic amino acids). Another six are polar but uncharged; these differ from one another in how polar they are. Five more are polar and capable of ionizing to a charged form; under typical cell conditions some of these five are acids, others bases. The remaining three have special chemical properties that play important roles in forming links between protein chains or forming kinks in their shape.

Polypeptides

A protein, as just mentioned, is composed of a long chain of amino acids linked end to end by peptide bonds. The general term for chains of this kind is **polypeptide.** Proteins are therefore long, complex polypeptides. The sequence of amino acids that make up a particular polypeptide chain is termed its **primary structure** (Figure 3-22). Because the R groups that distinguish the various amino acids play no role in the peptide backbone of proteins, a protein can be composed of any sequence of amino acids. A protein made up of 100 amino acids linked together in a chain might have any of 20^{100} different amino acid sequences. That would be the number 20 followed by 100 zeros, an enormous figure indeed. The great variability possible in the sequence of amino acids is per-

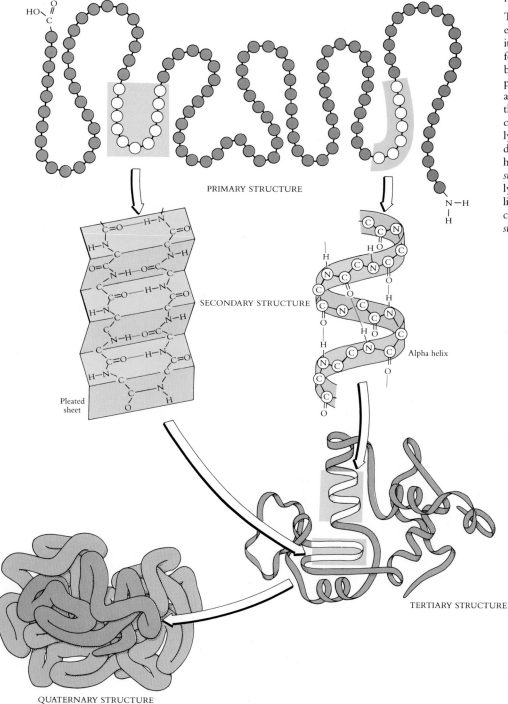

PRIMARY STRUCTURE

SECONDARY STRUCTURE

Pleated sheet

Alpha helix

TERTIARY STRUCTURE

QUATERNARY STRUCTURE

FIGURE 3-22

The amino acid sequence of the enzyme protein lysozyme, called its *primary structure,* encourages the formation of hydrogen bonds between nearby amino acids, producing coils called *alpha helices* and fold-backs called *pleated sheets;* these coils and fold-backs are called the *secondary structure.* The lysozyme protein assumes a three-dimensional shape like a cupped hand; this is called its *tertiary structure.* Many proteins (not lysozyme) aggregate in clusters like the one illustrated here; such clustering is called the *quaternary structure* of the protein.

haps the most important property of proteins, permitting great diversity in the kinds and therefore functions of specific proteins.

Each amino acid of a polypeptide interacts with its neighbors, forming hydrogen bonds. Because of these near-neighbor interactions, polypeptide chains tend to fold spontaneously into sheets or wrap into coils. The form that a region of a polypeptide assumes is called its local **secondary structure.**

The three-dimensional shape, or **tertiary structure,** of a protein depends heavily on its secondary structure. Proteins made up largely of sheets often form fibers that have a structural function (Figure 3-23), while proteins that have regions forming coils frequently fold into globular shapes. The shape of a globular protein is very sensitive to the order and nature of amino acids in the sequence. A change in the identity of a single amino acid can have very subtle, or profound, effects. We shall show in Chapter 7 it is because globular proteins can assume so many different shapes, that they are such effective biological catalysts.

FIGURE 3-23

Some of the more common structural proteins.
A   Collagen—strings of a tennis racket.
B   Fibrin—electron micrograph of a blood clot.
C   Keratin—a peacock feather.
D   Silk—a spider's web.
E   Hair—a woman's hair.

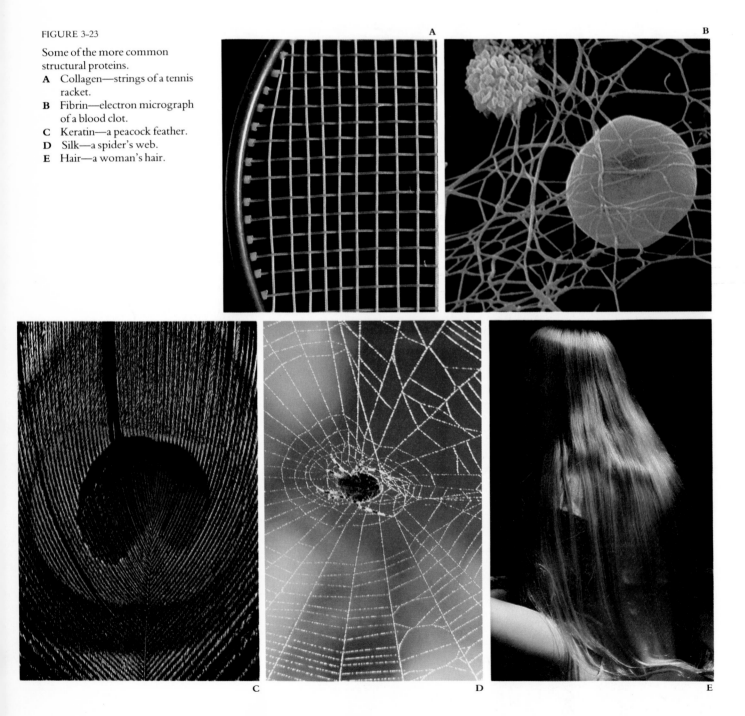

When two protein chains associate to form a functional unit, the chains are termed subunits. Hemoglobin is a protein composed of four subunits. The subunits need not all be the same, although they can be. In hemoglobin molecules, there are two identical subunits of one kind and two identical subunits of a second kind. For proteins that consist of subunits, the way these subunits are assembled into a whole is called the **quaternary structure.** Proteins are discussed in detail in Chapter 7.

> **The shape that proteins assume is determined by the sequence of amino acids in the polymer. Because different amino acid R groups have different chemical properties, the shape of a protein may be altered by a single amino acid change.**

NUCLEIC ACIDS

All organisms store the information specifying the structures of their proteins in nucleic acids. *Nucleic acids* are long polymers of repeating subunits called *nucleotides.* Each nucleotide, the basic repeating unit, is a composite molecule made up of three smaller building blocks (Figure 3-24):

1. A five-carbon sugar
2. A phosphate group (PO_4)
3. An organic nitrogen–containing base

In the formation of a nucleic acid chain, the individual sugars are linked together in a line by the phosphate groups. The phosphate group of one sugar binds to the hydroxyl group of another, forming an —O—P—O bond. This bond is called a **phosphodiester** bond. A nucleic acid is simply a chain of five-carbon sugars (called ribose sugars) linked by phosphodiester bonds, with an organic base protruding from each sugar. Each of the repeating phosphate-sugar-base links in the chain is a nucleotide.

Organisms encode the information specifying the amino acid sequence of their proteins as sequences of nucleotides in the nucleic acid called DNA (deoxyribonucleic acid). This encoded information is used in the everyday metabolism of the organism as well as being stored and passed on to the organism's descendants.

How does the structure of DNA permit it to store hereditary information? If DNA were a simple, monotonously repeating polymer, it could not encode the message of life. Imagine trying to write a story using only the letter E and no spaces or punctuation. All you could ever say is "EEEEEEE . . ." You need more than one letter to write. We use 26 letters in the English alphabet. The Chinese use thousands of individual characters to convey the same messages. You do not need so many individual symbols, of course, if the individual "letters" are grouped together into words. Morse code, which is used to transmit messages by telegraph, employs only two elements ("dot" and "dash"), as do most modern computers (0 and 1). Nucleic acids can encode information because they contain more than one kind of organic base. Each sugar link in a nucleic acid chain can have any one of four different organic bases attached to it. Just as in the English language, the sequence of letters encodes the information. In nucleic acids there are not 26 letters, as in English, but only four letters, the four organic bases that occur in nucleic acids. (As you will see shortly, one of the four bases is present in different versions in the two principal forms of nucleic acid.)

> **Organisms store and use hereditary information by encoding the sequence of the amino acids of each of their proteins as a sequence of nucleotides in nucleic acids.**

Organisms store hereditary information in two forms of nucleic acid. One form, **deoxyribonucleic acid (DNA),** provides the basic storage vehicle or master plan. The other form, **ribonucleic acid (RNA),** is similar in structure and is made as a template copy of portions of the DNA. This copy passes out into the rest of the cell, where it provides a blueprint specifying the amino acid sequence of proteins.

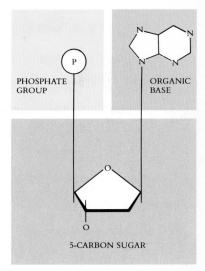

FIGURE 3-24

The nucleotide subunits of DNA and RNA have a composite structure; each is made up of three elements: a five-carbon sugar, an organic base, and a phosphate group.

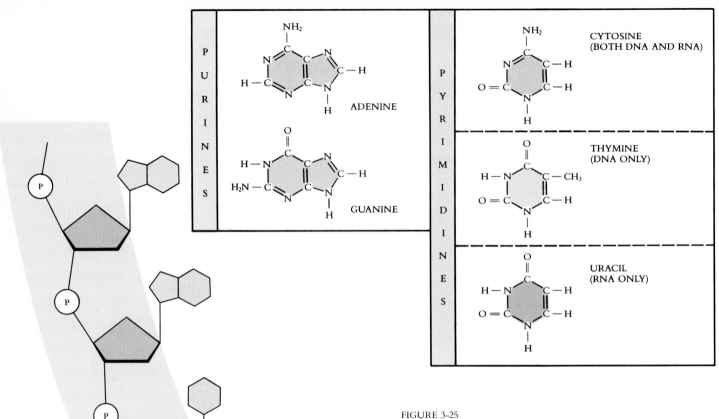

PURINES

NH₂

ADENINE

GUANINE

PYRIMIDINES

CYTOSINE
(BOTH DNA AND RNA)

THYMINE
(DNA ONLY)

URACIL
(RNA ONLY)

BASES

BACKBONE

FIGURE 3-25

The five nucleotide bases of nucleic acids. In DNA
the nucleotide thymine replaces the nucleotide
uracil found in RNA. In a nucleotide chain
nucleotides are linked to one another via
phosphodiester bonds, as shown on the left,
which is one half of a DNA helix.

Two of the four organic bases that make up DNA and RNA (Figure 3-25), adenine and guanine, are large, double-ring compounds called **purines.** The other organic bases that occur in these molecules, cytosine (in both DNA and RNA), thymine
(in DNA only), and uracil (in RNA only), are smaller, single-ring compounds called
pyrimidines. In discussing the sequence of bases in RNA and DNA, the organic bases
are usually referred to by their first initials: A, G, C, T, and U.

DNA chains in organisms exist not as single chains folded into complex shapes,
as in proteins, but rather as double chains (Figure 3-26). Two of the polymers wind
around each other like the outside and inside rails of a circular staircase. Such a winding
shape is called a **helix,** and one composed of two molecules winding about one another
like DNA is called a **double helix.** The steps of the helical staircase are hydrogen bonds
between the bases in one polymer chain and those opposite them in the other chain.
These hydrogen bonds hold the two chains together as a duplex. Details of the structure of DNA, and how it interacts with RNA in the production of proteins, will be presented in Chapter 13.

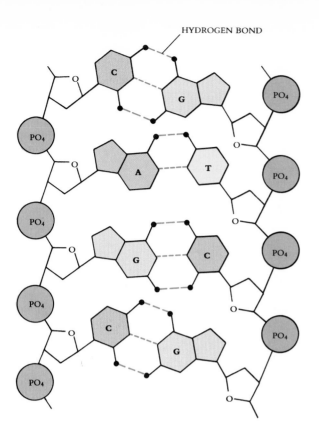

HYDROGEN BOND

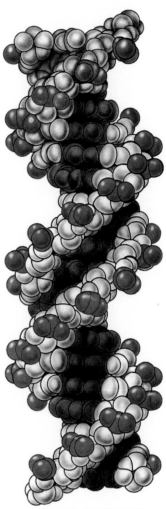

DNA DOUBLE HELIX

FIGURE 3-26

Hydrogen bond formation between the organic bases, called base pairing, causes the two chains of a DNA duplex to bind to each other. The molecule is not a straight chain, but rather a graceful double helix, rising like a circular staircase.

SUMMARY

1. The smallest stable particles of matter are protons, neutrons, and electrons, which associate to form atoms. The core (nucleus) of an atom is composed of protons and neutrons; electrons orbit around the core in a cloud.

2. The chemical behavior of an atom is largely determined by the distribution of its electrons, particularly the number of electrons in its outermost level. There is a strong tendency for atoms to have a fully populated outer level. Electrons are lost or gained until this condition is reached.

3. A molecule is a stable collection of atoms. The forces holding atoms together in a molecule are called chemical bonds.

4. The force of a chemical bond can result from the attraction of opposite charges, as in an ionic bond, or from the sharing of one or more pairs of electrons, as in a co-valent bond.

5. The chemistry of life is the chemistry of water. In water molecules the oxygen atom has a stronger attraction for the electrons shared between oxygen and the hydrogen atoms. As a result the oxygen atom is electron-rich (partial negative charge), and the hydrogen atoms are electron-poor (partial positive charge). This charge separation is like that of a magnet with positive and negative poles. Water is termed a "polar" molecule.

6. A hydrogen bond is formed by the attraction of the partial positive charge of one hydrogen atom of a water molecule with the partial negative charge of the oxygen atom of another. Water molecules tend to form the maximum number of hydrogen bonds, and tend to exclude nonpolar molecules.

7. Organisms store energy in carbon-hydrogen (C—H) bonds. The most important of the energy-storing carbohydrates is glucose, a six-carbon sugar.

8. Excess energy resources may be stored in complex sugar polymers called starches, especially in plants. Glycogen, a comparable storage polymer that is frequent in animals, is characterized by complex branching.

9. Fats are molecules containing many more C—H bonds than carbohydrates, providing more efficient energy storage. Fats are efficient storage molecules.

10. Proteins are linear polymers of amino acids. Because the 20 amino acids that occur in proteins have side groups with very different chemical properties, the function and shape of a protein is critically affected by its particular sequence of amino acids.

11. Hereditary information is stored as a sequence of nucleotides in a linear nucleotide polymer called deoxyribonucleic acid, or DNA. DNA is a double helix. A second form of nucleic acid, ribonucleic acid, or RNA, is similar in structure and is made as a template copy of portions of the DNA. In the cell, RNA provides a blueprint specifying the amino acid sequence of proteins.

REVIEW

1. Which molecule is most important to the evolution of life?

2. The three subatomic particles that make up atoms are _____, _____, and _____. Which two make up the nucleus of an atom?

3. If you came across a mass of krypton, it would not be a glowing green rock as depicted in some popular literature but rather a _____, because krypton is a noble element.

4. Plants store glucose as _____. Animals store glucose in long, highly branched chains called _____.

5. The nucleic acid _____ is the molecule that encodes the information specifying the amino acid sequences for proteins and thus contains the information necessary to make organisms.

SELF-QUIZ

1. Which three chemical forces act to allow atoms to combine to form molecules?
   (a) the tendency of electrons to occur in pairs
   (b) the tendency of electrons to reverse charge
   (c) the tendency of atoms to balance positive and negative charges
   (d) the tendency of the outer shell of the electrons to be full
   (e) the tendency of large masses to attract smaller ones

2. To satisfy the octet rule, a carbon atom must gain access to _____ additional electrons.
   (a) one   (d) four
   (b) two   (e) eight
   (c) three

3. Which is an example of a polar molecule that is soluble in water?
   (a) water   (d) fats
   (b) oxygen gas   (e) cellulose
   (c) ammonia

4. The primary structure of a protein is:
   (a) the sequence of amino acids that make up the chain
   (b) usually a helical coil
   (c) the shape of the protein
   (d) influenced by the amino acid sequence
   (e) when two proteins associate to form a functional unit

5. A nucleotide, the repeating subunit that makes up a nucleic acid, is composed of which *three* smaller building blocks?
   (a) a glucose
   (b) a fatty acid
   (c) a ribose
   (d) a phosphate group
   (e) a nitrogen-containing base

PROBLEMS

1. What is the molecular weight of water? How much would a mole of water weigh?

2. How many different three-base DNA sequences are possible?

THOUGHT QUESTIONS

1. Carbon (atomic number 6) and silicon (atomic number 14) both have four vacancies in their outer energy levels. Ammonia is even more polar than water. Why do you suppose life evolved in the form of organisms made up of carbon chains in water solution rather than ones of silicon in ammonia?

2. Carbon atoms can share four electron pairs to form molecules. Why doesn't carbon form a bimolecular gas, the way hydrogen (one pair of shared electrons), oxygen (two pairs of shared electrons), and nitrogen (three pairs of shared electrons) do?

FOR FURTHER READING

DOOLITTLE, R.: "Proteins," *Scientific American,* October 1985, pages 88–99. A good general description of protein primary, secondary, and tertiary structure, with emphasis on protein evolution.

FRIEDEN, E.: "The Chemical Elements of Life," *Scientific American,* July 1972, pages 52–64. A nice introduction to the diversity of atoms in living organisms, with emphasis on the diversity of trace elements.

KARPLUS, M., and A. McCAMMON: "The Dynamics of Proteins," *Scientific American,* April 1986, pages 42–51. Proteins are not fixed in their shape like car parts, but rather are flexible. This article explains why flexibility is critical to protein function.

SHARON, N.: "Carbohydrates," *Scientific American,* November 1980, pages 90–116. An overview of the structures of carbohydrates and the diverse roles they assume in organisms.

FIGURE 4-1   One can imagine that the earth looked something like this area in Yellowstone National Park before the origin of life. The brownish streaks, however, are masses of bacteria, indicating that the scene more closely resembles conditions that occurred billions of years ago soon after the origin of life.

THE ORIGIN OF LIFE

Chapter 4

Overview

Life originated on earth more than 3.5 billion years ago, within 1 billion years after our planet's formation. We do not know how life originated, although the evidence is consistent with the hypothesis that it evolved from nonliving materials. All organisms are composed of one or more cells, which are the basic units of life. The cells of bacteria have little internal organization and were the first kind of cells to appear on earth. Later a new kind of cell evolved with a more elaborate internal organization; organisms that have such cells are called eukaryotes. Eventually multicellular organisms originated independently from a number of different groups of eukaryotes.

For Review

Here are some important terms and concepts that you will encounter in this chapter. If you are not familiar with them, you should review them before proceeding.

Properties of life (Chapter 1)

Amino acids (Chapter 3)

Lipids (Chapter 3)

When we look around us and see a world teeming with life, it is difficult to imagine that there was a time when no life existed on earth, when no grass grew and no fish swam in the sea. However, the earth is much older than life. By studying how radioactive isotopes decay, scientists have determined that the earth was formed about 4.5 billion years ago. At first the earth was molten, but soon a thin crust of rock, the shell on which we live, solidified over the hot core. Early earth was a land of molten rock and violent volcanic activity. The oldest rocks that have survived on earth are about 3.9 billion years in age. These ancient rocks contain no definite traces of life, or at least none that can be recognized with our current level of technology.

Today the world is very different, and life exists in every crack and crevice (Figure 4–1). Except for a blast furnace, you would be hard pressed to think of a place anywhere on the surface of the earth where life does not exist in profusion. Where did all of this life come from?

It is not easy to answer this question. You cannot go back in time and see for yourself how life originated, nor are there any witnesses. We can learn something about what must have happened by studying the rocks of the earth, but the record of events is incomplete and often silent. Perhaps the most fundamental of these issues is the nature of the agency or force that led to the origin of life. In principle, there are at least three possibilities:

1. *Extraterrestrial origin.* Life may not have originated on earth at all but instead may have been carried to it, perhaps as an extraterrestrial infection of spores originating on a planet of a distant star. How life came to exist on *that* planet is a question we cannot hope to answer soon.
2. *Special creation.* Life forms may have been put on earth by supernatural or divine forces. This viewpoint, common to most Western religions, is the oldest hypothesis and is widely accepted. It forms the basis of the "scientific creationism" viewpoint discussed in Chapter 2.
3. *Evolution.* Life may have evolved from inanimate matter, with associations among molecules becoming more and more complex. In this view the force leading to life was selection; changes in molecules that increased their stability caused the molecules to persist longer.

In this book we deal only with the third possibility and thus attempt to understand whether the forces of evolution could have led to the origin of life—and, if so, how the process might have occurred. This is not to say that the third possibility is the correct one. Any one of the three possibilities listed might be true. Nor do any of the possibilities for the origin of life preclude the existence of a Divine Being, who might have acted in various ways. However, we are limiting the scope of our inquiry to scientific matters. Of these three possibilities, only the third permits the construction of testable hypotheses. Therefore evolution provides the only *scientific* explanation, that is, one that could potentially be disproved by experiment, by obtaining and analyzing actual information.

In our search for this understanding, we must look back to the time before life appeared, when the earth was just starting to cool. We must go back at least that far because there are fossils of simple living things called bacteria in rocks that are about 3.5 billion years old. Because of the existence of these fossils, we know that life had originated by the end of the first billion years of our planet's history. In attempting to determine how the first organisms originated, we must first consider the mode of origin of organic molecules, which are the building blocks of organisms. Then we shall consider how organic molecules might have become organized into living cells.

THE ORIGIN OF ORGANIC MOLECULES: CARBON POLYMERS

Scientists who study the conditions of the primitive earth are called geochemists. Geochemists believe that as the primitive earth cooled and its rocky crust formed, many gases were released from the molten core. It was a time of volcanoes, blasting enormous amounts of material skyward. These gases formed a cloud around the earth and were held as an atmosphere by the earth's gravity. The atmosphere we breathe now is very different from what it used to be; it has been changed by the activities of organisms, as we shall see later. Despite the changes, however, geochemists have been able to learn what the early atmosphere must have been like by studying the gases released by volcanoes and by deep sea vents in the earth's crust. Geochemists do not all agree on the exact composition of this original atmosphere, but they do agree that it was composed principally of nitrogen gas. It also contained significant amounts of carbon dioxide and water. It is probable, although not certain, that compounds in which hydrogen atoms were bonded to other light elements such as sulfur, nitrogen, and carbon were also present in the atmosphere of the early earth. These compounds would have been hydrogen sulfide (H_2S), ammonia (NH_3), and methane (CH_4).

The atmosphere of the early earth was probably rich in hydrogen, although there is debate on this point. We refer to such an atmosphere as a reducing one because of the ample availability of hydrogen atoms and associated electrons (in chemistry the donation of electrons to a molecule is called **reduction,** and the removal of electrons is **oxidation**). Little if any oxygen gas was present. In such a reducing atmosphere it does not take much energy to form the carbon-rich molecules from which life evolved. Later the earth's atmosphere changed as living organisms began to carry out photosynthesis, which involves harnessing the energy in sunlight to split water molecules and form complex carbon-containing molecules, giving off gaseous oxygen molecules in

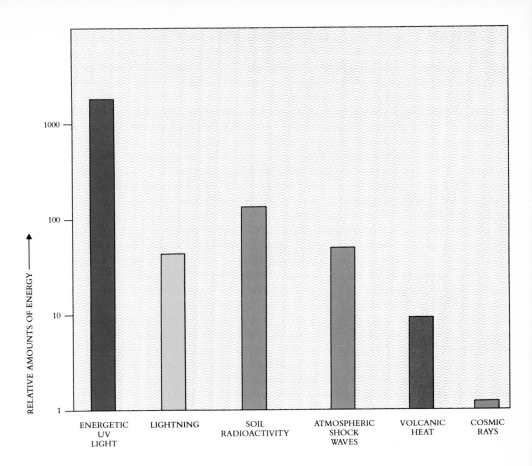

RELATIVE AMOUNTS OF ENERGY →

| ENERGETIC UV LIGHT | LIGHTNING | SOIL RADIOACTIVITY | ATMOSPHERIC SHOCK WAVES | VOLCANIC HEAT | COSMIC RAYS |

the process. Our atmosphere is now approximately 21% oxygen. In the oxidizing atmosphere that exists today, the spontaneous formation of complex carbon-containing molecules cannot occur.

Therefore, the first step in the evolution of life probably occurred in a reducing atmosphere that was devoid of gaseous oxygen and thus was very different from the atmosphere that exists now. Those were violent times, and the earth was awash with energy (Figure 4-2): solar radiation, lightning from intense electrical storms, violent volcanic eruptions, and heat from radioactive decay. Living on earth today, shielded from the effects of solar ultraviolet radiation by a layer of ozone gas (O_3) in the upper atmosphere, most humans cannot imagine the enormous flux of ultraviolet energy to which the early earth's surface was exposed. Subjected to ultraviolet energy and to other sources of energy as well (Figure 4-3), the gases of the early earth's atmosphere underwent chemical reactions with each other and formed a complex assemblage of molecules. In the covalent bonds of these molecules, some of the abundant energy present in the atmosphere was captured as chemical energy.

What kinds of molecules might have been produced? One way to answer this question is to repeat the process: (1) assemble an atmosphere similar to the one thought to exist on the early earth; (2) place this atmosphere over liquid water, which was present on the surface of the cooling earth; (3) exclude gaseous oxygen from the atmosphere since none was present in the atmosphere of the early earth; (4) maintain this mixture at a temperature somewhat below 100° C; and (5) bombard it with energy in the form of electrical sparks. When Harold C. Urey and his student Stanley L. Miller performed this experiment in 1953, they found that within 1 week 15% of the carbon that was originally present as methane gas had been converted into more complex carbon-based molecules.

Among the first substances produced in the Miller-Urey experiments (Figure 4-4) were molecules derived from the breakdown of methane, including formaldehyde and hydrogen cyanide. These molecules then combined to form more complex molecules containing carbon-carbon bonds, including the amino acids glycine and alanine.

FIGURE 4-2 (Left)

Before life evolved, the simple molecules in the earth's atmosphere combined to form more complex molecules. The energy that drove some of these chemical reactions came from lightning and other forms of geothermal energy.

FIGURE 4-3 (Right)

Sources of energy for the synthesis of complex molecules in the atmosphere of the primitive earth. Of the ultraviolet radiation, only the very short wavelengths (less than 100 nanometers) would have been effective in promoting chemical reactions. Electrical discharges are thought to have been more common on the primitive earth than they are now.

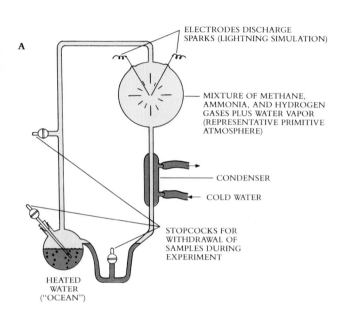

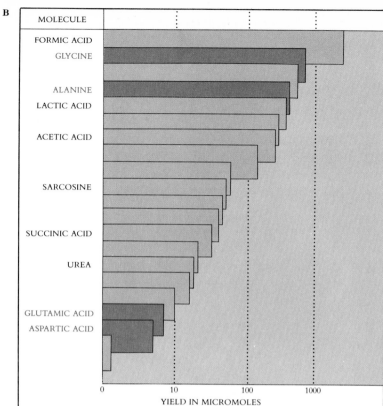

MOLECULE			
FORMIC ACID			
GLYCINE			
ALANINE			
LACTIC ACID			
ACETIC ACID			
SARCOSINE			
SUCCINIC ACID			
UREA			
GLUTAMIC ACID			
ASPARTIC ACID			
	0 10 100 1000		
	YIELD IN MICROMOLES		

FIGURE 4-4

A The Miller-Urey experiment. The apparatus consisted of a closed tube connecting two chambers. The upper chamber contained a mixture of gases thought to resemble the earth's atmosphere. Through this mixture electrodes discharged sparks, simulating lightning. Condensers then cooled the gases, causing water droplets to form, which passed to the second heated chamber, the "ocean." Any complex molecules formed in the atmosphere chamber would be carried dissolved in these droplets to the ocean chamber, from which samples were withdrawn for analysis.
B Some of the 20 most common complex molecules detected in the original Miller-Urey experiments are indicated. Among these 20 molecules are four amino acids, the names of which are shown in blue type.

Amino acids are important in understanding the origin of life because they are the basic building blocks of proteins, which are one of the major kinds of molecules of which organisms are composed. About 50% of the dry weight of each cell in your body consists of amino acids—alone or linked together into protein chains.

In later experiments, more than 30 different complex carbon-containing molecules were identified, including the amino acids glycine, alanine, glutamic acid, valine, proline, and aspartic acid. The production of amino acids indicates that proteins could have formed under conditions similar to those that probably existed on the early earth. Other biologically important molecules were also formed in these later experiments, including the purine base adenine, which is a constituent of both DNA and RNA. Thus at least some of the key molecules from which life evolved were created in the atmosphere of the early earth as a by-product of its birth.

Among the molecules that form spontaneously under conditions thought to be similar to those of the primitive earth are some of those that form the building blocks of organisms.

As the earth cooled, much of the water vapor present in the atmosphere condensed into liquid water and accumulated in the ever-expanding oceans. Judging from the results of the Miller-Urey experiments, we presume that the water droplets carried nucleotides, amino acids, and other compounds produced by chemical reactions in the atmosphere.

The primitive oceans must not have been pleasant places. It is odd to think of life originating from a dilute, hot, smelly soup of ammonia, formaldehyde, formic acid, cyanide, methane, hydrogen sulfide, and organic hydrocarbons. Yet within such an ocean arose the organisms from which all later life forms are derived. One cannot escape a certain curiosity about the earliest steps that eventually led to the origin of all living things on earth, including ourselves. How did organisms evolve from complex molecules? What is the "origin of life"?

ORIGIN OF THE FIRST CELLS

Many different kinds of molecules gather together or aggregate in water, much as people from the same foreign country tend to aggregate together in a large city. Sometimes the aggregation of molecules of one kind form a cluster big enough to see. If you shake up a bottle of oil-and-vinegar salad dressing, you can see this happen—small spherical bubbles of oil appear, grow in size, and fuse with one another. Small **microspheres** of this sort, only 1 to 2 micrometers in diameter, form spontaneously from lipid molecules when they are suspended in water. These little microspheres are called **coacervates** (see Figure 1-7). Simple coacervate microdrops that formed in the primeval soup by aggregation of lipids or of proteins were probably the first step in the evolution of cellular organization. Coacervates have several remarkably cell-like properties:

1. Coacervates form an outer boundary that has two layers and thus resembles a biological membrane.
2. Coacervates grow by accumulating more subunit molecules from the surrounding medium.
3. Coacervates form budlike projections and divide by pinching in two, as do bacteria.
4. Coacervates may contain amino acids and use them to facilitate several different kinds of chemical reactions that are mainly found in living cells.

A process of chemical evolution involving coacervate microdrops of this sort may have taken place before the origin of life. The early oceans must have contained untold numbers of these microdrops—billions in a spoonful, each one forming spontaneously, persisting for a while, and then dispersing. Some of the droplets would by chance have contained amino acids with side groups that were better able than the others to catalyze growth-promoting reactions. These droplets would have survived longer than the others because the persistence of both protein and lipid coacervates is greatly increased when they carry out metabolic reactions such as glucose degradation (breakdown) and when they are actively growing.

Over millions of years the complex microdrops that were better able to incorporate molecules and energy from the lifeless oceans of the early earth would have tended to persist more than the others. (This is the chemical equivalent of natural selection, discussed in Chapter 2.) Also favored would have been those microdrops that could use these molecules to expand in size and eventually grow large enough to break into "daughter" microdrops with features similar to those of their "parent" microdrop. The daughter microdrops would have been able to grow by utilizing the favorable combination of characteristics acquired from their parents. Some of the microdrops, by chance rearrangement of their parts, acquired the means to facilitate the transfer of this ability from parent to offspring. Those microdrops gained the property of heredity, and life had begun.

THE EARLIEST CELLS

The fossils that we have found in ancient rocks (Figure 4-5) represent an obvious progression from simple to complex organisms during the vast period of time that began no more than a billion years after the origin of the earth. Living things may have been present closer to the time of the origin of the earth, but rocks in which their fossils might have been preserved are unknown.

What do we know about these early life forms? We have learned from studying early microfossils that for most of the history of life, organisms were very simple. Like the bacteria living today, they were small (1 to 2 micrometers in diameter), single-celled creatures with little evidence of internal structure and no external appendages.

We call these simple organisms **prokaryotes,** from the Greek words for "before" and "kernel" or nucleus. The name reflects their lack of a **nucleus,** which is a spherical organelle (structure) within cells that evolved later in other kinds of organisms. We re-

FIGURE 4-5

Fossil bacteria.
A A scanning electron micrograph of one of the oldest fossils yet discovered, a bacterium from South African rocks 3.4 billion years old.
B Cross-sections through fossil bacteria from the Bitter Springs Formation of Australia, in which the cell walls are clearly visible. These fossils are about 850 million years old.

fer to the prokaryotes collectively as **bacteria.** More complex living forms did not appear until about 1.5 billion years ago. Therefore for at least 2 billion years—nearly half the age of the earth—bacteria were the only organisms that existed.

A Living Fossil

Most organisms living today resemble one another fundamentally, having the same kinds of membranes and hereditary systems and many similar aspects of metabolism. However, not all living organisms are exactly the same in these respects. If we look carefully in uncommon environments, we occasionally encounter organisms that are quite unusual, differing in form and metabolism from most other living things. Sheltered from evolutionary alteration in unchanging habitats that resemble those of earlier times, these living relics are the surviving representatives of the first ages of life on earth. In those ancient times biochemical diversity was the rule, and living things did not resemble each other in their metabolic features as closely as they do today. In places such as the oxygenless depths of the Black Sea or the boiling waters of hot springs (see Figure 4-1), we can still find bacteria living without oxygen and displaying a bewildering array of metabolic strategies. Some of these bacteria have shapes similar to those of the fossils of bacteria that lived 2 or 3 billion years ago.

The search for primitive and unusual bacteria can lead biologists to strange places, such as the ruins of an ancient stable near Harlech Castle in Wales. A stable existed on that site continuously for some 700 years in ancient times, and the soil beneath the stable is quite unusual. Seven centuries of urine and manure have rendered it rich in ammonia, raising the pH so high that little could be expected to live in it. There, however, was found a curious bacterium called *Kakabekia umbellata* shaped like a miniature umbrella.

This unusual prokaryote was alive and grew quite well when it was provided with concentrated ammonium hydroxide—a much stronger version of the same chemical solution one uses to strip wax from a kitchen floor. However, *Kakabekia* was not a novel life form. Bacteria with such a shape were already well known to biologists—but as ancient fossils! Its distinctive form is common in an outcrop of Gunflint chert in Canada. This ancient rock is about 2 billion years old and teems with well-preserved fossil bacteria. However, *Kakabekia* does not appear later in the fossil record. After an apparent absence of 2 billion years, it was found alive in the alkaline soil of the stable in Wales (Figure 4-6). Perhaps at this particular place, the conditions in the soil closely resemble those of a much earlier time, when global environments were very different from what they are today.

FIGURE 4-6

Kakabekia, an unusual bacterium from ammonia-rich soil in Wales. Fossils of similar bacteria have been found in rock nearly 2 billion years old from Ontario, Canada.

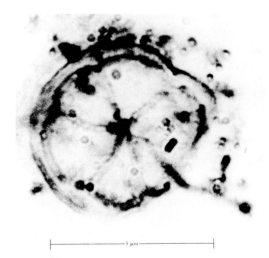

Methane-Producing Bacteria

What were other early bacteria like? Perhaps the most primitive ones that still exist today are the methane-producing bacteria. These organisms are typically simple in form and are able to grow only in an oxygen-free environment. For this reason they are said to grow "without air," or **anaerobically** (Greek *an,* without + *aer,* air + *bios,* life), and are poisoned by oxygen. The methane-producing bacteria convert CO_2 and H_2 into methane gas (CH_4). They resemble all other bacteria in that they possess hereditary machinery based on DNA, a cell membrane composed of lipid molecules, an exterior cell wall, and a metabolism based on an energy-carrying molecule called ATP. However, the resemblance ends there.

When the details of membrane and cell wall structure of the methane-producing bacteria are examined, they prove to be different from those of all other bacteria. There are also major differences in some of the fundamental biochemical processes of metabolism that are the same in other bacteria. The methane-producing bacteria are survivors from an earlier time when there was considerable variation in the mechanisms of cell wall and membrane synthesis, in reading the hereditary information, and in energy metabolism. They appear to represent a road not taken, a side branch of evolution.

Photosynthetic Bacteria

One additional kind of bacteria deserves mention here: those that have the ability to capture the energy of light and transform it into the energy of chemical bonds within cells. These bacteria are photosynthetic, like plants. The pigments used to capture light energy vary greatly in different groups; when these bacteria are massed, they often color the earth, water, or other areas where they grow with characteristic hues (see Figure 4-1).

One of the groups of photosynthetic bacteria that is very important in the history of life on earth is the **cyanobacteria,** sometimes called "blue-green algae." They have the same kind of chlorophyll pigment that is most abundant in plants, plus other pigments that are blue or red. Cyanobacteria produce oxygen as a result of their photosynthetic activities, and when they appeared at least 3 billion years ago, they played the decisive role in increasing the concentration of free oxygen in the earth's atmosphere from below 1% to the current level of 21%. Certain cyanobacteria are also responsible for the accumulation of massive limestone deposits.

The Origin of Modern Bacteria

The early stages of the history of life on earth seem to have been rife with evolutionary metabolic experimentation. Novelty abounded, and many biochemical possibilities were apparently represented among the organisms alive at that time. From the array of different early living forms, representing a variety of biochemical strategies, a very few became the ancestors of the great majority of organisms that are alive today. A few of the other "evolutionary experiments" such as *Kakabekia* and the methane-producing bacteria have survived locally or in unusual habitats, but most others became extinct millions or even billions of years ago.

> **Most organisms now living are descendants of a few lines of early bacteria. Many other diverse forms have not survived.**

Modern bacteria, for the most part, seem to have stemmed from a tough, simple little cell; its hallmark was adaptability. For at least 2 billion years, bacteria were the only form of life on earth (see Figure 4-6). All of the eukaryotes, including animals, plants, fungi, and protists, are their descendants.

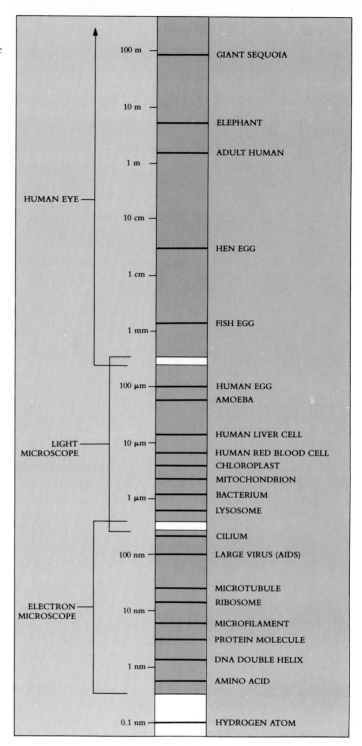

FIGURE 4-7

The size of things, from nanometers to meters. Bacteria are generally 1 to 2 micrometers (μm) thick, and human cells are typically orders of magnitude larger. The scale goes from nanometers (nm) to micrometers (μm) to millimeters (mm) to centimeters (cm) and finally meters (m).

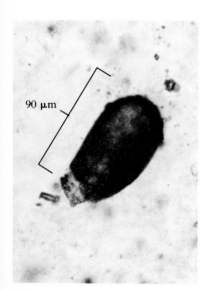

90 μm

FIGURE 4-8

Fossil unicellular eukaryote about 800 million years old. All life was unicellular until about the past 700 million years.

THE APPEARANCE OF EUKARYOTIC CELLS

All fossils that are more than 1.5 billion years old are generally similar to one another structurally. They are small, simple cells (Figure 4-7); most measure 0.5 to 2 micrometers in diameter, and none are more than about 6 micrometers thick.

In rocks about 1.5 billion years old we begin to see for the first time microfossils that are noticeably different in appearance from the earlier simpler forms. These cells are much larger than bacteria and have internal membranes and thicker walls (Figure 4–8). Cells more than 10 micrometers in diameter rapidly increased in abundance. Some fossil cells 1.4 billion years old are as much as 60 micrometers in diameter. Others, 1.5

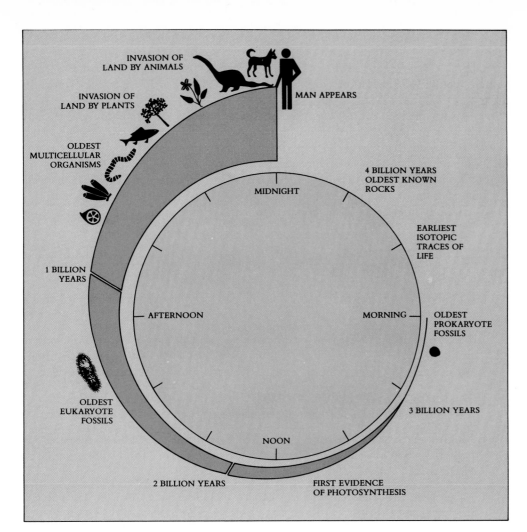

FIGURE 4-9

The clock of biological time. A billion seconds ago it was 1957, and most students using this text had not yet been born. A billion minutes ago Jesus was alive and walking in Galilee. A billion hours ago the first human had not been born. A billion days ago no biped walked on earth. A billion months ago the first dinosaurs had not yet been born. A billion years ago no creature had ever walked on the face of the earth.

billion years old, contain what appear to be small, membrane-bound structures. Many of these fossils have elaborate shapes, and some exhibit highly branched filaments, tetrahedral configurations, or spines.

These early fossil traces mark a major event in the evolution of life. A new kind of organism had appeared. These new cells are called **eukaryotes,** from the Greek words for "true" and "nucleus," since they possess an internal chamber called the cell nucleus. All organisms other than the bacteria are eukaryotes, and they rapidly evolved to produce all of the diverse organisms that inhabit the earth today, including ourselves (Figure 4-9). In the next chapter we shall explore in detail the structure of eukaryotes in relation to the factors involved in their origin.

For at least the first 2 billion years of life on earth, all organisms were bacteria. About 1.5 billion years ago, the first eukaryotes appeared.

IS THERE LIFE ON OTHER WORLDS?

The life forms that evolved on earth closely reflect the nature of this planet and its history. If the earth were farther from the sun, it would be colder, and chemical processes would be greatly slowed down. For example, water would be a solid, and many carbon compounds would be brittle. If the earth were closer to the sun, it would be warmer, chemical bonds would be less stable, and few carbon compounds would persist. Apparently the evolution of a carbon-based life form is possible only within the narrow range of temperatures that exists on earth, and this range of temperature is directly related to the distance from the sun.

The size of the earth has also played an important role because it has permitted a gaseous atmosphere. If the earth were smaller, it would not have a sufficient gravitational pull to hold an atmosphere. If it were larger, it might hold such a dense atmosphere that all solar radiation would be absorbed before it reached the surface of the earth.

Has life evolved on other worlds? In the universe there are undoubtedly many worlds with physical characteristics like those of our planet. The universe contains some 10^{20} stars with physical characteristics that resemble those of our sun; at least 10% of these stars are thought to have planetary systems. If only 1 in 10,000 planets is the right size and at the right distance from its star to duplicate the conditions in which life originated on earth, the "life experiment" will have been repeated 10^{15} times (that is, a million billion times). It seems likely that we are not alone.

SUMMARY

1. Of the many theories of how life might have evolved, only evolution provides a testable explanation.

2. The experimental re-creation of atmospheres, energy sources, and temperatures similar to those thought to have existed on the primitive earth leads to the spontaneous formation of amino acids and other biologically significant molecules.

3. The first cells are thought to have arisen by a process in which aggregations of molecules that were more stable persisted longer.

4. Microscopic fossils of bacteria are found continuously in the fossil record as far back as 3.5 billion years in the oldest rocks that are suitable for the preservation of organisms.

5. Bacteria were the only life forms on earth for 2 billion years or more. The first eukaryotes appear in the fossil record about 1.5 billion years ago. All organisms other than bacteria are descendants of these first eukaryotes.

6. Bacteria are metabolically, but not structurally, diverse. Some unusual ones, which are probably of ancient origin, survive in unusual habitats that may resemble those of the early earth.

REVIEW

1. Which of the possible explanations for the origin of life permits the construction of testable hypotheses?

2. Fossils of bacteria have been found in rocks that are about _____ billion years old.

3. The Miller–Urey experiment demonstrated that a number of biologically important molecules could be produced from a mixture simulating the earth's early atmosphere, using _____ as an energy source. Name another energy source that could have been used.

4. Small _____, which form spontaneously in a mixture of lipids with water, display a number of remarkably cell-like properties.

5. Eukaryotic cells, that is, cells that contain a true nucleus and are found in complex organisms like yourself, first appeared about _____ years ago.

SELF-QUIZ

1. According to geochemists, which of the following was *not* a major component of the earth's atmosphere when life began?
   (a) nitrogen gas   (d) hydrogen
   (b) methane   (e) oxygen gas
   (c) carbon dioxide

2. In which two locations would you *not* look for living organisms that might resemble some of the early prokaryotic organisms?
   (a) in the hot springs in Yellowstone National Park
   (b) in the depths of the Black Sea
   (c) on some of the older items in your refrigerator
   (d) in the floor of an old stable that has accumulated urine and manure for centuries
   (e) in the Gunflint chert, an ancient rock found in Canada

3. Organisms more complex than bacteria begin appearing as fossils in rocks that are about _____ years old.
   (a) 4.5 billion   (d) 600 million
   (b) 3.5 billion   (e) 6000
   (c) 1.5 billion

4. The methane-producing bacteria resemble other prokaryotic organisms in that they (choose three)
   (a) can survive in the presence of oxygen gas
   (b) use DNA as their hereditary information molecule
   (c) have a cell membrane composed of lipid molecules
   (d) use the molecule ATP to drive chemical reactions that require energy
   (e) show the same biochemical processes when synthesizing cell walls and cell membranes

5. Currently, the most likely place in our solar system on which life might occur is Europa, a moon of _____.
   (a) Mercury
   (b) Venus
   (c) Mars
   (d) Jupiter
   (e) Uranus

THOUGHT QUESTIONS

1. In Fred Hoyle's science fiction novel *The Black Cloud,* the earth is approached by a large interstellar cloud of gas. As the cloud orients around the sun, scientists discover that the cloud is feeding, absorbing the sun's energy through the excitation of electrons in the outer energy levels of cloud molecules, a process similar to the photosynthesis that occurs on earth. Different portions of the cloud are isolated from each other by associations of ions created by this excitation. Electron currents pass between these sectors, much as they do on the surface of the human brain, and endow the cloud with self-awareness, memory, and the ability to think. Using electricity produced by static discharges, the cloud is able to communicate with human beings and to describe its history, as well as to maintain a protective barrier around itself. The cloud tells our scientists that it once was smaller, having originated as a small extrusion from an ancestral cloud, but has grown by absorbing molecules and energy from stars like our sun, on which it has been grazing. Eventually the cloud moves off in search of other stars. Is the cloud alive? Which of its features would you consider important in deciding whether the cloud is alive or not?

2. If 1 in 10 of the stars that are like our sun has planets, if 1 in 10,000 of these planets is capable of supporting life, and if 1 of each million life-supporting planets evolves an intelligent life form, how many planets in the universe support intelligent life? Can you think of any objections to this estimate?

FOR FURTHER READING

DICKERSON, R.: "Chemical Evolution and the Origin of Life," *Scientific American,* September 1978, vol. 239 (3), pages 62–78. A lucid exposition of the chemical changes thought to have occurred during the evolution of life.

HOROWITZ, N.: "The Search for Life on Mars," *Scientific American,* November 1977, vol. 237 (5), pages 52–61. A fascinating account of what it means to search for alien life and the difficulty in knowing when you've found it.

MARGULIS, L., and D. SAGAN: *Microcosmos: Four Billion Years of Evolution from Our Microbial Ancestors,* Summit Books, Simon & Schuster, Inc., New York, 1986. In a beautifully written essay, this mother-son team outlines the evolution of life on earth, showing how all the features that we see today are derived from the early evolution of bacteria. Highly recommended.

SCHOPF, J.W.: "The Evolution of the Earliest Cells," *Scientific American,* 1978, vol. 239 (3), pages 110–139. How the earliest cells gave rise to the oxygen present in the earth's atmosphere today.

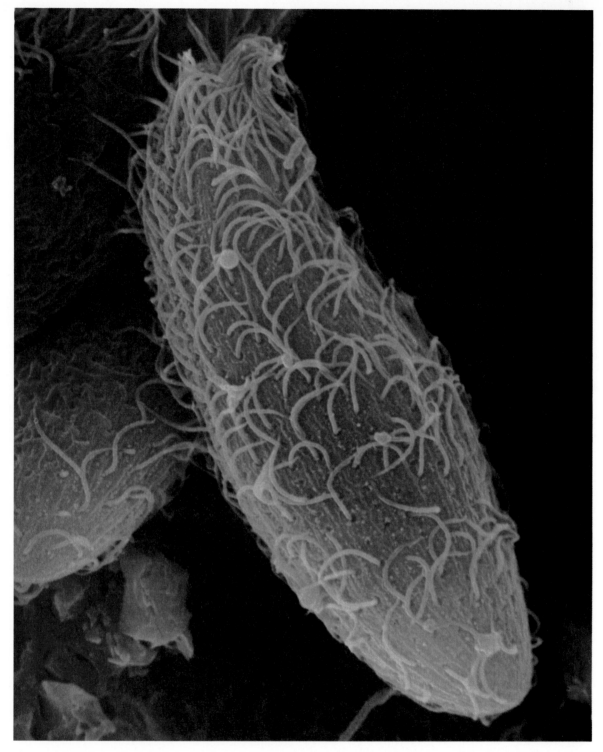

FIGURE 5-1   Scanning electron micrograph (×1000) of the single–celled protist *Dileptus anser*. The hairlike projections that cover the surface are cilia, which the organism undulates to propel itself through the water.

CELLS

Chapter 5

Overview

Unlike the simple cells of bacteria, eukaryotic cells exhibit a considerably greater degree of internal organization, with a dynamic system of membranes forming internal compartments. Some of these compartments are relatively permanent, such as the nucleus that isolates the hereditary apparatus from the rest of the cell. Others are transient, such as the lysosomes that contain digestive enzymes. The partitioning of the cytoplasm into functional compartments is the most distinctive feature of the eukaryotic cell.

For Review

Here are some important terms and concepts that you will encounter in this chapter. If you are not familiar with them, you should review them before proceeding.

Proteins   (Chapter 3)

Lipids (Chapter 3)

Distinction between prokaryotic and eukaryotic cells (Chapter 4)

Evolution of eukaryotes (Chapter 4)

All organisms are composed of cells. Some are composed of a single cell (Figure 5-1), and some, like us, are composed of many cells. The gossamer wing of a butterfly is a thin sheet of cells, and so is the glistening layer covering your eyes. The hamburger you eat is composed of cells, whose contents will soon become part of your cells. Your eyelashes and fingernails, orange juice, and the wood in your pencil—all were produced by or consist of cells. Cells are so much a part of life as we know it that we cannot imagine an organism that is not cellular in nature. In this chapter we will look more closely at cells and learn something of their internal structure. In the following chapters we will focus on cells in action, on how they communicate with their environment, grow, and reproduce.

GENERAL CHARACTERISTICS OF CELLS

Before launching into a detailed examination of cell structure, it is useful to first gain an overview of what to expect, of what we would find on the inside of a typical cell. What is a cell like? A bacterial cell, with its prokaryotic organization, is like a blimp; it has an outer framework that supports an inner membrane bag with a uniform interior. A eukaryotic cell is more like a submarine. A submarine has an outer hull open to the sea and a watertight inner pressure hull; eukaryotic cells also have a porous outer wall covering a membrane that regulates the passage of water and dissolved substances. Within the submarine is a central control room; the control room of a eukaryotic cell is a central compartment called the **nucleus.** The power to drive a submarine comes from the engine room; similarly, a eukaryotic cell's power comes from internal bacteria-like inclusions called **mitochondria.** The rooms of a submarine are divided into watertight compartments; a eukaryotic cell is also divided into separate rooms by a winding membrane system called the **endoplasmic reticulum.** A spinning propeller drives the submarine through the water; motile cells are driven by **flagella,** whip-like structures that undulate rapidly, driving the cell through the medium in which it is swimming.

A typical cell, then, is composed of three elements:

1. A membrane surrounds the cell, isolating it from the outside world. Chapter 6 describes the many passageways and communication channels that span these membranes. They provide the only connection between the cell and the outside world.

2. The nuclear region directs the activities of the cell. In bacteria the genetic material is mostly included in a single, closed, circular molecule of DNA, which resides in a central portion of the cell, unbounded by membranes. In eukaryotes, by contrast, a membrane, the **nuclear membrane,** surrounds the nucleus, which contains the DNA.

3. A semifluid matrix called the **cytoplasm** occupies the volume between the nuclear region and the cell membrane. In bacteria the cytoplasm contains the chemical wealth of the cell, the sugars and amino acids and proteins with which the cell carries out its everyday activities of growth and reproduction. In addition to these elements, the cytoplasm of a eukaryotic cell also contains numerous organized structures called **organelles** or "little organs". Many of these organelles are created by the membranes of the endoplasmic reticulum, which close off compartments within which different activities take place. The cytoplasm of eukaryotic cells also contains organelles that look like bacteria; these organelles, called **mitochondria,** provide power.

All cells share this architecture. In different broad classes of cells, however, the general plan is modified in various ways. For example, the cells of most kinds of organisms—plants, bacteria, fungi, protists—possess an outer cell wall that provides structural strength; animal cells do not. The cells of the majority of organisms possess a single nucleus, whereas the cells of fungi and some other groups have several nuclei. Most cells derive all their power from the kind of organelles called mitochondria, whereas plant cells contain a second kind of bacteria-like powerhouse, **chloroplasts,** in addition to their mitochondria. As we shall see, these differences are relatively minor compared with the many ways in which all cells resemble one another.

> **A cell is a membrane-bound unit containing hereditary machinery and other components, including enzymes, by virtue of which it is able to metabolize substances, to grow, and to reproduce.**

CELL SIZE

Sometimes important things seem so obvious that they are overlooked. In studying cells, for example, it is important that we do not overlook one of their most striking traits—their very small size. Cells are not like shoeboxes, big and easy to study. Instead, they are much smaller, so small that you cannot see a single one of your body's cells with the naked eye. Your body contains about 100 trillion cells. If each cell were the size of a shoebox and they were lined up end-to-end, the line would extend to Mars and back, over 500 million kilometers.

The Cell Theory

It is because cells are so small that they were not observed until microscopes were invented in the mid-seventeenth century. Cells were first described by Robert Hooke in

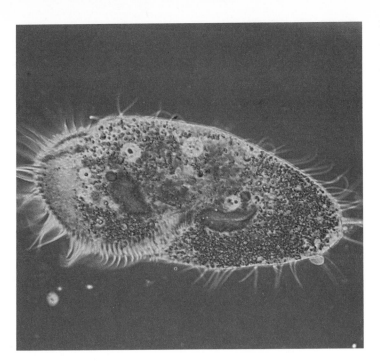

FIGURE 5-2

Among the "animalicules" that can be seen with a microscope are individuals of *Paramecium,* which dash this way and that, engulfing smaller organisms, and clearly are vibrantly alive.

FIGURE 5-3

Although the marine green alga *Acetabularia,* a protist, is a large organism with clearly differentiated parts, such as the stalks and elaborate "hats" visible here, individuals are actually single cells (each with many nuclei), several centimeters tall.

1665. Using a microscope he had built to examine a thin slice of cork, Hooke observed a honeycomb of tiny empty compartments similar to that shown in a photograph taken through a microscope of his time. He called the compartments in the cork *cellulae,* using the Latin word for a small room. His term has come down to us as **cells.** The first living cells were observed by the Dutch naturalist Antonie van Leeuwenhoek a few years later. Van Leeuwenhoek called the tiny organisms that he observed "animalicules"—little animals (Figure 5-2). For another century and a half, however, the general importance of cells was not appreciated by biologists. In 1838 the German Matthias Schleiden, after a careful study of plant tissues, made the first statement of what we now call the cell theory. Schleiden stated that all plants "are aggregates of fully individualized, independent, separate beings, namely the cells themselves." The following year, Theodor Schwann reported that all animal tissues are also composed of individual cells.

The cell theory in its modern form includes four principles:

1. All organisms are composed of one or more cells, within which the life processes of metabolism and heredity occur.
2. Cells are the smallest living things, the basic unit of organization of all organisms.
3. Although life evolved spontaneously in the hydrogen-rich environment of the early earth, biologists have concluded that additional cells are not originating at present. Rather, life on earth represents a continuous line of descent from those early cells.
4. Cells arise only by division of a previously existing cell.

All the organisms on earth are cells or aggregates of cells, and all of us are descendants from the first cells.

Why Aren't Cells Larger?

Cells are not all the same size. Individual cells of the marine alga *Acetabularia,* for example, are up to 5 centimeters long (Figure 5-3). The cells of your body, in contrast, are typically from 5 to 20 micrometers in diameter. If a typical cell in your body were again the size of a shoebox, an *Acetabularia* cell to the same scale would be about 2 kilometers high!

Cells are so small that you cannot see them with the naked eye. Most eukaryotic cells are between 10 and 30 micrometers in diameter. Why can't we see such small objects? Because when two objects are closer together than about 100 micrometers, the two light beams fall on the same "detector" cell at the rear of the eye. Only when two dots are farther apart than 100 micrometers will the beams fall on different cells, and only then can your eye tell that they are two objects and not one.

Robert Hooke and Antonie van Leeuwenhoek were able to see very small cells by magnifying their size, so that the cells appeared larger than the 100 micrometer limit imposed by the structure of the human eye. Hooke and van Leeuwenhoek accomplished this with simple microscopes, which magnified images of cells by bending light through a glass lens. To understand how such a single-lens microscope is able to magnify an image, examine Figure 5-A, *1*. The size of the image that falls on the screen of detector cells lining the back of your eye depends on how close the object is to your eye—the closer the object, the bigger the picture. Your eye, however, is unable to comfortably focus on an object closer than about 25 centimeters (Figure 5-A, *2, top*), because it is limited by the size and thickness of its lens. What Hooke and van Leeuwenhoek did was help the eye out by interposing a glass lens between the object and the eye (Figure 5-A, *2, bottom*). The glass lens added additional focusing power, producing an image of the close-up object on the back of the eye. Because the object is closer, however, the image on the back of the eye is bigger than it would have been had the object been 25 centimeters away from the eye. It is as big as a *much larger* object placed 25 centimeters

away would have appeared without the lens. You perceive the object as magnified or bigger.

The microscope used by van Leeuwenhoek consists of (1) a plate with a single lens, (2) a mounting pin that holds the specimen to be observed, (3) a focusing screw that moves the specimen nearer or farther from the eye, and (4) a specimen-centering screw.

Van Leeuwenhoek's microscope, although simple in construction, is very powerful. One of van Leeuwenhoek's original specimens, a thin slice of cork, was recently discovered among his papers. In the image of that section obtained with van Leeuwenhoek's own microscope, the magnification is 266 times, as good as many modern microscopes. The finest structures visible are less than 1 micrometer (1000 nanometers) in thickness.

Modern microscopes use two magnifying lenses (and a variety of correcting lenses) that act like back-to-back eyes. The first lens focuses the image of the object on the second lens. The image is then magnified again by the second lens, which focuses it on the back of the eye. Microscopes that magnify in stages by using several lenses are called **compound microscopes.** The finest structures visible with modern compound microscopes are about 200 nanometers in thickness. A contemporary light micrograph is shown in Figure 5-B, *1*.

Compound light microscopes are not powerful enough to resolve many structures within cells. A membrane, for example, is only 5 nanometers thick. Why not just add another magnifying lens to the microscope and so increase the resolving power? This approach doesn't work because when two objects are closer than a few hundred nanometers, the light beams

of the two images start to overlap. A light beam vibrates like a vibrating string, and the only way two beams can get closer together and still be resolved is if the "wave length" is shorter.

One way to do this is by using a beam of electrons rather than a light beam. Electrons have a much shorter wavelength, and a microscope employing electron beams has 400 times the resolving power of a light microscope. **Transmission electron microscopes** today are capable of resolving objects only 0.2 nanometer apart—just five times the diameter of a hydrogen atom. The specimen is prepared as a very thin section, and those areas that transmit more electrons (that is, those that are less dense) show up as bright areas in the micrographs (Figure 5-B, *2*).

Transmission electron microscopes receive their name because the electrons used to visualize the specimens are *transmitted* or passed through the material. A second kind of electron microscope, the **scanning electron microscope,** beams the electrons onto the surface of the specimen in the form of a fine probe, which passes back and forth rapidly. In images made with a scanning electron microscope, depressed areas and cracks in the specimen appear dark, while elevated areas such as ridges appear light. The electrons that are reflected back from the surface of the specimen, together with other electrons that the specimen itself emits as a result of the bombardment, are amplified and transmitted to a television screen, where the image can be viewed and photographed. Scanning electron microscopy yields striking three-dimensional images and has proved to be very useful in understanding many biological and physical phenomena (Figure 5-B, *3*).

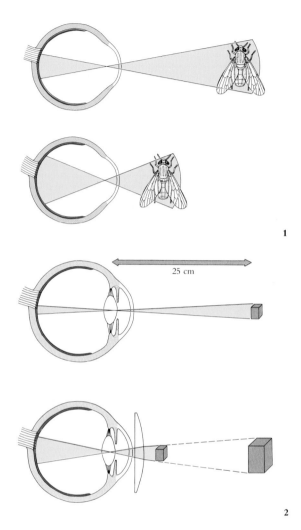

FIGURE 5-A

1 The closer an object is to the eye, the larger the image that falls on the back of that eye. The rear surface of the eye is covered with cells called photoreceptors, and when light falls on these cells they send signals to the vision centers of the brain.

2 (top) The eye will not comfortably focus an object closer than about 25 centimeters, because the lens of the eye must change shape to focus and cannot exceed this limit.

2 (bottom) The lens aids the eye in focusing the close object. Because the object is closer, it produces a larger image on the back of the eye, and so appears "larger." A much larger object would have been required to produce an image of the same size without the lens.

FIGURE 5-B

Three micrographs of the same cell, a fossil pollen grain of an extinct species of plant, *Integricorpus amicus*.

1 Light micrograph, $1600\times$.

2 Transmission electron micrograph, $2600\times$.

3 Scanning electron micrograph, $2000\times$.

Why are our bodies made up of so many tiny cells? The centralized control of each cell is essential in maintaining it as a distinct, functional unit. Substances travel through a cell by diffusion, which is a relatively slow process; if a cell were too large, it could not function efficiently. Thus the nucleus must send commands to all parts of the cell: molecules that direct the synthesis of certain enzymes, the entry of ions from the exterior, the assembly of organelles. These molecules must pass by diffusion from the nucleus to all parts of the cell, and it takes them a very long time to reach the periphery of a large cell. For this reason, an organism made up of relatively small cells is at an advantage over one composed of larger cells.

The advantage of small cell size is also seen in terms of what is called the **surface-to-volume ratio.** As cell size increases, the volume grows much more rapidly than does the surface area. For a round cell, surface area increases as the square of diameter, whereas volume increases as the cube. Thus a cell with 10 times greater diameter would have 10 squared or 100 times the surface area but 10 cubed or 1000 times the volume. A cell's surface provides its only opportunity to interact with the environment, and large cells have far less surface per unit volume than do small ones. All substances must enter and exit from a cell via the **plasma membrane,** a structure of fundamental importance that will be discussed later in this chapter and in more detail in Chapter 6. This membrane plays a key role in controlling cell function, something that is more effectively done when cells are relatively small.

We have many small cells rather than few large ones because small cells can be commanded more efficiently and have a greater opportunity to communicate with their environment.

THE STRUCTURE OF SIMPLE CELLS: BACTERIA

Bacteria are the simplest cellular organisms. Over 2500 species that have been given names are considered to be distinct, but doubtless many times that number actually exist and have not yet been described properly. Although these species are diverse in form (Figure 5-4), their organization is fundamentally similar: small cells about 1 to 10 micrometers thick, enclosed within a membrane and encased within a rigid cell wall, with no distinct interior compartments (Figure 5-5). Sometimes the cells of bacteria adhere in chains or masses, but fundamentally the individual cells are separate from one another.

FIGURE 5-4

Bacterial cells have several different shapes.
A A rod-shaped bacterium, *Pseudomonas,* a type associated with many plant diseases.
B *Streptococcus,* a more or less spherical bacterium in which the individuals adhere in chains.
C *Spirillum,* a spiral bacterium. This large bacterium has a tuft of flagella at each end.

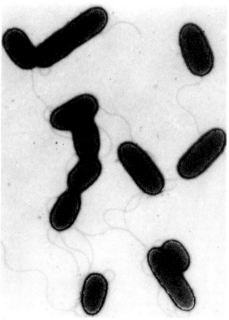

A

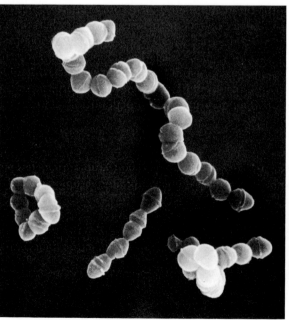

B

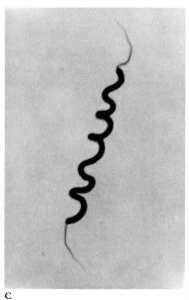

C

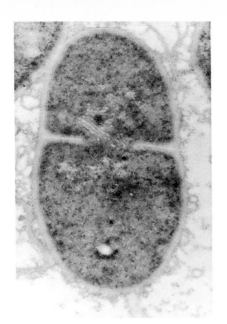

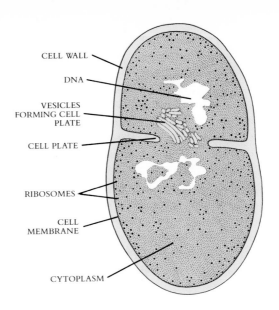

FIGURE 5-5

The structure of a bacterial cell (Corynebacterium; × 48,700).

CELL WALL

DNA

VESICLES FORMING CELL PLATE

CELL PLATE

RIBOSOMES

CELL MEMBRANE

CYTOPLASM

Compared with the other kinds of cells that have evolved from them, prokaryotic cells (bacteria) are smaller and lack interior organization.

Cell Walls

Bacteria are encased by a strong **cell wall,** in which a carbohydrate matrix (a polymer of sugars) is cross-linked by short peptide units. No eukaryotes possess cell walls with a chemical composition of this kind. Bacteria are commonly classified by differences in their cell walls as gram positive and gram negative. The name refers to the Danish microbiologist Hans Christian Gram, who developed a staining process that distinguishes the two classes of bacteria as a way to detect the presence of certain disease-causing bacteria. **Gram-positive** bacteria have a single, thick cell wall that retains the **gram stain** within the cell, causing the stained cells to appear purple under the microscope. More complex cell walls have evolved in other groups of bacteria. In them, the cell wall is thinner and it does not retain the Gram stain; such bacteria are called **gram negative.** Bacteria are susceptible to different kinds of antibiotics depending on the structure of their cell walls.

Simple Interior Organization

If you were to look at an electron micrograph of a thin section of a bacterial cell (see Figure 5-5), you would be struck by its simple organization. There are no internal compartments bounded by membranes or membrane-bounded organelles—the kinds of distinct structures that are so characteristic of eukaryotic cells. The entire cytoplasm of a bacterial cell is one unit, with no internal support structure; thus the strength of the cell comes primarily from its rigid wall.

The external membrane of bacterial cells often intrudes into the interior of the cell, where it may play an important role. When the cells of bacteria divide, for example, the circular, closed DNA molecule first replicates, and the two DNA molecules that result may attach to the cell membrane at different points. Their attachment may then assist in each one of the two identical units of DNA being included in a different one of the cells that results from the division. In some photosynthetic bacteria, the cell membrane is often extensively folded, with the folds extending into the cell's interior (Figure 5-6). On these folded membranes are located the bacterial pigments connected with photosynthesis.

FIGURE 5-6

Electron micrograph of a photosynthetic bacterial cell, *Prochloron,* showing the extensive folded photosynthetic membranes. The cellular DNA is located in the clear area in the central region of the cell.

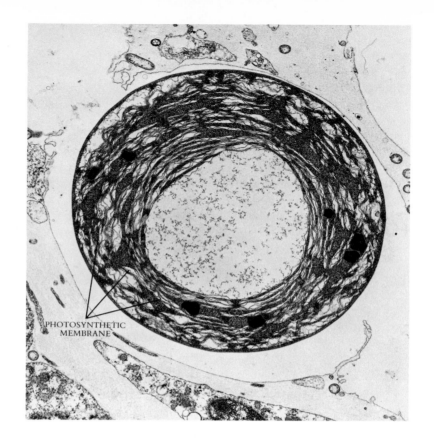

PHOTOSYNTHETIC MEMBRANE

Since there are no membrane-bounded compartments within a bacterial cell, however, both the DNA and the enzymes within such a cell have access to all parts of the cell. Reactions are not compartmentalized as they are in eukaryotic cells, and the whole bacterium operates very much as a single unit.

Bacteria are encased by an exterior wall composed of carbohydrates cross-linked by short peptides. They lack interior compartments.

A COMPARISON OF BACTERIA AND EUKARYOTIC CELLS

Eukaryotic cells (Figure 5-7) are far more complex than prokaryotic ones. Compared with their bacterial ancestors, eukaryotic cells exhibit significant morphological differences:

1. Their DNA is packaged tightly into compact units called **chromosomes,** which contain both DNA and proteins. These eukaryotic chromosomes are located within a separate organelle, the nucleus.
2. The interiors of eukaryotic cells are subdivided into membrane-bounded compartments, which permits one biochemical process to proceed independently of others that may be going on at the same time. The compartmentalization of biochemical activities in eukaryotes serves to increase the efficiency of the various processes.
3. The cells of animals and some protists lack cell walls.
4. The mature cells of plants often contain large fluid-filled internal sacs called **central vacuoles** that are not present in animal cells (Figure 5-8).

The interiors of eukaryotic cells are subdivided by membranes in a complex way. The DNA of the cell is present within the nucleus and associated with protein in units called chromosomes. Most eukaryotes possess cell walls, but they are lacking in animals and some single-celled organisms.

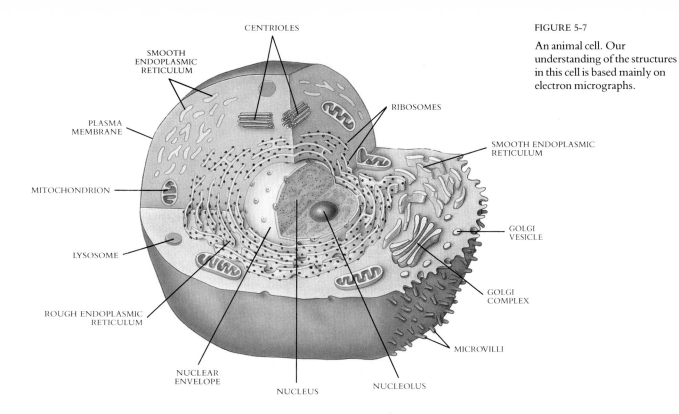

CENTRIOLES

SMOOTH ENDOPLASMIC RETICULUM

PLASMA MEMBRANE

MITOCHONDRION

LYSOSOME

ROUGH ENDOPLASMIC RETICULUM

NUCLEAR ENVELOPE

NUCLEUS

NUCLEOLUS

RIBOSOMES

SMOOTH ENDOPLASMIC RETICULUM

GOLGI VESICLE

GOLGI COMPLEX

MICROVILLI

FIGURE 5-7

An animal cell. Our understanding of the structures in this cell is based mainly on electron micrographs.

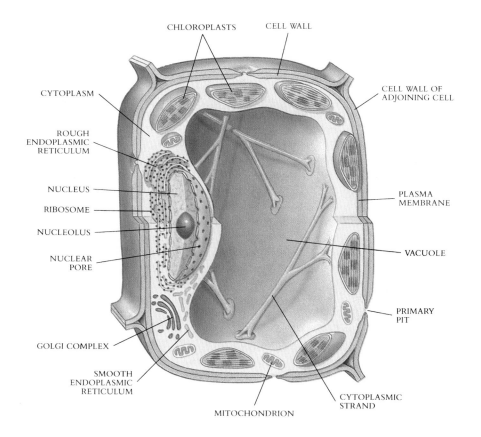

CHLOROPLASTS

CELL WALL

CYTOPLASM

ROUGH ENDOPLASMIC RETICULUM

NUCLEUS

RIBOSOME

NUCLEOLUS

NUCLEAR PORE

GOLGI COMPLEX

SMOOTH ENDOPLASMIC RETICULUM

MITOCHONDRION

CYTOPLASMIC STRAND

CELL WALL OF ADJOINING CELL

PLASMA MEMBRANE

VACUOLE

PRIMARY PIT

FIGURE 5-8

Most mature plant cells contain large central vacuoles, which occupy a major portion of the internal volume of the cell. The cytoplasm occupies a thin layer between vacuole and cell membrane. Within it are all of the cell's mitochondria and other organelles. The illustration shows the relative proportions of the different parts of a plant cell.

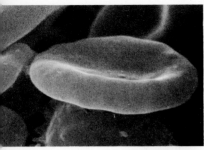

A

B

FIGURE 5-9

A A red blood cell. It is shaped like a flattened disk with a depressed center—not unlike a doughnut with a hole that doesn't go all the way through. **B** This electron micrograph of a thin section of a red blood cell (× 200,000) shows clearly the double nature of the plasma membrane, which is indicated by arrows. The membrane consists of two layers of phospholipid molecules with their hydrophobic tails pointing inward. The phospholipid molecules themselves are electron-dense, and therefore dark in this micrograph, while the area where the tails are concentrated is electron-transparent.

THE INTERIOR OF EUKARYOTIC CELLS: AN OVERVIEW

Although eukaryotic cells are diverse in form and function, they share in common a basic architecture. They all are bounded by a membrane called the **plasma membrane,** they all contain a supporting matrix of protein called a **cytoskeleton,** and they all possess numerous organelles. The organelles are of two general kinds: (class 1) membranes or organelles derived from membranes and (class 2) bacteria-like organelles.

Class 1	**Class 2**
endoplasmic reticulum	mitochondria
nucleus	chloroplasts
Golgi bodies	
lysosomes	
microbodies	

BOUNDARY OF THE EUKARYOTIC CELL: THE PLASMA MEMBRANE

The **plasma membrane** of a eukaryotic cell is a double layer of lipid about 9 nanometers thick (Figure 5-9). This membrane envelops the cell, and nothing can enter or leave the cell without passing across it. In fact, very little does cross the lipid layer itself. However, traversing the lipid layer are a variety of proteins that control the interactions of the cell with its environment.

1. *Channels.* Some of these proteins, called channels, act as doors that admit specific molecules to the cell; for example, membranes possess specific channels for sodium ions and others for glucose.
2. *Receptors.* Other proteins that cross the membrane serve to transmit information rather than molecules. These proteins, called receptors, induce changes within the cell when they come in contact with particular molecules on the cell surface. Some hormones induce changes in cells by binding to such surface receptors.
3. *Markers.* A third class of proteins embedded within the membrane serves to identify a cell as being of a particular type. This is very important in a multicellular individual, since cells must be able to recognize one another for tissues to form and function correctly.

Chapter 6 presents a more detailed look at the structure of cell membranes and how they function.

THE ENDOPLASMIC RETICULUM

As viewed with a light microscope, the interiors of eukaryotic cells exhibit a relatively featureless matrix, within which various organelles are embedded. With the advent of electron microscopes, however, a very striking difference becomes evident: the interiors of eukaryotic cells are seen to be packed with membranes. So thin that they are not visible with the relatively low resolving power of light microscopes, these membranes fill the cell, dividing it into compartments, channeling the transport of molecules through the interior of the cell, and providing the surfaces on which enzymes act. This system of internal compartments created by these membranes in eukaryotic cells constitutes the most fundamental distinction between eukaryotes and prokaryotes.

The extensive system of internal membranes that exists within the cells of eukaryotic organisms is called the **endoplasmic reticulum,** often abbreviated ER (Figure 5-10). The term *endoplasmic* means "within the cytoplasm," and the term *reticulum* comes from a Latin word that means "a little net." Like the plasma membrane, the ER is composed of a double layer of lipid, with various enzymes attached to its surface. The ER, weaving in sheets through the interior of the cell, creates a series of channels and interconnections between its membranes that isolates some spaces as membrane-enclosed bags called **vesicles.**

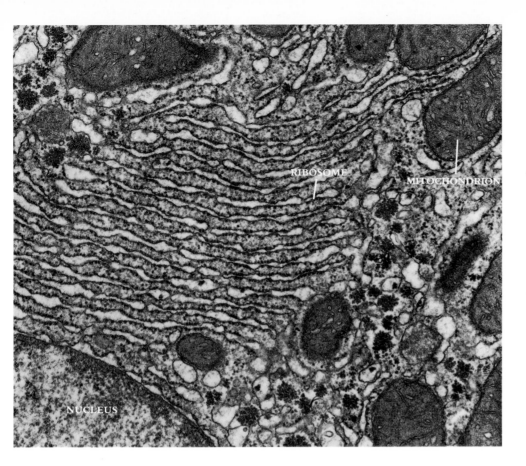

NUCLEUS

RIBOSOME

MITOCHONDRION

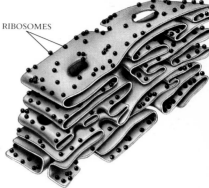

RIBOSOMES

FIGURE 5-10

Rough endoplasmic reticulum. The electron micrograph is of a rat liver cell, rich in ER-associated ribosomes. In the drawing you can see that the ribosomes are associated with only one side of the rough ER; the other side bounds a separate compartment within the cell into which the ribosomes extrude newly made proteins destined for secretion. Figure 5-15 shows how the endoplasmic reticulum functions in a living cell.

Rough ER

The surface of the ER is the place where the cell manufactures proteins intended for export from the cell, such as enzymes secreted from the cell surface. The manufacture is carried out by **ribosomes,** large molecular aggregates of protein and ribonucleic acid (RNA) that translate RNA copies of genes into protein. As new proteins are made on the surface of the ER, they are passed out across the ER membrane into the vesicle-forming system called the Golgi complex (discussed below). They then travel within vesicles to the inner surface of the cell, where they are released outside the cell in which they were produced. From the time the protein is first synthesized on the ER-bound ribosome and crosses into these channels, it is, in a sense, already located outside the cell.

The surfaces of those regions of the ER devoted to the synthesis of such transported proteins are heavily studded with ribosomes (see Figure 5-10). These membrane surfaces appear pebbly, like the surface of sandpaper, when viewed with an electron microscope. Because of this "rocky beach" appearance, the regions of ER that are rich in bound ribosomes are often termed **rough ER.** Regions of the ER in which bound ribosomes are relatively scarce are correspondingly called **smooth ER.**

Smooth ER

Many of the cell's enzymes cannot function when floating free in the cytoplasm; they are active only when they are associated with a membrane. ER membranes contain embedded within them many such enzymes. Enzymes anchored within the ER, for example, catalyze the synthesis of a variety of carbohydrates and lipids. In cells that carry out extensive lipid synthesis, such as the cells of the testicles, smooth ER is particularly abundant. Intestinal cells, which synthesize triglycerides, and the cells of the brain are also rich in smooth ER. In the liver, enzymes embedded within the smooth ER are involved in a variety of detoxification processes. Drugs such as amphetamines,

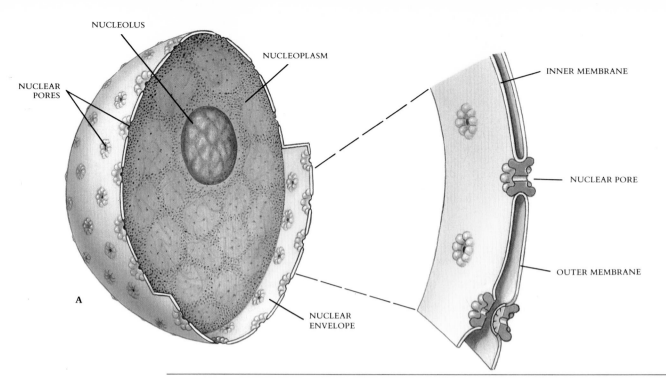

NUCLEOLUS

NUCLEAR PORES

NUCLEOPLASM

INNER MEMBRANE

NUCLEAR PORE

OUTER MEMBRANE

NUCLEAR ENVELOPE

A

FIGURE 5-11

The nucleus is composed of a double membrane called a nuclear envelope enclosing a fluid-filled interior **(A).** Within the fluid, the chromosomes are located. **B** was prepared for electron microscopy by the freeze-etch method, in which the cell is fractured and a micrograph taken of the resulting surface. In cross section **(C),** the individual nuclear pores *(P)* are seen to extend through the two membrane layers of the envelope; the dark material within the pore is protein, which acts to control access through the pore. *N* = nucleus, *C* = cytoplasm.

morphine, codeine, and phenobarbital are detoxified in the liver by components of the smooth ER.

> **The endoplasmic reticulum (ER) is an extensive system of membranes that divides the interior of eukaryotic cells into compartments and channels. Rough ER is devoted to the synthesis and transport of proteins across the membrane, whereas smooth ER is devoted to organizing the synthesis of lipids and other biosynthetic activities.**

THE NUCLEUS

The largest and most easily seen of the organelles within an eukaryotic cell is the **nucleus** (Figure 5-11), which was first described by the English botanist Robert Brown in 1831. The word "nucleus" is derived from the Greek word for a kernel or nut. In fact, nuclei are rather spherical in shape and bear some resemblance to nuts. In animal cells the nucleus is typically located in the central region. In some cells the nucleus seems to be cradled in this position by a network of gossamer-fine filaments. The nucleus is the repository of the genetic information that directs all the activities of a living cell. Some kinds of cells, such as mature red blood cells, discard their nuclei during the course of development. The development of these cells terminates when their nuclei are lost. They lose all ability to grow, change, and divide and become merely passive vessels.

The Nuclear Envelope

The surface of the nucleus is bounded by *two* layers of membrane, the outer membrane of the **nuclear envelope** (Figure 5-11). Many researchers believe that ER is the source of the nuclear envelope, which seems to be continuous with it. Scattered over the surface of the nuclear envelope, like craters on the moon, are shallow depressions called **nuclear pores.** These pores, 50 to 80 nanometers apart, form at locations where the two membrane layers of the nuclear envelope pinch together. The nuclear pores are not empty openings in the same sense that the hole in a doughnut is, with nothing filling the actual hole. Rather, nuclear pores contain embedded within them many proteins that act as molecular channels, permitting certain molecules to pass into and out of the

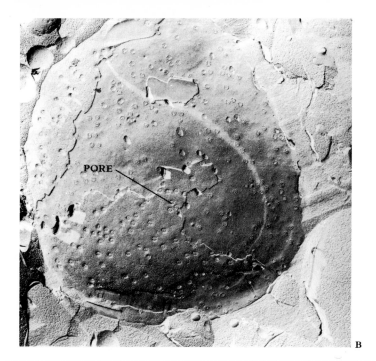

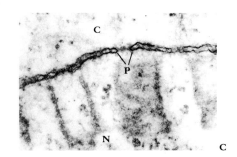

B

nucleus. Passage is restricted primarily to two kinds of molecules: (1) proteins moving into the nucleus, where they will be incorporated into nuclear structures or catalyze nuclear activities, and (2) RNA and protein-RNA complexes formed in the nucleus and subsequently exported to the cytoplasm.

> **The nucleus of eukaryotic cells is a vesicle that contains the cell's hereditary apparatus and isolates it from the rest of the cell.**

Chromosomes

Both in bacteria and in eukaryotes, all the hereditary information specifying cell structure and function is encoded in DNA. Unlike bacterial DNA, however, the DNA of eukaryotes is divided into several segments and associated with protein and RNA to form **chromosomes** (Figure 5-12). Association with protein enables eukaryotic DNA to wind up into a highly condensed form during cell division. Under a light microscope these condensed chromosomes are readily seen in dividing cells as densely staining rods. After cell division, eukaryotic chromosomes uncoil and can no longer be distinguished individually with a light microscope. Uncoiling the chromosomes into a more extended form permits the enzymes that make RNA copies of DNA to gain access to the DNA molecule. Only by means of these RNA copies can the hereditary information be used to direct the synthesis of enzymes.

> **A distinctive feature of eukaryotes is the organization of their DNA into chromosomes. Chromosomes can be condensed into compact structures when a eukaryotic cell divides and later unraveled so that the information that the chromosomes contain may be used.**

The Nucleolus

To make proteins a cell employs a special organelle, the **ribosome,** that reads the RNA copy of a DNA gene and uses the information it finds there to direct the synthesis of a protein. Ribosomes are made up of several special forms of RNA bound within a complex of several dozen different proteins. When a lot of proteins are being made, a cell

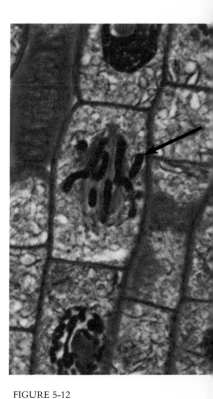

FIGURE 5-12

Eukaryotic chromosomes *(arrow),* such as those in this onion root tip cell, exist as several individual units, each of which carries only a portion of the genetic information. This contrasts markedly with the situation in bacteria, in which the principal DNA molecule exists as a single closed circle, not complexed with proteins.

FIGURE 5-13

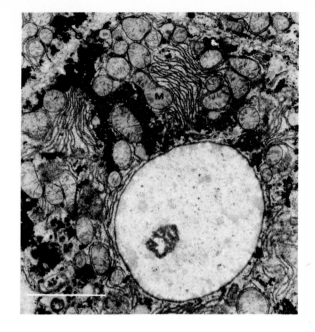

FIGURE 5-13

Electron micrograph of the interior of a rat liver cell, magnified about 6000 times. A single large nucleus occupies the lower center of the micrograph. The electron–dense area in the center of the nucleus is the nucleolus, the area where the major components of the ribosomes are produced. In the nucleoplasm around the nucleolus can be seen partly formed ribosomes. The nucleus is surrounded by a double nuclear envelope, penetrated here and there by pores. A comparison of this micrograph with Figure 5-6 makes the contrast between bacterial and eukaryotic cell organization apparent.

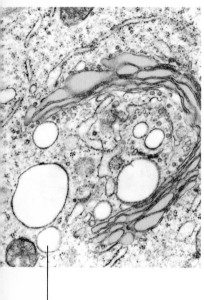

VESICLES

FIGURE 5-14

A Golgi body in cross section. Within the Golgi body the membrane (color) pinches together at its terminus to produce membrane-bounded vesicles. These vesicles contain whatever enzymes or other substances have been transported through the endoplasmic reticulum to the Golgi body.

needs a lot of ribosomes to handle the work load, and therefore it needs to be able to make large numbers of ribosomes quickly. To do this, many thousands of copies of the portion of the DNA encoding the RNA components of ribosomes, called **ribosomal RNA** (rRNA), are clustered together on the chromosome. Copying RNA molecules from the cluster rapidly generates large numbers of the molecules needed to produce ribosomes.

At any given moment many rRNA molecules dangle from the chromosome at the sites of these clusters of RNA genes. The proteins that will later form part of the ribosome complex bind to the dangling rRNA molecules. These areas where ribosomes are being assembled on the chromosomes are easily visible within the nucleus as one or more dark-staining regions, called **nucleoli** (singular, *nucleolus*; Figure 5-13). The nucleoli can be seen under the light microscope even when the chromosomes are extended, unlike the rest of the chromosomes, which are visible only when condensed. Because they are visible in nondividing cells, early scientists thought that the nucleoli were distinct cellular structures. We now know that this is not true. Nucleoli are, in fact, aggregations of rRNA and some ribosomal proteins that are transported into the nucleus from the rough ER and accumulate at those regions on the chromosomes where very active synthesis of rRNA is taking place.

THE GOLGI COMPLEX

At various locations in the cytoplasm are flattened stacks of membranes called **Golgi bodies** (Figure 5-14). These structures are named for Camillo Golgi, the nineteenth-century Italian physician who first called attention to them. Animal cells contain 10 to 20 Golgi bodies each (they are especially abundant in glandular cells, which are manufacturing the substances that they secrete), whereas plant cells may contain several hundred. Collectively the Golgi bodies are referred to as the **Golgi complex.**

Golgi bodies function in the collection, packaging, and distribution of molecules synthesized in the cell (Figure 5-15). The proteins and lipids that are manufactured on the rough and smooth ER membranes are transported through the channels of the ER, or as vesicles budded off from it, into the Golgi bodies. Within the Golgi bodies, many of these molecules are bound to polysaccharides, forming compound molecules. Among these are **glycoproteins,** consisting of a polysaccharide bound to a protein, and **glycolipids,** consisting of a polysaccharide bound to a lipid. The newly formed glycoproteins and glycolipids collect at the ends of the membranous folds of the Golgi bod-

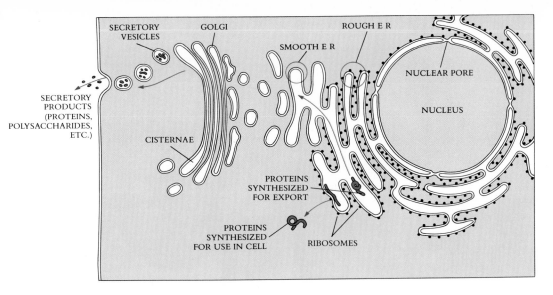

FIGURE 5-15

How proteins are secreted across membranes. Proteins targeted for other compartments within a cell than the ones in which they are made must pass across internal cell membranes. In this internal transport process, the proteins are first assembled by ribosomes on the rough ER, which extrude them across the endoplasmic reticulum into the channels which the ER creates within the cell. The proteins move through these channels to a zone of smooth ER, where some of them are chemically modified by enzymes embedded within the smooth ER membrane. They then pass through more channels to a Golgi body, which is the end of the channel. There they are encapsulated within vesicles formed by the pinching together of membranes at the end of the Golgi body. These secretory vesicles pass through the cytoplasm to the surface of the membrane encasing the compartment, or in some cases to the interior surface of the plasma membrane. There the vesicles fuse with the membrane, releasing their contents to the other side.

ies; these folds are given the special name **cisternae** (Latin, "collecting vessels"). Intermittently in such regions, the membranes of the cisternae push together, pinching off small membrane-bound vesicles containing the glycoprotein and glycolipid molecules. These vesicles then move to other locations in the cell, distributing the newly synthesized molecules within them to their appropriate destinations.

> **The Golgi complex is the delivery system of the eukaryotic cell. It collects, modifies, packages, and distributes molecules that are synthesized at one location within the cell and used at another.**

MICROBODIES

Microbodies, thought to be derived from endoplasmic reticulum, are organelles that carry a set of enzymes active in converting fat to carbohydrate, and another set that destroys harmful peroxides (H_2O_2). Similar sets of enzymes are found in the microbodies of plants, animals, fungi, and protists, although traditionally animal microbodies are called **peroxisomes** and plant microbodies are called **glyoxysomes.**

LYSOSOMES

Lysosomes provide an impressive example of the metabolic compartmentalization achieved by the activity of the Golgi complex. **Lysosomes** (Figure 5-16) are vesicles that contain in a concentrated mix the digestive enzymes of the cell, enzymes that catalyze the rapid breakdown of proteins, nucleic acids, lipids, and carbohydrates. In other words, lysosomes contain the hydrolytic enzymes responsible for the breakdown of essentially every major macromolecule in the cell.

Lysosomes digest worn-out cellular components, making way for newly formed ones while recycling the materials locked up in the old ones. This is perhaps their most important function in the cell. Cells can persist for a long time only if their component parts are constantly renewed. Otherwise the ravages of use and accident chip away at the metabolic capabilities of the cell, slowly degrading its ability to survive. Cells age for the same reason that people do, because of a failure to renew themselves. Lysosomes destroy the organelles of eukaryotic cells and recycle their component proteins and other molecules at a fairly constant rate throughout the life of the cell. Mitochondria, for example, are replaced in some tissues every 10 days, with lysosomes digesting the old mitochondria as new ones are produced.

If the contents of lysosomes are able to digest whole cells, it might have occurred to you to ask "What prevents lysosomes from digesting *themselves*?" They don't be-

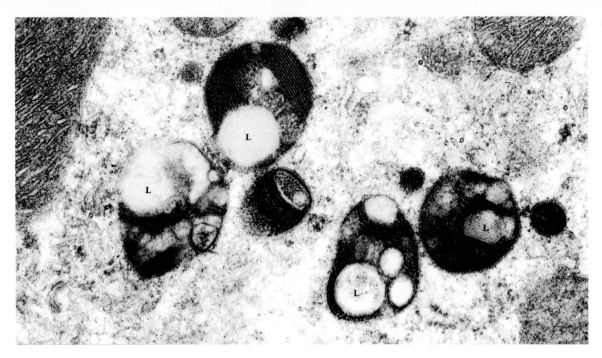

FIGURE 5-16

Lysosomes *(L)* are digestive organelles within many eukaryotic cells. These lysosomes are within the cytoplasm of mouse kidney cells.

cause lysosomes that are not actively engaged in digestive activities keep their battery of hydrolytic enzymes inactive by maintaining a low internal pH; they expend energy to pump protons into their interiors, the excessive protons creating an acid pH. At such acid pH's, the hydrolytic enzymes are not active. A lysosome in such a "holding pattern" is called a **primary lysosome.** It is not until a primary lysosome fuses with a food vacuole or other organelle that pH rises and the arsenal of hydrolytic enzymes is activated (a **secondary lysosome).** Maintaining an acid pH requires energy, of course, which is the reason that metabolically inactive eukaryotic cells die. Without the input of energy necessary to maintain a low lysosomal pH, hydrolytic digestive enzymes become active, lysosomal membranes are digested from within and disintegrate, and the digestive enzymes of the lysosomes pour out into the cytoplasm and destroy it. Bacteria, in contrast, do not possess lysosomes and do not die when they are metabolically inactive. They are simply able to remain quiescent until altered conditions restore their metabolic activity, a property that greatly heightens their ability to persist under unfavorable environmental conditions. For us, the very process that repairs the ravages of time to our cells may also lead to their destruction. An absolute dependency on a constant supply of energy is the price that we pay for our long lives.

> **Lysosomes are vesicles, formed by the Golgi complex, which contain digestive enzymes. The isolation of these enzymes in lysosomes protects the rest of the cell from their digestive activity.**

ENERGY-PRODUCING ORGANELLES

Most of the membrane-bound structures within eukaryotic cells are derived from the ER, but there are important exceptions. Eukaryotic cells also contain complex cell-like organelles that most biologists believe were derived from ancient **symbiotic bacteria. Symbiosis,** the living together in close association of two or more organisms, is one of the most prominent features of life on earth. An organism that is symbiotic within another is called an **endosymbiont.** The major endosymbionts that occur in eukaryotic cells are **mitochondria,** which occur in all but a very few eukaryotic organisms, and **chloroplasts,** which occur in algae and plants.

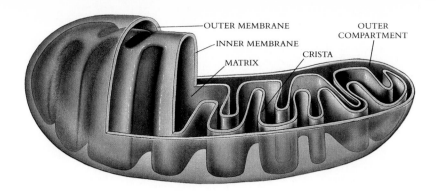

OUTER MEMBRANE

INNER MEMBRANE

MATRIX

CRISTA

OUTER COMPARTMENT

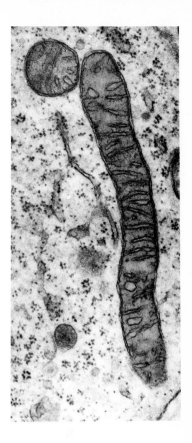

FIGURE 5-17

Mitochondria in longitudinal and cross section. These organelles are thought to have evolved from bacteria that long ago took up residence within the ancestors of present-day eukaryotes. Plant mitochondria tend to be shorter and thicker than the animal mitochondrion shown here.

Mitochondria

The mitochondria that occur in all but a very few eukaryotic cells are thought by most biologists to have originated as symbiotic, **aerobic** (oxygen-requiring) bacteria. The theory of the symbiotic origin of mitochondria has had a controversial history, and a few biologists still do not accept it. The evidence supporting the theory is so extensive, however, that in this book we will treat it as established. We will present the evidence as we proceed.

According to this theory, the bacteria that became mitochondria were engulfed by ancestral eukaryotic cells early in their evolutionary history. Before they had acquired these bacteria, the host cells were unable to carry out the metabolic reactions necessary for living in an atmosphere that contained increasing amounts of oxygen. Metabolic reactions requiring oxygen are collectively called **oxidative metabolism,** a process that the engulfed bacteria were able to carry out. The bacteria became mitochondria.

Mitochondria (singular, *mitochondrion*) are tubular or sausage-shaped organelles 1 to 3 micrometers long (Figure 5-17); thus they are about the same size as most bacteria. Mitochondria are bounded by two membranes. The outer membrane is smooth and was apparently derived from the ER of the host cell, whereas the inner one is folded into numerous contiguous layers called **cristae.** These cristae resemble the folded membranes that occur in various groups of bacteria. The cristae partition the mitochondrion into two compartments, an inner **matrix** and an **outer compartment.** On the surfaces of the membranes, and also submerged within them, are the proteins that carry out oxidative metabolism.

During the billion and a half years in which mitochondria have existed as endosymbionts in eukaryotic cells, most of their genes have been transferred to the chromosomes of their host cells. For example, the genes that produce the enzymes involved with the oxidative metabolism characteristic of mitochondria are located in the cell nucleus. Mitochondria still have their own genome, however, contained in a circular, closed molecule of DNA like those found in bacteria. On this mitochondrial DNA are located several genes that produce some of the proteins that are essential for the mitochondria's role as the site of oxidative metabolism. All of these genes are copied into RNA within the mitochondrion and used there to make proteins. In this process the mitochondria use small RNA molecules and ribosomal components that are also encoded within the mitochondrial DNA. These ribosomes are smaller than those of eukaryotes in general, resembling bacterial ribosomes in size and structure.

A eukaryotic cell does not produce brand new mitochondria each time the cell itself divides. Instead, the ones it has divide in two, doubling the number, and are partitioned between the new cells. Thus all mitochondria within a eukaryotic cell are produced by the division of existing mitochondria, just as all bacteria are produced from existing bacteria by cell division. Mitochondria divide by simple fission, splitting into two just as bacterial cells do, and apparently replicate and partition their circular DNA molecule in much the same way as do bacteria. Mitochondrial reproduction is not autonomous (self-governed), however, as is bacterial reproduction. Most of the

FIGURE 5-18

Chloroplast structure. The inner membrane of a chloroplast is fused to form stacks of closed vesicles called *thylakoids*. Within these thylakoids, photosynthesis takes place. Thylakoids are typically stacked one on top of the other in columns called *grana*.

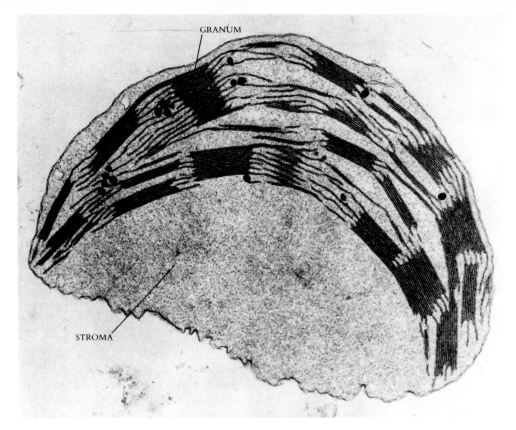

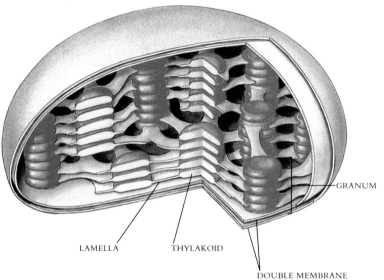

components required for mitochondrial division are encoded as genes within the eukaryotic nucleus and translated into proteins by the cytoplasmic ribosomes of the cell itself. Mitochondrial replication is thus impossible without nuclear participation, and mitochondria cannot be grown in a cell-free culture.

Chloroplasts

Symbiotic events similar to those postulated for the origin of mitochondria also seem to have been involved in the origin of chloroplasts, which are characteristic of photosynthetic eukaryotes (algae and plants). Chloroplasts, which apparently were derived from symbiotic photosynthetic bacteria, confer the ability to perform photosynthesis on these eukaryotes. The advantage that chloroplasts bring to the organisms that possess them is, therefore, obvious: these organisms can manufacture their own food. The photosynthetic endosymbionts that are thought to have given rise to chloroplasts were **anaerobic** (requiring no oxygen), photosynthetic bacteria.

> **Mitochondria apparently originated as endosymbiontic aerobic bacteria, whereas chloroplasts seem to have originated as endosymbiontic anaerobic, photosynthetic bacteria.**

The chloroplast body is bounded, like the mitochondrion, by two membranes that resemble those of mitochondria (Figure 5-18) and apparently were derived in a similar fashion. Chloroplasts are larger than mitochondria and their inner membranes have a more complex organization. They have a larger circular DNA molecule than do mitochondria, but many of the genes that specify chloroplast components are located in the cell nucleus so that the transfer of genetic material has been a part of their history also. Some components of the photosynthetic process are synthesized entirely within the chloroplast, which includes the specific RNA and protein components necessary to accomplish this. Photosynthetic cells typically contain from one to several hundred chloroplasts, depending on the organism involved or, in the case of multicellular photosynthetic organisms, the kind of cell involved. As is true of mitochondria, chloroplasts cannot be grown in a cell-free culture.

> **Both mitochondria and chloroplasts have transferred the bulk of their genomes to the host chromosomes but retain certain specific genes related to their functions. Neither kind of organelle can be maintained in a cell-free culture.**

THE CYTOSKELETON

The cytoplasm of all eukaryotic cells is crisscrossed by a network of protein fibers that support the shape of the cell and anchor organelles such as the nucleus to fixed locations (Figure 5-19). This network, called the **cytoskeleton,** cannot be seen with an ordinary microscope because the fibers are single chains of protein, much too fine for microscopes to resolve. The fibers of the cytoskeleton are a dynamic system, constantly being formed and disassembled. Individual fibers form by **polymerization,** a process in which identical protein subunits are attracted to one another chemically and spontaneously assemble into long chains. Fibers are disassembled in the same way, by the removal from one end of first one subunit and then another.

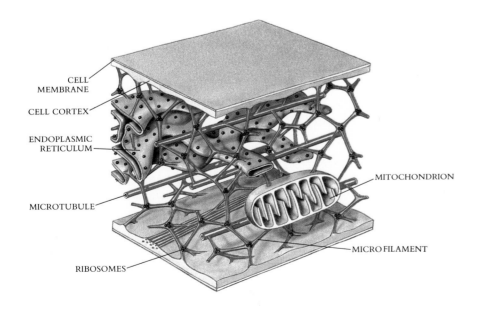

FIGURE 5-19

The cytoskeleton. In this diagrammatic cross section of a eukaryotic cell the mitochondria, ribosomes, and endoplasmic reticulum are all supported by a fine network of filaments, through which pass microtubules linking various portions of the cell together.

CELL MEMBRANE

CELL CORTEX

ENDOPLASMIC RETICULUM

MICROTUBULE

RIBOSOMES

MITOCHONDRION

MICROFILAMENT

Cells from plants and animals contain three different types of cytoskeleton fibers, each formed from a different kind of subunit (Figure 5-20):

Actin filaments. Actin filaments (also called microfilaments) are long protein fibers about 7 nanometers in diameter. Each fiber is composed of two chains of protein loosely wrapped around one another like two strands of pearls (Figure 5-20, A). Each "pearl" of a filament is a ball-shaped molecule of a protein called **actin,** which is the size of a small enzyme. Actin molecules, if left alone, will spontaneously form these filaments even in a test tube; a cell regulates the rate of their formation by means of other proteins that act as switches, turning on polymerization only when appropriate.

Microtubules. Microtubules are hollow tubes about 25 nanometers in diameter. Each microtubule is a chain of proteins wrapped round and round in a tight spiral (Figure 5-20, B). The basic protein subunit of a microtubule is a molecule of a protein a little larger than actin called **tubulin.** Like actin filaments, microtubules form spontaneously, but in a cell microtubules form only around specialized structures called **organizing centers** that provide a base from which they can grow.

Intermediate fibers. The most durable element of the cytoskeleton is a system of tough durable protein fibers, each a rope of threadlike protein molecules wrapped around one another like the strands of a cable (Figure 5-20, C). These fibers are characteristically 8 to 10 nanometers in diameter, intermediate in size between actin filaments and microtubules; this is why they are called intermediate filaments. Once formed, intermediate filaments are very stable and do not usually break down. The basic protein subunit of an intermediate fiber in most cells is a protein called **vimentin,** although some cells employ other fibrous proteins instead. Skin cells, for example, form their intermediate fibers from a protein called **keratin.** When skin cells die, the intermediate filaments of their cytoskeleton persist; hair and nails are formed in this way.

Both actin filaments and intermediate fibers are anchored to proteins embedded within the plasma membrane and provide the cell with mechanical support. Intermediate fibers act as intracellular tendons, preventing excessive stretching of cells, whereas actin filaments play a major role in determining the shape of cells. Because actin filaments can form and dissolve so readily, the shape of an animal cell can change quickly. If you look at the surface of an animal cell under a microscope, you will find it alive with motion, projections shooting outward from the surface and then retracting only to shoot out elsewhere moments later (Figure 5-21).

The cytoskeleton is responsible not only for the cell's shape, but it also provides a scaffold on which the enzymes and other macromolecules are located in defined areas of the cytoplasm. Many of the enzymes involved in cell metabolism, for example, bind to actin filaments, as do ribosomes that carry out protein synthesis. By anchoring particular enzymes near one another, the cytoskeleton serves, like the ER, to organize the cell's activities.

Actin filaments and microtubules also play important roles in cell movement. Pairs of microtubules are cross-linked at numerous positions by molecules of a protein called **dynein.** The shifting positions of these cross-links determine the relative motion of the microtubules. As an example of the results of microtubule movement, when we study cell reproduction in Chapter 9, we will see that chromosomes move to opposite sides of dividing cells because they are attached to shortening microtubules, and that the cell pinches into two because a belt of actin filaments contracts like a pursestring. Your own muscles use actin filaments to contract their cytoskeleton. Indeed, all cell motion is tied to these same processes. The fluttering of an eyelash, the flight of an eagle, and the awkward crawling of a baby all depend on the movements of actin filaments in the cytoskeletons of muscle cells.

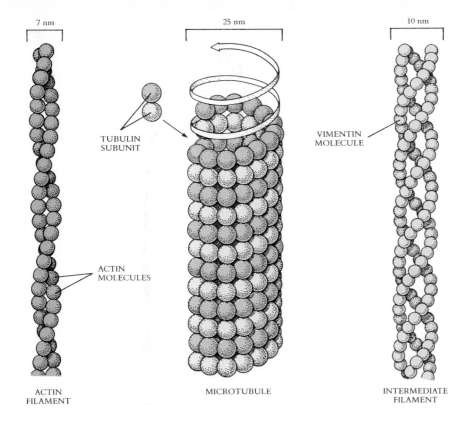

7 nm

25 nm

10 nm

TUBULIN
SUBUNIT

VIMENTIN
MOLECULE

ACTIN
MOLECULES

ACTIN
FILAMENT

MICROTUBULE

INTERMEDIATE
FILAMENT

FIGURE 5-20

A comparison of the molecules that make up the cytoskeleton. **A** Actin filaments. In the micrograph, actin filaments parallel the cell surface membrane in bundles known as stress fibers, which may have a contractile function. **B** Microtubules. Each microtubule is composed of a spiral array of tubulin subunits. The microtubules visible in this micrograph radiate from an area near the nucleus (most heavily stained region). Microtubules act to organize metabolism and intracellular transport in the nondividing cell. **C** Intermediate filaments. It is not known how the individual subunits are arranged in an intermediate filament, but the best evidence suggests that three subunits are wound together in a coil, interrupted by uncoiled regions. Intermediate filaments, like microtubules, extend throughout the cytoplasm. In a skin cell, such as the one shown, intermediate filaments form thick, wavy bundles that probably provide structural reinforcement.

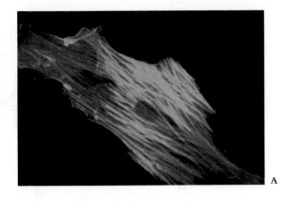

A

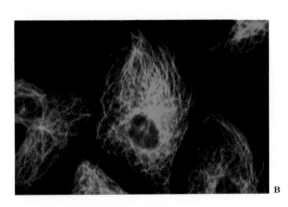

B

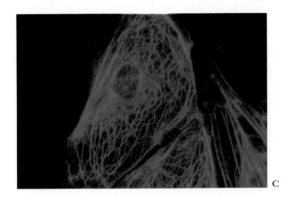

C

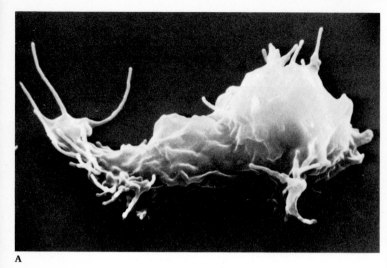

A

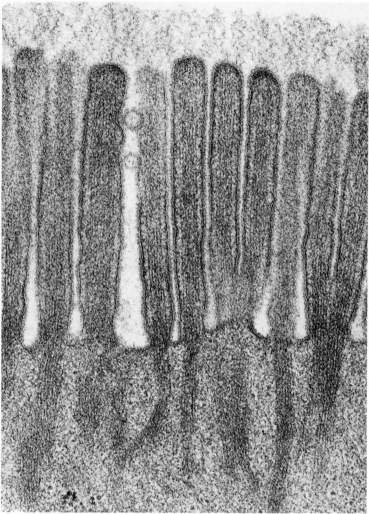

FIGURE 5-21

The surfaces of animal cells are in constant motion.

A This amoeba, a single-celled protist, is advancing toward you, its advancing edges extending projections outward. The moving edges have been said to resemble the ruffled edges of a woman's skirt.

B Animal cells often produce projections. This figure shows fingerlike projections, called microvilli, in the cells lining the human intestine. In such cells, a bundle of actin filaments is anchored at the tip. This bundle extends downward through the microvillus into the interior of the cell. The contraction of the microfilament below the cell surface shortens the protruding microvillus as the actin fibers attached to its tip are drawn inward. Since such contractions of the microfilament base can be quite rapid, microvilli can change their length quickly. They often appear to pop up almost instantaneously and to disappear just as quickly.

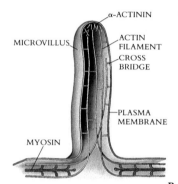

α-ACTININ

MICROVILLUS

ACTIN FILAMENT

CROSS BRIDGE

PLASMA MEMBRANE

MYOSIN

B

FLAGELLA

Flagella (singular, *flagellum*) are fine, long, threadlike organelles protruding from the surface of cells; they are used in locomotion and feeding. The flagella of bacteria are long protein fibers that are so efficient that the bacteria possessing them can move about 20 cell diameters per second. Imagine trying to run 20 body lengths per second. The bacteria swim by rotating their flagella (Figure 5-22). One or more flagella trail behind each swimming bacterial cell, depending on the species of bacterium. Each has a motion like a propeller caused by a complex rotary "motor" embedded within the cell wall and membrane. This rotary motion is virtually unique to bacteria; only a very few eukaryotes have organs that truly rotate.

Eukaryotic cells have a completely different kind of flagellum, based on a kind of cable made up of microtubules. Their flagella consist of a circle of nine microtubule pairs surrounding two central ones; they are called **9 + 2 flagella** (Figure 5-23). Completely different from bacterial flagella, this complex microtubular apparatus evolved very early in the history of the eukaryotes. Although the cells of many multicellular and some unicellular eukaryotes today no longer exhibit 9 + 2 flagella and are nonmotile (unmoving), the 9 + 2 arrangement of microtubules can still be found within them. The apparatus has merely been put to other uses. Like mitochondria, the 9 + 2 flagellum appears to be a fundamental component of the eukaryotic endowment.

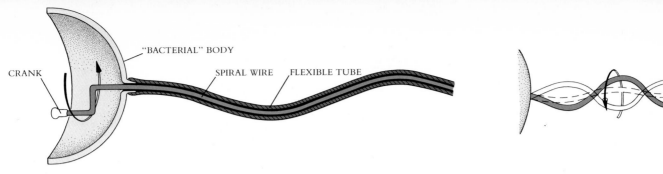

CRANK — "BACTERIAL" BODY — SPIRAL WIRE — FLEXIBLE TUBE

FIGURE 5-22

Bacteria swim by rotating their flagella. The photograph is of *Vibrio cholerae*, the microbe that causes the serious disease cholera. The unsheathed core visible at the top of the micrograph is composed of a single crystal of the protein *flagellin*. In intact flagella this core is surrounded by a flexible sheath. Imagine that you are standing inside the *Vibrio* cell, turning the flagellum like a crank. You would create a spiral wave that travels down the flagellum, just as if you were turning a wire within a flexible tube. The bacterium employs this kind of rotary motion when it swims.

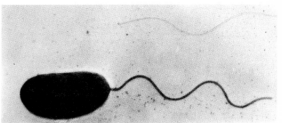

TABLE 5-1   EUKARYOTIC CELL STRUCTURES AND THEIR FUNCTIONS

STRUCTURE	DESCRIPTION	FUNCTION
STRUCTURAL ELEMENTS		
Cell wall	Outer layer of cellulose or chitin, or absent	Protection; support
Plasma membrane	Lipid bilayer in which proteins are embedded	Regulates what passes in and out of cell; cell-to-cell recognition
Cytoskeleton	Network of protein filaments	Structural support; cell movement
Flagella (cilia)	Cellular extensions with 9 + 2 arrangement of pairs of microtubules	Motility or moving fluids over surfaces
ORGANELLES		
Endoplasmic reticulum	Network of internal membranes	Forms compartments and vesicles
Ribosomes	Small, complex assemblies of protein and RNA, often bound to ER	Sites of protein synthesis
Nucleus	Spherical structure bounded by double membrane; contains chromosomes	Control center of cell; directs protein synthesis and cell reproduction
Chromosomes	Long threads of DNA associated with protein	Contain hereditary information
Nucleolus	Site on chromosome of rRNA synthesis	Assembles ribosomes
Golgi complex	Stacks of flattened vesicles	Packages proteins for export from cell; forms microbodies
Microbodies	Vesicles containing collections of oxidative and other enzymes	Isolate particular chemical activities from rest of cell
Lysosomes	Microbodies containing digestive enzymes	Digest worn-out mitochondria and cell debris; play role in cell death
ENERGY-PRODUCING ORGANELLES		
Mitochondria	Bacteria-like elements with inner membrane highly folded	Power plant of the cell; site of oxidative metabolism
Chloroplasts	Bacteria-like elements with vesicles containing chlorophyll	In plant cells; site of photosynthesis

Eukaryotic flagella are not difficult to see. If you were to examine a motile, single-celled eukaryote under a microscope, you would immediately notice the flagella protruding from its cell surface. The constant undulating motion of these flagella, caused by the movement of pairs of microtubules past one another, propels the cell through its environment and requires a constant input of energy into the flagellum. When examined carefully, each flagellum proves to be an outward projection of the interior of the cell, containing cytoplasm and enclosed by the cell membrane. Its microtubules are derived from a **basal body** situated just below the point at which the flagellum protrudes from the surface of the cell membrane. The arrangement of flagella differs greatly in different eukaryotes. If the flagella are numerous and organized in dense rows (see Figure 5-1), they are called **cilia** (singular, *cilium*), but cilia do not differ at all from flagella in structure.

In many multicellular organisms cilia carry out tasks far removed from their original functions of propelling cells through water. In several kinds of human tissues, for example, the beating of rows of cilia moves water over the tissue surface. The 9 + 2 arrangement of microtubules occurs in the sensory hairs of the human ear, where the bending of these hairs by pressure constitutes the initial sensory input of hearing. Throughout the evolution of eukaryotes, 9 + 2 flagella have been an element of central importance.

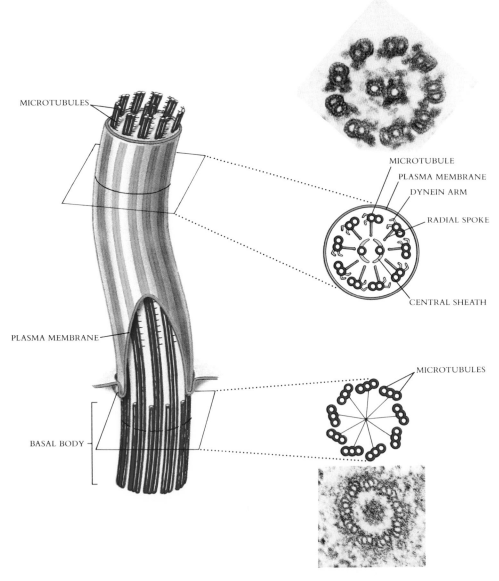

FIGURE 5-23

A eukaryotic flagellum springs directly from a basal body, is composed of nine pairs of microtubules, and has two microtubules in its core connected to the outer ring of paired microtubules by dynein arms.

TABLE 5-2   A COMPARISON OF BACTERIAL, ANIMAL, AND PLANT CELLS★

	BACTERIUM	ANIMAL	PLANT
EXTERIOR STRUCTURE			
Cell wall	Present (protein-polysaccharide)	Absent	Present (cellulose)
Cell membrane	Present	Present	Present
Flagella	May be present (1 strand)	May be present	Absent except in sperm of a few species
INTERIOR STRUCTURE			
ER	Absent	Usually present	Usually present
Microtubules	Absent	Present	Present
Centrioles	Absent	Present	Absent
Golgi bodies	Absent	Present	Present
ORGANELLES			
Nucleus	Absent	Present	Present
Mitochondria	Absent	Present	Present
Chloroplasts	Absent	Absent	Present
Chromosomes	A single circle of naked DNA	Multiple units, DNA associated with protein	Multiple units, DNA associated with protein
Ribosomes	Present	Present	Present
Lysosomes	Absent	Usually present	Equivalent structures called "spherosomes"
Vacuoles	Absent	Absent or small	Usually a large single vacuole in mature cell

★The structures found in the cells of the other two kingdoms of eukaryotic organisms, Protista and Fungi, are diverse but share the basic features of eukaryotic cells; most have cell walls and centrioles, and chloroplasts are present only in the photosynthetic protists known as algae.

AN OVERVIEW OF CELL STRUCTURE

The structure of eukaryotic cells is much more complicated and diverse than the structure of bacterial cells (Table 5-2). The most distinctive difference between these two fundamentally distinct cell types is the extensive subdivision of the interior of eukaryotic cells by membranes. The most visible of these membrane-bound compartments, the nucleus, gives eukaryotes their name. There are no equivalent membrane-bound compartments within prokaryotic cells. The membranes of some photosynthetic bacteria are extensively folded inwardly, but they do not isolate any one portion of the cell from any other portion. A molecule can travel unimpeded from any location in a bacterial cell to any other location.

In the following chapters we will consider the consequences of these structural differences and how they influence the metabolism and biochemistry of eukaryotes. We shall see that the metabolic processes that occur within eukaryotic cells differ from those of bacteria and that the differences, like the structural ones discussed in this chapter, are substantial.

SUMMARY

1. The cell is the smallest unit of life. It is composed of a nuclear region that contains the hereditary apparatus within a larger volume called the cytoplasm, which performs the day-to-day functions of the cell under the supervision of the nuclear region. In all cells the cytoplasm is bounded by a lipid membrane.

2. Bacteria, which have prokaryotic cell structure, do not have membrane-bound organelles within their cells. Their DNA is mostly included in a single, closed circular molecule.

3. The endoplasmic reticulum is a series of membranes that subdivides the interior of eukaryotic cells into separate compartments. It is the most distinctive feature of eukaryotic cells and one of the most important, since it makes it possible for the cells to carry out different metabolic functions separately.

4. The nucleus is the largest of the closed compartments created by internal membranes. Within it, the cell's DNA is separated from the rest of the cytoplasm.

5. A third key membrane-associated organelle is the Golgi complex, which serves as a cellular "express package service," packaging molecules within special membrane vesicles and transporting them to various locations in the cell.

6. One class of vesicle created by the Golgi complex consists of the lysosomes, vesicles containing high concentrations of digestive enzymes. A cell constantly expends energy to prevent digestive damage to these vesicles, and when this preventive activity ceases, the vesicles soon burst, digesting and killing the cell.

7. Not all the eukaryotic cell's internal organelles are created by internal membrane systems. Others appear to have been derived from symbiotic bacteria. By far the most widespread of these are mitochondria and chloroplasts.

8. Many, but not all, of the genes that originally were present in mitochondrial and chloroplast DNA seem to have been transferred to, or had their functions taken over by, the DNA in the chromosomes of the host cell. Both classes of organelles, however, have retained the genes necessary to create their particular distinctive structures. Without these structures, mitochondria would be unable to function in aerobic respiration, and chloroplasts would be unable to function in photosynthesis.

REVIEW

1. The photosynthetic pigments of photosynthetic bacteria are located on extensively convoluted infoldings of the _____.

2. The proteins embedded in eukaryotic plasma membranes perform three general classes of functions. They are: (1) _____, (2) _____, and (3) _____.

3. _____ are the "garbage collectors/recycling centers" of eukaryotic cells, digesting worn-out cellular components and recycling their components. These "garbage collector" organelles are formed by what other organelle?

4. Which two organelles are thought to have originated as endosymbionts in an early cell?

5. DNA in eukaryotic cells occurs in the _____, the _____, and the _____.

SELF-QUIZ

1. Which of the following statements is *not* included in what we call the cell theory?
    (a) All cells are so small that they cannot be seen with the naked eye.
    (b) All organisms are composed of one or more cells.
    (c) Cells are the smallest living things.
    (d) The hereditary information is contained within cells.
    (e) The cells that we see today are descended from early cells produced billions of years ago and are not produced anew at the present time.

2. Which of the following are components of all eukaryotic cells (choose three)?
    (a) DNA and associated proteins found in a membrane-bound structure called the nucleus
    (b) cytoplasm that contains numerous membrane-bound organelles
    (c) an intracellular matrix of protein called a cytoskeleton
    (d) gram-negative or gram-positive staining cell walls
    (e) chloroplasts

3. The DNA of eukaryotic cells is fragmented into several pieces called _____, unlike the DNA of bacteria, which is mostly contained in a single circular molecule.
    (a) ribosomes   (d) chromosomes
    (b) peroxisomes   (e) chloroplasts
    (c) RNA

4. Arrange in order the sequence of membrane-bound structures through which a protein that is going to be secreted by a cell must pass.
    (a) Golgi body
    (b) secretory vesicle
    (c) rough ER
    (d) smooth ER

5. Which of the following properties are exclusive to microtubules that help make up the cytoskeleton of eukaryotic cells (choose three)?
    (a) will form spontaneously if the subunits are present
    (b) are made of protein subunits called tubulin
    (c) are anchored to proteins in the plasma membrane
    (d) are hollow tubes about 25 nanometers in diameter
    (e) play an important role in cell movement

PROBLEM

1. A cubical cell that is about 5 micrometers across contains what percentage of the volume of a cubical cell that is 20 micrometers across?

THOUGHT QUESTIONS

1. Mitochondria are thought to be the evolutionary descendants of living cells, probably of symbiotic aerobic bacteria. Are mitochondria alive?

2. Some cells are very much larger than others. What would you expect the relationship to be between cell size and level of cell activity?

3. How does the Golgi complex know where to send what vesicle?

FOR FURTHER READING

DE DUVE, C.: *A Guided Tour of the Living Cell,* vols. 1 and 2, Scientific American Books, New York, 1986. This review has excellent illustrations and presents a wide variety of the techniques used to study cells.

MARGULIS, L.: *Symbiosis in Cell Evolution,* W.H. Freeman Co., San Francisco, 1980. A broad-ranging exposition of the theory that mitochondria, chloroplasts, and other organelles were acquired by eukaryotes from symbiotes, written by the chief proponent of the idea.

PORTER, K., and J. TUCKER: "The Ground Substance of the Living Cell," *Scientific American,* March 1981, pages 56-67. A lucid account of how the authors used high-resolution electron microscopy to detect the network of microfilaments within eukaryotic cells.

ROTHMAN, J.E.: "The Golgi Apparatus: Two Organelles in Tandem," *Science,* vol. 213, pages 1212-1219, 1981. A current view of how the Golgi complex operates.

WEBER, K., and M. OSBORN: "The Molecules of the Cell Matrix," *Scientific American,* October 1985, pages 110-120. An up-to-date description of the many different molecules that make up the cell cytoskeleton.

FIGURE 6-1   This beautiful foraminifera, less than a centimeter in diameter, is a marine protist, a single-celled creature of the sea. Protected like a porcupine by a shield of minute needles of calcium carbonate (the material limestone is made of), this cell carries out the same essential life processes that you do, ingesting and metabolizing food, moving from one place to another, and reproducing. And, as in your own body, all of these activities require that substances enter and leave the cell. The traffic across cell membranes is one of life's most fundamental processes.

HOW CELLS INTERACT WITH THE ENVIRONMENT

Overview

Every cell is enveloped within a liquid layer of lipid, a fluid shell that isolates its interior. Anchored within this liquid are a host of proteins that move about on the surface like ships on a lake. These proteins provide passages across the lipid layer for molecules and information and are the cell's only connection with the outer world. The combination of lipid shell and embedded proteins is called a biological membrane.

For Review   *Here are some important terms and concepts that you will encounter in this chapter. If you are not familiar with them, you should review them before proceeding.*

Polar nature of water (Chapter 3)

Hydrogen bonds (Chapter 3)

Structure of fat molecules (Chapter 3)

Types of membrane proteins (Chapter 5)

Among a cell's most important activities are its transactions with the environment, a give-and-take that never ceases. Cells are constantly feasting on food they encounter, ingesting molecules and sometimes entire cells. They dump their wastes back into the environment together with many other kinds of molecules. Cells continuously garner information about the world around them, responding to a host of chemical clues and often passing on messages to other cells. This constant interplay with the environment is a fundamental characteristic of all cells. Without it, life could not persist (Figure 6-1).

Imagine, now, that you were to coat a living cell in plastic, giving it a rock-hard, impermeable shell. All the cell's transactions with the environment would stop. No molecules could pass in or out, nor could the cell learn anything about the molecules around it. The cell might as well be a rock. Life in any meaningful sense would cease—unless, of course, you allowed for doors and windows in the shell.

That is, in fact, what living cells do. Every cell is encased within a lipid membrane, an impermeable shell through which no water-soluble molecules and little information (data about its surroundings) can pass, but the shell contains doors and windows made of protein. Molecules pass in and out of a cell through these passageways, and information passes in and out through the windows. A cell interacts with the world through a delicate skin of protein molecules embedded in a thin sheet of lipid. We call this assembly of lipid and protein a **plasmalemma,** or **plasma membrane.** The structure and function of this membrane are the subject of this chapter.

THE LIPID FOUNDATION OF MEMBRANES

The membranes that encase all living cells are sheets only a few molecules thick; it would take more than 10,000 of these sheets piled on one another to equal the thickness of this sheet of paper. These thin sheets are not simple in structure, however, like a soap bubble. Rather, they are made up of diverse collections of proteins enmeshed in a lipid framework, like corks bobbing on the surface of a pond.

Phospholipids

The lipid layer that forms the foundation of cell membranes is composed of molecules called **phospholipids.** Like the fat molecules you studied in Chapter 3, a phospholipid has a backbone derived from a three-carbon molecule called glycerol, with long chains of carbon atoms called fatty acids attached to this backbone. A fat molecule has three such chains, one attached to each carbon of the backbone; because these chains are nonpolar (do not form hydrogen bonds with water), the fat molecule is insoluble in water. A phospholipid, by contrast, has only two such chains attached to its backbone (Figure 6-2). The third position is occupied instead by a highly polar organic alcohol that readily forms hydrogen bonds with water. Because this alcohol is attached by a phosphate group, the molecule is called a *phospho*lipid.

One end of a phospholipid molecule is therefore strongly nonpolar (water insoluble), whereas the other end is extremely polar (water soluble). The two nonpolar fatty acids extend in one direction, roughly parallel to each other, and the polar alcohol group points in the other direction. Because of this structure, phospholipids are often diagrammed as a (polar) ball with two dangling (nonpolar) tails (Figure 6-3).

Phospholipids Form Bilayer Sheets

Imagine what happens when a collection of phospholipid molecules is placed in water. The long nonpolar tails of phospholipid molecules are pushed away by the water molecules that surround them, shouldered aside as the water molecules seek partners that can form hydrogen bonds. Water molecules always tend to form the maximum number of such bonds; the long nonpolar chains that can't form hydrogen bonds get in the way, like too many chaperones at a party. The best way to rescue the party is to put all the chaperones together in a separate room, and that is what water molecules do—they shove all the long, nonpolar tails of the lipid molecules together, out of the way. The polar heads of the phospholipids are "welcomed," however, because they form good hydrogen bonds with water. What happens is that every phospholipid molecule orients so that its polar head faces water and its nonpolar tails face away. Because there are *two* layers with the tails facing each other, no tails are ever in contact with water. The structure that results is called a **lipid bilayer** (see Figure 5-9). Lipid bilayers form spontaneously, driven by the forceful way in which water tends to form hydrogen bonds.

> **The basic foundation of biological membranes is a lipid bilayer that forms spontaneously. In such a layer, the nonpolar tails of phospholipid molecules point inward, forming a nonpolar zone in the interior of the bilayer.**

Lipid bilayer sheets of this sort are the foundation of all biological membranes. Because the interior of the bilayer is completely nonpolar, it repels any water-soluble molecules that attempt to pass through it, just as a layer of oil stops the passage of a drop of water (that's why ducks don't get wet). This barrier to the passage of water-soluble molecules is the key biological property of the lipid bilayer. It means that a cell, if fully encased within a pure lipid bilayer, would be completely impermeable to water-soluble molecules such as sugars, amino acids, and proteins. No cell, however, is so imprisoned. In addition to the phospholipid molecules that make up the lipid bilayer, the membranes of every cell also contain proteins that extend across the lipid bilayer, providing passages across the membrane.

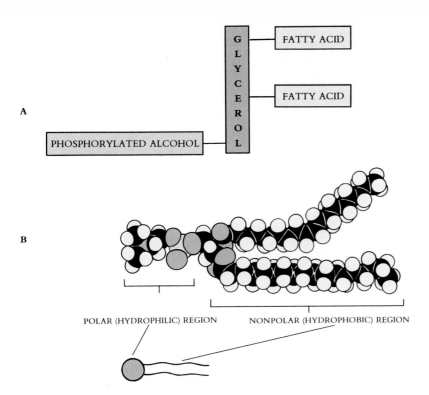

FIGURE 6-2

A A phospholipid is a composite molecule similar to a triglyceride, except in this case only two fatty acids are bound to the glycerol backbone, the third position being occupied by another kind of molecule called a phosphorylated alcohol.

B Since the phosphorylated alcohol usually extends from one end of the molecule, and the two fatty acid chains from the other, phospholipids are often diagrammed as a polar ball with two nonpolar tails.

POLAR (HYDROPHILIC) REGION   NONPOLAR (HYDROPHOBIC) REGION

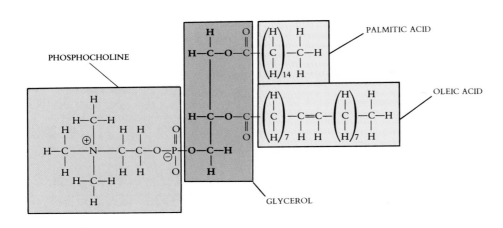

FIGURE 6-3

Molecular model of one of the most common membrane phospholipids, lecithin.

The Lipid Bilayer Is a Fluid

A lipid bilayer is very stable because the hunger of water for hydrogen bonding never stops. Although water continually urges phospholipid molecules into this orientation, it is indifferent to where individual phospholipid molecules are located. Water forms just as many hydrogen bonds when a particular phospholipid molecule is located here as there. As a result, individual lipid molecules are free to move about within the membrane (Figure 6-4). Because the individual molecules are free to move about, the lipid bilayer is not a solid like a rubber balloon, but rather a liquid like the "shell" of a soap bubble. The bilayer itself is a fluid, with the viscosity of olive oil. Just as the surface tension holds the soap bubble together, even though it is made of a liquid, so the hydrogen bonding of water holds the membrane together.

Some membranes are more fluid than others, however. The tails of individual phospholipid molecules attract one another when they line up close together, stiffening the membrane because aligned molecules must pull apart from one another before they

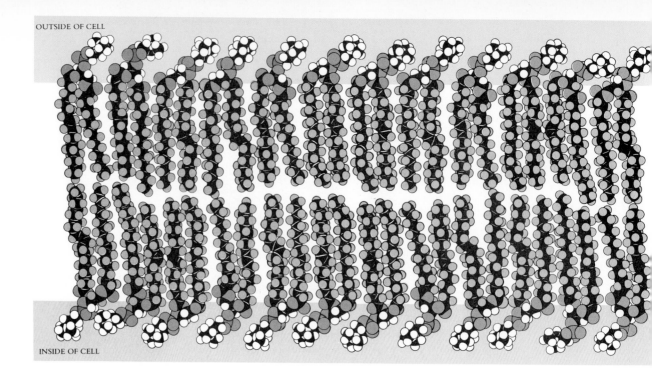

OUTSIDE OF CELL

INSIDE OF CELL

FIGURE 6-4

This diagram illustrates how the long nonpolar tails of the phospholipids orient toward one another. Because some of the tails contain double bonds, which introduce kinks in their shape, the tails do not align perfectly and the membrane is "fluid"—individual phospholipid molecules can move from one place to another in the membrane.

can move about in the membrane. The more alignment, the less fluid the membrane is. Some phospholipids have tails that don't align well because they contain one or more C═C double bonds, which introduce kinks in the tail. Membranes containing phospholipids of these sorts are more fluid than those that lack them. Sometimes membranes contain other short lipids like cholesterol, which prevent the phospholipid tails from coming into contact with one another and thus make the membrane more fluid.

ARCHITECTURE OF THE PLASMA MEMBRANE

A eukaryotic cell contains many membranes, and they are not all identical. They all share, however, the same fundamental architecture. Cell membranes are assembled from four components:

1. *A lipid bilayer foundation.* Every cell membrane has as its basic foundation a phospholipid bilayer. The other components of the membrane are enmeshed within the bilayer, which provides a flexible matrix and, at the same time, imposes a barrier to permeability.

2. *Transmembrane proteins.* A major component of every membrane is a collection of proteins that float within the lipid bilayer like icebergs on the sea (Figure 6-5). These proteins provide channels into the cell through which molecules and information pass. Membrane proteins are not fixed in position like headstones in a graveyard. Instead, they move about like boats on a lake. Some membranes are crowded with proteins, side by side, just as some lakes are so crowded with boats you can hardly see the water. In other membranes, the proteins are more sparsely distributed.

3. *Network of supporting fibers.* Membranes are structurally supported by proteins that reinforce the membrane's shape. This is why a red blood cell is shaped like a doughnut rather than being irregular. The plasma membrane is held in that shape by a scaffold of protein on its inner surface (Figure 6-5). Membranes use networks of other proteins to control the lateral movements of some key membrane proteins, anchoring them to specific sites so that they do not simply drift away. Unanchored proteins have been observed to move as much as 10 micrometers in 1 minute.

4. *Exterior glycolipids.* A thicket of carbohydrate and lipid extends from the cell surface. These chains act as cell identity markers. Different cell types exhibit different kinds of carbohydrate chains on their surfaces.

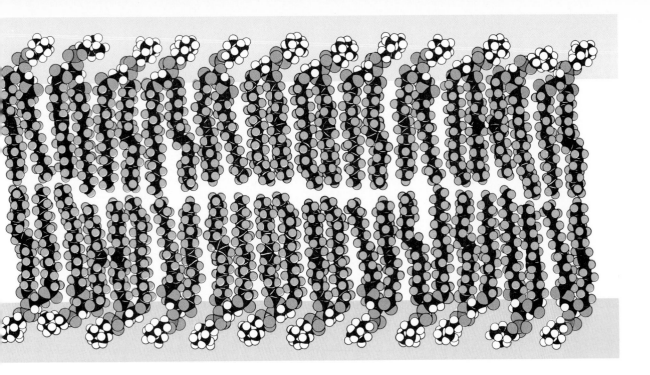

A membrane, then, is a sheet of lipid and protein that is supported by other proteins and to which carbohydrates are attached (Table 6-1). The key functional proteins act as passages through the membrane, extending all the way across the bilayer. How do these transmembrane proteins manage to span the membrane, rather than just floating on the surface, in the way that a drop of water floats on oil? The part of the protein that traverses the lipid bilayer is specially constructed, a spiral helix of nonpolar amino acids (Figure 6-6). Because water responds to nonpolar amino acids much as it does to nonpolar lipid chains, the nonpolar helical spiral is held within the interior of the lipid bilayer by the strong tendency of water to avoid contact with these amino acids. Although the polar ends of the protein protrude from both sides of the membrane, the protein itself is locked into the membrane by its nonpolar helical segment.

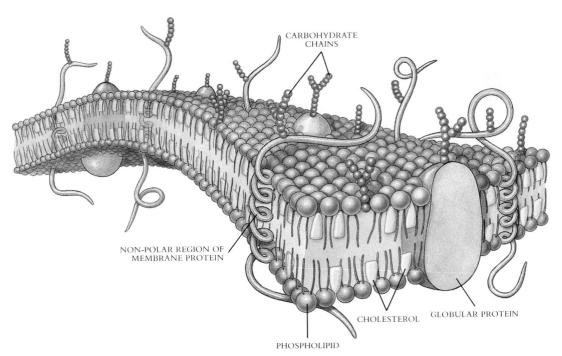

CARBOHYDRATE
CHAINS

NON-POLAR REGION OF
MEMBRANE PROTEIN

CHOLESTEROL   GLOBULAR PROTEIN

PHOSPHOLIPID

FIGURE 6-5

Within the lipid bilayer of animal cells, cholesterol molecules are inserted, adding fluidity. A variety of proteins protrude through the membrane, nonpolar regions of the proteins serving to tether them to the membrane's nonpolar interior. The three principal classes of membrane protein are channels, receptors, and cell surface markers. Carbohydrate chains (strings of sugar molecules) are often bound to these proteins, and to lipids in the membrane itself as well. These chains serve as distinctive identification tags, unique to particular types of cells.

TABLE 6–1 COMPONENTS OF THE CELL MEMBRANE

COMPONENT	COMPOSITION	FUNCTION	HOW IT WORKS	EXAMPLE
Lipid foundation	Phospholipid bilayer	Permeability barrier matrix for proteins	Water-soluble molecules excluded from nonpolar interior of bilayer	Bilayer of cell is impermeable to water-soluble molecules
Transmembrane proteins	Single-coil channels	Transport of large molecules across membrane	Carrier "flip-flops"	Glycophorin channel for sugar transport
	Multicoil channels	Transport of small molecules across membrane	Create a tunnel that acts as a passage	Photoreceptor
	Receptors	Transmit information into cell	Bind to cell surface portion of protein; alter portion within cell, inducing activity	Peptide hormones; neurotransmitters
Cell surface markers	Glycoprotein	"Self"-recognition	Shape of protein/carbohydrate chain characteristic of individual	Major histocompatibility complex protein recognized by immune system
	Glycolipid	Tissue recognition	Shape of carbohydrate chain is characteristic of tissue	A, B, O blood group markers
Interior protein network	Spectrin	Determines shape of cell	Forms supporting scaffold beneath membrane, anchored to both membrane and cytoskeleton	Red blood cell
	Clathrins	Anchor certain proteins to specific sites	Form network above membrane to which proteins are anchored	Localization of low-density lipoprotein receptor within coated pits

FIGURE 6-6

All membrane proteins are anchored within the lipid bilayer by nonpolar segments. In all cases studied to date, these segments have proved to be helical in secondary structure. Two general classes of membrane protein occur: proteins that traverse the membrane only once (receptors and some channels are of this sort) and proteins that traverse the membrane many times, creating a hollow "pipe" through the bilayer, as illustrated here. Many channels are of this sort.

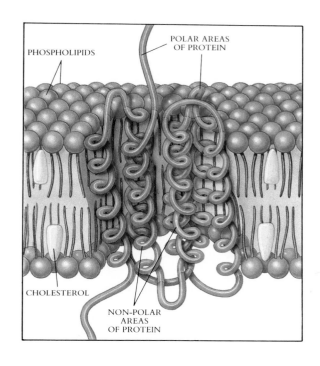

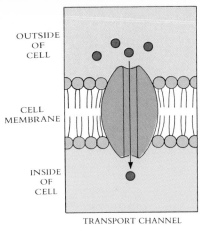

OUTSIDE OF CELL

CELL MEMBRANE

INSIDE OF CELL

TRANSPORT CHANNEL

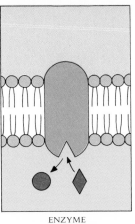

ENZYME

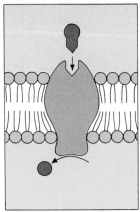

RECEPTOR SITE

FIGURE 6-7

Functions of plasma membrane proteins.

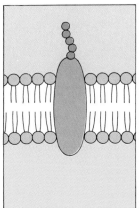

CELL IDENTITY MARKER

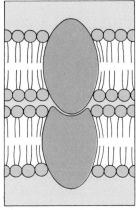

CELL ADHESION

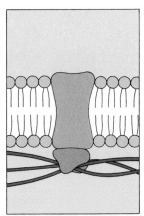

ATTACHMENT OF CYTOSKELETON

HOW A CELL'S MEMBRANES REGULATE INTERACTIONS WITH ITS ENVIRONMENT

The cell membrane, then, is a complex assembly of proteins floating like loosely anchored ships on a lipid sea. This is probably not the sort of cell surface you or I would have designed. It is not strong, and its surface components are not precisely located. It is, however, a design of enormous flexibility, permitting a broad range of interactions with the environment. A cell membrane is like a rack that can hold many different tools. A list of all the different kinds of proteins in one cell's plasma membrane would run to many pages. With these tools, the cell can interact with its environment in many different ways (Figure 6-7), admitting a particular molecule here, sensing the presence of a hormonal signal there. Like the tools of a busy factory, a cell's membrane proteins are in a state of high activity. If the proteins of a membrane made noise when they worked, like the tools in a factory do, a cell would be in a state of constant uproar.

We shall mention only six of the many ways in which a cell's membranes regulate its interactions with the environment:

1. Passage of water   Membranes are freely permeable to water, but the spontaneous movement of water into and out of cells sometimes presents problems.
2. Passage of bulk material   Cells sometimes engulf large hunks of other cells or gulp liquids.
3. Selective transport of molecules   Membranes are very picky about which molecules they allow to enter or leave the cell.
4. Reception of information   Membranes can identify chemical messages with exquisite sensitivity.
5. Expression of cell identity   Membranes carry name tags that tell other cells who they are.
6. Physical connection with other cells   In forming tissues, membranes make special connections with each other.

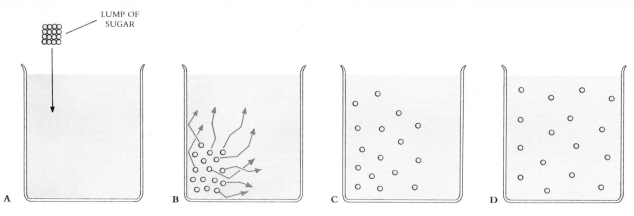

FIGURE 6-8

Diffusion. If a lump of sugar is dropped into a beaker of water, its molecules dissolve **(A)** and diffuse **(B, C)**. Eventually, diffusion results in a even distribution of sugar molecules throughout the water **(D)**.

THE PASSAGE OF WATER INTO AND OUT OF CELLS

Molecules dissolved in liquid are in constant motion, moving about randomly. This random motion causes a net movement of molecules toward zones where the concentration of the molecules is lower, a process called **diffusion** (Figure 6-8). Driven by random motion, molecules always "explore" the space around them, diffusing out until they fill it uniformly. A simple experiment will demonstrate this. Take a small jar, fill it with ink to the brim, cap it, place it at the bottom of a full bucket, and remove the cap. The ink molecules will slowly diffuse out until there is a uniform concentration in the bucket and the jar.

> **Diffusion is the net movement of molecules to regions of lower concentration as a result of random spontaneous molecular motions. Diffusion tends to distribute molecules uniformly.**

The cytoplasm of a cell consists of molecules such as sugars, amino acids, and ions dissolved in water. The mixture of these molecules and water is called a **solution.** Water, the most common of the molecules in the mixture, is called the **solvent,** and the other kinds of molecules dissolved in the water are called **solutes.**

Because of diffusion, both solvent and solute molecules in a cell will move from regions where their concentration is greater to a region where their concentration is less. When two regions are separated by a membrane, however, what happens depends on whether or not the molecule can pass freely through that membrane; most kinds of solutes that occur in cells cannot do so. Sugars, amino acids, and other solutes, for example, are water soluble and not lipid soluble, and so are imprisoned within the

FIGURE 6-9

In this experiment a 3% salt solution is employed to demonstrate osmosis. Within cells, metabolites have the same effect.
A The end of the tube containing the 3% salt solution is closed by stretching a differentially permeable membrane across its face that will pass water molecules but not salt molecules.
B When this tube is immersed in a beaker of distilled water, the salt cannot equalize its concentration in the tube and beaker by passing into the beaker, since it cannot cross the membrane; water can achieve the same effect, however, by passing into the tube from the beaker. The added water will cause the salt solution to rise in the tube.
C Water will continue to enter the tube from the beaker until the weight of the column of water in the tube exerts a downward force equal to the force drawing water molecules upward into the tube. This force is referred to as the osmotic pressure generated by the 3% salt solution in distilled water (see Figure 6-11).

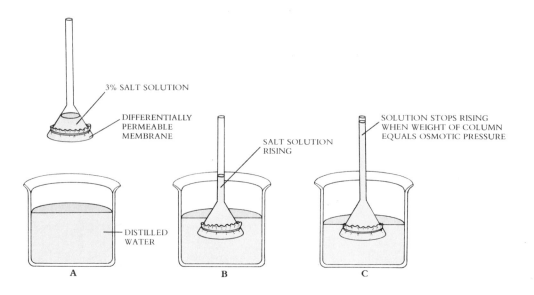

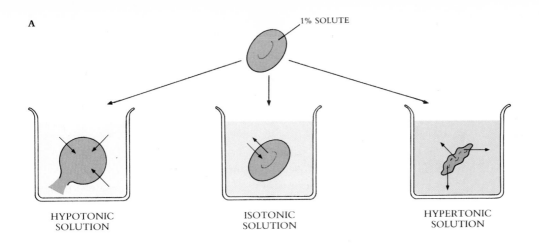

1% SOLUTE

HYPOTONIC
SOLUTION

ISOTONIC
SOLUTION

HYPERTONIC
SOLUTION

CELL WALL
CYTOPLASM
VACUOLE

HYPOTONIC SOLUTION

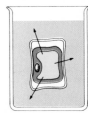

ISOTONIC SOLUTION

HYPERTONIC SOLUTION

cell; they are unable to cross the lipid bilayer of the membrane. Water molecules, in contrast, can pass through slight imperfections in the sheet of lipid molecules and so diffuse across the membrane into the cell. Water molecules stream into the cell across the membrane, thus diluting the high concentration of solutes within the cell so that they match more and more closely their lower concentrations in the outside solution. This form of net water movement into a cell is called **osmosis** (Figure 6-9).

> **Osmosis is the diffusion of water across a membrane that permits the free passage of water but not that of one or more solutes.**

The fluid content of a cell immersed in pure water is said to be **hypertonic** (Greek *hyper,* more than) with respect to its surrounding solution because it has a higher concentration of solutes than the water does. The surrounding solution, which has a lower concentration of solutes than the cell, is said to be **hypotonic** (Greek *hypo,* less than) with respect to the cell. A cell with the same concentration of solutes as its environment is said to be **isotonic** (Greek *iso,* the same) (Figure 6-10).

Intuitively you might think that as new water molecules diffuse inward, the pressure of the cytoplasm pushing out against the cell wall, called **hydrostatic pressure**, would build up, and this is indeed what happens. As water molecules continue to diffuse inward toward the area of lower *water* concentration (the concentration of the water is lower than outside the cell because of the dissolved solutes in it), the hydrostatic water pressure within the cell increases. We refer to pressure of this kind as **osmotic pressure** (Figure 6-11). Because osmotic pressure opposes the movement of

FIGURE 6-10

Osmosis is the movement of water across a membrane that is impermeable to solutes in the direction of lower water concentration, where dissolved solutes are more abundant.
A When the outer solution is hypotonic with respect to the cell, water will move in; when it is hypertonic, water will move out.
B In plant cells, the large central vacuole contains a high concentration of solutes, so that water tends to diffuse inward, causing the cells to swell outward against their rigid cell walls. However, if a plant cell is immersed in a high-solute (hypertonic) solution, water will leave the cell, causing the cytoplasm to shrink and pull in from the cell wall.

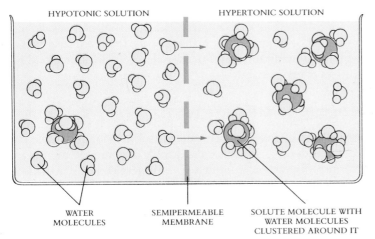

HYPOTONIC SOLUTION   HYPERTONIC SOLUTION

WATER
MOLECULES

SEMIPERMEABLE
MEMBRANE

SOLUTE MOLECULE WITH
WATER MOLECULES
CLUSTERED AROUND IT

FIGURE 6-11

How solutes create osmotic pressure. The addition of solutes to one side of a membrane does not change the overall concentration of water (the number of water molecules per unit volume), but it *does* reduce the number of water molecules that can move freely. Water molecules that are clustered about a solute molecule are not free to diffuse across the membrane. Water moves by osmosis from a hypotonic solution to a hypertonic solution because the hypotonic solution has a greater concentration of unbound water molecules that are free to cross the membrane.

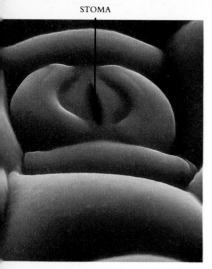

FIGURE 6-12

The cells surrounding this opening in a *Tradescantia* leaf are swollen by osmosis. The opening is called a stoma, and the swollen cells are called guard cells. When relaxed, these guard cells rest against one another, closing the opening. When swollen by osmosis from the surrounding cells, as they are here, the cells assume a rigid bowed shape, creating an opening between them.

water inward, such diffusion will not continue indefinitely. The cell will eventually reach an equilibrium—a point at which the osmotic force driving water inward is counterbalanced exactly by the hydrostatic pressure driving water out. In practice, the hydrostatic pressure at equilibrium is typically so high that an unsupported cell membrane cannot withstand it, and such an unsupported cell, suspended in water, will burst like an overinflated balloon. Cells whose membranes are surrounded by cell walls, in contrast, can withstand high internal hydrostatic pressures (Figure 6-12).

> **Within the closed volume of a cell that is hypertonic to its surroundings, the movement of water inward, which tends to lower the relative concentration difference of water, will at the same time increase the internal hydrostatic pressure. The net movement of water stops when an equilibrium condition is reached or the cell bursts.**

BULK PASSAGE INTO THE CELL

The lipid nature of biological membranes raises a second problem for growing cells. The metabolites required by cells as food are for the most part polar molecules that will not pass across the hydrophobic barrier interposed by a lipid bilayer. How then are organisms able to get food molecules into their cells? Particularly among single-celled eukaryotes, the dynamic cytoskeleton is employed to extend the cell membrane outward toward food particles such as bacteria. The membrane encircles and engulfs a food particle. Its edges eventually meet on the other side of the particle where, because of the fluid nature of the lipid bilayer, the membranes fuse together, forming a vesicle around the fluid. This process is called **endocytosis.** Endocytosis involves the incorporation of a portion of the exterior medium into the cytoplasm of the cell by capturing it within a vesicle (Figure 6-13, *A* to *D*).

If the material that is brought into the cell contains an organism (Figure 6-13, *C* and *D*) or some other fragment of organic matter, that particular kind of endocytosis is call **phagocytosis** (Greek *phagein,* to eat + *cytos,* cell). If the material brought into the cell is liquid and contains dissolved molecules, the endocytosis is referred to as **pinocytosis** (Greek *pinein,* to drink). Pinocytosis (Figure 6-13, *B*) is common among the cells of multicellular animals. Human egg cells, for example, are "nursed" by surrounding cells that secrete nutrients that the maturing egg cell takes up by pinocytosis.

> **Phagocytosis is a process in which cells literally engulf organisms or fragments of organisms, enfolding them within vesicles.**

Virtually all eukaryotic cells are constantly carrying out endocytosis, trapping extracellular fluid in vesicles and ingesting it. Rates of endocytosis vary from one cell type to another but can be surprisingly large. Some types of white blood cell, for example, ingest 25% of their cell volume each hour.

The reverse of endocytosis is **exocytosis,** the extrusion of material from a cell by discharging it from vesicles at the cell surface (Figure 6-13, *F*). In plants, vesicle discharge constitutes a major means of exporting the materials used in the construction of the cell wall through the plasma membrane. In animals, many cells are specialized for secretion using the mechanism of exocytosis.

SELECTIVE TRANSPORT OF MOLECULES

Endocytosis is an awkward means of governing entrance to the interior of the cell. It is expensive from an energy standpoint, since a considerable amount of membrane movement is involved. More important, endocytosis is not selective. Particularly when the process is pinocytotic, it is difficult for the cell to discriminate between different solutes, admitting some kinds of molecules into the cell while excluding others.

No cell, however, is limited to endocytosis. As we have seen, cell membranes are studded with proteins, which act as channels across the membrane. Because a given

A

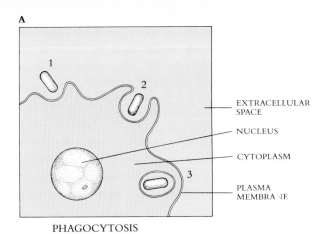

EXTRACELLULAR
SPACE

NUCLEUS

CYTOPLASM

PLASMA
MEMBRANE

PHAGOCYTOSIS

B

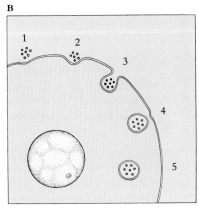

PINOCYTOSIS

FIGURE 6-13

Endocytosis and exocytosis.
Phagocytosis **(A)** is the ingestion by cells of other cells or
large fragments of cells, while
pinocytosis **(B)** is the ingestion by
cells of dissolved molecules. Both
phagocytosis and pinocytosis are
forms of endocytosis.
C Phagocytosis in action. The
large egg–shaped protist *Didinium
nasutum* has just begun eating the
smaller protist *Paramecium*.
D Its meal is practically over.
E and **F** Exocytosis. Proteins and
other molecules are secreted from
cells in small packets called
vesicles, whose membranes fuse
with the cell membrane, releasing
their contents to the cell surface.

C

D

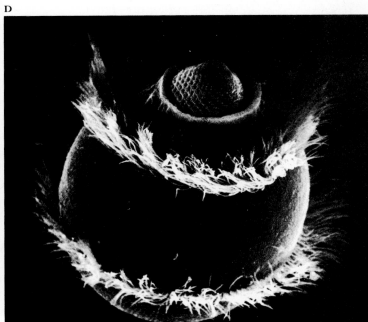

E

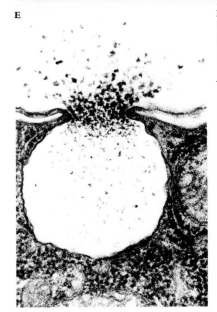

F

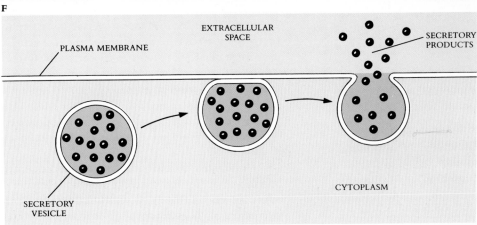

PLASMA MEMBRANE

EXTRACELLULAR
SPACE

SECRETORY
PRODUCTS

SECRETORY
VESICLE

CYTOPLASM

channel will transport only certain kinds of molecules, the cell membrane is **selectively permeable,** that is, it is permeable to some molecules and not to others.

> **Selective permeability allows the passage across a membrane of some solutes but not others. Selective permeability of cell membranes is the result of specific protein channels extending across the membrane; some molecules can pass through a specific kind of channel, but others cannot do so.**

The Importance of Selective Permeability

The most important property of any cell, in general, is that the cell constitutes an isolated compartment within which certain molecules can be concentrated and brought together in particular combinations. This essential isolation depends on allowing molecules to enter the cell selectively. Your home is private in the same sense—inoperable if no one including you can enter, but not private if everyone in the neighborhood is free to wander through at all times. The solution adopted by cells is the same one that most homeowners adopt: there are doors with keys, and only those possessing the proper keys can enter or leave. The channels through cell membranes are the doors to cells. These doors are not open to any molecule that presents itself; only particular molecules can pass through a given kind of door. A cell is able to control the entry and exit of many kinds of molecules by possessing many different kinds of doors. These channels through its membrane are among the most important functional feature of any cell.

Facilitated Diffusion

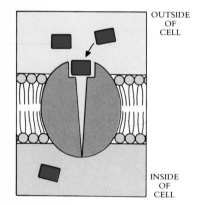

Some of the most important channels in the cell membrane are highly selective, facilitating *only* the passage of specific molecules or ions, but in either direction. An example is provided by the channel of vertebrate red blood cell membranes that transports negatively charged ions, or **anions.** This channel plays a key role in the oxygen-transporting function of these cells. The anion channels of the red blood cell membrane readily pass chloride ion (Cl^-) or carbonate ion (HCO_3^-) across the red blood cell membrane. We can easily demonstrate that the ions are moving by diffusion through the channels in the red blood cell membrane. If there is more Cl^- ion within the cell, we will see that the net movement is outward, whereas if there is more Cl^- ion outside the cell, then the net movement will be into the cell. Since the ion movement is always toward the direction of lower ion concentration, the transport process is one of diffusion.

Are these channels simply holes in the membrane, somehow specific for these anions? Not at all. If we repeat our hypothetical experiment, progressively increasing the concentration of Cl^- ion outside the cell above that inside the cell, the rate of movement of Cl^- ion into the cell increases only up to a certain point, after which it levels off and will proceed no faster despite increases in the concentration of exterior Cl^- ion. The reason that the diffusion rate will increase no further is that the Cl^- ions are being transported across the membrane by a "carrier," and all available carriers are in use: we have saturated the capacity of the carrier system. The transport of these anions, then, is a diffusion process facilitated by a "carrier." Transport processes of this kind are called **facilitated diffusion** (Figure 6-14).

> **Facilitated diffusion is the transport of molecules across a membrane by a carrier protein in the direction of lowest concentration.**

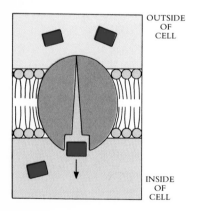

FIGURE 6-14

Facilitated diffusion is a carrier-mediated transport process.

Facilitated diffusion provides the cell with a ready means of preventing the buildup of unwanted molecules within the cell, or of gleaning from the external medium molecules that are present there in high concentration. Facilitated diffusion has two essential characteristics: (1) it is *specific,* with only certain molecules being able to traverse a given channel, and (2) it is *passive,* the direction of net movement being determined by the relative concentrations of the transported molecule inside and outside the membrane.

Active Transport

There are many molecules that the cell admits across its membrane but that are maintained within the cell at a concentration different from that of the surrounding medium. In all such cases, the cell must expend energy to maintain the concentration difference. The kind of transport that requires the expenditure of energy is called **active transport.** Active transport may maintain molecules at a higher concentration inside the cell than outside it by expending energy to pump more molecules in than would enter by diffusion, or it may maintain molecules at a lower concentration by expending energy to pump them out actively.

> **Active transport is the transport of a solute across a membrane to a region of higher concentration by the expenditure of chemical energy.**

Active transport is one of the most important functions of any cell. It is by active transport that a cell is able to concentrate metabolites. Without it, the cells of your body would be unable to harvest glucose molecules—a major source of energy—from the blood, since glucose concentration is often higher in the cells already than it is in the blood from which it is extracted. Imagine how difficult it would be to survive as a beggar if you could only obtain money from those who had less than you did! Active transport permits a cell, by expending energy, to take up additional molecules of a substance that is already present in its cytoplasm in concentrations higher than in the cell's environment.

There are many molecules that a cell takes up or eliminates against a concentration gradient. Some, like sugars and amino acids, are simple metabolites that the cell extracts from its surroundings and adds to its internal stockpile. Others are ions, such as sodium and potassium, that play a critical role in functions such as the conduction of nerve impulses. Still others are the nucleotides that the cell uses to synthesize DNA. These many different kinds of molecules enter and leave cells by way of a wide variety of different kinds of selectively permeable transport channels. Some of them are permeable to one or a few sugars, others to a certain size of amino acid, and still others to a specific ion or nucleotide. You might suspect that active transport occurs at each of these channels, but you would be wrong. There is *one* major active transport channel in cell membranes that transports sodium and potassium ions; all the others work by tying their activity to this all-important channel. The many channels in the membrane that the cell uses to concentrate metabolites and ions are called **coupled channels.** We shall discuss the sodium-potassium channel first, and then these coupled channels.

The Sodium-Potassium Pump

More than one third of all the energy expended by a cell that is not actively dividing is used to transport sodium (Na^+) and potassium (K^+) ions actively. The remarkable channel by which these two ions are transported across the cell membrane is referred to as the **sodium-potassium pump.** Most animal cells have a low internal concentration of Na^+ ions, relative to their surroundings, and a high internal concentration of K^+ ions. They maintain these concentration differences by actively pumping Na^+ ions out of the cell and K^+ ions in. The transport of these ions is carried out by a highly specific transmembrane protein channel. Passage through this channel entails changes in the shapes of the proteins within it.

The sodium-potassium pump is an active transport process, transporting Na^+ and K^+ ions from areas of low concentration to areas of high concentration. This transport into a zone of higher concentration is just the opposite of that which occurs spontaneously in diffusion; it is achieved only by the constant expenditure of metabolic energy. The energy used in the process is obtained from a molecule called **adenosine triphosphate (ATP),** the functioning of which will be explained in Chapter 7. Some membranes contain large numbers of $Na^+ - K^+$ channels, whereas others have few. The changes in protein shape that go on within an individual channel are very rapid. Each channel is capable of transporting as many as 300 Na^+ ions per second when working full tilt.

CHLORIDE CHANNELS AND CYSTIC FIBROSIS

Cystic fibrosis is a fatal disease of human beings in which affected individuals secrete thick mucus that clogs the airways of the lungs. These same secretions block the ducts of the pancreas and liver so that the few patients who do not die of lung disease die of liver failure. Cystic fibrosis is usually thought of as a children's disease, since few affected individuals live long enough to become adults. There is no known cure.

Cystic fibrosis is a genetic disease resulting from a defect in a single gene that is passed down from parent to child. It is the most common fatal genetic disease of Caucasians. One in 20 individuals possesses at least one copy of the defective gene. Most carriers are not afflicted with the disease; only those children who inherit two copies of the defective gene, one from each parent, succumb to cystic fibrosis—about 1 in 1800 Caucasian children.

Cystic fibrosis has proved to be difficult to study. Many organs are affected, and until recently it was impossible to identify the nature of the defective gene responsible for the disease. In 1985 the first clear clue was obtained. An investigator named Paul Quinton seized on a commonly observed characteristic of cystic fibrosis patients, that their sweat is abnormally salty, and performed the following experiment. He isolated a sweat duct from a small piece of skin and placed it in a solution of salt (NaCl) that was three times as concentrated as the NaCl inside the duct. He then monitored the movement of ions. Diffusion tends to drive both the Na^+ and Cl^- ions into the duct because of the higher outer ion concentrations. In skin isolated from normal individuals Na^+ ions indeed enter the duct, transported by the sodium-potassium pump; Cl^- ions follow, passing

through a passive channel. Both ions cross the membrane easily. In skin isolated from individuals with cystic fibrosis, the sodium-potassium pump transports Na^+ ions into the ducts, but no Cl^- ions enter. The passive chloride channels are not functioning in these individuals.

It appears that cystic fibrosis results from a defective channel within membranes, one that transports Cl^- ions across the membranes of normal individuals but not affected persons. It was learned in 1986 that the genetic defect is the result of an alteration in a protein regulating the activity of the channel, rather than in the transmembrane protein itself. The defective gene was isolated in 1987. Now that scientists have finally identified the primary cause of the disease they can embark on the road to find a cure.

FIGURE 6-15

A coupled channel. The sodium-potassium pump keeps the Na^+ ion concentration higher outside the cell than inside. There is thus a strong tendency for Na^+ ions to diffuse back in through the coupled channel—but their passage requires the simultaneous transport of a sugar molecule as well. The diffusion gradient driving Na^+ entry is so great that sugar molecules are pulled in too, even against a sugar concentration gradient.

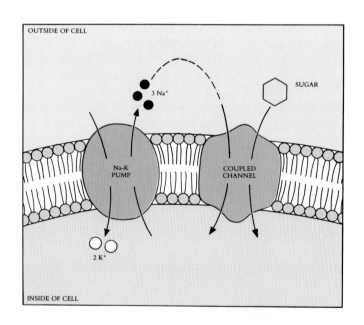

Coupled Channels

The accumulation of many amino acids and sugars by cells is also driven against a concentration gradient: the molecules are harvested from a surrounding medium in which their concentration is much lower than it is inside the cell. The active transport of these molecules across the cell membrane takes place by coupling them with Na^+ ions, which pass simultaneously through a channel by facilitated diffusion. This process is driven by the very low concentration of Na^+ ions within the cell, maintained by active transport of these ions outward by the sodium-potassium pump. The special transmembrane protein of the **coupled channel** allows Na^+ back into the cell only when the sugar or amino acid is also bound to the protein's exterior surface (Figure 6-15). As a result, the facilitated diffusion inward of the Na^+ ion also results in the importation into the cell of the sugar or amino acid. In this process, the sugar or amino acid is literally dragged along osmotically because the Na^+ ions are so much less concentrated within the cell than outside. Thus the transport of the sugar or amino acid inward to an area of its higher concentration, against the concentration gradient, occurs as a direct consequence of the sodium-potassium pump's activity.

The Proton Pump

The sodium-potassium pump is of central importance because it drives the uptake of so many different molecules. A second channel of equal importance in the life of the cell is the **proton pump.**

The proton pump involves two special transmembrane protein channels: the first pumps protons (H^+ ions) out of the cell, using energy derived from energy-rich molecules or from photosynthesis to power the active transport. This creates a proton gradient, in which the concentration of protons is higher outside the membrane than inside. As a result, diffusion acts as a force to drive protons back across the membrane toward a zone of lower proton concentration. The plasma membrane, however, is impermeable to protons, so the only way protons can diffuse back into the cell is through a second channel, which couples the transport of protons to the production of ATP. This process occurs within the cell and not primarily at the cell membrane. The net result is the expenditure of energy derived from metabolism or photosynthesis and the production of ATP. This mechanism, called **chemiosmosis,** is responsible for the production of almost all the ATP you harvest from food that you eat and for all the ATP produced by photosynthesis. ATP provides the cell with a usable energy source that it can employ in its many activities.

Transfer Between Cell Compartments

Relatively little is known about how molecules pass from one compartment of a cell to another. One process thought to be important is receptor-mediated endocytosis.

Receptor-Mediated Endocytosis

Within most eukaryotic cells there are vesicles that transport molecules from one place to another. In electron micrographs, some of these vesicles appear like membrane-bounded balloons coated with bristles on their interior (cytoplasmic) surfaces, as if they had a 2-day growth of beard. These coated vesicles appear to form by the pinching in of local regions of membrane coated with a network of proteins called coated pits (Figure 6-16). Coated pits act like molecular mousetraps, closing over to form an internal vesicle when the right molecule enters the pit. The trigger that releases the mousetrap is a protein called a receptor, embedded within the pit, that detects the presence of a particular target molecule and reacts by initiating endocytosis of the coated pit. By doing so, it traps the target molecule within the new vesicle. This process is called **receptor-mediated endocytosis.** It is quite specific, so that transport across the membrane occurs *only* when the correct molecule is present and positioned for transport.

The mechanisms for transport across cell membranes that we have considered in this chapter are summarized in Table 6-2.

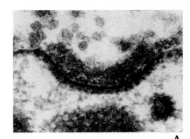

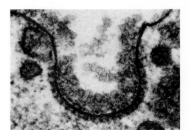

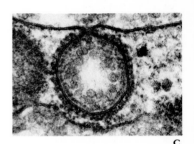

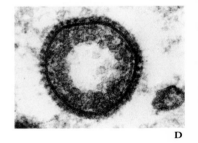

FIGURE 6-16

The stages in the formation of a coated vesicle are shown in these electron micrographs.

A A coated pit appears in the plasma membrane of a developing egg cell, covered with a layer of proteins. A coating of protein molecules can be seen just beneath the pit, on the interior side of the membrane. When an appropriate collection of molecules gathers in the coated pit, the pit deepens **(B),** the outer membrane of the cell closes in behind the pit **(C),** and the pit buds off to form a coated vesicle, which carries the molecules into the cell **(D).**

TABLE 6-2   MECHANISMS FOR TRANSPORT ACROSS CELL MEMBRANES

PROCESS	PASSAGE THROUGH MEMBRANE	HOW IT WORKS	EXAMPLE
NONSPECIFIC PROCESS			
Diffusion	Imperfections in lipid bilayer	Random molecular motion produces net migration of molecules toward region of lower concentration.	Movement of oxygen into cells
Osmosis	Imperfections in lipid bilayer	Diffusion of water across differentially permeable membrane.	Movement of water into cells placed in distilled water
Endocytosis			
Phagocytosis	Membrane vesicle	Particle is engulfed by membrane, which folds around it, forming a vesicle.	Ingestion of bacteria by white blood cells
Pinocytosis	Membrane vesicle	Fluid droplets are engulfed by membrane, which forms vesicles around them.	Nursing of human egg cells
Exocytosis	Membrane vesicle	Vesicles fuse with plasma membrane and eject contents.	Secretion of mucus
SPECIFIC PROCESS			
Facilitated diffusion	Protein channel	Molecule binds to carrier protein in membrane and is transported across; net movement is in direction of lowest concentration.	Movement of glucose into cells
Active transport			
Na-K pump	Protein channel	Carrier expends consumed energy to export Na ions against a concentration gradient.	Coupled uptake of many molecules into cells, against a concentration gradient
Proton pump	Protein channel	Carrier expends consumed energy to export protons against a concentration gradient.	Chemiosmotic generation of ATP
Carrier-mediated endocytosis	Membrane vesicle	Endocytosis triggered by a specific receptor.	Cholesterol uptake
Secretory synthesis	Protein channel	Ribosomes secrete newly made protein through channel in endoplasmic reticulum.	Passage of proteins into mitochondria

RECEPTION OF INFORMATION

So far in this chapter we have focused on membrane proteins that act as channels—doors across a membrane through which only particular molecules may pass. But cells also interact with their environments in many ways that do not involve the passage of molecules across membranes. A second major class of transaction that cells carry out with their environments involves information. In examining receptor-mediated endocytosis, we encounter for the first time a second general class of membrane proteins, **cell surface receptors.** Recall that the receptor, a protein embedded within the membrane of a coated pit, binds specifically to particles but does *not* itself provide a transport channel for the particle; what the receptor transmits into the cell is information.

In general, a cell surface receptor is an information-transmitting protein that extends across a cell membrane. The end of the receptor protein exposed on the cell surface has a shape that fits to specific hormones or other "signal" molecules, and when such molecules encounter the receptor on the cell surface, they bind to it. This binding produces a change in the shape of the other end of the receptor protein, the end protruding into the interior of the cell, and this change in shape in turn causes a change in cell activity in one of several ways.

Cell surface receptors play a very important role in the lives of multicellular animals. Among these receptors are, for example, the signals that pass from one nerve to another; the protein hormones such as adrenaline and insulin, which your body uses to regulate its metabolic level; and the growth factors such as epidermal growth factor, which regulate development. All of these substances act by binding to specific cell surface receptors. The antibodies that your body uses to defend itself against infection are themselves free forms of receptor proteins. Without receptor proteins, the cells of your body would be "blind," unable to detect the wealth of chemical signals that the tissues of your body use to communicate with one another.

EXPRESSION OF CELL IDENTITY

In addition to passing molecules across its membranes and acquiring information about its surroundings, a third major class of cell interaction with the environment involves conveying information *to* the environment. To understand why this is important, let us consider multicellular animals like ourselves. One of their fundamental properties is the development and maintenance of highly specialized groups of cells called **tissues.** Your blood is a tissue, and so is your muscle. What is remarkable about having tissues is that each cell within a tissue performs the functions of a member of that tissue and not some other one, even though all the cells of the body have the same genetic complement of DNA and are derived from a single cell at conception. How does a cell "know" to what tissue it belongs? In the course of development in human beings and other vertebrates, some cells move over others, as if they were seeking particular collections of cells with which to develop. How do they sense where they are? Every day of your adult life your immune system inspects the cells of your body, looking for cells infected by viruses. How does it recognize them? When foreign tissue is transplanted into your body, your immune system rejects it. How does it know that the transplanted tissue is foreign?

The answer to all these questions is the same: during development, every cell type in your body is given a banner proclaiming its identity, a set of proteins called **cell surface markers** that are unique to it alone. Cell surface markers are the tools a cell uses to signal to the environment what kind of cell it is.

Some cell surface markers are proteins anchored in plasma membranes. The immune system uses such marker proteins to identify "self," for example. All the cells of a given individual have the same "self" marker, called a **major histocompatibility complex (MHC)** protein. Because practically every individual makes a different version of the MHC marker protein, these proteins serve as distinctive markers for that individual.

Other cell surface markers are **glycolipids,** lipids with carbohydrate tails. These are the cell surface markers that differentiate the various organs and tissues of the vertebrate body. The markers on the surfaces of red blood cells that distinguish different blood types, such as A, B, and O, are glycolipids. During the course of development, the cell population of glycolipids changes dramatically as the cells divide and differentiate.

PHYSICAL CONNECTIONS BETWEEN CELLS

So far we have seen how cells pass molecules to and from the environment, how they acquire information from the environment, and how they convey information about their identity to the environment. A fourth major class of interaction with the environment concerns physical interactions with other cells. Most of the cells of multicellular organisms are in contact with other cells, usually as members of organized tissues in organs such as lungs, heart, or gut. The immediate environment of many of these cells is the mass of other cells clustered around it. The nature of the physical connections between a cell and the other cells of a tissue in large measure determines that cell's contribution to what the tissue will be like.

FIGURE 6-17

Desmosomes are simple points of attachment between animal cells. They do not contain channels connecting the interiors of the two attached cells.

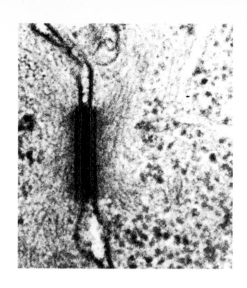

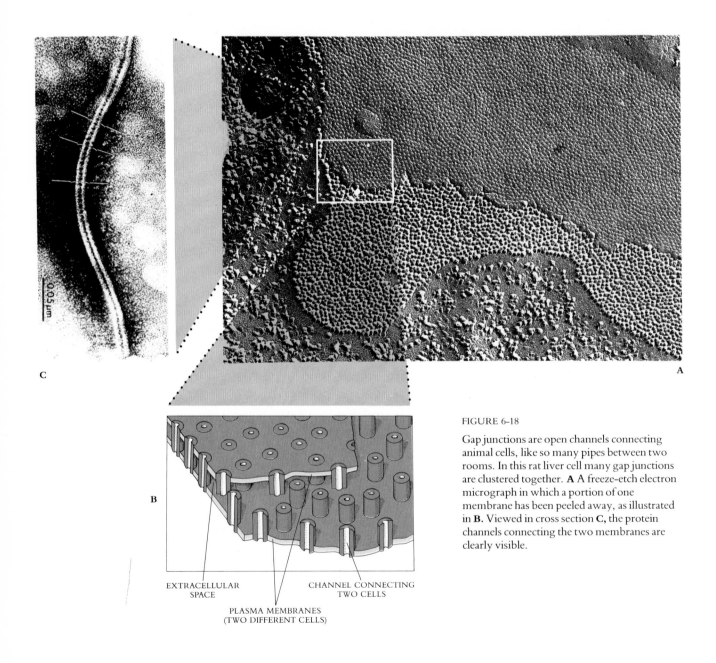

C

B

A

EXTRACELLULAR SPACE

CHANNEL CONNECTING TWO CELLS

PLASMA MEMBRANES (TWO DIFFERENT CELLS)

FIGURE 6-18

Gap junctions are open channels connecting animal cells, like so many pipes between two rooms. In this rat liver cell many gap junctions are clustered together. **A** A freeze–etch electron micrograph in which a portion of one membrane has been peeled away, as illustrated in **B.** Viewed in cross section **C,** the protein channels connecting the two membranes are clearly visible.

The locations on the cell surface where cells of a tissue adhere to one another are called **cell junctions.** There are three general classes in animals:

1. *Adhering junctions.* Adhering junctions hold cells together as if they were welds constructed of protein. They are called **desmosomes** (Figure 6–17) and are common in sheets of tissue that are subject to severe stress, such as skin and heart muscle.

2. *Organizing junctions.* Organizing cell junctions partition the plasma membrane into separate compartments. Many tissues, such as the cells that line the gut, form sheets only one cell thick. In the gut, one surface of this sheet of cells faces the digestive tract and the other faces the blood. For the tissue to function in food absorption from gut to bloodstream, the membrane channels on the cell membranes facing the digestive tract must pump food molecules inward; those facing the bloodstream must pump food molecules outward. It is very important that the two kinds of protein channels used for these two tasks be prevented from mixing within the fluid membrane. To prevent this mixing, the cells of the gut wall are joined with what are known as **tight junctions.** Tight junctions are belts of protein that girdle each cell like the belt around a pair of jeans. These protein belts act like fences, preventing any membrane proteins afloat in the lipid bilayer from drifting across the boundary from one side of the cell to the other. The belts of the different cells in the gut wall are all aligned the same way. In the gut the belts of adjacent cells are shoved tightly together, and there is no space between them through which leakage could occur. Therefore the two sides of the gut wall remain functionally distinct.

3. *Communicating junctions.* Communicating cell junctions of this kind pass small molecules from one cell to another. Sometimes the cells of tissue sheets are connected beneath their tight junction belts by open channels called **gap junctions** (Figure 6–18). Such junctions are passageways large enough to permit small molecules, such as sugar molecules and amino acids, to pass from one cell to another, while preventing the passage of larger molecules, such as proteins.

In plants the plasma membranes of adjacent cells come together through pairs of holes in the walls; the cytoplasmic connections that extend through such holes are called **plasmodesmata.** Animal cells, in contrast, are not enclosed in cell walls. The types of connections between cells are summarized in Table 6–3.

TABLE 6-3   TYPES OF INTERCELLULAR CONNECTIONS

TYPE	NAME	CHARACTERISTIC SPECIALIZATIONS IN CELL MEMBRANE	WIDTH OF INTERCELLULAR SPACE	FUNCTION
Adhering junctions	Desmosomes	Buttonlike welds joining opposing cell membranes	Normal size 24 nm	Hold cells tightly together
Organizing junctions	Tight junctions	Belts of protein that isolate parts of plasma membrane	Intercellular space disappears as the two adjacent membranes are adjacent	Form a barrier separating surfaces of cell
Communicating junctions	Gap junctions	Channels or pores through the two cell membranes and across the intercellular space	Intercellular space greatly narrowed to 2 nm	Provide for electrical communication between cells and for flow of ions and small molecules

HOW A CELL COMMUNICATES WITH THE OUTSIDE WORLD

Every cell is a prisoner of its lipid envelope, unable to communicate with the outside world except by means of the proteins that traverse its lipid shell. Anchored within the lipid layer by nonpolar segments, these proteins are the "senses" of the cell. They detect the presence of other molecules, often initiating responses within the cell. They provide doors into the cell through which food molecules, ions, and other molecules may pass, but like protective doormen they are very picky about whom they admit. They form the shape of a cell and bind one cell to another. They provide a cell with its identity, doing so by means of surface name tags that other cells can read. This diverse collection of proteins, together with the lipid shell within which they are embedded, constitute the cell's membrane system, a system that is among the cell's most fundamental features.

SUMMARY

1. Every cell is encased within a bilayer sheet of phospholipid, which exists as a liquid.

2. Because cells contain significant concentrations of sugars, amino acids, and other solutes, water tends to diffuse into them. As it does so, a hydrostatic pressure builds that will rupture cells lacking a wall or other means of support.

3. Lipid bilayers are selectively permeable and do not permit the diffusion of water-soluble molecules into the cell. These molecules gain entry by crossing one of a variety of transmembrane proteins embedded within the membrane, the proteins acting as transport channels.

4. Some channels involve carriers that transport molecules across the membrane much like cars of a train carry passengers across a bridge. This process is called facilitated diffusion because net movement is always in the direction of lowest concentration but cannot proceed any faster than the capacity of the number of carriers.

5. Some channels transport molecules against a concentration gradient by expending energy. The two most important ones transport sodium ions and protons. These channels create very low concentrations of sodium ions and protons (hydrogen ions) within the cell. These ions can diffuse back into the cell only through a second set of special channels, where their passage is often coupled to transport of another molecule inward or to the synthesis of ATP.

6. Passage across membranes from one cell compartment to another takes place when a particular molecule triggers a receptor in the membrane, initiating a form of endocytosis.

7. Many proteins embedded within the plasma membrane transmit information into the cell rather than transporting molecules. These proteins, called receptors, initiate chemical activity inside the cell in response to the binding of specific molecules on the cell surface.

8. Both proteins and glycolipids are used by cells as identification markers. These cell surface recognition markers permit cells of a given tissue to identify one another and also provide your body with a means of identifying foreign cells.

9. There are three general classes of cell junctions (connections between cells) in animal cells: (1) adhering junctions, or desmosomes, that hold cells together; (2) organizing junctions that partition the plasma membrane into separate compartments; and (3) communicating junctions that pass small molecules from one cell to another. In plants, the cell walls have openings that allow protoplasmic connections, called plasmodesmata, to connect adjacent cells.

REVIEW

1. What three components are necessary to make a phospholipid?

2. Which class of cell surface markers function in "self" versus "nonself" recognition?

3. What physical process allows you to smell perfume when you are near someone wearing it?

4. What mechanism is responsible for the production of almost all the energy that you obtain from food?

5. Give an example of a susbstance that acts by binding to cell surface receptors.

SELF-QUIZ

1. Phospholipids spontaneously form a lipid bilayer if placed in water. Why?
   (a) because phospholipids, like fats, are repelled by the polar nature of water
   (b) because phospholipids are polar molecules, so they mix easily with water
   (c) for the same reason that coacervates form spontaneously
   (d) they do not form a lipid bilayer spontaneously; proteins are needed to stabilize the phospholipids
   (e) because part of the molecule, the alcohol end, is polar and has affinity for the water, whereas the two nonpolar fatty acids are repelled by water and thus associate with the fatty acid tails of the other phospholipids

2. If you put an animal cell into a hypotonic solution, it will pop. However, if you put a plant cell into the same hypotonic solution, it will almost never pop. Why?
   (a) because the plant cell wall can withstand considerable hydrostatic pressure
   (b) because photosynthesis disrupts osmosis
   (c) because plant cells react oppositely from animal cells with regard to hypotonic solutions
   (d) because plant cells contain more water to begin with
   (e) because plant cells produce molecules that change the hypotonic solution into an isotonic one

3. How many of the following means of moving things into or out of cells require an expenditure of energy?
   (a) diffusion
   (b) osmosis
   (c) facilitated diffusion
   (d) active transport
   (e) sodium-potassium pump

4. How many of the following can move molecules *against* a concentration gradient?
   (a) diffusion
   (b) osmosis
   (c) facilitated diffusion
   (d) active transport
   (e) sodium-potassium pump

5. How are large proteins transported into cells?
   (a) diffusion
   (b) facilitated diffusion
   (c) active transport
   (d) the proton pump
   (e) receptor-mediated endocytosis

THOUGHT QUESTIONS

1. When a hypertonic cell is placed in water solution, water molecules move rapidly into the cell. If the introduced cell is hypotonic, however, water molecules leave the cell and enter the surrounding solution. How does an individual water molecule *know* what solutes are on the other side of the membrane?

2. Cells maintain many internal metabolite molecules at high concentrations by coupling their transport into the cell to the transport of sodium and potassium ions by the sodium-potassium pump. What happens to all the potassium ions that are constantly being pumped into the cell?

3. Why is a lipid bilayer membrane freely permeable to water, which is quite polar, when it is not freely permeable to ammonia, which is also polar and about the same molecular size?

FOR FURTHER READING

BRETSCHER, M.S.: "The Molecules of the Cell Membrane," *Scientific American*, October 1985, pages 100-108. A good description of the structure of the cell membrane and of how transmembrane proteins are anchored within the lipid bilayer.

HAKOMORI, S.: "Glycosphingolipids," *Scientific American*, May 1986, pages 44-53. Carbohydrate chains attached to lipid molecules play an important role in cell-to-cell recognition, an area of intensive present-day research.

STEAHLIN, L.A., and B.E. HULL: "Junctions Between Living Cells," *Scientific American*, May 1978, pages 140-152. Many passages may exist between adjacent cells that play important roles in the biology of organisms.

UNWIN, N., and R. HENDERSON: "The Structure of Proteins in Biological Membranes," *Scientific American*, February 1984, pages 78-94. A lucid account of how proteins are anchored within membranes by means of nonpolar segments.

ENERGY

FIGURE 7-1   Even a mother's love requires energy. These Japanese monkeys are huddled together sharing their body warmth, which they generate by converting body fat to ATP and then expending ATP to generate heat. Although their fur provides them with excellent insulation, they are constantly losing heat to the cold around them, and would not long survive without being able to generate heat to replace that which is lost.

ENERGY AND METABOLISM Chapter 7

Overview

The life processes of every cell are driven by energy. From cell growth and movement to the transport of molecules across cell membranes, all of a cell's activities require energy. Just as a car is powered by burning gasoline, so our bodies burn chemical fuel to obtain energy. Where is the energy in a chocolate bar? It is in the electrons, which spin in energetic orbitals about their atomic nuclei. Your cells strip these electrons away and use them to power their lives. Your every thought is fueled by the energy of electrons.

For Review

Here are some important terms and concepts that you will encounter in this chapter. If you are not familiar with them, you should review them before proceeding.

Mole (Chapter 3)

Nature of chemical bond (Chapter 3)

Nucleotides (Chapter 3)

Protein structure (Chapter 3)

Proton pump (Chapter 6)

A well-known story tells of an extravagant man who purchased an expensive car, drove it until it ran out of gas and sputtered to a stop—and then went out and bought another! You and I and every other living creature are like this man's car—we cannot run without fuel. Fuel is a source of energy, and energy drives everything that we do. When you whistle or walk down the street, you use energy. When you plan or hope or dream as you stand in the shower or sleep in bed, you use energy. In fact, it takes more energy to power your brain when you are sound asleep than it does to power the 60-watt bulb that lights your bedroom.

You obtain your energy from the food you eat. If you stopped eating, you would soon begin to lose weight as your body used up its stored energy. If you were not supplied with some outside source of energy, you would eventually die. The same is true of all other living things. Deprived of a source of energy, life stops. Why does this happen? Why cannot life simply continue? The reason is that each of the significant properties by which we define life—growth, reproduction, and heredity—uses energy. And once this energy has been used, it is dissipated as heat and cannot be used again. To keep life going, more energy must be supplied, like putting more logs on a fire (Figure 7-1).

Life can be viewed as a constant flow of energy, which is channeled by organisms to do the work of living. This chapter will focus on energy—what it is and how organisms capture, store, and use it. In the following two chapters we shall explore in more detail the energy-capturing and energy-using engines of cells, a network of chemical reactions that are the highway system for the energy of your body. This living chemistry—the total of all the chemical reactions that an organism performs—is called **metabolism.**

WHAT IS ENERGY?

Energy is defined as the ability to bring about change or, more generally, as the capacity to do work. Instinctively we all know something about energy. It is "work" such as the force of a falling boulder, the pull of a locomotive, or the swift dash of a horse; it is also "heat" such as the blast from an explosion or a warming fire. Energy can exist in many forms: as mechanical force, heat, sound, an electrical current, light, radioactivity, the pull of a magnet. All are able to create change, to do work.

Energy exists in two states. Some energy is actively engaged in doing work such as driving a speeding bullet or lifting a brick. This form of energy is called **kinetic energy,** or energy of motion. Other energy is not actively doing work but has the capacity to do so, just as a boulder perched on a hilltop has the capacity to roll downhill. This form of energy is called **potential energy,** or stored energy. Much of the work performed by living organisms involves the transformation of potential energy to kinetic energy (Figure 7-2).

> **Energy is the capacity to bring about change. It can exist in many forms: some are actively engaged in doing work, and others store the energy for later use.**

Because energy exists in many forms, there are many ways to measure it. The most convenient way is in terms of heat, because all other forms of energy can be converted into heat. Indeed, the study of energy is called **thermodynamics,** that is, "heat changes." The unit of heat most commonly employed in biology is the **kilocalorie (kcal),** which is equal to 1000 calories. A calorie is the heat required to raise the temperature of 1 gram of water 1 degree (from 14.5° to 15.5° C). It is important not to confuse the term calorie with a term often encountered in diets and discussions of nutrition, the Calorie (with a capital C), which is in fact another term for kilocalorie.

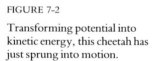

FIGURE 7-2

Transforming potential into kinetic energy, this cheetah has just sprung into motion.

THE LAWS OF THERMODYNAMICS

All of the changes in energy that take place in the universe, from nuclear explosions to the buzzing of a bee, are governed by two laws called the **laws of thermodynamics.** The **first law of thermodynamics** concerns amounts of energy. It states that energy can change from one form to another and can transform from potential energy to kinetic energy, but it can never be lost. Nor can any new energy be made. The total amount of energy in the universe remains constant.

> **The first law of thermodynamics states that energy cannot be created or destroyed; it can only undergo conversion from one form to another.**

The ground squirrel that you see gnawing on a nut in Figure 7-3 is busy acquiring energy. He is not creating new energy but rather transferring the potential energy stored in the nut's tissues to his own body, where it will fuel running, digging, and all his other daily activities. Where is the potential energy in the nut's tissue? It is stored in chemical bonds. Recall the nature of covalent chemical bonds, which are dicussed in Chapter 3. A covalent bond is created when two atomic nuclei share electrons, and breaking such a bond requires energy to pull the nuclei apart. Indeed, the strength of a covalent bond is measured by the amount of energy required to break it. For example, it takes 98.8 kilocalories of energy to break a mole (the atomic weight of a substance expressed in grams) of carbon-hydrogen (C—H) bonds.

What happens to the energy after the squirrel uses it? Some of it is transferred to other forms of potential energy—stored as fat, for example. Another portion accomplishes mechanical work such as bending blades of grass and running. Almost half is dissipated to the environment as heat, where it speeds up the random motions of molecules. This energy is not lost but rather converted to a nonuseful form, random molecular motion.

The **second law of thermodynamics** concerns this transformation of potential energy to the kinetic energy of random molecular motion, that is, to heat. It states that all objects in the universe tend to become more disordered and that the disorder in the universe is continuously increasing. The idea behind this law is easy to understand and part of everyone's experience. For example, it is much more likely that a stack of six soft-drink cans will tumble over than that six cans will spontaneously leap one onto another and form a stack. Stated simply, disorder is more likely than order. This is true of a child's room, of the desk where you study, of a waiting crowd of people—and of molecules.

> **The second law of thermodynamics states that disorder in the universe constantly increases. Energy spontaneously converts to less organized forms.**

As energy is transferred from one molecule to another, some always leaks away as kinetic energy of motion, thus increasing the random motion of molecules. At normal temperatures all molecules dance about randomly, and this added energy increases the pace of that dance. We refer to this form of kinetic energy as heat energy.

> **Heat is the energy of random molecular motion.**

With every transfer of energy, more potential energy is dissipated as heat. Although heat can be harnessed to do work when there is a gradient (that is how a steam engine works), it is generally not a useful source of energy for biological systems. Thus from a biological point of view the amount of useful energy in a system dissipates progressively. Although the total amount of energy does not change, the amount of **useful energy** available to do work decreases as progressively more energy is degraded to heat.

When energy becomes so randomized and uniform in a system that it is no longer available to do the work, the energy lost to disorder is referred to as **entropy.** Entropy is a measure of the disorder of a system. Sometimes the second law of thermodynamics

FIGURE 7-3

This ground squirrel is eating a nut. From the nut, and other food like it, the squirrel obtains all the energy it uses to run about, to grow, and to reproduce.

FIGURE 7-4

All the energy that powers life is captured from sunlight by photosynthetic organisms such as the plants that clothe these slopes in northern South America.

is stated simply as "entropy increases." When the universe was formed about 14 billion years ago, it had all the potential energy it will ever have. It has been becoming progressively more disordered ever since, with every energy exchange frittering away useful energy and increasing entropy. Someday all the energy will be random and uniform in distribution; the universe will have wound down like an abandoned clock. No stars will shine, no waves will break upon a beach. But this final state will not happen soon. Scientists speculate it will occur perhaps 100 billion years from now. Our species has been around for less than 1 million years, and life on earth has existed for about 3.5 billion years. Clearly many more immediate problems face us than the final end predicted by the second law of thermodynamics.

The first law of thermodynamics tells us that the universe as a whole is a closed system: energy doesn't come in or go out. The earth, however, is not a closed system; it is constantly receiving energy from the sun. It has been estimated that every year the earth receives in excess of 13×10^{23} calories of energy from the sun, which is equal to 2 million trillion calories per second. Much of this energy heats up the oceans and continents, but some is captured by photosynthetic organisms: plants (Figure 7-4), algae (photosynthetic protists), and photosynthetic bacteria. In photosynthesis, energy acquired from sunlight is converted to chemical energy, combining small molecules (water and carbon dioxide) into ones that are more complex (sugars). The energy is stored as potential energy in the bonds of the sugar molecules. This energy can then be shifted to other molecules by forming different chemical bonds or can be converted into motion, light, electricity—and heat. As each shift and conversion takes place, more energy is dissipated as heat. Energy continuously flows through the biological world, new energy from the sun constantly flowing in to replace the energy that is dissipated as heat.

Life converts energy from the sun to other forms of energy that drive life processes. The energy is never lost, but as it is used, more and more is converted to heat energy, a form of energy that is not useful in performing work.

OXIDATION-REDUCTION

The flow of energy into the biological world comes from the sun, which shines a constant beam of light on our earth and its moon. Life exists on earth because it is possible to capture some of that continual flow of energy and to transform it into chemical energy that can be transferred from one organism to another and used to create cattle, and fleas, and you. Where is the energy in sunlight, and how is it captured? To understand the answers to these questions, we need to look more closely at the atoms on which the sunlight shines. As described in Chapter 3, an atom is composed of a central nucleus surrounded by one or more orbiting electrons, and different electrons possess different amounts of energy, depending on how far from the nucleus they are and how strongly they are attracted to it. Light (and other forms of energy) can boost an electron to a higher energy level. The effect of this is to store the added energy as potential energy, that is, chemical energy that the atom can later release by dropping the electron back to its original energy level.

Energy stored in chemical bonds can be transferred to new chemical bonds, with the electrons shifting from one energy level to another. In some (but not all) of these chemical reactions, electrons actually pass from one atom or molecule to another. This class of chemical reaction is called an **oxidation-reduction reaction.** Oxidation-reduction (or redox) reactions are critically important to the flow of energy through living systems.

When an atom or molecule loses an electron, it is **oxidized,** and the process by which this occurs is called **oxidation.** The name reflects the fact that in biological systems oxygen, which strongly attracts electrons, is the most frequent electron acceptor (Figure 7-5).

When an atom or molecule gains an electron, it is **reduced,** and the process is called **reduction.** Oxidation and reduction always take place together because every electron that is lost by one atom (oxidation) is gained by some other atom (reduction) (Figure 7-6).

Oxidation is the loss of an electron; reduction is the gain of one.

Oxidation-reduction reactions play a key role in energy flow through biological systems because the electrons that pass from one atom to another carry with them their potential energy of position (that is, they maintain their distance from the nucleus). Energy that originally entered the system when light boosted electrons in a photosynthetic pigment to higher energy levels is passed from one molecule to another by these electrons. Until the electrons return to their original low-energy level, they continue to store potential energy.

Atoms can store potential energy by means of electons that orbit at higher-than-usual energy levels. When such an energetic electron is removed from one atom (oxidation) and donated to another (reduction), it carries the energy with it and orbits the second atom's nucleus at the higher energy level.

In biological systems electrons often do not travel alone from one atom to another but rather in the company of a proton. Recall that a proton and an electron together make up a hydrogen atom. Thus oxidation-reduction in a chemical reaction usually involves the removal of hydrogen atoms from one molecule and the gain of hydrogen atoms by another molecule. For example, in photosynthesis, hydrogen atoms are transferred from water to carbon dioxide, reducing the carbon dioxide to form glucose:

$$6CO_2 \ + \ 6H_2O \ + \ \text{energy} \rightarrow C_6H_{12}O_6 \ + \ 6O_2$$

| CARBON DIOXIDE | WATER | | GLUCOSE | OXYGEN |

This is a **redox** reaction, that is, one in which electrons move to higher energy levels.

FIGURE 7-5

The oxidation of foodstuffs that takes place during metabolism leads to the same chemical end point that burning it would have done—it is converted to CO_2 and water. Organisms, however, siphon off much of the energy, while burning disperses the chemical bond energy as heat.

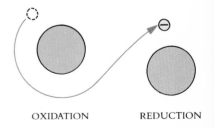

OXIDATION   REDUCTION

FIGURE 7-6

Oxidation is the loss of an electron; reduction is the gain of one.

The reduction of carbon dioxide to form a mole of glucose stores 686 kilocalories of energy in the chemical bonds of the glucose.

The energy stored in a glucose molecule is released in a process called **cellular respiration,** in which the glucose is oxidized. Hydrogen atoms are lost by glucose and gained by oxygen:

$$C_6H_{12}O_6 + 6O_2 \rightarrow 6CO_2 + 6H_2O + \text{energy}$$

The oxidation of a mole of glucose releases 686 kilocalories of energy, the same amount that was stored in making it.

ACTIVATION ENERGY

As the laws of thermodynamics predict, all chemical reactions tend to proceed spontaneously toward a state of maximum disorder and minimum energy. Reactions in which the products contain less energy than the reactants release the excess usable energy (technically called "free energy"). These reactions occur spontaneously and are called **exergonic.** In contrast, reactions in which the products contain more energy than the reactants require an input of usable energy from an outside source before they can proceed. These reactions do not occur spontaneously and are called **endergonic** (Figure 7-7).

> **Any reaction producing products that contain less free energy than the original reactants tends to proceed spontaneously.**

If all chemical reactions that release free energy tend to occur spontaneously, it is reasonable to ask why all such reactions have not already occurred. Clearly they have not. For example, when gasoline is ignited, the resulting chemical reaction proceeds with a net release of free energy. So why doesn't all the gasoline within all the automobiles of the world and beneath all filling stations just burn up right now? Because most reactions require an input of energy to get started, like the heat from the flame of

FIGURE 7-7

In an *endergonic* reaction **(A),** the products of the reaction contain more energy than the reactants, so the extra energy must be supplied in order for the reaction to proceed. In an *exergonic* reaction **(B),** the products contain less energy than the reactants, and the excess energy is released.

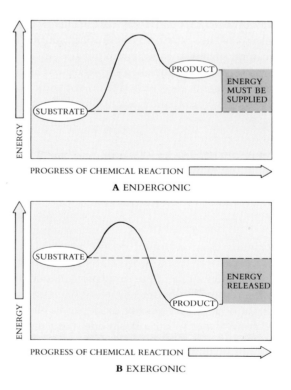

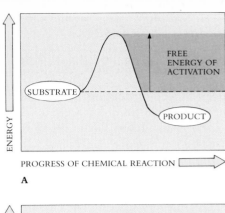

A

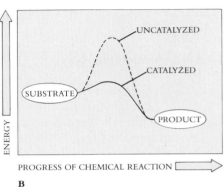

B

FIGURE 7-8

A chemical reaction proceeds because the products of the reaction possess less free energy than the reactants so that there is a net release of energy.
A Before a reaction occurs, energy must be supplied to destabilize existing chemical bonds. This energy is called the energy of activation.
B Enzymes are able to catalyze particular reactions because they lower the amount of activation energy required to initiate the reaction.

a match. Before it is possible to form new chemical bonds with less energy, it is necessary to break the existing bonds, and this requires extra energy, called *activation energy,* to destabilize existing chemical bonds and initiate a chemical reaction (Figure 7-8, *A*).

The speed, or reaction rate, of an exergonic reaction does not depend on how much energy the reaction releases but rather on the amount of activation energy that is required for the reaction to begin. Reactions that involve larger activation energies tend to proceed more slowly since fewer molecules succeed in overcoming the initial energy hurdle. Activation energies, however, are not fixed constants. Putting stress on particular chemical bonds can make them easier to break. The process of influencing chemical bonds in a way that lowers activation energies is called **catalysis,** and substances that perform catalysis are **catalysts.** Catalysts cannot violate the basic laws of thermodynamics; for example, they cannot make an endergonic reaction proceed spontaneously. Only exergonic reactions proceed spontaneously, and catalysis cannot change that. What catalysts *can* do is make a reaction rate very much faster.

> **The speed of a reaction depends on the activation energy necessary to initiate it. Catalysts reduce the amount of activation energy required and thus speed reactions.**

ENZYMES

The chemistry of living things, metabolism, is organized by controlling the points at which catalysis takes place. Therefore life is a process regulated by **enzymes,** which are agents that perform catalysis in living organisms. Enzymes are globular proteins whose shapes are specialized to form temporary associations with the molecules that are reacting. By putting stress on particular chemical bonds, an enzyme lowers the amount of activation energy required for new bonds to form (Figure 7-8, *B*). The reaction thus proceeds much faster than it otherwise would. Because the enzyme itself is not changed, it can be used over and over.

To understand how an enzyme works, consider the joining of carbon dioxide (CO_2) and water (H_2O) to form carbonic acid (H_2CO_3):

$$CO_2 \;+\; H_2O \;\leftrightarrows\; H_2CO_3$$
$$\text{CARBON DIOXIDE} \qquad \text{WATER} \qquad \text{CARBONIC ACID}$$

This reaction may proceed in either direction, but in the absence of an enzyme it is very slow because there is an appreciable need for activation energy. Only about 200 molecules of carbonic acid form in 1 hour. Given the speed at which events usually occur within cells, this reaction rate is like a snail racing in the Indianapolis 500. Cells overcome this problem by employing an enzyme within their cytoplasm called carbonic anhydrase (enzymes are usually given names that end in -*ase*), which speeds the reaction dramatically. In the presence of carbonic anhydrase, an estimated 600,000 molecules of carbonic acid form every second! The enzyme has speeded the reaction rate about 10 million times.

Cells employ proteins called enzymes as catalysts.

Thousands of different kinds of enzymes have been described, each catalyzing a different chemical reaction. By facilitating particular chemical reactions, the enzymes in a cell determine the course of metabolism in that cell, much as traffic lights determine the flow of traffic in a city. Not all types of cells contain the same enzymes, which is why there is more than one type of cell. For example, the chemical reactions within a red blood cell are very different from those going on within a nerve cell because the cytoplasm and membranes of red blood cells contain a different array of enzymes.

FIGURE 7-9

How the enzyme lysozyme works. The tertiary structure of lysozyme, diagrammed as a ribbon in **A** and shown in a three-dimensional model in **B** and **C**, forms a groove through the middle of the protein. This groove fits the shape of the chains of sugars that make up bacterial cell walls. When such a chain of sugars, indicated in yellow in **C**, slides into the groove, its entry induces the protein to alter its shape slightly to embrace the substrate more intimately. This *induced fit* positions a glutamic acid in the protein right next to the bond between two adjacent sugars, and the glutamic acid "steals" an electron from the bond, causing it to break.

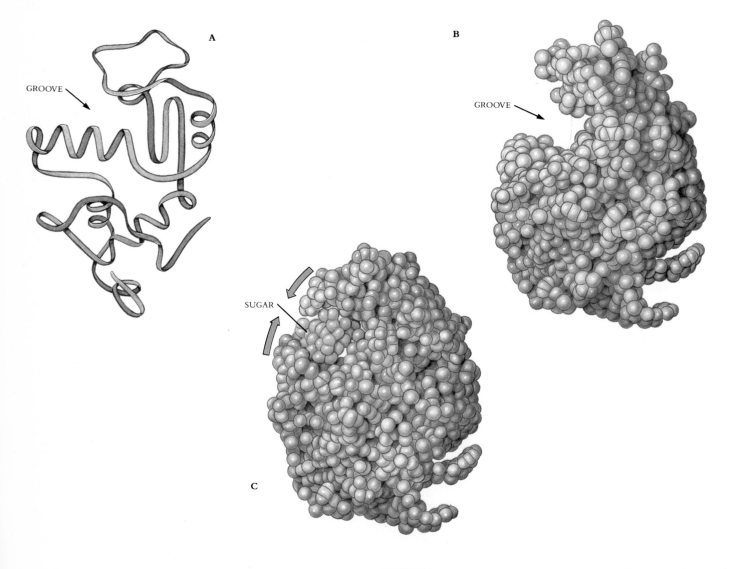

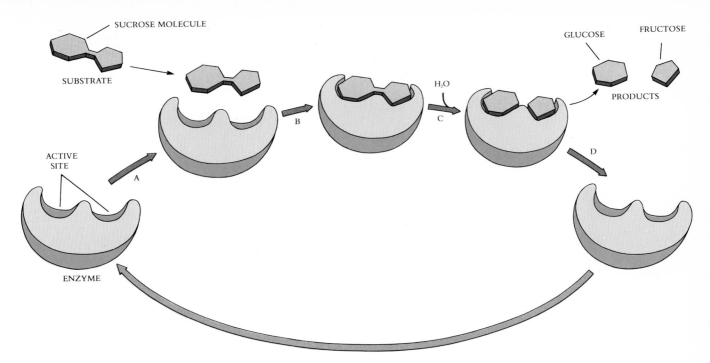

SUCROSE MOLECULE

SUBSTRATE

ACTIVE SITE

ENZYME

H₂O

GLUCOSE

FRUCTOSE

PRODUCTS

A

B

C

D

How Enzymes Work

Enzymes are globular proteins with one or more pockets or clefts on their surface, which resemble deep creases in a prune (Figure 7-9). These surface depressions are called **active sites.** They are the locations at which catalysis occurs. In order for catalysis to occur, the molecule on which the enzyme acts, called a **substrate,** must fit precisely into the surface depression so that many of its atoms nudge up against atoms of the enzyme. The substrate fits into the active site of the enzyme like a foot into a tight-fitting shoe, in very close contact. Proteins are not rigid, however, and in some cases the binding of substrate may induce the protein to adjust its shape slightly, allowing a better fit.

When a substrate molecule has bound to the active site of an enzyme, amino acid side groups of the enzyme are placed against certain bonds of the substrate, just as when you sit in a chair certain parts of your body press against the seat (Figure 7-10). These amino acid side groups chemically interact with the substrate, usually by stressing or distorting a particular bond, and consequently lower the amount of activation energy needed to break the bond.

Enzymes typically catalyze only one or a few different chemical reactions because they are very "picky" in their choice of substrate. The active site of each different kind of enzyme is shaped so that only a certain substrate molecule will fit into it.

Factors Affecting Enzyme Activity

The activity of an enzyme is affected by any change in conditions that alters its three-dimensional shape, including changes in temperature or pH or the binding to the protein of specific chemicals that regulate the enzyme's activity.

Temperature

The shape of a protein is determined by hydrogen bonds that hold its arms in particular positions and also by the tendency of noncharged ("nonpolar") segments of the protein to avoid water. Chemists call interactions of this second kind **hydrophobic,** or water-hating, interactions. Both hydrogen bonds and hydrophobic interactions are easily disrupted by slight changes in temperature. Most human enzymes function best within a relatively narrow temperature range between 35° and 40° C (close to body temperature). Below this temperature level, the bonds that determine protein shape are not

FIGURE 7-10

The catalytic cycle of an enzyme. Enzymes increase the speed with which chemical reactions occur, but are not altered themselves as they do this. In the reaction illustrated here, the enzyme sucrase is splitting the sugar sucrose (the sugar present in most candy) into its two parts, the simpler sugars glucose and fructose.

A First, the sucrose substrate binds to the active site of the enzyme, fitting into a depression in the enzyme surface.

B The binding of sucrose induces the sucrase molecule to alter its shape, fitting more tightly around the sucrose molecule.

C The active site, now in close proximity to the bond between the glucose and fructose parts of sucrose, breaks the bond.

D The enzyme releases the resulting glucose and fructose fragments, the products of the reaction, and is then ready to bind another molecule of sucrose and run through the catalytic cycle once again.

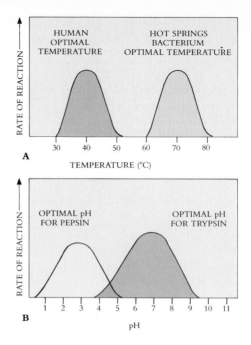

FIGURE 7-11

The activity of an enzyme is influenced both by temperature **(A)** and pH **(B)**. Human enzymes tend to work best at temperatures of about 40° C, and within a pH range of 6 to 8.

flexible enough to permit the induced change in shape that is necessary for catalysis; above this temperature, the bonds are too weak to hold the protein's arms in the proper position. Bacteria that live in hot springs, on the other hand, have proteins with stronger bonding between their arms and therefore can function at temperatures of 70° C or higher (Figure 7-11, *A*).

pH

A third kind of bond that acts to hold the arms of proteins in position is the bond that forms between oppositely charged amino acids such as glutamic acid (−) and lysine (+). These bonds are sensitive to hydrogen ion concentration. The more hydrogen ions available in the solution, the fewer negative charges and the more positive charges that occur. For this reason, most enzymes have a pH optimum just as they have a temperature optimum; this optimum usually lies in the range of pH 6 to 8. Proteins that are able to function in very acidic environments have amino acid sequences that maintain their ionic and hydrogen bonds even in the presence of high levels of hydrogen ion (Figure 7-11, B). For example, the enzyme pepsin digests proteins in your stomach at pH = 2, a very acidic level.

The Regulation of Enzyme Activity

The activity of an enzyme is sensitive not only to temperature and pH but also to the presence of specific chemicals that bind to the enzyme and cause changes in its shape. By means of these specific chemicals a cell is able to regulate which enzymes are active and which are inactive at a particular time. When the binding of the chemical alters the shape of the protein and thus shuts off enzyme activity, the chemical is called an **inhibitor;** when the change in the enzyme's shape is necessary for catalysis to occur, the chemical is called an **activator.**

The change in shape that occurs when an activator or inhibitor binds to an enzyme is called an **allosteric** change (Greek *allos,* other + *steros,* shape). Enzymes usually have special binding sites for the activator and inhibitor molecules that affect them, and these binding sites are different from their active sites. For example, the enzyme catalyzing the first step in a series of chemical reactions, or **biochemical pathway,** often has an inhibitor-binding site to which the molecule produced by the last step in the series binds. As the amount of this molecule builds up in the cell, it begins to bind to the initial enzyme in the biochemical pathway, thus inhibiting the activity of that enzyme. By this process the pathway is shut down when it is no longer needed. Such **end-product inhibition** is a good example of the way many enzyme-catalyzed processes within cells are self-regulating.

> **The activity of enzymes is regulated by allosteric changes in enzyme shape; these changes result when specific small molecules bind to the enzyme, molecules that are not substrates of that enzyme.**

Coenzymes

Often enzymes use additional chemical components called **cofactors** as tools to aid the catalysis. For example, many enzymes have metal ions locked into their active sites, and these ions help draw electrons from substrate molecules. The enzyme carboxypeptidase chops up proteins by employing a zinc ion to draw electrons away from the bonds being broken. Many of the trace elements such as molybdenum and manganese, which are necessary for your health, also utilize metal ions in this way. When the cofactor is a nonprotein organic molecule, it is called a **coenzyme.** Many of the vitamins that your body requires are parts of coenzymes.

> **Many enzymes employ metal ions or organic molecules to facilitate their activity; these are called cofactors. Cofactors that are nonprotein organic molecules are called coenzymes.**

In many enzyme-catalyzed oxidation–reduction reactions, energy-bearing electrons are passed from the active site of the enzyme to a coenzyme that serves as the electron acceptor. The coenzyme then carries the electrons to a different enzyme that is catalyzing another reaction, releases the electrons (and the energy they bear) to that reaction, and then returns to the original enzyme for another load of electrons. In most cases the electrons are paired with protons as hydrogen atoms. Just as armored cars transport cash around a city, so coenzymes shuttle energy, in the form of hydrogen atoms, from one place to another in a cell.

One of the most important coenzymes is the hydrogen acceptor **nicotinamide adenine dinucleotide,** usually referred to by the abbreviation NAD^+ (Figure 7-12).

FIGURE 7-12

The chemical structure of NAD^+. Nicotinamide adenine dinucleotide (NAD^+) is a composite molecule consisting of two nucleotides bound together. In a nucleotide, one or more phosphate groups is attached to a five-carbon sugar, and an organic base to the sugar's other end. The two nucleotides that make up NAD^+ are nicotinamide monophosphate (NMP) and adenine monophosphate (AMP), joined head-to-tail by their phosphate groups. The two nucleotides serve different functions in the NAD^+ molecule: AMP acts as a core, providing a shape recognized by many enzymes; NMP is the active part of the molecule, contributing a site that readily gains electrons (is reduced).

A   BASIC STRUCTURE OF A NUCLEOTIDE

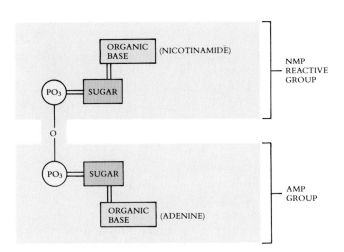

B   BASIC STRUCTURE OF NICOTINAMIDE
ADENINE DINUCLEOTIDE (NAD^+)

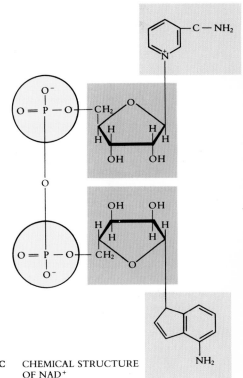

C   CHEMICAL STRUCTURE
OF NAD^+

When NAD$^+$ acquires a hydrogen atom from the active site of an enzyme, it becomes reduced as NADH. The energetic electron of the hydrogen atom is then carried by the NADH molecule, like money in your wallet. For example, the oxidation of foodstuffs in your body, from which you get the energy to drive your life, takes place by the cell's stripping of electrons from food molecules and donating them to NAD$^+$, thus forming a wealth of NADH. This wealth is the principal energy income of your cells. However, much of the NADH is eventually converted to another currency.

THE ENERGY CURRENCY OF ORGANISMS: ATP

The chief energy currency of all cells is a molecule called **adenosine triphosphate,** or **ATP.** Just as the bulk of the energy that plants harvest during photosynthesis is channeled into production of ATP, so is most of the NADH that soaks up the energy resulting from the oxidation of your food. The energy stored in the chemical bonds of fat and starch is converted to ATP, as is the energy carried by the sugars circulating in your blood. Cells then use their supply of ATP to drive active transport across membranes, to power movement, to provide activation energy for chemical reactions, to grow—almost every energy-requiring process that cells perform is powered by ATP. Because ATP plays this central role in *all* organisms it is clear that its role as the major energy currency of cells evolved early in the history of life.

Each ATP molecule (Figure 7-13) is composed of three subunits. The first subunit is a five-carbon sugar called **ribose,** which serves as the backbone to which the other two subunits are attached. The second subunit is **adenine,** an organic molecule composed of two carbon-nitrogen rings. Each of the nitrogen atoms in the ring has an unshared pair of electrons that weakly attract hydrogen ions. Adenine therefore acts as a chemical base and is usually referred to as a **nitrogenous base.** As described in Chapter 3, adenine plays another major role in the cell: it is one of the four nitrogenous bases that are the principal components of the genetic material DNA.

The third subunit is a **triphosphate group** (three phosphate groups linked in a chain). The covalent bonds linking these three phosphates are usually indicated by a squiggle ($\sim$) and are sometimes called **high-energy bonds.** When one is broken,

FIGURE 7-13

ATP is the primary energy currency of the cell. Like NAD$^+$, it has a core of AMP, a shape recognized by many enzymes. In this case, however, the reactive group added to the end of the AMP phosphate group is not another nucleotide but rather a chain of two additional phosphate groups. The bonds connecting these two phosphate groups to each other and to AMP are high energy ($\sim$) bonds. When the outer $\sim$ bond of ATP is cleaved, yielding adenosine diphosphate (ADP) and phosphate (P), 7.3 kcal of energy are released per mole of ATP. Similarly, when the second bond is cleaved, yielding AMP and either P or P$\sim$P, 7.3 kcal of energy are released per mole of ATP. Most energy exchanges in cells involve cleavage of only the outermost of the two bonds, converting ATP into ADP + P.

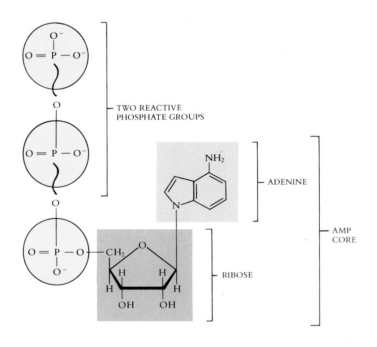

FIGURE 7-14

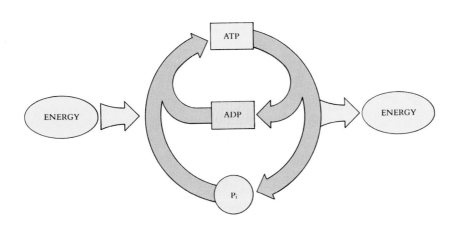

slightly more than 7 kilocalories of energy are released per mole of ATP. These phosphate bonds possess what a chemist would call "high transfer potential," that is, they are bonds that have a low activation energy and are broken easily, which releases their energy. In a typical energy transaction, only the outermost of the two high-energy bonds is broken, breaking off the phosphate group on the end. When this happens, ATP becomes **adenosine diphosphate (ADP)** and 7 kilocalories of energy are expended per mole of ATP.

Cells use ATP to drive endergonic reactions, reactions whose products possess more energy than their substrates. Such reactions will not proceed unless the reactants are supplied with the necessary energy, any more than a boulder will roll uphill. However, as long as the cleavage of ATP's terminal high-energy bond is more exergonic than the other reaction is endergonic, the overall energy change of the two "coupled" reactions is exergonic and the reaction will proceed. Because almost all endergonic reactions in the cell require less than 7 kilocalories of activation energy per mole, ATP is able to power all of the cell's activities. Thus, although the high-energy bond of ATP is not highly energetic in an absolute sense—for example, it is not nearly as strong as a carbon-hydrogen bond, it is more energetic than the activation energies of almost all energy-requiring cell activities. It is for this reason that ATP is able to serve as a universal energy donor.

ATP is the universal energy currency of all cells.

Cells contain a pool of ATP, ADP, and phosphate. ATP is constantly being cleaved into ADP plus phosphate to drive the endergonic, energy-requiring processes of the cell. An individual on a typical diet of 2000 Calories/day goes through about 125 pounds of ATP a day. Cells, however, do not maintain large stockpiles of ATP, just as most people do not carry large amounts of cash with them. Instead, cells are constantly recycling their ADP, withdrawing from their energy reserves to rebuild more ATP. Using the energy derived from foodstuffs and from stored fats or starches (or, in the case of plants, from photosynthesis), ADP and phosphate are recombined to form ATP, with 7 kilocalories of energy per mole contributed to each newly formed high-energy bond. If you could mark every ATP molecule in your body at one instant in time, and then watch them, they would be gone in a flash. Most cells maintain a particular molecule of ATP for only a few seconds before using it (Figure 7-14).

SUMMARY

1. Energy is the capacity to bring about change, to do work. Kinetic energy is actively engaged in doing work, whereas potential energy has the capacity to do work. Most energy transformations in living things involve conversion of potential energy to kinetic energy.

2. The first law of thermodynamics states that the amount of energy in the universe is constant and that energy cannot be destroyed or created. However, energy can be converted from one form to another.

3. The second law of thermodynamics says that disorder in the universe tends to increase. As a result, energy spontaneously converts to less organized forms.

4. The least organized form of energy is heat, which is random molecular motion. Heat energy can be used to do work when heat gradients exist, as in a steam engine, but cannot accomplish work in cells.

5. A redox reaction is one in which an electron is taken from one atom (oxidation) and donated to another (reduction). The electrons in redox reactions often travel in association with protons, as hydrogen atoms.

6. If the electron transferred in a redox reaction is an energetic electron, then the energy is transferred with the electron. The electron moves to a high-energy orbital in the reduced atom. Many energy transfers in cells take place by means of redox reactions.

7. Any chemical reaction whose products contain less free energy than the original reactants will tend to proceed spontaneously.

8. The speed of a reaction depends on the amount of activation energy required to break existing bonds. Catalysis is the process of lowering the amount of activation energies needed by stressing chemical bonds. Enzymes are the catalysts of cells.

9. Cells contain many different enzymes, each of which catalyzes a different reaction. A given enzyme is specific because its active site fits only one or a few potential substrate molecules.

10. Cells focus all of their energy resources on the manufacture of ATP from ADP and phosphate, which requires the cell to supply 7 kilocalories of energy obtained from photosynthesis or from electrons stripped from foodstuffs to form 1 mole of ATP. Cells then use this ATP to drive endergonic reactions.

REVIEW

1. Energy is defined as the capacity to do _____.

2. When energy is converted from one form to another (an example is the conversion of chemical energy to mechanical energy in a car), a large percentage is lost as heat. Does this contradict the first law of thermodynamics?

3. The earth is constantly receiving energy. Where does this energy come from?

4. What cellular process releases the chemical energy stored in glucose for use by organisms?

5. The agents that perform catalysis in the human body are called _____. What class of organic molecules are they? (See Chapter 3).

SELF-QUIZ

1. Energy can exist in many forms. Which three of the following are forms of energy?
   - (a) heat
   - (b) sound
   - (c) smell
   - (d) diffusion
   - (e) light

2. When energy becomes so randomized and uniform in a system that it is no longer available to do work, it is referred to as
   - (a) heat
   - (b) entropy
   - (c) enthalpy
   - (d) light
   - (e) chemical energy

3. Light energy is captured and utilized by _____. Because of this ability to utilize light energy to produce chemical energy, these organisms allow most of the rest of life on earth to exist.
   - (a) animals generally
   - (b) mammals
   - (c) plants
   - (d) fungi
   - (e) most bacteria

4. Chemical reduction is
   - (a) the gain of a proton
   - (b) the loss of a proton
   - (c) the gain of an electron
   - (d) the loss of an electron
   - (e) the exchange of an electron

5. The molecule that is used by all cells to do work, that is, to drive chemical reactions by moving ions or molecules against a concentration gradient, is
   - (a) ATP
   - (b) NAD^+
   - (c) NMP
   - (d) glucose
   - (e) water

PROBLEMS

1. If you eat a 1500-Calorie lunch consisting of a hamburger and some french fries, how many kilocalories have you consumed?

2. If you eat a mole of glucose, how many Calories do you consume?

THOUGHT QUESTIONS

1. Oxidation-reduction reactions occur between a wide variety of molecules. Why do you suppose reactions involving hydrogen and oxygen are the ones of paramount importance in biological systems?

2. On earth there are many more bears than there are lions and tigers. Why do you suppose this is so?

3. On summer nights in some parts of the country, one can often see fireflies glowing briefly in the dark. Do you suppose this light requires energy? If so, where does it come from?

FOR FURTHER READING

HINKLE, P.C., and R.E. McCARTY: "How Cells Make ATP," *Scientific American,* March 1978, pages 104–123. Describes how cells use electrons stripped from foodstuffs to make ATP.

LEHNINGER, A.L.: "How Cells Transform Energy," *Scientific American,* September 1961, pages 62-73. Written more than 25 years ago, this article still provides one of the clearest expositions of how cells channel energy through metabolism.

FIGURE 8-1   All of the energy that drives our metabolism comes from plants or from animals that eat plants. This winter wheat, rich with the products of months of capturing energy from the sun through the process of photosynthesis, will soon become flour and provide the energy for human activity.

Chapter 8

Overview

All organisms fuel their metabolism with energy from ATP. The simplest processes for generating ATP, and the ones that evolved first, involve the rearrangement of chemical bonds. Later, organisms evolved far more efficient means of carrying out this process: plants, algae, and some bacteria obtain ATP by using energetic electrons to drive proton pumps, obtaining the electrons both by photosynthesis and by oxidizing sugars and fats. Animals, fungi, most protists, and most bacteria are heterotrophs; they use only the second oxidative process, which occurs within the symbiotically derived mitochondria that exist in all but a very few eukaryotes.

For Review   *Here are some important terms and concepts that you will encounter in this chapter. If you are not familiar with them, you should review them before proceeding.*

Glucose (Chapter 3)

Chemiosmosis (Chapter 6)

Oxidation-reduction (Chapter 7)

ATP (Chapter 7)

Exergonic and endergonic reactions (Chapter 7)

Life processes are driven by energy. All the activities that organisms perform—the swimming of bacteria, the purring of a cat, your reading these words—use energy. Even though the ways that organisms use energy are many and varied, all of life's energy ultimately has the same beginning: the sun. Plants, algae, and some bacteria harvest the energy of sunlight by the process of photosynthesis, thus converting it to chemical energy (Figure 8-1). These organisms, along with a few others that use chemical energy in a similar way, are called **autotrophs,** or self-feeders. All organisms ultimately live on the energy produced by autotrophs. Organisms that do not have the ability to produce their own food are called **heterotrophs,** which means fed by others. At least 95% of the kinds of organisms on earth—all animals, all fungi, and most protists and bacteria—are heterotrophs; most of them live by feeding on the chemical energy fixed by photosynthesis. All of us, plants and bacteria, you and I, share the same ultimate dependency on the sun. We are all children of light.

FIGURE 8-2

How the cell drives endergonic reactions. The conversion of compound A to compound B requires the input of 4 kcal of energy per mole of compound A, and so cannot proceed unless this extra energy is provided from some other source. In cells, this extra energy is acquired from ATP. The cleavage of ATP to ADP + P yields 7.3 kcal of energy per mole of ATP, which more than makes up for the shortfall of the endergonic reaction. The two reactions, taken together, actually release 3.3 kcal of energy per mole of compound A. When an endergonic reaction is driven by a second energy-releasing reaction, the two reactions are said to be "coupled."

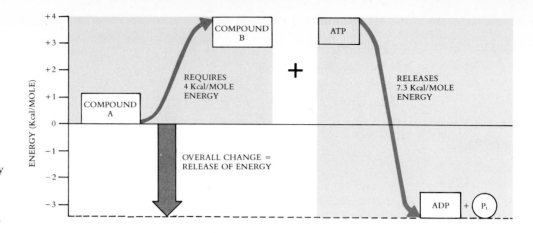

In this chapter we shall discuss the processes that cells use to derive energy from organic molecules and to convert that energy to ATP. We shall consider photosynthesis in detail in Chapter 9. We treat the conversion of chemical energy to ATP first because all organisms—both photosynthesizers and the heterotrophs that feed on them—are capable of transforming energy in this way. In contrast, less than 1 in 20 of the kinds of organisms on earth is capable of photosynthesis. As you will see, though, the two processes have much in common.

USING CHEMICAL ENERGY TO DRIVE METABOLISM

Your body contains more than 2000 different kinds of enzymes. These enzymes catalyze a bewildering variety of reactions. Many of the reactions occur in sequences called **biochemical pathways,** which are the organizational units of metabolism and the pathways of energy and materials in the cell. Just as the many metal parts of an automobile are organized into distinct subassemblies such as the carburetor, transmission, and brakes, so the many enzyme-catalyzed reactions of an organism are organized into biochemical pathways.

In biochemical pathways, exergonic reactions—that is, those reactions that involve a release of free energy—occur spontaneously; endergonic reactions—those reactions that require the addition of energy—do not. To drive endergonic reactions, all organisms use the same mechanism: they *couple* the reactions to the energy-yielding splitting of an energy-rich molecule such as ATP (Figure 8-2).

> **Chemical energy powers metabolism by driving endergonic reactions. Chemical energy is used to create ATP, and the splitting of ATP is coupled to these reactions, providing the necessary energy.**

HOW CELLS MAKE ATP: AN OVERVIEW

ATP is the energy currency of all living organisms. How do organisms make ATP? They do so in one of two ways:

1. *Substrate-level phosphorylation.* Because the formation of ATP from ADP plus inorganic phosphate (P_i) requires an input of free energy, ATP formation is endergonic, that is, it does not occur spontaneously. However, when coupled with an exergonic reaction that has a very strong tendency to occur, the synthesis of ATP from ADP plus P_i does take place. The reaction occurs because the release of energy from the exergonic reaction is greater than the input of energy required to drive the synthesis of ATP. The generation of ATP by coupling strongly exergonic reactions with the synthesis of ATP from ADP and P_i is called substrate level phosphorylation (Figure 8-3).

2. *Chemiosmotic generation of ATP.* All organisms possess transmembrane channels that function in pumping protons out of cells (see Chapter 6). Proton-pumping channels use a flow of excited electrons to induce a shape change in the transmembrane protein, which in turn causes protons to pass outward. As the proton concentration outside the membrane rises higher than that inside, the outer protons are driven inward by diffusion, passing back in through special proton channels that use their passage to induce the formation of ATP from ADP plus P_i. Because the chemical formation of ATP is powered by a diffusion force similar to osmosis, this process is called chemiosmosis (Figure 8-4).

The harvesting of chemical energy takes place in one or both of two stages:
1. **Substrate-level phosphorylation, which involves a reshuffling of chemical bonds to couple ATP formation to a highly exergonic reaction.**
2. **The transport of electrons to a membrane where they drive a proton pump and thus power the chemiosmotic synthesis of ATP.**

Substrate-level phosphorylation was probably the first of the two ATP-forming mechanisms to evolve. The process of glycolysis, which is the most basic of all ATP-generating processes and is present in every living cell, employs this mechanism.

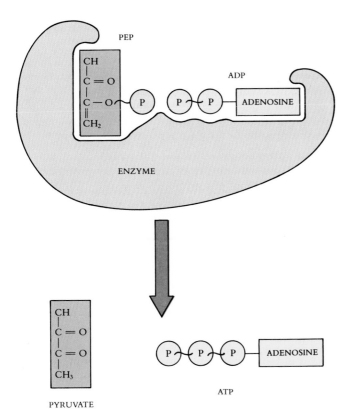

FIGURE 8-3

Substrate-level phosphorylation. Some ATP is made as a result of the direct transfer of a phosphate group from an enzymatic substrate to ADP. The reaction takes place because the phosphate bond of the substrate has higher energy than the phosphate bonds of ATP.

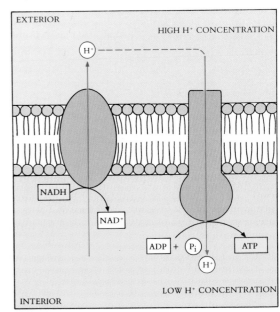

FIGURE 8-4

Chemiosmotic synthesis of ATP. Hydrogen ions are moved from one side of a membrane to the other, generating a proton gradient across the membrane. When protons move back across the membrane through special channels, their passage drives the synthesis of ATP.

However, most of the ATP that organisms make is produced by chemiosmosis. The electrons that drive proton-pumping channels involved in chemiosmosis are obtained by organisms from two sources. (1) In photosynthetic organisms, light energy boosts electrons to higher energy levels, and these electrons are channeled to proton pumps. (2) In all organisms, photosynthetic and nonphotosynthetic alike, high-energy electrons are extracted from chemical bonds and carried by coenzymes to proton pumps.

The second of these two chemiosmotic processes, the extraction of electrons from chemical bonds, is an oxidation-reduction process. Oxidation is defined as the removal of electrons (see Chapter 7). When electrons are taken away from the chemical bonds of food molecules to drive proton pumps, the food molecules are being oxidized chemically. This electron-harvesting process is given a special name: **cellular respiration.** Because cellular respiration uses oxygen, it is also known as **oxidative respiration.** This is the term we will use in this text. Oxidative respiration is the oxidation of food molecules to obtain energy. Do not confuse this with the breathing of oxygen gas that your body carries out, which is also called respiration.

The principal pathway of oxidative respiration in the body is called the **citric acid cycle,** after the six-carbon citric acid molecule formed in its first step. It is also called the **Krebs cycle,** after the biochemist Sir Hans Krebs, who discovered it. In the citric acid cycle, energetic electrons are stripped from food molecules and donated to NAD^+ molecules to form NADH molecules; in this process the electrons are accompanied by protons and thus travel as hydrogen atoms. These NADH molecules then donate the energetic electrons to proton-pumping channels to drive the chemiosmotic synthesis of ATP.

> **All cells make ATP by substrate-level phosphorylation and by oxidative respiration; some organisms also make ATP by photosynthesis.**

THE FATE OF A CANDY BAR

When you metabolize food and thus obtain the ATP that powers your life, you employ both substrate-level phosphorylation and oxidative respiration. To better understand what goes on, it is instructive to follow the fate of something you eat and to see what happens to it. Therefore let us eat a chocolate bar (Figure 8-5).

A chocolate bar, like many of the things that we consume, is a complex mixture of sugars, lipids, proteins, and other molecules. The first thing that happens in its journey toward ATP production is that the complex molecules are broken down to simple ones. Disaccharides such as sucrose are split into simple sugars, either glucose or sugars that are converted to glucose; proteins are split up into amino acids; and complex lipids are broken into smaller bits. These initial steps usually yield no usable energy, but they serve to assemble the energy wealth of a diverse array of complex molecules into a small number of simple molecules such as glucose.

For simplicity let us assume that our chocolate bar is entirely degraded to molecules of the six-carbon sugar glucose. Glucose occupies a central place in metabolism since many different foodstuffs are converted to it. Glucose is where the making of ATP begins.

The first stage of extracting energy from glucose is a 10-reaction biochemical pathway called **glycolysis.** In glycolysis, ATP is generated in two ways. For each glucose molecule, two ATP molecules are expended in preparing the glucose molecule, and four ATP molecules are formed by substrate-level phosphorylation, for a net yield of two ATP molecules. In addition, four electrons are harvested and used to form ATP molecules by oxidative respiration. The total yield of ATP molecules is small, however. When the glycolytic process is completed, the two molecules of **pyruvate** that are left still contain most of the energy that was present in the original glucose molecule.

The second stage of extracting energy from glucose, after glycolysis, is a cycle of nine reactions, the citric acid cycle. The pyruvate left over from glycolysis is first converted to a two-carbon molecule, which feeds into the cycle. In the cycle two more ATP molecules are extracted by substrate-level phosphorylation, and a large number

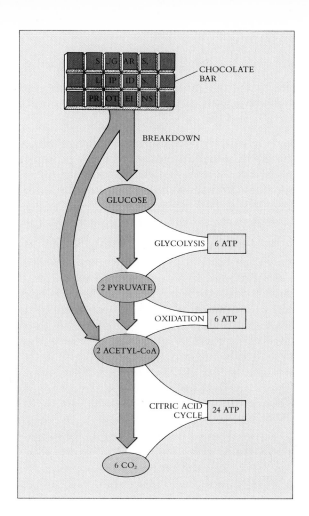

FIGURE 8-5

The fate of a chocolate bar. The bar is composed of sugar, chocolate, and other lipids and fats, protein, and many other molecules. This diverse collection is broken down to simple molecules: the sugars to glucose molecules and the proteins, fats, and other lipids to two-carbon molecules called acetyl-CoA. These breakdowns produce little or no energy, but they prepare the way for three major energy-producing processes. The first to occur is glycolysis, a process that in your body converts glucose to two molecules of pyruvate. The second process to occur is the oxidation of this pyruvate to two molecules of acetyl-CoA. The third process is the oxidation of acetyl-CoA molecules in the citric acid cycle.

of electrons are removed and donated to electron carriers. It is the harvesting of these electrons that leads to the greatest amount of ATP formation.

When the cycle is completed, the six-carbon glucose molecule has been divided into six molecules of CO_2, and 36 ATP molecules have been generated, four by substrate-level phosphorylation and 32 by oxidative respiration. This is a very good yield. Each ATP molecule represents the capture of 7.3 kilocalories of energy per mole of glucose; therefore 36 ATP molecules represents a total capture of $7.3 \times 36 = 263$ kilocalories per mole. The total energy content of the chemical bonds of glucose is only 686 kilocalories per mole; so we have succeeded in harvesting 38% of the available energy. By contrast, a car converts only about 25% of the energy in gasoline into useful energy.

This brief overview gives some sense of how cells organize their production of ATP from a food source such as a candy bar. We shall now examine glycolysis and the citric acid cycle as processes and study the ways in which they direct the flow of energy.

GLYCOLYSIS

Among the many simple molecules that are available as a consequence of degradation, the metabolism of primitive organisms focused on the simple six-carbon sugar glucose, undoubtedly in part because glucose was a major constituent of the carbohydrates of the cell. Glucose molecules can be dismantled in many ways, but primitive organisms evolved the ability to do it in a way that includes reactions that release enough free energy to drive the synthesis of ATP in coupled reactions. The process involves a sequence of 10 reactions that convert glucose into two three-carbon molecules of pyruvate. For each molecule of glucose that passes through this transformation, the cell acquires two ATP molecules. The overall process is called glycolysis.

An Overview of Glycolysis

Glycolysis consists of two very different processes, one wedded to the other:

1. Glucose is converted to two molecules of the three-carbon compound **glyceraldehyde 3-phosphate** (G3P), with the expenditure of ATP.
2. ATP is generated from G3P.

A **catabolic process** is one in which complex molecules are broken down into simpler ones; in contrast, an **anabolic process** is one in which more complex molecules are built up. Perhaps the original catabolic process involved only the ATP-yielding breakdown of G3P; the generation of G3P from glucose may have evolved later when alternative sources of G3P were depleted. Like many biochemical pathways, glycolysis may have evolved backwards, the last stages in the process being the most ancient.

The 10 reactions of glycolysis apparently proceed in four stages (Figure 8-6):

Stage a: Three reactions change glucose into a compound that can readily be split into three-carbon phosphorylated units. Two of these reactions require the cleavage of an ATP molecule, so that this stage, called **glucose mobilization,** requires the investment by the cell of two ATP molecules.

Stage b: The second stage is **cleavage,** in which the six-carbon product of the first stage is split into two three-carbon molecules. One is G3P, and the other is converted to G3P by another reaction.

Stage c: The third stage is **oxidation,** in which a hydrogen atom carrying a pair of electrons is removed from G3P and donated to NAD^+. NAD^+ acts as an electron carrier in the cell, in this case accepting the proton and two electrons from G3P to form NADH. Note that NAD^+ is an ion and that *both* electrons in the new covalent bond come from G3P.

Stage d: The final stage, called **ATP generation,** is a series of four reactions that convert G3P into another three-carbon molecule, **pyruvate,** and in the process generate two ATP molecules.

Because each glucose molecule is split into *two* G3P molecules, the overall net reaction sequence yields two ATP molecules and two molecules of pyruvate:

$$
\begin{array}{ll}
-\ 2\text{ATP} & \text{Stage a} \\
+\ 2(2\text{ATP}) & \text{Stage d} \\
\hline
+\ 2\text{ATP} &
\end{array}
$$

This is not a great amount of energy. When you consider that the free energy of the total oxidation of glucose to CO_2 and H_2O is -686 kilocalories per mole (the sign is negative to indicate that the products of the oxidation have less energy), and that ATP's high-energy bonds have an energy content of -7.3 kilocalories per mole apiece, then the efficiency with which glycolysis harvests the chemical energy of glucose is only 14.6/686, or 2% Although far from ideal in terms of the amount of energy that it releases, glycolysis does generate ATP; and for more than a billion years during the long, anaerobic first stages of life on earth, this reaction sequence was the only way for heterotrophic organisms to generate ATP from organic molecules.

> **The glycolytic reaction sequence generates a small amount of ATP by reshuffling the bonds of glucose molecules. Glycolysis is a very inefficient process, capturing only about 2% of the available chemical energy of glucose.**

The Universality of the Glycolytic Sequence

The glycolytic reaction sequence is thought to have been among the earliest of all biochemical processes to evolve. It uses no molecular oxygen and therefore occurs readily in an anaerobic environment. All of its reactions occur free in the cytoplasm; none is associated with any organelle or membrane structure. Every living creature is capable of

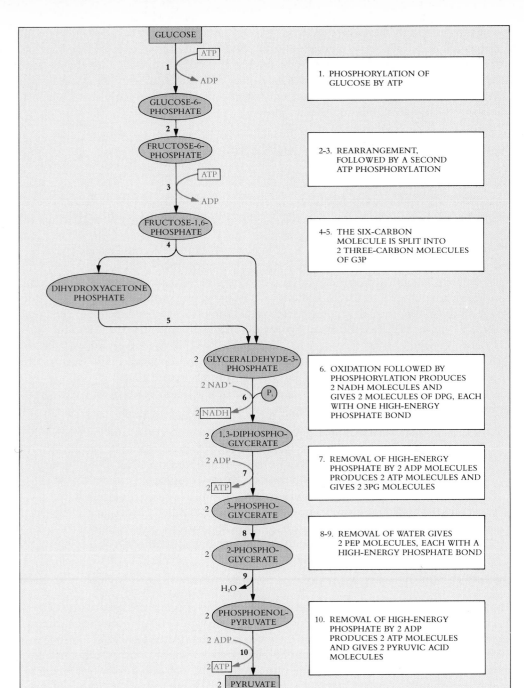

FIGURE 8-6

The glycolytic pathway. The first five reactions convert a molecule of glucose into two molecules of G3P. This process is endergonic and requires the expenditure of two ATP molecules to drive it. The second five reactions convert G3P molecules into pyruvate molecules and generate four molecules of ATP for each two molecules of G3P. These reactions also generate two molecules of NADH, which as you will see is further oxidized in your body to produce four more molecules of ATP. Subtracting the two ATP molecules expended in driving the inital endergonic reactions, the net yield in your body is thus six ATP molecules for each molecule of glucose. When cells are forced to operate without oxygen, the two molecules of NADH cannot be used to produce ATP, and the net yield is then only two molecules of ATP per molecule of glucose.

Within the figure:

GLUCOSE

1. PHOSPHORYLATION OF GLUCOSE BY ATP

GLUCOSE-6-PHOSPHATE

FRUCTOSE-6-PHOSPHATE

2-3. REARRANGEMENT, FOLLOWED BY A SECOND ATP PHOSPHORYLATION

FRUCTOSE-1,6-PHOSPHATE

4-5. THE SIX-CARBON MOLECULE IS SPLIT INTO 2 THREE-CARBON MOLECULES OF G3P

DIHYDROXYACETONE PHOSPHATE

2 GLYCERALDEHYDE-3-PHOSPHATE

2 NAD⁺   2 NADH   Pᵢ

6. OXIDATION FOLLOWED BY PHOSPHORYLATION PRODUCES 2 NADH MOLECULES AND GIVES 2 MOLECULES OF DPG, EACH WITH ONE HIGH-ENERGY PHOSPHATE BOND

2 1,3-DIPHOSPHO-GLYCERATE

2 ADP   2 ATP

7. REMOVAL OF HIGH-ENERGY PHOSPHATE BY 2 ADP MOLECULES PRODUCES 2 ATP MOLECULES AND GIVES 2 3PG MOLECULES

2 3-PHOSPHO-GLYCERATE

8-9. REMOVAL OF WATER GIVES 2 PEP MOLECULES, EACH WITH A HIGH-ENERGY PHOSPHATE BOND

2 2-PHOSPHO-GLYCERATE

H_2O

2 PHOSPHOENOL-PYRUVATE

2 ADP   2 ATP

10. REMOVAL OF HIGH-ENERGY PHOSPHATE BY 2 ADP PRODUCES 2 ATP MOLECULES AND GIVES 2 PYRUVIC ACID MOLECULES

2 PYRUVATE

---

carrying out the glycolytic sequence. However, most present-day organisms are able to extract considerably more energy from glucose molecules than glycolysis does. For example, of the 36 ATP molecules you obtain from each glucose molecule that you metabolize, only 2 of these are obtained by glycolysis. Why then is glycolysis still maintained even through its energy yield is comparatively meager?

This simple question has an important answer: evolution is an incremental process. Change occurs during evolution by improving on past successes. In catabolic metabolism, glycolysis satisfied the one essential evolutionary requirement: it was an improvement. Cells that could not carry out glycolysis were at a competitive disadvantage. By studying the metabolism of contemporary organisms, we can see that only those cells that were capable of glycolysis survived the early competition of life. Later improvements in catabolic metabolism built on this success. Glycolysis was

FIGURE 8-7

The conversion of pyruvate to ethanol takes place naturally in grapes left to ferment on vines, as well as in fermentation vats of crushed grapes. The process is carried out by yeasts, and when their conversion increases the ethanol concentration to about 12%, the toxic effects of the alcohol kill the yeasts. What is left is wine. Often wine will be aged in casks to allow time for more complex chemical reactions to occur that subtly alter the flavor of the wine. The casks must be airtight to prevent bacterial contamination, because bacteria will further ferment the ethanol to acetic acid (vinegar).

not discarded during the course of evolution but rather was used as the starting point for the further extraction of chemical energy. Nature did not, so to speak, go back to the drawing board and design a different and better metabolism from scratch. Rather, metabolism evolved as one layer of reactions was added to another, just as successive layers of paint can be found in an old apartment. We all carry glycolysis with us—a metabolic memory of our evolutionary past.

THE NEED TO CLOSE THE METABOLIC CIRCLE

Inspect for a moment the net reaction of the glycolytic sequence:

$$\text{Glucose} + 2\text{ ADP} + 2\text{ P}_i + 2\text{ NAD}^+ \rightarrow$$
$$2\text{ pyruvate} + 2\text{ ATP} + 2\text{ NADH} + 2\text{H}^+ + 2\text{H}_2\text{O}$$

You can see that three changes occur in glycolysis: (1) glucose is converted to pyruvate (2) ADP is converted to ATP and (3) NAD^+ is converted to NADH.

As long as foodstuffs that can be converted to glucose are available, a cell can continually churn out ATP to drive its activities, except for one problem: the NAD^+. As a result of glycolysis, the cell accumulates NADH molecules at the expense of the pool of NAD^+ molecules. A cell does not contain a large amount of NAD^+, and for glycolysis to continue, it is necessary to recycle the NADH that it produces back to NAD^+. Some other home must be found for the hydrogen atom taken from G3P, some other molecule that will accept the hydrogen and be reduced.

What happens after glycolysis depends critically on the fate of this hydrogen atom. One of two things generally happens:

1. *Oxidative respiration.* Oxygen is an excellent electron acceptor, and in the presence of oxygen gas the hydrogen atom taken from G3P can be donated to oxygen, forming water. This is what happens in your body. Because air is rich in oxygen, this process is referred to as **aerobic metabolism.**

2. *Fermentation.* When oxygen is not available, another organic molecule must accept the hydrogen atom instead. Such a process is called fermentation (Figure 8-7). This process, which is what happens when bacteria grow without oxygen, is referred to as **anaerobic metabolism,** or metabolism without oxygen.

OXIDATIVE RESPIRATION

In all aerobic organisms, the oxidation of glucose, which began in Stage c of glycolysis, is continued where glycolysis leaves off—with pyruvate. The evolution of this new biochemical process was conservative, as is almost always the case in evolution; the new process is simply tacked onto the old one. In eukaryotic organisms, oxidative respiration takes place exclusively in the mitochondria, which apparently originated as symbiotic bacteria. The oxidation of pyruvate takes place in two stages: the oxidation of pyruvate to form an intermediate product, acetyl–CoA, and the later oxidation of the acetyl–CoA. We will consider them in turn.

The Oxidation of Pyruvate

The first stage of oxidative respiration is a single oxidative reaction in which one of the three carbons of pyruvate is split off, departing as CO_2 (a reaction of this kind is called by chemists a **decarboxylation**) and leaving behind two remnants (Figure 8–8): (1) a pair of electrons and their associated hydrogen, which reduce NAD^+ to NADH; and (2) a two–carbon fragment called an **acetyl group.**

This reaction is very complex, involving three intermediate stages, and is catalyzed by an assembly of enzymes called a **multi-enzyme complex.** Such multi-enzyme complexes serve to organize a series of reactions so that the chemical intermediates do not diffuse away or undergo other reactions. Within such a complex, component polypeptides pass the reacting substrate molecule from one enzyme to the next in line without ever letting go of it. The complex of enzymes called **pyruvate dehydrogenase** that removes the CO_2 from pyruvate is one of the largest enzymes known—it contains 48 polypeptide chains. In the course of the reaction the two-carbon acetyl fragment removed from pyruvate is added to a cofactor, a carrier molecule called coenzyme A (CoA), forming a compound called **acetyl-CoA.**

$$Pyruvate + NAD^+ + CoA \rightarrow Acetyl\text{-}CoA + NADH + CO_2$$

This reaction produces a molecule of NADH, which is later used to produce ATP. Of far greater significance than the reduction of NAD^+ to NADH, however, is the residual fragment acetyl-CoA (Figure 8-9). Acetyl-CoA is important because it is produced not only by the oxidation of the product of glycolysis, as previously outlined, but also by the metabolic breakdown of proteins, fats, and other lipids. Acetyl-CoA thus provides a single focus for the many catabolic processes of the eukaryotic cell, all the resources of which are channeled into this single metabolite. Acetyl-CoA is the currency of oxidative metabolism.

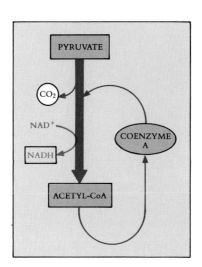

FIGURE 8-8

The oxidation of pyruvate. This complex reaction involves the reduction of NAD^+ to NADH and is thus a significant source of metabolic energy. Its product, acetyl-CoA, is the starting material for the citric acid cycle.

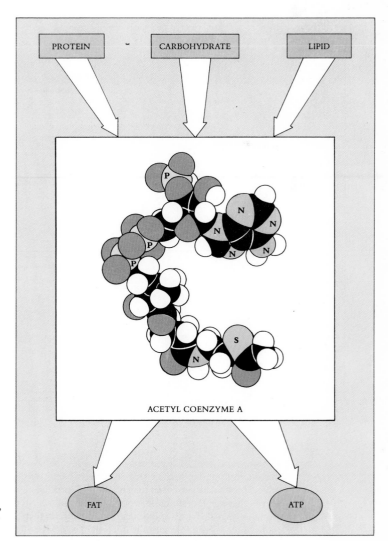

ACETYL COENZYME A

FIGURE 8-9

Acetyl-CoA contains a high-energy bond similar to that in ATP. Acetyl-CoA is the central molecule of energy metabolism. Almost all of the molecules that you use as foodstuffs are converted to acetyl-CoA when you metabolize them; the acetyl-CoA is then channeled into fat synthesis or into ATP production, depending on your body's energy requirements.

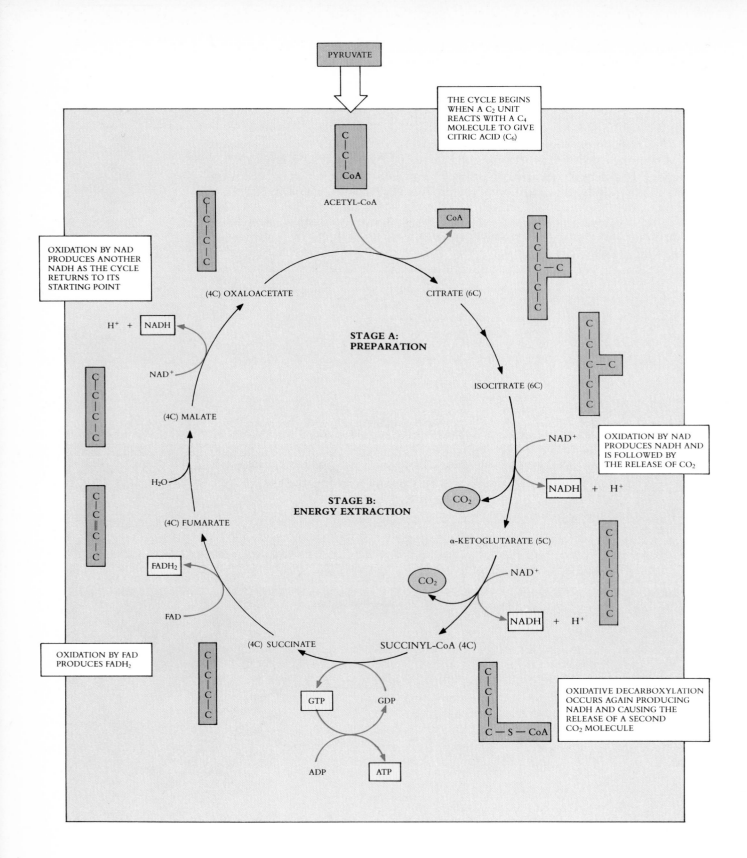

PYRUVATE

THE CYCLE BEGINS
WHEN A C_2 UNIT
REACTS WITH A C_4
MOLECULE TO GIVE
CITRIC ACID (C_6)

ACETYL-CoA

CoA

STAGE A:
PREPARATION

(4C) OXALOACETATE

CITRATE (6C)

OXIDATION BY NAD
PRODUCES ANOTHER
NADH AS THE CYCLE
RETURNS TO ITS
STARTING POINT

H^+ + NADH

NAD^+

ISOCITRATE (6C)

(4C) MALATE

NAD^+

OXIDATION BY NAD
PRODUCES NADH AND
IS FOLLOWED BY
THE RELEASE OF CO_2

H_2O

STAGE B:
ENERGY EXTRACTION

CO_2

NADH + H^+

(4C) FUMARATE

α-KETOGLUTARATE (5C)

$FADH_2$

CO_2

NAD^+

FAD

NADH + H^+

OXIDATION BY FAD
PRODUCES $FADH_2$

(4C) SUCCINATE

SUCCINYL-CoA (4C)

GTP   GDP

OXIDATIVE DECARBOXYLATION
OCCURS AGAIN PRODUCING
NADH AND CAUSING THE
RELEASE OF A SECOND
CO_2 MOLECULE

ADP   ATP

FIGURE 8-10

The citric acid cycle. The reactions of this sequence have been the focus of intense research, and a great deal is known about them. In stage A of the cycle, a two-carbon acetyl-CoA molecule is joined to a molecule of oxaloacetate and then rearranged, producing a molecule of isocitrate. This doesn't yield any energy. In stage B of the cycle, isocitrate is oxidized, re-forming oxaloacetate. In this process, two carbon atoms are expelled as CO_2 and energy is extracted in the form of three NADH molecules, one ATP molecule, and one molecule of $FADH_2$ (flavin adenine dinucleotide), which is an alternative electron carrier like NADH.

Although acetyl-CoA is formed by many catabolic processes in the cell, only a limited number of processes use acetyl-CoA. Most acetyl-CoA is either directed toward energy storage (it is used in lipid synthesis), or it is oxidized to produce ATP. Which of these two processes occurs depends on the amount of ATP in the cell. When ATP levels are high, the oxidative pathway is inhibited and acetyl-CoA is channeled into fatty acid biosynthesis, which is why people get fat when they eat too much. When ATP levels are low, the oxidative pathway is stimulated and acetyl-CoA flows into energy-producing oxidative metabolism.

The Oxidation of Acetyl-CoA

The oxidation of acetyl-CoA begins with the binding of the acetyl group to a four-carbon carbohydrate. The resulting six-carbon molecule is then passed through a series of electron-yielding oxidation reactions, during which two CO_2 molecules are split off, regenerating the four-carbon carbohydrate, which is then free to bind another acetyl group. The process is a continuous cyclical flow of carbon. In each turn of the cycle a new acetyl group comes in to replace the two CO_2 molecules that are lost, and more electrons are extracted.

The Reactions of the Citric Acid Cycle

The citric acid cycle, which oxidizes acetyl-CoA, consists of nine reactions, diagrammed in Figure 8-10. The cycle has two stages:

Stage a: Three **preparation reactions** set the scene. In the first reaction acetyl-CoA joins the cycle, and in the other two reactions chemical groups are rearranged.

Stage b: It is in the second stage that **energy extraction** occurs. Four of the six reactions are oxidations in which electrons are removed, and one generates an ATP equivalent directly by substrate-level phosphorylation.

Together the nine reactions constitute a cycle that begins and ends with oxaloacetate. At every turn of the cycle, acetyl-CoA enters and is oxidized to CO_2 and H_2O, and the electrons are channeled off to drive proton pumps that generate ATP.

The Products of the Citric Acid Cycle

During the oxidation of glucose, many electrons and some ATP have been generated. The extracted electrons are temporarily housed within NADH molecules. In one reaction the extracted electrons are not energetic enough to reduce NAD^+, and a different coenzyme called reduced **flavin adenine dinucleotide (FADH$_2$)** is used to carry these less energetic electrons. Let us count the number of molecules of ATP and of electron carriers that we have generated, starting with glucose (Table 8-1).

TABLE 8-1   THE OUTPUT OF AEROBIC METABOLISM

METABOLIC PROCESS	SUBSTRATE-LEVEL PHOSPHORYLATION	OXIDATION	
Glycolysis	2 ATP	2 NADH	
Decarboxylation of pyruvate ($\times 2$)		2 NADH	
Citric acid cycle ($\times 2$)	2 ATP	6 NADH	2 FADH$_2$
TOTAL	4 ATP	10 NADH	2 FADH$_2$

In the process of aerobic respiration the glucose molecule has been consumed entirely. Its six carbons were first split into three-carbon units during glycolysis. One of the carbons of each three-carbon unit was then lost as CO_2 in the conversion of pyruvate to acetyl-CoA, and the other two were lost during the oxidations of the citric acid cycle. All that is left to mark the passing of the glucose molecule is its energy, which is preserved in four ATP molecules and the reduced state of 12 electron carriers.

The oxidative metabolism of one molecule of glucose proceeds in three stages: glycolysis (which can be anaerobic), the oxidation of pyruvate, and the citric acid cycle. Both glycolysis and the citric acid cycle produce two ATP molecules by substrate-level phosphorylation, coupling ATP production to a reaction that involves a large release of free energy. All three processes involve oxidation reactions, that is, reactions that produce electrons. The electrons are donated to a carrier molecule, NAD^+ (or in one instance FAD^{++}).

The systematic oxidation of the pyruvate remaining after glycolysis generates two ATP molecules, which is as many as glycolysis produced. More important, this process harvests many energized electrons, which can then be directed to the chemiosmotic synthesis of ATP.

USING THE ELECTRONS GENERATED BY THE CITRIC ACID CYCLE TO MAKE ATP

The NADH and $FADH_2$ molecules formed during glycolysis and the subsequent oxidation of pyruvate each contain a pair of electrons gained when NADH was formed from NAD^+ and when $FADH_2$ was formed from FAD^+. The NADH molecules carry their electrons to the cell membrane (the $FADH_2$ is already attached to it), and there they transfer them to a complex membrane-embedded protein called **NADH dehydrogenase.** The electrons are then passed on to a series of respiratory proteins called **cytochromes** and other carrier molecules (Figure 8-11), one after the other, losing much of their energy in the process by driving several transmembrane proton pumps. This series of membrane-associated electron carriers is collectively called the **electron transport chain** (Figure 8-12). At the terminal step of the electron transport chain, the electrons are passed to the **cytochrome c oxidase complex,** which uses four of the electrons to reduce a molecule of oxygen gas and form water:

$$O_2 + 4H^+ + 4e^- \rightarrow 2H_2O$$

The electron transport chain puts the electrons harvested from the oxidation of glucose to work driving proton-pumping channels. The ultimate acceptor of the electrons harvested from pyruvate is oxygen gas, which is reduced to form water.

FIGURE 8-11

The structure of cytochrome c. Cytochromes are respiratory proteins that contain "heme" groups, complex iron-containing carbon rings often found in molecules that transport electrons. In the molecular model of cytochrome c, the heme group is indicated in red. Heme is an iron-containing pigment that has a complex ring structure with many alternating single and double bonds.

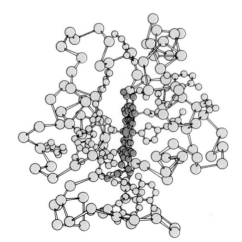

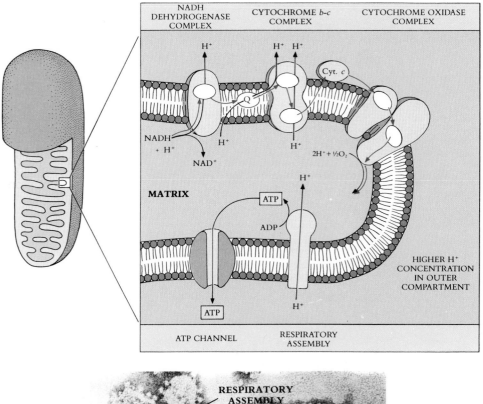

NADH DEHYDROGENASE COMPLEX · CYTOCHROME *b-c* COMPLEX · CYTOCHROME OXIDASE COMPLEX

MATRIX

ATP CHANNEL · RESPIRATORY ASSEMBLY

RESPIRATORY ASSEMBLY

FIGURE 8-12

The electron transport chain is composed of a series of electron carrier proteins embedded within the inner mitochondrial membrane. Electrons are passed from one carrier protein to another. Three of the carriers act as proton pumps. Upon accepting electrons, a protein carrier changes its shape, pumps a proton out into the outer compartment of the mitochondrion, and passes the electrons along to the next protein carrier in the chain. After reaching the third such proton pump, the electron is donated to a protein, which uses four electrons and four protons to reduce a molecule of oxygen to water. The electron transport chain creates a gradient in proton concentration, the outer mitochondrial compartment having many more protons than the inner matrix. The resulting diffusion force causes protons to re-enter the matrix, passing through special channels called **respiratory assemblies.** In the electron micrograph, the respiratory assemblies, magnified 320,000 times, resemble clusters of lollipops protruding into the matrix. Each passage of a proton back into the matrix through a respiratory assembly is coupled to the synthesis of a molecule of ATP within the matrix.

It is the availability of a plentiful electron acceptor (that is, an oxidized molecule) that makes oxidative respiration possible. In the absence of such a molecule, oxidative respiration is not possible. The electron transport chain used in aerobic respiration is similar to, and probably evolved from, the one that is employed in aerobic photosynthesis.

Thus the final product of oxidative metabolism is water, which is not an impressive result in itself. Recall, however, what happened in the process of forming that water. The electrons contributed by NADH molecules passed down the electron transport chain, activating three proton-pumping channels. Similarly, the passage of electrons contributed by $FADH_2$, which entered the chain later, activated two of these channels. The electrons harvested from the citric acid cycle have in this way been used to pump a large number of protons out across the membrane, and *that* result is the payoff of oxidative respiration.

In eukaryotes, oxidative metabolism takes place within the mitochondria, which are present in virtually all eukaryotic cells. The internal compartment or **matrix** of a mitochondrion contains within it the enzymes that perform the reactions of the citric acid cycle. Electrons harvested there by oxidative respiration are used to pump protons out of the matrix into the outer compartments, the space between the two mitochondrial membranes. As a high concentration of protons builds up, protons cross the inner membrane back into the matrix, driven by diffusion. The only way they can get in is through special channels, which are visible in electron micrographs as projections on the inner surface of the membrane (see Figure 8-12). The proton channel is made up of four polypeptide chains that traverse the membrane. At the inner boundary of the

FIGURE 8-13

A summary of proton movement in mitochondria during chemiosmosis. **A** The electron transport system utilizes energetic electrons from NADH to drive the pumping of electrons out of the inner matrix compartment. The result is a deficit of protons in the matrix and an excess in the outer compartment.
B Protons re-enter the matrix, driven by diffusion, through special channels; their passage is coupled to the synthesis of ATP from ADP + phosphate ion (P_i). ATP molecules then diffuse back out of the matrix and outer compartment through passive channels (red arrow).

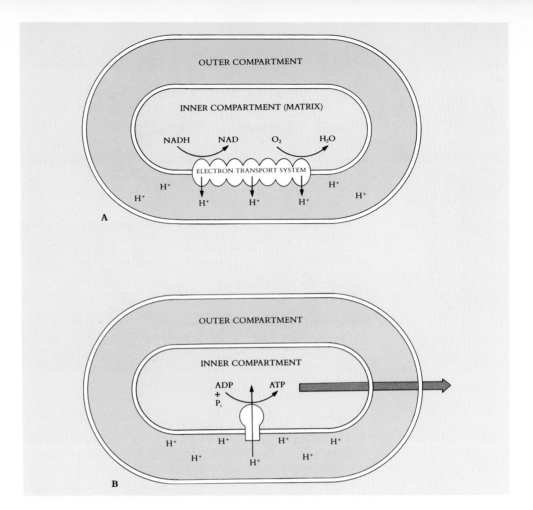

membrane, the channel is linked by a stalk of several proteins to a large protein complex composed of five polypeptide chains. This protein complex synthesizes ATP within the matrix when protons travel inward through the channel (Figure 8-13). The ATP then passes out of the mitochondrion by facilitated diffusion and into the cell's cytoplasm.

> **The electrons harvested from glucose and transported to the mitochondrial membrane by NADH drive protons out across the inner membrane. The return of the protons by diffusion generates ATP.**

AN OVERVIEW OF GLUCOSE CATABOLISM: THE BALANCE SHEET

How much metabolic energy does the chemiosmotic synthesis of ATP produce? One ATP molecule is generated chemiosmotically for each activation of a proton pump by the electron transport chain. Thus each NADH molecule that the citric acid cycle produces ultimately causes the production of three ATP molecules, since its electrons activate three pumps. Each $FADH_2$, which activates two of the pumps, leads to the production of two ATP molecules. However, because of an evolutionary development that we shall discuss later, eukaryotes carry out glycolysis in their cytoplasm and the citric acid cycle within their mitochondria. This separation of the two processes within the cell requires that the electrons of the NADH created during glycolysis be transported across the mitochondrial membrane, consuming one ATP molecule per NADH. Thus each glycolytic NADH produces only two ATP molecules in the final tally, instead of three.

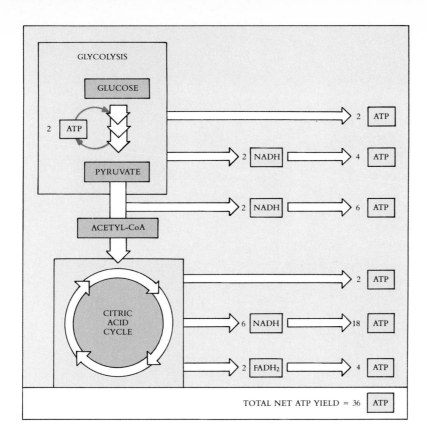

FIGURE 8-14

An overview of the energy extracted from the oxidation of glucose.

The overall reaction for the catabolic metabolism of glucose is

$$C_6H_{12}O_6 + 10NAD^+ + 2FAD^+ + 36ADP + 36P_i + 14H^+ + 6O_2 \rightarrow$$
(glucose) $\qquad\qquad\qquad 6CO_2 + 36ATP + 6H_2O + 10NADH + 2FADH_2$

As seen in Table 8-1, 4 of the 36 ATP molecules result from substrate-level phosphorylation; the remaining 32 ATP molecules result from chemiosmotic phosphorylation.

The overall efficiency of glucose catabolism is very high: the aerobic oxidation (combustion) of glucose yields 36 ATP molecules (a total of $-7.3 \times 36 = -263$ kilocalories per mole) (Figure 8-14). The aerobic oxidation of glucose thus has an efficiency of 263/686, or 38%. Compared with the two ATP molecules generated by glycolysis, aerobic oxidation is 18 times more efficient.

> **Oxidative respiration is 18 times more efficient than glycolysis at converting the chemical energy of glucose into ATP. It produces 36 molecules of ATP from each glucose molecule consumed, compared with the 2 ATP molecules that are produced by glycolysis.**

Oxidative metabolism is so efficient at extracting energy from photosynthetically produced molecules that its development opened up an entirely new avenue of evolution. For the first time it became feasible for heterotrophic organisms, which derive their metabolic energy exclusively from the oxidative breakdown of other organisms, to evolve. As long as some organisms produced energy by photosynthesis, heterotrophs could exist solely by feeding on these autotrophs. The high efficiency of oxidative respiration was one of the key conditions that fostered the evolution of nonphotosynthetic heterotrophic organisms. An overview of the oxidative respiration process is presented in Figure 8-15.

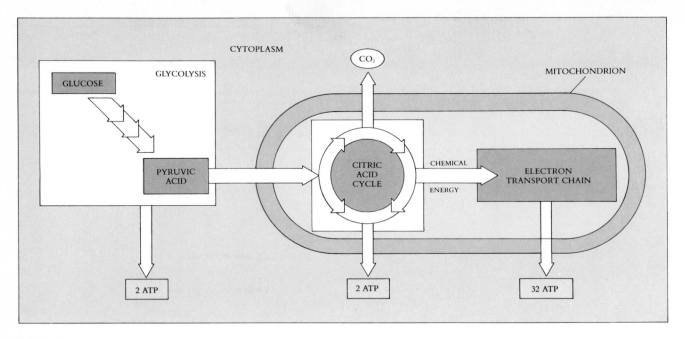

CYTOPLASM

GLYCOLYSIS

GLUCOSE

CO_2

MITOCHONDRION

PYRUVIC ACID

CITRIC ACID CYCLE

CHEMICAL

ENERGY

ELECTRON TRANSPORT CHAIN

2 ATP

2 ATP

32 ATP

FIGURE 8-15

An overview of oxidative respiration. In a eukaryotic cell, the citric acid cycle and the electron transport chain are located within the mitochondria, and glycolysis occurs in the cytoplasm. In the course of glycolysis, each glucose molecule is broken down into two molecules of pyruvic acid. This pyruvic acid then enters a mitochondrion, where it enters the citric acid cycle and is broken down to carbon dioxide. Chemical energy produced by the citric acid cycle is then transferred to the electron transport chain, which is located in the membranes of the cristae of the mitochondrion. Most of the ATP that results from respiration is formed at this stage.

FERMENTATION

In the absence of oxygen, oxidative respiration cannot occur because of the lack of an electron acceptor. In such a situation a cell must rely on glycolysis to produce ATP. In the absence of oxygen, cells donate the hydrogen atoms generated by glycolysis to organic molecules during a process called **fermentation.**

Bacteria carry out many different sorts of fermentations, each employing some form of carbohydrate molecule to accept the hydrogen atom from NADH and thus reform NAD^+:

$$\text{Carbohydrate} + \text{NADH} \rightarrow \text{Reduced carbohydrate} + NAD^+$$

More than dozen fermentation processes have evolved among bacteria, each process using a different carbohydrate as the hydrogen acceptor. Often the resulting reduced compound is an organic acid such as acetic acid, butyric acid, propionic acid, or lactic acid. In other organisms such as yeasts, the reduced compound is an alcohol.

Of these various bacterial fermentations, only a few occur among eukaryotes (Figure 8-16). In one fermentation process, which occurs in the single-celled fungi called yeasts (Figure 8-17), the carbohydrate that accepts the hydrogen from NADH is pyruvate, the end product of the glycolytic process. Yeasts remove a terminal CO_2 group from pyruvate, producing a toxic two-carbon molecule called **acetaldehyde,** and CO_2. Because of this CO_2 production, bread with yeast rises and "unleavened" bread—that is, bread without yeast—does not. Acetaldehyde then accepts the hydrogen from NADH, producing NAD^+ and ethyl alcohol, also called ethanol.

This particular type of fermentation has been of great interest to human beings, since it is the source of the ethyl alcohol in wine and beer. However, ethyl alcohol is an undesirable by-product from the yeast's standpoint since it becomes toxic to the yeast when it reaches high levels. That is why natural wine contains only about 12% alcohol—12% is the amount it takes to kill the yeast fermenting the sugars.

Most multicellular animals regenerate NAD^+ without decarboxylation. The processes that they use involve the production of by-products that are less toxic than alcohol. For example, your muscle cells use an enzyme called **lactate dehydrogenase** to add the hydrogen of the NADH produced by glycolysis back to the pyruvate that is the end product of glycolysis, converting pyruvate plus NADH into lactic acid plus NAD^+.

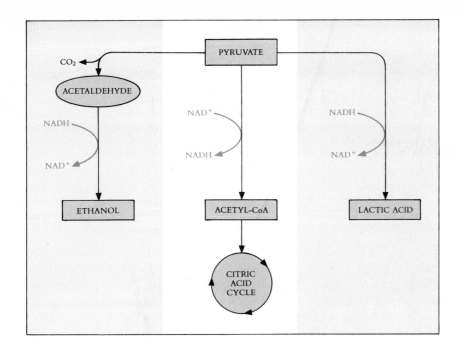

FIGURE 8-16

The metabolic fate of pyruvate, the product of glycolysis. In the presence of oxygen, pyruvate is oxidized to acetyl-CoA and enters the citric acid cycle. In the absence of oxygen, pyruvate instead is reduced, accepting the electrons extracted during glycolysis and carried by NADH. When pyruvate is reduced directly, as it is in your muscles, the product is lactic acid; when a CO_2 is first removed from pyruvate and the remainder is reduced, as in yeasts, the product is ethanol, also known as ethyl alcohol.

This process closes the metabolic circle, allowing glycolysis to continue for as long as the glucose holds out. Blood circulation removes excess lactic acid from muscles. When the lactic acid cannot be removed as fast as it is produced, your muscles cease to work well. Subjectively they feel tired or leaden. Try raising and lowering your arm rapidly a hundred times and you will soon experience this sensation. A more efficient circulatory system developed through training will let you run longer before the accumulation of lactic acid becomes a problem—this is why people train before they run marathon races—but there is always a point at which lactic acid production exceeds lactic acid removal. It is essentially because of this limit that the world record for running a mile is just under 4 minutes and not significantly less.

In fermentations, which are anaerobic processes, the electron generated in the glycolytic breakdown of glucose is donated to an oxidized organic molecule. In contrast, in aerobic metabolism such electrons are transferred to oxygen, generating ATP in the process.

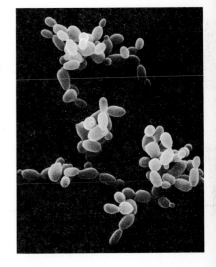

FIGURE 8-17

Scanning electron micrograph of a yeast.

SUMMARY

1. Metabolism is driven by the application of energy to endergonic reactions. Those reactions that require a net input of free energy are coupled to the cleavage of ATP, which involves a large release of free energy.

2. Organisms acquire ATP from photosynthesis or by harvesting the chemical energy in the bonds of organic molecules.

3. Energy can be harvested in two ways: substrate-level phosphorylation, in which some reactions involving a large decrease in free energy are coupled to ATP formation; or oxidative respiration, in which electrons are used to drive proton pumps, resulting in the chemiosmotic synthesis of ATP.

4. Glycolysis harvests chemical energy in the first of these two ways, by rearranging the chemical bonds of glucose to form two molecules of pyruvate and two molecules of ATP.

5. Organisms living in anaerobic environments require a mechanism to dispose of the electron and the associated hydrogen produced in the oxidation step of glycolysis. They are donated to one of a number of carbohydrates in a process called fermentation.

6. Aerobic organisms direct the electron to the mitochondrial membrane, where it drives a proton pump. Pyruvate is further oxidized, yielding many additional electrons, all of which are also channeled to the mitochondrial membrane to drive proton pumps.

7. The excess protons in the outer mitochondrial compartment created by the proton pumps diffuse back across the inner mitochondrial membrane through special channels, and the passage of these protons drives the production of ATP.

8. In total, the oxidation of glucose results in the net production of 36 ATP molecules, all but 4 of them produced chemiosmotically.

9. Within all eukaryotic cells, oxidative respiration of pyruvate takes place within the matrix of mitochondria. The mitochondria act as closed osmotic compartments from which protons are pumped and into which protons pass by diffusion to create ATP. The ATP then leaves the mitochondrion by facilitated diffusion.

REVIEW

1. Proton-pumping channels are responsible for the chemiosmotic generation of _____.

2. Give, in order, the three stages of the oxidative metabolism of glucose.

3. The end result of glycolysis is to break a glucose molecule up into two molecules of _____ and generate a net gain of two ATP molecules.

4. In eukaryotic cells, oxidative metabolism takes place in which organelle?

5. What is the net production of ATP from a molecule of glucose?

SELF-QUIZ

1. How do all cells drive thermodynamically unfavorable reactions?
   (a) by using the energy released from splitting ATP
   (b) by using light energy
   (c) these reactions occur only in warm-blooded organisms, and heat is used as the energy source
   (d) thermodynamically unfavorable reactions do not occur
   (e) the information is coded in the DNA

2. In glycolysis, in the absence of oxygen, ATP is generated by:
   (a) electron transport chain
   (b) substrate-level phosphorylation
   (c) chemiosmotic generation of ATP
   (d) citric acid cycle
   (e) photosynthesis

3. When oxygen is available as an electron acceptor, what happens after glycolysis?
   (a) glycolysis occurs agains
   (b) NADH is produced
   (c) fermentation
   (d) pyruvate is formed
   (e) oxidative respiration

4. By the end of oxidative metabolism, all of the carbons from glucose ($C_6H_{12}O_6$) are gone. Where did they go?
   (a) carbon dioxide
   (b) NADH
   (c) to make pyruvate
   (d) ATP
   (e) water

5. What is the purpose of oxidative metabolism?
   (a) carbon dioxide production
   (b) ATP production
   (c) water production
   (d) NAD^+ production
   (e) glucose utilization

PROBLEM

1. Thirty-six ATP molecules are produced per molecule of glucose. How many ATP molecules are produced from a mole of glucose? As a helpful hint, remember that there are 6.023×10^{23} molecules in a mole.

THOUGHT QUESTIONS

1. If you poke a hole in a mitochondrion, can it still perform oxidative respiration? Can fragments of mitochondria perform oxidative respiration? Explain.

2. Why have eukaryotic cells not dispensed with mitochondria, by placing the mitochondrial genes in the nucleus and incorporating the mitochondrial enzyme ATPase and the proton pump into the cell's smooth endoplasmic reticulum?

FOR FURTHER READING

DICKERSON, R.: "Cytochrome *c* and the Evolution of Energy Metabolism," *Scientific American,* March 1980, pages 136-154. A superb description of how the metabolism of modern organisms evolved.

HINKLE, P., and R. McCARTY: "How Cells Make ATP," *Scientific American,* March 1978, pages 104–125. A good summary of oxidative respiration, with a clear account of the events that happen at the mitochondrial membrane.

LEVINE, M., H. MUIRHEAD, D. STAMMERS, and D. STUART: "Structure of Pyruvate Kinase and Similarities with Other Enzymes: Possible Implications for Protein Taxonomy and Evolution," *Nature,* vol. 271, pages 626-630, 1978. An advanced article that recounts how the enzymes of oxidative metabolism may have evolved. Well worth the effort.

FIGURE 9-1   These corn plants, growing vigorously in the August sun, capture enough energy each day from sunlight to power your body for hours.

PHOTOSYNTHESIS

Chapter 9

Overview

We all depend on the process of photosynthesis, which is the means by which the energy that ultimately builds our bodies is captured from sunlight. Photosynthesis is one of the oldest and most fundamental of life processes: the major types of photosynthesis first evolved billions of years ago among the bacteria. The first of these to evolve apparently used the energy obtained from sunlight to split hydrogen sulfide, generating sulfur as a by-product. Later, in the cyanobacteria a system evolved in which water is split in the same way, yielding oxygen gas as a by-product. The outcome of this form of photosynthesis, performed for almost 3 billion years, has been the production of an atmosphere rich in oxygen; this atmosphere has set the stage for the evolution of all complex forms of life on earth, including ourselves.

For Review

Here are some important terms and concepts that you will encounter in this chapter. If you are not familiar with them, you should review them before proceeding.

Electron energy levels (Chapter 3)

Chemiosmosis (Chapters 6 and 8)

Oxidation-reduction (Chapter 7)

Glycolysis (Chapter 8)

For all its size and diversity, our universe might never have spawned life except for one characteristic of overriding importance: it is awash in energy. Everywhere in the universe matter is continually being converted to energy by thermonuclear processes. The energy resulting from those processes streams in all directions from the stars, including our sun. The total amount of radiant energy that reaches the earth from the sun each day is the equivalent of about 1 million Hiroshima–sized atomic bombs. Approximately one third of this energy is immediately radiated back into space, and most of the remainder is absorbed by the earth and converted to heat. Less than 1% of the energy that reaches the earth is captured in the process of photosynthesis to provide the energy that drives all the activities of life on earth (Figure 9-1). In this chapter we shall discuss how photosynthesis works and outline the different major kinds of photosynthesis.

THE BIOPHYSICS OF LIGHT

Where is the energy in light? What is there about sunlight that a plant can use to create chemical bonds? To answer these questions we need to begin by considering the physical nature of light itself. Perhaps the best place to start is in a laboratory in Germany in 1887 where a curious experiment was performed. A young physicist named Heinrich Hertz was attempting to verify a mathematical theory that predicted the existence of electromagnetic waves. To see whether such waves existed, Hertz constructed a spark generator in his laboratory–a machine composed of two shiny metal spheres standing near each other on slender rods. When a very high static electric charge built up on one sphere, sparks would jump across to the other sphere.

After Hertz had constructed this system, he proceeded to investigate whether the sparking would create invisible electromagnetic waves, as predicted by the mathematical theory. On the other side of the room he placed on an insulating stand a thin metal hoop that was not quite a closed circle. When he turned on the spark generator across the room, tiny sparks could be seen crossing the gap in the hoop! This was the first demonstration of radio waves. But Hertz noted a curious side effect as well. When light was shone on the ends of the hoop, the sparks crossed the gap more readily. This unexpected facilitation, called the **photoelectric effect,** puzzled investigators for many years.

Especially perplexing was the fact that the strength of the photoelectric effect depended not only on the brightness of the light shining on the gap in the hoop but also on its wavelength. Short wavelengths were much more effective than long wavelengths in producing the photoelectric effect. This effect was finally explained by Albert Einstein as a natural consequence of the physical nature of light: the light was literally blasting electrons from the metal surface at the ends of the hoop, creating positive ions and thus facilitating the passage of the electronic spark induced by the radio waves. Light consists of units of energy called **photons,** and some of these photons were being absorbed by the metal atoms of the hoop. In this process some of the electrons of the metal atoms were being boosted into higher energy levels and so ejected from the metal atoms into the gap.

All photons do not possess the same amount of energy. Some contain a great deal of energy, others far less. The energy content of light is inversely proportional to the wavelength of the light: short-wavelength light contains photons of higher energy than photons of long-wavelength light. For this reason the photoelectric effect was more pronounced with light of a shorter wavelength: the metal atoms of the hoop were being bombarded with higher-energy photons. Sunlight contains photons of many different energy levels, only some of which our eyes perceive as visible light. The highest-energy photons, which occur at the short-wavelength end of the **electromagnetic spectrum** (Figure 9-2), are gamma rays with wavelengths of less than 1 nanometer; the lowest-energy photons with wavelengths of thousands of meters are radio waves.

Sunlight contains a significant amount of ultraviolet light, which possesses considerably more energy than visible light because ultraviolet light has a shorter wavelength. Ultraviolet light is thought to have been an important source of energy on the primitive earth when life originated, before the oxygen-rich atmosphere developed. Today's atmosphere contains ozone (derived from oxygen gas), which absorbs most of the ultraviolet-energy photons in sunlight; but anyone who has been sunburned knows that considerable quantities of ultraviolet-energy photons manage to penetrate the atmosphere. As we shall learn in a later chapter, mysterious "holes" have recently been discovered in the ozone layer; these holes, which apparently have been caused by human activities, threaten to cause an enormous increase in the incidence of skin cancers in human beings throughout the world.

As described in Chapter 3, electrons occupy distinct energy levels in their orbits around the atomic nucleus. Boosting an electron to a different energy level requires just the right amount of energy—no more and no less; similarly, when climbing a ladder, you must raise your foot just so far to climb a rung, not 1 centimeter more or less. Therefore specific atoms can absorb only certain photons of light—those that corre-

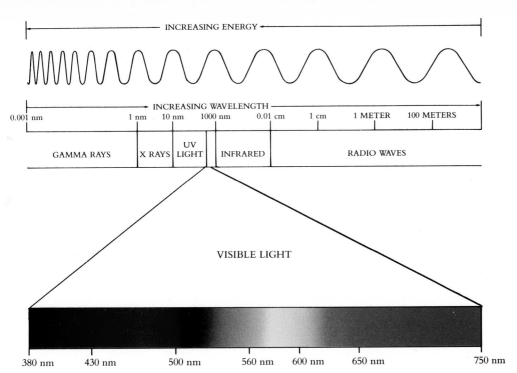

FIGURE 9-2

The electromagnetic spectrum. Light is a form of electromagnetic energy and is conveniently thought of as a wave. The shorter the wavelength of light, the greater the energy. Visible light represents only a small part of the electromagnetic spectrum, that between 380 and 750 nanometers.

spond to available energy levels. A given atom or molecule has a characteristic range, or **absorption spectrum,** of photons that it is capable of absorbing, depending on the electron energy levels that are available in it.

CAPTURING LIGHT ENERGY IN CHEMICAL BONDS

The energy of light is "captured" by a molecule that absorbs it because the photon of energy is used to boost an electron of the molecule to a higher energy level. Molecules that absorb light are called **pigments,** and organisms have evolved a variety of such pigments. Two general sorts are of interest here:

1. *Carotenoids.* Carotenoids consist of carbon rings linked to chains in which single and double bonds alternate. Carotenoids can absorb photons of a wide range of different energies, although not always with high efficiency. A typical carotenoid is **beta-carotene** (Figure 9-3). In beta-carotene two carbon rings are linked by a chain of 18 carbon atoms connected alternately by single and double bonds. Splitting a molecule of beta-carotene into equal halves results in the production of two molecules of **vitamin A.** When vitamin A is subsequently oxidized, **retinal,** the pigment used in human vision, is produced. Thus the claim that eating carrots (rich in beta-carotene) improves vision is based on fact. When retinal absorbs a photon of light, the resulting electron excitation causes a change in the shape of a pigment located in the membrane of certain cells in the eye and thus triggers a nerve impulse. Retinal absorbs photons that produce light ranging from violet (380 nanometers) to red (750 nanometers), and so determines the range of colors that we can see—**visible light.** Some other organisms use different light-absorbing pigments for vision and thus "see" a different portion of the electromagnetic spectrum. For example, most insects have eye pigments that absorb at lower wavelengths than retinal. As a result, bees can perceive ultraviolet-light, which is produced by photons with a shorter wavelength than violet but cannot see red, which is produced by a photon with a relatively long wavelength.

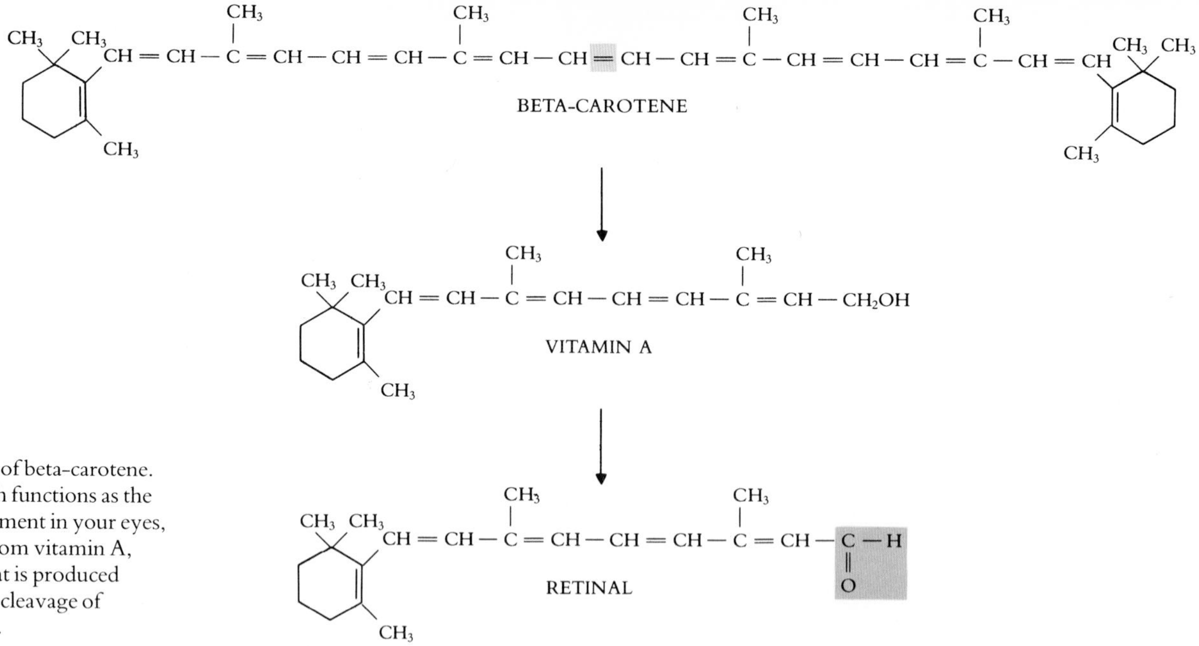

BETA-CAROTENE

VITAMIN A

RETINAL

FIGURE 9-3

The structure of beta-carotene. Retinal, which functions as the key visual pigment in your eyes, is produced from vitamin A, a molecule that is produced following the cleavage of beta-carotene.

2. *Chlorophylls.* Other biological pigments called chlorophylls absorb photons by means of an excitation process analogous to the photoelectric effect. These pigments use a metal atom (magnesium), which lies at the center of a complex ring structure called a **porphyrin ring.** This ring structure consists of alternating single and double carbon bonds. Photons absorbed by the pigment molecule excite electrons of the magnesium atom, which are then channeled away through the carbon-bond system. Outside the porphyrin ring, several small side groups are attached, which alter the absorption properties of the pigment in different kinds of chlorophyll.

Unlike retinal, the different kinds of chlorophyll absorb only photons of a narrow energy range. As you can see from their absorption spectra (Figure 9-4), the two kinds of chlorophyll that occur in plants—chlorophylls *a* and *b*—absorb primarily violet-blue and red light, respectively. The light between wavelengths of 500 and 600 nanometers is not absorbed by chlorophyll pigments and therefore is reflected by plants. The light reflected from a chlorophyll-containing plant has had all of its middle-energy photons absorbed by the chlorophyll except those in the 500 to 600 nanometer range. When these photons are subsequently absorbed by the retinal in our eyes, we perceive them as green.

A pigment is a molecule that absorbs light. The wavelengths absorbed by a particular pigment depend on the energy levels available in that molecule to which light-excited electrons can be boosted.

All plants and algae and all but one primitive group of photosynthetic bacteria use chlorophylls as their primary light-gatherers. Why don't these photosynthetic organisms use a pigment like retinal, which has a very broad absorption spectrum and can harvest light in the 500- to 600-nanometer wavelength range as well as at other wavelengths? The most likely hypothesis involves photoefficiency. Retinal absorbs a broad spectrum of light wavelengths but does so with relatively low efficiency; in contrast, chlorophyll absorbs in only two narrow bands, violet-blue and red, but does so with very high efficiency. Therefore by using chlorophyll plants and most other photosynthetic organisms achieve far higher overall photon capture rates than would be possible with a pigment that allows a broader but less efficient spectrum of absorption.

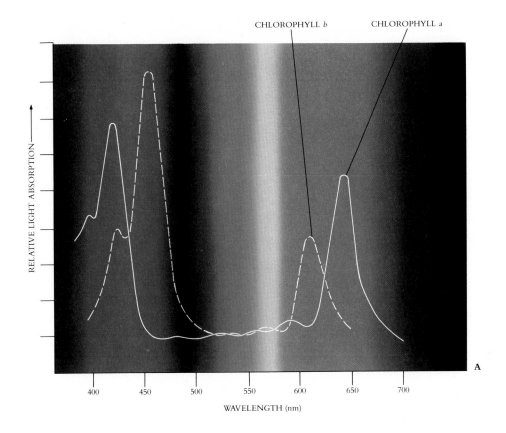

CHLOROPHYLL *b* CHLOROPHYLL *a*

RELATIVE LIGHT ABSORPTION

WAVELENGTH (nm)

A

CHLOROPHYLL *a*

CHLOROPHYLL *b*

B

FIGURE 9-4

The chlorophylls absorb predominantly violet-blue and red light in two narrow bands of the spectrum.

A Absorption spectra for chlorophyll a and chlorophyll b.

B The structures of chlorophyll a and chlorophyll b. The only difference between the two chlorophyll molecules is the substitution of a —CHO group in chlorophyll b for a —CH₃ group in cholorophyll a.

FIGURE 9-5

Overview of the three processes in photosynthesis.

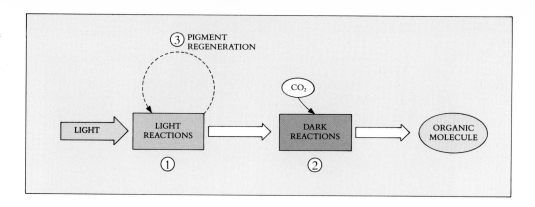

AN OVERVIEW OF PHOTOSYNTHESIS

Photosynthesis is a single term describing a complex series of events that involves three different kinds of chemical processes (Figure 9-5). The first process that occurs is the chemiosmotic generation of ATP by electrons, using energy captured from sunlight. The reactions involved in this process are called the **light reactions** of photosynthesis since the resultant synthesis of ATP takes place only in the presence of light. Second, the light reactions are followed by a series of enzyme-catalyzed reactions that use this newly generated ATP to drive the formation of organic molecules from atmospheric carbon dioxide. These reactions are called the **dark reactions** of photosynthesis because as long as ATP is available, they occur as readily in the absence of light as in its presence. Third, the pigment that absorbed the light in the first place is rejuvenated and made ready to initiate another light reaction.

Absorbing Light Energy

The light reactions occur on **photosynthetic membranes.** In photosynthetic bacteria these membranes are the cell membranes; in plants and algae all photosynthesis is carried out by the evolutionary descendants of photosynthetic bacteria, the **chloroplasts,** and the photosynthetic membranes are contained within the chloroplasts.

The light reactions occur in three stages:

1. A photon of light is captured by a pigment. The result of this **primary photoevent** is the excitation of an electron within the pigment.
2. The excited electron is shuttled along a series of electron-carrier molecules embedded within the photosynthetic membrane to a transmembrane proton-pumping channel, where the arrival of the electron induces the transport of a proton inward across the membrane. The electron is passed on to an acceptor.
3. The later exit of protons drives the chemiosmotic synthesis of ATP, just as it does in aerobic respiration.

Fixing Carbon

The chemical events that use the ATP generated by the light reactions take place readily in the dark. However, building an organism requires not only ATP but also a way of making organic molecules from carbon dioxide (CO_2). Organic molecules are "reduced" compared with CO_2 since they contain many C—H bonds (electrons and the associated hydrogens). To build an organism a source of hydrogens and the electrons to bind them to carbon are needed. Photosynthesis uses the energy of light to extract hydrogen atoms **(reducing power)** from water in a way that we shall explain later. In the dark reactions overall, atmospheric CO_2 is incorporated into carbon-containing molecules through a process called **carbon fixation.**

Replenishing the Pigment

The third kind of chemical event that occurs during continuous photosynthesis involves the electron that was stripped from the chlorophyll at the beginning of the light reactions. This electron must be returned to the pigment or another source of electrons must be used to replenish the supply of electrons in the pigment. Otherwise continual electron removal would cause the pigment to become deficient in electrons (bleached), and the pigment could then no longer trap photon energy by electron excitation. As we shall see, various organisms have evolved different approaches to solving this problem during the course of their evolutionary history.

Photosynthesis involves three processes: the use of light-ejected electrons to drive the chemiosmotic synthesis of ATP; the use of the ATP to fix carbon; and the replenishment of the photosynthetic pigment.

The overall process of photosynthesis may be summarized by a simple oxidation-reduction equation:

$$6CO_2 + 12H_2\star O \xrightarrow{\text{light}} C_6H_{12}O_6 + 6H_2O + 6\star O_2$$

ATMOSPHERIC CARBON DIOXIDE   WATER VAPOR   SUGAR   WATER   OXYGEN GAS FROM THE ORIGINAL WATER MOLECULES

Although H_2O appears on both sides of the equation, they are *not* the same water molecules. This can be demonstrated by carrying out photosynthesis with water vapor in which the oxygen atom is a heavy isotope. The majority of the heavy oxygen atoms (indicated in the equation by the symbol $\star O$) end up in oxygen gas and not in water.

HOW LIGHT DRIVES CHEMISTRY: THE LIGHT REACTIONS

Photosynthesis in plants, algae, and those bacteria in which it occurs is the result of a long evolutionary process. As this evolution progressed, new reactions were added to older ones, thus making the overall series of reactions more complex. Much of the evolution has centered on the light reactions, which apparently have changed considerably since they first evolved.

Evolution of the Photocenter

In all but the most primitive bacteria, light is captured by a network of chlorophyll pigments working together. Each chlorophyll molecule within the network, which is called a **photocenter,** is capable of capturing photons efficiently. Pigment molecules are held on a protein matrix within the photocenter, and their arrangement permits the channeling of excitation energy from anywhere in the array to a central point. The assembly of chlorophyll molecules thus acts collectively as a sensitive "antenna" to capture and focus photon energy. Its mode of operation is similar to the way a magnifying glass focusing light can generate enough heat energy at the point of focus to burn paper. Similarly, the photocenter channels the excitation energy gathered by any one of the pigment molecules to one called P_{700} (Figure 9-6), which is associated with a kind of membrane-bound protein called **ferredoxin.** The photocenter thus funnels to the ferredoxin many more electrons than would otherwise be possible.

When light of the proper wavelength strikes any pigment molecule of the photocenter, the resulting excitation passes from one chlorophyll molecule to another. The excited electron does not transfer physically from one pigment molecule to the next. Instead, the pigment passes the energy along to an adjacent molecule of the photocenter, after which the pigment's electron returns to the low-energy level it had before the photon was absorbed. A crude analogy to this form of energy transfer exists

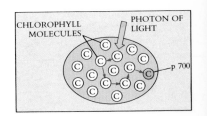

FIGURE 9-6

When light of the proper wavelength strikes any chlorophyll molecule within a photocenter, the light is absorbed by that pigment molecule. The excitation energy is then transferred from one molecule to another within the cluster of chlorophyll molecules until it encounters the P_{700} molecule of chlorophyll. P_{700} channels the energy out of the photocenter to the energy-harvesting machinery.

in the initial "break" in a game of pool. If the cue ball squarely hits the point of the triangular array of 15 pool balls, the two balls at the far corners of the triangle fly off and none of the central balls move at all. The energy is transferred through the central balls to the most distant ones. The protein matrix of the photocenters in which the molecules of chlorophyll are embedded serves as a sort of scaffold, holding individual pigment molecules in orientations that are optimal for energy transfer. In this way the process channels excitation energy to the membrane-bound ferredoxin in the form of electrons.

> **A photocenter, which is an array of pigment molecules, acts as a light antenna, directing photon energy captured by any of its members toward a single pigment molecule and thus amplifying the light-gathering powers of the individual pigment molecules.**

Where Does the Electron Go?

Photocenters probably evolved more than 3 billion years ago in bacteria similar to the group called green sulfur bacteria that exists today. In these bacteria the absorption of a photon of light by the photocenter results in the transmission of an electron from the pigment P_{700} to ferredoxin. As in many redox processes, the electron is accompanied by a proton, traveling as a hydrogen atom. Where does the proton come from? It is extracted by green sulfur bacteria from hydrogen sulfide (H_2S) through a process that produces elemental sulfur as a residual by-product.

The ejection of an electron from P_{700} and its donation to ferredoxin leaves P_{700} short one electron. Before the photocenter of the green sulfur bacteria can function again, the electron must be returned. These bacteria channel the electron back to the pigment through an electron-transport system (see Chapter 8) in which the electron's passage drives a proton pump and thus promotes the chemiosmotic synthesis of ATP. Therefore the path of the electron originally extracted from P_{700} is a circle (chemists call the process **cyclic photophosphorylation**) (Figure 9-7). However, the process is not a true circle. The electron that left P_{700} was a high-energy-level electron, boosted to its high-energy level by the absorption of a photon of energy; but the electron that returns has only as much energy left as it had before the photon absorption. The difference in the energy of that electron is the photosynthetic payoff, that is, the energy that is used to drive the proton pump.

> **In bacteria the electron ejected from the pigment by light travels a circular path in which light-excited electrons pass a proton pump and then return to the photocenter where they originated.**

The light reactions of plant photosynthetic systems evolved by adding to the simple cyclic photophosphorylation process of the green sulfur bacteria. Just as glycolysis is retained as a fundamental component of the respiratory metabolism of all organisms, so cyclic photophosphorylation remains a fundamental component of photosynthesis in the chloroplasts of all plants, algae, and most photosynthetic bacteria.

LIGHT REACTIONS OF PLANTS

For more than 1 billion years, cyclic photophosphorylation was the only form of photosynthetic light reaction that organisms used. However, it has a fundamental limitation: it is geared only toward energy production and not toward biosynthesis. To understand this point, consider for a moment that the ultimate point of photosynthesis is *not* to generate ATP but rather to fix carbon—to incorporate atmospheric carbon dioxide into new carbon compounds. Because the molecules (sugars) produced during carbon fixation are more reduced (have more hydrogen atoms) than their precursor (CO_2), a source of hydrogens must be provided. Cyclic photophosphorylation does not provide hydrogen atoms, and thus bacteria that use this process must scavenge hydrogens from other sources—a very inefficient undertaking.

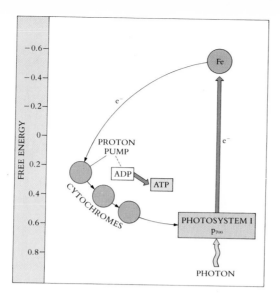

FIGURE 9-7

The path of an electron in bacterial forms of photosynthesis. Photosynthesis probably first evolved in organisms similar to the living green sulfur bacteria, in which it utilizes a single photocenter (called photosystem I in this figure). When an electron was ejected from this photocenter by a photon of light, it was passed to ferredoxin (Fe); from there it passed to a series of three cytochromes embedded within the photosynthetic membrane. The electron's passage serves to pump protons out across the membrane, creating a scarcity of protons within. From the proton pump, the electron passes back to the photocenter (photosystem I). Its passage is thus a circle, the electron returning to the photocenter from which it was initially ejected.

The Advent of Photosystem II

At some point after the appearance of the green sulfur bacteria, other kinds of bacteria evolved an improved version of the photocenter that solved the reducing power problem in a neat and simple way. These bacteria grafted a more powerful second photosystem onto the original one, using a new form of chlorophyll called chlorophyll *a*. This great evolutionary advance took place when the cyanobacteria originated, no less than 2.8 billion years ago.

In this second photosystem called **photosystem II,** molecules of chlorophyll *a* are arranged with a different geometry in the photocenter so that more of the shorter-wavelength photons of higher energy are absorbed than in the more ancient bacterial photosystem (called **photosystem I** in plants). As in the bacterial photosystem, energy is transmitted from one pigment molecule to another within the photocenter until it encounters a particular pigment molecule that is positioned near a strong membrane-bound electron acceptor. In photosystem II the absorption peak of this pigment molecule is 680 nanometers; therefore the molecule is called P_{680}.

How the Two Photosystems Work Together

Plants and algae, as contrasted with most photosynthetic bacteria, use both photosystems—a two-stage photocenter (Figure 9-8). The new photosystem acts first. When a photon jolts an electron from photosystem II, the excited electron is donated to an electron transport chain, which passes it along to photosystem I. In its journey to photosystem I, each electron drives a proton pump and thus generates an ATP molecule chemiosmotically.

When the electron reaches photosystem I, it has already expended its excitation energy in driving the proton pump and thus contains only the same amount of energy as the other electrons of this photosystem. However, its arrival does give the photosystem an electron that it can afford to lose. Photosystem I now absorbs a photon, boosting one of its pigment electrons to a high-energy level. The electron is then channeled to ferredoxin and is used to generate reducing power. In plants and algae ferre-

FIGURE 9-8

The path of an electron in a chloroplast. Plants and algae that carry out oxygen-forming photosynthesis employ two photocenters. First, a photon of light ejects an electron from photosystem II. This ejected electron is replaced with one obtained from water by a strongly electron-seeking protein called Z. In the process, a proton is pumped out across the membrane, contributing chemiosmotically to the production of a molecule of ATP. The ejected electron passes along the same chain of cytochromes employed in early forms of photosynthesis. Just as then, passage of the electron drives a proton pump, causing protons to be passed out across the membrane. From the proton pump, the electron passes to photosystem I. When this photosystem absorbs a photon of light, it ejects an electron to ferredoxin (Fe), just as it does in early forms of photosynthesis, but this time the electron does not follow a circular path and return to photosystem I. Instead, the electron passes to a soluble form of ferredoxin, which utilizes it to drive the formation of the electron carrier NADPH. For this reason, chemists call this form of photosynthesis **noncyclic photophosphorylation.**

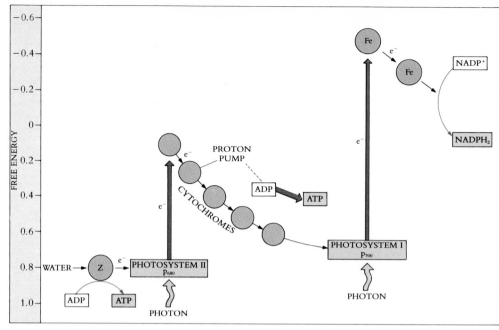

doxin contributes two electrons to reduce **nicotine adenine dinucleotide phosphate (NADP⁺),** generating NADPH. By using this molecule instead of the NAD used in oxidative respiration, plants and algae keep the flow of electrons in the two processes separate (Figure 9-9).

> **Plants and algae use a two-stage photocenter. First, a photon is absorbed by photosystem II, which passes an electron to photosystem I. During this process the electron uses its photon-contributed energy to drive a proton pump and thus generate a molecule of ATP. Then another photon is absorbed, this time by photosystem I, which also passes on a photon-energized electron. This second electron is channeled away to provide reducing power.**

Thus the energy produced in the first photoevent is spent in ATP synthesis; the energy of the second photoevent creates reducing power. These two processes together comprise the light reactions of eukaryotic photocenters.

The Formation of Oxygen Gas

By now you may wonder about the fate of P_{680}, the pigment of photosystem II that started the photosynthetic process by donating an electron. How does it make up for this loss if the electron is not returned but is instead expended to synthesize NADPH? As you might expect, an electron is obtained from another source. The loss of the excited electron from photosystem II converts P_{680} into a powerful oxidant (electron-seeker), which obtains the required electron from a protein called Z. The removal of this electron from Z renders it a strong electron-acceptor in turn, and Z obtains electrons from water to funnel to P_{680}. Z catalyzes a complex series of reactions in which water is split into electrons (passed to P_{680}), H^+ ions, and OH^- radicals. The OH^- radicals are collected and reassembled as water and oxygen gas. The hydrogen ions (protons) are exported across the membrane, thus augmenting the gradient in proton concentration that was established during the passage of electrons to photosystem I.

> **The electrons and associated protons that oxygen-forming photosynthesis employs to form reduced organic molecules are obtained from water. The leftover oxygen atoms of the water molecules combine to form oxygen gas.**

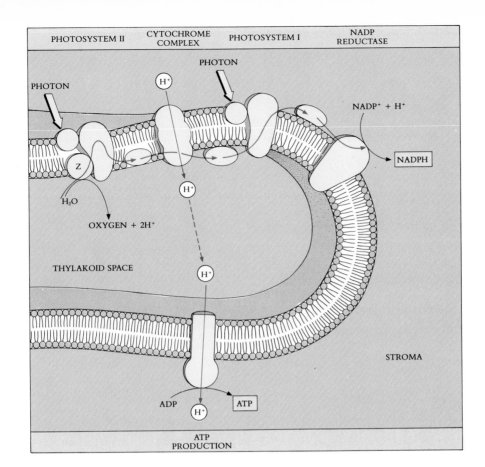

FIGURE 9-9

The photosynthetic electron transport system. When a photon of light strikes a pigment molecule in photosystem II, the energy is channeled to P_{680}, which reacts by donating an electron. This electron passes along a chain of membrane-bound cytochrome electron carriers to a proton pump, using the energy supplied by the electron to transport a proton across the membrane into the thylakoid. The resulting proton gradient drives the chemiosmotic synthesis of ATP. The electron then passes to photosystem I. When photosystem I absorbs a photon of light, P_{700} passes a high-energy electron to membrane-bound ferredoxin [Fe], which passes it to free ferredoxin. Free ferredoxin donates this electron to drive nitrogen fixation and NADPH generation. The loss of an electron at the start of this sequence by P_{680} causes it to obtain one from protein Z, which makes up the loss by capturing an electron from water. Water is split into a proton $[H^+]$, an electron $[e^-]$ that is passed to P_{680} by Z, and an OH^- radical; OH^- radicals are then reassembled into water and oxygen gas.

The use of a two-stage photocenter containing both photosystem I and photosystem II thus solves in a simple way the evolutionary problem of how to obtain reducing power for biosynthetic reactions. Even though the cyclic photophosphorylation of green sulfur bacteria provides ATP by cyclic photophosphorylation, it does not provide a ready means of generating NADPH. Therefore organisms that use this form of photosynthesis must make NADPH in a roundabout way and expend a lot of ATP to do so.

Comparing Plant and Bacterial Light Reactions

It is useful to compare the two-stage P_{680}/P_{700} photocenter with the P_{700} photocenter from which it evolved. The removal of an electron from P_{700} yields enough energy to extract hydrogen from H_2S (78 kilocalories) but not from H_2O (118 kilocalories). By contrast, the removal of an electron from P_{680} yields considerably more energy, and that energy is adequate to split water molecules, producing gaseous oxygen as a by-product. In cyanobacteria, algae, and plants, all of which use this double photocenter, there is no cyclic flow of electrons. Instead, electrons and associated hydrogen atoms are extracted continually from water and are eventually used to reduce $NADP^+$ to NADPH. As a result of stripping hydrogens from water, oxygen gas is continuously generated as a product of the reaction. This photosynthetic process generates all of the oxygen in the air that we breathe (Figure 9-10).

Every oxygen molecule in the air you breathe was once split from a water molecule by an organism carrying out oxygen-forming photosynthesis.

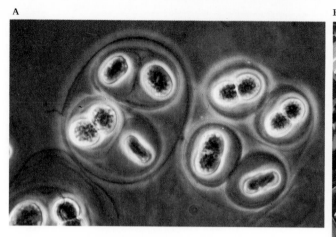

FIGURE 9-10

All of the oxygen we breathe, indeed all the oxygen in the earth's atmosphere, has been generated by photosynthesis. In the roughly 3 billion years between the advent of the cyanobacteria **(A)** and the cultivation of maize **(B),** enough oxygen has accumulated to account for 21% of the atmosphere.

Accessory Pigments

As we saw earlier in this chapter, a simple chemical modification converts chlorophyll *a* into chlorophyll *b*. Chlorophyll *b* has an absorption spectrum shifted toward the green wavelengths. It acts as an **accessory pigment** within the photocenter of plants, one that is able to absorb photons that chlorophyll *a* cannot. By doing this, chlorophyll *b* greatly increases the percentage of the photons of sunlight that a plant can harvest.

HOW THE PRODUCTS OF PHOTOSYNTHESIS ARE USED TO BUILD ORGANIC MOLECULES FROM CO_2

The preceding section was concerned with the light reactions of photosynthesis. These reactions use light energy to produce metabolic energy in the form of ATP and to produce reducing power in the form of NADPH. But this is only half the story. Photosynthetic organisms employ the ATP and NADPH produced by the light reactions to build organic molecules from atmospheric carbon dioxide. This later phase of photosynthesis, which comprises the so-called dark reactions, is carried out by a series of enzymes that are present in the chloroplasts of plants and algae and in the cells of many photosynthetic bacteria.

Carbon fixation depends on the presence of a molecule to which CO_2 can be attached. The cell produces such a molecule by reassembling the bonds of two of the intermediates of glycolysis—fructose 6-phosphate (F6P) and glyceraldehyde 3-phosphate (G3P)—to form a five-carbon sugar, **ribulose 1,5-bisphosphate (RuBP).**

The dark reactions of photosynthesis form a cycle, a circle of enzyme-catalyzed steps, just as the citric acid cycle is a circular biochemical pathway. In this cycle a carbon atom from atmospheric CO_2 is added to ribulose 1,5-bisphosphate (Figure 9-11), and the products run backwards through the glycolytic sequence to form fructose 6-phosphate molecules, some of which are used to reconstitute RuBP. The remainder enters the cell's metabolism as newly fixed carbon in glucose. This cycle of reactions is called the **Calvin cycle,** after its discoverer Melvin Calvin of the University of California, Berkeley. The full Calvin cycle is diagrammed in Figure 9-12.

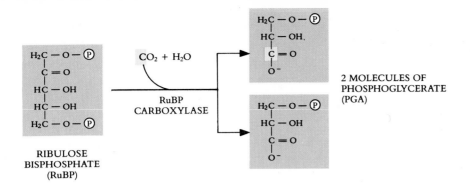

RIBULOSE
BISPHOSPHATE
(RuBP)

RuBP
CARBOXYLASE

$CO_2 + H_2O$

2 MOLECULES OF
PHOSPHOGLYCERATE
(PGA)

FIGURE 9-11

Melvin Calvin and his co-workers at the University of California–Berkeley worked out the first step of what later became known as the Calvin cycle by exposing photosynthesizing algae to radioactive carbon dioxide ($^{14}CO_2$). Following the fate of the radioactive carbon atom (shown in color in these diagrams), they found that it is first bound to a molecule of ribulose bisphospate (RuBP), which splits immediately, forming two molecules of phosphoglycerate (PGA), one of which contains the radioactive carbon atom.

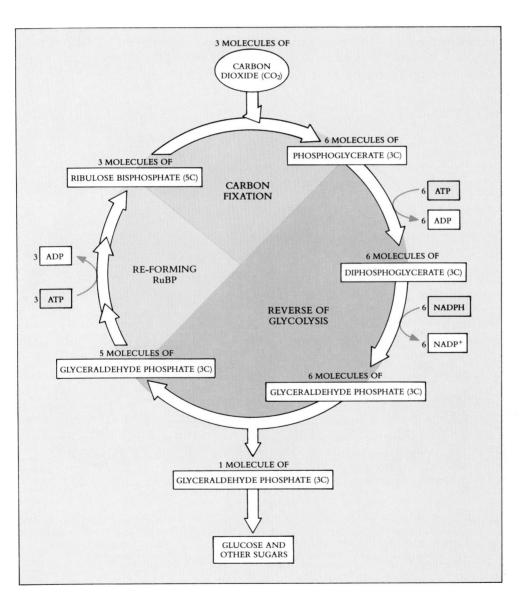

FIGURE 9-12

The Calvin cycle. For every three molecules of CO_2 that enter the cycle, one molecule of the 3-carbon compound glyceraldehyde phosphate is produced. Notice that the process requires energy, supplied as ATP and NADPH. This energy is generated by the light reactions, which generate ATP and NADPH.

THE CHLOROPLAST AS A PHOTOSYNTHETIC MACHINE

In eukaryotes all photosynthesis takes place in chloroplasts (Figure 9-13). As described in Chapter 5, the internal membranes of chloroplasts are organized into flattened sacs called thylakoids, which are stacked on top of one another in arrangements called grana (Figure 9-14). The photosynthetic pigments are bound to proteins embedded within the membranes of the thylakoids.

Each thylakoid is a closed compartment into which protons can be pumped. The thylakoid membrane is impermeable to protons and to most molecules; therefore transit across it occurs almost exclusively through transmembrane channels. The diffusion of protons from the thylakoid interior takes place at distinctive ATP-synthesizing proton channels. This is similar to the process in mitochondria, except that in mitochondria the diffusion gradient is in the opposite direction—protons diffuse *into* rather than out of a mitochondrion.) These channels protrude as knobs on the external surface of the thylakoid membrane, from which ATP is released into the surrounding fluid. This fluid matrix inside the chloroplast, within which the thylakoids are embedded, is called the **stroma.** It contains the Calvin-cycle enzymes that catalyze the dark reactions of carbon fixation, using the ATP and NADPH that the photosynthetic activity of the thylakoids produces (Figure 9-15). Within the chloroplast the thylakoid membrane pumps protons from the stroma into the thylakoid compartment. As hydrogen ions pass back through the membrane by way of enzymes (ATPases), phosphorylation of ADP occurs on the stroma side of the membrane (Figure 9-16).

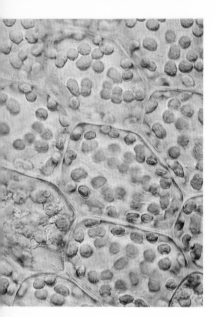

FIGURE 9-13

These moss cells are densely packed with bright-green chloroplasts.

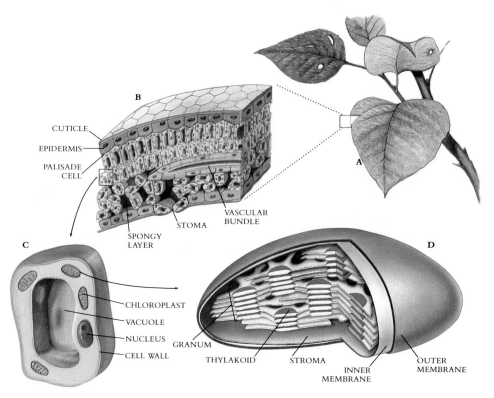

FIGURE 9-14

Journey into a leaf. **(A)** The leaf shown is from a plant that carries out the Calvin cycle. **(B)** In cross section, the leaf possesses a thick palisade layer whose cells **(C)** are rich in chloroplasts **(D).**

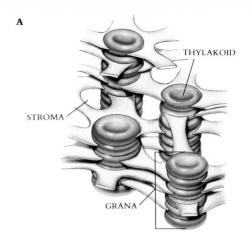

A

THYLAKOID

STROMA

GRANA

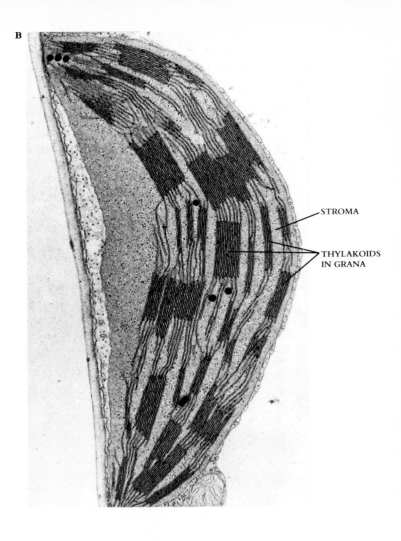

B

STROMA

THYLAKOIDS IN GRANA

FIGURE 9-15

Internal structure of a chloroplast.
A The basic photosynthetic unit is the thylakoid, a flattened, membranous sac. These sacs are stacked into columns called grana (singular, granum).
B In the micrograph you can see that a single chloroplast may contain many such grana. The interior matrix of a chloroplast, which is fluid, is called the stroma. The stroma contains the Calvin cycle enzymes, which catalyze carbon fixation by using ATP and NADPH generated on the exterior surface of the thylakoids by photosynthetic electron transport and photophosphorylation.

FIGURE 9-16

Chemiosmosis in a chloroplast. In this process, the thylakoid membrane pumps protons from the stroma into the thylakoid space. The phosphorylation of ADP occurs on the stroma side of the membrane as hydrogen ions pass across it through molecules of the enzyme ATPase, which are embedded in the membrane.

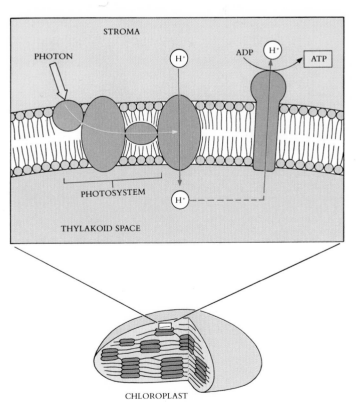

STROMA

PHOTON

H⁺

ADP

H⁺

ATP

PHOTOSYSTEM

H⁺

H⁺

THYLAKOID SPACE

CHLOROPLAST

A LOOK BACK

A cell's metabolism betrays its evolutionary past perhaps more than any other phase of its life. This is particularly true of cells that perform photosynthesis. The two-stage photocenter of modern plants (Figure 9-17) has as its second stage a photosystem that first evolved millions of years earlier in anaerobic bacteria and used hydrogen sulfide rather than water as a source of reducing hydrogen. The Calvin cycle uses part of the ancient glycolytic pathway, run in reverse, to produce glucose. The principal chlorophyll pigments of plants are but simple modifications of bacterial chlorophylls employed by anaerobic photosynthetic bacteria for billions of years before there were any plants. Thus by looking at a plant today, we see its history as well.

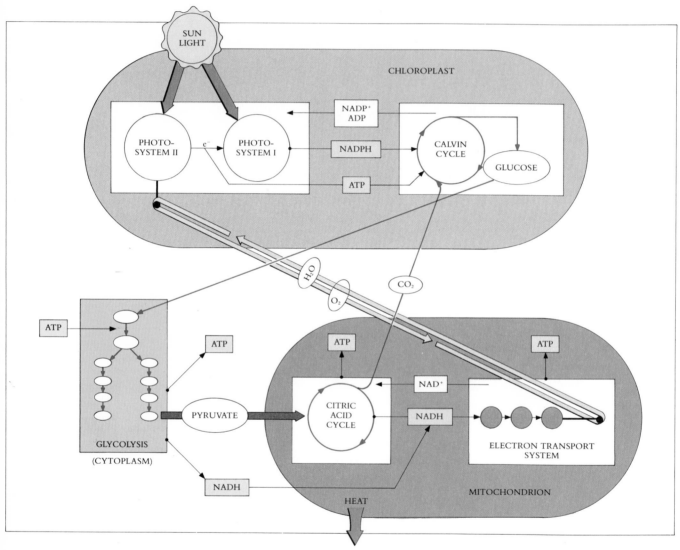

FIGURE 9-17

The metabolic machine. Within the chloroplast, sunlight drives the production of glucose, using up CO_2 and H_2O while generating O_2. This O_2 is converted back to water in the mitochondrion by accepting electrons harvested from glucose molecules after the electrons have been used by the electron transport system to drive the synthesis of ATP. Water and oxygen gas thus cycle between chloroplasts and mitochondria within a plant cell, as do glucose and CO_2. Cells without chloroplasts, such as your own cells, require an outside source of glucose and oxygen, and generate CO_2 and water.

In Chapters 29 through 31, we shall examine plants in detail; photosynthesis is only one aspect of plant biology, although an important one. We have treated photosynthesis here because photosynthesis evolved long before plants did and because all organisms depend directly or indirectly on photosynthesis for the energy that powers their lives.

SUMMARY

1. Light consists of energy packets called photons; the shorter the wavelength of light, the more energy in its photons.

2. When light strikes a pigment, photons are absorbed by boosting an electron to a higher energy level, thus causing an excited state. Most biological photopigments are carotenoids, such as the retinal of human eyes, or chlorophylls such as those that make grass green.

3. Photosynthesis seems to have evolved in organisms similar to the green sulfur bacteria, which use a network of chlorophyll molecules (a photocenter) to channel photon excitation energy to one pigment molecule, referred to as P_{700}. P_{700} then donates an electron to an electron transport chain, which drives a proton pump and returns the electron to P_{700} in a process called cyclic photophosphorylation.

4. The descendants of these bacteria developed a two-stage photocenter in which a new photosystem called photosystem II was grafted onto the old one. The new photosystem employed a new pigment, chlorophyll a, and was able to generate enough energy to use H_2O rather than H_2S as a hydrogen source.

5. In organisms with the two-stage photocenter, light is first absorbed by photosystem II, which jolts an electron out of one chlorophyll molecule, P_{680}. This has two effects: (1) the absence of the high-energy electron causes photosystem II to seek another electron actively, which results in the eventual splitting of water to obtain it with O_2 as the by-product; and (2) the high-energy electron is passed to photosystem I, driving a proton pump in the process and thus bringing about the chemiosmotic synthesis of ATP.

6. When the spent electron arrives at P_{700} from P_{680}, the P_{700} pigment absorbs a photon of light, boosting one of its electrons out. This electron is directed to ferredoxin, where it is used to drive the synthesis of NADPH from $NADP^+$, thus providing reducing power.

7. The ATP and reducing power produced by the light reactions are used to fix carbon in a series of dark reactions called the Calvin cycle. In this process ribulose 1,5-bisphosphate is carboxylated and the products run backwards through a series of reactions also found in the glycolytic sequence to form fructose 6-phosphate molecules, some of which are used to reconstitute RuBP. The remainder enters the cell's metabolism as newly fixed carbon in glucose.

REVIEW

1. Light energy is "captured" and used to boost an electron to a higher energy level in molecules called _____.

2. What is the point of photosynthesis? Is it to generate ATP or to fix carbon?

3. _____ is used as an electron carrier in the anabolic process of photosynthesis instead of _____, which is used in catabolic oxidative respiration.

4. All of the oxygen that you breathe has been produced by the splitting of water during _____.

5. The basic photosynthetic unit of a chloroplast is the _____, a flattened membranous sac that occurs in stacked columns called grana.

SELF-QUIZ

1. Carbon fixation requires the expenditure of ATP molecules. How is this ATP generated?
   (a) by chemiosmotic synthesis during the light reactions
   (b) by replenishment of the photosynthetic pigment
   (c) by formation of glucose during the dark reactions
   (d) by breaking the covalent bonds in carbon dioxide
   (e) none of the above

2. When P_{700} receives photon-induced excitation energy, the excitation energy is channeled to membrane-bound ferredoxin. What is the form of this excitation energy?
   (a) light energy   (d) electrons
   (b) protons   (e) sex
   (c) neutrons

3. Plants and algae that carry out oxygen-producing photosynthesis employ both photosystem I and photosystem II. This means that _____ and _____ are generated, both of which are required to form organic molecules from atmospheric carbon dioxide.
   (a) water   (d) NADH
   (b) ATP   (e) NADPH
   (c) ADP

4. The rate of photorespiration in most plants increases at higher temperatures. Some plants have evolved a somewhat round-about system to deal with this problem. This series of reactions is called
   (a) the C_3 pathway
   (b) the C_4 pathway
   (c) the Calvin cycle
   (d) photosystem II
   (e) cyclic photophosphorylation

5. A principal difference between a carotenoid and a chlorophyll is that
   (a) chlorophylls use a metal atom to snare electrons, whereas carotenoids do not
   (b) chlorophylls absorb light over a much broader range than do carotenoids
   (c) chlorophylls absorb light at much lower efficiency
   (d) carotenoids contain carbon ring molecules with alternating single and double carbon bonds, whereas chlorophylls do not
   (e) all of the above

THOUGHT QUESTIONS

1. What is the advantage of having many pigment molecules in each photocenter for every P_{700}? Why not couple *every* pigment molecule directly to an electron acceptor?

2. Why are plants that consume 30 ATP molecules to produce one molecule of glucose (rather than the usual 18 molecules of ATP per glucose molecule) favored in hot climates but not in cold climates? What role does temperature play?

FOR FURTHER READING

BJORKMAN, O., and J. BERRY: "High Efficiency Photosynthesis," *Scientific American,* October 1973, pages 80-93. A description of C_4 photosynthesis and other strategies carried out by plants that live in Death Valley, where the problem posed by photorespiration is acute.

GOVINDJEE, and R. GOVINDJEE: "The Absorption of Light in Photosynthesis," *Scientific American,* December 1974, pages 69-87. A good account of how a photocenter works.

MILLER, K.: "The Photosynthetic Membrane," *Scientific American,* October 1979, pages 100-113. Describes the way in which the key components of the photosynthetic apparatus are embedded in membranes.

GENETICS

Part Four

FIGURE 10-1 Just hatching from its egg, peering out at the world it is about to enter, this baby anaconda is the product of many cell divisions since it was first initiated by the fusion of egg and sperm. Many more cell divisions lie ahead as it grows into a 6-foot adult snake.

HOW CELLS REPRODUCE Chapter 10

Overview

Your body consists of some hundred trillion cells, all derived from a single cell at the start of your life as a fertilized egg. Many millions of successful cell divisions occurred while your body was reaching its present form. In each of these, the genetic material of the dividing cells was equally partitioned between their derivatives. To accomplish this, growing eukaryotic cells attach microtubules to each replicated chromosome and pull sister chromosomes to opposite poles of the cell, in a process called mitosis. In the production of their gametes, eggs and sperm, eukaryotic organisms employ another distinctive process of cell division, called meiosis.

For Review

Here are some important terms and concepts that you will encounter in this chapter. If you are not familiar with them, you should review them before proceeding.

Bacterial cell division (Chapter 5)

Chromosomes (Chapter 5)

Nuclear envelope (Chapter 5)

Microtubules (Chapter 5)

Microtubule-organizing center (Chapter 5)

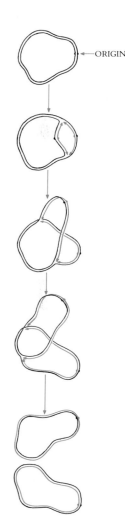

All living organisms grow and reproduce. Bacteria too small to see, alligators, the weeds growing on your lawn—from the smallest of creatures to the largest, all organisms produce offspring like themselves and pass on to them the hereditary information that makes them what they are (Figure 10-1). In this chapter, we begin our consideration of heredity with an examination of how cells reproduce themselves. The ways in which cell reproduction is achieved, and their biological consequences, have changed significantly during the evolution of life on earth.

CELL DIVISION IN BACTERIA

Among the bacteria, the process of cell division is simple. The genetic information, or **genome,** exists in bacteria as a single, circular, deoxyribonucleic acid (DNA) molecule, attached at one point to the interior surface of the cell membrane. This genome is replicated early in the life of the cell. At a special site on the chromosome called the **replication origin,** a battery of more than 22 different enzymes goes to work and starts to make a complete copy of the DNA molecule (Figure 10-2). When these enzymes have proceeded all the way around the circle of the DNA molecule, the cell then possesses two copies of the genome, attached side-by-side to the interior cell membrane.

FIGURE 10-2

The circular DNA molecule that constitutes the genome of a bacterium initiates replication at a single site, moving out in both directions. When the two moving replication points meet on the far side of the molecule, its replication has been completed.

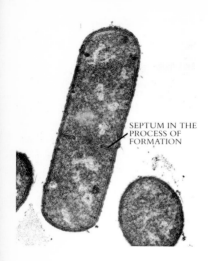

SEPTUM IN THE PROCESS OF FORMATION

FIGURE 10-3

Bacteria divide by a process of simple cell fission. Note here the septum forming between the two daughter cells.

The growth of a bacterial cell to an appropriate size induces the onset of cell division. First, new plasma membrane and cell wall materials are laid down in the zone between the attachment sites of the two "daughter" DNA genomes. This begins the process of binary fission. **Binary fission** is the division of a cell into two equal or nearly equal halves. As new material is added in the zone between the attachment sites, the growing plasma membrane pushes inward (invaginates), and the cell is progressively constricted in two (Figure 10-3). Initiating the constriction at a position between the membrane attachment sites of the two daughter DNA genomes provides a simple and effective mechanism for ensuring that each of the two new cells will contain one of the two identical genomes. Eventually the invaginating circle of membrane reaches all the way into the cell center, pinching the cell in two. A new cell wall forms around the new membrane, and what was originally one cell is now two.

Bacteria divide by binary fission, a process in which a cell pinches in two. The point where the constriction begins is located between where the two replicas of the chromosome are bound to the cell membrane, ensuring that one copy will end up in each daughter cell.

CELL DIVISION AMONG EUKARYOTES

The evolution of the eukaryotes introduced several additional factors into the process of cell division. Eukaryotic cells are much larger than bacteria, and they contain genomes with much larger quantities of DNA (Figure 10-4). This DNA, however, is located in several individual chromosomes, rather than in a single, circular molecule. In these chromosomes, the DNA is associated with proteins and wound into tightly condensed coils (Figure 10-5). The eukaryotic chromosome is a structure with complex organization, which contrasts strongly with the single, circular DNA molecule that plays the same role in bacteria.

FIGURE 10-4

A human chromosome contains an enormous amount of DNA. The dark element at the bottom of the photograph is the protein matrix of a single chromosome. All of the surrounding material is the DNA of that chromosome.

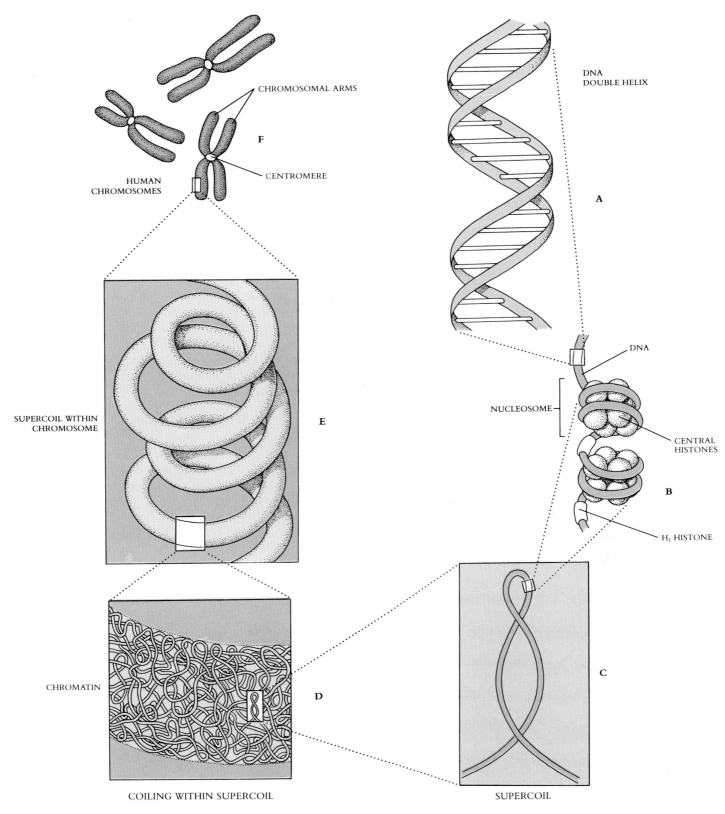

FIGURE 10-5

Levels of chromosomal organization. The DNA duplex is wound twice around aggregates of histones to form nucleosomes, and the string of nucleosomes is further coiled in a series of condensation stages. During cell division, chromosomes are fully condensed, as pictured here; at other times, much of the coiling is relaxed and chromosomes are not visible as discrete entities under a light microscope.

The Structure of Eukaryotic Chromosomes

In the century since their discovery, we have learned a great deal about chromosomes, their structure, and how they function. Eukaryotic chromosomes are composed of a complex of DNA, ribonucleic acid (RNA), and protein. This complex is called **chromatin.** Most eukaryotic chromosomes are about 60% protein and 40% DNA. A significant amount of RNA is also associated with chromosomes, because they are the sites of RNA synthesis. The DNA of a chromosome exists as one very long, double-stranded fiber, a **duplex,** which extends unbroken through the entire length of the chromosome. A typical human chromosome contains about half a billion (5×10^8) nucleotides in its DNA fiber. The amount of information one chromosome contains, therefore, would fill about 2000 printed books of 1000 pages each, assuming that the nucleotides were "words" and that each page had about 500 of them on it. If the strand of DNA from a single chromosome were laid out in a straight line, it would be about 5 centimeters (2 inches) long. This is much too long to fit into a cell. In the cell, however, the DNA is coiled, thus fitting into a much smaller space than would be possible if it were not.

How is the coiling of this long DNA fiber achieved? If we gently disrupt a eukaryotic nucleus and examine the DNA with an electron microscope, we find that it resembles a string of beads. Every 200 nucleotides, the DNA duplex is coiled about a complex of **histones,** which are small, very basic polypeptides, rich in the amino acids arginine and lysine. Eight of these histones form the core of an assembly called a **nucleosome** (Figure 10-6). Because so many of their amino acids are basic, histones are very positively charged. The DNA duplex, which is negatively charged, is strongly attracted to the histones and wraps tightly around the histone core of each nucleosome. The core thus acts as a "form" that promotes and guides the coiling of the DNA. Further coiling of the DNA occurs when the string of nucleosomes wraps up into higher-order coils called **supercoils** (see Figure 10-5).

Condensed portions of the chromatin are called **heterochromatin.** Some remain condensed permanently, so that their genes are never expressed. The remainder of the chromosome, called **euchromatin,** is not condensed except during cell division, when the movement of the chromosomes is made easier by the compact packaging that occurs at that stage. At all other times, the euchromatin is present in an open configuration, and its genes are active.

Chromosomes may differ widely from one another in appearance. They vary in such features as the location of a constricted segment called the **kinetochore**, which binds sister chromosomes together (see Figure 10-5, *F*); the relative length of the two arms (regions on either side of the kinetochore); size; staining properties; and the position of constricted regions along the arms. The particular array of chromosomes that an individual possesses, called its **karyotype** (Figure 10-7), may differ greatly among species, or sometimes even among individuals in a species. Karyotypes of individuals are often examined to detect genetic abnormalities, such as those arising from extra or lost chromosomes. The human birth defect known as Down syndrome, for example, is associated with a duplication of a particular chromosome, which can be recognized in photographs of the chromosome complement. In this way, the presence of the defect can be detected in samples taken from embryonic or other cells.

FIGURE 10-6

Nucleosomes are regions in which the DNA duplex is wound around a group of eight histones. In this electron micrograph of rat liver DNA, the nucleosomes exhibit the characteristic "beads on a string" structure that results.

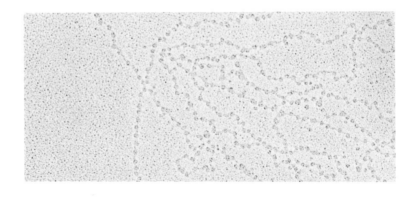

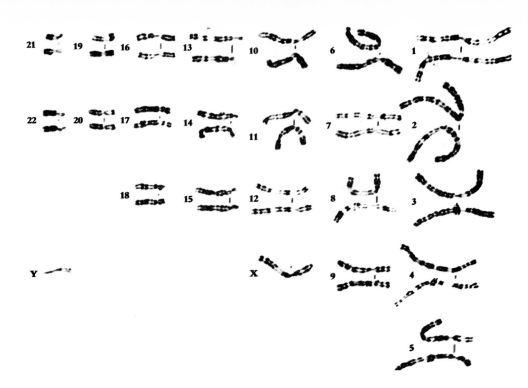

FIGURE 10-7

A human karyotype. The individual chromosomes that make up the 23 pairs can be seen to differ widely in size and in the position of the centromere, which is indicated in each case by a vertical black line between the pairs. In this preparation, they have been specially stained to indicate further differences in their composition and to distinguish them clearly from one another.

The Cell Cycle

The profound change in genome organization that occurred during the evolutionary transition from bacteria (a single circle of naked DNA) to eukaryotes (several segments of DNA packaged with protein) required radical changes in the way cells divide to partition two replicas of the genome accurately, one into each of two daughter cells. The processes that occur during the division of eukaryotic cells are conveniently diagrammed as a cell cycle:

$$G_1 \rightarrow S \rightarrow G_2 \rightarrow M \rightarrow C$$

G_1 phase	The growth phase of the cell. For many organisms, this occupies the major portion of the cell's life span.
S phase	The phase in which a replica of the genome is synthesized.
G_2 phase	The stage in which preparations are made for genomic separation. This includes the replication of mitochrondia and other organelles, chromosome condensation, and the synthesis of microtubules.
M phase	The phase in which the microtubular apparatus is assembled, binds to the chromosomes, and moves the sister chromosomes apart. This stage, called **mitosis,** is the essential step in the separation of the two daughter genomes.
C phase	The phase in which the cell itself divides, creating two daughter cells. This phase is called **cytokinesis.**

MITOSIS

Two phases of the eukaryotic cell division cycle that have received close attention from biologists are those that accomplish the physical division of the cell and its contents: M phase (mitosis) and C phase (cytokinesis). The time devoted to these two phases often represents only a small part of the cell cycle (Figure 10-8). We shall discuss them in turn.

The M phase of cell division, mitosis, has received more attention from biologists than any other aspect of the eukaryotic cell cycle. Biologists have long been fascinated by the intricate movements of the chromosomes as they separate. We shall describe the process as it occurs in animals and plants. Among fungi and some protists, different forms of mitosis occur, but the process varies little among different kinds of animals and plants.

FIGURE 10-8

The cell cycle of liver cells
grown in culture. During the S
phase, the chromosomes replicate;
they then consist of two identical
copies called sister chromatids
with a single centromere.

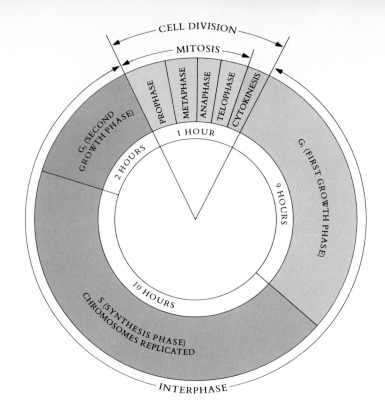

Traditionally, mitosis is subdivided into four stages (Figure 10-9): prophase, metaphase, anaphase, and telophase. Such a subdivision is convenient, but the process is actually continuous, with the stages flowing smoothly one into another.

Preparing the Scene: Interphase

Before the initiation of mitosis, events have occurred in the preceding **interphase** (that is, the G_1, S, and G_2 phases) that are essential for the successful completion of the mitotic process:

1. During S phase, each chromosome replicates to produce two daughter copies, called **sister chromatids.** The two copies remain attached to each other at a point of constriction, the centromere. The **centromere** is a specific DNA sequence of about 220 nucleotides, to which is bound a disk of protein called a **kinetochore.** The centromere occurs at a specific site on any given chromosome. At the completion of S phase, each replicated chromosome consists of two sister chromatids joined at the centromere sequences.

2. After the chromosomes are replicated in S phase, they remain fully extended and uncoiled and are not visible under a light microscope. In G_2 phase, the chromosomes begin the long process of **condensation,** coiling into more and more tightly compacted bodies.

3. During G_2 phase, the cells begin to assemble the machinery that they will later use to move the chromosomes to opposite poles of the cell. In animal cells, nuclear microtubule-organizing centers called **centrioles** replicate; plants and fungi lack visible centrioles. All cells undertake an extensive synthesis of tubulin, the protein of which microtubules are formed.

A consideration of some relationships will make the following discussion of mitosis simpler to understand. Each of the cells in your body except your **gametes** (eggs or sperm) is **diploid;** that is, each contains 46 chromosomes, consisting of **two** nearly identical copies of each of the basic set of 23 chromosomes. The basic **single** set of 23 chromosomes is called the **haploid complement** and is present in all of your gametes. The two nearly identical copies of each of the 23 different kinds of chromosomes are called homologous chromosomes or **homologues** (Greek *homologia,* agreement). In the S phase that precedes cell division, each of the two homologues replicates, producing in each case two identical copies called sister chromatids that remain joined together at the centromere. Thus, at the end of S phase, entering mitosis, a body cell contains a total of 46 replicated chromosomes, each composed of two sister chromatids joined by one centromere. How many centromeres does the cell contain? Forty-six. How many copies of the basic set of 23 chromosomes does it contain? Four, with a total of 92 chromosome copies (23 basic set × 2 homologues × 2 sister chromatids). The cell is said to contain 46 chromosomes and not 92 because, by convention, one counts the number of centromeres.

> **Interphase is that portion of the cell cycle in which the condensed chromosomes are not visible under a light microscope. It includes the G_1, S, and G_2 phases. In the G_2 phase, the cell mobilizes its resources for cell division.**

Formation of the Mitotic Apparatus: Prophase

When the chromosome condensation initiated in G_2 phase reaches the point at which individual condensed chromosomes first become visible with a light microscope, the first stage of mitosis, **prophase,** has begun. This condensation process continues throughout prophase, so chromosomes that start prophase as tiny threads may appear quite bulky before its conclusion. Ribosomal RNA (rRNA) synthesis ceases when that portion of the chromosome bearing the rRNA genes is condensed, with the result that the nucleolus, which was previously conspicuous, disappears.

> **Prophase is the stage of mitosis characterized by the condensation of the chromosomes.**

While the chromosomes are beginning to condense in prophase, another series of equally important events is also occurring: the assembly of the microtubular apparatus, which will be employed to separate the sister chromatids. Early in prophase in animal cells, the two centriole pairs start to move apart, forming between them an axis of microtubules referred to as **spindle fibers** (Figure 10-10). The centrioles continue to move apart until they reach the opposite poles of the cell, with a bridge of microtubules called the **spindle apparatus** extending between them. In plant cells, a similar bridge of microtubular spindle fibers forms between opposite poles of the cell (see Figure 10-9), although the microtubule-organizing center is not visible with a light microscope, and centrioles are absent.

During the formation of the spindle apparatus, the nuclear envelope breaks down, and its components are reabsorbed into the endoplasmic reticulum. The microtubular spindle thus extends all the way across the cell from one pole to the other. Its position determines the plane in which the cell will later divide, a plane that passes through the center of the nucleus at right angles to the spindle.

> **During prophase, the nuclear envelope is reabsorbed, and a network of microtubules called the spindle forms between opposite poles of the cell. The position of the spindle determines the plane in which the cell will divide.**

In animal mitosis, when the centrioles reach the poles of the cell, they radiate an array of microtubules outward, thus bracing the centrioles against the cell membrane. This arrangement of microtubules is called an **aster.** The function of the aster is not well known but is probably mechanical, acting to stiffen the point of microtubular attachment during the later contraction of the spindle. Plant cells, which have rigid cell walls, do not form asters.

As prophase continues, a second group of microtubules appears to grow out from the individual centromeres to the poles of the spindles. Two such microtubules extend from each chromosome, connecting the kinetochore of each of the sister chromatids to the two poles of the spindle. The two microtubules attached to a centromere both continue to grow until each has made contact with one of the poles of the cell. Because microtubules extending from each pole attach to their side of the centromere, the

A INTERPHASE

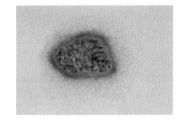

B PROPHASE

C METAPHASE

D ANAPHASE

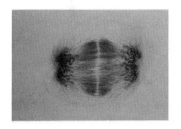

E EARLY TELOPHASE

F LATE TELOPHASE

FIGURE 10-9

The stages of mitosis in a plant cell. The chromosomes of the African blood lily, *Haemanthus katharinae,* are stained blue, and microtubules are stained red.

The stages of mitosis in an animal cell. In this series of micrographs, the chromosomes of a kangaroo have been stained blue and the spindle fibers stained brown by binding to them antibodies directed against tubulin. The mechanical role of microtubules in chromosome movement is clearly illustrated.

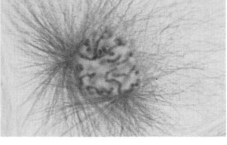

PROPHASE

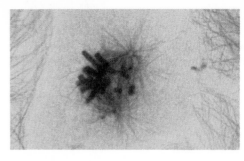

PROMETAPHASE

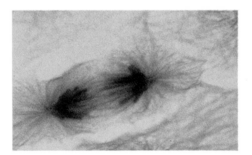

METAPHASE

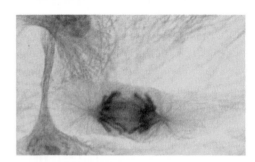

EARLY ANAPHASE

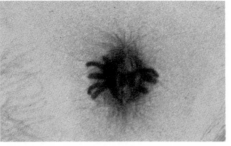

LATE ANAPHASE

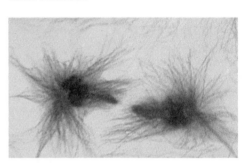

TELOPHASE

effect is to attach one sister chromatid to one pole, and the other sister chromatid to the other pole.

> **At the end of prophase, the centromere joining each pair of sister chromatids is attached by microtubules to the poles of the spindle.**

Separation of Sister Chromatids: Metaphase

The second phase of mitosis, **metaphase,** begins when the chromatid pairs align in the center of the cell. Viewed with a light microscope, the chromosomes appear to be lined up along the inner circumference of the cell in a circle perpendicular to the axis of the spindle. The three-dimensional arrangement resembles a line inscribing a great circle on the periphery of a sphere, just as the equator girdles the earth (Figure 10-11). An imaginary plane passing through this circle is called the **metaphase plate.** The metaphase plate is not a physical structure, but rather an indication about where the future axis of cell division will occur. Positioned by the microtubules attached to their kinetochores, all the chromosomes line up on the metaphase plate, with their centromeres neatly arrayed in a circle, each equidistant from the two poles of the cell.

> **Metaphase is the stage of mitosis characterized by the alignment of the chromosomes in a ring along the inner circumference of the cell. Each chromosome is drawn to that position by the microtubules extending from it to the two poles of the spindle.**

Each centromere possesses two kinetochores, one attached to the centromere region of each sister chromatid. A microtubule is attached to each kinetochore, extending to the opposite poles. This arrangement is critical to the process of mitosis. Any mistake in this positioning of the microtubules is disastrous. The attachment of the two microtubules to the same pole, for example, leads to a failure of the sister chromatids to separate, so that they end up in the same daughter cell.

At the end of metaphase, the centromeres divide. Each centromere splits in two, freeing the two sister chromatids from their attachment to one another. Centromere replication is simultaneous for all the chromosomes; the mechanism that achieves this synchrony is not known.

At the end of metaphase, the kinetochores separate, freeing the sister chromatids to be drawn in the next phase to opposite poles of the spindle by the microtubules attached to their half of the centromere.

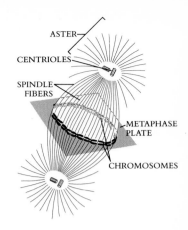

FIGURE 10-11

The metaphase plate is a great circle perpendicular to the axis of the spindle, around the edge of which the chromosomes array themselves during metaphase.

Separation of the Chromatids: Anaphase

Of all the stages of mitosis, anaphase is the most beautiful to watch and also the shortest. No longer tugged in two directions at once by opposing microtubules, like a rope in a tug-of-war, each sister chromatid now rapidly moves toward the pole to which its microtubule is attached. Two forms of movement take place simultaneously, each driven by microtubules:

1. *The poles move apart.* The microtubular spindle fibers slide past one another. Because the two members of each microtubule pair are physically anchored to opposite poles, their sliding past one another pushes the poles apart. Because the chromosomes are attached to these poles, they also move apart. In this process, the cell, if it is bounded by a flexible membrane, becomes visibly elongated. The mechanism of microtubular sliding is based on adenosine triphosphate (ATP)–driven changes in the shape of proteins bridging pairs of microtubules.

2. *The centromeres move toward the poles.* The microtubules attached to the centromeres shorten. This shortening process is not a contraction, since the microtubules do not get any thicker. Instead, tubulin subunits are continuously removed from the polar ends of the microtubules by the organizing center. As more and more subunits are removed, the progressive disassembly of the chromosome-bearing microtubule renders it shorter and shorter, pulling the chromosome ever closer to the pole of the cell.

Anaphase is the stage of mitosis characterized by the physical separation of sister chromatids. The poles of the cell are pushed apart by microtubular sliding, and the sister chromatids are drawn to opposite poles by the shortening of the microtubules attached to them.

Re-Formation of Nuclei: Telophase

The separation of sister chromatids achieved in anaphase completes the accurate partitioning of the replicated genome. This partitioning is the essential element of mitosis. With the play complete, the only tasks that remain in telophase are to dismantle the stage and remove the props. The spindle apparatus is disassembled, with the microtubules broken back down into tubulin monomers ready for use in constructing the cytoskeleton of the new cell. The nuclear envelope re-forms around each set of sister chromatids, which are now chromosomes in their own right and begin to uncoil into the more extended form that permits gene expression. One of the early genes to regain expression is the rRNA gene, resulting in the reappearance of the nucleolus.

Telophase is the stage of mitosis during which the mitotic apparatus assembled during prophase is disassembled, the nuclear envelope reestablished, and the normal use of the genes present in the chromosomes reinitiated.

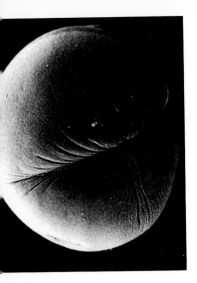

FIGURE 10-12

A cleavage furrow around a dividing sea urchin egg.

CYTOKINESIS

At the end of telophase, mitosis is complete. The eukaryotic cell has partitioned its replicated genome into two nuclei, which are positioned at opposite ends of the cell. While this process has been going on, the cytoplasmic organelles, such as the mitochondria and, if present, the chloroplasts, have also been reassorted to the areas that will separate and become the daughter cells. Their replication also occurs before cytokinesis, often in the S or G_2 stage. At this point, however, the process of cell division is still not complete. The division of the cell proper has not yet even begun. The stage of the cell cycle at which cell division actually occurs is called **cytokinesis.** Cytokinesis generally involves the cleavage of the cell into roughly equal halves.

> **Cytokinesis is the physical division of the cytoplasm of a eukaryotic cell into two daughter cells.**

In animal cells and in the cells of all other eukaryotes that lack cell walls, cytokinesis is achieved by pinching the cell in two with a contracting belt of microfilaments. As the contraction proceeds, a **cleavage furrow** becomes evident around the circumference of the cell (Figure 10-12) where the cytoplasm is being progressively pinched inward by the decreasing diameter of the microfilament belt. As contraction proceeds, the furrow deepens until it eventually extends all the way into the residual spindle, and the cell is literally pinched into two.

Plant cells possess a rigid cell wall, one far too strong to be deformed by microfilament contraction. A different approach to cell division has therefore evolved in plants. Plants assemble membrane components in their interior, at right angles to the spindle (Figure 10-13). This expanding partition is called a **cell plate.** It continues to grow outward until it reaches the interior surface of the cell membrane and fuses with it, at which point it has effectively divided the cell into two. Cellulose is then laid down on the new membranes, creating two new cells. The space between the two new cells becomes impregnated with pectins and is called a **middle lamella.**

Figure 10-14 shows a comparison between mitosis in plants and animals.

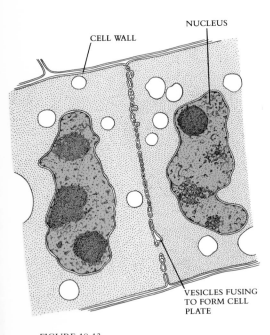

FIGURE 10-13

A cell plate forms between daughter nuclei in the dividing plant cell by vesicle fusion; once the plate is complete, there are two cells.

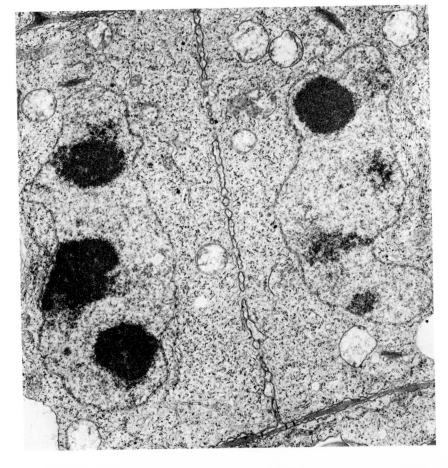

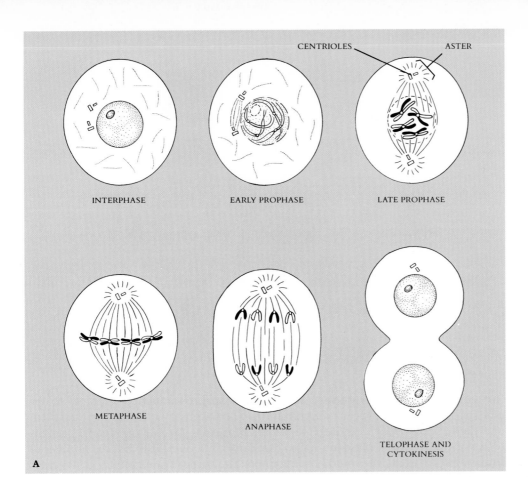

CENTRIOLES   ASTER

INTERPHASE   EARLY PROPHASE   LATE PROPHASE

METAPHASE   ANAPHASE   TELOPHASE AND CYTOKINESIS

A

FIGURE 10-14

A comparison of the stages of mitosis in an animal **(A)** and in a plant **(B)**.

CELL WALL

INTERPHASE   EARLY PROPHASE   LATE PROPHASE

METAPHASE   ANAPHASE   TELOPHASE AND CYTOKINESIS   CELL PLATE

B

FIGURE 10-15

Electron micrograph of a portion of a synaptonemal complex of a cup fungus, *Neotiella rutilans.* Synaptonemal complexes are characteristic of chromosome pairing in meiosis.

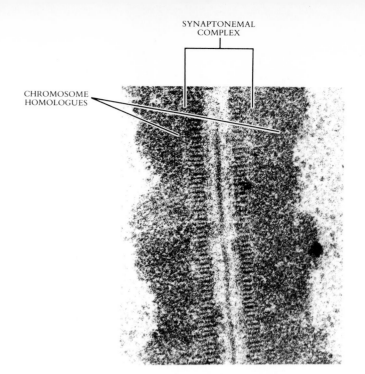

SYNAPTONEMAL
COMPLEX

CHROMOSOME
HOMOLOGUES

MEIOSIS

Most animals and plants reproduce sexually. In sexual reproduction, gametes of opposite sexes unite in the process of fertilization. The fusion of haploid gametes to form a new diploid cell, called a zygote, is known as **syngamy.** As you can imagine, cell fusion is not a process that can continue to occur over and over. If it did, the number of chromosomes in each cell would become impossibly large. For example, in 10 generations the 46 chromosomes present in each of your cells would have increased to more than 47,000. Even early investigators realized that there must be some mechanism during the course of gamete formation to reduce the number of chromosomes. If a special **reduction division** occurred in which cells are formed with half the number of chromosomes characteristic of most cells of that species, then the later fusion of these cells would allow a stable chromosome number. The mature individuals of each successive generation would have the same number of chromosomes. Investigators soon observed the expected reduction division process, called **meiosis.**

Only a year after chromosomes were first discovered by Walther Fleming in 1882, the Belgian biologist Pierre Joseph van Beneden was studying the chromosomes of *Ascaris,* a large roundworm that is parasitic in human beings and other vertebrate animals. van Beneden found, to his surprise, that the number of chromosomes was different in different types of cells. Specifically, he observed that eggs and sperm each contained two chromosomes, whereas the cells of the young embryo contained four chromosomes, as do all the cells in the body of a mature individual of *Ascaris.* From his observations, van Beneden was able to outline the basic process of meiosis: the gametes, eggs and sperm, each contain a single basic complement of chromosomes (we say that the egg and sperm are haploid); whereas the zygote resulting from the fusion of egg and sperm, and all the cells of the adult that it becomes, contains two copies of each chromosome (we say that the zygote is diploid).

Meiosis should be distinguished clearly from mitosis. An animal such as yourself contains two kinds of cells: **somatic** cells, which make up your body, and **germline** cells, which produce your gametes. Both the somatic cells and the gamete-producing cells are usually diploid. When a somatic cell undergoes mitosis, it divides to form two diploid daughter cells exactly like it; when a germline cell undergoes meiosis, however, it produces haploid cells with half the diploid number of chromosomes. In animals, only the germline cells undergo meiosis.

Meiosis is a process of nuclear division in which the number of chromosomes in cells is halved during gamete formation.

THE STAGES OF MEIOSIS

Although meiosis is a continuous process, we can describe it most easily by dividing it into arbitrary stages, just as is done in describing the process of mitosis. The two forms of nuclear division have much in common. Meiosis in a diploid organism consists, in a sense, of two rounds of mitosis, during the course of which two unique events occur:

1. In an early stage of the first of the two nuclear divisions, the two nearly identical versions of each chromosome, called **homologues,** pair with each other all along their length. Each chromosome has replicated during the S phase that precedes cell division and thus consists of two sister chromatids bound together by a single centromere. Thus a total of four chromatids have joined together. While the four chromatids are held together, they may exchange portions of DNA strands in a process called **crossing-over.** If the exchange takes place between chromatids of the different homologues, the two homologues will be held together, thus forming a pair of chromosomes that separates only later in meiosis. When the two homologues do separate, their chromatids may have exchanged genetic material, as we shall discuss in more detail later.
2. The second meiotic division is identical to a normal mitotic division, except that *the chromosomes do not replicate between the two divisions.*

> **Two important properties distinguish meiosis from mitosis:**
> 1. **In meiosis, the homologous chromosomes pair lengthwise, and their chromatids may exchange genetic material.**
> 2. **The sister chromatids of each homologue do not separate from one another in the first nuclear division during meiosis, and the chromosomes do not replicate between the two nuclear divisions.**

The two stages of meiosis are called meiosis I and meiosis II. Each stage is further subdivided into prophase, metaphase, anaphase, and telophase, just as in mitosis. In meiosis, however, prophase is much more complicated than it is in mitosis.

The First Meiotic Division
Prophase I

In prophase I, individual chromosomes first become visible under a light microscope, as their DNA coils more and more tightly. Since the chromosomes (DNA) already have replicated before the onset of meiosis, each of these threads actually consists of two sister chromatids joined at their centromeres. The two homologous chromosomes then line up side-by-side, a process that is called **synapsis.** A lattice of protein and RNA is laid down between the chromatids of the two homologous chromosomes. This lattice holds the chromatids in precise relation with one another, each gene located directly across from its corresponding sister on the homologue. The effect is similar to zipping up a zipper. The resulting complex is called a **synaptonemal complex** (Figure 10-15). Within it, the DNA duplexes of each chromatid unwind, and single strands of DNA pair with their complementary number *from the other homologue.* How many single strands of DNA are present in the synaptonemal complex? Eight. Count them: 2 per DNA duplex molecule × 2 sister chromatids per homologue × 2 homologues.

> **Synapsis is the close pairing of homologous chromosomes that occurs early in prophase I of meiosis. During synapsis, a molecular latticework called the synaptonemal complex aligns the genes of the two homologous chromosomes side-by-side. As a result, a DNA strand of one homologue can pair with the corresponding complementary DNA strand of the other.**

The process of synapsis initiates a complex series of events, called crossing-over, in which DNA is exchanged between the two homologous chromosomes. Once the process of crossing-over is complete, the synaptonemal complex breaks down. At that point, the nuclear envelope dissolves, and the chromatids begin to move apart from

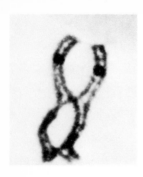

FIGURE 10-16

Chiasmata.

one another. A total of four chromatids exist for each kind of chromosome at this point: two homologous chromosomes, each replicated and so present twice as sister chromatids. The four chromatids cannot separate from one another completely, however, because they are held together in two ways:

1. The two sister chromatids of each homologue, created by DNA replication in S phase, are held together at the centromere.
2. The paired homologues are held together at the points where crossing-over occurred within the synaptonemal complex.

The points of crossing-over, where portions of chromosomes have been exchanged, can often be seen under a light microscope as X-shaped structures known as **chiasmata** (singular *chiasma*, the Greek word for a cross) (Figure 10-16). The presence of a chiasma indicates that two of the four chromatids of paired homologous chromosomes have exchanged parts, one participant from each homologue. The X-shaped chiasmata soon begin to move out to the end of the respective chromosome arm as the chromosomes separate. The process somewhat resembles moving a small ring down two strands of rope.

> In prophase I, individual DNA strands of the two homologues pair with one another. Crossing-over occurs between the DNA-paired strands, creating the chromosomal configurations known as chiasmata. These exchanges serve to lock the two homologues together.

Metaphase I

In the second phase of the first meiotic division, the nuclear envelope disperses and the microtubules form a spindle, just as in mitosis. A crucial difference exists, however, between this metaphase and that of mitosis: in meiotic metaphase I, one of the kinetochores of each centromere is inaccessible to microtubules, because the homologous chromosomes are held tightly together by chiasmata (Figure 10-17). In mitosis, on the other hand, *both* kinetochores of a centromere are bound by microtubules (and the chromatids therefore separate from one another when anaphase occurs). The chromosomes line up double-file in meiosis and single-file in mitosis.

Since microtubules bind only to one kinetochore of each centromere, the centromere of one homologue becomes attached to microtubules extending to one pole, whereas the centromere of the other homologue becomes attached to microtubules extending to the other pole. Each joined pair of homologues then lines up on the meta-

FIGURE 10-17

A key mechanical element necessary for reduction division in meiosis is the presence of chiasmata. These chiasmata are created by crossing-over during prophase I, when the chromosome homologues were close to one another within the synaptonemal complex. The chiasmata hold the two pairs of sister chromatids together; consequently, the spindle microtubules are able to bind only one face of each kinetochore. Later, when these microtubules shorten by sliding past one another, the terminal chiasmata break, and the two pairs of sister chromatids are drawn to opposite poles. However, no microtubule-driven separation of the individual chromatids occurs in the pairs of sister chromatids. In mitosis, by contrast, microtubules attach to both faces of each kinetochore; when they shorten, the pair of sister chromatids is split and the two individual chromatids are drawn to opposite poles.

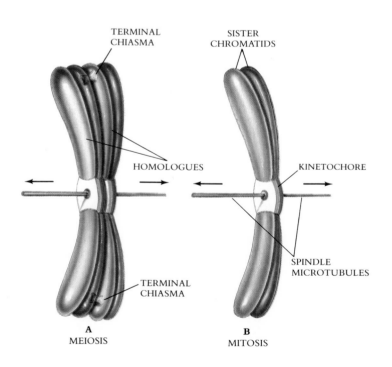

TERMINAL CHIASMA

SISTER CHROMATIDS

HOMOLOGUES

KINETOCHORE

TERMINAL CHIASMA

SPINDLE MICROTUBULES

A
MEIOSIS

B
MITOSIS

phase plate. For each pair of homologues, the orientation on the spindle axis is random; which homologue is oriented toward which pole is a matter of chance.

Anaphase I

After spindle attachment is complete, the microtubules attached to chromosome homologues begin to slide past one another and shorten. As this occurs, the microtubules break the chiasmata apart and pull the centromeres toward the two poles, dragging the chromosomes along with them. Because the microtubules are attached to only one of the kinetochores of each centromere, the individual centromeres are not pulled apart as they are in mitosis. Instead, the entire centromere proceeds to one pole, taking both sister chromatids with it. When the shortening of the spindle fibers is complete, each pole has a complete complement of chromosomes, consisting of one member of each homologous pair. Because the orientation of each pair of homologous chromosomes on the metaphase plate is random, the chromosome that a pole receives from each pair of homologues is also random with respect to all other chromosome pairs.

The two similar chromosomes of homologous pairs orient randomly on the metaphase plate.

FIGURE 10-18

The stages of meiosis.

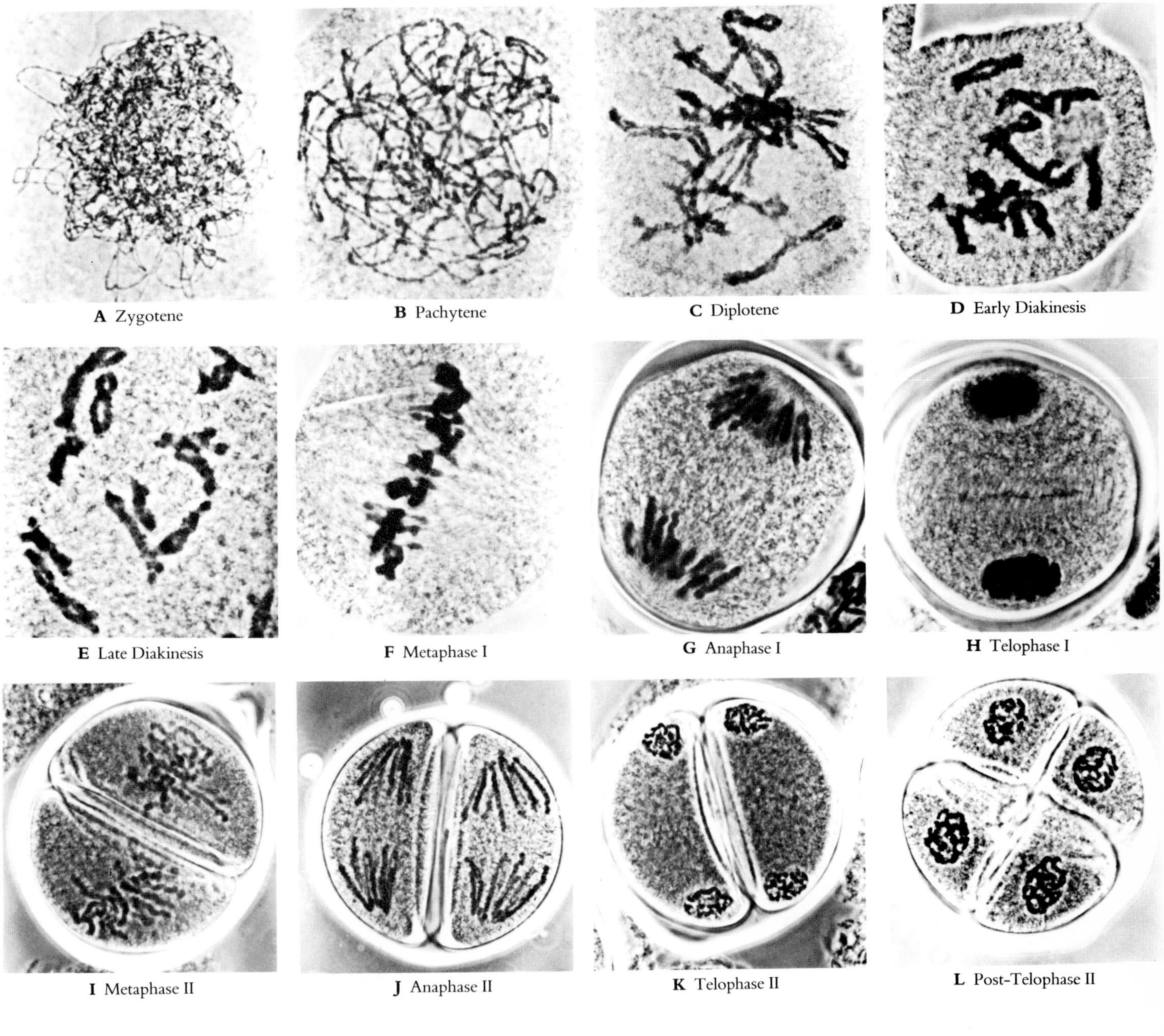

A Zygotene

B Pachytene

C Diplotene

D Early Diakinesis

E Late Diakinesis

F Metaphase I

G Anaphase I

H Telophase I

I Metaphase II

J Anaphase II

K Telophase II

L Post-Telophase II

Telophase I

At the completion of anaphase I, each pole has a complete complement of chromosomes, one member of each pair of homologues. There are half as many chromosomes and half as many centromeres as were present in the cell in which meiosis began. Each of these chromosomes replicated itself before meiosis began and thus contains two copies (sister chromatids) of itself attached at the centromere. These copies are not identical, however, because of the crossing-over that occurred in prophase I. The stage at which the two complements of chromosomes gather together at their respective poles to form two chromosome clusters is called telophase I. After an interval of variable length, the second meiotic division, meiosis II, occurs.

The first meiotic division is traditionally divided into four stages:
Prophase I   Homologous chromosomes pair and exchange segments.
Metaphase I Homologous chromosomes align on a central plane.
Anaphase I   Homologous chromosomes move toward opposite poles.
Telophase I Individual chromosomes gather together at the two poles.

The Second Meiotic Division

Meiosis II is simply a mitotic division, involving the products of meiosis I. At the completion of anaphase I, each pole has a haploid complement of chromosomes, each of which is composed of two sister chromatids attached at the centromere. Meiosis II occurs to separate these sister chromatids. Because of crossing-over in the first phase of meiosis, however, *these sister chromatids are not identical to one another in terms of base sequences of the two DNA molecules.* At both poles of the original cell, these two complements of chromosomes now divide mitotically. Spindle fibers bind each of the two kinetochores at the centromere region, which separate and move to opposite ends of each polar region. The result of this mitotic division is four haploid complements of chromosomes. At this point, the nuclei are reorganized, and nuclear envelopes form around each of the four haploid complements of chromosomes (Figure 10-18). The cells that contain these haploid nuclei may function directly as gametes, as they do in animals, or they may divide again mitotically, as they do in plants, fungi, and many protists, producing greater numbers of gametes.

Each of the four haploid products of meiosis contains a basic set of
chromosomes. These haploid cells may function directly as gametes, as they
do in animals, or may continue to divide by mitosis, as they do in plants,
fungi, and many protists.

THE IMPORTANCE OF MEIOTIC RECOMBINATION

The reassortment of genetic material that occurs during meiosis (Figure 10-19) is the principal factor that has made possible the evolution of eukaryotic organisms, in all their bewildering diversity, over the past 1.5 billion years. Meiosis represents an enormous advance in the ability of organisms to generate genetic variability. To understand why this is true, recall that most organisms have more than one chromosome. Human beings, for example, have 23 different pairs of homologous chromosomes. Each human gamete receives one of the two copies of each of the 23 different chromosomes, but which copy of a particular chromosome it receives is random. For example, the copy of chromosome number 14 that a particular human gamete receives has no influence on which copy of chromosome number 5 that it will receive. Each of the 23 pairs of chromosomes goes through meiosis independently of all the others, so there are 2^{23} (more than 8 million) different possibilities for the kinds of gametes that can be produced, and no two of them are alike. The subsequent fertilization of two gametes thus creates a unique individual, a new combination of the 23 chromosomes that probably has never occurred before and probably will never occur again.

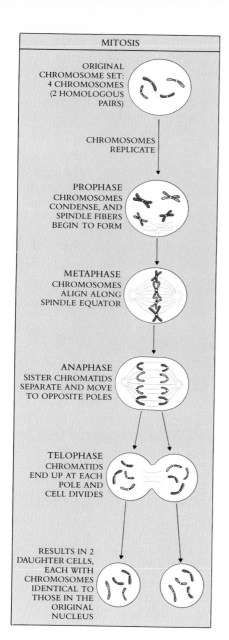

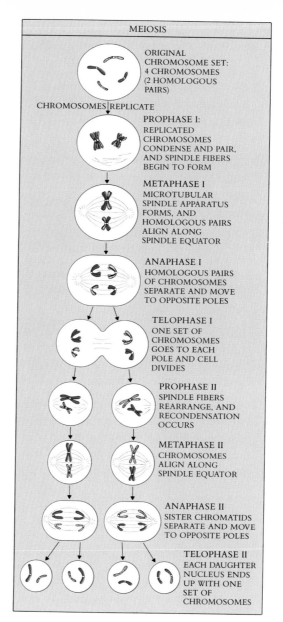

MITOSIS

ORIGINAL CHROMOSOME SET: 4 CHROMOSOMES (2 HOMOLOGOUS PAIRS)

CHROMOSOMES REPLICATE

PROPHASE
CHROMOSOMES CONDENSE, AND SPINDLE FIBERS BEGIN TO FORM

METAPHASE
CHROMOSOMES ALIGN ALONG SPINDLE EQUATOR

ANAPHASE
SISTER CHROMATIDS SEPARATE AND MOVE TO OPPOSITE POLES

TELOPHASE
CHROMATIDS END UP AT EACH POLE AND CELL DIVIDES

RESULTS IN 2 DAUGHTER CELLS, EACH WITH CHROMOSOMES IDENTICAL TO THOSE IN THE ORIGINAL NUCLEUS

MEIOSIS

ORIGINAL CHROMOSOME SET: 4 CHROMOSOMES (2 HOMOLOGOUS PAIRS)

CHROMOSOMES REPLICATE

PROPHASE I:
REPLICATED CHROMOSOMES CONDENSE AND PAIR, AND SPINDLE FIBERS BEGIN TO FORM

METAPHASE I
MICROTUBULAR SPINDLE APPARATUS FORMS, AND HOMOLOGOUS PAIRS ALIGN ALONG SPINDLE EQUATOR

ANAPHASE I
HOMOLOGOUS PAIRS OF CHROMOSOMES SEPARATE AND MOVE TO OPPOSITE POLES

TELOPHASE I
ONE SET OF CHROMOSOMES GOES TO EACH POLE AND CELL DIVIDES

PROPHASE II
SPINDLE FIBERS REARRANGE, AND RECONDENSATION OCCURS

METAPHASE II
CHROMOSOMES ALIGN ALONG SPINDLE EQUATOR

ANAPHASE II
SISTER CHROMATIDS SEPARATE AND MOVE TO OPPOSITE POLES

TELOPHASE II
EACH DAUGHTER NUCLEUS ENDS UP WITH ONE SET OF CHROMOSOMES

FIGURE 10-19

A comparison of meiosis with mitosis. Meiosis involves two serial nuclear divisions with no DNA replication between them and so produces four daughter cells, each with half the original amount of DNA. Crossing-over occurs in prophase I of meiosis.

Even this process, however, does not fully explain the diversity of the gametes that results from meiosis. As you will recall, pairs of homologous chromosomes exchange physical segments during meiotic prophase I (if they did not, the pairs would not remain associated in metaphase I, and meiosis would not proceed normally). The exchange that occurs as a result of this crossing-over adds even more recombination to the random assortment of chromosomes that occurs later in meiosis. Thus, the number of possible combinations that can occur among gametes is virtually unlimited.

SUMMARY

1. Bacterial cells divide by simple binary fission. The two replicated circular DNA molecules attach to the plasma membrane at different points, and fission is initiated between the points to which they are attached.

2. The cells of eukaryotes contain much more DNA than do those of bacteria. Eukaryotic DNA is complexed with histones and divided among several chromosomes. Some of the DNA is permanently condensed into heterochromatin, whereas the rest is condensed only during cell division.

3. DNA replication is completed before mitosis begins. Immediately before the onset of mitosis, condensation of the chromosomes and synthesis of microtubules begin. This period preceding mitosis is called interphase.

4. The first stage of mitosis is prophase, during which the mitotic apparatus forms. At the end of prophase, the nuclear envelope disassembles, and microtubules attach each pair of sister chromatids to the two poles of the cell.

5. The second stage of mitosis is metaphase, during which the chromosomes align along the periphery of a plane cutting through the center of the cell at right angles to the spindle axis. At the end of metaphase, the centromeres joining each pair of sister chromatids replicate, freeing each chromatid to be pulled to one of the poles of the cell by the microtubules attached to it.

6. The third stage of mitosis is anaphase, during which the chromatids physically separate, moving to opposite poles of the cell.

7. The fourth and final stage of mitosis is telophase, during which the mitotic apparatus is disassembled, the nuclear envelope re-forms, and the chromosomes uncoil.

8. Following mitosis, most cells undergo cytoplasmic cleavage, or cytokinesis. In cells without a cell wall, the cell body is pinched in two by a belt of microtubules drawing inward around its midsection. In plant cells and other cells with a cell wall, an expanding cell plate forms along the spindle midline.

9. Meiosis is a special form of nuclear division that occurs during gamete formation in most eukaryotes. In meiosis, there is a single replication of the chromosomes and two chromosome separations. Meiosis results in the formation of four nuclei, each with half of the original number of chromosomes. The cells that include these nuclei may serve as gametes.

10. Meiosis involves a pair of serial nuclear divisions. The two unique characteristics of meiosis are synapsis, the intimate pairing of homologous chromosomes, and the lack of chromosome replication before the second nuclear division.

11. Crossing-over is an essential element of meiosis. The crossing-over between homologues that occurs during synapsis binds the two homologous chromosomes together. As a result, the spindle fibers are able to bind to only one kinetochore of each homologous chromosome's centromere, since the other kinetochore is covered by the opposite homologue. For this reason, they do not pull the sister chromatids apart. Rather, their contraction pulls the paired homologous chromosomes apart, ultimately breaking the terminal chiasmata that link them.

12. At the end of meiosis I, one member of each pair of homologues is present at each of the two poles of the dividing nucleus. These chromosomes already consist of two chromatids, which differ from one another as a result of the crossing-over that occurred when the chromosomes were paired with their homologues. No further replication occurs before the next nuclear division, which is a normal mitosis that occurs at each of the two poles. The sister chromatids simply separate from one another. This results in the formation of four clusters of chromosomes, each with a complement of half the number of chromosomes that was present initially. Because each daughter nucleus has one copy of each chromosome, any or all of the clusters can function as a gamete.

13. Meiotic recombination is the principal factor that has made possible the evolution of eukaryotic organisms. Because of meiosis and crossing-over, the number of possible genetic combinations that can occur among gametes is virtually unlimited.

REVIEW

1. After the naked DNA duplex is replicated and separated in bacteria, the cell divides by a simple process called _____, which is made easier by invaginations of the growing plasma membrane.

2. Which phase of the cell cycle is the essential step in the separation of the two daughter genomes?

3. What determines the plane in which a cell will divide?

4. The process by which homologous chromosomes come to line up side-by-side during prophase I of meiosis is called _____.

5. Meiotic recombination greatly enhances the ability of organisms to generate genetic _____.

SELF-QUIZ

1. Eukaryotic DNA is coiled around _____, which are the first order of DNA packing.
   (a) nucleosomes
   (b) histones
   (c) basic amino acids
   (d) chromosomes
   (e) supercoils

2. Which of the following does *not* occur during interphase?
   (a) each chromosome replicates
   (b) the nucleolus disappears
   (c) the sister chromosomes are attached to each other at the centromere
   (d) the chromosomes begin the process of condensation
   (e) centrioles replicate

3. The stage of the cell cycle during which the cell actually divides to form two cells is called
   (a) G_1
   (b) S
   (c) G_2
   (d) mitosis
   (e) cytokinesis

4. Sexual reproduction in plants and animals involves the production of gametes through a special type of reduction division called _____, followed by the union of haploid gametes, which is called _____.
   (a) mitosis
   (b) syngamy
   (c) cytokinesis
   (d) chromatin
   (e) meiosis

5. Which of the following is *not* true of crossing-over?
   (a) it occurs during prophase I
   (b) it occurs during prophase II
   (c) it occurs between homologues
   (d) it is seen as **X**-shaped structures called chiasmata
   (e) exchange of DNA occurs between homologous chromosomes at the points of cross-over

PROBLEMS

1. A typical human genome contains about 5 billion (5×10^9) nucleotides; each nucleotide of the sugar phosphate backbone of DNA takes up about 0.34 nanometer (1000 nanometers = 1 micrometer; 1000 micrometers = 1 millimeter; 10 millimeters = 1 centimeter). If a typical human genome were spread out as a linear molecule, how long would it be? Answer in centimeters.

2. Human beings have 23 pairs of chromosomes, 22 pairs that play no role in sex determination, and an XX (female) or XY (male) pair. Ignoring the effects of crossing-over, what proportion of your eggs or sperm contain all of the chromosomes you received from your mother?

THOUGHT QUESTIONS

1. In what way does the short generation time in bacteria produce an effect similar to that of the extensive recombination that occurs as a result of meiosis in eukaryotes?

2. Which do you think evolved first, meiosis or mitosis? What do you think may have been some of the stages in the evolution of one from the other?

FOR FURTHER READING

KORNBERG, R.D., and A. KLUG: "The Nucleosome," *Scientific American,* February 1981, pages 52-64. This article describes how the DNA superhelix is wound on a series of protein spools.

MAZIA, D.: "The Cell Cycle," *Scientific American,* January 1974, pages 54-64. An account by one of those responsible for the development of the cell cycle concept.

SLOBODA, R.D.: "The Role of Microtubules in Cell Structure and Cell Division," *American Scientist,* vol. 68, pages 290-298, 1980. A good description of how the spindle apparatus works.

FIGURE 11-1   Human beings are extremely diverse in appearance. The differences between us are partly inherited and partly arise as a result of the environmental factors that we encounter during the course of our lives.

MENDELIAN GENETICS

Chapter 11

Overview

Why do members of a family tend to resemble one another more than they do members of other families? We have puzzled about heredity since before words were written down. Little progress in solving the puzzle was made, however, until Gregor Mendel conducted some simple but significant experiments in his monastery garden a century ago. What emerged from this obscure garden was nothing less than the key to understanding heredity, the pivotal piece of the jigsaw puzzle. Mendel developed a model, a set of rules that accurately predicted patterns of heredity. When these rules became widely known, investigators avidly sought and soon found the physical mechanisms responsible. We now know that inherited traits are specified by genes, which are portions of the DNA molecules of chromosomes. Mendel's construction of his model is one of the greatest intellectual accomplishments in the history of science.

For Review

Here are some important terms and concepts that you will encounter in this chapter. If you are not familiar with them, you should review them before proceeding.

DNA (Chapter 3)

Chromosomes (Chapters 5, 10)

Meiosis (Chapter 10)

Homologues (Chapter 10)

Chiasmata and crossing-over (Chapter 10)

Some variation in physical characteristics is evident in all of us. Groups of people from different parts of the world often have distinct appearances. Within any of these groups, individual families often differ greatly from one another. Look around you in class; rarely will any two students resemble one another closely. Even our brothers and sisters are not exactly like us, unless of course they are our identical twins. Any group of people includes individuals of very different appearance (Figure 11-1). The reasons for these differences and similarities have always fascinated us. And human beings are not unique in this respect. Great differences in appearance also often exist within other species. Dogs, for example, are born in many varieties and breeds of every size and form. All of these are still dogs, however, able to breed with one another and produce puppies (see Figure 17-2).

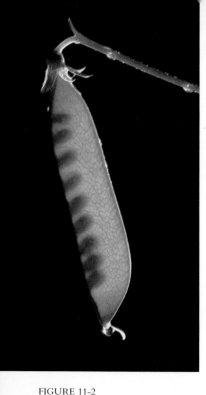

Variation is not surprising in itself. Differences in diet during development can have great effects on adult appearance, as can variation in the environments that different individuals experience. Many arctic mammals, for example, develop white fur when they are exposed to the cold of winter and dark fur during the warm summer months. One remarkable property in some patterns of variation, however, has especially fascinated and puzzled us: some of the differences that we observe between individuals are inherited, and thus passed down from parent to child.

As far back as there is a written record, such patterns of resemblance among the members of particular families have been noted and commented on. Some features shared by family members are unusual, such as the protruding lower lip of the Austrian royal family, the Hapsburgs, which is evident in pictures and descriptions of that family since the thirteenth century. Other characteristics are more familiar, such as the common occurrence of red-haired children within the families of red-haired parents. Such inherited features, the building blocks of evolution, are the focus of this chapter.

EARLY IDEAS ABOUT HEREDITY: THE ROAD TO MENDEL

Our understanding of heredity starts with a series of English gentleman farmers who were trying to improve varieties of agricultural plants. They carried out matings, called **crosses,** between individual plants, selecting the most desirable of the offspring of each cross. The offspring of a cross between dissimilar parents are often called **hybrids.** As long ago as 200 years, these farmers had learned that not all offspring of a cross received the same selection of inherited traits from the parents. The alternatives of a trait were "segregating" among the progeny; some progency exhibited one alternative, whereas others exhibited another. In their crosses, the farmers found that some alternative forms of a trait would appear more often among the progeny than others. They also discovered that the less frequent forms would disappear in one generation, only to reappear unchanged in the next. In one such series of experiments in the 1790s, T. A. Knight crossed two true-breeding varieties of the garden pea, *Pisum sativum* (Figure 11-2). One of these varieties had purple flowers; the other, white flowers. All the progeny of the cross had purple flowers. Among the offspring of these hybrids, however, some plants had purple flowers, and others, occurring less frequently, had white ones. Just as had happened in earlier studies, a character trait from one of the parents (white flowers) was hidden in one generation, only to reappear in the next.

In these deceptively simple results were the makings of a scientific revolution. It took another century, however, before the process of **segregation of alternative traits** was understood. Why did it take so long? One reason was that some characteristics appeared to be **blended** in the offspring, so that it was not clear that individual, distinct factors were involved. Another problem was that early workers did not quantify their results, and a numerical record of results proved to be crucial to understanding this process. For example, both Knight and later experimenters who carried out other crosses with pea plants noted that some traits had a "stronger tendency" to appear than others, but they did not record the numbers of the different classes of progeny. Science was young then, and it was not obvious that the numbers were important. They were. They almost always are.

MENDEL AND THE GARDEN PEA

The first quantitative studies of inheritance were carried out by an Austrian monk, Gregor Mendel (Figure 11-3). Born in 1822 to peasant parents, Mendel was educated in a monastery and went on to study science and mathematics at the University of Vienna, where he failed his examinations for a teaching certificate. Returning to the monastery, where he spent the rest of his life and eventually became abbot, Mendel initiated a series of experiments on plant heredity in its garden. The results of these experiments would ultimately change our views of heredity irrevocably.

For his experiments, Mendel chose the garden pea, the same plant that Knight and many others had studied earlier. The choice was a good one for several reasons:

1. Many earlier investigators had produced "hybrid" peas by crossing different varieties. From their results, Mendel knew that he could expect to observe the segregation of alternative traits among the progeny of crosses. That is, Mendel knew that some of the progeny would exhibit the alternative characteristic of one variety, whereas other progeny individuals would exhibit the alternative characteristic of the other variety.

2. Many varieties of peas were available. Mendel initially examined 32 varieties. Then, for further study, he selected lines that differed with respect to seven easily distinguishable traits, such as smooth versus wrinkled seeds, and purple versus white flowers (a characteristic that Knight had studied 60 years earlier).

3. Pea plants are small, easy to grow, produce large numbers of offspring, and mature quickly. Thus one can conduct experiments involving many plants and can obtain the results relatively soon.

4. The sexual organs of the pea are enclosed within the flower (Figure 11-4). The flowers of peas, as with those of most flowering plants, are bisexual, containing both the structures in which male gametes are produced (the anthers) and those in which female gametes are produced (the carpels). **Self-fertilization** takes place automatically within an individual flower if it is not disturbed. As a result of this process, the offspring of garden peas are the progeny of a single individual. One can let self-fertilization occur within an individual flower, or one can remove its anthers before fertilization, introduce pollen (produced in the anthers and giving rise to the gametes) from a strain with alternative characteristics, and thus perform an experimental cross. **Self-pollination** leads directly to self-fertilization in peas.

MENDEL'S EXPERIMENTAL DESIGN

Mendel usually conducted his experiments in three stages:

1. He first allowed pea plants of a given variety to produce progeny by self-fertilization for several generations. Mendel was thus able to ensure that the characteristics that he was studying were true-breeding, transmitted regularly from generation to generation. Pea plants with white flowers, for example, produced only plants with white flowers among their progeny, regardless of the number of generations studied.

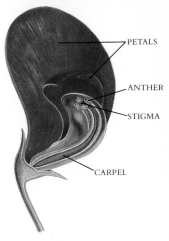

FIGURE 11-4

In a pea flower, the petals enclose the anthers, where male gametes are produced, and the carpels, where female gametes are produced. Self-fertilization takes place automatically unless the flower is disturbed.

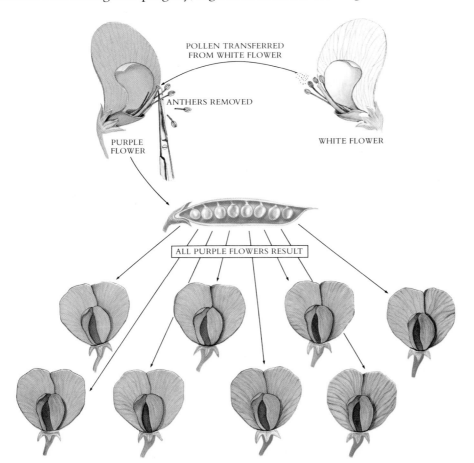

FIGURE 11-5

How Mendel conducted his experiments. In this cross of two varieties with different-colored flowers, Mendel pushed aside the petals of a white flower and cut off the anthers, where the male gametes are produced and enclosed in pollen grains, their units of dispersal. He then collected pollen from a purple flower (he needed only to open the petals and take out the anthers) and placed that pollen onto the female parts of the castrated white flower, where cross-fertilization took place. All of the seeds in the pod that resulted from this pollination were hybrids, with a white-flowered female parent and a purple-flowered male parent. Planting these seeds, Mendel observed what kinds of plants they produced. In this instance, all of them had purple flowers.

2. Mendel then conducted crosses between varieties exhibiting alternative traits (Figure 11-5). For example, he removed the male parts from a flower on a plant that produced white flowers and fertilized it with pollen from a purple-flowered plant. He would also carry out the reciprocal cross by reversing the procedure, using pollen from a white-flowered individual to fertilize a flower on a pea plant that produced purple flowers.

3. Finally, Mendel permitted the "hybrid" offspring produced by these crosses to self-pollinate for several generations. By doing so, he allowed the alternative traits to segregate among the progeny. That was the same experimental design that Knight and others had used much earlier. But Mendel added a new element: he counted the numbers of offspring in each class and in each succeeding generation. No one had ever done that before. The quantitative results that Mendel obtained were essential in helping him, and us, understand the process of heredity.

WHAT MENDEL FOUND

When Mendel crossed two contrasting varieties, such as purple-flowered plants with white-flowered plants, the hybrid offspring that he obtained did not have an intermediate flower color. Instead, the hybrid offspring always resembled one of the parents. It is customary to refer to these hybrid offspring as the **first filial,** or F_1, generation. Thus, in a cross of white-flowered with purple-flowered plants, the F_1 offspring all had purple flowers, just as Knight and others had reported earlier. Mendel referred to the trait expressed in the F_1 plants as **dominant** and to the alternative trait, which was not expressed in the F_1 plants, as **recessive.** For each pair of contrasting traits that Mendel examined, one proved to be dominant and the other, recessive.

After individual F_1 plants had been allowed to mature and self-pollinate, Mendel collected and planted the fertilized seed from each plant to see what the offspring in this **second filial,** or F_2, generation would look like. He found, just as Knight had earlier, that some F_2 plants exhibited the recessive trait. Latent in the F_1 generation, the recessive alternative reappeared among some F_2 individuals.

It was at this stage that Mendel instituted his radical change in experimental design. He counted the numbers of each type among the F_2 progeny. Mendel was investigating whether the proportions of the F_2 types would provide some clue about the mechanism of heredity. For example, he examined a total of 929 F_2 individuals in the cross between purple-flowered F_1 plants just described. Of these F_2 plants, 705 had purple flowers and 224 had white flowers. Almost exactly one quarter of the F_2 individuals (24.1%) exhibited white flowers, the recessive trait.

Mendel examined each of the seven pairs of contrasting traits in this way (Figure 11-6). The numerical result was always the same: three fourths of the F_2 individuals exhibited the dominant trait, and one fourth displayed the recessive trait. The ratio of dominant to recessive among the F_2 plants was always 3:1.

Mendel went on to examine how the F_2 plants behaved in later generations. He found that the one quarter that were recessive were always true-breeding. In the cross of white-flowered with purple-flowered plants just described, for example, the white-flowered F_2 individuals reliably produced white-flowered offspring when allowed to self-fertilize. By contrast, only one third of the dominant F_2 individuals (one quarter of the entire offspring) were true-breeding, whereas two thirds were not. This last class of plants produced dominant and recessive F_3 individuals in the ratio 3:1. This result suggested that, for the entire sample, the 3:1 ratio Mendel observed in the F_2 generation was really a disguised 1:2:1 ratio, with one-quarter true-breeding dominant individuals to one-half not-true-breeding dominant individuals to one-quarter true-breeding recessive individuals.

HOW MENDEL INTERPRETED HIS RESULTS

From these experiments, Mendel was able to learn four things about the nature of heredity. First, the traits that he studied did not produce intermediate types when crossed. Instead, alternatives were inherited intact as distinct characteristics that were

TRAIT	DOMINANT VS RECESSIVE	F_2 GENERATION RESULTS DOMINANT FORM	RECESSIVE FORM	RATIO
FLOWER COLOR	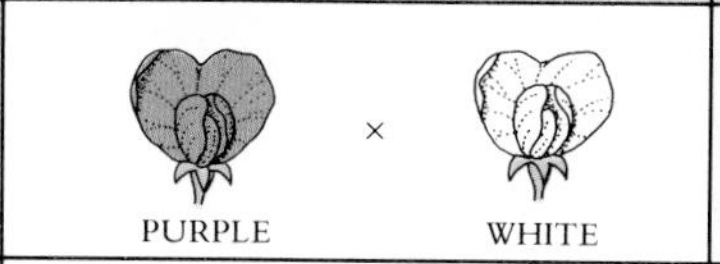 PURPLE × WHITE	705	224	3.15:1
SEED COLOR	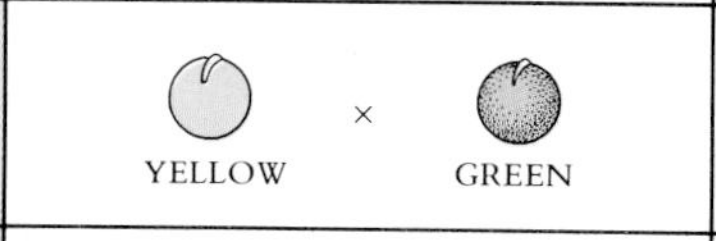 YELLOW × GREEN	6022	2001	3.01:1
SEED SHAPE	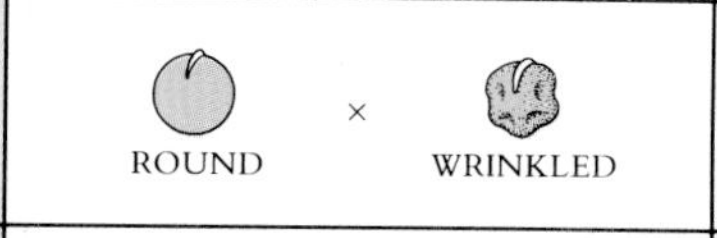 ROUND × WRINKLED	5474	1850	2.96:1
POD COLOR	 GREEN × YELLOW	428	152	2.82:1
POD SHAPE	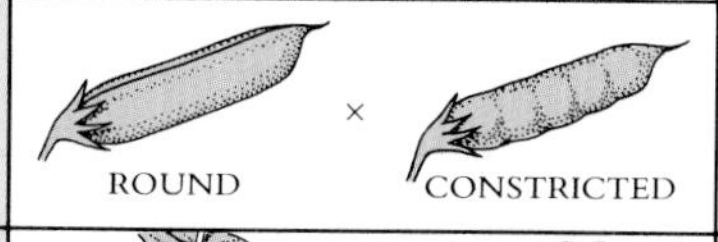 ROUND × CONSTRICTED	882	299	2.95:1
FLOWER POSITION	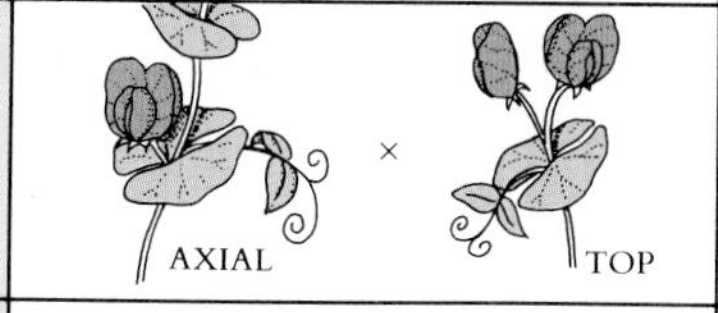 AXIAL × TOP	651	207	3.14:1
PLANT HEIGHT	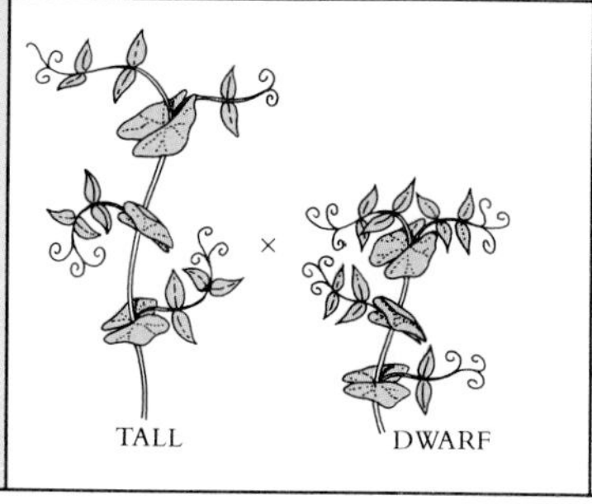 TALL × DWARF	787	277	2.84:1

FIGURE 11-6

The seven pairs of contrasting traits studied by Mendel in the garden pea. In each instance, one quarter of the F_2 generation individuals exhibited the recessive trait, whereas three quarters did not. The ratio of individuals with contrasting traits in the F_2 generation, in this case 3:1, is known as a "Mendelian ratio." Every one of the seven traits that Mendel studied yielded results very close to a theoretical 3:1 ratio.

either seen or not seen in a particular generation. Second, for each pair of traits that Mendel examined, one alternative was not expressed in the F_1 hybrids, although it reappeared in some F_2 individuals. *The "invisible" trait must therefore have been latent (present but not expressed) in the F_1 individuals.* Third, the pairs of alternative traits that Mendel examined segregated among the progeny of a particular cross, some individuals exhibiting one trait, some the other. Fourth, pairs of alternative traits were expressed in the F_2 generation in the ratio of three-fourths dominant to one-fourth recessive. This characteristic 3:1 segregation ratio is often referred to as the **Mendelian ratio.**

To explain these results, Mendel proposed a simple model. It has become one of the most famous models in the history of science, containing simple assumptions and making clear predictions. The model has the following five elements, which are re-phrased in modern terms when appropriate:

1. Parents do not transmit their physiological traits or form directly to their offspring. Rather, they transmit distinct information about the traits, what Mendel called **factors.** These factors later act in the offspring to produce the trait. In modern terms, the particular alternative trait that an individual will express is *encoded* within the factors that it receives from its parents.

2. In diploid organisms, with respect to each trait, each parent contains two factors, which may or may not be the same. If the factors are different, the individual is said to be **heterozygous** for that characteristic. If they are the same, the individual is **homozygous.**

FIGURE 11-7

An African albino. This boy lacks all melanin pigment. His albinism is caused by a recessive allele for which he is homozygous. One in every 130 persons carries a copy of this allele, but albino children are not common. It is unlikely that *both* parents of a family would be individuals that carry the allele. Even in the rare instances when such a marriage does take place, only a quarter of the children would be expected to be double-recessive albino individuals. About one newborn child in 17,000 is an albino.

3. The alternative forms of a factor, leading to alternative character traits, are called **alleles.** In modern terminology, we call Mendel's factors **genes.** We now know that one of Mendel's "factors," a gene, is composed of a sequence of nucleotides in a deoxyribonucleic acid (DNA) molecule. The position on a chromosome where a gene is located is often called a **locus.**

4. The two alleles, one contributed by the male and one by the female gamete, remain distinct; Mendel said they were "uncontaminated": alleles do not blend with one another or become altered in any other way. Thus, when this individual matures and produces its own gametes, these gametes include equal proportions of the elements that the individual received from its two parents.

5. The presence of a particular element does not ensure that the encoded trait will actually be expressed in the individual that carries the particular allele. In heterozygous individuals, only one (the dominant) allele achieves expression; the other (the recessive) allele is present but unexpressed. Modern geneticists refer to the collection of alleles that an individual contains, not all of which are expressed in the appearance or characteristics of that individual, as its **genotype.** The physical appearance of an individual is its **phenotype.**

Many traits in human beings exhibit dominant or recessive inheritance in a manner similar to the traits Mendel studied in peas. Albinism, for example, is a rare recessive condition that is characterized by the complete absence of the pigment melanin (Figure 11-7). Table 11-1 lists a few of the many human traits that are known to be inherited as recessive or dominant alleles.

The phenotype of an individual is the observable outward manifestation of the genes that it carries. The genotype, or genetic constitution of the organism, is the blueprint. The phenotype is the realized outcome.

TABLE 11-1   SOME DOMINANT AND RECESSIVE HUMAN TRAITS

TRAIT	PHENOTYPE
RECESSIVE TRAITS	
Common baldness	M-shaped hairline receding with age
Albinism	Lack of melanin pigmentation
Alkaptonuria	Inability to metabolize homogentisic acid
Red-green color blindnesses	Inability to detect red or green wavelengths of light
Cystic fibrosis	Abnormal gland secretion, leading to liver degeneration and lung failure
Duchenne's muscular dystrophy	Wasting away of muscles during childhood
Hemophilia	Inability of blood to clot
Sickle cell anemia	Defective hemoglobin that collapses red blood cells
DOMINANT TRAITS	
Middigital hair	Presence of hair on middle segment of fingers
Brachydactyly	Short fingers
Huntington's disease	Degeneration of nervous system, starting in middle age
Phenylthiocarbamide (phenylthiourea) tasting	Ability to taste PTC as bitter
Camptodactyly	Inability to straighten the little finger
Hypercholesterolemia (the most common human Mendelian disorder, affecting one in every 500 persons)	Elevated levels of blood cholesterol and high risk of heart attack
Polydactyly	Extra fingers and toes

These five elements constitute Mendel's model of the hereditary process. Does Mendel's model predict the sort of result that he actually obtained? Let us see.

The F₁ Generation

Consider again Mendel's cross of purple-flowered with white-flowered plants. Let us assign the symbol w to the recessive allele, associated with the production of white flowers. The symbol W refers to the dominant allele, associated with the production of purple flowers. By convention, genetic traits are usually assigned a letter symbol referring to their less common state, in this case the letter "W" for white flower color. The recessive allele (white flower color) is written in lower case, as w; the alternative dominant allele (purple flower color) is assigned the same symbol in upper case, W.

Using this system, the genotype of an individual that is true-breeding for the recessive white-flowered trait would be designated ww. In such an individual, both copies of the allele specify white flowers. Similarly, the genotype of a true-breeding purple-flowered individual would be labeled WW. A heterozygote would be designated Ww (the dominant allele is usually written first). Using these conventions and a times sign ($\times$) to denote a cross between two breeding lines, Mendel's original cross can be symbolized as $ww \times WW$ (Figure 11-8). Since the white-flowered parent can produce only w-bearing gametes and the purple-flowered parent can produce only W-bearing gametes, you can see that the union of an egg and a sperm from these parents can produce only heterozygous Ww offspring in the F₁ generation. Because the W allele is dominant, all these F₁ individuals are expected to have purple flowers. The w allele is present in these heterozygous individuals, but it is not phenotypically expressed.

The F₂ Generation

When F₁ individuals are allowed to self-fertilize, the W and w alleles segregate at random during gamete formation. This occurs because of the properties of meiosis, discussed in detail in Chapter 10. The orientation of the centromeres of the individual chromosomes, on which the genes are located, line up on the meiotic metaphase I plate in random orientation. Thus either chromosome, with the particular alleles that it bears, may pass to either pole. This orientation is also random with respect to all the other chromosomes. In addition, because of crossing-over, the particular *collection* of alleles of different genes that goes to a particular pole cannot be predicted accurately, even if the different genes in question are located on the same chromosome.

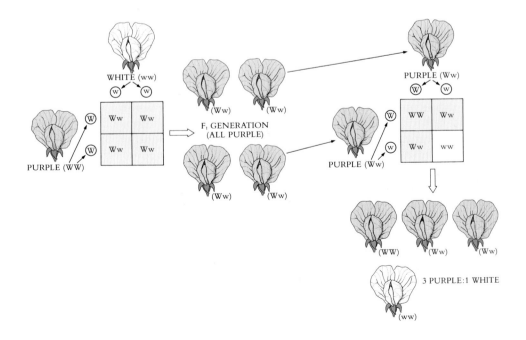

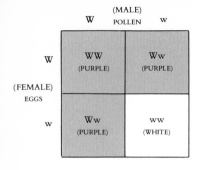

	(MALE)	
	W POLLEN w	
W	**WW** (PURPLE)	**Ww** (PURPLE)
(FEMALE) EGGS		
w	**Ww** (PURPLE)	ww (WHITE)

FIGURE 11-9

A Punnett square. One arrays the different possible types of female gametes according to the alleles they carry along one axis of a square, and the different possible types of male gametes along the other. Each potentially different kind of zygote then can be represented as the intersection of a vertical (female gamete) and horizontal (male gamete) line.

The subsequent union of these randomly formed gametes at fertilization to form F_2 individuals is also random and is not influenced by which alleles the individual gametes carry. What will the F_2 individuals look like? The possibilities may be visualized in a simple diagram called a **Punnett square** (Figure 11-9), named after its originator, the English geneticist Reginald Crundall Punnett. Mendel's model, analyzed in terms of a Punnett square, clearly predicts that in the F_2 generation one should observe three-fourths purple-flowered plants and one-fourth white-flowered plants, a phenotypic ratio of 3:1.

Further Generations

As one can see from Figure 11-9, there are really three kinds of F_2 individuals. Of them, ¼ are true-breeding ww white-flowered individuals, ½ are heterozygous Ww purple-flowered individuals, and ¼ are true-breeding WW purple-flowered individuals. The 3:1 phenotypic ratio is the expression of an underlying 1:2:1 genotypic ratio, in which the heterozygotes are phenotypically indistinguishable from the homozygous purple-flowered (dominant) individuals.

THE TEST CROSS

To test his model further, Mendel devised a simple and powerful procedure called the **test cross.** Consider a purple-flowered individual: is it homozygous or heterozygous? It is impossible to tell simply by looking at it. To learn its genotype, Mendel crossed it with a homozygous recessive individual. His model predicted totally different results for homozygous and heterozygous test plants (Figure 11-10):

> *Alternative 1*
> Test individual is homozygous. *WW × ww:* all offspring have purple flowers *(Ww)*.
>
> *Alternative 2*
> Test individual is heterozygous. *Ww × ww:* one half of offspring have white flowers *(ww)*.

To test his model, Mendel crossed heterozygous F_1 individuals back to the parent homozygous for the recessive trait. In such a cross, a *Ww* F_1 individual is crossed with its *ww* parent. Mendel predicted that the dominant and recessive trait would appear in a 1:1 ratio:

Gametes of homozygous recessive parent

	w
Gametes of F_1 individual *W* *w*	*Ww* *ww*

For each pair of alleles that he investigated, Mendel observed phenotypic test-cross ratios very close to 1:1, just as his model predicted.

Mendel's model thus accounted in a neat and satisfying way for the segregation ratios that he had observed. Its central conclusions—that alleles do not blend in heterozygotes, segregate in heterozygous individuals, and have an equal probability of being included in either gamete—have since been verified in countless other organisms. These points are usually referred to as **Mendel's First Law,** or the **Law of Segregation.** As you will see later in this chapter, the segregational behavior of alleles has a simple physical basis, but it was unknown to Mendel.

> **In modern terms, Mendel's First Law states that:**
> 1. **Alleles do not blend in heterozygotes.**
> 2. **When gametes are formed in heterozygous diploid individuals, the alleles segregate from one another.**
> 3. **Each gamete has an equal probability of possessing either member of an allele pair.**

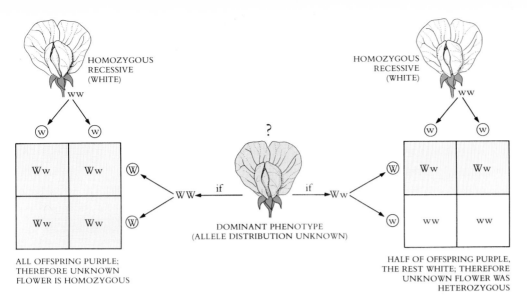

HOMOZYGOUS RECESSIVE (WHITE)
ww

Ww | Ww
Ww | Ww

ALL OFFSPRING PURPLE; THEREFORE UNKNOWN FLOWER IS HOMOZYGOUS

WW ← if

DOMINANT PHENOTYPE (ALLELE DISTRIBUTION UNKNOWN)

if → Ww

HOMOZYGOUS RECESSIVE (WHITE)
ww

Ww | Ww
ww | ww

HALF OF OFFSPRING PURPLE, THE REST WHITE; THEREFORE UNKNOWN FLOWER WAS HETEROZYGOUS

FIGURE 11-10

A test cross. Mendel used this to determine whether an individual exhibiting a dominant phenotype, such as purple flowers, is homozygous or heterozygous for the dominant allele. Both such individuals would be expected to exhibit purple flowers. Mendel crossed the individual in question to a plant he knew to be double recessive, in this case a plant with white flowers. If the purple-flowered individual being tested was homozygous dominant, all the progeny would have been heterozygotes with purple flowers. If the purple-flowered individual being tested was heterozygous for the gene controlling flower color, half of its progeny would have been homozygous for the recessive allele and therefore white-flowered. As usual, large progenies are needed to approach these theoretical proportions, and small progenies may vary widely based on chance alone.

INDEPENDENT ASSORTMENT

After Mendel had demonstrated that different alleles of a given gene segregate independently of one another in crosses, it seemed logical to ask whether different *genes* also segregated independently of one another. For example, would the particular alleles of a gene for seed shape that a gamete possessed have any influence on which allele the gamete had for a gene affecting seed color?

Mendel set out to answer this question in a straightforward way. He first established a series of true-breeding lines of peas that differed from one another with respect to two of the seven pairs of characteristics that he had studied. Second, he crossed contrasting pairs of the true-breeding lines. In a cross involving different seed-shape alleles (round, *W*, and wrinkled, *w*) and different seed-color alleles (yellow, *G*, and green, *g*), all the F_1 individuals were identical, each being heterozygous for both seed shape *(Ww)* and seed color *(Gg)*. The F_1 individuals of such a cross are said to be dihybrid individuals. A **dihybrid** is an individual heterozygous for two genes.

The third step in Mendel's analysis was to allow the dihybrid individuals to self-fertilize. If the segregation of alleles affecting seed shape and that of alleles affecting seed color were independent, the probability that a particular pair of seed-shape alleles would occur together with a particular pair of seed-color alleles would be simply the product of the two individual probabilities that each pair would occur separately. Thus, the probability of an individual with wrinkled, green seeds appearing in the F_2 generation would be equal to the probability of observing an individual with wrinkled seeds (¼) times the probability of observing an individual with green seeds (¼), or ¹⁄₁₆.

Since the genes concerned with seed shape and those concerned with seed color are each represented by a pair of alleles in the dihybrid individuals, four types of gametes are expected: *WG, Wg, wG, wg*. Thus, in the F_2 generation, there are 16 possible combinations of alleles, each of them equally probable (Figure 11-11). Of the 16 combinations, 9 possess at least one dominant allele for each gene (the second allele is usually signified with a dash, *W–G–*, which means that the second allele at the w locus and the second allele at the g locus may be either dominant or recessive) and thus should have round, yellow seeds. Three possess at least one dominant *W* allele but are double recessive for color *(W–gg)*. Three other combinations possess at least one dominant G allele but are double recessive for shape *(wwG–)*. One combination among the 16 is double recessive for both genes *(wwgg)*. The hypothesis that color and shape genes assort independently thus predicts that the F_2 generation of this dihybrid cross will display this ratio: nine individuals with round, yellow seeds to three individuals with round, green seeds to three individuals with wrinkled, yellow seeds to one with wrinkled, green seeds—a 9:3:3:1 ratio.

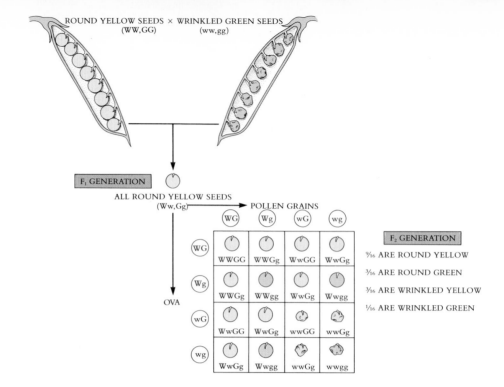

ROUND YELLOW SEEDS × WRINKLED GREEN SEEDS
(WW,GG)   (ww,gg)

F₁ GENERATION

ALL ROUND YELLOW SEEDS
(Ww,Gg)

POLLEN GRAINS

WG   Wg   wG   wg

OVA

WG — WWGG   WWGg   WwGG   WwGg
Wg — WWGg   WWgg   WwGg   Wwgg
wG — WwGG   WwGg   wwGG   wwGg
wg — WwGg   Wwgg   wwGg   wwgg

F₂ GENERATION

9/16 ARE ROUND YELLOW
3/16 ARE ROUND GREEN
3/16 ARE WRINKLED YELLOW
1/16 ARE WRINKLED GREEN

FIGURE 11-11

A dihybrid cross. The two pairs of contrasting traits employed by Mendel in this cross were round *(W)* versus wrinkled *(w)* seeds and yellow *(G)* versus green *(g)* seeds *(photos)*. In the F₁ generation, all seeds were round and yellow *(Ww, Gg)*. Each such individual is able to make four different kinds of gametes, its *W* allele paired in a gamete with either *G* or *g,* and its *w* allele paired with either *G* or *g.* If the male parent contributes four kinds of gametes and the female parent contributes four kinds of gametes, 16 (4 × 4) different kinds of zygotes are possible. As you can see in the Punnett square, nine of these possibilities produce round, yellow seeds; that is, they have at least one dominant allele for both traits and thus exhibit both dominant phenotypes. Three of the possible zygotes produce plants with round, green seeds; they have the dominant *W* allele but are homozygous *gg.* Three other zygotes produce plants with yellow, wrinkled seeds; they have at least one dominant *G* allele but are homozygous *ww.* Only 1 of the 16 possible combinations is a zygote that would give rise to a plant with wrinkled, green seeds, that is, one that is both homozygous *ww* and *gg.* The ratio of the four possible combinations of phenotypes is thus expected to be 9:3:3:1, the ratio that Mendel found repeatedly in such crosses.

What did Mendel actually observe? He examined a total of 556 seeds from dihybrid plants that had been allowed to self-fertilize, and he obtained the following results:

315	Round, yellow	*W–G–*
108	Round, green	*W–gg*
101	Wrinkled, yellow	*wwG–*
32	Wrinkled, green	*wwgg*

This is very close to a theoretical 9:3:3:1 ratio, which would have been 313:104:104:35. Thus, the two genes appeared to assort completely independently of one another. Note that this independent assortment of different genes in no way contradicts the independent segregation of alleles. Round and wrinkled seeds occur approximately in the ratio 3:1 (423:133), as do yellow and green seeds (416:140). Mendel obtained similar results for other pairs of traits.

Mendel's observation is often referred to as **Mendel's Second Law** or the **Law of Independent Assortment.** As you will see, genes that assort independently of one another, as did Mendel's genes, often do so because the genes are located on different chromosomes, which segregate from one another during the process of meiosis (Figure 11-12). A modern restatement of Mendel's Second Law is that *genes located on different chromosomes assort independently during meiosis.*

Mendel's Second Law of Heredity states that genes located on different chromosomes assort independently of one another.

Mendel's original paper describing his experiments, published in 1866, remains charming and interesting reading. His explanations are clear, and the logic of his arguments is presented in lucid detail. Unfortunately, Mendel failed to arouse much interest in his findings, which were published in the journal of the local natural history society. Only 115 copies of the journal were sent out, in addition to 40 reprints, which Mendel distributed himself. Although Mendel's results did not receive much notice during his lifetime, in 1900, 16 years after his death, three different investigators independently rediscovered his pioneering paper. They came across it while they were searching the literature in preparation for publishing their own findings, which were similar to those Mendel had quietly presented more than three decades earlier.

SOME KEY TERMS IN GENETICS

- **allele** One of two or more alternative states of a gene.
- **diploid** Having two sets of chromosomes, referred to as *homologues*. Animals, the dominant phase in the life cycle of plants, and some protists are diploid.
- **dominant allele** An allele that dictates the appearance of heterozygotes. One allele is said to be dominant over another if an individual heterozygous for that allele has the same appearance as an individual homozygous for it.
- **gene** The basic unit of heredity. A sequence of DNA nucleotides on a chromosome that encodes a polypeptide or RNA molecule and so determines the nature of an individual's inherited traits.
- **genotype** The total set of genes present in the cells of an organism. This term is often also used to refer to the set of alleles at a single gene locus.
- **haploid** Having only one set of chromosomes. Gametes, certain protists, and certain stages in the life cycle of plants are haploid.
- **heterozygote** A diploid individual carrying two different alleles of a gene on its two homologous chromosomes. Most human beings are heterozygous for many genes.
- **homozygote** A diploid individual whose two copies of a gene are the same. An individual carrying identical alleles on both homologous chromosomes is said to be *homozygous* for that gene.
- **locus** The location of a gene on a chromosome.
- **phenotype** The realized expression of the genotype. The phenotype is the observable expression of a trait (affecting an individual's structure, physiology, or behavior) that results from the biological activity of proteins or RNA molecules transcribed from the DNA.
- **recessive allele** An allele whose phenotypic effect is masked in heterozygotes by the presence of a dominant allele.

FIGURE 11-12

Independent assortment occurs because the orientation of chromosomes on the metaphase plate is random. Many different kinds of gametes are possible, based on the number 2 raised to a power equal to the haploid number of chromosomes. In this hypothetical three-chromosome cell, four of the eight possible orientations (2^3) are illustrated. Each orientation results in gametes with different combinations of parental chromosomes.

CHROMOSOMES: THE VEHICLES OF MENDELIAN INHERITANCE

Chromosomes are not the only kinds of organelles that segregate regularly when eukaryotic cells divide. Centrioles also divide and segregate in a regular fashion, as do the mitochondria and chloroplasts in the cytoplasm. In the early years of this century, it was not obvious that chromosomes were the vehicles for the information of heredity. The German geneticist Carl Correns first suggested a central role for chromosomes in 1900 in one of the papers announcing the rediscovery of Mendel's work. Soon after, observations that similar chromosomes paired with one another in the process of meiosis led directly to the **chromosomal theory of inheritance,** first formulated by the American Walter Sutton in 1902. Sutton argued as follows:

1. Reproduction involves the initial union of only two cells, egg and sperm. If Mendel's model is correct, then these two gametes must make equal hereditary contributions. Sperm, however, contain little cytoplasm. Therefore, the hereditary material must reside within the nuclei of the gametes.
2. Chromosomes segregate during meiosis in a fashion similar to that exhibited by the alleles Mendel studied.
3. Gametes have a copy of one member of each pair of homologous chromosomes; diploid individuals have a copy of both members of the pair. Similarly, Mendel found that gametes have one allele of a gene and that diploid individuals have two.
4. During meiosis, each pair of homologous chromosomes orients on the metaphase plate independently of any other pair. This independent assortment of chromosomes is a process similar to the independent assortment of alleles that Mendel studied.

FIGURE 11-13

White-eyed (mutant) and red-eyed (normal) *Drosophila*. The white-eyed defect in eye color is hereditary, the result of a mutation in a gene located on the sex-determining X chromosome. By studying this mutation, Morgan first demonstrated that genes are on chromosomes.

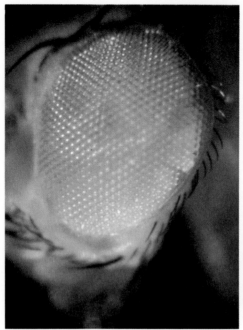 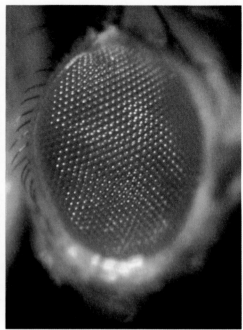

The proof that the genes were located on chromosomes was provided by a single small fly. In 1910, the American geneticist Thomas Hunt Morgan, studying the fly *Drosophila melanogaster,* detected a **mutant** fly, a male fly that differed strikingly from normal flies of the same species. In this fly, the eyes were white instead of the normal red (Figure 11-13).

Morgan immediately set out to determine if this new trait would be inherited in a Mendelian fashion. He first crossed the mutant male to a normal female to see if either red or white eyes were dominant. All F_1 progeny had red eyes, and Morgan therefore concluded that red eye color was dominant over white. Following the experimental procedure that Mendel had established long ago, Morgan then crossed flies from the F_1 generation with each other. Eye color did indeed segregate among the F_2 progeny, as predicted by Mendel's theory. Of 4252 F_2 progeny that Morgan examined, 782 had white eyes—an imperfect 3:1 ratio, but one that nevertheless provided clear evidence of segregation. Something was strange about Morgan's result, however, something that was totally unpredicted by Mendel's theory: all the white-eyed F_2 flies were males!

How could this strange result be explained? Perhaps it was not possible to be a white-eyed female fly; such individuals might not be viable for some unknown reason. To test this idea, Morgan test-crossed one of the red-eyed F_1 female progeny back to the original white-eyed male and obtained white-eyed and red-eyed males and females. So a female could have white eyes. Why then were there no white-eyed females among the progeny of the original cross?

The solution to this puzzle proved to involve sex. In *Drosophila,* the sex of an individual is influenced by the number of copies of a particular chromosome, the **X chromosome,** that an individual possesses. An individual with two X chromosomes is a female. An individual with only one X chromosome, which pairs in meiosis with a large, dissimilar partner called the **Y chromosome,** is a male. The female thus produces only X gametes, whereas the male produces both X and Y gametes. When fertilization involves an X sperm, the result is an XX zygote, which develops into a female. When fertilization involves a Y sperm, the result is an XY zygote, which develops into a male.

The solution to Morgan's puzzle lies in the fact that in *Drosophila* the white-eye trait resides on the X chromosome and is absent from the Y chromosome. We now know that the Y chromosome carries almost no functional genes. A trait that is determined by a factor on the X chromosome is said to be **sex linked** or **X-linked.** Knowing

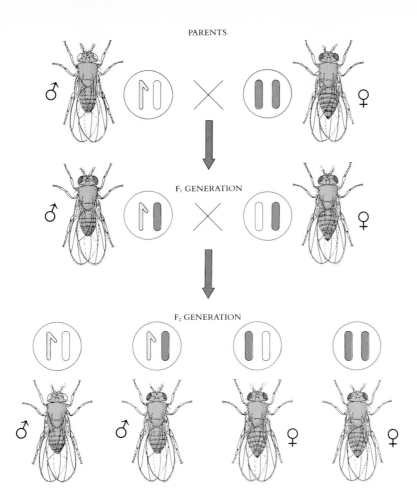

PARENTS

F₁ GENERATION

F₂ GENERATION

FIGURE 11-14

Morgan's experiment demonstrating the chromosomal basis of sex linkage in *Drosophila*. The white-eyed mutant male fly was crossed to a normal female. The F₁-generation flies all exhibited red eyes, as expected for flies heterozygous for a recessive white-eye allele. In the F₂ generation, every progeny fly obtains one sex chromosome from its father. If that chromosome is an X chromosome, then the progeny fly will be female. Because the F₁ father's X chromosome carries the normal red-eye allele, which is dominant, all of these female progeny flies will be red-eyed. If, on the other hand, the progeny fly obtains the Y chromosome of its father, it will be a male. Because the Y chromosome carries no genes, the X chromosome contributed by the F₁ mother will exhibit whatever allele it bears. Since half of them bear the white allele (the F₁ mother is heterozygous), half the male progeny will exhibit white eyes. This is just the result Morgan observed: all the white-eyed F₂-generation flies were male.

that the white-eye trait is recessive to the red-eye trait, we can now see that Morgan's result was a natural consequence of the Mendelian assortment of chromosomes (Figure 11-14).

Morgan's experiment is one of the most important in the history of genetics, because it presented the first clear evidence that Sutton was right and that the factors determining Mendelian traits do indeed reside on the chromosomes. The segregation of the white-eye trait, evident in the eye color of the flies, has a one-to-one correspondence with the segregation of the X chromosome, evident from the sexes of the flies.

The white-eye trait behaves exactly as if it were located on an X chromosome, which is indeed the case. The gene that specifies eye color in *Drosophila* is carried through meiosis as part of an X chromosome. In other words, Mendelian traits such as eye color in *Drosophila* assort independently because chromosomes do. When Mendel observed the segregation of alleles in pea plants, he was observing a reflection of the meiotic segregation of chromosomes.

Mendelian traits assort independently because they are determined by genes located on chromosomes, which also assort independently in meiosis.

CROSSING-OVER

Morgan's results led to the general acceptance of Sutton's chromosomal theory of inheritance. Scientists then attempted to find out why there are more different kinds of genes that assort independently than there are chromosomes. Curt Stern, an American geneticist, published in 1931 the experimental proof that the formation of chiasmata between paired chromosomes (see Chapter 10) also provided a means by which genes

FIGURE 11-15

Curt Stern's experiment demonstrating that a physical exchange of chromosomal arms occurs during crossing-over. Stern monitored crossing-over between two genes, recessive carnation eye color *(car)* and dominant Bar-shaped eye *(B)*, on chromosomes with physical pecularities that he could see under a microscope. Stern separated out those F_2-generation flies in which crossing-over had occurred between *car* and *B* producing new combinations of these alleles. The original F_1 female had Bar-shaped, carnation-colored eyes, whereas the F_2-generation flies in which crossing-over had occurred exhibited carnation-colored eyes of normal shape and Bar-shaped eyes of normal color. Stern then examined the chromosomes of these flies. All showed new combinations of the chromosomal abnormalities as well. When he examined the F_2 flies with parental combinations of eye traits, they all proved to possess the parental combinations of chromosomal abnormalities. The result is that whenever genes recombine through crossing-over, chromosomes recombine as well. The recombination of genes reflects a physical exchange of chromosome arms.

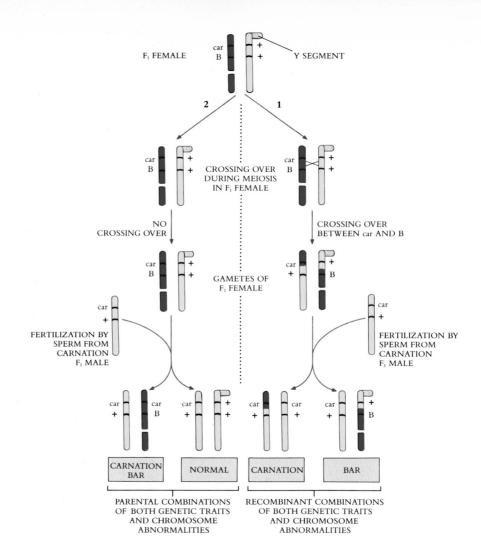

on the same chromosome could reassort into different combinations than those that occur originally. Stern's experiment is diagrammed in Figure 11-15.

Stern studied two sex-linked traits in strains of *Drosophila* whose X chromosomes were visibly abnormal at both ends. He first examined many flies and identified those in which an exchange had occurred with respect to the two eye traits. Stern then studied the chromosomes of these flies to see if their X chromosomes had exchanged arms. He found that all the individuals that had expressed nonparental combinations of the parental traits also possessed chromosomes whose abnormal ends could be seen to have exchanged (Figure 11-16). The conclusion was clear: genetic exchanges of traits on a chromosome, such as eye color, involves crossing-over, the physical exchange of chromosome arms.

The patterns of assortment of genes that actually occur indicate an important fact: in general, if genes are located relatively far apart on chromosomes, crossing-over is more likely to occur between them than if they are located close together. If the genes are located far enough apart, they segregate independently.

Genetic Maps

Because crossing-over occurs more often between two genes relatively far apart than between another set of two genes relatively close to each other, the frequency of crossing-over can be used to map the relative positions of genes on chromosomes. In a cross, the proportion of progeny derived from a gamete in which an exchange has occurred between two genes is a measure of how frequently cross-over events occur between them, and thus of the distance separating them.

FIGURE 11-16

Crossing-over occurs during prophase of meiosis I, when the two homologues are aligned together within the synaptonemal complex. The exchange creates a chiasma **(A).** For clarity, the chiasma has been allowed to break apart immediately after crossing-over **(B),** although in reality many chiasmata persist into metaphase. As you can see, the two homologues have exchanged arms at the point of crossing-over. When the two homologues are drawn to opposite poles **(C),** they produce gametes with new groups of alleles.

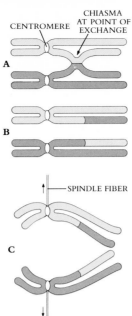

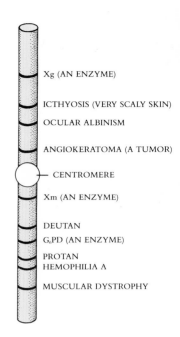

FIGURE 11-17

A recombination map of the human X chromosome. Deuteranopia *(deutan)* and protanopia *(protan)* are two forms of color blindness, *XM* is a protein found in the blood, and *Xg* is a cell-surface protein of red blood cells. At least 93 human genes have been shown to be located on the X chromosome, and more preliminary evidence indicates that almost as many others also belong on this chromosome.

A **genetic map** is a diagram showing the relative positions of genes. Genetic maps can be constructed by ordering fragments of DNA, by studing chromosomal alterations, or by performing crosses and seeing how frequently crossing-over occurs between pairs of genes. When the frequencies of crossing-over events in crosses are used to construct a genetic map, it is called a **recombination map.** Distances are measured in terms of the frequency of crossing over. On such a map, one "map unit" is defined as the distance within which a cross-over event is expected to occur, on the average, in one of 100 gametes. This unit, 1% recombination, is now called a **centimorgan,** in honor of Thomas Hunt Morgan.

In constructing a genetic map, one simultaneously monitors recombination among three or more genes located on the same chromosome. When genes are close enough together on a chromosome that they do not assort independently, they are said to be **linked** to one another. Figure 11-17 shows a genetic map for the human X chromosome.

GENETICS

After many centuries of speculation about heredity, the puzzle was finally solved in the space of a few generations. Guided by the work of Mendel and a generation of investigators determined to explain his results, the basic outline of how heredity works soon became clear. Hereditary traits are specified by genes, which are integral parts of chromosomes, and the movements of chromosomes during meiosis produce the patterns of segregation and independent assortment that Mendel reported. Two of the most important discoveries were that (1) chromosomes exchange genes during meiosis, and (2) genes located far apart on chromosomes are more likely to have an exchange occur between them. These findings allowed investigators to learn how genes are distributed on chromosomes long before we knew enough to isolate and study them.

This core of knowledge, this basic outline of heredity, has led to a long chain of investigation and questions. What is the physical nature of a gene? How is information encoded within genes? How do genes change, and why? How does a gene create a phenotype? The people that ask such questions are called geneticists, and the body of what they have learned and are learning is called genetics. Genetics is one of the most active subdisciplines of biology. The next few chapters will answer the questions just posed and many others.

SUMMARY

1. Knight and others noted the basic facts of heredity a century before Mendel. They found that alternative traits segregate in crosses and may mask each other's appearance. Mendel, however, was the first to quantify his data, counting the numbers of each alternative type among the progeny of crosses.

2. From counting progeny classes, Mendel learned that the alternatives that were masked in hybrids appeared only 25% of the time when they subsequently segregated in the F_2 generation. This finding, which led directly to Mendel's model of heredity, is usually referred to as the Mendelian ratio of 3:1, the ratio of dominant to recessive traits.

3. Mendel deduced from the 3:1 ratio that traits are specified by what he called discrete "factors," which do not blend. We refer to Mendel's factors as genes and to his alternative factors as alleles. Mendel deduced that pea plants contain two factors for each feature that he studied; we now know this is because they are diploid. When the two factors are not the same (when the plant is heterozygous), one factor, which Mendel described as dominant, determines the appearance (phenotype) of the individual.

4. When two heterozygous individuals mate, an individual offspring has a 50% (random) chance of obtaining either allele from the father, and a 50% chance of obtaining either allele from the mother. Thus, the probability of obtaining two dominant alleles, of being homozygous dominant, is $0.5 \times 0.5, = 0.25$, or 25%. Similarly, the probability of being homozygous recessive is 25%. The rest of the progeny, one half, are heterozygotes. Because the appearance of heterozygotes is specified by the dominant allele, the progeny thus appear as ¾ dominant, ¼ recessive, a ratio of 3:1 dominant to recessive.

5. When two genes are located on different chromosomes, the alleles included in an individual gamete are selected at random. The allele selected for one gene has no influence on which allele is selected for the other gene. Such genes are said to assort independently.

6. Thomas Hunt Morgan provided the first clear evidence that genes reside on chromosomes. Morgan demonstrated that the segregation of the white-eye trait in *Drosophila* flies was associated with the segregation of the X chromosome, the one responsible for sex determination.

7. Curt Stern reported the first evidence that crossing-over occurs between chromosomes. He showed that when two Mendelian traits exchange during a cross, visible abnormalities on the ends of the chromosomes bearing the traits also exchange.

8. The frequency of crossing-over between genes can be used to construct genetic maps. Such maps are representations of the physical locations of genes on chromosomes, inferred from the degree of crossing-over between particular pairs of genes.

REVIEW (Genetics problems appear in Appendix B)

1. The set of alleles that an individual contains is called its _____, whereas the physical appearance of that individual is its _____.

2. An individual that is homozygous for the recessive white-flower trait will produce gametes of what type?

3. A purple-flowered individual can have one of two genotypes. How would you unambiguously determine the genotype of this individual?

4. Which of Mendel's laws is demonstrated by the 9:3:3:1 ratio observed in the F_2 generation of a dihybrid cross?

5. Because _____ occurs more frequently between genes far from each other on a chromosome and less frequently between genes close together, this information can be used to determine the relative position of genes on a chromosome.

6. In a cross of purple- and white-flowered plants, the F_1 offspring observed by Mendel were all purple. A cross of two of these purple plants yielded F_2 purple- and white-flowered plants in the ratio of _____.

7. When two heterozygous individuals are crossed, the percentage of the progeny that exhibit the recessive trait is _____.

8. The exchange of segments of chromosomes during meiosis is called _____.

9. In Sweden, many more people possess blue eyes than eyes of all other colors combined. It follows that the allele determining blue eyes that occurs in Sweden is dominant to most other eye color alleles. True or false?

10. The closer two genes are on a chromosome, the less likely that crossing-over will occur between them. True or false?

11. When two heterozygous individuals are crossed, the proportion of the progeny that are true-breeding for the dominant trait is _____.

12. Under what circumstances does Mendel's second law (independent assortment) not hold?

SELF-QUIZ

1. Several people had carried out plant crosses before Mendel, but they are not credited with being the founder of the science of genetics. What did Mendel do that was different?
   (a) he used garden peas
   (b) he did specific crosses
   (c) he kept track of the numbers of individuals of the different types produced from his crosses
   (d) he removed the male parts from flowers so that self-fertilization could not occur
   (e) both a and b

2. In all of Mendel's crosses, the F_2 plants displayed a 3:1 ratio of dominant to recessive traits. Of those showing the dominant character, what proportion were pure-breeding?
   (a) ¼   (d) ⅔
   (b) ⅓   (e) all
   (c) ½

3. Which of the following is *not* one of the five elements of Mendel's model?
   (a) Parents transmit discrete "factors" of information that will act in the offspring to produce specific traits.
   (b) The presence of any particular element does not guarantee its expression.
   (c) The two elements contributed, one by each parent, do not influence each other.
   (d) All copies of "factors" are identical.
   (e) Each parent, with respect to each trait, contains two factors, one of which will be packaged at random into the gametes.

4. The white color of a homozygous recessive plant is its
   (a) phenotype   (b) genotype

5. Of the following crosses, which is a test cross?
   (a) $WW \times WW$ (c) $Ww \times ww$
   (b) $WW \times Ww$ (d) $Ww \times W$

6. Morgan's experiments on white-eyed *Drosophila* clearly showed that
   (a) white-eyed flies see better than normal flies
   (b) chromosomes are the carriers of genetic information
   (c) only female flies can have white eyes
   (d) diploid individuals have two copies of each trait
   (e) because eggs carry not only the genetic factors but large amounts of cytoplasm, the female has a greater influence on the genotype of the offspring

7. Of the seven genes that Mendel studied, only two are located by themselves on a particular chromosome. The others are on a chromosome with another of Mendel's traits. In view of this, how could Mendel have come to the conclusion that genes assort independently of one another?
   (a) the data are not clear on this point
   (b) they did not assort independently
   (c) Mendel erred
   (d) the traits were far enough apart that, because of the high probability of crossing-over between them, they behaved as if they were on different chromosomes
   (e) Mendel knew that the traits were on the same chromosome and took this into account

8. What is "independent assortment"?
   (a) Different genes on the same chromosome segregate independently of one another in genetic crosses.
   (b) Different alleles of the same gene segregate independently in genetic crosses.
   (c) Different genes on different chromosomes segregate independently of one another in genetic crosses.

9. How many alleles of a given gene are possible?
   (a) one   (d) four
   (b) two   (e) any number
   (c) three

FOR FURTHER READING

BLIXT, S.: "Why Didn't Gregor Mendel Find Linkage?" *Nature*, vol. 256, page 206, 1975. An interesting re-examination of the pea strains first studied by Mendel.

MORGAN, T.H.: "Sex-limited Inheritance in *Drosophila*," *Science*, vol. 32, pages 120–122, 1910. Morgan's original account of his famous analysis of the inheritance of the white-eye trait.

SUTTON, W.S.: "The Chromosomes of Heredity," *Biological Bulletin*, vol. 4, pages 213–251, 1903. The original statement of the chromosomal theory of heredity.

FIGURE 12-1   Czar Nicholas II of Russia, the last of the czars, and his wife Alexandra with their children, Olga, Tatiana, Maria, Anastasia, and Alexis. Alexandra was a carrier of the genetic disease hemophilia, a disorder in which the blood does not clot properly. She passed it on to her son Alexis. Since the type of hemophilia they had is caused by a mutant allele located on the X chromosome, it is not expressed in the heterozygous condition in women, who have two such chromosomes. It is not known which of her four daughters might have received the hemophilia allele from her, since none lived to bear children.

HUMAN GENETICS

Chapter 12

Overview

The principles of genetics apply not only to pea plants and *Drosophila,* but also to you. How closely you resemble your father or mother was largely established before your birth by the chromosomes that you received from them, just as meiosis in peas determined the segregation of Mendel's traits. Many of the alleles segregating within human populations, however, demand more serious concern than the color of a pea. Some of the most devastating human disorders result from alleles specifying defective forms of proteins with important functions in our bodies. These disorders are passed from parent to child. By studying human heredity, we are beginning to learn how to limit the misery that these disorders bring to so many human families.

For Review

Here are some important terms and concepts that you will encounter in this chapter. If you are not familiar with them, you should review them before proceeding.

Cell-surface proteins (Chapter 6)

Transport channels (Chapter 6)

Cholesterol uptake (Chapter 6)

Karyotypes (Chapter 10)

Sex linkage (Chapter 11)

Dominant and recessive traits (Chapter 11)

Biologists often investigate heredity by studying odd creatures such as the *Drosophila* fly, organisms that we may care about very little. The knowledge that biologists acquire by studying other organisms, however, often tells us a great deal about ourselves. Nowhere is this more true than in the study of genetics, the study of the ways in which individual traits are transmitted from one generation to the next. When biologists studying pea plants and *Drosophila* established that patterns of heredity reflect the segregation of chromosomes in meiosis, their discovery opened the door to the study of human heredity. By studying human chromosomes, it became possible to learn why certain hereditary disorders seem to leap generations and why, for example, others affect only males. Scientists are more and more able to predict which hereditary disorders parents might expect to pass on to their children, and with what probabilities (Figure 12-1).

Why do we devote a special chapter to human heredity? Because we are interested in ourselves. Although we humans pass our genes to the next generation in much the same way that other organisms do, we naturally have a special curiosity about ourselves. Most of us know someone, whether a relative or a friend's relative, who suffers from a condition that might be hereditary. If a member of your family has had a stroke, for example, it is difficult not to worry about your own future health, knowing that the propensity to suffer strokes can be hereditary. Few women have babies without some worrying about the possibility that their child may have some defect. From such standpoints, we are all human geneticists, interested in what the laws of genetics reveal about ourselves, our families, our friends, and our future children.

HUMAN CHROMOSOMES

Although chromosomes were discovered more than a century ago, the exact number of chromosomes that humans possess was not established accurately until 1956. At that time, appropriate techniques were developed that allowed investigators to determine accurately the number, shape, and form of human chromosomes and those of other mammals. Before then, the number of chromosomes characteristic of human beings had been known only approximately. We now know, however, that each typical human cell has 46 chromosomes.

How do biologists examine human chromosomes? They collect a blood sample, add chemicals that induce the blood cells in the sample to divide, and then add other chemicals that stop cell division at metaphase. Metaphase is the stage of mitosis when the chromosomes are most condensed and thus most easily distinguished from one another. After arresting the process of cell division in the sample at this stage, the biologist breaks the cells to spread out their protoplast and the chromosomes within it, then stains and examines the chromosomes. To facilitate the examination of the particular array of chromosomes, or karyotype, that a person possesses, the chromosomes are often photographed, and then the outlines of each chromosome are cut out, like paper dolls, and arranged in order (Figure 12-2).

Conventionally, the 23 different kinds of human chromosomes are arranged into seven groups, each characterized by a different size, shape, and appearance. These groups are designated A through G. Of the 23 pairs, 22 are perfectly matched in both males and females and are called **autosomes.** The remaining pair consists of two unlike members in males; in females, it consists of two similar members. The chromosomes that constitute this pair are called the **sex chromosomes.** Just as with *Drosophila,* females are designated XX and males XY. This indicates that the male pair of sex chromosomes contains one member (the Y chromosome) that bears few functional genes and thus differs in this respect from all other chromosomes (Figure 12-3).

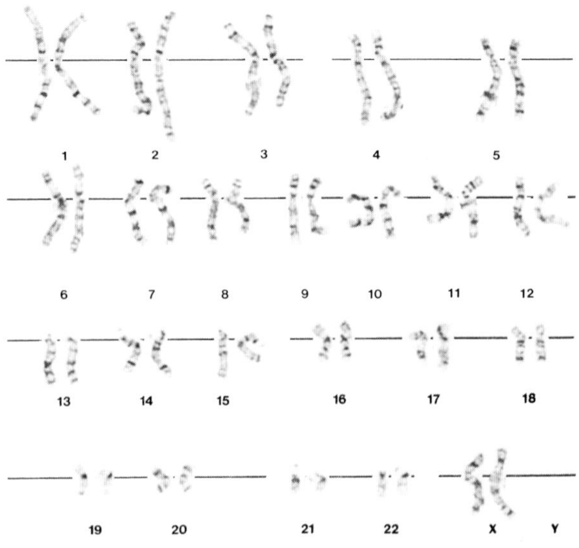

FIGURE 12-2

Human beings normally have 46 chromosomes in their somatic (body) cells, including 22 pairs of autosomes and two sex chromosomes. In a karyotype, such as the one shown here, the two members of each chromosome pair are grouped together, and the autosomes are arranged in order of descending size. Those chromosomes in which the centromere is near the middle (metacentric chromosomes) are placed first within a given size class; chromosomes with the centromere located nearer one end are placed after them. A normal woman has two X chromosomes, as shown, and a normal man has an X chromosome and a much smaller Y chromosome.

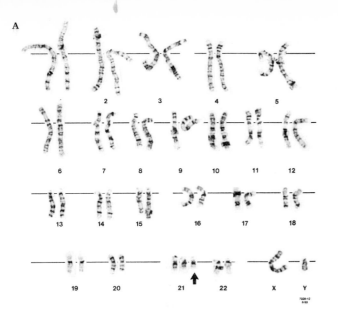

A

2   3   4   5

6   7   8   9   10   11   12

13   14   15   16   17   18

19   20   21   22   X   Y

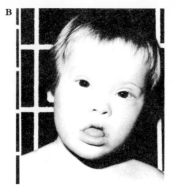

B

Down Syndrome

The karyotype seen in Figure 12-2 is characteristic of a very high proportion of all fe-male human cells. Almost everyone has the same karyotype for the same reason that al-most all cars have engines, transmissions, and wheels: other arrangements don't work well. Human individuals who have lost even one autosome (called **monosomics**) do not survive development. In all but a few cases, those who have gained an extra auto-some (called **trisomics**) also do not survive. Five of the smallest chromosomes, those numbered 13, 15, 18, 21, and 22, can be present in human beings as three copies and still allow the individual to survive for a time. Individuals with an extra chromosome 13, 15, or 18, however, undergo severe developmental defects, and infants with such a ge-netic constitution die within 3 months. In contrast, individuals who have an extra copy of chromosome 21 (Figure 12-3), and more rarely those who have an extra copy of chromosome 22, usually do survive to become adults. In such individuals, however, the maturation of the skeletal system is delayed, so that generally they are short and have poor muscle tone. Their mental development is also affected, and children with trisomy 21 or trisomy 22 are always mentally retarded. The developmental defect pro-duced by trisomy 21 was first described in 1866 by J. Langdon Down; thus it is called *Down syndrome* (formerly "Down's syndrome").

Down syndrome occurs frequently in all human racial groups, with an approxi-mate incidence of 1 in every 750 children. Similar conditions also appear in chimpan-zees and other related primates. In humans the defect is associated with a particular small portion of chromosome 21. When this chromosomal segment is present in three copies instead of two, Down syndrome results. In 97% of the human cases examined, all of chromosome 21 is present in three copies. In the other 3%, a small portion con-taining the critical segment has been added to another chromosome, in addition to the normal two copies of chromosome 21.

The developmental role of the genes whose duplication produces Down syn-drome is not known in detail, although clues are beginning to emerge from current re-search. When human cancer-causing genes (see Chapter 15) were localized on the hu-man chromosomes in 1985, one of them turned out to be located on chromosome 21, at precisely the location of the segment associated with Down syndrome. Cancer is in-deed more common in children with Down syndrome. The incidence of leukemia, for example, is 11 times higher in children with Down syndrome than in children of the same age without Down syndrome. Recently it has been demonstrated that a gene for Alzheimer's disease, a form of inherited senility, is also located on chromosome 21 near the genes governing Down syndrome, and it turns out that Down syndrome children indeed commonly get Alzheimer's disease.

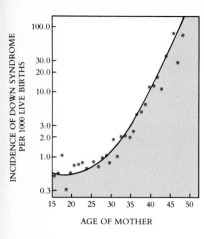

INCIDENCE OF DOWN SYNDROME
PER 1000 LIVE BIRTHS

AGE OF MOTHER

FIGURE 12-4

The graph shows the correlation between the age of the mother and the incidence of Down syndrome. The primary nondisjunction of chromosome 21 may result in Down syndrome if the gamete that receives the extra copy of this chromosome functions to produce a zygote. For unknown reasons, this nondisjunction event almost always occurs during the production of the female gamete, or egg, instead of the sperm. Its frequency increases sharply as a woman grows older.

How does Down syndrome arise? In humans, it almost exclusively results from **primary nondisjunction** (failure of chromosomes to separate in meiosis) of chromosome 21 in the meiotic event that leads to the formation of the egg. The cause of these primary nondisjunctions is not known, but as with cancer, their incidence increases with age (Figure 12-4). In mothers less than 20 years of age, the incidence of Down syndrome is only about 1 per 1700 births; in mothers 20 to 30 years old, the risk is only slightly greater, about 1 per 1400. In mothers 30 to 35 years old, however, the risk doubles, to 1 per 750. In mothers older than 45, the risk is as high as 1 in 16 births.

Primary nondisjunctions occur much more often in women than men because all the eggs that a woman will ever produce have begun their development, to the point of prophase of the first meiotic division, by the time the woman is born. In males, the development of sperm is initiated continuously. A much greater chance thus exists for problems of various kinds, including primary nondisjunction, to accumulate over time in the gametes of women than in those of men. For this reason, the age of the mother is more critical than that of the father when a couple contemplates childbirth.

Sex Chromosomes

As you have seen, humans possess 22 pairs of autosomes that are the same in both sexes and one pair of sex chromosomes that differs between males and females. Females are designated XX, since they have two similar sex chromosomes, and males are designated XY, since their sex chromosomes differ. The difference between them can be seen readily in Figure 12-3. The Y chromosome is highly condensed, and few of its genes are expressed. For this reason, recessive alleles that are present on the single X chromosome of males have no counterpart on the Y chromosome, at least no active counterpart. The characteristics of such alleles are often expressed, just as if they were present in a homozygous condition in a female. The Y chromosome is not completely inert genetically, however; it does possess some active genes. For example, the genes determining whether a zygote will develop into a male are located on the Y chromosome: any individual containing at least one Y chromosome is a male, and any individual containing no Y chromosome is a female. The number of X chromosomes an individual possesses has no effect on its expression of the features characteristic of males and females.

Because a female has two copies of the X chromosome and a male only one, you might think that female cells would produce twice as much of the proteins encoded by genes on the X chromosome. This does not occur, however, because in females one of the X chromosomes is inactivated shortly after sex determination, early in embryonic development. Such inactivation is called **Lyonization.** The inactivated chromosome can be seen as a deeply staining body, the **Barr body,** which remains attached to the nuclear membrane.

Individuals that lose a copy of the X chromosome or gain an extra one are not subject to the severe developmental abnormalities usually associated with similar changes in autosomes. When primary nondisjunction results in the acquisition of an extra X chromosome in a particular individual or in a female with only one X chromosome, the resulting individuals may become mature, as discussed next, although they have somewhat abnormal features.

The X Chromosome

Failure of the X chromosome to separate leads to some gametes that are XX and others that have no sex chromosome (designated O). If the XX gamete joins an X gamete to form a XXX zygote, the zygote that results develops into a female individual with one functional X chromosome and two Barr bodies. She is sterile but usually normal in other respects. If the XX gamete instead joins a Y gamete, the effects are more serious. The resulting XXY zygote develops into a sterile male who has many female body characteristics and, in some cases, diminished mental capacity. This condition, called **Klinefelter's syndrome,** occurs in about one out of every 500 male births.

The other gamete produced when the X chromosomes fail to separate, O, lacks any X chromosome. If this O gamete fuses with a Y gamete, the resulting OY zygote

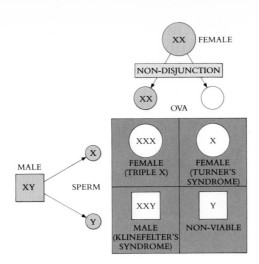

FIGURE 12-5

How nondisjunction can result in abnormalities in the number of sex chromosomes.

is inviable and fails to develop further. Human beings cannot survive without any of the genes on the X chromosome. If, on the other hand, the O gamete fuses with an X gamete to form an XO zygote, the result is a sterile female of short stature, a webbed neck, and sex organs that resemble those of an infant. The mental abilities of an XO individual are in the low-normal range. This condition, called **Turner's syndrome,** occurs roughly once in every 2000 female births. Figure 12-5 diagrams the ways in which nondisjunction can result in abnormalities in the number of sex chromosomes.

The Y Chromosome

The Y chromosome may also fail to separate in the anaphase II stage of meiosis. This leads to the formation of YY gametes and viable XYY zygotes, which develop into fertile males of normal appearance. The frequency of XYY among newborn males is about one per 1000. Interestingly, the frequency of XYY males in penal and mental institutions has been reported to be approximately 2% (that is, 20 per 1000), which is about 10 times the frequency of such individuals in the population at large. This observation has led to the suggestion that XYY males are inherently antisocial, a suggestion that has proved highly controversial. The observation is confirmed in some studies but not in others. In any case, the great majority of XYY males do not appear to develop patterns of antisocial behavior.

PATTERNS OF INHERITANCE

Imagine that you were trying to learn about an inherited trait that was present in your family. How would you find out if the trait is dominant or recessive, how many genes contribute to it, and how likely you might be to transmit it to your future children? If you were studying such a trait in *Drosophila,* you could conduct crosses, occasionally squashing a fly to examine its chromosomes. Studying your own heredity requires a more indirect approach.

To study heredity in humans, biologists look at the results of crosses that have already been made—they study family histories, called **pedigrees.** By studying which relatives exhibit a trait, it is often possible to determine whether the gene producing the trait is sex linked or autosomal and if the trait's phenotype is dominant or recessive. In many cases, one can infer which individuals are homozygous and which are heterozygous for the allele specifying the trait.

A Pedigree: Albinism

Albino individuals lack all pigmentation, so that their hair and skin are white. In the United States, about one Caucasian person in 38,000 and one black person in 22,000 are albinos. In the pedigree of albinism presented in Figure 12-6, each symbol represents one individual in the family history, with the circles representing females and the

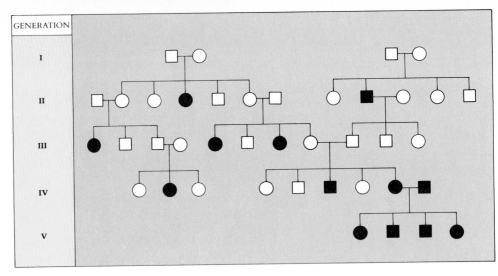

FIGURE 12-6

A pedigree typical of albinism. In pedigrees, males are conventionally shown as squares, females as circles, and marriages as horizontal lines connecting them, with the offspring shown below. The individuals who exhibit the trait being considered are indicated by solid symbols.

squares, males. In such a pedigree, individuals that exhibit a trait being studied, in this case albinism, are indicated by black solid symbols. Marriages are represented by horizontal lines connecting a circle and a square, from which a cluster of vertical lines indicate the progeny, arranged in order of birth.

How does one analyze the pedigree in Figure 12-6?

1. First, let us inquire whether albinism is sex linked or autosomal. If the trait is sex-linked, it may be expressed only in males, whereas if it is autosomal, it will appear in both sexes equally. In Figure 12-6, the proportion of affected males (5/13, or 39%) is similar to the proportion of affected females (8/21, or 37%). Thus we can conclude that the trait is autosomal rather than sex-linked.

2. Second, let us ask whether albinism is dominant or recessive. If the trait is dominant, every albino child will have an albino parent; if recessive, an albino child's parents can appear normal (both parents may be heterozygous). In Figure 12-6, parents of most of the albino children do not exhibit the trait, which indicates that albinism is recessive. Four children in one family *do* have albino parents because the allele is very common among the Hopi Indians, from which this pedigree was derived. Thus, homozygous individuals, such as these albino parents, are present among the Hopis in sufficient numbers that they sometimes marry. Note that in this marriage *both* parents are albino and *all* four children are albino, as you would expect if the trait were recessive and both parents were therefore homozygous for this allele.

3. Third, let us see if the trait is determined by a single gene or by several. If the trait is determined by a single gene, then albinos born to heterozygous parents should occur in families in 3:1 proportions, reflecting Mendelian segregation in a cross. Thus, you would expect that about 25% of the children should be albinos. If the trait were determined by several genes, on the other hand, the proportion of albinos would be much lower, only a few percent. In this case, 9/34 (you don't count the four children of the marriage between two homozygous individuals, since this is not a cross between heterozygotes), or 27% of the children are albinos, strongly suggesting that only one gene is segregating in these crosses.

This is how pedigree analysis is done. Looking at the pattern of inheritance of albinism, we were able to learn that albinism is an autosomal recessive trait controlled by a single gene. Other traits are studied in a similar way.

In pedigree analysis, one acquires information about the phenotypes of family members in order to infer the genetic nature of a trait from the pattern of its inheritance.

MULTIPLE ALLELES

Mendel studied pairs of contrasting traits in pea plants. His plants were either tall or short, and their flowers were purple or white. Similarly, Morgan's *Drosophila* flies had eyes that were either white or red (the normal color). Many human genes also exhibit two alternative alleles. An individual, for example, is either albino or pigmented. Remember, however, that many genes possess more than two possible alleles. One such gene encodes an enzyme that adds sugar molecules to lipids on the surface of our body's blood cells. These sugars act as recognition markers in our immune system and are called **cell surface antigens.** The gene encoding the enzyme is designated I and possesses three common alleles: (1) allele B, which adds the sugar galactose; (2) allele A, which adds a modified form of the sugar, galactosamine; and (3) allele O, which does not add a sugar.

> **Many genes possess multiple alleles, several of which may be common within populations.**

When more than one allele occurs, which allele is dominant? Often, no one allele is. Instead, each allele has its own effect. Thus, an individual heterozygous for the A and B alleles of the I gene produces both forms of the enzyme and adds both galactose and galactosamine to lipids on the cell surface. The cell surfaces of this individual's blood cells thus possess antigens with both kinds of sugar attached to them. Because both alleles are expressed simultaneously in heterozygotes, the A and B alleles are said to be **co-dominant.** Either is dominant over the O allele, because in heterozygotes the A or B allele leads to sugar addition and the O allele does not.

ABO Blood Groups

Different combinations of the three possible I gene alleles occur in different individuals, since each person possesses two copies of the chromosome bearing the I gene and may be homozygous for any allele or heterozygous for any two. The different combinations of the three alleles produce four different phenotypes:

1. Persons who add only galactosamine are called **type A** individuals. They are either AA homozygotes or AO heterozygotes.
2. Persons who add only galactose are called **type B** individuals. They are either BB homozygotes or BO heterozygotes.
3. Persons who add both sugars are called **type AB** individuals. They are, as we have seen, AB heterozygotes.
4. Persons who add neither sugar are called **type O** individuals. They are OO homozygotes.

These four different cell-surface phenotypes are called the ABO blood groups, or, less often, the Landsteiner blood groups, after the man who first described them. As Landsteiner first noted, your immune system can tell the difference between these four phenotypes. If a type A individual receives a transfusion of type B blood, the recipient's immune system will recognize that the type B blood cells possess a "foreign" antigen (galactose) and attack the donated blood cells. If the donated blood is type AB, this will also happen. However, if the donated blood is type O, no attack will occur, since no foreign galactose antigens are present on the surfaces of blood cells produced by the type O donor. In general, any individual's immune system will tolerate a transfusion of type O blood, and so type O individuals are called **universal donors.** Because neither galactose nor galactosamine are foreign to type AB individuals (they add both to their red blood cells), an AB individual may receive any type of blood and is called a **universal recipient.** Figure 12-7 shows the combinations in which agglutination of blood cells will occur.

In human populations, some of the ABO blood group phenotypes are more common than others (Table 12-1). In general, type O individuals are the most common, and type AB individuals the least common. Human populations, however,

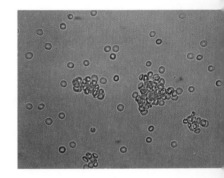

BLOOD TYPES AND AGGLUTINATION

RECIPIENT (SERUM TYPE)

BLOOD DONOR (BLOOD CELL TYPE)	A	B	AB	O
A	⊖	+	⊖	+
B	+	⊖	⊖	+
AB	+	+	⊖	+
O	⊖	⊖	⊖	⊖

+ = AGGLUTINATION

⊖ = NO REACTION

FIGURE 12-7

Type A red blood cells agglutinating in type O serum. The unclumped red blood cells are type O cells. The diagram shows the combinations in which agglutination will occur.

Human beings, like all other species, have differentiated in their characteristics as they have spread throughout the world. The local populations in one area often look impressively different from those that live elsewhere. For example, northern Europeans often have blond hair, fair skin, and blue eyes, whereas Africans often have black hair, dark skin, and brown eyes. In addition, other features, such as the blood groups discussed in this chapter, differ in proportion from area to area. These traits may play a role in adapting the particular populations to their environments. Thus, the blood groups may be associated with immunity to diseases characteristic of certain geographical areas, and dark skin shields the body from the damaging effects of ultraviolet radiation, which is much stronger in the tropics than in temperate regions.

All human beings, however, are capable of mating and producing fertile offspring. The reasons that they do or do not choose to associate with one another are purely psychological and behavioral (cultural). The number of races into which the human species might logically be divided, as well as which should be given names, has long been a point of contention. Some contemporary anthropologists divide people into as many as 30 races, others as few as three: Caucasoid, Negroid, and Oriental. American Indians, Bushmen, and Aborigines are particularly distinctive units that are sometimes regarded as distinct races.

The problem with classifying people or other organisms into races is that the characteristics used to define the races are usually not well correlated with one another. The way one determines the race to which a given group of people should be assigned is therefore always somewhat arbitrary. In human beings, it is simply not possible to delimit clearly defined races that can be recognized by a particular combination of characteristics. Variation patterns are somewhat correlated with geographical distribution, but they do not form clearly defined units that we can classify. Different groups of people have constantly intermingled and interbred with one another during the entire course of history. Today, the differences between human "races" are tending to break down rapidly as large numbers of individuals are constantly moving over the face of the globe, mixing with one another and recombining their characteristics.

In any event, individual differences within what might be considered races are greater than the differences between such units. This is a sound biological basis for dealing with each human being on his or her own merits, and not as a member of a particular "race."

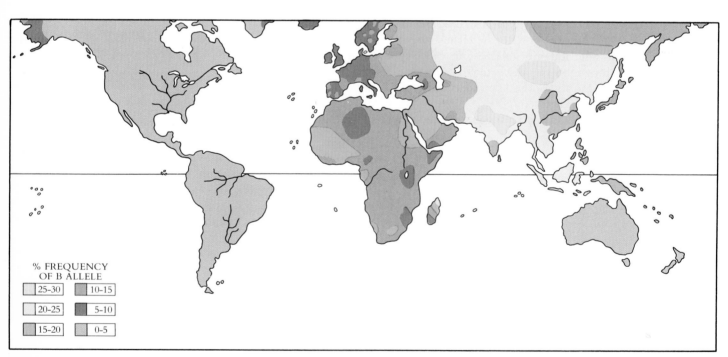

% FREQUENCY
OF B ALLELE

25–30	10–15
20–25	5–10
15–20	0–5

FIGURE 12-8

The frequency with which the B allele of the ABO blood group occurs throughout the world. By studying the distribution pattern of such alleles, biologists can often trace past migrations. The B allele is thought to have spread out from central Asia by migration.

TABLE 12-1   DISTRIBUTION OF ABO BLOOD GROUPS IN SOME HUMAN
             POPULATIONS

POPULATION	PHENOTYPE FREQUENCY (%)			
	A	B	AB	O
U.S. whites	39.7	10.6	3.4	46.3
U.S. blacks	26.5	20.1	4.3	49.1
African (Bantu)	25.0	19.7	3.7	51.7
Amerindians (Navaho)	30.6	0.2	0.0	69.1
Amerindians (Ecuador)	4.0	1.5	0.1	99.4
Japanese	38.4	21.9	9.7	30.1
Russians	34.6	24.2	7.2	34.0
French	45.6	8.3	3.3	42.7

differ greatly from one another. Among North American Indians, for example, the frequency of type A individuals is 31%, whereas among South American Indians, it is only 4%. Figure 12-8 illustrates the frequency with which B alleles occur in different parts of the world. This sort of genetic variation is an important property of genes in human and most other populations.

GENETIC DISORDERS

Most human genes vary. That is, among all living people, some individuals possess one allele of a particular gene, others another. However, most individuals possess the same allele as other people at each of most of their genes; most variant alleles are rare, and only a few occur frequently in human populations. Indeed, most human genes appear to possess only *one* common allele. Why only one allele? Probably because the proteins encoded by most genes must function in a very precise fashion in order for the many complex processes of development and the regulation of bodily functions to occur properly. Alternative alleles may lead to the production of proteins that do not function properly, if at all. Chapter 15 discusses the process, called mutation, that is responsible for the production of such alternative alleles. Here we need only note that the mutation process involves making random changes in genes. Changing a gene randomly rarely improves the functioning of its encoded protein, any more than randomly changing a wire in a computer is likely to improve the computer's functioning. On the other hand, variation is the material on which evolution, the progressive change and adaptation of life on earth, is based, so that changes in genes are, in general, quite necessary. Genetic disorders and other unfavorable conditions may be seen as a necessary consequence of this relationship.

You might expect that the alternative alleles with detrimental effects would be rare in human populations, and usually, but not always, they are. Sometimes a detrimental allele is common. This can happen in small, isolated communities, where an individual that happens to carry the allele is one of the few members. In many cases, we don't know why the allele is common. Whatever the reason that it is well represented, however, a common allele that results in unfavorable characteristics can can have disastrous effects on the group of humans in whom it occurs. When such an allele is recessive, as they usually are, two seemingly normal people who are heterozygous can produce children homozygous for the recessive allele who cannot avoid the detrimental effect of the mutant allele. This tragedy strikes many families. Learning how to avoid it is one of the principal goals of human genetics.

When a detrimental allele occurs at a significant frequency in human populations, the harmful effect that it produces is called a **genetic disorder.** Table 12-2 lists some of the most important human genetic disorders. We know a great deal about some of them, but much less about many others.

TABLE 12-2   SOME IMPORTANT GENETIC DISORDERS

DISORDER	SYMPTOM	DEFECT	DOMINANT/ RECESSIVE	FREQUENCY AMONG HUMAN BIRTHS
Cystic fibrosis	Mucus clogging lungs, liver, and pancreas	Failure of chloride ion transport mechanism	Recessive	$\frac{1}{1800}$ (whites)
Sickle cell anemia	Poor blood circulation	Abnormal hemoglobin molecules	Recessive	$\frac{1}{1600}$ (U.S. blacks)
Tay-Sachs disease	Deterioration of central nervous system while person is young	Defective form of enzyme hexosaminidase A	Recessive	$\frac{1}{1600}$ (Jews)
Phenylketonuria	Failure of brain to develop in infants	Defective form of enzyme phenylalanine hydroxylase	Recessive	$\frac{1}{18,000}$
Hemophilia (Royal)	Failure of blood to clot	Defective form of blood clotting factor IX	Sex-linked recessive	$\frac{1}{7000}$
Huntington's disease	Gradual deterioration of brain tissue in middle age	Production of an inhibitor of brain cell metabolism	Dominant	$\frac{1}{10,000}$
Muscular dystrophy (Duchenne)	Wasting away of muscles	Degradation of myelin coating of nerves stimulating muscles	Sex-linked recessive	$\frac{1}{10,000}$
Hypercholesterolemia	Excessive cholesterol levels in blood, leading to heart disease	Abnormal form of cholesterol cell-surface receptor	Dominant	$\frac{1}{500}$

Cystic Fibrosis

Cystic fibrosis is the most common fatal genetic disorder among whites (Figure 12-9). As we saw in the boxed essay in Chapter 6, the affected individuals secrete a thick mucus that clogs the airways of their lungs and the passages of their pancreas and liver. Among whites, about 1 in 20 individuals has a copy of the defective gene, but shows no symptoms; the double-recessive individuals account for about 1 in 1800 white children. These individuals inevitably die.

What causes cystic fibrosis? We have learned the answer only recently, as partly discussed in Chapter 6. Cystic fibrosis is a defect in how cells regulate the transport of chloride ions across their membranes. The transmembrane channels that normally transport chloride ions into cells do not function in individuals afflicted by this disorder. Consequently, water is prevented from passing from the bloodstream into the passages of the lung. As a result, the mucus that is a normal component of the lung's inner surface becomes too thick.

In 1986, investigators examined isolated bits of membrane from cells of patients with cystic fibrosis and found that the chloride-transport channels in these bits of membrane functioned perfectly. Rather than in the channels themselves, the problem lies in the way the activity of these channels is regulated by the living cell. Apparently, one of the proteins that controls channel activity is defective, so that the channels in the isolated bits of membrane were unaffected and functioned normally. Cystic fibrosis occurs when an individual is homozygous for the allele encoding the defective version of the protein regulating the chloride-transport channel. This allele is recessive to the normal-functioning version of the regulating protein. Thus, the chloride channels of heterozygous individuals function normally, and such persons do not develop cystic fibrosis. The defective regulatory protein responsible for the disorder was identified in 1987 and is being actively studied.

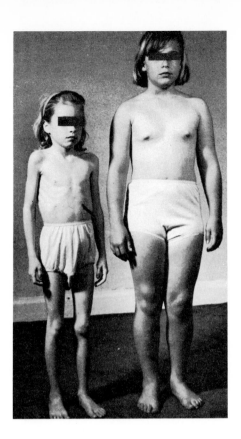

FIGURE 12-9

Two girls of the same age. The girl on the left has cyctic fibrosis. In affected individuals, the cell membrane proteins that normally transport chloride ion do not function. As a result, water does not pass from the bloodstream to the tissues of the lungs as it should, and the mucus that normally lines the insides of the lungs thickens, making breathing difficult. Few affected children live to be adults.

Sickle Cell Anemia

Sickle cell anemia is a heritable disorder in which the affected individuals are unable to transport oxygen to their tissues properly because the molecules within red blood cells that carry oxygen, molecules of the protein **hemoglobin,** are defective. When oxygen is scarce, these defective hemoglobin molecules become insoluble and combine with one another, forming stiff, rodlike structures. Surprisingly, the hemoglobin that occurs in such defective red blood cells differs from that occurring in normal red blood cells in only one out of a total of about 300 amino acid molecules. In the defective hemoglobin, one molecule of valine occurs in place of the glutamic acid that is located in the same position in normal hemoglobin. Red blood cells that contain large proportions of such defective molecules become sickle shaped and stiff; normal red blood cells are disk shaped and much more flexible (Figure 12-10). Because of their stiffness and irregular shape, the sickle-shaped red blood cells are able to move through the smallest blood vessels only with great difficulty. For the same reason, they also tend to accumulate in the blood vessels, forming clots. As a result, people who have large proportions of sickle-shaped red blood cells tend to have intermittent illness and a far shorter life span.

Individuals homozygous for the sickle cell allele show the characteristics just mentioned; those who are heterozygous for the allele are generally indistinguishable from normal persons. In the blood of people who are heterozygous for this trait, however, some of the red cells show the sickling characteristic when they are exposed to low levels of oxygen. The allele responsible for the sickle cell characteristic is particularly common among blacks. In the United States, for example, about 9% of blacks are heterozygous for this allele, and about 0.2% are homozygous and therefore have sickle cell anemia. In some groups of people in Africa, up to 45% of the individuals are heterozygous for this allele.

People who are homozygous for the sickle cell allele almost never reproduce because they usually die too young. This fact raises the question as to why the sickle cell allele has not been eliminated from all populations, rather than being maintained at high levels in some. The answer has proved much easier to find than has the parallel

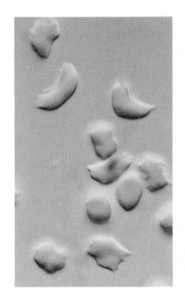

FIGURE 12-10

When they are exposed to low oxygen concentrations, many of the red blood cells of people who have the sickle cell trait (that is, who are heterozygous for the sickle cell allele) become sickle-shaped and distorted. In individuals who are homozygous for this trait, many of the red blood cells normally have such shapes.

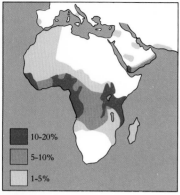

FREQUENCY OF SICKLE CELL GENE

10-20%

5-10%

1-5%

DISTRIBUTION OF *FALCIPARUM* MALARIA

FIGURE 12-11

Frequency of the sickle cell allele (top), and distribution of falciparum malaria (bottom). Falciparum malaria is one of the most devastating kinds of the disease; its distribution in Africa is closely correlated with that of the allele for the sickle cell characteristic.

question of why cystic fibrosis is so common. People who are heterozygous for the sickle cell allele are much less susceptible to falciparum malaria, which is one of the leading causes of illness and death, especially among young children, in the areas where the allele is common. In addition, for reasons not understood, women who are heterozygous for this allele are more fertile than are those who lack it. Consequently, even though most people who are homozygous recessive and therefore have sickle cell anemia die before they have any children, the sickle cell allele is maintained at high levels in populations where falciparum malaria is common (Figure 12-11).

Tay-Sachs Disease

Tay-Sachs disease is an incurable hereditary disorder in which the brain deteriorates. Affected children appear normal at birth and usually do not develop symptoms until about the eighth month, when signs of mental deterioration become evident. Within a year after birth, affected children are blind; they rarely live past their fifth year (Figure 12-12).

Tay-Sachs disease is a rare disorder in most human populations, occurring in 1 in 300,000 births. However, it has a high incidence among Jews of Eastern and Central Europe (Ashkenazim) and among American Jews (90% of whom are descendants of Eastern and Central European ancestors). Approximately 1 in 3600 such Jewish infants have this genetic disorder. Because it is a recessive condition, most people that carry the defective allele do not themselves develop the characteristic symptoms. An estimated 1 in 28 individuals in these Jewish populations is a heterozygous carrier of the allele.

What is responsible for the condition known as Tay-Sachs disease? Homozygous individuals lack an enzyme necessary to break down a special class of lipid called **gangliosides,** which occur within the lysosomes of brain cells. As a result, the lysosomes fill with gangliosides, swell, and eventually burst, releasing oxidative enzymes that kill the brain cell. There is no known cure for this condition.

Phenylketonuria

Many other hereditary disorders are not as common as cystic fibrosis in whites, sickle-cell anemia in blacks, or Tay-Sach's disease in Central and Eastern European Jews. Even though occurring less frequently, however, some additional genetic disorders do affect significant numbers of people. A good example of a relatively infrequent genetic disorder is **phenylketonuria (PKU),** a hereditary condition in which the affected individuals are unable to break down the amino acid phenylalanine. In such individuals, phenylalanine is instead converted to other chemicals that accumulate in the bloodstream. Although not harmful to an adult, these abnormal derivatives of phenylalanine are very harmful to infants, since they interfere with the development of brain cells. An infant with this disorder suffers severe mental retardation, and affected individuals rarely live beyond 30 years. When it is detected early enough, however, PKU can be treated nutritionally, and individuals with this genetic constitution can then develop and mature normally.

PKU is a recessive disorder caused by a mutant allele of the gene encoding the enzyme that normally breaks down phenylalanine. Only individuals homozygous for the mutant allele (in the United States, about 1 in every 15,000 infants) develop the disorder.

Hemophilia

Hemophilia is a hereditary condition in which the blood is slow to clot or does not clot at all. The reason that you do not bleed to death when you cut your finger is that the blood in the immediate area of the cut is converted to a solid gel that seals the cut, in a process similar to that which occurs in a puncture-proof tire. Such a blood clot forms from the polymerization of protein fibers circulating in the blood, in the same way that gelatin hardens. A variety of proteins are involved in this process, and all of them must

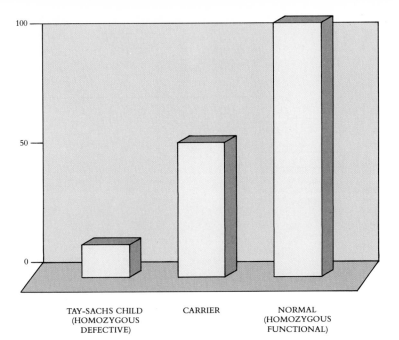

PERCENT OF
NORMAL ENZYME
FUNCTION

100

50

0

TAY-SACHS CHILD
(HOMOZYGOUS
DEFECTIVE)

CARRIER

NORMAL
(HOMOZYGOUS
FUNCTIONAL)

FIGURE 12-12

Tay-Sachs disease is a genetic disorder in which an enzyme critical to lipid metabolism does not function. This leads to harmful accumulations of fatty acids in the lysosomes of brain cells. Homozygous individuals typically have less than 10% of normal activities of the enzyme hexosaminidase A (left), whereas heterozygous individuals have about 50% of normal levels, enough to prevent deterioration of the central nervous system.

function properly for a blood clot to form. A mutation causing the loss of activity of any of these factors leads to a form of hemophilia.

Hemophilias are recessive disorders, expressed only when an individual does not possess at least one copy of the gene that is normal and so cannot produce one of the proteins necessary for clotting. Individuals homozygous for a mutant allele do not produce any active version of the affected clotting protein and thus cannot clot blood. Most of the dozen protein-clotting genes are on autosomes, but two (designated VIII and IX) are known to be located on the X chromosome. In the case of these particular protein-clotting genes, any male who inherits a mutant allele will develop hemophilia, because his other sex chromosome is the inactive Y, and thus he lacks a normal allele of the protein-clotting gene.

A mutation in factor IX occurred in one of the parents of Queen Victoria of England (1819–1901) (Figure 12-13). In the five generations since Queen Victoria, 10 of her male descendants have had hemophilia. The British royal family escaped the disorder, often called the **Royal hemophilia,** because Queen Victoria's son King Edward

FIGURE 12-13

Queen Victoria of England in 1894, surrounded by some of her descendants. Of Victoria's four daughters who lived to bear children, two, Alice and Beatrice, were carriers of Royal hemophilia. Two of Alice's daughters are standing behind Victoria (wearing feathered boas); to Victoria's right is Princess Irene of Prussia; to her left, Alexandra, who would soon become Tsarina of Russia (see Figure 12-1). Both Irene and Alexandra were also carriers of hemophilia.

FIGURE 12-14

The Royal hemophilia pedigree. From Queen Victoria's daughter Alice, the disorder was introduced into the Russian and Austrian royal houses, and from her daughter Beatrice, it was introduced into the Spanish royal house. Victoria's son Leopold, himself a victim, also transmitted the disorder in a third line of descent.

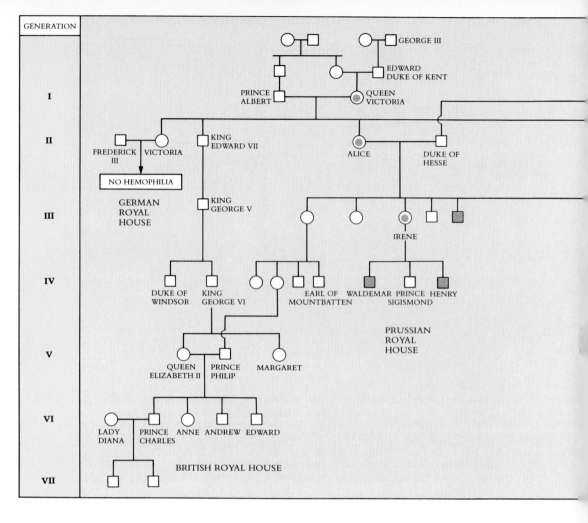

VII did not inherit the defective factor IX allele. Three of Victoria's nine children did receive the defective allele, however, and carried it by marriage into many of the royal families of Europe (Figure 12-14). It is still being transmitted to future generations among these family lines, except in Russia, where the five children of Alexandra, Victoria's granddaughter, were killed in the turbulent times soon after the Russian Revolution (see Figure 12-1).

Huntington's Disease

Not all hereditary disorders are recessive. Huntington's disease is a hereditary condition caused by a dominant allele that causes progressive deterioration of brain cells. (Figure 12-15). This disorder killed folksinger and songwriter Woody Guthrie. Perhaps 1 in 10,000 individuals develop the disorder. Because Huntington's disease is a dominant condition, every individual that carries an allele expresses it. You might then ask why the genetic disorder doesn't die out. The answer is that symptoms of Huntington's disease do not usually develop until the individuals are more than 30 years old, by which time most of them have already had children. Thus, the allele is transmitted before the lethal condition develops.

GENETIC COUNSELING

Although most genetic disorders cannot yet be cured, we are learning much about them, and progress toward successful therapy is being made in many cases. In the absence of a cure, however, the only recourse is to try to avoid producing children subject to these conditions. The process of identifying parents at risk for producing children with genetic defects, and of assessing the genetic state of early embryos, is called **genetic counseling.**

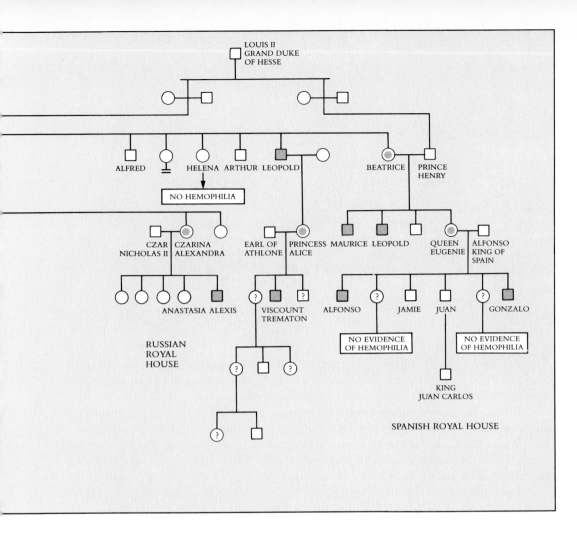

LOUIS II
GRAND DUKE
OF HESSE

ALFRED   HELENA   ARTHUR   LEOPOLD   BEATRICE   PRINCE HENRY

NO HEMOPHILIA

CZAR NICHOLAS II   CZARINA ALEXANDRA   EARL OF ATHLONE   PRINCESS ALICE   MAURICE   LEOPOLD   QUEEN EUGENIE   ALFONSO KING OF SPAIN

ANASTASIA   ALEXIS   VISCOUNT TREMATON   ALFONSO   JAMIE   JUAN   GONZALO

RUSSIAN ROYAL HOUSE

NO EVIDENCE OF HEMOPHILIA   NO EVIDENCE OF HEMOPHILIA

KING JUAN CARLOS

SPANISH ROYAL HOUSE

You might ask, "If the genetic defect is a recessive allele, how do potential parents *know* they carry the allele?" Pedigree analysis is often employed as an aid in genetic counseling. If one of your relatives has been afflicted with a recessive genetic disorder such as cystic fibrosis, there is a possibility that you are a carrier of the trait; in other words, you may carry the recessive allele in the heterozygous state. By analyzing your pedigree, it is often possible to estimate the likelihood that you are a carrier. When a couple is expecting a child and pedigree analysis indicates that both parents have a significant probability of being heterozygous carriers of a recessive allele responsible for a serious genetic disorder, the pregnancy is said to be a **high-risk pregnancy.** In such a pregnancy, there is a significant probability that the child will exhibit the clinical disorder.

FIGURE 12-15

Huntington's disease is a dominant genetic disorder that involves progressive deterioration of brain cells. The folksinger Woody Guthrie was a victim of Huntington's disease. Deterioration is usually not evident until later in life (left); less than 20% of affected individuals exhibit symptoms before age 30. It is because of this late age of onset that the disorder persists despite its dominant and fatal nature.

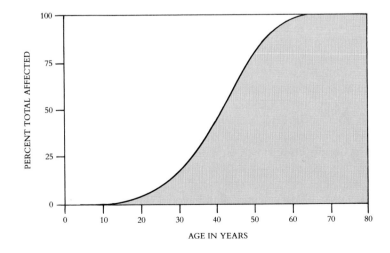

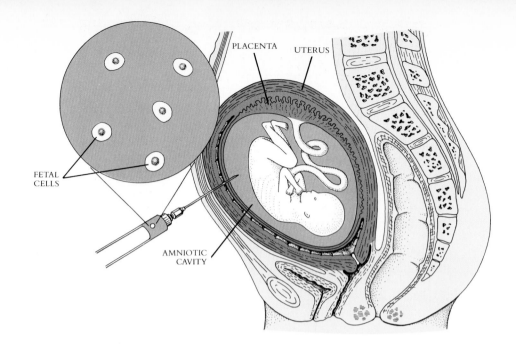

FIGURE 12-16

Amniocentesis. A needle is inserted into the amniotic cavity, and a sample of amniotic fluid, containing some free cells derived from the embryo, is withdrawn into a syringe. The fetal cells are then grown in tissue culture so that their karyotype and many of their metabolic functions can be examined.

PLACENTA   UTERUS

FETAL CELLS

AMNIOTIC CAVITY

Another class of high-risk pregnancies involves mothers who are more than 35 years old. As discussed earlier, the frequency of birth of infants with Down syndrome increases dramatically among the pregnancies of older women (see Figure 12-4).

When a pregnancy is diagnosed as being high risk, many women elect to undergo **amniocentesis,** a procedure that permits the prenatal diagnosis of many genetic disorders (Figure 12-16). In the fourth month of pregnancy, a small sample of amniotic fluid is removed by means of a sterile hypodermic needle. When the needle is inserted into the expanded uterus of the mother, its position and that of the fetus are usually observed simultaneously by means of a technique called **ultrasound** (Figure 12-17). Sound waves allow an image of the fetus to be seen, as do x rays, but the sound waves are not damaging to the mother or to the fetus. Since ultrasound allows the position of the fetus to be determined, the person withdrawing the amniotic fluid can avoid damaging the fetus. In addition, the fetus can be examined for the presence of major abnormalities. The amniotic fluid, which bathes the fetus, contains free-floating cells derived from the fetus. Once removed, these cells can be grown as tissue cultures in the laboratory. Studying these tissue cultures, genetic counselers can test for many of the most common genetic disorders:

1. *Enzyme activity tests.* In many cases it is possible to test directly for the proper functioning of the enzymes involved in genetic disorders; the lack of proper activity signals the presence of the disorder. Thus, the lack of the enzyme responsible for breaking down phenylalanine signals PKU, the absence of the enzyme responsible for the breakdown of gangliosides indicates Tay-Sachs disease, and so forth.

2. *Association with genetic markers.* For sickle cell hemoglobin, Huntington's disease, and one form of muscular dystrophy (a condition characterized by weakened muscles that do not function normally), investigators have searched and found other mutations on the same chromosome that, by chance, occurred at about the same time as the disorder-causing mutation. By testing for the presence of the second mutation, an investigator can identify individuals with a high probability of possessing the other disorder-causing defect. Identifying such mutations in the first place is a little like searching for a needle in a haystack, but persistent efforts have proved successful in these cases. The associated mutations are usually detected as alterations in DNA sequences, recognized through the presence of enzymes that cut strands of DNA at particular places. Such enzymes, called **restriction enzymes,** are discussed in Chapter 15.

3. *Identification of heterozygotes.* As demonstrated in Figure 12-12, for example, heterozygous individuals can often also be detected during genetic counseling and can plan accordingly.

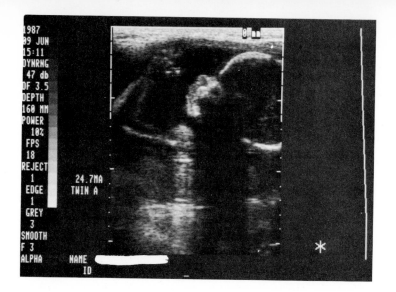

FIGURE 12-17

The appearance of a fetus as detected with ultrasound. During the fourth month of pregnancy, when amniocentesis is normally performed, the fetus usually moves about actively.

Genetic Therapy

When the analysis of amniotic fluid indicates a severe disorder in the fetus, the parents may consider terminating the pregnancy by means of therapeutic abortion. In some instances, other options are available. If PKU is diagnosed, for example, it is possible to avoid the defects of the disorder by placing the mother on a low-phenylalanine diet. This provides her and her unborn baby with enough phenylalanine to make proteins, but not enough to lead to buildup. After birth, the child is maintained on a low-phenylalanine diet until 6 years of age. At that age, the child's brain is fully developed and PKU is no longer a potential health problem.

Listed below are some of the organizations studying the major genetic disorders. Their support is helping to find new and better ways to manage the individual conditions and may eventually lead to the detection of ways to alleviate them completely.

Cystic Fibrosis Foundation
3379 Peachtree Rd., N.E.
Atlanta, GA 30326

National Hemophilia Foundation
25 West 39th St.
New York, NY 10018

Committee to Combat Huntington's Disease, Inc.
250 West 57th St., Suite 2016
New York, NY 10107

Muscular Dystrophy Association
810 Seventh Ave.
New York, NY 10019

National Association of Sickle Cell Disease, Inc.
945 South Western Ave.
Los Angeles, CA 90006

National Tay-Sachs and Allied Diseases Association
122 East 42nd St.
New York, NY 10017

We do not yet have the knowledge that would allow us to correct undesirable genetic conditions directly, but doing so is theoretically quite possible. As you will see in Chapter 15, biologists are making rapid progress in learning how to insert genes from one organism into another, and there is no reason why these techniques would not work to transfer genes from one human cell to another. Such transfers probably would have to be carried out **in vitro** (in a test tube) with embryos recently fertilized there, in order for the gene transfer to affect all the cells of the new individual. So far, such a process has never been carried out. As we learn more, it seems increasingly likely that it will be.

SUMMARY

1. Human body cells contain 46 chromosomes: 44 autosomes and 2 sex chromosomes. The autosomes form 22 pairs of homologous chromosomes in meiosis. The two sex chromosomes may be similar in size and appearance, as occurs in females, where they are designated XX. The Y chromosome, on the other hand, may be much smaller than the other X chromosome. XY individuals are male because the genes that initiate the development of male characteristics are located on the Y chromosome.

2. In humans, the loss of an autosome is invariably fatal. Gaining an extra autosome, which leads to a condition called trisomy, is also fatal, with few exceptions. The chromosomes that can be present in an extra copy and that still allow that individual to survive and become an adult are chromosomes 21 and 22. Individuals with an extra copy of chromosome 21 (three copies in all) are retarded and are referred to as having Down syndrome. As occurs in all trisomic persons, this trisomy arises from the chromosomes of that particular pair not separating during meiosis, usually in mothers over 35 years of age.

3. Patterns of inheritance observed in family histories, called pedigrees, can be used to determine the mode of inheritance of a particular trait. By such analysis, one can often determine the mode of inheritance of a particular trait. By such analysis, one can often determine if a trait is associated with a dominant or a recessive allele, if the gene determining the trait is located on the X chromosome (sex linked), and if the trait is specified by more than one gene.

4. Many human genes possess more than two common alleles. An example is the ABO blood group gene. This gene encodes an enzyme that adds sugars to the surfaces of blood cells. The A and B alleles add different sugars, and the O allele adds none. Human populations vary in the proportions of these three alleles that they possess.

5. Genetic disorders are often caused by alleles that encode abnormal proteins; the effects of these proteins lead to serious health problems. Some genetic disorders are relatively common in human populations, whereas others are rare. Many of the most important genetic disorders are associated with recessive alleles, the functioning of which may lead to the production of defective versions of enzymes that normally perform critical functions. Because such traits are determined by recessive alleles and therefore are expressed only in homozygotes, the alleles are not eliminated from the human population, even though their effects in homozygotes may be lethal. Dominant alleles that lead to severe genetic disorders are less common. In some of those occurring more frequently, the expression of the alleles does not occur until after the affected individuals reach their reproductive years.

6. Parents who suspect their children may express a genetic disorder or Down syndrome may elect to undergo amniocentesis. In this procedure, a sample of fetal cells obtained from amniotic fluid is used to establish a tissue culture, which can then be checked for the presence of a wide variety of genetic disorders.

7. For a few genetic disorders, such as phenylketonuria, it is possible to initiate therapy if the disorder is diagnosed during pregnancy and thus avoid the detrimental effects. For most genetic disorders, however, we do not know how to accomplish such a result. The direct transfer of normal human genes to the chromosomes of individuals who would otherwise suffer from a particular genetic disorder may prove possible in the future, although this has not yet been accomplished. Meanwhile, although no cures exist for any genetic disorder, many of the conditions can be managed with increasingly positive results.

REVIEW (Human Genetics problems appear in Appendix B)

1. What is the human diploid number of chromosomes? How many kinds of autosomes do you have?

2. The inactive X chromosome seen in the cells of human females is called a _____.

3. What percentage of whites possess at least one copy of the allele for cystic fibrosis?

4. Nondisjunction of an X chromosome during meiosis can lead to an individual with an XXX genotype or to an individual with an XXY genotype. The XXY genotype leads to a condition called _____ and occurs in about 1 in 500 male births.

5. _____ is a procedure that permits the prenatal diagnosis of many genetic disorders.

SELF-QUIZ

1. Down syndrome results from having three copies of chromosome
   (a) 13　　(c) 21　　(e) both c and d
   (b) 18　　(d) 22

2. Why are sex-linked traits expressed much more frequently in males than in females?
   (a) because males only have one copy of the X chromosome
   (b) because males only have one copy of the Y chromosome
   (c) because males are more developmentally fragile than females
   (d) sex-linked traits are carried on the X chromosome and therefore occur with equal frequency in males and females
   (e) these males lack an autosome

3. Which of the following genetic diseases is sex linked?
   (a) Royal hemophilia
   (b) Tay-Sachs disease
   (c) cystic fibrosis
   (d) hypercholesterolemia
   (e) none of the above

4. A detailed pedigree of a family with a history of a particular genetic disorder can help determine
   (a) whether the disorder is sex-linked
   (b) whether the disorder is autosomal
   (c) whether the disorder is dominant
   (d) whether the disorder is determined by a single gene or by several genes
   (e) all of the above

5. The sickle cell trait is fatal when homozygous, yet the allele is retained at fairly high levels in some areas of the world. Why?
   (a) because dominant alleles are difficult to purge from a population even if they are detrimental
   (b) because it has no effect on heterozygous individuals
   (c) because heterozygous individuals are resistant to falciparum malaria
   (d) because individuals homozygous for the trait generally live past their reproductive years
   (e) none of the above

THOUGHT QUESTIONS

1. As you can see in Table 12-1, both North American and South American Indians exhibit very low frequencies of ABO blood group allele B. Can you think of a reason why they should differ from other human populations in this regard?

2. Schizophrenia is a mental disorder in which a split occurs between thoughts and feelings and contact is lost with the environment. To assess whether schizophrenia is hereditary, twins who had been reared apart were studied; all they had in common was their being twins. Two kinds of twins were compared: those who were identical (monozygotic) and those who were not (dizygotic). Monozygotic twins are genetically identical, whereas dizygotic twins are normal brothers and/or sisters who happen to be born at the same time. Investigators asked if twins reared apart develop schizophrenia more often if they are genetically identical. Here is what they found, using data collected over 40 years: of 289 monozygotic twins studied, in 51% of cases both twins develop the disorder if one does; of 398 dizygotic twins studied, in 10% of cases both develop the disorder if one does. Do you think that these results suggest that schizophrenia is a hereditary disorder?

FOR FURTHER READING

FRIEDMANN, T.: "Prenatal Diagnosis of Genetic Disease," *Scientific American*, November 1971. A discussion of amniocentesis and diagnostic tests used to identify genetic disorders.

GOULD, S.J.: "Dr. Down's Syndrome," *Natural History*, vol. 89, pages 142-148, 1980. An account of the history of Down syndrome, a relatively common chromosomal abnormality that results in severe mental retardation.

HOOK, E.B.: "Behavioral Implications of the Human XYY Genotype," *Science*, vol. 179, pages 139-150, 1973. A review of some of the conflicting data on this controversial subject.

FIGURE 13-1   A key breakthrough in our understanding of heredity occurred in 1953, when James Watson, a young American postdoctoral student (he is the one peering up as if afraid their homemade model of the DNA molecule will topple over), and the English scientist Francis Crick (pointing) deduced the structure of DNA, the molecule that stores the hereditary information.

THE MECHANISM OF HEREDITY

Chapter 13

Overview

Our knowledge of Mendelian genetics permits us to predict patterns of inheritance but tells us nothing about *how* the information on chromosomes acts to determine whether the eyes of a fly are red or white or whether our eyes are brown or blue. Like a black box containing a machine we can't see, our chromosomes somehow manage to direct our development, to determine how tall we are, what color our hair is, and everything else about us. Using experiments, scientists have learned to look into that black box and study the machine within—the mechanism of heredity.

For Review   *Here are some important terms and concepts that you will encounter in this chapter. If you are not familiar with them, you should review them before proceeding.*

Structure of deoxyribonucleic acid (DNA) (Chapter 3)

How amino acid sequence determines protein shape (Chapter 3)

Ribosomes (Chapter 5)

The realization, which was reached at the beginning of this century, that patterns of heredity can be explained by the segregation of chromosomes during meiosis was one of the most important advances in human thought. Not only did it lead directly to development of the science of genetics and thus to great progress in agriculture and medicine, but it also profoundly influenced the way we think about ourselves. By removing the mystery from the process of heredity, it made the biological nature of human beings seem much more approachable. It also raised the question that was to occupy biologists for more than half a century: what is the exact nature of the connection between hereditary traits and the chromosomes?

We are now able to answer this question. We understand in considerable detail the mechanism by which the information on the chromosomes is converted into organisms with eyes, arms, and inquiring minds. Our understanding was not acquired all at once—deduced in a single flash of insight—but rather was developed slowly over many years by a succession of investigators. How they developed this understanding and what they learned are the subjects of this chapter (Figure 13-1).

AN EXPERIMENTAL JOURNEY

Several generations of biologists have worked to solve the riddle of heredity, applying to the problem creative thinking and rigorous experimentation. In this chapter we shall retrace their experimental journey. We shall not dwell for long at any one stage or worry overly about technical details. Instead we shall concentrate on tracing the major ideas and how the results of specific experiments have changed these ideas. The unraveling of the mechanism of heredity constitutes one of our greatest intellectual journeys.

We shall focus on a series of five questions that illustrates how scientific progress builds in stages, each advance depending on and elaborating the results that precede it. As in a "who done it" mystery or a Congressional hearing, what we have learned from answering each question has led naturally and inevitably to the questions that follow it:

1. Where do cells store hereditary information?
2. What component of the chromosomes contains the hereditary information?
3. How is the information in DNA reproduced so accurately?
4. What is the unit of hereditary information?
5. How do genes interact to produce a phenotype?

WHERE DO CELLS STORE HEREDITARY INFORMATION?

Perhaps the most basic question that we can ask about hereditary information is where in the cell it is stored. Of the many approaches that we might take to answer this question, let us start with a simple one: cut a cell into pieces and see which of the pieces can express hereditary information. For this experiment we need a single-celled organism that is large enough to be operated on conveniently and that is differentiated enough that the pieces can be told apart.

An elegant experiment of this sort was performed by Danish biologist Joachim Hammerling in 1943. Hammerling chose the large unicellular green alga *Acetabularia* as his experimental subject (Figure 13-2). Individuals of this genus have distinct foot, stalk, and cap regions, all of which are differentiated parts of a single cell. The nucleus of this cell is located in the foot. As a preliminary experiment, Hammerling tried amputating the caps or feet of individual cells. He found that when the cap was amputated, a new one regenerated from the remaining portions of the cell. However, when the foot was amputated and discarded, no new foot was regenerated. Hammerling concluded that the hereditary information resided within the foot, or basal portion, of *Acetabularia*.

To test this hypothesis, Hammerling selected individuals from two species of the genus in which the caps looked very different from each other: *Acetabularia mediterranea*, which has a disk-shaped cap, and *Acetabularia crenulata*, which has a branched, flowerlike cap. Hammerling cut the stalk and cap away from an individual of *A. mediterranea*; to the remaining foot he grafted a stalk cut from a cell of *A. crenulata* (Figure 13-3). The cap that formed looked something like the flower-shaped cap that is characteristic of *A. crenulata*, although it was not exactly the same.

FIGURE 13-2

The marine green alga *Acetabularia* has been the subject of many elegant experiments in developmental biology. Although *Acetabularia* is a large organism with clearly differentiated parts, such as the stalks and elaborate caps visible here, individuals are actually single cells.

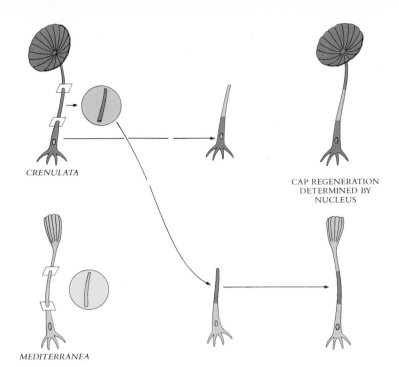

CRENULATA

CAP REGENERATION
DETERMINED BY
NUCLEUS

MEDITERRANEA

FIGURE 13-3

Hammerling's *Acetabularia* reciprocal graft experiment. To the foot of each species he grafted a stalk of the other. In each case, the cap that eventually developed was dictated by the foot and not the stalk. Although the first cap to develop was sometimes intermediate in appearance, when it was cut away and another cap allowed to form, that cap was always the one characteristic of the species from which the foot of the grafted cell was derived.

Hammerling then cut off this regenerated cap and found that a disk-shaped cap exactly like that of *A. mediterranea* formed in the second regeneration and in every regeneration thereafter. This experiment strengthened Hammerling's earlier conclusion that the instructions that specify the kind of cap that is produced are stored in the foot of the cell—probably in the nucleus—and that these instructions must pass from the foot through the stalk to the cap. In Hammerling's regeneration experiment the initial flower-shaped cap was formed as a result of the instructions that were already present in the transplanted stalk when it was excised from the original *A. crenulata* cell. In contrast, all later caps used new information derived from the foot of the *A. mediterranea* cell onto which the stalk had been grafted. In some unknown fashion the original instructions that had been present in the stalk were eventually "used up."

Hammerling's experiments identified the nucleus as the likely repository of the hereditary information, but to prove definitely that this was the case, isolated nuclei had to be transplanted. Such an experiment was performed in 1952 by the American embryologists Robert Briggs and Thomas King. Using a glass pipette drawn to a fine tip, and working with a microscope (Figure 13-4), Briggs and King removed the nucleus from a frog egg. Without the nucleus the egg would not develop. They then replaced the absent nucleus with one that they had isolated from a cell of a more advanced frog embryo of a different species (Figure 13-5). The implanting of this nucleus ultimately caused an adult frog to develop from the egg. Clearly the nucleus was directing the frog's development.

In eukaryotic cells, the hereditary information is stored in the nucleus.

Can each and every one of the many nuclei of a multicellular organism direct the development of an entire adult individual? The Briggs and King experiment did not answer this question definitively since the nuclei that they took from more advanced frog embryos often caused the eggs to which they were transplanted to develop abnormally. However, when John Gurdon at Yale University transplanted nuclei isolated from developed tadpole tissue into eggs from which the nuclei had been removed, the eggs developed normally. Clearly the nuclei within cells of tadpoles do indeed retain a full set of genetic instructions.

However, tadpoles are not adults. Have the nuclei in the cells of *adult* animals had any of their hereditary information removed during the course of their development? This question has proved difficult to answer since animal development is so complex.

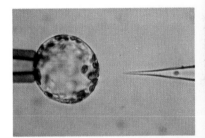

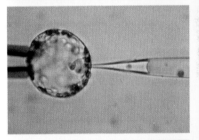

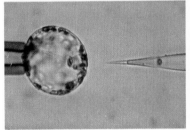

FIGURE 13-4

Working under a microscope, it is possible to pierce a cell with the fine tip of a glass micropipette without rupturing the cell. These micrographs show microinjection into the protoplast of a cell from the leaf of a tobacco plant (the protoplast is about 50 micrometers in diameter).

FIGURE 13-5

Briggs and King's nuclear transplant experiment. Two strains of frog were used that differed from each other in the number of nucleoli their cells possessed. The nucleus was removed from an egg of one strain, either by sucking the egg nucleus up into a micropipette or more simply by destroying it with UV light. Briggs and King then injected into this anucleate egg a nucleus obtained from a differentiated cell of the other species—in this case, a cell isolated from the intestine of a tadpole of that species. The hybrid egg was then allowed to develop. One of three results was obtained in individual experiments: (1) no growth occurred, perhaps reflecting damage to the egg cell during the nuclear-transplant operation; (2) normal growth and development occurred up to the early embryo stage, but subsequent development was not normal, and the embryo did not survive; or (3) normal growth and development occurred, eventually leading to the development of an adult frog. That frog is of the species that contributed the nucleus, and not of the species that contributed the egg. Only a few experiments gave this third result, but they served to clearly establish that the nucleus directs frog development.

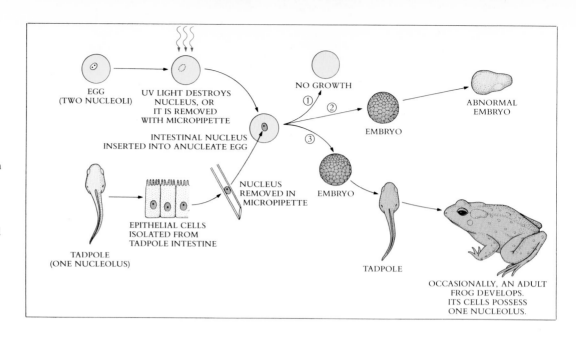

However, in plants a simple experiment did yield a clear-cut answer. At Cornell University in 1958, plant physiologist F.C. Steward let fragments of fully developed carrot tissue (bits of conducting tissue called **phloem**) swirl around in a rotating flask containing liquid growth medium. As individual cells broke away and tumbled through the liquid, Steward observed that these cells often divided and differentiated into multicellular roots. If these roots were then immobilized by placing them in a gel, they grew into entire plants that could be transplanted to soil and developed normally into maturity (see Chapter 30 and Figure 30-12). Steward's experiment confirms that in plants, at least some of the cells present in adult individuals do contain a full complement of hereditary information.

> **With rare exceptions the nucleus of all adult cells of multicellular organisms contains a full complement of genetic information.**

WHAT COMPONENT OF THE CHROMOSOMES CONTAINS THE HEREDITARY INFORMATION?

The identification of the nucleus as the source of hereditary information focused attention on the chromosomes, which were already suspected to be the vehicles of Mendelian inheritance. Specifically, biologists wondered how the actual hereditary information was arranged in the chromosomes. It was known that chromosomes contain both protein and DNA. On which of these was the hereditary information written?

During a brief period of 10 years or so after World War II, a flurry of investigators addressed this issue and resolved it clearly. Here we shall describe three very different kinds of experiments, each of which yields a clear answer in a simple and elegant manner.

The Griffith-Avery Experiments

As early as 1928 a British microbiologist named Frederick Griffith made a series of unexpected observations while experimenting with pathogenic (disease-causing) bacteria. When Griffith infected mice with a virulent strain of *Pneumococcus* bacteria, the mice died of blood poisoning, but when he infected similar mice with a strain of *Pneumococcus* that lacked a polysaccharide coat like the one possessed by the virulent strain, the mice showed no ill effects. The coat was apparently necessary for successful infection.

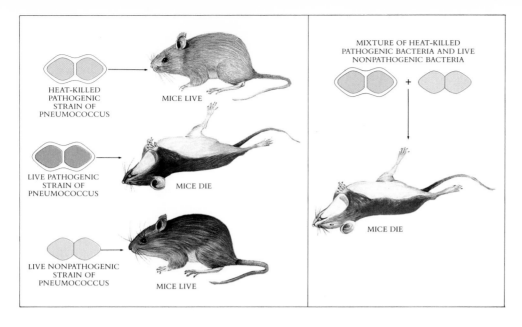

FIGURE 13-6

Griffith's discovery of transformation. The pathogenic bacterium *Pneumococcus* will kill many of the mice into which it is injected, but only if the bacterial cells are covered with a polysaccharide coat. Living bacteria without such polysaccharide coats do not harm the mice, since the coat is necessary for successful infection. However, it is not the coat itself that is the agent of disease: when Griffith injected dead bacteria possessing polysaccharide coats, the mice were not harmed. However, if he injected a mixture of dead bacteria with polysaccharide coats (harmless) and living mutant bacteria without such coats (harmless), many of the mice died. Griffith concluded that the live cells had been "transformed" by the dead ones—that the genetic information specifying the polysaccharide coat had passed from the dead cells to the living ones.

As a **control experiment,** that is, an experiment performed to make sure other factors were not responsible for the effect observed, Griffith injected normal (having a polysaccharide coat) but heat-killed bacteria into the mice to see whether the polysaccharide coat itself had a toxic effect. The mice remained healthy. As a final control, Griffith blended his two ineffective preparations—living mutant bacteria that lacked a coat and normal dead bacteria with intact coats—and injected the mixture into healthy mice (Figure 13-6). Unexpectedly, the injected mice developed disease symptoms, and many of them died. The blood of the dead mice was found to contain high levels of normal virulent *Pneumococcus* bacteria. Somehow the information specifying the polysaccharide coat had passed from the dead bacteria to the living but coatless bacteria in the control mixture, **transforming** then into normal virulent bacteria that infected and killed the mice.

Not until 1944 was the agent responsible for transforming *Pneumococcus* discovered. In a series of experiments, Oswald Avery and his co-workers characterized what he referred to as the "transforming principle." Its properties resembled those of DNA rather than protein: the activity of the transforming principle was not affected by protein-destroying enzymes but was lost completely in the presence of a DNA-degrading enzyme called DNAse.

When the transforming principle was purified, it indeed consisted mainly of DNA. Later it was shown that all but trace amounts of protein (0.02%) could be removed without reducing the transforming activity. The conclusion seemed inescapable: DNA is the hereditary material in bacteria. It has since proved possible to use purified DNA to change the genetic characteristics of eukaryotic cells in tissue culture, and even to inject pure DNA into fertilized *Drosophila* eggs and thereby alter the genetic characteristics of the resulting adult.

The Hershey-Chase Experiment

Avery's results were not widely appreciated at first; many biologists preferred to believe that proteins were the depository of the hereditary information. However, another convincing experiment was soon performed which was difficult to ignore. The experiment involved viruses. Viruses consist of either RNA (ribonucleic acid) or DNA and are covered with a protein coat (viruses are described in more detail in Chapter 25). The investigators focused on bacterial viruses, or **bacteriophages,** and performed an experiment analogous to the transplant experiments previously described. When a bacteriophage infects a bacterial cell, it first binds to the cell's outer surface and then injects its hereditary information into the cell. There, the hereditary information directs the production of thousands of new virus particles within the cell. The host bacterial cell eventually falls apart, or **lyses,** releasing the newly made viruses.

In 1952 Alfred Hershey and Martha Chase set out to identify the material injected into the bacterial cell at the start of an infection. Using a strain of bacteriophage known as T2, which contains DNA rather than RNA, they designed an experiment to determine whether the genetic material was DNA or protein (Figure 13-7). Hershey and Chase labeled the DNA of one preparation of T2 bacteriophage with a radioactive iso-

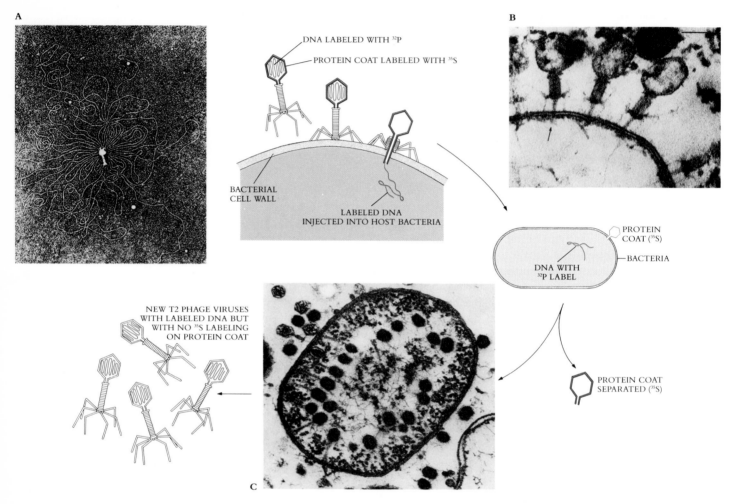

FIGURE 13-7

The Hershey-Chase experiment. The T2 bacterial viruses that they employed have a simple structure: they are composed of a protein envelope within which DNA is packaged.

A DNA is released from a single virus particle. Hershey and Chase labeled the T2 virus particles with either of two radioisotopes: in one experiment, the protein bodies were labeled with ^{35}S (sulfur occurs in protein in the amino acids cysteine and methionine, but does not occur in DNA), and in another experiment the DNA molecules were labeled with ^{32}P (phosphorus occurs in the phosphate groups of DNA, but does not occur in proteins). These T2 particles were then allowed to infect bacterial cells. Each virus body binds to the outside of the cell but does not enter; instead, it injects its DNA into the cell.

B Individual DNA strands enter the cell from virus particles bound to its surface. Within the cell, the injected DNA commandeers the machinery of the cell and directs the synthesis of all the parts necessary to make new viruses.

C New virus particles are assembled from parts within an infected cell. Eventually these new viruses will rupture the cell and be released into the surroundings. When Hershey and Chase used a Waring blender to knock the virus particles off the bacterial cells after the initial injection of virus DNA, and separated the bacterial cells (with the injected viral DNA) from the liquid medium (which had the dislodged virus bodies in it), they found that, when ^{35}S-labeled virus was used, the bulk of the ^{35}S label (and thus the virus protein) was now in the medium; when ^{32}P-labeled virus was used, the ^{32}P (and thus the DNA) was present in the interiors of the bacterial cells. It follows that it was the DNA that was responsible for directing the production of new viruses.

tope of phosphorus (^{32}P) and at the same time labeled the protein coats of another preparation with a radioactive isotope of sulfur (^{35}S). This was technically simple since phosphorus is found only in nucleic acids, not in proteins, and sulfur is found only in proteins (in the amino acid cysteine) and not in nucleic acids. Since the radioactive ^{32}P and ^{35}S isotopes emit particles of very different energies when they decay, they are easily told apart. The labeled viruses were permitted to infect two growing cultures of bacteria. The bacterial cells were then agitated violently in a blender to shake the protein coats of the infecting viruses loose from the bacterial surface to which they were attached. Centrifuging the cells from solution, Hershey and Chase found that the ^{35}S label was now predominantly in solution with the dissociated virus particles, whereas the ^{32}P label had transferred to the interior of the cells. The viruses later released from the infected bacteria contained the ^{32}P label. It follows that the hereditary information injected into the bacteria that specified the new generation of virus particles was contained in DNA and not in protein.

The Fraenkel-Conrat Experiment

Objections continued to be raised regarding the hypothesis that DNA is the genetic material because some RNA viruses contain *no* DNA and yet manage to reproduce quite satisfactorily. What is the genetic material in this case?

In 1957 Heinz Fraenkel-Conrat and co-workers isolated **tobacco mosaic virus,** or **TMV,** from tobacco leaves. From ribgrass *(Plantago),* a common weed, they isolated a second, similar virus called **Holmes ribgrass virus,** or **HRV.** In both TMV and HRV the viruses consist of protein and a single strand of RNA. After isolating these viruses, the scientists chemically dissociated each of them by separating their protein from their RNA. By putting the protein component of one virus with the RNA of another, Fraenkel-Conrat and associates were able to reconstitute hybrid virus particles. A hybrid results from the crossing or combining of two genetically unlike species.

The payoff of the experiment was in the next step. To choose betwen protein and RNA as the genetic material of the viruses, Fraenkel-Conrat infected healthy tobacco plants with a hybrid virus composed of TMV protein capsules and HRV RNA, being careful not to include any nonhybrid virus particles (Figure 13-8). The tobacco leaves that were infected with the reconstituted hybrid virus particles developed the sort of lesions that are characteristic of HRV and that normally form on infected ribgrass. Clearly the hereditary properties of the virus are determined by the nucleic acid in its core and not by the protein in its coat.

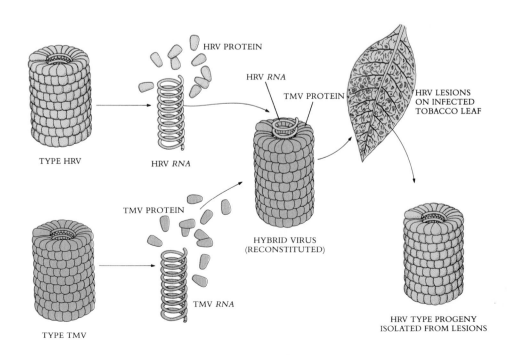

TYPE HRV   HRV PROTEIN   HRV *RNA*   HRV *RNA*   TMV PROTEIN   TMV PROTEIN   TMV *RNA*   TYPE TMV   HYBRID VIRUS (RECONSTITUTED)   HRV LESIONS ON INFECTED TOBACCO LEAF   HRV TYPE PROGENY ISOLATED FROM LESIONS

FIGURE 13-8

Fraenkel-Conrat's virus-reconstitution experiment. Both TMV and HRV are plant RNA viruses that infect tobacco plants, causing lesions on the leaves. Because the two viruses produce different kinds of wounds on the plant, the source of any particular infection can be identified. In this experiment, TMV and HRV were both dissociated into protein and RNA, and the protein and RNA separated from one another. Then hybrid virus particles were produced by mixing the HRV RNA and the TMV protein and allowing virus particles to form from these ingredients. When the reconstituted virus particles were painted onto tobacco leaves, lesions developed—of the HRV type. From these lesions normal HRV virus particles could be isolated in great numbers; no TMV viruses could be isolated from the lesions. It follows that the RNA (HRV) and not the protein (TMV) contains the information necessary to specify the production of the viruses.

Later studies have shown that many virus particles contain RNA rather than the DNA found universally in cellular organisms. Some of these RNA viruses, called **retroviruses,** make DNA copies of themselves when they infect a cell, copies that can then be inserted into the nuclear DNA as though they are genes that occur normally in the cellular organism. The virus lives within the cells of affected individuals as though it were a part of the cells' genes. An outbreak results whenever the virus begins to reproduce RNA copies of itself. The virus responsible for AIDS is a retrovirus.

DNA is the genetic material for all cellular organisms and most viruses, although some viruses use RNA.

HOW IS THE INFORMATION IN DNA REPRODUCED SO ACCURATELY?

As it became clear that the DNA molecule was the repository of the hereditary information, investigators began to puzzle over how such a seemingly simple molecule could carry out such a complex function. As discussed in Chapter 3, a DNA molecule is simply a repeating chain of identical five-carbon sugars linked together head to tail. One of four ring-shaped organic bases—adenine (A), guanine (G), thymine (T), and cytosine (C)—protrudes from each of the sugars.

It was at first difficult to understand how DNA plays such a complex role in heredity, because DNA was initially thought to be a simple repeating polymer such as AGTC AGTC AGTC. However, by the late 1940s, careful chemical analyses by Erwin Chargaff and his colleagues at Columbia University had revealed that the ratio of bases in different DNAs varies widely and that the sequences are not as simple as had originally been thought. Chargaff did observe an important underlying regularity: the amount of adenine present in all DNA molecules is equal to the amount of thymine, and the amount of guanine equals the amount of cytosine. However, the amount of adenine plus thymine often differs greatly from the amount of guanine plus cytosine.

Although the significance of the regularities pointed out by Chargaff was not immediately obvious, it soon became clear. Two British chemists, Rosalind Franklin and Maurice Wilkens, had performed x-ray crystallographic analysis of fibers of DNA. In this process the DNA molecule is bombarded with an x-ray beam. When individual x-rays encounter atoms, their path is bent or diffracted; the pattern created by the sum total of all these diffractions can be captured on a piece of photographic film. Such a pattern resembles the ripples created on a smooth lake by a rock tossed into it. By careful analysis of the diffraction pattern, it is possible to develop a three-dimensional image of the molecule. The diffraction patterns that Franklin obtained suggested that the DNA molecule was a helical coil with repeating elements of 2.0 and 3.4 nanometers.

Learning informally of Franklin's results in 1953, James Watson and Francis Crick, two young investigators at Cambridge University, quickly worked out the probable structure of the DNA molecule (see Figure 13-1). They analyzed the problem deductively. They first built models of the nucleotides and then determined how these nucleotides could be assembled into a molecule that fit what they knew about the structure of DNA. They tried various possibilities, first assembling molecules with three strands of nucleotides wound around one another to stabilize the helical shape. None of these early efforts proved satisfactory. They finally hit on the idea that the molecule might be a simple double helix in which the bases of two strands point inward toward each other. If the larger kind of base, a **purine** (adenine or guanine), is always paired with the smaller kind, a **pyrimidine** (cytosine or thymine), the diameter of the duplex stays the same—2.0 nanometers. Because hydrogen bonds develop between the two strands, the helical form is stabilized, turning once every 3.4 nanometers.

It immediately became apparent why Chargaff had obtained the results that he had: since the purine adenine (A) will not form proper hydrogen bonds in this structure with cytosine (C) but will with thymine (T), every A is paired to a T (Figure 13-9). Similarly, the purine guanine (G) will not form proper hydrogen bonds with thymine but will with cytosine, so that every G is paired with a C.

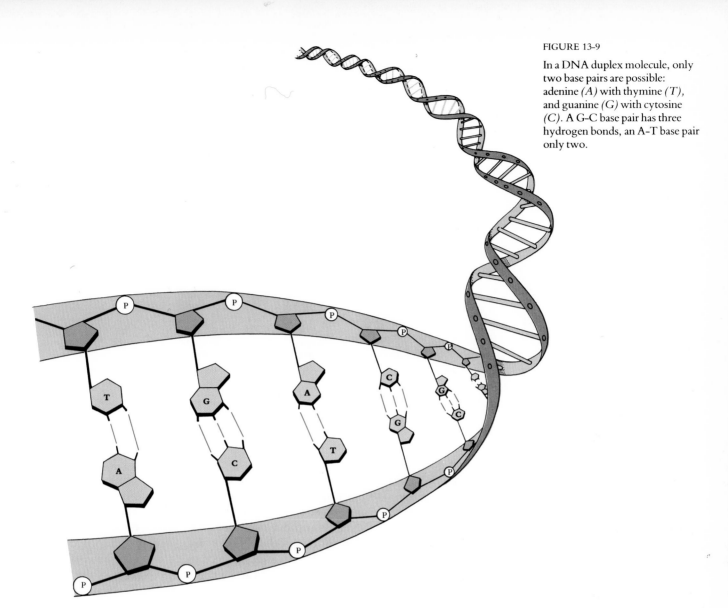

FIGURE 13-9

In a DNA duplex molecule, only two base pairs are possible: adenine *(A)* with thymine *(T)*, and guanine *(G)* with cytosine *(C)*. A G-C base pair has three hydrogen bonds, an A-T base pair only two.

The Watson–Crick model suggested that the basis for copying the genetic information is **complementarity.** One chain of the DNA molecule may have any conceivable base sequence, but this sequence completely determines the sequence of its partner in the duplex (the sugar and phosphate side chains remain the same in both copies). For example, if the sequence of one chain is ATTGCAT, the sequence of its partner in the duplex *must* be the complement TAACGTA. Each base of the chain in the duplex is the complement of the corresponding base of the other. To copy (replicate) the DNA molecule, one needs only to "unzip" it and construct a new complementary chain alongside each naked single strand. This process occurs in cell division.

The form of DNA replication suggested by the Watson–Crick model is called **semiconservative** because after one round of replication, the physical integrity of each of the two original complementary strands is conserved but the original duplex is not conserved. This prediction of the Watson–Crick model was tested in 1958 by Matthew Meselson and Frank Stahl, then of the California Institute of Technology. These two scientists grew bacteria for several generations in a medium containing the heavy isotope of nitrogen (^{15}N), so that the DNA of the bacteria eventually became denser than normal. They then transferred the growing cells to a new medium containing the normal lighter isotope ^{14}N and harvested the DNA at various points in time (Figure 13-10).

FIGURE 13-10

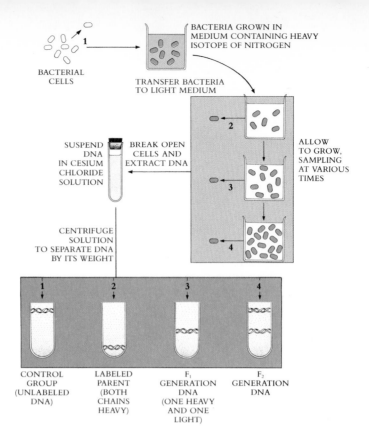

The Meselson and Stahl experiment. Bacterial cells were grown for several generations in a medium containing heavy nitrogen isotopes, and then transferred to a new medium containing only the normal lighter nitrogen isotope. At various times thereafter, samples were taken, and the DNA centrifuged in a cesium chloride solution. Because the cesium ion is so massive, the cesium chloride tends to settle in the rapidly spinning tube, establishing a gradient of cesium concentration. DNA molecules sink in the gradient until they reach a place where the cesium concentration has the same density as the density of the DNA; the DNA then "floats" at that position. Because DNA built with heavy nitrogen isotopes is denser than normal DNA, it sinks to a lower position on the cesium gradient. What Meselson and Stahl found was that after one generation in "light" medium, a single band of intermediate density halfway between heavy and light was obtained (one strand of each duplex was labeled, the other was not); after a second cell division, two bands were obtained, one intermediate (one of the two strands labeled) and one light (neither strand labeled). Meselson and Stahl concluded that replication of the DNA duplex involves building new molecules by separation of strands and assembly of new partners on these templates.

At first all the DNA that the bacteria manufactured was heavy. But as the new DNA that was being formed incorporated the lighter nitrogen isotope, the DNA density fell. After one round of DNA replication was complete, the density of the bacterial DNA had decreased to a value exactly intermediate between all–light-isotope DNA and all–heavy-isotope DNA. After another round of replication, two density classes were observed: one intermediate and the other light, corresponding to DNA that included none of the heavy isotope. These results showed that after one round of replication, each daughter DNA duplex possessed one of the labeled strands of the parent molecule. When this hybrid duplex replicated, it contributed one heavy strand to form another hybrid duplex and one light strand to form a light duplex. Meselson and Stahl's experiment thus clearly confirmed the prediction of the Watson-Crick model that DNA replicates in a semiconservative manner.

> The basis for the great accuracy of DNA replication is complementarity. A DNA molecule is a duplex, containing two strands that consist of complementary sequences of bases, so either strand can be used to reconstruct the other.

WHAT IS THE UNIT OF HEREDITARY INFORMATION?

It is one thing to demonstrate that hereditary information resides within replicating molecules of DNA, but quite another to understand what sort of information is stored there. Since the time of Mendel, geneticists had puzzled over this question. Mendelian traits in peas, involving colors and shapes, were the end result of complex processes. What kind of change in the hereditary information would change a Mendelian trait?

The first answer to this question came soon after Mendel's experiments, but its significance was not readily appreciated. In 1902 a British physician named Archibald Garrod, who was working with one of the early Mendelian geneticists, his countryman William Bateson, noted that certain diseases among his patients were prevalent in particular families. Indeed, if several generations within such families were examined, some of these disorders appeared to be controlled by simple recessive alleles. Garrod concluded that these disorders were Mendelian traits that had resulted from past changes in the hereditary information and had been passed down from an ancestor to the affected families.

Garrod proceeded to examine several of these disorders in detail. In one such disorder, **alkaptonuria,** the patients passed urine that rapidly turned black when exposed to air. Such urine contained homogentisic acid (alkapton), which air oxidized. In normal individuals, homogentisic acid is broken down into simpler substances, but the affected patients were unable to carry out that breakdown. With considerable insight, Garrod concluded that the patients suffering from alkaptonuria lacked the enzyme necessary to catalyze this breakdown and, more generally, that many inherited disorders might reflect enzyme deficiencies.

From Garrod's finding it is but a short leap of intuition to guess that the information encoded within the DNA of chromosomes is used to specify specific enzymes. However, this point was not actually established for nearly four decades. Then in 1941 a series of experiments by the Stanford University geneticists George Beadle and Edward Tatum finally provided definitive evidence. Beadle and Tatum deliberately set out to create Mendelian mutations in the chromosomes; they then studied the effects of these mutations on the organism.

One of the reasons that Beadle and Tatum's experiments produced clear-cut results—and one of the characteristics of most successful laboratory experiments in biology—is that the researchers made an excellent choice of experimental organism. They chose the bread mold *Neurospora,* a fungus that can readily be grown in the laboratory on a **defined medium** (that is, a medium that contains only known substances such as glucose and sodium chloride, rather than some uncharacterized cell extract such as ground-up yeasts). They first allowed various strains of the fungus to grow on a

FIGURE 13-11

Beadle and Tatum's procedure for isolating nutritional mutations in *Neurospora,* a fungus that grows well on a medium that contains the nutrients it needs. In Beadle and Tatum's experiments, individual irradiated strains of *Neurospora* were crossed with one another. These immediately underwent meiosis and then mitosis to produce eight spores of each with a single nucleus. Such spores were grown on complete medium, which contains all of the necessary amino acids and vitamins that the fungus normally manufactures and requires to grow. Once the colonies were established, the individual spores were taken and tested to see if they would grow on minimal medium, which lacks the amino acids and vitamins that the fungus normally manufactures. Any strains that would not grow on minimal medium, but would grow on complete medium, contained one or more mutations in the genes that are necessary to produce one of the substances in the complete but not the minimal medium. To find out which one, the line was tested for its ability to grow on minimal medium supplemented with particular substances. The mutation illustrated here is an arginine mutant, a cell line that has lost the ability to produce arginine. It will not grow on minimal medium but will grow on minimal medium to which only arginine has been added.

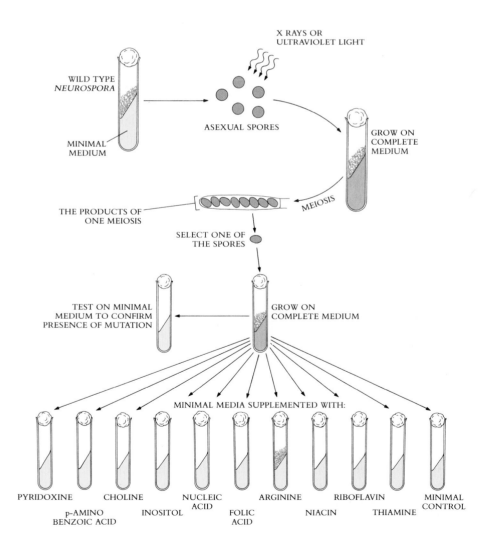

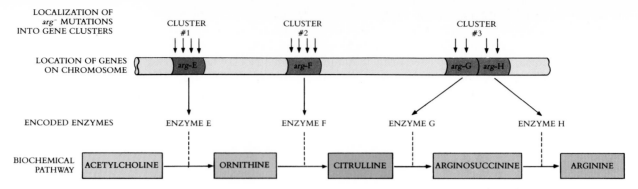

FIGURE 13-12

The chromosomal locations of the many arginine mutations isolated by Beadle and Tatum cluster around three locations, corresponding to the locations of the genes encoding the enzymes that carry out arginine biosynthesis.

complete medium, which is a medium that contains all necessary nutrients and supplies those nutrients to the growing organism, whether or not individual strains can manufacture the metabolites for themselves. In this way the investigators were able to preserve strains that had experienced changes in the nucleotide sequence of their DNA, even if the changes destroyed the ability to make one or more of the compounds that the fungus needed for normal growth. Damage of this sort is called **mutation,** and the strains that have lost the ability to use one or more compounds for normal development are called **mutant strains.**

The next step was to test the progeny of the fungi growing on complete medium to see whether any mutations leading to metabolic deficiency were present in any of the fungus strains. Beadle and Tatum did this by attempting to grow subdivisions of individual fungal strains on a **minimal medium** that contained only sugar, ammonia, salts, a few vitamins, and water. A cell that had lost the ability to make a necessary metabolite would not grow on such a medium. Using this approach, Beadle and Tatum succeeded in identifying and isolating many deficient mutants.

To determine the nature of each deficiency, Beadle and Tatum tried adding various chemicals to the minimal medium to find one that would make it possible for a given strain to grow (Figure 13-11). In this way they were able to pinpoint the nature of the biochemical problems that many of the mutants had. Many of the mutants proved unable to synthesize a particular vitamin or amino acid. For example, the addition of arginine permitted the growth of a group of mutant strains, called *arg* mutants. When the chromosomal position of each mutant *arg* gene was located, they were found to cluster in three areas (Figure 13-12). Since the genes that Beadle and Tatum were studying acted sequentially in different biochemical pathways, the scientists were able to understand the nature of these pathways by discovering the particular places at which the pathway was blocked.

For each enzyme in the arginine biosynthetic pathway (recall that enzymes are proteins that speed up specific chemical reactions, and that a particular enzyme is associated with each step in a biochemical pathway), Beadle and Tatum were able to isolate a mutant strain that apparently had a defective form of that enzyme. The mutation was always located at *one* of a few specific chromosomal sites, a different site for the particular gene associated with each enzyme. Thus each of the mutants that Beadle and Tatum examined could be explained in terms of a mutation in one (and only one) gene that was located at a single site on one chromosome. The geneticists concluded that genes produce their effects by specifying the structure of enzymes. They called this relationship the **one gene–one enzyme hypothesis.**

Genetic traits are expressed largely as a result of the activities of enzymes. Organisms store hereditary information by encoding the structures of enzymes in their chromosomes.

What sort of information must a gene encode in order to specify a protein? For some time this was not at all clear since protein structure seemed to be impossibly complex. For example, it was not evident whether or not a particular kind of protein had a consistent sequence of amino acids that would be the same in one individual molecule as it was in another of that same kind of protein. However, the picture changed in 1953, the same year in which Watson and Crick unraveled the structure of DNA. In that same year the great English biochemist Frederick Sanger, after many years of work, announced the complete sequence of amino acids in the protein insulin. Sanger's achievement was extremely significant because it demonstrated for the first time that proteins consist of definable sequences of amino acids. For any given form of insulin, each molecule has the same amino acid sequence as every other, and this sequence can be learned and written down. All enzymes and other proteins are strings of amino acids arranged in a certain definite order. Therefore the information necessary to specify an enzyme is an ordered list of amino acids.

In 1956 Sanger's pioneering work was followed by Vernon Ingram's analysis of the molecular basis of sickle cell anemia, a protein defect that is inherited as a Mendelian disorder among humans (Chapter 12). By analyzing the structure of normal and sickle cell hemoglobin, which is an oxygen-carrying blood protein, Ingram, working at Cambridge University, showed that sickle cell anemia is caused by the change of the amino acid at a single position in the protein from glutamic acid to valine. Hemoglobin consists of two pairs of identical parts called alpha and beta chains, which are produced by different genes and are combined after they are formed. Consequently, the whole hemoglobin molecule is a **tetramer,** or four-part molecule (Figure 13-13). The alleles

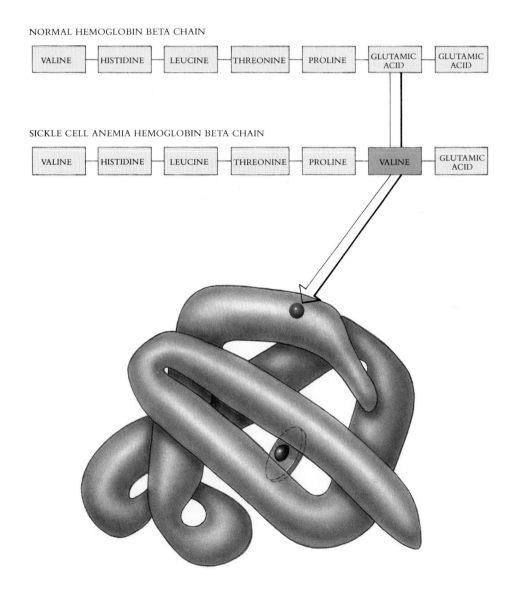

NORMAL HEMOGLOBIN BETA CHAIN

VALINE — HISTIDINE — LEUCINE — THREONINE — PROLINE — GLUTAMIC ACID — GLUTAMIC ACID

SICKLE CELL ANEMIA HEMOGLOBIN BETA CHAIN

VALINE — HISTIDINE — LEUCINE — THREONINE — PROLINE — VALINE — GLUTAMIC ACID

FIGURE 13-13

Sickle cell hemoglobin is produced by a recessive allele of the gene encoding the beta chain of hemoglobin. It represents a single amino acid change from glutamic acid to valine at the sixth position in the chain; in the folded beta chain molecule, this position contacts the alpha chain, and the amino acid change causes the hemoglobins to aggregate into long chains, which alters the shape of the cell. Hemoglobin aggregated in this way within sickled cells does a very poor job of transporting oxygen, so that individuals homozygous for the sickle cell hemoglobin allele, which can synthesize only this kind of hemoglobin, suffer from severe anemia and may die if they are not treated.

of the gene encoding the beta chain of the sickle cell hemoglobin differ from the normal form of the same gene only in their specification of this one amino acid.

These experiments and other related ones have finally brought us to a clear understanding of what the unit of heredity is: it is the information that encodes the amino acid sequence of an enzyme or other protein or of a polypeptide that makes up part of such a protein. We call the sequence of nucleotides that encode this information a **gene.** Although most genes encode polypeptides, there are also genes devoted to the production of special forms of RNA, many of which play important roles in protein synthesis.

> **The amino acid sequence of a particular protein is specified by a corresponding sequence of nucleotides in the DNA. The protein may consist of one or more polypeptides. The nucleotide sequences that specify such polypeptides are called genes.**

HOW DO GENES INTERACT TO PRODUCE A PHENOTYPE?

Many of the enzymes in cells participate in biochemical pathways, and when an investigator is studying the inheritance of two genes specifying enzymes in the same pathway, it may be difficult to sort out what is going on. For example, consider the standard pattern of Mendelian inheritance expected of genes at two loci assorting independently of each other. When individuals that are heterozygous for the two different genes mate (a dihybrid cross), four different phenotypes are possible among the progeny: (1) the dominant phenotype of both genes is displayed, (2) one of the dominant phenotypes is displayed, (3) the other dominant phenotype is displayed, or (4) neither dominant phenotype is displayed. Mendelian assortment predicts that these four possibilities will occur in the proportions 9:3:3:1. Sometimes, however, it is not possible for an investigator to identify successfully each of the four possible phenotypic classes because two or more of the classes look alike.

An example is provided by the corn, *Zea mays*. Some commercial varieties exhibit a purple pigment called anthocyanin in their seed coats, whereas others do not. In 1918 the pioneering geneticist R.A. Emerson crossed two pure-breeding corn varieties, neither of which typically exhibit any anthocyanin pigment, and obtained the surprising result that all of the F_1 plants produced purple seeds. When crossed, the two white varieties, which had never been observed to make the pigment, produced progeny that uniformly made the pigment.

When two of these pigment-producing F_1 plants were crossed to produce an F_2 generation, 56% were pigment producers and 44% were not. What was going on? Emerson correctly deduced that *two* genes were involved in the pigment-producing process, and that the second cross had been a dihybrid cross such as described by Mendel. Mendel predicts nine possible different genotypes, formed in 16 different ways in equal proportions ($9 + 3 + 3 + 1 = 16$), so that Emerson's result corresponded to $0.56 \times 16 = 9$ and $0.44 \times 16 = 7$—that is, he had observed a **modified ratio** of 9:7 instead of the usual 9:3:3:1 ratio.

In this case the pigment anthocyanin is produced from a colorless molecule by two enzymes that work one after the other—in other words, the pigment is the product of a two-step biochemical pathway:

To produce the pigment, a plant must possess at least one good copy of each enzyme. The dominant alleles encode functional enzymes, and the recessive alleles encode defective nonfunctional ones. Of the 16 genotypes predicted by random assortment, 9 contain at least one dominant allele of both genes (C–D–)—these are the purple

FIGURE 13-14

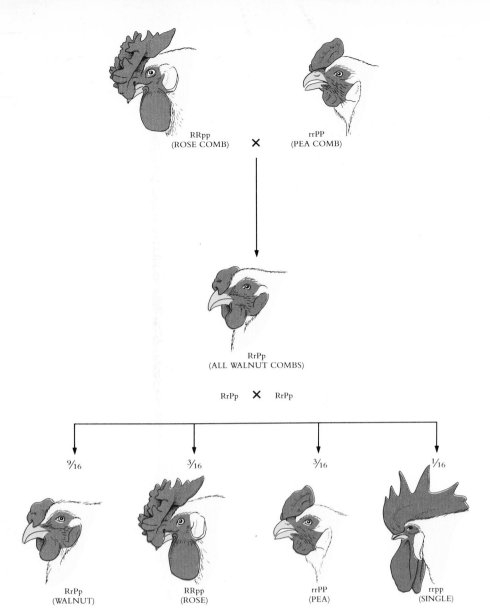

RRpp
(ROSE COMB) × rrPP
(PEA COMB)

RrPp
(ALL WALNUT COMBS)

RrPp × RrPp

9/16 — RrPp (WALNUT)

3/16 — RRpp (ROSE)

3/16 — rrPP (PEA)

1/16 — rrpp (SINGLE)

In many pure-breeding lines of chickens, the individuals have distinctive combs. Two kinds of combs are "rose," a large irregular comb, and "pea," a more compact solid one. When chickens with these two different kinds of combs are crossed, the resulting hybrid individuals have a new and very different kind of comb, rather than one of those characteristic of the parents. This new type of comb is called "walnut," which is what it looks like. When chickens with walnut combs are crossed with one another, they produce some offspring with walnut combs, others with rose or pea combs, and others with still another comb type, called "single." When crossed to one another, chickens with single combs produce pure lines of individuals with single combs, as in Leghorn chickens. How can these results be explained? There is not one gene segregating here, but two. The walnut combs of the first F_1 generation occur in individuals that are heterozygous for both rose and pea genes, while individuals with single combs are homozygous recessive for both of these genes. In a dihybrid cross, as you might expect, the four comb types are produced in 9:3:3:1 proportions.

progeny—and 7 genotypes (C–dd, ccDd, ccdd) do not. The 9:7 ratio that Emerson observed resulted from the pooling of the three nonpigmented phenotypic classes that lack dominant alleles at both loci (3 + 3 + 1 = 7).

Few phenotypes are the result of the action of only one gene. Most traits reflect the functioning of many genes that act sequentially or jointly. When one gene is present in mutant form, it is impossible to judge whether the others are functioning properly or not. Such interactions between genes are the basis of the phenomenon we call **epistasis.** Epistasis is an interaction between the products of two genes in which one of the products modifies or masks the phenotypic expression of the other. Epistatic interactions between genes often make the interpretation of particular phenotypes very difficult (Figure 13-14).

The ability of one gene to mask the phenotypic expression of another is called epistasis.

Many other cases of modified Mendelian ratios have been reported in which the phenotypic classes do not reflect the underlying 9:3:3:1 genotypic ratio. Sometimes, for example, an organism will require that at least one of the two genes under study encode a functioning product in order for the individual to survive—individuals that are homozygous recessive at both loci die. In this case the observed phenotypic ratio would be 9:3:3:0 or 3:1:1. It is important to realize that cases such as these represent a limitation in our ability to see what is going on—not a failure of Mendelian assortment. The 9:3:3:1 assortment ratio occurs among the gametes; the difficulty arises in knowing which phenotypes correspond to which gametes.

THE GENE CONCEPT

Fortunately the concept of a gene as a particular sequence of nucleotides in a DNA molecule encoding a polypeptide has the virtue that it is very concrete. Consider, for example, changes in the nucleotide sequence of genes that sometimes occur by accident. We call such changes mutations. Although a diploid individual possesses no more than two alleles at one time, this does not mean that only two allele alternatives are possible for a given gene. Any of the nucleotides are potentially subject to change. Almost all genes that have been studied exhibit several different alleles among the individuals of a population, any one or two of which may be carried by one individual. Now imagine that you have encountered two mutations in a survey of a group of individuals. Using this concept of a gene, we can say simply that the two mutations represent different alleles of the same gene if the mutations change nucleotides within the same coding sequence of the DNA; if they alter different coding sequences, then they represent mutations of different genes.

But what if you don't know the DNA sequence of the gene(s) you are studying? For example, imagine that you encounter two white-eyed flies. Are these two individuals mutant for the same gene? As we saw when considering modified Mendelian ratios, different genotypes can produce individuals that look the same; therefore you cannot be sure by looking at the flies that they represent the same class of mutation. How do you decide? By carrying out a cross, or by transferring the flies' DNA in other ways that we shall discuss in Chapters 14 and 15, you can create a heterozygous individual, that is, one that contains one copy of each gene (or allele). Then you ask whether this heterozygous individual has the mutant phenotype. If the two mutations are in the same gene, then neither chromosome of the heterozygote bears a functional copy of the gene, and the individual will appear mutant. On the other hand, if the two mutations are in different genes, then one chromosome will bear a functional copy of one gene and the other chromosome will bear a functional copy of the other; the individual will have one good copy of each gene and therefore will not appear mutant. In this case the two mutants are said to **complement** each other. The absence of complementation in heterozygotes allows us to tell different genes from one another.

Mutations that do not complement one another affect the same gene.

The difficulty of inferring genotypes from an examination of phenotypes is not the only problem that the practicing geneticist encounters in trying to understand how particular traits are inherited. Among others are the following:

Continuous variation. When multiple genes act jointly to determine a trait such as height or weight, the contribution resulting from the segregation of one particular gene is difficult to monitor, just as it is difficult to follow the flight of one bee within a swarm. Because all of the participating genes play a role in determining the phenotype and because many are segregating independently of one another, a gentle gradation in degree of difference is seen when many individuals are examined (Figure 13-15).

Pleiotropy. An individual genetic alteration will often have more than one effect on the phenotype. Such an alteration is said to be **pleiotropic.** Thus when the pioneering French geneticist Lucien Cuenot studied yellow fur (a dominant trait) in mice, he was unable to obtain a pure-breeding homozygous yellow strain by crossing individual yellow mice with one another—individuals that were homozygous for the yellow allele died. The yellow allele was pleiotropic:

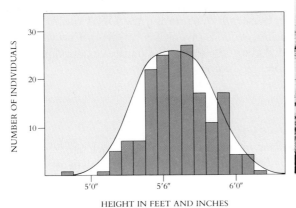

HEIGHT IN FEET AND INCHES

one effect was yellow color, but another effect was a lethal developmental defect. Thus a pleiotropic gene alteration may be dominant with respect to one phenotypic consequence (yellow fur) and recessive with respect to another (lethal developmental defect). Therefore the way in which mutations alter phenotypes makes it difficult to unravel the underlying genetic basis of the change in phenotypes. By examining the characteristics of organisms, we are also studying the consequences of the action of products made by genes, and these products often perform other functions about which we are ignorant.

Incomplete dominance. Not all alternative alleles are fully dominant or recessive in heterozygotes in the way that Mendel's alternative alleles were. Sometimes heterozygous individuals are intermediate between parents, and in other cases they resemble one parent but not precisely. For example, a cross between a red-flowering snapdragon and a white-flowering snapdragon produces heterozygotes that are an intermediate pink color (Figure 13-16).

Environmental effects. The degree to which many alleles are expressed depends on the environment. Early in this century, for example, scientists wondered whether the more compact plants that grew along the seashore differed genetically from the taller individuals of the same species that they encountered inland. Later experimental studies demonstrated that such differences were largely genetically determined, although every gardener knows that plants may vary greatly in appearance depending on where they grow. We shall further consider environmental effects of this sort in Chapter 17.

FIGURE 13-15

Variation in height among students of the 1914 class of the Connecticut Agricultural College. Because many genes contribute to height, and tend to segregate independently of one another, there are many possible combinations. The cumulative contribution of different combinations of alleles to height form a continuous spectrum of possible heights—a random distribution, in which the extremes are much rarer than the intermediate values.

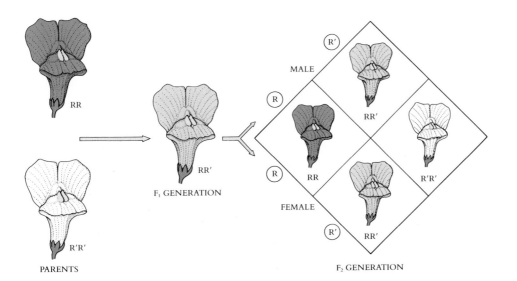

PARENTS

F₁ GENERATION

MALE

FEMALE

F₂ GENERATION

FIGURE 13-16

A cross between a red-flowered snapdragon, which has the genotype RR, and a white-flowered one (R′R′). Neither allele is dominant, and the heterozygotes have pink flowers and the genotype RR′.

All of these potential complications make the job of the Mendelian geneticist more difficult. However, our study of genes need not be restricted to looking at phenotypes, and more precise relationships can be seen by examining the DNA molecules themselves. In Chapter 14 we shall examine genes as physical entities—distinct segments of DNA. What geneticists have learned about the molecular nature of genes, particularly within the last 10 years, is revolutionizing biology.

SUMMARY

1. Eukaryotic cells store hereditary information within the nucleus. When the nucleus is transplanted, the hereditary specifications of an organism are also transplanted.

2. In viruses, bacteria, and eukaryotes, the hereditary information resides in nucleic acids. The transfer of pure nucleic acid can lead to the transfer of hereditary traits. In all cellular organisms the genetic material is DNA, although in some viruses the genetic material is RNA.

3. During cell division the hereditary message is duplicated with great accuracy. The mechanism that achieves this high degree of accuracy is complementarity: the DNA molecule is organized as a double helix with two strands that consist of complementary corresponding bases. By the addition of complementary bases, either one of the strands can be used to re-form the helix.

4. Most hereditary traits reflect the actions of enzymes. The traits are hereditary because the information necessary to specify these enzymes is encoded within the DNA.

5. Sometimes Mendelian assortment does not produce the expected $9:3:3:1$ phenotypic ratio in a dihybrid cross, because alleles of some genes interfere with our ability to differentiate alleles of others. Thus whether an individual is dominant or recessive at a step late in a pathway does not affect the phenotype if the pathway is blocked earlier by alteration of another gene. The ability of one gene to mask the phenotypic expression of another is called epistasis.

6. Two mutations can be assumed to affect the same gene if the two mutations do not complement one another in heterozygotes. Complementation provides an operational definition of a gene.

REVIEW

1. Where do cells store their hereditary information?

2. The transforming principle in the Griffith-Avery experiments proved to be _____.

3. Erwin Chargaff observed that the amount of adenine in all DNA molecules is equal to the amount of _____, and that the amount of guanine is equal to the amount of _____.

4. Genetic traits are expressed largely as a result of the activities of _____, which are necessary for biosynthetic processes to take place.

5. The ability of one gene to mask the phenotypic expression of another is called _____.

SELF-QUIZ

1. The Hershey-Chase experiment demonstrated that
   (a) virus DNA injected into bacterial cells is apparently the factor involved in directing the production of new virus particles
   (b) virus protein injected into bacterial cells is apparently the factor involved in directing the production of new virus particles
   (c) ^{32}P labeled protein is injected into bacterial cells by viruses
   (d) the transforming principle is the DNA and not the polysaccharide coat
   (e) RNA is the genetic material of some viruses

2. The Watson-Crick model of DNA structure suggested that the basis for the faithful copying of the genetic material is complementarity. This means that if you know the base sequence of one strand is AATTCG, then the sequence of the other strand must be
   (a) AATTCG
   (b) TTGGAC
   (c) TTAACG
   (d) TTAAGC
   (e) do not have enough information

3. The sequence of DNA nucleotides that encodes the amino acid sequence of an enzyme is
   (a) a gene   (d) a phenotype
   (b) a polypeptide   (e) a dominant trait
   (c) a protein

4. If you wanted to know whether two different white-eyed flies have the same or different mutations resulting in white eyes, what would be the simplest and first step in trying to find out?
   (a) transfer the gene to bacteria
   (b) sequence the gene
   (c) do backcrosses
   (d) cross the flies to see whether the F_1 has white or red (wild type) eyes
   (e) none of the above

5. Characteristics such as height in humans that exhibit continuous variation are generally controlled by
   (a) a single dominant gene
   (b) a recessive gene
   (c) epistatic interactions
   (d) pleiotropy
   (e) multiple genes

PROBLEM

1. You are presented with two test tubes, each containing purified DNA, and told that one of the tubes contains human double-stranded DNA and the other contains virus single-stranded DNA. You analyze the base composition of the two preparations, with the following results:

   Tube 1: 22.1% A : 27.9% C : 29.7% G : 22.1% T
   Tube 2: 31.3% A : 31.3% C : 18.7% G : 18.7% T

   Which of the two tubes contains single-stranded DNA?

FOR FURTHER READING

CRICK, F.H.C.: "The Double Helix: A Personal View," *Nature,* vol. 248, page 766, 1974. Francis Crick's own recollections of the hectic days when he and James Watson deduced that the structure of DNA is a double helix.

JUDSON, H.F.: *The Eighth Day of Creation,* Simon and Schuster, New York, 1979. The definitive historical account of the experimental unraveling of the mechanisms of heredity, based on personal interviews with the participants. Although you might find some of the science heavy going, this book is full of the feel of how science is really conducted.

LAWN, R., and G. VEHAR: "The Molecular Genetics of Hemophilia," *Scientific American,* March 1986, pages 48-54. A fine description of the way in which modern molecular techniques have permitted the isolation of the gene responsible for this famous genetic disorder, which is discussed in Chapter 12.

WATSON, J.D., and F.H.C. CRICK: "A Structure for Deoxyribose Nucleic Acid," *Nature,* vol. 171, page 737, 1953. The original report of the double helical structure of DNA. Only one page long, this paper marks the birth of molecular genetics.

WATSON, J.D.: "The Double Helix," Atheneum Publishing Co., Inc., New York, 1968. A lively, often irreverent account of what it was like to discover the structure of DNA, recounted by someone in a position to know.

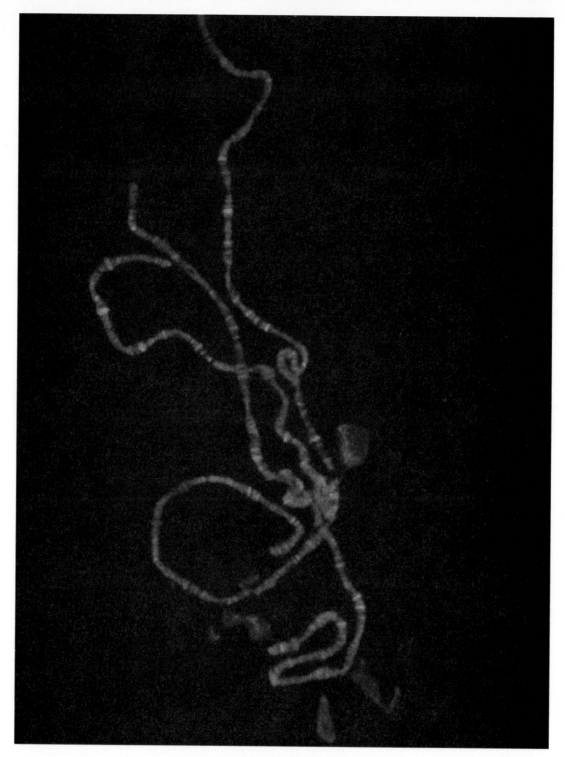

FIGURE 14-1   In these giant chromosomes of the fly *Drosophila melanogaster*, the individual strands of DNA are replicated many times in parallel. For this reason, the activity of individual genes can readily be detected. In this micrograph, active genes appear as brighter bands. At any one time, only a fraction of a chromosome's many genes are actively being transcribed to produce RNA molecules.

GENES AND HOW THEY WORK Chapter 14

Overview

What is a gene? Obtaining the answer to this question has been perhaps the greatest advance in biology in this century. We now know that genes are strings of DNA nucleotides arranged in groups that specify the structure of a polypeptide, sometimes of an entire protein. Special proteins bind to the DNA at the beginning of genes and then move through them, making an RNA copy of the gene as they go. It is this RNA copy that cells use to produce the polypeptides specified by the genetic information. Cells often control when a gene is "turned on" by controlling production of these RNA copies.

For Review

Here are some important terms and concepts that you will encounter in this chapter. If you are not familiar with them, you should review them before proceeding.

Ribosomes (Chapter 5)

Structure of DNA and RNA (Chapters 3, 13)

Enzymes and enzyme activity (Chapter 7)

Structure of eukaryotic chromosomes (Chapter 10)

Each of us bears in every cell of our bodies more information than there is in a telephone book for a large city. This information is the hereditary information, the instructions that specify what you are like. These instructions dictate that you will have arms and not fins, hair and not feathers, two eyes and not one. The color of your eyes, the texture of your fingernails, whether you dream in color—all of the many traits that you receive from your parents are recorded in every cell of your body. Biologists learned by experiment that long DNA molecules, which in eukaryotes associate with proteins to form chromosomes (Figure 14–1), contain this information. The information is arrayed in little blocks like entries in a dictionary. Each block is a gene specifying a particular polypeptide. Some polypeptides are entire proteins; other proteins are formed of two or more gene products. Proteins constitute the tools of heredity. Many proteins are enzymes that carry out reactions within cells: what you are is a result of what they do. The essence of gene function is the ability of a cell to use information in its DNA to bring about the production of particular polypeptides. In this way genes affect what that cell will be like. In this chapter we examine how this happens.

SMALL SUBUNIT

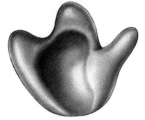

LARGE SUBUNIT

RIBOSOME

FIGURE 14-2

A ribosome is composed of two subunits. The smaller subunit fits into a depression on the surface of the larger one.

CELLS USE RNA TO MAKE PROTEIN

If you wanted to find out how a cell uses its DNA to direct the production of particular proteins, what is the first thing you would do? Perhaps the simplest question you might ask is "Where in the cell are proteins made?" You can answer this question by placing cells for a short time in a medium containing radioactive amino acids. The cells will take up the radioactively labeled amino acids for the short time that they are exposed to them. When investigators looked to see where in the cells these radioactive proteins first appeared, they found that proteins assembled not in the nucleus, where the DNA is, but rather in the cytoplasm, on large protein aggregates called ribosomes (Figure 14-2). These little polypeptide-making factories were very complex, containing more than 50 different proteins. They also contained a very different sort of molecule, RNA. As you will recall from Chapter 3, RNA is very similar to DNA (Figure 14-3); its presence in ribosomes hints that RNA molecules play an important role in polypeptide synthesis.

A cell contains many kinds of RNA, but there are three major classes:

Ribosomal RNA. The class of RNA that is found in ribosomes together with characteristic proteins is called **ribosomal RNA** or **rRNA.** During polypeptide synthesis, rRNA molecules provide the site on the ribosome where the polypeptide is assembled.

Transfer RNA. A second class of RNA, called **transfer RNA,** or **tRNA,** is much smaller. Cells contain more than 60 different kinds of tRNA molecules, which float free in the cytoplasm. During polypeptide synthesis, tRNA molecules transport amino acids to the ribosome for use in building the polypeptide, and position each amino acid at the correct place on the lengthening polypeptide chain.

Messenger RNA. A third class of RNA is **messenger RNA** or **mRNA.** Each mRNA molecule is a long single strand of RNA that passes from the nucleus to the cytoplasm. During polypeptide synthesis, mRNA molecules carry information from the chromosomes to the ribosomes to direct which polypeptide is assembled.

These three kinds of RNA act together with the ribosome proteins and a special class of enzymes and other proteins. These molecules constitute a system that reads the genetic message and produces the polypeptide specified by the particular message. RNA molecules are the apparatus that a cell uses to translate its hereditary information.

FIGURE 14-3

RNA is very similar to DNA, but there are two important differences in the nucleotides. The first is in the sugar components; in place of deoxyribose, RNA contains ribose, which has an additional oxygen atom. Second, RNA contains the pyrimidine uracil (U) instead of thymine (T). Thymine has a methane group ($-CH_3$) in place of $-H$ that is characteristic of uracil in the upper right hand corner of the ring as shown here. An additional difference between these two classes of nucleic acids is that RNA does not have a regular helical structure, and is usually single-stranded.

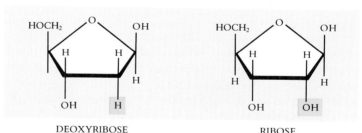

DEOXYRIBOSE   RIBOSE

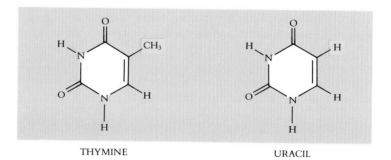

THYMINE   URACIL

You can think of this information as a message written in the code specified by the sequence of nucleotides in the DNA. The cell's polypeptide-producing apparatus reads the message one gene at a time, translating the genetic code of each gene into a particular polypeptide. Biologists have learned to read this code, and in so doing have learned much about what genes are, and how they work to dictate what a protein will be like and when it will be made.

AN OVERVIEW OF GENE EXPRESSION

The hereditary apparatus works in much the same way in all organisms. A copy of each active gene is made, and at a ribosome the copy directs the step-by-step assembly of a chain of amino acids. There are many minor differences in the details of gene expression between bacteria and eukaryotes, and a single major difference that we shall discuss later in this chapter. The basic apparatus used in gene expression, however, appears to be the same in all organisms. It apparently has persisted virtually unchanged since very early in the history of life. The process of gene expression occurs in two phases, transcription and translation.

Transcription

In the first stage of gene expression, an RNA copy of the gene is made (Figures 14-1 and 14-4). Like all classes of RNA, the copy is formed on a DNA template. The process that occurs at this stage of polypeptide synthesis is called **transcription.** The messenger RNA molecule is said to have been **transcribed** from the DNA. Transcription begins when a special enzyme, an **RNA polymerase,** binds to a particular sequence of nucleotides on one of the DNA strands, a sequence located at the edge of a gene. Starting at that end of the gene, the RNA polymerase assembles a single strand of mRNA with a nucleotide sequence complementary to that of the DNA strand it has bound. **Complementarity** refers to the way in which the two single strands of DNA that form a double helix relate to one another, with A (adenine) pairing with T (thymine), and G (guanine) pairing with C (cytosine). An RNA strand complementary to a DNA strand has the same relationships, but with U (uracil) in place of thymine.

The RNA polymerase moves along the strand into the gene, encounters each DNA nucleotide in turn, and adds the corresponding complementary RNA nucleotide to the growing mRNA strand. When the enzyme arrives at a special "stop" signal at the far edge of the gene, it disengages from the DNA and releases the newly assembled mRNA chain. This chain is complementary to the DNA strand from which the polymerase assembled it. It is an **RNA transcript** (copy), called the **primary mRNA transcript,** of the DNA nucleotide sequence of the gene.

FIGURE 14-4

The process of RNA transcription, schematically represented. First, the enzyme RNA polymerase attaches to the DNA, which opens up at that point. One of the strands of DNA then functions as a template, on which nucleotide building blocks are assembled into RNA. The hydrogen bonds between the two strands of DNA re-form as the RNA polymerase moves along the DNA molecule.

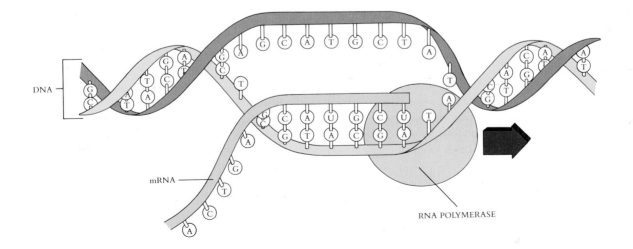

DNA

mRNA

RNA POLYMERASE

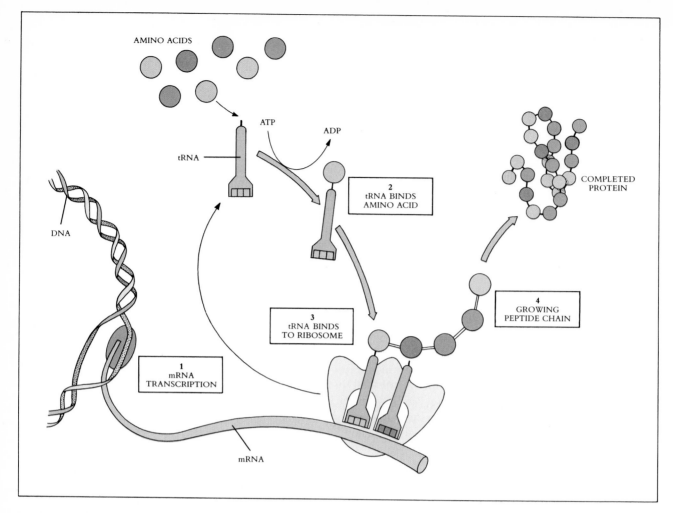

FIGURE 14-5

An overview of protein synthesis. mRNA molecules are transcribed from DNA genes encoding proteins *(1)*, and in turn are translated by tRNAs. Each of these tRNAs has a specific three-nucleotide sequence, or **anticodon,** that fits a particular mRNA three-nucleotide sequence or **codon**. The shape of a particular tRNA is such that it attaches to a particular kind of amino acid *(2)*. As a ribosome moves along the mRNA strand, a tRNA coupled with a particular amino acid fits into place *(3)*, enzymatically coupling its amino acid to a growing polypeptide chain *(4)*. The process links together a series of amino acids corresponding to the sequence of nucleotides in the DNA template on which the mRNA was initially transcribed.

Translation

The second stage of gene expression is the synthesis of a polypeptide, using the information contained on an mRNA molecule to direct the choice of amino acids. This process of mRNA-directed polypeptide synthesis is called **translation,** because nucleotide-sequence information is translated into amino acid–sequence information. Translation begins when an rRNA molecule within the ribosomes binds to one end of an mRNA transcript. Once it has bound to the mRNA molecule, a ribosome moves down the mRNA molecule in increments of three nucleotides. At each step, the ribosome adds an amino acid to a growing polypeptide chain. It continues to do this until it encounters a "stop" signal that indicates the end of the polypeptide. The ribosome then disengages from the mRNA and releases the newly assembled polypeptide. An overview of protein synthesis is presented in Figure 14–5.

> **The information encoded in genes is expressed in two stages: transcription, in which a polymerase enzyme assembles an mRNA molecule whose nucleotide sequence is complementary to the gene's template DNA strand; and translation, in which a ribosome assembles a polypeptide, using the mRNA to specify the amino acids.**

THE GENETIC CODE

Protein synthesis occurs on the ribosomes, where the initial portion of the mRNA transcribed from a gene binds to an rRNA molecule interwoven in the ribosome (Figure 14–6). The mRNA lies on the ribosome in such a way that only a three-nucleotide portion of the mRNA molecule, called a **codon,** is exposed at the polypeptide-making

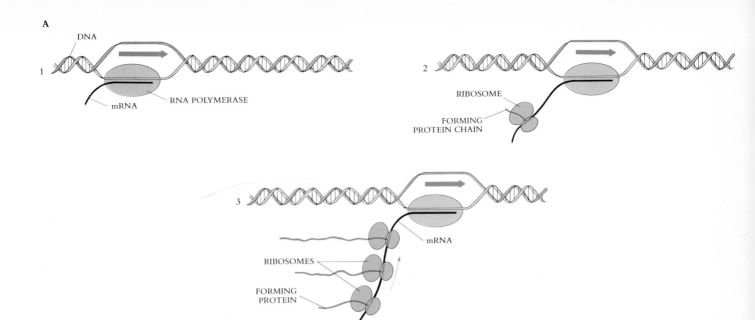

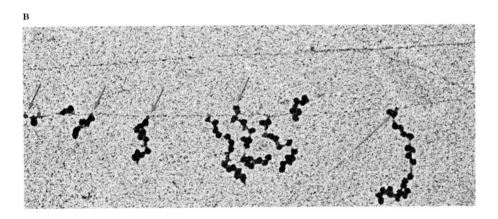

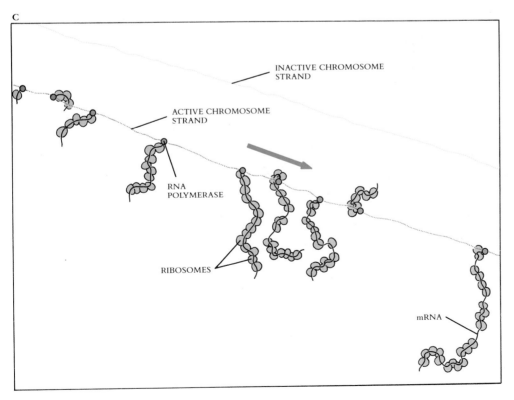

A As soon as the polymerase begins to transcribe a mRNA molecule for the DNA, a series of ribosomes bind one after another to the mRNA's ribosome recognition site (which is the part of the mRNA that is transcribed from the DNA first). Each of these ribosomes then begins in turn to translate the coding sequence into polypeptide.

FIGURE 14-6

Bacteria have no nucleus and hence no membrane barrier between DNA and cytoplasm.
A As soon as the polymerase begins to transcribe a mRNA molecule for the DNA, a series of ribosomes bind one after another to the mRNA's ribosome recognition site (which is the part of the mRNA that is transcribed from the DNA first). Each of these ribosomes then begins in turn to translate the coding sequence into polypeptide.
B In the electron micrograph of genes being transcribed in the common colon bacterium, *Escherichia coli,* you can see every stage of the process.
C From each mRNA molecule dangling from the DNA, a series of ribosomes is assembling polypeptides, one seeming to follow the next down the mRNA.

site. As each bit of the mRNA message is exposed in turn, a molecule of tRNA with the complementary three-nucleotide sequence, or **anticodon,** binds to it (Figure 14-7). Because this tRNA molecule carries a particular amino acid, only that amino acid and no other is added to the polypeptide in that position. Protein synthesis occurs as a series of tRNA molecules bind one after another to the exposed portion of the mRNA molecule as it moves through the ribosome. Each of these tRNA molecules has attached to it an amino acid, and the amino acids that they bring to the ribosome are added, one after another, to the end of a growing polypeptide chain.

How does a particular tRNA molecule come to possess the specific amino acid that it does, and not just any amino acid? The correct amino acid is placed on each tRNA molecule by a collection of 20 enzymes called **activating enzymes.** There is one activating enzyme for each of the 20 common amino acids. An activating enzyme binds the amino acid that it recognizes to a tRNA molecule (Figure 14-8).

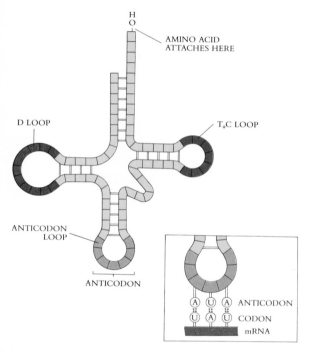

FIGURE 14-7

Structure of a tRNA molecule. The T$_\psi$C loop and the D loop function in binding to the ribosomes during polypeptide synthesis. The activating enzyme adds an amino acid to the free single-stranded −OH end. The third loop contains the anticodon sequence. It is the complementarity between the codons of mRNA and the anticodons of tRNA that makes possible the precise control of polypeptide synthesis.

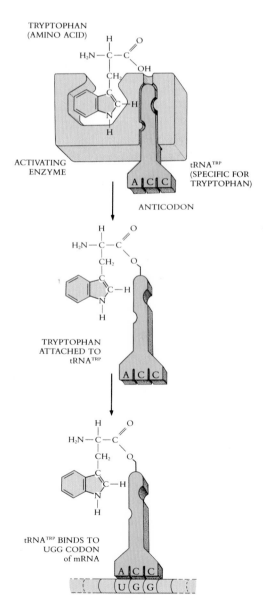

FIGURE 14-8

Activating enzymes are the elements within cells that "read" the genetic code. Each kind of activating enzyme recognizes and binds a specific amino acid such as tryptophan, on the one hand, and also recognizes and binds the tRNA molecules with anticodons specifying that amino acid, such as ACC for tryptophan on the other hand.

If you consider the nucleotide sequence of mRNA to be a coded message, then the 20 activating enzymes are the code books of the cell, the instructions for decoding the message. The code word recognized by an activating enzyme is three nucleotides long. Since there are four different kinds of nucleotides in mRNA (cytosine, guanine, adenine, and uracil instead of thymine), there are 4^3, or 64, different three-letter code words (codons) possible. Some of the activating enzymes recognize only one tRNA molecule, corresponding to one of the code words. Others recognize two, three, four, or six different tRNA molecules, each containing a different anticodon. The base sequences of the tRNA anticodons are complementary to the associated sequences of mRNA. The list of different mRNA codons specific for each activating enzyme is called the **genetic code** (Table 14-1). Each activating enzyme is signified by the amino acid that it adds; for example, the activating enzyme that adds leucine to its tRNA is designated leucine.

With a few minor exceptions, the genetic code is the same in all organisms, a given codon corresponding to the same amino acid. Three of the 64 codons (UAA, UAG, and UGA) do not correspond to triplets that are recognized by any activating enzyme. These three codons, called **nonsense codons,** serve as "stop" signals in the mRNA message, marking the end of a polypeptide. The "start" signal, which marks the beginning of a polypeptide amino acid sequence within an mRNA message, is the codon AUG. This triplet also encodes the amino acid methionine. The ribosome uses the first AUG that it encounters in the mRNA message to signal the start of its translation.

TABLE 14-1   THE GENETIC CODE

FIRST LETTER	SECOND LETTER				THIRD LETTER
	U	C	A	G	
U	Phenylalanine	Serine	Tyrosine	Cysteine	U
	Phenylalanine	Serine	Tyrosine	Cysteine	C
	Leucine	Serine	Stop	Stop	A
	Leucine	Serine	Stop	Tryptophan	G
C	Leucine	Proline	Histidine	Arginine	U
	Leucine	Proline	Histidine	Arginine	C
	Leucine	Proline	Glutamine	Arginine	A
	Leucine	Proline	Glutamine	Arginine	G
A	Isoleucine	Threonine	Asparagine	Serine	U
	Isoleucine	Threonine	Asparagine	Serine	C
	Isoleucine	Threonine	Lysine	Arginine	A
	(Start); Methionine	Threonine	Lysine	Arginine	G
G	Valine	Alanine	Aspartate	Glycine	U
	Valine	Alanine	Aspartate	Glycine	C
	Valine	Alanine	Glutamate	Glycine	A
	Valine	Alanine	Glutamate	Glycine	G

A codon consists of three nucleotides read in the sequence shown above. For example ACU codes threonine. The first letter, A, is read in the first column; the second letter, C, from the second letter columns; and the third letter, U, from the third letter column. Each of the codons is recognized by a corresponding anticodon sequence on a tRNA molecule. Some tRNA molecules recognize more than one codon sequence but always for the same amino acid. Most amino acids are encoded by more than one codon. For example, threonine is encoded by four codons (ACU, ACC, ACA, and ACG), which differ from one another only in the third position.

> All organisms possess a battery of 20 enzymes, called activating enzymes. Each activating enzyme carries one of the 20 common amino acids and recognizes a particular set of anticodons, three-base nucleotide sequences in tRNA molecules. The mRNA codons specific for each of the 20 activating enzymes and their associated amino acids constitute the genetic code.

THE MECHANISM OF PROTEIN SYNTHESIS

The initial portion of an mRNA molecule, called the **leader region,** is responsible for the first event in polypeptide synthesis, the formation of an **initiation complex** (Figure 14–9). First, a methionine-carrying tRNA, *met* **tRNA,** binds to the ribosome. Special proteins called **initiation factors** position the *met* tRNA on the ribosomal surface. Proper positioning of this first amino acid is critical, because it determines the reading frame (the groups of three) with which the nucleotide sequence will be translated into a polypeptide. This initiation complex, guided by another initiation factor, then binds to mRNA. It is important that the complex bind to the beginning of a gene, so that all of the gene is translated. In bacteria, the beginning of each gene is marked by a sequence that is complementary to one of the rRNA molecules on the ribosome. This ensures that genes are read from the beginning. Each mRNA binds to the ribosomes that read it by base-pairing between the sequence at its beginning and the complementary sequence on the rRNA which is a part of the ribosome.

After the initiation complex has formed, synthesis of the polypeptide proceeds as follows (Figure 14–10):

1. The ribosome exposes the codon on the mRNA immediately adjacent to the initiating AUG codon, positioning it for interaction with another incoming tRNA molecule. When a tRNA molecule with the appropriate anticodon appears, this new incoming tRNA binds to the mRNA molecule at its exposed codon position. Special proteins called **elongation factors** (so called because they aid in making the mRNA molecule longer) help to position the incoming tRNA. Binding the incoming tRNA to the ribosome in this manner places the amino acid at the other end of the incoming tRNA molecule directly adjacent to the initial methionine, which is dangling from the initiating tRNA molecule still bound to the ribosome.

2. The two amino acids undergo a chemical reaction, in which the initial methionine is released from its tRNA and is attached instead by a peptide bond to the adjacent incoming amino acid. The abandoned tRNA falls from its site on the ribosome, leaving that site vacant.

3. In a process called **elongation** (see Figure 14–10), the ribosome now moves along the mRNA molecule a distance corresponding to three nucleotides. The ribosome is guided by elongation factors. This movement repositions the growing chain (at this point containing two amino acids) and exposes the next codon of the mRNA. This is the same situation that existed in step 1. When a tRNA molecule appears that recognizes this next codon, the anticodon of the incoming tRNA binds this codon, placing a new amino acid adjacent to the growing chain. The growing chain transfers to the incoming amino acid, as in step 2, and the elongation process continues (Figure 14–11).

4. When a chain-terminating nonsense codon is encountered, no tRNA exists to bind to it. Instead, it is recognized by special **release factors,** proteins that release the newly-made polypeptide.

> Protein synthesis is carried out by ribosomes, which bind to sites at one end of the mRNA and then move down the mRNA in increments of three nucleotides. At each step of the ribosome's progress, it exposes a three-nucleotide sequence to binding by a tRNA molecule with the complementary nucleotide sequence. The amino acid carried by that particular tRNA molecule is added to the end of the growing polypeptide chain.

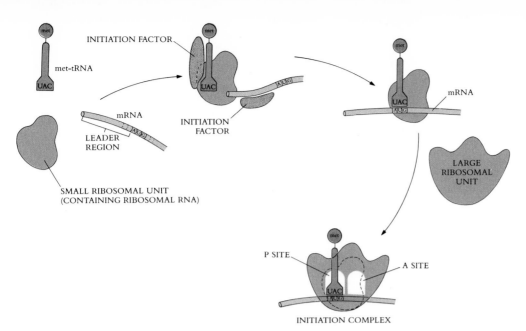

FIGURE 14-9

Formation of the initiation complex. Proteins called initiation factors play key roles in positioning the small ribosomal subunit and the *met* tRNA molecule at the beginning of the mRNA message. When the *met* tRNA is positioned over the first AUG codon sequence on the mRNA, the large ribosomal subunit binds, forming the A and P sites, and polypeptide synthesis begins.

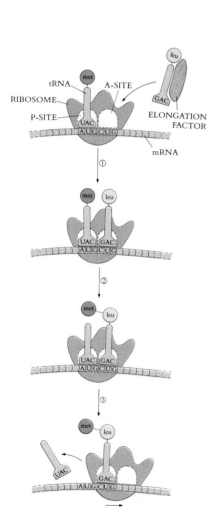

FIGURE 14-10

How polypeptide synthesis proceeds. **1** In the elongation process, the A site is occupied by the tRNA with an anticodon complementary to the mRNA codon exposed in the A site. **2** The growing polypeptide chain (in this illustration the first amino acid, methionine) is transferred to the incoming amino acid (in this illustration, leucine), **3,** the tRNA to which it was previously bound falls away, and the ribosome moves three nucleotides to the right.

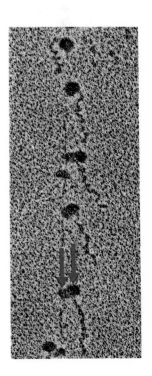

FIGURE 14-11

These ribosomes are reading along an mRNA molecule of the fly *Chironomus tentans* from bottom to top, assembling polypeptides that dangle behind them like the tail of a tadpole. Clearly visible are the two subunits (arrows) of each ribosome translating the mRNA.

Every nucleotide in the transcribed portion of a bacterial gene participates in an amino acid–specifying codon. The order of amino acids in the protein is the same as the order of the codons in the gene. For many years it was assumed that all organisms behaved in this logical way. In the late 1970s, however, biologists were amazed to discover that this relationship, with which they were familiar, did not apply to eukaryotes. Instead, eukaryotic genes are encoded in segments that are cut out from several locations along the transcribed mRNA and are later stitched together to form the mRNA that is eventually translated in the cytoplasm. With the benefit of hindsight, it is easy to design an experiment to reveal this unexpected mode of gene organization:

1. Isolate the mRNA corresponding to a particular gene. Much of the mRNA of red blood cells, for example, is related to the production of the hemoglobin and ovalbumin proteins, making it easy to purify the mRNAs from the genes related to these proteins.

2. Using an enzyme called **reverse transcriptase,** make a DNA version of the mRNA that has been isolated. Such a version of a gene is called **"copy" DNA (cDNA).**

3. Using genetic engineering techniques (Chapter 15), isolate from the nuclear DNA the portion that corresponds to one of the actual hemoglobin genes. This procedure is called "cloning."

4. Mix single-strand forms of this hemoglobin cDNA and nuclear DNA, and let them pair with each other **(hybridize)** and form a duplex.

When this experiment was performed and the resulting duplex DNA molecules were examined with an electron microscope, the hybridized DNA did not appear as a single duplex. Instead, unpaired loops were observed (Figure 14-A). In a related example, it was found that there are seven different sites in the ovalbumin gene at which the nuclear version contains long nucleotide sequences that are not present in the cytoplasmic cDNA version. The conclusion is inescapable: nucleotide sequences are removed from within the gene transcript before the cytoplasmic mRNA is translated into protein. These internal noncoding sequences are introns.

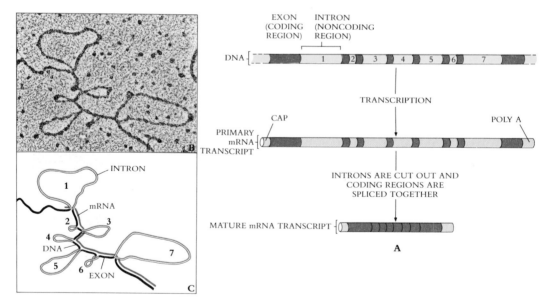

FIGURE 14-A

A The ovalbumin gene and its primary transcript contain seven segments not present in the mRNA version, which the ribosomes use to direct the synthesis of protein. These intron segments are removed by enzymes that cut out the introns and splice together the exons.

B The seven loops are the seven introns represented in the schematic drawing, **C,** of the DNA and primary mRNA transcript.

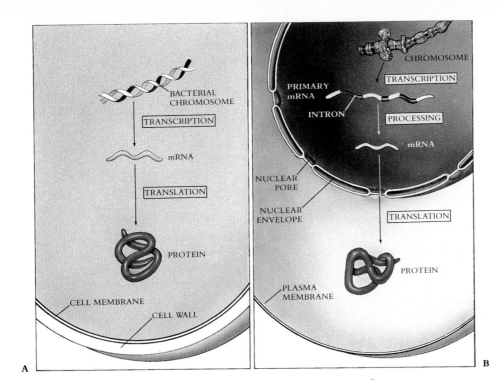

FIGURE 14-12
Genes are used very differently in prokaryotes and eukaryotes.
A The genes on a bacterial chromosome are transcribed into mRNA, which is translated directly. In other words, the sequence of DNA nucleotides in bacteria corresponds to the sequence of amino acids in the encoded polypeptide.
B The genes of a eukaryote are typically very different, containing long stretches of nucleotides called introns, which do not correspond to amino acids within the encoded polypeptide but rather serve some other function. These introns are removed from the mRNA transcript of the gene before the mRNA is used to direct the synthesis of the encoded polypeptide.

PROTEIN SYNTHESIS IN EUKARYOTES

Protein synthesis occurs in a similar way in both bacteria and eukaryotes, although there are differences. Because eukaryotes have a nucleus, their mRNA molecules must cross a membrane before they are utilized. This is not the case in bacteria. Eukaryotic mRNA molecules rarely contain copies of more than one gene; bacterial mRNA molecules often do. The ribosomes of eukaryotes are slightly larger than those of bacteria. One important difference is that, unlike bacterial genes, most eukaryotic genes are much larger than they need to be. They contain long stretches of nucleotides that are cut out of the mRNA transcript before it is used in polypeptide synthesis. Because these sequences are removed from the mRNA transcript before it is used, they are not translated into polypeptide. Intervening (lying between) the polypeptide-specifying portions of the gene, they are called **introns** (Figure 14–12). The remaining segments of the gene, the nucleotide sequences that encode the amino-acid sequence of the polypeptide, are called **exons.** Exons are scattered among larger noncoding intron sequences. Despite these differences, eukaryotes and bacteria make polypeptides in much the same way.

In a typical human gene the nontranslated (intron) portion of a gene is 10 to 30 times larger than the coding (exon) portion. For example, even though only 432 nucleotides are required to encode the 144 amino acids of human hemoglobin, there are actually 1356 nucleotides in the primary mRNA transcript of the hemoglobin gene.

REGULATING GENE EXPRESSION

Cells rarely have energy to waste. How rapidly a cell grows, how much it moves about—all of a cell's living activities require energy, which is usually in short supply. For this reason, evolution has strongly favored those cells which make particular proteins only when the proteins are useful, thus avoiding wasteful unnecessary synthesis. For example, there is little point for a cell to produce an enzyme when its substrate, the target of its activity, is not present in the cell. Much energy can be saved if production of the enzyme is delayed until the appropriate substrate is encountered and the enzyme's activity will be useful to the cell. From a broader perspective, the growth and development of many organisms, including human beings, entails a long series of biochemical steps, each step delicately tuned to achieve a precise effect. Specific enzyme activities are called into play and used to bring about a particular change. Once this change occurs, those particular enzyme activities cease, lest they disrupt other activities that follow. Many genes are transcribed in a carefully prescribed order, each gene for a specified period of time. The hereditary message is played like a piece of music on a grand organ, in which the genes are the notes and the hereditary information that regulates their expression is the score.

An organism controls the expression of its genes largely by controlling when the transcription of individual genes begins. Most genes have special nucleotide sequences called **regulatory sites.** These sites act as points of control. These sequences are recognized by specific regulatory proteins in the cell which bind to the sites.

Sometimes the site at which the regulatory protein binds to the DNA is located between the site at which the transcription enzyme RNA polymerase binds and the beginning edge of the gene that the polymerase will transcribe. If that is the case, the presence of the regulatory protein bound to its site blocks the movement of the polymerase toward the gene. To understand this more clearly, imagine that you are shooting a cue ball at the eight ball on a pool table, and that someone places a brick on the table between the cue ball and eight ball. This brick is like the regulatory protein that binds to the DNA: its placement blocks movement of the cue ball to the eight ball, just as placement of the regulatory protein between polymerase and gene blocks movement of the polymerase to the gene. This process of blocking transcription is called **repression.** The regulatory protein responsible for the blockage is called a **repressor protein** (Figure 14-13).

In other situations, the binding of a regulatory protein to the DNA is needed to start the transcription of genes. The "turning on" of the transcription of specific genes is called **activation;** the regulatory protein whose binding turns on transcription is called an **activator protein.** Activation can be achieved by several mechanisms. In some cases, the activator protein's binding promotes the unwinding of the DNA duplex. This aids production of an mRNA transcript of a gene, because the RNA polymerase, although it can bind to a double-stranded DNA duplex, cannot produce an mRNA transcript from such a duplex. mRNA is transcribed from a single strand of the duplex.

The cell controls the transcription of particular genes by influencing the shape of the regulatory proteins. Regulatory proteins have binding sites not only for DNA but

FIGURE 14-13

An example of repression. The regulatory protein controlling transcription of the genes encoding lactose-utilizing enzymes binds to the DNA between the RNA polymerase binding site and the gene to be transcribed. The large whitish sphere bound to the DNA strand at the upper right is the regulatory protein known as the *lac* repressor. The relationship between such a repressor and a DNA molecule is shown in the diagram. Since the repressor protein fills the groove of the DNA double helix, the RNA polymerase cannot attach there, and the repressor blocks transcription.

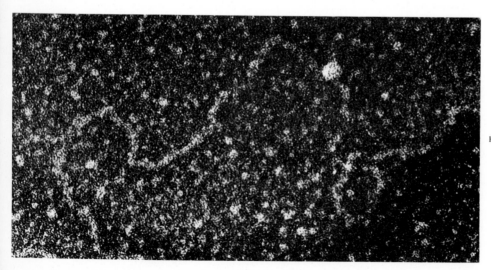

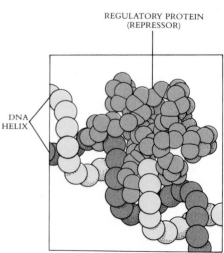

REGULATORY PROTEIN
(REPRESSOR)

DNA
HELIX

also for specific small molecules within the cell. The binding of one of these small molecules can change the shape of a regulatory protein and thus destroy or enhance its ability to bind DNA. Sometimes the protein in its new shape may no longer recognize the regulatory site on the gene. In other cases, the reshaped regulatory protein may recognize a regulatory site that it had previously ignored.

The cell thus uses the presence of particular "signal" molecules in the cell to stop particular regulatory proteins or to mobilize them for action. These regulatory proteins, in turn, repress or activate the transcription of particular genes. The pattern of metabolites in the cell sets "on/off" protein regulatory switches, and by doing so achieves a proper configuration of gene expression.

Organisms control the expression of their hereditary information by selectively inhibiting the transcription of some genes and facilitating the transcription of others. Control over transcription is exercised by modifying the shape of regulatory proteins and thus influencing their tendency to bind to sites on the DNA that influence the initiation of transcription.

THE ARCHITECTURE OF A GENE

Next we shall examine in more detail a single cluster of genes, to see how the structure of these genes achieves the precise and timed production of the proteins they encode. The cluster of three genes we shall examine encodes three proteins used by bacteria to obtain energy from the sugar lactose. These proteins include two enzymes and a membrane-bound transport protein (a permease). Researchers have found that this cluster is typical of how genes are organized in bacteria (Figure 14–14). There are four different regions within the cluster:

1. Three coding sequences specify the three lactose-utilizing enzymes. All three sequences are transcribed onto the same piece of mRNA, and are part of an operational unit called an **operon.** An operon consists of one or more structural genes and the associated regulatory elements, the operator and the promoter, described below. This particular operon is called the *lac* **operon,** because the three genes are all involved in lactose utilization. Such a pattern of clustered coding sequences is common among bacteria but is not known in higher eukaryotes.
2. Upstream (opposite the direction the RNA polymerase moves in transcribing) from the three coding sequences is the binding site for the bacterial ribosome. This nucleotide sequence is located within an initial untranslated portion of the mRNA, sometimes called a **leader region.** Each mRNA molecule transcribed from the cluster is composed of the leader region and the three coding sequences, transcribed in order. The cluster and the leader region are referred to as a **transcription unit.**
3. Upstream from the transcription unit is a specific DNA nucleotide sequence which the polymerase recognizes and to which it binds. Such polymerase-recognition sites are called **promoters,** because they promote transcription.
4. Between the promoter site and the transcription unit is a regulatory site, the **operator,** where a repressor protein binds to block transcription. Upstream from the promoter is another regulatory site, **CAP,** where an activator protein binds. This facilitates the unwinding of the DNA duplex and so enables the polymerase to bind to the nearby promoter.

Genes encoding enzymes have regulatory regions. The segment that is transcribed into mRNA is called a transcription unit and consists of the elements that are involved in the translation of the mRNA: the ribosome-binding site and the coding sequences. In front of the transcription unit on the DNA are the elements involved in regulating its transcription: binding sites for the polymerase and for regulatory proteins.

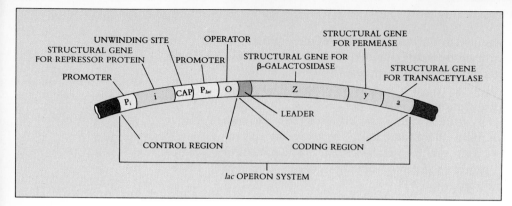

The *lac* region of the *Escherichia coli* chromosome. The left-most segment (P*i*) is a site where RNA polymerase binds, called a **promoter site.** Immediately downstream of it (that is, in the direction that the polymerase moves after it binds and begins transcription) is the gene *i*, which encodes a regulatory protein (the repressor protein). The RNA polymerase reads this sequence, producing the corresponding mRNA, and then dissociates from the DNA. Just to the left of the *i* gene is an unwinding site called CAP, a second RNA polymerase binding site called P*lac*, and a site called the **operator** to which the repressor protein can bind. Just downstream of these sites are three genes encoding the two enzymes and the permease that are involved in the metabolism of lactose. Because they encode the primary structure (amino acid sequence) of enzymes, they are called structural genes.

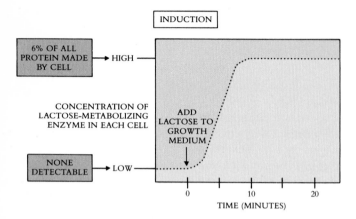

FIGURE 14-15

How the *lac* operon of the common colon bacterium, *Escherichia coli*, works. Bacterial cells growing in the absence of the sugar lactose do not contain measurable amounts of the enzyme beta-galactosidase, which splits lactose into glucose and galactose, sugars which the bacteria can utilize as food. When lactose is added to a solution containing a growing population of bacteria, the bacteria begin to produce large amounts of the enzyme. Within 10 minutes after lactose is added, 6% of all the protein made by a typical cell will be the enzyme beta-galactosidase.

The regulatory region of the *lac* operon controls production of the transcription unit through a series of complex steps. A special activator protein stimulates transcription of the *lac* operon when the cell is low in energy. As a result, the enzymes needed for the metabolism of lactose are produced only when the cell requires the energy that lactose would provide. The repressor protein blocks the binding of RNA polymerase in the promoter region under most circumstances. Cells in this condition are said to be **repressed** with respect to *lac* operon transcription.

Like the activator protein, the *lac* repressor protein is capable of changes in shape. When lactose binds to the repressor protein, the protein assumes a different shape, one that does not recognize the operator sequence. If the cell contains much lactose, therefore, the *lac* repressor proteins become inactive. For this reason, adding lactose to a growing bacterial culture causes a burst of synthesis of the lactose-utilizing enzymes (Figure 14-15). Transcription of the enzymes is said to have been *induced* by the lactose. This element of the control system ensures that the *lac* operon is transcribed only in the presence of lactose.

The *lac* operon is thus controlled at two levels: the lactose-utilizing enzymes are not produced unless lactose is available; even if lactose is available, these enzymes are not produced unless the cell needs the energy. Similarly precise control mechanisms are known in eukaryotes, but this example illustrates the complex nature of cellular control of protein synthesis.

HOW DNA REPLICATES

The complementary nature of the DNA duplex provides a ready means of duplicating the molecule. If one were to unzip the molecule, one would need only to assemble the appropriate complementary nucleotides on the exposed single strands in order to form two daughter duplexes of the same sequence. The density label experiments of Meselson and Stahl (Chapter 13) demonstrate that this is indeed what happens. When a DNA molecule replicates, the double-stranded DNA molecule separates at one end, forming a **replication fork** (Figure 14-16), and each separated strand serves as a template for the synthesis of a new complementary strand. Indeed, electron micrographs reveal Y-shaped DNA molecules at the point of replication, just as the model predicts.

To the surprise of those investigating the way in which DNA molecules replicate, it turned out that the two new daughter strands are synthesized on their templates in different ways. One strand is built by simply adding nucleotides to its end. This strand grows inward toward the junction of the Y as the duplex unzips. This strand ends with an —OH group and is called the "3-prime," or 3′, end. The enzyme that catalyzes this process is called **DNA polymerase.**

However, when investigators searched for a corresponding enzyme that added nucleotides to the other strand (which ends with a —PO₄ group and is called the 5′ end), they were unable to find one. Nor has anyone ever found one. DNA polymerases add only to the 3′ ends of DNA strands.

How does the polymerase build the 5′ strand? Along this strand, the chain is also formed in the 3′ direction, the polymerase jumping ahead and filling in backward. The DNA polymerase starts a burst of synthesis at the point of the replication fork and moves outward, adding nucleotides to the 3′ end of a short new chain until this new segment fills in a gap of 1000 to 2000 nucleotides between the replication fork and the end of the growing chain to which the previous segment was added. The short new chain is then added to the growing chain, and the polymerase jumps ahead again to fill in another gap. In effect, it copies the template strand in segments about 1000 nucleotides long and stitches each new fragment to the end of the growing chain. This mode of replication is referred to as **discontinuous synthesis.** If one looks carefully at electron micrographs showing DNA replication in progress, one of the daughter strands behind the polymerase appears single-stranded for about 1000 nucleotides.

A DNA molecule copies (replicates) itself by separating its two strands and using each as a template to assemble a new complementary strand, thus forming two daughter duplexes. The two strands are assembled in different directions.

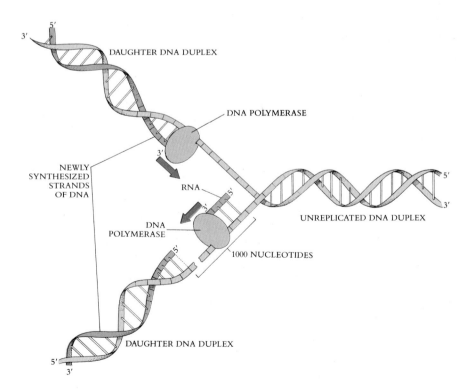

DAUGHTER DNA DUPLEX

DNA POLYMERASE

NEWLY SYNTHESIZED STRANDS OF DNA

RNA

DNA POLYMERASE

UNREPLICATED DNA DUPLEX

1000 NUCLEOTIDES

DAUGHTER DNA DUPLEX

FIGURE 14-16

A DNA replication fork. The colored upper left-hand strand is being copied from left to right by a DNA polymerase that is adding nucleotides to the growing 3′ end. The colored lower left-hand strand is being copied from right to left by a DNA polymerase, which is adding nucleotides to the 3′ end of a fragment; when the polymerase reaches the 5′ end of the previous fragment, it binds this new fragment to it. It then goes back and starts a new fragment about 1000 bases ahead of where it started last time. Each fragment is started with a few bases of RNA, which are later removed.

SUMMARY

1. The expression of hereditary information in all organisms takes place in two stages. First, an mRNA molecule with a nucleotide sequence that is complementary to a particular segment of the DNA is synthesized by the enzyme RNA polymerase. Second, an amino acid chain is assembled by a ribosome, which uses the mRNA sequence to direct its choice of amino acids. The first process is called transcription, the second translation.

2. The coding sequence of a gene is read in increments of three nucleotides from the mRNA by a ribosome. The ribosome positions the three-nucleotide segments of the message so that a tRNA molecule with the complementary base sequence can bind to it. Attached to the other end of the tRNA molecule is an amino acid, which is added to the end of the growing polypeptide chain.

3. The actual decoding of the genetic message is carried out by a family of 20 activating enzymes, each of which binds the amino acid that it recognizes to a tRNA molecule. The three-base sequence, or anticodon, of the tRNA molecule is complementary to the corresponding codon of an mRNA molecule.

4. In eukaryotes, the mRNA transcript is transported from the nucleus to the cytoplasm before it is translated.

5. A single mRNA molecule is called a transcription unit. The unit contains both the coding sequence, which directs the ribosome's choice of amino acids, and a leader region to which the RNA polymerase binds.

6. Cells control transcription largely by regulating the ability of RNA polymerase molecules to initiate transcription. In some instances, a nucleotide sequence located between the promoter and the transcription unit acts as a binding site for a repressor protein. When bound to the site, this repressor protein prevents the read-through of the RNA polymerase, just as a tree that has fallen across a road blocks traffic.

7. In other instances, nucleotide sequences located near the promoter site act as binding sites for activator proteins. The binding of the activator protein to the promoter site causes the unwinding of the DNA duplex. Since RNA polymerase can translate only a single strand of DNA, this unwinding greatly speeds the translation process.

8. Most eukaryotic genes contain additional sequences called introns embedded within the coding sequences of their transcription units. These introns are removed from the transcript before it is translated. Their function appears to be regulatory. They are thought to influence the availability of transcripts for translation. Consequently, the control of gene expression may be exercised at both the transcription and translation steps in eukaryotes.

REVIEW

1. Strings of DNA nucleotides arranged in groups that specify the structure of a protein are called _____.

2. After mRNA is transcribed from the DNA, it is _____ into a polypeptide at a ribosome.

3. How many nucleotides does it take to specify an amino acid?

4. A codon is part of a(n) _____ molecule that specifies a particular amino acid. An anticodon is the corresponding portion of a(n) _____ molecule.

5. A regulatory protein responsible for blockage of DNA transcription is called a _____.

SELF-QUIZ

1. RNA is a nucleic acid like DNA, but there are important differences. Choose these three differences from the following list.
   (a) The sugar in RNA is ribose instead of deoxyribose.
   (b) RNA contains uracil instead of thymine.
   (c) RNA is a single-stranded molecule.
   (d) RNA is a double-stranded molecule.
   (e) RNA is necessary for protein synthesis.

2. What is the first step in RNA transcription?
   (a) Nucleotides are assembled into RNA.
   (b) The DNA strands open up.
   (c) The hydrogen bonds of the DNA re-form.
   (d) RNA polymerase attaches to the DNA.
   (e) The DNA replicates.

3. Which amino acid is specified by the codon CUC? (Consult Table 14-1)
   (a) lysine   (d) proline
   (b) alanine   (e) leucine
   (c) glutamate

4. Two key differences between bacterial and eukaryotic mRNA are
   (a) bacterial mRNA molecules do not need ribosomes for transcription
   (b) eukaryotic mRNA molecules rarely contain more than one gene, whereas bacterial mRNAs often do
   (c) ribosomes of eukaryotes are much larger than those of bacteria
   (d) the primary mRNA transcripts of eukaryotes contain introns that are cut out before translation
   (e) in bacterial mRNA the nucleotide sequence is not arranged in codons

5. In the *lac* region of the bacterium *Escherichia coli's* chromosome, where does the repressor protein bind?
   (a) at the activator protein site
   (b) at the promoter site
   (c) at the operator site
   (d) at the transcription unit
   (e) at the permease gene site

PROBLEM

You are provided with a sample of aardvark DNA, obtained at great risk of life. As part of your investigation of this DNA, you transcribe messenger RNA from the DNA and purify it. You then separate the two strands of the DNA and analyze the base composition of each strand, and of the mRNA. You obtain the following results:

	A	G	C	T	U
DNA strand 1	19.1	26.0	31.0	23.9	0
DNA strand 2	24.2	30.8	25.7	19.3	0
mRNA	19.0	25.9	30.8	0	24.3

Which strand of the DNA is the "sense" strand that serves as the template for mRNA synthesis?

FOR FURTHER READING

DARNELL, J.: "The Processing of RNA," *Scientific American,* October 1983, pages 90-102. An up-to-date description of how intron sequences are removed from messenger RNA transcripts before they are utilized to direct the assembly of polypeptides.

DICKERSON, R.E.: "The DNA Helix and How It is Read," *Scientific American,* December 1983, pages 94-112. X-ray analysis of DNA molecules of different sequence shows that sequence differences create subtle differences in shape; these may influence DNA site recognition by proteins. A very clear exposition of a complex subject.

LAKE, J.: "The Ribosome," *Scientific American,* August 1981, pages 84-91. Recent studies show that ribosomes have unexpectedly complex shapes, providing a different view of polypeptide synthesis than had previously been accepted.

FIGURE 15-1   Genetic variation occurs in all natural populations. Except for identical twins, no two individuals are ever alike. Even though these penguins all seem the same to us, they are as different from one another as you are from the other students in your class. The reason that all penguins—and all people—are different is that their genes are constantly being altered by natural processes.

HOW GENES CHANGE

Chapter 15

Overview

The evolution of life on earth depends critically on the existence of genetic variation in nature. Only when heritable differences exist can one life form replace another. In this chapter, we consider the mechanisms responsible for creating genetic change. Broadly speaking, there are two such factors—mutation and recombination—and each occurs in many different forms.

For Review   *Here are some important terms and concepts that you will encounter in this chapter. If you are not familiar with them, you should review them before proceeding.*

Cancer (Chapter 1)

AIDS (Chapter 1)

Independent assortment (Chapters 10 and 11)

Crossing-over (Chapter 11)

The chromosomes of your body are remarkably good at copying themselves. During the course of your development, when your body grew from one to about a hundred trillion cells, the total length of DNA that was copied, and copied accurately, is staggering. If the total DNA in that first cell were stretched out, with the DNA content of all 46 chromosomes lined up, one after the other, the DNA would extend about as high as an adult person. After growth to adulthood, all of the DNA in your adult body, stretched end to end, would extend about 200 billion kilometers! To avoid the accumulation of mistakes during this extended replication process, each cell has a mechanism by which the copied DNA is checked for errors. This is important, since it ensures that the gametes your body will make as an adult carry unchanged the same genes that were originally contributed to the first cell by your parents.

Some changes do occur, however; random chemical accidents alter the DNA. Such changes in the genetic message are rare. Typically, a particular gene is altered in only one out of a million gametes. Limited as that might seem, this steady trickle of change is the very stuff of evolution. Every single difference between the genetic message specifying you and the one specifying your cat, or the fleas on your cat, arose as a result of genetic change (Figure 15-1).

In this chapter we shall consider those factors that produce changes in genes. Broadly speaking, two sorts of processes—mutation and recombination—are important agents of genetic change. We shall consider them in turn.

MUTATION

A change in the genetic message of a cell is referred to as a **mutation.** Some mutational changes affect the message itself, producing alterations in the sequence of DNA nucleotides. These alterations in the coding sequence typically are called **point mutations,** and involve only one or a few nucleotides. Other classes of mutation involve changes in the way the genetic message is organized. In both bacteria and eukaryotes, individual genes may move from one place on the chromosomes to another by a process called **transposition.** When a particular gene moves to a different location, there is often an alteration in its expression or in that of the neighboring genes. In eukaryotes, large segments of chromosomes may change their relative location or undergo duplication. Such **chromosomal rearrangement** often has drastic effects on the expression of the genetic message.

As you saw in Chapter 10, the bodies of animals are usually partitioned into two cell classes: those cells destined to form gametes **(germline cells)** and those which form the rest of the body **(somatic cells).** Cells of either kind can undergo mutation. When a mutation occurs within a germline cell, the mutational change is passed along to later generations as part of the hereditary endowment contributed by the gametes derived from that cell. When a mutation occurs within a somatic cell, such as one of the cells lining your lung, the mutation is not passed along to later generations of organisms, since the mutation does not affect the gametes. Mutation of a somatic cell **(somatic mutation)** may, however, have drastic effects on the individual organism in which it occurs, since it *is* passed on to all the cells of the organism that are descended from that mutant cell. Thus if a mutant lung cell divides, the line of cells derived from it all carry the mutation. This is what occurs in many forms of lung cancer.

In plants, on the other hand, growth is continuous, as you will see in Chapter 29. Certain groups of cells—the meristems—divide continuously throughout the life of the organism. Mutations that affect meristematic cells may be expressed later either in the gametes or in the body cells of the plant, depending on the eventual fate of that particular cell line.

All organisms are subject to point mutations. Such mutations result either from chemical or physical damage to the DNA, or from spontaneous pairing errors that occur during DNA replication. The first class of mutation is of particular practical importance, because modern industrial societies produce and release into the environment many chemicals capable of damaging DNA, chemicals we call **mutagens.**

> **Point mutations are changes in the hereditary message of an organism. They may result from physical or chemical damage to the DNA or from spontaneous errors during DNA replication.**

DNA Damage

Although there are many different ways in which a DNA duplex can be damaged, three are of major importance:

1. ionizing radiation
2. ultraviolet radiation
3. chemical mutagens

Ionizing Radiation

High-energy radiation such as x rays and gamma rays are highly mutagenic. When such radiation reaches a cell, it is absorbed by the atoms that it encounters, imparting energy to the electrons in their outer shells and causing these electrons to be ejected from the atoms. The ejected electrons leave behind ionized atoms with unpaired electrons. Each of these ionized atoms is called a **free radical.** Because most of a cell's atoms reside in water molecules and not in DNA, the great majority of free radicals are created from the water molecules that are encountered by **ionizing radiation** and not from the DNA.

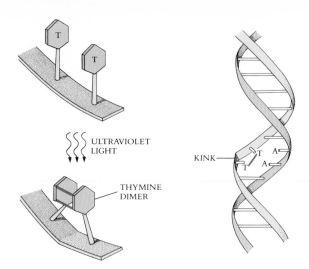

FIGURE 15-2

Ultraviolet (UV) radiation is absorbed by pyrimidine bases, particularly thymine. When two thymines are adjacent to one another in a DNA strand, the absorption of UV radiation can cause the formation of a covalent bond between them—a thymine dimer. Such a dimer introduces a "kink" into the double helix, which prevents replication of the duplex by DNA polymerase.

Most of the damage that the DNA suffers is thus indirect. It occurs because free radicals are highly reactive chemically, reacting violently with the other molecules of the cell, including DNA. The impact of free radicals on a chromosome is like that of shrapnel from a grenade blast on a human body.

The locations on the DNA where damage occurs are random, and the damage is often severe. Cells can heal some of this damage. Thus nucleotides in which chemical changes have occurred can be excised from the molecule, and if one of the bonds that links the nucleotides in one of the double helix strands is broken, it can be repaired. This **mutational repair** is not always accurate, however, and some of the mistakes are incorporated into the genetic message.

Much of the radiation emitted by nuclear reactions is high-energy radiation. Radioactive material destroys organisms by the damage this radiation does to their cells and also does more long-range damage to organisms that survive exposure, by creating mutations in their DNA. Leukemia is a rare form of blood cancer; exposure to high-energy radiation causes mutations that greatly increase its incidence. The proportion of survivors of the atomic blast at Hiroshima who later developed leukemia depended on how far away from the center of the blast they were; the farther away, the lower the dose of radiation they received and the less likely they were to develop leukemia.

Ultraviolet Radiation

Ultraviolet (UV) radiation, the component of sunlight that leads to suntan (and sunburn), is much lower in energy than are x rays. When molecules absorb UV radiation, the radiation does not impart enough energy to cause the molecules to eject electrons; consequently, free radicals are not formed. The only molecules that are capable of absorbing UV radiation, in fact, are certain organic ring compounds such as the DNA bases.

When DNA bases absorb UV energy, the electrons in their outer shells become reactive. If one of the nucleotides on either side of an absorbing pyrimidine is also a pyrimidine, a double covalent bond is formed between the two pyrimidines. The resulting cross-link between adjacent bases of the DNA strand is called a **pyrimidine dimer** (Figure 15-2). If such a cross-link is left unrepaired, it can potentially block DNA replication, in which case the damage would be lethal. What actually happens in most cases, however, is that cellular **UV repair systems** either (1) cleave the bond that links the adjacent pyrimidines, or (2) excise the entire pyrimidine dimer from the strand and fill in the gap (Figure 15-3), using the other strand as a template. In those rare instances in which a pyrimidine dimer does escape such cleavage or excision, the replicating polymerase simply fails to replicate the portion of the strand that includes the pyrimidine dimer, skipping ahead and leaving the problem area to be filled in later. This filling-in process is often error-prone, however, and it may create mutational changes in the base sequence of the gap region.

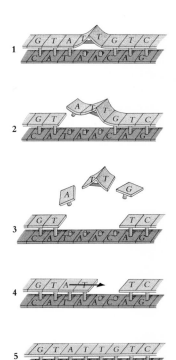

FIGURE 15-3

Thymine dimers are repaired by excising out the troublesome dimer as well as a short run of nucleotides on either side of it, and then filling in the gap using the other strand as a template.

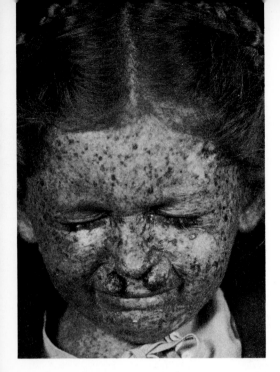

FIGURE 15-4

Phenotypic effects of the inherited genetic disorder xeroderma pigmentosum, in which those who have the disease develop extensive malignant skin tumors after exposure to sunlight. Xeroderma pigmentosum is characteristic of individuals who have a particular gene present in the homozygous recessive state; such individuals are less efficient in the repair of DNA damage by exposure to ultraviolet light.

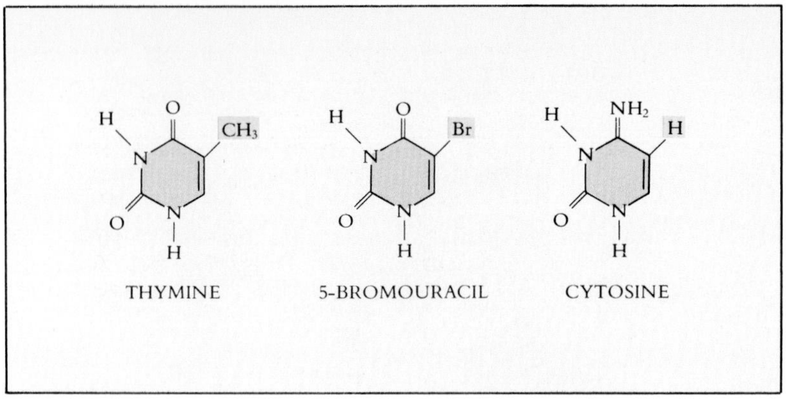

FIGURE 15-5

Mutations may be caused by chemicals that look like DNA bases. For example, DNA polymerase cannot distinguish between thymine and 5-bromouracil, which are very similar in shape. Once incorporated into a DNA molecule, however, 5-bromouracil tends to rearrange to a form which resembles cytosine and pairs with guanine. When this happens, what was originally an A–T base pair becomes a G–C base pair.

Without a mechanism for repairing the damage to DNA caused by UV radiation, sunlight would wreak havoc on the cells of your skin. There is a rare hereditary disorder among humans called **xeroderma pigmentosum** (Figure 15-4), in which just this problem occurs. Individuals homozygous for a mutation that destroys the ability of the body's cells to repair UV damage develop extensive skin tumors after exposure to sunlight. Because of the many different proteins involved in excision and repair of pyrimidine dimers, mutations in as many as six different genes can cause the disease.

Chemical Mutagens

Many mutations result from the direct chemical modification of the DNA bases. The chemicals that act on DNA fall into three classes: (1) chemicals that look like DNA nucleotides but pair incorrectly when they are incorporated into DNA (Figure 15-5); (2) chemicals that remove the amino group from adenine or cytosine, causing them to mispair; and (3) chemicals that add hydrocarbon groups to nucleotide bases, also causing them to mispair. This last group includes many particularly potent mutagens that are commonly used in laboratories, as well as compounds that are sometimes released into the environment, such as mustard gas.

> The three major sources of mutational damage to DNA are (1) high-energy radiation, such as x rays, which physically break the DNA strands; (2) low-energy radiation, such as UV light, which creates DNA cross-links whose removal often leads to errors in base selection; and (3) chemicals that modify DNA bases and thus alter their base-pairing behavior.

MUTATION AND CANCER

Among the most important chemical mutagens are those that cause cancer. Cancer is a growth disorder of cells. It starts when an apparently normal cell begins to grow in an uncontrolled and invasive way (Figure 15-6). The result is a ball of cells, called a **tumor,** that constantly expands in size. When this ball remains a hard mass, it is called a **sarcoma** (if connective tissue, such as muscle, is involved) or a **carcinoma** (if epithelial tissue, such as skin, is involved). If cells leave the mass and spread throughout the body (Figure 15-7), forming new tumors at distant sites, the spreading cells are called **metastases.** Cancer is perhaps the most deadly and frightening of human diseases. Of the children born in 1987, one third will contract cancer; one quarter of the male children and fully one third of the females will someday die of cancer. Not surprisingly, a great deal of effort has been expended to learn the cause of this disease.

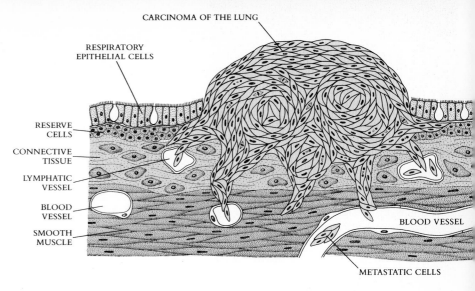

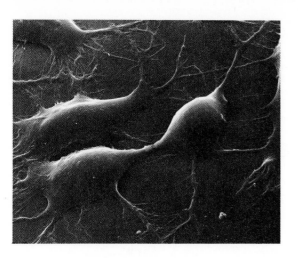

FIGURE 15-6

Cancer cells. These mouse nerve cells, growing over one another on a culture dish, were isolated from a tumor.

FIGURE 15-7

Portrait of a cancer. The ball of cells is a carcinoma, developing from epithelial cells lining the interior surface of a human lung. As the mass of cells grows, it invades surrounding tissues, eventually penetrating into lymphatic vessels and blood vessels, both of which are plentiful within the lung. These vessels carry metastatic cancer cells throughout the body, where they lodge and grow, forming new centers of cancerous tissue.

Some cancers appear to be caused by viruses. A sarcoma (connective tissue) cancer of chickens, for example, is caused by a small virus called a **retrovirus.** A retrovirus is a virus that packages its hereditary information as RNA rather than DNA, copying the RNA back to DNA in order to replicate. When the chicken sarcoma virus was compared with related viruses that do not cause cancer, it was found to contain an extra gene usually found not in viruses but in chickens (Figure 15-8). This gene, which activates cell division, is usually turned off in chickens, but when it was incorporated into the virus the gene did not respond to the "shutoff" controls in the cell and remained active. The cancer is thus the result of inappropriate activity of a growth-regulating gene.

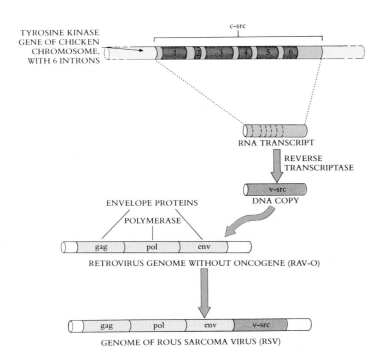

FIGURE 15-8

Structure of the Rous sarcoma virus (RSV) retrovirus. The virus contains only a few virus genes, encoding the virus protein envelope (called the *env* and *gag* genes), and reverse transcriptase, which produces a DNA copy of its RNA genome (called *pol*). In addition, it contains a gene, *src* (for sarcoma), that it picked up from chicken DNA during some previous infection. The RAV-0 virus shown here lacks this *src* gene, but is identical to the RSV retrovirus in all other respects. RSV causes cancer in chickens, RAV-0 does not.

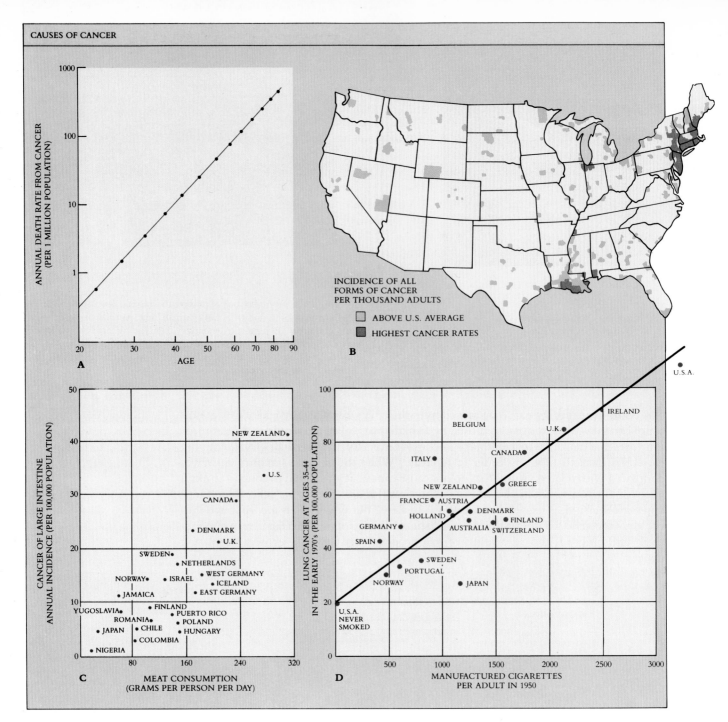

FIGURE 15-9

Potential agents in causing cancer.

A The annual death rate in the United States from cancer (all forms) is much higher among older segments of the population. When the annual death rate from cancer is plotted on a logarithm scale as a function of age, a straight line is obtained. This exponential rise suggests that several independent events are required to give rise to cancer.

B The incidence of cancer per 1000 people is not uniform throughout the United States. Rather, it is centered in cities and in the Mississippi Delta. This suggests that pollution and pesticide runoff may contribute to cancer.

C One of the most common deadly cancers in the United States, that of the large intestine, is not at all common in many other countries, such as Japan. Its incidence appears to be related to the amount of meat an average individual consumes—a high meat diet relative to fiber slows passage of food through the intestine, prolonging exposure of the intestinal wall to the oxidation by-products of digestion.

D The biggest killer among cancers is lung cancer, and the most important environmental agent producing lung cancer is cigarettes. It takes 20 years for the full effect to be seen. When levels of lung cancer in many countries are compared, the incidence of lung cancer among adult males between 40 and 50 years of age is strongly correlated with the cigarette consumption in that country 20 years earlier.

Most human cancers, however, are not associated with viruses. An intensive effort to isolate factors causing cancers in humans has led to the conclusion that most, if not all, human cancers are the result of mutation. Observe, for example, the environmental factors illustrated in Figure 15-9. As we noted earlier, the likelihood that a person who was living in Hiroshima, Japan, in 1945 will contract leukemia is clearly related to the amount of radiation his or her body has absorbed. The likelihood that you, living in the United States, will develop cancer depends very much on where you live, how much meat you eat, and whether or not you smoke. The likelihood that a nonsmoker living in the Southwest who eats less than 80 grams of meat per day will develop cancer is far less than that of a smoker living in New York City and consuming the U.S. average of 280 grams of meat per day. Clearly, how you live has a great deal to do with whether or not you will develop cancer.

Why should *how* you live affect the likelihood of getting cancer? Investigators have searched intensively for environmental factors that are potential agents in causing the disease. Many have been found, including ionizing radiation such as x rays and a variety of chemicals. What do these cancer-causing agents, or **carcinogens,** have in common? They are all potent mutagens, producers of mutations. This observation led to the suspicion that cancer might be caused, at least in part, by the creation of mutations.

How does mutation lead to cancer? The answer to this question has come from studying tumors to see what is different about them. To do this, investigators have used a technique called **transfection.** Transfection consists of (1) the isolation of the nuclear DNA from human tumor cells; (2) its cleavage into random fragments by the use of enzymes we shall discuss later in this chapter; and (3) the testing of the fragments individually for the ability of any particular fragment to induce cancer in the cells that take it up.

By using transfection techniques, researchers found that a single specific gene isolated from a cancer cell is all that is needed to transform normally dividing cells in tissue culture into cancerous ones. The cancerous cells differed from the normal ones only with respect to this one gene. In some cases, the cancer-inducing gene identified by transfection proved to be the same as one of the cancer-causing genes carried by a cancer-causing virus. Cancer-causing genes are often called **oncogenes,** from the Greek word for cancer, *oncos.*

Oncogenes are normal genes gone wrong. By identifying oncogenes through transfection and then isolating them, investigators have been able to compare them with their normal counterparts. In this way, researchers have been able to study why normal genes were converted into cancer-causing ones. The analysis of a number of oncogenes associated with cancers of various tissues has led to the following conclusions:

1. The initiation of many cancers involves changes in cellular activities that occur at the inner surface of the plasma membrane. In a normal cell these activities are associated with the turning-on of cell division (Figure 15-10).
2. The difference between a normal gene encoding one of the proteins that carry out cell division and a cancer-inducing oncogene need only be a single point mutation in the DNA. In a human bladder carcinoma, for example, a single nucleotide alteration from G to T converts a glycine in a plasma-membrane protein called *ras* into a valine in the cancer-causing protein. There is no other difference between the normal and cancer-inducing forms of the *ras* gene.
3. The mutation of a gene such as *ras* to an oncogene form in different tissues can lead to different forms of cancer. There are probably no more than a few dozen different genes whose mutation can lead to cancer, and there may be far fewer than that.
4. The onset of many cancers involves the action of two or more different oncogenes. The initiation of cancer may require changes at both the plasma membrane and in the nucleus (Figure 15-11). This may be the reason why most cancers occur in people over 40 years old. It is as if human cells accumulate mutational changes, and time is required for several such mutations to occur in the same cells.

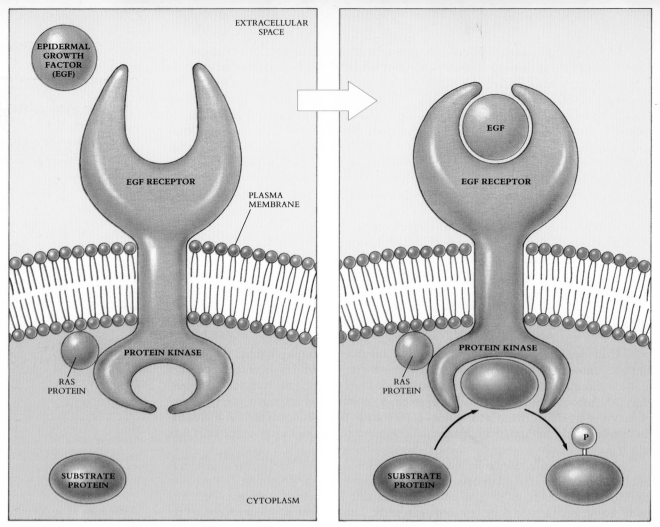

EPIDERMAL
GROWTH
FACTOR
(EGF)

EGF RECEPTOR

EGF

EGF RECEPTOR

PLASMA
MEMBRANE

PROTEIN KINASE

PROTEIN KINASE

RAS
PROTEIN

RAS
PROTEIN

P

SUBSTRATE
PROTEIN

SUBSTRATE
PROTEIN

CYTOPLASM

FIGURE 15-10

Mutations that lead to cancer often involve proteins of the plasma membrane. The ones portrayed here, all of which have been implicated in cancers, are associated with cell division. In a normal cell, cell division is triggered by a protein called *epidermal growth factor* (EGF), which binds to a receptor protein on the exterior surface of the cell and adds a phosphate group to one or more of its tyrosine amino acids. This addition of a phosphate alters the shape of the portion of the receptor protruding into the cell, initiating a signal that passes to the cell nucleus and initiates cell division. The level of EGF necessary to start this process is affected by another plasma-membrane protein, *ras*.

The emerging picture of cancer is one that involves aborted regulation of the genes that normally signal the onset of cell proliferation. Cancer seems to occur when several of the controls that cells normally impose on their own growth and division become inoperative. In most human cancers the controls become inoperative because of mutation of cellular genes that have regulatory functions. It is because smoking and improper diet increase the frequency of mutations that they lead to a higher incidence of cancer.

Cancer is a growth disease of cells in which the controls that normally restrict cell proliferation do not operate. The critical factor in initiating cancer seems to be the inappropriate activation of one or more proteins that regulate cell division, transforming cells to a state of cancerous growth.

It is difficult to avoid the conclusion that the way to avoid cancer is to avoid mutagens. As noted in Chapter 1, for example, few things can contribute more to your health than avoiding smoking.

TRANSPOSITION

Many mutations occur not because of chemical changes in the DNA, but rather because a foreign bit of DNA is injected into the middle of a gene, destroying the ability of the gene to function. These moving bits of DNA are called **transposable elements,** or **transposons** (Figure 15-12), and their movement is referred to as **transposition.** Transposition is a form of gene transfer that occurs both in bacteria and in eukaryotes. Transposons, unlike other genes, don't stay put on chromosomes. Instead, every once

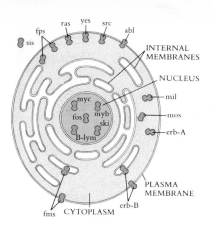

FIGURE 15-11

Sites of action in the cell of the proteins encoded by 16 of the 30 known cancer-causing genes.

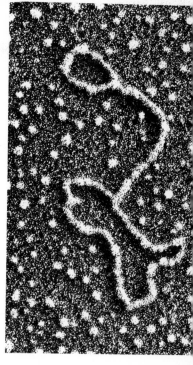

FIGURE 15-12

A transposable genetic element (transposon). Transposons form a characteristic stem-and-loop structure, as seen in this electron micrograph. The structure arises from the "inverted repeat" nature of the nucleotide sequences at the two ends of the transposon DNA.

in a while, often after many generations in one location, a transposon will abruptly move to a new position on the chromosomes, with the location of its new residence apparently chosen at random. They move about the chromosomes like so many Mexican jumping beans. Sometimes genes are carried along with transposons. These genes then become nomadic as well, carried along as parts of randomly moving genetic elements like passengers in a car with no driver.

Transposition is a one-way gene transfer to a random location on the chromosome. Genes move because they are associated with mobile genetic elements called transposons.

Many transposons exist in multiple copies scattered about on the chromosomes. In *Drosophila,* for example, more than 30 different transposons are known, most of them located at some 20 to 40 different sites throughout the genome. In all, the known transposons in a *Drosophila* cell account for perhaps 5% of its total DNA. Fewer different transposons are thought to occur in humans, but some of them are repeated many thousands of times.

The Impact of Transposition

The transposition of any portion of a chromosome is relatively rare, although such transposition is perhaps 10 times more likely than is the occurrence of a random mutational change. The transposition of a particular mobile element occurs perhaps once in every 100,000 cell generations. There are many elements in most cells, however, and many generations to consider. Viewed over long periods of time, transposition has had enormous evolutionary impact. Some of the ways in which this has happened are as follows:

1. *Mutation.* The insertion of a mobile element within a gene often destroys the gene's function. Such a loss of gene function is called **insertional inactivation.** It is thought that a significant fraction of the spontaneous mutations observed in nature has in fact resulted from this effect.
2. *Gene mobilization.* Evolution sometimes favors gene mobilization, the bringing together in one place of genes that are usually located at different positions on the chromosomes. In bacteria, for example, a number of different genes encode enzymes that make the bacteria resistant to one or more antibiotics, such as penicillin. The administration of several antibiotic drugs simultaneously to sick people was a common medical practice some years ago. Unfortunately, this simultaneous exposure to many antibiotics favors the persistence of segments of DNA that have managed to acquire several resistance genes as a result of transposition. Bacteria in which such resistance genes are linked are able to resist or destroy a variety of antibiotic drugs and are thus immune to all of them simultaneously.

3. *Rapid recombination.* More and more examples are being discovered of novel functions made possible by the ability of organisms to use transposition to move specific genes to random locations. A particularly important instance is the gene mobilization practiced by African **trypanosomes,** the protozoa that are responsible for sleeping sickness, a serious disease of human beings and other animals (see Chapter 26). The surface polysaccharides of these organisms determine their infective properties. Humans and other animals that are targets of trypanosome infection defend against these organisms by producing defensive proteins called antibodies that are directed against these surface polysaccharides. Trypanosomes defeat these defenses with transposition. The trypanosome genome contains a cluster of several thousand different polysaccharide-encoding sequences. Because the cluster lacks a promoter, none of these sequences can be transcribed as they are. The necessary promoter is located at a second site, called the **transcription site.** This site is within a transposon that periodically jumps randomly from one position to another within the cluster. These sporadic changes in which sequence is transcribed result in the appearance of a new surface polysaccharide before the immune system of the host has completely mobilized an attack against the old one.

MUTATIONS ALTERING CHROMOSOME ORGANIZATION

Chromosomes undergo several different kinds of gross physical alterations that have significant effects on the locations of their genes. The three most important are **deletions,** in which a portion of a chromosome is lost; **translocations,** in which a segment of one chromosome is transposed to another nonhomologous chromosome; and **inversions,** in which the orientation of a portion of a chromosome becomes reversed.

Deletions

Deletions are the loss of chromosomal material. Most deletions are very harmful because they halve the number of gene copies within a diploid genome and thus seriously affect the level of transcription. Deletions are one of the most serious kinds of mutation, since the loss of a chromosomal segment cannot be repaired. Ironically, most deletions appear to be produced by a cell's own mutation-repairing mechanisms working too well: when chromosomes pair, sequences may sometimes misalign, looping out a portion of one strand. Usually such misaligned pairing, which is called **slipped mispairing** (Figure 15-13), is only transitory, and the strands revert quickly to the normal arrangement. If the error-correcting system of the cell encounters such a mispairing before it reverts, however, it will attempt to "correct" it, usually by cutting out the loop. This results in a deletion of several hundred nucleotides from one of the strands. Some chemicals specifically promote the occurrence of deletions by stabilizing the loops that are produced by slipped mispairing, thus increasing the length of time during which the loops are vulnerable to being cut out.

FIGURE 15-13

Slipped mispairing occurs when a sequence is present in more than one copy on a chromosome, and the copies on homologous chromosomes pair out of register. The loop that is produced by this mistake in pairing is sometimes excised by the cell's repair enzymes, producing a short deletion and often altering the codon that is being "read" during the process of protein synthesis. Any chemical that tends to stabilize the loops increases the chance that this will happen.

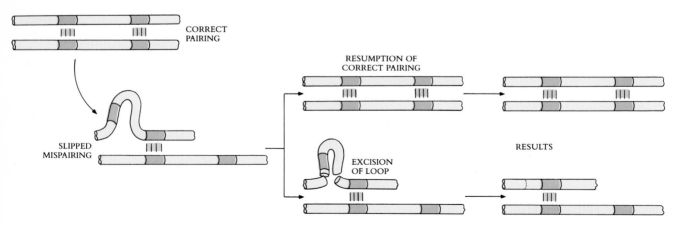

CORRECT PAIRING

SLIPPED MISPAIRING

RESUMPTION OF CORRECT PAIRING

EXCISION OF LOOP

RESULTS

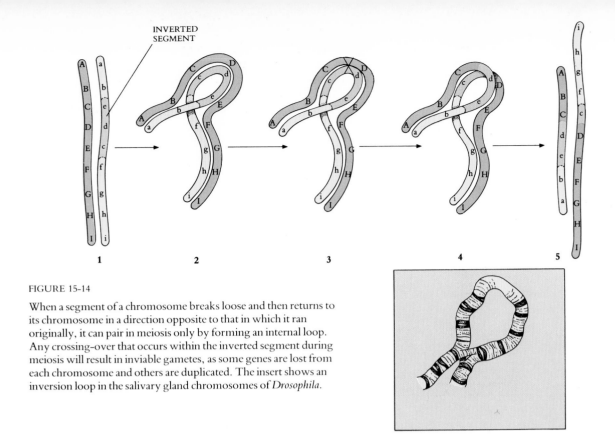

1   2   3   4   5

FIGURE 15-14

When a segment of a chromosome breaks loose and then returns to
its chromosome in a direction opposite to that in which it ran
originally, it can pair in meiosis only by forming an internal loop.
Any crossing-over that occurs within the inverted segment during
meiosis will result in inviable gametes, as some genes are lost from
each chromosome and others are duplicated. The insert shows an
inversion loop in the salivary gland chromosomes of *Drosophila*.

**Spontaneous errors in DNA replication occur very rarely. They result from
transient changes in the conformation of nucleotides and also from the
accidental mispairing of nonidentical but similar sequences.**

Chromosomal Rearrangement

The movement of an entire segment of a chromosome to a new location on the same
chromosome or to a different chromosome is called **translocation.** Translocations often have important effects on gene expression. If a gene is moved to a new site near a
patch of heterochromatin (a highly condensed and genetically inactive portion of the
chromosome), for example, it frequently becomes inactive.

   Inversion is the flipping around of a sequence, back to front. Inversions usually
do not alter gene expression but are important because of their effect on recombination. Recombination within a region of the chromosome that is inverted on one member of a homologous pair and not on the other (Figure 15-14) leads to serious problems
during meiosis, because none of the gametes produced after such a crossover event will
have a complete set of genes.

   Other chromosomal alterations change the number of gene copies that an individual possesses. Whole chromosomes may be gained or lost. An individual who has
gained or lost a whole chromosome is said to be an **aneuploid.** Whole diploid sets of
chromosomes may be multiplied; an individual, tissue, or cell that has one set of chromosomes is said to be **haploid,** one with two sets is **diploid**; any with three or more entire sets of chromosomes is said to be **polyploid.** A portion of a chromosome may also
be duplicated, a condition that also results in gene imbalance and thus is harmful in
many cases.

GENE TRANSFER

Genes are not fixed in their locations on chromosomes, like words engraved in granite.
They move around, although at a very low frequency. Some genes move because they
are part of **plasmids,** which are fragments of DNA separate from the main circular
molecule that occurs in bacteria; we shall discuss these later. Some plasmids, called

FIGURE 15-15

Electron micrograph of plasmids in a bacterium.

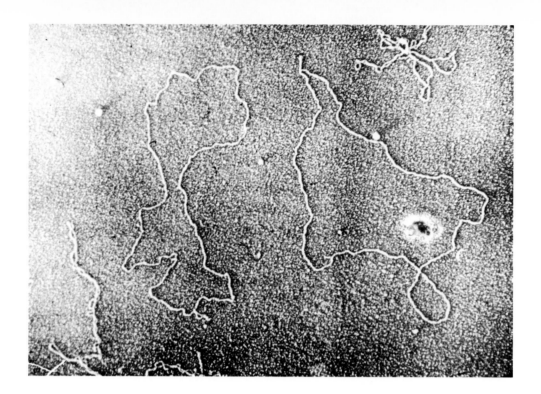

transfer plasmids, are able to enter and leave the main bacterial chromosome, but only at specific places where a nucleotide sequence occurs that is also present on the plasmid DNA. Other genes move as part of transposons, which migrate from one chromosomal position to another at random. Transposons occur both in bacteria and in eukaryotes. Both of these gene-transfer processes—plasmid movement and transposon movement—probably represent versions of the earliest form of genetic recombination that evolved, one that is still responsible for much of the genetic recombination that occurs today.

Plasmids

In bacteria with plasmids, about 5% of the total DNA is found outside the main chromosome in small circular molecules of DNA (Figure 15-15). Some plasmids are very small, containing only one or a few genes; others are quite complex and contain many genes.

To understand how plasmids arise, consider a hypothetical stretch of DNA along a bacterial chromosome, a stretch that contains two copies of a repeated nucleotide sequence. Because this sequence occurs twice on the chromosome, it is possible for the chromosome to form a transient "loop" in such a way that the two nucleotide sequences base-pair with each other and create a transient double duplex. All cells have enzymes, called **recombination enzymes,** that can recognize such double duplexes. These enzymes can cause the two duplexes to exchange strands. The result of such an exchange, as you can see in Figure 15-16, is to free the loop from the rest of the chromosome. The resulting free loop is a transfer plasmid. The DNA that makes up the plasmid corresponds to the genes between the two duplicated sequences. These genes are no longer present in the main chromosome but reside only on the plasmid.

A transfer plasmid that was created by recombination can reenter a chromosome the same way that it left. Sometimes the region of the plasmid DNA involved in the original exchange, called the **recognition site,** aligns with its mate on the main chromosome; if a second recombination event occurs anywhere in the common nucleotide sequence while they are aligned, it will result in the re-incorporation of the plasmid sequence into the chromosome. The plasmid is then said to have been **integrated** into the chromosome. This integration can occur at any site where the shared

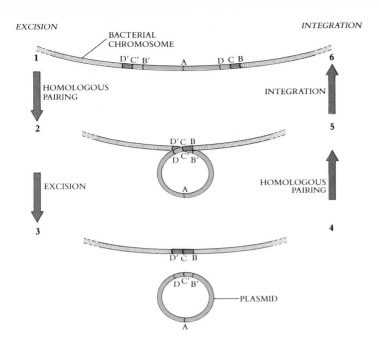

EXCISION   INTEGRATION

FIGURE 15-16

Integration and excision of a plasmid. Because the ends of the two sequences on the bacterial chromosome are the same—D′, C′, B′, and D, C, B—it is possible for the two ends to pair. The prime marks (as in D′) are used merely to indicate a different origin for the elements in the sequence. Steps 1 to 3 show the sequence of events if the strands exchange during the pairing. The result is excision of the loop and a free circle of DNA—a plasmid. Steps 4 to 6 show the sequence followed when a plasmid integrates itself into a bacterial chromosome.

sequence (or any other shared sequence) exists, so that plasmids sometimes integrate at positions different from those at which they arose. If they are re-incorporated into the chromosome at a new position, they transport their genes with them to the new location.

The locations to which genes are transported by plasmids correspond to the positions of DNA sequences on the main chromosome that also occur in the plasmids.

Gene Transfer Among Bacteria

A startling discovery that Joshua Lederberg and Edward Tatum made in the 1950s was that bacteria can pass plasmids from one cell to another. The plasmid that was studied by Lederberg and Tatum was a fragment of the chromosome of the bacterium *Escherichia coli*. It was given the name "F," for fertility factor, since only cells containing that plasmid could act as plasmid donors. The **F plasmid** contains a DNA replication origin and several special genes that promote its transfer to other cells. These genes encode protein subunits that assemble on the surface of the bacterial cell, forming a hollow tube called a **pilus,** whose construction resembles that of a brick smokestack protruding from a roof.

When the pilus of one cell makes contact with the surface of another cell that lacks pili, and therefore does not contain the F plasmid, the pilus initiates a series of changes within the first cell that causes a copy of the plasmid to be transferred across the pilus and into the other cell (Figure 15-17).

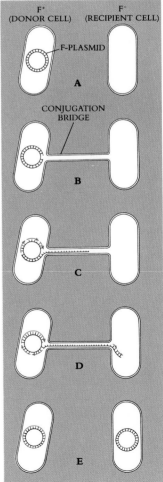

FIGURE 15-17

Gene transfer between bacteria. The long pilus connecting the two cells is called a **conjugation bridge.** Across it the F plasmid replicates a copy of its chromosome. One strand is passed across, the other serving as a template to build a replacement; when the single strand enters the recipient cell, it serves as a template to assemble a double-stranded duplex. When the process is complete, both cells contain a complete copy of the circular plasmid DNA molecule.

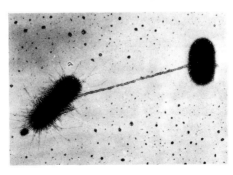

FIGURE 15-18

It takes over an hour for the entire bacterial chromosome to pass across the conjugation bridge from donor to recipient cell. Because this bridge is fragile, it is often broken before all of the chromosome has passed across. It is possible for an investigator to break all the conjugation bridges in a cell suspension by agitating the suspension rapidly in a blender. By conducting parallel experiments in which the blender is turned on at different times after the start of conjugation, investigators have been able to locate the positions of various genes on the bacterial chromosome. The closer genes are to the origin of replication, the sooner the blender must be turned on to block their transfer.

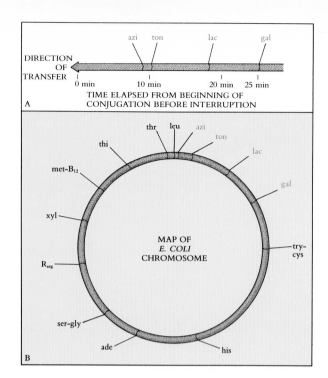

The F plasmid does not have to be free to carry out transfers of this kind. It will transfer just as readily if it is integrated into the bacterial chromosome. In this case, however, the cell begins to transfer a copy of *the entire bacterial chromosome* across to the recipient cell (Figure 15-18). Transfer proceeds just as if the bacterial chromosome were simply a small plasmid. The transfer of bacterial genes in this way is called **conjugation.**

> **Plasmids such as the F plasmid transfer bacterial genes from one individual bacterium to another by integrating the entire circular bacterial chromosome into the small plasmid circular DNA molecule. When a copy of the plasmid DNA molecule is replicated across to another cell, the integrated bacterial DNA is replicated across as well.**

GENETIC ENGINEERING

Not all genetic recombination occurs by accident. In 1980 geneticists succeeded for the first time in introducing a human gene, the one that encodes the protein interferon, into a bacterial cell. Interferon is a rare protein, difficult to purify in any appreciable amounts, that increases human resistance to viral infection. It may prove to be a useful therapy against cancer. This possibility has been difficult to explore, however, since the purification of the substantial amounts of interferon that are required for widespread clinical testing would, until recently, have been prohibitively expensive. A cheap way to produce interferon was needed, and introducing the gene responsible for its production into a bacterial cell made this possible.

The bacterial cell that had acquired the human interferon gene proceeded to produce large amounts of interferon and to grow and divide. Soon there were many millions of bacterial cells in the culture, all of them descendants of the original bacterial cell that had the human interferon gene, and all of them producing interferon. This procedure of introducing a gene into another cell, called **gene cloning,** had succeeded in making every cell in the culture a miniature factory for the production of human interferon. This interferon experiment and others like it marked the birth of a set of new technologies collectively known as **genetic engineering.**

Genetic engineering is based on the ability to cut up DNA into recognizable pieces and to rearrange these pieces in different ways. In the experiment just described, the gene segment carrying the interferon gene was inserted into a plasmid, which brought the inserted gene in with it when it infected the bacterial cell. Most other genetic engineering approaches have used the same general strategy of carrying the gene of interest into the target cell by first incorporating it into an infective plasmid or virus.

The success of the initial step in a genetic engineering experiment is the key to the whole procedure. As you might expect, success depends on being able to cut up the source DNA (human DNA in the interferon experiment, for example) and the plasmid DNA in such a way that the desired fragment of source DNA can be spliced permanently into the plasmid genome. This cutting is performed by a special kind of enzyme called a **restriction endonuclease.** These restriction enzymes have the ability to recognize and cleave specific sequences of nucleotides in a DNA molecule. Bacteria use them as a part of their defense against viruses, cutting up the invading viruses as they enter the bacterial cells and thus rendering them inoperative. Restriction enzymes are the basic tools of genetic engineering.

Restriction Enzymes

The DNA sequences that restriction enzymes recognize are typically four to six nucleotides long and are symmetrical. Their symmetry is of a special kind, called **twofold rotational symmetry** (Figure 15-19). Such symmetry means that when the break occurs in the middle of the sequence recognized by the restriction enzyme, it runs off in the same order in both directions from the break. The nucleotides at one end of the recognition sequence are complementary to those at the other end, so that the two strands of the duplex have the same nucleotide sequence running in opposite directions for the length of the recognition sequence. This has two consequences:

1. Because the same recognition sequence occurs on both strands of the DNA duplex (running in opposite directions), the restriction enzyme is able to recognize and cleave both strands of the duplex, effectively cutting the DNA duplex in half.

2. Because the position of the bond cleaved by a particular restriction enzyme is typically not in the center of the recognition sequence to which it binds, and because the sequence is running in opposite directions on the two strands, the sites at which the two strands of a duplex are cut are offset from one another. Take your two hands, one palm up and the other palm down, and fit the little fingers and ring fingers together tip-to-tip—offset like that. After cleavage, the two fragments of DNA duplex each possess a short single strand a few nucleotides long dangling from the end. Look carefully at the diagram (Figure 15-20). *The two single-stranded tails are complementary to one another.*

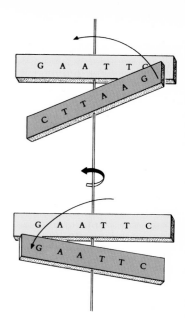

FIGURE 15-19

The nucleotide sequences recognized by restriction enzymes have twofold rotational symmetry. Thus if the sequence C-T-T-A-A-G is rotated through the plane of the paper as illustrated here, it becomes its complementary sequence, G-A-A-T-T-C. The importance of twofold rotational symmetry is that, because the two DNA strands are read in opposite directions, both strands have the same sequence.

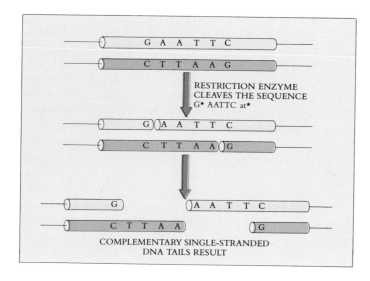

FIGURE 15-20

A restriction enzyme cleaves the sequence G-A-A-T-T-C at the position indicated by ★. Because the same sequence occurs on both strands, both are cut. The position of ★ is not the same on the two strands, however, with the result that single-stranded tails are produced. Because of the twofold rotational symmetry of the sequence, the single-stranded tails are complementary to each other—they have the same sequence of nucleotides—and they will therefore associate and form a double helix.

Every cleavage by a given kind of restriction enzyme takes place at the same recognition sequence. By chance, this sequence will probably occur somewhere in any given sample of DNA, so that a restriction endonuclease will cut DNA from any source into fragments. Each of these fragments will have the dangling sets of complementary nucleotides (sometimes called "sticky ends") characteristic of that endonuclease. Because the two single-stranded ends produced at a cleavage site are complementary, they can pair with each other. Once they have done so, the two strands can then be joined back together with the aid of a sealing enzyme called a **ligase,** which re-forms the phosphodiester bonds. This latter property makes restriction endonucleases the invaluable tools of the genetic engineer: *any* two fragments produced by the same restriction enzyme can be joined together. Fragments of elephant DNA and colon bacterium DNA cleaved by the same bacterial restriction enzyme can be joined to one another just as readily as can two bacterial fragments, because they have the same complementary nucleotide sequences at their ends.

A restriction enzyme cleaves DNA at specific sites, generating in each case two fragments whose ends have one strand of the duplex longer than the other. Because the tailing strands of the two cleavage fragments are complementary in nucleotide sequence, *any* pair of fragments produced by the same enzyme, from any DNA source, can be joined together.

PUTTING GENES INTO PLASMIDS

The first transfer of a gene from one organism to another was carried out by the American geneticists Stanley Cohen and Herbert Boyer in 1973. Cohen and Boyer used a restriction endonuclease to cut up a large bacterial plasmid, and from the resulting fragments they isolated one fragment 9000 nucleotides long, which contained both the sequence necessary for replicating the plasmid—the **replication origin**—and a gene that conferred resistance to an antibiotic, tetracycline.

Because both ends of this fragment were cut by the same restriction enzyme (called *Escherichia coli* restriction endonuclease number 1, or EcoR1), they can join together to form a circle, a small plasmid that Cohen dubbed pSC101 (plasmid of Stanley Cohen, isolate number 101). Cohen and Boyer used the same restriction enzyme, EcoR1, to cut up DNA that had been isolated from an adult amphibian, the African clawed toad, *Xenopus laevis.* They then mixed the toad DNA fragments with opened-circle molecules of pSC101, allowed bacterial cells to take up DNA from the mixture, and selected for bacterial cells that had become resistant to tetracycline (Figure 15-21). From among these pSC101-containing cells they were able to isolate ones containing the toad ribosomal RNA gene. These versions of pSC101 had the toad gene spliced in at the EcoR1 site. Instead of joining to one another, the two ends of the pSC101 plasmid had joined to the two ends of the toad DNA fragment that contained the ribosomal RNA gene.

pSC101 containing the toad ribosomal RNA gene is a true **chimera,** an organism that contains a mixture of features from different distinct organisms. In ancient Greek mythology, a chimera was an imaginary she-monster represented as vomiting flames, with a lion's head, goat's body, and dragon's or serpent's tail. Although dragons are difficult to obtain (!), it would be possible today to create a chimera that included genetic material from lions, goats, and snakes. The toad-bacteria chimera that Boyer and Cohen produced had features that never existed in nature and probably would never have evolved there. It is a form of **recombinant DNA,** a DNA molecule created in the laboratory by molecular geneticists by joining together bits of several genomes into a novel combination.

The first recombinant genome produced by genetic engineering was a bacterial plasmid into which an amphibian ribosomal RNA gene was inserted in 1973.

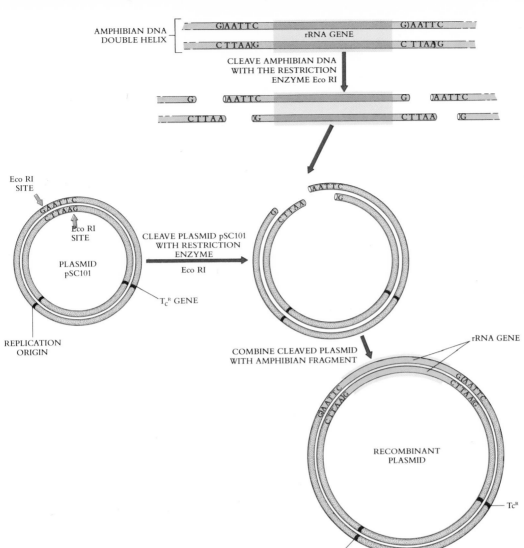

FIGURE 15-21

The first successful genetic engineering experiment. Cohen and Boyer inserted an amphibian gene encoding rRNA into a bacterial plasmid called pSC101 (the hundred and first plasmid isolated by Stanley Cohen). pSC101 contains a single site cleaved by the restriction enzyme EcoR1. The rRNA encoding region was inserted into pSC101 at that site by cleaving the rRNA region with EcoR1 and then allowing the complementary sequences to come together to form a new double helix.

There has been considerable discussion about the potential danger of inadvertently creating an undesirable life form in the course of recombinant DNA experiments. What if someone fragmented the DNA of a cancer cell and then incorporated the fragments at random into viruses that are propagated within bacterial cells? Might there not be a danger that among the resulting bacteria there could be one capable of constituting an infective form of cancer?

Even though most recombinant DNA experiments are not dangerous, such concerns are real and need to be taken seriously. Both scientists and individual governments monitor these experiments to detect and forestall any hazard. Experimenters have gone to considerable lengths to establish appropriate experimental safeguards. The bacteria used in many recombinant DNA experiments, for example, are unable to live outside of laboratory conditions; many of them are obligate anaerobes, organisms poisoned by oxygen. Decidedly dangerous experiments, such as the random experiment involving oncogenes that was described above, are prohibited.

RECENT PROGRESS IN GENETIC ENGINEERING

The 1980s have seen an explosion of interest in applying genetic engineering techniques to practical human problems. Perhaps the most obvious commercial application and the one that was seized on first was to introduce human genes that encode clinically important proteins into bacteria. Because they can be grown cheaply in bulk,

FIGURE 15-22

These two mice are genetically identical except that the large one has one extra gene—the gene encoding a potent rat growth hormone not normally present in mice. The gene was added to the mouse genome by human genetic engineers and is now a stable part of the mouse's genetic endowment.

bacteria that incorporate human genes can produce the human proteins that they specify in large amounts. This method has been used to produce several forms of human interferon; to place rat growth hormone genes into mice (Figure 15-22); and to manufacture many commercially valuable nonhuman enzymes. Ultimately it will prove possible to alter the genetic constitution of human beings, an area of ethical, moral, and practical concern: who will make the decisions about what it is proper and desirable to do?

A second major area of genetic engineering activity is the manipulation of the genes of key crop plants. In plants, the primary experimental difficulty has been identifying a suitable vector medium (carrier) for introducing genes into target organisms. Plants do not possess the many plasmids that bacteria do, so the choice of potential vectors is therefore limited.

The most successful results obtained thus far with plant systems have involved a bacterial plasmid called Ti, which infects plants and causes crown-gall tumors to develop. Part of this Ti plasmid integrates into the plant DNA; it has proved possible to attach other genes to this portion of the plasmid. Attempts are currently under way to use such techniques to introduce genes that will improve the resistance of crop plants to disease, frost, and other forms of stress; improve their nutritional balance and protein content; and confer herbicide resistance on them. Recently, for example, a bacterial gene endowing resistance to the herbicide Roundup has been made to work in plants (Figure 15-23). This is of real interest to farmers, since Roundup is a particularly powerful and broad-spectrum herbicide that will kill any green plant. A crop resistant to Roundup would never have to be weeded if the field were simply treated with the herbicide. A more long-range goal is to increase yield and plant size in crop plants; however, we do not yet know, for the most part, which genes are responsible for these complex characteristics. It has also proved difficult to introduce the genes responsible for capturing atmospheric nitrogen (nitrogen-fixing genes) from bacteria into plants. These genes enable bacteria to change nitrogen into a form in which it can be used by other organisms, but the genes do not seem to function properly in their new eukaryotic environment. Furthermore, it is possible that they might require so much energy to function that the new plant would not be productive. Experiments in many related areas are, however, being pursued actively.

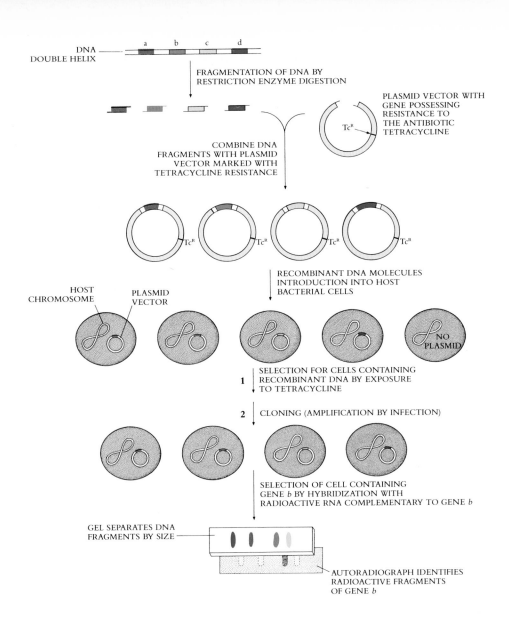

DNA DOUBLE HELIX

a b c d

FRAGMENTATION OF DNA BY RESTRICTION ENZYME DIGESTION

PLASMID VECTOR WITH GENE POSSESSING RESISTANCE TO THE ANTIBIOTIC TETRACYCLINE

Tc^R

COMBINE DNA FRAGMENTS WITH PLASMID VECTOR MARKED WITH TETRACYCLINE RESISTANCE

Tc^R Tc^R Tc^R Tc^R

RECOMBINANT DNA MOLECULES INTRODUCTION INTO HOST BACTERIAL CELLS

HOST CHROMOSOME   PLASMID VECTOR

NO PLASMID

1 SELECTION FOR CELLS CONTAINING RECOMBINANT DNA BY EXPOSURE TO TETRACYCLINE

2 CLONING (AMPLIFICATION BY INFECTION)

SELECTION OF CELL CONTAINING GENE *b* BY HYBRIDIZATION WITH RADIOACTIVE RNA COMPLEMENTARY TO GENE *b*

GEL SEPARATES DNA FRAGMENTS BY SIZE

AUTORADIOGRAPH IDENTIFIES RADIOACTIVE FRAGMENTS OF GENE *b*

FIGURE 15-23

A generalized scheme for cloning a gene. Imagine that you wish to clone the gene conferring resistance to the herbicide Roundup from the bacterial strain in which it was discovered. You would first fragment the bacterial DNA with a restriction enzyme. Then you would incorporate the fragments into a plasmid containing a resistance gene, and introduce the plasmid population into bacterial cells, adjusting concentrations so that on the average one plasmid enters each cell. You would then expose the cells to the antibiotic tetracycline, killing any cell that does not now harbor a plasmid. The surviving cells are put onto solid growth medium, such as agar, so that each cell is held apart from one another and forms a separate colony. A portion of each colony is then checked for the presence of the Roundup-resistant gene. One way in which such checks are carried out is to check each colony for the presence of a DNA fragment that will hybridize to a radioactive "probe" (form a new double helix with it). The probe—a section of DNA that contained the sequence for which you were searching— would be a fragment of mRNA or DNA known to contain the sequence, but labeled with one or more radioactive atoms so that double helices containing the probe could be identified. Such DNA, with a new sequence of nucleotides, would be recombinant DNA.

Even more practical, and of immediate applicability, is the genetic alteration of bacteria by these methods. For example, ice forms on plants largely because the crystals develop around individuals of bacteria, such as *Pseudomonas syringiae,* that coat the leaves. Mutant forms of this bacterium are known in nature and have also been produced experimentally, forms that lack the gene that makes possible the formation of ice crystals. Theoretically, if this bacteria were substituted for the predominant ones in the fields, the onset of frost would be delayed, in effect, for the crops, and their productivity could be enhanced. Genetically engineered forms of *P. syringiae* were first tested in a field in California in April 1987. Other bacteria have been developed that produce toxins specific to particular insect and other pests. If they are introduced into the fields, the application of large amounts of ecologically harmful insecticides can be avoided.

THE IMPORTANCE OF GENETIC CHANGE

All evolution begins with genetic change, with the processes described in this chapter: mutation, which creates new alleles, the raw materials for change; recombination, which shuffles and sorts the changes; transposition, which alters gene location; and chromosomal rearrangement, which alters the arrangement of entire chromosomes.

Some changes produce alterations in an organism that enable it to leave more off-spring, and these changes are preserved as the genetic endowment of future generations. Other changes reduce the ability of an organism to leave offspring. These changes tend to be lost, since the organisms that carry them contribute fewer members to future generations than those that lack them.

Evolution, the subject of the next section, can be viewed as the selection of particular combinations of alleles from a pool of different alternatives. The rate of evolution is ultimately limited by the rate at which these different alternatives are generated. Genetic change provides the raw material for evolution.

SUMMARY

1. A mutation is any change in the hereditary message. Changes in one or a few nucleotides are called point mutations. They may arise as a result of physical damage from ionizing radiation or mistakes made during the correction of chemical damage caused by the absorption of ultraviolet light. Occasionally changes occur spontaneously as a result of errors in pairing that take place during the course of DNA replication.

2. Cancer is a growth disease of cells, in which the regulatory controls that usually restrain cell division don't work.

3. The testing of DNA fragments from tumor cells to identify those fragments capable of inducing cancer has led to the isolation of a variety of cancer-causing genes. In every case, the cancer-causing gene is a normal gene whose product has a role in cell proliferation, but that becomes active when it should not be.

4. In some cases, the initiation of a cancer results from a single nucleotide mutation in a normal gene. In other cases, a normal gene is acquired by a transposable retrovirus and transcribed at the high rate characteristic of the virus.

5. The best way to avoid cancer is to avoid those things that cause mutation, notably smoking.

6. Transposable elements may integrate themselves and any genes that are associated with them, anywhere on the chromosomes. The mechanism that they employ to move from one chromosomal location to another is not well understood.

7. Bacterial genes may be transferred by association with plasmids, small circles of DNA that regularly move from cell to cell by conjugation. Because the plasmid enters the bacterial genome by a reciprocal exchange of strands in a region of sequence similarity, it may carry genes only to certain locations on the bacterial chromosome. At these sites, nucleotide sequences occur that are also represented on the plasmid.

8. Genetic engineering is the isolation of specific genes and their transfer to new genomes. The key to genetic engineering technology is a special class of enzymes called restriction endonucleases, which cleave DNA molecules into fragments. These fragments have short one-strand tails that are complementary in nucleotide sequence to each other.

9. Because the sequences of the tails of the two cleavage products are complementary, they can pair with one another and rejoin together, whatever their source. For this reason, DNA fragments from very different genomes can be combined.

10. The first combination of fragments from different genomes (recombinant DNA) was achieved by Stanley Cohen and Herbert Boyer in 1973. They inserted an amphibian ribosomal RNA gene into a bacterial plasmid.

REVIEW

1. Evolution depends on the existence of genetic variation in nature. Two important agents of this genetic variation are _____ and recombination.

2. A chemical capable of damaging DNA is called a _____.

3. Retroviruses use _____ instead of DNA as their hereditary information molecule.

4. Cancer-causing genes are called _____, from the Greek word for cancer.

5. The three major types of chromosomal alterations are _____, _____, and _____.

SELF-QUIZ

1. Eukaryotes, unlike bacteria, can repair double-stranded breaks in their DNA by using
   (a) ionizing radiation
   (b) pyrimidine dimers
   (c) the synaptonemal complex to line up homologous chromosomes
   (d) neither can repair double-stranded breaks
   (e) both can repair double-stranded breaks

2. The three major sources of DNA mutations are
   (a) ionizing radiation
   (b) ultraviolet radiation
   (c) retroviruses
   (d) transposition
   (e) chemical mutagens

3. Which of the following is *not* associated with elevated cancer risk?
   (a) high-fiber diet
   (b) cigarettes
   (c) atomic radiation
   (d) diet high in meat
   (e) living in the Mississippi delta

4. The evolutionary impact of transposition is enormous. Three reasons for this are
   (a) transposons cause mutation
   (b) transposition often causes cell death
   (c) retroviruses are responsible for some diseases
   (d) transposition can bring together genes that are actually located on different parts of a chromosome—as in bacterial plasmids
   (e) transposition can cause rapid recombination of genes

5. Restriction enzymes are an invaluable tool for genetic engineering because
   (a) they cut both strands of DNA at specific sites
   (b) "sticky" or complementary ends of the cleaved DNA are produced
   (c) "sticky" ends of DNA from very different DNA sources can be easily joined together
   (d) they are tools of the meat-tenderizing industry
   (e) a, b, and c

THOUGHT QUESTIONS

1. In medical research, mice are often used as model systems in which to study the immune system and other physiological systems that are important to human health. Many medical centers maintain large colonies of mice for these studies. In one such colony under your supervision, a hairless mouse is born. What minimal evidence would you accept that this variant represents a genetic mutation?

2. An American couple both work in an atomic energy plant, and both are exposed daily to low-level background radiation. After several years, the couple has a child, and this child proves to be affected by *Duchenne muscular dystrophy*, a sex-linked recessive genetic defect, one in which the mutant locus is located on the X chromosome. Both parents are normal, as are the grandparents. The couple sue the plant, claiming that the abnormality in their child was the direct result of radiation-induced mutation of their gametes, radiation from which the company should have protected them. Before reaching a decision, the judge insists on knowing the sex of the child. Which sex would be more likely to result in an award of damages, and why?

3. The evidence associating lung cancer with smoking is overwhelming and, as you have learned in this chapter, we now know in considerable detail the mechanism whereby smoking induces cancer—it introduces powerful mutagens into the lungs, which cause mutations to occur; when a growth regulating gene is mutated by chance, cancer results. The process is no more mysterious than the fact that death results from shooting shotguns at random in crowded football stadiums. In light of this, why do you suppose that cigarette smoking is still legal?

4. The American Cancer Society has estimated that one quarter of the male children and one third of the female children born in 1987 will someday die of cancer. Their prediction implies that boys are only 75% as likely as girls to die of cancer. Can you think of any reason(s) why so many more females than males are expected to die of cancer?

FOR FURTHER READING

BISHOP, J.M.: "Oncogenes," *Scientific American*, March 1982, pages 82-92. The story of the chicken sarcoma gene and how it causes cancer, by the man who first identified the protein product of the gene. This article is particularly good at showing the chain of reasoning that underlies a major scientific advance.

CAIRNS, J.: "The Treatment of Diseases and the War Against Cancer," *Scientific American*, November 1985, pages 51-59. There are many different kinds of cancer, and this article does an unusually good job of describing the different forms and the progress that has been made in treating some of them.

CHILTON, M.: "A Vector for Introducing New Genes Into Plants," *Scientific American*, June 1983, pages 50-60. A description of the only successful plant genetic engineering vector developed to date, important because of its agricultural implications.

FEDEROFF, N.: "Transposable Genetic Elements in Maize," *Scientific American*, June 1984, pages 85-98. An excellent introduction to the studies for which Barbara McClintock was awarded the 1984 Nobel Prize in Physiology or Medicine, rephrased in molecular terms that were not available when she made her important discoveries.

KELLER, E.F.: *A Feeling for the Organism*, W.H. Freeman & Co., San Francisco, 1983. A well-written biography of 1984 Nobel Prize winner Barbara McClintock, which will give you a feeling for how discoveries are made in science.

STAHL, F.W.: "Genetic Recombination," *Scientific American*, February 1987, pages 90-101. A beautifully illustrated account of our growing appreciation of the way in which chromosomes trade parts during the course of meiosis.

WEINBERG, R.: "The Molecular Basis for Cancer," *Scientific American*, November 1983, pages 126-144. This important paper by a developer of the transfection approach describes how transfection experiments yielded our first look at a cancer-causing gene.

EVOLUTION

Part Five

FIGURE 16-1   In evolutionary terms, this orangutan is a relatively close relative of ours. The possibility that this might be true has disturbed many people since Darwin first established a logical explanation for evolution, namely natural selection, in 1859. Although the theory of evolution by natural selection is now accepted by almost all biologists and indeed forms the foundation of modern biology, it still stirs heated debate in nonscientific circles.

THE EVIDENCE FOR EVOLUTION

Chapter 16

Overview

The theory of evolution by natural selection, which many find controversial, is the corner-stone of modern biology. That evolution has occurred is regarded as a demonstrated fact by biologists, most of whom also agree with Darwin's explanation of natural selection as the way in which evolution occurs. During the period of well over a century that has passed since Darwin first suggested this explanation for evolution, we have learned a great deal about the mechanisms operating to bring about evolution.

For Review

Here are some important terms and concepts that you will encounter in this chapter. If you are not familiar with them, you should review them before proceeding.

Darwin's theory of natural selection (Chapter 2)

Scientific creationism (Chapter 2)

Levels of organization (Chapter 2)

Origin of life (Chapter 4)

Alleles (Chapter 11)

Heterozygosity (Chapter 11)

ABO blood groups (Chapter 12)

Of all the major ideas of biology, evolution is perhaps the best known to the general public. This is not because the theory of evolution by natural selection is well under-stood by the average person, but rather because many people mistakenly believe that evolution represents a challenge to their religious beliefs (Figure 16-1). In Chapter 2 we reviewed the current form of this controversy, an attempt to require the teaching of the set of religious dogmas known as "scientific creationism" in public-school science classes. Similar highly publicized criticisms of evolution have occurred ever since the time of Darwin, and it is likely that others will occur in the future. For this reason it is important that, during the course of your study of biology, you address the issue squarely. Just what *is* the evidence for evolution?

In this chapter, we look first at how biologists currently view the evolutionary process, and we ask "What factors are thought to bring about evolutionary change?" We then look at individual cases in which evolutionary change in natural populations can be seen to be adaptive, a key tenet of Darwin's theory. Finally, we review the evidence that supports the general validity of Darwin's proposal.

EVOLUTION AND ADAPTATION

The genetic machinery of cells ensures that the hereditary instructions that specify the nature and characteristics of each organism will be faithfully followed. If you plant a radish seed in your garden, you get a radish and not a turnip. However, as you saw in the previous chapter, mutation and recombination alter genetic instructions. For this reason, few radishes are exactly alike, and no other human being is exactly like you (unless you have an identical twin). Often the differences between individuals have an important bearing on their relative chances to bear offspring. It was insights of this kind that led Darwin to formulate his theory of natural selection, which you studied in Chapter 2. Darwin agreed with many earlier philosophers and naturalists, who deduced that the many kinds of organisms we see around us in the world were produced by a process of evolution. Unlike his predecessors, however, Darwin proposed a mechanism that is capable of accounting for the process of evolution—natural selection (Table 16-1). He proposed that new kinds of organisms evolve from older ones because they occur in different environments, in which natural selection favors those genetic differences which better suit an organism to survive and reproduce (Figure 16-2).

When the word "evolution" is mentioned, it is difficult not to conjure up images of dinosaurs, of woolly mammoths frozen in blocks of ice, or of Darwin confronting a monkey. Traces of ancient forms of life, now extinct, survive as fossils that help us piece together the evolutionary story. With such a background, we usually think of evolution as meaning changes in the kinds of animals and plants on earth, changes that take place over long periods of time, with new forms replacing old ones. This kind of evolution is called **macroevolution.**

Much of the focus of Darwin's theory, however, is directed not at the way in which new species are formed from old ones, but rather at the way in which changes occur within species. **Natural selection,** the explanation that Darwin proposed for evolutionary change, is the process whereby some individuals in a population—those with favored characteristics—produce more surviving offspring than others that lack these characteristics. As a result, the population gradually will come to include more and more individuals with the favored characteristics, generation after generation, assuming that the characteristics have a genetic basis. In this way, the populations evolve. Change of this sort within populations is called **microevolution.** Natural selection is the process by which microevolutionary change occurs. The result of the process is

---

TABLE 16-1   THE LOGICAL ARGUMENT FOR NATURAL SELECTION

Fact 1: All species have such great potential fertility that their population size would increase exponentially if all individuals born would again reproduce successfully.

Fact 2: Except for minor annual fluctuations and occasional major fluctuations, populations normally display stability.

Fact 3: Natural resources are limited. In a stable environment they remain relatively constant.

Inference 1: Since more individuals are produced than can be supported by the available resources but population size remains stable, it means that there must be a fierce struggle for existence among individuals of a population, resulting in the survival of a part, often a very small part, of the progeny of each generation.

Fact 4: No two individuals are exactly the same; rather, every population displays enormous variability.

Fact 5: Much of this variation is heritable.

Inference 2: Survival in the struggle for existence is not random but depends in part on the hereditary constitution of the surviving individuals. This unequal survival constitutes the process of *NATURAL SELECTION.*

Inference 3: Over the generations this process of natural selection will lead to a continuing gradual change of populations, that is, *EVOLUTION,* and to the production of new species.

From Mayr, E. 1982. The growth of biological thought. Harvard University Press, Cambridge. (Emphasis added.)

FIGURE 16-2

Two peppered moths, *Biston betularia,* resting on a piece of bark. The dark moth is more easily seen on this light bark, which shows no evidence of industrial pollution. This chapter will deal with the ways in which such differences arise, are maintained, and change in frequency.

called **adaptation,** the incorporation of features that promote the likelihood of survival and reproduction by an organism in a particular environment.

The essence of Darwin's explanation of evolution is that progressive adaptation by natural selection is responsible for evolutionary changes *within* a species, changes that, when they accumulate, lead to the creation of new kinds of organisms, new species. Darwin's theory thus has two elements: **adaptation,** a microevolutionary process, and **species formation,** a macroevolutionary process. Evolution is said to involve the creation of novelty by adaptation and the partitioning of novel changes into groups by species formation. This chapter considers the role of adaptation in detail; the next chapter discusses how species form.

VARIATION IN NATURE

In considering Darwin's theory that evolution is a progressive series of adaptive changes brought about by natural selection, it is best to start by looking at the raw material available for the selective process—the genetic variation present among individuals from which natural selection chooses the best-suited alleles.

As you saw in Chapter 12, a group of individuals living together, a so-called "natural population," can contain among its members a great deal of genetic variation. How much? Biologists have looked at many different genes in an effort to answer this question.

1. *Blood groups.* Chemical analysis has revealed the existence of more than 30 different blood group genes in human beings, in addition to the ABO locus. At least a third of these genes are routinely found to be present in several alternative allelic forms in human populations. In addition to these, more than 45 additional variable genes are also known to encode proteins in human blood cells and plasma, but these are not considered to define blood groups. Thus there are more than 75 genetically variable genes in this one system alone.

2. *Enzymes.* Alternative alleles of the genes specifying particular enzymes are easy to distinguish. The differences in their nucleotide sequences alter the ways in which the proteins specified by these alleles behave in simple physical tests. One of the most popular of these is to measure how fast the alternative proteins migrate in an electric field (a process called **electrophoresis**). A great deal of variation is found at enzyme-specifying loci (Table 16-2; Figure 16-3). Among human beings, about 5% of the loci of a typical individual are heterozygous. That is, if you picked an individual at random, and from one of his or her cells selected a gene at random, the chances are 1 in 20 (5%) that the individual would be heterozygous for that gene.

TABLE 16-2   LEVELS OF GENETIC VARIATION AT ENZYME-ENCODING GENES

GROUP	PERCENTAGE OF GENES HETEROZYGOUS IN AN AVERAGE INDIVIDUAL
INVERTEBRATES	
Drosophila	15%
Other insects	15%
Marine invertebrates	12%
Land snails	15%
VERTEBRATES	
Fishes	8%
Amphibians	8%
Reptiles	5%
Birds	4%
Mammals	5%
PLANTS	
Self-pollinating	3%
Outcrossing	8%

Considering the entire human genome, it is fair to say that almost all individuals are different from one another; this is true of almost all other organisms as well. In nature, genetic variation is the rule.

The Hardy-Weinberg Principle

Genetic variation within natural populations was a puzzle to Darwin and his contemporaries. The way in which meiosis produces genetic segregation among the progeny of a hybrid had not yet been discovered. The theories of the day predicted that dominant alleles should eventually drive recessive ones out of a population, thus eliminating any genetic variation. Selection, they thought, should favor an optimal form.

The solution to this problem was developed independently and published almost simultaneously in 1908 by G.H. Hardy, an English mathematician, and G. Weinberg, a German physician. They pointed out that in a large population in which there is random mating, and in the absence of forces that change the proportions of the alleles at a given locus (which will be discussed below), the original proportions of the genotypes will remain constant from generation to generation. Because these proportions don't change, the genotypes are said to be in **Hardy-Weinberg equilibrium.**

In algebraic terms the Hardy-Weinberg principle is written as an equation. Its form is what is known as a binomial expansion. For a gene with two alternative alleles, which we will call *A* and *a,* the equation looks like this:

$$(p + q)^2 = p^2 + 2pq + q^2$$

(INDIVIDUALS HOMOZYGOUS FOR ALLELE *A*)   (INDIVIDUALS HETEROZYGOUS WITH ALLELES *A* + *a*)   (INDIVIDUALS HOMOZYGOUS FOR ALLELE *a*)

In statistics, **frequency** is defined as the proportion of individuals falling in a certain category in relation to the total number of individuals being considered. Thus, in a population of 10 cats, with 7 white and 3 black cats, the respective frequencies would be 0.7 (or 70%) and 0.3 (or 30%). In the algebraic terms of this equation, the letter p designates the frequency of one allele, the letter q the frequency of the alternative allele. By convention, the more common of the two alleles is designated $p,$ the rarer allele $q.$ Because there are only two alleles, $p + q$ must always equal 1.

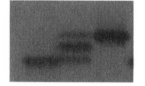

FIGURE 16-3

Gel electrophoresis of protein samples from three individuals of the common killifish, *Fundulus heteroclitus.* The gel was stained to detect activity of the enzyme glucose phosphate isomerase. Proteins have migrated from the bottom of the gel (as shown here) towards the top, with the sample from each fish filling a separate lane. The fish from which the sample in the left lane was obtained is homozygous for one allele of this enzyme, and the fish on the right is homozygous for a second, alternative allele. The fish represented by the middle sample is a heterozygote individual that has a copy of both of these alleles.

This equation is extraordinarily useful to biologists, because it permits an investigator to predict readily what proportion of a population will be homozygous for a given allele, and what proportion will be heterozygous, simply from a knowledge of that allele's frequency. As an example, consider the recessive allele that is responsible for cystic fibrosis. It is present in North American whites at a frequency of about 22 per 1000 individuals, or 0.022. What proportion of North American whites is expected to express this trait? The frequency of double recessive individuals (q^2) is expected to be 0.022×0.022, or 1 in every 2000 individuals. What proportion can be expected to be heterozygous carriers? If the frequency of the recessive allele q is 0.022, then the frequency of the dominant allele p must be $1 - 0.022$, or 0.978. The frequency of heterozygous individuals ($2pq$) is thus expected to be $2 \times 0.978 \times 0.022$, or 43 in every 1000 individuals.

How valid are these calculated predictions? For many genes, they prove to be very accurate. Most human populations are large and effectively randomly mating and so are similar to the "ideal" population envisioned by Hardy and Weinberg. As you will see, however, for some genes the calculated predictions do *not* match the actual values. The reasons they do not do so tell us a great deal about evolution.

> **The Hardy-Weinberg principle states that in a large population mating at random and in the absence of forces that would change their proportions, genetic segregation will not alter the proportion of alleles at a given locus.**

WHY DO ALLELE FREQUENCIES CHANGE?

According to the Hardy-Weinberg principle, the proportions of homozygotes and heterozygotes in a large random-mating population will remain constant, as long as the individual allele frequencies do not change. This reservation tacked onto the end of the statement is important. In fact, it is the key to the importance of the Hardy-Weinberg principle for biology, because individual allele frequencies *are* changing all the time in natural populations, with some alleles becoming more common than others. The Hardy-Weinberg principle establishes a convenient baseline against which to measure such changes. By looking at how various factors alter the proportions of homozygotes and heterozygotes in populations, we can identify those forces that are changing the allele frequencies within the population.

Many factors can alter allele frequencies. Only five, however, alter the proportions of homozygotes and heterozygotes enough to produce significant deviations from the proportions predicted by the Hardy-Weinberg principle. These are **mutation, migration** (including both immigration into and emigration out of a given population), **genetic drift** (random loss of alleles, which is more likely in small populations), **nonrandom mating,** and **selection.** Of these, only the last factor, selection, produces adaptive evolutionary change; the others are important in establishing the genetic constitution of individual populations.

> **Five factors can bring about a deviation from the proportions of homozygotes and heterozygotes predicted by the Hardy-Weinberg principle: mutation, migration, genetic drift, nonrandom mating, and selection.**

Mutation

Mutation from one allele to another obviously can change the proportions of particular alleles in a population. Mutation rates are generally so low, however, that they cannot alter Hardy-Weinberg proportions of common alleles very much. Many genes mutate about 1 to 10 times per 100,000 cell divisions, although of course a mutation in *some* gene of an individual will occur much more frequently than that. Since most environments are constantly changing, populations rarely last long enough to accumulate differences in allele frequency produced this slowly. Nonetheless, as you saw in Chapter 15, mutation is the ultimate source of genetic variation and thus makes evolution possible.

FIGURE 16-4

The yellowish-green cloud around these Monterey pines, *Pinus radiata*, is pollen, being dispersed by the wind. The male gametes within the pollen reach the egg cells of the pine passively in this way. In genetic terms, such dispersal is a form of migration.

Migration

Migration is defined in genetic terms as the movement of individuals from one population into another. It can be a powerful force in upsetting the genetic stability of natural populations. Sometimes migration is obvious, as when an animal moves from one place to another. If the characteristics of the newly arrived animal differ from those of the animals already there, the genetic composition of the receiving population may be altered, assuming the newly arrived individual or individuals are well-enough adapted to survive in the new area and to mate successfully. Other important kinds of migration are not as obvious to the observer. These subtler movements include, for example, the drifting of gametes or immature stages of marine animals or plants from one place to another (Figure 16-4). The male gametes of flowering plants are often carried great distances by insects and other animals that visit their flowers. Their seeds may also be blown in the wind or carried by animals or other agents to new populations far from their place of origin. No matter how it occurs, migration can alter the genetic characteristics of populations and prevent them from maintaining Hardy-Weinberg equilibrium. The evolutionary role of migration is more difficult to assess, however, and depends heavily on the selective forces prevailing at the different places where the species occurs.

Genetic Drift

In small populations the individual alleles of a given gene are all represented in a few individuals. If by chance anything happens to these individuals, the alleles they carry may be lost from the population. The same sort of chance is involved in rolling dice or in flipping a coin. In small populations the frequencies of alleles may be changed drastically by chance alone. Because so few individuals are involved, many of the alleles originally present in the population may be lost from it. Since this process appears to occur randomly, as if allele frequencies were drifting, it is known as **genetic drift.** A series of small populations that are isolated from one another may come to differ strongly as a result of genetic drift.

> **Genetic drift is the random loss of alleles from loci as a result of accident. It is of special significance in small populations.**

Sometimes one or a few individuals are dispersed and become the founders of a new, isolated population at some distance from their place of origin. When this occurs, the alleles that they carry are of special significance. Even if these alleles are rare in the source population, in their new area they will be a significant fraction of the whole population's genetic endowment. This effect—by which rare alleles and combinations of alleles may be enhanced in the new populations—is called the **founder principle.** It is a particularly important factor in the evolution of organisms on distant oceanic islands, such as the Hawaiian Islands. Most of the kinds of organisms that occur in such areas have probably been derived from one or a few initial "founders." In a similar way, isolated human populations are often dominated by the genetic features that were characteristic of their founders, if only a few individuals were involved initially.

Nonrandom Mating

Individuals with certain genotypes sometimes mate with one another more commonly than would be expected on a random basis. **Inbreeding** (mating with relatives), a type of nonrandom mating that is characteristic of many groups of organisms, will cause the frequencies of particular genotypes to differ greatly from those predicted by the Hardy-Weinberg principle. Inbreeding does not change the frequency of the alleles, but rather the proportion of individuals that are homozygous. In inbred populations there are more homozygous individuals than predicted by the Hardy-Weinberg principle. It is for this reason that populations of self-fertilizing plants consist primarily of homozygous individuals, while "outcrossing" plants, which interbreed with individuals different from themselves, have a higher proportion of heterozygous individuals (Figure 16-5).

Because inbreeding increases the proportion of homozygous individuals in a population, it tends to promote the occurrence of double-recessive combinations. It is for this reason that marriage between relatives is discouraged—it greatly increases the

FIGURE 16-5

This wild bee *(Ptilothrix)*, loading the pollen baskets on its hind legs with pollen from the desert "poppy" *(Kalstroemia)* in eastern Arizona, illustrates the way that flowering plants have co-opted animals to the task of dispersing their pollen precisely from plant to plant. The pollen that the bee has gathered will be used to provision the cell in which its larva will reach maturity. Other pollen grains, adhering to the hairy body of the bee, will have a good chance of reaching the female parts of other flowers of this same species. Outcrossing—breeding with individuals other than yourself and your close relatives—is as important for plants as it is for all eukaryotic organisms. It is a major factor in making possible their adaptation to the environment.

FIGURE 16-6

Inbreeding often has detrimental effects. In Japan it is common for first cousins to marry, and the Japanese population as a result is more homozygous than that of the United States, where such marriages are generally forbidden. Because of this increased homozygosity, recessive alleles tend to be expressed more often in Japan. For each of the five genetic disorders illustrated here, children of parents who are first cousins are far more likely to be homozygous and express the trait in Japan than in the Untied States—even though four of the five recessive alleles are more common in the United States!

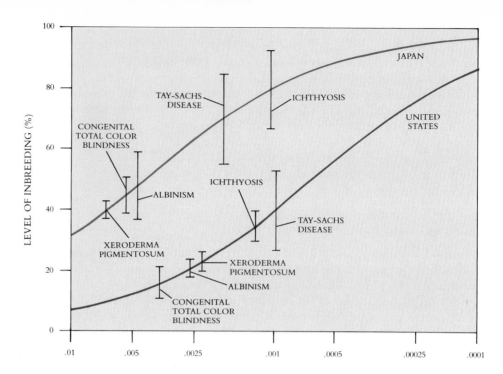

FREQUENCY OF RECESSIVE ALLELE

possibility of producing children homozygous for an allele associated with one or more of the genetic disorders discussed in Chapter 12. A dramatic example of this can be seen in Figure 16-6, which compares the incidence of five important genetic diseases in the United States, where inbreeding is rare, with the incidence in Japan, where it is more common. Even though alleles specifying some disorders, such as Tay-Sachs disease, are rarer in Japan, there is a much greater likelihood that a marriage between first cousins will produce affected children there because the Japanese population is more homozygous in general.

Selection

As Darwin pointed out, some individuals leave behind more progeny than others, and the rate at which they do so is affected by their inherited characteristics. We refer to this process as **selection** and speak both of **artificial selection** and of **natural selection.** In artificial selection the breeder selects for the desired characteristics, such as those of the dogs shown in Figure 17-2. In natural selection, the environment plays this role, with conditions in nature determining which kinds of individuals in a population are the most fit and so affecting the proportions of genes among individuals of future populations. This is the key point in Darwin's proposal that evolution occurs because of natural selection: the environment imposes those conditions that determine the results of selection and thus the direction of evolution.

Like mutation, migration, and genetic drift, selection causes deviations from Hardy-Weinberg proportions by directly altering the frequencies of the alleles. Darwin argued that greater reproduction of particular genotypes, which is how he defined selection, is the primary force that shapes the pattern of life on earth.

Although selection is perhaps the most powerful of the five principal agents of genetic change, there are limits to what it can accomplish. These limits arise because alternative alleles may interact in different ways with other genes. These interactions tend to set limits on how much a phenotype can be altered. For example, selecting for large clutch size in barnyard chickens eventually leads to eggs with thinner shells that break more often than before. Because of the limits imposed by gene interactions, strong selection is apt to result in rapid change initially, but the change soon comes to a halt as the interactions between genes increase. For this reason, we do not have pigs as large as cows, chickens that lay twice as many eggs as the best layers do now, or corn with an ear at the base of every leaf.

There is a second factor that limits what selection can accomplish: selection acts only on phenotypes. Only those characteristics that are expressed in the body of an organism can affect the ability of that organism to produce progeny. For this reason, selection does not operate efficiently on rare recessive alleles, simply because they do not often come together as homozygotes and there is no way of selecting them unless they do come together. For example, when a recessive allele a is present at a frequency q equal to 0.1, 10% of the alleles for that particular gene will be a, but only one out of a hundred individuals (q^2) will be double recessive and so display the phenotype associated with this allele. For lower allele frequencies, the effect is even more dramatic: if the frequency in the population of the recessive allele $q = 0.01$, the frequency of homozygotes in that population will be only 1 in 10,000.

> **Selection, the differential reproduction of genotypes as a result of the way that their phenotypic characteristics lead to reproductive success in the environment, is a powerful mechanism for producing deviations from Hardy-Weinberg equilibrium.**

What this means is that selection against undesirable genetic traits in humans or domesticated animals is very difficult unless the heterozygotes can also be detected. For example, if for a particular undesirable recessive allele r, $q = 0.01$, and none of the homozygotes for this allele were allowed to breed, it would take 1000 generations, or about 25,000 years in humans, to lower the allelic frequency by half, to 0.005. At this point, after 25,000 years of work, the frequency of homozygotes would still be 1 in 40,000, or 25% of what it was initially. This is the basic reason that few geneticists advocate **eugenics,** the science that deals with efforts to change the genetic characteristics of human beings by artificial selection. Aside from the moral implications, such efforts are essentially doomed to failure by the sheer difficulty of producing the desired results within any plausible human time frame.

Which Force is Most Important in Evolution?

Any of the five forces just discussed can cause genetic variation to occur in a population. Individual alleles may make different contributions to the **fitness,** or reproductive capacity, of the organisms in which they occur (selection); alternatively, alleles may be maintained in different frequencies by genetic drift or by migration from other populations. Mutation may affect the frequencies of very rare alleles. In practice, which of these effects is the most important in evolution? This is an important question to answer, since among the five factors only selection would be expected to produce adaptive evolutionary change. The other effects are random in direction, and so "neutral" in their evolutionary effect. Although many alleles in natural populations have been examined in an effort to evaluate these relationships precisely in individual cases, precise answers have proved to be difficult to obtain. This is because assigning a precise role to the contribution that a particular allele makes to the fitness of an organism in nature is so difficult. A population may maintain particular alleles at high frequencies in natural populations because the enzymes they produce make different contributions to individual fitness, or simply because of the structure and dynamics of the population.

THE EVIDENCE THAT NATURAL SELECTION PRODUCES MICROEVOLUTIONARY CHANGE

Although it is not clear how much of the wealth of genetic variation that we see in nature is present because it is being maintained by natural selection, it is abundantly clear that at least some of it is. When microevolutionary changes are produced by natural selection, the process is called **adaptation.** To illustrate this key point, that at least some of the genetic variation in nature is adaptive, let us examine a few well-studied cases.

Sickle Cell Anemia

Sickle cell anemia, a hereditary disorder affecting hemoglobin molecules in the blood, was first reported by a Chicago physician, James B. Herrick, in 1910. A West Indian Negro student exhibited symptoms of severe anemia that appeared related to abnormal red blood cells: "The shape of the red cells was very irregular, but what especially attracted attention was the large number of thin, elongated, sickle-shaped and crescent-shaped forms." The disease was soon found to be relatively common among American blacks. In Chapter 12 we noted that this disorder, which affects roughly 2 American blacks out of every 1000, is caused by a recessive allele. Using the Hardy-Weinberg equation, you can calculate that the frequency with which the sickle cell allele occurs in the American black population; this frequency is the square root of 0.002, or approximately 0.045. In contrast, the frequency of the allele among American whites is only about 0.001.

Sickle cell anemia is usually fatal. The disease occurs because of a single amino acid change, repeated in the two beta chains of the hemoglobin molecule. In this change a valine is substituted for the usual glutamic acid at a location on the surface of the protein near the oxygen-binding site. Unlike glutamic acid, valine is nonpolar (water-hating), and its presence on the surface of the molecule creates a "sticky" patch that will attempt to escape from the polar water environment by binding to another similar patch. As long as oxygen is bound to the hemoglobin molecule, there is no problem, since the oxygen atoms shield the critical area of the surface. When oxygen levels fall, however, such as after exercise or at high altitudes, then oxygen is not so readily bound to hemoglobin, and the exposed sticky patch binds to similar patches on other molecules, eventually producing long fibrous clumps. The result is a deformed, "sickle-shaped" red blood cell.

Individuals who are heterozygous for the valine-specifying allele (designated allele S) are said to possess sickle cell trait. They produce some sickle-shaped red blood cells, but only 2% of the level seen in homozygous individuals.

The average incidence of the S allele in West Africa is about 0.12, a far higher value than that found among American blacks. From the Hardy-Weinberg principle you can calculate that one in five Central African individuals is heterozygous at the S allele, and one in a hundred develops the fatal form of the disorder. People who are homozygous for the sickle cell allele almost never reproduce, because they usually die before they reach reproductive age. Why is the S allele not eliminated from Central Africa by selection, rather than being maintained at such high levels? Because people who are heterozygous for the sickle cell allele are much less susceptible to malaria, which is one of the leading causes of illness and death, especially among young children, in the areas where the allele is common. In addition, for reasons that are not understood, women who are heterozygous for this allele are more fertile than are those who lack it. Consequently, even though most people who are homozygous recessive die before they have children, the sickle cell allele is maintained at high levels in these populations because of its role in resistance to malaria in heterozygotes and its association with increased fertility in female heterozygotes.

As Darwin's theory predicts, it is the environment that acts to maintain the sickle cell allele at high frequency. In this case the characteristic of the environment of Central Africa that is exercising selection is the presence of malaria. For the people living in areas where malaria is frequent, maintaining a certain level of the sickle cell allele in the populations has adaptive value (see Figure 12-11). Among American blacks, many of whom have lived for some 15 generations in a country where malaria has been relatively rare in most areas and is now essentially absent, the environment does not place a premium upon resistance to malaria, so there is no adaptive value to counterbalance the ill effects of the disease. In this nonmalaria environment, selection is acting to eliminate the S allele.

Peppered Moths and Industrial Melanism

The peppered moth, *Biston betularia,* is a European moth that rests on the trunks of trees during the day (see Figure 16-2). Until the mid-nineteenth century, almost every individual of this species that was captured had light-colored wings. From that time on, individuals with dark-colored wings increased in frequency in the moth populations near industrialized centers until they made up almost 100% of these populations. The black individuals have a dominant allele that was present in populations before 1850, although very rare then. Biologists soon noticed that in industrialized regions where the dark moths were common, the tree trunks were darkened, almost black by the soot of pollution, and the dark moths were much less conspicuous resting on them than were the light moths. In addition, the air pollution that was spreading in the industrialized regions had killed many of the light-colored lichens that had occurred on the trunks of the trees earlier, thus making these trunks even darker than they would have been otherwise.

Can Darwin's theory explain the increase in frequency of the dark allele? Why was it an advantage for the dark moths to be less conspicuous? Although initially there was no evidence, the ecologist H.B.D. Kettlewell hypothesized that the moths were eaten by birds while they were resting on the trunks of the trees during the day. He tested the hypothesis by rearing populations of peppered moths in which dark and light individuals were evenly mixed. Kettlewell then released these populations into two sets of woods: one, near Birmingham, was heavily polluted; the other, in Dorset, was unpolluted. Kettlewell set up rings of traps around the woods to see how many of both kinds of moths survived. To be able to evaluate his results, he had marked the moths that he had released with a dot of paint on the underside of their wings, where it could not be seen by the birds.

In the polluted area near Birmingham, Kettlewell trapped 19% of the light moths, but 40% of the dark ones. This indicated that the dark moths had had a far better chance of surviving in these polluted woods, where the tree trunks were dark. In the relatively unpolluted Dorset woods, Kettlewell recovered 12.5% of the light moths but only 6% of the dark ones. These results indicated that where the trunks of the trees were still light-colored, the light moths had a much better chance of survival than the dark ones. He later solidified his argument by placing hidden blinds in the woods and actually filming birds eating the moths. The birds that Kettlewell observed sometimes actually passed right over or next to a moth that was of the "correct" color and thus well-concealed.

Industrial melanism is a phrase used to describe the evolutionary process in which initially light-colored organisms become dark as a result of natural selection. The process, which is common among moths that rest on tree trunks, takes place because the dark organisms are better concealed from their predators in habitats that have been darkened by soot and other forms of industrial pollution.

Dozens of other species of moths have changed in the same way as the peppered moth in industrialized areas throughout Eurasia and North America, with dark forms becoming more common from the mid-nineteenth century onward as industrialization spread. In the second half of the twentieth century, with the widespread implementation of pollution controls, the trends are being reversed, not only for the peppered moth in many areas in England, but also for many other species of moths throughout the northern continents. Such examples provide some of the best documented instances of changes in allelic frequencies of natural populations in response to selective forces.

Multiple Drug Resistance

Following the widespread introduction of antibiotics after World War II, individuals began to come down with bacterial illnesses that were difficult to treat. The first problems arose when bacteria such as *Neisseria,* which causes gonorrhea, developed strains that were resistant to the drugs commonly used to treat them. Soon the problem was compounded by the appearance of strains of bacteria that were simultaneously resistant to a wide variety of different antibiotics. As you saw in Chapter 15, this was because alleles conferring resistance to various antibiotics were concentrated on their plasmids and transmitted in this way. The multiple resistance arose because several different kinds of antibiotics were routinely administered simultaneously, and natural selection

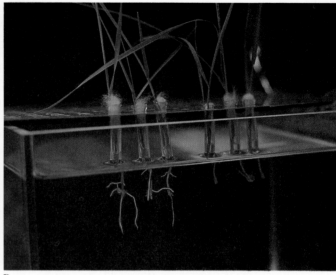

A

B

FIGURE 16-7

A Mine at Drws-y-Coed, Wales, showing soil contaminated by lead-rich tailings.
B Bent grass, *Agrostis tenuis,* tolerant and nontolerant strains growing in 0.5 millimolar copper, one of the metals that contaminate the mine tailings.

was swift in the new environment that was created in this way. As a result of the lesson we learned in this way, we now administer antibiotics only when necessary, and one at a time.

Lead Tolerance

A.D. Bradshaw and other investigators have studied the grasses that grow on the tailings, or refuse, around lead mines in Wales (Figure 16-7). These tailings, areas in which the soil is rich in lead, are almost bare of plants, but a few kinds, including bent grass, *Agrostis tenuis,* was found there. Bradshaw compared the growth patterns of plants of *Agrostis* taken from nearby pastures and areas where lead is not abundant in the soil with those of plants from the mine tailings. He grew plants taken from the two different kinds of soils side-by-side in samples of both soil types.

In normal pasture soil, the plants from the mine soil were smaller and grew more slowly than the ordinary pasture plants, but they did survive. In the altered environment of lead-rich mine soil, the plants that Bradshaw had collected there grew well. In complete contrast, the pasture plants were, with a few exceptions, unable to grow in the mine soil, most of them dying within a few months. The exceptions, however, were significant: in one sample of 60 plants from a pasture, 3 showed some ability to grow in the lead-rich soil. Such plants were undoubtedly the kind from which those that could grow in the mine soil had been selected originally. In this way a race of bent grass that was able to grow well in lead-rich mine soil evolved, the altered environment selecting in favor of individuals tolerant of high levels of lead.

Since plants able to grow on lead-rich soil are found in association with mines less than a century old, the populations of *Agrostis tenuis* clearly are able to evolve—change their genotypic frequencies—quickly when the environment demands it. Similarly rapid changes have now been documented for other populations of organisms; they tell us a great deal about the process of evolution.

An Overview of Adaptation

These four case histories, of sickle cell anemia, industrial melanism, multiple drug resistance, and lead tolerance, are among the best-documented cases of adaptation and provide clear evidence that microevolutionary changes can be produced by natural selection. They are typical of many other situations, all of which share the same fundamental characteristic: changes occur in the frequencies of alleles in populations, and these changes alter the characteristics of the population to make it better adapted to the environment in which it is living. In every case, it is the *environment* that dictates the direction and extent of the change. Just as Darwin's theory demands, it is the nature of the environment that leads to natural selection and so determines the direction of evolutionary change.

THE EVIDENCE OF MACROEVOLUTION

Adaptation within natural populations such as those we have just described constitutes strong evidence that Darwin was right in arguing that selection could bring about genetic change within populations. The examples we reviewed offer direct and compelling evidence of microevolutionary change due to selection. Because we can see the evolution as it occurs, we know that Darwin was correct. What of macroevolution, however? What is the evidence that macroevolution has led to the diversity of life on earth?

In Chapter 2 we outlined the evidence that Darwin presented in favor of the hypothesis that macroevolution had occurred, a view that had become generally accepted by the time he carried out his studies, but for which he was the first to provide an explanation of the mechanism. It is summarized in Table 2-1. A great deal of additional evidence has accumulated since then (Figure 16-8), much of it far stronger than that available to Darwin and his contemporaries. Among the many lines of available evidence, we shall review seven.

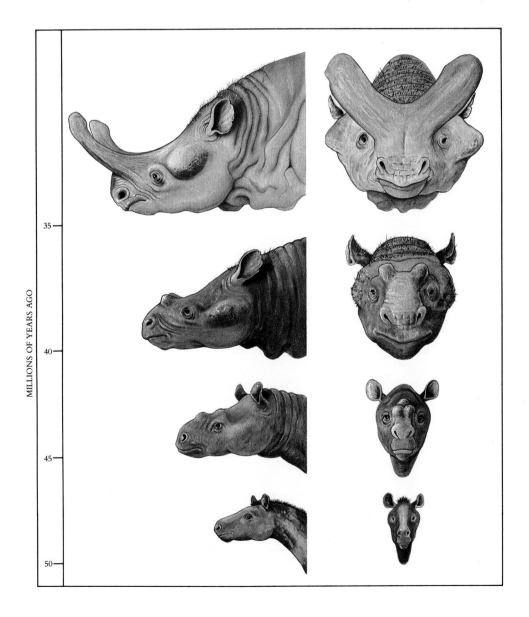

FIGURE 16-8

Evolution in a group of hoofed mammals known as titanotheres between the Early Eocene Epoch (about 50 million years ago) and the Early Oligocene Epoch (about 36 million years ago). During this period of time, the small, bony protuberances that began to appear by the Middle Eocene (about 45 million years ago) evolved into relatively large, blunt horns.

MILLIONS OF YEARS AGO

In considering the evolution of major groups of organisms, we sometimes find that the differences among them are so great that it is difficult to trace the lines of descent. Especially complex features, such as the eye, appear so unlike any possible earlier structure from which they may have been derived that the question of their origin greatly puzzled early students of evolution.

When we consider eyes in more detail, however, we find that they have in fact evolved many times, as shown by the fundamental differences between the eyes found in different groups of organisms. Some eyes consist of a single photoreceptor cell, and others are complex, like those of the vertebrates, with focusing lenses and color sensitivity. Zoologists, analyzing these differences, have calculated that eyes have evolved independently, and through

intermediate stages, at least 38 different times in various groups of animals (Figure 16-A).

In other words, not all of the structures we call eyes are homologous with one another. By analyzing and understanding their similarities and differences, biologists can more precisely interpret the relationships among the group in which they occur and more accurately understand the outlines of their evolution. Thus the difficult problem of how eyes evolved, when it is solved, assists us in interpreting broad evolutionary patterns of great interest.

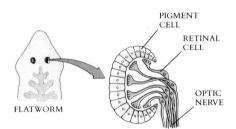

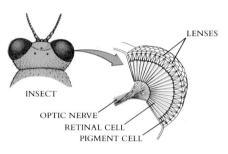

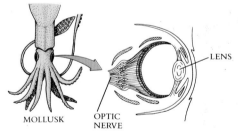

FIGURE 16-A

Although they are superficially similar, the eyes that occur in four different phyla of animals differ greatly in structure and are not homologous with one another. Each has evolved separately and, despite the apparent structural complexity, has done so from simpler structures.

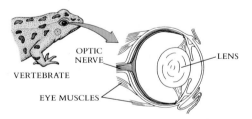

The Fossil Record

The most direct evidence of macroevolution is to be found in the fossil record; we now have a far more complete understanding of this record than was available in Darwin's time. Fossils are created when organisms become buried in sediment, the calcium in bone and other hard tissue is mineralized, and the sediment eventually is converted to rock. Sedimentary layers of rock reveal a history of life on earth in the fossils they contain.

By dating the rocks in which fossils occur, we can get a very accurate idea of how old the fossils are. In Darwin's day rocks were dated by their position with respect to one another; rocks in deeper strata are generally older. Knowing the relative position of sedimentary rocks and the rates of erosion of different kinds of sedimentary rocks in different environments, geologists of the nineteenth century had derived a fairly accurate idea of the relative ages of rocks.

More recently, much more accurate ways of dating rocks have been derived, and these provide dates that are absolute, rather than relative. Nowadays, rocks are dated by measuring the degree to which certain radioisotopes that they contain have decayed; the older the rock, the more its isotopes have decayed. Because radioactive isotopes decay at a constant rate that is not altered by temperature or pressure, the isotopes in a rock act as an internal clock, measuring the time since the rock was formed. For fossils less than 30,000 years old, the decay of carbon-14 (with a half-life—the time it takes for half of a sample of given size to change—of 5568 years) is used; for older fossils, investigators examine the decay of radioactive potassium-40 into argon and calcium (half-life of 1.3 billion years). For very old fossils, a third isotope measure, the decay of uranium-238 into lead (half-life 4.5 billion years) is used. An investigator has only to measure the proportion of uranium-238 to lead in the rock in order to estimate the rock's age.

When fossils are arrayed according to their age, from oldest to youngest, they provide evidence of progressive evolutionary change in the direction of greater complexity. Among the hoofed mammals illustrated in Figure 16-8, for example, small bony bumps on the nose can be seen to change progressively until they become large blunt horns. In the evolution of horses (see Figure 2-11), the number of toes on the front foot is gradually reduced from four to one. About 200 million years ago, oysters underwent a change from small curved shells to larger, flatter ones, progressively flatter fossils being seen in the fossil record over a period of 12 million years (Figure 16-9). A host of other examples are known, all illustrating a record of progressive change. The demonstration of this progressive change is one of the strongest lines of evidence that evolution has occurred.

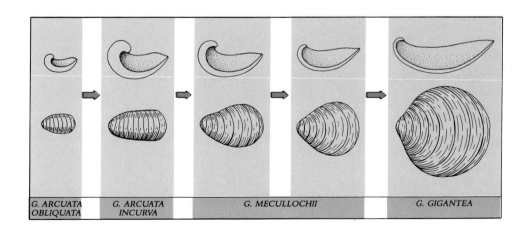

G. ARCUATA OBLIQUATA   G. ARCUATA INCURVA   G. MECULLOCHII   G. GIGANTEA

FIGURE 16-9

The apparently gradual evolution of a group of coiled oysters during a portion of the Early Jurassic Period that lasted about 12 million years. During this interval, the shell became larger, thinner, and flatter. These animals rested on the ocean floor, and it may be that the larger, flatter shells were more stable against potentially disruptive water movements.

FIGURE 16-10

This graph shows the estimated number of nucleotide substitutions in the gene for cytochrome *c* plotted against the estimated time since the divergence of various pairs of different organisms. The fact that the relationship can be plotted as a straight line suggests that there has been an approximately constant rate of substitution of nucleotides at the different positions in this gene per million years.

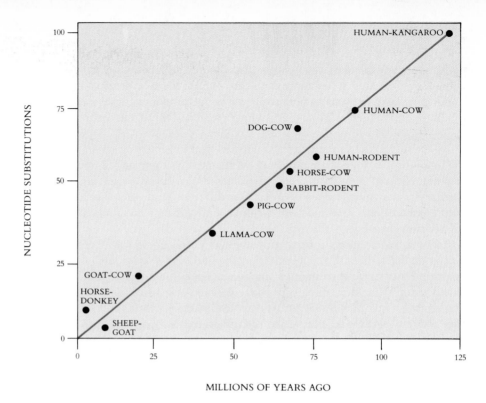

The Molecular Record

If you think about it, Darwin's theory implies that a record of evolutionary changes is present in the cells of each of us, in our DNA. According to evolutionary theory, every evolutionary change involves the substitution of new versions of genes for old ones, the new arising from the old by mutation and coming to predominance because of favorable selection. Thus a series of evolutionary changes involves a progressive accumulation of genetic change in the DNA. Organisms that are more distantly related will accumulate a greater number of evolutionary differences. This is indeed what is seen when DNA sequences are compared between various organisms: for example, the longer the time since the organisms diverged as judged by the age of their fossils, the greater the number of differences in the nucleotide sequence of the gene for cytochrome *c,* a protein you may recall from Chapter 8 that plays a key role in oxidative metabolism (Figure 16-10). The same regular pattern of change is seen in hemoglobins and many other proteins. Again we see that evolutionary history involves a pattern of progressive change.

Some genes, such as the ones specifying the protein hemoglobin, have been particularly well studied, and the entire time course of their evolution can be laid out with confidence by tracing the origin of particular substitutions in their nucleotide sequences (Figure 16-11). The pattern of descent that is obtained is called a **phylogenetic tree.** It represents the evolutionary history of the gene. You should note that the progressive changes seen in the hemoglobin molecule produce a tree that reflects precisely the evolutionary relationships predicted by a study of anatomy. Whales, dolphins, and porpoises cluster together, as do the primates and the hooved animals. The pattern of progressive change seen in the molecular record constitutes very strong direct evidence for macroevolution.

Homology

A third demonstration of the process of macroevolution lies in the fact that many organisms exhibit structures that appear to have been derived from a common ancestral form. The forelimbs of all mammals, for example, contain the same pattern of bones, although they now carry out a variety of different functions (see Figure 2-12). All ver-

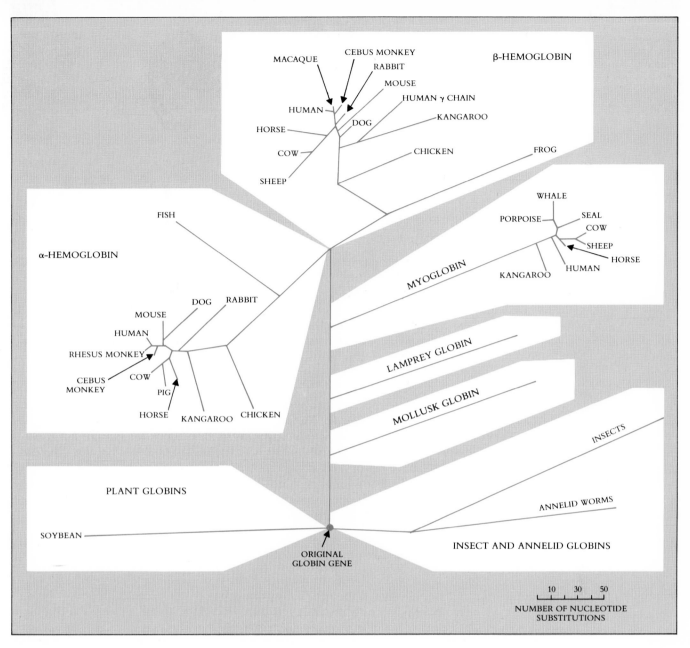

FIGURE 16-11

The evolution of a number of kinds of organisms as reconstructed from substitutions in the globin gene. The length of the various lines is relative to the number of nucleotide substitutions.

tebrates have the same pattern of bones, muscles, nerves, blood circulation, and organs, the pattern becoming gradually more complex as one moves from the fishes to amphibians to reptiles to mammals. It is difficult to avoid the conclusion that progressive change is taking place.

Development

In many cases the evolutionary history of an organism can be seen to unfold during its development, with the embryo exhibiting characteristics of the embryos of its ancestors. For example, early in their development human babies possess gill slits like those of a fish, and later exhibit a tail, the vestige of which we carry to adulthood as the coccyx at the end of our spine. Human embryos even possess a fine fur (called lanugo) during the fifth month of development. These relict developmental forms argue strongly that our development has evolved, with new instructions being layered on top of old ones and the overall developmental program getting progressively longer. Some vertebrate embryos are shown in Figure 16-12.

FIGURE 16-12

The embryos of various groups of vertebrate animals, showing the primitive features that all share early in their development, such as gills and a tail.

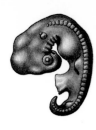
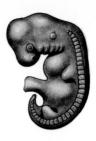

FISH   REPTILE   BIRD   HUMAN

Vestigial Structures

Many organisms possess structures with no apparent function that resemble the structures of presumed ancestors. You, for example, possess a complete set of muscles for wiggling your ears, just as a coyote does. Figure 16-13, *A* illustrates the skeleton of a baleen whale, a representative of the group that contains the largest living mammals. It contains a pelvic bone just as other mammals do, even though such a bone serves no function in the whale. Another example is the human vermiform appendix, a hollow, wormlike appendage of the cecum, the sac in which the large intestine begins. The vermiform appendix apparently is vestigial and represents the degenerate terminal part of the cecum. Although some suggestions have been made, it is difficult to assign any current function to the vermiform appendix. In many respects it is a dangerous organ: quite often it becomes infected, leading to an inflammation called **appendicitis.** If it is not removed surgically when inflamed, the vermiform appendix may burst, allowing the contents of the gut to come in contact with the lining of the body cavity. This condition can be fatal if unchecked. It is difficult to understand vestigial structures such as these in any way other than as evolutionary relics, holdovers from the evolutionary past. They argue strongly for the common ancestry of the members of the groups that share them, regardless of how different they have become over time.

Parallel Adaptation

Different geographical areas sometimes exhibit plant and animal communities of similar appearance, even though the individual plants and animals that make up these communities may be only distantly related to one other. It is difficult to explain these similarities as resulting from coincidence. In the best-known case, the continent of Australia separated from the other continents more than 50 million years ago, before

FIGURE 16-13

A vestigial feature in the skeleton of a baleen whale, a representative of the group of mammals that contains the largest living species. The enlargement shows the pelvic bones, which resemble those of other mammals, but are only weakly developed in the whale and have no apparent function.

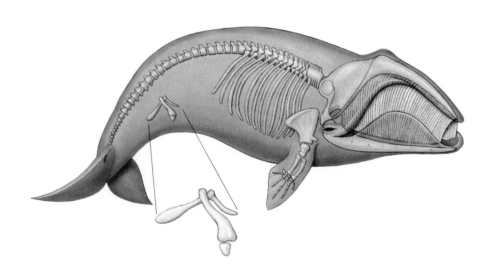

PLACENTAL MAMMALS   **MARSUPIAL MAMMALS**

FIGURE 16-14

The parallel adaptation of marsupials in Australia and placental mammals in the rest of the world.

MOLE

MARSUPIAL MOLE

ANTEATER

NUMBAT (ANTEATER)

MOUSE

MARSUPIAL MOUSE

LEMUR

SPOTTED CUSCUS

FLYING SQUIRREL

FLYING PHALANGER

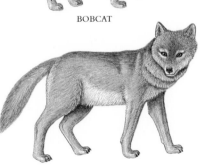

BOBCAT

TASMANIAN "TIGER CAT"

WOLF

TASMANIAN WOLF

placental mammals—the group that dominates throughout most of the world—are thought to have arrived in the area. Today only a few recently introduced mammals in Australia are placental. The bulk of Australian mammals are marsupials, members of a group in which the young are born in a very immature condition and held in a pouch until they are ready to emerge into the outside world. Marsupials are thought to have evolved earlier than placental mammals, and they probably arrived in Australia before its separation from Antarctica. What are the Australian marsupials like? To an astonishing degree, they resemble the placental mammals that are present on the other continents (Figure 16-14). The similarity of some of the individual members of these two sets of mammals, in which specific kinds have similar habits and find their food in similar ways, argues strongly that they have evolved similarly in different, isolated areas as a result of similar selective forces.

Patterns of Distribution

Darwin was the first to present evidence that the animals and plants living on oceanic islands resemble most closely the forms of the nearest continent—a relationship that would not make sense if they were all specially created. This kind of relationship, which has been observed many times since Darwin with the increasing exploration of the earth's surface, strongly suggests that the island forms evolved from individuals that came to the islands from the adjacent mainland at some time in the past. In many cases the island forms are *not* identical to those that still occur on the nearby continents. The Galapagos finch of Figure 2-6 has a very different beak than its South American relative, for example. In the absence of evolution, there seems to be no logical explanation of why individual kinds of plants and animals were clearly related to, but have diverged in their features from, other kinds of plants and animals that occur on the adjacent mainlands. As Darwin pointed out, this relationship provides strong evidence that macroevolution has occurred.

TABLE 16-3   EXAMPLES OF THE EVIDENCE FOR EVOLUTION

The fossil record	When fossils are arrayed in the order of their age, a progressive series of changes is seen
The molecular record	The longer organisms have been separated according to the fossil record, the more differences are seen in their DNAs
Homology	All vertebrates contain a similar pattern of organs, suggesting that they are related to one another
Development	During development, humans exhibit characteristics of other vertebrates, which suggests that humans are related to the other forms
Vestigial structures	Many vertebrates contain structures that have no function but that resemble functional structures of other vertebrates; this suggests that the structures are inherited from a common ancestor
Parallel adaptation	The marsupials in Australia closely resemble the placental mammals of the rest of the world, which suggests that parallel selection has occurred
Patterns of distribution	Inhabitants of ocean islands resemble forms of the nearest mainland but show some differences, which suggests that they have evolved from mainland migrants

In sum, the evidence for macroevolution is overwhelming (Table 16-3). Almost all biologists would agree both that (1) macroevolutionary changes have occurred, and (2) microevolutionary changes result from natural selection. In the next chapter we shall consider Darwin's proposal that microevolutionary changes have led directly to macroevolutionary ones, the key argument in his theory that evolution occurs by natural selection.

SUMMARY

1. Macroevolution describes the grand outlines of evolution, which is based on the evolution of species. Microevolution, also called adaptation, refers to the evolutionary process itself. Adaptation leads to species formation and thus ultimately to macroevolution.

2. The Hardy-Weinberg model illustrates the fact that in large populations with random mating, allele and genotype frequencies—and consequently phenotype frequencies—will remain constant indefinitely, provided that selection, net mutation or migration in one direction, inbreeding, and genetic drift do not occur.

3. Fitness is the tendency of some organisms to leave more offspring than do competing ones. The genetic traits possessed by the fit individuals will appear in greater proportions among members of succeeding generations. This process is called selection.

4. Alleles at enzyme-specifying loci are abundant in natural populations and provide clear examples of simple genetic differences. Population geneticists are interested in learning whether the alleles are maintained by selection or are adaptively neutral and simply maintained by the operation of the genetic systems in the populations where they occur.

5. There is clear evidence of microevolutionary change in natural populations. For example, the disease called sickle cell anemia occurs when an altered hemoglobin molecule, associated with a particular allele, is well represented in the red blood cells. If this allele is present in homozygous form, it is almost invariably lethal; if it is present in heterozygous form, it not only does not produce anemia, but it also confers resistance to malaria and increases fertility of heterozygous females. For these reasons the allele has reached high frequencies in certain African populations in areas where malaria occurs frequently.

6. The peppered moth *Biston betularia* and the grass *Agrostis tenuis* provide other examples of microevolutionary changes in natural populations. Changes in both kinds of organisms have resulted from adaptation to conditions associated with human activities.

7. Two direct lines of evidence argue that macroevolution has occurred: (1) the fossil record, which exhibits a record of progressive change correlated with age, and (2) the molecular record, which exhibits a record of accumulated changes, the amount of change correlated with age as determined in the fossil record.

8. Several indirect lines of evidence argue that macroevolution has occurred, including progressive changes in homologous structures, relict and vestigial developmental structures, parallel patterns of evolution, and patterns of biogeographical distribution.

REVIEW

1. Genetic change within populations, which is the result of adaptation, is called _____.

2. Small, randomly mating populations of a species that are isolated from one other can come to differ greatly in their characteristics. In the absence of net migration or mutation, this may occur as a result of _____ or of _____.

3. Which of the five mechanisms for producing deviations from Hardy-Weinberg allele frequencies is the most important in evolution?

4. _____ _____ is a term used to describe the evolutionary process in which a population of predominantly light-colored individuals is gradually dominated by dark-colored individuals as a result of natural selection. The best-studied example of this is the peppered moth, *Biston betularia*.

5. What kind of evidence for evolution is illustrated by the striking similarity of the fossil marsupial sabertooth and the placental sabertooth?

SELF-QUIZ

1. Approximately what percentage of your loci are heterozygous?
   (a) 1%   (d) 15%
   (b) 5%   (e) 25%
   (c) 10%

2. Which of the following is *not* a factor that causes change in the proportions of homozygous and heterozygous individuals in a population?
   (a) mutation
   (b) migration
   (c) genetic drift
   (d) random mating
   (e) selection against a specific allele

3. If you came across a population of plants and discovered a surprisingly high level of homozygosity, what would you predict about their mating system?
   (a) their pollen is dispersed by wind
   (b) their pollen is spread by insects that visit their flowers
   (c) they probably reproduce asexually
   (d) they are predominantly outcrossing
   (e) they are predominantly self-fertilizing

4. The sickle cell (S) allele has a relatively high frequency (0.12) in West Africa, even though individuals homozygous for this allele usually die before they reach reproductive age. Why has this allele persisted in the population in high frequencies when there appears to be such strong natural selection against it?
   (a) because heterozygous individuals are resistant to malaria
   (b) because individuals homozygous for this characteristic are resistant to malaria
   (c) because females heterozygous for this allele are more fertile than those who lack it
   (d) because females homozygous for this allele are more fertile than those who lack it
   (e) both a and c

5. Successful adaptation is defined simply as
   (a) an increase in fitness
   (b) moving to a new location
   (c) producing offspring
   (d) evolving new traits
   (e) living longer

PROBLEM

1. In a large, randomly mating population with no forces acting to change gene frequencies, the frequency of homozygous recessive individuals for the character extra-long eyelashes is 90 per 1000, or 0.09. What percentage of the population carries this very desirable trait but displays the dominant phenotype, short eyelashes? Would the frequency of the extra-long-lash allele increase, decrease, or remain the same if long-lashed individuals preferentially mated with each other and no one else?

THOUGHT QUESTIONS

1. The North American human population is similar to the ideal Hardy-Weinberg population in that it is very large (more than 270 million people in the United States and Canada alone) and generally randomly mating. Although mutation occurs, it does not lead by itself to great changes in allele frequencies. Migration from Latin American and Asian countries occurs at relatively high levels—perhaps 1% per year—however. The following data were obtained in 1976 by the geneticist A. E. Mourant about relative numbers of individuals bearing the two alleles of what is known as the MN blood group:

	MM	**MN**	**NN**	**Total**
Observed number of individuals	1787	3037	1305	6129

   Do these data suggest that migration or some other factor is acting to perturb the Hardy-Weinberg proportions of the three genotypes?

2. In Central Africa in addition to the A and S alleles discussed in this chapter, there exists in low frequency a third hemoglobin allele called C. Individuals heterozygous for C and the normal allele A are susceptible to malaria just as AA homozygotes are, but CC individuals are resistant to malaria—and do *not* develop anemia. Assuming that the Bantu people entered Central Africa relatively recently from a land where malaria is not common (we think this is what happened), and that among the original settlers both C and S alleles were rare, can you suggest a reason that CC individuals have not become predominant?

3. Will a dominant allele that is lethal be removed from a large population as a result of natural selection? What factors might prevent this from happening?

FOR FURTHER READING

GOULD, S.J.: "The Panda's Thumb: More Reflections in Natural History," W. W. Norton & Company, New York, 1980. A collection of well-written and witty essays about evolution from *The Journal of Natural History*, which can be consulted for many others.

SIBLEY, C., and J. ALQUIST: "Reconstructing Bird Phylogenies by Comparing DNAs," *Scientific American*, February 1986, pages 82-92. A good introduction to the way in which biologists are beginning to use the tools of molecular biology to answer questions about evolution.

SIMPSON, G.G.: "Fossils and the History of Life," *Scientific American Library*, New York, 1983. A short and beautifully illustrated account of how fossils are used to learn about life's evolutionary past, by a master of the field.

FIGURE 17-1   Along the course followed by this mountain stream in the Sierra Nevada of California, you can see a diverse array of environments, from the dry, windswept rock of the mountain face to the boggy meadows bordering the stream in the foreground. As populations adapt to each of these different environments, they come to diverge more and more from one another. Eventually the populations may become so different that they can no longer interbreed, forming new species.

HOW SPECIES FORM

Chapter 17

Overview

To establish that Darwin's theory of evolution is correct, it is not enough to show that evolution has occurred, that the dinosaurs' ancestors were fishes, and that our ancestors were apelike primates. To be sure, it is necessary to *see* evolution in action. Although the replacement of one kind of organism with another usually occurs too slowly to observe in one human lifetime, less drastic changes within individual species can be observed. Many of these changes, when examined by biologists, turn out to reflect the selection of better-adapted kinds of organisms, just as Darwin proposed. The key issue, then, is the evidence that such changes *within* species lead to differences *between* species, to the formation of *new* species.

For Review

Here are some important terms and concepts that you will encounter in this chapter. If you are not familiar with them, you should review them before proceeding.

Darwin's studies (Chapters 2 and 16)

Meiosis (Chapter 10)

Selection (Chapter 16)

Adaptation in populations (Chapter 16)

If he were alive today, Darwin would be amazed at the quantity of the evidence that has accumulated to support his theory that natural selection is the primary mechanism responsible for evolution. He would not have needed convincing; the case he built in *Origin of Species* is a very strong one. The case for evolution now rests solidly on three pillars of evidence, one establishing that natural selection is the agent of microevolutionary change, a second that microevolutionary change leads to macroevolutionary change, and a third that macroevolution has indeed occurred.

In the preceding chapter, we considered the evidence that macroevolution has occurred, and that selection is the agent of microevolutionary change, that is, change within populations. We saw that natural selection drives evolution by favoring genetic variations that better enable organisms to survive and reproduce. Different environments demand different responses from the organisms that survive in them, and there is a continuing process of change in relation to these environments (Figure 17-1). This is the key lesson of the previous chapter: evolution is not blind; instead, it is directed by the environment. Just as a football coach has his team try a variety of plays but keeps in the team's game plan only those plays that work, so a population of organisms keeps only those changes that "work." The population doesn't decide which changes to keep, any more than the football coach does. The pattern of success determines the outcome.

In this chapter we shall consider the third body of evidence on which Darwin's theory rests: the evidence that microevolutionary change leads to macroevolutionary change, that adaptive changes *within* a species convert to differences *between* species. This third point is the crux of Darwin's evolutionary argument.

THE NATURE OF SPECIES

How do adaptive changes in populations lead to the origin of new species? Darwin was extremely interested in this question because he considered species to be the fundamental units of evolution. As we consider this question, it helps to start by carefully examining what the concept "species" means and how this concept has changed through the years.

John Ray, an English clergyman and scientist, was one of the first to propose a definition of the category "species." In about 1700 he pointed out how a species could be recognized: all the individuals that belonged to it could breed with one another and produce progeny that were still members of that species. Even if two different-looking individuals appeared among the progeny of a single mating, they were still considered to belong to the same species. All dogs were one species, because all dogs, no matter how different they appear, can breed with one another and produce offspring (Figure 17-2). Similarly, all cabbages, including such diverse forms as cauliflower, rutabaga, and brussels sprouts, are interfertile members of the same species; their characteristics were obtained by artificial selection. Dogs, however, are not the same species as cats. Dogs and cats cannot mate and produce little cogs and dats. Similarly, trout are not the same species as goldfish, nor are monkeys the same species as humans, and so forth.

Informally, people had always recognized species. Indeed, the word "species" is simply Latin for "kind." However, with Ray the species began to be regarded as an important biological unit that could be catalogued and understood. With other scientists of his time, Ray believed that species did not change. This view held, for the most part, until it was shattered by Darwin in 1859. Darwin considered the fact that species existed to be important evidence in support of his theory of evolution. Darwin explained the relative constancy of species by saying that each had its own distinctive role in nature, a role that we in modern terms call a **niche.**

As Mendelian genetics became widely accepted early in this century, there was a growing desire to define *species* more precisely. The definition that began to emerge in the 1930s was stated by the American evolutionist Ernst Mayr as follows: species were "groups of actually or potentially interbreeding natural populations which are reproductively isolated from other such groups." In other words, hybrids between species occur rarely in nature, but individuals that belong to the same species are able to interbreed freely. This early definition has proved to be too simple. In fact, there are essentially no barriers to hybridization between the species in many groups of organisms. For example, among trees, some groups of mammals, and fishes generally, separate species of particular groups, such as oaks, *are* able to form fertile hybrids with one another, even though they may not do so in nature, or may do so only rarely, or under particular circumstances.

FIGURE 17-2

All of the different breeds of dogs are members of the same species, *Canis familiaris,* and all can breed successfully with one another. The great differences between breeds of dogs were produced by artificial selection carried out by dog breeders.

FIGURE 17-3

The mule, a generally sterile hybrid between a female horse (mare) and a male donkey. By any standard, the horse and donkey are distinct species; mules, of course, cannot interbreed with either of them. The combination of the characteristics of horses and donkeys in the mule makes it a highly desirable work animal. Mules are always obtained by repeating the same interspecific cross.

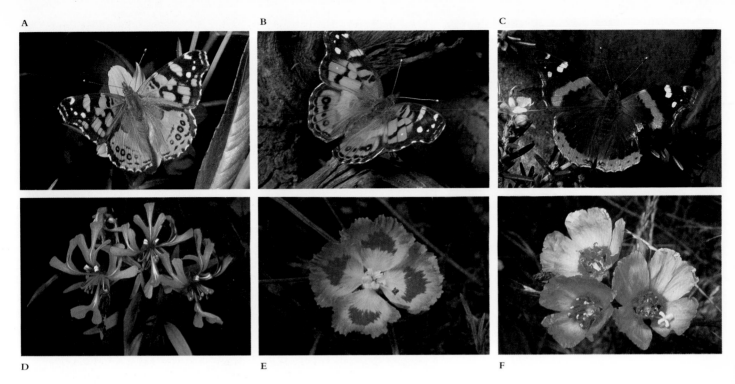

In other groups of organisms, hybrids do not form between species of a particular **genus** (plural **genera,** a group of related species), or such hybrids are generally sterile (Figure 17-3). In some groups of organisms, even local populations are sometimes not capable of interbreeding with one another. This pattern occurs, for example, in many groups of annual plants. For still other kinds of plants and animals, we do not know whether the species are capable of forming hybrids or not.

In practice, scientists usually recognize species primarily because they differ from one another in their features (Figure 17-4), not by the presence of breeding barriers. For these reasons modern biologists define a **species** as a group of organisms that is unlike other such groups of organisms and does not normally interbreed extensively with those other groups in nature, although they may do so under artificial conditions.

> **Species are groups of organisms that differ in one or more characteristics and do not interbreed extensively in nature even if they occur together.**

THE DIVERGENCE OF POPULATIONS

The path from microevolutionary change within a species to the divergence of that species into two separate species involves the division of a group of similar individuals into two groups that usually are less similar. The new species usually do not interbreed with one another, either because they are incapable of doing so or because the nature of their occurrence prevents them from doing so. We refer to the separation of a species into dissimilar groups as **divergence,** and to the mechanisms that prevent successful mating as **reproductive isolating mechanisms.** We shall consider divergence first.

Every species is composed of local populations, groups of individuals that live together in a particular place and are more likely to meet each other than members of other populations. The size of an **effective breeding population,** those members of a local population that actually interbreed with one another, is often extremely small (Figure 17-5), with little exchange of individuals, and thus of genetic material, between different local populations. Because of this, local populations tend to adjust individually to the demands of their particular environment. As they do so, their characteristics change.

FIGURE 17-4

Distinct species in one genus of animals and one of plants. Butterflies of the genus Vanessa.
A American painted lady, *Vanessa virginiensis;*
B Western painted lady, *Vanessa annabela;*
C Red admiral, *Vanessa atlanta.* Annual plants of the genus *Clarkia,* all from California.
D *Clarkia concinna;*
E *Clarkia speciosa;*
F *Clarkia rubicunda.* This genus, which has about 45 species in and near California, and one in areas of Chile and Argentina with a similar climate, was discovered on the Lewis and Clark Expedition and named in honor of George Rogers Clark.

A

B

FIGURE 17-5

Jasper Ridge is a biological preserve in the foothills above Stanford University, just south of San Francisco, California. The checkerspot butterfly *Euphydryas editha* **(A)** occurs here, seemingly as one continuous population throughout the open grassland **(B),** which occurs as an "island" surrounded by oak forest and chaparral. Extensive studies by Paul Ehrlich and his colleagues over many years have shown that the butterflies actually do not fly around much, but exist as a series of discontinuous populations. These local populations, separated by dashed lines in the maps **(C),** are free to respond to the selective forces that are characteristic of their particular parts of the ridge. Changes in these populations, which have remained essentially constant in overall distribution for more than 25 years, are shown in the maps. Illustrated are the locations of first capture of individual butterflies in 1961, 1962, and 1963.

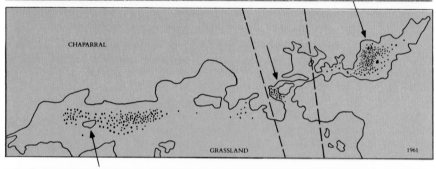

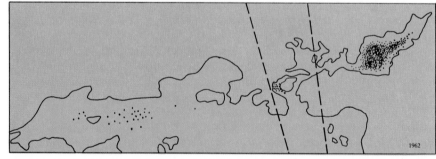

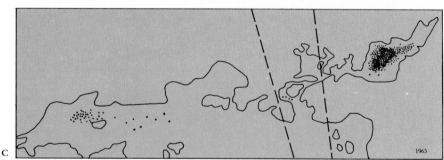

C

The rates at which local populations change depend primarily on the strength of the selective forces that their local environments impose. If these forces are strong, the local populations will change rapidly. Because the environments of local populations can be very different, selection can cause local populations to adapt quite differently. Microevolution, the adaptive molding of a population, thus leads directly to divergence between local populations. If the divergence proceeds far enough, the result is the formation of separate species from what began as subpopulations of the same species. Thus microevolution, if it proceeds far enough, produces macroevolution.

The characteristics of local populations tend to diverge if their local environments differ. Since there is often only a limited exchange of genetic material between local populations, even if they are geographically close, all populations tend to become increasingly divergent from one another in their characteristics over time.

Divergence itself can produce new species, groups of individuals that are markedly different from one another. The new species may become so carefully attuned to their local environments that they can no longer survive in other places and so do not come into contact with other species. Many tree species are like this, as are almost all fish species. In these situations, the formation of species does not involve the creation of reproductive isolating mechanisms. The focused nature of their adaptation keeps them apart.

REPRODUCTIVE ISOLATING MECHANISMS

Once microevolutionary adaptation produces divergence, how do the species that result maintain their identity? In some cases, geographical isolation prevents the isolated species from exchanging genes, so they stay different. In other cases, two sorts of changes may occur to preserve the difference even if the isolated species come back into contact.

1. Isolated populations may come to exploit different resources in different ways. Because they now approach the environment differently, the populations will remain distinct even if in the future they migrate back into contact with one another. This sort of divergence is common: local populations may come to occupy different habitats—the kinds of places where they live—to carry out different stages of their life cycle at different times, or to practice different feeding habits. They may also come to behave differently, and thus tend not to mate with one another, or simply not to come into contact with one another. Because of any of these changes, the populations become reproductively isolated from each other. The factors involved in this sort of reproductive isolation, in which the formation of zygotes is prevented, are called **prezygotic factors. A zygote** is the first cell formed after the fusion of egg and sperm in a eukaryotic, sexual organism. The zygote divides by mitosis and ultimately gives rise to an adult organism.
2. Sometimes isolated populations can no longer interbreed, not because they approach their environment differently, but because their chromosome organizations or physiological makeups are no longer compatible. Individuals from two isolated populations may mate if they meet, but reproduction is not successful. If the genes of the two kinds of organisms do not function together harmoniously in development, a very complex process, then the hybrid individuals will not be viable. The factors involved in this sort of reproductive isolation, which prevents the proper functioning of zygotes, are termed **postzygotic factors.**

Local populations of organisms tend to become increasingly different from one another in all characteristics. If their patterns of adaptation to the environment differ so greatly that they cannot exist in the same habitat, they have developed prezygotic factors leading to reproductive isolation. Alternatively, they may develop genetic or physiological incompatibility, because of postzygotic reproductive factors.

Prezygotic Mechanisms

Geographical isolation. Most species of most genera simply do not exist together in the same places: their geographical ranges differ. Species are generally adapted to different kinds of climates, or to different habitats. If this is the case, there is no possibility of natural hybridization between them. They may, however, hybridize if they are brought together in zoos, parks, or botanical gardens.

As an example of geographical isolation, the English oak, *Quercus robur,* occurs throughout areas of Europe that have a relatively mild, oceanic climate. In its characteristics, the English oak is quite similar to the valley oak, *Quercus lobata,* of California, and quite different from the scrub oak, *Quercus dumosa,* also of California and adjacent Baja California (Figure 17-6). All of these species can hybridize with one another and form fertile hybrids. The English oak does not hybridize with the others in nature, however, simply because its geographical range does not overlap with theirs. Similarly, although lions, *(Panthera leo)* and tigers *(Panthera tigris)* do not now occur together in nature, they do mate and produce hybrids in zoos. The hybrids in which the tiger is the father, called "tiglons," are viable and fertile; less is known about "ligers," hybrids in which the lion is the father.

Ecological isolation. Even if two species occur in the same area, they may occur in different habitats and thus may not hybridize with one another. If, on the other hand, they do hybridize with one another, the hybrids may not be well represented in the overall population, because they may not be as fit in the habitat of either of their parents. In the latter case, one would speak of a postzygotic isolating mechanism that will be discussed next.

For example, in India the ranges of lions and tigers overlapped until about 150 years ago. Even when they did, however, there were no records of natural hybrids. Lions stayed mainly in the open grassland and hunted in groups called **prides;** tigers tended to be solitary creatures of the forest. Because of their ecological differences, lions and tigers rarely came into direct contact with one another, even though their ranges overlapped over thousands of square kilometers.

Similar situations occur among plants. We have already mentioned two species of oaks that occur in California (Figure 17-7). Valley oak, a graceful tree, can be as tall as 35 meters and occurs in the fertile soils of open grassland on gentle slopes and valley floors in central and southern California. In contrast, scrub oak is an evergreen shrub, usually only 1 to 3 meters tall, which often forms the kind of dense scrub known as chaparral. The scrub oak is found on steep slopes, in less fertile soils. Hybrids between these very different oaks do occur, and they are fully fertile, but they are rare. The sharply distinct habitats of their parents limit their occurrence together, and there is no intermediate habitat where the hybrids might flourish.

Seasonal isolation. Two species of wild lettuce, *Lactuca graminifolia* and *L. canadensis,* grow together along roadsides throughout the southeastern United States. Hybrids between these two species can easily be made experimentally and are completely fertile. Such hybrids are rare in nature, however, because *L. graminifolia* flowers in early spring and *L. canadensis* flowers in summer. When their blooming periods occasionally overlap, the two species do form hybrids, which may even be abundant locally.

Many species of birds and amphibians that are closely related have different breeding seasons. Differences of this kind prevent hybridization between such species. For example, five species of frogs of the genus *Rana* occur together in most of the eastern United States. The peak time of breeding is different for each of them, so hybrids are rare. In insects such as termites and ants, mating occurs when winged, reproductive

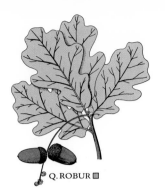

Q. DUMOSA ☐

Q. ROBUR ☐

Q. LOBATA ☐

FIGURE 17-6

Leaf and acorn characteristics of three species of white oaks, English oak, *Quercus robur;* valley oak, *Q. lobata;* and scrub oak, *Q. dumosa.* The range of *Q. robur,* which is in Europe, does not overlap with those of the other two species, which are in North America. Although it is interfertile with them, hybridization cannot occur under natural circumstances.

A

B

FIGURE 17-7

A *Quercus lobata* in Yosemite Valley.
B *Quercus dumosa,* a shrub, in the Coast Ranges south of San Francisco. The ranges of these two very distinct oak species, which are nevertheless interfertile, are shown in Figure 17-6, along with drawings illustrating details of their characteristics.

individuals swarm from the nest. Species often differ in their swarming times, which eliminates the possibility of hybridization between them.

   Behavioral isolation. In Chapter 43, we will consider the often elaborate courtship and mating rituals of some animals. Related species of organisms, such as birds, often differ in their mating rituals, which tends to keep the species distinct even if they do occur in the same places. Indeed, much animal communication is related to the selection of mates.

   In the Hawaiian Islands there are more than 500 species of flies of the genus *Drosophila.* This is one of the most remarkable concentrations of species in a single animal genus anywhere. Many of these flies differ greatly from other species of *Drosophila,* exhibiting characteristics that can only be described as bizarre. The genus occurs throughout the world, but nowhere are the flies more diverse in their external appearance or behavior than in Hawaii.

   The Hawaiian species of *Drosophila* are long-lived and often very large compared with their relatives on the mainland. The females are more uniform than the males, which are often bizarre and highly distinctive. The males display complex territorial behavior and elaborate courtship rituals. Some of these are shown in Figure 17-8.

A

B

FIGURE 17-8

Males and females of two closely related species of Hawaiian flies of the genus *Drosophila*, illustrating activities associated with territorial defense and courtship.

A *Drosophila silvestris*. After approaching the female from the rear, the male has lunged forward while vibrating his wings with his head under the wings of the female, and raised his forelegs up and over the female's abdomen. Specialized hairs on the dorsal surface of one of the leg segments are then "drummed" over the dorsal surface of female's abdomen as part of the final stages of courtship.

B and **C** *Drosophila heteroneura*. In this species, the males have heads that are greatly expanded laterally.

B A male with extended wings approaching and displaying towards a female in typical courtship posture.

C Two males have locked antennae as part of the aggressive behavior involved in the defense of their territories. Groups of males, called *leks,* form in defended mating grounds.

D A stage in the courtship of *Drosophila clavisetae*. In this group of species, the males raise their abdomens up over their backs and spray a chemical signal over the female; this signal seems to play an important role in the mating process.

The patterns of mating behavior among Hawaiian species of *Drosophila* are of great importance in maintaining the distinctiveness of the individual species. Despite the great differences between them, which are so evident in Figure 17-8, *Drosophila heteroneura* and *D. silvestris,* for example, are very closely related. Hybrids between them are fully fertile. They occur together over a wide area on the island of Hawaii, but hybridization has been observed at only one locality. The very different and complex behavioral characteristics of these flies obviously play the major role in maintaining their distinctiveness.

Mechanical isolation. There are structural differences between some related species of animals that prevent mating. Besides such obvious features as size, the structure of the male and female copulatory organs may be so incompatible that mating cannot occur. In many insect and other arthropod groups, the sexual organs, particularly those of the male, are so diverse that they are used as a primary basis for classification. This diversity in structure is generally presumed to have some importance in maintaining differences between species.

Similarly, the flowers of related species of plants often differ significantly in their proportions and structures. Some of these differences limit the transfer of pollen from one plant species to another, and thus decrease the frequency of hybridization between them.

Prevention of gamete fusion. In animals that simply shed their gametes into water, the eggs and sperm derived from different species may not attract one another. Many hybrid combinations involving land animals are not realized because the sperm of one species may function so poorly within the reproductive tract of another that fertilization never takes place. The growth of pollen tubes may be impeded in hybrids between different species of plants. In both plants and animals the operation of such mechanisms may prevent the union of gametes even after successful mating. The prevention of fusion between gametes is the last kind of prezygotic isolating mechanism possible before hybrids are formed.

Prezygotic mechanisms lead to reproductive isolation by preventing the formation of hybrid zygotes. The principal mechanisms are geographical, ecological, seasonal, behavioral, and mechanical isolation, and prevention of gamete fusion.

Postzygotic Mechanisms

All of the factors that we have discussed so far tend to prevent hybridization. If hybridization does occur, however, and zygotes are produced, there are still many factors that may prevent those zygotes from developing into normal, functional, fertile F_1 individuals. Development in any species is a complex process. In hybrids the genetic complements of two species may be so different that they cannot function together normally in embryonic development. For example, hybridization between sheep and goats usually produces embryos that die in the earliest developmental stages. In other instances, the development of sex organs in hybrids may be abnormal, the chromosomes derived from the respective parents may not pair properly, or hybrid fertility may be lower than normal.

Even if hybrid progeny survive the embryo stage, they may not develop normally. If the hybrids are weaker than their parents, they will almost certainly be eliminated in nature. Even if they are vigorous and strong, as in the case of the mule—a hybrid of the horse and the donkey (see Figure 17-3)—they may still be sterile and thus incapable of contributing to succeeding generations.

> **Postzygotic mechanisms are those in which hybrid zygotes develop abnormally or fail to develop entirely, or the hybrids cannot become established in nature.**

Most species are separated by combinations of prezygotic and postzygotic isolating factors, similar to the ones we have just discussed. For example, two related species may occur in different habitats, produce their gametes at different times of the year, have different behavioral patterns, and produce inviable embryos even if hybridization does take place. Such patterns, in which more than one factor functions to limit the frequency of hybrids between two species, either form while the original populations are isolated from one another or may be strengthened when they come into contact. For example, if the hybrids do not complete their development beyond the embryo stage, then any factor that limits the hybridization that produces the zygotes (prezygotic mechanisms) would be an advantage. Individuals that form hybrids that do not function well in nature waste reproductive energy by doing so, and are less fit than individuals that do not form such hybrids.

As a result of human population pressures, thousands of kinds of plants, animals, and microorganisms all over the world are facing extinction. One of these is the dusky seaside sparrow. A population greater than 6,000 of these canary-sized birds used to inhabit the cordgrass flats along the St. Johns River and on Merritt Island in Brevard County, central Florida, but the last survivor (Figure 17-A) died in June, 1987. It had been living in a large cage in a secluded area on Discovery Island, an accredited zoo especially concerned with the preservation of selected species of birds, in Disney World near Orlando. The wild birds were driven near extinction by mosquito-abatement efforts near the Kennedy Space Center—heavy spraying with insecticides and then flooding the marshes—and other intensive development in the region, followed by wildfires. By the late 1960s, it was evident that the sparrows faced extinction. The last known nesting in the wild occurred in 1975, and the six remaining birds were captured in 1980.

The last few dusky seaside sparrows that had been captured in nature were systematically backcrossed with members of the Gulf Coast subspecies of the seaside sparrow, however. So while the dusky seaside sparrow is now extinct, much of its genetic constitution has been preserved by these backcrossing efforts. The future release of these backcross progeny may serve to introduce a new species to the region with much of the evolutionary heritage of the one we have lost.

FIGURE 17-A

One of the three dusky seaside sparrows surviving in 1985, a male. The last six birds were all captured and backcrossed to a related subspecies in captivity at Discovery Island. (© Walt Disney Productions.)

THE EVIDENCE THAT DIVERGENCE LEADS TO SPECIES FORMATION

The best evidence that microevolutionary divergence leads to species formation has been gathered by biologists who study the nature of local populations. These biologists have been able to demonstrate the existence of populations at all levels of differentiation. If the processes resulting in the formation of these differentiated populations were to continue, they would eventually lead to the formation of distinct species. Because many intermediate stages exist in nature and have been studied, we have a good understanding of the processes that lead to the evolution of species.

In many species, the individuals that occur in one part of its geographical range have different features from those that occur elsewhere. Such groups of distinctive individuals are informally called **races,** groups of individuals with distinctive characteristics, often occurring in different places. Darwin considered the existence of races to be an important demonstration of the processes that led to the formation of species, just as we do today. Sometimes races are named and given the rank of **subspecies,** which is simply a formal way of designating them. Races may change over time into the clusters of species that we will consider next.

Ecological Races

Distinct races occur in many kinds of plants and animals, and even in humans. The features that distinguish them, however, may or may not be highly correlated with one another, so their distinctiveness is usually relative. Nonetheless, by studying almost any species with a wide distribution, individual differences can be found, and these may be considered characteristics of races. How this is done depends on the person doing the classifying, but such patterns of differentiation undeniably occur in virtually any feature one cares to study. As seen in our consideration of human races in Chapter 12, the various features by which races have been distinguished are often not well correlated with one another, but they may nonetheless appear very distinct.

Ecotypes in Plants

Ecological races were first studied in detail in plants. As every gardener knows, the same plants may differ greatly in appearance, depending on the places where they are grown (Figure 17-9). This is true even for genetically identical parts of the same individual plant, **clones.** The part of an individual plant that is most in the sun often produces leaves that are unlike the leaves the same plant produces in the shade. For example, shade leaves are usually thinner, broader, and have more internal air spaces than do sun leaves. Thus it seemed to many biologists in the nineteenth century that environmental factors, rather than genetic differences, might account for many differences between races and even between species of plants. As a result, it became important to determine whether this conclusion was accurate.

In the 1920s and 1930s, the Swedish botanist Gote Turesson performed experiments that were designed to test whether differences between races of plants were largely determined genetically or by environmental factors. Turesson observed that many plants had distinctive races that grew in different habitats. These races differed from one another in characteristics such as height, leaf size and shape, degree of hairiness, flowering time, and branching pattern. Turesson dug up individual plants representing these races and cultivated them together in his experimental garden in Lund, Sweden. In nearly every case he found that the unique features of individual races were maintained when the plants were grown in a common environment. Most of the characteristics that he observed, therefore, had a genetic basis; a few of them were environmental. Turesson called the ecological races that he studied and showed to have a genetic basis **ecotypes.**

Ultimately, the studies that were first put on a scientific basis by Turesson and continued by others led to the conclusion that most differences between individuals, populations, races, and species of plants are genetically fixed. Despite the fact that plants can change their characteristics in relation to the environments in which they grow, the differences between them are usually fixed genetically in the course of their evolution.

FIGURE 17-9

Two forms of Bishop pine, *Pinus muricata*, in Marin County, just north of San Francisco, California. The individual on the left is growing in a flat field behind Inverness Ridge, where it is protected from the ocean winds; the one on the right, growing in an exposed place, has been dwarfed and shaped by the winds. Without experimental study, it would not be possible to determine whether these two individuals differ genetically or not.

Ecological Races in Animals

Similar ecological races are, of course, also found in animals. The differences between animal populations may be **morphological** (involving the appearance) or **physiological** (involving how the body functions), and they produce races that have the same basis as do ecotypes in plants. The differences between animal races may be striking. For example, the larger races of some bird species may often consist of individuals that weigh three or four times as much as individuals of the smaller races. Races may differ in their tolerance to different temperatures, in the speed of their larval development, in their behavioral characteristics—in short, in virtually any feature that can be measured and studied. Their features are almost always genetically determined.

Clusters of Species

One of the most visible manifestations of species formation is the existence in certain locations of groups of closely related species. These species often have evolved relatively recently from a common ancestor. Such clusters are particularly impressive on groups of islands, in series of lakes, or in other sharply discontinuous habitats. The existence of these clusters makes sense only in the context of their arising by microevolutionary divergence from an ancestral form occupying diverse habitats. One of the best-known clusters of species is a group of birds found on the Galapagos Islands, where it was studied by Charles Darwin.

Thirteen species of Darwin's finches occur on the Galapagos Islands. One species lives on Cocos Island, which lies about 1000 kilometers to the north. This group of birds provides one of the most striking and best-studied examples of species formation.

The Galapagos Islands are a remarkable natural laboratory of evolution. The islands are all relatively young in geological terms, and they have never been connected with the adjacent mainland of South America or with any other source area. The lowlands of the Galapagos Islands are covered with thorn scrub. At higher elevations, which are attained only on the larger islands, there are moist, dense forests. All of the organisms that occur on these islands have reached them by crossing the sea, as a result of chance dispersal in the water, by wind, or by transport via another organism.

On oceanic islands there is often a disproportionate representation of certain groups of organisms. For example, in addition to the 13 species of Darwin's finches on the Galapagos, there are only 7 other species of land birds. Perhaps the ancestor of Darwin's finches reached these islands earlier than the ancestors of the other birds. In that case, all the habitats where birds occur on the mainland would have been unoccupied, and the ancestor of Darwin's finches would have been able to take advantage of them all. As new arrivals moved into these vacant niches, adopting new lifestyles, they were subjected to diverse sets of selective pressures. Under these circumstances, the ancestral finches rapidly split into a series of diverse populations, and some of these eventually became species.

The descendants of the original finches that reached the Galapagos Islands now occupy many different kinds of habitats (Figure 17-10). These habitats encompass a variety of niches comparable to those occupied by several distinct groups of birds on the mainland (Figure 17-11). Among the 13 species of Darwin's finches that inhabit the Galapagos, there are three main groups:

1. *Ground finches,* which feed on seeds of different sizes. The size of their bills is related to the size of the seeds on which the birds feed.
2. *Tree finches,* which have bills suitable for feeding on insects. One of the tree finches has a parrotlike beak and feeds on buds and fruit in trees. Another has a chisel-like beak with which it carries around a twig or cactus spine, which it uses to probe for insects in crevices. It is an extraordinary example of a bird that uses a tool.
3. The *warbler finch,* an unusual bird that plays the same ecological role in the Galapagos woods that warblers play on the mainland, searching continually over the leaves and branches for insects.

FIGURE 17-10

Darwin's finches.
A Medium ground finch, *Geospiza fortis,* on Daphne Island. Note the large, cone-shaped bill, adapted for seed eating, and presumably similar to that of the common ancestor of the group.
B Warbler finch, *Certhidea olivacea,* on Santa Cruz Island. This species is the most distinctive of Darwin's finches.

A

B

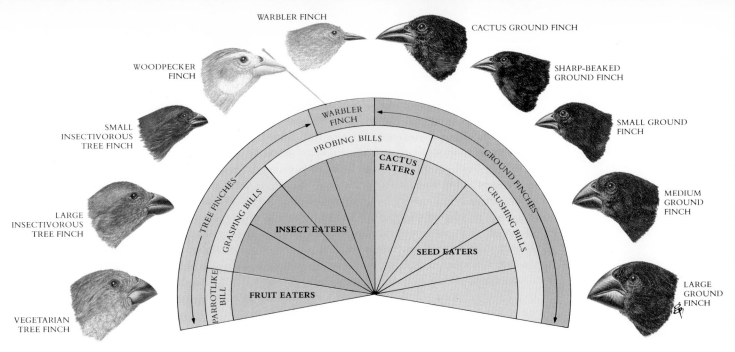

The evolution of Darwin's finches on the Galapagos Islands and Cocos Island provides a classical example of species formation. It illustrates how adaptation to local conditions produces the divergence that is the heart of species formation. By the same sorts of processes, species are originating continuously in all groups of organisms. Another example of evolutionary radiation on a group of islands is provided by the Hawaiian species of *Drosophila* discussed earlier (see Figure 17-8).

The evolution of organisms on islands differs from that in mainland areas only in the degree of opportunity afforded for *rapid* evolution. Because evolution on islands is often rapid, and therefore recent, Darwin and other investigators have found the study of plants and animals on islands especially informative.

DOES EVOLUTION OCCUR IN SPURTS?

One aspect of species formation that is now under active discussion by biologists concerns the degree to which the rate of macroevolution, as judged by the fossil record, is periodically accelerated and then reduced. In 1972 Niles Eldredge of the American Museum of Natural History in New York and Stephen Jay Gould of Harvard University proposed that it was the norm of evolution to proceed in spurts. They described the process of macroevolution as consisting of long periods in which evolutionary forces are in equilibrium and little change occurs, interrupted by occasional brief periods of rapid evolutionary change. Their proposal is called the hypothesis of **punctuated equilibria.** It suggests that the fossil record should be interrupted or discontinuous, a point on which there is considerable disagreement. Eldredge and Gould contrasted their hypothesis of punctuated equilibria with that of **gradualism,** or gradual evolutionary change, a view they ascribe to Darwin. Scientists are actively studying the fossil record and evidence from the living members of groups of organisms to test this hypothesis.

THE EVIDENCE FOR EVOLUTION: AN OVERVIEW

In this and the preceding chapter we have considered the body of evidence that supports the three central tenets of evolution. The evidence that macroevolution has occurred is so strong as to be beyond further question. Although scientists are currently engaged in lively discussions about the rate at which macroevolution proceeds, and about whether it proceeds gradually or in spurts, there is essentially no disagreement among practicing biologists about the fact of macroevolutionary change.

The evidence for microevolution depends partly on observations of evolutionary change now in progress. In many cases it has been possible to observe evolutionary

FIGURE 17-11

Ten species of Darwin's finches from Indefatigable Island, one of the Galapagos Islands, showing differences in bills and feeding habits. The bills of several of these species resemble those of different distinct families of birds on the mainland, a condition that presumably arose when the finches evolved new species in habitats where other kinds of birds normally occur on the mainland, and which are not available to finches there. The woodpecker finch uses cactus spines to probe in crevices of bark and rotten wood for food. All of these birds are thought to have been derived from a single common ancestor.

change within populations of a single species, in which selection has proven to be a powerful agent of adaptive change, just as Darwin proposed in 1859. Many factors cause natural populations to change progressively over time, including both environmentally guided selection on the one hand and random factors such as genetic drift on the other hand. Among the many changes that occur, those resulting from selection lead to microevolution, the adaptive evolution of the species. As populations change, the accumulation of different adaptive characteristics eventually leads to the formation of different species, macroevolution.

The evidence that microevolutionary change leads to macroevolutionary change is extensive. Biologists have observed the stages of the species-forming process in many different kinds of plants, animals, and microorganisms. These stages are indicated by such features as genetic variation in individual populations, genetic differences between different populations, divergence of populations from one another in response to different adaptive pressures, ecological races, progressively stronger reproductive isolation, and clusters of related species, each species occupying an ecologically distinct, but often adjacent, habitat. These patterns are so frequently observed in nature that there can be no doubt of the way in which species normally originate. Microevolution drives macroevolution, just as Darwin proposed.

SUMMARY

1. The theory of evolution is based on three lines of evidence: that macroevolution occurs, that natural selection is responsible for the adaptive nature of microevolution, and that microevolutionary change (evolution within a species) leads to macroevolutionary change (evolution of new species).

2. Species are kinds of organisms that differ from one another in one or more characteristics and that do not normally hybridize freely with other species when they come into contact in nature. They often cannot hybridize with one another at all. Individuals within a given species, on the other hand, usually are able to interbreed freely.

3. Populations change as they adjust to the demands of their environments. Even populations of a given species that are close to one another geographically are normally effectively isolated. Such populations are free to diverge in ways that are responsive to the needs of their particular environment.

4. In general, if the selective forces bearing on populations differ greatly, the populations will diverge rapidly; if these selective forces are similar, the populations will diverge slowly.

5. Among the factors that separate populations, and species, from one another are geographical, ecological, seasonal, behavioral, and mechanical isolation, as well as factors that inhibit the fusion of gametes or the normal development of the hybrid organism. Reproductive isolation between species arises as a normal by-product of the progressive differentiation of populations.

6. The evidence that microevolutionary change leads to macroevolutionary change is that it is possible to observe the intermediate stages of the process directly.

7. Ecological races and subspecies differentiate within species but often still intergrade with one another. The differences between races, in both plants and animals, are mostly determined by genetic factors.

8. Clusters of species arise when the differentiation of a series of populations proceeds further, past race formation. On islands such differentiation is often rapid, because numerous open habitats are available. In many continental areas, differentiation is not as rapid, but there are local situations, such as those where many different kinds of habitats are developing close to one another, where differentiation may be rapid.

REVIEW

1. Groups of organisms that differ from each other in one or more characteristics and do not hybridize extensively in nature are called _____.

2. The adaptive molding of local populations of a species, which leads to divergence of those populations, is called _____. This process, if it proceeds far enough, results in _____.

3. Lions and tigers will mate if brought together, and the hybrids are fertile. Which isolating mechanism is responsible for the lack of hybrids between these species in nature?

4. Mechanisms such as the prevention of gamete fusion so that hybrids are not formed are called _____ mechanisms. These mechanisms can be contrasted with_____ mechanisms, which act after fertilization has occurred.

5. Is the evolution of Darwin's finches on the Galapagos Islands an example of microevolution or of macroevolution?

SELF-QUIZ

1. Choose the phrase(s) which best complete this statement. Local populations of a species
   (a) respond to environmental conditions in their particular habitat
   (b) tend to mate more frequently with other individuals in their population
   (c) are groups of individuals that live in a particular place
   (d) can be genetically different from other populations of the species
   (e) all of the above

2. Which is *not* a prezygotic isolating mechanism?
   (a) geographical isolation   (d) hybrid sterility
   (b) ecological isolation   (e) mechanical isolation
   (c) seasonal isolation

3. Complete this sentence with the most appropriate choice. Gote Turesson performed a series of experiments to determine what caused the observed differences between races of plants. When plants were grown in a common garden, he found that
   (a) plants maintained their differences, and he concluded that the differences had a genetic basis
   (b) plants maintained their differences, and he concluded that the differences were environmental

4. On the Galapagos Islands, the descendants of a South American finch have produced a cluster of species. Darwin's finches fill a variety of niches. Which niche is *not* filled by any of these finches?
   (a) seed eater
   (b) cactus eater
   (c) fish eater
   (d) insect eater
   (e) fruit eater

5. In continental areas, as on islands, evolution sometimes results in clusters of species. In what kind of continental area would you expect to find species clusters?
   (a) an area with an array of diverse habitats
   (b) an arid area
   (c) an alpine area
   (d) a tropical area
   (e) a grassland

THOUGHT QUESTION

1. In the fall of 1986 the Supreme Court of the United States heard arguments for and against a Louisiana law requiring that "creation science" be taught in public schools on an equal footing with evolution. One of the arguments advanced in support of the law was that evolution is as much a religion as creationism, reflecting the beliefs of scientists and their faith in a particular world view rather than the certain knowledge of objective reality that they claim. In June of 1987, the decision of the Court was announced: by 7 to 2, the justices struck down the law, rejecting the "creation science" argument. How would you have voted?

FOR FURTHER READING

GOULD, S.J.: *Ever Since Darwin,* W.W. Norton & Company, New York, 1977. An entertaining collection of essays on evolution and Darwinism.

HITCHING, F.H.: *The Neck of the Giraffe or Where Darwin Went* Chaucer Press (Pan), London, 1982. An entertaining and informal presentation of all the arguments currently being advanced *against* Darwin's theory.

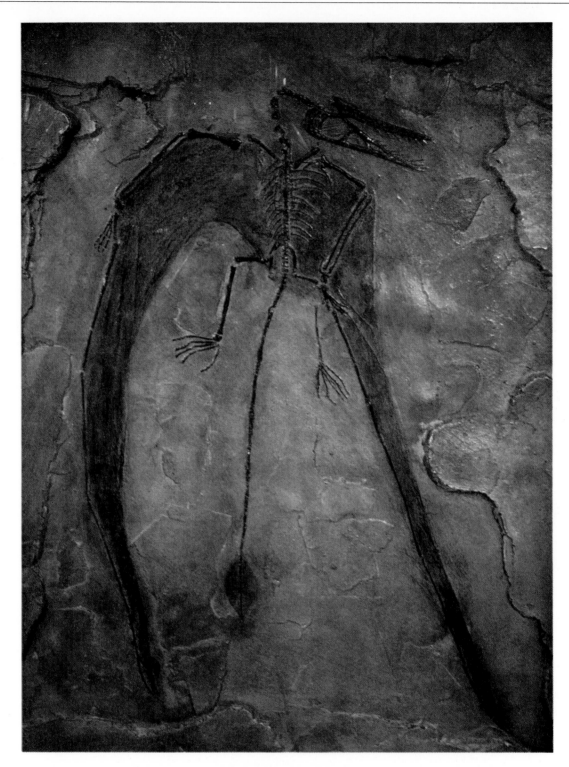

FIGURE 18-1   Flight has evolved three separate times among the vertebrates. Birds and bats, which evolved independently from reptiles and mammals, respectively, are still with us. Pterosaurs, a diverse group of flying reptiles that lived for more than 100 million years, became extinct with the dinosaurs about 65 million years ago. The wing membrane stretches along the greatly extended fourth finger of this fossil pterosaur, a member of the genus *Rhamphorhychus*.

THE EVOLUTION OF LIFE ON EARTH

Chapter 18

Overview

The central lesson of evolution is that the many kinds of plants, animals, and microorganisms now living on earth are the end result of a long history of success and failure, as organisms have for billions of years striven to survive and reproduce. Although much of what has happened is hidden from us now, traces of the evolutionary journey to the present remain frozen in rock as fossils. By studying fossils, we have been able to get a fairly clear picture of how life has evolved on earth since the first cells formed more than 3 billion years ago.

For Review

Here are some important terms and concepts that you will encounter in this chapter. If you are not familiar with them, you should review them before proceeding.

Kingdoms (Chapter 2)

The origin of life (Chapter 4)

Radioisotope dating of fossils (Chapter 16)

The origin of species (Chapter 17)

All of the many kinds of organisms now living on earth represent only a small fraction of the total that have lived. Dinosaurs were once the dominant land vertebrates (Figure 18-1), but today no dinosaurs walk the earth. The trilobites, sea organisms related to the living horseshoe crabs, dominated the ocean floor in the Cambrian Period more than 500 million years ago (Figure 18-2), but they are gone now; so are the ammonites, octopuses with shells that were one of the dominant forms of ocean life in the Cretaceous Period 100 million years ago. As children, few of us escape fascination with dinosaurs; they appear so different that it is as if they inhabited a different world. And yet dinosaurs are the ancestors of the turkey you eat at Thanksgiving, and relatives of the ancestors of the rattlesnakes and turtles of the modern world. We know a great deal about dinosaurs and about the many other forms of life that have preceded us in the evolutionary parade of life on earth, because we have found their fossils. A fossil allows us to visualize the body of an organism that lived in the past, because certain structures are mineralized and so preserved in rock. We don't have fossils of all the kinds of organisms that have lived before us, but we have found enough to piece together a fairly complete picture of the past. We will review that picture in this chapter. First, a few words about fossils.

FIGURE 18-2

A A reconstruction of a community of marine organisms in the Cambrian Period, 500 to 550 million years ago. The swimming animals, which have jointed legs and large, compound eyes, are trilobites, early members of the arthropod phylum. On the sea floor is a colony of sponges, members of another ancient animal phylum.
B A fossil trilobite. Although this group of animals disappeared nearly 250 million years ago, reconstructions such as that shown at the left can be made on the basis of fossils such as this.

A

B

THE DISCOVERY OF A LIVING COELACANTH

New discoveries relating to the history of life on earth are not always made in the fossil record. Thus in 1938 scientists were surprised by the announcement that a trawler fishing in the Indian Ocean off the coast of South Africa had landed a large, very strange fish (Figure 18-A). The specimen, which was about 2 meters long, became rotten before it was seen by an ichthyologist (a scientist who specializes in the study of fish) and was skinned so that some of it could be saved. Nevertheless, as soon as J.L.B. Smith of the University of Grahamstown in South Africa saw it, he recognized it as a member of an ancient group of "lobe-finned" fishes (fishes in which the fins more closely resemble the limbs of amphibians and other land vertebrates than is usual). These fishes had been described from fossils more than a century earlier and had been thought to have been extinct for about 70 million years! They were called coelacanths, and the newly discovered living fish was dubbed *Latimeria chalumnae*.

Coelacanths are well represented in the fossil record for more than 300 million years, from about 390 million years to about 70 million years ago. Until *Latimeria* was discovered alive, it was assumed that the group, which

was somewhat similar to the ancestors of the terrestrial vertebrates, had become extinct long ago. Here, however, was indisputable evidence that one of their descendants was still alive, swimming the warm waters of the western Indian Ocean.

Because the first specimen had been skinned, scientists had no opportunity at first to learn about is internal parts, features that were of great interest in terms of its relationship to terrestrial vertebrates and other fishes. Leaflets and posters were distributed among the fishing communities of southern and eastern Africa, offering rewards for another specimen of this remarkable fish. It was not until 1952, however,

that a second coelacanth was brought to the attention of scientists. It was landed in the Comoro Islands, about 3000 kilometers northeast of the place where the first specimen was caught. Coelacanths were landed occasionally in the Comoros and were well known to local fishermen.

Living mostly at depths of 150 to 300 meters in the sea, *Latimeria* is a very strange animal. Its features mark it as a member of the evolutionary line that gave rise to the terrestrial tetrapods. By studying the dozens of specimens that have been landed since 1952, scientists have been able to shed additional light on the nature of this ancient and archaic group of vertebrates.

FIGURE 18-A
The living coelacanth, *Latimeria*.

FOSSILS

Only a tiny fraction of the organisms living at any one time are preserved as fossils. Most of the fossils that we do have are preserved in **sedimentary** rocks. Sedimentary rocks are formed of particles of other rocks that are weathered off, deposited at the bottom of bodies of water, and then hardened. During the formation of sedimentary rocks, dead organisms are sometimes washed down to the bottom along with the particles, and if they are buried before their decay is complete, they may form fossils. The actual parts of the fossil organisms are almost always replaced by minerals. A fossil rarely contains any of the material that made up the body of the organism originally; rather it is a mineralized replica of that body or its part.

Fossils provide concrete evidence about the history of particular groups of organisms. They are found mainly in sedimentary rocks.

Dates are usually assigned to fossils based on the layers of sedimentary rock in which they occur and the other kinds of fossils that are associated with them. Direct methods of dating fossils, by determining the percentages of radioactive isotopes of certain elements that are present in the rocks, are also available, as mentioned in Chapter 16.

THE EARLY HISTORY OF LIFE ON EARTH

The earth itself is about 4.6 billion years old, and the oldest rocks that have persisted in recognizable form are about 3.8 billion years old. For many years, scientists believed that there were no fossils in such ancient rocks, but we now know that the fossils were simply too small to be seen. The earliest fossils found so far, all of them bacteria, are about 3.5 billion years in age.

Massive limestone deposits called **stromatolites** (Figure 18-3) became frequent in the fossil record about 2.8 billion years ago. Produced by cyanobacteria, stromatolites were abundant in virtually all freshwater and marine communities until about 1.6 billion years ago. Today stromatolites are still being formed, but only under conditions of high salinity, aridity, and high light intensities. About 2 billion years ago many different kinds of bacteria existed, including single, rounded cells; filaments apparently divided by cross-walls; tubular structures; branching filaments; and several unusual forms that do not fit well into any of these categories. For most of the time in which life has existed on earth, the only organisms in existence were bacteria. Fossils that definitely seem to represent unicellular protists, the first eukaryotes, first appear about 1.5 billion years ago.

FIGURE 18-3

A This diagram shows how aggregations of the mineral calcium carbonate build up around massive colonies of cyanobacteria. Such deposits are known as stromatolites.
B Stromatolites in the intertidal zone at Shark Bay, Western Australia. The largest structures are about 1.5 meters across. These stromatolites formed during a time of slightly higher sea level perhaps 1000 to 2000 years ago.

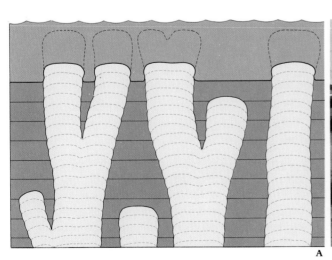

A

B

The oldest fossils are bacteria; they date from about 3.5 billion years ago. The oldest eukaryotic fossils, unicellular protists, are from about 1.5 billion years ago.

We find the first known fossils of multicellular organisms in rocks about 630 million years old from southern Australia. Their appearance marks the onset of a major period of the earth's history, **Phanerozoic time.** Within Phanerozoic time, we begin to speak in terms of geological eras, periods, epochs, and ages. For many years the oldest known fossils were those from the Cambrian Period (590 to 505 million years ago), the first period of the Paleozoic Era. The geological eras, with their dates in millions of years before the present, are as follows:

1. *Paleozoic Era.* 590 to 248 million years ago. The name of this era is derived from the Greek words *paleos,* "old," + *zoos,* "life." Until the discoveries outlined above were made, this was the oldest period from which fossils were known.
2. *Mesozoic Era.* 248 to 65 million years ago. Greek *mesos,* "middle."
3. *Cenozoic Era.* 65 million years ago to the present. Greek *coenos,* "recent."

All of the strata older than the Cambrian Period, and thus older than the Paleozoic Era, are classified as Precambrian. To earlier scientists there seemed to be no fossils at all in Precambrian rocks, and their absence was regarded as a great mystery. Now, however, we know that fossil organisms were not detected simply because they were unicellular and therefore so small that they were difficult to observe. With the evolution of external skeletons (that is, hard shells) visible fossils became abundant. Those multicellular animals that lived before the evolution of external skeletons in many cases were not well-preserved. The evolution of nearly all of the larger groups of organisms occurred during the Cambrian Period.

Multicellular fossils first appear about 630 million years ago, with traces of multicellular animals preserved in rocks up to 700 million years old.

THE PALEOZOIC ERA

The six periods of the Paleozoic Era, together with the approximate times that they began and ended (in million of years before the present) are:

	START	END
Cambrian	590	505
Ordovician	505	438
Silurian	438	408
Devonian	408	360
Carboniferous	360	286
Permian	286	248

We present these periods at this point so that they may provide a point of reference for the upcoming discussion.

Origins of Major Groups of Organisms

As we just mentioned, the Paleozoic Era was the time of origin and early diversification for virtually all of the major groups of organisms that survive at the present time, except for the plants. Its beginning, the Cambrian Period (590 to 505 million years ago), was the period when multicellular animals became more diverse more rapidly (in geological terms) than was ever the case later. **Phyla** (singular **phylum**), the major groups into which kingdoms are divided (such as mollusks, sponges, or flatworms), appeared

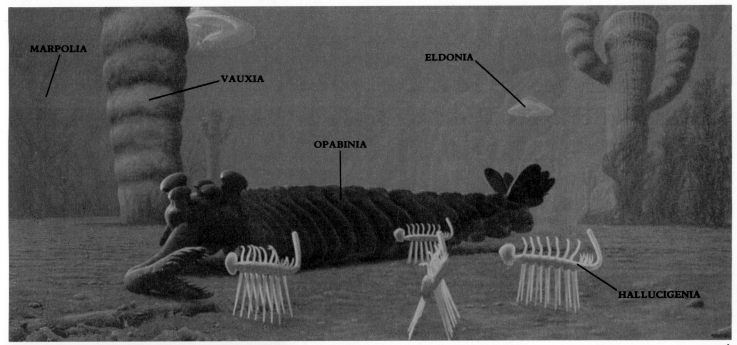

MARPOLIA

VAUXIA

ELDONIA

OPABINIA

HALLUCIGENIA

A

B

FIGURE 18-4

A A shallow seafloor during the Cambrian Period, about 530 million years ago. Fossils of the organisms shown here, together with well over a hundred others, are found together in the Burgess Shale, a formation that was formed from a fine-grained mud in the sea but has been uplifted to high elevations in the Rocky Mountains of British Columbia. *Opabinia*, which was about 12 to 15 centimeters in length, had a jointed grasping organ unlike anything seen in living animals. *Hallucigenia* is a very peculiar organism that also has no known relationships among living organisms. *Eldonia* is a cnidarian, *Vaxia* is sponge, and *Marpolia* is a multicellular alga. These three organisms represent living phyla.

B Fossil of *Hallucigenia sparsa,* shown in the reconstruction. This specimen is 12.5 millimeters long. The relationships of this bizarre animal are obscure. The animal seems to have been supported on seven pairs of elongate spines; its trunk bore seven long tentacles and an additional group of tentacles near the rear end. *Hallucigenia* may have been a scavenger on the bottom of the sea.

mainly at this time and exclusively in the sea. In contrast, plants originated on land, although they were derived from aquatic ancestors, the green algae. The basic record of the diversification of animal life on earth is a marine record, and the fossils that we have from the Paleozoic Era all originated in the sea.

Many of the multicellular animals that occurred in the early Paleozoic Era have no living relatives. In studying their remains in the rocks, one senses that this was a period of "experimentation" with different body forms and ways of life (Figure 18–4). Multicellular organisms seem to have diversified along many new evolutionary pathways, some of which ultimately led to the contemporary phyla of animals, and others to extinction. The trilobites (see Figure 18–2), for example, appear to be the ancestor of at least one living group, the horseshoe crabs, whereas the ammonites, abundant 100 million years ago, have no surviving descendants.

The early Paleozoic Era was a time of extensive diversification for marine animals. Many new kinds of animals appeared, and some of them have persisted to the present.

Until the close of the Cambrian Period, all of the animals in the sea fed on unicellular organisms that floated freely in the water. During the following period, the Ordovician, true predators appeared, and the diversity of animals became even more extensive. The first corals originated during the Ordovician Period also, and began to change the structure of marine communities permanently. By the end of this period, practically every mode of life that has ever existed had already evolved: for example, bottom feeders, scavengers, carnivores, colonies, and drifting multicellular forms. With all of the adaptive zones in which these animals occur having been filled, the later opportunities for the additional evolution of novel forms have been much more limited.

The Invasion of the Land

Only a few phyla of organisms have invaded the land successfully; most others have remained exclusively marine. The first organisms that colonized the land were the plants, and they did so near the close of the Silurian Period, about 410 million years ago. The features of plants evolved in relation to their colonization of the land and were those that made them best suited for such colonization. Among these features are cuticle and bark, water-resistant outer coverings with specialized openings for gas exchange penetrating them (see Chapter 29); tissues for conducting water and nutrients; and drought-resistant pollen grains and spores. The ancestors of plants were specialized members of a group of photosynthetic protists known as the green algae. Although most algae are aquatic, the immediate ancestors of plants might themselves have been semiterrestrial. It was, however, with the plants themselves that the occupation of the land truly began (Figure 18-5, A).

The second major invasion of the land, and perhaps the most successful, was by the **arthropods,** a phylum of hard-shelled animals with jointed legs and a segmented body (Figure 18-5, B). Arthropods were originally marine organisms; the trilobites shown in Figure 18-2 were members of this phylum, for example. Among the descendants of the first arthropods to invade the land are the insects, which have wings. This second invasion of the land occurred at about the same time as the evolution of the plants, about 410 million years ago. The body plan of arthropods has proved so well adapted to life on land that insects and other classes of this phylum now represent a large majority of all species of organisms. Among the features that are important to the success of the arthropods on land are their drought-resistant cuticles and efficient structures for conserving water and exchanging gases with the atmosphere (see Chapter 28). It seems certain that the plants colonized the land before arthropods did, since plants would have been essential sources of food and shelter for the first arthropods that emerged from the sea.

The third major invasion of the land was by the vertebrates. Vertebrates are members of a phylum of animals called the **chordates** (see Chapter 19), and they include our own ancestors. The first of the vertebrates were the amphibians, represented today by such animals as frogs, toads, and salamanders. The earliest amphibians

FIGURE 18-5

A This fragment of an unknown plant from the Silurian Period, about 423 million years ago, shows the earliest evidence of an apparently well-organized conducting strand, consisting of banded, elongated conducting cells surrounded by unbanded, smooth-walled cells. It cannot be determined from the material found so far whether this organism was terrestrial or aquatic, or whether it was actually a plant, a large alga, or a transitional form.
B A mite (phylum Arthropoda) from the Devonian Period, about 376 to 379 million years ago, of what is now New York State. Land animals appeared about 40 million years earlier than the time this mite lived, but they were initially very scarce.

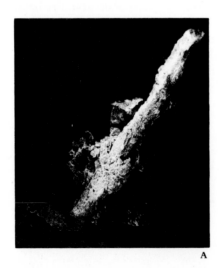

A

B

FIGURE 18-6

Reconstruction of *Ichthyostega,* one of the first amphibians with efficient limbs for crawling on land, an improved olfactory sense associated with a lengthened snout, and a relatively advanced ear structure for picking up airborne sounds. Despite these features, *Ichthyostega,* which lived about 350 million years ago, was still quite fishlike in overall appearance, and it represents a very early amphibian.

known are from the end of the Devonian Period, appearing just over 360 million years ago (Figure 18-6). Among their descendants on land are the reptiles. Different groups of reptiles, in turn, ultimately became the ancestors of the birds and the mammals, as we shall discuss further in Chapter 19. The air-breathing lungs of vertebrates; their scales, fur, and feathers, which regulate heat loss; efficient circulatory and waste-removal systems; and internal fertilization admirably fit them to life on land.

The fact that all three of these major groups of organisms—plants, arthropods, and vertebrates—colonized the land within a few tens of millions of years of one another is probably related to the development of suitable environmental conditions, such as the formation of a layer of ozone in the atmosphere, which blocked ultraviolet radiation. Ozone (O_3) forms in equilibrium with oxygen (O_2) and thus was not abundant until the activities of photosynthetic bacteria had elevated the level of oxygen in the atmosphere sufficiently. These conditions have allowed the existence of multicellular organisms in terrestrial habitats for more than 400 million years, about a tenth of the age of the earth.

In addition to the plants, arthropods, and vertebrates, a fourth large phylum that has colonized the land consists of the fungi, which are actually a distinct kingdom of organisms. The success of the fungi on land, where they probably have been present for as long as any group of organisms, might be related to the structure of their cell walls, which are rich in chitin, a drought-resistant substance that also forms the external skeletons of arthropods.

> **Plants and arthropods colonized the land about 410 million years ago; amphibians, the first terrestrial vertebrates, arrived about 50 million years later. Fungi may also have colonized the land at about the same time as plants. Earlier, there were no terrestrial, multicellular organisms.**

Mass Extinctions

One of the most prominent features of the history of life on earth has been the periodic occurrence of major episodes of extinction. During the course of geological time there have been five such events. In each of them a large proportion of the organisms on earth at that time became extinct. Four of these events occurred during the Paleozoic Era, the first of them near the end of the Cambrian Period (about 505 million years ago). At that time most of the existing families of trilobites (see Figure 18-2) became extinct. A second major extinction marked the close of the Ordovician Period, about 438 million years ago, and a third occurred at the close of the Devonian Period, about 360 million years ago.

The fourth and most drastic extinction event in the history of life on earth happened during the last 10 million years of the Permian Period, which ended the Paleozoic Era, about 238 to 248 million years ago. It is estimated that approximately 96% of all species of marine animals that were living at that time may have become extinct!

The last of the trilobites and many other groups of organisms disappeared forever. The fifth and most recent major extinction event occurred at the close of the Mesozoic Era, 65 million years ago. It is familiar to most of us as the time when dinosaurs became extinct.

> **In the history of life on earth there have been five major extinction events. The extinction event that took place at the end of the Permian Period, when about 96% of the species of marine animals went extinct, was the most drastic.**

Large-scale extinction events other than the five major ones summarized above also have occurred. One surprising correlation that was reported in 1983 by J. John Sepkoski and David M. Raup of the University of Chicago is that such events appear to occur regularly every 26 to 28 million years. The most recent such event occurred about 11 million years ago. The search for the sort of cosmic event, such as periodic comet showers, that could cause major events of extinction to occur with such regularity is now under way, as other scientists attempt to test the validity of the correlation made by Sepkoski and Raup. It has been hypothesized that the major extinction event that ended the Mesozoic Era is correlated with the impact of a large meteorite, as we shall discuss below. Changes in the relative positions of the continents and oceans also have played an important role in extinctions on a regional scale. Numerous other hypotheses have been advanced in explanation of major extinction events, but there is no general agreement concerning which explanation is true. The extinctions almost certainly had multiple causes. Within our own lives, the activities of human beings are bringing about an episode of mass extinction fully comparable to anything that has occurred in the past (see Chapter 23).

THE MESOZOIC ERA

The Mesozoic Era has traditionally been divided into three periods. The approximate times that they began and ended, in millions of years before the present, are:

	START	END
Triassic	248	213
Jurassic	213	144
Cretaceous	144	65

The Mesozoic Era, which began about 248 million years ago and ended about 65 million years ago, was a time very different from the present and one of intensive evolution of terrestrial plants and animals. The major evolutionary lines on land had been established during the mid-Paleozoic Era, but the evolutionary expansion of these lines—a radiation that led to the establishment of the major groups of living organisms today—took place in the Mesozoic Era. In tracing the evolution of these lines, we need to first consider for a moment events that happened just before the Mesozoic Era, in the Permian Period (286 to 248 million years ago), a time of drought and extensive glaciation that concluded the Paleozoic Era.

The Permian Period, the last period of the Paleozoic Era, ended with the greatest wave of extinction in the history of life on earth. In the sea, only about 4% of the species that were present earlier in the Permian Period survived to occur in the following Triassic Period. The Mesozoic and Paleozoic Eras were initially recognized as distinct from each other because of the effects of this major extinction event: the marine animals of the Paleozoic Era can be recognized instantly as different from those of the Mesozoic Era. The few kinds of marine organisms that survived into the Mesozoic Era, including gastropods (a group of mollusks that includes snails and their relatives) and bivalves (mollusks such as oysters and clams), crustaceans (a group of arthropods that includes crabs, shrimp, and lobsters), fishes (aquatic chordates), and echinoderms (a phylum that includes starfish and sea urchins) began to evolve rapidly during the Mesozoic Era, producing many new kinds of organisms. Some of these had ways of living that were radically different from those of their ancestors. For example, the first efficient burrowers appeared among the echinoderms.

Both on land and in the sea, the number of species of almost all groups of organisms has been climbing steadily since the Permian extinction 250 million years ago and is now at an all-time high. Even though the evolutionary radiation of marine organisms during this period has been spectacular, the story of the evolution of life on land during the Mesozoic Era is of even greater interest for us, for we are products of that history. Many of the significant events in the history of the vertebrates took place during the Mesozoic Era. Chapter 19 is devoted to describing the history of the vertebrates.

The History of Plants

The earliest known fossil plants are from the late Silurian Period, about 410 million years ago. By the close of the Paleozoic Era, plants had become abundant and diverse. Shrubs and then trees evolved and came to form forests, a kind of community that was abundant by the Carboniferous Period (360 to 286 million years ago). These Carboniferous forests in turn formed many of the great coal deposits that we are consuming now. Much of the land was low and swampy at this time and provided excellent conditions for the preservation of plant remains. In these coal deposits, we have a relatively complete record of the horsetails, ferns, and primitive seed-bearing plants that made up these ancient forests (Figure 18-7).

The Permian Period (286 to 248 million years ago), the last part of the Paleozoic Era, was cool and dry. The swamps of the preceding Carboniferous Period, which existed at a time of worldwide moist and warm climates, largely disappeared. The Permian Period seems to have been one of ecological stress, during which many kinds of new life forms originated. One of these groups was the conifers, a group of seed-bearing plants that is represented today by pines, spruces, firs, and similar trees and shrubs. Today the descendants of these conifers still form extensive forests in many temperate and subtropical areas. Seed-bearing plants with featherlike leaves, similar to the living group called cycads, were abundant in the Mesozoic Era and helped to give that period

FIGURE 18-7

Reconstruction of a Carboniferous Period forest, in which early vascular plants were dominant. All of the evolutionary lines illustrated here have become extinct, although other forms from the same phyla still exist. This reconstruction was made by the Field Museum of Natural History in Chicago.

its nickname, "the age of dinosaurs and cycads." Ultimately, however, the flowering plants, which apparently originated during the second half of the Mesozoic Era, became the dominant group of plants on the land.

The oldest fossils definitely known to be flowering plants are from the early Cretaceous Period, about 127 million years ago. It seems likely that the group actually originated somewhat earlier, but no one is certain how much earlier. Like the mammals, the flowering plants were for a long time a minor group; they became dominant, however, during the last part of the Cretaceous Period and have been more abundant than any other group of plants for about 100 million years.

The evolution of the flowering plants, which began in the Mesozoic Era, has continued strongly to the present. Today there are about 240,000 species of this large group, which greatly outnumbers all other kinds of plants. As the flowering plants became more diverse, so did the insects, which had feeding habits that were closely linked with the characteristics of the flowering plants (Figure 18-8); the two groups have evolved together. Indeed, all groups of terrestrial organisms, including mammals, birds, and fungi, have characteristics that are largely related to those of the flowering plants. These groups now dominate life on the land, literally making our world look the way it does.

The earliest fossil plants, about 410 million years old, evolved into others that formed extensive forests within 50 million years of their appearance. Conifers evolved during the Permian Period (286 to 248 million years ago). Together with the flowering plants, which appeared in the fossil record about 127 million years ago, conifers have come to dominate the modern landscape.

The Movement of Continents

Alfred Wegener, a German scientist, in 1915 proposed the idea that Africa and South America had once been joined as part of a giant land mass he called **Pangaea.** Wegener's theory was not accepted by the great majority of geologists and biologists of his day, but evidence gathered over the last 20 years has proven him right. We now know that, about 200 million years ago in the early Jurassic Period, the major continents were all together in one huge supercontinent (Figure 18-9).

We have learned that the earth's crust and associated upper mantle, which together are about 100 to 150 kilometers thick, are divided into plates. There are seven enormous plates and a series of smaller ones that lie between them. The continents are

FIGURE 18-8

A blister beetle, *Pyrota concinna,* eating the petals of a daisy in Zacatecas, Mexico. Beetles were among the first visitor to the flowers of the early flowering plants, spreading their pollen and thus bringing about cross-fertilization as they flew from flower to flower. The evolution of the flowering plants has been closely intertwined with that of the insects for more than 100 million years.

FIGURE 18-9

The positions of the continents at 200 million, 135 million, and 65 million years ago, and at present.

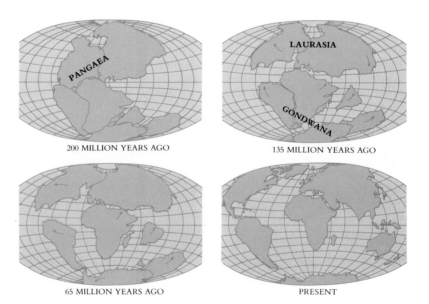

200 MILLION YEARS AGO

135 MILLION YEARS AGO

65 MILLION YEARS AGO

PRESENT

carried along on the surface of these plates as they move, like boxes on a conveyor belt. Earthquakes generally are caused by the relative movement of these plates. Thus an earthquake along the San Andreas Fault in California, such as the one that destroyed San Francisco in 1906, results from the relative movement of two gigantic segments of the earth's crust and mantle. Most mountains are thrust up by plate movements, as the light rocks pile up where plates collide. For example, the peaks of the Himalayas have been carried to the highest elevations on earth as a result of the collision of the Indian subcontinent with Asia, a process that has been under way for tens of millions of years and is still continuing today.

The continents have gradually moved apart from their positions 200 million years ago as parts of Pangaea. For example, the opening of the South Atlantic Ocean began about 125 to 130 million years ago, with Africa and South America directly connected earlier. South America has moved slowly toward North America, with which it was first connected by a land bridge, the Isthmus of Panama, between 3.1 and 3.6 million years ago. When the bridge was complete, many South American plants and animals such as the opossum (Figure 18-10) and the armadillo migrated overland into North America. At the same time, numerous North American plants and animals such as oaks, deer, and bears moved into South America for the first time. Changes on a greater or lesser scale occurred in the positions of all the continents, playing a major role in the patterns of distribution of organisms that we see today.

Australia provides a striking example of the way in which continental movements have affected the nature and distribution of organisms. Until about 53 million years ago, Australia was joined with a much warmer Antarctica, as was South America. Marsupials, which are best represented in Australia and South America at present, seem clearly to have moved overland between these continents by way of Antarctica when this was still possible. (Recently, a fossil mammal and some fossil plants have been discovered in Antarctica, and many exciting discoveries clearly lie in the future.) A land connection between Australia and South America seems to have persisted until about 38 million years ago. As the two continents moved farther from Antarctica, the land bridges connecting the three continents were eventually lost, and overland migration between them became impossible. From then on, as Australia moved northward toward the tropical islands fringing Southeast Asia, its peculiar plants and animals evolved in isolation.

One important effect of the separation of Australia and South America from Antarctica was the creation of a belt of open sea all the way around the world. This meant that the southern oceans, driven by the spinning of the earth to circle the globe, could complete the journey without moving through tropical regions and thus without being warmed by tropical temperatures. The very cold circulation that resulted is called the Circumantarctic Current. Its formation led directly to the development of the Antarctic ice sheet, which reached its full size by about 10 million years ago. Ultimately the cold temperatures associated with the formation of that enormous mass of ice triggered the onset of widespread continental glaciation in the Northern Hemisphere during the past few million years.

The movements of continents over the past 200 million years have profoundly affected the distribution of organisms on earth. Some of the major events include the linkage of South America with North America about 3.1 to 3.6 million years ago and the separation of Australia and South America from Antarctica, which triggered the formation of the southern and ultimately the northern ice sheets.

The Extinction of the Dinosaurs

Everyone is generally familiar with the disappearance of the dinosaurs, an event of global importance that took place about 65 million years ago at the end of the Cretaceous Period. Less discussed, but actually of more fundamental importance, was the disappearance of many other kinds of organisms that took place at about the same time. Among the **plankton**—free-drifting protists and other organisms that are still abun-

FIGURE 18-10

The North American opossum, *Didelphis virginiana*. Marsupials, which once lived in Europe, Africa, and Antarctica, are now best represented in Australia and South America. They formerly occurred in North America but became extinct about 40 million years ago. This particular marsupial, the opossum, is a mammal that migrated into North America from South America once the Isthmus of Panama had been uplifted above the sea, an event that occurred several million years ago. Armadillos entered North America in a similar fashion at about the same time.

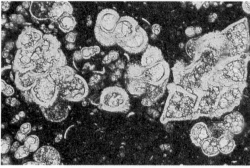
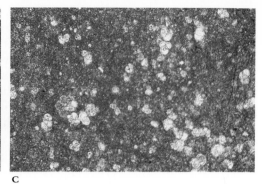

A   **B**   **C**

FIGURE 18-11

Evidence of the abrupt change that occurred at the end of the Cretaceous Period, 65 million years ago, in the vicinity of Gubbio, northern Italy.
A The white limestone below was deposited under the sea in the closing years of the Cretaceous Period, and the red limestone above was deposited during the first years of the Tertiary Period. They are separated by a layer of clay about 1 centimeter thick, in which iridium is abundant. The most abundant fossils in these sediments are members of the single-celled group of protists called forams. They were planktonic, floating freely in the sea.
B Large, ornamented, diverse forams, in sediments deposited during the late Cretaceous Period.
C Small, relatively unornamented, much less diverse forams deposited a few million years later, during the early years of the Tertiary Period. They are about one sixth the diameter of those that existed before the formation of the iridium layer.

dant in the sea—many of the larger (but still microscopic) forms suddenly disappeared about 65 million years ago, and a much lower number of smaller ones took their place. The same rapid changes occurred in at least some nonplanktonic marine animal groups such as bivalves. The ammonites, a large and diverse group of relatives of octopuses with shells, abruptly disappeared. In 1980 a group of distinguished scientists headed by physicist Luis W. Alvarez of the University of California, Berkeley, presented a controversial hypothesis about the reasons for this drastic change.

Alvarez and his associates discovered that the usually rare element iridium was abundant in a thin layer that marked the end of the Cretaceous Period not only in the strata in Italy, where the period had first been defined, but also in many other parts of the world at the same time (Figure 18-11). Iridium is rare on earth but common in meteorites. Alvarez and his colleagues proposed that, if a large meteorite, or asteroid, had struck the surface of the earth then, a dense cloud would have been thrown up. The cloud would have been rich in iridium, and as its particles settled, the iridium would have been incorporated in the layers of sedimentary rock that were being deposited at that time. By darkening the world, the cloud would have greatly slowed or temporarily halted photosynthesis and driven many kinds of organisms to extinction (Figure 18-12).

Calculations have shown that the cloud produced by a meteorite about 20 kilometers in diameter would certainly have caused these effects to occur. Daytime conditions would have resembled those on a moonless night—below the amount of light required for photosynthesis—for several months. In biological communities like the marine plankton, which are based directly on the continuous production of food by photosynthesis, such an effect would have had serious disruptive effects and could have produced the sudden change seen in the fossil record. Disruption of photosynthesis may also have been responsible for the extinction of certain other kinds of organisms—but is it reasonable to assume that it would have done away with the dinosaurs?

The Alvarez hypothesis has been controversial. It is not clear, for example, that the dinosaurs became extinct suddenly, as they should have if driven to extinction by a meteorite collision. Also, it is not clear that other kinds of animals and plants show the kinds of patterns that would have been predicted from these effects. Whether or not a meteorite impact caused widespread extinction 65 million years ago is a hypothesis that is still under active consideration.

> **The occurrence of a worldwide layer rich in iridium 65 million years old suggests that a giant meteorite struck the earth at that time and threw up a huge cloud. The role of this cloud in the extinction of various kinds of organisms is under active investigation.**

THE CENOZOIC ERA: THE WORLD WE KNOW

We conclude this chapter with a brief account of some of the major evolutionary changes that have occurred during the past 65 million years, changes that have resulted in the conditions we now experience. The relatively warm and moist climates of the early Cenozoic Era have gradually given way to today's climates. By making the es-

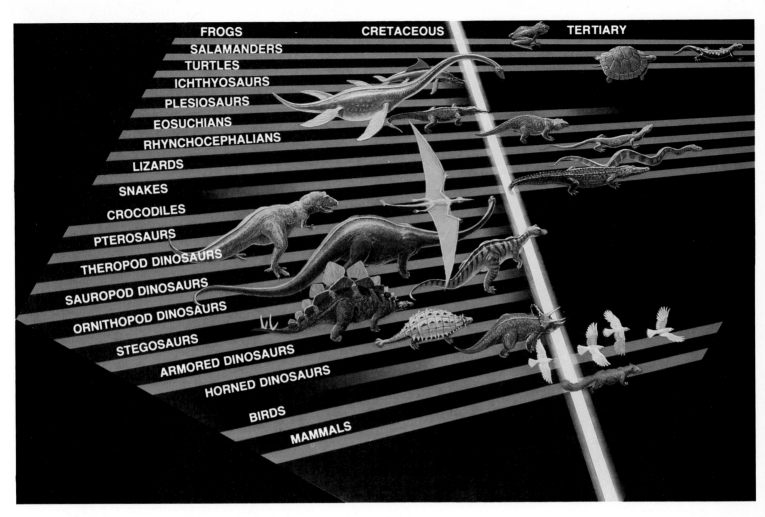

FIGURE 18-12

Extinction of the dinosaurs. For reasons that are still debated, the dinosaurs became extinct 65 million years ago, in a major extinction event that also eliminated all the great marine reptiles (plesiosaurs and ichthyosaurs), as well as the largest of the primitive land mammals. The birds and smaller mammals survived and went on to occupy the aerial and terrestrial niches left vacant by the dinosaurs. Crocodiles, small lizards, and turtles also survived, but reptiles never again achieved the diversity of the Cretaceous Period.

tablishment of the Circumantarctic Current possible, the final separation of South America and Antarctica about 27 million years ago set the stage for worldwide glaciation. The ice mass that has been formed as a result of this glaciation has made the climate cooler near the poles, warmer near the equator, and drier in the middle latitudes than ever before.

In general, forests covered most of the land area of the continents, except for Antarctica, until about 15 million years ago, when they began to recede rapidly. During the time when these forests were receding and modern plant communities were appearing, some of the continents were approaching one another again after having been widely separated during most of the Cenozoic Era. The organisms in Australia and South America, particularly, evolved in isolation from all those in the rest of the world, mainly during the Cenozoic Era. Evolution in isolated regions of this kind has been responsible for the distinctive characteristics of the groups of plants and animals found in different regions of the world. During the past several million years, the formation of extensive deserts in northern Africa, the Middle East, and India made migration very difficult for the organisms of tropical forests between Africa and Asia. This formation of desert barriers, in turn, has provided further opportunities for evolution in isolation. In general, the overall character of the Cenozoic Era has been set by a deteriorating climate, sharp differences in habitats even within small areas, and the regional evolution of distinct groups of plants and animals. These factors have enhanced the opportunities for the rapid formation of many new species by the processes outlined in Chapter 17.

Throughout the 65 million years of the Cenozoic Era, the world climate has deteriorated steadily, and the distributions of organisms have become more and more regional in character.

SUMMARY

1. Fossils provide a record of life in the past. They occur largely in sedimentary rocks, with the actual organic remains gradually being replaced by minerals.

2. The earth originated about 4.6 billion years ago, and the oldest rocks are about 3.8 billion years old. Fossil bacteria about 3.5 billion years old are the oldest direct evidence of life on earth. Stromatolites, massive deposits of limestone formed by cyanobacteria, appear in the fossil record starting about 2.8 billion years ago.

3. The first unicellular eukaryotes appeared about 1.5 billion years ago; all earlier forms of life were bacteria. Multicellular animals appeared about 700 million years ago; the earliest, soft-bodied forms are poorly represented in the fossil record.

4. During Phanerozoic time, which started about 630 million years ago, fossils of multicellular organisms become frequent. Phanerozoic time is divided into three eras: the Paleozoic Era, 590 to 248 million years ago; the Mesozoic Era, 248 to 65 million years ago; and the Cenozoic Era, 65 million years ago to the present.

5. All of the phyla of organisms except plants seem to have evolved during the Cambrian Period (590 to 505 million years ago).

6. The evolution of hard skeletons, including shells and similar structures, about 570 million years ago seems to have been correlated with the appearance of many diverse modes of life among animals.

7. Plants and terrestrial arthropods appeared about 410 million years ago; terrestrial vertebrates (amphibians) appeared about 360 million years ago. These groups of organisms, together with the fungi, have dominated life on the land since then.

8. The Paleozoic Era ended with the Permian Period (286 to 248 million years ago), a cold and dry period during which conifers and reptiles first became evident.

10. The Mesozoic Era was a time when the outlines of life on earth as we know it were established. Flowering plants had become dominant by the end of the era; insects, mammals, birds, and other groups had begun to evolve in relation to the diversity of these plants.

11. There have been five major episodes of mass extinction during the history of life on earth. The most drastic was that at the end of the Permian Period, about 248 million years ago, when some 96% of marine animals became extinct. At the end of the Cretaceous Period, 65 million years ago, the dinosaurs and many other kinds of organisms disappeared.

12. The diversity of living things, measured in terms of species number, has been increasing steadily for the past 250 million years and is at an all-time high now. The wide separation of the continents and their arrangement into novel configurations probably has played a role in this trend.

REVIEW

1. The giant land mass that was formed when all the continents united is called _____.

2. In biological classification, major groups into which kingdoms are divided are called _____.

3. The oldest fossils, which are bacteria, are about _____ years old.

4. What four major groups of organisms colonized the land from the sea?

5. The most drastic extinction event in the history of life on earth happened at the end of the _____ period.

SELF-QUIZ

1. Place the following periods and eras in order, oldest first.
   - (a) Permian
   - (d) Cenozoic
   - (b) Cambrian
   - (e) Devonian
   - (c) Cretaceous

2. Which of the following invaded land *first?*
   - (a) fungi
   - (d) vertebrates
   - (b) arthropods
   - (e) we don't know
   - (c) plants

3. A stromatolite is a
   - (a) colony of fossil cyanobacteria
   - (b) limestone deposit
   - (c) kind of fossil coral reef
   - (d) coelacanth
   - (e) mass of compacted seashells

4. Match the organism with the period in which it arose.
   - (a) amphibians
   - (1) Cambrian
   - (b) chordates
   - (2) Permian
   - (c) conifers
   - (3) Devonian
   - (d) first plants
   - (4) Cretaceous
   - (e) flowering plants
   - (5) Silurian

5. During the extinction that marked the end of the Cretaceous Period, how many of the following forms did *not* become extinct?
   - (a) flowering plants
   - (d) mammals
   - (b) smaller plankton
   - (e) fishes
   - (c) ammonites

THOUGHT QUESTIONS

1. Dinosaurs and mammals both lived throughout the Mesozoic Era, a period of more than 150 million years; all this time the dinosaurs were the dominant form, mammals being a minor group. Both mammals and small reptiles survived the Cretaceous extinction. Why do you suppose reptiles did not go on to become dominant again, rather than mammals?

2. There seems to be no iridium layer preserved in the rocks from four of the five major periods when mass extinction occurred. What are some of the ways in which one could account for these four extinction events?

3. Separation of Australia and South America from Antarctica led to the formation of a huge sheet of ice over Antarctica. Do you think that a land bridge connecting Alaska to Siberia would have had a similar effect, forming a huge Arctic ice mass over the North Pole?

FOR FURTHER READING

LEWIN, R.: *Thread of Life: The Smithsonian Looks at Evolution,* Smithsonian Books, Washington, D.C., 1982. A beautifully illustrated chronicle of the history of life on earth.

McMENAMIN, M.A.S.: "The Emergence of Animals," *Scientific American,* April 1987, pages 94–102. Why did the animals of 570 million years ago become diverse so rapidly, taking up many new ways of life?

RAUP, D.M.: *The Nemesis Affair,* W.W. Norton Co., New York, 1985. An account by one of the co-discoverers of the 26 million–year extinction cycle of how they discovered it and how it was treated in the media. Fascinating reading.

RICHARDSON, J.: "Brachiopods," *Scientific American,* September 1986, pages 100–106. A fascinating group of marine organisms, well represented in the fossil record, illustrates many of the kinds of changes that have occurred during the history of life on earth.

VIDAL, G.: "The Oldest Eukaryotic Cells," *Scientific American,* February 1984, pages 48–57. An up-to-date review of the evidence for the origin of eukaryotic cells; the author concludes that this event occurred about 1.4 billion years ago.

WARD, P.: "The Extinction of the Ammonites," *Scientific American,* October 1983, pages 136–146. An account of the complex fossils of these mollusks, with interesting ideas on the possible causes of their extinction.

FIGURE 19-1   These footprints were made in Africa 3.7 million years ago. A mother and child walking on the beach might leave such tracks. But these tracks are not human. Preserved in volcanic ash, these tracks record the passage of two individuals of the genus *Australopithecus,* the group from which our genus, *Homo,* evolved. The tracks continue 24 meters before disappearing; by looking at them we journey into our past.

HOW WE EVOLVED: VERTEBRATE EVOLUTION

Chapter 19

Overview

The vertebrates originated in the sea as jawless fishes. The bony fishes, which evolved from them, are the most plentiful kind of vertebrates today. The first vertebrate land-dwellers were amphibians, but they are not truly terrestrial because they still require frequent access to water. The first true terrestrial vertebrates were the reptiles, which gave rise independently to the birds and the mammals, including human beings. Human beings are biologically similar to the apes and, together with them, make up an evolutionary line that has existed for at least 36 million years. *Australopithecus,* a genus that includes the direct ancestors of humans, appeared in Africa perhaps 5 million years ago. Ultimately *Australopithecus* gave rise to our genus, *Homo,* also in Africa. The genus *Homo* has existed for about 2 million years; *Homo sapiens,* our species, has existed for perhaps 500,000 years.

For Review

Here are some important terms and concepts that you will encounter in this chapter. If you are not familiar with them, you should review them before proceeding.

Theory of evolution (Chapter 16)

How species evolve (Chapter 17)

Major features of evolutionary history (Chapter 18)

Origin of mammals (Chapter 18)

The first vertebrates were jawless fishes that swam in the seas 500 million years ago. From them evolved first the sharks, with light skeletons made of cartilage, and then the bony fishes, which, unlike sharks, do not have to swim constantly in order to breathe. The first vertebrates to walk on land were the amphibians, which evolved from bony fishes about 350 million years ago. The reptiles were probably derived from the amphibians, but they may have originated directly among the kinds of fishes that also gave rise to the amphibians. From different groups of reptiles, in turn, were derived the two most successful terrestrial groups of vertebrates, the mammals and birds. In this chapter we shall review the evolutionary journey from jawless fishes to mammals like ourselves (Figure 19-1).

Figure 19-2 shows the evolutionary relationship between the seven major groups, or **classes,** of vertebrates that have living representatives. Phyla are divided into classes, which in turn are divided into smaller categories, **orders,** such as rodents, carnivores, and bats among mammals. These orders in turn consist of one or more

FIGURE 19-2

Diagram of the evolutionary relationships of the seven classes of vertebrates that include living representatives, and some of their extinct relatives.

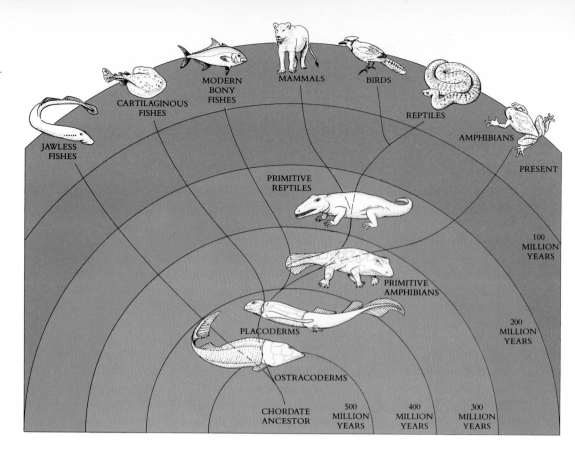

families, such as cats and weasels among the carnivores. In turn, families include **genera** and **species.** Such a system of classification, in which successively smaller units of classification are included within one another, like boxes within boxes, is called a **hierarchical** system. It will be illustrated further in Chapter 24.

GENERAL CHARACTERISTICS OF THE VERTEBRATES

The approximately 42,500 species of chordates (phylum Chordata), including birds, reptiles, amphibians, fishes, and mammals, are distinguished by three principal features: (1) a single, dorsal, hollow **nerve cord,** or main trunk, with which the nerves that reach the different parts of the body are connected; (2) a rod-shaped **notochord,** which forms between the nerve cord and the developing gut (which becomes the stomach and intestines) in the early embryo; and (3) **pharyngeal slits.** The **pharynx** is a muscular tube that connects the mouth cavity and the esophagus. It serves as the gateway to the digestive tract and to the windpipe, or **trachea.**

With the exception of two groups of relatively small marine animals, the tunicates and lancelets, which will be considered in Chapter 28, all chordates are vertebrates. Vertebrates differ from other chordates in usually having a **vertebral column,** a tube of hollow bones called vertebrae, which encloses the dorsal nerve cord like a sleeve and protects it. In addition, vertebrates (except for the agnathans, as you will see) have a distinct and well–differentiated head; as a result, they are sometimes called the **craniate** chordates (Greek *kranion,* skull). Most vertebrates have a bony skeleton, although the living members of two of the classes of fishes, Agnatha (lampreys and hagfishes) and Chondrichthyes (sharks and rays), have a cartilaginous one. In vertebrates the notochord becomes surrounded and then replaced during the course of embryological development by the vertebral column. All of the features of chordates, however, are evident in their embryos, even among the most advanced vertebrates (Figure 19-3).

Vertebrates are a group of chordates characterized by a vertebral column surrounding a dorsal nerve cord.

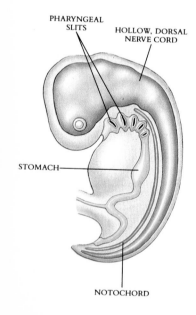

FIGURE 19-3

Embryos reveal some of the principal distinguishing features of the chordates, even though these features may be lost during the course of development.

JAWLESS FISHES

The first vertebrates to evolve were jawless fishes, members of the class Agnatha, about 540 to 550 million years ago, in the mid–Cambrian Period. For more than 100 million years, they were the *only* vertebrates. Many of the major groups of agnathans are extinct, and the living ones, the lampreys and hagfishes, belong to a single, fairly uniform group. Both are naked, eel-like fishes. In the living agnathans, the notochord persists throughout the life of the animal, and the body is supported by an internal skeleton made of cartilage. We know from their fossils, however, that the ancestral agnathans had bony skeletons and therefore that skeletons of this kind were lost during the course of evolution of the living ones.

Lampreys have round mouths that function like suction cups (Figure 19-4). Using these specialized mouths, lampreys attach themselves to bony fishes. Once attached, they rasp through the skin of the fish with their tongues, which are covered with sharp spines, sucking out the blood of the fish through the hole that they make. Sometimes lampreys can be so abundant as to constitute a serious threat to commercial fisheries.

> **Agnathans were the first vertebrates; in fact, for more than 100 million years they were the *only* vertebrates. All agnathans have cartilaginous skeletons and lack jaws. The living agnathans are the parasitic lampreys and hagfishes.**

THE APPEARANCE OF JAWED FISHES
The Evolution of Jaws

Jaws first developed among vertebrates that lived about 410 million years ago, toward the close of the Silurian Period. The biting jaws characteristic of modern vertebrates were produced by the evolutionary modification of one or more of the gill arches, originally the areas between the gill slits (Figure 19-5). Jaws allowed ancient fishes to become proficient predators, to defend themselves, and to collect food from places and in ways that had not been possible earlier. Fishers were able to bite and chew their food instead of sucking it or filtering it in like all of the more primitive chordates did. Because the early jawed fishes outcompeted their jawless ancestors, only about 63 species of agnathans have survived to the present. Over the same period of time, fishes have become dominant throughout the waters of the world. Of the approximately 42,500 species of living chordates, about half are fishes.

Sharks and Rays

The first jawed fishes to evolve were members of the class Chondrichthyes: sharks, dogfish sharks, skates, and rays (Figure 19-6). Many hundreds of extinct species of the class Chondrichthyes are known from the fossil record, but there are only about 850 living ones. These fishes are mostly scavengers and carnivores, and the internal skeleton of the living forms is composed of cartilage—a soft, light, and elastic material. Such skeletons are lighter and more buoyant than those of the early agnathans that the Chondrichthyes replaced. In addition to the evolutionary advance represented by such a skeleton, the Chondrichthyes were the first vertebrates to develop efficient fins. With these strong advantages, they soon largely replaced agnathans throught the world.

> **The sharks and other Chondrichthyes were the first jawed vertebrates and the first to develop efficient fins. They evolved from agnathans that had bony skeletons, but soon developed cartilaginous ones.**

The skin of Chondrichthyes is covered everywhere with small, pointed **denticles,** similar in structure to the teeth of other vertebrates. As a result, the skin of these fishes has a rough, sandpaper-like texture. Their teeth, which are abundant in the fossil record, are enlarged versions of these denticles.

FIGURE 19-4

Lampreys attach themselves to fishes by means of their suckerlike mouths, rasp a hole in the body cavity, and suck out the blood and other fluids within.

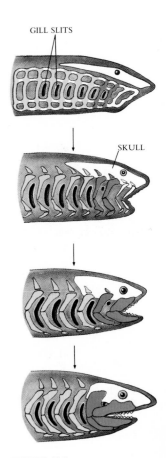

GILL SLITS

SKULL

FIGURE 19-5

Jaws evolved from the anterior gill arches of ancient, jawless fishes.

A   **B**   **C**

FIGURE 19-6

Members of the class
Chondrichthyes, which are
mainly predators or scavengers,
are constantly and gracefully in
motion. As they move, they
create a flow of water over the
gills, extracting oxygen from this
water. Three representatives of
the class are shown here.
A Blue shark.
B Diamond sting ray.
C Manta ray.

Sharks drive themselves through the water by sinuous motions of the whole body and by their thrashing tails. Such motion tends to drive the sharks downward, but this tendency is corrected by their two spreading pectoral fins. In the skates and rays, these pectoral fins have become enlarged and are undulated when these fishes move, which gives these animals a very characteristic appearance.

For their supply of oxygen, many sharks (and those bony fishes that swim constantly) depend on a constant stream of water that is forced over the gills and brings dissolved oxygen with it; others are able to pump water through their gills while they are stationary. Fishes that obtain their oxygen by moving constantly can literally drown if they are prevented from swimming. Drowned sharks are often found trapped in the nets used to protect Australian beaches, for example.

BONY FISHES

The vast majority of the more than 18,000 known species of fishes are bony fishes (class Osteichthyes). Although bones are heavier than cartilaginous skeletons, bony fishes are still buoyant because they possess a **swim bladder** (Figure 19-7). A swim bladder is a gas-filled sac that allows the fishes to regulate their buoyant density; for this reason, they can remain suspended at any depth in the water. Swim bladders apparently evolved as outpocketings of the pharynx, specialized for respiration, in fishes of the Devonian Period. Stationary bony fishes do not drown because they use muscles to draw water through their gills, by way of their gill slits. These gill slits are simply the pharyngeal slits that are characteristic of all chordates, at least when they are embryos.

Bony fishes, which comprise about half of the species of living vertebrates, clearly evolved in fresh water, as can be seen from the places where early fossils of the class are found and the characteristics of these ancient fishes. Bony fishes apparently entered the sea only after a number of their distinctive evolutionary lines had appeared. Bony fishes are abundant both in the modern seas and in fresh water (Figure 19-8). Many kinds of fishes spend a portion of their lives in fresh water and another portion in the sea.

FIGURE 19-7

Diagram of a swim bladder. This
structure, which evolved as an
outpocketing of the pharynx, is
used by the bony fishes to control
their buoyancy in water.

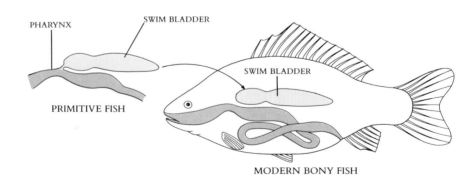

FIGURE 19-8

The bony fishes (class
Osteichthyes) are extremely
diverse. Four examples are
shown here.
A Koran angelfish, *Pomacanthus
semicircularis,* in Fiji; one of the
many striking fishes that lives
around coral reefs in tropical seas.
B Puffer. Puffers, which are also
marine fishes, avoid being eaten
by inflating themselves when
attacked, thus projecting their
spines outward.
C A sea horse. An example of
the bizarre forms of some fishes.
Sea horses move slowly and are
difficult to see among the
marine algae and sea grasses where
they live.
D Moray eel, Rangiroa, French
Polynesia.

**Bony fishes (class Osteichthyes) possess swim bladders, which enable them to
increase their buoyancy and thus to offset the greater weight of a bony
skeleton. They are the most successful of the vertebrates, accounting for
about half of all living vertebrate species.**

THE INVASION OF THE LAND

About 350 million years ago a member of the large group of bony fishes called the lobe-
finned fishes began the evolutionary journey onto land. Today, the lobe-finned fishes
are represented by the lungfishes, freshwater fishes with species in Australia, Africa,
and South America; and by the **coelacanth,** a unique marine fish that is found in the
warm waters of the western Indian Ocean (see boxed essay, page 346). The lobe-finned
fishes, which were to become the ancestors of all **tetrapods,** or terrestrial, basically
four-limbed vertebrates, perhaps began to explore an existence on land at the margins
of freshwater ponds or swamps. In these animals, which ultimately became the earliest
amphibians, primitive lungs gradually developed into the kinds of efficient air-breathing organs seen in modern tetrapods. These lungs were originally supplementary organs, making it possible to breathe air directly when this was necessary; lungs of this
sort still occur today in the lungfishes.

In these early terrestrial vertebrates, efficient locomotion on land was gradually
made possible by the evolution of strong skeletal supports in the **thoracic** portion of the
body, that part just behind the head (Figure 19-9). Such supports provided a more rigid
base for the limbs, which were derived from the kinds of fins found in the lobe-finned

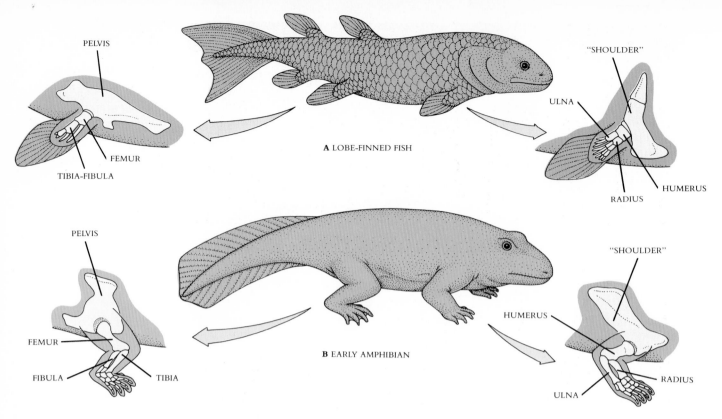

PELVIS

FEMUR

TIBIA-FIBULA

A LOBE-FINNED FISH

"SHOULDER"

ULNA

HUMERUS

RADIUS

PELVIS

FEMUR

FIBULA

TIBIA

B EARLY AMPHIBIAN

"SHOULDER"

HUMERUS

RADIUS

ULNA

FIGURE 19-9

A comparison between the limbs of a lobe-finned fish and those of a primitive amphibian.
A A lobe-finned fish. Some of the animals of this appearance could probably move out onto land.
B A primitive amphibian. As illustrated by their skeletal structure, the legs of such an animal could clearly function on land much better than the fins of the lobe-finned fish.

fishes. These two evolutionary advances—that is, the abilities to use gaseous oxygen for respiration and to move from one place to another on land—became the basis of the success and the later evolutionary radiation of vertebrates on land.

AMPHIBIANS

The first land vertebrates were amphibians, members of a class that first became abundant about 300 million years ago during the Carboniferous Period. The early amphibians had rather fishlike bodies; short, stubby legs; and lungs (see Figure 18-6). Ultimately these amphibians probably gave rise to all of the other tetrapods, although there is a remote possibility that the reptiles evolved directly from the fishes; in that case the amphibians would be an evolutionary "dead end." Even so, the amphibians constitute the group of tetrapods most like the ones that originally evolved from the fishes. As a group, amphibians, unlike reptiles, birds, and mammals, depend on the availability of water during their early stages of development. Many amphibians live in moist places even when they are mature. There are about 3150 species in this class. Members of this class were the dominant vertebrates on land for about 100 million years, until they were gradually replaced in their dominant role by the reptiles.

The two most familiar orders of amphibians are (1) Caudata, which have tails as adults, and (2) Anura, which lack tails as adults. There are of course other differences between the members of these two orders, but the presence or absence of a tail is the most obvious one. Caudata are the salamanders, mud puppies, and newts, which comprise a total of about 350 species; Anura, the frogs and toads, comprise about 2800 species (Figure 19-10). Amphibians possess the first true lungs, but these organs are relatively inefficient. Respiration also takes place through their thin, moist, glandular skin, which lacks scales in both the orders Caudata and Anura, and through the lining of their mouth. The constant loss of water through the skins of amphibians is one of the reasons that these animals must remain, for the most part, in moist habitats.

Amphibians lay their eggs, which lack water-retaining external membranes and shells and dry out rapidly, directly in water or in moist places. They also lack specialized organs for copulation. In Caudata, packets of sperm are taken into the body of the

A

B

C

D

FIGURE 19-10

Some of the kinds of amphibians, representing the diversity of the class Amphibia.

A Giant toad, *Bufo marinus*.
B Red-eyed tree frog, *Agalychnis callidryas*.
C Tiger salamander, *Ambystoma tigrinum*.
D Tennessee cave salamander, *Gyrinophilus palleucus*. This species and a number of other salamanders remain permanently larval in form throughout their lives, but become sexually mature and breed.

female after the male deposits them, and fertilization is internal; in Anura, it is external. Anuran larvae are tadpoles, which usually live in the water, where they feed on minute algae. Their adults, which are highly specialized for jumping and very different from the larvae in appearance, are carnivorous. Young salamanders are carnivorous like the adults and look like small versions of them. Many salamanders swim efficiently and return to water to breed.

> **Amphibians are terrestrial but still depend on a moist environment; their eggs are laid in water, and the development of their larvae takes place there.**

Although amphibians appear to us as primitive, they are in fact members of a successful group—one that has survived over 300 million years. The amphibians evolved long before the dinosaurs and have, thus far, outlasted them by 65 million years.

REPTILES

Reptiles (class Reptilia) have a dry skin, covered with scales, that efficiently retards water loss. As a result, the members of this large class, consisting of nearly 6000 living species, are independent of free water, unlike their ancestors. Nonetheless, of the four orders of reptiles that have living representatives (Figure 19-11), the Crocodilia (crocodiles, alligators, and caimans) live in water, as do most of the members of the order Chelonia (turtles and tortoises). Members of a third, and by far the largest, order, Squamata (lizards, snakes, and other reptiles) are almost entirely terrestrial. Snakes evolved from lizards through the loss of their limbs and other evolutionary modifications.

FIGURE 19-11

Representatives of the major groups of reptiles (class Reptilia).
A River crocodile, *Crocodilus acutus*. The crocodiles and their relatives resemble birds and mammals in having a completely four-chambered heart; all other living reptiles have a three-chambered one. Crocodiles, like birds, are related to dinosaurs, rather than to any of the other living reptiles.
B Red-bellied turtles, *Pseudemys rubriventris*. This attractive turtle occurs frequently in the northeastern United States.
C An Australian skink, *Sphenomorphus*. Some burrowing lizards lack legs, and the snakes evolved from one line of legless lizards.
D Smooth greensnake, *Opheodrys vernalis*.

CRETACEOUS   JURASSIC

The first reptiles appeared during the age of dominance of the amphibians, the Carboniferous Period, about 300 million years ago. Immediately following the Carboniferous Period, the Permian Period (286 to 248 million years ago) was a period of widespread glaciation and drought. The special adaptations of reptiles, including their water-resistant skin and more efficient lungs, probably gave them an advantage as arid climates became widespread, and over the next 50 million years reptiles gradually replaced amphibians. The dinosaurs, by far the best-known group of fossil organisms, appeared near the start of the Mesozoic Era, at least 225 million years ago. They dominated the land for most of the following 180 million years, until the start of the Tertiary Period (Figure 19-12).

Reptiles were the first truly terrestrial vertebrates. They have more efficient lungs than amphibians and so do not need to breathe through their skin, which is dry and efficiently retards water loss.

One of the most critical adaptations of reptiles in relation to their life on land is the evolution of the **amniotic egg,** which protects the embryo from drying out, nourishes it, and enables it to develop outside of water (Figure 19-13). Amniotic eggs, characteristic of reptiles and birds (and of the very few egg-laying mammals) retain their own water. They contain a large **yolk,** the primary food supply for the embryo, and abundant **albumen,** or egg white, which provides additional nutrients and water. An egg membrane, called the **amnion,** surrounds the developing embryo, enclosing a liquid-filled space within which the embryo develops. Blood vessels grow out of the embryo through the membranes of the egg to its surface, where they take in oxygen and release carbon dioxide. The egg is more easily permeable to these gases than to water. In most reptiles, the egg shell is leathery. In addition to the amniotic egg, another adaptation of reptiles to a terrestrial existence is the presence of copulatory organs, which the male inserts into the female, thus protecting the eggs and sperm from drying out before their fusion.

The key to successful invasion of the land by vertebrates, as it was by arthropods, was the elaboration of ways to avoid desiccation in dry air. In addition to a dry skin that retained water, a second pivotal innovation of reptiles was the watertight amniotic egg, which contains embryonic nutrients and is permeable to gases but not to water.

FIGURE 19-12

Some of the remarkable diversity of dinosaurs, as shown in a reconstruction from the Peabody Museum, Yale University. This painting covers a span of approximately 320 million years during the later Paleozoic and Mesozoic Eras, ending 65 million years ago at the end of the Cretaceous Period. Throughout this vast period of time, the remarkable increase in structural complexity and overall diversity of the dinosaurs can be seen—until they abruptly became extinct, giving way to the dominant mammals of the Cenozoic Era. Flowering plants can be seen for the first time at the left-hand side of the illustration. The names of the geological periods are given along the bottom of the painting.

TRIASSIC　　　　　　　　　　PERMIAN　　　　　　　　CARBONIFEROUS

FIGURE 19-13

The amniotic egg is perhaps the most important feature that allows reptiles to live in a wide variety of terrestrial habitats.

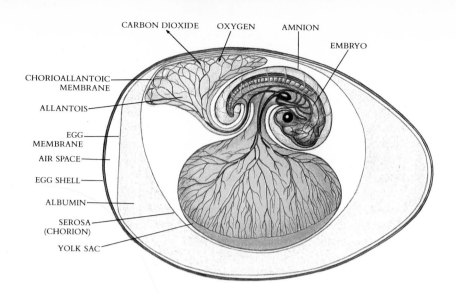

TEMPERATURE CONTROL IN LAND ANIMALS

Reptiles, like amphibians and fishes but unlike birds and mammals, are **ectothermic** (from the Greek words *ectos,* "outside" + *thermos,* "heat"), regulating their body temperatures by taking in heat from the environment. They have, however, developed a wide array of behavioral mechanisms that enable them to control their internal temperatures with remarkable precision. Even though ectothermic animals are called "cold-blooded," they often may actually be able to maintain bodily temperatures much warmer than their surroundings, as a result of their behavior. Several scientists have postulated that at least some dinosaurs, like the birds and mammals, were **endothermic** (Greek *endos,* "inside"), capable of regulating their body temperatures internally. Endothermic animals are also sometimes called "warm-blooded" and **homeotherms** (Greek *homios,* similar).

The fact that birds (class Aves) and mammals (class Mammalia) are endothermic makes them unique among living organisms. Consequently, the members of these two classes are able to remain active at night, even if temperatures are cool. They are also able to live at higher elevations and farther north and south than most reptiles and amphibians. Endotherms can maintain their internal organs within a very narrow range of temperature more or less independently of external conditions, an adaptation that makes possible the more efficient maintenance and functioning of these animals as compared with the ectotherms. About 80% of the calories in the food of endotherms, however, goes to maintain their body temperature; a reptile that is the same size as a mammal can therefore subsist on about a tenth of the amount of food required by the mammal. For this reason, reptiles are able to live in places such as some extreme deserts where food may be too scarce to support many birds and mammals.

BIRDS

There are approximately 9000 species of living birds (class Aves; Figure 19-14). Birds were derived from reptilian ancestors during the Mesozoic Era; *Archaeopteryx* (Figure 19-15) is one of the earliest known birds, from about 150 million years ago. In many ways, it is little more than a reptile with feathers; presumably its power of flight was somewhat limited, and it may have been somewhat of a glider.

The modern descendants of *Archaeopteryx* and other ancient birds have become true masters of the air. In them, the wings are forearms, modified in the course of evolution. The scaly skin of birds has feathers, flexible, strong organs that can be replaced and that make up an excellent airfoil—a surface, such as a wing or rudder, that is designed to obtain resistance on its surfaces from the air through which it moves—for flying. The membranes that make up the flying surface of a bat or one of the extinct groups of flying reptiles can be severely damaged by a single rip. In contrast, the feath-

FIGURE 19-14

The birds (class Aves) are a large and very successful group of about 9000 species, more than any other class of vertebrates except the bony fishes. These photographs show some of their diversity.

A Ostrich, *Struthio camelus.* The group of wingless birds that includes the ostrich (Africa), rhea (South America), emu (Australia), and kiwi (New Zealand)—the ratite birds—seems to represent an evolutionary line distinct from all other birds. They migrated between their now widely separated southern lands before these lands drifted apart.

B A bee-eater, *Merops,* in Tanzania. The bee-eaters nest communally, with many adult individuals participating in the care of the young.

C Purple finch, *Carpodacus purpureus,* one of the very large group of birds called passerines. Darwin's finches and their relatives, shown in Figures 17-11 and 17-12, are passerines, as are sparrows, warblers, chickadees, and many other familiar birds.

D Northern saw-whet owl, *Aegolius acadius.* Owls, hawks, eagles, and similar predatory birds are called raptors.

E A pair of wood ducks, *Aix sponsa,* showing the marked difference between males and females.

FIGURE 19-15

Artist's reconstruction of *Archaeopteryx,* an early bird about the size of a crow. Closely related to its ancestors among the bipedal dinosaurs, *Archaeopteryx* lived in the forests of central Europe 150 million years ago. The teeth and long, jointed tail are features not found in any modern birds. Discovered in 1862, *Archaeopteryx* was cited by Darwin in support of his theory of evolution. The true feather colors of *Archaeopteryx* are not known.

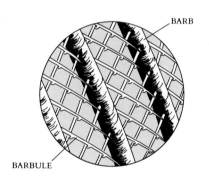

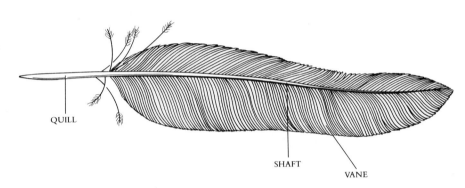

BARB

QUILL

BARBULE

SHAFT

VANE

FIGURE 19-16

A feather, showing the way in which the vanes—secondary branches—are linked together by microscopic barbules.

ers of birds are individually replaceable. Even after the loss of many individual feathers, the bird can keep flying. Several hundred microscopic hooks along the sides of the individual barbules of a feather attach the barbules to one another. These hooks unite the feather so that in a marvelous way, unique to birds, a feather is perfectly adapted for flight (Figure 19-16). Overall, flight in birds is made possible by light, hollow bones, by the replacement of scales with feathers, and by the development of highly efficient lungs that supply the large amounts of oxygen necessary to sustain muscle contraction during prolonged flight.

In birds, the eggs are highly resistant to drying out. Fertilization is direct, although copulatory organs are found only in a few groups, including ducks and geese.

Birds evolved from reptiles and, with the bats, are one of the two living vertebrate groups that have achieved a full mastery of the air.

MAMMALS

The first known mammals are from the early Mesozoic Era, about 200 million years ago. They were clearly derived from reptiles; in fact, a number of the early fossil organisms are difficult to classify satisfactorily either as reptiles or as mammals, and a number of transitional forms are known. The first mammals were probably small; fed on insects, as suggested by their teeth; and, judging from their large eyes, may have been nocturnal. Mammals remained secondary to the reptiles, both in absolute numbers and in numbers of species, until about 65 million years ago, following the start of the Tertiary Period. After the extinction of the dinosaurs at that time (see Chapter 18), mammals became abundant.

FIGURE 19-17

Monotremes (class Mammalia).
A A duck-billed platypus, *Onithorhynchus anatinus*, at the edge of a stream in Australia. The "duck bill" and webbed feet of this unique mammal can be seen easily in this photograph.
B Spiny echidna, *Tachyglossus aculeatus*.

Characteristics of Mammals

There are about 4500 species of living mammals (class Mammalia), including human beings (see Figures 19-17, 19-18, 19-19, and other figures in Chapters 18 and 19). Like birds, mammals are homeotherms. Their skin is covered with hair during at least some stage of their life cycle. Like birds and crocodiles, mammals have a four-chambered heart with complete double circulation: separate systems for circulating oxygen-rich and oxygen-poor blood. Mammals have copulatory organs like their reptilian ancestors, and fertilization is internal. Mammals nourish their young with milk, a nutritious substance produced in the mammary glands of the mother. Their locomotion is advanced over that of the reptiles, which in turn is advanced over that of the amphibians; the legs of mammals are positioned much farther under the body than those of reptiles and are suspended from limb girdles, which permit greater leg mobility.

> **Mammals nourish their young with milk and are covered with hair rather than scales or feathers.**

Monotremes

In all but a few mammals, the young are not enclosed in eggs when they are born. The **monotremes,** or egg-laying mammals, consisting of the duck-billed platypus and the echidna, or spiny anteater (Figure 19-17), are the only exceptions to this rule. These animals occur only in Australia and New Guinea, and they are not known from the fossil record. In echidnas, the eggs are transferred to a special pouch, where they are held until the young hatch.

Marsupials

The **marsupials** are mammals in which the young are born early in their development, sometimes as soon as 8 days after fertilization, and are retained in a pouch, or **marsupium.** Living marsupials are found only in Australia, where they are abundant and diverse (see Figures 16-14 and 19-18), and in North and South America. Ancient fossil marsupials, 40 to 100 million years old, are found in North America, where they became extinct until their geologically recent reintroduction, which occurred after North and South America were linked 3 to 4 million years ago. The opossum, a familiar mammal in North America, arrived from the south in just this way (see Figure 18-10).

FIGURE 19-18

A marsupial (class Mammalia). Kangaroo with young in its pouch.

B

FIGURE 19-19

Placental mammals (class Mammalia).
A Snow leopard, *Uncia*, a cat (order Carnivora).
B Starnosed mole, *Condylura cristata*, a burrowing insectivore (order Insectivora).
C White-tailed deer, *Odocoileus virginianus*, abundant in eastern and central temperate North America (order Artiodactyla).
D Orca, *Orcinus orca*, a carnivorous whale (order Cetacea).

A

Placental Mammals

Most modern mammals are **placental mammals.** The first organ to form during the course of their embryonic development is the placenta. The **placenta** is a specialized organ, held within the womb in the mother, across which she supplies the offspring with food, water, and oxygen and through which she removes wastes. Both fetal and maternal blood vessels are abundant in the placenta, and substances can thus be exchanged efficiently between the bloodstreams of the mother and her offspring. In placental mammals, unlike marsupials, the young undergo a considerable period of development before they are born.

> **Some mammals (monotremes) lay eggs; others (marsupials) give birth to embryos that continue their development in pouches (as do the monotremes); and still others (placental mammals) nourish their developing embryos within the body of the mother by means of a placenta until development is almost complete.**

Placental mammals are extraordinarily diverse. Representatives of several orders of placental mammals are shown in Figure 19-19. One evolutionary line of mammals, the bats, has joined insects and birds as the only group of living animals that truly flies; another, the Cetacea (whales, dolphins, and porpoises), has reverted to an aquatic habitat like that from which the ancestors of mammals came hundreds of millions of years ago.

THE EVOLUTION OF PRIMATES

The small Mesozoic mammals that ultimately gave rise to the primates were rare for much of their history. They were rare during the first 10 million years of the Cenozoic Era (Table 19-1), when other kinds of mammals appeared and became diverse very

C   D

TABLE 19-1   SUBDIVISIONS OF THE CENOZOIC ERA

TIME	START (MILLIONS OF YEARS AGO)	END (MILLIONS OF YEARS AGO)
TERTIARY PERIOD	65	2
Paleocene Epoch	65	55
Eocene Epoch	55	38
Oligocene Epoch	38	25
Miocene Epoch	25	5
Pliocene Epoch	5	2
QUATERNARY PERIOD	2	Present
Pleistocene Epoch	2	0.01
Holocene Epoch	0.01	Present

rapidly. All of them were probably **nocturnal** (active primarily at night), judging from their large eyes and small size. The earliest fossils that resemble any living primates belonged to the group known as **prosimians.** By the end of the Eocene Epoch (38 million years ago), prosimians were abundant in North America and Eurasia and were probably present in Africa also. Their descendants now live only in tropical Asia and adjacent islands, in Africa, and especially on the island of Madagascar, which is about twice the size of the state of Arizona and lies in the Indian Ocean about 325 kilometers off the east coast of Africa (Figure 19-20).

Anthropoids

The ancestors of monkeys, apes, and humans appear for the first time in the fossil record about 36 million years ago in the early Oligocene Epoch. These **anthropoid** primates—monkeys, apes, and human beings and their direct ancestors—seem to have replaced prosimians rather rapidly. Monkeys and apes are almost all **diurnal,** that is active during the day. They are agile animals that feed mainly on fruits and leaves, and most of them live in trees. Monkeys and apes differ markedly in these respects from the smaller, mostly nocturnal prosimians. Their well-developed **binocular** vision (vision in which both eyes are located at the front of the head and function together to provide a three-dimensional view of the object being examined) has evolved in relation to a reduction in their snouts and is a great advantage in leaping through the treetops. This evolutionary trend, in turn, resulted in the flat faces that are characteristic of most members of the order.

FIGURE 19-20

Prosimians.
A Ringtail lemur, *Lemur catta.*
All living lemurs are restricted to the island of Madagascar.
B Tarsier, *Tarsius syrichta,* tropical Asia. Note the binocular vision
and large eyes of the tarsier, adaptations to nocturnal living in the
trees.

The anthropoid primates have larger brains, and their braincase forms a larger portion of the head than it does in the prosimians and other mammals. The thumb of anthropoids is opposable: it stands out at an angle to the other digits and can be bent back against them, as when the animal grasps an object. Some of the advanced anthropoids have developed an extraordinary ability to manipulate objects by using this opposable thumb. Monkeys and apes live in groups in which complex social interaction occurs. In addition, they tend to care for their young for prolonged periods. Such a pattern of care makes possible the prolonged period of learning that is characteristic of these animals. The maternal care given the young of anthropoids seems to be related in some way to their large brains and the long periods that are necessary for their development.

> Contemporary primates are characterized by the expansion and elaboration of their brains; the shortening of their snouts, together with well-developed binocular vision; the free mobility of the fingers and toes, with the forefinger and thumb opposable; and an increasing complexity and quantity of social behavior.

The Evolution of Hominoids

The apes, together with the **hominids** (human beings and their direct ancestors), make up a group called the **hominoids.** The earliest known fossils of hominoids are from the Early Miocene Epoch (20 to 25 million years ago) and occur in the Old World. The living apes are usually classified as members of two families, the Pongidae (chimpanzees, gorillas, and orangutans) and the Hylobatidae (gibbons) (Figure 19-21). The living apes are confined to relatively small areas in Africa and Asia; no apes ever occurred in the New World.

FIGURE 19-21

A Gorilla, *Gorilla gorilla*.
B Mueller gibbon, *Hylobates muelleri*.
C Chimpanzee, *Pan troglodytus*.
D Orangutan, *Pongo pygmaeus*.

Apes have larger brains than monkeys do. Apes exhibit the most adaptable behavior of all mammals except for human beings. All apes lack tails. With the exception of the gibbons, which are relatively small, all apes are larger than any surviving monkey.

Although the fossil evidence is still being accumulated, biochemical studies have told us a great deal about our relationships to the apes and their relationships to one another (Figure 19-22). Based on the rate with which differences seem to accumulate in the proteins they have studied, investigators have calculated that the evolutionary line leading to gibbons diverged from that leading to the other apes about 10 million years ago; the line leading to orangutans split off about 8 million years ago; and the split between hominids and the line leading to chimpanzees and gorillas occurred about 4 million years ago. The time of divergence between chimpanzees and gorillas appears to have been a little less than 3.2 million years ago, as determined by the same methods. The relationship between chimpanzees, gorillas, and human beings is so close that if they were members of any other group of organisms, they would almost certainly be classified as members of the same genus.

> **Molecular evidence suggests that chimpanzees and gorillas diverged from one another a little less than 3.2 million years ago, hominids from that line about 5 million years ago, orangutans from all of them about 8 million years ago, and gibbons from the rest about 10 million years ago. The absolute timing of these events depends on the interpretation of the fossil record and on the relationship of molecular patterns to time.**

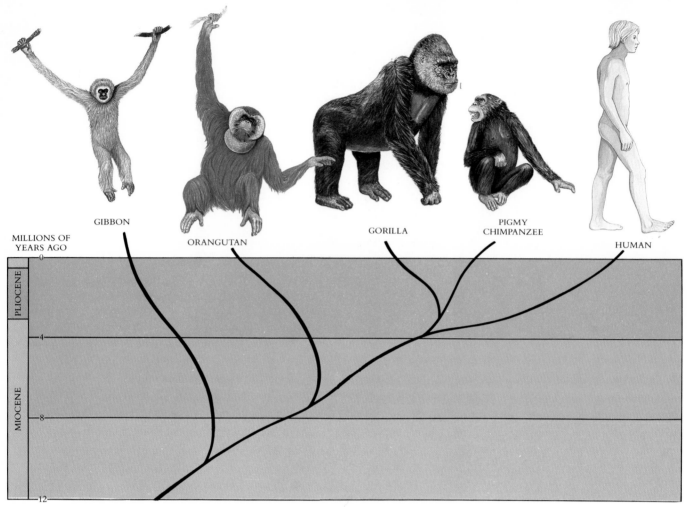

MILLIONS OF
YEARS AGO

GIBBON

ORANGUTAN

GORILLA

PIGMY
CHIMPANZEE

HUMAN

PLIOCENE

MIOCENE

0

4

8

12

FIGURE 19-22

The evolution of living
hominoids, the apes and
human beings.

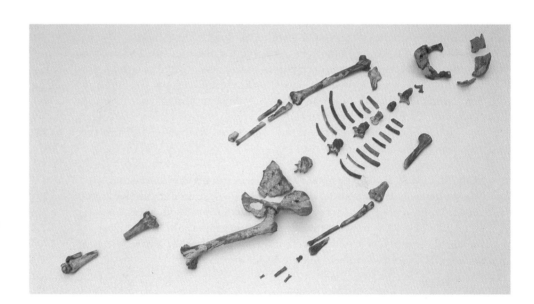

FIGURE 19-23

"Lucy," from Ethiopia, the most
complete skeleton of
Australopithecus discovered so far.
Many more complete skeletons
are needed for a better
understanding of our ancestors.

THE APPEARANCE OF HOMINIDS
The Australopithecines

The two critical steps in the evolution of humans were the evolution of bipedalism and the enlargement of the brain. The earliest definite hominids, which share many of the characteristics that we regard as distinctively human, are assigned to the genus *Australopithecus* (Figure 19-23). Among their human features were bipedal locomotion and rounded jaws. The australopithecines lived on the ground in the open grasslands of eastern and southern Africa, weighed upwards of 20 kilograms, and were up to 1.3 meters tall. Their brains were about the size of those of present-day gorillas, about 350 to 450 cubic centimeters in volume, as compared with an average volume of about 1450 cubic centimeters in modern humans. Individuals of *Australopithecus* appear to have eaten many different kinds of plant and animal food, judging from the characteristics of their teeth and jaws. Although there is no general agreement on this point, there may have been four or more species of the genus *Australopithecus*. Their characteristics indicate clearly that this genus was ancestral to our genus, *Homo*.

The oldest known fossils that may represent *Australopithecus* are 5 to 5.5 million years old. They thus date from close to the postulated time of divergence of the hominid and chimpanzee-gorilla evolutionary lines. Other fossils of *Australopithecus* are commonly found from 4 million years ago onward at a number of localities in East and South Africa.

The first definite hominids, *Australopithecus*, appeared about 5 million years ago in Africa. They lived on the ground, weighed less than 20 kilograms, and were less than 1.3 meters tall. They disappeared about 1.3 million years ago.

The Use of Tools: *Homo habilis*

Starting about 2 million years ago, tools appear in the sedimentary beds of a wide area of East and South Africa. Such tools have been associated with the fossils of the first known humans, the earliest members of the genus *Homo*. The fossils and tools were discovered in the Olduvai Gorge, Tanzania. Because these people used tools, they were named *Homo habilis* (Latin *habilis,* able). *Homo habilis* appeared in the fossil record about 2 million years ago and persisted for about half a million years (Figure 19-24).

Judged from its physical characteristics, *Homo habilis* was certainly derived from *Australopithecus*. As the use of tools was developed in its earliest populations, the characteristics of the genus *Homo* may have begun to emerge. Judged from the structure of its hands, *Homo habilis* regularly climbed trees, like *Australopithecus,* although it mainly stayed on the ground and walked erect on its two legs. Skeletons found in 1987 indicate that *H. habilis* was small in stature, like *Australopithecus.*

Homo habilis is the first member of our own genus and the first primate to use tools consistently. This species appeared in East Africa about 2 million years ago and survived for about 500,000 years. In this species, which certainly evolved from Australopithecus, fully developed human characteristics first emerge.

Homo erectus

All of the early evolution of the genus *Homo* seems to have taken place in Africa. There, fossils belonging to the second and only other extinct species of *Homo, Homo erectus,* are widespread and abundant from 1.7 million until about 500,000 years ago. By 1 million years ago, *Homo erectus* had migrated into Asia and Europe. These large humans, who were about the size of our species, *Homo sapiens,* had brains that were roughly twice as large as those of their ancestors, prominent brow ridges, rounded jaws, and massive teeth. *Homo erectus* survived in Asia until about 250,000 years ago, much longer than in Africa or in Europe.

FIGURE 19-24

The path of human evolution, showing the many early branches that are known from fossil evidence to have existed. *Australopithecus robustus* and *A. boisei* seem to represent evolutionary dead ends, with no living descendants. Whether *A. africans* was ancestral to *Homo habilis* or to *A. robustus* is currently in dispute, as is whether Neanderthal man is an ancestor of modern man or a parallel line that did not survive.

The wide distribution of tools made by *Homo erectus* suggests that their widely dispersed bands may have been in communication with one another. The first evidence of the use of fire by humans occurs at the campsites of this species in the Rift Valley of Kenya at least 1.4 million years ago, and fire was characteristically associated with populations of *Homo erectus* from that time onward.

Homo erectus, the second species of our genus to evolve, appeared in Africa about 1.7 million years ago and migrated throughout Eurasia by 1 million years ago. It disappeared in Africa about 500,000 years ago and in Asia about 250,000 years ago. Homo erectus used fire, starting at least 1.4 million years ago, and made characteristic stone tools.

Modern Humans: *Homo sapiens*

We recognize the first appearance of our own species about half a million years ago in Europe because of our larger brains, reduced teeth, and enlargement of the rear of our skulls as compared with those of of *Homo erectus*. *Homo sapiens* appeared in Europe as *H. erectus* was becoming rarer. It appears likely, in fact, that *Homo sapiens* represents a derivative of *H. erectus* with certain improved, "modern" features. Ultimately, our species seems to have replaced its ancestor, starting in Europe.

Neanderthals

Populations of *Homo sapiens* of the sort that are called Neanderthals were abundant in Europe and western Asia between about 70,000 and about 32,000 years ago, judging from the fossil record. Such fossils were first discovered in the valley of the Neander River in German in 1856, and the name given to humans of this sort is derived from that of the valley.

Compared with ourselves, Neanderthals were powerfully built, short, and stocky. Their skulls were massive, with protruding faces, projecting noses, and rather heavy bony ridges over the brows. Their brains were even larger than those of modern humans, a fact that may have been related to their heavy, large bodies. The Neanderthals made diverse tools, including scrapers, borers, spearheads, and hand axes. Some of the Neanderthal tools were used for scraping hides, which they used for clothing. They lived in hutlike structures or in caves. Neanderthals took care of their injured and sick and commonly buried their dead, often placing food and weapons, and perhaps even flowers, with the bodies. Such attention to the dead suggests strongly that they believed in a life after death. For the first time, the kinds of thought processes that are characteristic of modern *Homo sapiens,* including symbolic thought, are evident in these acts.

> **Our species, *Homo sapiens,* may be as much as 500,000 years old; it has been abundant for the past 150,000 years. Neanderthals were abundant in Europe and western Asia from about 70,000 to about 32,000 years ago.**

Cro-Magnons

About 34,000 years ago the European Neanderthals were abruptly replaced by people of essentially modern character, the so-called Cro-Magnons. We can only speculate about why this sudden replacement occurred, but it was complete all over Europe in a short period of time. There is some evidence that the Cro-Magnons came from Africa, where fossils of essentially modern aspect, but as much as 100,000 years old, have been found. They seem to have replaced the Neanderthals completely in southwest Asia by 40,000 years ago, and then spread across Europe during the next few thousand years, coexisting and possibly even interbreeding with the Neanderthals for several thousand years.

The Cro-Magnons used sophisticated stone tools that rapidly became more diverse and thus more useful for many purposes. They also made a wide variety of tools out of bone, ivory, and antler, materials that had not been used earlier. The Cro-Magnons were hunters who killed game by using complex tools. In addition, their evident social organization suggests that they may have been the first people to have fully modern language capabilities. Soon after Cro-Magnon people appeared, they began to make what were apparently ritual paintings in caves (Figure 19-25). The culture associated with these impressive and mysterious works of art existed during a period that was cooler than the present. Such conditions prevailed at the time of the last great expansion of continental ice. During that period, grasslands inhabited by large herds of grazing animals occurred across Europe. The animals that occurred in such habitats are often depicted in the cave paintings.

Humans of modern appearance eventually spread across Siberia to the New World, which they reached at least 12,000 to 13,000 years ago, after the ice had begun to retreat at the end of the last glacial period. As people spread throughout the world, they were apparently responsible for the extinction of many populations of animals

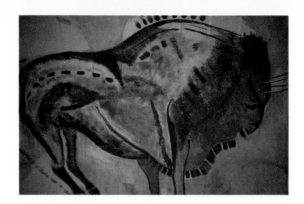

and plants, a process that has been accelerated recently. Human remains are often associated with the bones of the large animals they hunted, and these animals often disappeared promptly from different parts of the world shortly after humans arrived for the first time. By the end of the Pleistocene Epoch, about 10,000 years ago, there were only about 10 million people throughout the entire world (compared with more than 5 billion now).

SUMMARY

1. Chordates are characterized by a single, dorsal, hollow nerve cord; a flexible rod, the notochord, which forms on the dorsal side of the gut in the early embryo; and pharyngeal slits.

2. Vertebrates comprise most groups of the phylum Chordata. They differ from other chordates in that they usually possess a vertebral column, a distinct and well-differentiated head, and a bony skeleton.

3. Members of the class Agnatha differ from the other vertebrates in that they lack jaws. Once abundant and diverse, they are represented among the living vertebrates only by the lampreys, which attach themselves to other fishes and suck their blood, and the hagfishes.

4. The two classes of fishes other than Agnatha consist of animals that have jaws, as do the members of the other four classes of vertebrates. Jawed fishes constitute about half of the estimated 42,500 species of vertebrates and are dominant in fresh and salt waters everywhere. Of the two classes of jawed fishes, the Chondrichthyes, or cartilaginous fishes, consist of about 850 species of sharks, rays, and skates; the Osteichthyes, or bony fishes, consist of about 18,000 species.

5. The first land vertebrates were the amphibians, one of the four classes of tetrapods (basically four-limbed vertebrates). Amphibians depend on water and lay their eggs in moist places. In many species the larvae, and in some the adults too, live in water. The amphibians probably gave rise to the reptiles, which otherwise may have been derived directly from the fishes.

6. The reptiles were the first vertebrates that were fully adapted to terrestrial habitats. Amniotic eggs, which evolved in this group but are also characteristic of the birds and the very few egg-laying mammals, represent a significant adaptation to the dry conditions that are widespread on land.

7. The birds and mammals were derived from the reptiles and are now the dominant groups of vertebrates on land in many areas. The members of these two classes have independently become endothermic (homeothermic), capable of regulating their own body temperatures. With few exceptions, all other living animals are ectothermic; their body temperatures are set by external influences.

8. Primates first appear 55 to 38 million years ago, with the first anthropoid primates (a group that now includes monkeys, apes, and humans) appearing about 30 million years ago. Primates have proportionately large brains; binocular vision; and five digits, including an opposable thumb; and they exhibit complex social interactions.

9. The apes, which appeared 20 to 25 million years ago, gave rise to the gibbons, orangutans, and then to the African apes (chimpanzees and gorillas) and hominids (humans and their direct ancestors), starting more than 10 million years ago. Apes and hominids collectively are termed hominoids.

10. The earliest hominids, family Hominidae, belong to the genus *Australopithecus*. The members of this genus were the direct ancestors of humans. They appeared in Africa about 5 million years ago and were small, up to about 1.3 meters tall and about 20 kilograms in weight, with brains about 450 cubic centimeters in volume.

11. About 2 million years ago, with an enlargement of brain size that was perhaps associated with the increased use of tools, the genus *Australopithecus* gave rise to humans belonging to the genus *Homo*. The first species of this genus, *Homo habilis*, appeared in Africa about 2 million years ago. It became extinct about 1.5 million years ago.

12. The second species of *Homo, Homo erectus*, appeared in Africa at least 1.7 million years ago. *Homo erectus* migrated from Africa to Eurasia about 1 million years ago. They were replaced by our species, *Homo sapiens*, about 500,000 years ago in Africa and about 250,000 years ago in Asia.

REVIEW

1. One of the pivotal innovations that led to the success of reptiles is the watertight egg, called a(n) _____ egg.

2. No mammals lay eggs. True or false?

3. Of the five classes of vertebrates, which has the most species?

4. The first hominids appeared about 5 million years ago. They belonged to the genus _____.

5. We are members of the genus *Homo*. How many species have existed in this genus?

SELF-QUIZ

1. Which of the following are *not* distinguishing features of chordates?
   (a) dorsal nerve cord   (d) gills
   (b) spine   (e) tail
   (c) notochord

2. Arrange the following animals in the order in which they are thought to have evolved.
   (a) amphibians   (d) reptiles
   (b) bony fishes   (e) sharks
   (c) lampreys

3. Arrange the following hominids in order of their evolutionary distance from modern humans, the most distant first.
   (a) *Australopithecus*   (d) *Homo habilis*
   (b) Cro-Magnons   (e) *Neanderthals*
   (c) *Homo erectus*

4. Arrange the following animals in order of their evolutionary distance from humans, the most distant first.
   (a) gibbons   (d) monkeys
   (b) gorillas   (e) orangutans
   (c) prosimians

5. Arrange the following animals in the order of their diversity (number of species), the one with the greatest number of species first.
   (a) amphibians   (d) mammals
   (b) birds   (e) reptiles
   (c) bony fishes

THOUGHT QUESTIONS

1. Of the 42,500 species of living chordates, twice as many (about 19,000 fishes) live in the sea as on the surface of the land (about 6000 reptiles and 4500 mammals). Why do you think there are so many more species of chordates in the sea than on land?

2. Our species, *Homo sapiens,* evolved from *Homo erectus* less than 1 million years ago. Do you think that evolution of the genus *Homo* is over, or might another species of man evolve within the next million years? Do you think this would involve the extinction of *H. sapiens?*

3. Flying reptiles were a very diverse group during the age of dinosaurs. However, although the present age might be considered the age of mammals, bats (the only flying mammals) exhibit far less diversity than dinosaurs did. Can you suggest an explanation for this?

4. Studies of the DNA of primates have revealed that humans differ from gorillas in only 1.4% of the DNA nucleotide sequences, and from chimpanzees in only 1.2%. This degree of genetic similarity (about 1%) is the same as is usually seen among "sibling" species (that is, species that have only recently evolved from a common ancestor). Yet humans are assigned not only to a different genus but to a different family! Do you think this is legitimate, or are humans just a rather unusual kind of African ape?

FOR FURTHER READING

BAKKER, R.T.: *The Dinosaur Heresies,* William Morrow & Co., Inc., New York, 1986. A fine discussion of the controversy about whether dinosaurs could control their own temperatures internally; well-written and beautifully illustrated.

EASTMAN, J., and A. DE VRIES: "Antarctic Fishes," *Scientific American,"* November 1986, pages 106-114. Did you ever wonder why fishes don't freeze in the Antarctic seas, even though ice forms in the oceans in which they swim? These biologists showed that the fishes produce their own potent antifreeze.

EISENBERG, J.F.: *The Mammalian Radiations: An Analysis of Trends in Evolution, Adaptation, and Behavior,* University of Chicago Press, Chicago, 1981. A marvelous, biologically oriented account of the mammals.

FOSSEY, D.: *Gorillas in the Mist,* Houghton Mifflin Co., Boston, 1983. The late Dian Fossey devoted many years to a study of mountain gorillas, and this book summarizes her findings.

HORNER, J.R.: "The Nesting Behavior of Dinosaurs," *Scientific American,* April 1984, pages 130-137. The members of a group of dinosaurs named the hadrosaurs that lived in Montana 80 million years ago appear to have built nests and to have exhibited social behavior. An interesting example of paleontological detective work.

JOHANSON, D., AND M. EDEY,: *Lucy: The Beginnings of Humankind,* Simon and Schuster, New York, 1981. The account of an expedition to the Afar region of Ethiopia and the significance of the findings of early fossils of *Australopithecus* to an understanding of early human evolution.

LEAKEY, R.E.: *The Making of Mankind,* E.P. Dutton & Co. Inc., New York, 1981. A beautifully illustrated and well-written book on human evolution during the past several million years, based on much personal research and reflection.

MORRELL, V.: "Announcing the Birth of a Heresy," *Discover,* vol. 8(3), March 1987, pages 26-50. Engaging article that presents new evidence that dinosaurs were active, warm-blooded, and gregarious. Read also the following page (51) by Stephen Jay Gould.

O'BRIEN, J.: "The Ancestry of the Giant Panda," *Scientific American,* November 1987, pages 102-107. A clear account of how modern molecular techniques are being used to solve a famous evolutionary puzzle: is the panda a raccoon or a bear?

OSTROM, J.: "Bird Flight: How Did it Begin," *American Scientist,* vol. 67, pages 46-56, 1979. Fascinating account of theories of the origin of feathers and flight.

PILBEAM, D.: "The Descent of Hominoids and Hominids," *Scientific American,* March 1984, pages 84-96. Good digest of contemporary research on human evolution.

ECOLOGY

Part Six

FIGURE 20-1   Each plant in this forest competes with all the individuals around it for light, soil nutrients, and moisture. The competition for available resources is fierce, especially during the relatively brief period in spring before the trees have so many leaves that photosynthesis is limited.

HOW SPECIES INTERACT WITH THEIR ENVIRONMENT

Chapter 20

Overview

No species escapes selection. Often this selection results from competition with other individuals for something an organism requires, such as food or shelter. One of the strongest forces that shapes the nature of relationships within populations of a species, as well as between species, is competition for limiting resources. For each species, individuals reproduce and populations get bigger until the number of individuals is as large as the habitat is able to support. How the populations of a species grow is a key component of its evolutionary success or failure. This is as true of humans as of any other species. As our world grows more crowded with humans, we compete ever more ferociously with the rest of life, with African antelopes, prairie grasses, and all the other organisms that share the earth's resources. This sort of competition is the chief sculptor of ecological relationships. It determines the future of all species and will determine the future of ours.

For Review

Here are some important terms and concepts that you will encounter in this chapter. If you are not familiar with them, you should review them before proceeding.

Human population growth (Chapter 1)

Interactions between genes (Chapter 13)

The Hardy-Weinberg principle (Chapter 16)

Selection (Chapter 16)

The nature of species (Chapter 17)

Stepping back and viewing the broad panorama of evolution, as we have done in the preceding section, reveals the unity that underlies the diversity of life. All of us, camels and elephants and humans, share a common history of adaptation and progressive change. It is important, however, not to limit our view to looking backward. Evolution is still going on. Right now, all the organisms on earth are involved in a day-to-day struggle to survive and to reproduce. Each living thing on earth is interacting with its physical environment and with the organisms around it in a test of its own solution to the problem posed by living (Figure 20-1). We call the study of the interactions of organisms with one another and with their physical environment **ecology**. A study of ecology quickly reveals that evolution is not a simple chain of events, one leading to the next in straight-line order. Rather, evolution is more like the lives we ourselves lead, in which a single action can have many diverse consequences, some of them not easily predictable. Ecologists study these relationships and the ways in which they determine the outcome of biological change.

In this chapter we consider how populations, members of the same species that live together in the same place, interact with each other and with their environments. It is here that evolution and genetics meet, here that the kind of individual specified by an organism's genes is tested against the demands of its environment. The testing process is selection, which we began to discuss in Chapter 16. The outcome of selection is, as you have seen, dictated by the nature of the environment, both the physical environment of temperature and moisture and light and the biological environment of other organisms. After considering how selection operates within natural populations, we then focus on the interactions between species and their environments, both physical and biological.

SELECTION IN ACTION

Selection operates in natural populations of a species something like skill does in football games: in any individual game, it is difficult to predict the winner, since chance can play an important role in the outcome, but over a long season the teams with the most skillful players usually win the most games. In nature, too, those individuals best suited to their environment tend to win the evolutionary game by leaving the most offspring,

FIGURE 20-2

Schematic representation of the three different kinds of natural selection acting on a trait, such as height, that varies in a population. In this diagram, dots represent individuals that do not contribute to the next generation. The curves represent measurements for the trait taken on each individual in the population. All three kinds of natural selection have the same starting point, and the three series show the way that selection alters the distribution of the characteristic as time passes, moving to the right.

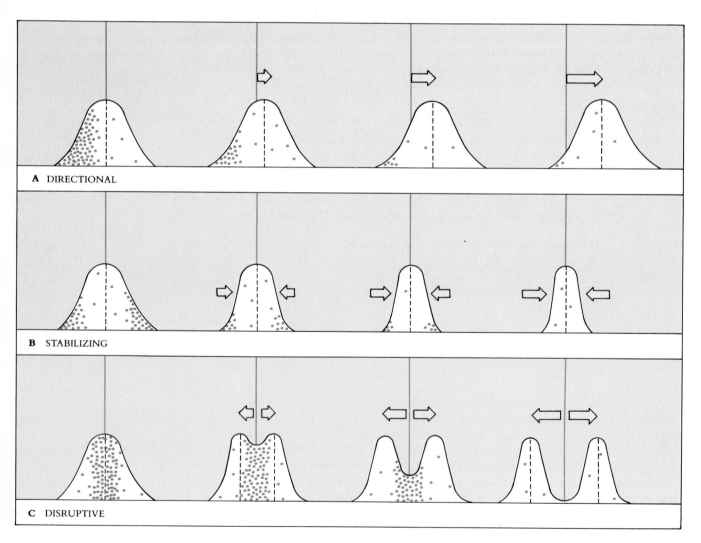

A DIRECTIONAL

B STABILIZING

C DISRUPTIVE

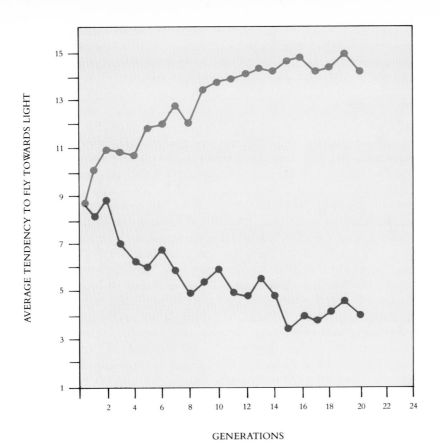

GENERATIONS

FIGURE 20-3

In generation after generation (expressed along the bottom of the graph), a population of the fly *Drosophila* was subjected to strong selection. The hollow circles represent the average results for each generation when the individual flies were scored for their tendency to fly toward light more strongly than usual, and the solid circles represent the same kind of measurement when the selection was *against* the tendency of the flies to fly toward light. When flies that tended strongly to fly to light were selected and used as the parents for the next generation, their offspring had a greater and greater tendency in successive generations to fly toward light, indicating that the characteristic has a genetic basis. The opposite results were obtained by selecting flies that tended *not* to fly toward light. The scale along the left-hand axis is an arbitrary measure of tendency to fly toward light, averaged for all the flies in the population.

although chance can play a major role in the life of any one individual. Selection, you can see, is a statistical concept, just as betting is. Although you cannot predict the fate of any one individual, or any one coin toss, you can often predict which kind of individual will tend to become more common in populations of a species.

To visualize how selection operates to increase the frequency of particular traits in a population, it is simplest to consider traits that are specified by single genes. As you saw in Chapter 16, selection against dominant alleles can proceed rapidly and always leads eventually to the elimination of the allele. Selection against recessive alleles is far slower and soon reaches limits in its ability to change the proportion of recessive alleles in the population. This relationship puts real limits on our ability to change populations for traits controlled by recessive alleles, even if the genetic basis of the trait is simple.

Forms of Selection

In nature many traits, perhaps most, are affected by more than one gene. The interactions between genes are typically complex, as you saw in Chapter 13. For example, alleles of many different genes play a role in determining human height (see Figure 13-15). In such cases, selection operates on all the genes, influencing most strongly those that make the greatest contribution to the phenotype. How selection changes the population depends on which genotypes are favored.

Directional Selection

When selection acts to eliminate one extreme from an array of phenotypes (Figure 20-2, *A*), the genes promoting this extreme become less frequent in the population. Thus in the *Drosophila* population illustrated in Figure 20-3, the elimination by the investigator of flies that move toward light causes the population to contain fewer individuals with alleles promoting such behavior. The result is that if you were to pick an individual at random from the new fly population, there is a lesser chance it would spontaneously move toward light than if you had selected a fly from the old population. The population has been changed by selection in the direction of lower light attraction. This form of selection is called **directional selection.**

Stabilizing Selection

When selection acts to eliminate *both* extremes from an array of phenotypes (Figure 20-2, *B*), the result is to increase the frequency of the intermediate type, which is already the most common. In effect, selection is operating to prevent change away from this middle range of values. In a classic study carried out after an "uncommonly severe storm of snow, rain, and sleet" on February 1, 1898, 136 starving English sparrows were collected and brought to the laboratory of H.C. Bumpus in Brown University at Providence, Rhode Island. Of these, 64 died and 72 survived. Bumpus took standard measurements on all the birds. He found that among males, surviving birds tended to be bigger, as one might expect from the action of directional selection. However, among females, Bumpus observed a different result. The females that perished were not smaller, on the average, than those that survived, but among them were many more individuals that had extreme measurements—measurements that were unusual for the population as a whole. Selection had acted most strongly against these individuals. When selection acts in this way, the population contains fewer individuals with alleles promoting extreme types. Selection has not changed the most common phenotype of the population, but rather made it even more common by eliminating extremes. Many examples similar to Bumpus' female sparrows can be given. In human beings, for example, infants with intermediate weight at birth have the highest survival rate (Figure 20-4). In ducks and chickens, eggs of intermediate weight have the highest hatching success. This form of selection is called **stabilizing selection.**

FIGURE 20-4

Stabilizing selection for birth weight in human beings. The distribution of birth weights is presented as a histogram, the percentage of individuals with each birth weight expressed along the bottom axis. The percentage of babies of each birth weight that died is indicated by the dots, joined by a line. As you can see, the death rate among babies is lowest at an intermediate birth weight; both smaller and larger babies have a greater tendency to die at or near birth than those around the optimum weight, which is between 7 and 8 pounds.

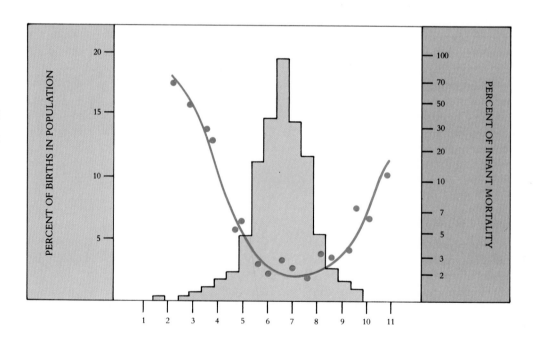

BIRTH WEIGHT IN POUNDS

Disruptive Selection

In some situations, selection acts to eliminate rather than favor the intermediate type (Figure 20-2, *C*). A clear example is provided by the different color patterns of the African butterfly *Papilio dardanus*. In different parts of Africa, the color pattern of this butterfly is dramatically different, in each instance being a close copy of some other species that birds don't like to eat (*Papilio dardanus* is said to be a "mimic"). Any intermediate patterns that do not look like distasteful butterflies would be readily detected and eaten by birds, and so any intermediate patterns that occur are selected against. In this case, selection is acting to eliminate the intermediate phenotypes, in effect partitioning the population into homozygous groups. This form of selection is called **disruptive selection.**

The Limits to Selection

What limits the power of selection to change a population? Why can't agriculturalists produce a 2-ton cow or corn plants that have an ear inside every leaf? First, because there is a limit to what genes can specify, and second, because extreme phenotypes tend to have other undesirable side effects. For example, commercial breeders in the United States carried out an intensive breeding program with turkeys, starting in 1944, to obtain larger and more meaty birds. The program was very successful over the course of 20 years but eventually had to be abandoned, because as larger and larger birds were obtained, they also tended to be much less fertile. Their eggs did not hatch as well, and there were fewer of them. Most complex phenotypic characteristics result from the integrated contributions of many genes. Since these genes also tend to have other effects, selecting for one set of phenotypic characteristics tends to disturb a number of others. As the strength of the selection is increased, so is the disturbance.

THE CONCEPT OF POPULATION

Selection acts on individuals. In the everyday struggle to survive and leave offspring, it is individual organisms that must survive storms, find food for their offspring, escape from predators, and otherwise cope with the demands of living. Individuals, however, do not evolve—species do. The alleles that an individual carries are determined when the egg and sperm that give rise to that individual fuse in the process of fertilization; they do not change during its life. Rather, evolution acts to alter which kinds of individuals are common in populations—which alleles are frequent and which are rare. Selection, by acting on individuals, tends to weed out the less successful alleles and, in this way, changes the genetic makeup of the population and produces evolutionary change. It is populations that are the basic units of evolutionary change, the building blocks of evolution. If we wish to understand the significance of how an individual interacts with other organisms, both of different species and members of its own species, then we must consider all the members of its population, fellow travelers on its evolutionary journey.

A **population** consists of the individuals of a given species that occur together at one place and at one time. This is a flexible definition that allows us to speak in similar terms of the world's human population, of the population of protists in the gut of an individual termite, or of the deer that inhabit a wood.

Every population has characteristic features such as size (the number of individuals that it contains), density (how crowded its members are), dispersion (whether its members are evenly or unevenly spaced), sex ratio (proportion of males to females), and demography (how many of its members are young, middle-aged, and old). Some populations are small, and many of these are now in danger of extinction (Figure 20-5). Each population occupies a particular place and plays a particular role in the web of life of which it is a part. That role, about which we will have more to say later, is defined as the **niche** of that population.

A population consists of the individuals of a species that occur together at one place and one time. The role of a population in nature is called its niche.

FIGURE 20-5

These animals exist in very small populations, and are therefore in danger of extinction unless the activities that are threatening them are brought under control.

A The Sumatran rhinoceros, *Didermoceros sumatrensis,* reduced to a few hundred individuals on the island of Sumatra and on the adjacent Asian mainland. Poaching to obtain the horns, which are in great demand for medicinal uses, and destruction of habitat are responsible for the decline of this species.

B Mediterranean monk seal, *Monachus monachus.* Fewer than 500 individuals of this species are believed to exist; they are confined to remote cliffbound coasts and islands in the Mediterranean. The seals have a low birth and survival rate, and are killed by fishermen.

C The kagu, *Rhynochetus jubatus,* the only member of a family of birds that is restricted to the island of New Caledonia, in the southwestern Pacific Ocean. It has been brought to the brink of extinction by introduced dogs that have run wild on the island, and by habitat destruction.

A

HOW POPULATIONS GROW

One of the most important characteristics of any population is whether or not it is growing and, if so, how rapidly. From observing nature in ways that range from the informal to the highly sophisticated, we know that most populations tend to remain relatively constant in number, regardless of how many offspring the individuals produce. This relationship, clearly pointed out by Thomas Malthus near the end of the eighteenth century, was one of the fundamental observations that led Charles Darwin to develop his theory of evolution by natural selection. If populations did not tend to

FIGURE 20-6

A scene in Kokee State Park, Kauai, Hawaii. Two vines that are not native to Hawaii, namely the blackberry *Rubus penetrans,* from temperate eastern North America, and the passion flower *Passiflora molissima,* from South America, have grown over and overwhelmed the native plants. Such uncontrolled growth is characteristic of many introduced populations, particularly on islands.

B

C

remain constant in number, they would eventually become impossibly large. Darwin calculated, for example, that if elephants had one offspring per year for 10 years, elephants would soon cover the face of the earth many times over. The fact that the earth is not covered with elephants reflects the fact that nature is acting to hold down the size of natural populations. To understand how nature operates to limit population growth, we first consider situations in which populations do grow very rapidly in size for a time (Figure 20–6), and we then ask what brings their growth to a halt.

Biotic Potential

The rate at which a population will increase when there are no limits of any sort on its rate of growth is called its **innate capacity for increase,** or **biotic potential.** This theoretical rate is almost always impossible to calculate, however, because there usually are limits to growth. What biologists actually tend to calculate is the **actual rate of population increase,** which is defined as the difference between the birth rate and the death rate per given number of individuals per unit of time. The actual rate of increase, r, may also be affected by the movement of individuals into or out of the area. For example, the actual rate of increase for the human population of the United States during the closing decades of the twentieth century is more than half made up of immigrants to the country, less than half from the growth in numbers of the people already living in it.

The innate capacity of growth for any population is constant, determined largely by the physiology of the organism. The actual growth rate, on the other hand, depends on the death rate as well as the birth rate and so changes as the population increases in size. In general, as the population grows more crowded and begins to exhaust its resources, the death rate rises. If you were to keep track of the number of individuals in a population from its beginning, you would see that the population would grow rapidly at first, the number of individuals increasing exponentially (2, 4, 8, 16, 32, 64, 128, . . .). Soon, however, the rate of increase would slow as the death rate began to rise. Eventually, just as many individuals would be dying as being born. The early, rapid phase of population growth (Figure 20–7) lasts only for a short period, usually when an organism reaches a new habitat in which it has abundant resources. In this connection, one may think of such examples as dandelions reaching the fields, lawns, and meadows of North America from Europe for the first time; of algae colonizing a newly formed pond; or of the first terrestrial immigrants that arrive on an island recently thrust up from the sea.

TIME (HOURS)	NUMBER OF INDIVIDUALS
10	1,048,576
9½	524,288
9	262,144
8½	131,072
8	65,536
7½	32,768
7	16,384
6½	8,912
6	4,096
5½	2,048
5	1,024
4½	512
4	256
3½	128
3	64
2½	32
2	16
1½	8
1	4
½	2
0	1

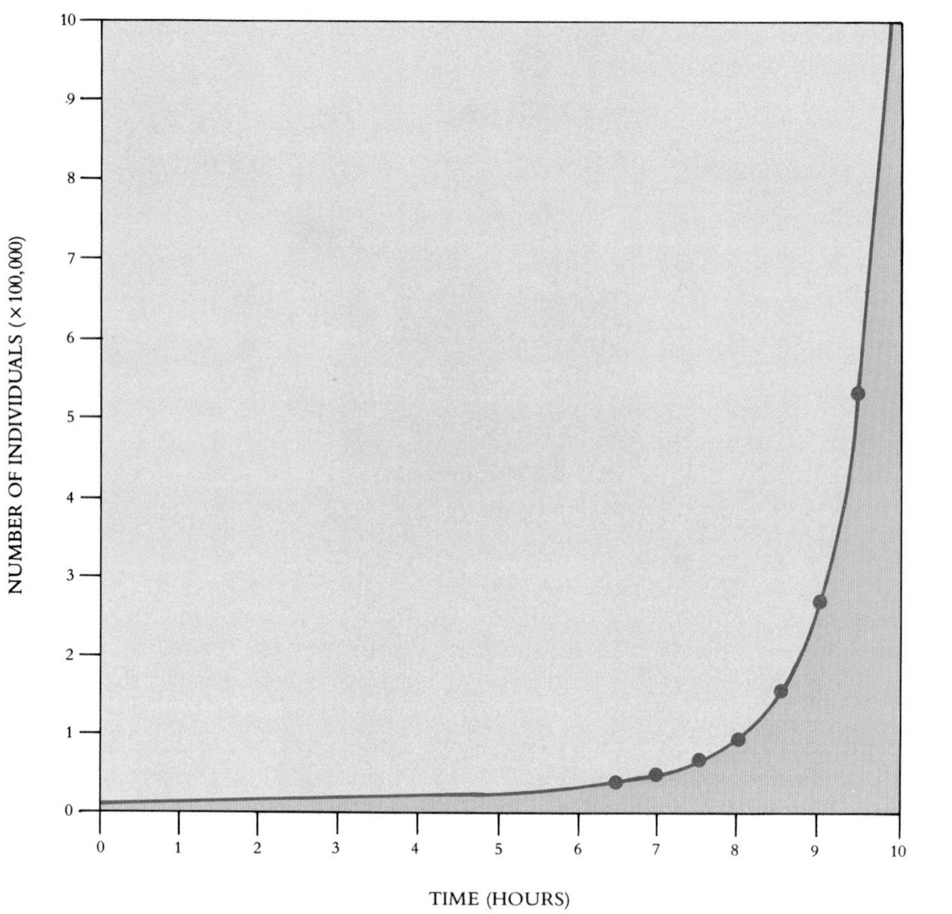

FIGURE 20-7

Exponential growth in a population of bacteria whose individuals are dividing every half hour.

FIGURE 20-8

The sigmoid growth curve, characteristic of biological populations, was named because it resembles the shape of the Greek letter sigma (σ). Such a curve begins with a period of exponential growth like that shown in Figure 20-7; when the population approaches its environmental limits (K), the growth begins to slow down, and it finally stabilizes, fluctuating around the maximum number of individuals that the environment will hold.

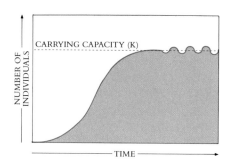

Carrying Capacity

No matter how rapidly new populations grow, they eventually reach some environmental limit imposed by shortages of an important factor such as space, light, water, or nutrients. A population ultimately stabilizes at a certain size, called the **carrying capacity** of the particular place where it lives. The carrying capacity, symbolized by K, is the

FIGURE 20-9

Migratory locusts, *Locusta migratoria,* a legendary plague of large areas of Africa and Eurasia. When the population density reaches a certain level, the locusts change in appearance and behavior and take off as a swarm.

number of individuals that can be supported at that place indefinitely. As the carrying capacity is approached, the rate of growth of the population slows down greatly, because, in effect, there is less "room" for each additional individual. Graphically, this relationship is the S-shaped **sigmoid growth curve,** characteristic of biological populations (Figure 20-8).

> **The size at which a population stabilizes in a particular place is defined as the carrying capacity of that place for that species.**

Processes such as competition for resources, emigration, and accumulation of toxic waste products all increase as a population approaches its carrying capacity for a particular habitat. The resources for which the members of the population must increasingly compete may include food, shelter, light, mating sites, or any other factor necessary for them to carry out their life cycle and reproduce.

Density-Dependent and Density-Independent Effects

Effects such as those we have just discussed regulate the growth of populations and are called **density-dependent effects.** Among animals, they are often accompanied by hormonal changes that may bring about alterations in behavior that directly affect the ultimate size of the population. One striking example occurs in migratory locusts (short-"horned" grasshoppers), which, when they are crowded, produce hormones that cause them to enter a new, migratory phase: the locusts take off as a swarm and fly long distances to new habitats (Figure 20-9). In contrast, **density-independent effects** are caused by factors, such as the weather and physical disruption of the habitat, that operate regardless of population size.

> **Density-dependent effects are controlled by factors that come into play particularly when the population size is larger; density-independent effects are controlled by factors that operate regardless of population size.**

FIGURE 20-10

Fishing on Georges Bank, off the coast of New England in the northeastern United States. Intensive fishing here has driven the populations of commercially valuable fish species such as cod and halibut below the point of optimal yield. As increasingly sophisticated methods are used for fishing, the danger of such consequences increases.

FIGURE 20-11

The common dandelion (*Taraxacum officinale*) is an *r* strategist; it produces numerous seeds by means of which the plants reproduce rapidly, as every gardener knows. In the dandelion, the seeds are produced asexually, so that all of the daughter individuals are genetically identical to their parents.

Modern agricultural practice makes use of the characteristics of the sigmoid growth curve. Early in the history of a population, resources are not yet limiting the growth of individuals, and net productivity is highest. For best yields, this is the best time to "harvest" a population. Commercial fisheries, for example, attempt to operate so that fish populations are always harvested in the steep, rapidly growing parts of the curve. The point of **optimal yield** (maximum sustainable catch from the population) lies partway up the sigmoid curve. Harvesting the population of an economically desirable species near this point will result in much better yields than can be obtained either when the population approaches the carrying capacity of its habitat or when it is very small (Figure 20-10). Overharvesting a population that is smaller than this critical size can destroy its productivity for many years, as evidently happened to the Peruvian anchovy fishery after the population had been depressed by the 1972 El Niño weather disturbance. Although it is often difficult to determine the population levels of commercially valuable species that are most suitable for long-term, productive harvesting, such estimates are the subject of much study because they are so important from both a commercial and an ecological point of view. In forestry, the harvesting of mature trees for plywood, pulp, or similar products tends to exploit the upper part of the sigmoid curve, for example, and it is very important to understand the characteristics of the populations for proper management.

> In natural systems that are exploited by humans, such as agricultural systems and fisheries, the aim is to exploit the population of the early, most productive part of the rising portion of the sigmoid growth curve.

r Strategists and *K* Strategists

Many species, such as annual plants, a number of insects, and bacteria have very fast rates of population growth. An individual bacterium, for example, can reproduce in less than 20 minutes. Growth that is this rapid cannot be controlled effectively by reducing population sizes. In such species, small surviving populations will soon enter an exponential pattern of growth and regain original sizes. In contrast, a comparable reduction in population size among relatively slow-breeding organisms, such as whales, rhinoceroses, California redwoods, or most tropical rain forest trees, can lead directly to extinction.

Slow-growing populations of organisms tend to be limited in number by the carrying capacity of the environment and are sometimes called **K strategists.** Such

FIGURE 20-12

A female humpbacked whale *(Megaptera)*, swimming with its calf in the waters off Hawaii, a *K* strategist. Most whales have one calf at a time. Roger Payne, who has contributed a great deal to our understanding of whales, has written: "In our long sad history, we have brought hundreds of species to extinction. Unless we change our priorities, we will soon eradicate hundreds of thousands more. There is a curious fact, however: among all of the species we have destroyed, there is not one that occurred worldwide. . . . The closest that we have ever come to taking that fatally insane step is with the right whale—a species that once occurred off the western and eastern shores of every continent. We came so close to destroying that species that it really is something of a miracle that it survived."

organisms, including the whales and redwoods just mentioned, usually live in habitats that are fairly stable and predictable. In contrast, species whose populations are characterized by very rapid growth followed by sudden crashes in population size tend to live in unpredictable and rapidly changing environments. These organisms have a high intrinsic rate of increase, or *r*, and are called **r strategists.** Many organisms are neither "pure" *r* strategists nor "pure" *K* strategists. Rather, their reproductive strategies lie somewhere between these two extremes or change from one extreme to the other under certain environmental circumstances.

In general, *r* strategists have many offspring (Figure 20-11). Their offspring are small, mature rapidly, and receive little or no parental care. Many offspring are produced as a result of each reproductive event. Such organisms include dandelions, cockroaches, aphids, and mice.

In contrast, the *K* strategists tend to have few offspring. These offspring are large, mature slowly, and often receive intensive parental care. This group consists of organisms such as coconut palms, whooping cranes, whales, and humans (Figure 20-12). Many of them are in danger of extinction; some of these were shown in Figure 20-5.

Human Populations

Human beings use their culture to avoid or postpone the effects of most limitations on their population size, which is currently increasing at the rate of about 1.7% per year. Some 85 million people are added annually to a total population that already numbers well over 5 billion people. That's three every second! At this rate, the world population will double in just over 40 years. As we shall discuss in Chapter 23, such a rate of growth has potential consequences for our future that are extremely grave. By reducing the population sizes of most other species as our own populations grow so rapidly, we are driving to extinction those relatively large and slow-breeding species that we view as desirable, while favoring those weed and pest species that compete directly with our own enormous populations for food and other critical resources.

> **The rapidly growing human population is tending to exterminate *K* strategists, such as whales, lions, and forest trees, while favoring *r* strategists, including houseflies, cockroaches, and dandelions.**

MORTALITY AND SURVIVORSHIP

A population's intrinsic rate of increase depends on the ages of the organisms in it and the reproductive performance of the individuals in the various age groups. Very young and very old individuals, for example, are not as reproductively active as the rest of the population. The speed with which a population can grow depends critically on how many of its members are reproductive, and thus on the **age distribution** of the popula-

FIGURE 20-13

Survivorship curves for the oyster, for a microscopic freshwater animal called the hydra, and for human beings. Each population is assumed to have started with 1000 individuals. Humans have a Type I life cycle, the hydra Type II, and oysters Type III. The shapes of the respective curves are determined by the percentages of individuals in populations of these organisms that die at different ages.

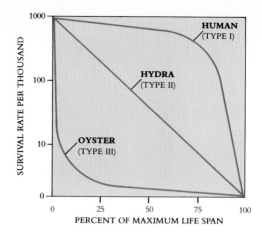

tion. A population with many reproductive individuals will grow much faster than one with few reproductive individuals. Age distributions differ greatly from species to species and even, to some extent, from place to place within a given species.

One way to express the age distribution of a population is the survivorship curve. **Survivorship** is defined as the percentage of an original population that is living at a given age. Samples of different kinds of survivorship curves are shown in Figure 20-13. In the hydra, individuals are equally likely to die at any age, as indicated by the straight survivorship curve. Oysters, on the other hand, produce vast numbers of offspring, only a few of which live to reproduce. Once they become established and grow into reproductive individuals, their rate of death, or **mortality,** is extremely low. Even though human infants are susceptible to death at relatively high rates, the highest mortality rates in people occur later in life, in their postreproductive years.

By convention, life cycles are designated as type I, the type found in humans, where a large proportion of the individuals appear to reach their physiologically determined maximum age; type II, the situation found in hydra, in which the mortality rate remains more or less constant at all ages; and type III, the kind of life cycle found in oysters, in which mortality is high in the early stages but then declines. Many animals and protist populations in nature probably have survivorship curves that lie somewhere between those characteristic of type II and type III, and many plant populations, with high mortality at the seed and seedling stages, are probably closer to type III. Humans have probably approached type I more and more closely through the years, with the birth rate remaining relatively constant or declining somewhat but the death rate dropping markedly.

> In type I survivorship curves, a large number of individuals reach their maximum theoretical age. In type II curves, mortality is largely independent of age. In type III curves, mortality is highest during the young stages. These curves are convenient ways to describe the different kinds of life histories.

DEMOGRAPHY

Demography is the statistical study of population size. Its name comes from two Greek words, *demos,* the people (the same root that we see in the word "democracy"), and *graphein,* to write. It therefore literally means writing about, or describing, people. Demography is the science that enables us to predict whether populations will remain the same size in the future or how they may change. It considers how different patterns of age distribution in populations alter their size through time.

An example of the age distribution of men and women is shown in Figure 20-14. We can predict the future sizes of a population by multiplying the number of women in each age class by the average number of female offspring produced by a woman alive at that age and then adding these totals for each age group to see whether the new number will exceed, equal, or be less than the number of women in the sample under consid-

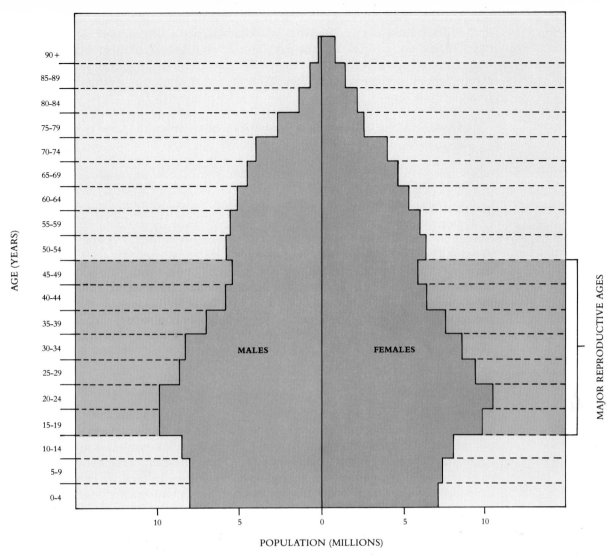

FIGURE 20-14

The age distribution of human males and females in the population in 1980 in the United States.

eration. By such means, the future growth trends of the human population as a whole and of individual countries and regions can be determined. Such calculations form much of the basis for the considerations that we shall present in Chapter 23.

FACTORS LIMITING THE DISTRIBUTION OF SPECIES

Populations that grow faster do not necessarily occupy larger areas. Most organisms, however rapidly their population is growing, are found only within certain geographical areas, referred to as the **range** of the species. A species occurs within its range and not elsewhere either because physical barriers prevent its spread (the ocean that surrounds a Pacific island, for example), or because selection acts against individuals that attempt to survive beyond the range of their species. The selective factors that limit the distributions of organisms can be grouped for convenience under three headings: **climatic, biotic,** and **edaphic.** Of these, climatic factors are perhaps the most important in determining the distribution of organisms. However, edaphic factors (those pertaining to the soil) interact with climate in complex ways and may also play a key role in delineating distributions (Figure 20-15). Biotic, or biological, factors may likewise be significant: the kinds of biological communities that occur in particular places greatly influence the limits of distribution of organisms, both members of that community and others.

Climatic, biotic, and edaphic factors are often intricately linked. For example, California redwoods *(Sequoia sempervirens)* directly affect the local "microclimate" that occurs around their bases and thus the distribution of many other kinds of organisms. In addition, certain communities may constitute barriers to the migration of many species of plants, animals, and microorganisms that do not occur in them. Plants and animals that need normal amounts of light cannot occur within, or even in shaded places at the edges of, a redwood forest; for them, the forest may be as inhospitable an environment as the sea.

Climatic, biotic, and edaphic factors interact to determine the limits of distribution of particular kinds of organisms.

Climatic Factors

The distribution of a particular species of organism at the margins of its range is generally determined by the continually changing climates in these areas, as well as by the genetic variability of its population. Also of importance is the dispersibility of the organism. The conditions that occur in extremely hot, cold, wet, or otherwise extraordinary years constantly act to eliminate populations of organisms at the limits of their distribution. On the other hand, if appropriate genetic combinations are represented in these marginal populations or are achieved by recombination, and if the dispersal mechanisms of the species are adequate, the species may be able to evolve and to extend its range into new areas, reaching pockets of favorable habitat beyond the current margins of their ranges.

Edaphic Factors

The nutrients present in the soil play significant roles in determining the ranges of plant species (see Figure 20-15). For example, the old, weathered soils of Australia are low in phosphates and in molybdenum, an essential element whose supply is generally adequate everywhere else in the world. The native plant communities of Australia are able to grow on these very poor soils. The plants characteristic of these communities often have leathery, evergreen leaves that appear to play an important role in conserving phosphorus and other nutrients. When we attempt to replace such plant communities

FIGURE 20-15

This aerial view of Jasper Ridge, in the Coast Ranges of California above Stanford University, clearly illustrates the difference between the vegetation on soil derived from serpentine rock, which is rich in magnesium and poor in calcium, and sandstone, which has quantities of these elements more suitable for the growth of most plants. The soil that formed from serpentine is dominated by the golden-flowered herb *Lasthenia californica;* that derived from sandstone is dominated by grasses. Jasper Ridge was the locality for Paul Ehrlich's important studies of butterfly populations, illustrated in Figure 17-5.

FIGURE 20-16

The fynbos plant community, shown here in the Cape Peninsula of South Africa, grows on highly infertile soils. Many of the plants that grow in the fynbos retain scarce nutrients such as phosphorus in their evergreen leaves.

with agricultural crops, however, great amounts of fertilizer often must be applied. Communities similar to those of Australia also occur on relatively infertile soils in South Africa (Figure 20-16).

Deciduous plants, whose leaves are shed at the end of each growing season, do not normally conserve nutrients as effectively as evergreen plants. Certain species of oaks and other deciduous trees and shrubs are exceptions to this generality. These plants hold their dead leaves through most or all of the winter; during this time, nutrients are extracted from these leaves for use mainly during the period of active growth in the spring. Because deciduous plants do not conserve nutrients as well as evergreen ones, they are not as successful or well represented on the more sterile soils of Australia and similar regions.

Biotic Factors

Biotic factors often limit the distribution of organisms in quite evident ways. For instance, a species of herbivore cannot spread beyond the limits of the plant on which that species feeds unless new, genetically differentiated populations of individuals that can feed successfully on other kinds of plants evolve in the original population. Similarly, no carnivore can live apart from the animals that make up its prey. A plant that is pollinated by a particular insect does not normally set seed beyond the range of that insect, although it may be physiologically able to do so. In all these ways, organisms affect one another's ranges; they have done so ever since bacteria began to change the nature of the earth's atmosphere and its soils several billion years ago.

SUMMARY

1. Selection alters the distribution of alleles within populations so that individuals that are better suited to their environments increase in frequency within populations. Both physical and biological factors are important agents of selection.

2. The rate of growth of any population is defined as the difference between the birth rate and the death rate per individual per unit of time. The actual rate is affected both by emigration from the population and by immigration into it.

3. Most populations exhibit a sigmoid growth curve, which implies a relatively slow start in growth, a rapid increase, and then a leveling off when the carrying capacity of the environment of the species is reached. Populations can be harvested most effectively when they are in the rapid growth phase.

4. *r* strategists have large broods and rapid rates of population growth; *K* strategists are limited in size by the carrying capacity of their environments. They tend to have fewer offspring, with slower rates of population growth.

5. Survivorship curves are used to describe the characteristics of growth in different kinds of populations. Type I populations are those in which a large proportion of the individuals approach their physiologically determined limits of age. Type II populations have a constant mortality throughout their lives. Type III populations have very high mortality in their early stages of growth, but an individual surviving beyond that point is likely to live a very long time.

6. Climatic, biotic, and edaphic factors limit the distributions of organisms. Among these, climatic factors are the most important, but the others are significant as well.

REVIEW

1. The three principal kinds of selection are _____, _____, and _____.

2. Selection against a dominant allele alters allele frequencies faster than selection against a recessive allele. True or false?

3. Are populations with higher proportions of children expected to grow rapidly or more slowly than populations with lower proportions of children and a greater proportion of individuals of childbearing age?

4. Does the biotic potential of a population depend on the population's size?

5. Is a slow-growing population of organisms more likely to be a *K* reproductive strategist or an *r* reproductive strategist?

SELF-QUIZ

1. Which one of the following is *least* characteristic of a species that is an *r* reproductive strategist?
   (a) small size of offspring
   (b) large numbers of offspring per litter
   (c) rapid maturation
   (d) extensive parental care
   (e) all are characteristic

2. What is the difference between a type II and a type III survivorship curve?
   (a) In type II, mortality is higher in the early stages than it is in type III.
   (b) In type III, mortality is the same in early stages as it is in type II.
   (c) In type III, mortality is higher in the early stages than it is in type II.
   (d) In type II, a large proportion of individuals survive to the maximum age, whereas in type III they do not.
   (e) none of the above

3. For a recessive allele, selection acts
   (a) most quickly when the allele frequency is low
   (b) most quickly when the allele frequency is high
   (c) at a rate that does not depend on the allele's frequency

4. Arrange these processes in the order of the speed with which they will cause a population to evolve to a new form, starting with the fastest.
   (a) selection for a rare recessive allele
   (b) selection for a dominant allele
   (c) selection for a common recessive allele
   (d) stabilizing selection

5. How many of the following are edaphic factors?
   (a) average daily temperature
   (b) distribution of food source
   (c) soil nutrients
   (d) average rainfall
   (e) biological diversity

PROBLEMS

1. The frequency of the recessive allele that, when homozygous, leads to cystic fibrosis is about 0.024 in North American whites. Of every 100,000 births, 55 individuals are expected to exhibit the disease. How many are expected to be carriers (heterozygous for the cystic fibrosis allele)?

2. If the frequency of individuals homozygous for the recessive allele leading to albinism (lack of pigmentation) were reduced from one in one thousand to one in ten thousand, what would be the expected change in the proportion of individuals heterozygous for the allele?

THOUGHT QUESTIONS

1. Many species that are K reproductive strategists, such as whooping cranes and whales, are in danger of extinction. Can you think of any r reproductive strategists that are in danger of extinction? In light of this, why do you think K reproductive strategists are so common?

2. Representative plants from two commercial lines of corn were crossed, and the progeny then crossed with themselves for eight successive generations. The "hybrid" corn in the F_1 generation was much more vigorous than in either commercial line, although this effect falls off in later generations. Why do you suppose the hybrid corn is so much more vigorous? Why has selection not operated to endow the commercial lines with this vigor if it can be achieved in a single generation?

3. In the classic study of sparrow mortality during a winter storm, Bumpus reported that the largest male birds had the best chance of survival (directional selection), whereas most extreme female birds (largest and smallest) were selected against (stabilizing selection). Can you suggest an explanation why the results for male and female birds were different?

FOR FURTHER READING

BEDDINGTON, J.R., and R.M. MAY: "The Harvesting of Interacting Species in a Natural Ecosystem," *Scientific American,* November 1982, pages 62-69. The authors analyze the changing populations of whales and other animals that feed on the krill (shrimp) populations in the Antarctic Ocean as an example of the problems of using a biological resource without destroying it.

MYERS, J., AND OTHERS: "Conservation Strategy for Migratory Species," *American Scientist,* vol. 75, pages 19-26, 1983. Shorebirds, which often migrate great distances, illustrate the need for international cooperation in conservation.

FIGURE 21-1   These topi and zebras graze together in great herds on the Serengeti plains of East Africa. They share available food and water, and are subjected to many of the same dangers, yet they live together without apparent conflict. Evolution often favors cooperation.

HOW SPECIES INTERACT
WITH ONE ANOTHER

Chapter 21

Overview

Communities are bound together by the interactions of the organisms that make them up; the species that occur together in a community evolve together. Competition between two species limits their coexistence in the community: if they are competing for the same resources, one will win and the other must change or be driven to extinction. Predation and other forms of interaction among populations play important roles in limiting population size. We ourselves are part of a network of interactions that occur among the many organisms that inhabit earth. This web of ecological interactions is the very fabric of ecology. In large measure it dictates what the world is like and what sorts of evolutionary changes are likely to occur.

For Review

Here are some important terms and concepts that you will encounter in this chapter. If you are not familiar with them, you should review them before proceeding.

Symbiotic origin of mitochondria (Chapter 5)

Quantitative inheritance (Chapter 11)

Selection (Chapter 16)

The Hardy-Weinberg principle (Chapter 16)

The nature of species (Chapter 17)

Population growth (Chapter 20)

Evolution grants no free rides. Every individual organism on earth competes for its place within the biosphere. The organisms best able to meet the challenges of their environment leave the most offspring, and therefore it is their characteristics that multiply in nature. The environment imposes two challenges: (1) the challenge to cope with the demands of the physical environment, and (2) the challenge to compete with other organisms for the use of the environment. In Chapter 20 we focused on populations that consist of only one species, and we considered the factors that determine how rapidly such a population grows. Then we examined how this growth leads to increased competition for limited resources among the members of the population. In this chapter we shall focus on the ways that selective forces have shaped the interactions between *different* species. The competition that occurs between individuals of different species has produced some of the most interesting ecological relationships (Figure 21-1).

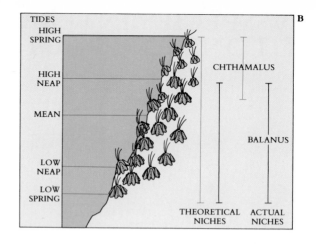

FIGURE 21-2

Balanus balanoides, one of the two competing species of barnacles that were studied by J.H. Connell along the coasts of Scotland. The microscopic larvae of barnacles are dispersed widely, but they settle down as adults. The distribution of the two species with respect to the different tides at Connell's study site is shown in **B**.

COMPETITION BETWEEN SPECIES

Two organisms that both use a resource that is in short supply are said to compete with each other, even if they are different species. Competition between different species is called **interspecific competition.** This sort of competition is, as you might expect, strongest between organisms that use the same resources: plants compete largely with other plants, **herbivores** (plant-eaters) with other herbivores, and **carnivores** (meat-eaters) with other carnivores.

More than 50 years ago the Soviet ecologist G.F. Gause formulated what he called the **principle of competitive exclusion.** This principle states that if two species are competing with each other for the *same* limited resource, then one of the species will be able to use that resource more efficiently than the other, and therefore the former will eventually eliminate the latter. Gause was able to demonstrate these relationships experimentally, often using different species of protists that are easy to maintain in culture. His valuable studies formed the theoretical basis for studies of competition.

> The principle of competitive exclusion states that if two species are competing with each other for the same limited resource in the same place, one of them will be able to use the resource more efficiently than the other and eventually will drive that second species to extinction in the area where they are competing.

Competition in Nature

Examples demonstrating the principle of competitive exclusion can be found and studied, both by observation and by experiment, in nature. For example, the competitive interactions between two species of barnacles that grow together on the same rocks along the coast of Scotland have been investigated by J.H. Connell of the University of California, Santa Barbara. Barnacles, which are marine crustaceans (Chapter 28), have free-swimming larvae that settle down, cement themselves to rocks, and then remain permanently attached at that point. Of the two species Connell studied, *Chthamalus stellatus* lives in shallower water where it is often exposed to air by tidal action, whereas *Balanus balanoides* occurs lower down where it is rarely exposed to the atmosphere (Figure 21-2). In this deeper zone *Balanus* could always outcompete *Chthamalus* by crowding it off the rocks, undercutting it, and replacing it even where it had begun to grow. The actual area that each species occupies is called its **actual niche.**

However, when Connell removed *Balanus* from the area, he found that *Chthamalus* was easily able to occupy the deeper zone, indicating that only the presence of *Balanus* there set lower limits on its occurrence. The total areas that *Chthamalus* could occupy in the absence of *Balanus* is called its **theoretical niche.** In other words, the niche of an organism is limited by the properties of the other organisms with which it occurs, as well as by physical factors in its environment (see Chapter 20). Even if *Chthamalus* was removed from the area, *Balanus* could not move into shallower water; thus its upper limits seem to be set by its physiological characteristics.

> **The theoretical niche of an organism is the niche that it might occupy if competitors were not present. The actual niche is the niche that it does occupy under natural circumstances.**

When thinking about the concept of niche, remember that it is simply the role an organism plays in its community. A niche may be described in such terms as space, food, temperature, appropriate conditions for mating, and requirements for moisture. A full portrait of an organism's niche also includes its behavior and the ways in which this behavior changes at different seasons and different times of the day. All of these factors taken together determine where different kinds of organisms are able to occur. Over time, of course, the characteristics of organisms evolve, and their niches—and possibilities of the organism occurring in different areas—change accordingly.

COEVOLUTION

Competition leads to selection. This is true when two organisms share the same resource, and it is also true when one organism is the resource of the other. These sorts of selective pressures have led to major evolutionary changes, usually involving two-way changes in which both competing organisms evolve. In response to such pressures, plants, herbivores, and carnivores have changed and adjusted to one another continually over millions of years (Figure 21-3). For example, many of the features of flowering plants have evolved in relation to the dispersal by insects of the plant's gametes to other members of the same species. Insects in turn have evolved a number of special traits that enable them to obtain food efficiently from the flowers of the plants they visit.

These two-way interactions, which involve the long-term mutual evolutionary adjustment of the characteristics of different organisms in relation to one another, are examples of **coevolution.** Coevolutionary interactions take many forms. Here we shall discuss some examples under the general headings of predator-prey interactions and symbiosis.

> **Coevolution is a term that describes the long-term evolutionary adjustment of one group of organisms to another. It is an important factor in determining the structure of communities.**

PREDATOR-PREY INTERACTIONS
Predation

Sometimes one species is an important component of another's niche. This is particularly true when one species feeds on another. We call the consumption of one species by another **predation.** When predation continues over long periods of time, the characteristics of both predator and prey are often changed, and coevolution between them takes place.

As an example of the way populations of predators and prey may achieve a balance, we shall consider the introduction of prickly pear cactus *(Opuntia)* into Australia and its later control by a South American moth, *Cactoblastis,* that was introduced for that purpose. When it was first introduced, the cactus overran wide areas and became so abundant that these areas were useless for grazing cattle (Figure 21-4). The introduction of the moth changed the situation dramatically, however. Its larvae feed on the

FIGURE 21-3

Several examples of the coevolution between species are evident here. The color of the crab spider, a carnivore, has changed during the course of its evolution to resemble that of the flowers where it hides, awaiting its prey, which comes to the flowers seeking food. The prey in this case is not a bee, but a harmless fly that takes advantage, in an ecological sense, of the fact that predators such as birds avoid bees, because they sting; although the fly cannot sting, it also is avoided. Both the flowers of the sunflower relative shown here (family Asteraceae) and many features of the fly, such as its mouthparts, have coevolved: the fly is able to obtain its food efficiently, and the plant takes advantage of the visits of such flies, and other insects, to spread its pollen from place to place.

A   B

cactus pads and rapidly destroy the plants. Within relatively few years, the moth had
reduced the cactus to the status of a rare species in many regions where it was formerly
abundant. It is now unusual to find an individual of *Cactoblastis* in the area, but the moth
is still present and evidently keeps the cactus in check. As the cactus and the moth continue to coexist in this kind of balance, the characteristics that determine their mutual
relationships will change, much like the characteristics of the populations considered in
Chapter 17. As they do, coevolution occurs.

Diseases display similar relationships. If the disease-causing organisms kill all of
their hosts, they eliminate their source of food. For this reason, those strains that are
less virulent but can spread efficiently from host to host will be favored by natural selection and will survive. Again, the characteristics of both host populations and those
of the disease-causing organisms will change over time.

The history of the viral disease myxomatosis, which was introduced into Australia and New Zealand to control rabbits, provides a good example of this principle.
The rabbits were imported to these countries as a convenient source of meat, but they
soon ran wild and began to destroy the plants over wide areas of the countryside (Figure 21-5). When the myxomatosis virus was introduced to control them, most of the
rabbits soon died. The most virulent strains of myxomatosis disappeared with the dead

FIGURE 21-5

Rabbits crowding around a water
hole in Australia. Imported from
Europe as a source of meat, rabbits
ran wild in Australia, New
Zealand, and elsewhere, and
eventually were controlled by a
lethal virus that caused the disease
myxomatosis. The disease did not
kill all the rabbits, however, but
eventually reached an equilibrium
state, in which sublethal strains
could spread from rabbit to rabbit.
If the virus had killed all the
rabbits, it would itself have gone
extinct. Putting it differently,
those strains that do not immediately kill their hosts have a
selective advantage over those
that do.

rabbits. Later, less lethal strains of the virus became apparent in the remaining rabbit populations. At the same time, strains of rabbits that were resistant to the disease began to appear. The relationship between both populations has reached an equilibrium in which both can coexist indefinitely.

A disease that always kills its host will die with it; a disease that produces sublethal effects has the opportunity to spread to another host.

Plant Defenses Against Herbivores

Of all the ecological interactions that occur on earth, none is more common than the predator-prey interactions between plants and the animals that eat them. All of the energy that animals use ultimately comes from photosynthetic organisms. At least 2 million species of herbivores feed on some 250,000 species of plants. The nature of life on earth has been largely determined by the ways in which plants avoid being eaten and by the ways herbivores succeed in eating them.

The most obvious ways in which terrestrial plants limit the activities of herbivores are morphological defenses. For example, thorns, spines, and prickles play an important role in discouraging browsers, although some browsers have learned to overcome these obstacles (Figure 21-6). Simple strengthening of the plant parts by the deposition of the mineral silica constitutes an important element in the protective system of some plants such as grasses (Figure 21-7). By depositing enough silica in the cells, the plants simply become too tough to eat.

Best known and perhaps most important in the defenses of plants against herbivores are the **secondary chemical compounds,** so called to distinguish them from **primary compounds,** which are normal chemicals used in metabolism. Virtually all plants and apparently many algae as well contain secondary compounds that play such a defensive role. For decades botanists and chemists tended to regard these substances as waste products that the plants produced, but this notion has now been discredited.

Secondary compounds, that is, chemicals that are not involved in primary metabolic processes, play the dominant role in protecting plants from being eaten by herbivores.

Many different kinds of chemical compounds, some extraordinarily complex, are involved in plant defenses. As a rule, different kinds of compounds are characteristic of particular groups of plants. For example, the mustard family (Brassicaceae; also called Cruciferae) is characterized by **mustard oils,** the substances that give the pungent aromas and tastes to such plants as mustard, cabbage, watercress, radish, and horseradish. The same tastes that we enjoy signal the presence of chemicals that are toxic to many groups of insects.

The Evolution of Herbivores

Associated with each family or other group of plants that is protected by a particular kind of secondary compound are certain groups of herbivores that are able to feed on them, often as the herbivore's exclusive food source. In general, those herbivores that feed on a restricted array of plant families, perhaps only one, feed on plant groups in which certain kinds of secondary compounds are well represented. For example, the larvae of cabbage butterflies (family Pieridae) feed almost exclusively on plants of the mustard and caper families, as well as on a few other small families that are also characterized by the presence of mustard oils. At some point the cabbage butterflies developed the ability to break down the mustard oils and thus feed on these mustards and capers without harming themselves. To those herbivores that feed on plants that are poisonous to most animals, the poisons become signals of the presence of food. Once the ancestors of cabbage butterflies acquired the ability to feed on these plants, they had registered an important evolutionary "breakthrough." They were able to use a new resource without having to compete for it with other herbivores.

FIGURE 21-6

Gerenuk, an antelope that grazes successfully on some of the spiny shrubs and trees of the African savanna that other browsers do not use as food.

FIGURE 21-7

These zebras graze selectively on the grass species that make up this East African savanna, depending in part on the degree to which the grass' leaves are reinforced with silica. In general, the more silica in the leaf cells, the less likely zebras and other grazing animals are to eat that particular kind of grass.

FIGURE 21-8

Butterflies that feed on plants that are toxic to most insects. All stages of the life cycle of the monarch butterfly *(Danaus plexippus)* are protected from birds and other predators by the poisonous chemicals that occur in the milkweeds and dogbanes on which they feed as larvae. Both the caterpillars **(A)** and the adult butterflies **(B)** "advertise" their poisonous nature with warning coloration.

Chemical Defenses in Animals

Some groups of animals that feed on plants rich in secondary compounds receive an extra benefit that is of great ecological importance. When the caterpillars of monarch butterflies (Figure 21-8) feed on plants of the milkweed family, they do not break down the poisonous cardiac glycosides that protect these plants from most herbivores. Instead, they store them in structures called fat bodies. As a result, these butterflies are themselves protected from predators. The cardiac glycosides in the leaves that the caterpillars eat are concentrated, stored, and passed through the chrysalis stage to the adult and even to the eggs; thus all stages are protected from predators. A bird that eats a monarch butterfuly quickly regurgitates it and thenceforth avoids the conspicuous orange-and-black pattern that characterizes the adult monarch (Figure 21-9). The active principles in drugs such as opium, marijuana, and belladonna are all secondary compounds that are used by the plants in which they occur to defend themselves against herbivores, but they also have significant physiological effects on vertebrates.

Many animals manufacture the chemicals that they use in their defense. In fact, animals manufacture and use a startling array of substances to perform an incredible variety of defensive functions. Certain frogs and many arthropods such as bees, wasps, scorpions, and spiders have chemicals that they use to defend themselves and to kill their prey. Frogs of the family Dendrobatidae, for example, secrete toxins in the mucus that covers their skin (Figure 21-10). These toxins are so powerful that a few micrograms will kill a person if they are injected into the bloodstream, and they are used by some of the native peoples of northern South America and Central America to poison the darts with which they hunt.

A

B

FIGURE 21-9

A A cage-reared bluejay, which had never seen a monarch butterfly before, eating one.
B The same bird a few minutes later, regurgitating the butterfly. Such a bird is not likely to attempt to feed on orange-and-black insects again.

FIGURE 21-10

The arrow-poison frogs of the family Dendrobatidae, like this individual of *Dendrobates lehmanni,* are abundant in the forests of Latin America and extremely poisonous to vertebrates. American native people employ extracts from frogs of this family to poison the darts that they use in their blowguns to kill birds and mammals.

FIGURE 21-11

Warning coloration is displayed by the spotted skunk, *Spilogale putorius* **(A)**; the poisonous gila monster, *Heleoderma suspectus*, a member of the only genus of poisonous lizards in the world **(B)**; and the red-and-black African grasshopper *Phymatus morbillosus* **(C)**, which feeds on succulent spurges, *Euphorbia*, plants that are highly poisonous.

Warning and Protective Coloration

Like the monarch (see Figure 21-8), other insects in North America that feed regularly on plants of the milkweed family are, in general, brightly colored. Among them are longhorn beetles, whose larvae feed on the roots of the milkweed plants, and bright red milkweed bugs. In some parts of the world there are bright red grasshoppers and other very obvious insects. These herbivores clearly are "advertising" their poisonous nature by their bright colors, or **warning coloration.** If an animal or plant is poisonous, it becomes a selective advantage for it to make itself obvious so that predators avoid it (see Figure 21-9). Therefore the more brightly colored individuals in a population of such a species will tend to avoid being eaten and leave more progeny than other individuals that are less conspicuous. Warning coloration is characteristic of animals that have effective defense systems, including not only poisons but also stings, bites, and other means of repelling predators. Some examples of animals with warning coloration are shown in Figure 21-11.

Warning coloration serves to keep potential predators away from poisonous or otherwise dangerous prey.

The opposite of advertising is camouflage, or deliberate concealment. Plant-eating insects and other herbivores that are not protected by particular chemical defenses are often **cryptically colored**—that is, colored so as to blend in with their surroundings and thus to be hidden from predators (Figure 21-12).

FIGURE 21-12

A striking example of cryptic coloration. This nighthawk is almost invisible against a background of pebbles.

B

C

D

A

FIGURE 21-13

Butterflies of the genus *Limenitis* feed on plants that are not toxic; the butterflies are therefore not protected by chemicals. Different species of *Limenitis*, however, have become Batesian mimics of other poisonous butterflies, and are therefore avoided by birds and other predators.
A and **B** The viceroy butterfly, *Limenitis archippus*. The larvae of the viceroy, which feed on willows and other nontoxic plants, are cryptically colored and thus concealed from birds and other predators.
C and **D** The monarch, *Danaus plexippus,* toxic to predators, is the model for the viceroy.

Mimicry

During the course of their evolution, many unprotected species have come to resemble distasteful ones that exhibit warning coloration. Provided that the unprotected animals are present in numbers that are low relative to those of the species that they resemble, they too will be avoided by predators. If they are too common, of course, predators will have as much chance of finding them as finding a protected individual, and thus the warning coloration ceases to protect either species. A pattern of resemblance of this kind is called **Batesian mimicry,** after the British naturalist H. W. Bates, who first called it to general attention in the 1860s. Bates also carried out the first scientific studies of this phenomenon, which has been of interest to naturalists ever since.

> **In Batesian mimicry, unprotected species resemble others that are distasteful and exhibit warning coloration. If the unprotected organisms are relatively scarce, they will be avoided by predators.**

Many of the best-known examples of Batesian mimicry occur among the butterflies and moths. One mimic among North American butterflies is the viceroy, *Limenitis archippus* (Figure 21-13). This butterfly, which resembles the poisonous monarch, ranges from central Canada south through much of the United States east of the Sierra Nevada and Cascade Range into Mexico.

Another kind of mimicry, **Muellerian mimicry,** was named for the German biologist Fritz Mueller, who first described it in 1878. In Muellerian mimicry several unrelated but protected animal species come to resemble one another. For example, a number of different kinds of stinging wasps have black-and-yellow–striped abdomens, but they may not all be descended from a common ancestor (Figure 21-14). In general, yellow-and-black and bright red tend to be common color patterns that warn predators that rely on vision that animals with such coloration are to be avoided.

> **Muellerian mimicry is a phenomenon in which two or more unrelated but protected species resemble one another, thus achieving a kind of group defense.**

SYMBIOSIS

Symbiotic relationships are those in which two kinds of organisms live together. All symbiotic relationships provide the potential for coevolution between the organisms involved in them, and in many instances the results of this coevolution are fascinating. The major kinds of symbiotic relationships include (1) **commensalism,** in which one of the species benefits and the other species neither benefits nor is harmed; (2) **mutualism,** in which both participating species benefit; and (3) **parasitism,** in which one species benefits but the other is harmed.

Examples of symbiosis include lichens, which are associations of certain fungi and green algae (cyanobacteria) (see Chapter 26); mycorrhizae, which are associations of fungi and the roots of plants (Chapter 26); and the associations of bacteria and root

A

B

C

D

FIGURE 21-14

The familiar yellow-and-black stripes of the yellowjackets and other wasps, such as *Vespula arenaria* (**A**), form the basis for large Batesian and Muellerian mimicry complexes. Here we see Batesian mimics representing three separate orders of insects, all rarer than yellowjackets and with patterns of behavior similar to those of the dangerous wasps that they resemble: (**B**) a flower fly, *Chrysotoxum;* (**C**) a longhorn beetle; (**D**) a sesiid moth, *Aegeria rutilans,* whose caterpillars are a pest on strawberries. None of these mimics stings or is poisonous, yet all are conspicuous members of the communities where they occur, flying about actively and remaining in full view at all times. They would be easy prey if they were not protected by their resemblance to yellowjacket wasps.

nodules that occur in legumes and certain other plants, which enable the host plants to fix atmospheric nitrogen and enable the bacteria to obtain carbohydrates (Chapter 31).

> **Symbiotic relationships are those in which two or more kinds of organisms live together.**

Commensalism

In nature the individuals of one species often grow attached to those of another. For example, birds nest in trees, or orchids grow on the branches of other plants. The host plant is usually unharmed, while the organism that grows or nests on it benefits. Similarly, various marine animals such as barnacles or sea anemones grow on other, often actively moving sea animals and are thus carried passively from place to place (Figure 21-15). These "passengers" presumably gain more protection from predation than they would if they were fixed in one place, and they also reach new sources of food. The increased water circulation that such animals receive as their host moves around may be of great importance, particularly if the passengers obtain their food by straining water to obtain the small organisms in it.

The best-known examples of commensalism involve the relationship between certain small tropical fishes and marine animals called sea anemones (Figure 21-16). These fishes have evolved the ability to live among the tentacles of the sea anemones, even though these tentacles quickly paralyze other fishes that touch them (see Chapter 28). The anemone fishes feed on the leftovers from the meals of the host anemone, remaining uninjured under remarkable circumstances. Other fishes live in similar associations with larger fishes or with other marine animals.

FIGURE 21-15

The barnacles that are evident on the back of this blowing gray whale, surfacing in San Ignacio Lagoon, Baja California, are carried from place to place with the whale, straining their food from the water as the whale moves about. Unlike most barnacles, which are anchored to fixed substrates, these animals have continuous access to fresh sources of the small, free-floating organisms on which they feed.

FIGURE 21-16

Fishes, such as this individual of *Amphiprion perideraion* in Guam, often form symbiotic associations with sea anemones, gaining protection by remaining among their tentacles and gleaning scraps from their food.

Mutualism

Examples of mutualism abound in nature and are of fundamental importance in determining the structure of biological communities. A particularly striking tropical example involves certain Latin American species of the plant genus *Acacia* and the ants that are associated with them. In these species the leaf parts called stipules are modified as paired, hollow thorns; consequently, these particular species are called bull's horn acacias. The thorns are inhabited by stinging ants of the genus *Pseudomyrmex*, which do not occur anywhere else.

At the tips of the leaflets of these acacias are unique, protein-rich bodies called Beltian bodies—after Thomas Belt, a nineteenth-century British naturalist who first wrote about them based on his observations in Nicaragua. Beltian bodies do not occur in species of *Acacia* that are not inhabited by ants, and their role is clear: they serve as a primary food for the ants. In addition, the plants secrete nectar from glands near the bases of the leaves. The ants eat this nectar also, and feed it and the Beltian bodies to their larvae as well (Figure 21-17).

Apparently this association is beneficial to the ants, and one can readily see why they inhabit acacias of this group. They and their larvae are protected within the swollen thorns, and the trees provide a ready source of a balanced diet, including the sugar-rich nectar and the protein-rich Beltian bodies. What, if anything, do the ants do for the plants? This question had fascinated observers for nearly a century until it was answered by Daniel Janzen, then a graduate student at the University of California, in a beautifully conceived and executed series of field experiments.

Whenever any herbivore lands on the branches or leaves of an acacia that is inhabited by such ants, the ants immediately attack and devour it. Thus the ants protect the acacias from being eaten, and the herbivore also provides additional food for the ants, which continually patrol the branches. Related species of acacias that do not have the special features of the bull's horn acacias and are not protected by ants have bitter-tasting substances in their leaves that the bull's horn acacias lack. Evidently these bitter-tasting substances protect the acacias in which they occur, in an evolutionary alternative to the way that the ants protect the species that they inhabit.

The ants that live in the bull's horn acacias also assist their hosts in competing with other plants. The ants cut away any branches of other plants that touch the bull's horn acacia in which they are living and thus create, in effect, a tunnel of light through which the acacia can grow, even in the lush deciduous forests of lowland Central America. Without the ants, as Janzen showed experimentally by poisoning the ant col-

FIGURE 21-17

A Ants of the genus *Pseudomyrmex* live within the inflated, spinelike stipules of the bull's-horn acacias, a unique group of this large genus of plants that occurs from Mexico to northern South America.
B Nectaries at the base of the leaves provide a sugar-rich fluid that the ants harvest.
C Beltian bodies are unique, protein-rich structures borne at the ends of the leaflets; the ants harvest them for food.
D As shown here, the ants cut away much of the surrounding vegetation that crowds the acacias. They also attack most kinds of herbivores that threaten the acacias.

A

B

C

D

A

B

FIGURE 21-18

A parasitic flowering plant and a close, nonparasitic relative; both are members of the plant family Convolvulaceae, the morning glory family.
A Morning glory *(Ipomoea)*.
B Dodder *(Cuscuta)* is a parasite. It has lost its chlorophyll and its leaves in the course of its evolution, and, like animals in general, is heterotrophic—unable to manufacture its own food; instead, it obtains its food from the host plants on which it grows.

onies that inhabited individual plants, the acacia is unable to compete successfully in this habitat. Finally, the ants bring organic material into their nests, and the part that they do not consume, together with their excretions, provides the acacias with an abundant source of nitrogen, which is an essential nutrient.

Janzen first determined these relationships by making careful observations of what was going on. Once he had formulated his hypotheses about the nature of the system, he poisoned the ant colonies in some of the plants and studied the fates of those particular plants that were growing in the same habitats as others that had healthy ant colonies. Many similar experiments will be needed before we can begin to understand the functioning of temperate ecosystems better, much less the vastly more complex ones of the tropics.

Parasitism

The concept of parasitism seems obvious, but individual instances are often surprisingly difficult to distinguish from predation and from various other kinds of symbiosis. Many instances of parasitism are well known. Among the vertebrates instances involve members of many different phyla of animals and protists. Invertebrates also have series of parasites that live within their bodies. However, bacteria and viruses are often not considered parasites, even though they fit our definition. Lice, which are insects that live on the bodies of birds and mammals, are normally considered parasites, but mosquitoes are not, even though they draw food from the same birds and mammals in a similar manner. Although the mosquitoes may fly away when they have finished feeding, they may be extremely closely associated with their hosts ecologically and in every other sense.

Some flowering plants, too, are parasitic on other plants (Figure 21-18), and a few are serious pests of crops. A number of examples of parasites will be considered in Chapters 25, 26, and 28 in relation to the groups of organisms in which they occur.

Internal parasitism is generally marked by much more extreme specialization than external parasitism, as shown by the many different examples of protist and invertebrate parasites that infect humans. The more closely the life of the parasite is linked with that of its host, the more its morphology and behavior are likely to have been modified in the process (Figure 21-19). Of course, the same is true of symbiotic relationships of all sorts. Conditions within the body of another organism are very different from those encountered outside and are apt to be much more constant in every way. Consequently, the structure of the parasite is often simplified, and unnecessary armaments and structures are lost as it evolves.

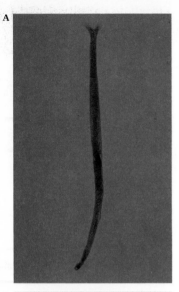

FIGURE 21-19

Parasitic animals.
A A male hookworm, *Ancyclostoma duodenale*. These parasites of humans live in the intestine and range in size from 8 to 10 mm for males and 10 to 13 mm for females.
B The human flea, *Pulex irritans*. Fleas are flattened from side to side, and slip easily through hair; their ancestors were insects that were larger, more brightly colored, not flattened, and had wings. The structural and behavioral modifications of fleas have come about in relation to a parasitic way of life.

Despite these qualifications and inconsistencies in terminology, the general meaning of the term **parasite** is clear. Parasitism may be regarded as a special form of predation in which the predator is much smaller than the prey and remains closely associated with it. Parasitism is harmful to the prey organism but beneficial to the parasite.

SUMMARY

1. Each species plays a specific role in its ecosystem. This role is called its niche.

2. Gause's principle of competitive exclusion states that if two species compete with each other for the same limited resources, then one will be able to use them more efficiently than the other and will drive its less efficient competitor to change or extinction. In terms of the niche concept, this principle can be restated: no two species can occupy the same niche indefinitely.

3. An organism's theoretical niche is the total niche that the organism would occupy in the absence of competition. The actual niche of an organism is a description of its actual role in nature.

4. Predator-prey relationships are of crucial importance in limiting population size. They appear to maintain a kind of balance, because if the prey is completely eliminated, the predator will die also.

5. Coevolution is the process by which a species adjusts to other kinds of organisms by genetic change over long periods of time. It is a stepwise process that ultimately involves mutual adjustment by all of the interacting groups of organisms.

6. Plants are often protected from herbivores, fungi, and other agents by chemicals that the plants manufacture. Such chemicals, which are not part of the primary metabolism of the plant, are called secondary compounds.

7. Particular classes of secondary compounds are usually characteristic of closely related families of plants. The herbivores that can feed on such plants either break the secondary compounds down or store them in their bodies. If they do the latter, the chemicals may in turn protect the herbivores from their predators.

8. Batesian mimicry is a situation in which a palatable or nontoxic organism resembles another kind of organism that is distasteful or toxic. Muellerian mimicry occurs when several toxic or dangerous kinds of organisms resemble one another.

9. Symbiotic relationships are those in which two kinds of organisms live together. There are three principal kinds of symbiotic relationships: in commensalism, one benefits and the other is unaffected; in mutualism, each organism benefits; and in parasitism, one benefits but the other is harmed.

REVIEW

1. Competition between different species is called _____ competition.

2. The niche that an organism might occupy if competitors were not present is called its _____ niche.

3. Chemicals that are not involved in primary metabolic processes and that often play a role in defenses against predation are called _____ compounds.

4. Poisonous animals tend to be _____ colored.

5. Mutualism, commensalism, and parasitism are all _____ relationships, in which two kinds of organisms live together.

SELF-QUIZ

1. If *Balanus* barnacles are removed from an area deep on rocks of the Pacific seacoast, *Chthamalus* barnacles occupy the empty zone; if *Chthamalus* barnacles are removed from a shallower area higher on the same rocks, *Balanus* barnacles do not occupy the empty zone. Which species of barnacle has the larger theoretical niche, *Chthamalus* or *Balanus?*

2. Grazing tends to decrease the variety of species that make up a community. True or false?

3. If two species live together with one benefiting and the other neither benefiting nor being harmed, their relationship is described as
   (a) mutualism   (d) predation
   (b) commensalism   (e) competition
   (c) parasitism

4. In which form of mimicry do unprotected species resemble other species that are distasteful and that exhibit warning coloration?
   (a) Mullerian mimicry
   (b) aposematic mimicry
   (c) Batesian mimicry
   (d) Commensal mimicry
   (d) none of the above

5. A lichen is a symbiotic association between _____ and _____ .

PROBLEM

1. Imagine that you sowed two kinds of grass seed, Kentucky bluegrass and fescue, in various mixtures in different areas of your yard. In those places where the majority of the grass seed that you planted was Kentucky bluegrass, the bluegrass germinated more quickly and filled that segment of the yard with small bluegrass plants; in those place where the majority of seed was fescue, the bluegrass still predominated early because of its more rapid germination. However, fescue grows more quickly than bluegrass, and by the end of the summer the entire lawn was fescue—no bluegrass was to be seen. Can you imagine a density of bluegrass and fescue that would result in a mixture of grass on the lawn?

THOUGHT QUESTIONS

1. How is it possible for ecological generalists—that is, those organisms that feed on a wide range of foods—to coexist with specialists in natural communities, as they do throughout the world? Why don't all species simply become generalists or specialists?

2. What kinds of ecosystems are most resistant to introduced species? Why?

FOR FURTHER READING

BEDDINGTON, J.R., and R.M. MAY: "The Harvesting of Interacting Species in a Natural Ecosystem," *Scientific American,* November 1982, pages 62-69. The authors analyze the changing populations of whales and other animals that feed on the krill (shrimp) populations in the Antarctic Ocean as an example of using a biological resource without destroying it.

HUTCHINSON, G.E.: *The Ecological Theater and the Evolutionary Play,* Yale University Press, New Haven, Conn., 1965. This series of beautifully written essays, by one of the great masters of ecology, is an outstanding introduction to many of its most interesting concepts.

JACKSON, J.B.C., and T.P. HUGHES: "Adaptive Strategies of Coral-Reef Invertebrates," *American Scientist,* vol.73, pages 265-274, 1985. Coral-reef environments, which are regularly disturbed by storms and predation, provide an excellent example of a complex, balanced marine community.

MORSE, D.H.: "Milkweeds and Their Visitors," *Scientific American,* July 1985, pages 112-119. Nectar-feeders, herbivores, predators, and parasites gather on milkweed plants, forming a model ecological community, which the author has studied in detail.

OWEN, D.: *Camouflage and Mimicry,* University of Chicago Press, 1980. A short, beautifully illustrated book. Many of the photographs are based on the author's field experience in West Africa.

SCHOENER, T.W.: "The Controversy over Interspecific Competition," *American Scientist,* vol. 70, pages 586-595, 1982. Interesting discussion about the interplay between competition and predation in structuring natural communities.

FIGURE 22-1   Unlike the moon, all of the earth's surface, even its deserts, is teeming with life, although it may not always appear so. The same ecological principles apply to the organization of all of the earth's communities, both on land and in the sea, although the details differ greatly.

COMMUNITIES AND ECOSYSTEMS   Chapter 22

Overview

The earth provides living organisms with much more than a place to stand or swim. Many chemicals cycle between our bodies and the physical environment around us. We live in a delicate balance with our physical environment, one easily disturbed by human activities. The collection of organisms that live in a place, and all the physical aspects of the environment that affect how they live, operate as a fundamental biological unit, the ecosystem. Much of the effort of ecologists focuses on improving our understanding of how ecosystems function.

For Review

Here are some important terms and concepts that you will encounter in this chapter. If you are not familiar with them, you should review them before proceeding.

Nitrogen (Chapter 3)

Metabolic energy (Chapter 7)

Respiration (Chapter 8)

Major features of evolution (Chapter 16)

Ecology is concerned with the most complex level of biological integration. It attempts to tell us why particular kinds of organisms can be found living in one place and not in another. In previous chapters we have dealt with the physical and biological variables that govern the distribution of organisms and the factors that maintain the numbers of particular kinds of organisms at certain levels. In this chapter we focus on principles that govern the functioning of assemblages of organisms.

In ecological terms, populations of different organisms that live together in a particular place are called **communities.** A community, together with the nonliving factors with which it interacts, is called an **ecosystem.** An ecosystem regulates the flow of energy and the cycling of the essential elements on which the lives of its constituent plants, animals, and other organisms depend. Communities and ecosystems exist both on land and in the sea (Figure 22-1).

> **A community is the interacting set of different kinds of organisms that occur together at a particular place. An ecosystem is that set of organisms, together with the nonliving factors with which it interacts.**

ECOSYSTEMS

Ecosystems are the most complex level of biological organization. They are systems in which there is a regulated transfer of energy and an orderly, controlled cycling of nutrients. The individual organisms and populations of organisms in an ecosystem act as part of an integrated whole, adjust over time to their role in the ecosystem, and relate to one another in complex ways that we only partly understand. Despite their differences, all ecosystems are governed by the same principles and restricted by the same limitations. The earth is a closed system with respect to chemicals, but an open one in terms of energy. That is, no new chemicals are being added from outside, but energy is constantly being added from the sun. Ecosystems function to regulate the capture and expenditure of that energy and the cycling of those chemicals. As you will see in this chapter, all organisms, including human beings, depend on the activities of a few other organisms to recycle the basic components of life: plants, algae, and some bacteria in the case of carbon, and other bacteria in the case of nitrogen.

THE CYCLING OF NUTRIENTS IN ECOSYSTEMS

All of the substances that occur in organisms, including water, carbon, nitrogen, and oxygen, as well as a number of others that are ultimately derived from the weathering of rocks, cycle through ecosystems. These cycles are geological ones that involve the biologically controlled cycling of chemicals and therefore are called **biogeochemical cycles.** Among the elements that are cycled are some that come from rocks, including phosphorus, potassium, sulfur, magnesium, calcium, sodium, iron, cobalt, and all the other elements that are essential for plant growth (see Chapter 31). All organisms require carbon, hydrogen, oxygen, nitrogen, phosphorus, and sulfur in relatively large quantities; the other elements are required in smaller amounts.

We speak of the **cycling** of materials in ecosystems because these materials are incorporated from the atmosphere or from weathered rock first into the bodies of organisms; they then sometimes pass from these organisms into the bodies of other organisms that feed on these primary ones, and ultimately are returned to the nonliving world. When this occurs, the nutrients may possibly be incorporated again into the bodies of other organisms. Some examples will help to clarify the ways in which different cycles function.

The Water Cycle

All life directly depends on the presence of water. In fact, the bodies of most organisms consist mainly of this substance. Water is the source of the hydrogen ions whose movements generate ATP in organisms, and for that reason alone, it is indispensable to their functioning. Thus the **water cycle** (Figure 22-2) is the most familiar of all biogeochemical cycles. The oceans cover nearly three fourths of the earth's surface. From their surface, water evaporates into the atmosphere, a process that is powered by energy from the sun. This water eventually precipitates back to earth and passes into surface and subsurface bodies of fresh water. Most of it falls directly into the oceans, but some falls onto the earth. Only about 2% of all the water on earth, however, is fixed in any form—frozen, held in the soil, or incorporated into the bodies of organisms. All of the rest is free water, circulating between the atmosphere and the earth. Regardless of where this water is held temporarily, however, it eventually returns to the atmosphere and the oceans.

Much less obvious than the surface waters, which we see in streams, lakes, and ponds, is the groundwater, which occurs in permeable, saturated, underground layers of rock, sand, and gravel called **aquifers.** In many areas, groundwater is the most important reservoir of water; for example, it amounts to more than 96% of all fresh water in the United States. It flows much more slowly than surface water, anywhere from a few millimeters to as much as a meter or so per day. In the United States, groundwater provides about a quarter of the water used for all purposes and provides about half of the population with drinking water. Three quarters of the country's cities and most of

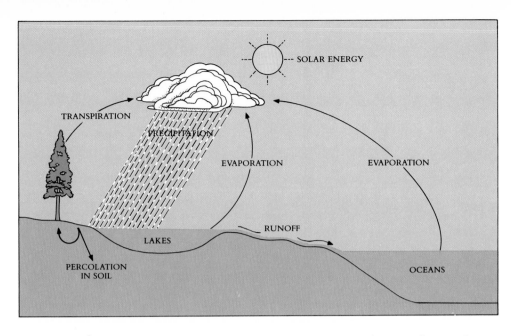

FIGURE 22-2

The water cycle.

its rural areas rely at least in part on groundwater reserves. The use of groundwater throughout the world is growing much more rapidly than is the use of surface water.

> **Some 96% of the fresh water in the United States consists of groundwater. This groundwater, which already provides a quarter of all the water used in this country, will become even more important in the future.**

The Carbon Cycle

The **carbon cycle** is based on carbon dioxide, which makes up only about 0.03% of the atmosphere. The worldwide synthesis of organic compounds from carbon dioxide and water results in the fixation of about a tenth of the roughly 700 billion metric tons of carbon dioxide in the atmosphere each year (Figure 22-3). This enormous amount of biological activity takes place as a result of the combined activities of photosynthetic bacteria, algae, and plants. All heterotrophic organisms—including the nonphotosynthetic bacteria and protists, the animals, and a relatively few plants, such as dodder (see Figure 21-18,*B*), that have lost the ability to photosynthesize—obtain their carbon indirectly from the organisms that fix it. As a result of oxidative respiration and the ultimate decomposition of their bodies, organisms release carbon dioxide to the atmosphere again. Once there, it can be reincorporated into the bodies of other organisms.

> **About a tenth of the estimated 700 billion metric tons of carbon dioxide in the atmosphere is fixed annually by the process of photosynthesis.**

In addition to the carbon dioxide in the atmosphere, approximately 1000 billion metric tons are dissolved in the ocean; more than half of this quantity is in the upper layers, where photosynthesis takes place. The fossil fuels, primarily oil and coal, contain more than 5 trillion additional metric toms of carbon, and between 600 billion and 1 trillion metric tons are locked up in living organisms at any one time.

The release of the carbon in fossil fuels as carbon dioxide, a process that is proceeding rapidly as a result of human activities, currently appears to be changing global climates and may do so even more rapidly in the future. Before widespread industrialization, the concentration of carbon dioxide in the atmosphere was about 260 to 280 parts per million (ppm). During the 25-year period starting in 1958, this concentration increased from 315 ppm to more than 340 ppm and is continuing to rise rapidly. Such increased levels of carbon dioxide in the atmosphere profoundly affect the way in

FIGURE 22-3

The carbon cycle.

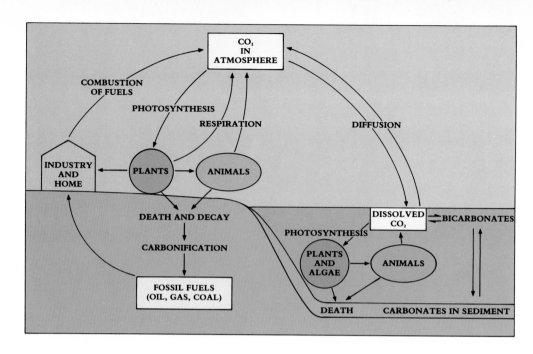

which the earth is heated by the sun, primarily because carbon dioxide traps the longer wavelengths of infrared light, or heat, and prevents them from radiating into space. By doing so, it creates what is known as a greenhouse effect, which acts to raise the overall temperature of the earth and its atmosphere. The rise in global carbon dioxide concentrations is being monitored very carefully, because the potential consequences of a warmer earth are very serious.

The Nitrogen Cycle

Nitrogen gas constitutes nearly 80% of the earth's atmosphere by volume, but the total amount of fixed nitrogen in the soil, oceans, and the bodies of organisms is only about 0.03% of that figure. The nitrogen in the atmosphere cannot be used by most organisms, because few of them possess the special enzyme system necessary to break the very strong triple bond of nitrogen gas (N_2). Those that do have the enzymes, including several genera of bacteria, carry out a process called **nitrogen fixation.** Once nitrogen has been "fixed," it cycles within biological systems; all living organisms depend on nitrogen fixation. Without it, they would be unable to synthesize proteins, nucleic acids, and other necessary nitrogen-containing compounds. Nitrogen fixation is the means by which a very small fraction of the enormous reservoir of nitrogen that exists in the earth's atmosphere is made available for biological processess (Figure 22-4).

Although dozens of genera of bacteria have the ability to fix nitrogen, only a few genera that form symbiotic associations with plants usually fix enough nitrogen to be of major significance. Plants that regularly form symbiotic associations with nitrogen-fixing bacteria, such as the legumes (members of the plant family Fabaceae, which includes peas, beans, alfalfa, mesquite, and many other well-known plants), can grow in soils that have such low amounts of nitrogen that they are unsuitable for most other plants. It is estimated that a legume crop may add as much as 300 to 350 kilograms of nitrogen per hectare per year, whereas all other sources may add about 15 additional kilograms. The overwhelming importance of these symbiotic associations is therefore obvious.

> **Although nitrogen gas constitutes nearly 80% of the earth's atmosphere, it becomes available to organisms only through the activities of a few genera of bacteria, especially those which live symbiotically on the roots of legume and certain other plants.**

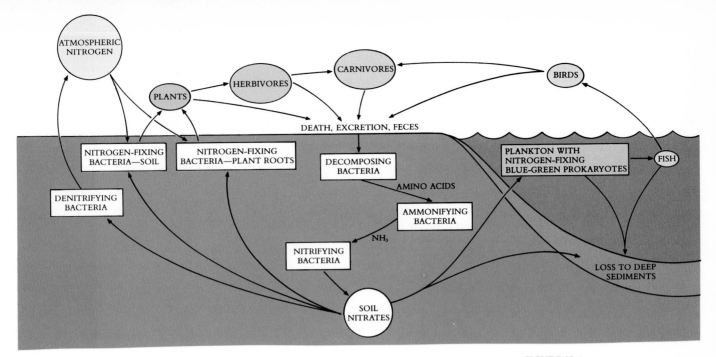

FIGURE 22-4

The nitrogen cycle.

The Phosphorus Cycle

In all biogeochemical cycles other than those involving water, carbon, oxygen, and nitrogen, the reservoir of the nutrient exists in mineral form, rather than in the atmosphere. The **phosphorus cycle** (Figure 22-5) is presented here as a representative example of all other mineral cycles, because of the critical role that phosphorus plays in plant nutrition worldwide.

FIGURE 22-5

The phosphorus cycle.

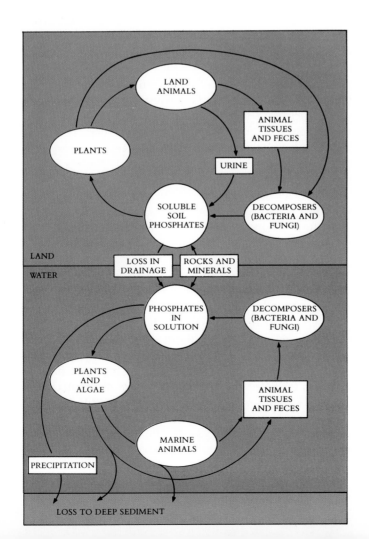

FIGURE 22-6

Pelicans roosting on a small island in the Gulf of California, Mexico. Sea birds bring up phosphorus from the deepest layers of the sea by eating fish and other marine animals and depositing their remains as guano on the rookeries.

FIGURE 22-7

The tropical rain forest has a large biomass and a high net productivity, even though it grows on soils that are often low in some of the nutrients that are essential for plant growth. How it does this is described in the text on p. 435.

Phosphates—charged phosphorus ions—exist in the soil only in small amounts, because they are relatively insoluble and are present only in certain kinds of rocks. If the phosphates are not lost to the sea by way of rivers and streams, they may be absorbed by plants and incorporated into compounds such as ATP, nucleic acids, and membrane proteins. Fungi associated with plants roots—the associations are called **mycorrhizae**—facilitate the transfer of phosphates from the soil to plants, a very important association (see Chapter 26). Other plants that grow on very poor soils do not form mycorrhizae, but may instead have clusters of very fine roots that perform a similar function.

When animals or plants die, slough off parts, or, in the case of animals, excrete waste products, the phosphorus may be returned to the soil or water and then cycle again through other organisms. When phosphates are lost to the deep sea, they may be recycled through the activities of sea birds that eat fishes and other animals that feed in deep waters (Figure 22-6).

THE FLOW OF ENERGY IN ECOSYSTEMS

An ecosystem includes two different kinds of living components, autotrophic and heterotrophic ones. The autotrophic components, consisting of plants, algae, and some bacteria, are able to capture light energy and manufacture their own food. To support themselves, the heterotrophic ones, including animals, fungi, most protists and bacteria, and nongreen plants, must obtain organic molecules that have been synthesized by autotrophs.

Once energy enters an ecosystem, mainly after it is captured as a result of photosynthesis, it is slowly released as metabolic processes proceed. The autotrophs that first acquire this energy provide all of the energy that heterotrophs use. Ecosystems, as well as the organisms that make them up, can in one sense be viewed as systems that have become adapted through time to delay the release of the energy obtained from the sun back into space. Energy flows one way through an ecosystem, while the nutrients are continually recycled.

Looking at an ecosystem as a whole, we can speak of its **net primary productivity,** which we define as the total amount of energy that is converted to organic compounds in a given area per unit of time. The **net productivity** of the ecosystem is the total amount of energy fixed by photosynthesis per unit of time, minus that which is expended by the metabolic activities of the organisms in the community. The total weight of all of the organisms living in the ecosystem, called its **biomass,** increases as a result of its net production. Some ecosystems—for example, a cornfield or a cattail swamp—have a very high net primary productivity. Others, such as tropical rain forests (Figure 22-7), also have a relatively high net primary productivity, but a rain forest has a much larger biomass than a cornfield; consequently, the net primary productivity of rain forest is much lower in relation to its total biomass. In such communities as sugar cane fields, coral reefs, and estuaries, the net primary productivity per square meter per year may range from roughly 3500 to 9000 grams. The productivity of marshlands and tropical forests is somewhat less, and that of deserts is about 200 grams.

Trophic Levels

Green plants, the primary producers of an ecosystem, generally capture about 1% of the energy that falls on their leaves, converting it to food energy. In especially productive systems, this percentage may be a little higher. When these green plants are consumed by other organisms, usually only about 10% of the plant's accumulated energy is actually converted into the bodies of the organisms that consume them.

Among these consumers, several levels may be recognized. The **primary consumers,** or herbivores, feed directly on the green plants. **Secondary consumers,** carnivores and the parasites of animals, feed directly or indirectly on the herbivores. **Decomposers** break down the organic matter accumulated in the bodies of other organisms. Another, more general, term that includes decomposers is **detritivores.** Detritivores are organisms that live on the refuse of an ecosystem—not only on dead

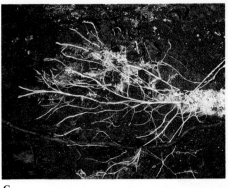

A   **B**   **C**

organisms but on the cast-off parts of organisms. They include large scavengers, such as crabs, vultures, and jackals, as well as decomposers such as bacteria and fungi (Figure 22-8).

All of these levels, and usually additional ones, are represented in any fairly complicated ecosystem. They are called **trophic levels,** from the Greek word *trophos,* which means "feeder." Organisms from each of these levels, feeding on one another, make up a series that is called a **food chain.** The length and complexity of food chains vary greatly. In real life, it is rather rare for a given kind of organism to feed on only one other kind of organism. Usually, each will feed on two or more other kinds and in turn will be fed on by several other kinds of organisms. When diagrammed, the relationship appears as a series of branching lines, rather than as one straight line; it is called a **food web** (Figure 22-9).

A certain amount of the energy that is ingested and retained by the organisms at a given trophic level goes toward heat production. A great deal of the energy is used for digestion and work, and usually 40% or less goes toward growth and reproduction. An invertebrate such as a worm or insect typically uses about a quarter of this 40% for growth; in other words, about 10% of the food that an invertebrate eats is turned into its own body, and thus into potential food for its predators. Although the comparable figure varies from approximately 5% in carnivores to nearly 20% for herbivores generally, 10% is a good average value for the amount of organic matter that is present at each step in a food chain, or each successive trophic level, and that reaches the next level.

A plant fixes about 1% of the sun's energy that falls on its green parts. The successive members of a food chain, in turn, process about 10% of the available energy in the organisms on which they feed into their own bodies.

FIGURE 22-8

A   The grass in the East African savannas is a primary producer, capturing energy from the sun. Grazing mammals obtain their food from the grass, and may in turn be consumed by predators.
B   This crab, *Gecarcinus quadratus,* photographed on the beach at Mazatlan, Mexico, is a detritivore, playing the same role that vultures and similar animals do in other ecosystems.
C   Fungi, such as the one shown here growing through the soil in Costa Rica, are, along with the bacteria, the primary decomposers of terrestrial ecosystems.

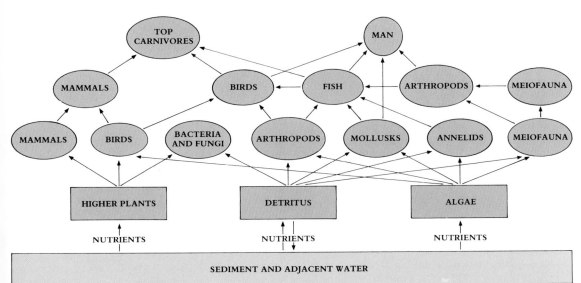

FIGURE 22-9

The food web in a salt marsh, showing the complex interrelationships between organisms. Each level of rectangles, ovals, and circles represents a trophic level feeding on, or gaining energy from, the layer below. Annelids, mollusks, and arthropods are groups of animals that will be discussed in Chapter 28. **Meiofauna** is a collective term for all the very small animals that live between grains of sand.

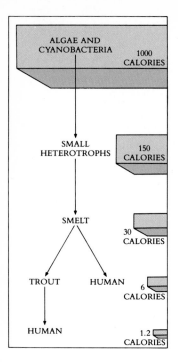

FIGURE 22-10

In Lake Cayuga, autotrophic plankton fix the energy of the sun, heterotrophic plankton feed on them, and both are consumed by smelt. The smelt are eaten by trout, with about a tenfold loss in fixed energy; for humans, the amount of biomass available in smelt is at least ten times greater than that available in trout, which they prefer to eat.

Lamont Cole of Cornell University studied the flow of energy in a freshwater ecosystem in Lake Cayuga in upstate New York. He calculated that about 150 calories of each 1000 calories of energy that were fixed by algae and cyanobacteria were transferred into the bodies of small heterotrophs (Figure 22-10). Of these, about 30 calories were incorporated into the bodies of smelt, the principal secondary consumers of the system. If human beings eat the smelt, they gain about 6 calories from each 1000 calories that originally entered the system. If, on the other hand, trout eat the smelt and we eat the trout, we gain only about 1.2 calories from each original 1000.

Relationships of this kind make it clear that organisms, including people, that subsist on an all-plant diet obviously have more food and energy available to them than do carnivores. Such considerations will become increasingly important in the future, not only for the efficient management of fisheries, but also in an effort to maximize the yield of food for a hungry and increasingly overcrowded world.

Food chains generally consist of only three or four steps. The loss of energy at each step is so great that very little of the original energy remains in the stystem as usable energy after it has been incorporated successively into the bodies of organisms at four trophic levels. There are generally far more individuals at the lower trophic levels of any ecosystem than at the higher ones. Similarly, the biomass of the primary producers present in a given ecosystem is greater than that of the primary consumers, with successive trophic levels having a lower and lower biomass, and correspondingly less potential energy. Larger animals characteristically are members of the higher levels; to some extent, they *must* be larger in order to capture enough prey to support themselves.

These relationships, if shown diagrammatically, appear as pyramids (Figure 22-11). We may therefore speak of "pyramids of biomass," "pyramids of energy," "pyramids of number," and so forth, as characteristic of ecosystems. Occasionally the pyramids are inverted. For example, in ecosystems consisting of organisms suspended in fresh or salt water, the autotrophic organisms may reproduce rapidly, thus being ca-

FIGURE 22-11

Pyramids of numbers, biomass, and energy. In the aquatic community, biomass is greater in the zooplankton (heterotrophic plankton) than in the phytoplankton (autotrophic plankton) because of rapid turnover among the organisms engaged in primary production. This is an unusual situation among ecosystems generally.

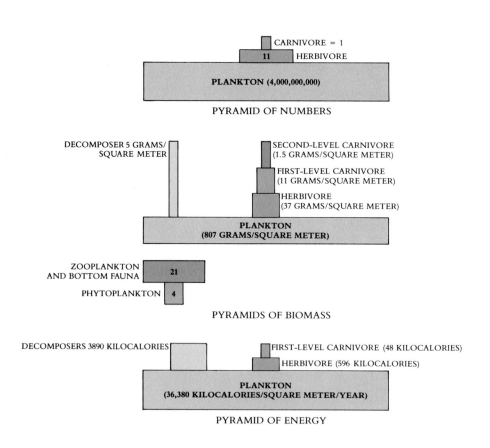

FIGURE 22-12

A pond near Tallahassee, Florida, with abundant aquatic vegetation. As the process of succession proceeds, the pond will eventually be filled and then gradually the area where it once existed will become indistinguishable from the surrounding vegetation.

pable of supporting a population of heterotrophs that is larger in biomass and more numerous than themselves. Pyramids of energy, on the other hand, cannot be inverted because of the necessary loss of energy at each step.

ECOLOGICAL SUCCESSION

Even when the climate of a given area remains stable year after year, ecosystems have a tendency to change from simple to complex in a process that is known as **succession.** This process is familiar to anyone who has seen a vacant lot or cleared woods slowly but surely become occupied with larger and larger plants and more and more different kinds of them, or seen a pond become filled with vegetation that encroaches from the sides and gradually turns it into dry land (Figure 22-12).

Succession is continuous and worldwide in scope. If a wooded area is cleared, and the clearing is left alone, plants will slowly reclaim the area. Eventually, the traces of the clearing will disappear, and the whole area will again be woods. This kind of succession, which occurs in areas that have been disturbed and that were originally occupied by living organisms, is called **secondary succession.** Human beings are often responsible for initiating secondary succession throughout the regions of the world that they inhabit. Secondary succession may also take place after fire has burned off an area, for example, or after the eruption of a volcano (Figure 22-13).

Primary succession, in contrast to secondary succession, occurs on some bare, lifeless substrate, such as rocks or open water, where organisms gradually occupy the area and change its nature. On bare rocks, for example, cyanobacteria, algae, or lichens (associations of fungi and algae discussed in Chapter 26) may grow first. Acidic secretions from the lichens help to break down the stone and form small pockets of soil. Mosses may then colonize these pockets of soil, eventually followed by ferns and the seedlings of other plants. Over many thousands of years, or even longer, the rocks may be completely broken down, and the vegetation over an area where there was once a rock outcrop may be just like that of the surrounding grassland or forest.

A   B

FIGURE 22-13

Mount St. Helens in the state of Washington erupted violently on May 18, 1980; the lateral blast devastated more than 600 square kilometers of forest and recreation lands within 15 minutes, as seen in **A,** which shows an area near Clearwater Creek 4 months after the eruption. Five years later, succession was underway at the same spot **(B),** with shrubs such as blackberries, blueberries, and dogwoods following the first plants that became established immediately after the blast.

Similarly, a lake may gradually accumulate organic matter and fill in (see Figure 22-12). Plants standing along the edges of the lake, such as cattails and rushes, and those growing submerged, such as pondweeds, together with other organisms, may contribute to the formation of a rich organic soil. As this process continues, the lake may increasingly be filled in with terrestrial vegetation. Eventually, the area where the lake once stood, like the rock outcrop we just described, may become an indistinguishable part of the surrounding vegetation.

Primary succession takes place in areas that are originally open, such as dry rock faces or new ponds. Secondary succession, in contrast, takes place in areas that have been disturbed after having been occupied by living things earlier.

Ponds and bare rocks in the same region often come to feature the same kind of vegetation as one another—the vegetation that is characteristic of the region as a whole. This relationship led the American ecologist F.E. Clements, at about the turn of the century, to propose the concept of **climax vegetation.** The term refers to the fact that this characteristic vegetation type is thought to be controlled by the *climate* of that region. However, with an increasing realization that the climate keeps changing, the process of succession is so slow, and the nature of a region's vegetation is being determined to a greater and greater extent by human activities, ecologists no longer consider the term "climax vegetation" to be as useful as they once did.

One of the most interesting features of succession is that the organisms involved in its early stages often are participants in symbiotic systems. Lichens, so important in the early colonization of rocks, are symbiotic associations of cyanobacteria or algae and fungi. This association appears to enable the lichens to withstand harsh conditions of moisture and temperature. Legumes and other plants that harbor nitrogen-fixing bacteria in nodules on their roots are often among the first colonists of relatively infertile soils. Trees with mycorrhizae are apparently more resistant to conditions of moisture stress and perhaps other environmental extremes than are those trees that have no mycorrhizae. It is believed that the first land plants had mycorrhizae and that this association was highly beneficial to them in the sterile, open habitats that they would have encountered.

Symbiotic associations, such as lichens, legumes with their associated nitrogen-fixing bacteria, and plants with mycorrhizal fungi, appear to play an unusually important role in early successional communities.

As any ecosystem matures, there is an increase in total biomass but a decrease in net productivity. The earlier successional stages are more productive than the later ones. Agricultural systems are examples of early successional stages in which the process is intentionally not allowed to go to completion and the net productivity is high (Figure 22-14). There are more kinds of species in mature ecosystems than in immature

FIGURE 22-14

Intensively cultivated areas are kept permanently at an early successional stage, as shown here in this newly cut hay field. Agriculture expends energy to arrest succession at an early stage, and thus to enhance productivity.

ones, and the number of heterotrophic species increases even more rapidly than the number of autotrophic species. This progression is related to the decreasing net productivity of increasingly mature ecosystems and to the fact that mature ecosystems have a greater ability to regulate the cycling of nutrients than do disturbed and immature ones. It seems that the plants and animals that appear in the later stages of succession may be more specialized, in general, than those that exist in the earlier stages. The late-successional species fit together into more complex communities and have much narrower ecological requirements, or niches.

> **Communities at early successional stages have a lower total biomass, higher net productivity, fewer species, many fewer heterotrophic species, and less capacity to regulate the cycling of nutrients than do communities at later successional stages.**

THE IMPACT OF CLIMATE ON ECOSYSTEMS

If the climate on earth were the same everywhere, there would be few different kinds of ecosystems, just oceans and lakes at different depths and land at different altitudes above sea level. The world contains a great diversity of ecosystems because its climate in fact varies a great deal. On a given day Miami, Florida, and Bangor, Maine, often have very different weather. There is no mystery about this. The tropics are warmer than the temperate regions because the sun's rays arrive almost perpendicular at regions near the equator, while near the poles their angle of incidence spreads them out over a much greater area (Figure 22-15, *A*), providing less energy per unit area. This simple fact, that because the earth is a sphere some parts of it receive more energy from the sun than others, is responsible for many of the major climatic differences that occur over the earth's surface, and thus indirectly for much of the diversity of ecosystems.

The earth's annual orbit around the sun and its daily rotation around its own axis are both important in determining world climate (Figure 22-15, *B*). Because of the daily cycle, the climate at a given latitude is relatively constant, with a constant mixing of climates and temperatures at that latitude. Because of the annual cycle and the inclination of the earth's axis at approximately 23.5 degrees from its plane of revolution around the sun, there is a progression of seasons away from the equator in all parts of the earth. One of the poles is closer to the sun than the other at all times, because the angle and direction of the earth's inclination are maintained as it rotates around the sun.

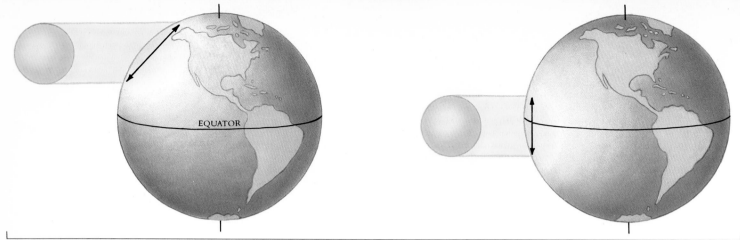

A

FIGURE 22-15

Relationships between the earth and the sun are critical in determining the nature and distribution of life on earth.
A A beam of solar energy striking the earth in middle latitudes is spread over a wider area of the earth's surface than a similar beam striking the earth near the equator.
B The rotation of the earth around the sun has a profound effect on climate. In the northern and southern hemispheres, temperatures change in an annual cycle because the earth is slightly tilted on its axis in relation to its pathway around the sun.

B

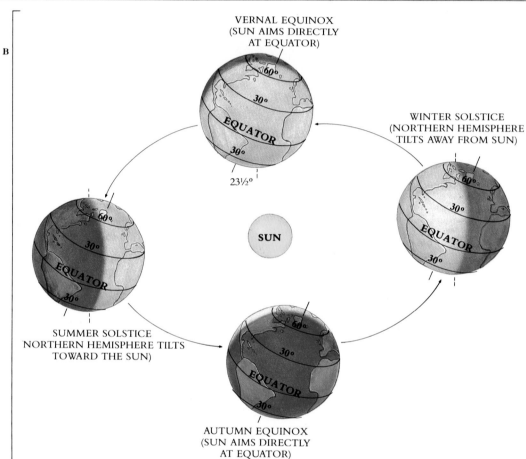

Major Circulation Patterns

Near the equator, warm air rises and flows toward the poles (Figure 22-16, *A*). As it rises, this warm air loses most of its moisture, which is why it rains so much in the tropics. This region of rising air is one of low pressure, the doldrums, which draws air from both north and south of the equator. When the air masses that have risen reach about 20 to 30 degrees north and south latitude, the air, now cooler, sinks and becomes reheated, producing a zone of decreased precipitation—warm air holds more moisture than cooler air. Consequently, all of the great deserts of the world lie near 20 to 30 degrees north or 20 to 30 degrees south (see Figure 22-23). Air at these latitudes is still warmer than it is in the polar regions, and it continues to flow toward the poles. It rises again at about 60 degrees north and south latitude and flows back toward the equator. Another air mass that rises here descends near the poles, producing a zone of very low temperature. These interactions, together with the rotation of the earth, result in the major patterns of atmospheric circulation (Figure 22-16, *B*).

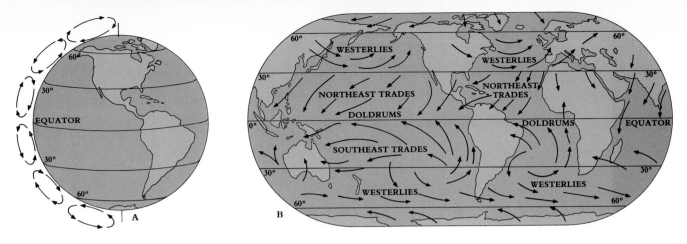

Patterns of Circulation in the Ocean

Patterns of circulation in the ocean are determined by the major patterns of atmospheric circulation just discussed, but they are modified by the location of the land masses around which and against which the ocean currents must flow. The circulation is dominated by huge surface **gyrals** (Figure 22-17), circular patterns that move around the subtropical oceans at about 30 degrees north latitude and 30 degrees south latitude. These gyrals move clockwise in the Northern Hemisphere and counterclockwise in the Southern Hemisphere. They profoundly affect life not only in the oceans but also on coastal lands by the ways in which they redistribute heat. For example, the Gulf Stream, in the North Atlantic, swings away from North America near Cape Hatteras, North Carolina, and reaches Europe near the southern British Isles. Because of the Gulf Stream, western Europe is much warmer and thus more temperate than is eastern North America at the same latitudes.

In South America the Humboldt Current carries cold water northward up the west coast and helps to make possible an abundance of marine life that supports the fisheries of Peru and northern Chile. Marine birds, which feed on these organisms, are responsible for the commercially important guano deposits of these countries. These deposits, like those shown in Figure 22-6, are rich in phosphorus, which is brought up from the ocean depths by the upwelling of cold water that occurs from the generally mountainous slopes that border the Pacific. When the coastal waters are not as cold as usual, a devastating phenomenon called "El Niño" occurs, which affects the vitality of both marine and terrestrial populations of animals and plants worldwide.

FIGURE 22-16

General patterns of atmospheric circulation.
A The pattern of air movement out from and back to the earth's surface.
B The major wind currents across the face of the earth.

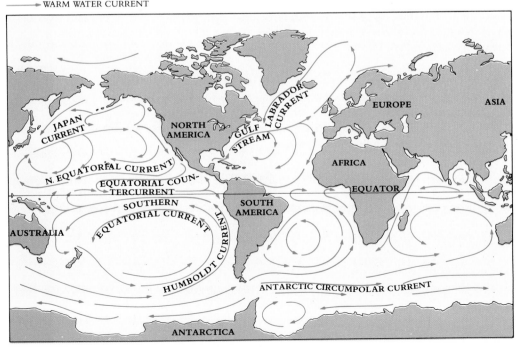

FIGURE 22-17

The circulation in the oceans moves in great surface spiral patterns called gyrals and profoundly affects the climate on adjacent lands.

FIGURE 22-18

Fishes and many other kinds of animals find food and shelter among the kelp beds that occur in the coastal water of temperate regions.

THE OCEANS

Nearly three quarters of the earth's surface is covered by ocean. The seas have an *average* depth of more than 3 kilometers, and they are, for the most part, cold and dark. Heterotrophic organisms are found even at the greatest ocean depths, which reach nearly 11 kilometers in the Marianas Trench of the western Pacific Ocean, but photosynthetic organisms are confined to the upper few hundred meters of water (Figure 22-18). Organisms that live below this level obtain almost all of their food indirectly, as a result of photosynthetic activities that occur above; these activities result in organic debris that drifts downward.

Many fewer species live in the sea than on land. Probably more than 90% of all species of organisms occur on land, including the great majority of members of a few large groups—especially insects, mites, fungi, and plants. Each of these groups has marine representatives, but these comprise only a very small fraction of the total number of species. On land the barriers between habitats are sharper, and variations in elevation, parent rock, degree of exposure, and other factors have all been crucial to the evolution of the milions of species of terrestrial organisms. On the other hand, most phyla originated in the sea, and almost all are represented there now, whereas only a few phyla occur on land.

Although representatives of almost every phylum occur in the sea, only a relatively few phyla occur on land. However, the few phyla that are terrestrial have many species: an estimated 90% of living species of organisms are terrestrial. This is because the boundaries between different habitats are sharper on land than they are in the sea.

The marine environment consists of three major kinds of habitats: (1) the **neritic zone,** the zone of shallow waters along the coasts of the continents; (2) the **surface layers** of the open sea; and (3) the **abyssal zone,** the deep-water areas of the oceans.

The Neritic Zone

The neritic zone of shallow water is small in area, but it is inhabited by very large numbers of species compared with other parts of the ocean (Figure 22-19). The intense and

FIGURE 22-19

Tidepools—depressions in coastal rock shelves in which water remains after the tide goes out—are often occupied by rich and varied communities of organisms. They are subject to exposure and drying out, depending on how far the tide goes out.

sometimes violent interaction between sea and land in this zone gives a selective advantage to well-secured organisms that can withstand being washed away by the continual beating of the waves. Part of this zone, the **intertidal,** or **littoral,** region is exposed to the air whenever the tides recede.

Because of the way in which it gives access to the land, the intertidal zone must have been home for the ancestors of the first land organisms. The complex structures necessary to anchor them and protect them from drying out in this turbulent zone seem in some cases to have made possible their success on land. The world's great fisheries also occur on banks in the coastal zones, where nutrients, derived from the land, are often more abundant than in the open ocean.

The Surface Zone

Drifting freely in the upper, better-illuminated waters of the ocean is a diverse biological community, primarily consisting of microscopic organisms called the **plankton.** Fishes and other larger organisms that swim in these same waters constitute the **nekton,** whose members feed mainly on plankton. Together, the organisms that make up the plankton and the nekton provide all of the food for those that live below. Some of the members of the plankton, including the algae and some bacteria, are photosynthetic. Collectively these organisms account for about 40% of all the photosynthesis that takes place on earth, even more by some calculations. Most of the plankton occurs in the top 100 meters of the sea, the zone into which light from the surface penetrates freely. Perhaps half of the total photosynthesis in this zone is carried out by organisms less than 10 micrometers in diameter, including cyanobacteria and the smallest algae.

The Abyssal Zone

In the deep waters of the sea, below the top 300 meters, occur some of the most bizarre organisms found on earth. Many of these animals have some form of bioluminescence (Figure 22-20), by means of which they communicate with one another or attract their prey. In the mud of the ocean floor, or along rifts from which warm water issues, live similar assemblages of peculiar creatures (Figure 22-21). Bacteria are apparently rather frequent in the deeper layers of the sea and as decomposers are as important in this zone as they are on land and in freshwater habitats.

FIGURE 22-20

The luminous spot below the eye of this deep-sea fish results from the presence of a symbiotic colony of luminous bacteria. Similar luminous signals are a common feature of deepsea animals that move about.

FIGURE 22-21

These giant beardworms are members of a small phylum of animals, the Pogonophora. They are living along warm-water vents in fissures along the Galapagos Trench, deep in the Pacific Ocean. Water jets from these fissures at a scalding 350° C, but it soon cools to the 2° C temperature of the surrounding water. Hydrogen sulfide also emerges from these vents in abundance. Bacteria use the hydrogen sulfide as a source of energy; these bacteria in turn make possible the existence of the diverse community of animals— including these remarkable beardworms—which was discovered in 1977.

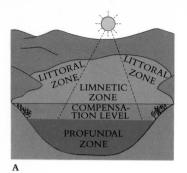

A

FIGURE 22-22

The pattern of stratification in a large pond or lake in temperate regions **(A)** is upset in the spring and fall overturns. Of the three layers of water shown in **(B)** at the far right, the hypolimnion consists of the densest water, at 4° C; the epilimnion consists of warmer water that is less dense; and the thermocline is the the zone of abrupt change in temperature that lies between them. If you have dived into a pond in temperate regions in the summer, you have experienced the existence of these layers directly.

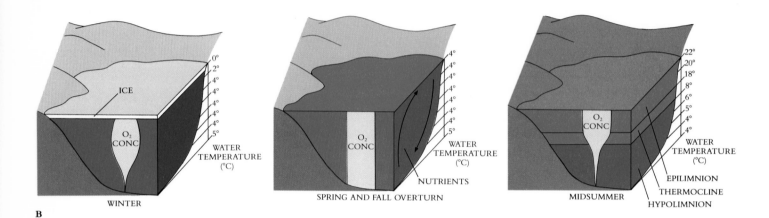

B

FRESH WATER

Freshwater habitats are distinct from both marine and terrestrial ones, but they are very limited in area. Inland lakes cover some 1.8% of the earth's surface, and running water about 0.3%. All freshwater habitats are strongly connected with terrestrial ones, with marshes and swamps constituting intermediate habitats. In addition, a large amount of organic and inorganic material continuously enters bodies of fresh water from communities growing on the land nearby.

Ponds and lakes, like the ocean, have three zones in which organisms occur: a **littoral zone;** a **limnetic zone,** inhabited by plankton and other organisms that live in open water; and a **profundal zone,** below the limits of effective light penetration. **Thermal stratification**—the formation of layers of water with different temperatures—is characteristic of the larger lakes in temperate regions (Figure 22-22). Since water at about 4° C is denser (heavier per unit volume) than water that is either warmer or colder, water at this temperature sinks to the bottom of ponds and lakes. In winter it sinks beneath cooler water, which freezes at the surface at 0° C. Below the ice, the water remains between 0° C and 4° C, and plants and animals survive there. In spring, as the ice melts, the surface water is warmed to 4° C and again sinks below the cooler water, bringing that water to the top with nutrients from the lower regions of the lake. This process is known as the **spring overturn** and has a counterpart in the fall.

BIOMES

Biomes are climatically defined assemblages of organisms that have a characteristic appearance and are distributed over a wide area on land. Biomes are classified in several ways, but, for our purposes, we shall group them into seven categories: (1) tropical forests; (2) savannas; (3) deserts; (4) grasslands; (5) temperate deciduous forests; (6) taiga; and (7) tundra. These biomes differ remarkably from one another because they have evolved in regions with very different climates. These and some of the related com-

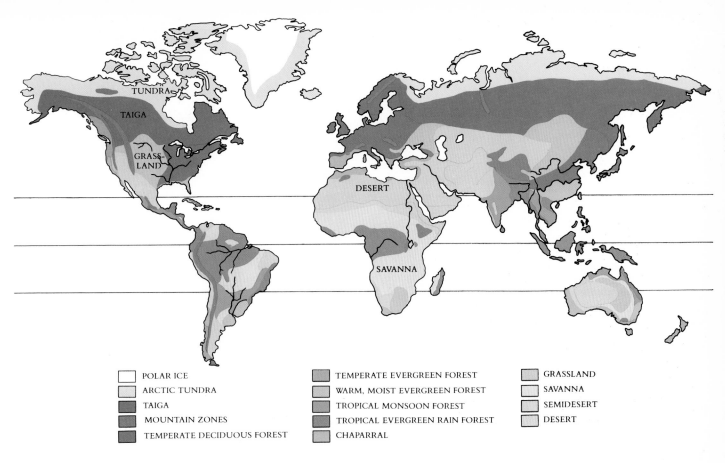

☐ POLAR ICE	☐ TEMPERATE EVERGREEN FOREST	☐ GRASSLAND
☐ ARCTIC TUNDRA	☐ WARM, MOIST EVERGREEN FOREST	☐ SAVANNA
☐ TAIGA	☐ TROPICAL MONSOON FOREST	☐ SEMIDESERT
☐ MOUNTAIN ZONES	☐ TROPICAL EVERGREEN RAIN FOREST	☐ DESERT
☐ TEMPERATE DECIDUOUS FOREST	☐ CHAPARRAL	

FIGURE 22-23

The distribution of biomes.

munities are shown in Figure 22-23. We shall now examine some of the distinctive features of the seven biomes we have chosen to discuss in more detail.

Tropical Rain Forests

The **tropical rain forests** are the richest biome in terms of number of species, probably containing at least half of the species of terrestrial organisms—more than 2 million species! The rainfall in areas where tropical rain forests occur is generally 200 to 450 centimeters per year, with little difference in its distribution from season to season. The communities that make up tropical forests are rich in species and diverse, so that each kind of animal, plant, or microorganism is often represented in a given area by very few individuals. For example, there are seldom fewer than 40 species of tree per hectare; this is four or five times as many as are typical of temperate forests. In a single square mile of tropical forest in Peru or Brazil, there may be 1200 or more species of butterflies—twice the total number found in the United States and Canada combined! The ways of life of tropical organisms are often specialized and highly unusual (Figure 22-24).

Tropical rain forests have a high net productivity, even though they exist mainly on quite infertile soils. Most of the nutrients are held within the plants themselves and are rapidly recycled when the plants die or when parts, such as leaves, are lost. Most of the roots of the tall trees spread through the top 1 or 2 centimeters of the soil, extracting the nutrients promptly from fallen leaves and other plant parts and recycling them back to the trees and other plants from which they have fallen. When tropical rain forests are cut and burned, an abundant supply of nutrients runs out into the soil. This temporary fertilization makes possible the cultivation of crops for a few years, but then, as nutrients are depleted, the people must move on; if there are too many people in a given region, the forest is destroyed (see Chapter 23).

FIGURE 22-24

Characteristic animals and plants of the tropical rain forest.

A Huge trees, many with buttresses like the ones that are prominent in this photograph, dominate tropical rainforests.

B An orchid growing on a tree trunk. More plants and animals occur in the canopy layers of a tropical rain forest, high above the ground, than anywhere else in this richest of all biological communities.

C A three-toed sloth *(Bradypus infuscatus)*, common mammal of the rain forest, characteristically hanging underneath a branch in Panama.

D Bats are the most abundant group of mammals in the tropics. These are white tent bats *(Ectophylla albicollis)* in Costa Rica.

E Butterflies are far more abundant in the tropics than elsewhere. This is *Chlorippe kallina,* in Brazil.

F A leaflike katydid from the forest canopy in Brazil. The markings on its wings resemble holes bitten in a genuine leaf by other insects, making the camouflage more nearly perfect.

A

B

C

D

E

F

There are extensive tropical rain forests in South America, particularly in and around the Amazon Basin; in Africa, particularly in portions of West Africa; and in Southeast Asia. Human populations in the countries that contain the world's rain forests, already a majority of the global population, are destroying them rapidly. Consequently, few of these forests will be left in an undisturbed condition anywhere in the world by the first part of the next century. The destruction of tropical rain forests as an untouched biome will be accompanied by the extinction of a major proportion of all plant, animal, and microorganism species on earth—perhaps as much as a quarter of the total species—during the lifetime of many of us.

Savannas

In areas of reduced annual precipitation or prolonged annual dry seasons, open tropical and subtropical **deciduous forests**—forests in which most of the trees and shrubs lose their leaves at some season of the year—give way to the kind of open grassland called **savannas** (Figure 22-25). The huge herds of grazing mammals, with their associated predators, that inhabit the savannas of Africa are well known and spectacular. Such animal communities occurred in North America during the Pleistocene Epoch but have persisted mainly in Africa. On a global scale, the savanna biome is transitional between tropical rain forest and desert. Generally 90 to 150 centimeters of rain fall each year in savannas. There is a wider fluctuation in temperature here during the year than in the tropical rain forests, and there is seasonal drought. These factors have led to the evolution of an open landscape, often with widely spaced trees; many of the animals and plants are active only during the rainy season. Savannas have often been converted to agricultural purposes throughout the world and provide most of the agricultural products for many tropical and subtropical countries.

Desert

Less than 25 centimeters of annual precipitation usually falls in the world's desert areas—such a low amount that water is the predominant controlling variable for most biological processes and is also highly variable in quantity both during a given year and between years. In desert regions the vegetation is characteristically sparse (see Figure 2-19). Recall that such regions occur around 20° to 30° north and south latitude, where the warm air that rises near the equator falls and precipitation is limited. Deserts are most extensive in the interiors of continents, especially in Africa (the Sahara Desert), Eurasia, and Australia. Less than 5% of North America is desert. The organisms that occur in deserts are often bizarre in appearance or way of life.

FIGURE 22-25

The herds of grazing mammals that inhabit the African savannas are one of the world's greatest sights.

Desert survival depends on water conservation by structural, behavioral, or physiological adaptations. Plants and animals may restrict their activity to favorable times of the year, when water is present; they must also avoid high temperatures. Most desert vertebrates live in deep, cool, and sometimes even somewhat moist burrows, and some of them that are active over a greater portion of the year emerge from these burrows only at night, when temperatures are relatively cool. Some, like camels, can drink large quantities of water when it is available and can then safely withstand the loss of much of it. Many animals simply migrate to or through the desert, where they exploit food that may be abundant seasonally; when the food disappears, the animals move on to more favorable areas.

Grasslands

Temperate grasslands once covered much of the interior of North America (Figure 22-26), and they were widespread in Eurasia and South America as well. Such grasslands are often highly productive when they are converted to agriculture, and many of the rich agricultural lands in the United States and southern Canada were originally occupied by **prairies,** another name for temperate grasslands. The roots of perennial grasses characteristically penetrate far into the soil, and grassland soils tend to be deep and fertile. In North America the prairies were once inhabited by huge herds of buffalo and pronghorns, which had wolves, bears, and other predators, as well as various groups of American Indians, hunting them for food and clothing. These are almost all gone now, with most of the prairies having been converted to the richest agricultural region on earth, stretching across a wide region of the north-central United States and adjacent Canada.

Temperate Deciduous Forests

In areas of the Northern Hemisphere with relatively warm summers, relatively cold winters, and sufficient precipitation, **temperate deciduous forest** occurs (Figure 22-27). This biome covers very large areas, particularly over much of the eastern United

FIGURE 22-26

Tall-grass prairie stretched over thousands of square kilometers in the interior of North America when Europeans first came to the area. In North America, tall-grass prairies occur towards the east, where the precipitation is more abundant, and short-grass prairies westward, toward the Rocky Mountains, where the climate is drier. Over much of its former area, tall-grass prairie has been replaced by cultivated fields.

FIGURE 22-27

The leaves of the trees in the temperate deciduous forest often change in color before they fall.

States and Canada and an extensive region in Eurasia. It is the region of deer, bears, beavers, and raccoons—the familiar animals of the temperate regions. Many of the plants flower before the trees form their leaves in the early spring. In temperate deciduous forests the annual precipitation generally ranges from about 75 to about 250 centimeters, well distributed throughout the year but generally unavailable to animals and plants in the winter—because it is usually frozen. Because the deciduous forests represent the remnants of more extensive forests that stretched across North America and Eurasia a few million years ago, these remaining areas—especially those in eastern Asia and eastern North America—share animals and plants that were once more widespread. Alligators, for example, are found only in China and in the southeastern United States.

The **chaparral** of California and adjacent regions (Figure 22-28) is historically derived from deciduous forests. It consists of evergreen, often spiny shrubs and low trees that form extensive communities in summer-dry regions. The forests of conifers that are so impressive in western North America also are remnants of mixed forests that existed in the past, related both to the present-day deciduous forests and to the taiga.

Taiga

Taiga is the northern forest of coniferous trees, primarily spruce, hemlock, and fir, that extends across vast areas of Eurasia and North America (Figure 22-29). Here the winters are long and cold, and most of the limited amount of precipitation falls in the summer. Many large mammals live in the taiga, including elk, moose, deer, and such carnivores as wolves, bear, lynx, and wolverines. Traditionally, much fur trapping has gone on in this region, which is also important in lumber production. Because of the latitude where taiga occurs, the days are short in winter (as little as 6 hours) and correspondingly long in summer. During the summer, plants may grow rapidly, and crops often attain a large size in a surprisingly short time. Marshes, lakes, and ponds are common here and are often fringed by willows or birches. Most of the trees in the taiga tend to occur in dense stands of one or a few species.

FIGURE 22-28

In the Coast Ranges in California, the boundary between the evergreen shrub association known as chaparral—in which the trees and shrubs retain their leaves all year—and grassland is often sharp, as shown in this photograph taken along the western edge of the Santa Clara Valley near Morgan Hill. The oak trees shown in the lower left of the photograph represent the edge of an oak woodland community that occurs in the same region.

A

B

FIGURE 22-29

A Taiga, here a forest dominated by spruces and alders, in Alaska.
B Moose are one of the characteristic animals of the taiga.

A

B

C

Tundra

Farthest north in Eurasia, North America, and their associated islands, between the taiga and the permanent ice, occurs the open, often boggy, grassland community known as the **tundra** (Figure 22-30). This is an enormous biome, extremely uniform in appearance, that covers a fifth of the earth's land surface. Trees are small and are mostly confined to the margins of streams and lakes; in general the tundra's appearance is like that of some parts of the prairies.

Annual precipitation in the tundra is very low, usually less than 25 centimeters, and the water is unavailable for most of the year because it is frozen. During the brief Arctic summers, water sits on frozen ground, and the surface of the tundra is often extremely boggy then. **Permafrost,** or permanent ice, usually exists within a meter of the surface.

THE FATE OF THE EARTH

Now that you have finished your survey of the biomes and the corresponding communities that occur in the sea, you have gained some appreciation of the different kinds of climate under which these areas have evolved over tens or hundreds of millions of years. In just the last several hundred years, in contrast, human populations have grown over most of the land surface of the globe. Having already converted many

FIGURE 22-30

A Tundra in Mt. McKinley National Park, Alaska.
B Caribou live in large herds that migrate across the tundra. Sometimes called reindeer, they have been domesticated by the Lapps in northern Scandanavia as a source of meat, milk, and hides. Following the Chernobyl disaster in the summer of 1986, most of these reindeer became radioactive and therefore unfit for human consumption, as a result of radioactive particles settling out of the atmosphere and becoming concentrated in the lichens and plants consumed by the reindeer.
C Arctic ground squirrels live in burrows and consume plant material actively during the short Arctic summers.

biomes to agriculture, urban areas, or other uses, human populations are attacking those biomes—such as the tropical rain forests—that are less suitable for exploitation and about which we know much less. The human population doubled from 1950 to 1987, and most of that growth has occurred in the warmer portions of the globe. There are now so many of us that we must manage the whole earth as a single system, if those who follow us are to find a stable ecological situation, and one in which they can live out their lives in peace and relative prosperity. In the next chapter, we shall examine some of the factors that will determine our success or failure in this great enterprise.

SUMMARY

1. Populations of different organisms that live together in a particular place are called communities. A community, together with the nonliving components of its environment, is called an ecosystem. Ecosystems regulate the flow of energy, ultimately derived from the sun and the cycling of nutrients.

2. Only 2% of the water on earth is fixed in any way; the rest is free. In the United States, 96% of the fresh water is groundwater.

3. About 10% of the roughly 700 billion metric tons of free carbon dioxide in the atmosphere is fixed each year through photosynthesis. An additional trillion metric tons of carbon dioxide is dissolved in the ocean, and five times as much carbon as represented in it is locked up as coal, oil, and gas. About as much carbon exists in living organisms at any one time as in the atmosphere.

4. Plants convert about 1% of the energy that falls on their leaves to food energy. The herbivores that eat the plants, and the other animals that eat the herbivores, constitute a set of trophic levels. At each of these levels, only about 10% of the energy fixed in the food is fixed in the body of the animal that eats that food. For this reason, food chains are always relatively short.

5. Primary succession takes place in areas that are originally bare, like rocks or open water. Secondary succession takes place in areas where the communities of organisms that existed initially have been disturbed.

6. Both sorts of succession lead ultimately to the formation of climax communities, whose nature is controlled primarily by the climate of the area concerned, although the human influence on many of these communities is increasing. Such communities have more total biomass, less net productivity, more species, many more heterotrophic species, and a higher capacity of regulating the cycling of nutrients within them than do the earlier successional stages.

7. Ocean communities occur in three major kinds of environments: the neritic zone, the surface layers, and the abyssal zone. The neritic zone, which lies along the coasts, is small in area but very productive and rich in species. The surface layers are the habitat of plankton (drifting organisms), and nekton (actively swimming ones).

8. Freshwater habitats make up only about 2.1% of the earth's surface; most of them are ponds and lakes. As autumn changes to winter, cooler water forms and sinks, since water is most dense at 4° C, mixing the water of the lake. When winter changes to spring, warmer water sinks, and a similar mixing occurs.

9. Biomes are major terrestrial assemblages of plants, animals, and microorganisms that occur together in similar habitats; biomes are defined largely by climate. We discuss seven major biomes, or terrestrial communities, in this book: (1) tropical forests; (2) savannas; (3) deserts; (4) grasslands; (5) temperate deciduous forests; (6) taiga; and (7) tundra.

REVIEW

1. The gas _____ makes up nearly four fifths of the earth's atmosphere.

2. The total weight of all the organisms living in an ecosystem is called the _____ of that ecosystem.

3. The biome that is transitional between tropical rain forest and desert is _____ .

4. What does an ecosystem possess that a community does not?

5. The total amount of energy that an area converts to organic compounds over a period of time is called its _____ .

SELF-QUIZ

1. In which of the following cycles does the reservoir of the nutrient exist in a mineral form?
   (a) carbon cycle
   (b) phosphorus cycle
   (c) nitrogen cycle
   (d) two of the above
   (e) all of the above

2. Herbivores are
   (a) primary consumers
   (b) secondary consumers
   (c) decomposers
   (d) detritivores
   (e) parasites

3. Communities at early successional stages have _____ relative to the more advanced communities of later stages.
   (a) higher biomass
   (b) lower productivity
   (c) more species
   (d) less regulated nutrient cycling
   (e) none of the above

4. Of the 700 billion metric tons of carbon dioxide in the atmosphere, how much of the carbon is fixed into organic compounds by photosynthesis each year?
   (a) 0.03%   (d) 5%
   (b) 10%   (e) all of it
   (c) less than 0.1%

5. When a green plant is consumed by another organism, how much of the energy of the plant is converted into the body of the animal that consumes it?
   (a) none of it   (d) 10% of it
   (b) all of it   (e) ⅓ of it
   (c) 1% of it

THOUGHT QUESTIONS

1. How could you increase the net primary productivity of a desert?

2. Why does the productivity of an ecosystem increase as it becomes more mature?

3. At what successional stage would you characterize a field of wheat? What does this imply as to its stability and productivity?

FOR FURTHER READING

ATTENBOROUGH, D.: *The Living Planet: A Portrait of the Earth,* William Collins Sons & Co., Ltd., and British Broadcasting Corporation, London, 1984. A beautifully written and illustrated account of the biomes.

LAWS, R.: "Antarctica: A Convergence of Life," *New Scientist,* September 1983, pages 608-616. A dynamic and beautifullly illustrated view of one of the most productive oceans on earth.

McNAUGHTON, S.J.: "Grazing Lawns: Animals in Herds, Plant Form, and Coevolution," *American Naturalist,* vol. 124, pages 863-886, 1984. The fascinating story of the ways in which differences in grasses and in the populations of animals that depend on them help to determine the structure of the savanna biome.

MAY, R.M.: "The Evolution of Ecological Systems," *Scientific American,* September 1978, pages 160-175. An important essay on the interplay between ecology and evolution, by one of the masters of both fields.

PERRY, D.R.: "The Canopy of the Tropical Rain Forest," *Scientific American,* November 1984, pages 138-147. This article describes efforts to explore the richest and least understood biological community on earth.

SISSON, R.: "Tide Pools: Windows Between Land and Sea," *National Geographic,* vol. 169(2), pages 252-259, 1986. A beautifully illustrated tour of a California tide pool, alive with organisms.

FIGURE 23-1 The girl gazing from this page faces an uncertain future. She is a refugee, an Afghani. The whims of war have destroyed her home, her family, all that is familiar to her. Her expression carries a message about our own future. The message is that our future, like hers, is now on us, and that the problems which humanity faces in living on an increasingly unstable, overcrowded, and polluted earth are no longer hypothetical, a dilemma for our children, but are with us today and demanding a solution.

THE FUTURE OF THE BIOSPHERE Chapter 23

Overview

By applying knowledge about the characteristics of ecosystems and organisms to enhance and stabilize food production, biologists have an important contribution to make in feeding the rapidly growing world population. The development of agriculture by the roughly 5 million people who lived in the world about 11,000 years ago led through a series of stages to the very large human population—more than 5 *billion* people—that exists now. Through the improvement of existing crops and domesticated animals and the development of new crops, biologists will play the key role in solving the problems generated by the world's expanding human population. Biological literacy will be an essential component of human progess in the future.

For Review

Here are some important terms and concepts that you will encounter in this chapter. If you are not familiar with them, you should review them before proceeding.

Overpopulation (Chapter 1)

Human evolution (Chapter 19)

Population growth characteristics (Chapter 20)

Characteristics of biomes (Chapter 22)

Cycling of minerals and flow of energy (Chapter 22)

The human race is growing rapidly and affecting the global ecosystem to a profound and unprecedented degree. In preparing for the future, we must use wisely our biological knowledge in managing the life-giving systems on which our collective prosperity—and even our survival—depends. In this chapter, we consider the implications of the explosive growth of the human population, our growing need for food and other resources, and the ways in which we may be able to apply our biological knowledge to the solution of the problems presented by the finite resources of the world on which we live (Figure 23-1).

THE POPULATION EXPLOSION

Only about 5 million people lived throughout the entire world when agriculture was first developed about 11,000 years ago. However, given the new and much more dependable source of food that became available as a result of agriculture, human populations began to grow rapidly, climbing to an estimated 130 million people by the time

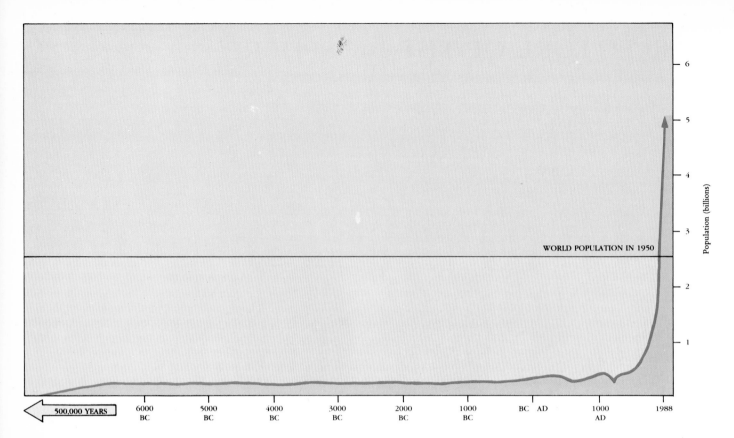

500,000 YEARS | 6000 BC | 5000 BC | 4000 BC | 3000 BC | 2000 BC | 1000 BC | BC AD | 1000 AD | 1988

WORLD POPULATION IN 1950

Population (billions)

FIGURE 23-2

Growth curve of the human population.

FIGURE 23-3

One of the most critical human inventions was that of agriculture. Here, in southern India, the people still thresh sorghum using primitive tools whose designs are probably thousands of years old. By producing abundant supplies of food, agriculture made possible the rapid expansion of human culture.

of Christ—more than a 25-fold increase during the roughly 9000 years that followed the development of agriculture (Figure 23-2). The global population 2000 years ago was about half that of the United States and Canada at present. As human populations grew in the Middle East, their agricultural herds became tremendously destructive to the pastures and slopes on which they fed, often disrupting the control mechanisms of these ecosystems and causing erosion and flooding. Similar effects were repeated in the New World when the same animals were introduced there. Nonetheless, thanks to the increased production of food made possible by agriculture (Figure 23-3), large numbers of people were able to live together in permanent settlements for the first time. As a result, villages and towns developed in all areas where agriculture was practiced by about 3000 BC, some 5000 years ago.

The world population increased some 25-fold during the 9000 years from the development of agriculture to the time of Christ, when it amounted to an estimated 130 million people.

The ability to live together on a long-term basis in relatively large settlements made possible, in turn, the specialization of professions in these centers. This was the necessary condition for the development of modern culture. Such advances as the development of metal tools and utensils, for example, could not originate until after the towns were settled.

By 1650, the world population had reached 500 million, with many people living in substantial urban centers in which many important innovations were taking place. In other words, there were about four times as many people in 1650 as there had been at the time of Christ. The Renaissance in Europe, with its renewed interest in science, ultimately led to the establishment of industry in the seventeenth century, and to the Industrial Revolution of the late eighteenth and early nineteenth centuries. Centers of cultural innovation arose in Europe and elsewhere, making possible in turn still more increases in world population.

The Present Situation

For the past 300 years, and probably for much longer, the human birth rate (as a global average) has remained relatively constant at about 30 births per year per 1000 people. It may be lower now than it has been historically, at about 27 births per year per 1000 people, but this difference is probably not significant. However, with the spread of better sanitation, coupled with modern medical techniques, the death rate has fallen steadily, to an estimated 1987 level of about 11 deaths per 1000 people per year. The difference between these two figures amounts to an annual, worldwide increase in human population of approximately 1.7%. Such a rate of increase may seem relatively small, but it would lead to a doubling of the world population from its present level of more than 5 billion people in only 41 years!

The world population reached an unprecedented 5 billion people by early 1987, and the *annual* increase in population amounts to about 85 million people, a number substantially larger than the total population of either Britain, France, or Germany. At this rate, more than 232,000 people are added to the world population each day, or more than 160 every minute! The world population is expected to rise much higher—to well over 6 billion people by the end of the century—and then to add about 1.7 billion more during the following 20 years. As recently as 1950, the total world population was only about 2.5 billion people.

In view of the limited resources available to the human population and the necessity of our learning how to manage these resources well, the first and most necessary step toward global prosperity is to stabilize the population. One of the surest signs of the pressue we are putting on the environment is our utilization of about 40% of the total global photosynthetic productivity. Given that statistic, doubling our population in about 41 years certainly poses extraordinarily severe problems. The facts virtually demand restraint in population growth: if and when we develop the technology that would allow greater numbers of people to inhabit the earth in a stable condition, it will be possible to increase our numbers to whatever level might be appropriate at that time.

In the late 1980s, the global human population of more than 5 billion people was growing at a rate of approximately 1.7% annually. At that rate, it will reach well over 6 billion people by 2000, and about 7.8 billion by 2020.

By the year 2000, about 60% of the people in the world will be living in countries that are at least partly tropical or subtropical. An additional 20% will be living in China, and the remaining 20%—one in five—in the so-called developed countries: Europe, the Soviet Union, Japan, the United States, Canada, Australia, and New Zealand. While the populations of the developed countries are growing at an annual rate of only about 0.6%, those of the less developed, mostly tropical countries (excluding China) are growing at an annual rate estimated in 1987 to be about 2.4%. For every person living in an industrialized country like the United States in 1950, there were two people living elsewhere; by 2020, just 70 years later, there will be five.

Since the people living in industrialized countries control at least 80% of the world's wealth and material goods and enjoy a standard of living perhaps 20 times better than those found in many developing countries, the changing ratios of total population in each of these sectors should be a matter of special concern for everyone. To provide a few examples, the infant mortality rate in industrial countries in 1986 was 11 per thousand, while in developing countries it was 99 per thousand; the respective life expectancies at birth were 73 years versus 56 years, and even lower in some developing countries where many people are malnourished.

As you learned in Chapter 20, the age structure of a population determines how fast the population will grow. For this reason, it is essential in predicting the future growth patterns of a population to know what proportion of its individuals have not yet reached childbearing age. In developed countries such as the United States, about a fifth of the population consists of people under 15 years of age; in developing countries

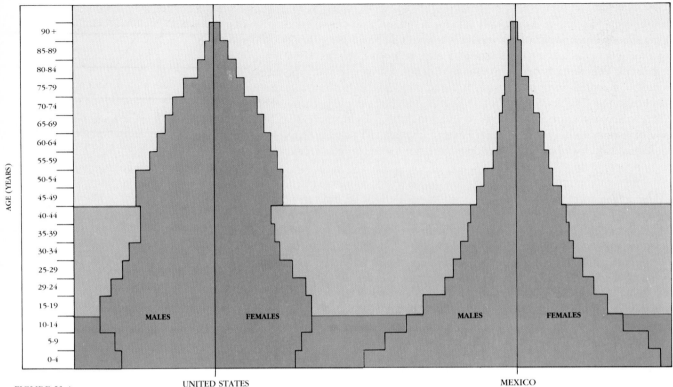

AGE (YEARS)

90 +
85-89
80-84
75-79
70-74
65-69
60-64
55-59
50-54
45-49
40-44
35-39
30-34
25-29
29-24
15-19
10-14
5-9
0-4

MALES   FEMALES   MALES   FEMALES

UNITED STATES   MEXICO

FIGURE 23-4

The age structure of the United States and Mexico in 1977. Dark green indicates the pre-reproductive years, blue the reproductive years. The rate of population growth in Mexico, already rapid, will increase considerably in the future because so much of its population will be entering into their reproductive years. The "bulge" in the population structure in the United States *(BB)* indicates the "Baby Boomer" generation. Males are indicated on the left in each graph, females on the right.

such as Mexico, the proportion is typically about twice as high (Figure 23-4). Thus, even if the policies that most tropical and subtropical countries have established to limit their own population growth are carried out consistently for decades, the populations of these countries will continue to grow well into the twenty-first century, and the developed countries will constitute a smaller and smaller proportion of the world's population. For example, if India, with a mid-1986 population of 785 million (with 39% under 15 years old) managed to reach a simple replacement reproductive rate by the year 2000, its population still would not stop growing until the mid-twenty-first century (Figure 23-5). By 2020, India will have a population of nearly 1.2 billion people and will still be growing rapidly, judged from the country's age structure and current patterns of growth. As a result of such worldwide trends, the proportion of the world's population living in the developed countries, a third in 1950, will have fallen to a sixth by 2020; the proportion living in the tropics will have risen from 45% to 64% in the same 70-year period.

The proportion of people living in the developed countries is falling—from a third of the total world population in 1950 to a projected sixth in 2020—while the proportion of people living in countries that are at least partly tropical or subtropical is rising from 45% to 64% during the same period.

Most countries are devoting considerable attention to slowing the growth rate of their populations, and there are genuine signs of progress. It is estimated that the world population may stabilize by the close of the twenty-first century at about 10.5 billion people. No one knows whether the world can support so many people indefinitely, but we must assume that we will be able to find the ways to enable it to do so. Among the tasks that we must accomplish is the development of new ways to feed, clothe, and shelter adequately a population two to four times larger than the present one. The quality of life that will be available for our children in the next century will depend to a large extent on our ability to achieve this goal.

FIGURE 23-5

With its human population growth outstripping its resources, India faces an uncertain future. At present rates of growth, India will pass China in total population during the first half of the twenty-first century. This scene was photographed in Ranchi, Bihar State, India, where it is increasingly difficult for the city administrators to collect taxes and pay for the basic services they provide.

FOOD AND POPULATION

Even though experts estimate that enough food is produced in the world to provide an adequate diet for everyone in it, the distribution of this food is so unequal that large numbers of people live in hunger. In the United States, the Soviet Union, Japan, and Europe, there is, on the average, a quarter to a third *more* food available per person than the United Nations Food and Agriculture Organization (FAO) set in the late 1970s as the calorie intake necessary to maintain moderate physical activity. In contrast, such countries as Angola, Bangladesh, Bolivia, Ecuador, Haiti, India, Kenya, and Zimbabwe had 10% to 15% *less* than this minimum available per person and, for the most part, little cash with which to purchase more. Worldwide, 300 million to 400 million people each consume less than 80% of the U.N.-recommended standards, a diet insufficient to prevent stunted growth and serious health risks. Only the United States, Canada, Argentina, and one or two other temperate-zone countries have emerged as consistent exporters of food.

Of the approximately 2.7 billion people living in the tropics in the mid-1980s, the World Bank estimated that more than 1 billion are living in absolute poverty. These people cannot reasonably expect to provide adequate food for themselves and their families on a consistent basis. Among the malnourished people mentioned above, UNICEF estimates that about 35 million starve to death every year in the tropics, including more than 14 million children under the age of five (Figure 23-6). This amounts to nearly 100,000 people, and 40,000 babies, each day. Worse, many millions of additional children exist only in a state of lethargy, their mental capacities often permanently impaired by their lack of access to adequate quantities of food.

About 40% of the approximately 2.7 billion people living in or near the tropics in the late 1980s existed in a state of absolute poverty; many of them were malnourished.

In the developed countries, about 72% of the population was living in urban centers by the late 1980s—more than twice the proportion in the less developed countries; and a far smaller proportion of the total population engaged in farming in the developed countries than did in the developing countries. One of the most alarming trends taking place in tropical countries, however, is a massive movement to urban centers.

FIGURE 23-6

Consuelo Andena, 16, holding her 1-year-old baby, Carolina, in the malnutrition ward of the Women and Children's Hospital in Tegucigalpa, Honduras, in 1983. Such tragic scenes are common throughout the tropics, where more than 40,000 babies perish every day from starvation.

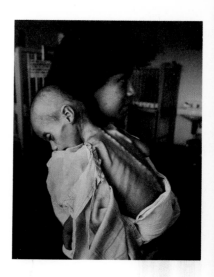

FIGURE 23-7
Mexico City.

For example, Mexico City, the largest city in the world, is plagued by smog, traffic, inadequate waste disposal, and other problems—and it will have a population of more than 30 million by the end of the century (Figure 23-7). The prospects of supplying adequate food, water, and sanitation to these people, in a country whose 1987 population (about 85 million) is expected to double in the next 27 years, are almost unimaginable. It is no wonder that there is such massive emigration out of Mexico and countries that are similarly troubled.

Consider what such rapid population growth means to a country. For example, Brazil, which had about 146 million people in mid–1987, along with the world's largest international debt and one of the highest rates of inflation anywhere, was then growing in population at about 2.3% per year and was estimated to double in population size in 30 years. Thus Brazil would need to double all of its food production, water supplies, fuel, housing, educational and medical facilities, and so on, during the next 30 years *just to stay even* (Figure 23-8). Even if Brazil does perform this unprecedented economic miracle, then its people will be only as well off in 30 years as they are now—no better off. If the miracle does not occur, then more than the currently estimated 40% of the people of Brazil who are living in absolute poverty will be in that condition.

The populations of the less developed countries are growing so rapidly that an unprecedented effort will be required to keep the proportion of hungry, poor people—now amounting to a fifth of the total world population—at its current level while their populations double during an average period of 29 years.

A

B

C

FIGURE 23-8

A Manioc (also called cassava; *Manihot esculenta*) is a widespread and very important tropical root crop, comparable in its use to potatoes in temperate regions. It is often grown on marginal lands that are fertile for a few years following the clearing of the forests.
B A Brazilian woman is preparing the starchy tubers to eat.
C In Brazil, manioc is being used extensively for the production of ethyl alcohol, which is mixed with gasoline to make a fuel (gasohol) for motors that is suitable for cars, buses, and trucks and helps to make Brazil more independent of foreign sources of oil. On the other hand, by locking up large tracts of land—as much as one sixth of the arable land in the country has at times been scheduled for the production of manioc for gasohol—land that might have produced food for the poor is being devoted to another purpose. Such are the trade-offs that must be made when weighing scarce resources; the decisions rarely favor the poor, however.

The considerations about population growth that we have just presented indicate clearly why a stable population structure is as necessary for us as it is for the other organisms that share this planet with us. Although we have the ingenuity and ability to increase the supply of resources available to us, we urgently need to do so; meanwhile, we must also find ways to reduce the widespread poverty, malnutrition, disease, and starvation that haunt the lives of fully a quarter of our fellow human beings. That is the great task of biology.

THE FUTURE OF AGRICULTURE

What are the prospects for increased agricultural productivity in the future, and how can we improve them? When the development of agriculture began about 11,000 years ago, there were about 5 million people in the whole world. They took up agricultural practices independently in the Middle East, in eastern Asia, and at one or more centers in North and South America, with different kinds of plants and animals domesticated in each area. As a result of these early efforts, nearly all of the major crops in world commerce were discovered; most of them have been cultivated for hundreds or even thousands of years. Only a few, including rubber and oil palms, have entered widespread cultivation since 1800. One key feature for which nearly all of our important crops were first selected was ease of growth by relatively simple methods, often by the women who stayed behind in camps while the men were away hunting.

How many plants do we use at present? Just three species—rice, wheat, and corn—supply more than half of all human energy requirements; only about 150 kinds of plants are used extensively; and only about 5000 have ever been used for food. It is estimated that there may be tens of thousands of additional kinds of plants, among the world total of some 250,000 species, that could be used for human food if their properties were fully explored and they were brought into cultivation. There are many uses for plants other than food, too. For example, oral contraceptives for many years were produced from Mexican yams; the muscle relaxants used in surgery worldwide came from curare, an Amazonian vine used traditionally to poison darts used in hunting; and the cure for Hodgkin's disease was developed in the early 1970s from the rosy periwinkle, a widely cultivated native of Madagascar.

The reasons for which we select new crops now are often very different from those that appealed to our ancestors, living as they did in small groups around the foothills of the Near East or on the temperate slopes of the mountains in Mexico. Standards of cultivation have changed, and we use many products from plants—oils, drugs, and other chemicals, for example—that often would not have led to their cultivation earlier. In a time when human activities threaten to drive many of the world's plants and other organisms to extinction during our lifetimes and those of our children, we need to search systematically for new and useful crops that fit the multiple needs of modern society in ways that would not have been considered earlier. We must begin more diligent efforts to find and bring into cultivation more kinds of useful plants before they are gone forever.

> **Only about 150 kinds of plants, out of the roughly 250,000 known, play an important role in international trade at present. Many more could be developed by a careful search for new crops.**

THE PROSPECTS FOR MORE FOOD

The most rapid growth of human populations is taking place in the tropics and subtropics, and a majority of the world population of people already lives there. If people everywhere are to be fed adequately, the increases in food production must take place primarily in the tropics and subtropics. Even though it is sometimes thought that the productive soils of temperate regions could simply be used more efficiently to produce enough food to feed everyone in the world, this is not the case, at least for the foreseeable future: exports from the developed world provided only about 8% of the total food consumed in the developing world in the mid-1980s, yet these exports constituted half of the total food exports from the United States and other major food-producing countries. The agricultural lands in the United States are being cultivated so intensively that a considerable amount of topsoil is already being lost, and there is genuine concern for the future. Drastic increases in U.S. productivity do not seem possible in the near future.

During the 1950s and 1960s, the so-called Green Revolution took place. This revolution depended on the development of new, improved strains of wheat and rice at international centers that were organized by groups such as the Rockefeller and Ford foundations, with the assistance of many other governments and agencies. As a result of the efforts made at these centers, the production of wheat in Mexico, for example, increased nearly 10-fold between 1950 and 1970, and Mexico temporarily became an exporter of wheat rather than an importer (Figure 23-9). During the same decades, food production in India was largely able to outstrip even a population growth of approximately 2.3% annually, and China became self-sufficient in food.

Despite these and similar success stories, the Green Revolution has had its limitations. The agricultural techniques necessary to raise Green Revolution crops require the expenditure of large amounts of energy, which often are not readily available in the tropics and subtropics. Because of the extensive use of fertilizers, pesticides, herbicides, and machinery, it requires about 10 times as much energy per calorie to produce rice in the United States as it does in the Philippines, and more than 1000 times as much energy to produce each calorie of energy in wheat in the United States as it does by traditional farming methods in India. Energy prices are often held at artificially low levels in developing countries, so that it may actually be more expensive for a poor rural farmer to grow an equivalent amount of grain as for a large-scale farmer using developed-world technology. In this respect, the introduction of Green Revolution methods in some regions has actually worsened poverty for many of the people and lessened their access to food, fuel, and other commodities on which they depend.

The Green Revolution has depended for its success on improved versions of current crops grown on lands that were already under cultivation. By now, however, almost all of the land that can be cultivated by available methods is *already* cultivated. Each hectare of cultivated land in the tropics supported approximately three people in 1975; it will need to support approximately twice as many in 2000—and it will need to

FIGURE 23-9

Improved strains such as this dwarf wheat helped to make Mexico an exporter, rather than an importer, of wheat during some years in the 1960s and 1970s. The improved strains of wheat were developed by CIMMYT, the International Center for the Improvement of Maize and Wheat, an important research institution located in Mexico. CIMMYT is one of approximately 14 such institutions located throughout the world, all striving to improve some aspect of third-world agriculture.

do so year after year without the use of abundant fertilizers or other methods that are not economically feasible. Many of the trends that are taking place in tropical countries, such as the loss of agriculturally productive land to urban sprawl or the deterioration of soil quality because of intensive irrigation, actually run counter to the need to protect the currently cultivated land and to keep it in prime condition. Furthermore, the increased use of pesticides and herbicides may have undesirable side effects, and erosion tends to accompany the intensification of agriculture throughout the world.

The Green Revolution has resulted in increased food production in many parts of the world through the use of improved crop strains. These strains often depend on increased inputs of fertilizers, water, pesticides, and herbicides, as well as the greater use of machinery—means that are not normally available to the poor.

Certain nutritional problems have also arisen in connection with the Green Revolution. The overconcentration on cereal crops has tended to lower the production of other nutritionally important plants, including legumes, oilseeds, and vegetables of all kinds. Correcting this imbalance remains a serious problem for Third World planners. The linked sets of cereals and legumes—such as rice and beans—that our Stone Age ancestors domesticated and used are as nutritionally important to us as they were to them. The reason that legumes and cereals are often combined is that together they provide a balanced set of the amino acids required by humans for proper growth. In the modern world, we are tending to neglect such combinations. The varied strains of crops that are grown on small farms may also be driven out by fewer kinds of modern strains, which produce a better yield if large-scale inputs of chemicals and the use of machinery are possible (Figure 23-10). Despite these short-term advantages, the loss of the unique, traditional strains of crops currently cultivated by small, rural farmers throughout the world may ultimately prevent the particular crop plants from being able to grow in less favorable habitats or to withstand important diseases in the longer run (Figure 23-11). **Monoculture**—the exclusive cultivation of a single crop over wide areas—is an efficient way to use certain kinds of soils, but it carries the risk of an entire crop being destroyed with the appearance of a single pest species or disease. Clearly, mixed cropping systems will need to be employed widely if all the different kinds of areas that can be used for crops are to be brought into production.

Biological diversity made possible the development of agriculture in the first place; genetic diversity still enables crops to adapt to new situations. The preservation of plant and other genetic material in so-called gene banks, either artificial—as in botanical gardens, zoos, or seed banks—or natural (in reserves) is a matter of high priority, especially in view of the very high rates of extinction that we shall review later in this chapter.

In improving the world food supply, the most promising strategy is to improve the productivity of crops that are already being grown on lands that are under cultivation. There are relatively few parts of the world where additional land exists that can be brought into cultivation using currently available technology. The land that is already cultivated and the crops that are already known to be highly productive are our most important resource for the immediate future. In the development of these crops, biologists have a crucial role to play. Traditional methods of plant breeding and selection must be applied fully and to many crops of importance in the tropics and subtropics, in addition to wheat, corn, and rice. The techniques of genetic engineering (see Chapter 15) are being applied extensively to crop plants, and to some extent animals, and these techniques offer great promise for the future. For example, it will be possible to produce plants that are resistant to specific herbicides, which can therefore be applied much more effectively to crops for weed control. It will soon be possible, through the methods of genetic engineering, to produce new strains of plants that will grow successfully in areas where the particular crop plant could not grow before. Eventually it will also be possible to introduce desirable characteristics into important crop plants, such as the ability to fix nitrogen, carry out C_4 photosynthesis, or produce substances that deter pests and diseases in abundance. The ability to transfer genes between organ-

FIGURE 23-10

A traditional farm in Thailand, with coconuts, *Leucaena* (a fast-growing, nitrogen-fixing legume that provides food and firewood), sugar palms, litchi nuts, and the nitrogen-fixing water fern *Azolla* growing in ponds. Land crabs are frequent in such farms in Thailand, and are harvested as a source of protein. Farms of this type, with their trees and mixed crops, are extremely suitable for many sites in the tropics, but the overall tendency is to try to replace them with large-scale agriculture modeled after that which is successful in temperate regions.

FIGURE 23-11

Corn *(Zea mays)* and a wild relative.

A The genetic uniformity of hybrid corn, enhanced by selection over decades, is clearly demonstrated by this crop in Illinois, one of the most productive areas for corn in the world.

B Corn is an annual grass, but this field in Mexico is dominated by a wild perennial relative, *Zea diploperennis,* which was discovered in 1977. The two species can be crossed to produce fertile hybrids, and *Z. diploperennis* is resistant to the seven main types of virus diseases that damage corn as a crop. It has also been used to develop a new kind of perennial corn, which shows promise for cultivation in some of the infertile soils of subtropical areas.

C Spikelets of *Z. diploperennis.*

D This commercially important and scientifically interesting relative of corn might never have been found at all, because it occurs only in one field, which is about the size of a football field. Logging operations are widespread in the Sierra de Manantlán, Jalisco, Mexico, where it lives, and its single population could easily have been destroyed by the rapidly increasing human population pressures of the area.

A

isms, which became a practical possibility for the first time only in 1973, will be of great importance in the improvement of crop plants before the twentieth century draws to a close.

The oceans were once regarded as an inexhaustible source of food, but overexploitation of their resources is actually limiting the world catch from year to year, while these catches are costing more in terms of energy. It has been estimated that the mismanagement of fisheries, mainly through overfishing, local pollution, and the destruction of fish breeding and feeding grounds, has already lowered the catch of fish in the sea by about 20% from its maximum levels. The decline in the numbers of whales in the oceans of the world is a tragic and well-known example of the way fisheries have been and are being destroyed (see Figure 20-12).

The development of new kinds of food, such as microorganisms grown in culture in nutrient solutions, should definitely be pursued. For example, the photosynthetic, nitrogen-fixing cyanobacterium *Spirulina* is being used in this way, with experimentation under way in several countries to see whether it can be developed into a commercial food source (Figure 23-12). Masses of this bacterium are used traditionally as food in Africa, Mexico, and other regions. *Spirulina* thrives in very alkaline water and has a higher protein content than soybeans; the ponds in which it grows are 10 times more productive, on the average, than wheat fields. Ultimately the enrichment of human food through protein-rich concentrates of microorganisms will provide important nutritional supplements to our diets. There are, however, psychological barriers that must be overcome to persuade people to eat such foods, and the processes required to produce them tend to be energy-expensive.

THE TROPICS

More than half of the world's population was living in the tropics and subtropics by the 1980s, and this percentage is increasing rapidly. The well-being of these people should be a matter of concern to everyone, including inhabitants of the developed countries, such as ourselves. Although this is clearly true on humanitarian grounds alone, it also can easily be justified economically and in terms of the self-interest of the people living in developed countries. Not only do we obtain many commodities, including foods, minerals, and other goods, from the tropics, but as many as one in six manufacturing

B

C

D

A

B

C

FIGURE 23-12

Spirulina, a cyanobacterium that is being grown increasingly as a source of protein.
A Individuals of *Spirulina*, each about 250 micrometers long.
B A mat of *Spirulina* on a conveyor belt in a processing plant in Mexico.
C Ponds for the production of *Spirulina* in Thailand.

jobs in the developed world depends on exports to the tropics. In addition, as much as half of the food exports of developed countries go to the tropics. If this world commerce is to continue, the tropical countries must be able to develop sustainable agricultural systems and other sorts of productive industries so they can more adequately feed and otherwise provide for their rapidly growing populations.

Many people in the tropics engage in **shifting agriculture**—clearing and cultivating a patch of forest, growing crops for a few years, and then moving on (Figure 23-13). Such agricultural systems, more colorfully called "slash-and-burn" cultivation, are effective when the human populations that practice them are relatively small. Indeed, they still seem to be working well in such areas as Zaire, where about 35 million people are spread out over a very large area. In these systems, the minerals in the trees run out into the soil, and crops like manioc (tapioca, cassava; see Figure 23-8) and beans can be cultivated successfully for a few years. The fertility of the soil is then exhausted, and the cultivator must move on and clear another patch of forest.

Provided that the forests have decades or even centuries to recover—depending on the nature of the forest being cut—shifting agriculture can be practiced indefinitely without permanent harmful effects. There is renewed interest in the ways in which people have traditionally cultivated these forests, and some of the discoveries that have been made—for example, the extensive irrigated terraces of the Mayas—certainly may have applicability for meeting contemporary needs. With the rapidly growing human populations of the tropics, however, the pressures on the forests have been intensified, and they often do not have sufficient time to recover. Nor is there time for the development of new, more appropriate technologies or the re-introduction of traditional ones. The greater the population of a given area, the more often a given patch of forest is cultivated and the less completely it recovers. Firewood gathering is also hastening the demise of many tropical forests; about 1.5 billion people worldwide—a third of the global population—depend on firewood as fuel and are cutting the local supplies faster than the trees can regenerate themselves.

Shifting agriculture is one major factor in the destruction of tropical forests, but a number of other factors are significant also. We can illustrate these factors by reference to the tropical rain forest, biologically the richest of the tropical biomes. In the late 1980s, it is estimated that about 6 million square kilometers of tropical rain forest still exist in a relatively undisturbed form. This area, about three-quarters the size of the United States exclusive of Alaska, represents about half of the original extent of tropi-

FIGURE 23-13

A family has cleared a small patch of forest in the northern Amazon Basin of Brazil; they will be able to grow crops on it for a few years, and then will need to move on to another part of the original forest. Such activities can be sustained indefinitely if population levels are low enough, but that will rarely be the case in the closing years of the twentieth century.

A

B

FIGURE 23-14

A This palm is the best individual of its species, *Hyophorbe verschaffeltii,* known in nature. It is shown here in forest remnants on the island of Rodrigues, in the western Indian Ocean, the same island where that famous flightless pigeon, the dodo, became extinct more than three centuries ago. Who knows whether this palm, or one of the hundreds of other species of palms threatened with extinction around the world, holds the economic promise of oil palm, which has become the basis of a multi-billion-dollar industry over the past half-century, and is now grown on thousands of square kilometers throughout the tropics?
B *Hyophorbe verschaffeltii* is not likely to become extinct, however, because it is preserved in cultivation on the campus of the University of Mauritius, on the island of Mauritius. The palms are the second most important family of plants in the world economically, being surpassed only by the grasses.

cal rain forest. From it, about 100,000 square kilometers were being clear cut per year, and another 100,000 square kilometers severely disturbed by shifting cultivation, firewood gathering, and allied practices. The total area of tropical rain forest destroyed—and therefore permanently removed from the world total—amounted to an area about the size of Kansas annually. At such a rate, all of the tropical rain forest in the world would be gone in about 30 years, but in many regions, the rate of destruction is much more rapid.

Only about 1 tree is now being planted in the tropics for every 10 that are cut. Even though the firewood, lumber, and pasture needs of tropical countries could easily be met on lands that have already been cut over, the irresistible urge exists to attack the standing forests and "develop" them, thus reducing them to the status of a non-renewable resource like oil or other minerals. International economic pressures are making this trend even stronger and thus contributing significantly to global instability now and in the future.

As a result of such overexploitation, experts predict, there will be little undisturbed tropical forest left anywhere in the world by the early twenty-first century. Many areas now occupied by forest will still be tree-covered, but these trees will represent only a small percentage of those that now grow in these areas. Many species of tropical plants, animals, and microorganisms can reproduce only under the conditions in which they live in the original forest. Consequently they are threatened with extinction or, at the very least, with exclusion from large areas. In fact, a fifth or even more of the species on earth may become extinct during the next 30 or 40 years, amounting to a million or more species. Viewing the situation in another way, several species per day are probably becoming extinct now (see Figure 20-5). Many of these species that are becoming extinct inhabit ecologically devastated islands such as St. Helena in the south Atlantic and Rodrigues (Figure 23-14) in the western Indian Ocean, or similarly devastated areas on continents, such as the Atlantic forests of Brazil or the lowlands of western Ecuador. By early in the next century, the rate could easily reach several species per *hour;* and it would continue to climb for at least another 50 years. Overall, this would amount to an extinction event that has been unparalleled for at least 65 million years, since the end of the age of dinosaurs. The number of species in danger of extinction during our lives is far greater than the number that became extinct at that distant time.

FIGURE 23-15

A When tropical evergreen forests are removed, the ecological consequences can be disastrous—despite the fact that the undisturbed forests have one of the highest rates of net primary productivity of any plant community in the world. These fires are destroying rain forest in Brazil, which is being cleared for cattle pasture.
B The consequences of deforestation can be seen on these mid-elevation slopes in Ecuador, which now support only low-grade pastures where highly productive forest, which protected the watersheds of the area, grew in the 1970s.
C As time goes by, the consequences of tropical deforestation may become even more severe, as shown here by this extensive erosion in an area of Tanzania from which forest has been removed.

A

The prospect of such an extinction event is of major concern. Only one out of every six tropical organisms has even been given a scientific name, so many species that are about to become extinct will never have been seen by any human being. As these species disappear, so does our opportunity to learn about them, not only scientifically, but also in terms of their possible benefits for ourselves. We have an intrinsic curiosity about the plants, animals, and microorganisms with which we share this planet, and we would like to understand them better individually. Each is unique, and with its loss we lose forever the chance to use it for any purpose whatever. The fact that our entire supply of food is based primarily on 20 kinds of plants, out of the quarter million kinds that are available, should give us pause to consider what it means to be living in a generation during which a high proportion of the remainder are being lost permanently. Many of them would surely be of great use to us if we knew about their properties (see Figure 23-11).

What biologists can do in the face of this crisis is to help design intelligent plans for finding those organisms most likely to be of use and saving them from extinction. They must also participate in sound, globally based schemes to preserve as much as possible of the biological diversity of life on earth so that the options for our descendants may be as wide as possible. With the loss of tropical forest, and of biological communities throughout the world, we are permanently losing many opportunities not only for knowledge, but for increased prosperity—whether we realize it or not. It is biologists who must understand this message and inform their fellow citizens of its importance to them.

Very little undisturbed tropical forest will be left anywhere in the world by the early twenty-first century. Primarily as a result of this destruction, a fifth or more of the estimated 5 million species of plants, animals, and microorganisms on earth may become extinct during the next several decades, many of them without ever having been discovered or considered in terms of their potential usefulness to human beings.

Viewing the consequences of uncontrolled deforestation in the tropics and subtropics in another way, tropical forests are complex, productive ecosystems that work

C

B

well in the areas where they have evolved. The sad truth is that we do not know, for the most part, how to replace them with other productive ecosystems that will support human beings. When we cut a forest or open a prairie in the North Temperate Zone, we provide the basis for a farm that we know can be worked for generations. In the tropics, for the most part, we simply do not know how to engage in continuous agriculture in most areas that are not now under cultivation. When we clear a tropical forest, we engage in a one-time consumption of natural resources that will not be available again (Figure 23-15). The complex ecosystems that have been built up over billions of years are now being dismantled, in almost complete ignorance, by the human species.

What biologists must do is to learn more about the construction of sustainable agricultural ecosystems that will meet human needs in tropical and subtropical regions. The ecological principles that we have been reviewing in the last three chapters are universal principles. The undisturbed tropical rain forest has one of the highest rates of net primary productivity of any plant community on earth, and it is therefore logical to assume that it can be harvested for human purposes in a sustainable, intelligent way. Simply passively allowing it to be consumed is something that biologists and concerned individuals should attempt to avoid. Sound development of the tropics, an urgent matter in view of the very large numbers of humans who live in tropical countries, must be based on sound biology, as well as on achieving stable human population levels and alleviating the problems of poverty and widespread malnutrition. Biologists must address themselves increasingly to problems that have traditionally been the concern of agronomists, animal breeders, and other agriculturists, and must apply the results of their research to the creation of a stable global ecosystem.

POLLUTION

The activities of human being are now so extensive that our waste products are threatening the continued productivity—and in some cases even the continued existence—of not only many natural communities but also of some of the communities that we have created for our own benefit. The by-products of our industry and the chemicals that

FIGURE 23-16

Smog, associated with a temperature inversion, in the Los Angeles Basin.

we use in agriculture pass into various kinds of ecosystems and cause various kinds of disturbances. These disruptions depend on the nature of these ecosystems and the ways in which the chemicals circulate through them.

Air pollution is one of the most familiar forms of pollution with which we must contend. Much of it arises from the use of fossil fuels such as coal and oil without adequate steps being taken to remove the by-products that result from their combustion. Automobiles, which run on gasoline, an oil derivative, contribute heavily to the smog that hangs over cities such as Los Angeles (Figure 23-16) and causes damage to the natural forests far beyond the city limits. Such pollution makes living in places like Los Angeles and other major urban centers with similar problems relatively unhealthy for anyone who suffers from lung or heart problems. Many plants, too, are highly susceptible to damage from air pollution, which causes significant losses in natural vegetation, around homes and factories, and in agricultural situations.

Acid rain, better called acid precipitation, is a particular result of air pollution that has perplexed scientists and public officials throughout the industrial world since the late 1970s. "Acid rain" is a blanket term for the process whereby industrial pollutants such as nitric and sulfuric acids—introduced into the upper atmosphere by factory smokestacks hundreds of feet tall—are spread over wide areas by the prevailing winds and then fall to earth with the precipitation. The pH of unpolluted, natural precipitation often lies between 5.0 and 5.6. At present, however, the pH in some areas of the United States, Canada, and Europe often approaches 4.0—perhaps 10 times lower than normal (recall that pH is calculated on a log scale); in some cases it may even approach 2.0!

In both Europe and North America, tens of thousands of lakes are dying biologically as a result of acid precipitation (Figure 23-17). At pH levels below 5.0, many fish species and other aquatic animals die, and any lowering of the natural pH of the water may have other detrimental effects. In southern Sweden and elsewhere, groundwater is now regularly found to have a pH between 4.0 and 6.0, its acidity resulting from the acid precipitation that is slowly filtering down into the underground reservoirs, thus threatening the water supplies of future generations. It is also estimated that at least 3.5 million hectares of forest in the Northern Hemisphere are being affected by acid precipitation, and the scope of the problem is clearly growing. Unfortunately, since air pollution does not respect national boundaries, the problem of alleviating acid preci-

FIGURE 23-17

Twin Pond in the Adirondacks of upstate New York, one of the many lakes in the region in which the levels of acidity in the lake have killed the fishes, amphibians, and most of the other kinds of animals and plants that once occurred there.

piation has a very significant political dimension and has resulted in a great deal of strife between nations and even between different regions within individual countries.

Many related kinds of pollution have arisen as a result of human industry. For example, the polymers known as plastics are produced in abundance but are essentially impervious to being broken down by known natural forces. Even though scientists are trying to develop strains of bacteria that can decompose plastics, their efforts have been largely unsuccessful. Consequently, virtually all of the polymers in the plastic items that have been produced since the 1950s, made almost entirely from fossil fuels, are still with us. Collectively, these plastics constitute a new form of pollution of the world ecosystem for which there is, as yet, no known solution.

Water pollution is another very serious problem that exists on a global scale. There is simply not enough water available to dispose of the diverse substances that today's enormous human population produces continuously. Despite the implementation of ever-improved methods of sewage treatment throughout the world, our lakes, streams, and groundwater are becoming increasingly polluted. For example, household detergents, which contain phosphates, may flow into lakes and lead to the overgrowth of algae, which may in turn change the overall pattern of ecological relationships of the lake and the quality of its water.

Widespread agriculture, carried out increasingly by modern methods, also causes very large amounts of many new kinds of chemicals to be introduced into the global ecosystem. These include pesticides, herbicides, and fertilizers. Developed countries like the United States now attempt to monitor the side effects of these chemicals carefully. Unfortunately, however, large quantities of many toxic chemicals that were manufactured in the past still circulate in the ecosystems of these nations.

For example, the chlorinated hydrocarbons, a class of compounds that includes DDT, chlordane, lindane, and dieldrin, have all been banned for normal use in the United States, where they were once used very widely. These molecules break down very slowly and accumulate in animal fat. Furthermore, as they pass through a food chain, they are increasingly concentrated. DDT, for example, caused serious problems by leading to the production of very thin, fragile eggshells in many bird species in the United States and elsewhere until the late 1960s, when it was banned in time to save the birds from extinction. Chlorinated hydrocarbons have many other undesirable side effects, several of which are still poorly understood.

Obviously, a "back to nature" approach—one that ignores the important contributions made to our standard of living by the intelligent use of chemicals—will not allow us to care adequately for the needs of the current world population. Even less would such a backward approach allow us to feed the additional billions of people who will join us during the next few decades. On the other hand, it is essential that we use our technology as intelligently as possible and with due regard for the protection of the productive capacity of all parts of the earth, on which we all depend.

FIGURE 23-18

By understanding the ecological relationships within forests like this Venezuelan one, biologists will help to lay the foundation for a stable and peaceful world.

WHAT BIOLOGISTS HAVE TO CONTRIBUTE

Our human population is at an unprecedented size and is straining the productive capacity of the global ecosystem more than it ever has. Of the more than 5 billion people living in the late 1980s, 1 of every 5 is living in absolute poverty, 1 of every 10 is malnourished, and a hundred thousand are starving to death each day. The productive agricultural soils of temperate regions are rapidly being lost to erosion; our water supplies are becoming increasingly contaminated and restricted; and species of plants, animals, and microorganisms are being lost to extinction at a rate that has not occurred for the past 65 million years. The social and economic consequences of these biological problems are increasingly familiar, manifesting themselves as governmental instability, inflation, massive international debts, war, and huge numbers of displaced people and immigrants moving throughout the world.

The development of appropriate solutions to these massive problems must rest partly on the shoulders of politicians, economists, bankers, engineers—many different kinds of people. The basis for solving them permanently, however, and achieving a stable, productive world, must be biological (Figure 23-18). The energy that bombards the earth comes in essentially inexhaustible amounts from the sun; living systems capture that energy and fix it in molecules that can be utilized by all organisms to sustain their life processes. We and all other living beings depend on the proper utilization of that renewable energy and of the other materials on which our civilization depends. We must come to understand better the processes involved and to develop new systems that will work efficiently in areas where they are not available now. That is our challenge for the future.

It is clear that a scientific education has become necessary for everyone, so that we may understand the basis for our continued existence on earth and the steps that we will need to take to improve the quality of our lives. Biology should play a major part in that education and is of critical importance in improving the standard of living for our fellow human beings. Biological literacy is no longer a luxury for intelligent human beings who want to play a constructive role in improving the world. It has become a necessity.

SUMMARY

1. Agriculture was developed in several centers, starting about 11,000 years ago. The growing herds of domesticated animals caused widespread ecological destruction in the Old World, and the same phenomenon began to occur in the New World when the animals were introduced there nearly 500 years ago.

2. About 5 million people lived throughout the world when agriculture was first developed. With the availability of this new steady source of food, the population grew rapidly, reaching about 130 million at the time of Christ and 500 million by 1650. Agriculture made possible the development of villages, towns, and finally cities. Cultural specialization occurred in these centers, resulting in the development of writing and, ultimately, of science.

3. By the late 1980s, the world population was more than 5 billion people, was growing at a rate of 1.7% per year, and was estimated to double within approximately 40 years. It is growing much more rapidly in the tropics and subtropics than in developed countries, so the relative proportions of people in these two areas are changing rapidly. By 2020, nearly two thirds of the people in the world are expected to live in the tropics and subtropics, and only a sixth in the developed countries—half the proportion that lived there in 1950.

4. In the 1980s, the World Bank estimated that about 1 billion of the more than 2.5 billion people living in the tropics and subtropics existed in absolute poverty. Some 35 million people a year were starving to death, about 14 million of them children under 5 years old. With the populations of tropical countries doubling

about every 29 years, the task of feeding all these extra people even as well as is being done now will be extraordinarily difficult.

5. More food must be produced for the inhabitants of the tropics and subtropics. It can be produced in appropriate quantity only in the countries where it is needed. As a result, a great deal of research into the properties of sustainable agricultural systems in the tropics and subtropics is required, and strict attention must be paid to the needs of the rural poor if stable systems are to be put in place.

6. The Green Revolution resulted from a large-scale effort on the part of agricultural research institutes to increase the productivity of wheat and corn (maize) in the subtropics. It was enormously successful in increasing the food production of such countries as Mexico, Pakistan, and India, although it had relatively little effect on the well-being of the rural poor and tended to be energy-expensive.

7. The destruction or disturbance of virtually all of the forests of the tropics during the next several decades will bring about the extinction of as much as 20% of the species of plants, animals, and microorganisms on earth. Many of these will never have been seen by a scientist, and their possible contribution to human welfare will be lost forever.

8. Pollution of air, water, and the other life-supporting systems of the biosphere is a serious by-product of human population growth, and it must be brought under control.

9. The application of the principles of biology to the human condition has never been more necessary than it is now. Only the full attention of society and many talented individuals to the solution of these grave problems will make it possible for our children and grandchildren to enjoy the same benefits that we have.

REVIEW

1. Will a country with a greater proportion of its population below childbearing age grow faster or slower in the future than a country that has a greater proportion of its population within the childbearing ages?

2. During the next 30 to 40 years, what percentage of the world's species are expected to become extinct?

3. When agriculture first developed about 11,000 years ago, there were approximately _____ people in the world.

4. Three major factors that are bringing about the destruction of the tropical forests are _____ , _____ , and _____ .

5. At what annual rate is the world population growing? _____ .

SELF-QUIZ

1. The present population of the world is about
   (a) 500 million   (d) 5 billion
   (b) 100 billion   (e) 50 billion
   (c) 10 billion

2. Which of the following crops have entered cultivation on an international scale since 1800?
   (a) wheat   (d) peanuts
   (b) manioc   (e) soybeans
   (c) rubber

3. We obtain more than half of our food energy from how many different kinds of plants?
   (a) 200   (d) 12
   (b) 1800   (e) 20
   (c) 3

4. The population of developing countries is expected to double in approximately _____ years.
   (a) 10   (d) 40
   (b) 20   (e) 50
   (c) 30

5. What is the main problem with the "shifting cultivation" practiced in the tropics?
   (a) Soil in the tropics is ruined when it is plowed and exposed to oxygen.
   (b) The natural trees are the principal repository of nutrients, and when crops are removed the soil becomes nutrient-poor.
   (c) Burning of trees chars the soil so that it is unable to support crops.
   (d) It does not permit the intensive use of fertilizers upon which the Green Revolution depends.
   (e) Two of the above.

THOUGHT QUESTIONS

1. Can we ever produce enough food and other materials so that population growth will not be a matter of concern? How?

2. Decreased birth rates have ultimately followed decreased death rates in many countries, for example, Germany and Great Britain. Do you think this will eventually occur worldwide and solve our population problems?

3. Some have argued that attempts by the United States to promote lowering of the birth rate in the underdeveloped countries of the tropics is no more than economic imperialism, and that it is in the best interests of these countries that their populations grow as rapidly as possible. How would you respond to this argument?

FOR FURTHER READING

BATIE, S.S., and R.G. HEALY: "The Future of American Agriculture," *Scientific American,* February 1983, pages 45-53. An interesting analysis of the ecological and economic factors involved.

BROWN, L. (ed.): "State of the World, 1987," W.W Norton & Co., New York, 1987. Perhaps the most important recent book on the ecological problems faced by an overcrowded world. Easy to read and authoritative; highly recommended.

EHRLICH, P.R.: "Humankind's War Against *Homo sapiens,*" *Defenders,* November 1985, pages 4-12. An impassioned plea to stop the wave of extinctions now resulting from human activities.

HINMAN, C.W.: "Potential New Crops," *Scientific American,* July 1986, pages 33-37. A well-written review of present efforts to develop new commercial crops.

HUXLEY, A.: *Green Inheritance,* Anchor Press/Doubleday, Garden City, N.J., 1985. An outstanding account of the many ways in which we depend on plants and should therefore be concerned with their preservation.

POWER, J.F., and R.F. FOLLETT: "Monoculture," *Scientific American,* March 1987, pages 78-87. Growing one crop repeatedly on the same land has advantages, but it may not be good in the long run.

VIETMEYER, N.D.: "Lesser-Known Plants of Potential Use in Agriculture and Forestry," *Science,* vol. 232, pages 1379-1384, 1986. An excellent, brief survey of little-known but extremely useful plants that humans could use to improve food and firewood supplies in the future.

VITOUSEK, P., AND OTHERS: "Human Appropriation of the Products of Photosynthesis,'" *BioScience,* vol. 36, pages 368-372, 1986. Nearly 40% of terrestrial photosynthesis productivity is consumed as a result of human activities now, leaving little room for future growth.

BIOLOGICAL DIVERSITY Part Seven

FIGURE 24-1   This weevil, a kind of beetle, is dead, lying on a leaf in the tropical rain forest of Costa Rica. It has been killed by the fungus whose fruiting structure now rises above it, scattering spores to the four winds. The filaments of the fungus grew through the body of the weevil and eventually killed it, converting its body into the body of the fungus. There are more than 100,000 named species of fungi and 40,000 named species of weevils, by far the largest family of the animal kingdom. Both groups inflict billions of dollars worth of damage on crops annually.

THE FIVE KINGDOMS OF LIFE Chapter 24

Overview

There are at least 5 million different kinds of organisms living on earth, of which about 1.5 million have been catalogued. In order to be able to discuss and study them, scientists group them in categories and give those categories particular names. Every organism is assigned to a species, and the species are grouped in genera, which consist of one or more species. The names of organisms are written with the name of the genus first, followed by a distinctive word that, together with the name of the genus, makes up the name of that particular species. In turn, genera are grouped into families, families into orders, orders into classes, classes into phyla (called divisions in some plants, fungi, and algae), and phyla into kingdoms. Each of these more inclusive groups has its own distinctive set of properties, so that the name of an organism, which indicates its position in the system, tells us a great deal about that organism.

For Review   *Here are some important terms and concepts that you will encounter in this chapter. If you are not familiar with them, you should review them before proceeding.*

The five kingdoms of life (Chapter 2)

Prokaryotic versus eukaryotic cells (Chapter 4)

Definition of a species (Chapter 17)

How species form (Chapter 17)

Biological classification (Chapter 18)

All living organisms have a great deal in common. They are all composed of one or more cells, carry out metabolism based on energy transfer via ATP, encode hereditary information in DNA, and have evolved from simpler forms. In addition, they evolve, becoming progressively more complex, and occur together in populations which have certain properties of their own. These populations make up communities and ecosystems, which provide the overall structure of life on earth. So far in this text, we have stressed these common themes, considering the family of general principles that apply to all organisms. Now it is time to focus on the *differences* between organisms, to consider the diversity of the biological world and the properties of the individual organisms that make it up. For the rest of this text, we will examine the parade of life on earth, from the simplest microbes to elephants and bears and redwood trees. Like going to the zoo, examining biological diversity is interesting, challenging, and a great deal of fun (Figure 24–1).

THE CLASSIFICATION OF ORGANISMS

Millions of different kinds of organisms exist; to talk about them and study them, it is necessary that they have names. Going back as early as we can trace, organisms have been grouped in basic units, such as oaks, cats, and horses. Eventually, these units, which were often given the same names used by the Romans and Greeks, began to be called **genera** (singular, **genus**). Starting in the Middle Ages, these names were written in Latin, the language of scholarship at that time, or given a Latin form. Thus oaks were assigned to the genus *Quercus,* cats to *Felis,* and horses to *Equus*—names that the Romans applied to these groups of organisms. For genera that were not known in antiquity, of course, new names had to be invented.

> **The names of genera were the basic points of reference in classification systems, and these names eventually came to be written in Latin.**

The Polynomial System

Before the 1750s, scholars usually added a series of additional descriptive terms to the name of the genus when they wanted to designate a particular species. (Both the singular and plural of the word "species" are the same.) These phrases, starting with the name of the genus, made up what came to be known as **polynomials,** strings of Latin words and phrases consisting of up to 12, 15, or even more words. Not only were such polynomials cumbersome, but they also could be altered at will by later authors. As a result, a given organism really did not have a single name that was its alone. Instead, its names were a series of different descriptive phrases that could be related to one another only by scholars. This was a burdensome and imprecise system of naming.

The Binomial System

The simplified system of naming plants, animals, and microorganisms that has been standard for more than two centuries stems from the works of the Swedish biologist Carl Linnaeus (1707-1778; Figure 24-2). Linnaeus' ambition, like that of many of his predecessors, was to catalog all the kinds of organisms and minerals. In the 1750s he produced several major works that, like his earlier books, employed the polynomial system, which had been well established for nearly a century as the basic system for naming organisms.

FIGURE 24-3

A Willow oak, *Quercus phellos.*
B Red oak, *Quercus rubra.*
Although they are clearly oaks *(Quercus),* these two species differ sharply in the shapes and sizes of their leaves and in many other features, including their overall geographical distributions.

QUERCUS PHELLOS
(WILLOW OAK)

A

QUERCUS RUBRA
(RED OAK)

B

As a kind of shorthand, however, Linnaeus also included a two-part name for each species, and it is these two-part names, or **binomials,** that have become our standard way of designating them. For example, he designated the willow oak *Quercus phellos* and the red oak *Quercus rubra* (Figure 24-3), even though he also included the polynomial names for these species. The convenience of the short, new names proposed by Linnaeus was so great that they were immediately accepted generally.

In Linnaeus' binomial system the name of a species consists of two parts, the first of which is the generic name.

By agreement, the scientific names of organisms are the same throughout the world and thus provide a standard and precise way of communicating about organisms, whether the language of a particular scholar is Chinese, Arabic, Russian, or English. In contrast, the names that are given locally often differ greatly from place to place; thus a "bear" in Australia may be either a koala bear (a marsupial) or one of the large carnivores (placental mammals) more usually known by this name. For general communication, it is important to have one standard set of names. The system of naming plants, animals, and microorganisms established by Carl Linnaeus has served the science of biology well for nearly 250 years.

WHAT IS A SPECIES?

In Chapter 17 we reviewed the nature of species and saw that there are no absolute criteria that can be applied to the definition of this category. Individuals that belong to a given species—for example, dogs (see Figure 17-2)—may look very unlike one another. Nevertheless, they are generally capable of hybridizing with one another, and the different forms can appear in the progeny of a single mated pair. Species remain rel-

A

FIGURE 24-4

A Black-footed ferret, *Mustela nigripes.* One of the most elegant members of the weasel family, the black-footed ferret preys on prairie dogs on the plains of the north-central United States. Unfortunately, it is near extinction, with the 20 or so remaining individuals threatened by a viral disease called canine distemper. Foxes belong to the same family, Canidae, as dogs, coyotes, and wolves, but are unable to interbreed with them. The general similarity in form and characteristics of these mammals is obvious, however. Together with weasels (including the black-footed ferret), bears, cats, and other related mammals, all of the animals shown here belong to the order Carnivora, the carnivores.
B A dog, *Canis familiaris.* All dogs are capable of interbreeding with one another, regardless of how different they appear.
C Coyotes, *Canis latrans,* occur widely in North America.
D The gray wolf, *Canis lupus,* extends all around the Northern Hemisphere. Coyotes and wolves belong to the same genus as dogs, and often produce fertile hybrids with them where their ranges come together.

B

C

D

atively constant in their characteristics, can be distinguished from other species, and do not normally interbreed when they come together with other species in nature. For example, dogs are not capable of interbreeding with foxes, which, although they are generally similar, are members of another, completely distinct, group of mammals. In contrast, dogs can and do form fully or partly fertile hybrids with related species such as wolves and coyotes, which are also members of the genus *Canis* (Figure 24-4). The transfer of characteristics between these species has, in some areas, changed the characteristics of both of the interbreeding units.

The criteria just mentioned for species apply primarily to those that regularly **outcross**—interbreed with individuals other than themselves. In some groups of organisms, including bacteria and many eukaryotes, **asexual reproduction**—reproduction without sex—predominates. The species of these organisms clearly cannot be characterized in the same way as the species of outcrossing animals and plants; they do not interbreed with one another, much less with the individuals of other species. De-

FIGURE 24-5

The hierarchical system used in classifying an organism, in this case a dog.
Kingdom: Monera, Plantae, Fungi, Protista, ANIMALIA.
Phylum: Anthophyta, Mollusca, CHORDATA, etc.
Class: Amphibia, Reptilia, MAMMALIA, etc.
Order: Primates, Insectivora, CARNIVORA, etc.
Family: Mustelidae, Ursidae, CANIDAE, etc.
Genus: *Vulpes, Urocyon, CANIS, etc.*
Species: *Canis latrans, C. lupus, C.* FAMILIARIS, etc.

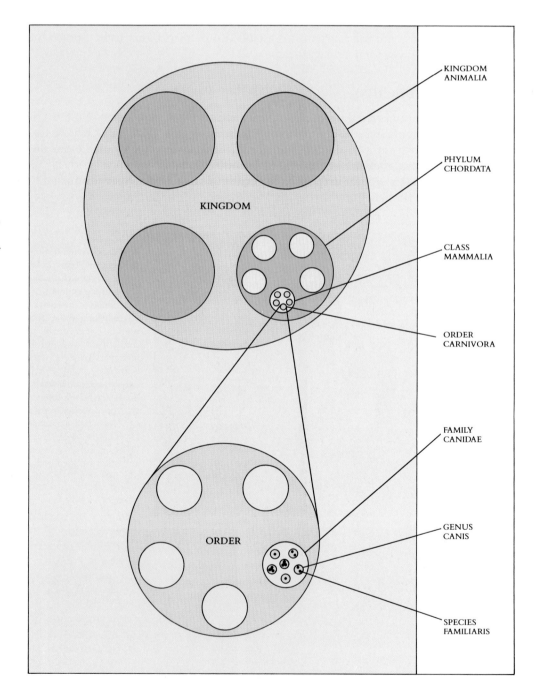

KINGDOM ANIMALIA

PHYLUM CHORDATA

KINGDOM

CLASS MAMMALIA

ORDER CARNIVORA

FAMILY CANIDAE

ORDER

GENUS CANIS

SPECIES FAMILIARIS

spite these difficulties, biologists generally agree on the kinds of units that they classify as species, although these units share no biological characteristics uniformly.

Species differ from one another in at least one characteristic and generally do not interbreed freely with one another where their ranges overlap in nature.

HOW MANY SPECIES ARE THERE?

Since the time of Linnaeus, about 1.5 million species have been named. This is a far greater number of organisms than Linnaeus suspected to exist when he was developing his system of classification in the eighteenth century. The actual number of species in the world, however, is undoubtedly much greater, judging from the very large numbers that are still being discovered. **Taxonomists**—scientists who study and classify different kinds of organisms—estimate that there are at least 5 million species of organisms on earth, and that at least two thirds of these—more than 3 million species—occur in the tropics. Considering that no more than 500,000 tropical species have been named, our knowledge of these organisms is very limited.

THE TAXONOMIC HIERARCHY

Biological systems of classification are **hierarchical,** with the basic units arranged like boxes within boxes (Figure 24-5). In such a system, grouping genera into the larger, more inclusive categories known as **families** is an effective way to remember and study their characteristics. Families of organisms such as "finches," "legumes," and "mammals" include many genera and have characteristics of their own. Thus, for example, the oaks *(Quercus),* beeches *(Fagus),* and chestnuts *(Castanea)* were grouped, along with other genera, into the beech family, Fagaceae, because of the many features they had in common. Similarly, the tree squirrels *(Sciurus),* Siberian and western North American chipmunks *(Eutamias),* and marmots *(Marmota)* were grouped with other related mammals in the family Sciuridae (Figure 24-6). If one knows that a genus belongs to a particular family, one immediately knows a great many more of its features.

The **taxonomic system** also includes several more inclusive units than families, as mentioned in connection with our survey of the history of life on earth in Chapter 18.

FIGURE 24-6

Three genera of the squirrel family, Sciuridae, which differ greatly in their characteristics and are adapted for different modes of life.

A

B

C

A The red squirrel, *Tamiasciurus hudsonicus,* an agile dweller in the trees.
B Townsend's chipmunk, *Eutamias townsendii,* a member of a genus of diurnal, brightly colored, very active, ground-dwelling squirrels.
C Yellow-bellied marmot, *Marmota flaviventris,* a large squirrel that lives in burrows. Yellow-bellied marmots are social, tolerant, and playful; they live in harems that consist of a single territorial male together with numerous females and their offspring. The flying squirrel shown in Figure 24-7 represents a fourth genus of the family Sciuridae, one that is nocturnal; the three species shown here are all diurnal (day-active).

TABLE 24-1   SAMPLE CLASSIFICATIONS OF THREE REPRESENTATIVE ORGANISMS

	HUMAN BEINGS	HONEY BEE	RED OAK
Kingdom	Animalia	Animalia	Plantae
Phylum (division)	Chordata	Arthropoda	Anthophyta
Class	Mammalia	Insecta	Dicotyledones
Order	Primates	Hymenoptera	Fagales
Family	Hominidae	Apidae	Fagaceae
Genus	*Homo*	*Apis*	*Quercus*
Species	*Homo sapiens*	*Apis mellifera*	*Quercus rubra*

They function in the same way as do the names of families, making possible more efficient communication about organisms. Families are grouped into **orders,** orders into **classes,** and classes into **phyla** (singular, **phylum**), which are called **divisions** among plants, fungi, and algae. The phyla or divisions are in turn grouped into **kingdoms,** the most inclusive units of classification. To remember the order of the taxonomic categories that are used to classify organisms, you may wish to memorize this phrase: "**K**indly **P**ay **C**ash **O**r **F**urnish **G**ood **S**ecurity" (kingdom–phylum–class–order–family–genus–species).

> **By convention, species are grouped into genera, genera into families, families into orders, orders into classes, and classes into phyla (or divisions). Phyla are the basic units within kingdoms. Such a system is hierarchical.**

Table 24–1 shows, respectively, how the human species, the honey bee, and the red oak are placed in taxonomic categories at the seven different hierarchical levels. The categories at the different levels may include many, a few, or only one item, depending on the nature of the relationships in the particular groups involved. Thus, for example, there is only one living genus of the family Hominidae, but, as we have mentioned, several living genera of Fagaceae. Each of the categories at every level implies to someone familiar with the system, or with access to the appropriate books, both a series of characteristics that pertain to that group and also a series of organisms that belong to it.

FIGURE 24-7

For obvious reasons, people have always considered plants and animals to be very distinct from one another. The southern flying squirrel, *Glaucomys volans,* left, shown here feeding on berries, appears completely unlike the ladyslipper orchid, *Cyrripedium calceolus,* right. Animals ingest their food, and most animals move about from place to place; plants manufacture their food through the process of photosynthesis and are stationary. These and other major differences have traditionally caused plants and animals to be regarded as distinct kingdoms, with their similarities being appreciated only in the past hundred years or so.

THE FIVE KINGDOMS OF ORGANISMS

During the past 50 years, biologists have recognized that the most fundamental distinction among groups of organisms is not that between animals and plants (Figure 24-7), but rather that which separates prokaryotes and eukaryotes. The prokaryotes, or bacteria (Figure 24-8), differ from all other organisms in their lack of membrane-bound organelles and microtubules, as well as in their possession of simple flagella. Their DNA is not associated with proteins and is not located in a nucleus. Bacteria do not undergo sexual recombination in the same sense that eukaryotic organisms do, although forms of genetic recombination occur occasionally in some bacteria. The cell walls of most bacteria contain muramic acid, which is only one of the many biochemical peculiarities by which they differ from all eukaryotes. Bacteria existed for at least 2.5 billion years before the appearance of eukaryotes. In recognition of their highly distinctive features, the bacteria have been assigned to a kingdom of their own, **Monera,** which will be treated in detail in the next chapter.

The features of eukaryotes contrast sharply with those of prokaryotes, as we stressed in Chapter 5 and will now review. All eukaryotes have microtubules in their cytoplasm, and their cells are characterized by classes of membrane-bound organelles. They have a clearly defined nucleus that is enclosed within a double membrane, the nuclear envelope. Their chromosomes are complex structures in which the DNA is characteristically associated with proteins; the chromosome divide and are distributed in a regular manner to the daughter cells as a result of the process of mitosis. When flagella and cilia are present in eukaryotes, they have a characteristic 9 + 2 structure that consists of microtubules. Mitochondria are among the complex organelles that occur in the cells of all eukaryotes. In addition, eukaryotes exhibit integrated multicellularity of a kind that is not found in bacteria, as well as true sexual reproduction.

Despite their often bewildering diversity of form, eukaryotes are much less diverse metabolically than bacteria. All eukaryotes are believed to have evolved from a common ancestor. Eukaryotes consist primarily of groups of organisms that are single-celled, together with several multicellular groups that were derived from single-celled ancestral eukaryotes (Figure 24-9).

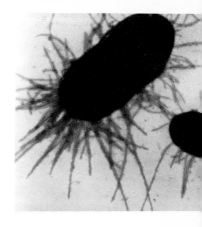

FIGURE 24-8

Bacteria, like these individuals of *Escherichia coli,* constitute the kingdom Monera. They are prokaryotes, a term that describes a pattern of cellular organization completely different from that of all other organisms. The slender projections extending from the surfaces of these bacteria are called pili.

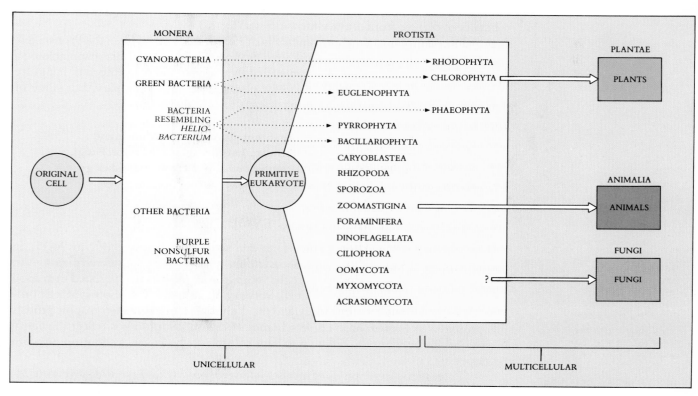

FIGURE 24-9

Diagram of the evolutionary relationships of the phyla of organisms. The solid lines indicate evolutionary relationships, the dotted ones symbiotic events.

A

B

C

FIGURE 24-10

Animals, plants, and fungi are the three multicellular kingdoms derived from members of the kingdom Protista.
A Tortoise beetle, family Chrysomelidae, in Costa Rica.
B Part of the complex leaf of a palm.
C A mushroom, Mill Valley, California.

THE KINGDOMS OF EUKARYOTIC ORGANISMS

The classification of organisms, especially eukaryotes, into kingdoms is clearly somewhat arbitrary, and no scheme has been universally accepted yet. Here we have adopted what is called the **five-kingdom system,** in which four of the kingdoms are eukaryotic organisms. The fifth kingdom, Monera, contains the bacteria, as we mentioned earlier. In the five-kingdom system of classification, the array of diverse eukaryotic phyla consisting predominantly of single-celled organisms is assigned to the kingdom Protista (see Chapter 26). Three major multicellular lines that have originated from this kingdom, each with a great many species, are assigned to separate kingdoms: Animalia, Plantae, and Fungi (Figure 24-10). Each of the three multicellular kingdoms—plants, animals, and fungi—has evolved from a different ancestor among the Protista. Although the fungi include a few single-celled forms, the yeasts, these have been derived from multicellular ancestral forms.

The members of the three multicellular kingdoms derived from the Protista differ in their mode of nutrition. Thus animals ingest their food, plants manufacture it, and fungi digest it by means of enzymes that they secrete and then absorb it. Fungi are filamentous and have no moveable cells; plants are mainly stationary, but some of them have moveable sperm; and many animals are highly motile.

> **We recognize five kingdoms of organisms. Two of them, Monera and Protista, are predominantly unicellular; the other three kingdoms, Fungi, Animalia, and Plantae, are multicellular and derived from individual groups of protists.**

FEATURES OF EUKARYOTIC EVOLUTION

Protists are very diverse both in their form and in their biochemistry (Figure 24-11). In older systems of classification, all photosynthetic protists were considered plants, even if they were unicellular, and so were the fungi. Those protists that ingested their food like the animals were considered small, very simple animals. The photosynthetic protists are called **algae** (singular, **alga**), and the traditional name for heterotrophic protists is **protozoa** or **protozoans.** These schemes of classification do not reflect the actual relationships between the different groups of protists, however, as do more modern ones.

Three of the larger phyla of protists—the red, brown, and green algae (Figure 24-12)—have attained multicellularity independently of the animals, plants, and fungi; there are some multicellular representatives in most of the other phyla of Protista also.

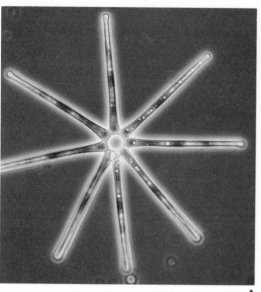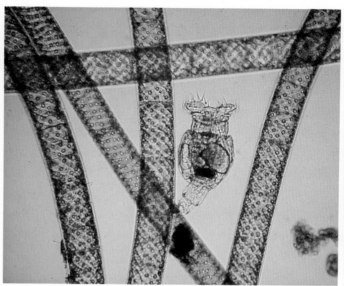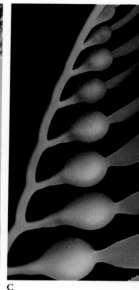

A   B   C

FIGURE 24-11

The members of the kingdom Protista are very diverse, as suggested by these photographs.

A A freshwater diatom, member of a mostly photosynthetic, unicellular group of organisms, that occurs widely in fresh and salt water characterized by silica shells.

B A green alga, *Spirogyra,* an organism with threadlike growth form and a single, ribbon-shaped spiral chloroplast, by means of which it manufactures its own food, with a rotifer, a small, motile animal that captures and ingests its food.

C A kelp, one of the brown algae, a member of a completely multicellular division (phylum) of the kingdom Protista. The individual kelp may be many meters in length. Brown, red, and green algae are divisions (phyla) of Protista not classified as plants, even though they are all photosynthetic, because they were derived independently from different ancestors among the Protista. Many green algae and a few red algae are unicellular.

A   C

FIGURE 24-12

Three phyla of algae have become multicellular independently and differ greatly in their photosynthetic pigments and structure.

A The brown algae (phylum Phaeophyta) are large, complex algae which often dominate north-temperate seacoasts.

B *Ulva* is a green alga (phylum Chlorophyta), a group that includes many unicellular organisms as well as a number that are multicellular.

C Some red algae (phylum Rhodophyta) are shown here growing on sponges. Members of this phylum occur at greater depths in the sea than do the members of either of the other phyla. Different groups of bacteria seem to have given rise to the chloroplasts of each of these phyla of algae as a result of ancient symbiotic events.

TABLE 24-2   CHARACTERISTICS OF THE FIVE KINGDOMS

KINGDOM	CELL TYPE	NUCLEAR ENVELOPE	MITO-CHON-DRIA	CHLORO-PLASTS	CELL WALL
Monera	Prokaryotic	Absent	Absent	None (photosynthetic membranes in some types)	Noncellulose (polysaccharide plus amino acids)
Protista	Eukaryotic	Present	Present	Present (some forms)	Present in some forms, various types
Fungi	Eukaryotic	Present	Present	Absent	Chitin and other noncellulose polysaccharides
Plantae	Eukaryotic	Present	Present	Present	Cellulose and other polysaccharides
Animalia	Eukaryotic	Present	Present	Absent	Absent

Two of the phyla of algae that were just mentioned are primarily multicellular; the brown algae (phylum Phaeophyta) and the red algae (phylum Rhodophyta). The green algae (phylum Chlorophyta) include many kinds of multicellular organisms and even larger numbers of unicellular ones. The red, brown, and green algae have become multicellular independently. In an evolutionary sense, the red and brown algae are certainly distinct from the plants, and green algae include the ancestor of plants. However, the green algae are basically aquatic and much simpler in form than the plants, so we will treat them as protists. All three of these phyla of algae are discussed in more detail in Chapter 26.

Fungi, plants, and animals are clearly defined, major evolutionary groups that have each had a single common ancestor different from the others. For this reason, we shall treat them as separate kingdoms and leave all remaining eukaryotic organisms as members of the kingdom Protista. Grouping all of these predominantly unicellular, eukaryotic organisms together allows us to compare them directly and to emphasize the similarities and differences between them. It should be kept in mind, however, that the kingdom Protista is much more diverse than any of the other three kingdoms of eukaryotic organisms. The characteristics of the five kingdoms are outlined in Table 24-2.

Of the four kingdoms of eukaryotes, the most diverse is Protista, the protists. Probably every phylum of Protista has at least some multicellular representatives included in it, and three of the groups—the red, brown, and green algae—are completely or largely multicellular.

MEANS OF GENETIC RECOMBINATION, IF PRESENT	MODE OF NUTRITION	MOTILITY	MULTI-CELLULARITY	NERVOUS SYSTEM
Conjugation, transduction, transformation	Autotrophic (chemosynthetic or photosynthetic or heterotrophic)	Bacterial flagella, gliding or nonmotile	Absent	None
Fertilization and meiosis	Photosynthetic or heterotrophic, or combination of these	9 + 2 cilia and flagella ameboid, contractile fibrils	Absent in most forms	Primitive mechanisms for conducting stimuli in some forms
Fertilization and meiosis	Absorption	Nonmotile	Present in most forms	None
Fertilization and meiosis	Photosynthetic, chlorophylls *a* and *b*	9 + 2 cilia and flagella in gametes of some forms, none in most forms	Present in all forms	None
Fertilization and meiosis	Digestion	9 + 2 cilia and flagella, contractile fibrils	Present in all forms	Present, often complex

SYMBIOSIS AND THE ORIGIN OF THE EUKARYOTIC PHYLA

As discussed in Chapter 5, two of the principal organelles of eukaryotic cells, mitochondria and chloroplasts, share a number of unusual characteristics with each other and with the bacteria from which they were apparently derived.

As you saw in Chapter 18, the first eukaryotes appeared about 1.5 billion years ago; their descendants soon become abundant and diverse. With a very few exceptions, all modern eukaryotes possess mitochondria, and it is therefore clear that these organelles were acquired early in the history of the group. Mitochondria are most similar to nonsulfur purple bacteria, the group that probably gave rise to them originally (see Chapter 5).

In contrast to mitochondria, which are relatively uniform from group to group, chloroplasts fall into three classes that are distinct in their biochemistry. Each of these three classes seems to have been derived from a different bacterial ancestor (see Chapter 25). Even today, so many bacteria and unicellular protists are symbiotic that the incorporation of smaller organisms with desirable features into cells appears not to be a difficult process. The symbiotic events that gave rise to the chloroplasts of different groups of protists seem to have taken place independently, and the same groups of bacteria to have been involved more than once.

> **Eukaryotic cells acquired chloroplasts by symbiosis not once, but at least several times during the evolution of different groups of protists.**

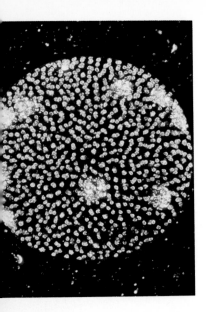

THE EVOLUTION OF MULTICELLULARITY AND SEXUALITY
Multicellularity

True **multicellularity,** a condition in which the activities of the individual cells are co-ordinated and the cells themselves are in contact, is a property of eukaryotes alone and one of their major characteristics. The cell walls of bacteria occasionally adhere to one another, and bacterial cells may also be held together within a common sheath. Consequently, some bacteria form filaments, sheets, or three-dimensional bodies, but the degree of integration of individual bacteria within these groups of cells is much more limited than is the case in multicellular eukaryotes.

Multicellularity evolved numerous times among the protists. The brown, green, and red algae all evolved multicellularity independently, as did animals, plants, and fungi. The evolution of multicellularity allowed organisms to deal with their environments in novel ways (Figure 24-13). Distinct types of cells, tissues, and organs can be differentiated within the complex bodies of multicellular organisms. With such functional division within its body, a multicellular organism can protect itself, move about, seek mates and prey, and carry out other activities on a scale and with a complexity that would have been impossible for its unicellular ancestors. With all of these advantages, it is not surprising that multicellularity has arisen independently so many times.

> **Multicellularity evolved numerous times among the protists. All of the complex differentiations that we associate with advanced life forms depend on multicellularity, which must have been highly advantageous to have evolved independently so often.**

Sexuality

The second major characteristic of eukaryotic organisms as a group, besides multicellularity, is sexuality. Although some interchange of genetic material does occur in bacteria (Chapters 5 and 25), it is certainly not a regular, predictable mechanism in the same sense that sex is in eukaryotes. The regular alternation between **syngamy**—the union of male and female gametes—and meiosis constitutes the **sexual cycle** that is characteristic of eukaryotes. It differs sharply from anything found in bacteria.

Some eukaryotes are haploid all their lives, but with few exceptions animals and plants are diploid at some stage of their lives. In the cells of animals and plants there are two sets of chromosomes, derived respectively from their male and female parents. These chromosomes segregate regularly by the process of meiosis; since meiosis involves crossing-over, no two products of a single meiotic event are ever identical. As a result, the offspring of sexual, eukaryotic organisms vary widely, thus providing the raw material for evolution.

> **Sexual organisms are able to evolve rapidly in relation to the demands of their environment because they produce variable progeny. Sexual reproduction, involving the regular alternation between syngamy and meiosis, is the process that makes their evolutionary adjustment possible.**

In many of the unicellular phyla of protists, sexual reproduction occurs only during times of stress. Meiosis may have evolved originally as a means of producing new, well-adapted forms that would increase the chances of survival during such times. The first eukaryotes were probably haploid, and diploids seem to have arisen on a number of separate occasions by the fusion of haploid cells, which then eventually divided by meiosis.

Eukaryotes are characterized by three major types of life cycles. In the simplest of these, the zygote is the only diploid cell. Such a life cycle is said to be characterized by **zygotic meiosis,** since the zygote immediately undergoes meiosis. In most animals the gametes are the only haploid cells. They exhibit **gametic meiosis;** meiosis produces the

FIGURE 24-13

Volvox (phylum Chlorophyta; green algae). Individual biflagellated (they have two flagella each), motile, unicellular green algae are united in *Volvox* as a hollow bowl of cells, which moves by means of the beating of the flagella of its individual cells. Some of these cells are specialized in form and function: a few, near the posterior (rear when moving) end of the colony, are reproductive cells, but most are relatively undifferentiated. Some species of *Volvox* have cytoplasmic connections between the cells, which function in the coordination of the activities of the colony. *Volvox* represents an unusual form of multicellularity, which evolved independently from that of all other multicellular organisms.

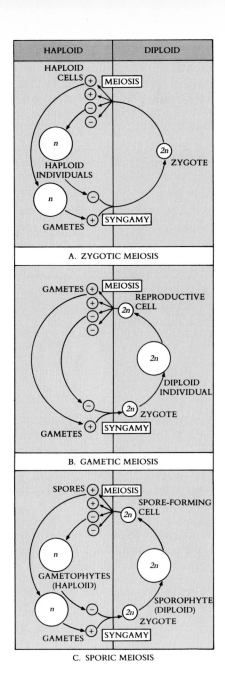

HAPLOID	DIPLOID

HAPLOID CELLS ⊕ MEIOSIS
⊕
⊖
⊖
n
HAPLOID INDIVIDUALS
2n ZYGOTE
n
⊖
GAMETES ⊕ SYNGAMY

A. ZYGOTIC MEIOSIS

GAMETES ⊕ MEIOSIS
⊕ 2n REPRODUCTIVE CELL
⊖
⊖
2n
DIPLOID INDIVIDUAL
⊖ 2n ZYGOTE
GAMETES ⊕ SYNGAMY

B. GAMETIC MEIOSIS

SPORES ⊕ MEIOSIS
⊕ 2n SPORE-FORMING CELL
⊖
⊖
n
GAMETOPHYTES (HAPLOID)
2n
SPOROPHYTE (DIPLOID)
n ⊖ 2n ZYGOTE
GAMETES ⊕ SYNGAMY

C. SPORIC MEIOSIS

FIGURE 24-14

Diagram of the three major kinds of life cycles in eukaryotes. Top to bottom: **(A)** zygotic meiosis, **(B)** gametic meiosis, and **(C)** sporic meiosis.

gametes are the only haploid cells. They exhibit **gametic meiosis;** meiosis produces the gametes, which fuse, giving rise to a zygote. In plants there is a regular alternation between a multicellular haploid phase and a multicellular diploid phase. The diploid phase produces spores that give rise to the haploid phase, and the haploid phase produces gametes that fuse to give the zygote. The zygote is the first cell of the multicellular diploid phase. This kind of life cycle is characterized by **alternation of generations** and has **sporic meiosis.** These three major types of eukaryotic life cycles are diagrammed in Figure 24–14.

> **In a life cycle characterized by zygotic meiosis, the zygote is the only diploid cell; in one characterized by gametic meiosis, the gametes are the only haploid cells. In plants there is an alternation of generations, in which both diploid and haploid cells divide by mitosis; their type of life cycle is characterized by sporic meiosis.**

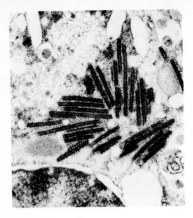

FIGURE 24-15

A kind of virus called GD7 replicating itself in a human white blood cell, or macrophage. Viruses are nonliving segments of DNA or RNA, ultimately derived from the genomes of either bacteria or eukaryotes; they have the ability to take over the protein-synthesizing machinery of their host cells. Many viruses manufacture protein coats around their nucleic acid cores.

ORGANISMS AND EVOLUTION

All of the groups we have discussed so far are clearly organisms, but there is another group that lies on the borderline between living and nonliving: the viruses. Viruses present a special classification problem. They are nonliving, but capable of replication. They are basically fragments of nucleic acids probably derived from ancient cells. They occur in both eukaryotes and bacteria. Most of them have the capacity to organize protein coats around themselves (Figure 24-15). Viruses are able to direct the machinery of their host cells to manufacture more virus material, but they cannot exist on their own. They cannot logically be placed in any of the kingdoms, since they are not organisms. Viruses are discussed in Chapter 25.

The remaining chapters of this section will present a detailed discussion of the diversity of the major groups of organisms, stressing their evolutionary relationships with one another. Multicellularity and sexuality, which we have just discussed, are major themes that underlie the evolutionary advances that we will examine. The colonization of the land, which was carried out by multicellular organisms complex enough to function in this harsh habitat, is perhaps the feature that contributed most to the definition of the major groups of organisms that we see today.

SUMMARY

1. Carl Linnaeus, an eighteenth-century Swedish scientist, developed the system of naming organisms that is still used today. In it, every species of organism has a binomial name, which consists of two words, the first of which is the name of the genus and the second of which, when combined with the name of the genus, designates the species.

2. In the hierarchical system of classification used to describe organisms, genera are grouped into families, families into orders, orders into classes, classes into phyla (called divisions in some groups), and phyla into kingdoms.

3. There are at least 5 million species of plants, animals, and microorganisms in the world. We have named only a third of them at most thus far, and only about 500,000 of the estimated 3.4 million or more species in the tropics.

4. The fundamental distinction among living organisms is that which separates the bacteria (kingdom Monera), which have a prokaryotic form of cellular organization, from the eukaryotes.

5. Most phyla of eukaryotes are placed in the diverse kingdom Protista. Among the phyla of Protista, virtually all have at least some multicellular representatives, and three—red algae (Rhodophyta), brown algae (Phaeophyta), and green algae (Chlorophyta)—are exclusively or largely multicellular.

6. Three groups of multicellular eukaryotes are so large and differ so sharply from their ancestors among the Protista that each is classified as a different kingdom. These are the animals (Animalia), plants (Plantae), and fungi (Fungi).

7. Viruses are not organisms and are not included in the classification of organisms. They are portions of the genomes of organisms.

8. True multicellularity and sexuality are both exclusively properties of the eukaryotes. Multicellularity confers a degree of protection from the environment and the ability to carry out a wider range of activities than is available to unicellular organisms. Sexuality permits extensive, orderly, genetic reproduction.

REVIEW

1. _____ _____ devised the binomial system of nomenclature.

2. About how many species of organisms have been named? About how many organisms are estimated to live on the earth?

3. In what way are eukaryotes less diverse than prokaryotes?

4. The three kingdoms that consists exclusively or nearly so of multicellular organisms have evolved from different members of which eukaryotic kingdom?

5. _____ present a special classification problem because they are nonliving, but capable of replication.

SELF-QUIZ

1. List the taxonomic categories from most inclusive to least inclusive.
   (a) order
   (b) kingdom
   (c) species
   (d) phylum or division
   (e) family
   (f) genus
   (g) class

2. To which (A) kingdom, (B) phylum, and (C) order do you belong?
   (a) Mammalia
   (b) Arthropoda
   (c) Anthophyta
   (d) Animalia
   (e) Chordata

3. Which kingdom contains only unicellular organisms?
   (a) Monera
   (b) Protista
   (c) Fungi
   (d) Plantae
   (e) Animalia

4. Which three of the following are characteristics shared by all eukaryotes?
   (a) their DNA is not associated with protein
   (b) they have microtubules in their cytoplasm
   (c) they possess membrane-bound organelles in their cytoplasm
   (d) they have a nuclear envelope
   (e) the cell walls of most of them contain muramic acid

5. Which of the following is not included in any kingdom?
   (a) viruses
   (b) bacteria
   (c) algae
   (d) oak trees
   (e) human beings

PROBLEM

1. Taxonomists think that our planet has at least 5 million species and that about two thirds of these occur in the tropics. No more than 500,000 tropical species have been named, so what is the minimum number of tropical species that are still to be named?

FOR FURTHER READING

MARGULIS, L., and K. V. SCHWARTZ: *Five Kingdoms: An Illustrated Guide to the Phyla of Life on Earth,* W.H. Freeman & Company, San Francisco, 1982. A marvelous account of the diversity of organisms, beautifully illustrated and presented in a logical fashion throughout.

MAYR, E.: "Biological Classification: Toward a Synthesis of Opposing Methodologies," *Science,* vol. 214, pages 510–516, 1981. A thoughtful consideration of alternative methods of classifying, with emphasis on genera and species.

ROSS, H.H.: *Biological Systematics,* Addison–Wesley Publishing Co., Inc., Reading, Mass., 1974. A nicely balanced overview of taxonomy.

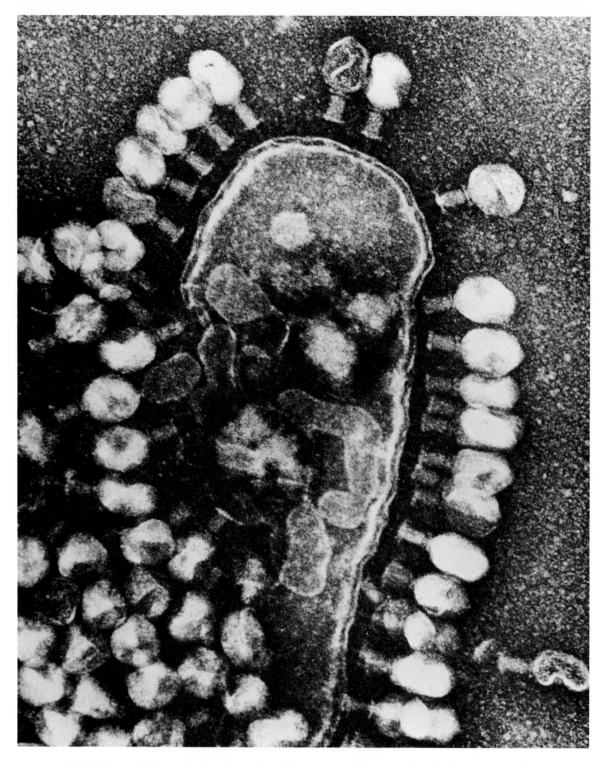

FIGURE 25-1   T4 bacteriophage (virus) particles surround the end of a cell of a common colon bacterium, *Escherichia coli*. The structure of the virus particles is complex. Each particle injects its DNA, a long, coiled molecule within the viral coat, into the bacterium, where the viral DNA directs the production of more virus material by using the machinery of the bacterial cell. Viruses are fragments of bacterial or eukaryotic genomes that have the ability to spread from host to host and to replicate themselves.

THE INVISIBLE WORLD: BACTERIA AND VIRUSES

Chapter 25

Overview

Invisible to us because they are so small, bacteria and viruses are known to everyone as agents of disease. The role of bacteria is, however, much broader: our life on earth would not be possible without them, since they perform many of the essential functions of ecosystems. Bacteria are the simplest, most ancient, and most abundant of organisms. A virus is a fragment of a nucleic acid, either DNA or RNA, that is able to reproduce within the cells of organisms. Viruses are not alive and are not organisms, but because of their disease-producing potential they are very important biological entities.

For Review

Here are some important terms and concepts that you will encounter in this chapter. If you are not familiar with them, you should review them before proceeding.

Definition of organism (Chapter 1)

Structure of bacterial cell (Chapter 5)

Bacterial photosynthesis (Chapter 9)

Bacterial cell division (Chapter 10)

Bacterial conjugation (Chapter 15)

Nitrogen cycle (Chapter 22)

Bacteria are the oldest, the structurally simplest, and the most abundant form of life on earth; they are the only organisms with prokaryotic cellular organization (Figure 25-1). Represented in the oldest rocks from which fossils have been obtained—rocks about 3.5 billion years old—bacteria were abundant for well over 2 billion years before eukaryotes appeared in the world. They were largely responsible for creating the properties of the atmosphere and the soil during the long ages in which they were the only form of life on earth. Bacteria are the only members of the kingdom Monera.

Bacteria are exceedingly abundant: in a single gram of fertile agricultural soil there may be 2.5 billion bacteria. In one hectare of wheat land in England, the weight of bacteria in the soil is approximately equal to that of 100 sheep! It is not surprising, then, that bacteria play a very important part in the web of life on earth. From a human perspective, they are both very useful and very worrisome. On the one hand, bacteria play key roles in the cycling of minerals within our ecosystems, while on the other hand bacteria are responsible for many of the most deadly of human diseases and the most devastating diseases of plants and other animals. Our constant companions, bacteria are present in everything we eat and on everything we touch. No other organisms except plants have as great an impact on human life, for better and for worse.

Although bacteria are the smallest and simplest organisms, they are not the smallest or simplest agents of disease. Viruses, which are segments of nucleic acid, are also often serious agents of disease. Like bacteria, viruses have had a major impact on human history, and still do. AIDS, a viral disease that has become prominent recently, is but one familiar example. In this chapter we consider first bacteria and then viruses (see Figure 25-1).

BACTERIAL STRUCTURE

We discussed bacterial cell structure in Chapters 4 and 5. The most distinctive characteristic of bacteria is that they lack the internal organization that is the hallmark of the eukaryotic cell. Bacterial cells lack both a nucleus and a system of endoplasmic reticulum. Although the interior of bacterial cells may sometimes contain membranes, such interior membranes do not subdivide the cells into separate compartments, as the endoplasmic reticulum does the interior cytoplasm of eukaryotic cells.

Bacterial cell walls consist of a network of polysaccharide molecules (a polymer of sugars) connected by short peptide cross-links (see Figure 5-5). Some species of bacteria are gram positive (see p. 83) with a cell wall about 15 to 80 nanometers thick. In the more common gram-negative bacteria, large molecules of **lipopolysaccharide**—a polysaccharide chain with lipids attached to it—are deposited in a lipid bilayer membrane over a thinner wall. Gram-negative bacteria are resistant to many antibiotics to which gram-positive ones, lacking the lipopolysaccharide layer, are susceptible. Beyond these layers a gelatinous layer, the **capsule,** often surrounds bacterial cells.

Bacteria are mostly simple in form, varying mainly from straight and rod-shaped **(bacilli)** or spherical **(cocci)** to long and spirally coiled **(spirilla)** (see Figure 5-4). Some bacteria change into stalked structures; grow long, branched filaments; or form erect structures that release **spores,** single-celled bodies that grow into new bacteria (Figure 25-2). Among the rod-shaped bacteria, some adhere end-to-end after they have divided, forming chains. Still other bacteria form different kinds of cell groups.

Bacteria, like all other living cells, divide. In the case of the bacteria, the mode of division is **binary fission.** In this process, reviewed in Chapter 10, an individual cell

FIGURE 25-2

Although no bacteria are truly multicellular, some adhere to one another, like *Stigmatella aurantiaca,* one of the gliding bacteria. The rod-shaped individuals move together, forming the composite spore-bearing structures shown here; millions of spores, basically individual bacteria, are eventually released from these structures.

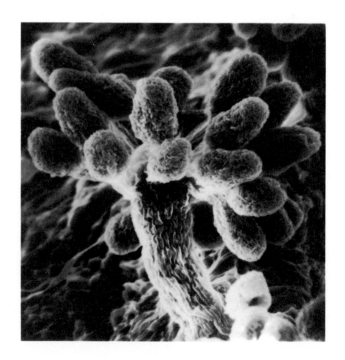

simply increases in size and divides in two. The cell membrane and cell wall grow inward and eventually divide the cell by forming a new wall from the outside toward the center of the old cell (see Figure 10-3).

Bacterial cells have a very simple structure. The cytoplasm contains no internal compartments or organelles and is bounded by a membrane encased within a cell wall composed of one or more layers of polysaccharide.

BACTERIAL ECOLOGY AND METABOLIC DIVERSITY

Bacteria occur in the widest possible range of habitats and play key ecological roles in virtually all of them. Some thrive in hot springs, for example, where the usual temperatures may range as high as 78° C; others have been recovered living beneath 430 meters of ice in Antarctica. Bacteria are abundant in groundwater, where they were once thought to be absent. Still other bacteria, capable of dividing only under high pressures, exist around deep-sea vents, where the water is at temperatures as high as 360° C. These bacteria use the energy they get from the conversion of hydrogen sulfide to sulfur to power their own growth and reproduction, and all other organisms in the vicinity of the deep-sea vents live directly or indirectly by consuming these bacteria (see Figure 22-21).

The reason that bacteria are able to play such a varied ecological role is their metabolic diversity. Different kinds of bacteria vary extensively from one another in almost all aspects of their metabolism, in marked contrast to eukaryotes, which are relatively uniform. All eukaryotes, for example, follow the same general patterns of respiration, glucose breakdown, photosynthesis, and synthesis of nucleic acids and proteins. Bacteria are far more diverse in their ways of carrying out each of these vital functions. In view of their abundance and metabolic versatility, it is not surprising that bacteria play a key ecological role in virtually all habitats on earth.

Photosynthetic Bacteria

Like plants, the photosynthetic bacteria contain chlorophyll, but it is not held within plastids. The photosynthetic processes carried out by different kinds of bacteria are diverse and reflect the long evolutionary path to green plant photosynthesis. One group of photosynthetic bacteria, the cyanobacteria (Figure 25-3) (also called "blue-green algae"), was discussed in Chapters 4 and 18 in connection with the evolution of life on earth. The plastids of the eukaryotic red algae are almost certainly derived from cyanobacterial ancestors. The structure of cyanobacteria is relatively simple. In contrast to those of the red algae, the plastids of the green algae, and therefore those of the plants, are more similar to bacteria of the genus *Prochloron* (see Figure 5-6). Still other groups of algae, including the dinoflagellates, golden-brown algae, and diatoms, have plastids with different characteristics, suggesting an origin from a different kind of bacterium.

Chemoautotrophic Bacteria

Chemoautotrophic bacteria derive the energy they use for their metabolism from the oxidation of chemical sources of energy such as the reduced gases ammonia (NH_3), methane (CH_4), or hydrogen sulfide (H_2S). Chemoautotrophs do not require sunlight; in the presence of one of the chemicals just mentioned, and with nitrogenous salts, oxygen, and carbon dioxide, they can manufacture all of their own amino acids and proteins. The chemoautotrophs appear to be a very ancient and primitive phylum of bacteria.

Heterotrophic Bacteria

Most bacteria are heterotrophs, obtaining their energy from organic material formed by other organisms. Once organic material is formed, it must be broken down if it is to be used again by organisms. Bacteria and fungi are the principal organisms that occupy

FIGURE 25-3

The cyanobacterium *Anabaena*, in which the individual cells adhere in filaments. The large, clear cell (at the upper right) is a heterocyst, an enlarged, specialized cell in which nitrogen fixation occurs. Cyanobacteria such as *Anabaena* exhibit some of the closest approaches to multicellularity among the bacteria.

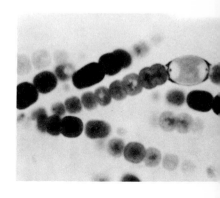

this trophic level, playing the leading role in breaking down the organic molecules that have been formed by biological processes. By doing so, bacteria and fungi make the nutrients in these molecules available once more for recycling. Decomposition is just as indispensable to the continuation of life on earth as is photosynthesis.

Nitrogen-Fixing Bacteria

Another group of bacteria of great ecological importance is the one that includes the nitrogen-fixing bacteria, such as *Rhizobium,* which live in nodules on the roots of legumes. Along with a few other genera that are not associated with legumes and live free in the soil, *Rhizobium,* which is the most important of all nitrogen-fixers, converts atmospheric nitrogen to a form in which it can be used by living organisms as part of the nitrogen cycle (see Figure 22-4). It is a member of a group of bacteria that are **aerobic** (require oxygen for growth), gram negative, and mostly motile by means of flagella.

BACTERIA AS PATHOGENS

Many costly diseases of plants are associated with particular bacteria; almost all kinds of plants are susceptible to one or more kinds of bacterial disease. The symptoms of these plant diseases vary, but they are commonly manifested as spots of various sizes on the stems, flowers, or fruits, or by wilting or local rotting. A familiar example of a bacterial disease is citrus canker, which broke out in Florida in August, 1984, and led to the destruction of more than 4 million citrus seedlings in 4 months in an effort to halt its spread. The effects of this disease, combined with recent freezes, are threatening the future of the $2.5 billion citrus industry in Florida.

Many human diseases are also caused by bacteria, including typhoid fever, dysentery, plague, cholera, typhus, leprosy, tetanus, bacterial pneumonia, whooping cough, and diphtheria. Enormous sums are spent annually in the effort to reduce the likelihood of these infections and their devastating effects.

Several genera of **pathogenic** (disease-causing) **bacteria** are of particular importance for humans. For example, members of the genus *Streptococcus* (see Figure 5-4, *B*) are associated with strep throat, scarlet fever, rheumatic fever, and other infections. Interestingly, the scarlet fever bacterium produces its characteristic and deadly **toxin** (poison) only if it is infected with the appropriate bacterial virus. Tuberculosis, another bacterial disease, is still a leading cause of death in humans. These diseases are mostly spread through the air.

An important air-borne microbe is the pathogenic genus *Staphylococcus,* which causes widespread hospital infections. Another disease associated with bacteria of this genus is toxic shock syndrome, which is characterized by fever, lowered blood pressure, vomiting, diarrhea, and a rash in which the skin peels. It is caused by certain strains of *S. aureus.* About 85% of the cases of toxic shock syndrome reported in the United States have occurred in menstruating women who are using tampons at the time the disease occurs, but both men and women can contract the disease. Superabsorbent tampons of a kind that are no longer manufactured contained fibers that provide an environment that enhances the production of the disease-causing toxin by the bacterium.

One of the more recently detected bacterial diseases of humans is legionellosis (Legionnaires' disease), which is believed to affect about 125,000 people in the United States annually. The disease develops into a severe form of pneumonia, which proves fatal to about 15% to 20% of its victims if it is untreated. Discovered in 1976, when a mysterious lung ailment proved fatal to 35 American Legion members attending a convention in Philadelphia, legionellosis is caused by small, flagellated, rod-shaped bacteria with pointed ends that have been given the name *Legionella* (Figure 25-4). These bacteria are common in water, preferring warm water at about 40° C to 50° C. In the human body, they destroy the monocytes, a type of white blood cell that normally plays a major defensive role against most microorganisms. The bacteria are gram negative and can be destroyed by treatment with the antibiotic erythromycin.

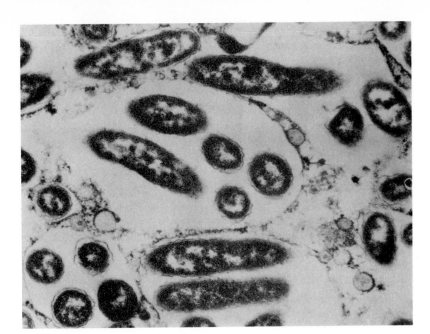

FIGURE 25-4

The dark, oval objects in this photograph are bacteria of the genus *Legionella,* the group of bacteria that causes the disease legionellosis.

A number of important bacterial diseases are sexually transmitted; they are called **venereal diseases.** Among the most common of these diseases are gonorrhea, caused by the bacterium *Neisseria gonorrhoeae,* and syphilis, caused by *Treponema pallidum,* a corkscrew-shaped bacterium of the group called the spirochetes. Originally both of these diseases were easily controlled with antibiotics such as penicillin, but the appearance of penicillin-resistant strains of gonorrhea makes the treatment of this disease much more difficult than formerly. Gonorrhea is much more common and less serious than syphilis, which can be fatal; gonorrhea has infected about 500 people per 100,000 in the United States each year during the 1980s, syphilis fewer than 20. Gonorrhea can be detected in men easily, because of the discharge of pus from the penis and burning sensations during urination; in women it is more difficult to detect, producing at most very mild symptoms. However, if untreated it can lead to infection and inflammation of the oviducts, which can cause their blockage and consequently sterility. Syphilis generally produces a hard, painful ulcer called a **chancre** within 3 weeks of infection, which soon disappears. From 2 to 4 months later, there is a generalized skin rash, following which the disease may become inactive or may ultimately produce damage to the nervous system or circulatory system. In syphilis, as in gonorrhea, antibiotic resistance has recently become a serious problem.

More common than either syphilis or gonorrhea are **chlamydial infections** caused by the bacterium *Chlamydia trachomatis.* These infections, which are usually relatively mild, are controllable with the antibiotic tetracycline. If they are left untreated, however, they can cause serious complications also.

One human bacterial disease—dental caries, or cavities—affects almost everyone. The disease arises in the film on our teeth, which is called **dental plaque;** this film consists largely of bacterial cells surrounded by a polysaccharide matrix. Most of the bacteria are filamentous cells, classified as *Leptotrichia buccalis,* which extend out perpendicular to the surface of the tooth, but many other bacterial species are present in plaque also. Tooth decay is caused by the bacteria that are present in the plaque, which persists especially in places that are difficult to reach with a toothbrush. Diets that are high in sugars are especially harmful to the teeth, because lactic acid bacteria (especially *Streptococcus sanguis* and *S. mutans*) ferment the sugars to lactic acid, a substance that causes the local loss of calcium from the teeth. Once the hard tissue has started to break down, the lysis of proteins in the matrix of the tooth enamel starts, and tooth decay begins in earnest. Fluoride makes the teeth more resistant to decay because it retards the loss of calcium. Since tooth decay is an infectious disease caused by bacteria, it can in principle be controlled by antibiotics; a number of studies are now under way that bear on this problem.

> **Bacteria are important disease-causing organisms in plants and animals, including humans. Among the many human diseases that they cause are scarlet fever, rheumatic fever, tuberculosis, typhoid, and legionellosis.**

VIRUSES

The simplest organisms living on earth today are bacteria, and we think that they closely resemble the first living organisms to evolve on earth. Even simpler than the bacteria, however, are the viruses. The earliest indirect observations of viruses, other than simply observations of their effects, were made near the end of the nineteenth century. At that time several groups of European scientists, working independently, concluded that the infectious agents associated with a plant disease known as tobacco mosaic and those associated with hoof-and-mouth disease in cattle were not bacteria. They reached this conclusion because these infectious units were not filtered out of solutions by the kinds of fine-pored porcelain filters that were routinely used to remove bacteria from various media. Investigating the properties of the filtered material, scientists found not only that viruses were much smaller than any known bacteria but also that viruses could reproduce themselves only within living host cells and therefore lacked some of the critical machinery by which living cells are able to reproduce themselves.

ARE VIRUSES ALIVE?

For many years after their discovery, viruses were regarded as very primitive forms of life, perhaps the ancestors of bacteria. We now know this view to be incorrect. The true nature of viruses first became evident in 1933 when Wendell Stanley, then of the Rockefeller Institute, prepared an extract of tobacco mosaic virus and purified it. Surprisingly, the purified virus precipitated (separated out of solution) in the form of crys-

FIGURE 25-5

Tobacco mosaic virus.
A Virus particles in a crystalline array.
B Diagram of tobacco mosaic virus structure. Experiments on this virus first began to reveal the chemical nature of viruses.

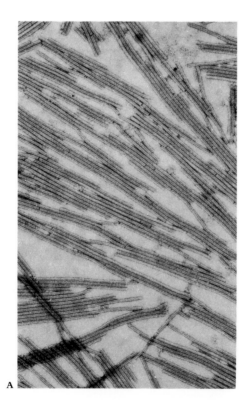

RNA COIL

PROTEIN COAT

A

B

THE ERADICATION OF SMALLPOX

One of the greatest triumphs of modern medicine has been the eradication of smallpox everywhere in the world. This dread disease killed millions of people, and when first introduced to new populations, as in the Spanish conquest of Mexico, it sometimes eliminated half or more of the total population. Fortunately there are no known hosts of the smallpox virus other than human beings, so when all of the people who were susceptible in all of the areas where smallpox still occurred were inoculated in the late 1970s, the disease was eliminated completely.

Officials of the World Health Organization, established in 1948 as a specialized agency of the United Nations, noted that smallpox had already been eliminated in North America and Europe. By 1959 the disease had been eliminated throughout the Western Hemisphere, except for five South American countries, and an intensive worldwide campaign was initiated. As late as 1967, smallpox was still endemic in 33 countries, and the campaign appeared to be faltering. Countries throughout the world, including the Soviet Union and the United States, manufactured and donated large quantities of vaccine, and improved methods of vaccination were developed.

Although reporting was poor, there were probably 10 to 15 million cases of smallpox worldwide in 1967. Attention was focused on areas in which cases had actually occurred. The last case of smallpox on the Indian subcontinent was contracted by a 3-year-old girl on October 16, 1975. By 1978, no more cases were reported in Somalia, the last country of the world in which the scourge of smallpox persisted, and none have been reported since, anywhere in the world (Figure 25-A).

FIGURE 25-A

Ali Maow Maalin, of Merka, Somalia, contracted the last known case of smallpox reported anywhere in the world in 1977. At the time, he was 23 years old.

---

tals (Figure 25-5, *A*). Stanley was able to show by this method that viruses can better be regarded as chemical matter than as living organisms, at least in any normal sense of the word "living." The purified crystals still retained the ability to reinfect healthy tobacco plants and so clearly *were* the virus itself, not merely a chemical derived from it. With Stanley's experiments, scientists began to understand the nature of viruses for the first time.

Within a few years other scientists were able to follow up on Stanley's discovery and demonstrate that tobacco mosaic virus consisted simply of an RNA molecule surrounded by a coat of protein molecules (Figure 25-5, *B*; also see Figure 13-8). Many plant viruses have a similar composition, but most other viruses have DNA in place of RNA. Nearly all viruses form a protein sheath, or **capsid,** around their nucleic acid core. In addition, many viruses form an **envelope,** rich in proteins, lipids, and glycoprotein molecules, around the capsid. The only exceptions are **viroids,** small, naked RNA molecules that infect some plant cells.

Most viruses form a protein sheath, or capsid, around their nucleic acid core, and a lipid-rich protein envelope around the capsid.

No viruses have the ability to grow or replicate on their own. Viruses reproduce only when they enter cells and utilize the cellular machinery of their hosts (see Figure 25-1). They are able to reproduce themselves in this way because they carry genes that are translated into proteins by the cell's genetic machinery, leading to the production of more viruses. Outside of its host cell, a virus is simply a fragment of nucleic acid en-

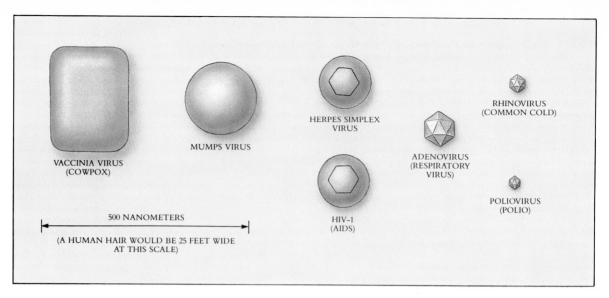

FIGURE 25-6

The diversity of viruses. At the same scale, a human hair would be more than 8 meters thick.

cased in protein, a helpless portion of a cellular genome. The fact that individual kinds of viruses contain only a single type of nucleic acid, either DNA or RNA, is one of the major reasons that they can reproduce only within living cells. All true organisms contain both DNA and RNA, and both are essential components of their genetic machinery. Viruses also lack ribosomes, as well as all of the enzymes necessary for protein synthesis and energy production.

In light of their simple chemical nature and total inability to exist independently from other organisms, earlier theories that viruses represent a kind of halfway house between life and nonlife have now largely been abandoned. Instead, viruses are now viewed as fragments of the genomes of organisms; they could not have existed independently of preexisting organisms.

As an analogy to a virus, consider a computer whose operation is directed by a set of instructions in a program, just as a cell is directed by DNA-encoded instructions. A new program can be introduced into the computer that will cause the computer to cease what it is doing and instead devote all of its energies to making copies of the introduced program. The new program is not itself a computer, however, and cannot make copies of itself when outside the computer, lying on the desk. The introduced program, like a virus, is simply a set of instructions.

Viruses are fragments of DNA or RNA that have become detached from the genomes of bacteria or eukaryotes and have the ability to replicate themselves within cells. Viruses are nonliving and are not organisms. Their genetic material consists of DNA or RNA, but not both.

THE DIVERSITY OF VIRUSES

Viruses occur in virtually every kind of organism that has been investigated for their presence. Viruses are almost always highly specific in the hosts they infect and do not reproduce anywhere else. In light of this observation, there should be nearly as many kinds of viruses as there are kinds of organisms—perhaps millions of them. Since there often is more than one kind of virus in a given organism, the actual number of kinds of viruses might even be much greater.

The smallest viruses are only about 17 nanometers in diameter, the largest ones up to 1000 nanometers (1 micrometer) in their greatest dimension; they vary greatly in appearance (Figure 25-6). The largest viruses, therefore, are just visible with a light microscope. Most viruses can be detected only by using the higher resolution of an electron microscope. Viruses are directly comparable with molecules in size, a hydrogen atom being about 0.1 nanometer in diameter and a large protein molecule being several hundred nanometers in its greatest dimension.

VIRUSES AND DISEASE

Depending on which genes a virus genomic fragment carries, it can often seriously disrupt the normal functioning of the cells that it infects. For thousands of years, diseases caused by viruses have been known and feared. Among them are smallpox, chickenpox, measles, German measles (rubella), mumps, influenza, colds, infectious hepatitis, yellow fever, polio, rabies, and AIDS, as well as many other diseases not as well known. One series of viral diseases that has been discussed a great deal in recent years is herpes, which includes one category of viruses (herpesvirus 1) associated with cold sores and fever blisters, another (herpesvirus 2) that causes venereal disease, a third (herpesvirus 3) that is responsible for chickenpox and sometimes shingles, and many others.

THE INFECTION CYCLE OF AN AIDS VIRUS

Viruses infect bacteria, plants, animals, and probably all other organisms. As an example of how a viral infection proceeds, we shall examine how the AIDS virus, discussed in Chapter 1, infects humans. Most other virus infections follow a similar course, although the details of entry and replication differ in individual cases. The infection cycle of the AIDS virus is outlined in Figure 25-7.

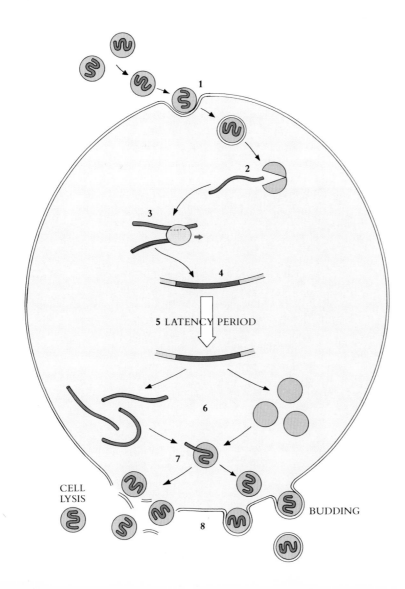

FIGURE 25-7

The infection cycle of an AIDS virus.
1   The HIV virus responsible for AIDS attaches to a CD4 receptor protein on the surface of a T4 lymphocyte cell, and enters the cell by endocytosis.
2   The viral RNA is released into the cell's cytoplasm.
3   A DNA copy is made of the virus RNA.
4   The DNA copy enters the cell's chromosomal DNA.
5   After a long period (typically 5 years), the virus genes initiate active transcription.
6   Both HIV RNA and HIV proteins are made.
7   Complete HIV virus particles are assembled.
8   Some cells are lysed, releasing free HIV, while others bud out HIV by exocytosis.

Attachment

When an AIDS virus is introduced into the human bloodstream, almost always sexually or through a contaminated hypodermic needle, the virus particle circulates throughout the body but does not infect most of the cells that it encounters.

When the AIDS virus encounters a certain kind of white blood cell called a T4 cell, however, it begins its assault by infecting these cells. Most other animal viruses are similarly narrow in their requirements. Polio goes only to certain spinal nerve cells, hepatitis to the liver, and rabies to the brain. How does a virus like AIDS recognize a specific kind of target cell like a T4 cell? Recall from Chapter 6 that every kind of cell in your body has a specific array of cell surface receptors, which are designed to bind particular hormones and growth factors. Cells also possess one or more kinds of cell surface markers that they use to identify themselves to other similar cells. The AIDS virus recognizes T4 cells because each AIDS particle possesses a glycoprotein on its surface that precisely fits a protein called CD4 on the T4 cell surface. Other viruses possess other glycoproteins that key in on other cell types. Because most human cells lack the CD4 surface marker, they are immune to AIDS infection. When the AIDS virus encounters a T4 cell, however, it is able to attach to the T4 cell surface, as these cells contain the proper CD4 marker on their cell surfaces.

Entry

After docking with a T4 cell, the AIDS virus penetrates the cell membrane. Like other animal viruses, the AIDS virus enters the cell by endocytosis, with the cell membrane folding inward to form a deep cavity around the virus particle. Plant viruses normally enter the cells of their hosts at points of injury, whereas bacterial viruses (called **bacteriophages,** or "bacteria eaters"; see Figure 25-1) shed their coats outside their host cell, injecting their nucleic acids through the host cell wall.

Replication

Once within the host cell, an AIDS virus particle sheds its protective coat. This leaves a single strand of virus RNA floating in the cytoplasm, along with a virus enzyme that was also within the virus shell. This enzyme, called **reverse transcriptase,** synthesizes a double strand of DNA complementary to the virus RNA. This double-stranded DNA

FIGURE 25-8

AIDS viruses released from infected T4 cells, blue in this photograph, soon spread over neighboring T4 cells, infecting them in turn. The individual AIDS particles are very small; more than 200 million would fit on the period at the end of this sentence.

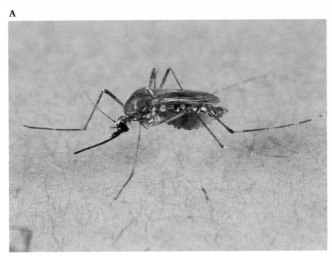

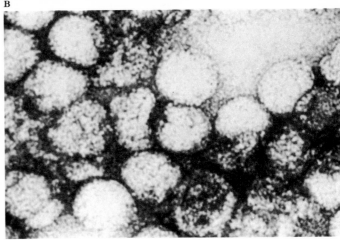

FIGURE 25-9

A *Aedes* mosquito.
B The virus that causes yellow fever. The building of the Panama Canal was possible only after the eradication of this mosquito, which has been a serious scourge of humankind for as long as we have records. The reservoir of the virus is in monkeys, from which the mosquitoes spread it to human beings. Many viruses, called arboviruses, are spread from host to host by arthropods (insects, mites, ticks, and their relatives).

then inserts itself into the chromosomes of the host T4 cell, where one of two things then happens: either the copy remains quiet ("latent"), its genes not being transcribed into virus proteins, or the copy becomes active, taking over part of the host cell machinery and directing it to produce many copies of the virus. In the latter case, the cell eventually dies, releasing thousands of new virus particles, which infect other T4 cells (Figure 25-8).

> **Some viruses have the ability to incorporate themselves into the chromosomes of their hosts; in this form, they are said to be latent if they do not begin replicating themselves immediately.**

The conditions that determine when a latent AIDS virus will become active are not well known. Infections by other microbes seem to trigger AIDS outbreaks, perhaps because the infected T4 cells are involved in the immune response. Many latent human viruses are known to be activated by external stimuli, such as ultraviolet radiation and some chemicals. This is precisely what happens when a fever blister develops on your lip because of the activation of a latent herpes virus, or when a cell bearing a latent retrovirus is suddenly converted into a rapidly dividing cancerous cell by a carcinogen, a process described in Chapter 15.

One of the most important viruses in human history has been the yellow fever virus (Figure 25-9). This virus is one of the arboviruses, or viruses that are spread by arthropods—*Aedes* mosquitoes in the case of yellow fever. In their modes of dispersal, as in their manner of infection, viruses are incredibly versatile. As shown by the recent history of AIDS, and by our increasing understanding of the role of viruses and virus-like particles in causing cancer (see Chapter 15), it is evident that many new viral diseases will appear among the increasingly crowded populations of human beings and other animals and plants in the future. The very best efforts of biological scientists will be required to deal with them effectively.

SIMPLE BUT VERSATILE ORGANISMS

For at least 2 billion years—probably more than half of the history of life on earth—bacteria were the only organisms in existence. They have survived to the present in rich diversity, however, by exploiting an amazingly diverse set of habitats, some of them unchanged since the beginnings of the evolution of the world as we know it. Ecologically, their metabolic activities are of fundamental importance. In a broader sense, bacteria also contribute directly to the functioning of all but a few eukaryotes, since they are the ancestors of mitochondria. Considering the high probability that all chloroplasts are symbiotic bacteria, it may be said that even photosynthesis, like chemosynthesis and nitrogen fixation, is exclusively a property of the bacteria.

Although bacteria are clearly very simple living organisms, viruses are best thought of as segments of genomes. We have learned a great deal about the structure of viruses, but there is doubtless much more to be discovered, probably including the existence of kinds of particles that are unsuspected at present. The nature of viruses suggests that new forms are evolving constantly. While this process continues, we clearly have a great deal to learn about the existing viruses.

SUMMARY

1. Bacteria are the only organisms with prokaryotic cellular organization. They are the oldest and simplest organisms but are metabolically much more diverse than all of the other forms of life on earth combined.

2. Bacteria have cell walls consisting of a network of polysaccharide molecules connected by short peptide cross-links. In the gram-negative bacteria, large molecules of lipopolysaccharide are deposited over this layer.

3. Bacteria are rod-shaped (bacilli), spherical (cocci), or spiral (spirilla). Bacilli or cocci may adhere in small groups or chains.

4. Bacteria and fungi break down organic compounds and make the substances in them available for other living organisms once again. Different groups of bacteria fix the energy from the sun in different ways, and some bacterial groups use chemical processes as a source of energy also.

5. Only bacteria are capable of fixing atmospheric nitrogen, thus making it available for their own metabolic activities and those of other organisms. Some of these bacteria live in nodules on the roots of plants, whereas others are free-living, mainly in the soil.

6. The most frequent form of reproduction in bacteria is by binary fission.

7. Viruses are fragments of bacterial or eukaryotic genomes that are able to replicate within cells by using the genetic machinery of those cells. They are not alive and are not organisms.

8. The cells to which a virus will attach are determined by proteins that make up the coats and envelopes of the virus.

REVIEW

1. Are viruses alive?

2. Chemoautotrophic bacteria derive their energy from the oxidation of chemicals such as ammonia, methane, or hydrogen sulfide. Do they need light energy for this process?

3. *Treponema pallidum,* the cause of syphilis, belongs to what phylum of bacteria?

4. Can viruses replicate outside of a cell?

5. To what kind of human cells does the AIDS virus attach?

SELF-QUIZ

1. The chloroplasts of green plants are similar to, and possibly derived from, bacteria of the genus
   (a) *Prochloron*
   (b) *Pseudomonas*
   (c) *Heliobacterium*
   (d) *Nitrosomonas*
   (e) *Streptococcus*

2. Which three of the following human diseases are caused by bacteria?
   (a) tetanus
   (b) AIDS
   (c) toxic shock syndrome
   (d) cholera
   (e) herpes

3. Polio viruses attack nerve cells, hepatitis attacks the liver, and AIDS attacks white blood cells. How does each virus know which cell to attack?
   (a) They enter all cells, but only certain cell types are vulnerable to each kind of virus.
   (b) They key in on certain cell surface molecules characteristic of each cell type.
   (c) The virus in each case is derived from that kind of cell.
   (d) The cell types practice virus-specific phagocytosis.
   (e) None of the above.

4. The genetic information of viruses is encoded in a molecule of
   (a) RNA
   (b) DNA
   (c) proteins
   (d) either RNA or DNA, but not both
   (e) either DNA or proteins, but not both

5. Most viruses are very specific about the cells that they infect. What kind of cells does the rabies virus infect?
   (a) T4 cells
   (b) liver cells
   (c) red blood cells
   (d) brain cells
   (e) spinal nerve cells

THOUGHT QUESTIONS

1. Why do we say that bacteria are alive and viruses are not?

2. Did the different groups of viruses arise from one another? Describe the process whereby they did evolve.

FOR FURTHER READING

GOODFIELD, J.: *Quest For the Killers,* Birkhauser Press, Boston, 1985. An illuminating account of how smallpox and four other diseases are being combated in the field by public health scientists.

HENDERSON, D.A.: "The Eradication of Smallpox," *Scientific American,* October 1976, pages 25-33. A description of the campaign that ultimately led to the worldwide elimination of this deadly disease.

HOGLE, J.M., M. CHOW, and D.J. FILMAN: "The Structure of Poliovirus," *Scientific American,* March 1987, pages 42-49. The new understanding of the structure of this virus, explained here, should suggest ways to thwart other viruses.

LAPPÉ, M.: *Germs That Won't Die,* Anchor Press/Doubleday, Garden City, N.Y., 1982. An interesting account of the author's theories about the overuse and misuse of antibiotics, and his suggestions about how they may be made more effective.

WOESE, C.R.: "Archaebacteria," *Scientific American,* June 1981, pages 98-122. Presents arguments that the archaebacteria represent an independent kingdom of organisms.

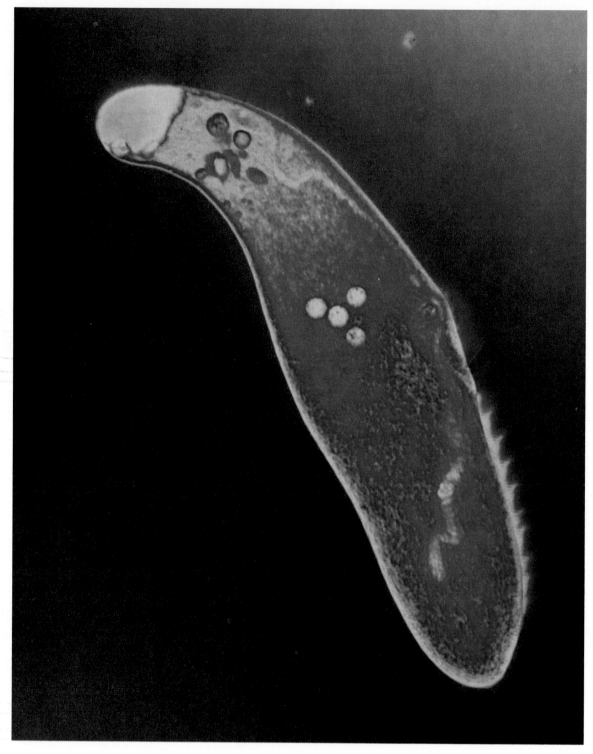

FIGURE 26-1   Although multicellularity has evolved many times among the protists, most protists are microscopic, single-celled organisms. This one is typical in having a complex interior organization. Protists are far more diverse than any other kingdom of life.

THE ORIGINS OF MULTICELLULARITY: PROTISTS AND FUNGI

Chapter 26

Overview

Most kinds of eukaryotes are protists, members of the very diverse kingdom known as Protista. Protists are typically unicellular organisms, but two phyla of algae, the red and brown algae, consist almost entirely of multicellular ones. A third phylum, the green algae, has many multicellular representatives. Multicellularity originated in each of these groups independently. Other protists gave rise to three of the five kingdoms we recognize as distinct: the plants, the animals, and the fungi. Plants manufacture their food, animals ingest it, and fungi absorb it after releasing enzymes into their surroundings. Along with the bacteria, fungi are the major decomposers of the biosphere, breaking down organic molecules and making the material in them available for recycling. This process is as necessary for the continuation of life on earth as photosynthesis.

For Review   *Here are some important terms and concepts that you will encounter in this chapter.*
If you are not familiar with them, you should review them before proceeding.

Symbiotic origin of mitochondria and chloroplasts
(Chapters 5 and 25)

Mitosis (Chapter 10)

Major features of evolutionary history (Chapter 18)

The five kingdoms of life (Chapter 24)

Life cycle patterns (Chapter 24)

The Protista consist of phyla that are primarily unicellular, although every group has at least a few multicellular representatives. Multicellularity probably has evolved independently in each of them. Two of the protist phyla, the red algae (phylum Rhodophyta) and the brown algae (Phaeophyta), consist largely of multicellular organisms and may attain great size (Figure 26-1). Another group, the green algae (Chlorophyta), are of special interest as the ancestors of the plants. Many other green algae are also multicellular (see Figures 24-12, *B,* and 24-13). Most of the protists are unicellular, however, and these members exhibit the greatest diversity.

FIGURE 26-2

The evolutionary radiation of the eukaryotes. Of the five kingdoms of living organisms, four are eukaryotes. Protista are by far the most diverse of these, and individual protists gave rise to the other three kingdoms. The ancestral groups of animals and plants are well established, but no living protist has the characteristics that we believe occurred in the ancestor of the fungi.

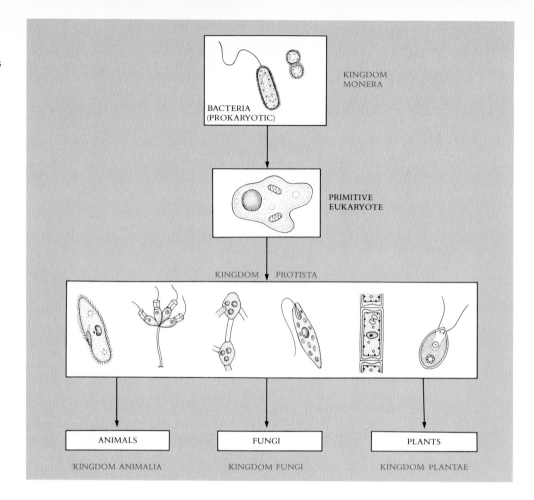

No transitional forms survive between the bacteria and eukaryotes. Because of their simple form, however, we are sure that the first eukaryotes were unicellular organisms similar to some protists. They rapidly differentiated into many distinct evolutionary lines and in turn gave rise to the major groups of eukaryotes—plants, animals, and fungi (Figure 26-2). Later chapters will discuss the plants and animals; this chapter covers the fungi. The fungi are a kingdom of diverse organisms that have a great impact on the functioning of the biosphere, decomposing organic materials and returning them to ecosystems. The fungi and those bacteria that perform similar functions are as indispensable as plants to the continuation of life on earth, playing a critical role in the carbon cycle.

THE PROTISTS

Of the four kingdoms of eukaryotes, the most diverse kingdom is Protista, the protists (Figure 26-3). Among the protists are the simplest eukaryotes, but many single-celled protists have a structure far more complex than the bacteria. Virtually all the diversity of the eukaryotes is found among the protists; in fact, the group includes all the eukaryotes that are not separated as animals, plants, and fungi. Some protists have chloroplasts and manufacture their own food; these are called the algae. Other phyla of protists ingest their food, as do the animals, and still others absorb their food, as do the fungi.

Despite their diversity, grouping all the protists together allows us to compare them directly to emphasize their similarities and differences. In reviewing them, however, keep in mind that the kingdom Protista is a much more diverse group than any of the other three kingdoms of eukaryotic organisms, although metabolically they are far less diverse than the bacteria.

A **B** **C**

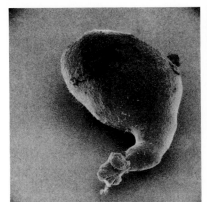  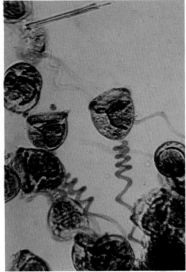

Symbiosis and the Origin of Eukaryotes

To unravel the evolutionary relationships of the protists, one must understand the history of their major symbiotic organelles, mitochondria and chloroplasts. Since these were derived in early protists from different groups of bacteria, the relationships of these organelles are different from the overall relationships of the cells within which they occur. One must keep both kinds of relationships in mind to appreciate the protists properly.

Mitochondria are thought to have originated from bacterial ancestors that had characteristics similar to the purple, nonsulfur bacteria. Mitochondria are characteristic of all living eukaryotes except for several zoomastigotes (see the following section) and *Pelomyxa palustris* (Figure 26-3, *A*), an unusual amoeba-like organism found on the muddy bottoms of freshwater ponds and placed in its own phylum. The existence of a few protists that lack mitochondria probably indicates that the initial stages of this group's evolution took place before mitochondria were acquired. This event must have taken place about the time that eukaryotes originated, approximately 1.5 billion years ago.

> **Mitochondria, which probably originated from symbiotic nonsulfur purple bacteria, are absent in two unusual groups of protists. These may represent a stage of evolution before ancestral eukaryotes acquired these organelles.**

The mitochondria that occur in all but a very few eukaryotes generally possess similar features. On the basis of their characteristics, it is difficult to be sure whether these mitochondria all originated from a single symbiotic event, or whether similar bacteria became symbiotic independently in different groups of early eukaryotes. For chloroplasts, however, the story is very different. As you learned in Chapter 25, three biochemically distinct classes of chloroplasts exist, each resembling a different bacterial ancestor. Thus the chloroplasts of red algae possess chlorophyll *a*, carotenoids, and an unusual class of accessory pigments called phycobilins. These chloroplasts were almost certainly derived from symbiotic cyanobacteria. In contrast, the green chloroplasts of plants and green algae seem to have been derived from bacterial ancestors similar to *Prochloron* (see Figure 5-6). Finally, the chloroplasts of several other groups of algae (brown algae, diatoms, and dinoflagellates), which have chlorophylls *a* and *c*, carotenoids, and distinctive yellowish-brown pigments, probably were derived from another group of bacteria. These relationships are outlined in Figure 24-9, which you should use as a guide to the relationships between the different groups discussed in this chapter.

> **Eukaryotic cells acquired chloroplasts by symbiosis not once, but at least three times during the evolution of different groups of protists.**

FIGURE 26-3

The kingdom Protista predominantly includes a great diversity of single-celled organisms, such as those shown here.
A *Pelomyxa palustris,* a unique, amoebalike protist, which lacks mitochondria and mitosis. *Pelomyxa* may represent a very early stage in the evolution of the eukaryotic cell. This species is classified as the only member of its own phylum, Caryoblastea.
B Still others have more unusual forms of multicellular construction, such as the multinucleate plasmodium of a plasmodial slime mold (phylum Myxomycota). Individual cells cannot be distinguished in such a plasmodium, but they retain their distinctness.
C *Vorticellia* (phylum Ciliophora), which is heterotrophic, feeds largely on bacteria and has a retractable stalk. This is an amazing degree of complexity for a single cell; it goes far beyond the complexity found in any bacterium.

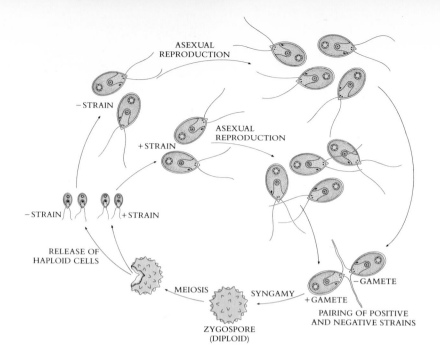

FIGURE 26-4

Life cycle of *Chlamydomonas* (phylum Chlorophyta). Individual cells of this microscopic, biflagellated alga, which are haploid, divide asexually, producing identical copies of themselves. At times, such haploid cells act as gametes, fusing to produce a zygote, as shown at the lower right-hand side of the diagram. The zygote develops a thick, resistant wall, becoming a zygospore. Within this diploid zygospore, meiosis takes place, ultimately resulting in the release of four haploid individuals. Because of the segregation during meiosis, two of these individuals are what are called the + strain, the other two the − strain. Only + and − individuals are capable of mating with one another when syngamy does take place, although both may divide asexually and reproduce themselves in that way also.

MAJOR GROUPS OF PROTISTS
Multicellular Protists

Three phyla of photosynthetic protists consist largely or entirely of multicellular organisms (see Figure 24-12). Among these, the green algae (phylum Chlorophyta) consist of about 7000 species; most are aquatic, but some are semiterrestrial in damp places. They are biochemically similar to the plants, which we suppose were derived from multicellular green algae. The growth form of the marine genus *Ulva* (see Figure 24-12, *B*) is platelike (the individuals have a thickness of two cells), and that of the freshwater genus *Volvox* (see Figure 24-13) is a globular sphere of flagellated cells. Other green algae, however, are filamentous, such as *Spirogyra* (see Figure 24-11, *B*), or unicellular and either flagellated or not. Flagellated unicellular organisms such as *Chlamydomonas* (Figure 26-4) apparently gave rise to colonial and eventually multicellular ones such as *Volvox;* their origin occurred separately from that of other multicellular green algae. *Acetabularia,* which we considered in Chapter 13 as a key experimental organism for the study of development, is a member of this phylum (see Figures 13-2 and 13-3). In *Acetabularia,* the individual cells are very large. Many of the green algae have cell walls composed totally or partly of cellulose.

The other two groups of multicellular, photosynthetic algae are primarily marine. The red algae (phylum Rhodophyta) consist of about 4000 species. Their chloroplasts are similar to those of cyanobacteria. An unusual feature of the red algae, which grow at greater depths in the sea than any other photosynthetic organism, is that they completely lack flagellated cells at any stage of their life cycle. (A representative red alga is shown in Figure 24-12, *C*.) Some members of this phylum, as well as some green and brown algae, are harvested as important sources of food in the Orient. Flagellated reproductive cells and the biochemistry of their chloroplasts distinguish the brown algae (phylum Phaeophyta) from the red algae. Approximately 1500 species of brown algae exist, some of which attain great size (up to more than 100 meters long) and complexity. The kelps (see Figures 22-18; 24-11, *C;* and 24-12, *A*) and rockweeds are one large group of brown algae, often commercially harvested for fertilizer or for their chemicals, such as sodium and potassium salts.

Unicellular Protists

Fewer multicellular representatives occur in the other phyla of protists. Most ciliates (phylum Ciliophora), of which *Paramecium* (see Figures 1-4 and 5-2) is a well-known example, are unicellular organisms with a very complex structure. They move about

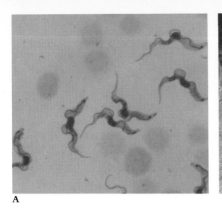

A

B

FIGURE 26-5

A *Trypanosoma,* the zoomastigote that causes sleeping sickness, among red blood cells. The nuclei (dark-staining bodies); the anterior (forward-projecting) flagella; and the undulating, changeable shape of the trypanosomes may be seen in this photograph.
B A tsetse fly, shown here sucking blood from a human arm, in Tanzania, East Africa.

by the beating of fine flagella, which occur in lines of cilia on their cells. *Vorticellia* (see Figure 26-3, *C*) is a particularly complex member of this phylum, which consists of at least 8000 primarily aquatic species. A third genus is *Tetrahymena.* An unusual mode of reproduction known as **conjugation** characterizes the members of the group, whereby nuclei are exchanged between individuals through tubes that connect them during the period of conjugation.

In another group of heterotrophic, primarily unicellular protists, the zoomastigotes (phylum Zoomastigina), the cells move by means of one or more longer flagella. There are thousands of species within this phylum. Many zoomastigotes are parasitic. For example, the serious and widespread diseases trypanosomiasis or "sleeping sickness" (Figure 26-5), East Coast fever, and Chagas' disease, which are especially prevalent in the tropics, are caused by trypanosomes, a particular group of zoomastigotes. The presence of these diseases has kept cattle out of much of Africa, thus complicating the problems of feeding the people.

Trypanosomes also have complicated life cycles. When they are ready to spread to a mammal from the flies that carry them around, trypanosomes acquire a thick coat of glycoprotein antigens that protects them from the host's antibodies. As mentioned in Chapter 15, trypanosomes can change this coat so rapidly that the immune systems of cattle and humans cannot mount a defense against it. A major effort is now being made to develop vaccines against trypanosomes, and some of the prototype vaccines are being tested in the field.

Other groups of zoomastigotes are equally interesting. For example, some that live in the guts of wood roaches and termites possess cellulase enzymes that allow these insects to digest cellulose and thus live on a diet of wood. The insects are unable to manufacture these enzymes themselves, but the presence of the zoomastigotes, which are always in their guts, enables them to live as if they did have this ability. Members of this phylum include the choanoflagellates, which have a characteristic collar at one end of the cell from which the single flagellum protrudes (Figure 26-6). The choanoflagellates are clearly the ancestors of the sponges (see Figure 28-3) and probably the ancestors of all other animals as well. The members of a satellite group of the zoomastigotes, the euglenoids (sometimes considered to be a separate phylum, Euglenophyta), often possess chloroplasts with characteristics resembling those of the plants.

Two other important groups of protists also possess chloroplasts, which are similar in biochemistry to one another and also to the brown algae: the diatoms (phylum Bacillariophyta) and the dinoflagellates (Dinoflagellata). Diatoms, of which there are about 11,500 living species and many others known only as fossils, have a very characteristic "shell" made of two parts, like the halves of a box (Figure 26-7). Chemically, the shell is opaline silica. Huge masses of diatoms have been deposited in certain locations, such as near Lompoc, California, where the strata are hundreds of meters thick. Although only about 1000 species of dinoflagellates exist, they are important as the causal agent of "red tides," during which millions of dinoflagellates, secreting toxic substances, color the sea red and poison fish and shellfish by the thousands. Other dinoflagellates become symbiotic in marine animals such as coral (Figure 26-8). The presence of these dinoflagellates allows the coral colony to produce its own food by photosynthesis. Corals are thus able to flourish in nutrient-poor tropical waters.

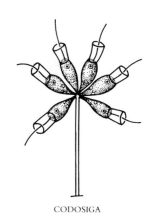

CODOSIGA

FIGURE 26-6

A choanoflagellate, *Codosiga,* a member of the group of zoomastigotes (phylum Zoomastigina) that certainly included the ancestor of sponges, one of the phyla of animals, judged from the unique structure of their cells, which is shared only with that group. The choanoflagellates also probably gave rise to the other animals. This particular choanoflagellate is colonial, living in groups with others of the same species.

CYCLOSPORINE: A MODERN "WONDER DRUG"

Fungi manufacture a remarkable arsenal of chemicals, which they employ in their interactions with the environment. The members of this kingdom display an intimate relationship with their environment, obtaining the nutrients that they require and protecting themselves solely through the materials that they secrete. Furthermore, the fungi have exceedingly diverse metabolisms. In view of these properties, it is not surprising that so many drugs and other useful substances are obtained from fungi or that they play such an important role in various industrial processes.

In 1970, microbiologists working at the Sandoz Corporation in Basel, Switzerland, isolated two new strains of Fungi Imperfecti from soil samples that they had obtained from Norway and Wisconsin. Both of these strains produced a substance that was later named cyclosporine (Figure 26-A), which was first studied because it controlled the growth of some other fungi and was relatively nontoxic for such a metabolically active substance. One of these fungi, *Tolypocladium inflatum,* was selected as the commercial source of cyclosporine because it grows in submerged culture, an important property for industrial production. Following these original discoveries, cyclosporine was found to be a cyclic molecule consisting of 11 amino acids, one of which is not known to occur anywhere else in nature.

Investigators soon discovered that (1) cyclosporine suppressed the type of immunity that causes organ transplants to be rejected and (2) it did not have the harmful side effects associated with the standard drugs used to suppress immune reactions. Cyclosporine causes these effects by inactivating T cells (see Chapter 42). Most importantly, the inactivation ceases when the drug is no longer being administered, so the body can return to normal. The other drugs used previously for this process killed bone marrow cells, which are the source of all blood cells. Loss of bone marrow cells causes intolerable side effects because of the patient's greatly increased susceptibility to disease and infection. Cyclosporine does not damage bone marrow cells.

Before cyclosporine became available in 1979, about one in three patients who had received liver transplants survived for a full year. With the new drug, the percentage of survival has doubled: up to 90% of kidney transplants now succeed, as opposed to no more than half just a few years ago. Heart transplants, which had essentially been abandoned before cyclosporine was available, are now being performed once again. No instances of heart rejection have occurred when the drug has been used. Cyclosporine is also extremely promising for application in diseases that involve autoimmunity. These diseases, which are known or suspected to include multiple sclerosis, juvenile diabetes, and lupus, have been extremely difficult to treat in the past. The drug also shows great potential in treating certain diseases caused by parasites, including schistosomiasis (cyclosporine kills the worms themselves), malaria, and trichinosis. The antiparasitic effects of cyclosporine do not depend on its ability to cause suppression of immune systems. For this reason, scientists are searching for derivative molecules that would be able to kill the parasites without causing immune system suppression, which is not usually desirable.

As research on cyclosporine continues actively and new applications are reported almost monthly, scientists are curious about how many additional "wonder drugs" may be among the remarkable chemical arsenal of the fungi. These organisms apparently have a virtually limitless potential for both harming and helping human beings.

FIGURE 26-A

1 Synthetic crystal of cyclosporine.
2 Conidia-bearing branches of *Tolypocladium inflatum,* ×4100.

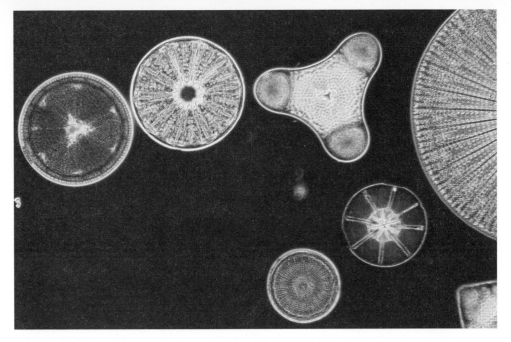

FIGURE 26-7

Several different kinds of diatoms (phylum Bacillariophyta). The diverse shapes of the silica shells of some members of this phylum are shown here.

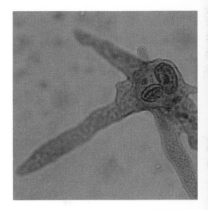

FIGURE 26-8

Free-living dinoflagellates (phylum Dinoflagellata) have two flagella, one lying in a shallow girdle around the center of the cell, which is armored with stiff cellulose plates, and the other directed backward. Many dinoflagellates, however, live symbiotically within marine animals, such as the coral tentacles shown here, where they are the abundant, tiny brownish objects.

The amoebas (phylum Rhizopoda) are heterotrophic protists that are abundant throughout the world in fresh and salt water, as well as in the soil. Many species are parasites of animals. For example, it is estimated that more than 10 million people in the United States are infected by amoebas, about 2 million of whom show symptoms of amebic dysentery. Amoebas lack cell walls, flagella, meiosis, and any form of sexuality but do carry out mitosis and possess mitochondria. They move from place to place by means of their **pseudopods,** from the Greek words for "false" and "feet" (Figure 26-9).

The well-known tropical disease malaria is caused by the members of another phylum of amoeba-like protists, the sporozoans (phylum Sporozoa). Sporozoans are nonmotile, spore-forming parasites of animals. Nearly 4000 species are known. Approximately 250 million people are affected by malaria at any one time, and 2 to 4 million of them die each year. The symptoms include chills, fever, sweating, an enlarged and tender spleen, confusion, and great thirst. Malaria kills most children under 5 years old who contract it. The disease is caused by sporozoans of the genus *Plasmodium,* which are carried by mosquitoes of the genus *Anopheles* (Figure 26-10). Once the *Plasmodium* organisms reach the bloodstream of humans or other mammals, they move to

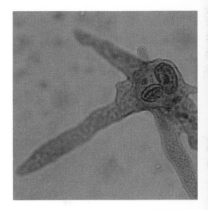

FIGURE 26-9

Amoeba proteus (phylum Rhizopoda). This is a relatively large amoeba that is commonly used in teaching and for research in cell biology. The projections are pseudopods; an amoeba moves simply by flowing into them. The nucleus of the amoeba is plainly visible.

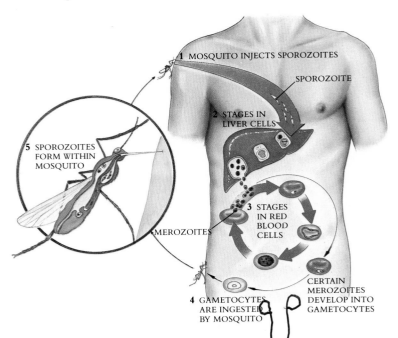

1 MOSQUITO INJECTS SPOROZOITES

SPOROZOITE

2 STAGES IN LIVER CELLS

5 SPOROZOITES FORM WITHIN MOSQUITO

MEROZOITES

3 STAGES IN RED BLOOD CELLS

CERTAIN MEROZOITES DEVELOP INTO GAMETOCYTES

4 GAMETOCYTES ARE INGESTED BY MOSQUITO

FIGURE 26-10

Life cycle of *Plasmodium* (phylum Sporozoa), the sporozoan that causes malaria. The stages in the life cycle are called sporozoites, merozoites, and gametocytes, and they play the roles indicated in this diagram.

FIGURE 26-11

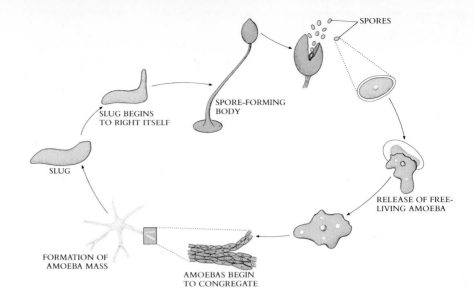

Life cycle of a cellular slime mold (phylum Acrasiomycota). The amoebas emerge from spores, congregate at certain stages of their life cycle, and then differentiate into stalks and spore-forming bodies. Individuals of this group feed on bacteria and other small organisms.

the liver, where they begin to divide. They then pass back into the bloodstream and invade the red blood cells, dividing rapidly within them and causing them to become enlarged and ultimately to rupture. This releases toxic substances throughout the body of the host, bringing about the well-known cycle of fever and chills that is characteristic of malaria. Malaria has proved difficult to control both because the mosquitoes have become resistant to insecticides and because the parasites have developed resistance to the chemicals, such as quinine, that are used to kill them.

Two diverse phyla of protists have historically, and incorrectly, been included among the fungi. These are the so-called plasmodial slime molds (phylum Myxomycota) and cellular slime molds (phylum Acrasiomycota), both of which have ameboid stages in their life cycles (Figure 26-11). There are fewer than 500 species of plasmodial slime molds and only a few dozen species of cellular slime molds.

THE MOST DIVERSE KINGDOM OF EUKARYOTES

You are probably convinced now of the diversity of the protists. The phyla we have considered in this chapter consist mostly of microscopic, unicellular organisms, but many multicellular ones have been derived from them. In their ways of life, the habitats where they occur, and the details of their life cycles, the protists are extraordinarily diverse.

Among the phyla that make up this great group, we are especially interested in those that apparently gave rise to the other exclusively or predominantly multicellular evolutionary lines that we call animals, plants, and fungi. The choanocytes, one of the groups of zoomastigotes (see Figure 26-6), appear certainly to be the ancestors of the sponges and probably all animals: the fundamental similarity between their unique structure and that of the collar cells of sponges (see Chapter 28) is simply too great to be explained in any other way. Green algae, which share with plants the same photosynthetic pigments, cell wall structure, and chief storage product (starch), certainly include the ancestors of plants, a kingdom that achieved its distinctive features when it invaded the land. The ancestors of the fungi are not known, since their features are distinct from those of all living groups of Protista.

THE FUNGI

Fungi are an ancient group of organisms at least 400 million, and possibly 800 million, years old. This distinct kingdom of organisms consists of approximately 100,000 named species, but **mycologists,** the scientists who study them, believe several times that many may exist. Although fungi have traditionally been included in the plant kingdom, they have no chlorophyll and resemble plants only in lacking mobility and _these are great at parties. They're fungis!_

A   **B**   **C**

D   **E**

FIGURE 26-12

Representatives of the two divisions (phyla) of fungi.
A A morel, _Morchella esculenta,_ a delicious edible ascomycete (division Ascomycota) that appears in early spring in the north-temperate woods, especially under oaks.
B _Amanita muscaria,_ the fly agaric, a poisonous basidiomycete (division Basidiomycota).
C A cup fungus (division Ascomycota) in the rain forest of the Amazon Basin.
D A shelf fungus, _Coriolus versicolor_ (division Basidiomycota), growing on a tree trunk. Basidia, the reproductive structures, line the underside of this fungus.
E A puffball (division Basidiomycota). Basidia form within puffballs, which have no external openings; puffballs release their basidiospores by rupturing when mature. All visible structures of fleshy fungi like the ones shown here arise from an extensive network of filaments (**hyphae**) that penetrate and interweave with the substrate on which they grow. (Figure 26-12, _A_ courtesy of Carolina Biological Supply Company.)

growing from the ends of somewhat linear bodies (Figure 26-12). Even these similarities prove to be misleading, however, when fungi are examined closely. Some plants have motile sperm with flagella; plants almost certainly originated from ancestors that had flagella. In contrast, no fungi ever have flagella, and no evidence indicates that their ancestors possessed flagella. Fungi are basically filamentous in their growth form, consisting of slender filaments, whereas plants are three-dimensional. Unlike plants, fungi obtain their food by secreting enzymes out of their bodies and onto or into their substrate. They then absorb into their bodies the materials that these enzymes make available.

Many fungi are harmful because they decay, rot, and spoil many different materials as they obtain food. They also can cause serious diseases in plants and animals, including human beings. Other fungi, however, are extremely useful. The manufacture of both bread and beer depends on the biochemical activities of yeasts, single-celled fungi that produce abundant quantities of ethanol and carbon dioxide. Both cheese and wine achieve their delicate flavors because of the metabolic processes of certain fungi, and other fungi make possible the manufacture of such Oriental delicacies as soy sauce and tofu. Vast industries depend on the biochemical manufacture of organic substances such as citric acid by using fungi in culture, and yeasts are now employed on a large scale to produce protein for the enrichment of animal food. Many antibiotics, including the first one widely used, penicillin, are derived from fungi. Other fungi are used to convert complex organic molecules into other molecules, such as in the synthesis of many commercially important steroids. Because so many are harmful and because so many are beneficial, fungi hold great importance for all of us.

Fungi absorb their food after digesting it externally by secreting enzymes. This unique mode of nutrition, combined with their filamentous growth form and complete lack of flagella, makes the members of this kingdom highly distinctive.

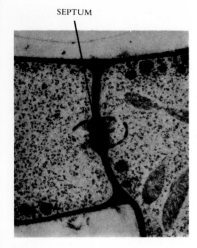

SEPTUM

FIGURE 26-13

Transmission electron micrograph of a section through a hypha of the basidiomycete *Laetisaria arvalis,* showing a septum. Septa of this kind, lined with a thick, barrel-shaped structure, are charactertistic of one phase of the life cycle of basidiomycetes.

FIGURE 26-14

Fungal mycelium growing through leaves on the forest floor in Maryland.

Fungal Ecology

Fungi, along with bacteria, play an essential role as decomposers in the biosphere. They break down organic materials and return the substances locked up in those molecules to circulation in the ecosystem. In this way, critical biological building blocks, such as compounds of carbon, nitrogen, and phosphorus, that have been incorporated into the bodies of living organisms are released and made available for other organisms. For the same reason, fungi often cause serious damage to materials that are used by human beings and are responsible for major human, animal, and plant diseases.

Fungal Structure

Fungi exist mainly as slender filaments, barely visible with the naked eye, which are called **hyphae** (singular, **hypha**). These hyphae may be divided into cells by cross walls called **septa** (singular **septum**). These septa rarely form a complete barrier, however, except for those separating the reproductive cells. Cytoplasm characteristically flows freely throughout the hyphae, passing right through the major pores in the septa (Figure 26-13). Because of this cytoplasmic streaming, proteins, which are synthesized throughout the hyphae, may be carried to the actively growing tips of the hyphae. As a result, the growth of fungal hyphae may be rapid with abundant food and water and a high enough temperature.

A mass of hyphae is called a **mycelium** (plural, **mycelia**). This term, like *mycologist,* is derived from the Greek word for fungus, *myketos.* The mycelia of fungi (Figure 26-14) constitute a system that may be many kilometers long, although concentrated in a much smaller area. This system grows through and penetrates the environment of the fungus, resulting in a unique relationship between a fungus and its environment. All parts of a fungus are metabolically active, continually interacting with the soil, wood, or other material in which the mycelia are growing.

> **Fungi exist primarily in the form of filamentous hyphae, which are completely divided by septa only when reproductive organs are formed. These hyphae surround and penetrate the substrate within which the fungi are growing.**

In two of the three divisions of fungi, structures composed of interwoven hyphae, such as mushrooms, puffballs, and morels, are formed at certain stages of the life cycle. These structures may expand rapidly because of cytoplasmic streaming and growth in their kilometers of hyphae. For this reason, mushrooms can force their way through tennis court surfaces or appear suddenly in your lawn.

The cell walls of fungi are not formed of cellulose as are those of the plants and many groups of protists. Other polysaccharides are typical constituents of fungal cell walls, however, and one of them, chitin, occurs especially frequently. Chitin is the same material that makes up the major portion of the hard shells, or **exoskeletons,** of arthropods, a group of animals that includes insects and crustaceans (see Chapter 28). Chitin is far more resistant to microbial degradation than is cellulose.

Mitosis in fungi differs from that found in other organisms. The nuclear envelope does not break down and re-form, and the spindle apparatus is formed within it. In addition, centrioles are lacking in all fungi. Overall, fungal features suggest that the kingdom originated from some unknown group of single-celled eukaryotes that lacked flagella. Certainly the fungi differ sharply from all other groups of living organisms.

Spores, always nonmotile, constitute a common means of reproduction among the fungi. They may be formed through either asexual or sexual processes. When the spores land in a suitable place, they germinate, giving rise to a new fungal hypha. Since the spores are very small, they may remain suspended in the air for long periods. Because of this property, fungal spores may be blown great distances from their place of origin, a factor explaining the extremely wide distributions of many kinds of fungi. Unfortunately, many fungi that cause diseases of plants and animals are spread rapidly and widely by such means.

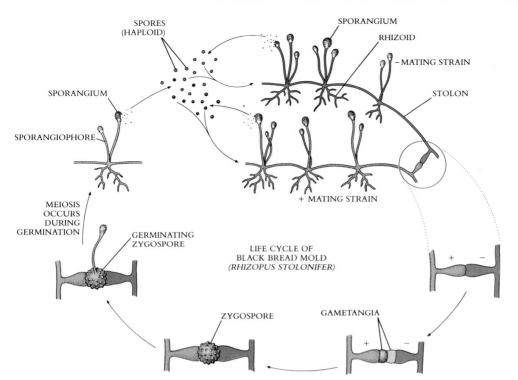

SPORES (HAPLOID)

SPORANGIUM

SPORANGIUM

RHIZOID

SPORANGIOPHORE

– MATING STRAIN

STOLON

MEIOSIS OCCURS DURING GERMINATION

GERMINATING ZYGOSPORE

+ MATING STRAIN

LIFE CYCLE OF BLACK BREAD MOLD (RHIZOPUS STOLONIFER)

ZYGOSPORE

GAMETANGIA

FIGURE 26-15

Life cycle of *Rhizopus,* a zygomycete. The hyphae grow over the surface of the bread or other material on which the fungus feeds, producing erect sporangium-bearing stalks in clumps. If both + and − strains are present in a colony, they may grow together and their nuclei may fuse, producing a zygote. This zygote, which is the only diploid cell of the life cycle, acquires a thick, black coat, and is then called a zygospore. Meiosis occurs during its germination, and normal, haploid hyphae grow from the resulting haploid cells.

MAJOR GROUPS OF FUNGI

Fungi comprise three divisions (phyla), representatives of which are shown in Figure 26-12. Of these, the **zygomycetes** (division Zygomycota) are the simplest in form and structure; they consist of only about 600 species. Their simple reproductive organs are called **gametangia** (singular, **gametangium**). When they fuse, they form hard, resistant spores called **zygospores,** which (1) contain the zygotes, (2) are characteristic of the whole group, and (3) differ greatly between the different species and genera. Spores are also formed without sexual reproduction taking place; these **asexual** spores are produced in bodies called **sporangia** (singular, **sporangium**). The life cycle of the black bread mold, *Rhizopus,* is shown in Figure 26-15.

The second division of fungi, the **ascomycetes** (division Ascomycota), includes more than 30,000 described species. Among them are such familiar and economically important fungi as cup fungi and morels (see Figure 26-12, *A* and *C*), truffles, and yeasts. Ascomycetes also include the organisms that cause many of the most serious plant diseases. The chestnut blight, *Endothia parasitica,* has almost exterminated the American chestnut throughout its native range, and Dutch elm disease, *Ceratocystis ulmi,* has decimated elms around the world. The characteristic reproductive structure of the ascomycetes is the **ascus** (plural, **asci**), a club-shaped element that is formed within a structure consisting of densely interwoven hyphae, the **ascocarp.** The hyphae of ascomycetes are haploid. Syngamy, which occurs in the young asci, is immediately followed by meiosis, producing four, eight, or more spores. Asexual reproduction by means of **conidia** (singular, **conidium**), which are multinucleate spores in fungi cut off at the ends of the hyphae, is characteristic of most ascomycetes. Asexual reproduction is also a common feature among the more than 25,000 additional species of Fungi Imperfecti, a group of fungi in which sexual reproduction is not known (Figure 26-16). Most Fungi Imperfecti would be classified as ascomycetes if their sexual structures were found. Yeasts, the only single-celled fungi, are primarily ascomycetes too.

Basidiomycetes, the third division of fungi (division Basidiomycota), with about 25,000 named species, are the most familiar fungi. These include not only the mushrooms, toadstools, puffballs (Figure 26-12, *E*), jelly fungi, and shelf fungi (Figure 26-12, *D*), but also many important plant pathogens among the groups called rusts and smuts (Figure 26-17). In place of the asci of the ascomycetes, the basidiomycetes form **basidia** (singular, **basidium**). At the apex of basidia, the spores that result from meiosis

FIGURE 26-16

Scanning electron micrograph of conidia (spores) of Fungi Imperfecti, ascomycetes in which sexual reproduction is unknown. The conidia are the round balls at the end of special hyphae called **conidiophores.** Shown here are the characteristic conidiophores of *Penicillium. Penicillium* and the closely related genus *Aspergillus* are among the most important fungi economically. They produce penicillin, give the characteristic flavors and aromas to cheeses such as Roquefort and Camembert, and are used in fermenting soy sauce and soy paste, among other applications.

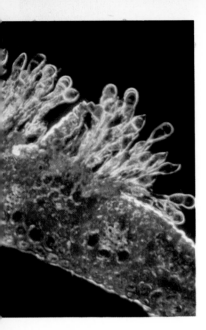

FIGURE 26-17

Wheat rust, *Puccinia graminis,* one of about 7000 species of rusts (division Basidiomycota), all of them plant pathogens. This species causes enormous economic losses to wheat wherever it grows, and is combatted largely by breeding resistant wheat varieties. Mutation and recombination in wheat rust constantly produce new virulent strains and make it necessary to replace the existing wheat varieties constantly. Wheat rust alternates between two different hosts, wheat and barberries, and needs both to complete its life cycle. The sexual stages of wheat rust take place only on barberries, and the eradication of these plants helps to control this disease.

are elevated (Figure 26-18); otherwise, the details of the life cycles are roughly similar to those of the ascomycetes. Asexual reproduction, however, is relatively uncommon.

LICHENS

A **lichen** is a symbiotic association between a fungus and a photosynthetic partner. Ascomycetes are the fungal partners in all but about 20 of the 25,000 described species of lichens (Figure 26-19); the exceptions, mostly tropical, are basidiomycetes. Most of the visible body of a lichen consists of its fungus, but within the tissues of that fungus are found either cyanobacteria or green algae (Figure 26-20). Specialized fungal hyphae penetrate the photosynthetic cells held within them and transfer nutrients directly to the fungal partner, which in turn protects its bacterial or algal cells from drying out.

The durable construction of the fungus, linked with the photosynthetic properties of its partner, has enabled lichens to invade the harshest of habitats: the tops of mountains, the farthest north and south latitudes, and dry, bare rock faces in the desert. In such desolate, exposed areas, lichens are often the first colonists, breaking down the rocks and setting the stage for the invasion of other organisms. Lichens with a cyanobacterium for a photosynthetic partner have a particular advantage, because they are able to fix atmospheric nitrogen and thus contribute it to their habitat for use by other pioneering organisms.

Lichens are often strikingly colored because of their pigments, which probably play a role in protecting the photosynthetic partner from the destructive action of the sun's ray. These same pigments may be extracted from the lichens and used as natural dyes, for example, in the traditional method of manufacturing Harris tweed (now, however, colored with synthetic dyes).

Lichens are symbiotic associations between a fungus, usually an ascomycete, and a photosynthetic partner, which may be either a green alga or a cyanobacterium.

FIGURE 26-18

Life cycle of a basidiomycete (division Basidiomycota). In primary mycelia, there is only one nucleus in each cell (the septa are perforated); in the secondary mycelia, which are formed by the fusion of primary mycelia (plasmogamy), there are two nuclei, one derived from each of the strains that gave rise to the secondary mycelia, within each cell. Secondary mycelia ultimately may become massed and interwoven, forming the basidiocarp.

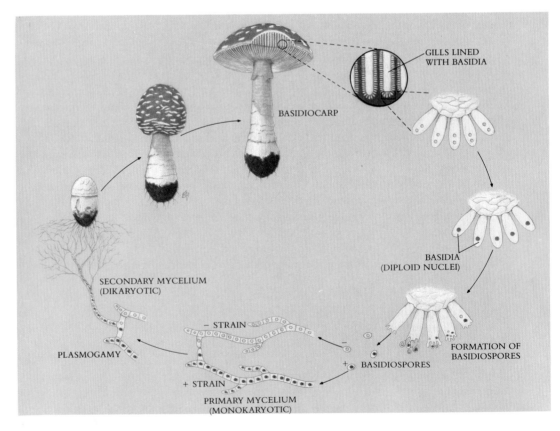

Lichens are able to survive in inhospitable habitats partly by being able to dry or freeze to a condition one might call suspended animation. Once the drought or cold has passed, the lichens recover quickly and resume their normal metabolic activities, including photosynthesis. The growth of lichens may be extremely slow in harsh environments; many relatively small ones appear to be thousands of years old and therefore are among the oldest living things on earth. Lichens are extremely sensitive to pollutants in the atmosphere and thus can be used as bioindicators of air quality. Many species are characteristically absent near major cities and other sources of pollution.

MYCORRHIZAE

The roots of about 80% of all kinds of plants normally are involved in symbiotic relationships with certain specific kinds of fungi; it has been estimated that these fungi probably amount to 15% of the total weight of the world's plant roots. Associations of this kind are termed **mycorrhizae,** from the Greek words for "fungus" and "roots". To a certain extent, the fungi involved in mycorrhizal association replace and have the same function as the fine projections from the epidermis, or outermost cell layer, of the terminal portions of the roots called **root hairs.** When mycorrhizae are present, they aid in the direct transfer of phosphorus, zinc, copper, and probably other nutrients from the soil to the roots of plants. The plant, on the other hand, supplies organic carbon to the symbiotic fungus (Figure 26-21).

Mycorrhizae are symbiotic associations between plants and fungi.

The earliest fossil plants often are found to have mycorrhizal roots. Such associations, which were common during the initial invasion of the land by plants, may have played an important role in allowing this invasion to occur. The soils available at such time would have been sterile and completely lacking in organic matter. Plants that form mycorrhizal associations are particularly successful in infertile soils today. When one considers this fact, and the fossil evidence, the suggestion that mycorrhizal associations were characteristic of the earliest plants seems reasonable.

FIGURE 26-19

Lichens on a fog-swept rock in coastal California.

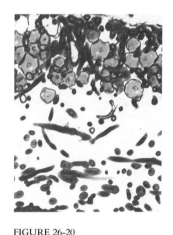

FIGURE 26-20

Trans-section of a lichen, showing the fungal hyphae, more densely packed into a protective layer on the top, and, especially, the bottom layer of the lichen. The green cells near the upper surface of the lichen are those of a green alga. Penetrated by fungal hyphae, they supply carbohydrates to the fungus.

FIGURE 26-21

Soybeans without endomycorrhizae (left) and with different strains of mycorrhizae (center and right).

SUMMARY

1. The kingdom Protista consists of the exclusively or predominantly unicellular phyla or eukaryotes, together with three phyla that include many multicellular organisms: the red algae, brown algae, and green algae. The phyla of protists are highly diverse.

2. *Pelomyxa palustris,* an amoeba-like organism, lacks both mitosis and mitochondria and apparently represents a very early stage in the evolution of the eukaryotes.

3. Chloroplasts originated several times among the protists. The process apparently involved the members of at least three different groups of bacteria: cyanobacteria (in the red algae); *Prochloron*-like organisms (in the green algae and euglenoids); and organisms derived from another bacterial group (in the diatoms, dinoflagellates, and brown algae).

4. The three major multicellular groups of eukaryotes, plants, animals, and fungi, all originated from protists. Because the three groups are so large and important, however, each is considered a distinct kingdom. The three are not related directly to one another. The plants originated from green algae; the sponges, and probably all animals, originated from the choanoflagellates, a group of zoomastigotes; and the ancestors of the fungi are unknown.

5. About 10 other phyla of protists are mentioned, including Rhizopoda, the amoebas; Ciliophora, the ciliates; Bacillariophyta, the diatoms; and Dinoflagellata, the dinoflagellates. Two phyla that have stages in their life histories in which individuals move like amoebae, and which have sometimes been grouped with the fungi, are Myxomycota, the plasmodial slime molds, and Acrasiomycota, the cellular slime molds.

6. One of the zoomastigotes causes sleeping sickness and several other primarily tropical diseases. A second kind includes the ancestors of the sponges and animals. A third zoomastigote, sometimes separated as the phylum Euglenophyta, contains chloroplasts that closely resemble those of the green algae and plants.

7. The malarial parasite, *Plasmodium,* is a member of the phylum Sporozoa. Carried by mosquitoes, it multiplies rapidly in the liver of humans and other primates. This results in the cyclical fevers characteristic of malaria because of the release of toxins into the bloodstream of the host.

8. The fungi are a distinct kingdom of eukaryotic organisms characterized by their filamentous growth form, lack of chlorophyll and motile cells, chitin-rich cell walls, and external digestion of food by the secretion of enzymes. Along with the bacteria, they are the decomposers of the biosphere.

9. Fungal filaments, called hyphae, collectively make up a mass called the mycelium. Mitosis in fungi occurs within the nuclear envelope.

10. Symbiotic systems involving fungi include lichens and mycorrhizae. The fungal partners in lichens are almost entirely ascomycetes, which derive their nutrients from green algae or cyanobacteria. Mycorrhizae are symbiotic associations between plants and fungi that are characteristic of the great majority of plants.

REVIEW

1. Four characteristics that differentiate plants from fungi are _____, _____, _____, and _____.

2. Flagella are common among the various phyla (divisions) of algae except one. Which phylum of algae lacks flagellated cells at any stage in the life cycle?

3. Which group of protists is believed to have given rise to animals and certainly gave rise to the sponges?

4. The slender filaments that make up the body of a fungus are called _____.

5. _____, a chemical isolated from a fungus, suppresses the response of the immune system to transplanted organs and has greatly increased the success rate of organ transplants since it became available in 1979.

SELF-QUIZ

1. Which kingdom of eukaryotes is most diverse?
   (a) Monera (d) Plantae
   (b) Protista (e) Animalia
   (c) Fungi

2. Which of the following groups of algae have chloroplasts identical to those found in plants?
   (a) red algae (d) *Prochloron*
   (b) brown algae (e) euglenoids
   (c) green algae

3. Which of the following eukaryotes do *not* carry out meiosis?
   (a) amoebas (d) *Prochloron*
   (b) fungi (e) two of the above
   (c) choanoflagellates

4. Two phyla of protists mentioned in this text have historically, and incorrectly, been placed in the kingdom Fungi. They are
   (a) plasmodial slime molds (Myxomycota)
   (b) bread molds (Zygomycota)
   (c) cellular slime molds (Acrasiomycota)
   (d) yeasts (Ascomycota)
   (e) smuts (Basidiomycota)

5. Mycorrhizae are
   (a) symbiotic associations between plants and fungi
   (b) an association found in about 80% of plants
   (c) an aid in mineral uptake by plants
   (d) all of the above
   (e) none of the above

THOUGHT QUESTIONS

1. If fungi have been so successful as to become an entire kingdom, why do you suppose the protist that first gave rise to the fungi has not persisted?

2. Both plants and animals have individuals that are very large. Whales weigh several tons, and a mature redwood tree even more. Why do you suppose there are no similarly large fungi?

FOR FURTHER READING

DONELSON, J.E., and M.J. TURNER: "How the Trypanosome Changes its Coat," *Scientific American,* February 1985, pages 44-51. Trypanosomes survive in the bloodstream by evading the immune system; they do this by switching on new genes and encoding new surface antigens.

GODSON, G.N.: "Molecular Approaches to Malaria Vaccines," *Scientific American,* May 1985, pages 52-59. This article chronicles recent advances in the important quest for an effective malaria vaccine.

KOSIKOWSKI, F.V.: "Cheese," *Scientific American,* May 1985, pages 88-99. A fascinating account of how more than 2000 varieties of cheese are made and how bacteria and fungi participate in the process.

LAMBRECHT, F.: "Trypanosomes and Hominid Evolution," *BioScience,* vol. 35, pages 640-646, 1985. This fascinating article charts the probable effects of sleeping sickness in determining the course of human history.

STROBEL, G.A., and G.N. LANIER: "Dutch Elm Disease," *Scientific American,* August 1981, pages 56-66. This deadly fungal infection of elms may be brought under control with the aid of fungicides, beetle traps, and bacteria that inhibit its growth.

WEBB, A.D.: "The Science of Making Wine," *American Scientist,* vol. 72, pages 360-367, 1984. This account traces the steps that have made this ancient practical art a modern science.

FIGURE 27-1   The extraordinary evolutionary radiation of the angiosperms, which has resulted in such strange structures as the insect-trapping leaves of these Malaysian pitcher plants *(Nepenthes albomarginata),* has made them the dominant plant group on land for the past 100 million years.

Chapter 27

Overview

Over a relatively short time, plants evolved into a very diverse group, creating the great forests and other plant communities where the vertebrates, insects, and fungi evolved and diversified. About 266,000 species of plants are now living. The two major groups of plants are the bryophytes and the vascular plants. Seeds that evolved among the vascular plants provide a means to protect young individuals. Flowers, which occur in only one group of living vascular plants, evolved to guide the activities of insects and other pollinators so that they would carry pollen rapidly and precisely from one flower to another of the same species. The flowering plants dominate every spot on earth except the great northern forests, the polar regions, the high mountains, and the driest deserts. Despite their overwhelming success, this group originated relatively recently.

For Review *Here are some important terms and concepts that you will encounter in this chapter. If you are not familiar with them, you should review them before proceeding.*

History of plants (Chapter 18)

Classification of organisms (Chapter 24)

Mycorrhizae (Chapter 26)

Of the five kingdoms of living organisms, the three we have discussed so far consist of organisms that are mostly small relative to us. Bacteria and most protists are unicellular, and fungi, although multicellular, are rarely as large as your fist. In the plant kingdom, however, we encounter many species that consist of large individuals. A tree can be 30 meters or more tall, with a mass of many metric tons. Plants are the dominant organisms of the terrestrial landscape, and we believe they were the first organisms to invade the land successfully from the sea, where life first evolved and flourished (Figure 27-1).

Despite all the diversity of life on earth, only three groups are responsible for virtually all of the fixation of carbon that occurs and thus for providing the sustenance of all living organisms. The algae and photosynthetic bacteria carry out most marine photosynthesis, whereas plants are the dominant photosynthetic organisms on land. Plants are multicellular eukaryotic organisms that have (1) cellulose-rich cell walls; (2) chloroplasts that contain chlorophylls *a* and *b,* together with carotenoids; and (3) starch as their primary carbohydrate food reserve. This chapter discusses the characteristics of the major plant groups (Figure 27-2), focusing on differences in their life cycles. Chapters 30 to 32 will treat specific aspects of the biology of plants in more detail.

A

B

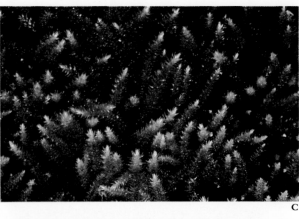

C

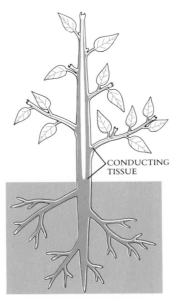

D

FIGURE 27-2

Representatives of four divisions of plants.

A Only flowering plants (division Anthophyta), such as this silk–cotton tree in the Peruvian Andes, add color to the landscape. Their flowers are specialized to attract insects and other animals.

B Maidenhair fern, *Adiantum pedatum* (division Pterophyta).

C A moss (division Bryophyta).

D A branch of Norway spruce, *Picea nigra* (division Coniferophyta), in the Alps. Seeds are produced in the large cones, pollen in the smaller ones.

CONDUCTING TISSUE

FIGURE 27-3

The conducting tissues of a vascular plant, shown diagrammatically in green. Water passes continuously in through the roots, up through the stems, and out through the leaves; at the same time, sugar molecules, manufactured as a result of photosynthesis in the leaves and green stems, pass through a parallel conducting system to all parts of the plant, where they are used for growth.

THE GREEN INVASION OF THE LAND

Plants, fungi, and insects are the only major groups of organisms that occur almost exclusively on land; several other phyla, including chordates and mollusks (Chapter 28), are also very well represented on land. The groups that occur almost exclusively in terrestrial habitats, however—the plants, fungi, and insects—all probably evolved here, whereas the chordates and mollusks originated in the water. Of the groups that did evolve on land, the ancestors of the plants were almost certainly the first to become terrestrial. As you have seen with fungi (Chapter 26) and will see with insects (Chapter 28), a major evolutionary challenge in the transition from an aquatic to a terrestrial habitat is **desiccation,** the tendency of organisms to lose water to the air. Various structures evolved to supply water to the different parts of the plant and at the same time conserve it.

As with insects, efficient plumbing systems helped the earliest vascular plants to achieve their ecological dominance of the land. Vascular plants are so named because, unlike the bryophytes, they have **vascular tissue.** The word *vascular* comes from the Latin *vasculum,* meaning a vessel or duct, and refers, in the case of plants, to their plumbing system. Vascular tissue consists of specialized strands of elongated cells that run from near the tips of a plant's roots through its stems and into its leaves. The vascular tissue conducts water with dissolved minerals (nutrients), which come in through the roots, and carbohydrates, which are manufactured in the green parts of the plant (Figure 27-3). Therefore, water and nutrients reach all parts of the plant, as do the carbohydrates that provide energy for the synthesis of its different structures. The bryophytes (mosses, liverworts, hornworts) do not have such specialized conducting systems, but it is not clear whether they never had them or whether they had them and lost them. In either case, bryophytes, being smaller than most vascular plants, do not have as great a need for transport within the plant.

Most plants also are well protected from drying out by their **cuticle,** an outer covering formed from a waxy substance called **cutin.** The cuticles that cover the exposed surfaces of plants are impermeable to water and thus provide a key barrier to water loss. Passages exist through the cuticle, however, in the form of specialized pores called **stomata** (singular, **stoma;** Figure 27-4) in the leaves and sometimes the green portions of the stems. Stomata allow carbon dioxide to pass into the plant bodies and allow water and oxygen to pass out of them; the cells that border stomata expand and contract,

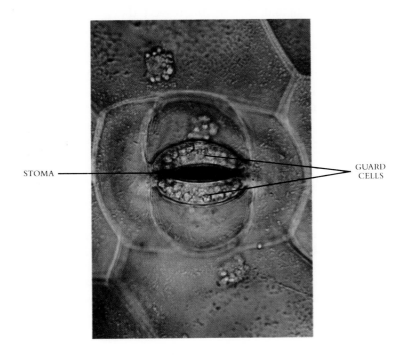

STOMA

GUARD CELLS

FIGURE 27-4

A stoma. The guard cells flanking the stoma contain chloroplasts, unlike the other epidermal cells. Water passes out through the stomata, and carbon dioxide enters by the same portals. The mechanism of opening and closing of stomata will be described in Chapter 31.

thus controlling these movements. The same stomata that allow carbon dioxide into the plant, however, also provide a way for water to leave. A vascular plant depends on the constant flow of water, in through the roots and out of the stomata.

> **Vascular plants are defined primarily by their possession of a specialized plumbing system, the vascular system. This system involves strands of elongated, specialized cells that transport water and dissolved nutrients, and other strands that transport carbohydrate molecules.**

In addition to their structural features, plants also developed a special relationship with fungi that evidently was a key factor in their successful occupation of terrestrial habitats. Mycorrhizae are characteristic of some 80% of all plants and are frequently seen in fossils of the earliest plants (see Chapter 26). They were probably critical to the ancestral plants as they adjusted to life on land.

Many other features developed gradually and aided the evolutionary success of plants on land. For example, in the first plants there was no fundamental difference between the above-ground and underground parts. Later, roots and shoots with specialized structures evolved, each suited to its particular environment (Chapter 29). **Leaves,** expanded areas of photosynthetically active tissue, evolved and diversified in relationship to the varied habitats that existed on land. Specializations in key reproductive features improved the methods by which plants protected their young and were dispersed from place to place.

The more specialized roots, stems, leaves, and reproductive features that evolved in plants were important factors in the overwhelming success of this group on land. An estimated 250,000 species of vascular plants and 16,000 additional species of bryophytes are now in existence. Today plants dominate every part of the terrestrial landscape, except the extreme polar regions and the highest mountaintops.

THE PLANT LIFE CYCLE
Alternation of Generations

Understanding the different kinds of life cycles that occur among plants provides an important key to understanding their evolutionary relationships. All plants exhibit **alternation of generations,** in which a diploid generation, or **sporophyte,** alternates with a haploid generation, or **gametophyte** (see Figure 24-14, C). *Sporophyte* literally means "spore plant," and *gametophyte* means "gamete plant"; these terms indicate the kinds of reproductive structures that the respective generations produce (Figure 27-5).

Most animals are diploid, and in this respect they resemble the sporophyte generation of a plant. Such animals, however, produce eggs and sperm, which fuse directly to form a zygote. In contrast, the sporophyte generation of a plant does not produce gametes as a result of meiosis. Instead, meiosis takes place in specialized cells, called **spore mother cells,** and results in the production of haploid **spores,** the first cells of the gametophyte generation. Spores do not fuse with one another as gametes do; instead, they divide by mitosis, producing a multicellular haploid individual, the gametophyte (Figure 27-6).

In turn, the gametes—eggs and sperm—eventually are produced by the gametophyte as a result of mitosis. They are haploid, as is the gametophyte that produces them. When they fuse to form a zygote, the first cell of the next sporophyte generation has come into existence. The zygote grows into a sporophyte, in which meiosis ultimately occurs.

> **Plant life cycles are marked by an alternation of generations of diploid sporophytes with haploid gametophytes. Sporophytes produce spores, which are haploid and result from meiosis; the spores grow into gametophytes. Gametophytes produce gametes, which result from mitosis. The fusion of gametes produces a zygote, the first cell of the sporophyte generation.**

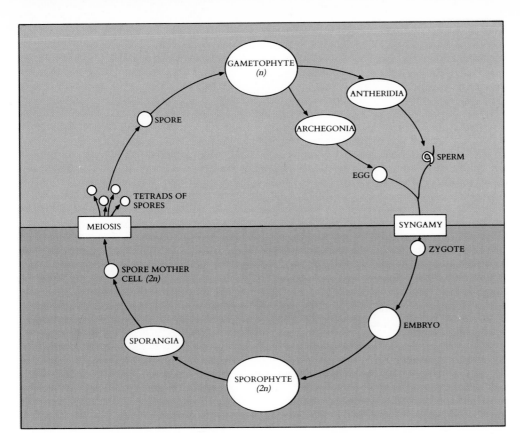

FIGURE 27-5

A generalized plant life cycle (sporic meiosis). In a life cycle of this kind, gametophytes, which are haploid (n), alternate with sporophytes, which are diploid (2n). Antheridia (male) and archegonia (female), which are the sex organs (gametangia), are produced by the gametophyte, and they in turn produce sperm and eggs, respectively. These ultimately come together in the process of syngamy to produce the first diploid cell of the sporophyte generation, the zygote. Meiosis takes place within the sporangia, the spore-producing organs of the sporophyte, resulting in the production of the spores, which are haploid and are the first cells of the gametophyte generation.

In some plants, including bryophytes and ferns, the gametophyte is green and free living; in others it is not green and is nutritionally dependent on the sporophyte. When you look at a moss or liverwort (bryophytes), you see largely gametophyte tissue; the sporophytes are usually smaller brown or yellow structures attached to or enclosed within the tissues of the gametophyte. In vascular plants, the gametophytes are always much smaller than the sporophytes and usually are nutritionally dependent on sporophytes and enclosed within their tissues. When you look at a vascular plant, what you see is a sporophyte, with rare exceptions.

The Specialization of Gametophytes

The gametes of the first plants differentiated within specialized organs called **gametangia** (singular, **gametangium**), in which the eggs and sperm were surrounded by a jacket of cells. Complex multicellular gametangia of this sort are still found in all the less specialized members of the plant kingdom, including the bryophytes and several divisions (phyla) of vascular plants. Eggs and sperm are formed within different kinds of gametangia. Those in which eggs are formed are called **archegonia** (singular, **archegonium**), and those in which sperm are formed are called **antheridia** (singular, **antheridium**). An archegonium produces only one egg; an antheridium produces many sperm. These structures usually look very different from one another (Figure 27-7). In some bryophytes and less advanced vascular plants, including the ferns, antheridia and archegonia exist together on the same gametophyte. In other members of these groups, and typically in bryophytes, the two kinds of gametangia are on separate gametophytes. In the more advanced vascular plants, including all of the vascular plants that form seeds, the gametangia have been lost during the course of evolution. The eggs or sperm develop directly from individual cells of the respective gametophyte. Some of the gametophytes bear only eggs, and others bear only sperm.

FIGURE 27-6

Diagram of the life cycle of a heterosporous vascular plant. Although for convenience both megaspores and microspores are shown here arising from a single meiotic event, each such event would actually result in the production of all megaspores or all microspores.

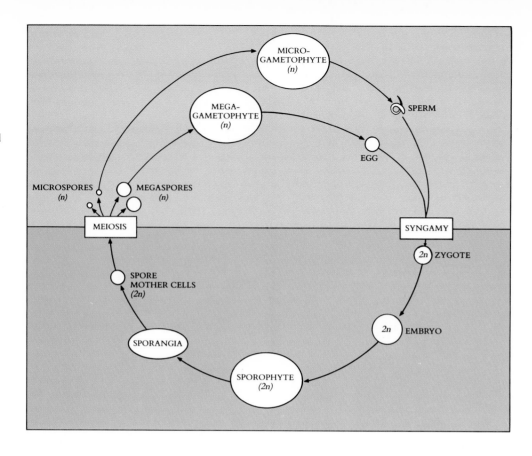

FIGURE 27-7

Gametangia of bryophytes.
A Trans-section through the archegonium of the liverwort *Marchantia.* A single egg differentiates within the lower, swollen portion of the archegonium.
B Trans-sections through a group of moss antheridia. The smaller cells in each of these elongate structures will give rise to sperm which, when liberated by the rupturing antheridium, swim through free water to the mouth of the archegonium.

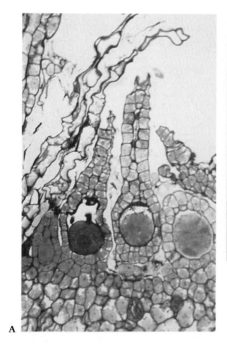

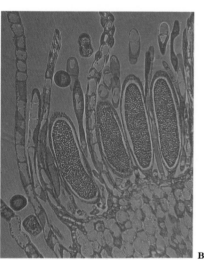

When one kind of gametophyte bears antheridia and another kind bears archegonia, the two kinds of gametophytes may look different from one another. If they do, the gametophytes that form antheridia are called **microgametophytes,** and those that form archegonia are called **megagametophytes.** These two formidable-looking terms literally mean no more than "small gametophytes" and "large gametophytes." In nearly all plants that do have two different kinds of gametophytes, the gametophytes arise from two different kinds of spores, **microspores** and **megaspores.** Plants that produce two different-looking spores are called **heterosporous;** those that produce only one kind of spore are **homosporous.**

Spores are formed as a result of meiosis in the sporophyte generation. Their differentiation occurs within specialized, multicellular structures called **sporangia** (singular, **sporangium**). If a plant forms both megaspores and microspores, each of these will be formed in a different kind of sporangium called, respectively, **megasporangia** and **microsporangia.**

As we describe the two major groups of plants, bryophytes and vascular plants, you will see a progressive reduction of the gametophyte from group to group. You will also see increasing specialization for life on the land, culminating with the remarkable structural adaptations of the flowering plants, the dominant plant group today.

MOSSES AND OTHER BRYOPHYTES

The bryophytes consist of mosses, liverworts, and hornworts (Figures 27-2, *C,* and 27-8). Most bryophytes are small: few exceed 2 centimeters in length. They are especially common in relatively moist places, both in the tropics and in temperate regions. In the Arctic and the Antarctic, bryophytes are the most abundant plants. Not only do they boast the most individuals in these harsh regions, but also the greatest number of species. Regardless of where they grow, however, bryophytes require free water at some time of the year to reproduce sexually, as well as for their growth and development.

FIGURE 27-8

Bryophytes.
A Hair-cup moss, *Polytrichum,* left. The leaves below belong to the gametophyte. Each of the yellowish-brown stalks, with the capsule at its summit, is a sporophyte. Although moss sporophytes may be green and carry out a limited amount of photosynthesis when they are immature, they are soon completely dependent on the gametophyte in a nutritional sense.
B A liverwort, *Marchantia.* The sporophytes are borne within the tissues of the umbrella-shaped structures that arise from the surface of the flat, green, creeping gametophyte. These particular structures develop archegonia within their tissues; another kind of similar structure, borne on different plants of *Marchantia,* produces the antheridia.

A   B

FIGURE 27-9

Moss life cycle. On the gametophytes, which are haploid, the sperm are released from antheridia. They then swim through free water to the archegonia and down their neck to the egg. Fertilization takes place there; the resulting zygote develops into a sporophyte, which is diploid. The sporophyte grows on the gametophyte and eventually produces spores as a result of meiosis. These spores are released and germinate, giving rise to gametophytes. The gametophytes initially are threadlike; they grow along the ground. Ultimately, buds form on them, from which leafy gametophytes arise.

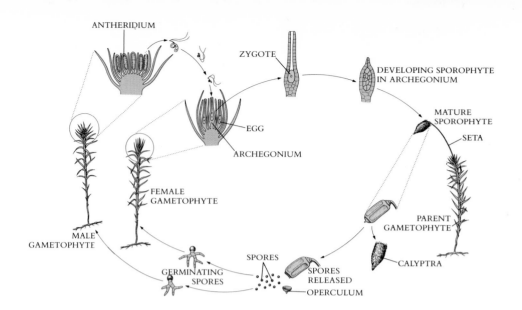

The largest class of bryophytes, and probably the one most familiar to you, is the mosses. Many small, tufted plants are mistakenly called "mosses"; for example, "Spanish moss" is actually a flowering plant, a relative of the pineapple. Even eliminating these imposters, however, leaves approximately 10,000 species of true mosses. They are found almost everywhere on earth. Figure 27-9 summarizes a moss life cycle.

Two major features distinguish bryophytes from vascular plants:

1. The general lack of specialized vascular tissues in bryophytes. Many moss sporophytes do have a central strand of somewhat specialized water-conducting tissue in their stems, a condition analogous to what occurs in vascular plant sporophytes. Food-conducting tissue, however, has been identified in only a few genera of bryophytes. Even if such tissues are present, their structures are much less complex than those found in the vascular plants. The stems and leaves of mosses and liverworts differ from those of vascular plants in that they lack vascular tissue.
2. The sporophytes of bryophytes are almost always smaller than, and derive their food from, the gametophytes.

Bryophytes are relatively simple plants. Taking all their features into account, we may view them as either (1) primitive plants or (2) plants that have became simplified during the course of their evolution. Which possibility accurately reflects the status of the bryophytes is not known.

> **Bryophytes lack the specialized vascular tissues of the vascular plants.**
> **Sporophytes of bryophytes photosynthesize only to a limited extent, if at all, and draw their food from the gametophytes on which they are produced.**

VASCULAR PLANTS

The first vascular plants appeared no later than the early Silurian Period, approximately 430 million years ago. The vascular plants are distinguished from the bryophytes by (1) their large, dominant, and nutritionally independent sporophytes; (2) efficient conducting tissues (see Figure 27-3); (3) specialized leaves, stems, and roots; (4) cuticles and stomata (see Figure 27-4); and (5) the evolution of seeds in some.

> **The vascular plants are characterized by their dominant sporophytes; efficient conducting tissues; specialized stems, leaves, and roots; cuticles and stomata; and seeds.**

Growth of the Vascular Plants

Early vascular plants were characterized by **primary growth,** growth that results from cell division at the tips of the stems and roots. Primary growth is also characteristic of herbaceous plants and of the young stems of all plants today (see Chapter 29). This sort of growth is like erecting a smokestack by continuing to add bricks to the top: it gets taller without the part that is already there becoming larger. Within stems that have resulted from primary growth are **vascular bundles.** These bundles of elongated cells function in conducting water with dissolved minerals (nutrients) and carbohydrates throughout the plant body. Vascular bundles, and the similar but more massive tissues that occur in woody plants, are a key characteristic of the vascular plants, not present in any other group.

  Secondary growth was an important early development in the evolution of vascular plants. In secondary growth, cell division takes place actively in regions around the plant's periphery as a result of repeated cell divisions that occur in a cylindrical zone. Secondary growth causes a plant to grow in diameter. Only after the evolution of secondary growth could vascular plants develop thick trunks and therefore grow tall. This evolutionary advance made possible the development of forests and, consequently, the domination of the land by plants. According to the fossil record, secondary growth had evolved independently in several different groups of vascular plants by the middle of the Devonian Period, approximately 380 million years ago.

> **Primary growth results from cell division at the tips of stems and roots, whereas secondary growth results from the division of a cylinder of cells around the plant's periphery.**

Conducting Systems of the Vascular Plants

Two types of conducting elements in the earliest plants have become characteristic of the vascular plants as a group. **Sieve elements** are soft-walled cells that conduct carbohydrates away from the areas where they are manufactured. **Tracheary elements** are hard-walled cells that transport water and dissolved minerals up from the roots. Both kinds of cells are elongated, and both occur in strands. Sieve elements are the characteristic cell types of a tissue called **phloem;** tracheary elements are characteristic of tissue called **xylem.** In primary tissues, which result from primary growth, these two types of tissue are often associated with one another in the same vascular strands.

> **Water and nutrients are carried in the xylem, which consists primarily of hard-walled cells called tracheary elements. Carbohydrates, in contrast, are carried in the phloem, which consists of soft-walled cells called sieve elements.**

What Is a Seed?

Seeds are a characteristic feature of some groups of vascular plants but not of others. A **seed** contains an embryo surrounded by a protective coat. This embryo's development has been temporarily arrested (Figure 27-10). Seeds are units by means of which plants, being rooted in the ground, are dispersed to new places. Many seeds have devices, such as the wings on the seeds of pines or maples or the plumes on the seeds of a dandelion, that help them to travel very efficiently.

  The seed is a crucial adaptation to life on land because it protects the embryonic plant from drying out or being eaten when it is at its most vulnerable stage. Most kinds of seeds have abundant food stored in them, either inside the embryo or in specialized storage tissue. The rapidly growing young plant uses the seed as a ready source of energy; the seed thus plays the same role as the yolk of an egg. The evolution of the seed was clearly a critical step in the domination of the land by plants.

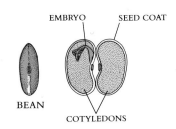

FIGURE 27-10

Diagram of a bean seed, with the tissues in it identified. The cotyledons are the first leaves; in such a seed, they are filled with stored food.

Seedless Vascular Plants

In four divisions of vascular plants, the living representatives do not form seeds. The most familiar seedless division is the ferns (see Figure 27-2, *B*), which include about 12,000 living species and make up division Pterophyta. The club mosses and their relatives (division Lycophyta) include four living genera with a total of about 1000 species. The horsetails, with a single genus *(Equisetum)* and 15 species, make up the division Sphenophyta. Another division, the Psilophyta, consists of two genera of tropical and south-temperate plants, the whisk ferns.

FIGURE 27-11

Fern life cycle. The gametophytes, which are haploid, grow on moist ground or similar places (see Figure 27-5). Eggs and sperm develop in archegonia and antheridia, respectively, on their lower surface. The sperm, when released, swim through free water to the mouth of the archegonium, entering and fertilizing the single egg. Following the fusion of egg and sperm to form a zygote—the first cell of the diploid sporophyte generation—the zygote starts to grow within the archegonium. Eventually, the sporophyte become much larger than the gametophyte—it is what we know as a fern plant. On its leaves occur clusters of sporangia, within which meiosis occurs and spores are formed. The release of these spores, which is explosive in many ferns, and their germination, lead to the development of new gametophytes.

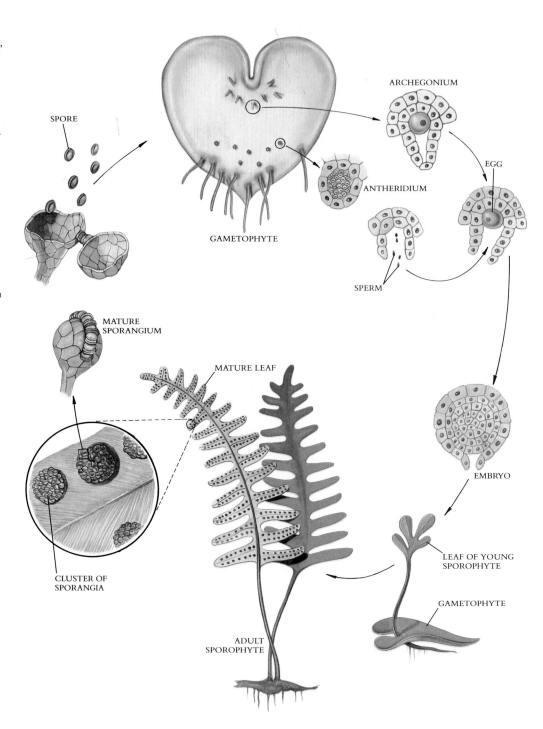

SPORE

ARCHEGONIUM

GAMETOPHYTE

ANTHERIDIUM

EGG

SPERM

MATURE SPORANGIUM

MATURE LEAF

EMBRYO

CLUSTER OF SPORANGIA

LEAF OF YOUNG SPOROPHYTE

GAMETOPHYTE

ADULT SPOROPHYTE

The life cycle of a fern (Figure 27-11) differs from that of a moss primarily in the much greater development, independence, and dominance of the fern's sporophyte, which is much more complex than that of a moss.

Gymnosperms

In the four divisions of living seed plants, the **ovules**—the structures that eventually become the seeds after fertilization—are not completely enclosed by the tissues of the parent individual on which they are carried at the time of pollination. These four divisions, the conifers, cycads, ginkgoes, and gnetophytes, are called **gymnosperms** (Greek *sperma,* seed, and *gymnos,* naked; in other words, naked-seeded plants). In fact, the seeds of gymnosperms often become enclosed by the tissues of their parents by the time they are mature, but their ovules are naked at the time of pollination.

Botanists generally agree that the four divisions of gymnosperms did not have an immediate common ancestor and are not directly related to one another. The four groups differ in many ways, and seeds may have evolved independently in each of them. For example, the cycads and ginkgoes (Figure 27-12, *A* and *B*) still have motile sperm, as do the vascular plants that do not form seeds. The cycads (division Cycadophyta) consist of 10 genera and about 100 species, widespread throughout the warmer portions of the world; one of the most familiar kinds is the sago palm. This is not a true palm, however, despite the palmlike or fernlike appearance of the cycad leaves; palms are flowering plants. The ginkgo *(Ginkgo biloba)* is the only living species of the division Ginkgophyta. Its seeds resemble small plums, with a fleshy, unpleasantly scented outer covering. The ginkgo is not known anywhere as a wild plant, but it was preserved in cultivation around the temples and in the gardens of Japan and China, where Europeans first encountered it several centuries ago. The division Gnetophyta has few species.

The four divisions of gymnosperms probably are not directly related to one another, and seeds may have evolved independently in each of them.

FIGURE 27-12

Representatives of three of the divisions of gymnosperms.
A An African cycad, *Encephalartos kosiensis;* division Cycadophyta. The cycads are seed plants with fernlike leaves.
B Maidenhair tree, *Ginkgo biloba,* the only living representative of the division Ginkgophyta, a group of plants that was abundant 200 million years ago. Among living seed plants, only the cycads and *Ginkgo* have swimming sperm.
C Engelmann spruce, *Picea engelmannii,* in the Rocky Mountains; division Coniferophyta.

A   B   C

FIGURE 27-13

Pine life cycle. In all seed plants, the gametophyte generation is greatly reduced. A germinating pollen grain is the mature microgametophyte of a pine; megagametophytes, in contrast, develop within the tissues of the ovule. When the pollen tube grows to the vicinity of the megagametophye, sperm are released, fertilizing the egg and producing a zygote there. The development of the zygote into an embryo takes place within the ovule, which matures into a seed. Eventually the seed falls from the cone and germinates, the embryo resuming growth and becoming a new pine tree.

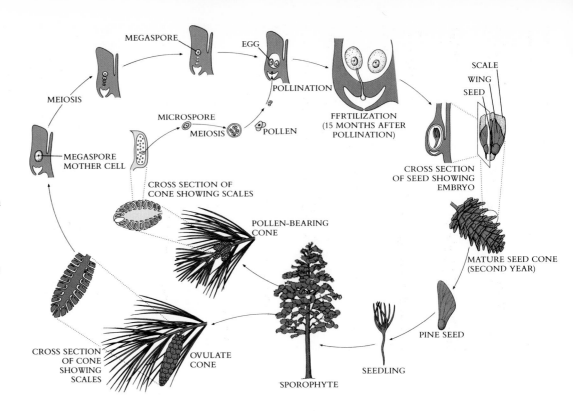

The most familiar of the divisions of gymnosperms is the division Coniferophyta, the conifers (Figures 27-2, *D,* and 27-12, *C*). With about 550 living species, the conifers include such tree groups as pines, spruces, firs, yews, redwoods, bald cypress, junipers, cedars, and others. Most conifers are evergreen, and they form vast forests in many of the temperate and cooler portions of the globe. In their life cycle (Figure 27-13), the pollen grains, which contain the male gametes, are carried by the wind to the vicinity of the ovules, the structures holding the female gametes. The ovules are carried on cone scales (in pines and their relatives) or similar structures. Within the seeds of conifers and other gymnosperms, stored food is provided for the embryo within tissue derived from the megagametophyte.

THE FLOWERING PLANTS

Flowering plants, or angiosperms (division Anthophyta), are the dominant photosynthetic organisms nearly everywhere on land. This great group of some 235,000 species includes our familiar trees (except conifers), shrubs, herbs, grasses, vegetables, and grains—in short, nearly all of the plants that we see every day. Virtually all of our food is derived, directly or indirectly, from the flowering plants; in fact, more than half comes from just three species: rice, corn, and wheat (see Chapter 23). In a sense, the remarkable evolutionary success of the angiosperms is the culmination of the plant line of evolution. Our earliest record of flowering plants is from approximately 123 million years ago, in the first part of the Cretaceous Period. Angiosperms, therefore, are by far the youngest of all the divisions of plants, despite their world dominance for most of the last hundred million years.

Angiosperms differ from other seed plants in their possession of flowers and fruits. The basic structure of the flower consists of four **whorls;** a whorl is a circle of parts present at a single level along the axis. The two outer whorls of the flower are mainly concerned with protecting the flower and attracting insects and other visitors to it; the inner ones contain the male and/or female gametophytes. The third whorl consists of the **stamens,** which produce pollen grains.

A

B

C

Several examples of the remarkable diversity of angiosperm flowers.
A Wild geranium, *Geranium*, a woodland plant, showing the 5 free petals, 10 stamens, and fused carpel.
B Tiger lily, *Lilium canadense*, with six free colored attractive flower parts (they cannot be separated in lilies into sepals and petals) and six free anthers. The carpels are fused together.
C Fragrant water lily, *Nymphaea odorata*, a flower with numerous, free, spirally arranged parts that intergrade with one another in form.

The innermost floral whorl, the **gynoecium,** consists of the **carpels.** In angiosperms, the ovules are enclosed within the carpels, and pollination is indirect. This contrasts with gymnosperms, the other seed-bearing plants, in which the ovules are carried on the surface of scales and the pollen is blown to them by the wind (Figure 27-14).

In many angiosperms, the pollen grains are carried from flower to flower by insects; in others the pollen is transported passively in the wind, as it is in most gymnosperms. Chapter 30 discusses the details of angiosperm reproduction. Briefly, however, the pollen reaches a specialized area of the carpel, called the **stigma,** where each pollen grain then forms a **pollen tube,** which grows through the tissue of the carpel and may ultimately reach the egg. Within each pollen grain are two sperm, which travel down the tube. One sperm fuses with the egg to form a zygote.

During meiosis, angiosperms produce eight haploid nuclei within an **embryo sac.** One of these nuclei becomes an egg. Two others, held together in a large cell at the center of the embryo sac, are called **polar nuclei.** The second sperm from the pollen grain fuses with these two polar nuclei, forming the **primary endosperm nucleus.** Since the primary endosperm nucleus is formed from the fusion of three haploid ($1n$) nuclei, it is triploid ($3n$), whereas the zygote is diploid ($2n$). The primary endosperm nucleus divides rapidly by mitosis, giving rise to a specialized kind of triploid nutritive tissue called the **endosperm,** one of the distinctive features of angiosperms. The endosperm either may be digested by the growing embryo or may be present in the mature seed to nourish the germinating seedling. The unique process whereby one sperm fuses with the egg while the other fuses with the polar nuclei is called **double fertilization.** This process is characteristic of angiosperms and is found nowhere else. Figure 27-15 outlines the angiosperm life cycle.

> **Double fertilization, a process that is unique to the angiosperms, occurs when one sperm nucleus fertilizes the egg and the second one fuses with the polar nuclei. These two events result in the formation of, respectively, the zygote and the primary endosperm nucleus. The latter divides to produce the endosperm, the nutritive tissue that occurs in the seeds of angiosperms.**

FIGURE 27-15

Angiosperm life cycle. As in the pine, the sporphyte is the dominant generation. Eggs form within the megagametophyte, or embryo sac, inside the ovules, which in turn are enclosed in the carpels—members of the inner whorl of the flower. The carpel is differentiated in most angiosperms into a slender portion, or style, ending in a stigma, the receptive surface on which the pollen grains germinate. The pollen grains, meanwhile, are formed within the sporangia of the anthers and complete their differentiation to their mature, three-celled stage in angiosperms either before or after grains are shed. Fertilization is distinctive in angiosperms, being a double process. A sperm and an egg come together, producing a zygote; at the same time, another sperm fuses with the two polar nuclei, producing the primary endosperm nucleus, which is triploid. Both the zygote and the primary endosperm nucleus divide mitotically, giving rise, respectively, to the embryo and the endosperm. The endosperm is the tissue, unique to angiosperms, that nourishes the embryo and young plant. Some plants, such as the one shown in Figure 27-11, transfer all of the stored food in the endosperm to the cotyledons before the seed is mature.

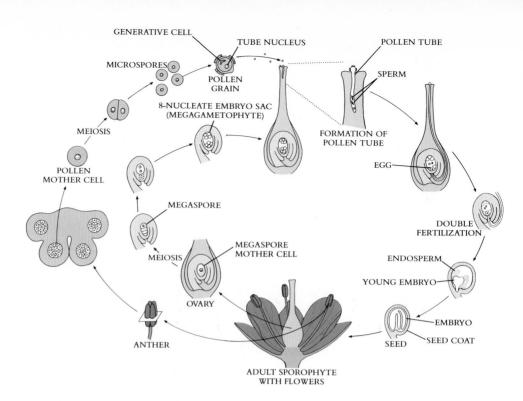

The two outer whorls of the angiosperm flower are (1) the outer **calyx,** the individual parts of which, the **sepals,** are often green and leaflike and surround the flower; and (2) the inner **corolla,** the individual parts of which are called **petals** (Figure 27-16). The petals are often colored and attractive to insects and other animals that visit the flowers and spread their pollen from flower to flower. Chapter 30 presents more details about how the features of flowers have evolved.

Monocots and Dicots

The two classes in the angiosperms, division Anthophyta, are (1) the Monocotyledones, or **monocots** (about 65,000 species) and (2) the Dicotyledones, or **dicots** (about 170,000 species). The monocots include lilies, grasses, cattails, palms, agaves, yuccas, pondweeds, orchids, and irises. The dicots include the great majority of familiar angiosperms of all kinds.

Monocots and dicots differ from one another in several features. For example, the venation in the leaves of monocots usually consists of parallel veins, whereas the leaves of dicots have netlike **(reticulate)** veins. In the flowers of dicots, the members of a given whorl are generally in fours or fives; in monocots, they are usually in threes (Figure 27-17). The embryos of dicots, as the name of the class implies, generally have two seedling leaves, or **cotyledons;** those of monocots have one cotyledon. The arrangement in monocots has been derived from that in dicots through the suppression of one of the cotyledons. Likewise, monocots share with the most primitive dicots a very similar kind of single-pored pollen.

> **The two classes of angiosperms are monocotyledons (monocots) and dicotyledons (dicots). Monocots have one cotyledon, or seedling leaf; usually parallel venation in their leaves; and flower parts often in threes. Dicots have two cotyledons; usually net (reticulate) venation; and flower parts in fours or fives. Several other anatomical differences also exist between the members of these classes.**

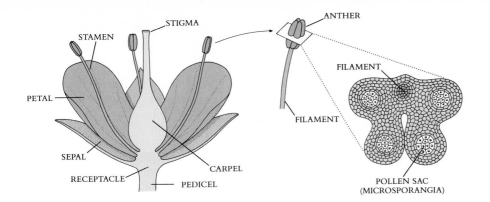

FIGURE 27-16

Simplified diagram of an angiosperm flower. The members of the outermost whorl of the flower, the sepals, make up the calyx, and the other whorls, working inward, are the corolla (made up of petals), androecium (anthers), and gynoecium (carpels). The microsporangia differentiate within the anthers, as do the megasporangia within the carpels.

A   B

FIGURE 27-17

The bright red flowers of fire-pink, *Silene virginica,* with their flower parts in fives, are typical of the dicots **(A);** those of *Trillium ozarkanum* variety *pusillum,* with flower parts in threes, are typical of the monocots **(B).**

Monocotyledons and dicotyledons have been distinct from one another from almost the time of the first appearance of angiosperms. Dicots are the more primitive of these two classes, and monocots were probably derived from early dicots before the middle of the Cretaceous Period.

A Very Successful Group

Vascular plants have come to dominate the land, all over the world. Many of the features of the flowering plants seem to be correlated with successful growth under arid and semiarid conditions, which have been spreading throughout the world during the history of this group. Flowers effectively ensure the transfer of gametes over substantial distances and therefore promote outcrossing, because insects, birds, and other animals are more precise in carrying their pollen from one plant to another than is the wind. Fruits, probably carried from place to place by animals, would have been especially effective in assisting the spread of some of these early flowering plants from one patch of favorable habitat to another. The tough, often leathery leaves of angiosperms, their efficient cuticles and stomata, and their specialized conducting elements were all important factors in survival and growth under arid conditions, just as they are in stressful conditions at present. The many natural insecticides produced by these plants (Chapter 21) were essential for their survival as well. As the early angiosperms evolved, all of these features that contributed to their success developed further, and the group continued to evolve more and more rapidly.

SUMMARY

1. The evolution of plants' conducting tissues, cuticle, stomata, and seeds has made them progressively less dependent on available water.

2. Plants all have an alternation of generations, in which gametophytes, which are haploid, alternate with sporophytes, which are diploid. The spores that sporophytes form as a result of meiosis grow into gametophytes, which produce gametes (sperm and eggs) as a result of mitosis.

3. The major groups of vascular plants are the bryophytes (division Bryophyta), including mosses, liverworts, and hornworts; and the vascular plants, which make up nine other divisions. Vascular plants have well-defined conducting strands, some specialized to conduct water and dissolved minerals and others specialized to conduct the food molecules that the plants manufacture.

4. The gametophytes of bryophytes and ferns are green and nutritionally independent. The sporophytes of bryophytes are mostly brown or straw-colored at maturity and nutritionally dependent, at least in part, on the gametophytes.

5. In all seed plants (gymnosperms and angiosperms) and in a few ferns, the gametophytes are either female (megagametophytes) or male (microgametophytes). Megagametophytes produce only eggs; microgametophytes produce only sperm. These eggs and sperm are produced, respectively, from megaspores, which are formed as a result of meiosis within megasporangia, and microspores, which are formed in a similar manner within microsporangia.

6. In gymnosperms, the ovules are exposed directly to pollen at the time of pollination. In angiosperms, the ovules are enclosed within a carpel, and a pollen tube grows through the carpel to the ovule.

7. The nutritive tissue in gymnosperm seeds is derived from the expanded, food-rich gametophyte. In angiosperm seeds, the nutritive tissue is unique and is formed from a cell, usually resulting from the fusion of the polar nuclei of the embryo sac with a sperm cell.

8. The pollen of gymnosperms is usually blown about by the wind. Pollen of angiosperms is often carried from flower to flower by various insects and other animals, which are called pollinators.

9. Angiosperms are at least 123 million years old and have been the dominant plant group in terrestrial habitats for about 100 million years.

10. Angiosperm flowers typically consist of four whorls. From the inside outward, these whorls are called the gynoecium (consisting of the carpels), androecium (stamens), corolla (petals), and calyx (sepals).

11. There are two classes of angiosperms. Monocotyledones, or monocots, include about 65,000 species, largely with parallel venation, flower parts in threes, and one cotyledon. Dicotyledones, or dicots, include approximately 170,000 species, largely with reticulate venation, flower parts in fours or fives, and two cotyledons.

12. Angiosperms were successful in terrestrial habitats because of their relatively drought-resistant vegetative features, including their vascular systems, cuticles, and stomata. Their flowers, fruits, and seeds, however, play an even more important role. Flowers make possible the precise transfer of pollen and therefore outcrossing, even when the stationary individual plants are widely separated. Fruits, with their complex adaptations, enable angiosperms to disperse widely.

REVIEW

1. Plant gametes fuse to form a diploid zygote that grows by mitosis to the _____ generation.

2. How do moss sperm get to the archegonium?

3. A sperm is haploid. Endosperm is _____.

4. The gymnosperms are seed plants; unlike the angiosperms, in which the seeds are contained within a fruit, the seeds of gymnosperms are _____.

5. Name the two classes of angiosperms.

SELF-QUIZ

1. What key feature probably allowed the ancestor of the plants to invade the land?
   - (a) leaves
   - (b) roots
   - (c) chlorophyll
   - (d) a vascular system
   - (e) an outer covering resistant to desiccation

2. Which of the following are found *only* in vascular plants?
   - (a) a waxy cuticle
   - (b) carbohydrate molecules
   - (c) a vascular system
   - (d) cell walls
   - (e) stomata

3. Which plants *lack* archegonia and antheridia?
   - (a) gametophytes
   - (b) bryophytes
   - (c) ferns
   - (d) advanced vascular plants
   - (e) none of the above

4. An example of a seedless vascular plant is
   - (a) a hornwort
   - (b) a club moss
   - (c) a pine tree
   - (d) a fruit tree
   - (e) a lily

5. Which of the following are unique characteristics of the angiosperms?
   - (a) double fertilization
   - (b) flowers
   - (c) fruits
   - (d) insect pollination
   - (e) vascular system

THOUGHT QUESTION

1. Why do mosses and ferns both require free water to complete their life cycles? At which stage of the life cycle is the water required? Do angiosperms also require free water to complete their life cycles? What are the reasons for the difference, if any?

FOR FURTHER READING

HEYWOOD, V.H. (editor): *Flowering Plants of the World*, Mayflower Books, Inc., New York, 1978. An outstanding guide to the families of flowering plants.

RICHARDSON, D.H.S.: *The Biology of Mosses*, John Wiley & Sons, Inc., New York, 1981. Excellent, concise account of the mosses.

TAYLOR, T.N.: *Paleobotany: An Introduction to Fossil Plant Biology*, McGraw-Hill Book Co., New York, 1981. An excellent survey of fossil plants.

FIGURE 28-1   Female lion and her cub, gazing over the plains of the Serengeti, in Tanzania, East Africa. Many of the animals of the Serengeti are becoming rare except in parks, as the expanding human population reduces their habitat.

Chapter 28

ANIMALS

Overview

In this chapter we shall trace the long evolutionary history of the animals. All of the major animal phyla first evolved in the sea during the Cambrian Period. All but two of them are still water-dwellers—only the phyla containing vertebrates and the arthropods (spiders and insects) are fully terrestrial. The earliest animals to evolve in the sea had no distinct tissues or organs. Later there evolved a succession of animals with well-defined tissues—first solid worms and then worms with progressively more complex body cavities. The next innovation in body design was segmentation, in which bodies were assembled from similar subunits, like the cars of a train. In the arthropods, jointed appendages evolved. A radical change in the organization of the embryo accompanied the evolution of the sea stars and their relatives, the echinoderms, and of our own phylum, the chordates.

For Review

Here are some important terms and concepts that you will encounter in this chapter. If you are not familiar with them, you should review them before proceeding.

Major features of evolutionary history (Chapter 18)

Vertebrate evolution (Chapter 19)

Classification (Chapter 24)

The five kingdoms of life (Chapter 24)

Choanoflagellates (Chapter 26)

One kingdom of life remains for us to discuss: the animal kingdom. Animals are the most familiar organisms to most people. We are animals, and so are fleas and worms and jellyfish. Animals are multicellular heterotrophs; there are no photosynthetic animals and no unicellular ones. The animal kingdom evolved from protists in water, and much of its evolution since then has been in the sea. Of the 35 phyla in the animal kingdom, only 2 have been overwhelmingly successful as land-dwellers: the arthropod phylum of spiders and insects (Figure 28-1) and our own chordate phylum, which includes the vertebrates. In this chapter we shall follow that evolutionary journey, tracing the path that has led from the protists to sponges and spiders and bears.

We shall start with the simplest animals, the sponges, and then talk about jellyfish, snails, and a variety of different worms. We shall then consider insects, the most successful of animal groups, followed by the sea stars and other echinoderms, which comprise the phylum most closely related to our own, and finish with chordates like

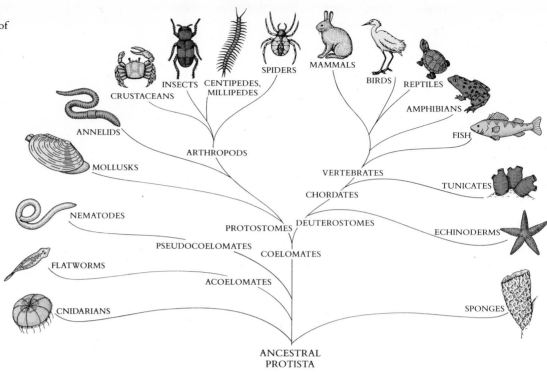

ourselves. The evolutionary relationships between the major groups of animals will be
discussed throughout this chapter and are presented visually in Figure 28-2.

SOME GENERAL FEATURES OF ANIMALS

Animals are extraordinarily diverse in form. They range in size from a few that are
smaller than many protists to others like the truly enormous whales and giant squids.
The cells in animals are exceedingly diverse in form and function and are of fundamental importance in making up their complex bodies, which function so well under such
a wide variety of circumstances. Except for the sponges, animal cells are organized into
tissues, which are groups of cells combined into a structural and functional unit. In
most animals the tissues are organized into **organs,** which are complex structures that
are made up by the composition of two or more kinds of tissues.

Most animals reproduce sexually. Their gametes—eggs and sperm—do not divide by mitosis. With few exceptions animals are diploid; their gametes are the only
haploid cells in their life cycles. The complex form of a given animal develops from a
zygote formed from the union of male and female gametes. In a characteristic process
of embryonic development, discussed in Chapter 42, the zygote first undergoes a series
of mitotic divisions and becomes a hollow ball of cells called the **blastula.** This developmental stage occurs in all animals. In most animals the blastula folds inward at one
point to form a hollow sac with an opening at one end called the **blastopore.** An embryo with a blastopore is called a **gastrula.** The subsequent growth and movement of
the cells of the gastrula produce the digestive system. The details of early embryonic
development differ widely from one phylum of animals to another, as you will see.
These details often provide important clues to the evolutionary relationships among
the phyla.

THE SPONGES—ANIMALS WITHOUT TISSUES

Sponges are the simplest of animals. The cells of a sponge are not organized into tissues. Sponges lack organs, and most sponges completely lack symmetry. The bodies
of sponges consist of little more than masses of cells embedded in a gelatinous matrix.
There is relatively little coordination among the cells—a sponge can pass through a fine

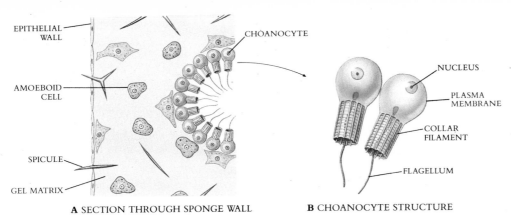

EPITHELIAL WALL

CHOANOCYTE

AMOEBOID CELL

SPICULE

GEL MATRIX

A SECTION THROUGH SPONGE WALL

NUCLEUS

PLASMA MEMBRANE

COLLAR FILAMENT

FLAGELLUM

B CHOANOCYTE STRUCTURE

FIGURE 28-3

A The microstructure of a sponge, as seen in a transection through the animal; note the choanocytes lining the pore.
B Details of two choanocytes.

silk mesh, with individual clumps of cells separating, and then reaggregate on the other side.

The body of a young sponge is shaped like a sac or vase. The body wall is covered on the outside by a layer of flattened cells called the **epithelial wall.** Facing into the internal cavity are specialized, flagellated cells called **choanocytes,** or collar cells (Figure 28-3, *A*). Between the choanocytes and epithelial wall is a gelatinous, protein-rich matrix in which occur various types of amoeboid cells, minute needles of calcium carbonate or silica called **spicules,** and fibers of a tough protein called **spongin.** The spicules and spongin may occur together, or only one of the two may be present. These elements not only strengthen the body of the sponge, but they may also deter predators.

The body of a sponge is perforated by tiny holes. The name of the phylum, Porifera, refers to this system of pores. The beating of the flagella of the many choanocytes that line the body cavity draws water in through the pores and drives it through the sponge, thus providing the means by which the sponge acquires food and oxygen and expels wastes. The flagellum of each choanocyte beats independently. In some sponges 1 cubic centimeter of tissue can propel more than 20 liters of water a day in and out of the sponge body! The movement of water through the pores and channels of sponges is a primitive form of the circulatory systems that occur in other, more complex animals.

> **Sponges are unique in the animal kingdom because they possess choanocytes, which are special flagellated cells whose beating drives water through the body cavity.**

The microscopic examination of individual choanocytes (Figure 28-3, *B*) reveals an important substructure: the base of each flagellum is surrounded by a collar of small, hairlike projections, which resemble a picket fence. The beating flagellum of the choanocyte draws water through the openings in the collar. Any food particles in the water are also drawn in and are trapped. The trapped particles pass directly through the plasma membrane, and are later digested either by the choanocyte itself or by a neighboring amoeboid cell.

Each choanocyte closely resembles a protist with a single flagellum and is exactly like the group of unicellular zoomastigotes called choanoflagellates (see Chapter 26, Figure 26-6). Based on their unique features and the close resemblance between choanoflagellates and the choanocytes of sponges, it seems almost certain that choanoflagellates are the ancestors of the sponges. They may also be the ancestors of the other animals as well, but it is difficult to be certain of a direct relationship between sponges and other groups.

There are about 10,000 species of marine sponges (Figure 28-4) and about 150 additional species that live in fresh water. In the sea, sponges are abundant at all depths. Although some sponges are tiny, no more than a few millimeters across, others, like the loggerhead sponges, may reach 2 meters or more in diameter. Although larval sponges are free-swimming, the adults are **sessile,** or anchored in place.

FIGURE 28-4

Diversity in sponges (phylum Porifera).
A A barrel sponge, a large sponge in which the form is somewhat organized.
B Red boring sponge *(Cliona delitrix),* a crustose (encrusting) sponge.

A

B

FIGURE 28-5

Representative cnidarians
(phylum Cnidaria).
A A soft coral.
B Yellow cup coral.
C A jellyfish, *Aurelia aurita.*
D A sea anemone.

A

B

C

D

CNIDARIANS: THE RADIALLY SYMMETRICAL ANIMALS

All animals other than sponges are called **eumetazoans,** or "true animals"; they have a definite shape and symmetry and nearly always distinct tissues. Three distinct cell layers form in the embryos of all eumetazoans: an outer **ectoderm,** an inner **endoderm,** and an in-between **mesoderm.** These layers ultimately differentiate into the tissues of the adult animal. In general the nervous system and outer covering layers called **integuments** develop from the ectoderm, the muscles and skeletal elements develop from the mesoderm, and the intestine and digestive organs develop from the endoderm.

> **All eumetazoans possess three embryonic tissues: ectoderm, mesoderm, and endoderm.**

Only the two most primitive of the eumetazoan phyla are **radially symmetrical,** with parts arranged around a central axis like the petals of a daisy. These two phyla are Cnidaria (pronounced ni–DAH–ree–ah), which includes jellyfish, hydra, sea anemones, and corals, and Ctenophora (pronounced Tea–NO–fo–rah), a minor phylum that includes the comb jellies. The bodies of all other eumetazoans are marked by a fundamental bilateral symmetry (see Figure 28-8).

All cnidarians (Figure 28-5) are carnivores that capture their prey, such as fishes and crustaceans, with the tentacles that ring their mouth. Their bodies, like those of all eumatazoans, consist of three layers and differ greatly from those of sponges. Cnidarians are basically gelatinous in construction. More than 9000 species are known.

There are two basic body forms among the cnidarians, **polyps** and **medusae** (Figure 28-6). Polyps are cylindrical, pipe-shaped animals that are usually attached to a rock. In polyps the mouth faces away from the rock on which the animal is growing and therefore is often directed upward. Many polyps build up a hard shell, an internal skeleton, or both. In contrast, most medusae are free-floating and are often umbrella-shaped. Their mouths usually point downward, and the tentacles hang down around the mouth. Medusae are commonly known as jellyfish because of their thick, gelatinous interior. Many cnidarians occur only as polyps, whereas others exist only as medusae; still others alternate between these two phases during the course of their life cycles.

Individual cnidarian species may either be medusae, which are floating, bell-shaped animals with the mouth directed downward, or polyps, which are anchored animals with the mouth directed upward. In some cnidarians these two forms alternate during the life cycle of the organism.

Nematocysts

The tentacles of the cnidarians bear stinging cells called **cnidocytes.** The name of the phylum Cnidaria refers to these cells, which are highly distinctive and occur in no other group of organisms. Within each cnidocyte there is a **nematocyst** (Figure 28-7), which is best thought of as a small but very powerful harpoon. Cnidarians use nematocysts to spear their prey and then draw the harpooned prey back to the tentacle containing the cnidocyte. To propel the harpoon the cnidocyte uses water pressure. Here's how it works: using transmembrane channels such as those described in Chapter 6, each cnidocyte builds up a very high internal concentration of ions. Because the cnidocyte's membrane is not permeable to water, this creates an intense osmotic pressure. If a flagellum-like trigger on the cnidocyte is touched, other transmembrane channels open, permitting water to rush in. The resulting hydrostatic pressure pushes the barbed filament of the nematocyst violently outward. Because the filament is shot from the nematocyst so forcefully, the barb can penetrate even the hard shell of crustaceans. Nematocyst discharge is one of the fastest cellular processes in nature. The entire process takes place in about 3 milliseconds, with a maximum velocity of 2 meters per second.

Cnidarians characteristically possess a specialized kind of cell called a cnidocyte. Each cnidocyte contains a nematocyst, which is like a harpoon and is used to attack prey. Nematocysts are found in no phylum other than Cnidaria.

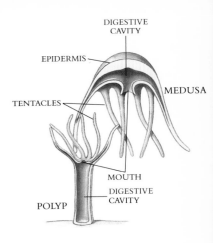

FIGURE 28-6

The two kinds of cnidarians, the medusa (above) and the polyp (below). These two phases alternate in the life cycles of many cnidarians, but a number—including the corals and sea anemones, for example—exist only as polyps.

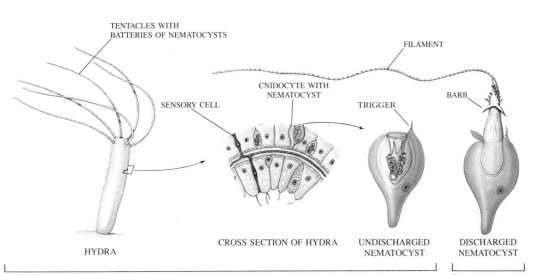

FIGURE 28-7

A Details of the structure of *Hydra,* showing the position of the cnidocytes and the structure of the nematocysts they contain.
B A discharged nematocyst from *Hydra.*

Extracellular Digestion

A major evolutionary innovation in the cnidarians, as compared with sponges, is the extracellular digestion of food—that is, digestion within a gut cavity rather than within individual cells. This evolutionary advance has been retained by all of the more advanced groups of animals. Cnidarians have a digestive cavity with only one opening (Figure 28-6). Digestive enzymes, primarily proteases, are released from cells lining the walls of the cavity and partially break down the food. Unlike the process in more advanced invertebrates, however, digestion is not completely extracellular: it only fragments food into small bits that are then engulfed by the cells lining the gut by the process of phagocytosis (Chapter 6).

What is important about the evolution of extracellular digestion that precedes the phagocytosis and intracellular digestion in cnidarians is that for the first time it became possible to digest an animal larger than oneself. Cnidarians simply tackle the job a little at a time.

> **Cnidarians are the most primitive animals that exhibit extracellular digestion.**

THE EVOLUTION OF BILATERAL SYMMETRY

Unlike a radially symmetrical animal, a **bilaterally symmetrical** animal (Figure 28-8) has a right half and a left half that are mirror images of each other. There are a top and a bottom, better known respectively as the **dorsal** and **ventral** portions of the animal. There is also a front or **anterior** end and a back or **posterior** end, and therefore right and

FIGURE 28-8

How bilateral (**A**) and radial (**B**) symmetry differ.

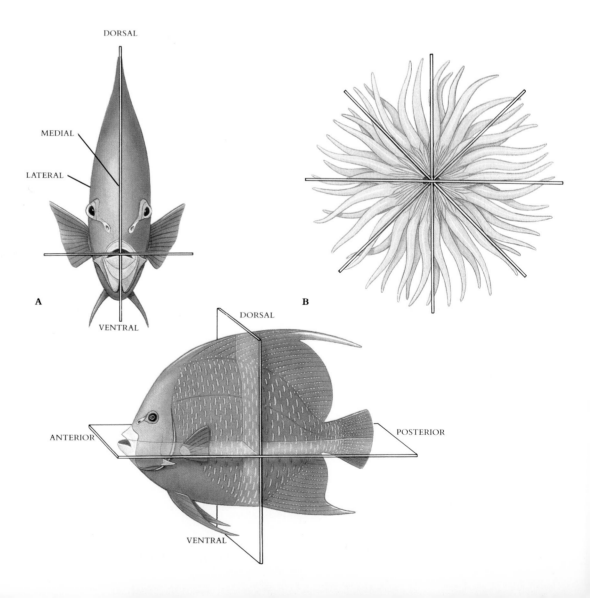

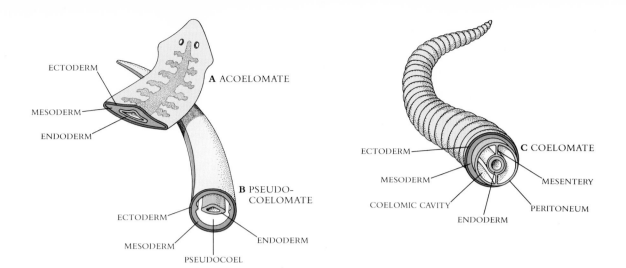

FIGURE 28-9

Three plans for construction of animal bodies, representing **(A)** acoelomate, **(B)** pseudocoelomate, and **(C)** coelomate patterns.

left sides. All eumetazoans other than cnidarians and ctenophores are bilaterally symmetrical (even sea stars, as you will see). The bilaterally symmetrical animals constitute a major advance because this symmetry allows different parts of the body to become specialized in different ways.

Three basic body plans occur in bilaterally symmetrical animals (Figure 28-9):

1. In the more advanced phyla, the tissues of the mesoderm open during development, leading to the formation of a particular kind of body called the **coelom.** The digestive, reproductive, and other internal organs develop within or around the margins of this coelom and are suspended within it by double layers of mesoderm known as **mesenteries.** Animals in which a coelom develops are called **coelomates.**

2. In another group of phyla, the body cavity develops between the mesoderm and the endoderm rather than within the mesoderm. Because of the way it originates, the body cavity of these animals lacks the characteristic epithelial lining derived from mesoderm that is found in a true coelom. An **epithelium** is a type of tissue that covers an exposed surface or lines a tube or cavity. The kind of cavity described is called a **pseudocoel,** and the animals in which it occurs are called **pseudocoelomates.**

3. Some bilaterally symmetrical animals have no body cavity at all, other than the digestive system. Animals that have this kind of a body plan are called **acoelomates.**

The body architecture of bilaterally symmetrical animals follows one of three patterns:
1. Coelomate, possessing a cavity bounded by mesoderm
2. Pseudocoelomate, possessing a cavity between endoderm and mesoderm
3. Acoelomate, possessing no body cavity
All vertebrates and the more advanced invertebrates are coelomates.

SOLID WORMS: THE ACOELOMATE PHYLA

Among bilaterally symmetrical animals, those with the simplest body plan are the acoelomates; they are solid worms that lack any internal cavity other than the digestive tract (see Figure 28-9, *A*). By far the largest phylum of acoelomates, with about 15,000 species, is Platyhelminthes, which includes the flatworms (Figure 28-10). These ribbon-shaped, soft-bodied animals are flattened from top to bottom, like a piece of tape or ribbon.

Although in structure they are among the simplest of all the bilaterally symmetrical animals, flatworms exhibit traces of many of the evolutionary trends that are so

FIGURE 28-10

A marine, free-living flatworm (phylum Platyhelminthes).

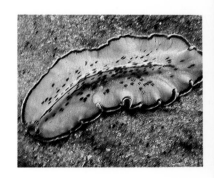

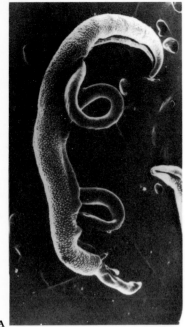

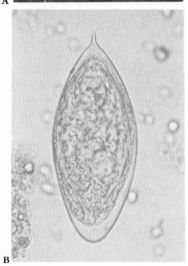

FIGURE 28-11

Schistosoma.
A Scanning electron micrograph of copulating male and female individuals. The thicker male has a canal that runs the length of its body, within which the female is held during insemination and egg-laying. The worms mate in the human bloodstream.
B An egg containing a mature larva.

highly developed among the members of more advanced phyla. The members of this phylum are the simplest animals in which organs occur. Flatworms have distinct bilateral symmetry and a definite head at the anterior end. The presence of organs, bilateral symmetry, and a distinct head is characteristic of all the more advanced phyla of animals.

Most species of flatworms are parasitic and occur within the bodies of members of almost every other animal phylum. Flatworms range in size from 1 millimeter or less to many meters long, as in some of the tapeworms.

The acoelomates, typified by the flatworms, are the most primitive bilaterally symmetrical animals and are the simplest animals in which true organs occur.

Flatworms lack circulatory systems, and most of them have a gut with only one opening. Therefore they excrete wastes directly into the gut and out through the mouth. To a lesser extent they also excrete wastes by means of specialized, bulblike cells lined with cilia that function primarily to regulate the water balance of the organism. The nervous systems of flatworms are simple; only the tiny swellings that occur near the leading (anterior) end of some flatworms resemble brains.

The free-living flatworms belong to the class Turbellaria, a large group with about 121 families. There are two classes of parasitic flatworms: flukes (class Trematoda) and tapeworms (class Cestoda). Among the best-known flukes are the blood flukes of the genus *Schistosoma* (Figure 28-11), which afflict some 200 to 300 million people (about 1 in 20 of the world's population) throughout tropical Asia, Africa, Latin America, and the Middle East. The eggs of *Schistosoma* leave the human body through the urine and the feces, and their larvae develop within the bodies of freshwater snails. Eventually these snails release a highly infectious stage of the worm's life cycle; the individuals of this stage burrow into human skin and eventually reach the intestine or the bladder, where they may live for up to 30 years causing inflammation, discomfort, and sometimes death. Tapeworms also alternate between various hosts; the juvenile beef tapeworm occurs in cattle, but the mature form of the tapeworm occurs in humans.

THE EVOLUTION OF A BODY CAVITY

The body organization of the other bilaterally symmetrical animals differs from the solid worms in an important way: all the other bilaterally symmetrical animals possess an internal body cavity. The evolution of an internal body cavity made possible a significant advance in animal architecture. Consider for a moment the limitations of a solid body: a solid worm has no internal circulatory or digestive system, and all of its internal organs are pressed on by muscles and are thus deformed by muscular activity. Even though flatworms do have digestive systems, they are subject to the same type of problems.

An internal body cavity circumvents these limitations, and its development was an important step in animal evolution. Perhaps the most important advantage of an internal body cavity is that the body's organs are located within a fluid-filled enclosure where they can function without having to resist pressures from the surrounding muscles. In addition, the fluid that fills the cavity may act as a circulatory system, freely transporting food, water, waste, and gases throughout the body. Without the free circulation made possible by such a system, every cell of an animal must be within a short distance of oxygen, water, and all of the other substances that it requires.

The digestive system of an animal, like its circulatory system, also functions much more efficiently within an internal body cavity such as a coelom than it can when embedded in other tissues. When the gut is suspended within such a cavity, food can pass through the gut freely and at a rate controlled by the animal because the gut does not open and close where the animal moves. By controlling the rate of passage of the food, the body allows it to be digested much more efficiently than would be possible otherwise. Waste removal is also carried out much more efficiently under such circumstances.

In addition, an internal body cavity provides space within which the gonads (ovaries and testes) can expand, allowing the accumulation of large numbers of eggs and sperm. Such accumulation helps to make possible all of the diverse modifications of breeding strategy that characterize the phyla of higher animals. Furthermore, large numbers of gametes can be released when the conditions are as favorable as possible for the survival of the young animals.

Nematodes

Nematodes are bilaterally symmetrical, cylindrical, unsegmented worms that possess a pseudocoelomate body plan (see Figure 28-9, B). They lack a defined circulatory system; the circulatory role is performed by the fluids that move within the pseudocoel. Nematodes, like all coelomates, have a complete, one-way digestive tract, which functions like an assembly line, the food being acted on in different ways in each section. First the food is broken down, then absorbed, then the wastes are treated and stored, and so on. The body of a nematode is covered by a flexible, thick cuticle, which is shed periodically as the individual grows. A layer of muscles extends beneath the epidermis along the length of the worm. These longitudinal muscles push both against the cuticle and against the pseudocoel, whipping the body from side to side as the nematode moves.

There are 12,000 recognized species in the phylum Nematoda, most of them microscopic animals that live in soil. Nematodes occur practically everywhere. It has been estimated that a spadeful of fertile soil contains, on the average, 1 million nematodes. So many new kinds of nematodes are found when environments are sampled that some scientists think there actually may be 500,000 or more species of this phylum, the great majority of which have never been collected or studied. Almost every species of plant and animal that has been studied has been found to have at least one parasitic species of nematode living in it, and nematodes are one of the most serious groups of pests on agricultural and horticultural crops throughout the world. Roundworms and hookworms are common nematode parasites of humans in the United States.

ADVENT OF THE COELOMATES

We have met two of the three bilaterally symmetrical body plans: the acoelomates, represented by the solid flatworms, and the pseudocoelomates, which are worms with a body cavity that develops between their mesoderm and endoderm. Even though both of these body plans have proven very successful, a third way of organizing the body has evolved and occurs in the bulk of the animal kingdom—the "higher" invertebrates and vertebrates. It involves the development of a coelom, which is a body cavity that originates within the mesoderm (Figure 28-9, C).

The Advantage of the Coelom

Both pseudocoelomate and coelomate animals possess a fluid-filled body cavity, a great improvement in body design compared with the less advanced solid worms. What then is the functional difference between a pseudocoel and a coelom, and why has the latter kind of body cavity been so much more overwhelmingly successful in evolutionary terms? In coelomates the body cavity develops not between endoderm and mesoderm but entirely within the mesoderm. This makes it easier for complex organ systems to develop. The evolutionary specialization of the internal organs of the coelomates has far exceeded that of the pseudocoelomates. For example, very few pseudocoelomates possess a true circulatory system, whereas many coelomates have such a system.

In addition, the presence of a coelom allows the digestive tract, by its coiling or folding within the coelom, to be longer than the animal itself. The longer passage allows for storage organs for undigested food, longer exposure to the enzymes for more complete digestion, and even storage and final processing of food remnants. Such an arrangement allows an animal to eat a great deal when it is safe to do so and then to hide

FIGURE 28-12

Patterns of embryonic development in coelomates, and of egg cleavage in protostomes and deuterostomes. The progressive division of cells during embryonic growth is called **cleavage.** In **spiral cleavage,** each new cell buds off at an angle oblique to the axis of the embryo; this kind of cleavage is characteristic of nearly all protostomes. In **radial cleavage,** which is characteristic of deuterostomes, the cells divide parallel to and at right angles to the polar axis. As a result, the pairs of cells that result from each division are positioned directly above and below one another. These differences are correlated with the other features of embryonic development illustrated here.

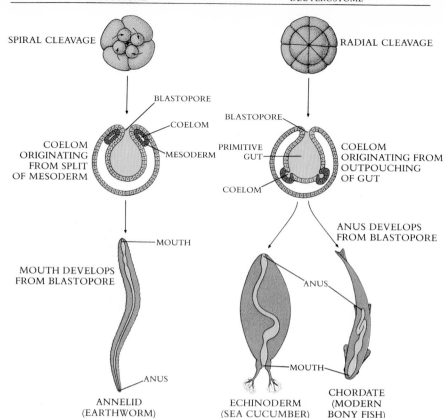

during the digestive process, thus limiting the animal's exposure to predators. The tube within the coelom architecture is also more flexible, thus allowing the animal greater freedom to move.

> **The evolution of the coelom was a major improvement in animal body architecture. It permitted the development of a closed circulatory system, provided a fluid environment within which digestive, sexual, and other organs could be suspended, and facilitated muscle-driven body movement.**

An Embryonic Revolution

There are two major branches of coelomate animals, representing two distinct evolutionary lines. In the first, which includes the mollusks, annelids, and arthropods as well as some smaller phyla, the mouth (stoma) develops from or near the blastopore (Figure 28-12). This pattern of embryonic development also occurs in all noncoelomate animals. An animal whose mouth develops in this way is called a **protostome.** If the animal has a distinct anus or anal pore, it develops later in another region of the embryo. The fact that this kind of developmental pattern is so widespread makes it virtually certain that it is the original one for animals as a whole and that it was characteristic of the common ancestor of all eumetazoan animals.

A second, very distinct pattern of embryological development occurs in the echinoderms, the chordates, and a few other small related phyla. In these animals the anus forms from or near the blastopore, and the mouth forms later on another part of the blastula (Figure 28-12). Animals in this group of phyla are called the **deuterostomes.** They are clearly related to one another by their shared pattern of embryonic development.

In protostomes the developmental fate of each cell in the embryo is fixed when that cell first appears. Even at the four-celled stage, each cell is different, and not one of them, if separated from the others, can develop into a complete animal because the chemicals that act as developmental signals are localized in different parts of the egg.

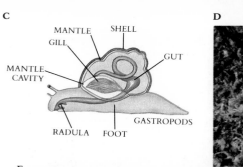

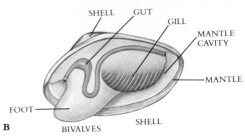

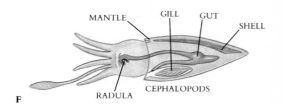

FIGURE 28-13

Mollusks.

A A bivalve, the scallop, *Chlamys hericia.* Note the blue eyes around the margin of the body.

B Body plan of a bivalve.

Gastropods.

C Body plan of a gastropod.

D The terrestrial snail *Allogona townsendiana.*

E A nudibranch, or marine slug, *Hermissenda crassicornis.*

Cephalopods.

F Body plan of a cephalopod.

G An octopus. Octopuses generally move slowly along the bottom of the sea.

H Pearly nautilus, *Nautilus pompilius,* from the Philippines.

Consequently, the cleavage divisions that occur after fertilization separate different signals into different daughter cells. On the other hand, in deuterostomes the first cleavage divisions of the fertilized embryo result in identical daughter cells, any one of which can, if separated, develop into a complete organism. The commitment of individual cells to developmental pathways occurs later.

We begin our discussion of the coelomate animals with the three major phyla of protostomes—the mollusks, annelids, and arthropods. These three phyla include such familiar animals as clams, snails, octopuses, earthworms, lobsters, spiders, and insects. In the members of these phyla we can observe all of the major advances that are associated with the evolution of the coelom. We shall then consider the two largest phyla of deuterostomes—the echinoderms and the chordates.

MOLLUSKS

The **mollusks** (phylum Mollusca) include the snails, clams, scallops, oysters, cuttlefish, octopuses, slugs, and many other familiar animals. Three classes of mollusks are representative of the phylum: (1) Gastropoda—snails, slugs, limpets, and their relatives; (2) Bivalvia—clams, oysters, scallops, and their relatives; and (3) Cephalopoda—squids, octopuses, cuttlefishes, and nautilus (Figure 28-13). The mollusks are the largest animal phylum, except for the arthropods, in terms of named species; there are at least 110,000, and probably at least that many more still to be discovered. Mollusks are

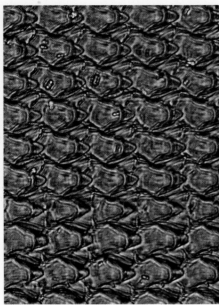

A

B

FIGURE 28-14

A The radula, or rasping tongue, of a snail.
B Enlargement of the rasping teeth on a radula.

widespread and often abundant in marine, freshwater, and terrestrial habitats. With about 35,000 species, the terrestrial mollusks far outnumber the roughly 20,000 species of terrestrial vertebrates.

The body of a mollusk is composed of a **visceral mass** that contains the body's organs, a muscular foot that is used in locomotion, and a heavy fold of tissue called a **mantle** wrapped around the visceral mass like a cape. Within the visceral mass are found the organs of digestion, excretion, and reproduction. The folds of the mantle enclose a cavity between the dorsal wall and the visceral mass; within this mantle cavity are the mollusk's gills, which comprise a specialized respiratory system of filamentous projections, rich in blood vessels, that capture oxygen and release carbon dioxide. Mollusk gills are very efficient. Many gilled mollusks extract 50% or more of the dissolved oxygen from the water that passes through the mantle cavity. In most members of this phylum the other surface of the mantle also secretes a protective **shell.** Many mollusks can withdraw for protection into their mantle cavity, which lies within the shell. In the squids and octopuses the mantle cavity has been modified to create the jet-propulsion system that enables the animals to move rapidly through the water.

The foot of a mollusk is muscular and may be adapted for locomotion, for attachment, for food capture (in squids and octopuses), or for various combinations of these functions. Some mollusks secrete mucus, forming a path that they glide along on their foot. In the cephalopods (squids and octopuses) the foot is divided into arms, which are also called tentacles.

One of the most characteristic features of mollusks is the **radula,** a rasping tonguelike organ that all members of the phylum have, except bivalves, in which it has almost certainly been lost during evolution. The radula consists primarily of chitin and is covered with rows of pointed, backward-curving teeth (Figure 28-14). It is used by some of the snails and their relatives—members of a group known as the gastropods— to scrape algae and other food materials off their substrates and then to convey this food to the digestive tract. Other gastropods are active predators; they use their radula to puncture their prey and extract food from it.

> Mollusks are the second largest phylum of animals in terms of named species. They characteristically have bodies with three distinct sections: head, visceral mass, and foot. All mollusks except the bivalves also possess a unique rasping tongue called a radula.

A

FIGURE 28-15

Annelids.
A Earthworms mating (class Oligochaeta).
B Shiny bristleworm, *Oenone fulgida*, another polychaete.
C Fan worm, *Sabella melanostigma*, a polychaete.
D A freshwater leech, with young leeches visible inside its body.

B

C

D

THE RISE OF SEGMENTATION

Unlike the mollusks, the other major phyla of protostomes, the annelids and the arthropods, have segmented bodies. Just as it is efficient for workers to construct a tunnel from a series of identical prefabricated parts, so these advanced protostome coelomates are "assembled" as a chain of identical segments, like the boxcars of a train. Segmentation underlies the organization of all advanced animals. In some adult arthropods the segments are fused, making it difficult to perceive the underlying segmentation, but it is usually apparent in embryological development. Even among vertebrates, the backbone and muscular areas are segmented.

> **Segmentation is a feature of the advanced coelomate phyla, notably the annelids and the arthropods, although it is not obvious in some of them.**

ANNELIDS

The **annelids** (phylum Annelida), one of the major animal phyla, are segmented worms (Figure 28-15). They are abundant in marine, freshwater, and terrestrial habitats throughout the world. Their segments are visible externally as a series of ringlike structures running the length of the body. Internally the segments are divided from one another by partitions. In each of the cylindrical segments of these animals, the digestive and excretory organs are repeated in tandem. Leeches are one of three classes of annelid worms (class Hirudinea). They are freshwater predators or bloodsuckers and were formerly used in medicine for blood-letting, which was thought to be beneficial. Cur-

FIGURE 28-16

An earthworm in external view and with a cross section showing the anatomy in a simplified form.

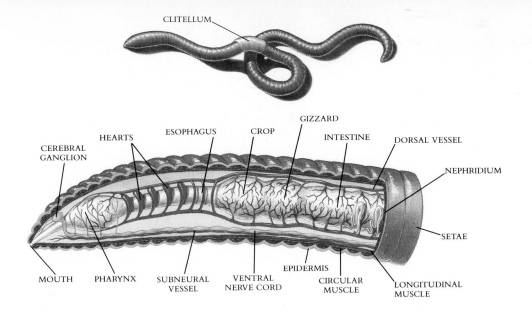

rently the enzymes that leeches use to prevent blood from clotting are extracted and used to treat certain conditions in which blood clots are a serious risk. The two other classes of annelids are Polychaeta, a largely marine group of about 8000 species, and Oligochaeta, the earthworms, with about 3100 species. Earthworms are often present in very large numbers and are extremely important in aerating and enriching the soil.

The basic body plan of the annelids (Figure 28-16) is a tube within a tube: the internal digestive tract is a tube suspended within the coelom, which is tube running from mouth to anus. The anterior segments have become modified, and a well-developed **cerebral ganglion,** or **brain,** is contained in one of them. The sensory organs are mainly concentrated near the anterior end of the worm's body. Some of these are sensitive to light, and elaborate eyes with lenses and retinas have evolved in certain members of the phylum. Separate nerve centers, or ganglia, are located in each segment but are interconnected by nerve cords. These nerve cords are responsible for the coordination of the worm's activities.

Annelids use their muscles to crawl, burrow, and swim. In each segment the muscles play against the fluid in the coelom. This fluid creates a hydrostatic (liquid-supported) skeleton that gives the segment rigidity, like an inflated balloon. Because each segment is separate, each is able to contract or expand independently. Therefore a long body can move in ways that are quite complex. For example, when an earthworm crawls on a flat surface, it lengthens some parts of its body while shortening others.

Each segment of an annelid typically possesses **setae** (singular, **seta**), which are bristles of chitin that help to anchor the worms during locomotion or when it is in its burrow. Because of the setae, annelids are often called "bristle worms." The setae are absent in all but one species of the leeches.

The annelids are characterized by serial segmentation. The body is composed of numerous similar segments, each with its own circulatory, excretory, and neural elements and each with its own array of setae.

ARTHROPODS

With the evolution of the first annelids, many of the major innovations of animal structure had already appeared: the division of tissues into three primary types (endoderm, mesoderm, and ectoderm), bilateral symmetry, coelomic body architecture, and segmentation. However, one further innovation remained. It marks the origin of the body plan that is characteristic of the most successful of all animal groups, the arthropods. This innovation was the development of jointed appendages.

The name "arthropod" comes from two Greek words, *arthros,* jointed, and *podes,* feet. All arthropods have jointed appendages. The numbers of these appendages are

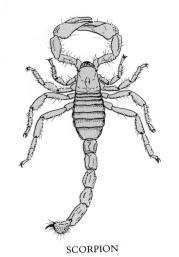

SCORPION

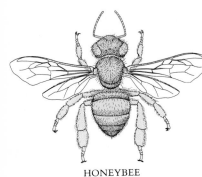

HONEYBEE

FIGURE 28-17

Evolution from many to few body segments among the arthropods; the scorpion and the honeybee are examples of arthropods with different numbers of body segments, as you can see by counting them.

A Roman Ilekys B

FIGURE 28-18

Some arthropods have a very tough exoskeleton, like this South American scarab beetle, *Dilobderus abderus* (order Coleoptera) **(A)**; others have a fragile exoskeleton, like the green darner dragonfly, *Anax junius* (order Odonata) **(B)**.

progressively reduced in the more advanced members of the phylum, and the nature of the appendages differs greatly in different subgroups. Thus individual appendages may be modified into antennae, mouthparts of various kinds, or legs.

Arthropod bodies are segmented like those of annelids, a phylum to which at least some of the arthropods are clearly related. The members of some classes of arthropods have many body segments. In others the segments have become fused together into functional groups such as the head or thorax of an insect (Figure 28-17). But even in these arthropods the original segments can be distinguished during the development of the larvae.

> **Arthropods, like annelid worms, are segmented, but in many arthropods the individual segments are fused into functional groups such as the head or thorax of an insect.**

The arthropods have a rigid external skeleton, or **exoskeleton,** in which chitin is an important element (Figure 28-18). The exoskeleton provides places for muscle attachment, protects the animal from predators and injury, and, most importantly, impedes water loss. As an individual outgrows its exoskeleton, that exoskeleton splits open and is shed, allowing the animal to increase in size. This process, which is known as **ecdysis,** is controlled by hormones. The eggs of arthropods develop into immature forms that may bear little or no resemblance to the adults of the same species; most members of this phylum change their characteristics as they develop from stage to stage during a process called **metamorphosis.**

Arthropods are by far the most successful of all life forms. Approximately a million animal species—about two thirds of all the named species on earth—are members of this gigantic phylum, and it is estimated that there are many millions more awaiting discovery. There are several times as many species of arthropods as there are of all other plants, animals, and microorganisms put together. A hectare of lowland tropical forest is estimated to be inhabited, on the average, by 41,000 species of insects. Many suburban gardens may have 1500 or more species of this gigantic class of organisms. The insects and other arthropods are abundant in all of the habitats on this planet, but they especially dominate the land, where along with the flowering plants and the vertebrates they determine the very structure of life. In terms of individuals, it has been estimated that approximately a billion billion insects are alive at any one time!

DIVERSITY OF THE ARTHROPODS

Many arthropods have jaws, or **mandibles,** formed by the modification of one of the pairs of anterior appendages (but not the one nearest the anterior end). These arthropods are called **mandibulates;** they include the crustaceans, insects, centipedes, millipedes, and a few other small groups. The remaining arthropods, which include the spiders, mites, scorpions, and a few other groups, lack mandibles. These animals are called the **chelicerates.** Their mouthparts, called **chelicerae** (Figure 28-19), evolved from the appendages nearest the anterior end of the animal. They often take the form

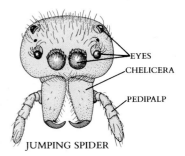

JUMPING SPIDER

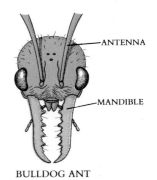

BULLDOG ANT

FIGURE 28-19

In the chelicerates, such as the jumping spider, the chelicerae are the foremost appendages of the body. Mandibulates, in contrast, such as the bulldog ant, have the antennae at the anterior (front) end of the body.

of pincers or fangs. The crustaceans seem to have evolved mandibles separately and are not thought to be directly related to the insects and other mandibulates.

Chelicerates

The fossil record of chelicerates goes back as far as that of any multicellular animals, about 630 million years. A major group of extinct arthropods, the trilobites, was also abundant then, and the living horseshoe crabs seem to be directly descended from them. By far the largest of the three classes of chelicerates is the largely terrestrial class Arachnida, with some 57,000 named species, including the spiders, ticks, mites, scorpions, and daddy longlegs (Figure 28-20). Scorpions are probably the most ancient group of terrestrial arthropods; they are known from as early as the Silurian Period, some 425 million years ago. There are about 35,000 named species of spiders (order Araneae). The order Acarai, the mites, is the largest in terms of number of species and the most diverse of the arachnids.

FIGURE 28-20

Arachnids are an extremely diverse group of chelicerate arthropods, as shown here by a few examples.

A The scorpion *Uroctonus mordas,* showing the characteristic pincers and segmented abdomen, ending in a sting, raised over the animal's back. White young scorpions cluster on this individual's back.
B An orb-weaving spider, the arrowhead spider *Micranthena,* in Peru.
C Tarantulas are large hunting spiders, powerful enough to prey on small lizards, mammals, and birds in addition to insects. They do not spin webs.
D The other genus of poisonous spiders of this area, the brown recluse, *Loxosceles reclusa.* Both species are common throughout temperate and subtropical North America, but bites are rare in humans.
E One of the two poisonous spiders in the United States and Canada, the black widow spider, *Latrodectus mactans.*

A

B

C

D

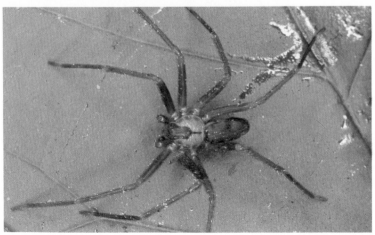

E

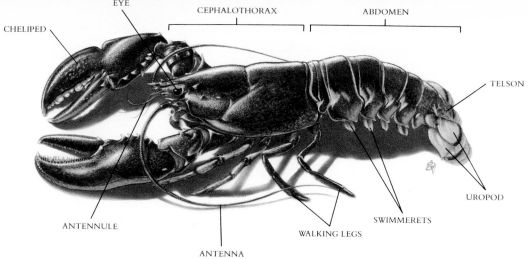

EYE

CHELIPED

CEPHALOTHORAX

ABDOMEN

TELSON

UROPOD

ANTENNULE

ANTENNA

WALKING LEGS

SWIMMERETS

FIGURE 28-21

The principal external features of a lobster, *Homarus americanus*. Some of the specialized terms used to describe crustaceans are indicated; for example, the head and thorax are fused together into a cephalothorax.

A

B

C

D

E

F

FIGURE 28-22

Crustaceans.
A Gooseneck barnacles, *Lepas anatifera*, feeding. Barnacles remain fixed in position when they are mature, kicking food into their mouths with their feet.
B A marine amphipod, *Eurius* (order Amphipoda), thriving in water at −1.9° C in Antarctica. Many amphipods, known as beach fleas or sand fleas, are terrestrial or semiterrestrial in their habits.
C Sowbugs, *Porcellio scaber,* representatives of the terrestrial isopods (order Isopoda).
D A copepod, member of an abundant group of marine and freshwater crustaceans (order Copepoda), most of which are a few millimeters long. Copepods are important components of the plankton.
E A freshwater crayfish, *Procambarus.*
F Sally lightfoot crab, *Grapsus grapsus*. This crab species is semi-terrestrial, ranging widely overland in search of food. Many crustaceans are important sources of food for humans.

Mandibulates: Crustaceans

The crustaceans (subphylum Crustacea, with a single class) are a large, diverse group of primarily aquatic organisms, including some 35,000 species of crabs, shrimps, lobsters (Figure 28-21), crayfish, barnacles, water fleas, pillbugs, and related groups (Figure 28-22). Often incredibly abundant in marine and freshwater habitats and playing a role of critical importance in virtually all aquatic ecosystems, crustaceans have been called "the insects of the water." Most crustaceans have two pairs of antennae, three pairs of chewing appendages, and various numbers of pairs of legs. Crustaceans differ from the insects—but resemble the centipedes and millipedes—in that they have legs on their abdomen as well as on their thorax. They are the only arthropods with two pairs of antennae.

A

B

C

FIGURE 28-23

Insects and some relatives.
A A soldier fly, *Ptecticus trivittatus* (order Diptera). Flies have only one pair of wings; during the course of their evolution, the second pair has been converted to halteres (balancers).
B An African blister beetle (order Coleoptera), which has spread its two tough, leathery forewings, exposing its delicate flying wings. More than 350,000 species of beetles—a fifth of all species of animals—have been described.
C A mayfly, *Hexagenia limbata* (order Ephemeroptera), immediately after molting to reach the adult form. Adult mayflies, the reproductive stage, do not feed and live for only a few hours or a few days; their nymphs, which are aquatic, live underwater for months or even years.
D A centipede, *Scolopendra*.
E A millipede, *Sigmoria*. Centipedes are active predators, whereas millipedes are sedentary herbivores. Millipedes and centipedes are closely related to insects; all of the other animals illustrated here are insects.
F Termites, *Macrotermes bellicosus* (order Isoptera). The large, sausage-shaped individual is a queen, specialized for laying eggs; the smaller individuals around it are nonreproductive workers. Only the reproductive males and females (queens) have wings, which are soon lost in the queens.
G Copulating grasshoppers (order Orthoptera).

Mandibulates: Insects and Their Relatives

The insects, class Insecta, are by far the largest group of arthropods, whether measured in terms of numbers of species or numbers of individuals; as such, they are the most abundant group of eukaryotes on earth. Insects live in every conceivable habitat on land and in fresh water, and a few have even invaded the sea. One scientist has calculated that there are probably about 300 million insects alive at any one time for each person on earth! More than 70% of all the named animal species are insects, and the actual proportion is undoubtedly much higher, because millions of additional forms await detection, classification, and naming. There are about 90,000 described species in the United States and Canada, and the actual number of species in this region probably approaches 125,000. A glimpse at the enormous diversity of insects is presented in Figure 28-23.

Insects have three body sections—the head, thorax, and abdomen; three pairs of legs, all attached to the thorax; and one pair of antennae (Figure 28-24). Most insects have compound eyes, which are composed of many independent visual units (Figure 28-25, *A*). The insect thorax consists of three segments, each of which has a pair of legs, although occasionally one or more of these pairs of legs is absent. The thorax is almost entirely filled with muscles that operate the legs and wings. Insects have two pairs of wings, which are attached to segments of the thorax, although in the flies and some other groups one or both of these pairs has been lost during the course of evolution. The wings of adult insects are solid sheets of chitin with strengthening veins made of chitin tubules. Like many other arthropods, insects have no single major respiratory organ. Their respiratory system consists of small, branched, cuticle-lined air ducts called **tracheae** (singular, **trachea;** Figure 28-25, *B*), which are a series of tubes that transmit oxygen throughout the body. The tracheae ultimately branch into very small **tracheoles**. Air passes into the tracheae through specialized openings called **spiracles,** which occur on many of the segments of the thorax and abdomen.

D

E

F

G

FIGURE 28-24

The external anatomy of a grasshopper (order Orthoptera), illustrating the major structural features of insects, the most numerous group of arthropods. The left wing has been removed to show the features of the abdomen.

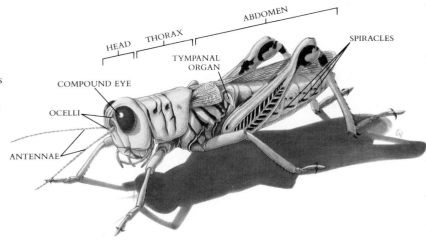

HEAD · THORAX · ABDOMEN · SPIRACLES · TYMPANAL ORGAN · COMPOUND EYE · OCELLI · ANTENNAE

A

B

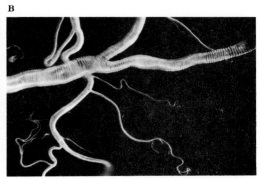

FIGURE 28-25

Two characteristic features of insects.
A The compound eyes found in insects are complex structures. Three ocelli—simple eyes—can be seen between the compound eyes of the robberfly (order Diptera).
B A portion of the tracheal system of a cockroach.

Also characteristic of insects are complex patterns of metamorphosis, in which certain stages give way to others. In some insects such as grasshoppers, cockroaches, and termites the stages grade from one to the other without abrupt changes; in others such as butterflies and moths, flies, and beetles, the changes are abrupt (Figure 28-26).

DEUTEROSTOMES

Two outwardly dissimilar large phyla, Echinodermata and Chordata, have a series of key embryological features that are very different from those of the other animal phyla (see Figure 28-12). Because it is extremely unlikely that these features evolved more than once, it is believed that both phyla share a common ancestry. They are the members of a group that we call the deuterostomes. Deuterostomes diverged from protostome ancestors more than 630 million years ago.

A

B

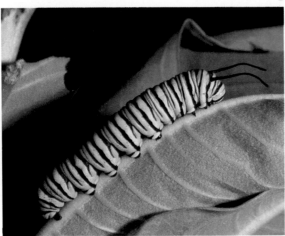

C

FIGURE 28-26

The life cycle of the monarch butterfly, *Danaus plexipus* (order Lepidoptera), illustrating complete metamorphosis.
A Egg.
B Larva (caterpillar) feeding on a leaf. The pseudopods, or false legs, that occur on the abdomen of larval butterflies and moths are not related to the true legs that occur in the adults.
C Larva preparing to shed its skin and become a chrysalis.
D Chrysalis.
E The colors of the wings of the adult butterfly can be seen through the thin outer skin of the chrysalis; the adult is nearly ready to emerge.
F to **H** Adult butterfly emerging.
I Adult monarch butterfly.

F

Deuterostomes, like protostomes, are coelomates. They differ fundamentally from protostomes, however, in the way the embryo grows: the blastopore of a deuterostome becomes the animal's anus, and the mouth develops at the other end; for a brief period of development, they have identical daughter cells, each of which, if separated from the others at an early enough stage, can develop into a complete organism; and whole groups of cells move around during the course of embryonic development to form new tissue associations.

ECHINODERMS

Echinoderms (phylum Echinodermata) are an ancient group of marine animals that are very well represented in the fossil record. The term *echinoderm* means "spine-skin," an apt description of many members of this phylum, which consists of about 6000 living

D

E

I

G

H

FIGURE 28-27

Echinoderms (phylum
Echinodermata).

A A sand dollar, *Echinarachnius
   parma* (class Echinoidea).
B Giant red sea urchin,
   Strongylocentrotus franciscanus
   (class Echinoidea).
C A sea star, *Oreaster occidentalis*
   (class Asteroidea).
D A sea cucumber, *Stichopus*
   (class Holothuroidea).
E A feather star (class
   Crinoidea).
F A brittle star, *Ophiothrix* (class
   Ophiuroidea).

A

B

C

D

E

F

species. Many of the most familiar animals seen along the seashore, such as the sea stars (starfish), brittle stars, sea urchins, sand dollars, and sea cucumbers, are echinoderms (Figure 28-27). All are bilaterally symmetrical as larvae but radially symmetrical as adults; thus they are said to be fundamentally bilaterally symmetrical. They certainly evolved from bilaterally symmetrical ancestors. Echinoderms are well represented not only in the shallow waters of the sea but also in its abyssal depths. All but a few of them are bottom-dwellers. The adults range from a few millimeters to more than 1 meter in diameter or length.

Basic Features of Echinoderms

Echinoderms have a five-part body plan corresponding to the arms of a sea star. As adults, these animals have no head or brain. Their nervous systems consist of central nerve rings from which branches arise. The animals are capable of complex response patterns, but there is no centralization of function. Apparently the centralization of the nervous system is not feasible in animals with radial symmetry.

The **water vascular system** of an echinoderm radiates from a ring canal, which encircles the animal's esophagus. Five radial canals extend into each of the five parts of the body and determine its basic symmetry (Figure 28-28). Water enters the water vascular system through a sievelike plate on the animal's surface, from which it flows through a tube to the ring canal. The five radial canals in turn extend out of the animal through short side branches into the many hollow tube feet. In some echinoderms each tube foot has a sucker at its end; in others, suckers are absent. At the base of each tube foot is a muscular sac, the **ampulla,** which contains fluid. When the sac contracts, the fluid is prevented from entering the radial canal and is forced into the tube foot, thus extending it. When extended, the foot attaches itself to the substrate. Longitudinal muscles can then shorten the foot, and the animal is pulled forward to a new position as water is forced back into the foot by the muscular sac. In a sea star any one of the arms may go first. By such action repeated in many tube feet, sea stars, sand dollars, and sea urchins slowly move along.

As adults, echinoderms are radially symmetrical animals whose bodies typically have a pattern that repeats itself five times. By means of a unique water vascular system connected to their tube feet, they are able to move about.

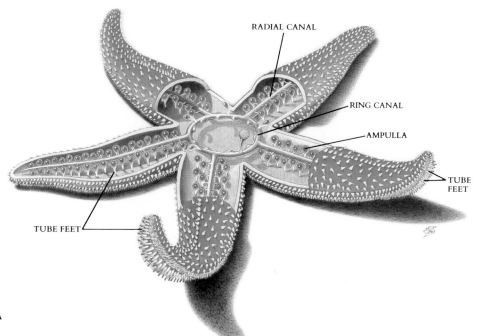

RADIAL CANAL

RING CANAL

AMPULLA

TUBE FEET

TUBE FEET

A

B

FIGURE 28-28

A The water vascular system of a sea star.
B The extended tube feet of a sea star. *Ludia magnifica.*

FIGURE 28-29

A human embryo. The segments called somites, which develop into muscle and vertebrae, are clearly distinct at this stage of development, reflecting the fundamentally segmented nature of all chordates.

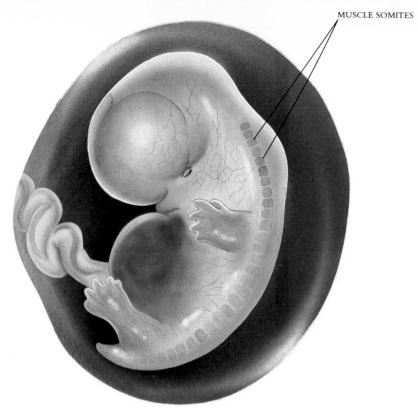

MUSCLE SOMITES

CHORDATES

The chordates (phylum Chordata) are the best understood and most familiar group of animals. There are some 42,500 species of chordates, a phylum that includes the birds, reptiles, amphibians, fishes, and mammals (and therefore our own species).

The three principal features of the chordates were presented in Chapter 19, where the evolution of vertebrates was discussed, but are reviewed here briefly. All chordates have a single, hollow nerve cord, a notochord, and pharyngeal slits (see Figure 19-3), and each of these features has played an important role in the evolution of the phylum. In the more advanced vertebrates the dorsal nerve cord becomes differentiated into the brain and spinal cord. The notochord, which persists throughout the life cycle of some of the invertebrate chordates, becomes surrounded and then is replaced by the vertebral column during the embryological development of vertebrates. Pharyngeal slits are present in the embryos of all vertebrates but are lost later in the development of the terrestrial vertebrates. However, the presence of these structures in all vertebrate embryos provides a clue to the aquatic ancestry of the group.

> **Chordates are characterized by a single, hollow dorsal nerve cord. At some point in the embryonic development of all chordates, the notochord, which is a flexible rod, forms dorsal to the gut, and slits are present in the pharynx. The notochord persists into the adult stage in the less advanced chordates.**

In addition to these three principal features, a number of other characteristics tend to distinguish the chordates. In their body plan, chordates are more or less segmented, and distinct blocks of muscles can be seen clearly in many less specialized forms (Figure 28-29). Chordates have an internal skeleton against which the muscles work, and either this skeleton or the notochord makes possible the extraordinary powers of locomotion that characterize the members of this group. Finally, chordates have a tail that extends beyond the anus, at least during their embryonic development; nearly all other animals have a terminal anus.

This great phylum is divided into three major groups. Two of them, the tunicates and the lancelets, are **acraniates,** that is, they lack a brain. The **craniate** chordates are the **vertebrates,** which we have already examined in detail in Chapter 19. We shall review the two other subphyla of chordates in more detail here.

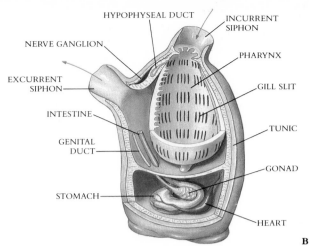

FIGURE 28-30

Tunicates (phylum Chordata, subphylum Tunicata).
A   A beautiful blue-and-gold tunicate.
B   Diagram of the structure of a tunicate.

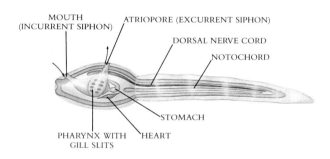

FIGURE 28-31

Diagram of the structure of a larval tunicate, which is very much like that of the postulated common ancestor of the chordates.

Tunicates

The tunicates (subphylum Tunicata) are a group of about 1250 species of marine animals. Most of them are **sessile**—remaining fixed in one spot—as adults (Figure 28-30). When they are adults, these animals lack visible signs of segmentation or of a body cavity. It would be very difficult to discern the evolutionary relationships of an adult tunicate by examining its features.

The tadpolelike larvae of tunicates, however, plainly exhibit all of the basic characteristics of chordates (notochord and nerve cord) and mark the tunicates as having the most primitive combination of features found in any chordate (Figure 28-31). They do not feed, and they have a poorly developed gut. The larvae remain free-swimming for no more than a few days; then they settle to the bottom and attach themselves to a suitable substrate by means of a sucker.

Lancelets

The lancelets (subphylum Cephalochordata; Figure 28-32) are scaleless, fishlike marine chordates a few centimeters long. They occur widely in shallow water throughout the oceans of the world. There are about 23 species of lancelets, which were given their English name because of their similarity to a lancet, a small, two-edged surgical knife. In the lancelets the notochord persists through the animal's life and runs the entire length of the dorsal nerve cord.

A

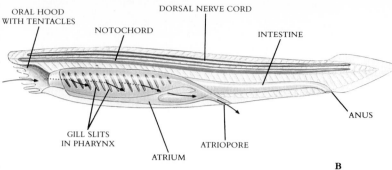

B

FIGURE 28-32

A Two lancelets, *Branchiostoma lanceolatum* (phylum Chordata, subphylum Cephalochordata), partly buried in shell gravel, with their anterior ends protruding. The muscle segments are clearly visible in this photograph; the numerous square, pale yellow objects along the side of the body are gonads, indicating that these are male lancelets.
B The structure of a lancelet, showing the path by which the water is pulled through the animal.

FIGURE 28-33

Embryonic development of a vertebrate. During the course of evolution, or of development, the flexible notochord is surrounded and eventually replaced by a cartilaginous or bony covering the centrum. The neural tube is protected by an arch above the centrum, and the vertebra may also have a hemal arch, which protects the major blood vessels below the centrum. The vertebral column is a strong, flexible rod against which the muscles pull when the animal swims or moves.

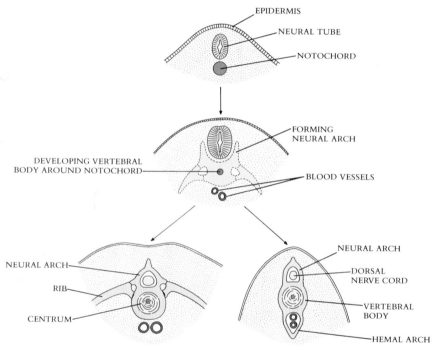

Vertebrates

Vertebrates (subphylum Vertebrata) differ from other chordates in that they usually possess a vertebral column, which replaces the notochord to a greater or lesser extent in adult individuals of different members of the subphylum. In addition, the vertebrates, or craniate chordates, have a distinct head with a skull and brain. The hollow dorsal nerve cord of most vertebrates is protected with a U-shaped groove formed by paired projections from the vertebral column (Figure 28-33). The seven classes of living vertebrates were discussed and illustrated in Chapter 19; three of them are fishes and four are tetrapods. The classes of fishes are Agnatha, the lampreys and hagfishes; Chondrichthyes, the cartilaginous fishes, sharks, skates, and rays; and Osteichthyes, the bony fishes, the dominant group of fishes today. The four classes of tetrapods are Amphibia, the amphibians, including salamanders, frogs, and toads; Reptilia, the reptiles; Aves, the birds; and Mammalia, the mammals.

SUMMARY

1. The animals, kingdom Animalia, comprise some 35 phyla and at least 4 million species. Animals are heterotrophic, multicellular organisms that ingest their food.

2. The sponges (phylum Porifera) are characterized by specialized, flagellated cells called choanocytes; by a lack of symmetry in the bodily organization of most species; and by a lack of tissues and organs.

3. The remaining animals have radially or bilaterally symmetrical body plans and distinct tissues. The cnidarians (phylum Cnidaria) are predominantly marine, radially symmetrical animals with unique stinging cells called cnidocytes, each of which contains a specialized harpoon apparatus, or nematocyst.

4. Bilaterally symmetrical animals include three major evolutionary lines: acoelomates, which lack a body cavity; pseudocoelomates, which develop a body cavity between the mesoderm and the endoderm; and coelomates, which develop a body cavity (coelom) within the mesoderm.

5. Acoelomates are the most primitive bilaterally symmetrical animals. They lack an internal cavity, except for the digestive system, and are the simplest animals that have organs, which are structures made up of two or more tissues. The most prominent phylum of acoelomates is the phylum Platyhelminthes, which includes the free-living flatworms and the parasitic flukes and tapeworms.

6. Pseudocoelomates, exemplified by the nematodes (phylum Nematoda), have a body cavity that develops between the mesoderm and the endoderm. This pseudocoel provides a place to which the muscles can attach, giving these worms enhanced powers of movement.

7. The two major evolutionary lines of coelomate animals—the protostomes and the deuterostomes—are represented among the oldest known fossils of multicellular animals, dating back some 630 million years.

8. In the protostomes the mouth develops from or near the blastopore, and the early divisions of the embryo are spiral. At early stages of development the fate of the individual cells is already determined, and they cannot develop individually into a whole animal. The major phyla of protostomes are Mollusca, Annelida, and Arthropoda.

9. In the deuterostomes the anus develops from or near the blastopore, and the mouth forms later on another part of the blastula. At early stages of development, each cell of the embryo can differentiate into a whole animal. The major phyla of deuterostomes are Echinodermata and Chordata.

10. Arthropods are the most successful of all animals in terms of numbers of individuals and numbers of species, as well as in terms of ecological diversification. Like the annelids, arthropods have segmented bodies; but in arthropods some of the segments have become fused during the course of evolution.

11. Insects, which have six legs, are the largest class of organisms, with at least 2 million and perhaps as many as 30 million species.

12. Echinoderms are exclusively marine deuterostomes that are radially symmetrical as adults. They have a unique water vascular system that includes tube feet, by means of which some echinoderms move.

13. Chordates are characterized by their single, hollow dorsal nerve cord; a notochord; and pharyngeal slits, at least during the embryonic stage. There are three subphyla: tunicates, lancelets, and vertebrates.

REVIEW

1. Of the 35 phyla of animals, name the two that are represented by large numbers of terrestrial species.

2. In all eumetazoans, which embryonic layer gives rise to the digestive organs?

3. Name the two groups of animals that have fundamentally distinct patterns of embryonic development.

4. Which animal phylum has the most species?

5. Animals that develop a body cavity between the mesoderm and the endoderm are called _____.

SELF-QUIZ

1. Choanocytes of sponges bear a striking resemblance to the _____, members of one group of zoomastigotes.
   (a) acoelomates
   (b) choanoflagellates
   (c) cnidarians
   (d) collar cells
   (e) none of the above

2. Where does a coelom originate?
   (a) in the endoderm
   (b) between the endoderm and the mesoderm
   (c) in the mesoderm
   (d) between the mesoderm and the ectoderm
   (e) in the ectoderm

3. Some coelomates are
   (a) protostomes
   (b) deuterostomes
   (c) spiders
   (d) human beings
   (e) all of the above

4. Which two phyla listed here are deuterostomes?
   (a) Platyhelminthes
   (b) Nematoda
   (c) Arthropoda
   (d) Echinodermata
   (e) Chordata

5. Three characteristics that distinguish the chordates from other animals are
   (a) a single, hollow, dorsal nerve cord
   (b) a notochord at some time in development
   (c) pharyngeal slits at some time in development
   (d) segmentation
   (e) a nervous system

THOUGHT QUESTIONS

1. Why do we believe that the ancestor of the deuterostomes was a protostome?

2. Why is it believed that echinoderms and chordates, which are so dissimilar, are members of the same evolutionary line?

FOR FURTHER READING

CAMERON, J.N.: "Molting in the Blue Crab," *Scientific American,* May 1985, pages 102-109. An interesting discussion of the way the blue crab molts; soft-shelled crabs are individuals caught right after molting.

GOSLINE, J.M., and M.E. DEMONT: "Jet-Propelled Swimming in Squids," *Scientific American,* January 1985, pages 97-103. For short distances a squid can jet through the water as rapidly as a fish, by contracting radial and circular muscles in its boneless mantle wall.

HADLEY, N.: "The Arthropod Cuticle," *Scientific American,* July 1986, pages 104-112. A major innovation aiding the success of insects as they conquered the terrestrial environment, the cuticle is far more than a simple waxy covering.

RICKETTS, E.F., J. CALVIN, and J.W. HEDGPETH: *Between Pacific Tides,* 5th ed., Stanford University Press, Stanford, Calif., 1985. An outstanding, ecologically oriented treatment of the marine organisms of the Pacific Coast of the United States.

YONGE, C.M.: "Giant Clams," *Scientific American,* April 1975, pages 96-105. These enormous undersea bivalves, although they are filter-feeders, probably owe their large size to the fact that photosynthetic dinoflagellates live within their tissues and nourish them.

PLANT BIOLOGY

Part Eight

FIGURE 29-1   Regardless of how modified they have become in the course of evolution, all plants have the same basic structure. In this barrel cactus *(Ferocactus diguetii),* photographed on an island in the Gulf of California, Mexico, the stem is greatly swollen and stores water efficiently; the photosynthetic cells lie mainly just beneath the surface of the thick, tough epidermal layers of the stem. Leaves are present only in seedlings.

THE STRUCTURE OF PLANTS Chapter 29

Overview

Plants, which cover the surface of the earth in bewildering variety, all possess the same fundamental architecture. Although roots and shoots differ in their basic structure, growth at the tips throughout the life of the individual is characteristic of both. All parts of plants have an outer covering, called dermal tissue, and ground tissue, within which is embedded vascular tissue that conducts water, nutrients, and food throughout the plant. Plants cannot move, but they adjust to their environment by growing and changing their form. Their light-gathering leaves are supported on a stem, which is connected to the water-gathering roots penetrating the soil below.

For Review

Here are some important terms and concepts that you will encounter in this chapter. If you are not familiar with them, you should review them before proceeding.

Cell structure (Chapter 5)

How cells divide (Chapter 10)

History of plants (Chapter 18)

Major groups of plants (Chapter 27)

Monocots and dicots (Chapter 27)

Plants dominate and give color to the living world, providing the basic structure for the terrestrial communities in which other organisms live. Only plants, algae, and a few kinds of bacteria can carry out photosynthesis (Figure 29-1), capturing the energy of the sun (which bombards the surface of the earth constantly in enormous quantities) and making it available for their own metabolic activities and those of all other organisms. The world is populated with millions of kinds of organisms, the great majority of them still unknown, but without the process of photosynthesis no life would be possible. We began our study of plants in Chapter 27, and in this section we consider their structure and function in more detail.

THE ORGANIZATION OF A PLANT

A plant never becomes an "adult" in the way an animal does. Rather, plants simply keep growing, adding new cells, tissues, and organs at the ends of their shoots and roots, becoming ever larger. Some large patches of prairie grasses are thought to be single individuals that have been growing in one place since the glaciers receded more than

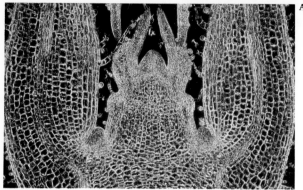

APICAL MERISTEM

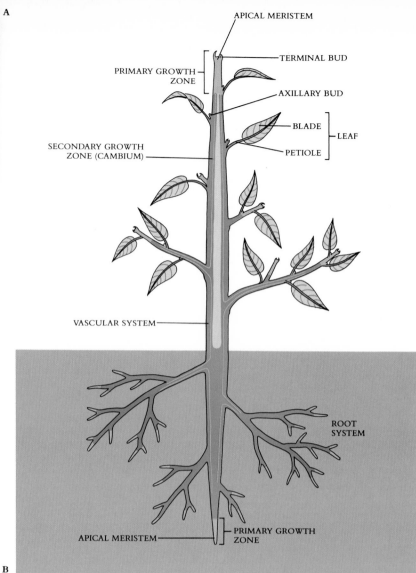

APICAL MERISTEM

TERMINAL BUD

AXILLARY BUD

BLADE — LEAF

PETIOLE

PRIMARY GROWTH ZONE

SECONDARY GROWTH ZONE (CAMBIUM)

VASCULAR SYSTEM

ROOT SYSTEM

APICAL MERISTEM

PRIMARY GROWTH ZONE

B

FIGURE 29-2

Some of the features of a plant. The terms in this illustration will be explained as we discuss different parts of the plant body, which consists of a shoot (stems and leaves) and root. Primary growth takes place as a result of the division of clusters of cells, the **apical meristems,** which are located at the ends of the roots and the stems; secondary growth takes place laterally, allowing the plant to enlarge in girth.

A Transection of a shoot apex in *Coleus.*

B A diagram of a vegetative plant body.

10,000 years ago. Trees attain great ages, and potatoes and many other crops are simply propagated over and over again as parts of a single clone plant, producing generation after generation of genetically identical individuals.

Compared with the other two main evolutionary lines, or kingdoms, of organisms that are fundamentally multicellular—animals and fungi—plants differ because they are green: they photosynthesize. Photosynthesis takes place largely in leaves, which are flattened organs that are arranged to capture the sun's light, and in young green stems (Figure 29-2). There are no leaves on the underground portions of a plant, obviously, and the growth patterns in the roots and shoots differ fundamentally. This chapter is devoted primarily to outlining these differences and to analyzing the way these organs, together with their specialized cell and tissue types and their appendages, form the plant body. Although the similarities between a cactus, an orchid, and a pine tree might not be obvious at first sight, plants have a fundamental unity of structure. This unity is reflected in the construction plan of their respective bodies; in the way they grow, produce, and transport their food; and in how they regulate their development.

A vascular plant is organized along a vertical axis, like a pipe. The part below ground is called the **root;** the part above ground is called the **shoot.** The root penetrates the soil and absorbs water and various ions, which are crucial for plant nutrition. It also anchors the plant. The shoot consists of stem and leaves. The **stem** serves as a framework for the positioning of the **leaves,** the structure in which most photosynthesis

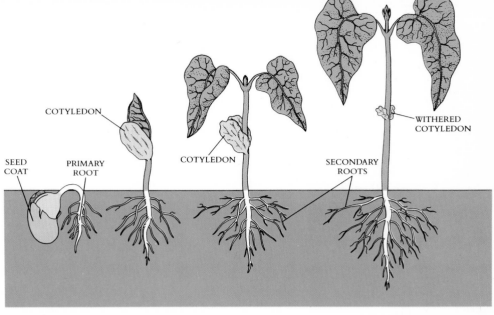

COTYLEDON

WITHERED
COTYLEDON

SEED
COAT

PRIMARY
ROOT

COTYLEDON

SECONDARY
ROOTS

takes place. The arrangement, size, and other characteristics of the leaves are of critical importance in the plant's production of food. Flowers, other reproductive organs, and ultimately fruits and seeds are formed on the shoot as well.

When an embryo is formed within a seed, it remains dormant for a time. Such an embryo consists of an axis, usually with one or two cotyledons, or embryonic leaves (Figure 29-3). In general, embryos of monocots, such as lilies or corn, have only one cotyledon, whereas embryos of dicots, such as beans or oaks, have two. The food in a mature seed may be stored either in endosperm (a common condition in monocots) or in the cotyledons (as in many dicots).

In an embryo, the **apical meristems**—regions of active cell division that occur at the tips of roots and shoots—differentiate early, thus establishing the growth pattern that will persist throughout the life of the plant. In the germination of a seed, the embryonic root may emerge first, anchoring the seedling in the soil, or the shoot may emerge at the same time or even earlier.

> **An apical meristem is a region of active cell division that occurs at or near the tips of the roots and shoots of plants.**

TISSUE TYPES IN PLANTS

The organs of a plant, the leaves, roots, and stem, are composed of different mixtures of tissues, just as your legs are composed of bone, muscle, and connective tissue. A tissue is a group of similar cells organized into a structural and functional unit, cells that are specialized in the same way. In plants there are three major tissue types: **vascular tissue,** which conducts water and dissolved minerals up the plant and conducts the products of photosynthesis throughout; **ground tissue,** in which the vascular tissue is embedded; and **dermal tissue,** the outer protective covering of the plant. Each major tissue type is composed of distinctive kinds of cells, whose structures are related to the functions of the tissues in which they occur. For example, the vascular tissue is composed of **xylem,** which conducts water and dissolved minerals, and **phloem,** which conducts carbohydrates, mostly sucrose, that the plant uses as food. Xylem and phloem are the two principal types of conducting tissue in a vascular plant.

> **The three major types of tissues in plants are (1) dermal tissue, which covers the outside of the plant; (2) ground tissue, which fills its interior; and (3) vascular tissue, which conducts water, minerals, carbohydrates, and other substances throughout the plant and is embedded in ground tissue.**

FIGURE 29-3

Seeds and stages of germination in the common bean, *Phaseolus vulgaris.* In the bean, as in most dicots, the food that is initially produced in the endosperm is absorbed by the embryo during the course of its development and primarily located in the cotyledons by the time the seed is mature.

A

B

FIGURE 29-4

Plants live for very different lengths of time. Desert annuals like those shown in **A** complete their entire life span in a few weeks, whereas trees—such as the giant redwood, *Sequoiadendron giganteum* **(B),** which occurs in scattered groves along the west slope of the Sierra Nevada in California—live 2000 years or more. When plants do live for more than a year, they accumulate wood; even in such plants, however, the stems that are produced during the current year are herbaceous.

TYPES OF MERISTEMS

Animals grow all over. When you are growing from a child into an adult, your torso grows at the same time your legs do. Imagine if instead you grew in only one place, your legs simply getting longer and longer. This is similar to the way a plant grows. Plants contain zones of unspecialized cells called **meristems,** whose only function is to divide. Every time one of these cells divides, one of its two offspring remains in the meristem, while the other goes on to differentiate into one of the three kinds of plant tissue, and ultimately to become part of the plant body.

Primary growth in plants is initiated by the apical meristems. The growth of these meristems results primary in the extension of the plant body. As it elongates, it forms what is known as the **primary plant body,** which is made up of the **primary tissues.** The primary plant body comprises the young, soft shoots of a tree or shrub, or the entire plant in some short-lived plants.

Secondary growth involves the activity of the **lateral meristems.** Lateral meristems are cylinders of meristematic tissue. The continued division of their cells results primarily in the thickening of the plant body. There are two kinds of lateral meristems: the **vascular cambium,** which gives rise to ultimately thick accumulations of secondary xylem and phloem, and the **cork cambium,** from which arise the outer layers of the bark in both roots and shoots. The tissues formed from the lateral meristems, comprising most of the bulk of trees and shrubs, are known collectively as the **secondary plant body,** and its tissues are known as **secondary tissues.** Figure 29-2 compares primary and secondary growth zones.

> **The primary plant body, which includes the young, soft shoots and roots, arises from the apical meristems. Once the lateral meristems begin to function, they produce the secondary plant body, which is characterized by thick accumulations of conducting tissue and the other cell types associated with it.**

The ways in which meristems function in the production of a mature plant determine its nature. A **woody plant,** for example a tree or shrub, is one in which secondary growth has been extensive. An **herbaceous plant** is one in which secondary growth has been limited. Herbaceous plants produce new shoots each year, either from underground portions of the plant or from seeds. If they complete their entire life cycle within a year, they are called **annuals;** if shoots are produced year after year, they are called **perennials.** Members of a less common class of plants, **biennials,** form a leafy shoot the first year and then go on to flower the second year. Two extremes are shown in Figure 29-4.

PLANT CELL TYPES
Ground Tissue

Parenchyma cells are the least specialized and the most common of all plant cell types (Figure 29-6, for example, shows parenchyma cells). They form masses in leaves, stems, and roots. Parenchyma cells, unlike some of the other cell types, are characteristically alive at maturity, with fully functional cytoplasm and a nucleus. They are, therefore, capable of further division. Most parenchyma cells have only **primary cell walls,** which are mostly cellulose that is laid down while the cells are still growing. **Secondary cell walls,** in contrast, are deposited between the cytoplasm and primary wall of a fully expanded cell.

Collenchyma cells, which are also living at maturity, form strands or continuous cylinders beneath the epidermis of stems or leaf stalks and along veins in leaves. They are usually elongated, with unevenly thickened primary walls (Figure 29-5), which are their distinguishing feature. Strands of collenchyma provide much of the support for plant organs in which secondary growth has not taken place.

Parenchyma cells, which are usually living at maturity, are the most common type of cells in the primary plant body. They lack secondary cell walls. Collenchyma cells, which are also living at maturity, are elongated cells with unevenly thickened primary walls. They provide much of the support for plants in which secondary growth has not taken place.

In contrast to parenchyma and collenchyma cells, **sclerenchyma** cells have tough, thick secondary walls; they usually do not contain living protoplasts when they are mature. There are two types of sclerenchyma: **fibers,** which are long, slender cells that usually form strands, and **sclereids,** which are variable in shape but often branched. Sclereids are sometimes called stone cells, because they make up the bulk of the stones of peaches and other "stone" fruits, as well as that of nut shells (Figure 29-6). Both fibers and sclereids are tough and thick-walled; they serve to strengthen the tissues in which they occur.

Flattened **epidermal cells,** which are often covered with a thick, waxy layer called the **cuticle,** cover all parts of the primary plant body. These cells are the most abundant kind in the epidermis, or skin, of a plant. They protect the plant and provide an effective barrier preventing water loss. A number of kinds of specialized cells occur among the epidermal cells, including guard cells, trichomes, and root hairs.

Guard cells are paired cells. With the opening that lies between them, they make up the **stomata** (singular, **stoma**) (see Figures 27-4 and 29-12), which occur frequently in the epidermis of leaves and occasionally on the outer parts of the shoot, such as stems or fruits. The passage of oxygen and carbon dioxide into and out of the leaves, as well as the loss of water from them, takes place almost exclusively through the stomata, which open and shut in response to external factors such as the supply of moisture (see Chapter 31).

Trichomes are outgrowths of the epidermis, superficially like the hairs of mammals, that vary greatly in form in different kinds of plants. They play a major role in controlling the loss of water from leaves and other plant parts and in regulating the temperature of plant parts. Similar to trichomes, but actually extensions of single epidermal cells, are **root hairs,** which occur in masses just behind the very ends of the roots (Figure 29-7). They keep the roots in intimate contact with the particles of soil, and are soon worn off as the root continues to grow.

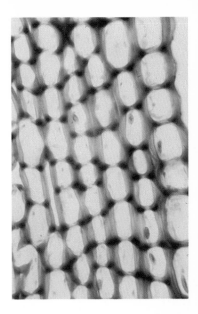

FIGURE 29-5

Collenchyma cells, with thickened side walls, from a young branch of elderberry *(Sambucus)*. In other kinds of collenchyma cells, the thickened areas may occur at the corners of the cells or in other kinds of strips.

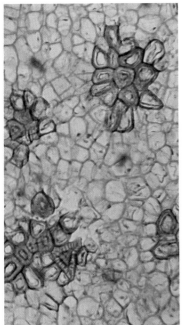

FIGURE 29-6

Clusters of thick-walled sclereids ("stone cells") in pulp of a pear; the surrounding, thin-walled cells are parenchyma. Such clusters of sclereids give pears their gritty texture.

FIGURE 29-7

A germinating seedling of radish, *Raphanus sativus,* showing the abundant, fine root hairs that form in back of the root apex.

FIGURE 29-8

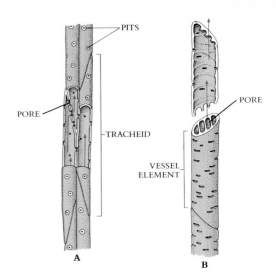

Conducting Cells

Vascular plants contain two kinds of conducting or vascular tissue. These are the xylem and the phloem.

Xylem

Xylem is the principal water-conducting tissue of plants. It forms a continuous system running throughout the plant body. Within this system, water passes from the roots up through the shoot in an unbroken stream; Chapter 31 discusses the way in which this stream is maintained. Dissolved minerals also are taken into plants through their roots, as a part of this stream of water. When the water reaches the leaves, much of it passes into the air as water vapor, mainly through the stomata.

The two principal types of conducting elements in the xylem are **tracheids** and **vessel elements** (Figure 29-8), both of which have thick secondary walls, are elongated, and have no living protoplast at maturity. In conducting elements composed of tracheids, water flows from tracheid to tracheid through openings called **pits** in the secondary walls. In contrast, vessel elements have not only pits but also definite openings or **perforations** in their end walls by which they are linked together, and through which water flows. A linked row of vessel elements forms a **vessel**. Primitive angiosperms have only tracheids, but the majority of living angiosperms have vessels. Vessels conduct water much more efficiently than do strands of tracheids. In addition to the conducting cells, xylem likewise includes fibers and parenchyma cells.

> The major types of conducting cells of the xylem are tracheids and vessel elements. Incorporated in the xylem, however, are also fibers and parenchyma cells.

Phloem

Phloem is the principal food-conducting tissue in the vascular plants. Different kinds of plants have one of two different kinds of phloem cells: **sieve cells** or **sieve-tube members.** Clusters of pores known as **sieve areas** occur on both kinds of cells and connect the protoplasts of adjoining sieve cells and sieve-tube members. Both cell types are living, but their nuclei are lost during the process of their maturation.

Angiosperms contain sieve-tube members, phloem cells in which the pores in some of the sieve areas are larger than those in the others; such sieve areas are called **sieve plates.** Sieve-tube members occur end-to-end, forming longitudinal series called **sieve tubes.** Other vascular plants contain sieve cells, phloem cells in which the pores in all of their sieve areas are roughly the same diameter. Specialized parenchyma cells known as **companion cells** occur regularly in association with sieve-tube members (Figure 29-9), but are lacking in plants with sieve cells. In an evolutionary sense, sieve-

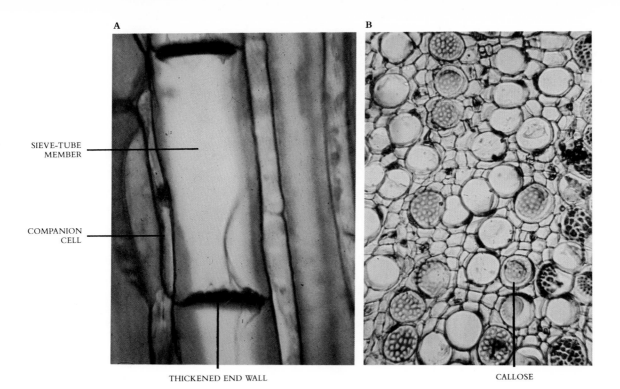

A

SIEVE-TUBE
MEMBER

COMPANION
CELL

THICKENED END WALL

B

CALLOSE

FIGURE 29-9

Phloem of squash *(Cucurbita)*.
A A sieve-tube member, connected with the cells above and below to form a sieve tube. Note the thickened end walls, which are at right angles to the sieve tube. The narrow cell with the nucleus below the sieve-tube member is a companion cell.
B Transection of phloem, showing a substance called callose deposited around the sieve areas in the end walls of the sieve tube elements. Callose is believed mostly to be deposited when a plant is injured, and not to be present in normally functioning phloem.

tube members clearly are advanced over sieve cells; they are more specialized and presumably more efficient.

> **The principal cell types in phloem are sieve cells or sieve-tube members. Sieve cells and sieve-tube members lose their nucleus in the course of becoming mature.**

SHOOTS

The shoot of a plant is technically that portion that lies above the cotyledons, and the root is the portion below the place where they are attached. In most plants, all of the above-ground parts are portions of the shoot. Leaves are usually the most prominent organs of the shoot and determine its appearance.

Leaves

Leaves, outgrowths of the shoot apex, are the light-capturing organs of most plants. The only exceptions to this are found in some plants, such as cacti, in which the stems are green and have largely taken over the function of photosynthesis for the plant (see Figure 29-1).

The apical meristems of stems and roots are capable of growing indefinitely under appropriate conditions. Leaves, in contrast, grow by means of **marginal meristems,** which flank their thick central portions. These marginal meristems grow outward and ultimately form the blade of the leaf, while the central portion becomes the midrib. Once a leaf is fully expanded, its marginal meristems cease to function: their growth is called **determinate,** whereas that of apical meristems is **indeterminate.**

Most leaves have a flattened portion, the **blade,** and a slender stalk, the **petiole** (see Figures 29-2 and 29-10). In addition, there may be two leaflike organs, the **stipules,** that flank the base of the petiole, where it joins the stem. Veins, usually consisting of both xylem and phloem, run through the leaves. In many monocots the veins are **parallel;** in most dicots the pattern is **net** or **reticulate venation** (Figure 29-11).

FIGURE 29-10

The leaves of different species of plants often differ greatly.

A The undivided, but lobed, leaves of a tulip tree, *Liriodendron tulipifera*.

B The divided leaves of the sensitive plant, *Mimosa pudica*. In both species, the leaves have a petiole, a slender stalk that connects the broad part to the stem.

A   **B**

A   **B**

FIGURE 29-11

The leaves of dicots, such as this African violet relative **(A)**, have net, or reticulate, venation; those of monocots, like this palm **(B)**, have parallel venation.

A typical leaf contains masses of parenchyma through which the vascular bundles, or veins, run. The masses of parenchyma that occur in leaves are called **mesophyll** ("middle-leaf"). Beneath the upper epidermis of a leaf, there are one or more layers of closely packed, column-like parenchyma cells called **palisade parenchyma** (Figure 29-12; see also Figure 9-14). The rest of the interior of a leaf, except for the veins, consists of a tissue called **spongy parenchyma.** Between the spongy parenchyma cells are large intercellular spaces, which function in gas exchange and in the passage of carbon dioxide from the atmosphere to the mesophyll cells. These intercellular spaces are connected, directly or indirectly, with the stomata.

The cells of the mesophyll, especially those near the leaf surface, are packed with chloroplasts. These cells constitute the plant's primary site of photosynthesis. Water and minerals are brought from the roots to the leaves in the xylem strands of the veins. Once the water reaches the ends of the veins in the leaves, it passes into the photosynthetic mesophyll cells. Because the surfaces of these cells border on the intercellular spaces of the interior of the leaf, and because water can pass through cell membranes and cell walls relatively easily, much of the water evaporates into the intercellular space and can then escape from the leaf through the open stomata. Thus a structure that allows absorption of the essential carbon dioxide also allows the escape of water and necessitates its constant replenishment. At the same time that water is leaving through the stomata, the products of photosynthesis are being transported from the leaves to all other parts of the plant through the phloem strands of the same veins.

> **The mesophyll in a leaf consists of two types of parenchyma cells, both packed with chloroplasts. Palisade parenchyma cells are columnar and closely packed together, whereas spongy parenchyma cells are loosely packed and separated by large intercellular spaces.**

Stems
Primary Growth

In the primary growth of a shoot, leaves first appear as **leaf primordia** (singular, **primordium**), or rudimentary young leaves, which cluster around the apical meristem, unfolding and growing as the stem itself elongates (see Figure 29-2, *B*). The places on the stem at which leaves form are called **nodes.** The portions of the stem between these attachment points are called the **internodes.** As the leaves expand to maturity, a **bud,** a tiny, undeveloped side-shoot, develops in the **axil** of each leaf, the angle between a branch or leaf and the stem from which it arises (Figure 29-2, *B,* where the buds are visible as densely staining masses at the base of the larger pair of leaf primordia). These buds, which have their own leaves, may elongate and form lateral branches, or they may remain small and dormant. A hormone diffusing downward from the terminal bud of the shoot continuously suppresses the expansion of the lateral buds (see Chapter 31). These buds begin to expand when the terminal bud is removed. Therefore, gardeners who wish to produce bushy plants or dense hedges crop off the tops of the plants, thus removing their terminal buds.

Within the soft, young stems, the strands of vascular tissue, xylem and phloem, either occur as a cylinder in the outer portion of the stem, as is common in dicots (Figure 29-13), or are scattered through it, as is common in monocots. At the stage when only primary growth has occurred, the inner portion of the ground tissue of a stem is called the **pith,** and the outer portion is the **cortex.**

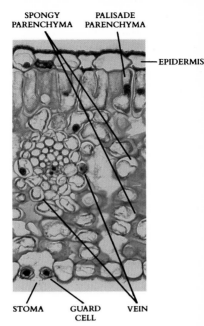

SPONGY PARENCHYMA PALISADE PARENCHYMA

EPIDERMIS

STOMA GUARD CELL VEIN

FIGURE 29-12

Cross section of a lily leaf, showing palisade and spongy parenchyma; a vascular bundle, or vein; and the epidermis, with paired guard cells flanking the stoma that is visible on the lower surface of the leaf, and the intercellular spaces that lie below the stomata.

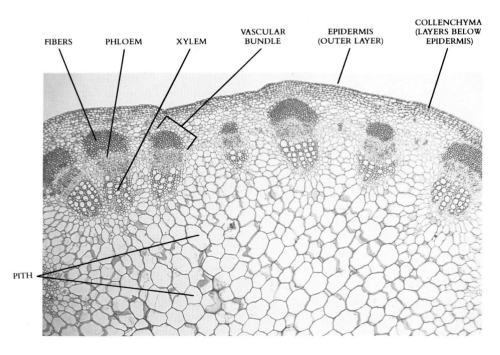

FIBERS PHLOEM XYLEM VASCULAR BUNDLE EPIDERMIS (OUTER LAYER) COLLENCHYMA (LAYERS BELOW EPIDERMIS)

PITH

FIGURE 29-13

Cross section of a part of a young stem in a dicot, the common sunflower, *Helianthus annuus.* In this species and most dicots, the vascular bundles are arranged around the outside of the stem, forming a ring in transection. In monocots, they are usually scattered throughout the pith.

FIGURE 29-14

An early stage in the differentiation of vascular cambium in an elderberry *(Sambucus canadensis)* stem. The primary vascular bundles have been separated by the division of cells in the young vascular cambium.

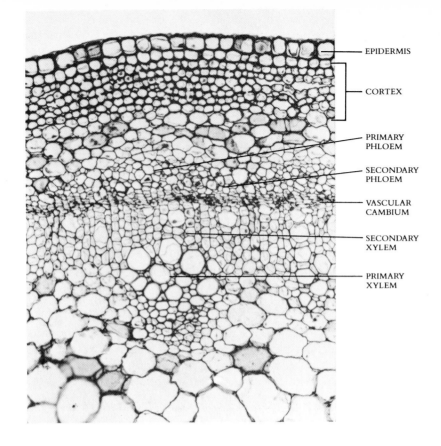

EPIDERMIS

CORTEX

PRIMARY PHLOEM

SECONDARY PHLOEM

VASCULAR CAMBIUM

SECONDARY XYLEM

PRIMARY XYLEM

FIGURE 29-15

Lenticels in the bark of bigtooth aspen *(Populus grandidentata)*. Oxygen reaches the layers of living tissue beneath the bark by way of such lenticels.

Secondary Growth

In stems, secondary growth is inititated by the differentiation of the **vascular cambium,** which consists of a thin cylinder of actively dividing cells that is located between the bark and the main stem in mature woody plants. The vascular cambium differentiates from parenchyma cells within the vascular bundles of the stem, between the primary xylem and the primary phloem (Figure 29-14). These begin to divide actively first, and then the vascular cambium grows into a cylinder when some of the parenchyma cells between the bundles also begin to divide. When it has been established, the vascular cambium consists of elongated, somewhat flattened cells with large vacuoles. The cells that divide from the vascular cambium outwardly, toward the bark, become secondary phloem; those that divide from it inwardly become secondary xylem. The cells of the vascular cambium also divide side-to-side, allowing the stem to get larger and larger as the tree or shrub becomes more mature.

While the vascular cambium is becoming established, a second kind of lateral cambium, the **cork cambium,** normally also develops in the outer layers of the stem. The cork cambium usually consists of plates of dividing cells that move deeper and deeper into the stem as they divide. Outwardly, the cork cambium splits off densely packed **cork cells;** they contain a fatty substance and are nearly impermeable to water. Cork cells are dead at maturity. Inwardly, the cork cambium divides to produce a dense layer of parenchyma cells. The cork, the cork cambium that produces it, and this dense layer of parenchyma cells make up a layer called the **periderm,** which is the outer protective covering of the plant. Oxygen reaches the layers of living cells under the bark through areas of loosely organized cells called **lenticels,** which are often easily identifiable on the outer surface of bark (Figure 29-15).

The periderm consists of cork to the outside, the cork cambium in the middle, and layers of parenchyma cells to the inside of a stem or root. Penetrated by areas of more loosely packed tissue known as lenticels, the periderm retards water loss from the secondary plant body.

Wood

Wood is one of the most useful, economically important, and beautiful products that we obtain from plants. From an anatomical point of view, **wood** is accumulated secondary xylem. As the secondary xylem ages, its cells become infiltrated with gums and resins, and the wood becomes darker. For this reason, the wood located nearer the central regions of a given trunk, called **heartwood,** is often darker in color and denser than the wood nearer the vascular cambium, which is still actively involved in transport within the plant (Figure 29-16). Commercially, wood is divided into hardwoods and softwoods. **Hardwoods** are the woods of dicots, regardless of how hard or soft they actually may be; **softwoods** are the woods of conifers.

Because of the way it is accumulated, wood often displays rings. In temperate regions, these rings are **annual rings** (Figure 29-17): they reflect the fact that the vascular cambium divides more actively when water is plentiful and temperatures are suitable for growth than when water is scarce and the weather is cold. The abrupt discontinuity between the layers of larger cells, with proportionately thinner walls, that form in the growing season (in most temperate regions, during the spring and early summer) and those that form later, is often very evident. For this reason, the annual rings in a tree trunk can be used to calculate the age of the tree.

Bark is a term used to refer to all of the tissues of a mature stem or root outside of the vascular cambium (Figure 29-18). Because the vascular cambium has the thinnest-walled cells that occur anywhere in a secondary plant body, it is the layer at which the bark breaks away from the accumulated secondary xylem. The inner layers of the bark are primarily phloem. Its outer layers consist of the periderm, and the very outermost ones are cork.

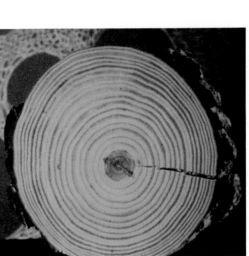

FIGURE 29-16

Dark heartwood and light sapwood in a section of a white poplar trunk.

FIGURE 29-17

Annual rings in a tree trunk.

FIGURE 29-18

The structure of bark.

FIGURE 29-19

Median longitudinal section of a
root tip in corn, *Zea mays,*
showing the differentiation of
epidermis, cortex, and column of
vascular tissues.

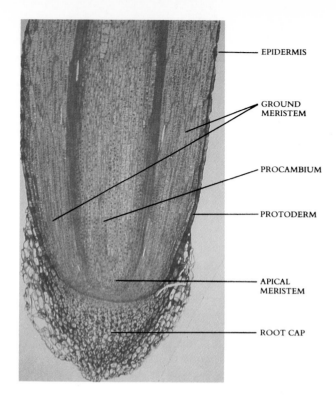

EPIDERMIS

GROUND
MERISTEM

PROCAMBIUM

PROTODERM

APICAL
MERISTEM

ROOT CAP

Roots

Roots have simpler patterns of organization and development than do stems (Figure
29-19). Although different patterns exist, we shall describe a kind of root that is found
in many dicots. There is no pith in the center of the vascular tissue of the root in most
dicots. Instead, these roots have a central column of xylem with radiating arms. Between these arms are located strands of primary phloem (Figure 29-20). Around the
column of vascular tissue, and forming its outer boundary, is a cylinder of cells one or
more cell layers thick called the **pericycle.** The outer layer of the root, as in the shoot, is
the **epidermis.** The mass of parenchyma in which the vascular tissue of the root is located is the cortex. Its innermost layer, the **endodermis,** consists of specialized cells that
regulate the flow of water between the vascular tissues and the outer portion of the
root.

 The apical meristem of the root divides and produces cells both inwardly, back
toward the body of the plant, and outwardly. Outward cell division results in the formation of a thimblelike mass of relatively unorganized cells, the **root cap,** which covers
and protects the root's apical meristem as it grows through the soil.

 The root elongates relatively rapidly just behind its tip. Above that zone are
formed abundant root hairs (see Figure 29-7). Virtually all absorption of water and
minerals from the soil takes place through the root hairs, which greatly increase the
surface area and therefore the absorptive powers of the root. In plants that have mycorrhizae (Chapter 26), the root hairs are often greatly reduced in number, and the fungal
filaments of the mycorrhizae play a role similar to that of the root hairs.

 One of the fundamental differences between roots and shoots has to do with the
nature of their branching. In stems, branching occurs from buds on the surface of the
stem; in roots, branching is initiated well back of the root apex, as a result of cell divisions in the pericycle. The lateral root primordia grow out through the cortex toward
the surface of the root (Figure 29-21), eventually breaking through and becoming established as lateral roots. Secondary growth in roots, both main roots and laterals, is
similar to that in stems, with the vascular cambium being initiated after the division of
cells located between the primary xylem and the primary phloem.

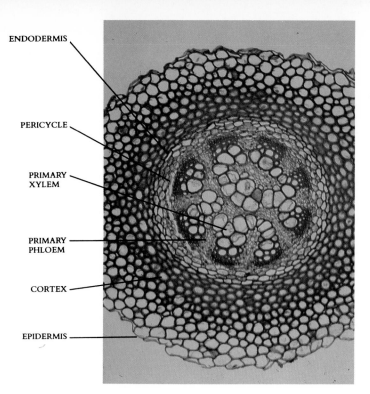

ENDODERMIS

PERICYCLE

PRIMARY
XYLEM

PRIMARY
PHLOEM

CORTEX

EPIDERMIS

FIGURE 29-20

Cross section through a root of a clubmoss, *Lycopodium,* which has an anatomical structure like that of a dicot. The central xylem and phloem, cortex, and epidermis are visible.

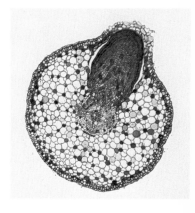

FIGURE 29-21

A lateral root growing out through the cortex of black willow, *Salix nigra.* The origin of lateral roots occurs beneath the surface of the main root, while the origin of lateral stems occurs at the surface.

SUMMARY

1. A plant body is basically an axis that has two parts, a root and a shoot. Within it, there are three principal tissue types: vascular tissue, ground tissue, and dermal tissue. Dermal tissue covers the outside of the plant; vascular tissue conducts substances through it; and ground tissue is the matrix in which the vascular tissue is embedded.

2. Plants grow in length by means of their apical meristems, zones of active cell division at the ends of the roots and the shoots.

3. Plants that complete their entire life cycle within a year are called annuals; those that require 2 years to reach maturity and then flower just once are called biennials; and those that flower year after year once they have reached maturity are known as perennials. Perennial plants may be soft, as are herbs, or else woody, as are trees and shrubs.

4. Parenchyma cells are the most abundant type of ground tissue in the plant body. They usually have only primary walls, which are composed mainly of cellulose and are laid down while the cells are still growing. Parenchyma cells are living at maturity.

5. Collenchyma cells often form strands or continuous cylinders in the plant body, for which they provide the chief source of strength. They may be recognized by the uneven thickenings of their primary cell walls.

6. Sclerenchyma cells have tough, thick walls with secondary thickening laid down after the cells have reached full size. Fibers, which are elongated strengthening cells, are sclerenchyma cells.

7. Tracheids and vessel elements are the principal conducting elements of the xylem. They have thick cell walls and lack protoplasts at maturity. Water reaches the leaves after entering the plant through the roots and passing upward via the xylem. Water vapor passes out of leaves by entering the intercellular spaces, evaporating, and then moving out through the stomata; carbon dioxide enters the plant by the same route.

8. Carbohydrates are conducted through the plant primarily in the phloem, whose elongated conducting cells are living but lack a nucleus. These conducting cells are called sieve cells and sieve-tube members.

9. The growth of leaves is determinate, and that of shoots and roots is indeterminate. Leaves, which are mainly flattened organs of the shoot specialized for photosynthesis, expand in size by means of marginal meristems.

10. Stems branch because of the growth of buds that form externally at the point where the leaves join the stem; roots branch by forming centers of cell division within their cortex. Young roots grow out through the cortex, eventually breaking through the surface of the root.

11. Secondary growth in both stems and roots takes place following the formation of lateral meristems known as vascular cambia. These cylinders of dividing cells form xylem internally and phloem externally. As a result of their activity, the girth of a plant increases.

12. Wood is accumulated secondary xylem; it often displays rings because it exhibits different rates of growth at different seasons.

13. The cork cambium forms in both roots and stems during the initial stages of establishment of the vascular cambium. It produces cork externally and a dense layer of parenchyma internally. The cork, cork cambium, and underlying parenchyma layers collectively are called the periderm. This layer, the outermost of bark, is perforated, in the stem, by areas of loosely organized tissue called lenticels.

REVIEW

1. Areas of active cell division that occur at the shoot and root tips are called _____.

2. Primary growth in plants in initiated by the _____, whereas secondary growth involves activity of the _____.

3. Epidermal cells are covered with a waxy _____, which retards water loss from the plant.

4. Which leaf cells are the primary site of photosynthesis?

5. The term _____ refers to all of the tissue of a woody stem outside the vascular cambium.

SELF-QUIZ

1. The vegetative organs of plants are (pick three)
   (a) flowers
   (b) stems
   (c) roots
   (d) fruits
   (e) leaves

2. Parenchyma cells
   (a) are alive at maturity
   (b) have only primary cell walls
   (c) are capable of further division
   (d) are found in both the primary and secondary plant body
   (e) all of the above

3. Xylem
   (a) is the principal water-conducting tissue of plants
   (b) conducts the products of photosynthesis
   (c) is made of a number of different cell types
   (d) is not found in roots
   (e) a and c

4. What goes on at the lenticels?
   (a) water absorption
   (b) gas exchange
   (c) photosynthesis
   (d) production of the periderm
   (e) a and c

5. Cells of the _____ regulate the flow of water laterally between the vascular tissues and the cell layers in the outer portions of the root.
   (a) periderm
   (b) endodermis
   (c) pericycle
   (d) pith
   (e) xylem

THOUGHT QUESTIONS

1. If you hammer a nail into the trunk of a tree 2 meters above the ground when the tree is 6 meters tall, how far above the ground will the nail be when the tree is 12 meters tall?

2. What is the difference between primary and secondary growth?

FOR FURTHER READING

GALSTON, A.W., P.J. DAVIS, and R.L. SATTER: *The Life of the Green Plant,* ed. 3, Prentice-Hall, Inc., Englewood Cliffs, N.J., 1980. Very well-written, physiologically oriented treatment of the structure and functioning of plants.

RAVEN, P.H., R.F. EVERT, and H. CURTIS: *The Biology of Plants,* 4th ed., Worth Publishers, Inc., New York, 1981. A comprehensive treatment of general botany, emphasizing structural botany.

FIGURE 30-1   The rich colors and textures of this flower head of a South African species of *Gazania,* a member of the sunflower family (Asteraceae), have evolved in response to the sensory perceptions and activities of insects. There are two types of flowers in the head; the outermost are extended into rays, and the inner ones, called disc flowers, are symmetrical.

FLOWERING PLANT REPRODUCTION

Chapter 30

Overview

The flowering plants dominate every spot on earth except the great northern forests, the polar regions, the high mountains, and the driest deserts. Among the features that have contributed to their success are their unique reproductive structures, the flower and the fruit. Flowers bring about the precise transfer of pollen by insects and other animals. Because they can "control" the activities of these animals, plants can exchange gametes with one another even though each plant is rooted in one place. Fruits play a role of extraordinary importance in the dispersal of angiosperms from place to place. Not only were both flowers and fruits important in making possible the early success of the flowering plants, but their evolution has produced most of the striking differences that we see among different members of the class.

For Review

Here are some important terms and concepts that you will encounter in this chapter. If you are not familiar with them, you should review them before proceeding.

Meiosis and mitosis (Chapter 10)

Evolution of major groups (Chapter 18)

Coevolution of insects and flowering plants (Chapter 21)

Angiosperm life cycle (Chapter 27)

Plants are, with insects, one of the few major groups of organisms to evolve entirely on land. As you learned in Chapter 18, insects and plants coevolved in conquering terrestrial environments, insects forming a key link in the reproduction of many of the flowering plants that now cover the face of our earth (Figure 30-1). Each one of us deals with plant reproduction every day without thinking about it. The bread we eat is the ground-up seed of a grass, wheat; the roses that a boy gives his girlfriend evolved as structures attractive to insects; and the honey that we put on bread is produced by bees from nectar, the bribe a flower uses to induce the bees to carry the flower's gametes to another plant. In this chapter we examine how flowering plants, or angiosperms, reproduce.

FIGURE 30-2

A bumblebee, *Bombus,* covered with pollen while visiting a flower. This bee will transfer large quantities of pollen to the next flower that it visits.

FLOWERS AND POLLINATION

As you saw in Chapter 27, all plant life cycles are characterized by an alternation of generations, which involves diploid sporophytes and haploid gametophytes. In plants, the haploid cells that result from meiosis *always* divide by mitosis to produce a gametophyte generation. In flowering plants, this gametophyte generation is very reduced—the size of a fleck of dust—and completely enclosed within the tissues of its parent sporophyte. The ovules of flowering plants, the structures that become the seeds when mature, are enclosed within carpels, and pollination is indirect. In contrast, gymnosperms have direct pollination, in which the ovules are exposed at the time pollen reaches them.

Unlike the reproductive organs of animals, the reproductive structures of plants are not permanent parts of the adult individual. Flowers and the reproductive organs of other plant groups develop seasonally, being produced at times of the year that are favorable for pollination. In a flower, the corolla and sometimes other floral parts are often brightly colored and attractive to insects and other animals that may visit the flower and thus carry its pollen passively from place to place. Those flowering plants in which the pollen is carried from place to place by wind or in which self-pollination predominates often have dull-colored or green corollas and other flower parts.

Pollination by Animals

In many angiosperms, the pollen grains are carried from flower to flower by insects and other animals that visit the flowers for food or other rewards (Figure 30-2), or are deceived into doing so because the characteristics of the flower suggest that they may offer such rewards. Successful pollination depends on the plants attracting insects and other animals regularly enough that these animals will carry pollen from one flower of that particular species to another. Flowers that are regularly visited by animals normally provide some kind of food reward, often in the form of a liquid called **nectar.** Nectar is rich not only in sugars but also in amino acids and other substances.

The relationship between such animals, known as **pollinators,** and the flowering plants has been one of the most important features of the evolution of both groups. By using insects to transfer pollen, the flowering plants can disperse their gametes on a regular and more or less controlled basis, despite the fact that they are anchored to their substrate.

For pollination by animals to be effective, it is necessary that a particular insect or other animal visit numerous plant individuals of the same species. The color and form of flowers have been shaped by evolution to promote such specialization. Yellow flowers, for example, are particularly attractive to bees, whereas red flowers attract birds but are not particularly noticed by insects (Figure 30-3). Some flowers have very long floral tubes with the nectar produced deep within them; only the long, slender beaks of hummingbirds or the long, coiled tongues of moths or butterflies can reach nectar sup-

FIGURE 30-3

Hummingbirds and flowers.
A Ruby-throated hummingbird extracting nectar from the flowers of a trumpet flower *(Campsis radicans).*
B Poinsettia *(Euphorbia pulcherrima)* flowers. The individual flowers, shown here, each have a large nectary at one side. The clusters of yellowish flowers are made more attractive to birds because of the large red leaves that surround them.

plies of this kind. Specialized flowers that attract certain kinds of animals consistently take advantage of these visits in spreading their pollen accurately from individual to individual. The animals that visit flowers in this way perform the same functions for the flowering plants that they do for themselves when they actively search out mates.

The most numerous insect-pollinated angiosperms are those pollinated by bees (see Figure 30-2), a large group of insects consisting of some 20,000 species. Bees are the most frequent, characteristic, and constant visitors to particular kinds of flowers today. On the other hand, they certainly did not pollinate the most primitive angiosperms, in which beetles (Figure 30-4) may have played a similar role; bees were not abundant and may not even have existed when angiosperms first appeared. The diversity of flowering plants that we see today is closely related to later specialization of angiosperms in relation to bees. The coevolutionary relationships between the members of these two groups are often tightly linked.

Pollination by Wind

In certain angiosperms, pollen is blown about by the wind, as it is in most gymnosperms, and reaches the stigmas passively. For such a system to operate efficiently, the individuals of a given plant species must grow relatively close together, because wind does not carry pollen very far or very precisely, compared with insects or other animals. Because gymnosperms such as spruces or pines grow in dense forests, they are very effectively pollinated by wind. Wind-pollinated angiosperms such as birches, alders, and ragweed (Figure 30-5) also tend to grow in dense stands. The flowers of wind-pollinated angiosperms are usually small, greenish, and odorless, and their petals are reduced in size or absent. They typically produce large quantities of pollen. Certainly the ancestors of the angiosperms were wind-pollinated, but whether the very first members of the group were pollinated by insects or by the wind cannot be determined with certainty. The association with insects, however, is an ancient one for the flowering plants. Wind-pollination seems certainly to have evolved secondarily in all angiosperms in which it occurs today; in other words, they all had insect-pollinated ancestors.

FIGURE 30-4

A soldier beetle, *Chauliognathus pennsylvanicus,* visiting a daisy flower (family Asteraceae). Beetles were abundant and diverse at the time the angiosperms first evolved; they still predominate among the visitors to the flowers of many relatively primitive angiosperms. Beetles very probably played a key role in the early evolution of flowers.

FIGURE 30-5

The flowers of a birch, *Betula.* Birches have two different kinds of flowers, pollen-producing (staminate) ones, which hang down in long, yellowish tassles, and ovule-bearing (pistillate) ones, which mature into characteristic conelike structures and occur in small reddish-brown clusters. Both are shown here.

Self-Pollination

In some angiosperms the pollen does not reach other individuals at all: instead, it is shed directly onto the stigma of the same flower, sometimes in bud. This results in self-pollination and inbreeding, with evolutionary consequences that will be discussed on page 587.

SEED FORMATION

The long chain of events between fertilization and maturity is called **development.** During development, cells become progressively more specialized, or **differentiated.** Development in seed plants results first in the production of an embryo, which remains dormant within a seed until the seed germinates. The first stage in the development of a plant zygote is active cell division. The **zygote** formed by union of sperm and egg divides repeatedly to form an organized mass of cells, the **embryo.** In angiosperms the differentiation of cell types within the embryo begins almost immediately after fertilization. By day 5 the principal tissue systems can be detected within the embryo mass, and within another day, the root and shoot apical meristems can be detected. Cell movement does not occur during the process of embryonic development in plants, as it does in animals. Instead, cells differentiate where they are formed, their positions determining in large measure their future developmental fates. In animals, specific cell movements play an important role in development.

Early in the development of an angiosperm embryo, a profoundly significant event occurs. The embryo simply stops developing. In many plants the development of the embryo is arrested soon after apical meristems and the first leaves, or cotyledons, are differentiated. The **integuments,** the coats surrounding the embryo, develop into a relatively impermeable seed coat, which encloses the quiescent embryo within the seed, together with a source of stored food.

Once a seed coat forms around the embryo, most of the embryo's metabolic activities cease; a mature seed contains only about 10% water. Under these conditions, the seed and the young plant within it are very stable. Germination (Figure 30-6) cannot take place until water and oxygen reach the embryo, a process that sometimes involves cracking the seed. Seeds of some plants have been known to remain viable for hundreds of years (Figure 30-7).

FIGURE 30-6

The germination of a seed, such as this garden pea *(Pisum sativum),* involves the fracture of the seed coat, from which the young shoot and root emerge.

FIGURE 30-7

A seedling grown from seeds of lotus, *Nelumbo nucifera,* recovered from the mud of a dry lake bed in Manchuria, northern China. The radiocarbon age of this seed indicates that it was formed in about 1515; another seed that was germinated was estimated to be at least a century older. The coin is included to give some idea of the size.

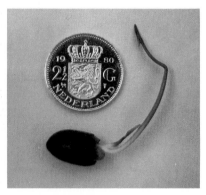

The role of environmental factors helps to ensure that the plant will germinate only under appropriate conditions. Sometimes the seeds are held within tough fruits that will not crack until, for example, they are exposed to the heat of a fire, a strategy that clearly results in the germination of a plant in an open habitat. The seeds of other plants will germinate only when inhibitory chemicals have been washed out of their seed coats, guaranteeing germination when sufficient water is available. Still other plants will germinate only after they pass through the intestines of birds or mammals, or are regurgitated by them. This process both weakens the seed coats and also ensures the dispersal of the plants involved.

While a seed's embryo is embedded in its coat, water and oxygen are largely excluded and the metabolism in the seed is slowed down greatly. To reactivate its metabolism, the embryo has only to be provided with water, oxygen, and a source of metabolic energy. The final release of arrested development, germination, is then cued to specific signals from the environment.

Seeds are clearly important adaptively in at least three respects:

1. Seeds permit plants to postpone development when conditions are unfavorable and to remain dormant until more advantageous conditions arise. Under conditions in which young plants might or might not become established, a plant can "afford" to have some seeds germinate, because others remain dormant.
2. By tying the reinitiation of development to environmental factors, seeds permit the course of embryo development to be synchronized with critical aspects of the plant's habitat.
3. Perhaps most importantly, the dispersal of seeds facilitates the migration and dispersal of genotypes into new habitats. The seed also offers maximum protection to the young plant at its most vulnerable stage of development.

FRUITS

Paralleling the evolution of flowers of the angiosperms, and nearly as spectacular, has been the evolution of their fruits. Besides the many ways that fruits can be formed, they exhibit a wide array of modes of specialization in relation to their dispersal.

Fruits that have fleshy coverings, often black, or bright blue, or red, are normally dispersed by birds and other vertebrates. Like the red flowers that we discussed in relation to pollination by birds, the red fruits signal an abundant food supply (Figure 30-8, A). By feeding on these fruits, the birds and other animals may carry seeds from place to place and thus transfer the plants from one suitable habitat to another.

A

B

FIGURE 30-8

Animal-dispersed fruits.
A The bright red berries of dogwood, *Cornus florida,* are highly attractive to birds, just as red flowers are. The birds carry the seeds they contain for great distances.
B Figs, *Ficus carica,* are whole clusters of flowers turned inside out. They are pollinated by tiny wasps, which enter through the hole in the end of the fig. When mature, figs are consumed, and the seeds of the individual tiny flowers are scattered about by birds and mammals.

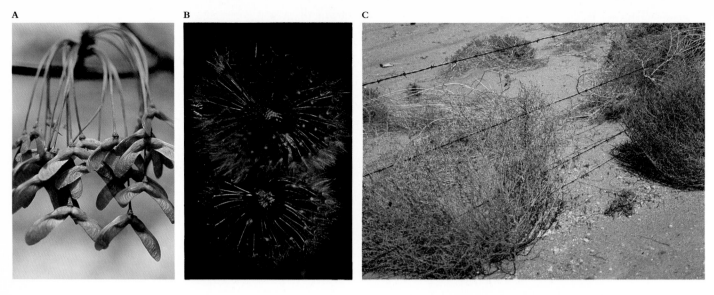

A   **B**   **C**

FIGURE 30-9

Wind-dispersed fruits.
A   The mature double fruits of maples, *Acer,* are blown away from their parent trees.
B   Dandelion, *Taraxacum officinale.* The "parachutes" disperse the fruits widely in the wind, much to the gardener's despair.
C   Tumbleweed, *Salsola,* in which the whole dead plant becomes a light, wind-blown structure that rolls about scattering seeds.

FIGURE 30-10

Coconuts, *Cocos nucifera,* sprouting on a sandy beach. One of the most useful plants for humans in the tropics, coconuts have become established even on the most distant islands by drifting on the waves.

Other kinds of specialized fruit dispersal have evolved many different times in the flowering plants. Fruits with hooked spines, like snakeroot, beggar ticks, or burdock (Figure 30-8, *B*) are characteristic of several genera of plants that occur in the northern deciduous woods. Such fruits are often spread from place to place by mammals, now including humans. In addition, mammals like squirrels disperse and bury seeds. Other fruits or seeds have wings and are blown about by the wind; the wings on the seeds of pines and those on the fruits of ashes or maples (Figure 30-9, *A*) play identical ecological roles. The dandelion (Figure 30-9, *B*) provides a familiar example of a kind of fruit that is dispersed by the wind; the dispersal of seeds of such plants as milkweeds, willows, and cottonwoods is similar.

Still other fruits, like the coconut and those of certain other plants that characteristically occur on or near beaches, are regularly spread from place to place by water (Figure 30-10). Dispersal of this sort is especially important in the colonization of distant island groups, such as the Hawaiian Islands. It has been calculated that the seeds of about 175 original flowering plant species must have reached Hawaii to have evolved into the roughly 970 native species found there today.

GERMINATION

The first step in germination of a seed occurs when it imbibes water. Because the seed is so dry at the start of germination, it takes up water with great force. Once this has occurred, metabolism within the seed resumes. Initially, metabolism may be anaerobic, but when the seed coat ruptures, aerobic metabolism takes over. At this point, it is important that oxygen be available to the developing embryo, because plants, which drown for the same reason people do, require oxygen for active growth. Few plants produce seeds that germinate successfully underwater, although some, like rice, have evolved a tolerance of anaerobic conditions.

Germination and early seedling growth require the mobilization of metabolic reserves stored in the starch grains of **amyloplasts** (plastids that are specialized to store starch) and protein bodies. Fats and oils are also important food reserves in some kinds of seeds. They can readily be digested during germination to produce glycerol and fatty acids, which yield energy through aerobic respiration and can also be converted to glucose. Any of these reserves may be stored in the embryo itself or in the endosperm, depending on the kind of plant.

How are the genes that transcribe the enzymes involved in the mobilization of food resources activated? Experimental studies have shown that, in the endosperm of the cereal grains at least, this occurs when hormones called **gibberellins** are synthesized by the embryo. These hormones initiate a burst of mRNA and protein synthesis. It is not known whether the gibberellins act directly on the DNA, or through chemical intermediates in the cytoplasm. DNA synthesis apparently does not occur during the early stages of seed germination, but becomes important when the **radicle,** or embryonic root, has grown out of the seed coats (see Figure 29-3).

> During early germination and seed establishment, the vital mobilization of the food reserves stored in the embryo or the endosperm is mediated by hormones.

GROWTH AND DIFFERENTIATION

Once a seed has germinated, the plant's further development depends on the activities of the meristematic tissues, which interact with the environment. As described in Chapter 29, the shoot and root apical meristems give rise to all of the other cells of the adult plant.

Differentiation, the formation of specialized tissues, can be considered to occur in five stages in plants (Figure 30-11):

STAGE 1   is the formation of the embryo by cell division of the zygote formed by the sperm's fertilization of the egg within the ovule.

STAGE 2   is the differentiation within the embryo of the apical meristems, which begins almost immediately in angiosperms. The apical meristems can be detected when there are as few as 40 cells in the growing embryo. Apical meristems are largely responsible for primary growth.

STAGE 3   is the differentiation from apical meristems of the cambiums, which are largely responsible for secondary growth.

STAGE 4   is the production of **primordia,** which are cells fully committed to becoming leaves, shoots, or roots. Leaf and shoot primordia develop directly from apical meristem cells, while root primordia develop from the root cambium, called the pericycle.

STAGE 5   is the production of fully differentiated tissues and structures, including xylem, phloem, leaves, shoots, and roots.

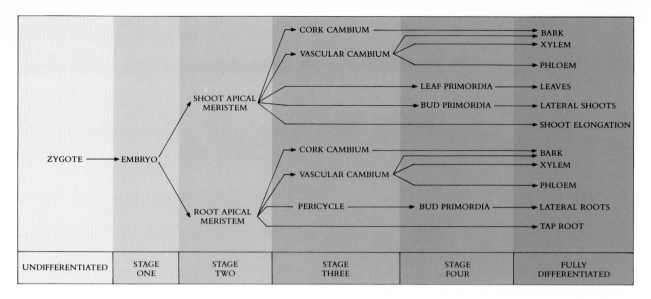

UNDIFFERENTIATED	STAGE ONE	STAGE TWO	STAGE THREE	STAGE FOUR	FULLY DIFFERENTIATED

FIGURE 30-11

Stages in plant differentiation.

FIGURE 30-12

The technique of isolating phloem tissue from carrots *(Daucus carota)*, as carried out in the laboratory of F.C. Steward at Cornell University. The discs of tissue were grown in a flask in which the medium was constantly agitated so as to bring a fresh supply of nutrients to the masses of **callus** (undifferentiated cells) that soon formed.

PIECES OF PHLOEM

TRANSVERSE SECTION OF CARROT ROOT

INDIVIDUAL CELLS ARE SEPARATED AND CULTURED IN MEDIUM PLUS COCONUT MILK

EMBRYOS

PLANTLET

MATURE PLANT

After a seed germinates, the pattern of growth and differentiation that was established in the embryo is repeated indefinitely, until the plant dies. Differentiation in plants, however, unlike that in animals, is largely reversible. Botanists first demonstrated in the 1950s that individual differentiated cells isolated from mature individuals could give rise to entire individuals. F.C. Steward, for example, was able to induce isolated bits of phloem taken from carrots to form new plants, plants that were normal in appearance and fully fertile (Figure 30-12). Regeneration of entire plants from differentiated tissue has since been carried out in many plants, including cotton, tomatoes, and cherries. These experiments clearly demonstrate that the original differentiated phloem tissue still contains all of the genetic potential needed for the differentiation of entire plants. No information is lost during plant tissue differentiation, and no irreversible steps are taken.

In plants, regeneration is not confined to the laboratory. Nature does it too. Asexual reproduction is a regeneration process in which differentiated root or stem cells form a new shoot with its own set of roots. As a result, entire plants may form from horizontal roots, and aboveground and underground runners such as **stolons** and **rhizomes.** For example, colonies of aspen, *Populus tremuloides,* often consist of a single individual that has given rise to a colony of genetically identical trees by producing new stems from its horizontal roots (Figure 30-13).

FIGURE 30-13

The contrast between the golden quaking aspens *(Populus tremuloides)* and the dark green Engelmann spruces *(Picea engelmannii),* which is evident in this autumn scene near Durango, Colorado, makes it possible to see the clones of the aspen, large colonies produced by single individuals spreading underground and sending up new shoots periodically. The aspens are deciduous, losing their leaves in the winter, whereas the spruces, like many other gymnosperms, are evergreen.

REPRODUCTIVE STRATEGIES OF PLANTS

Flowers, seeds, fruits, and asexual reproduction in plants can all be viewed as ways in which the plants increase or decrease the rate of recombination, a genetic process that has been one of the most powerful forces in the evolution of life on earth. By rapidly generating new combinations of alleles, recombination gives natural selection the raw material from which to select new and better-adapted phenotypes. Recombination is not always advantageous, however. Species that are well adapted to their environments have little to gain by shuffling their genes about, because most new combinations are less well-attuned to the plants' present environments. In plants, reproductive strategies that favor **outcrossing,** hybridization with another individual of the same species that is not a close relative and thus promote recombination, occur in some species, whereas different strategies that favor inbreeding and thus minimize recombination occur in other species.

Factors Promoting Outcrossing

Outcrossing is critically important for the adaptation and evolution of all eukaryotic organisms. One of the ways that outcrossing is enhanced in certain plant species is by the possession of separate male and female flowers. A flower that has only ovules, not pollen (see Figure 30-5), is called **pistillate;** functionally it is female. A flower that produces only pollen is functionally male; it is called **staminate.**

Staminate and pistillate flowers may occur on separate individuals in a given plant species, as for example in willows. Such plants, whose sporophytes produce either only ovules or only pollen, are called **dioecious** (from the Greek words for "two houses"). In other kinds of plants—oaks, birches, and ragweed (see Figure 30-5, *A* and *B*)—the two types of flowers may be produced on the same plant; such plants are called *monoecious* ("one house"). The separation of ovule-producing and pollen-producing flowers that occurs in dioecious and monoecious plants makes outcrossing even more likely than it would otherwise be. Most angiosperms are neither dioecious nor monoecious, however; each of their flowers includes both pollen-producing and ovule-producing structures, that is, stamens and carpels.

An even simpler way that many flowers promote outcrossing is through the physical separation of the anthers and the stigmas. If the flower is constructed in such a way that these organs do not come into contact with one another, there will naturally

THE FLOWERING OF BAMBOOS AND THE STARVATION OF PANDAS

In many Asian bamboo species, all the individuals of the species flower and set seed simultaneously. These cycles tend to occur at very long intervals, ranging from 3 years upward. Most of the cycles are between 15 and 60 years long. The most extreme example is the Chinese species *Phyllostachys bambusoides,* which seeded massively and throughout its range in 919, 1114, between 1716 and 1735, in 1833 to 1847, and in the late 1960s. The last event involved cultivated plants in widely separated areas, including England, Russia, and Alabama. *Phyllostachys bambusoides,* therefore, has a flowering cycle of about 120 years, set by an internal clock that runs more or less independently of environmental circumstances.

The bamboos with cycles of this kind spread mainly by the production of rhizomes, horizontal underground stems. When they do flower, nearly all individuals flower at once, set large quantities of seed, and die. Huge numbers of animals, including rats, pigs, and pheasants, often migrate into areas where bamboos are fruiting to feed on the seeds; humans also gather them for food. Apparently the plants put all of their energy into producing such vast amounts of seeds that large numbers of seedlings survive, even though most of the seeds are eaten.

A particular conservation problem that has attracted widespread notice in the 1980s is the relationship between the flowering of bamboos (primarily of the genus *Fargesia*) and the survival of the giant panda, a spectacular animal that still survives in a few mountain ranges in southwestern China. Humans have occupied so much of the former range of the panda that the wild population now consists of only a few thousand individuals, which live only in several widely separated areas. In these places, only one or a few species of bamboo provide virtually all of the pandas' food. The mass flowering of some of these bamboo species in 1984, which led to their death over large areas, drove many of the pandas to the brink of starvation, and only a massive international effort made it possible to rescue many of the animals. The predictable episodes of mass flowering in bamboos will need to be considered carefully in planning for the pandas' survival.

FIGURE 30-A

1   A panda in a large enclosure in a stand of the bamboo *Fargesia spathacea.* Bamboos of the genus *Fargesia* are the most important panda food in the Min Mountains of northern Sichuan Province, China, where this photograph was taken.

2   A Chinese scientist examining a stand of the bamboo *Fargesia nitida* that has flowered and dropped its leaves; the whole plant will soon be dead. This bamboo covers large areas of the Min Mountains and other mountain areas of southwestern China where pandas occur. It flowered extensively in the Min Mountains between 1974 and 1976, and again in 1982 and 1983; as a result, many pandas needed to be rescued and fed in captivity.

A

B

C

FIGURE 30-14

In the genus *Epilobium,* most species are self-pollinating, but a few are outcrossing. **A** and **B** *Epilobium angustifolium* is strongly outcrossing and is one of the first plants in which the process of pollination was studied (in the 1790s). In it, the anthers first shed pollen, then the style swings up into a similar position in the flower and its four lobes open; these flowers are functionally staminate at first and then become pistillate about 2 days later. The flowers open progressively up the stem, so that the lowest ones are visited by bees first. Working up the stem, the bees encounter pollen-shedding, staminate-phase flowers, become covered with pollen, and carry it passively to the lower, functionally pistillate flowers of another plant. Shown here are **(A)** flowers in the staminate phase and **(B)** flowers in the pistillate phase. The stigmas are the white, cross-shaped structures.
(C) *Epilobium ciliatum.* In this species, of a kind that is more common in the genus, the anthers shed their pollen directly on the large, creamy, undivided stigmas of the same flower, which are mature at the same time, thus bringing about self-pollination.

be a tendency for the pollen to be transferred to the stigma of another flower, rather than to the stigma of its own flower (Figure 30-14, *A* and *B*).

Another device that occurs widely in flowering plants and increases outcrossing is **genetic self-incompatibility.** In self-incompatible plants, the pollen from a given individual does not function on the stigmas of that individual, or the embryos resulting from self-fertilization do not function. Self-incompatibility is a mechanism that increases outcrossing even though the flowers of such plants may produce both pollen and ovules, and their stamens and stigmas may be mature at the same time.

Self-Pollination

Self-pollination is also very frequent among angiosperms. In fact, probably more than half of the angiosperms that occur in temperate regions self-pollinate regularly. Most of these have small, relatively inconspicuous flowers in which the pollen is shed directly onto the stigma, sometimes even before the bud opens (Figure 30-14, *C*). You might logically ask why there are so many self-pollinated plant species, if outcrossing is just as important for plants, in genetic terms, as it is for animals. There are two basic reasons for the frequent occurrence of self-pollinated angiosperms:

1. Self-pollination is advantageous in harsh climates, and as a result the plants in which it occurs do not need to be visited by animals to produce seed. As a consequence, self-pollinated plants can grow in areas where the kinds of insects or other animals that might visit them are absent or very scarce, as in the Arctic or at high elevations.

2. In genetic terms, self-pollination produces progenies that are more uniform than those that result from outcrossing. Such progenies may contain high proportions of individuals well adapted to particular habitats. Where the habitat occurs regularly, inbreeding is advantageous because it produces a greater proportion of well-adapted progeny than outcrossing does. Many successful weeds are self-pollinating, because the habitat of weeds has been made uniform and spread all over the world by human beings.

SUMMARY

1. The flowers of angiosperms make possible the precise transfer of pollen and thus enable even widely separated plants to outcross effectively. Pollen is transferred by insects, particularly bees, by birds and bats, and by the wind.

2. The change of a zygote into a mature individual, initiated immediately after fertilization, is called development. Development is a process of progressive specialization that results in differentiation, the production of highly individual tissues and structures. In plants, differentiation is fully reversible. Whole plants can be regenerated from cultures of single cells, as long as they retain a living protoplast and a nucleus.

3. Differentiation of angiosperm embryos ceases after the major organs have been laid down, and each now-dormant embryo is encased within a dry casing, becoming a seed that maintains a state of suspended development until it is broken or moistened.

4. The seeds of angiosperms remain within the carpel, which develops into a fruit. Many fruits are fleshy and often sweet; animals that consume fruit may carry the seeds for long distances before excreting them as solid waste. The seeds, not harmed by the animal digestive system, can then germinate in the new location.

5. Seeds, because of their rigid, relatively impermeable seed coat, germinate only when they receive water and appropriate environmental cues. In the germination of seeds, mobilization of the food reserves stored in the cotyledons and in the endosperm is critical. In cereal grains, this process is mediated by hormones known as gibberellins.

6. Outcrossing in different angiosperms is promoted by the separation of the pollen-producing and ovule-producing structures into different flowers, or even onto different individuals; by genetic self-incompatibility; and by the separation of the pollen and the stigmas within a given flower with respect to position or time of maturation.

7. Self-pollination occurs in more than half of the angiosperm species of temperate regions. It is most common among plants in harsh climates, or those that occur in widespread, uniform habitats, such as weeds.

REVIEW

1. Which type of cell division, mitosis or meiosis, produces gametes in plants?

2. _____ are the most numerous and constant pollinators of insect-pollinated flowers.

3. The first step in seed germination occurs when the seed _____

4. During the early germination of cereal grains, hormones called _____ are synthesized by the embryo; they mediate the mobilization of food reserves.

5. _____ is the hybridization of a plant with another individual of the same species that is not a close relative.

SELF-QUIZ

1. The flowers of angiosperms
   (a) are often colorful so that insects are attracted to them
   (b) usually offer nectar or other rewards to insects that visit them
   (c) are usually attractive only to a specific set of pollinators
   (d) are often green or dull-colored if self-pollination or wind-pollination predominates
   (e) all of the above

2. Why are seeds an important evolutionary improvement over spores?
   (a) because they contain little water
   (b) because they can remain dormant until conditions are right for germination
   (c) because they enhance dispersal and thus migration of genotypes
   (d) because seeds such as beans, corn, and rice are important food sources
   (e) both b and c

3. Fleshy fruits that are brightly colored are often dispersed by
   (a) insects   (d) birds
   (b) wind   (e) attaching to the fur of mammals
   (c) water

4. Leaf and shoot primordia develop from
   (a) the xylem
   (b) the epidermis
   (c) the apical meristem
   (d) the vascular cambium
   (e) none of the above

5. Which three are factors that promote outcrossing in flowering plants?
   (a) having dioecious plants
   (b) physical separation of stamens and carpels
   (c) inbreeding
   (d) genetic self-incompatability
   (e) asexual reproduction

THOUGHT QUESTIONS

1. Angiosperms usually produce both pollen and ovules within a single flower, or at least on a single plant. Why don't all angiosperms simply self-pollinate?

2. Self-pollination does occur within at least some species of a wide variety of different kinds of plants, usually those living in harsh environments, or among "weedy" species. Why do you suppose that animals coping with similar environments never developed similar reproductive strategies? Can you think of any animals that do self-fertilize?

3. Why is plant development so much more closely linked with environmental cues than animal development?

4. When you eat a tossed salad, what plant parts are you consuming?

FOR FURTHER READING

BATRA, S.W.T.: "Solitary Bees," *Scientific American,* February 1984, pages 120-127. Excellent article on these diverse and fascinating pollinators, some of which are of great commercial importance.

COOK, R.E.: "Clonal Plant Populations," *American Scientist,* vol. 71, pages 244-253, 1983. Excellent discussion on the role of asexual reproduction among plants in natural populations.

JANZEN, D.H.: "Why Bamboos Wait So Long to Flower," *Annual Reviews of Ecology and Systematics,* vol. 7, pages 347-391, 1976. A fascinating essay about the natural history of bamboos.

PETTITT, J., S. DUCKER, and B. KNOX: "Submarine Pollination," *Scientific American,* March 1981, pages 134-145. Sea grasses (marine angiosperms) flower underwater, shedding pollen that is carried from plant to plant by waves.

SCHALLER, G.B., HU JINCHU, PAN WENSHI, and ZHU JING: *The Giant Pandas of Wolong,* University of Chicago Press, Chicago, 1985. A fascinating account of the mutual adaptations of pandas and bamboo; this book offers the first glimpse of the life of pandas in their remote mountain home in western China.

SHEPARD, J.F.: The Regeneration of Potato Plants from Leaf-Cell Protoplasts," *Scientific American,* May 1982, pages 112-121. This article provides valuable insight into the way experiments in this area are designed.

FIGURE 31-1   Man-high, this cutleaf Silphium is a plant of the tall-grass prairie of the American West. With a sunflower-like bloom that is the size of a saucer, Silphium is one of the most striking of prairie wildflowers.

Overview

The body of a plant is basically a tube embedded in the ground and extending up into the light, where expanded surfaces, the leaves, capture the sun's energy and participate in gas exchange. The warming of the leaves by the sunlight increases evaporation from them, creating a suction that draws water into the plant through the roots and up the plant through the xylem to the leaves. Transport from the leaves and other photosynthetically active structures to the rest of the plant occurs through the phloem. Most of the nutrients critical to plant metabolism are accumulated by the roots and are then carried in the water stream throughout the plant. The growth of the plant itself is regulated by plant hormones. Partly through the mediation of such hormones, plants respond in complex ways to external stimuli such as touch, light, gravity, and day length.

For Review

Here are some important terms and concepts that you will encounter in this chapter. If you are not familiar with them, you should review them before proceeding.

Osmotic pressure (Chapter 6)

Transport of ions across membranes (Chapter 6)

Photosynthesis (Chapter 9)

Conducting tissues (Chapter 29)

Seed germination (Chapter 30)

When you first look at a tree, it does not appear vibrantly alive. It does not bound from one place to another like a gazelle, or growl, or respond to a caress. The tree's stolid appearance is deceiving, however; its internal structure is more complex than you might suspect. It has a plumbing system (Figure 31-1), as you do, which sends fluids from one part to another, and special organs for reproduction and the gathering of energy. Like you, it regulates its growth and the functioning of its organs with hormones, chemicals that act as messengers to coordinate the many activities of the body. In this chapter we focus on these activities, first discussing the movement of fluids within plants and the many substances that plant fluids transport. We then examine one class of these substances in more detail, the hormones that control how the plant grows.

FIGURE 31-2

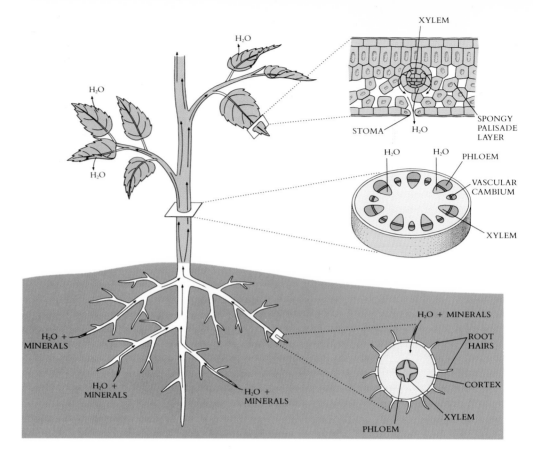

WATER MOVEMENT

Functionally, a plant is essentially a tube with its base embedded in the ground (Figure 31-2). At the base of the tube are roots, and at its top are leaves. For a plant to function, two kinds of transport processes must occur. First, the carbohydrate molecules that are produced in the leaves by photosynthesis must be carried to all of the other living cells of the plant. To accomplish this, liquid, with these carbohydrate molecules dissolved in it, must move both up and down the tube. Second, nutrients and water in the ground must be taken up by the roots and ferried to the leaves and other cells of the plant. In this process, liquid moves up the tube.

It is not unusual for many of the leaves of a large tree to be more than 10 stories off the ground. Did you ever wonder how a tree manages to raise water so high? To understand how this happens, imagine filling with water a long, hollow tube closed at one end and placing the tube, open end down, in a full bucket of water. Gravity acts (pushes) on the column of air over the bucket: the weight of the air (at sea level) exerts an amount of pressure that is defined as 1 atmosphere downward on the water in the bucket and thus presses the water up into the tube. Gravity also acts to pull the water down within the tube, however. The interaction of these two forces, both dependent on gravity, determines the level of water in the tube. At sea level the water rises to about 10.4 meters. If the tube is any higher, a vacuum will form in the upper, closed end of the tube and will fill with water vapor.

To illustrate how water rises higher than 10.4 meters in a plant, open the tube and blow across the upper end. The stream of relatively dry air will cause water molecules to evaporate from the surface of the water in the tube. Does the level of water in the tube fall? No. As water molecules are drawn from the top, they are replenished by new ones that are forced up from the bottom. This, in essence, is what happens in plants. The passage of air across leaf surfaces results in the loss of water by evaporation, creating a suction force at the open upper end of the "tube," while water is pushed up from below by atmospheric pressure. In addition to this principal factor, the **adhesion** of water molecules to the walls of the very narrow tubes that occur in plants (see Figure 31-1) also helps to maintain the flow of water to the tops of plants.

> **Water rises in a plant beyond the point at which it would be supported by atmospheric pressure (10.4 meters at sea level) because evaporation from its leaves produces a suction that pulls up on the entire water column all the way down to the roots.**

Why does a column of water in a tall tree not collapse simply because of its weight? The answer is that water has an inherent strength that arises from the cohesion of its molecules; that is, from their tendency to form hydrogen bonds with one another. Because of these hydrogen bonds, a column of water resists separation. This resistance, called tensile strength, varies inversely with the diameter of the column; that is, the smaller the diameter of the column, the greater the tensile strength. Therefore plants must have very narrow transporting vessels in order to take advantage of the tensile strength.

Transpiration

More than 90% of the water that is taken in by the roots of a plant is ultimately lost to the atmosphere, almost all of it from the leaves. It passes out primarily through the stomata in the form of water vapor. The process by which water leaves the plant is known as **transpiration.** First, the water passes into the pockets of air within the leaf by evaporating from the walls of the spongy mesophyll that lines the intercellular spaces (see Figure 29-12). These intercellular spaces open to the outside of the leaf by way of the stomata. The water that evaporates from the surfaces of the spongy mesophyll cells lining the intercellular spaces is continuously replenished from the tips of the veinlets in the leaves (see Figure 31-1). Since the strands of xylem conduct water within the plant in an unbroken stream all the way from the roots to the leaves, when a portion of the water vapor in the intercellular spaces passes out through the stomata, the supply of water vapor in these spaces is continually renewed.

Because they are constantly losing so much water to the atmosphere, and because the presence of this water is essential to their metabolic activities, growing plants depend on a continuous stream of water entering and leaving their bodies at all times. Water must always be available to their roots. Such structural features as the stomata, the cuticle, and the substomatal spaces have evolved in response to two contradictory requirements: to maintain this stream of water and to admit carbon dioxide into the plant. Before considering how plants resolve this problem, however, we must consider the absorption of water by the roots.

The Absorption of Water by Roots

Most of the water absorbed by the plant comes in through the root hairs, which collectively have an enormous surface area (see Figure 29-7). The root hairs are always **turgid**—plump and swollen with water—because they have a higher concentration of dissolved minerals than the water in the soil solution; water, therefore, tends to move into them steadily. Once it is inside the roots, the water passes inward to the conducting elements of the xylem.

Water is not the only substance that enters the roots by passing into the cells of the root hairs. The membranes of these root hair cells contain a variety of ion transport channels, which actively pump specific ions into them even against large concentration gradients. These ions, many of which are plant nutrients, are then transported throughout the plant as a component of the water flowing through the xylem.

At night, when the relative humidity may approach 100% at the leaf surface, there may be no transpiration from the leaves. Under these circumstances, the negative pressure component of the water potential (suction due to evaporation) becomes very

The continued existence and growth of plants depends on an adequate supply of water. Plants lack closed circulation systems, and only the continuous stream of water that flows through them keeps them healthy.

Even plants, however, can receive too much water. This occurs when the soil is flooded, a condition that can arise when rivers or streams overflow their banks or when rainfall is heavy, irrigation is excessive, or drainage is poor. Flooding rapidly depletes the available oxygen in the soil and blocks the normal reactions that take place in roots and make possible the transport of minerals and carbohydrates. Abnormal growth patterns may result, and the plants may ultimately "drown." Hormone levels change in flooded plants (ethylene, for example, often increases, whereas gibberellins and cytokinins usually decrease), and these changes may also contribute to the abnormal growth patterns. Flooding that involves moving water, which brings in new supplies of oxygen, is much less harmful than flooding involving standing water, which does not; flooding that occurs when a plant is dormant is much less harmful than flooding when the plant is growing actively.

Physical changes that may occur in the roots as a result of oxygen deprivation may halt the flow of water through the plant, paradoxically drying out the leaves, even though the roots of the same plant may be standing in water. Because of such stresses, the stomata of flooded plants often close. In some plants the closing of the stomata maintains the turgor of the leaves.

Many plants, of course, grow in places that are often flooded naturally; they have adapted to these conditions during the course of their evolution (Figure 31-A, *1*). One of the most frequent adaptations among such plants is the formation of **aerenchyma,** loose parenchymal tissue with large air spaces in it. Aerenchyma is very prominent in water lilies and many other aquatic plants. Oxygen may be transported from the parts of plants above the water to those below by way of passages in the aerenchyma. This supply of oxygen allows oxidative respiration to take place even in the submerged portions of the plant. Some plants normally form aerenchyma,

whereas others, subject to periodic flooding, can form it when necessary. Plants also respond to flooded conditions by forming larger lenticels, which facilitate gas exchange, and additional adventitious roots.

Plants such as mangroves (Figure 31-A, *2,* and 31-A, *3*), which are normally flooded with salt water, not only must provide a supply of oxygen for their submerged parts, but also must control their salt balance. The salt must be excluded, actively excreted, or diluted as it enters. The arching stilt roots of mangroves are connected to long, spongy, air-filled roots that emerge above the mud. These spongy air roots have large lenticels on their above-water portions through which oxygen enters; it is then transported to the submerged roots. In addition, the succulent leaves of mangroves contain large quantities of water, which dilutes the salt that reaches them. Many plants that grow in such conditions excrete large quantities of salt through more or less specialized glands.

FIGURE 31-A

Adaptations of plants to flooded conditions.
1 The "knees" of bald cypress *(Taxodium),* are formed wherever it grows in wet conditions.
2 The air roots of black mangrove *(Avicennia nitida),* like those of white mangrove, bring oxygen to the roots of the plants.
3 White mangrove *(Rhizophora),* with its stilt roots, is a familiar sight along shores throughout the world's tropics and subtropics.

1

3

small or nonexistent. At such times water does not travel upward in the xylem. Active transport of ions into the roots, however, continues to take place under these circumstances. It results in an increasingly high ion concentration within the cells, a concentration that causes water to be drawn into the root hair cells by osmosis. The result is the movement of water into the plant and up the xylem columns, which we call **root pressure.**

> Root pressure, which is active primarily at night, is caused by the continued, active accumulation of ions by the roots of a plant at times when transpiration from the leaves is very low or absent.

The Regulation of Transpiration Rate

The only way plants can control water loss on a short-term basis is to close their stomata. Many plants can do this when they are subjected to water stress. The stomata must be open at least part of the time, however, so that CO_2, which is necessary for photosynthesis, can enter the plant. In its pattern of opening or closing its stomata, a plant must respond to both the need to conserve water and the need to admit CO_2.

The stomata open and close because of changes in the water pressure of their guard cells (Figure 31-3). The guard cells of stomata are the only cells of the epidermis to possess chloroplasts. They stand out for this reason and because of their distinctive shape—they are thicker on the side next to the stomatal opening and thinner on their other sides and ends. When the guard cells are turgid (plump and swollen with water), they become bowed in shape, as do their thick inner walls, thus opening the stomata as wide as possible

The guard cells use ATP-powered ion transport channels through their membranes to concentrate ions actively. This concentration causes water to enter the guard cell osmotically. As a result, these cells accumulate water and become turgid, opening the stomata. The guard cells remain turgid only so long as the active transport channels pump ions, chiefly potassium (K^+), into the cells and so maintain the higher solute concentration there. Keeping the stomata open therefore requires a constant expenditure of ATP. When the active transport of ions into the guard cells ceases, the higher concentration of ions within the guard cells causes the ions to move out into the surrounding cells by diffusion. Ultimately, water leaves the guard cells also, which then become somewhat "limp" or deflated, and the stomata between them close.

> Stomata open when their guard cells become turgid. Their inner surfaces are thickest, and they bow inwardly when the pressure within the cells is high. Keeping the guard cells turgid requires a constant expenditure of ATP.

CARBOHYDRATE TRANSPORT

Most of the carbohydrates manufactured in the leaves and other green parts of the plant are moved through the phloem to other parts of the plant. This process, known as **translocation,** is responsible for the availability of suitable carbohydrate building blocks at the actively growing regions of the plant. The carbohydrates that are concentrated in storage organs such as tubers, often in the form of starch, are also converted into transportable molecules such as sucrose and are moved through the phloem. The liquid in the phloem contains 10% to 25% dissolved solid matter, almost all of which is sucrose.

The movement of substances in the phloem can be remarkably fast; rates of 50 to 100 centimeters per hour have been measured. The movement of water and dissolved nutrients within the sieve tubes is a passive process that does not require the expenditure of energy. The movement of materials transported in the phloem occurs because of hydrostatic pressure, which develops as a result of osmosis (Figure 31-4). The overall process by which this occurs is called **mass flow.** First, sucrose produced as a result of photosynthesis is actively loaded into the phloem tubes of the veinlets. This increases the solute concentration of the sieve tubes, so water passes into them by osmosis. An

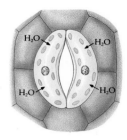
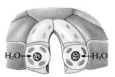

A GUARD CELLS OPEN

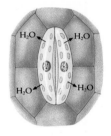

B GUARD CELLS CLOSED

FIGURE 31-3

A The thick inner side of each of the two yellow guard cells that make up a stoma bow outwardly, opening the stoma, when solute pressure is high within the guard cells.
B When the solute pressure is low within the guard cells, the guard cells become limp and close the stoma.

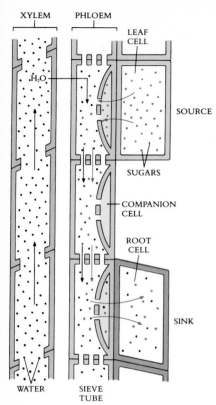

XYLEM PHLOEM

LEAF CELL

H₂O

SOURCE

SUGARS

COMPANION CELL

ROOT CELL

SINK

WATER   SIEVE TUBE

FIGURE 31-4

Diagram of mass flow. In this diagram, sucrose molecules are indicated by red dots, water molecules by black ones. Moving from the parenchyma cells of a leaf or another part of the plant into the conducting cells of the phloem, the sucrose molecules are then transported to other parts of the plant by mass flow and unloaded where they are required.

area where the sucrose is made is called a **source;** an area where it is being taken from the sieve tubes is called a **sink.** Sinks include the roots and other regions where the sucrose is being unloaded. There the solute concentration of the sieve tubes is decreased as the sucrose is removed. As a result of these processes, water moves in the sieve tubes from the areas where sucrose is being taken in to those areas where it is being withdrawn, and the sucrose moves passively with it.

PLANT NUTRIENTS
Essential Nutrients

Plants require a number of inorganic nutrients. Some of these are **macronutrients,** which the plants need in relatively large amounts, and others are **micronutrients,** those required in trace amounts. There are nine macronutrients: carbon, hydrogen, and oxygen (the three elements found in all organic compounds); and nitrogen, potassium, calcium, phosphorus, magnesium, and sulfur. Each of these nutrients approaches or exceeds 1% of the dry weight of a healthy plant. The seven micronutrient elements that in most plants constitute from less than one to several hundred parts per million by dry weight are iron, chlorine, copper, manganese, zinc, molybdenum, and boron.

> **The macronutrients are substances required by plants in relatively large amounts. Each of them approaches or exceeds 1% of a plant's dry weight. Micronutrients are substances that plants require in only trace amounts of one to several hundred parts per million.**

The six macronutrients other than carbon, hydrogen, and oxygen, and the seven micronutrients are involved in plant metabolism in many ways. Potassium ions regulate the turgor pressure of guard cells and therefore the rate at which the plant loses water and takes in carbon dioxide. Calcium is an essential component of the **middle lamellae,** structural elements laid down between plant cell walls, and it also helps to maintain the physical integrity of membranes. Magnesium is a part of the chlorophyll molecule. The presence of phosphorous in many key biological molecules such as nucleic acids and ATP has been explored in detail in the earlier chapters of this book. Nitrogen is an essential part of amino acids and proteins, of chlorophyll, and of the nucleotides that make up nucleic acids.

A number of ions are components of enzyme systems and serve as cofactors in essential biochemical reactions. Potassium, for example, which reaches higher concentrations in plants than any other elements except carbon and oxygen, affects the conformation of many proteins and probably affects at least 60 different enzymes while they are functioning. Zinc appears to play a similar role in the synthesis of the important plant growth hormone auxin. Plants with an inadequate supply of zinc display symptoms that derive mainly from a lack of cell elongation, apparently reflecting a shortage of auxin. When an essential nutrient is present in short supply, a plant will display deficiency symptoms characteristic of that nutrient. For this reason, a trained observer can often tell what chemicals should be supplied simply by observing the appearance of a plant.

Some kinds of plants have specific nutritional requirements that are not shared by others. Silica, for example, is essential for the growth of many grasses because it helps to retard their complete destruction by herbivores (see Figure 21-7), it is not required by plants in general. Cobalt is necessary for the normal growth of the nitrogen-fixing bacteria associated with the nodules of legumes and is therefore an essential element for the normal growth of these plants. Nickel seems to be essential for soybeans, and its role in the nutrition of other plants needs further investigation.

In general, the elements that animals require reach them through plants, which therefore form an indispensable link between animals and the reservoirs of chemicals in nature. Some elements that animals require, such as iodine, come by way of plants but are not required by the plants. Iodine is very rare in soils; a shortage of it in the human diet can lead to the condition known as goiter.

REGULATING PLANT GROWTH: PLANT HORMONES

Hormones are chemical substances produced in small, often minute, quantities in one part of an organism and then transported to another part of the organism, where they bring about physiological responses. The activity of hormones results from their ability to stimulate certain physiological processes and to inhibit others. How they act in a particular instance is influenced both by what the hormones themselves are and by how they affect the particular tissue that receives their message.

In animals, hormones are usually produced at definite sites, normally in organs that are solely concerned with their production. In plants, on the other hand, hormones are produced in tissues that are not specialized for that purpose but that carry out other, usually more obvious, functions. There are at least five major kinds of hormones in plants: auxin, cytokinins, gibberellins, ethylene, and abscisic acid. Other kinds of plant hormones certainly exist, but they are less well understood. The study of plant hormones, especially our attempts to understand how they produce their effects, is an active and important field of current research.

> **Five major kinds of plant hormones are reasonably well understood:**
> **auxins, cytokinins, gibberellins, ethylene, and abscisic acid.**

The Discovery of the First Plant Hormone

In his later years, the great evolutionist Charles Darwin became increasingly devoted to the study of plants. In 1881 he and his son Francis published a book called *The Power of Movement in Plants*. In this book, the Darwins reported their systematic experiments concerning the way in which growing plants bend toward light, a phenomenon known as **phototropism.** They made many observations in this field and also conducted experiments using young grass seedlings.

Charles and Francis Darwin found that the seedlings they were studying normally bent strongly toward a source of light, if the light came primarily from one side. However, if they covered the upper part of a seedling with a cylinder of metal foil so that no light reached its tip, the shoot would not bend (Figure 31-5). The Darwins obtained this result even though the region where the bending normally occurred was still exposed. Light reached this part of the seedling directly, but bending did not occur. If they covered the end of the shoot with a gelatin cap, which transmitted light, however, the shoot would bend as if it were not covered at all.

To explain this unexpected finding, the Darwins hypothesized that when the shoots were illuminated from one side, an "influence" arose in the uppermost part of the shoot, was then transmitted downward, and caused the shoot to bend. For some 30 years, the Darwins' perceptive experiments remained the sole source of information about this interesting phenomenon. Then a series of experiments was performed by several botanists, who demonstrated that the substance that caused the shoots to bend was a chemical. They cut off the tip of a grass seedling and then replaced it, but separated it from the rest of the seedling by a block of agar, a gelatinous medium often used in biological experiments. The seedling reacted as if there had been no change. Something was evidently passing from the tip of the seedling through the agar into the region where the bending occurred. The "something" was a plant hormone called **auxin.** We now know that auxin acts to regulate cell growth in plants.

How the Shoot Apex Uses Auxin to Control Growth

How auxin controls the growth of a plant was discovered by Frits Went, a Dutch plant physiologist, in 1926. Went obtained these results in the course of studies for his doctoral dissertation. Carrying the earlier experiments an important step farther, Went cut off the tips of grass seedlings that had been illuminated normally and set these tips on agar. He then took grass seedlings that had been grown in the dark and cut off their tips in a similar way. Finally, Went cut tiny blocks from the agar on which the tips of the light-grown seedlings had been placed and put them on the tops of the decapitated

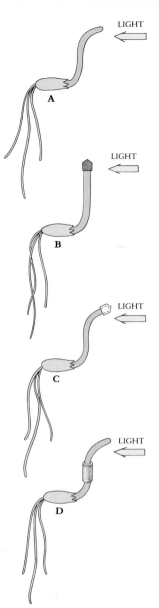

FIGURE 31-5

The Darwins' experiment. Young grass seedlings normally bend toward the light **(A).** When the tip of a seedling was covered by a lightproof collar **(B)** (but not when it was covered by a transparent one **[C]**), this bending did not occur. When the collar was placed below the tip **(D),** the characteristic light response still took place. From these experiments, the Darwins concluded that, in response to light, an "influence" that causes bending was transmitted from the tip of the seedling to the area below the tip, where bending normally occurs.

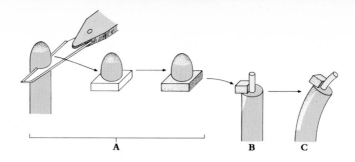

FIGURE 31-6

Frits Went's experiments.
A Went removed the tips of the grass seedlings and put them on agar.
B Blocks of the agar were then put on one side of the ends of other grass seedlings grown in the dark from which the tips had been removed.
C The seedlings bent away from the side on which the agar block was placed. Went concluded that the substance that he named auxin promoted the elongation of the cells, and that it accumulated on the side of a grass seedling away from the light.

dark-grown seedlings, but set off to one side (Figure 31-6). Even though these seedlings had not been exposed to the light themselves, they bent *away* from the side in which the agar blocks were placed.

As a result of his experiments, Went was able to show that the substance that had flowed into the agar from the tips of the light-grown grass seedlings enhanced cell elongation. This chemical messenger caused the tissues on the side of the seedling into which it flowed to grow more than those on the opposite side. He named the substance that he had discovered **auxin,** from the Greek word *auxein,* which means "to increase."

Went's experiments provided a basis for understanding the responses that the Darwins had obtained some 45 years earlier. The grass seedlings bent toward the light because the auxin contents on the two sides of the shoot differed. The side of the shoot that was in the shade had more auxin, and therefore its cells elongated more than those on the lighted side, bending the plant toward the light. Later experiments by other investigators showed that auxin in normal plants migrates from the illuminated side to the dark side in response to light and thus causes the plant to bend toward the light.

THE MAJOR PLANT HORMONES
Auxin

Only one form of auxin occurs in nature (Figure 31-7), indoleacetic acid (IAA). Indoleacetic acid is produced at the shoot apex, in the region of the apical meristem, and diffuses continuously downward. The term *auxin* is now used to refer both to the naturally occurring substance and to those related synthetic molecules that produce similar effects.

Auxin acts by increasing the plasticity of the plant cell wall. A more plastic wall will stretch more during active cell growth, while its protoplast is swelling. Since very low concentrations of auxin promote cell wall plasticity, it must be broken down rapidly to prevent its accumulation. Plants break auxin down by means of the enzyme **indoleacetic acid oxidase.** By controlling the levels of both IAA and IAA oxidase, plants can regulate their growth very precisely.

Auxin controls various plant responses in addition to those involved in phototropism. One of these is the suppression of lateral bud growth. How can auxin, a growth promoter, also inhibit growth? Apparently the cells around lateral buds produce the chemical ethylene under the influence of auxin. The ethylene, in turn, inhibits the growth of the lateral buds. When the terminal bud is removed, removing the source of auxin, the lateral buds grow, producing bushy plants. The number of flowers on an individual plant is also increased in this situation.

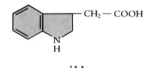

IAA
(INDOLEACETIC ACID)

FIGURE 31-7

Indoleacetic acid (IAA), the only known naturally occurring auxin.

> The only known naturally occurring auxin, indoleacetic acid (IAA), is produced in the apical meristems of shoots and diffuses downward, suppressing the growth of lateral buds. In young grass seedlings and other herbs, it plays a major role in stem elongation, migrating from the illuminated portions of the stem to the dark portions and thus causing the stems to grow toward the light.

Synthetic Auxins

Synthetic auxins are routinely used to control weeds. When they are used as herbicides, they are applied in higher concentrations than those at which IAA normally occurs in plants. One of the most important of the synthetic auxins used in this way is 2,4-dichlorophenoxyacetic acid, usually known as 2,4-D. It kills weeds in lawns without harming the grass because 2,4-D affects only broad-leaved dicots. When treated, the weeds literally "grow to death," rapidly depleting all metabolic reserves so that there is no source of energy to carry out transport or other essential functions.

Closely related to 2,4-D is the herbicide 2,4,5-T (2,4,5-trichlorophenoxyacetic acid), which is widely used to kill woody seedlings and weeds. Notorious as the "Agent Orange" of the Vietnam War, 2,4,5-T is easily contaminated with a by-product of its manufacture, **dioxin.** Dioxin, which is believed to be extremely toxic to people, is the subject of great environmental concern in the United States and elsewhere (Figure 31-8).

Cytokinins

A **cytokinin** is a plant hormone that, in combination with auxin, stimulates cell division in plants and determines the course of differentiation. Substances that have these properties are widespread both in bacteria and in eukaryotes. In vascular plants, most cytokinins seem to be produced in the roots and transported from there throughout the rest of the plant. Cytokinins seem to be necessary for mitosis and cell division to take place. They apparently work by influencing the synthesis or activation of proteins that are specifically required for mitosis.

The naturally occurring cytokinins all appear to be derivatives of the purine base adenine. In contrast to auxins, cytokinins *promote* the growth of lateral branches. Similarly, auxins promote the formation of lateral roots, whereas cytokinins inhibit it. As a consequence of these relationships, the balance between cytokinins and auxin determines, along with other factors, the appearance of a mature plant.

> **Cytokinins are plant hormones that, in combination with auxin, stimulate cell division and determine the course of differentiation.**

Gibberellins

Gibberellins are named for the fungus genus *Gibberella,* which causes a disease of rice in which the plants grow to be abnormally tall. This "foolish seedling disease" of rice was investigated in the 1920s by Japanese scientists, who found that if they grew the fungus in culture, they could obtain a chemical completely free of the fungus itself that would affect the rice plants in a way similar to the fungus. This substance, isolated in 1939 and chemically characterized in 1954, was the first of what proved to be a large class of naturally occurring plant hormones called the **gibberellins** (Figure 31-9).

Synthesized in the apical portions of both stems and roots, gibberellins have important effects on stem elongation in plants and play the leading role in controlling this process in mature trees and shrubs. In these plants the application of gibberellins characteristically promotes internode elongation, and this effect is enhanced if auxins are present also. Gibberellins are also involved with many other aspects of plant growth such as inducing flowering and hastening seed germination.

FIGURE 31-8

Personnel from the Missouri Department of Natural Resources and the U.S. Environmental Protection Agency taking earth samples in Times Beach, Missouri. These samples were then tested for dioxin. The entire community of Times Beach was abandoned in 1983, after it was found to be heavily contaminated with dioxin that had been an impurity in oil spread on the roads.

FIGURE 31-9

A Although more than 60 gibberellins have been isolated from natural sources, apparently only GA_1 is active in shoot elongation. All of the other gibberellins have a similar structure.

B Cabbage *(Brassica oleracea),* a biennial that is native to the seacoasts of Europe, will "bolt"—flower—when the heads are treated with gibberellin. Most biennials exhibit similar behavior. In nature, they usually flower during their second year of growth.

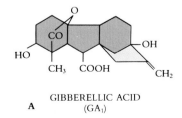

GIBBERELLIC ACID
(GA_1)

A   B

> Gibberellins are an important class of plant hormones. They are produced in the apical regions of shoots and roots and play the major role in controlling stem elongation for most plants, acting in concert with auxin and other plant hormones.

Ethylene

Long before its role as a plant hormone was appreciated, the simple hydrocarbon ethylene (H_2C—CH_2) was known to defoliate plants when it leaked from gas lights in street lamps. **Ethylene** is, however, a natural product of plant metabolism and appears to be the main factor in the formation of specialized layers of cells that precedes the dropping off of leaves from plants. We have already mentioned the way in which auxin, diffusing down from the apical meristem of the stem, may stimulate the production of ethylene in the tissues around the lateral buds and thus retard their growth. Ethylene also suppresses stem and root elongation, probably for similar reasons.

Ethylene is also produced in large quantities during a certain phase of the ripening of fruits, when their respiration is proceeding at its most rapid rate. At this phase of ripening, complex carbohydrates are broken down into simple sugars, cell walls become soft, and the volatile compounds associated with flavor and scent in the ripe fruits are produced. When ethylene is applied to fruits, it hastens their ripening. One of the first lines of evidence that led to the recognition of ethylene as a plant hormone was the observation that gases that came from oranges caused premature ripening in bananas. Such relationships have led to major commercial uses. For example, tomatoes are often picked green and then artificially ripened as desired by the application of ethylene. Ethylene is widely used to speed the ripening of lemons and oranges as well. Carbon dioxide produces effects in fruits opposite to those of ethylene, and fruits that are being shipped are often kept in an atmosphere of carbon dioxide if they are not intended to ripen yet.

> Ethylene, a simple gaseous hydrocarbon, is a naturally occurring plant hormone. It plays the key role in controlling the falling off of leaves, flowers, and fruits from the plants on which they form; auxin retards their tendency to fall.

Abscisic Acid

Abscisic acid is a naturally occurring plant hormone that is synthesized mainly in mature green leaves, in fruits, and in root caps. The hormone was given its name because applications of it stimulate leaves to age rapidly and fall off (the process of *abscission*), but there is little evidence that it plays an important natural role in this process. Abscisic acid suppresses the growth and elongation of buds and promotes aging, also counteracting some of the effects of the gibberellins (which stimulates the growth and elongation of buds) and auxin (which tends to retard aging).

TROPISMS

Tropisms, or responses to external stimuli, control the patterns of growth of plants and thus their appearance. Plants adjust to the conditions of their environment by growth responses. **Positive tropisms** are those in which the movement or reaction is in the direction of the source of the stimulus, whereas **negative tropisms** are those that result in movement or growth in the opposite direction. We shall consider three major classes of plant tropisms: phototropism, gravitropism, and thigmotropism.

Phototropism

We introduced **phototropism,** the bending of plants toward unidirectional sources of light, in our discussion of the action of auxin. In general, stems are positively phototropic, growing toward the light, whereas roots are negatively phototropic, growing away from it. The phototropic reactions of stems are clearly of adaptive value because they allow plants to capture greater amounts of light than would otherwise be possible. Auxin is involved in most, if not all, of the phototropic growth responses of plants.

Phototropisms are growth responses of plants to a unidirectional source of light. They are mostly, if not entirely, mediated by auxin.

Gravitropism

Another familiar plant response is **gravitropism**, formerly known as geotropism. This tropism causes stems to tend to grow upward and roots downward (Figure 31-10); both of these responses are clearly of adaptive significance. Stems that grow upward are apt to receive more light than those that do not; roots that grow downward are more apt to encounter a more favorable environment than those that do not. The phenomenon is now called gravitropism because it is clearly a response to gravity and not to the earth (prefix "geo") as such.

It seems likely that **amyloplasts,** starch-containing plastids, play an important role in the perception of gravity by plants. The amyloplasts are heavy capsules that contain large amounts of calcium and starch. In roots, the cells in which amyloplasts occur are apparently located in the central cells of the root cap; removing the root cap stops the root's responses to gravity in most cases. In shoots, on the other hand, gravity is clearly sensed along the whole length of the stem, probably by the functioning of similar amyloplasts in certain cells. Gravity causes the amyloplasts to fall to the lower side of a given cell. There they apparently set in motion a series of reactions that eventually causes the shoots and roots to bend. The amyloplasts reach the lower side of their cells within a minute, and the bending of the root or shoot may occur within as little as 10 minutes.

Gravitropism, the response of a plant to gravity, generally causes shoots to grow up and roots to grow down. The force of gravity apparently is sensed in special cells with amyloplasts, starch-containing plastids.

Thigmotropism

Still another response of plants that is commonly observed is **thigmotropism,** a name derived from the Greek root *thigma,* meaning "touch." Thigmotropism is defined as the response of plants to touch. The responses by which tendrils curl around and cling to stems or other objects are surprisingly rapid and clear (Figure 31-11); the ways in which twining plants such as bindweed coil around objects are analogous. This behavior is the result of rapid growth responses to touch. Specialized groups of cells in the epidermis appear to be concerned with thigmotropic reactions, but again, their exact mode of action is not well understood.

Thigmotropisms are growth responses of plants to touch.

TURGOR MOVEMENTS

Some kinds of plant movements are based on reversible changes in the **turgor pressure** (defined as the pressure within a cell resulting from the movement of water into the cell) of specific cells, rather than on differential growth or cell enlargement. One of the most familiar of these has to do with the changing patterns of leaf position that certain plants exhibit at night and in the day (Figure 31-12). For example, the attractively spotted leaves of the prayer plant *(Maranta)* spread horizontally during the day but become more or less vertical at night.

Turgor movements of plants are reversible and involve changes in the turgor pressure of specific cells. They allow plants to orient their leaves and flowers in different positions.

FIGURE 31-10

The stems of this fallen tree are growing straight up because they are negatively gravitropic (growing against the force of gravity) and also because they are positively phototropic (growing towards the light).

FIGURE 31-11

These tendrils coil around the stem because of their positive thigmotropism.

FIGURE 31-12

Wood sorrel *(Oxalis)* in the day with leaflets held horizontally **(A)**, and at night with leaflets held down **(B)**.

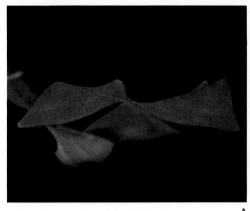
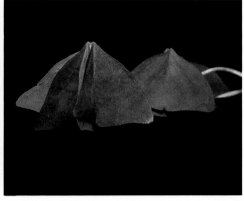

A

B

FIGURE 31-13

Chrysanthemum, a short-day plant. Millions of chrysanthemums are induced to flower by darkening them for part of the 24-hour cycle during times of year when the days are not short enough for them to flower naturally. This individual is being exhibited in Japan.

PHOTOPERIODISM

Essentially all eukaryotic organisms are affected by the cycle of night and day, and many features of plant growth and development are keyed to the changes in the proportions of light and dark in the daily 24-hour cycle. Such responses constitute **photoperiodism,** a mechanism by which organisms measure seasonal changes in relative day and night length. One of the most obvious of these photoperiodic reactions concerns the productions of flowers by angiosperms.

Day length changes with the seasons; the farther from the equator one is, the greater the variation. The flowering responses of plants fall into three basic categories in relation to day length. **Short-day plants** being to form flowers when the days become shorter than a critical length (Figure 31-13). **Long-day plants,** on the other hand, initiate flowers when the days become longer than a certain length. Thus many fall flowers are short-day plants, and many spring and early-summer flowers are long-day plants. Commercial plant growers use these responses to day length to bring plants into bloom when they are wanted for sale. In addition to the long-day and short-day plants, a number of kinds of plants are described as **day neutral;** they produce flowers whenever environmental conditions are suitable, without regard to day length. The complex chemical basis of these responses is now reasonably well understood.

> **Short-day plants start to form flowers when the days become shorter than a certain critical day length; long-day plants form flowers when the days become longer than a certain length. Day-neutral plants do not have specific day-length requirements for flowering.**

DORMANCY

Plants respond to their external environment largely by changes in growth rate. As you might imagine, the ability to stop growing altogether when conditions are not favorable is a critical factor in their survival.

In temperate regions, we generally associate dormancy with winter, when low temperatures and the unavailability of water because of freezing make it impossible for plants to grow. During this season the buds of deciduous trees and shrubs remain dormant, and the apical meristems remain well protected inside enfolding scales. Perennial herbs spend the winter underground as stout stems or roots packed with stored food. Many other kinds of plants, including most annuals, pass the winter as seeds.

In climates that are seasonally dry, dormancy will occur primarily during the dry season, whenever in the year it falls. In dry conditions, plants stay in a dormant condition by using strategies similar to those that the plants of temperate areas rely on in winter.

Annual plants occur frequently only in areas of seasonal drought. Seeds are ideal mechanisms for allowing annual plants to bypass the dry season when there is insufficient water for growth. When it rains, they can germinate and the plants can grow rapidly to take advantage of the relatively short period when water is available.

SUMMARY

1. Water flows through plants in a continuous column, driven mainly by transpiration through the stomata. The plant can control water loss primarily by closing its stomata. The cohesion of water molecules and their adhesion to the walls of the very narrow cell columns through which they pass are additional important factors in maintaining the flow of water to the tops of plants.

2. The stomata open when their guard cells are turgid and bow out, thus causing the thickened inner walls of these cells to bow away from the opening.

3. The movement of dissolved sucrose and other carbohydrates in the phloem does not require energy. Sucrose is loaded into the phloem near sites of photosynthesis and unloaded at the places where it is required.

4. Nutrients move primarily through plants as solutes in the water column in the xylem. Their selective admission into the plant and their subsequent movement between living cells require the expenditure of energy.

5. The nine macronutrients, substances that are each present at concentrations of 1% or more of a plant's dry weight, are carbon, hydrogen, oxygen, nitrogen, potassium, calcium, phosphorus, magnesium, and sulfur. In addition, there are seven known micronutrients, each present at concentrations of from one to several hundred parts per million of dry weight.

6. Hormones are chemical substances produced in small quantities in one part of an organism and transported to another part of the organism, where they bring about physiological responses. The tissues in which plant hormones are produced are not specialized particularly for that purpose, nor are there usually clearly defined receptor tissues or organs.

7. There are five major classes of naturally occurring plant hormones: auxins, cytokinins, gibberellins, ethylene, and abscisic acid. They often interact with one another in bringing about growth responses.

8. Auxins are produced at the tips of shoots and diffuse downward, suppressing the growth of lateral buds. In young grass seedlings and other herbs, they play a major role in promoting stem elongation.

9. Cytokinins are necessary for mitosis and cell division in plants. They promote the growth of lateral buds and inhibit the formation of lateral roots.

10. Gibberellins play the major role in stem elongation in most plants.

11. Tropisms in plants are growth responses to external stimuli. A phototropism is a response to light, gravitropism a response to gravity, and thigmotropism a response to touch.

12. The flowering responses of plants fall into two basic categories in relation to day length. Short-day plants begin to form flowers when the days become shorter than a given critical length; long-day plants do so when the days become longer than a certain length.

13. Dormancy is a necessary part of plant adaptation that allows a plant to bypass unfavorable seasons such as winter when the water may be frozen, or periods of drought. Dormancy also allows plants to survive in many areas where they would be unable to grow otherwise.

REVIEW

1. _____ are substances produced in minute quantities in one part of the plant and transported to other parts of the plant where they are physiologically active.

2. The active transport of ions into the root hairs results in a high concentration of ions in these cells, so that water is drawn into the cells by _____.

3. Name the six macronutrients other than carbon, hydrogen, and oxygen that are required for plant growth.

4. In contrast to auxins, cytokinins _____ the growth of lateral branches.

5. _____ is the response of roots to gravity, which causes them to grow downward.

SELF-QUIZ

1. Some 90% of the water taken in by the roots of plants is lost to the atmosphere. Which leaf structure accounts for the greatest portion of this water loss?
   (a) the cuticle (d) stomata
   (b) veinlets (e) mesophyll
   (c) transpiration

2. Water can rise to the top of a 20-foot tree primarily because of _____.

3. Keeping guard cells turgid and thus keeping the stomata open requires a constant expenditure of energy. What is the location at which the necessary ATPs are produced?
   (a) the roots
   (b) photosynthesis carried on in the guard cells themselves
   (c) mesophyll cells
   (d) rhizomes
   (e) all the cells of the plant

4. Plants bend toward the light because (pick two)
   (a) cells on the shaded side of the stem elongate more than those on the sunny side
   (b) cells on the sunny side of the stem elongate more than those on the shaded side
   (c) cells on the shaded side of the stem accumulate auxin
   (d) cells on the sunny side of the stem accumulate auxin
   (e) auxin suppresses lateral bud growth

5. Dormancy in plants is commonly associated with which two of the following?
   (a) warm temperatures
   (b) cold temperatures
   (c) lack of carbon dioxide
   (d) lack of nutrients
   (e) lack of available water

THOUGHT QUESTIONS

1. Why do gardeners often remove many of a plant's leaves after transplanting it?

2. If you grew a plant that initially weighed 200 grams but eventually weighed 50 kilograms, in a pot, would you expect the soil in the pot to change weight? If so, how much, and why?

3. When poinsettias are kept inside a house following the holiday season, they rarely bloom again. Why do you think this might be, and what might you do to get them to produce flowers a second time?

FOR FURTHER READING

EVANS, M., R. MOORE, and K. HASENSTEIN: "How Roots Respond to Gravity," *Scientific American,* December 1986, pages 112-120. Did you ever wonder why roots grow down and not up? These authors speculate that calcium ions settle in roots, where they activate proteins that promote growth.

SISLER, E.C., and S.F. YANG: "Ethylene, The Gaseous Plant Hormone," *BioScience,* vol. 33, pages 233-238, 1984. An up-to-date review of this important plant hormone.

TREWAVAS, A.: "How Do Plant Growth Substances Work?" *Plant, Cell, and Environment,* vol. 4, pages 203-228, 1981. A concise introduction to plant hormones.

ANIMAL BIOLOGY

Part Nine

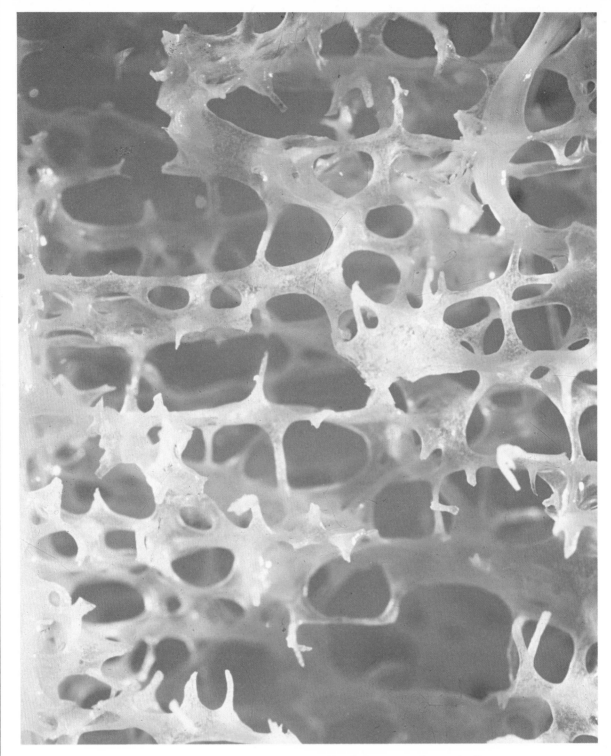

FIGURE 32-1   While we think of bone as hard and solid, the interiors of many bodies are composed of a delicate and surprising latticework. Bone, like most of the many tissues present in your body, is a dynamic structure, constantly renewing itself.

THE VERTEBRATE BODY

Overview

The three fundamental embryonic tissues—ectoderm, mesoderm, and endoderm—are determined early during the embryonic development of vertebrates and differentiate during the later course of development into several hundred kinds of cells, depending on the kind of animal. These cells make up the four major tissue types: epithelium, connective tissue, muscle, and nerve. Each of these tissue types can be recognized by its structure, function, and origin. The organs of the body are often composed of several different tissue types.

For Review

Here are some important terms and concepts that you will encounter in this chapter. If you are not familiar with them, you should review them before proceeding.

The sodium–potassium pump (Chapter 6)

The evolution of vertebrates (Chapter 19)

Embryonic development in animals (Chapter 28)

Basic structure of chordates (Chapters 19 and 28)

Coelom (Chapter 28)

Most of us have been to a zoo and have seen the many different kinds of animals there—long-necked giraffes, armor-plated rhinoceroses, seals that swim, and birds that fly and sing. Monkeys look at us, perhaps wondering why we are looking at them; snakes slither, squirrels dart about, and peacocks strut. All of these animals are vertebrates, and although they may seem very different from each other, they are in fact very much alike "under the skin." All vertebrates share the same basic body plan, with the same sorts of organs operating in much the same way. In this chapter we shall begin a detailed consideration of the biology of vertebrates and of the often intricate and fascinating structure of their bodies (Figure 32-1). We shall focus on human biology because the architecture of the human body provides a good focal point for discussing the structure and functioning of vertebrate bodies in general and because human biology is of particular importance to all of us. We each want to know how our body works and why it functions the way it does.

Not all vertebrates are the same, of course, and some of the differences are important. For example, a fish doesn't breathe the same way you do. So we shall not limit ourselves to human beings but rather shall describe the human animal in the context of vertebrate diversity. The differences reflect our evolutionary history and offer important lessons in why we function the way we do.

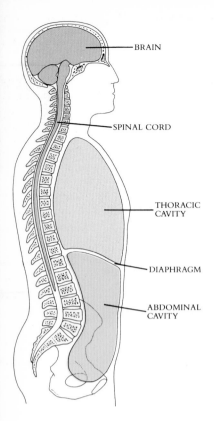

THE HUMAN ANIMAL

An incredible machine of great beauty, the human body has the same general body architecture that all vertebrates have (Figure 32-2). It includes a long tube that travels from one end of the body to the other, from mouth to anus; this tube is suspended within an internal body cavity called the coelom (see Chapter 28). In human beings the coelom is divided into two parts: (1) the thoracic cavity, which contains the heart and lungs; and (2) the abdominal cavity, which contains the stomach, intestines, and liver. The body of all vertebrates is supported by an internal scaffold or skeleton made up of jointed bones that grow as the body grows. A bony skull surrounds the brain; and a column of hollow bones, the vertebrae, surrounds the dorsal nerve cord, or spinal cord.

Human beings are mammals and, like all other mammals, are warm-blooded, regulating their internal temperature at a relatively constant value. Humans keep their temperature at about 98° F. Like all other mammals, human beings have hair rather than scales or feathers; and like all other mammals except monotremes, humans give birth to live young. Human development is a lengthy process; infants are nursed for a long time and mature slowly.

Levels of Organization

The bodies of vertebrates, like those of all other multicellular animals, are composed of different cell types. The bodies of adult vertebrates contain between 50 and several hundred different kinds of cells, depending on the kind of vertebrate and how finely you differentiate between cell types. Groups of similar cells are organized into **tissues,** which are structural and functional units.

Organs are body structures composed of several different tissues grouped together into a larger structural and functional unit. Your heart is an organ. It contains cardiac muscle tissue wrapped in connective tissue and wired with nerves. All of these tissues work together to pump blood through your body. An **organ system** is a group of organs that function together to carry out the principal activities of the body. For example, the digestive organ system is composed of individual organs concerned with the breaking up of food (teeth), the passage of food to the stomach (esophagus), the storage of food (stomach), the digestion and absorption of food (intestine), and the expulsion of solid residue (rectum). The human body contains 11 principal organ systems (Table 32-1), which will be presented later in this section.

TISSUES

Early in the development of any vertebrate, the growing mass of cells differentiates into three fundamental embryonic tissues; **endoderm, mesoderm,** and **ectoderm.** These three kinds of embryonic tissue in turn differentiate into the hundreds of different cell types that are characteristic of the adult vertebrate body. As you saw in Chapter 28, where they were reviewed in an evolutionary context, these diverse cell types are traditionally grouped on a functional basis into four basic types of tissues: **epithelium, connective tissue, muscle,** and **nerve** (Figure 32-3). Blood cells, for example, are one kind of connective tissue, and bone another.

Epithelium

Epithelial cells are the guards and protectors of the body. They cover its surface and determine which substances enter it and which do not. The organization of the vertebrate body is fundamentally tubular, with developmental derivatives of ectodermal cells covering the outside (skin), those of endodermal cells lining the hollow inner core (alimentary canal and gut), and those of mesodermal cells lining the body cavity (coelom).

FIGURE 32-2

Vertebrates have a dorsal central nervous system, consisting of a spinal cord and brain, enclosed in vertebrae and the skull. In mammals, a muscular diaphragm divides the coelom into the thoracic cavity and the abdominal cavity.

TABLE 32-1   THE MAJOR VERTEBRATE ORGAN SYSTEMS

SYSTEM	FUNCTIONS	COMPONENTS	DETAILED TREATMENT
Circulatory	Transports cells and materials throughout the body	Heart, blood vessels, blood, lymph, and lymph structures	Chapter 36
Digestive	Captures soluble nutrients from ingested food	Mouth, esophagus, stomach, intestines, liver, and pancreas	Chapter 34
Endocrine	Coordinates and integrates the activities of the body	Pituitary, adrenal, thyroid, and other ductless glands	Chapter 40
Urinary	Removes metabolic wastes from the bloodstream	Kidney, bladder, and associated ducts	Chapter 41
Immune	Removes foreign bodies from the bloodstream	Lymphocytes, macrophages, and antibodies	Chapter 37
Integumentary	Covers the body and protects it	Skin, hair, nails, and sweat glands	Chapter 32
Muscular	Produces body movement	Skeletal muscle, cardiac muscle, and smooth muscle	Chapter 33
Nervous	Receives stimuli, integrates information, and directs the body	Nerves, sense organs, brain, and spinal cord	Chapters 38 and 39
Reproductive	Carries out reproduction	Testes, ovaries, and associated reproductive structures	Chapter 42
Respiratory	Captures oxygen and exchanges gases	Lungs, trachea, and other air passageways	Chapter 35
Skeletal	Protects the body and provides support for locomotion and movement	Bones, cartilage, and ligaments	Chapter 33

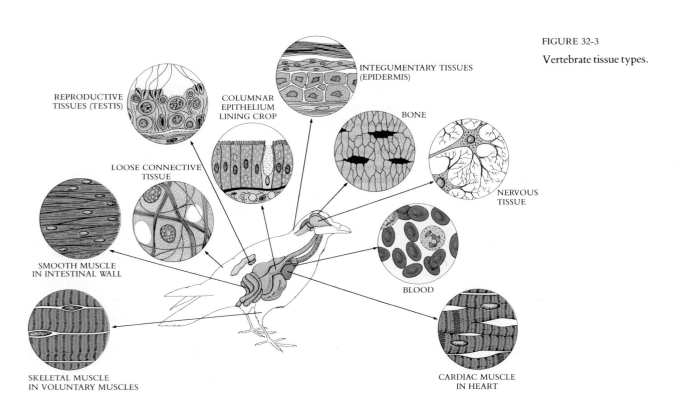

FIGURE 32-3

Vertebrate tissue types.

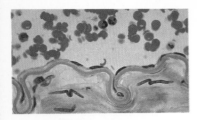

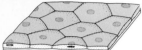

A SIMPLE SQUAMOUS

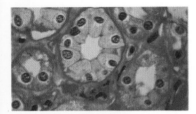

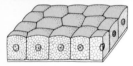

B SIMPLE CUBOIDAL

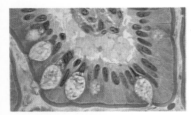

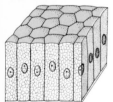

C COLUMNAR

FIGURE 32-4

Types of epithelial tissue based on shape.
A Squamous epithelium lines the artery seen here. The nuclei are characteristically flattened. The round cells above the epithelium are blood cells in the lumen, or hollow interior, of the artery.
B Cuboidal epithelium forms the walls of these kidney tubules, seen in cross section.
C Columnar epithelium forms the outer cell layer of this human ileum. Interspersed among the epithelial cells are goblet cells, which secrete mucus.

All of these kinds of **epidermal,** or "skin" cells are broadly similar in form and function, and they are collectively called the **epithelium.** The epithelial layers of the body function in four different ways:

1. They protect the tissues beneath them from dehydration and mechanical damage, a particularly important function in land-dwelling vertebrates.
2. They provide a selectively permeable barrier that can facilitate or impede the passage of materials into the tissue beneath. Because epithelium encases all of the body's surfaces, every substance that enters or leaves the body must cross an epithelial layer.
3. They provide sensory surfaces. Many sensory nerves end in epithelial cell layers.
4. They secrete materials. Most secretory glands are derived from invaginations of layers of epithelial cells that occur during embryonic development.

Layers of epithelial tissue are usually only one or a few cells thick. Individual epithelial cells possess a small amount of cytoplasm and have a relatively low metabolic rate. Few blood vessels pass through the epithelium; instead, the circulation of nutrients, gases, and wastes in epithelial tissue occurs by diffusion from the capillaries of neighboring tissue.

Epithelial tissues possess remarkable regenerative powers. The cells of epithelial layers are constantly being replaced throughout the life of the organism. For example, the liver, which is a gland formed of epithelial tissue, can readily regenerate substantial portions of tissue that have been surgically removed from it.

There are three general classes of epithelial tissue (Figure 32-4): **simple epithelium, stratified epithelium,** and **glands.**

The membranes that line the lungs and the major cavities of the body are composed of simple epithelial cells a single cell layer thick. The skin, or epidermis (Figure 32-5), is composed of more complex epithelial cells, several cell layers thick. The glands of the body are also composed of epithelial tissue. These body glands produce sweat, milk, saliva, digestive enzymes, and many different hormones.

The skin of the body and the lining of the respiratory and digestive tracts are composed of epithelium. Much of the internal lining of respiratory and digestive tracts is simple epithelium, only one cell thick, whereas external skin is composed of epithelium that is many cell layers thick.

FIGURE 32-5

For the most part, skin ages gradually, but on the face, the changes are more dramatic. At 19, Lorraine's face appears youthful and smooth. Forty years later, her production of skin oil is much less; her face appears less smooth and elastic, and begins to exhibit wrinkles.

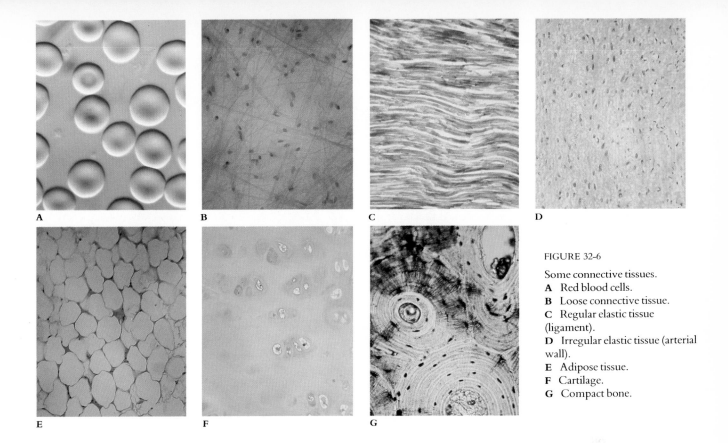

FIGURE 32-6

Some connective tissues.
A Red blood cells.
B Loose connective tissue.
C Regular elastic tissue (ligament).
D Irregular elastic tissue (arterial wall).
E Adipose tissue.
F Cartilage.
G Compact bone.

Connective Tissue

The cells of connective tissue provide the body with its structural building blocks and also with its most potent defenses. Unlike epidermal cells, the cells that make up loose connective tissue are not stacked tightly; instead, they are spaced well apart from one another. Connective tissue cells, which are derived from the mesoderm, fall into three categories: (1) the cells of the immune system, which acts to defend the body; (2) the cells of the skeletal system, which supports the body; and (3) the cells that store and distribute substances throughout the body (Figure 32-6).

The cells of the immune system are small and roam the body within the bloodstream (Figure 32-7); they are mobile hunters of invading microorganisms and foreign substances (see Chapter 37). There are three principal kinds of immune system cells: (1) **macrophages,** which are mobile, phagocytic cells able to engulf and digest invading bacteria, fungi, and other microorganisms, as well as cellular debris; (2) **lymphocytes,** or white blood cells, which either synthesize antibodies or attack virus-infected cells; and (3) **mast cells,** which synthesize the molecules involved in the body's response to **trauma** (injury), including histamine, which produces the inflammation response, and heparin, which prevents blood clotting. Allergies are the result of the stimulation of mast cells by other cells of the immune system.

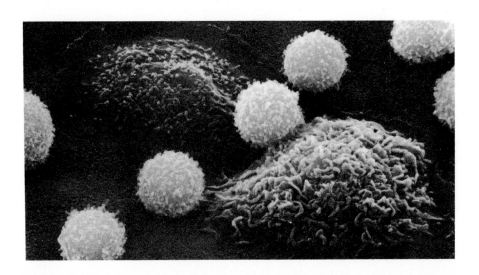

FIGURE 32-7

Lymphocytes and macrophages. The lymphocytes are small and spherical; the macrophages are larger and more irregular in form.

FIGURE 32-8

Collagen fibers. Each fiber is composed of many individual collagen strands and can be very strong.

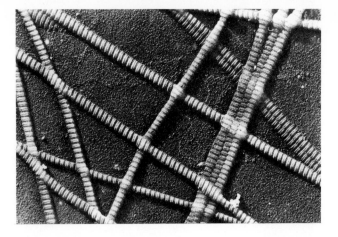

Three principal kinds of connective tissues occur in the skeletal system: fibroblasts, cartilage, and bone. The three are distinguished principally by the nature of the matrix that is laid down between the individual cells.

Fibroblasts. The most common kind of connective tissue in the vertebrate body consists of **fibroblasts,** which are flat, irregularly branching cells that secrete structurally strong proteins into the matrix between the cells. The most commonly secreted protein is **collagen** (Figure 32-8), which is the most abundant protein in vertebrate bodies. One quarter of all animal protein is collagen. Fibroblasts are active in wound healing. They multiply rapidly in wound tissue, forming granulated fibrous scar tissue that possesses a collagen matrix.

Cartilage. **Cartilage** is a specialized connective tissue in which the collagen matrix between cells is formed at positions of mechanical stress. In cartilage the fibers are laid down along the lines of stress in long, parallel arrays. The result of this process is a firm and flexible tissue that has great strength. Cartilage covers the ends of bones that come together in joints, such as the knee, ankle, and elbow.

Bone. **Bone** is a special form of cartilage in which the collagen fibers are coated with a calcium phosphate salt. Bone is more rigid than collagen and is strong but not brittle. The structure of bone and the way in which it is formed are discussed in Chapter 33.

The third general class of connective tissue is composed of cells that are specialized for the accumulation and transport of particular molecules. **Sequestering connective tissues** include pigment-containing cells and the fat cells of adipose tissue. The most important tissues of this class are red blood cells, called **erythrocytes** (Figure 32-9). There are about 5 billion erythrocytes in every milliliter of human blood. Erythrocytes act as the transporters of oxygen in the vertebrate body. During the process of their maturation in mammals, they lose their nucleus and mitochondria, and their endoplasmic reticulum dissolves. As a result of these processes, mammalian erythrocytes are relatively inactive metabolically, but they are not empty. Large amounts of the iron-containing protein hemoglobin are produced within the erythrocytes and remain in the mature cell. Hemoglobin is the principal carrier of oxygen and carbon dioxide in vertebrates, as it is in many other groups of animals. Each erythrocyte contains about 300 million molecules of hemoglobin.

The fluid intracellular matrix, or **plasma,** in which the erythrocytes move, is both the "banquet table" and the "refuse heap" of the vertebrate body. Practically every substance used by cells is dissolved in plasma. These substances include the sugars, lipids, and amino acids, which are the fuel of the body, as well as the products of metabolism. The plasma also contains inorganic salts such as the calcium used to form bone, fibrinogen from the liver, albumin, which gives the blood viscosity, and antibody proteins produced by lymphocytes. Every substance such as urea that is secreted or discarded by cells is also present in the plasma.

FIGURE 32-9

Blood cells. White blood cells, or leukocytes, are roughly spherical and have irregular surfaces with numerous extending pili. Red blood cells, or erythrocytes, are flattened spheres, typically with a depressed center.

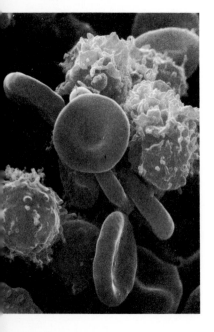

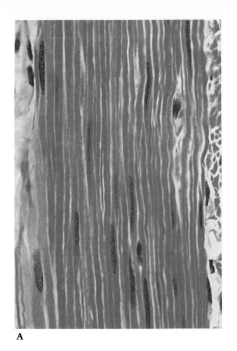

A

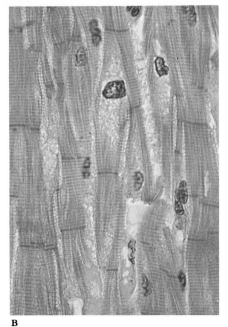

B

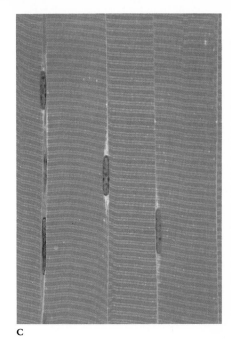

C

SMOOTH MUSCLE

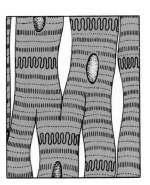

CARDIAC MUSCLE

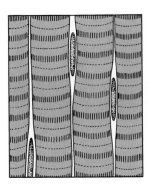

SKELETAL MUSCLE

Muscle

Muscle cells, formed from the mesoderm early in development, are the "workhorses" of the vertebrate animal. The distinguishing characteristic of muscle cells, the one that makes them unique, is the relative abundance of actin and myosin microfilaments within them. These microfilaments are present as a fine network in all eukaryotic cells (see Chapter 5), but they are far more abundant in muscle cells. In muscle cells the actin microfilaments are bunched together with thicker filaments of myosin into many thousands of fibers called **myofibrils.** The myofibrils shorten when the actin and myosin filaments slide past each other. The shortening of these myofibrils can cause the muscle cell to change shape. Because there are so many filaments all aligned the same way in muscle cells, a considerable force is generated when all of the myofibrils shorten at the same time. Vertebrates possess three different kinds of muscle cells (Figure 32-10): smooth muscle, striated muscle, and cardiac muscle.

Nerve

The fourth major class of vertebrate tissue is nervous tissue. It is composed of two kinds of cells: (1) **neurons,** which are specialized for the transmission of nerve impulses; and (2) **supporting cells,** which support and insulate the neurons. The supporting cells also assist in the propagation of the nerve impulse by playing an important part in the maintenance of the ionic composition of nerve tissue. They are also believed to supply the neurons with nutrients and other molecules.

FIGURE 32-10

Types of muscle.
A Smooth muscle cells are long and spindle shaped, with a single nucleus.
B Cardiac muscle cells, such as these from a human heart, also contain a single nucleus and are organized into long branching chains that interconnect, forming a lattice.
C Skeletal or striated muscle cells are formed by the fusion of several muscle cells, end to end, to form a long fiber with many nuclei.

TABLE 32-2   TYPES OF VERTEBRATE TISSUE

TISSUE	TYPICAL LOCATION	TISSUE FUNCTION	CHARACTERISTIC CELL TYPES
EPITHELIAL			
Simple epithelium			
Squamous	Lining of lungs, capillary walls, and blood vessels	Cells very thin; provides a thin layer across which diffusion can readily occur	Epithelial cells
Cuboidal	Lining of some glands and kidney tubules; covering of ovaries	Cells rich in specific transport channels; functions in secretion and specific absorption	Gland cells
Columnar	Surface lining of stomach, intestines, and parts of respiratory tract	Thicker cell layer; provides protection and functions in secretion and absorption	Epithelial cells
Stratified epithelium			
Squamous	Outer layer of skin; lining of mouth	Tough layer of cells; provides protection	Epithelial cells
Columnar	Lining of parts of respiratory tract	Functions in secretion of mucus; dense with cilia that aid in movement of mucus; provides protection	Gland cells; ciliated epithelial cells
MUSCLE			
Skeletal	Voluntary muscles of the body	Powers walking, lifting, talking, and all other voluntary movement	Sarcomeres
Smooth	Walls of blood vessels, stomach, and intestines	Powers rhythmic contractions not under conscious control but commanded by central nervous system	Smooth muscle cells
Cardiac	Walls of heart	Highly interconnected cells; promotes rapid spread of signals initiating contraction	Heart muscle cells

TABLE 32-2   TYPES OF VERTEBRATE TISSUE—cont'd

TISSUE	TYPICAL LOCATION	TISSUE FUNCTION	CHARACTERISTIC CELL TYPES
CONNECTIVE			
Vascular			
Blood	Circulatory system	Transports O_2, CO_2, nutrients, and wastes; functions as highway of immune system, and stabilizer of body temperature	Erythrocytes; lymphocytes
Connective tissue proper			
Loose	Beneath skin and other epithelia	Support; provides a fluid reservoir for epithelium	Fibroblasts
Dense	Tendons; sheath around muscles; kidney; liver; dermis of skin	Provides flexible strong connections	Fibroblasts
Elastic	Ligaments; large arteries; lung tissue; skin	Enables tissues to expand and then return to normal size	Fibroblasts
Adipose	Fat beneath skin and on surface of heart and other internal organs	Provides insulation, food storage, and support of breasts and kidneys	Fat cells
Cartilage	Spinal disks; knees and other joints; ear; nose; tracheal rings	Provides flexible support; functions in shock absorption and reduction of friction on load-bearing surfaces	Chondrocytes
Bone	Most of skeleton	Protects internal organs; provides rigid support for muscle attachment	Osteocytes
NERVE			
Sensory cells	Eyes; ears; surface of skin	Receives information about body's condition and about exterior world	Rods and cones; muscle stretch receptors
Signal-transmitting cells	Nerves	Transmits signals	Neurons
Information-processing cells	Brain and spinal cord	Integrates information	Interneurons

FIGURE 32-11

A human neuron. The cell body is at the upper left, with the axon extending up out of view. The branching network of fibers extending down from the cell body is made up of dendrites, which carry signals to the cell body.

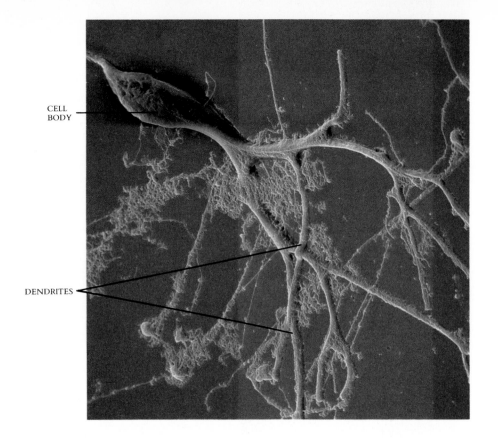

CELL BODY

DENDRITES

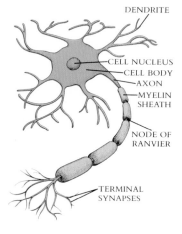

DENDRITE

CELL NUCLEUS
CELL BODY
AXON
MYELIN SHEATH

NODE OF RANVIER

TERMINAL SYNAPSES

FIGURE 32-12

Idealized structure of a vertebrate neuron. Many dendrites lead to the cell body, from which a single long axon extends. In some neurons, specialized for rapid signal conduction, the axons are encased at intervals within myelin sheaths. At its far end, an axon may terminate at one cell, or branch to several cells, or connect to several locations on one cell.

Neurons (Figure 32-11) are cells specialized to conduct signals rapidly throughout the body. Their membranes are rich in ion-transporting channels, which pump ions out of the cell and thus maintain a charge difference between the interior and exterior of the cell. A current is initiated when channels in a local area of the membrane open, permitting ions to reenter from the exterior and thus wiping out the charge difference—a process called **depolarization.** This depolarization in a local area of the membrane tends to open nearby channels of the neuron membrane, starting a wave that propagates down the nerve as a **nerve impulse.** The nature of nerve impulses varies, as described in more detail in Chapter 38, but all nerve impulses propagate along nerves as waves of membrane depolarization.

The cell body of a neuron contains within it the cell nucleus. From the cell body project two kinds of cytoplasmic elements that carry out the transmission functions of the neuron. The first of these elements consists of **dendrites,** which are threadlike protrusions. The dendrites act as antennae for the reception of nerve impulses from other cells or sensory systems. The second kind of cytoplasmic element that projects from the cell body of a neuron is the **axon.** An axon is a long tubular extension of the cell that provides a channel for the transmission of the nerve impulse away from the cell body (Figure 32-12). Because axons can be quite long, some nerve cells are very long indeed. For example, a single neuron that activates the muscles in your thumb may have its cell body in the spinal cord and possess an axon that extends all the way across your shoulder and down your arm to your thumb. Single neuron cells more than a meter in length are common.

The nerves of the vertebrate body appear as fine white threads when they are viewed with the naked eye, but are actually composed of bundles of axons and dendrites. Like a telephone trunk cable, these nerves include large numbers of independent communications channels—bundles of hundreds of axons and dendrites, each connecting a nerve cell to a muscle fiber. In addition, the nerve contains numerous supporting cells bunched around the axons. In the **brain** and **spinal cord,** which together make up the **central nervous system,** these supporting cells are called **glial cells.** The supporting cells associated with projecting axons and all other nerve cells that make up the **peripheral nervous system** are called **Schwann cells.**

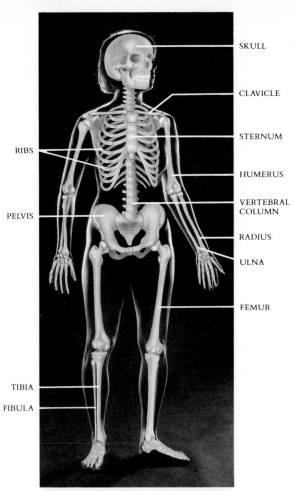

SKULL

CLAVICLE

STERNUM

HUMERUS

VERTEBRAL COLUMN

RADIUS

ULNA

FEMUR

RIBS

PELVIS

TIBIA

FIBULA

SKELETAL SYSTEM

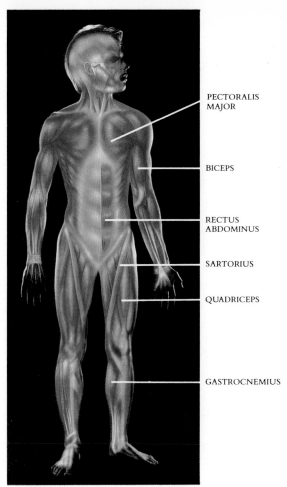

PECTORALIS MAJOR

BICEPS

RECTUS ABDOMINUS

SARTORIUS

QUADRICEPS

GASTROCNEMIUS

MUSCULAR SYSTEM

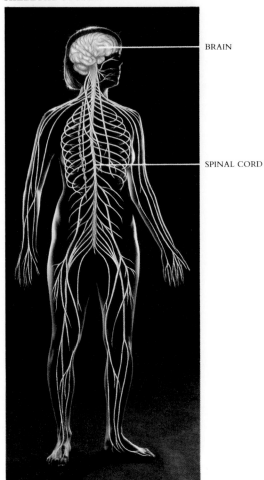

BRAIN

SPINAL CORD

NERVOUS SYSTEM

FIGURE 32-13

The major organ systems of vertebrates. Each organ of the body is composed of combinations of the four classes of tissues discussed in this chapter, assembled in various ways. The many organs that, working together, carry out the principal activities of the body are traditionally grouped together as organ systems. On this and the following two pages are presented diagrams of 10 of the major organ systems of the human body.

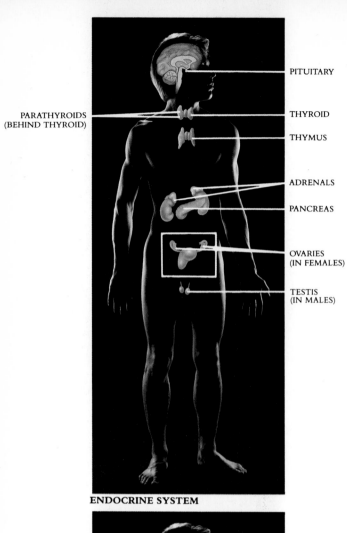

PITUITARY

PARATHYROIDS
(BEHIND THYROID)

THYROID

THYMUS

ADRENALS

PANCREAS

OVARIES
(IN FEMALES)

TESTIS
(IN MALES)

ENDOCRINE SYSTEM

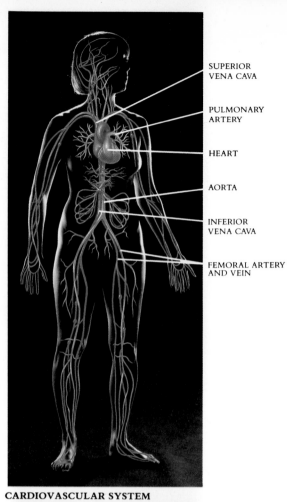

SUPERIOR
VENA CAVA

PULMONARY
ARTERY

HEART

AORTA

INFERIOR
VENA CAVA

FEMORAL ARTERY
AND VEIN

CARDIOVASCULAR SYSTEM

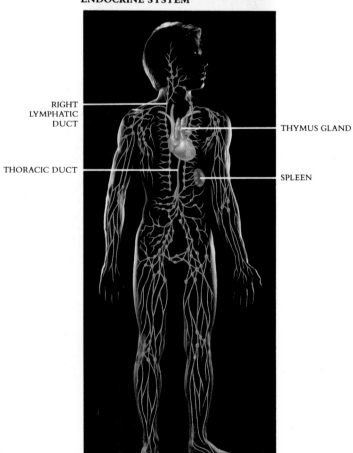

RIGHT
LYMPHATIC
DUCT

THORACIC DUCT

THYMUS GLAND

SPLEEN

LYMPHATIC SYSTEM

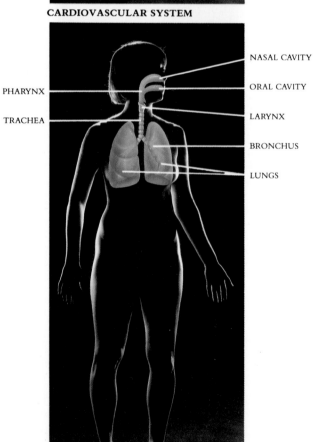

NASAL CAVITY

ORAL CAVITY

PHARYNX

LARYNX

TRACHEA

BRONCHUS

LUNGS

RESPIRATORY SYSTEM

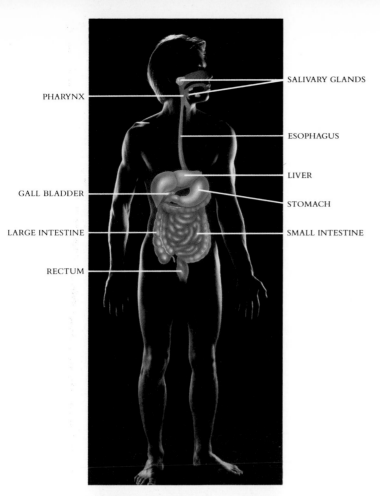

PHARYNX

SALIVARY GLANDS

ESOPHAGUS

LIVER

GALL BLADDER

STOMACH

LARGE INTESTINE

SMALL INTESTINE

RECTUM

DIGESTIVE SYSTEM

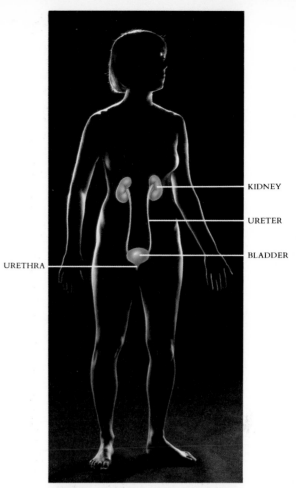

KIDNEY

URETER

BLADDER

URETHRA

URINARY SYSTEM

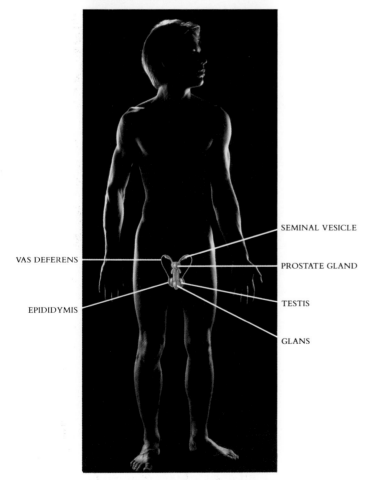

SEMINAL VESICLE

VAS DEFERENS

PROSTATE GLAND

TESTIS

EPIDIDYMIS

GLANS

REPRODUCTIVE SYSTEM—MALE

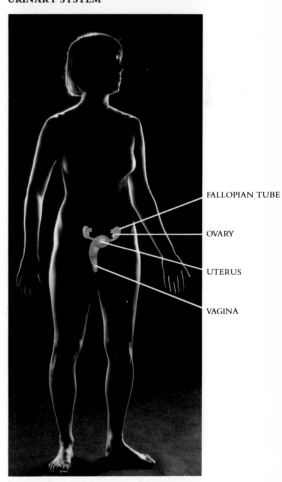

FALLOPIAN TUBE

OVARY

UTERUS

VAGINA

REPRODUCTIVE SYSTEM—FEMALE

SUMMARY

1. The four basic types of tissue are epithelium, connective tissue, muscle, and nerve.

2. Epithelium covers the surfaces of the body. The lining of the major body cavities is composed of epithelium, as are the exterior skin and the major gland systems.

3. Connective tissue supports the body mechanically and defensively. The structural connective tissues are fibroblasts (which secrete the structural protein collagen), cartilage, and bone. The defensive connective tissues are macrophages and other cells of the immune system that attack and destroy foreign bacteria and virus-infected cells. Other connective tissues are pigment-producing cells, fat cells, and blood cells.

4. The contraction of cells within muscle tissue provides the force for mechanical movement of the body.

5. Nerve cells provide the body with a means of rapid communication. There are two kinds of nerve cells—neurons and supporting cells. Neurons are specialized for the conduction of electrical impulses.

REVIEW

1. In human beings the coelom is divided into two parts. What are they?

2. Connective tissue cells are derived from which embryonic tissue?

3. _____ is the most abundant protein in vertebrate bodies.

4. Microfilaments made of actin and myosin are much more abundant in _____ cells than in any other eukaryotic cell type.

5. Neurons have two kinds of projections from the cell body, the _____ and the _____.

SELF-QUIZ

1. Which of the following is *not* one of the four basic types of tissues of the adult vertebrate body?
   (a) muscle
   (b) nerve
   (c) connective
   (d) mesoderm
   (e) epithelium

2. Which of the following is *not* a function of the epithelial layers?
   (a) to secrete material
   (b) to store and distribute substances throughout the body
   (c) to protect the tissues beneath from dehydration and mechanical damage
   (d) to provide a selectively permeable barrier
   (e) to provide sensory surfaces

3. Which of the following is *not* a cell type of the connective tissue?
   (a) lymphocytes
   (b) fat cells
   (c) columnar cells
   (d) erythrocytes
   (e) macrophages

4. Cartilage functions in many ways. Which of the following is *not* one of them?
   (a) makes up the hard external part of your ear
   (b) forms the ends of bones in knees and other joints
   (c) forms the nails on the tips of your fingers and toes
   (d) makes spinal disks firm and flexible
   (e) forms the firm tip of your nose

5. Which of the following is *not* a principal organ system of the human body?
   (a) circulatory system
   (b) urinary system
   (c) lymphatic system
   (d) secretory system
   (e) muscular system

THOUGHT QUESTIONS

1. Land was successfully invaded four times—by plants, fungi, arthropods, and vertebrates. Since bodies are far less buoyant in air than in water, each of these four groups evolved a characteristic hard substance to lend mechanical support. Describe and contrast these four substances, discussing their advantages and disadvantages. Can you imagine any other substance (such as plastic) that would have been superior to any of these?

2. Your body contains 206 bones. As you grow, all 206 of these bones must increase in size and maintain proper proportions with one another. How is the growth of all these bones coordinated?

FOR FURTHER READING

CAPLAN, A.: "Cartilage," *Scientific American,* October 1984, pages 84–97. An interesting account of the many roles played by cartilage in the vertebrate body.

CURREY, J.: *The Mechanical Adaptations of Bones,* Princeton University Press, Princeton, N.J., 1984. A functional analysis of why different bones are structured the way they are, with an unusually well-integrated evolutionary perspective.

NATIONAL GEOGRAPHIC SOCIETY: *The Incredible Machine,* National Geographic Society, Washington, D.C., 1986. A series of outstanding articles on the human body, focusing on its major organ systems. Beautifully illustrated and fun to read.

FIGURE 33-1   This hawk seems delicate and fragile as it glides overhead, but if it sights prey it may dive, falling as rapidly as a stone, only to rise back up in a flurry of wingbeats. Its flight reflects the complex integration of nerves and muscles—as does the way a fish swims, the way a racehorse runs, and the way you walk.

HOW ANIMALS MOVE

Chapter 33

Overview

One of the obvious differences between animals and plants is that most animals move about from one place to another and plants don't. Whether it is swimming, flying, or walking, movement almost always requires that an individual shift the position of a limb against a resistance. This is done by contracting muscles attached to the limb. In invertebrates, the muscles are attached to a somewhat brittle outer skeleton, which must become thicker as the animals become larger to perform its function as an anchor point. In vertebrates, on the other hand, the muscles are attached to bones, and muscle contraction moves one bone relative to another within a flexible skin that stretches to accommodate the movement.

For Review *Here are some important terms and concepts that you will encounter in this chapter. If you are not familiar with them, you should review them before proceeding.*

Actin filaments (Chapter 5)

ATP (Chapter 8)

Chitin (Chapter 26 and 28)

Vertebrate tissues (Chapter 32)

Some animals remain in one place all their adult lives, rooted like plants. The barnacles that encrust submerged rocks and the bottoms of ships are immobile, for example. They are an exception, however, to a very general rule, which is that animals move about from place to place, often with great speed (Figure 33-1). Many animals are creatures of constant activity, and few stay still for long. Of the three multicellular kingdoms that evolved from protists, only animals explore their environment in this active way; plants and fungi move only by growing, or as the passive passengers of wind and water. To move, all animals use the same basic mechanism, the contraction of muscles. In this chapter we examine how animals use muscles to achieve movement, with our focus on vertebrates.

FIGURE 33-2

An example of the action of muscles in a flea, a grasshopper, and a human being. Even though they differ by several orders of magnitude in size, all three jump to similar heights. In all three, antagonistic (opposing) muscle pairs control the movement of the legs. Muscles can exert force only by becoming shorter. In the flea and the grasshopper (**A** and **B**), the muscles are attached to the inside of the skeleton, whereas in the human **(C)**, the muscles are attached to the bones. In each case, the flexor muscles move the lower leg closer to the body and the extensor muscles move it away. Because the flea muscle is more massive for the size of the flea, it is capable of producing much greater acceleration .

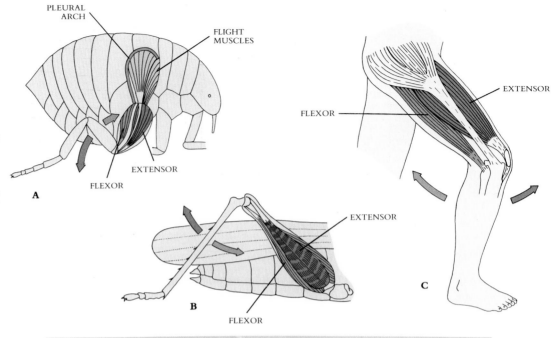

	A FLEA	**B** GRASSHOPPER	**C** MAN
BODY MASS	0.5 mg	3 g	70 kg
TYPICAL JUMP HEIGHT (cm)	20 cm	59 cm	60 cm
ACCELERATION (gravities)	245	15	1.5

THE MECHANICAL PROBLEMS POSED BY MOVEMENT

If you've ever tried to lift a large boulder or push down a big tree, you are familiar with the basic problem posed by movement, which is that gravity tends to hold objects in one place. The reason that you cannot toss a boulder with your little finger is that gravity is pulling down on the boulder far harder than you finger can push up. To move the boulder, you have to lift with a greater force than gravity is exerting. All motion must meet this simple requirement.

Organisms use the chemical energy of ATP to supply that force. By splitting an ATP molecule into ADP and P_i (inorganic phosphate), 7.3 kilocalories of energy per mole (the atomic weight of a substance—in this case ATP—expressed in grams) is made available to do the work of movement. Organisms apply this energy to alter the length of structural elements within the cytoskeletons of certain cells called muscle cells, causing the cells to shorten. When a lot of muscle cells shorten all at once, they can exert a great deal of force.

If this were all there is to movement, organisms would not move. Instead, they would simply pulsate as their muscles contracted and relaxed, contracted and relaxed, in futile cycles. For a muscle to produce movement, it must direct its force against another object. Most soft-bodied invertebrates, which have neither an internal nor an external skeleton, solve this problem by using the relative incompressibility of water as a kind of skeleton, directing the force of their muscles against the water. Those few soft-bodied invertebrates like slugs, which occur on land but have no shell or internal skeleton, move by attaching themselves to the land. Most animals, however, are able to move because the opposite ends of their muscles are attached to different parts of their bodies, so that, for example, muscle contraction results in the beating of a wing or the lifting of a leg (Figure 33-2).

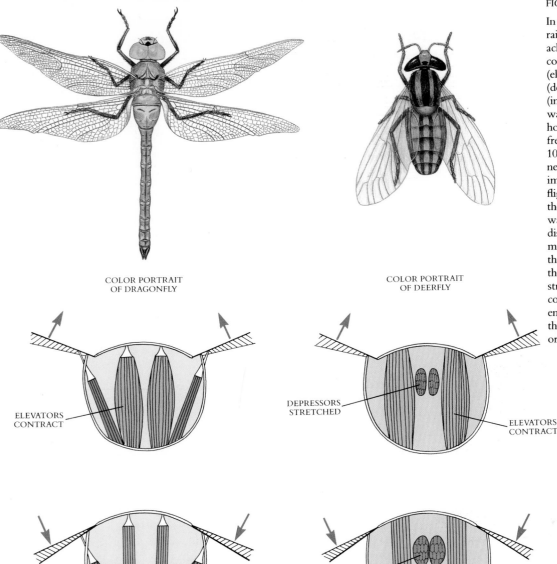

COLOR PORTRAIT
OF DRAGONFLY

COLOR PORTRAIT
OF DEERFLY

ELEVATORS
CONTRACT

DEPRESSORS
STRETCHED

ELEVATORS
CONTRACT

DEPRESSORS
CONTRACT

DEPRESSORS
CONTRACT

ELEVATORS
STRETCHED

A

B

In invertebrates the muscles are attached in various ways. Arthropods, for example, have muscles that are attached to the rigid chitin exoskeleton, enabling them to swim, to walk, and to fly (Figure 33-3). As long as an individual is small enough, this is an effective strategy. Chitin is relatively brittle, however, and the exoskeleton must be much thicker to bear the pull of the muscles in large insects than in small ones. In an insect the size of a human being, the exoskeleton would need to be so thick that the animal could hardly move. This relationship puts real limits on the size of insects.

In vertebrates, muscles are attached to an internal skeleton of bone, which is both rigid and flexible and yet able to bear far more weight than chitin. Instead of a rigid exterior skeleton, vertebrates have a soft, flexible exterior, which stretches to accommodate the movement of their bodies. Whenever you bend your arm, the skin covering the joint of your elbow stretches; if it didn't, it would tear. The flexibility of the skin is a necessary component of vertebrate movement, which otherwise is determined largely by muscles and their attachments to bones.

FIGURE 33-4

The structure of skin. Vertebrate skin is composed of three layers: an outer **epidermis**, a lower **dermis,** and a underlying layer of **subcutaneous tissue.** The epidermis of skin is from 10 to 30 cells thick—about as thick as this page. The outer cell layer, called the **stratum corneum,** is the one you see when you look at your arm or face. Cells from this layer are continuously worn off and sometimes injured by the friction and stress that accompanies many of the body's activities. They also lose moisture and dry out. The body deals with this damage not by repairing cells, but by replacing them. Cells from the stratum corneum are shed continuously and replaced by new cells produced deep within the epidermis. A cell normally lives in the stratum corneum for about a month. The dermis of skin is from 15 to 40 times thicker than the epidermis. The thick dermis provides structural support for the epidermis and a matrix for the many nerve endings, muscles, and specialized cells residing within skin. A fine network of blood vessels passes through it. The leather of a belt is animal dermis. The layer of subcutaneous tissue below the dermis is composed primarily of fat-rich cells. They act as shock absorbers and provide insulation to conserve body heat. This tissue varies greatly in thickness in different parts of the body. The eyelids have none, whereas the buttocks and thighs may have a lot. Embedded within the epidermis are millions of specialized cells called melanocytes, which produce a brownish pigment called melanin. All people have about the same number of melanocytes in their skin, but they differ in the amount of melanin that the individual melanocytes produce. The more melanin the melanocytes produce, the darker the skin. The result is a range of human skin tones from white and yellow to various shades of brown and black.

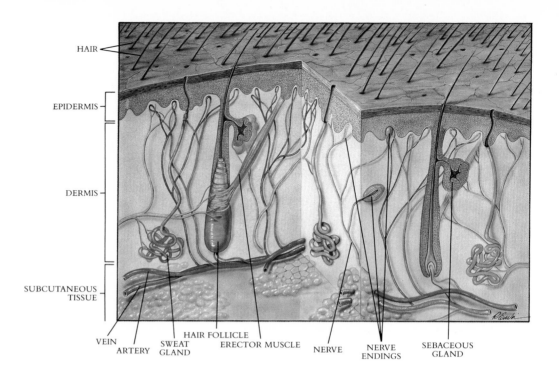

SKIN AND BONES
Skin

The skin of a vertebrate is far more than simply an elastic container of epithelial cells encasing the body's muscles, blood, and bones. Instead, it is a dynamic organ that has many functions:

1. Skin is a protective barrier. It keeps out microorganisms that would otherwise infect the body. Because skin is waterproof, it keeps the fluids of the body in and other fluids out; when you soak in a bathtub, your body absorbs little or no water. Skin cells also contain a pigment called melanin, which absorbs potentially damaging ultraviolet radiation from the sun.
2. Skin provides a sensory surface. Sensory nerve endings in skin act as your body's pressure gauge, telling you how gently to caress a loved one, how hard to hold a pencil. Other sensors embedded in skin detect pain, heat, and cold. Skin is the body's point of contact with the outside world.
3. Skin compensates for body movement. Skin stretches when you reach for something and contracts quickly when you stop reaching. It expands when you grow and shrinks when you lose weight.
4. Skin helps control the body's internal temperature. When the temperature is cold, the blood vessels in the skin contract so that less of the body's heat is lost to the surrounding air. When it is hot, these same vessels expand, and glands in the skin release sweat, whose evaporation cools the body surface.

Skin is the largest organ of the vertebrate body. In an adult human, 15% of the total body weight is skin. Much of the multifunctional role of skin reflects the fact that crammed in among its cells are many other specialized cells. One square centimeter of human skin contains 200 nerve endings, 10 hairs and muscles, 100 sweat glands, 15 oil glands, 3 blood vessels, 12 heat receptors, 2 cold receptors, and 25 pressure-sensing receptors (Figure 33-4).

Bone

Bone is a special form of cartilage in which the collagen fibers are coated with a calcium phosphate salt. The great advantage of bone over chitin as a structural material is that it is strong without being brittle. To understand the properties of bone, first consider those of fiberglass. Fiberglass is composed of glass fibers embedded in epoxy glue. The individual fibers are rigid, giving great strength, but they are brittle, like the cuticle of arthropods. The epoxy component of fiberglass, on the other hand, is flexible but weak. The composite, fiberglass, is both rigid and strong. When a fiber breaks because of stress and a crack starts to form, the crack runs into glue before it reaches another fiber. The glue distorts and reduces the concentration of the stress, and the adjacent fibers consequently are not exposed to the same high stress. In effect, the glue acts to spread the stress over many fibers.

The construction of bone is similar to that of fiberglass. Small, needle-shaped crystals of a calcium-containing mineral, hydroxyapatite, surround and impregnate collagen fibrils of bone. The fibrils are placed parallel to the axis of long bones and also parallel to the curved ends of bones in joints. As a result of the way these fibers are placed, no crack can penetrate far into bone without encountering a hard mass of hydroxyapatite crystals embedded in a collagenous matrix. Bone is more rigid than collagen, just as fiberglass is more rigid than epoxy. On the other hand, bone is more flexible and resistant to fracture than is hydroxyapatite—or chitin.

New bone is formed by cells called **osteoblasts,** which secrete the collagen fibers on which calcium is later deposited. Bone is laid down in thin, concentric layers called **lamellae,** like so many layers of paint on an old pipe. The lamellae are laid down as a series of tubes around narrow channels called **Haversian canals,** which run parallel to the length of the bone. The Haversian canals are interconnected and contain blood vessels and nerve cells. The blood vessels provide a lifeline to living bone-forming cells, while the nerves control the diameter of the blood vessels and thus the flow through them. When bone is first formed in the embryo, osteoblasts use the cartilage skeleton as a template for bone formation. Later, new bone is formed along lines of stress.

Bone is formed in two stages: first, collagen is laid down in a matrix of fibrils along lines of stress, and then calcium minerals impregnate the fibrils. These minerals provide rigidity, while the collagen provides flexibility.

The bones of the vertebrate skeleton are composed of two kinds of tissue. The ends and interiors of long bones are composed of an open lattice of bone called **spongy bone tissue,** or **marrow.** Within this lattice framework, most of the body's red blood cells are formed. Surrounding the spongy bone tissue at the core of bones are concentric layers of **compact bone tissue** (Figure 33-5) in which the collagen fibrils are laid down in a pattern that is far denser than the marrow. The compact bone tissue gives the bone the strength to withstand mechanical stress (Figure 33-6).

MUSCLE

Even though the cells of almost all eukaryotic organisms appear capable of shape changes mediated by actin microfilaments, many multicellular animals have evolved specialized cells devoted almost exclusively to this purpose. These cells contain far more microfilaments than occur in the fibers of single-celled eukaryotes. Such specialized animal cells are called **muscle cells.** As we mentioned in Chapter 32, vertebrates possess three different kinds of muscle cells: smooth muscle, skeletal muscle, and cardiac muscle.

Smooth Muscle

Smooth muscle was the earliest form of muscle to evolve, and it is found throughout the animal kingdom. **Smooth muscle** cells are long and spindle-shaped, each cell containing a single nucleus (see Figure 32-10, *A*). The interiors of smooth muscle cells are packed with actin-myosin myofibrils, but the individual myofibrils of a cell are not aligned into organized assemblies. Smooth muscle tissue is organized into sheets of cells. In some tissues the muscle cells contract only when they are stimulated by a nerve or hormone, and then all of the cells contract together as a unit. An example are the

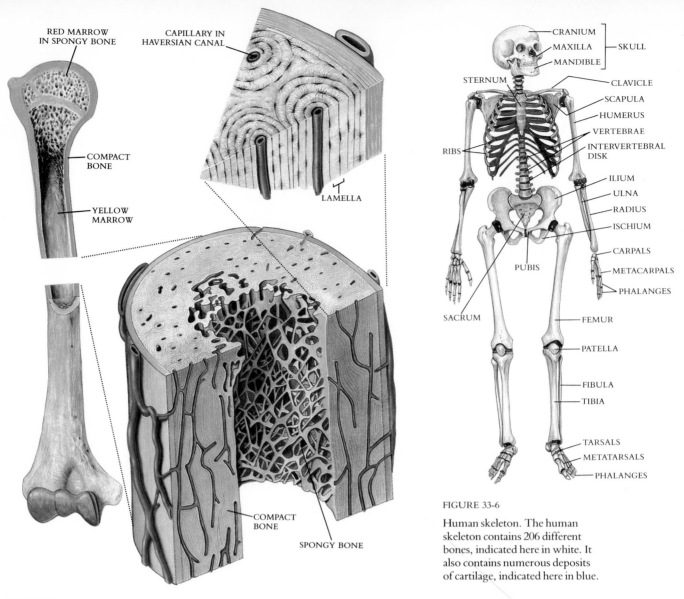

RED MARROW IN SPONGY BONE

CAPILLARY IN HAVERSIAN CANAL

COMPACT BONE

YELLOW MARROW

LAMELLA

COMPACT BONE

SPONGY BONE

CRANIUM

MAXILLA — SKULL

MANDIBLE

STERNUM

CLAVICLE

SCAPULA

HUMERUS

VERTEBRAE

INTERVERTEBRAL DISK

RIBS

ILIUM

ULNA

RADIUS

ISCHIUM

CARPALS

METACARPALS

PHALANGES

PUBIS

SACRUM

FEMUR

PATELLA

FIBULA

TIBIA

TARSALS

METATARSALS

PHALANGES

FIGURE 33-6

Human skeleton. The human skeleton contains 206 different bones, indicated here in white. It also contains numerous deposits of cartilage, indicated here in blue.

FIGURE 33-5

The structure of compact bone shown at three levels of detail. Some parts of bones are dense and compact, giving the bone strength. Other parts are spongy, with a more open lattice; it is here that most red blood cells are formed. The Haversian canals are evident in the enlargement.

muscles found lining the walls of many vertebrate blood vessels and those which make up the iris of the vertebrate eye. In other smooth muscle tissue, such as that found in the wall of the gut, the individual cells contract spontaneously, leading to a slow, steady contraction of the tissue.

Skeletal Muscle

Skeletal muscles are the muscles associated with the skeleton (Figure 33-7). They are also called **striated muscles** because they are obviously marked with lines (Latin *striae;* see Figure 32-10, *C*). Striated muscle cells are produced during development by the fusion of several cells at their ends to form a very long fiber. Each muscle cell or muscle fiber still contains all of the original nuclei pushed out to the periphery of the cytoplasm. Each striated muscle is a tissue made up of numerous individual muscle cells that act as a unit. These striated muscle cells represent a distinct improvement in muscle cell organization, as compared with that found in the smooth muscle cells. Imagine a large raft being towed upstream by many small canoes, each canoe bound to the raft by its own towline. This is analogous to the contraction of smooth muscle, each smooth muscle cell participating individually in the contraction of the muscle. Now imagine placing all the rowers in one galley where they row in concert, pulling the raft far more effectively. This is analogous to the contraction of striated muscle, in which numerous muscle cells pool their resources.

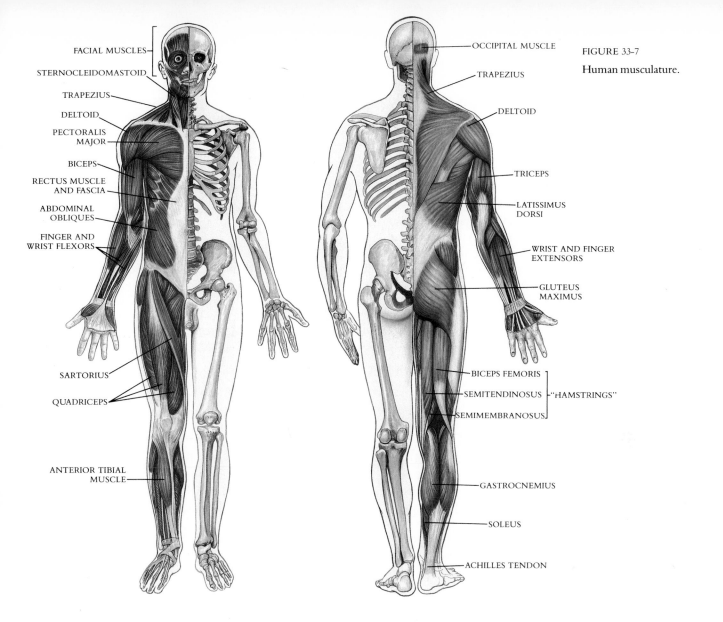

FIGURE 33-7

Human musculature.

FACIAL MUSCLES
STERNOCLEIDOMASTOID
TRAPEZIUS
DELTOID
PECTORALIS MAJOR
BICEPS
RECTUS MUSCLE AND FASCIA
ABDOMINAL OBLIQUES
FINGER AND WRIST FLEXORS
SARTORIUS
QUADRICEPS
ANTERIOR TIBIAL MUSCLE

OCCIPITAL MUSCLE
TRAPEZIUS
DELTOID
TRICEPS
LATISSIMUS DORSI
WRIST AND FINGER EXTENSORS
GLUTEUS MAXIMUS
BICEPS FEMORIS
SEMITENDINOSUS
SEMIMEMBRANOSUS
"HAMSTRINGS"
GASTROCNEMIUS
SOLEUS
ACHILLES TENDON

Cardiac Muscle

The vertebrate heart is composed of striated muscle fibers arranged very differently from the fibers of skeletal muscle. Instead of very long multinucleate cells running the length of the muscle, heart muscle is composed of chains of single cells, each with its own nucleus. These chains of cells are organized into fibers that branch and interconnect, forming a latticework. This lattice structure is critical to how heart muscle functions. Heart contraction is initiated at one location by the opening of transmembrane channels that admit ions into the muscle cells there, altering the charge of their membranes. This change in membrane charge is called depolarization. When two cardiac muscle fibers touch one another, their membranes make an electrical junction. As a result, the electrical depolarization of the initial fiber initiates a wave of contraction throughout the heart, with the wave of depolarization rapidly passing from one fiber to another across these junctions. For this reason, a mass of heart muscle tends to contract all at once, rather than gradually.

HOW CELLS MOVE

Cells move by expanding and contracting portions of their surfaces. This ability to alter surface relationships arises from dynamic changes in their cytoskeletons. The key elements are the fine microfilaments, elements that are only 6 nanometers thick.

ACTIN
MOLECULES

FIGURE 33-8

An actin filament.

The Structure of Microfilaments

Microfilaments in muscle cells, called **myofilaments,** are composed of the protein **actin** and three associated protein components. The major one that we are concerned with here is the protein **myosin.** The two additional proteins—**troponin** and **alpha-actinin**—are involved in skeletal muscle; we will have more to say about them later.

1. *Actin.* Actin microfilaments are one of the two major components of myofilaments. The individual actin proteins are the size of a small enzyme. Actin molecules polymerize to form thin filaments (Figure 33-8). The filaments consist of two strings of monomers wrapped around one another, like two strands of pearls loosely wound together. The result is a long, thin, helical filament with a diameter of about 6 nanometers.
2. *Myosin.* The other major protein associated with myofilaments, myosin, is a protein molecule more than 10 times longer than an individual actin molecule. Myosin has an unusual shape: one end of the molecule consists of a very long rod, whereas the other end consists of a double-headed globular region. In electron micrographs a myosin molecule looks like a two-headed snake. Like actin, myosin spontaneously forms into filaments (Figure 33-9).

How Myofilaments Contract

The contraction of myofilaments occurs when the heads of the myosin filaments change their shape. This causes the myosin fiber to slide past the actin filament, its globular heads (called **S-1 units**) "walking" step by step along the actin (Figure 33-10). Each step by an S-1 head uses a molecule of ATP.

How does microfilament sliding lead to cell movement? Myofilaments convert the sliding of fibers into motion by anchoring the actin, the ends of which are bound to the anchoring protein **alpha-actinin.** Alpha-actinin is found widely distributed on the interior surfaces of the plasma membranes of eukaryotic cells. Because the actin is not free to move with respect to the alpha-actinin to which it is bound, the zones between alpha-actinin anchors shorten when the microfilaments slide past myosin. Because the myofilament may be attached at both of its ends to membrane, its overall shortening moves the membranes to which the myofilament is attached.

Myofilaments are composed of a highly ordered complex of actin and myosin. The secret of muscle contraction lies in the way in which the actin and myosin fibers are combined. They *interdigitate.* The arrangement is diagrammed in Figure 33-11, with a myosin filament interposed between two pairs of actin filaments, its S-1 heads jutting out toward the actin filaments on each side.

> **The contraction of vertebrate muscles and many other kinds of cell movement in eukaryotes result from the movements of microfilaments within cells. The microfilaments are composed of long, parallel fibers of actin cross-connected by myosin. Their movement results from a ATP-driven shape change in myosin.**

FIGURE 33-9

The structure of myosin.
A Each myosin molecule is a coil of two chains wrapped around one another; at the end of each chain is a globular region referred to as the "head."
B Myosin molecules are usually combined into filaments, which are cables of myosin from which the heads protrude at regular intervals.

A

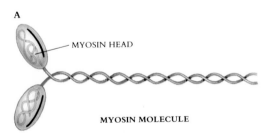

MYOSIN HEAD

MYOSIN MOLECULE

B

MYOSIN FILAMENT

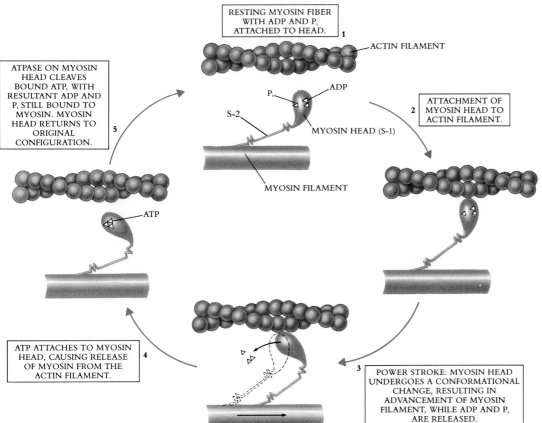

RESTING MYOSIN FIBER
WITH ADP AND P_i
ATTACHED TO HEAD. **1**

ACTIN FILAMENT

ATPASE ON MYOSIN
HEAD CLEAVES
BOUND ATP, WITH
RESULTANT ADP AND
P_i STILL BOUND TO
MYOSIN. MYOSIN
HEAD RETURNS TO
ORIGINAL
CONFIGURATION. **5**

P_i   ADP

S-2

MYOSIN HEAD (S-1)

2 ATTACHMENT OF
MYOSIN HEAD TO
ACTIN FILAMENT.

MYOSIN FILAMENT

ATP

ATP ATTACHES TO MYOSIN
HEAD, CAUSING RELEASE
OF MYOSIN FROM THE
ACTIN FILAMENT. **4**

3 POWER STROKE: MYOSIN HEAD
UNDERGOES A CONFORMATIONAL
CHANGE, RESULTING IN
ADVANCEMENT OF MYOSIN
FILAMENT, WHILE ADP AND P_i
ARE RELEASED.

FIGURE 33-10

The mechanisms of microfilament movement during muscle contraction. Myosin moves along actin (from left to right in this diagram) by first binding to it and then hunching forward as the result of a change in the shape of the myosin head. The splitting of ATP recocks the mechanism, returning the myosin head to its extended position.

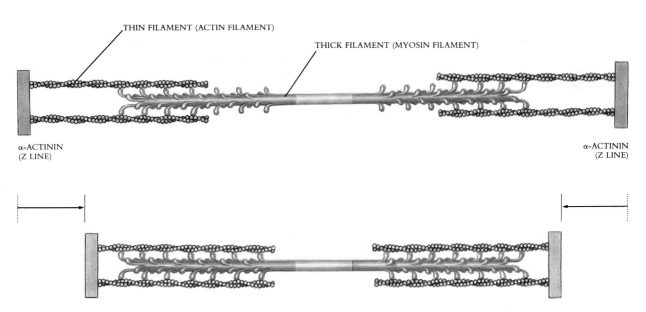

THIN FILAMENT (ACTIN FILAMENT)

THICK FILAMENT (MYOSIN FILAMENT)

α-ACTININ
(Z LINE)

α-ACTININ
(Z LINE)

FIGURE 33-11

The interaction of actin microfilaments and myosin filaments in vertebrate muscle. The heads on the two ends of the myosin filament are oriented in opposite directions, so that as the right-hand end of the myosin molecules "walks" along the actin filaments at the right, it pulls them and the attached alpha-actinin (a component called the Z line in muscles) leftward toward the center. The left-hand end of the same myosin molecule "walks" in a leftward direction, pulling its actin filaments and their attached Z line rightward toward the center. The result is contraction, with both Z lines moving toward the center.

FIGURE 33-12

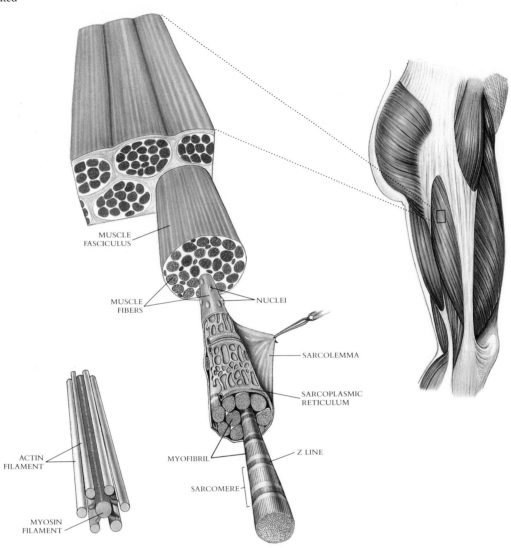

MUSCLE
FASCICULUS

MUSCLE
FIBERS

NUCLEI

SARCOLEMMA

SARCOPLASMIC
RETICULUM

ACTIN
FILAMENT

MYOFIBRIL

Z LINE

SARCOMERE

MYOSIN
FILAMENT

How Striated Muscle Contracts

As noted earlier, striated muscle cells are produced during development by the fusion
of several cells, end to end, to form a very long fiber (Figure 33–12). A single fiber will
typically run the entire length of a vertebrate muscle. Each cell, or muscle fiber, still
contains all of the original nuclei, pushed out to the periphery of the cytoplasm by a
central cable of 4 to 20 myofibrils. The cytoplasm in striated muscle is given a special
name, the **sarcoplasm.** The myofibrils that run down the center of a muscle cell are
highly organized to promote simultaneous contractions (see Figure 33–11):

1. A myofibril is made up of a long chain of contracting units called **sarcomeres,**
   lined up like the cars on a train.
2. Each sarcomere is composed of interdigitating filaments of actin and myosin.
   One collection of actin filaments is attached to the front and another to the back
   of the sarcomere, to plates of alpha–actinin referred to as **Z lines.** These front and
   back assemblies of actin filaments are not long enough to reach each other in the
   center of the sarcomere. They are joined to one another by interdigitating
   myosin filaments.
3. The sarcomere contracts when the heads of the myosin filaments change their
   shape. Since these heads are in contact with the actin filaments, the effect of their
   contraction is to pull the myosin along the actin.

4. The orientation of myosin is such that it moves along actin toward the Z line. Because both ends of the myosin filaments move in this manner simultaneously, the effect is to pull the two Z lines together, contracting the sarcomere.
5. Simultaneous contraction of all the sarcomeres of a myofibril results in an abrupt and forceful shortening of the myofibril. All of the myofibrils of a muscle fiber usually also contract in concert at the same time, producing a very strong contraction in the length of the muscle fiber cell.

The sarcomeres are lined up, in register with one another, all along the length of the stacked myofibrils (see Figure 32-10, *C*). This gives striated muscle, as viewed with a light microscope, the distinctive pattern of bands or striations that give it its name.

Muscle cells are rich in actin and myosin, which form myofibrils that are capable of contraction. Many muscle cells contracting in concert can exert considerable force.

How Nerves Signal Muscles to Contract

In vertebrate striated skeleton muscle, contraction is initiated by a nerve impulse. The nerve impulse arrives as a wave of depolarization along the nerve membrane. The nerve fiber is embedded in the surface of the muscle fiber, forming a **neuromuscular junction** (Figure 33-13). When such a wave of depolarization reaches the end of a neuron, at the point where the neuron attaches to a muscle (the **motor endplate**), it causes that membrane to release the chemical acetylcholine from the nerve cell into the junction. The acetylcholine passes across to the muscle membrane and opens the ion channels of that membrane, depolarizing it. Acetylcholine is called a **neurotransmitter,** because it transmits the nerve impulse across the junction.

How does the depolarization of the nerve fiber membrane cause the contraction of the muscle fibers? The endoplasmic reticulum of a striated muscle cell, which is called the **sarcoplasmic reticulum,** wraps around each myofibril like a sleeve (see Figure 33-12). As a result, the entire length of every myofibril is very close to the intracellular space bounded by the sarcoplasmic reticulum membrane system. Within the sarcoplasmic reticulum are embedded numerous calcium ion transport channels. In resting muscle, calcium ions are actively pumped through these channels, concentrating all of the calcium ions of that cell within the spaces of the sarcoplasmic reticulum. The depolarization of the muscle fiber membrane opens ion channels in the sarcoplasmic reticular membrane (Figure 33-14) and so causes the release of this concentrated calcium across the reticular membrane into the cytoplasm (sarcoplasm), where it is free to interact with myofibrils. This calcium acts as a trigger to initiate contraction of the myofibril. It does this in the following way:

1. In resting muscle, myosin filaments are not free to interact with actin, because the sites on actin where the myosin heads must make contact are not available. The heads are covered by a globular protein called troponin.
2. Troponin molecules are able to bind calcium ions, and when they do, the troponin molecules change their shape. As a result of this change in shape, the troponin is repositioned to a new location, where it does not interfere with myosin interaction. Only when this repositioning has occurred can contraction take place.

Thus the release of calcium by the nerve's stimulation of the sarcoplasmic reticulum releases the troponin trigger and results in the contraction of the myofibril.

LOOKING AHEAD

In the next chapter we shall discuss the source of the energy used in movement—digestion. Later, in Chapter 38, we present more of the details of the nervous system, which determines which muscles will move and when.

FIGURE 33-13

The way in which the branches of this neuron axon connect with a skeletal muscle is shown here. The thick central portion is the body of the axon, with small branches each ending in a motor end plate, where the nerve and muscle join to form a neuromuscular junction.

FIGURE 33-14

How acetylcholine acts to open Ca^{++} ion channels and so initiate muscle contraction. Embedded within the post-synaptic membrane of neuromuscular junctions are enzymes that convert ATP to a cyclic form, cyclic AMP. These enzymes (AC) are turned "off" unless the neurotrasmitter acetylcholine is bound to an adjacent membrane receptor (A). When an imcoming nerve impulse releases acetylcholine into the synaptic cleft, the acetylcholine binds receptor A, which activates AC to produce cyclic AMP. The cyclic AMP in turn activates an enzyme (protein kinase) that adds a phosphate group to the closed Ca^{++} ion channels, causing them to change their shape and open. The opening of the membrane's Ca^{++} ion channels causes Ca^{++} ions to flood into the sarcomere cell, releasing muscle contraction. The triggering of muscle contraction by a nerve impulse is not automatic—it depends upon the state of two "on/off" switches: 1. If an inhibitory neurotransmitter is bound to another adjacent receptor I, then AC remains shut off; 2. The sarcomere must have an adequate supply of energy; when it does, a GTP molecule (an alternative form of ATP formed when cell energy levels are high) is bound to the protein linking A to AC, and only then can A activate AC.

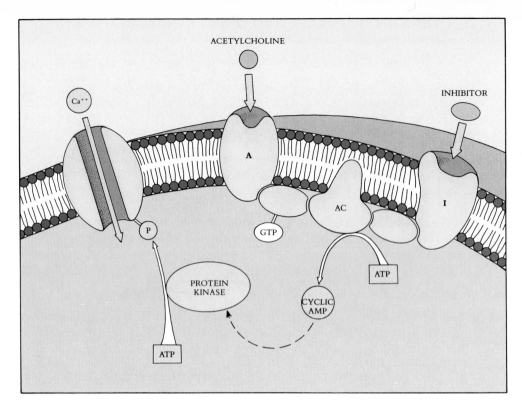

SUMMARY

1. In vertebrates, movement results from the contraction of muscles anchored to bones. When the limb containing the bone is pulled to a new position, the skin stretches to accommodate the change.

2. Skin is an epithelial tissue that covers the surface of the vetebrate body, protecting it while compensating for body movement.

3. Bone is a form of cartilage in which collagen fibers are impregnated with calcium salts. The salts act like the glass fibers in fiberglass, creating a material that is strong without being brittle.

4. There are three kinds of muscle: smooth muscle, which is organized into sheets and contracts spontaneously; skeletal muscle, which is organized into trunks of long fibers and contracts only when stimulated by a nerve; and cardiac muscle, in which the fibers are interconnected and may initiate contraction spontaneously.

5. Muscle cells contract as a result of shortening of myofilaments within the cystoskeleton. The myofilaments are composed of the proteins actin and myosin, together with two other proteins which together control the contraction.

6. In a myofilament the myosin is located between adjacent actin filaments. Changes in the shape of the ends of the myosin molecule, driven by the splitting of ATP molecules, cause the myosin molecule to move along the actin, producing contraction of the myofilament.

7. In vertebrate striated skeleton muscle, contraction is initiated by a nerve impulse. Acetylcholine passes across the neuromuscular junction from the nerve to the muscle, initiating the process that causes the muscle to contract.

REVIEW

1. The muscles of a beetle are attached to its _____, whereas your muscles are attached to your _____.

2. Vertebrate skin is composed of three layers. What are they?

3. Most of the red blood cells in your body are produced in the _____ of long bones such as the tibia.

4. The major components of myofibrils are _____ and _____, both of which spontaneously form into filaments.

5. When troponin molecules bind _____, they change shape and muscle contraction occurs.

SELF-QUIZ

1. Which of the following is *not* a function of skin?
   (a) controls the internal temperature
   (b) provides muscle attachment sites
   (c) provides a sensory surface
   (d) compensates for body movement
   (e) provides a protective barrier

2. Which is more likely to be found at the core of bones?
   (a) compact bone tissue
   (b) spongy bone tissue (marrow)
   (c) osteoblasts
   (d) myofilaments
   (e) nothing; the interior is hollow

3. Haversian canals
   (a) have bone lamellae laid down around them in concentric rings
   (b) run parallel to the long axes of bones
   (c) are interconnected
   (d) have blood vessels and nerves running through them
   (e) all of the above

4. When you lift your leg, what types of muscles are doing the lifting?
   (a) smooth muscles
   (b) striated muscles
   (c) cardiac muscles
   (d) a and b
   (e) all of the above

5. The source of energy for muscle contraction is
   (a) actin
   (b) myosin
   (c) sarcomeres
   (d) myofibrils
   (e) ATP

PROBLEM

1. If you weigh 60 kilograms, approximately how much does your skin weigh?

THOUGHT QUESTIONS

1. Myofilaments can contract forcefully, pulling membranes attached to the two ends toward one another. Myofilaments cannot expand, however, pushing membranes attached to the two ends of a myofilament apart from one another. Why is it that myofilaments can pull but not push?

2. Among long-distance runners and committed joggers, the long bones of the leg often develop "stress fractures," numerous fine cracks running parallel to one another along the lines of stress. In most instances, stress fractures occur when runners push themselves much farther than they are accustomed to running. Runners who train by gradually increasing the distances they run develop stress fractures rarely, if ever. What protects this second kind of runner?

FOR FURTHER READING

CARAFOLI, E., and J. PENNISTON: "The Calcium Signal," *Scientific American*, November 1985, pages 70-78. The release of calcium ion is the only known way in which the electricity of the nervous system is able to produce changes in the body. Nerves regulate all muscle contractions and hormone secretions by controlling the level of Ca^{++} ions.

COHEN, C.: "The Protein Switch of Muscle Contraction," *Scientific American*, November 1975, pages 36-45. How proteins associated with myofilaments interact with Ca^{++} ions to trigger contraction.

SCHMIDT-NIELSEN, K.: *Animal Physiology: Adaptation and Environment*, ed. 3, Cambridge University Press, New York, 1983. Chapter 11 presents an outstanding treatment of muscles and bones from an evolutionary perspective.

SMITH, D.: "The Flight Muscles of Insects," *Scientific American*, June 1965, pages 76-89. Some insects flap their wings more rapidly than nerves can carry successive impulses. This article explains how they accomplish this seemingly impossible feat.

FIGURE 34-1   Eating is something we all do. Some of us prefer hamburgers; others, like this lion, enjoy giraffes. Most of the substance of this giraffe will become part of the lion's body or be burned to supply it with energy. Eventually, the residue will be discarded. All the food that each of us consumes suffers the same fate, being converted to body tissue, energy, and refuse.

HOW ANIMALS DIGEST FOOD

Chapter 34

Overview

Digestion is the conversion of parts of organisms into small molecules that can be easily used as food by cells. In vertebrates, food is digested as it passes through a long, one-way digestive tract: the food is first pulverized in the mouth and then degraded into molecular fragments in the stomach. The digestion of these fragments to simple molecules is completed in the small intestine. The small molecules that are the product of digestion are then absorbed into the body through the walls of the small intestine, and the residual solids are concentrated in the large intestine and eliminated.

For Review

Here are some important terms and concepts that you will encounter in this chapter. If you are not familiar with them, you should review them before proceeding.

Acid (Chapter 3)

Lysosomes (Chapter 5)

Oxidative respiration (Chapter 8)

A vertebrate body is a complex colony of many cells. Like a city, it contains many individuals that carry out specialized functions. It has its own police (macrophages), its own construction workers (fibroblasts), and its own telephone company (the nervous system). Just as in a city, the many individual cells of the vertebrate body need to be provided with food that is trucked in from elsewhere (Figure 34-1). Among the cells of a vertebrate's body, there are no farmers; no vertebrate contains photosynthetic cells. Instead, all of an animal's cells are nourished with food that the animal obtains outside itself and transports to the individual cells. Many of the major organ systems of an animal are involved in this acquisition of energy. The digestive system acquires organic foodstuffs; the respiratory system acquires the oxygen necessary to metabolize the food; the circulatory system transports both food and oxygen to the individual cells of the body; and the excretory system rids the body of wastes produced by metabolism. In this chapter we consider the first of these activities, digestion.

FIGURE 34-2

The digestive systems that occur in various animal phyla are diverse, as shown here.

A The digestive system of a flatworm, in which there is only one opening to the outside. Most of the digestion in flatworms takes place after the food particles are incorporated into the cells that line the digestive cavity.

B The digestive system of an adult female mosquito. Most blood that the mosquito sucks from another animal is stored in the diverticulum for later use.

C The digestive system of a bird. A bird's crop plays a role similar to that of the diverticulum of a mosquito. In the gizzard, hard food is mixed and crushed with rocks and pebbles, often called brit, and thus prepared for more efficient digestion in the intestine.

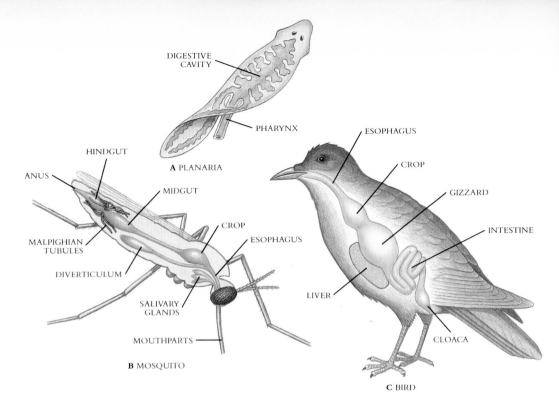

THE NATURE OF DIGESTION

Animals obtain the metabolic energy needed for growth and activity by degrading the chemical bonds of organic molecules. In Chapter 7 and 8 we considered these degradation processes in detail. They include the breakdown of sugar molecules in glycolysis and the oxidation of pyruvate in the citric acid cycle. What these processes have in common is that they act on simple molecules, on amino acids, lipids, sugars, and fragments of these molecules. Eating another organism, however, does not in itself provide a rich source of such molecules to an animal, because few organisms contain significant concentrations of free sugars and amino acids. Instead, the simple molecules are incorporated into long chains, into starches, fats, and proteins. Before an animal can obtain the energy from its food, it must degrade these molecules into the simple compounds from which they were built. This process is called **digestion.**

Like most of the body's other systems, digestion has become increasingly complex during the evolution of animals. Their protist ancestors, as you have seen, simply incorporated food particles within their bodies by the process of phagocytosis. Cnidarians break down the bodies of their prey by secreting enzymes into their central cavity and then incorporate the fragments into their cells, where digestion continues. Flatworms have a highly branched digestive system with only one opening (Figure 34-2, A). Because of its branching, such a digestive system has a greatly increased absorptive surface. Although some of the food particles are broken down by the extracellular secretion of enzymes, as in cnidarians, most are simply incorporated into the cells that line the flatworm's digestive tract and digested there.

Other, more complex animals have a more complete digestive tract, with a mouth and an anus. In them, there are various methods for breaking down the food mechanically, as by grinding in the gizzard of an earthworm or bird, or enzymatically. Many such animals (Figure 34-2, B and C) also have a crop, which can be filled with food when it is available; this stored food can be used later.

In vertebrates generally, digestion is carried out by two types of agent: by hydrochloric acid (HCl), which indiscriminately breaks up large proteins into smaller pieces, and by a variety of enzymes. Enzymes that break up proteins into amino acids are called **proteases;** enzymes that break up starches and other carbohydrates into sugars are called **amylases;** and those that break up lipids and fats into small segments are called **lipases.** Most digestive enzymes cannot tolerate high acid concentrations, so the vertebrate digestive process is carried out in two phases: acid digestion takes place first, in the stomach, and then the food moves to the small intestine, where the acid is neutral-

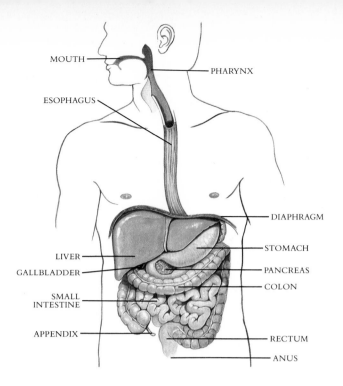

FIGURE 34-3

The human digestive system. Food passes from the mouth through the pharynx and esophagus to the stomach, where acids break up proteins and other molecules. From the stomach, food passes to the small intestine, where nutrients are further broken up and absorbed into the body's bloodstream, and then to the large intestine or colon, where water is resorbed and the residual material is compacted. Exit is through the rectum and out of the anus. Above the stomach is the liver, which plays many important roles in digestion; nested below the stomach is the pancreas, which secretes many of the digestive enzymes.

ized and the battery of digestive enzymes continues the digestive processes. The products of digestion then pass across the wall of the small intestine into the bloodstream. Figure 34-3 illustrates the organization of the human digestive system, which is representative of the kinds of digestive systems found in other vertebrates.

> **In the stomach, proteins are broken up into short fragments by acids; the fragments are then cleaved into individual amino acids by a variety of enzymes in the stomach and small intestine. Also in the small intestine, starches are digested by amylases and fats are digested by lipases.**

WHERE IT ALL BEGINS: THE MOUTH

The food of all vertebrates is taken in through their mouths, which in all groups except for the birds typically contain teeth (Figure 34-4). The teeth of different kinds of vertebrates are specialized in different ways, depending on whether they usually feed on animals or plants, and how they obtain what they eat. Human beings are omnivores, eating both plant and animal food regularly. As a result, our teeth are structurally

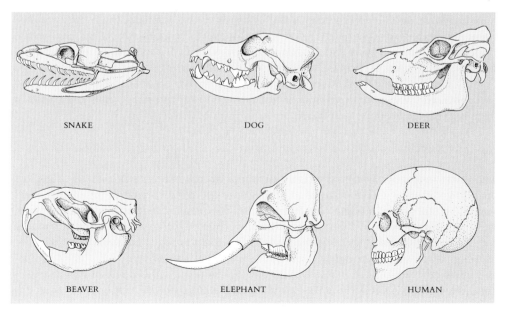

SNAKE   DOG   DEER

BEAVER   ELEPHANT   HUMAN

FIGURE 34-4

The arrangement of teeth in a variety of vertebrates. Teeth are in each case specialized for particular tasks. In the snake the teeth slope backward, to aid in retention of prey during swallowing. In carnivores such as the dog, "canine" teeth specialized for ripping food predominate. In herbivores such as deer, grinding teeth predominate. In the beaver the foreteeth are specialized as incisors—chisels. In the elephant, two of the upper front teeth are specialized as weapons. Human beings are omnivores—we consume both plant and animal food—and have a relatively broad-function mouth containing canine, chisel, and grinding teeth.

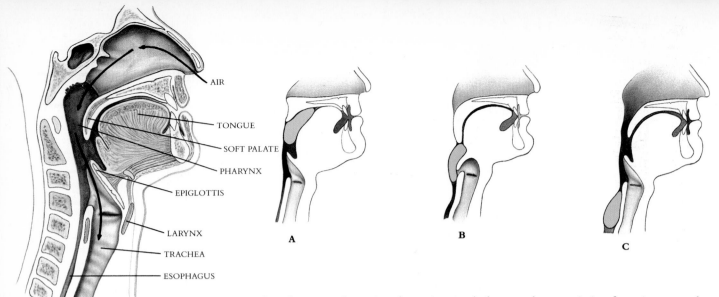

AIR

TONGUE

SOFT PALATE

PHARYNX

EPIGLOTTIS

LARYNX

TRACHEA

ESOPHAGUS

A   B   C

FIGURE 34-5

How human beings swallow. As food passes back past the rear of the mouth **(A)**, it presses the soft palate against the back wall of the pharynx, sealing off the nasal passage; as the food passes on down, a flap of tissue called the epiglottis folds down **(B)**, sealing the respiratory passage; after the food enters the esophagus, the soft palate relaxes and the epiglottis is raised **(C)**, opening the respiratory passage between the nasal cavity and the trachea.

intermediate between the pointed, cutting teeth that are characteristic of carnivores and the flat, grinding teeth characteristic of herbivores.

Within the vertebrate mouth, the tongue mixes the food with a mucus solution, the **saliva.** In humans, saliva is secreted into the mouth by three pairs of salivary glands, which are located in the mucosal lining of the mouth. The saliva moistens and lubricates the food so that it is swallowed more readily and does not abrade the tissue it passes on its way down through the esophagus. The saliva also contains the enzyme amylase, which initiates the breakdown of starch and other complex polysaccharides into smaller fragments. This action by amylase is the first of the many digestive processes that occur as the food passes through the digestive tract.

> Vertebrate teeth serve to shred animal tissue and to grind plant material. Saliva secreted into the mouth moistens the food, which aids its journey into the digestive system and begins the enzyme-catalyzed process of degradation.

THE JOURNEY OF FOOD TO THE STOMACH

After passing through the opening at the back of the mouth (Figure 34-5), the food enters a tube called the **esophagus,** which connects the pharynx to the stomach. No further digestion takes place in the esophagus. Its role is that of an escalator, moving food down toward the stomach. In adult human beings the esophagus is about 25 centimeters long, and its lower end opens into the stomach proper. The lower two thirds of the esophagus is enveloped in smooth muscle. Successive waves of contraction of these muscles move food down through the esophagus to the stomach. Such rhythmic sequences of waves of muscular contraction in the walls of a tube are called **peristalsis.** Because the movement of food through the esophagus is primarily caused by these peristaltic contractions, humans can swallow even if they are upside down.

The exit of food from the esophagus to the stomach is controlled by the **esophageal sphincter.** When this sphincter is contracted, it prevents the food in the stomach from moving back up the esophagus.

PRELIMINARY DIGESTION: THE STOMACH

The stomach (Figure 34-6) is a saclike portion of the digestive tract. In it, the digestive process is organized. The stomach collects ingested food, breaks it into small fragments, and feeds it in a controlled fashion into the primary digestive organ, the small intestine. In the lining of the stomach, the epithelium overlies a deep layer of connective tissue (Figure 34-7) called **mucosa,** below which are located a complex array of muscles, blood vessels, and nerves.

Food in the stomach is attacked by both acid and enzymes, both of which are secreted from deep depressions called **gastric pits** (Figure 34-8), which dot the upper ep-

FIGURE 34-6

The upper digestive tract of a human being. Food enters the stomach from the esophagus. The epithelial walls of the stomach are dotted with gastric pits, which contain glands that secrete acid and digestive enzymes. The entrance to the small intestine—the duodenum—is controlled by a band of muscle called the pyloric sphincter. The surface of the duodenum is covered with microscopic villi, which greatly increase its surface area and so aid in absorption of nutrients.

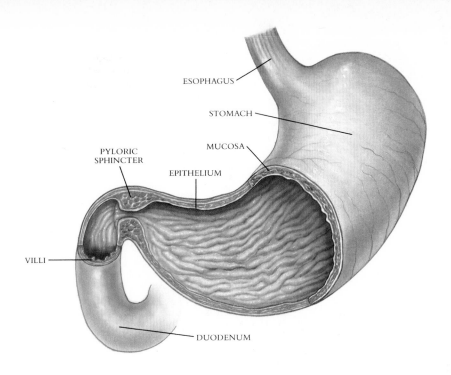

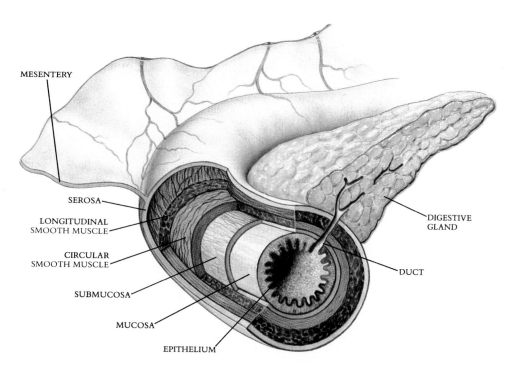

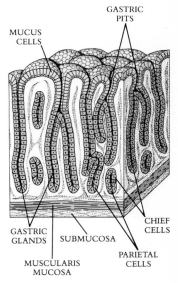

FIGURE 34-7

Organization of the digestive tract. Both the stomach and the intestinal tract have the same general organization, being tubes encased within successive layers of different tissues. The innermost layer of tissue is a thin cylinder of epithelium. This is encased by a thick layer of mucosa, a connective tissue rich in blood vessels, glands, and nerves. Surrounding the mucosa are layers of muscle tissue and an envelope of tough connective tissue called serosa, often connected to body structures by thin but tough sheets of tissue that hold the digestive tract in position within the body cavity.

FIGURE 34-8

Gastric pits are deep invaginations of the stomach epithelium down into the underlying mucosa, at the base of which parietal and chief cells secrete hydrochloric acid and a protein that is cleaved into the enzyme pepsin into the lumen—hollow space—of the pit.

ithelial surface of this organ. The human stomach secretes about 2 liters of hydrochloric acid every day, creating a very concentrated acid solution, the **gastric fluid.** The HCl breaks up connective tissue. It does this because the very low pH values, between 1.5 and 2.5, that are created by the HCl change the ionization of carboxyl and amino side groups of proteins. This process causes the folded proteins of connective tissue to open out and disrupts their associations with one another. Further digestion of protein is accomplished by the enzyme pepsin, which cleaves proteins into short polypeptides. Carbohydrates and fats are not digested in the stomach. Mucus secreted by the stomach lining protects the walls of the stomach from its own digestive juices.

> **In the stomach, concentrated acid breaks up connective tissue and protein into molecular fragments, which are further digested by pepsin into short polypeptides.**

The digestive tract exits from the stomach at a muscular constriction called the **pyloric sphincter** (see Figure 34–6). The pyloric sphincter is the gate to the small intestine, where the terminal stages of digestion occur. The capacity of the small intestine is limited, and its digestive processes take time. Consequently, efficient digestion requires that only relatively small portions of food be introduced from the stomach into the small intestine at any one time. When a small volume of food, which is by now highly acidic, passes into the small intestine, the acid introduced with the food acts as a signal, resulting in the closing of the pyloric sphincter. As time passes, the food is digested and the acid that entered the small intestine with it is neutralized. At a certain point in the process, the pH of the small intestine reaches a level that acts as a signal to the pyloric sphincter to open again. An additional small portion of food is introduced from the stomach into the small intestine, and the process continues.

TERMINAL DIGESTION AND ABSORPTION: THE SMALL INTESTINE

The small intestine is the true digestive vat of the vertebrate body. Within it, carbohydrates, proteins, and fats are broken down into simple sugars, amino acids, and fatty acids. Once these small molecules have been produced, they all pass across the epithelial wall of the small intestine into the bloodstream. Some of the enzymes necessary for these digestive processes are secreted by the cells of the intestinal wall. Most, however, are introduced into a short initial segment of the small intestine, called the **duodenum,** through a duct from a gland called the pancreas.

The **pancreas** secretes a number of different enzymes that break down carbohydrates, proteins, and fats in a variety of ways. This important gland also possesses specialized cells that secrete a base, **bicarbonate,** which neutralizes the acid derived from the stomach. Most of the digestive enzymes secreted by the pancreas will not work in acid solution, and this process of neutralizing the solution is therefore necessary for the enzymes to function.

Because fats are insoluble in water, they tend to enter the small intestine as small globules that are not attacked readily by the enzymes secreted by the pancreas. Before fats can be digested, they must be emulsified into droplets small enough to allow enzymes to break down the fat. This process is carried out by a collection of detergent molecules secreted by a second gland, the **liver.** The liver produces these detergent molecules, known as **bile salts,** and secretes them into the duodenum through a duct. These bile salts render the fats in the digestive tract soluble. Humans store excess secreted bile in the **gall bladder.**

Of the approximately 6–meter length of the human small intestine, only the duodenum region (the first 25 centimeters—about 4% of the total length) is actively involved in digestion. The rest of the small intestine is highly specialized to aid in the absorption of the products of digestion by the bloodstream. The epithelial wall of the small intestine is covered with fine fingerlike projections called **villi,** which are so small that it takes a microscope to see them (Figure 34–9). In turn, each of the epithelial cells

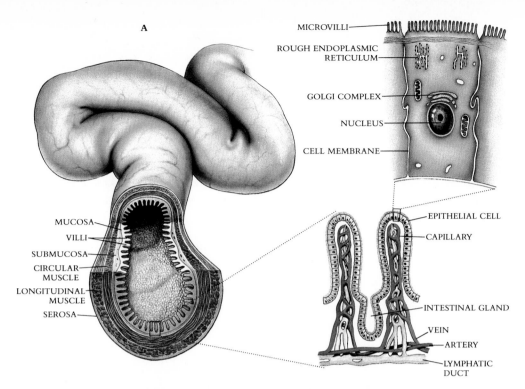

A

MICROVILLI

ROUGH ENDOPLASMIC RETICULUM

GOLGI COMPLEX

NUCLEUS

CELL MEMBRANE

MUCOSA
VILLI
SUBMUCOSA
CIRCULAR MUSCLE
LONGITUDINAL MUSCLE
SEROSA

EPITHELIAL CELL
CAPILLARY

INTESTINAL GLAND
VEIN
ARTERY
LYMPHATIC DUCT

FIGURE 34-9

A Cross section of the small intestine.
B Villi, shown in a scanning electron micrograph, are very densely clustered, giving the small intestine an enormous surface area, which is very important in efficient absorption of the digestion products.

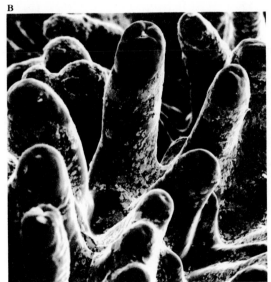

B

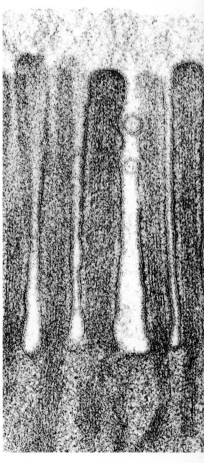

FIGURE 34-10

Intestinal microvilli.

covering the villi is covered on its outer surface by a field of cytoplasmic projections called **microvilli** (Figure 34-10). Both kinds of projections greatly increase the absorptive surface of the epithelium lining the small intestine. The average surface area of the small intestine of an adult human being is about 300 square meters, a greater area than the surface covered by many houses. The membranes of the epithelial cells contain carrier systems that actively transport sugars and amino acids across the membrane; fatty acids cross passively by diffusion.

> **Most digestion occurs in the first 25 centimeters of the 6-meter length of the small intestine, in a zone called the duodenum. The rest of the small intestine is devoted to the absorption of the products of digestion.**

The amount of material passing through the small intestine is startlingly large. An average human consumes about 800 grams of solid food and 1200 milliliters of water each day, for a total volume of about 2 liters. To this amount is added about 1.5 liters of fluid from the salivary glands, 2 liters from the gastric secretions of the stomach, 1.5 liters from the pancreas, 0.5 liter from the liver, and 1.5 liters of intestinal secretions. The total adds up to a remarkable 9 liters. However, although the flux is great, the *net* passage is small. Almost all of these fluids and solids are reabsorbed during their passage through the intestines, with about 8.5 liters passing across the walls of the small intestine and an additional 350 milliliters through the wall of the large intestine. Of the 800 grams of solid and 9 liters of liquid that enter the digestive tract in one day, only about 50 grams of solid and 100 milliliters of liquid leave the body as feces. The normal fluid absorption efficiency of the digestive tract thus approaches 99%, which is very high indeed.

The liver is the body's principal metabolic factory, turning foodstuffs arriving from the digestive tract in the bloodstream into substances that are used by the different cells of the body. The liver is the largest internal organ of the body. In an adult human being the liver weighs about 1.2 kilograms and is the size of a football. It carries out a wide variety of metabolic functions, many of which we shall discuss in later chapters of this book. It supplies quick energy, metabolizes alcohol, makes proteins, stores vitamins and minerals, regulates blood clotting, regulates the production of cholesterol, and detoxifies poisons. In addition, the liver also secretes cholesterol and the phospholipid lecithin, both of which play a role in rendering fat soluble.

CONCENTRATION OF SOLIDS: THE LARGE INTESTINE

The large intestine, or **colon,** is much shorter than the small intestine, occupying approximately the last meter of the intestinal tract. No digestion takes place within the large intestine, and only about 4% of the absorption of fluids by the body occurs there. The large intestine is not convoluted, lying instead in three relatively straight seg-

FIGURE 34-11

Although no vertebrate produces the enzyme cellulase, some are able to achieve the digestion of cellulose by using bacteria that live within their digestive tracts to produce this necessary enzyme. Cows and related mammals have a special digesting pouch called a rumen in which such bacteria live. It is located in front of the stomach and is one of four separate digestive chambers. Such mammals regurgitate and rechew the contents of their rumens ("chewing the cud"), which helps them to digest the cellulose in the food they consume far more efficiently than in other vertebrates, which lack a rumen.

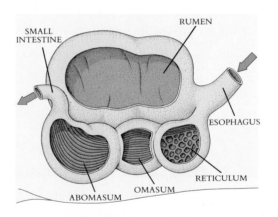

ments, and its inner surface does not possess villi. Consequently, the large intestine has less than one-thirtieth the absorptive surface area of the small intestine. Although sodium, vitamin K, and some other products of bacterial metabolism are absorbed across its wall, the primary function of the large intestine is to act as a refuse dump. Within it, undigested material, primarily bacterial fragments and cellulose (Figure 34-11), is compacted and stored. Many bacteria live and actively divide within the large intestine; the excess bacteria are incorporated into **feces.** Bacterial fermentation produces gas within the colon at a rate of about 500 milliliters per day. This rate increases greatly after the consumption of beans or other vegetable matter because the passage of undigested plant material into the large intestine provides material for fermentation.

> **The large intestine serves primarily to compact the solid refuse remaining after digestion of food, thus facilitating its elimination.**

Like all good things, digestion eventually comes to an end. In this case the end is a short extension of the large intestine called the **rectum.** Compacted solids within the colon pass into the rectum as a result of the peristaltic contractions of the muscles encasing the large intestine. From the rectum, the solid material passes out of the anus through two **anal sphincters.** The first of these is composed of smooth muscle; it opens involuntarily in response to a pressure-generated nerve signal from the rectum. The second sphincter, in contrast, is composed of striated muscle. It is subject to voluntary control from the brain, thus permitting a conscious decision to delay defecation.

NUTRITION

The ingestion of food by vertebrates serves two ends: it provides a source of energy, and it also provides raw materials that the animal is not able to manufacture for itself (Figure 34-12). The liver maintains a very constant level of glucose in the blood and stores several hours' reserve of glucose in the form of glycogen. Any intake of food in

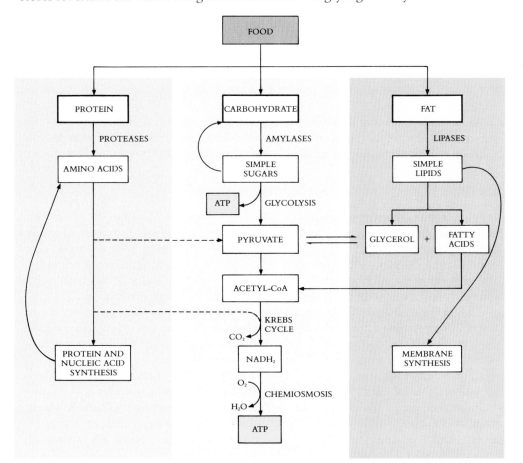

FIGURE 34-12

The fate of food. Most food can be broadly considered to consist of carbohydrate, fat, or protein. During digestion, protein is broken down into amino acids, carbohydrate is broken down into simple sugars, and fats are absorbed, often without degradation. The resulting sugars and lipids may be considered members of a common pool of metabolites, since each may be converted to the other. Amino acids form a distinct metabolic pool—although there is some interconversion of amino acids and sugars, the amount is relatively small. In general, amino acids are channeled into synthesis of nitrogen-containing molecules such as proteins, hormones, and nucleotides, and the carbohydrates and fats are used to store energy and assemble structural components of the cell. (Modified from A. Vander et al., *Human Physiology,* third edition, McGraw-Hill.)

DANGEROUS EATING HABITS

In the United States, serious eating disorders have become much more common since the mid-1970s. The most frequent of these are **anorexia nervosa,** a condition in which the afflicted person literally starves himself or herself; and **bulimia,** a condition in which individuals gorge themselves and then cause themselves to vomit, so that their weight stays constant. For reasons that we do not understand, 90% to 95% of those suffering from these eating disorders are female, and researchers estimate that 5% to 10% of the adolescent girls and young women in the United States suffer from eating disorders. As many as one of five female high school and college students may suffer from bulimia.

Those who suffer from anorexia are typically shy, well-behaved young women who feel embarrassed about their bodies. Eventually, they often begin to show signs of severe malnutrition and abnormally low temperatures and pulse rates. Although malfunction of the pituitary gland may

produce similar symptoms, anorexia nervosa is now understood to be a psychiatric disturbance that has severe psychic and psychological consequences. Its medical consequences may be severe enough to result in death.

Bulimia is usually accompanied by severe feelings of guilt, depression, and a feeling of anxiety and helplessness. The eating binges that are characteristic of bulimia often involve large quantities of carbohydrate-rich junk food. The accompanying medical consequences result from the frequent purging, which reduces the body's supply of potassium. Potassium plays an important role in regulating body fluids; its loss may lead to muscle weakness or paralysis, an irregular heartbeat, and kidney disease. Like anorexia nervosa, bulimia often leads to decreased resistance to infection.

Although both anorexia nervosa and bulimia are now believed to be psychiatric disorders, scientists are continuing to search actively for

physiological explanations for these serious conditions. Changes in the levels of certain hormones are associated with these conditions, but whether these changes cause or result from the conditions has not been demonstrated clearly. In treating these conditions, it is essential that the shame that is often associated with them be recognized as a problem, and that professional help be sought promptly. Both a psychiatrist or a psychologist and a physician should be involved, and the family as a whole must play a strong, supportive role for the treatment to be effective. Even with effective treatment, both conditions may persist for years; it is therefore essential that they be understood. Parents, teachers, and others should be alert to the existence of anorexia and bulimia and should provide the kind of education about food and diet that, coupled with a supportive environment, will lead to the early detection and cure of these eating disorders if they do occur.

---

excess of that required to maintain the glycogen reserve results in one of two consequences. Either the excess glucose is metabolized by the muscles and other cells of the body, or it is converted to fat and stored within fat cells. Fats represent a very efficient way to store energy. Thus we have the simple equation:

$$\text{FOOD} - \text{EXERCISE} = \text{FAT}$$

In wealthy countries such as those of North America and Europe, the obesity that results from chronic overeating and from imbalanced diets high in carbohydrates is a significant human health problem. In the United States, about 30% of middle-aged women and 15% of middle-aged men are classified as overweight, weighing at least 20% more than the average weight for their height (Figure 34-13). Being overweight is strongly correlated with coronary heart disease and many other disorders.

Over the course of their evolution, many vertebrates have lost the ability to synthesize different substances that nevertheless continue to play critical roles in their metabolism. Substances that an animal cannot manufacture for itself but that are necessary for its health must be obtained in other ways—in its diet. When essential organic substances are used in trace amounts, they are called **vitamins.** Human beings, monkeys, and guinea pigs, for example, have lost the ability to synthesize ascorbic acid (vitamin C) and will develop the disease **scurvy**—a disease characterized by weakness, spongy gums, and bleeding of the skin and mucus membranes, which can ultimately prove fatal—if vitamin C is not supplied in sufficient quantities in their diets. All other mam-

OBESITY STARTS HERE*			
*FOR ADULTS OF MEDIUM BUILD BETWEEN THE AGES OF 25 AND 59, INCLUDING CLOTHES AND ALLOWING FOR 1″ HEELS. (BASED ON THE 1983 METROPOLITAN LIFE TABLES.)			
MEN		WOMEN	
HEIGHT	WEIGHT	HEIGHT	WEIGHT
5′6″	174	5′2″	150
5′7″	178	5′3″	154
5′8″	181	5′4″	157
5′9″	185	5′5″	161
5′10″	188	5′6″	164
5′11″	192	5′7″	168
6′0″	196	5′8″	172
6′1″	200	5′9″	175
6′2″	205	5′10″	179

FIGURE 34-13

Obesity is usually characterized as the state of being more than 20% heavier than the average person of the same sex and height. For a variety of heights, these are the weight values at which obesity begins in Americans, using average 1985 weights.

mals, as far as is known, are able to synthesize ascorbic acid. Vitamin K is obtained by mammals from the bacteria that are symbiotic in their intestines, but birds must obtain it from their food. Human beings require at least 13 different vitamins (Table 34-1).

A vitamin is an organic substance that is required in minute quantities by an organism for growth and activity, but that the organism cannot synthesize.

Some of the substances that vertebrates are not able to synthesize are required in more than trace amounts. Many vertebrates, for example, require one or more of the 20 amino acids that are necessary to synthesize proteins. Human beings are unable to synthesize 8 of these 20 amino acids: lysine, tryptophan, threonine, methionine, phenylalanine, leucine, isoleucine, and valine. These amino acids, called **essential amino acids,** must be obtained by humans from proteins in the food they eat (Figure 34-14). All vertebrates have also lost the ability to synthesize certain polyunsaturated fats that provide backbones for fatty acid synthesis. Some essential substances that vertebrates do synthesize for themselves cannot be manufactured by the members of other animal groups. Some carnivorous insects, for example, are unable to synthesize the cholesterol that is required for the synthesis of steroid hormones; they must therefore obtain cholesterol in their diet. Other insects have the ability to convert other steroids, which they consume with their food, to cholesterol.

In addition to energy and those organic compounds that cannot be synthesized, the food that an animal consumes must also supply essential minerals such as calcium and phosphorus. It must also include a wide variety of **trace elements,** which are minerals that are required in very small amounts. Among the trace elements are iodine (a component of thyroid hormone), cobalt (a component of vitamin B_{12}), zinc and molybdenum (components of enzymes), manganese, and selenium. All of these, with the possible exception of selenium, are essential for plant growth also; they are obtained by the animals that require them either directly from plants or from animals that have eaten the plants.

Interestingly, one essential characteristic of food is simply bulk, its content of undigested fiber. The large intestine of humans, for example, has evolved as an organ adapted to process food that has a relatively high fiber content. Diets that are low in fiber, which are common in the United States, result in a slower passage of food through the colon than is desirable. This low dietary fiber content is thought to be associated with the levels of colon cancer in the United States, levels that are among the highest in the world.

TABLE 34-1   MAJOR VITAMINS

VITAMIN	FUNCTION	DIETARY SOURCE	RECOMMENDED DAILY ALLOWANCE (milligrams)	DEFICIENCY SYMPTOMS	SOLUBILITY
Vitamin A (retinol)	Used in making visual pigments, maintenance of epithelial tissues	Green vegetables, milk products, liver	1	Night blindness, flaky skin	Fat
B-COMPLEX VITAMINS					
B_1	Coenzyme in CO_2 removal during cellular respiration	Meat, grains, legumes	1.5	Beriberi, weakening of heart, edema	Water
B_2 (riboflavin)	Part of coenzymes FAD & FMN, which play metabolic roles	In many different kinds of foods	1.8	Inflammation and breakdown of skin, eye irritation	Water
B_3 (niacin)	Part of coenzymes NAD^+ and $NADP^+$	Liver, lean meats, grains	20	Pellagra, inflammation of nerves, mental disorders	Water
B_5 (pantothenic acid)	Part of coenzyme-A, a key connection between carbohydrate and fat metabolism	In many different kinds of foods	5 to 10	Rare: fatigue, loss of coordination	Water
B_6 (pyridoxine)	Coenzyme in many phases of amino acid metabolism	Cereals, vegetables, meats	2	Anemia, convulsions, irritability	Water
B_{12} (cyanocobalamin)	Coenzyme in the production of nucleic acids	Red meat, dairy products	0.003	Pernicious anemia	Water
Biotin	Coenzyme in fat synthesis and amino acid metabolism	Meat, vegetables	Minute	Rare: depression, nausea	Water
Folic acid	Coenzyme in amino acid and nucleic acid metabolism	Green vegetables, whole	0.4	Anemia, diarrhea	Water
Vitamin C	Important in forming collagen, cement of bone, teeth, connective tissue of blood vessels; may help maintain resistance to infection	Fruit, green leafy vegetables	45	Scurvy, breakdown of skin, blood vessels	Water
Vitamin D (calciferol)	Increases absorption of calcium and promotes bone formation	Dairy products, cod liver oil	0.01	Rickets, bone deformities	Fat
Vitamin E (tocopherol)	Protects fatty acids and cell membranes from oxidation	Margarine, seeds, green leafy vegetables	15	Rare	Fat
Vitamin K	Essential to blood clotting	Green leafy vegetables	0.03	Severe bleeding	Fat

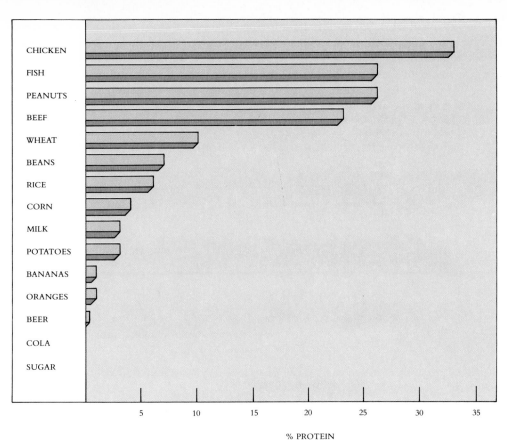

FIGURE 34-14

The protein content of a variety of common foods.

The vertebrate digestive systems that we have described in this chapter are much larger and more complex than those of the invertebrates discussed in Chapter 28. In the vertebrate systems, the working surfaces are convoluted and therefore much more extensive, and the different portions of the system are specialized for particular functions in the digestive process. Nutritional specialization in the mammals has likewise become complex, and has led to many of the problems that humans face in selecting an appropriate diet. In the next chapter, we shall consider the way in which vertebrates acquire oxygen, another critical component of their systems for energy use and release.

SUMMARY

1. Digestion is the rendering of parts of organisms into amino acids and sugars, which can be metabolized by heterotrophic organisms.

2. The digestive tract of vertebrates is one-way. The initial portion leads from a mouth through an esophagus to a stomach.

3. In most mammals, the stomach juices are concentrated acid, in which the protein-digesting enzyme pepsin is active.

4. Food passes from the stomach to the small intestine, where the pH is neutralized and a variety of enzymes synthesized in the pancreas act to complete digestion. Most digestion occurs in the first 25 centimeters of the small intestine, in a zone called the duodenum.

5. The products of digestion are absorbed across the walls of the small intestine, which possess numerous villi and so achieve a very great surface area. Amino acids and sugars are transported by specific transmembrane channels, whereas fatty acids, which are lipid soluble, cross passively across the membranes of the villi. In the process of digestion, fats need to be made soluble by detergent molecules secreted by the liver.

6. The large intestine has little digestive or absorptive activity; it functions principally to compact the refuse that is left over from digestion for easier elimination.

7. Vertebrates lack the enzymes necessary to synthesize many necessary compounds and must obtain these enzymes, which are called vitamins, from their diet. A number of trace elements must also be present in the diet.

REVIEW

1. _____ is the process of breaking down the macromolecules in food into simple compounds.

2. A band of muscle called the _____ controls the entrance to the duodenum from the stomach.

3. The very numerous _____ that cover the epithelial wall of the small intestine greatly increase the surface area for absorption of digested foods.

4. If food energy is not metabolized by muscle or other body cells, it is stored as _____.

5. Which vitamin do you get from your symbiotic intestinal bacteria?

SELF-QUIZ

1. Starchy foods such as potatoes begin being digested in the
   (a) mouth
   (b) esophagus
   (c) stomach
   (d) duodenum
   (e) small intestine

2. Protein-rich foods, such as a steak, are broken down in the stomach by the combined action of (pick two)
   (a) low pH
   (b) high pH
   (c) amylase
   (d) pepsin
   (e) bile

3. Most of the enzymes that complete the breakdown of food items into simple sugars, amino acids, and fatty acids are secreted by which of the following?
   (a) salivary glands
   (b) gastric pits
   (c) liver
   (d) pancreas
   (e) large intestine

4. The main function of the large intestine is
   (a) digestion
   (b) absorption
   (c) secretion
   (d) fermentation
   (e) compaction

5. Which of the following is an essential amino acid?
   (a) alanine
   (b) leucine
   (c) cystine
   (d) asparagine
   (e) proline

THOUGHT QUESTIONS

1. Human beings obtain vitamin K from symbiotic bacteria living in their gastrointestinal tract. Many bacteria also produce ascorbic acid, vitamin C. Can you suggest a reason why people have not evolved a symbiotic relationship with bacteria that would result in their obtaining bacterial vitamin C?

2. Many digestive enzymes are synthesized in the pancreas and released into the duodenum. Since this is the case, why do mammalian digestive systems go to the trouble of producing pepsin in the stomach? What does pepsin do that the others do not?

FOR FURTHER READING

DeGABRIELE, R.: "The Physiology of the Koala," *Scientific American,* July 1980. These Australian marsupials are adapted to a highly specific and unusual diet.

MOOG, F.: "The Lining of the Small Intestine," *Scientific American,* November 1981, pages 154–176. A clear description of the most important absorptive surface in the human body.

SCRIMSHAW, N.S., and V.R. YOUNG: "The Requirements of Human Nutrition," *Scientific American,* September 1976, pages 50–64. If you want to know what you should eat and what perhaps you should not, this article will point the way. A particularly good treatment of the important roles of trace elements in the human diet.

WOLIN, M.: "Fermentation in the Rumen and Human Large Intestine," *Science,* vol. 213, pages 1463–1468, 1981. The intestines of cows, humans, and other vertebrates are teeming with bacteria.

FIGURE 35-1   This bald eagle, like most humans, is a meat eater, and like us it must breathe in air to obtain the oxygen necessary to metabolize what it eats. That is what the hole in the beak is for.

HOW ANIMALS CAPTURE OXYGEN

Chapter 35

Overview

Oxygen enters the bodies of animals by diffusion into water. The evolution of respiratory mechanisms among the vertebrates has favored changes that maximize the rate of this diffusion. The most efficient aquatic mechanism to evolve is the gill of bony fishes; the most efficient aerial respiration mechanism is the two-cycle lung of birds. Both of these structures achieve high efficiency by countercurrent flow. In the vertebrates, hemoglobin plays a critical role in the transport of oxygen and carbon dioxide in the respiration process.

For Review   *Here are some important terms and concepts that you will encounter in this chapter. If you are not familiar with them, you should review them before proceeding.*

Chemistry of carbon dioxide (Chapter 3)

Diffusion (Chapter 6)

Oxidative respiration (Chapter 8)

Adaptation of vertebrates to terrestrial living (Chapters 19 and 28)

Red blood cells (Chapter 32)

Human beings, like all other animals, obtain carbon compounds by consuming other organisms and then metabolize these carbon compounds to obtain the energy they use to move, to grow, and to think (Figure 35-1). The biochemical mechanism of this **oxidative metabolism** was discussed in Chapter 8. Basically, animals obtain their energy by removing electrons from carbon compounds they eat and then using these electrons to drive a series of proton-pumping channels in the membrane to generate ATP. Afterwards, the electrons are donated to oxygen gas (O_2), which combines with hydrogen (H^+) to form water (H_2O). The carbon atoms that are left over after the electrons have been stripped off combine with oxygen and are released as carbon dioxide (CO_2). In most vertebrates, the water produced as a result of this metabolism is simply diluted into the much larger volume of the body's internal fluid: metabolism is in effect a process that utilizes oxygen and produces carbon dioxide. This final balance sheet determines one of the principal physiological challenges facing all animals—how to obtain oxygen and dispose of carbon dioxide. The uptake of oxygen and the release of carbon dioxide together are called **respiration.**

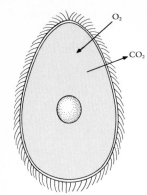

A SINGLE-CELLED ORGANISM

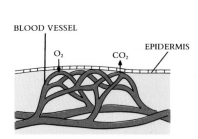

B TRANSCUTANEOUS RESPIRATION

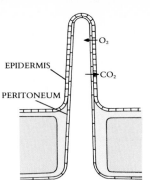

C PAPULA OF ECHINODERM

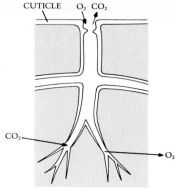

D INSECT TRACHEAL SYSTEM

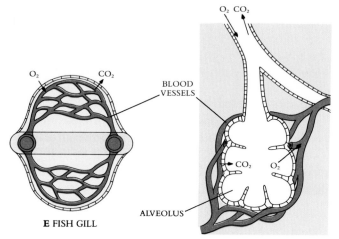

E FISH GILL

F LUNG WITH ALVEOLI

FIGURE 35-2

Gas exchange in animals may take place in a variety of ways.
A   Gases diffuse directly into single-celled organisms.
B   Amphibians and many other multicellular organisms respire across their skin.
C   Echinoderms have protruding papulae, which provide an increased respiratory surface.
D   Insect respire through spiracles, openings in their cuticle.
E   The gills of fishes provide a very large respiratory surface and employ countercurrent exchange.
F   Mammalian lungs provide a large respiratory surface but do not permit countercurrent exchange.

THE EVOLUTION OF RESPIRATION

Animals do not capture oxygen from their environments actively; to do so would require the expenditure of more energy than the animals would obtain by utilizing the captured oxygen. For similar reasons, a working mother doesn't pay the babysitter more than the mother's own salary. In all animals, the capture of oxygen and discharge of carbon dioxide are passive processes, the gases moving into and out of cells by diffusion. The force that drives the movement of the gases is the difference in oxygen concentration between the interior of the organism and its exterior environment.

Oxygen diffuses slowly. The levels of oxygen required by oxidative metabolism in most organisms cannot be obtained by diffusion alone over distances greater than about 0.5 millimeter. This factor severely limits the size of organisms that obtain their oxygen entirely by diffusion from the environment into the metabolizing cytoplasm. Single-celled protists are small enough that diffusion distance presents no problem; but as the size of an organism increases, the problem soon becomes significant.

Major changes in the mechanism of respiration have occurred during the evolution of animals (Figure 35-2). In general, these changes have tended to optimize the rate of diffusion by (1) increasing the surface area over which diffusion takes place; (2) decreasing the thickness of tissue through which the gas must pass to reach the interior of the organism; and (3) increasing the concentration difference between the organism and its environment.

Creating a Water Current

Most of the more primitive animal phyla possess no special respiratory organs. The sponges (phylum Porifera), cnidarians (phylum Cnidaria), many flatworms (phylum

Platyhelminthes) and roundworms (phylum Nematoda), and some annelids (phylum Annelida) all obtain their oxygen by diffusion directly from surrounding water. How do they overcome the limits imposed by diffusion? By beating with their cilia, these organisms create a water current, by means of which they continuously replace the water over the diffusion surface. Because of this continuous replenishment with water containing fresh oxygen, *the exterior oxygen concentration does not decrease as diffusion proceeds.* Although each oxygen gas molecule that passes into the organism has been removed from the surrounding volume of water, the exterior oxygen concentration does not fall, because a new volume of water with a higher oxygen concentration is constantly replacing the one from which the oxygen has been removed.

Increasing the Diffusion Surface

Most of the more advanced invertebrates (mollusks, arthropods, echinoderms), as well as the vertebrates, possess special respiratory organs that both increase the surface area available for diffusion and reduce the thickness of the tissue separating the internal fluid from the surroundings (see Figure 35-2). These organs are of two kinds: (1) those that facilitate exchange with water and (2) those that facilitate exchange with air. As a rough rule of thumb, aquatic respiratory organs increase the diffusion surface by extensions of tissue, called **gills,** that project from the body out into the water. Atmospheric respiratory organs, on the other hand, involve invaginations into the body; in terrestrial vertebrates, the respiratory organs are internal sacs called **lungs.**

Perhaps the simplest of the respiratory organs are the tracheae of arthropods (Chapter 28). Tracheae are extensive series of passages connecting the surface of the animal to all portions of its body (see Figure 35-2, *D*). Oxygen diffuses from these passages directly to the cells, without the intervention of an active circulation system. Piping air directly to the cells in this manner works very well in organisms such as insects, which have small bodies relative to those of vertebrates, since air must move only a relatively short distance within their bodies. This relationship, however, severely limits the potential body size of organisms that obtain their oxygen in this way.

THE GILL AS AN AQUEOUS RESPIRATORY MACHINE

By far the most successful aqueous respiratory organ, the gill, evolved among the bony fishes. In these animals, water passes through the mouth into two cavities, which are situated behind the mouth on each side of the head. From these cavities the water passes back out of the body. The gills hang like curtains between the mouth and the entrance to each of the cavities.

Many fishes that swim continuously, such as tuna, have practically immobile covers over their gill cavities. They swim with their mouths partly open, forcing water constantly over the gills, a process that amounts to a form of **ram ventilation.** Most bony fishes, however, have flexible gill covers that permit a pumping motion using muscles. In such gills, there is an uninterrupted one-way flow of water over the gills even when the fish is not swimming (Figure 35-3).

FIGURE 35-3

How a fish breathes. The gills are suspended between the mouth cavity and the outside, under a cover called the gill cover, which seals off the opening when pressed down. Breathing occurs in two stages.
A When the oral valve of the mouth is open, closing the gill cover increases the volume of the mouth cavity so that water is drawn in.
B When the oral valve is closed, opening the gill cover decreases the volume of the mouth cavity, forcing water out over the gills to the outside.

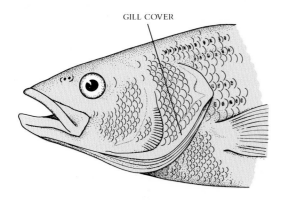

GILL COVER

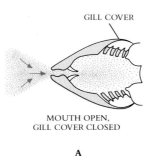

GILL COVER

MOUTH OPEN,
GILL COVER CLOSED

A

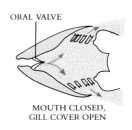

ORAL VALVE

MOUTH CLOSED,
GILL COVER OPEN

B

FIGURE 35-4

Structure of a fish gill. Several rows of gills lie beneath the gillcover on each side of a fish's head **(A).** A gill is composed of two rows of filaments **(B),** each of which bears rows of thin, disk-like lamellae **(C).** Water passes from the gill arch out over the filaments (from left to right in the diagram). Water always passes the lamellae in the same direction—which is opposite to the direction the blood circulates across the lamellae. The success of the gill's operation depends critically on the opposite orientation of blood and water flow.

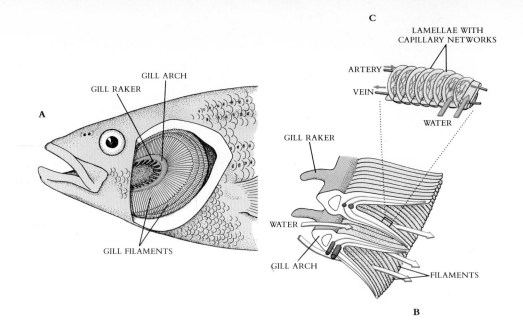

In addition to maintaining a high diffusion rate by providing a continuous flow of water, the gills of fishes are constructed in such a way that they actually maximize the concentration difference between tissue and environment. To understand how this is done, we must look carefully at the structure of a gill. Each gill is composed of thin, membranous gill filaments that project out into the flow of water (Figure 35-4). Within each filament are rows of thin, disklike lamellae arrayed parallel to the direction of water movement. Water flows over these lamellae from front to back. Here is the key: within each lamella, the blood circulation is arranged so that *blood is carried in the direction opposite to the movement of the water,* from the back of the lamella to the front.

Because the water flowing over the lamellae and the blood flowing within lamellae run in opposite directions, the concentration difference is maximized. At the back of the gill the least oxygenated blood meets the least oxygenated water and is able to remove oxygen from the water. By the time the blood reaches the front of the gill it has acquired a lot of oxygen, but it is able to acquire still more oxygen by diffusion from the water entering the gill. The reason that it can do this is because the new water entering the front of the lamellae is richer in oxygen than the water that has already flowed past the gills and lost some of its oxygen. This kind of **countercurrent flow** (Figure 35-5) ensures a continuous gradient of concentration, and diffusion continues to occur all along the gill.

If the flows of water and blood had been in the same direction, the concentration difference would have been high initially, as the oxygen-free blood met the new water that was entering. The concentration difference would have fallen rapidly, however, as the water lost oxygen to the blood. Since the blood oxygen concentration rises as the water oxygen falls, much of the oxygen in the water would remain there when the blood and water concentrations became equal and diffusion ceased. In countercurrent flow, in contrast, the blood oxygen level encountered by the water becomes lower and lower as the level of oxygen in the water falls. The result is that, in countercurrent flow, the blood can attain concentrations of oxygen as high as those which exist in the water entering the gills. Fish gills are the most efficient respiratory machines that occur among organisms. Gills are able to maximize the rate of diffusion in an oxygen-poor medium, obtaining as much as 85% of the available oxygen.

The gill is the most efficient of all respiratory organs. Its great efficiency derives from the countercurrent flow of water past the blood vessels of the gills.

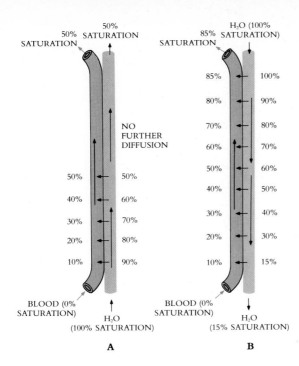

FIGURE 35-5

Countercurrent exchange.
A When blood and water flow in the same direction, dissolved oxygen gas can diffuse from the water into the blood rapidly at first, because of the large concentration difference (0% in blood versus 100% in water), but the difference decreases as more oxygen diffuses from water into blood, until finally the concentrations of oxygen in water and blood are equal. At this point there is no concentration difference to drive any further diffusion. In this example, blood can obtain no more than 50% dissolved oxygen in this fashion.
B When blood and water flow in opposite directions, the initial concentration difference between water and blood is not as great (0% in blood versus 15% in water) but is sufficient for diffusion to occur from water to blood. As more oxygen diffuses into the blood, raising its oxygen concentration, the blood encounters water with higher and higher oxygen concentrations; at every point, the oxygen concentration is higher in the water than in the blood, so that diffusion continues. In this example, the blood reaches an 85% saturation level for dissolved oxygen.

FROM AQUATIC TO ATMOSPHERIC BREATHING: THE LUNG

Although aquatic animals were the first to evolve, their descendants have successfully invaded the land many times. On land, the respiratory challenge is very different than the one that exists in water.

Water is relatively poor in dissolved oxygen; it contains only 5 to 10 milliliters of oxygen per liter of water. Air, in contrast, is rich in oxygen, containing about 210 milliliters of oxygen per liter of air. Not surprisingly, many members of otherwise aquatic groups utilize atmospheric air as a source of oxygen; these include many mollusks, crustaceans, and fishes.

When organisms first became fully terrestrial, the air became the source of their oxygen. An entirely new respiratory apparatus evolved, one that was based on internal passages rather than on gills. Why were gills not maintained in terrestrial organisms, in view of the fact that they are such superb oxygen-capturing mechanisms? Gills were lost for two principal reasons:

1. Air is less buoyant than water. Because the fine, membranous lamellae of gills lack structural strength, they must be supported by water in order to avoid collapsing on one another. A fish out of water, although awash in oxygen, soon suffocates because its gills collapse into a mass of tissue. This collapse greatly reduces the diffusion surface of the gill.

2. Water diffuses into air through the process of **evaporation.** Atmospheric air is rarely saturated with water vapor, except immediately after a rainstorm. Consequently, organisms that live in air are constantly losing water to the atmosphere. Gills would have provided an enormous surface for water loss.

Two main systems of internal oxygen exchange evolved among terrestrial organisms. One was the tracheae of insects mentioned earlier, and the other was the lung. Both systems sacrifice respiratory efficiency in order to maximize water retention. Insects prevent excessive water loss by closing the external openings of the tracheae whenever possible. They do this whenever body carbon dioxide levels are below a certain level.

Lungs, in contrast, minimize the effects of drying out by eliminating the one-way flow of oxygen that was such an effective means of increasing the efficiency of aquatic respiratory systems. In organisms that respire by means of lungs, the air moves into the lung through a tubular passage and then back out again via the same passage.

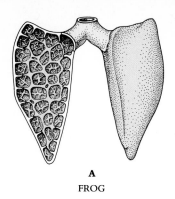

A
FROG

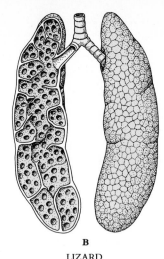

B
LIZARD

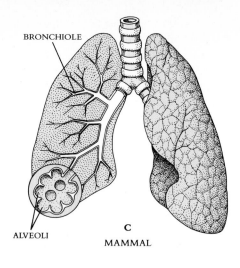

BRONCHIOLE

ALVEOLI

C
MAMMAL

FIGURE 35-6

Evolution of the vertebrate lung.
A The amphibian lung.
B The reptile lung.
C The mammalian lung. An enlarged group of alveoli is shown to represent these structures, which are very numerous.

When each breath is completed, the lung still contains a volume of air, the **residual volume.** In human beings this volume is about 1200 milliliters. Each inhalation adds from 500 milliliters (resting) to 3000 milliliters (exercising) of additional air. Each exhalation removes approximately the same volume as inhalation added, reducing the air volume in the lung once more to about 1200 milliliters. Because the diffusion surfaces of the lungs are not exposed to fully oxygenated air, but rather to a mixture of fresh and partly depleted air, the concentration difference is far from maximal, and the respiratory efficiency of lungs is much less than that of gills. Oxygen capture is lessened by this two-way flow of air—but so is water loss.

Evolution of the Lung

There is so much more oxygen in air than in water that low respiratory efficiency does not appear to have presented a critical problem to early land-dwellers; the amphibian lung is hardly more than a sac with a convoluted internal membrane (Figure 35-6, *A*). Because the inner membrane surface of the lung is convoluted, the surface area of amphibian lungs available for diffusion is great, although it is still not large enough to provide all the oxygen necessary for the organism. Consequently, amphibians obtain much of their oxygen by diffusion through their moist skin.

Reptiles are far more active than amphibians, and they have significantly greater metabolic demands for oxygen. The early reptiles could not rely on their skins for respiration; living fully on land, they are "watertight," avoiding desiccation by possessing a dry scaly skin. Again, as in aquatic organisms, the respiratory apparatus has changed in ways that tend to optimize respiratory efficiency. The lungs of reptiles possess many small chambers within their surface called **alveoli,** which are clustered together like a bunch of grapes (Figure 35-6, *B*). The alveoli greatly increase the diffusion surface of the lung.

The metabolic demands for oxygen became even greater with the evolution of birds and mammals, which unlike reptiles and amphibians maintain a constant body temperature by heating their bodies metabolically. The lungs of mammals contain many clusters of alveoli (Figure 35-6, *C*; see also Figure 35-8). Humans, for example, have about 300 million alveoli in their two lungs. The increase in the number of alveoli enlarged yet again the total diffusion surface of the lung. In humans, the total surface devoted to diffusion can be as much as 80 square meters, an area about 42 times the surface area of the body.

In amphibians, one-way flow through the respiratory organ was abandoned in favor of a saclike lung. Increases in efficiency among the reptiles and mammals have been achieved by enlargement of the lung's internal surface area.

There is a limit to the improvements that can be made by increases in the diffusion surface of the lung, a limit that is probably approached by the more active mammals. With the advent of birds, flying introduced respiratory demands that exceeded the capacity of a saclike lung. Many birds rapidly beat their wings for prolonged periods during flight; such rapid wing movement uses up a lot of energy quickly because it depends on the frequent contraction of wing muscles. Flying birds must carry out intensive oxidative respiration within their cells to replenish the ATP expended by contracting flight muscles. They thus require a great deal of oxygen, more oxygen than a saclike lung, even one with a large surface such as a mammalian lung, is capable of delivering. The lungs of birds cope with the demands of flight by employing a new respiratory mechanism, one that produces a significant improvement in respiratory efficiency.

An avian lung works like a two-cycle pump (Figure 35-7). When a bird inhales air, the air passes directly to a nondiffusing chamber called the **posterior air sac.** When the bird exhales, the air flows into a lung. On the following inhalation, the air passes from the lung to a second air sac, the **anterior air sac.** Finally, on the second exhalation, the air flows from the anterior air sac out of the body. What is the advantage of this complicated passage? It creates a unidirectional flow of air through the lungs! Thus there is no dead volume as in the mammalian lung, and the air passing through the lung of a bird is always fully oxygenated. Correspondingly, the flow of blood past the avian lung runs in the direction opposite to the air flow in the lung, as in the gills of a fish. The oxygenated blood leaving the lung can thus contain more oxygen than exhaled air, a capacity not achievable by mammalian lungs.

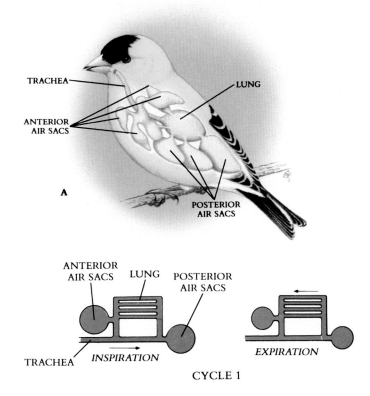

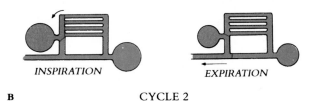

FIGURE 35-7

How a bird breathes.
A The respiratory system of a bird is composed of anterior air sacs, lungs, and posterior air sacs.
B Breathing occurs in two cycles. *Cycle one:* Air is drawn from the trachea into the posterior air sacs and then is exhaled through the lungs. *Cycle 2:* Air is drawn from the lungs into the anterior air sacs and then is exhaled through the trachea. Passage of air through the lungs is always in the same direction, from posterior to anterior (right to left in this diagram). Because blood circulates in the lung from anterior to posterior, the lung achieves countercurrent flow and thus is very efficient at picking up oxygen from the air.

FIGURE 35-8

The human respiratory system
(A). Air passes through the
nostrils or mouth via the trachea
into the bronchi, bronchioles, and
alveoli **(B),** where it comes into
contact with the blood vessels.

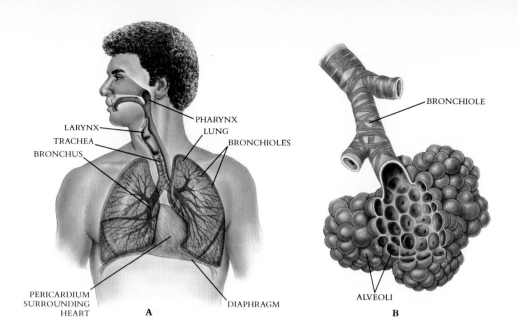

THE MECHANICS OF HUMAN BREATHING

Humans possess a pair of lungs located in the chest, or **thoracic,** cavity. The two lungs hang free within the cavity, being connected to the rest of the body only at the one position where the lung's blood vessels and air tube enter (Figure 35-8). This air tube is called a **bronchus.** It connects each lung to a long tube called the **trachea,** which passes upward in the body, past the voice box, or **larynx,** and opens into the rear of the mouth (Figure 35-9). Air normally enters through the nostrils, where it is warmed. In addition, the nostrils are lined with hair that filter out dust and other particles. As the air passes through the nasal cavity, an extensive array of cilia on its epithelial lining further filters the air and moistens it (Figure 35-10). The air then passes through the back of the mouth, crossing the path of food as it enters first the larynx and then the trachea. From there it passes down through the bronchus to the lungs.

The human respiratory apparatus is simple in structure, functioning as a one-cycle pump. The thoracic cavity is bounded on its sides by ribs, which are capable of flexing, and on the bottom by a thick layer of muscle, the **diaphragm** (see Figure 35-8), which separates the thoracic cavity from the abdominal cavity. Each lung is covered by a very thin, smooth membrane called the **pleural membrane.** A second pleural membrane marks the interior boundary of the thoracic cavity, into which the lungs hang. Within the cavity the weight of the lungs is supported by water, the **intrapleural fluid.**

The intrapleural fluid not only supports the lungs, but also plays another important role by permitting an even application of pressure to all parts of the lung. You can visualize the pleural membranes as a system of two balloons of different sizes, one nested inside the other, with the space between them completely filled with water. The inner balloon opens out to the atmosphere (Figure 35-11), so that two forces act on this inner balloon: air pressure from the atmosphere pushes it outward, and water pressure from the intrapleural fluid pushes it inward.

During **inhalation,** the walls of the chest cavity expand. The rib cage moves outward and upward, and the diaphragm moves downward by stretching taut. In effect, we have enlarged the outer balloon by pulling it in all directions. This expansion of the fluid space causes the fluid pressure to decrease to a level less than that of the air pressure within the inner balloon. As a result, the wall of the inner balloon is pushed out. As the inner balloon expands, its internal air pressure decreases, and air moves in from the atmosphere. During **exhalation,** the ribs and diaphragm return to their original resting position. In doing so, they exert pressure on the fluid. This pressure is transmitted uniformly by the fluid over the entire surface of the lung (the inner balloon), forcing air from the inner cavity back out to the atmosphere.

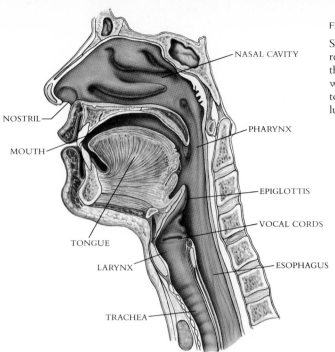

NASAL CAVITY

NOSTRIL

MOUTH

PHARYNX

EPIGLOTTIS

VOCAL CORDS

TONGUE

LARYNX

ESOPHAGUS

TRACHEA

FIGURE 35-9

Side view of the human upper respiratory tract. Most air enters through the nostrils, moving by way of the pharynx past the larynx to the trachea and ultimately to the lungs.

FIGURE 35-10

Respiratory cilia such as these line the trachea. They moisten the air and aid its passage.

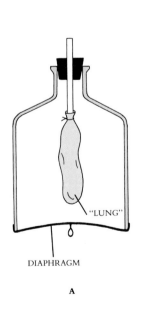

"LUNG"

DIAPHRAGM

A

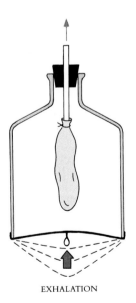

INHALATION

B

EXHALATION

C

FIGURE 35-11

This figure illustrates the way in which air is taken into and expelled from the lungs by means of diaphragm action. When the diaphragm is pulled down, as in the "lung" shown in **B,** the lungs expand; when it is relaxed, the lungs contract.

FIGURE 35-12

How a human breathes.
A Inhalation. The diaphragm and walls of the chest cavity expand, increasing its volume. As a result of the large volume, air is sucked in through the trachea. The ways in which the various muscles affect the process are indicated.
B Exhalation. The diaphragm and chest walls return to their normal positions, reducing the volume of the chest cavity and forcing air outward through the trachea. Muscles that contract during inhalation relax during exhalation.

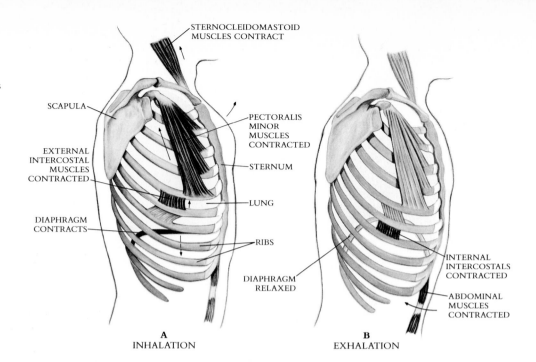

STERNOCLEIDOMASTOID MUSCLES CONTRACT

SCAPULA

PECTORALIS MINOR MUSCLES CONTRACTED

EXTERNAL INTERCOSTAL MUSCLES CONTRACTED

STERNUM

LUNG

DIAPHRAGM CONTRACTS

RIBS

DIAPHRAGM RELAXED

INTERNAL INTERCOSTALS CONTRACTED

ABDOMINAL MUSCLES CONTRACTED

A
INHALATION

B
EXHALATION

Breathing, the active cycle of inhalation and exhalation (Figure 35-12), thus depends on the regular contraction of the muscles surrounding the thoracic cavity. The rate of breathing is regulated by a respiratory center in the brainstem. Chemoreceptors in the arterial walls detect changes in carbon dioxide levels and transmit nerve impulses to the respiratory center, which sends appropriate signals to the muscles of the diaphragm and rib cage.

> **The active pumping of air in and out through the lungs is called breathing. The lungs function by suction during inhalation. The expansion of the chest cavity draws air into the lungs, and the return of the ribs to their resting position drives air from the lungs.**

HOW RESPIRATION WORKS: GAS TRANSPORT AND EXCHANGE

When oxygen has diffused from the air into the moist cells lining the inner surface of the lung, its journey has just begun. Passing from these cells into the bloodstream, the oxygen is carried throughout the body by the circulation system (to be described in Chapter 36). It has been estimated that it would take a molecule of oxygen 3 years to be transported from your lung to your toe if the transport depended only on diffusion, unassisted by a circulatory system.

Oxygen moves within the circulatory system on carrier proteins that bind dissolved molecules of oxygen. This binding occurs in the capillaries surrounding the alveoli of the lungs. The carrier proteins later release their oxygen molecules to metabolizing cells at distant locations in the body.

Hemoglobin and Gas Transport

The carrier protein that is used by all vertebrates is hemoglobin. **Hemoglobin** is a protein composed of four polypeptide subunits (see Figure 13-13); each of the four polypeptides is combined with a metal ion in such a way that oxygen can be bound reversibly to the ion. Hemoglobin is synthesized by erythrocytes (red blood cells) and remains within these cells, which circulate in the bloodstream like ships bearing cargo.

The higher the **partial pressure of oxygen**—the amount of air pressure caused by the oxygen content of the atmosphere—in the air within the lungs, the more hemoglobin in the blood that will combine with oxygen. At the oxygen partial pressures encountered in the blood supply of the lung, most hemoglobin molecules are saturated with oxygen. However, the gas carbon monoxide (CO), which is a component of automobile exhaust and a minor atmospheric component, binds to hemoglobin more readily than oxygen does. This is the reason that carbon monoxide poisoning can result in death. In more normal circumstances, however, carbon dioxide (CO_2) is the important gas, along with oxygen, that interacts with hemoglobin. In the presence of carbon dioxide, the hemoglobin molecule assumes a different shape, causing it to more easily give up its oxygen. This effect is of real importance, since CO_2 is produced by the tissues at the site of cell metabolism. For this reason, the blood unloads oxygen more readily to those tissues undergoing metabolism and generating CO_2.

At the same time that the red blood cells are unloading oxygen, they are also absorbing CO_2 from the tissue. Perhaps one fifth of the CO_2 that the blood absorbs is bound to hemoglobin. The remaining 80% of the carbon dioxide diffuses from the blood plasma into the cytoplasm of the red blood cells. There an enzyme, **carbonic anhydrase,** catalyzes the combination of CO_2 with water to form carbonic acid (H_2CO_3), which dissociates into bicarbonate (HCO_3^-) and hydrogen (H^+) ions. This process removes large amounts of CO_2 from the blood plasma, facilitating the diffusion of more CO_2 into it from the surrounding tissue. The facilitation is critical to CO_2 removal, since the difference in CO_2 concentration between blood and tissue is not large (only 5%).

The red blood cells carry their cargo of carbonate ions back to the lungs. The lower CO_2 concentration in the air inside the lungs causes the carbonic anhydrase reaction to proceed in the reverse direction, releasing gaseous CO_2, which diffuses outward from the blood into the alveoli. With the next exhalation, this CO_2 leaves the body. Hemoglobin has a greater affinity for O_2 than for CO_2 at low CO_2 concentrations. For this reason, the diffusion of CO_2 outward from the red blood cells causes the hemoglobin within these cells to release its bound CO_2 and take up oxygen instead. The red blood cells, with their newly bound oxygen, then start the next respiratory journey.

In the next chapter we shall consider in more detail the journey of the red blood cell. Traveling between the lungs, where they acquire oxygen and release CO_2, and respiring tissues, where they release oxygen and acquire CO_2, the body's red blood cells traverse a complex highway that passes to all parts of the body. Nor are red blood cells the only traffic on this highway. Just as the roads and sidewalks transport all the commerce of a city (although not all the information, since much of this passes over phone lines), so the circulatory systems of the vertebrate body transports all the material that moves from one part of the body to another (although not all the information, as much of this moves by way of the nerves).

SUMMARY

1. All animals obtain oxygen by diffusion. The evolution of respiratory mechanisms among animals has tended to favor changes that improve the rate of diffusion. These include changes that decrease the length of the path over which diffusion occurs, changes that increase the surface area over which diffusion occurs, and changes that maximize the difference in oxygen concentration between environment and tissue.

2. The most efficient aquatic respiratory organ is the gill of bony fishes. Fishes take water in through their mouths, move it past their gills, and pass it out of their bodies. This one-way flow is the secret to high respiratory efficiency, since it permits fishes to establish a countercurrent flow of blood: blood vessels are located within the gills in such a way that blood flows in a direction opposite that of water.

3. Gills will not work in air, since air is not buoyant enough to support their fine latticework of passages. That is the reason fishes drown in air even though there is much more oxygen available in air than there is in the water in which they normally live.

4. The amphibian lung is a simple sac with two-way flow in and out. In it, gaseous diffusion across the small internal surface area is supplemented by diffusion across the moist skin.

5. The further evolution of the lung in reptiles and mammals has involved no fundamental changes. Instead, there has been a progressive increase in the internal surface area of the lung, achieved by partitioning the inner surface into increasingly numerous chambers called alveoli. The lungs of humans and other mammals possess about 300 million alveoli, with a combined surface area that is about 42 times the body's surface area.

6. A fundamental change in the atmospheric lung is characteristic of the birds, which utilize a series of air chambers and two-cycle breathing to effect a one-way flow of air. Just as in the gill of bony fishes, the establishment of a one-way flow of the diffusing medium permits countercurrent flow, which is by far the most efficient diffusion mechanism.

7. Humans, like other terrestrial vertebrates, breathe by expanding the cavity within which the lungs hang. This action expands the lungs, sucking air inward.

8. In the respiration of terrestrial vertebrates, hemoglobin within the red blood cells binds the oxygen, which diffuses across the lung capillaries into the blood plasma. The circulating system carries these red blood cells to the respiring tissues of the body.

9. In respiring tissues, there is much less oxygen than in the blood and much more CO_2, due to the consumption of oxygen and generation of CO_2 by respiring cells. Hemoglobin responds to the higher CO_2 concentration by unloading its oxygen, which diffuses out of the red blood cells into the tissue. The CO_2 is absorbed into the red blood cells and carried to the lungs, where it is discharged.

REVIEW

1. The uptake of oxygen and the release of carbon dioxide together are called _____.

2. An increase in the carbon dioxide concentration of blood causes the same effect on hemoglobin's ability to combine with oxygen as does a decrease in the partial pressure of oxygen within the lungs. True or false?

3. The first lungs seem to have evolved from the _____ of fishes.

4. Although lungs are less efficient respiratory structures than gills, they have the great advantage to terrestrial organisms of minimizing _____.

5. The protein _____ carries oxygen from your lungs throughout your body via the circulatory system.

SELF-QUIZ

1. Which of the following phyla possess organisms with no special respiratory organs?
   (a) Porifera
   (b) Cnidaria
   (c) Platyhelminthes
   (d) Chordata
   (e) a, b, and c

2. Within the lamellae of gills, the blood circulation is arranged so that blood is carried in the opposite direction to the movement of water. The functional significance of this arrangement is that
   (a) it helps to maintain the temperature of the organism equal to the water temperature, thus enhancing diffusion
   (b) it results from a developmental constraint
   (c) it ensures a continuous gradient of concentration difference between the blood and the water, so that diffusion continues to occur all along the gill
   (d) it increases the surface area for diffusion
   (e) it allows some kinds of fishes to continue to get oxygen even if they are not moving

3. Arrange the following in the order in which air contacts them during breathing, starting from the mouth and nose:
   (a) bronchus
   (b) alveoli
   (c) larynx
   (d) hemoglobin
   (e) trachea

4. Which type of terrestrial organism has the most efficient type of lung?
   (a) terrestrial mollusks
   (b) amphibians
   (c) reptiles
   (d) birds
   (e) mammals

5. Air normally enters your body through the nostrils, which are filled with hairs. What is the function of these hairs?
   (a) they filter out dust and other particles
   (b) they are a noise-making device
   (c) they slow and regulate the passage of air
   (d) they moisten the air
   (e) they allow you to breathe while you eat

THOUGHT QUESTIONS

1. Often people who appear to have drowned can be revived, in some cases after being under water for as long as half an hour. In every case of full recovery after extended submergence, however, the person had been submerged in very cold water. Is this observation consistent with the fact that oxygen is twice as soluble in water at 0° C as it is at 30° C?

2. Can you think of a reason why a respiratory system has not evolved in which oxygen is actively transported across respiratory membranes, in place of the passive process of diffusion across these membranes that is employed universally?

3. If by accident your pleural membrane were punctured, would you be able to breathe?

FOR FURTHER READING

FEDER, M., and W. BURGGREN: "Skin Breathing in Vertebrates," *Scientific American,* November, 1985, pages 126–142. Many vertebrates do a significant portion of their breathing through their skin.

PERUTZ, M.F.: "Hemoglobin Structure and Respiratory Transport," *Scientific American,* December 1978, pages 92–125. An account of how hemoglobin changes its shape to facilitate oxygen binding and unloading, by the man who won a Nobel Prize for unraveling the structure of hemoglobin.

SCHMIDT-NIELSEN, K.: "How Birds Breathe," *Scientific American,* December 1971, pages 73-79. A fascinating account of the discovery of unidirectional flow in avian lungs, by a great comparative physiologist.

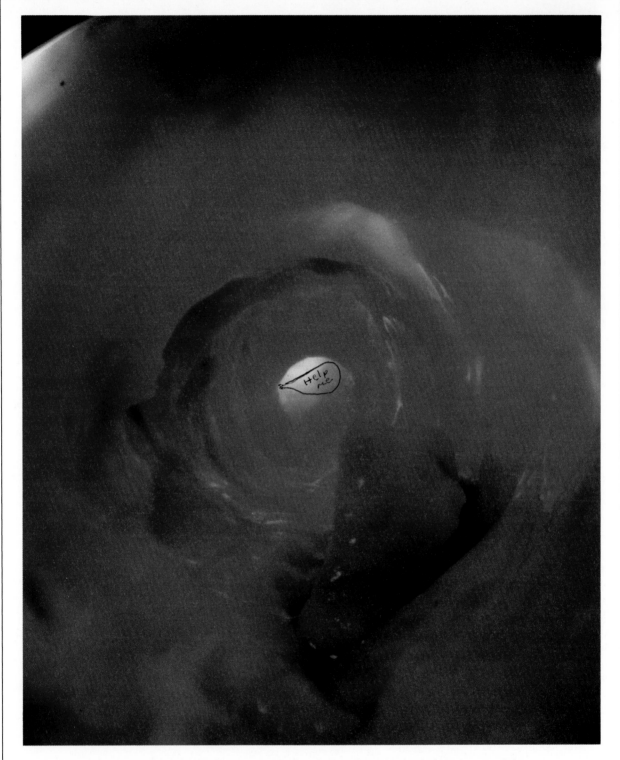

FIGURE 36-1   The interior of an artery caked with fatty deposits. Such deposits retard the circulation of blood. Fragments that break off can block arterial branches and produce strokes or heart attacks.

CIRCULATION

Chapter 36

Overview

In vertebrates, circulatory systems are the highways over which red blood cells carry oxygen to the tissues and remove carbon dioxide. The fluid of the blood also transports glucose and amino acids to the cells and carries away wastes. The composition of the blood is kept constant by the liver, which monitors and adjusts metabolite levels in the blood. The key to circulation in vertebrates is the organ that pumps the blood through the system, the heart.

For Review *Here are some important terms and concepts that you will encounter in this chapter. If you are not familiar with them, you should review them before proceeding.*

Glycogen (Chapter 3)

Gills (Chapters 19, 35)

Erythrocytes (Chapter 32)

How hemoglobin carries oxygen (Chapter 35)

Of the many tissues of the vertebrate body, few have the emotive impact of blood. In movies and in real life the sight of blood connotes violence, injury, and bodily harm. In literature, blood symbolizes the life force, which is not a bad analogy in real life either. This chapter is about blood, its circulation through the body, and the many ways in which it affects the body's functions (Figure 36-1). Although vertebrates possess many other organ systems that are necessary for life, it is the activities of the blood that bind them together into a functioning whole.

THE EVOLUTION OF CIRCULATORY SYSTEMS

The capture of nutrients and gases from the environment is one of the essential tasks that all living organisms must carry out. In animals, with their relatively large bodies with many elements, this function has become more complex. In less complex animals, such as roundworms, the fluid within the body cavity constitutes a primitive kind of circulatory system, one that permits materials to pass from one cell to another without leaving the organism. Within such simple systems, which are called **open circulatory systems,** there is no distinction between circulating fluid and body fluid generally. Arthropods have open circulatory systems (see Chapter 28), in which a muscled tube within the central body cavity forces the cavity fluid out through a network of interior channels and spaces. The fluid then flows back into the central cavity. Most animals, however, have **closed circulatory systems,** in which the circulatory system fluid

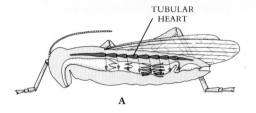
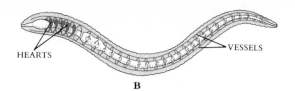

A

B

FIGURE 36-2

Open and closed circulatory systems.
A An open circulatory system in an insect. The tubular heart of insects pumps blood out to the tissues of the body, from which it seeps back into the heart, rather than traveling in vessels.
B A simple closed circulatory system in an annelid. The series of hearts push blood out through a system of vessels to the body's tissues, then back via a second vessel system.

is separated from the rest of the body's fluids and does not mix with them (Figure 36-2). In annelids, for example, two major tubes or vessels extend the length of the worm, with branches extending out from each tube to the muscles, skin, and digestive organs.

A great advantage of closed circulatory systems is that they permit regulation of fluid flow by means of muscle-driven changes in the diameters of the vessels. In other words, with a closed circulatory system the different parts of the body can maintain different circulation rates.

A closed circulatory system is characteristic of all vertebrates. It has four principal functions: (1) nutrient and waste transport; (2) oxygen and carbon dioxide transport; (3) temperature maintenance; and (4) hormone circulation. The third function, that of temperature maintenance, is associated with the passage of the blood throughout the body; it is important in all vertebrates.

THE CARDIOVASCULAR SYSTEM

The circulatory system of vertebrates is composed of three elements: (1) the **heart,** a muscular pump that we shall consider later in this chapter; (2) the blood vessels (Figure 36-3; see also Figure 36-14), a network of tubes that pass through the body; and (3) the blood, which circulates within these vessels. The plumbing of the closed circuit, the heart and vessels, is known collectively as the **cardiovascular system.** Blood moves within this cardiovascular system, leaving the heart through vessels known as **arteries.** From the arteries, the blood passes into a larger network of **arterioles,** or smaller arteries. From these, it eventually is forced through the **capillaries,** a fine latticework of very narrow tubes, which get their name from the Latin word *capillus,* "a hair." It is while passing through these capillaries that the blood exchanges gases and metabolites with

FIGURE 36-3

This ruptured tube is a blood vessel, an element of the human circulatory system. It is full of red blood cells, which move through these blood vessels transporting oxygen and carbon dioxide from one place to another in the body.

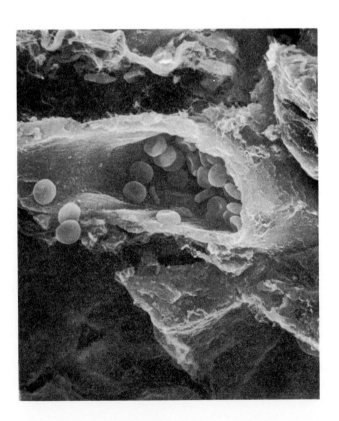

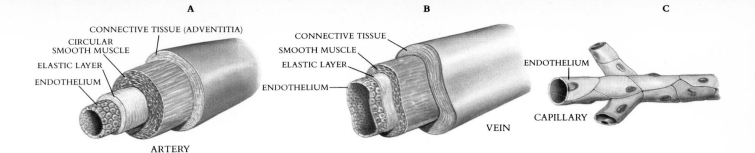

A **B** **C**

CONNECTIVE TISSUE (ADVENTITIA)
CIRCULAR
SMOOTH MUSCLE
ELASTIC LAYER
ENDOTHELIUM

ARTERY

CONNECTIVE TISSUE
SMOOTH MUSCLE
ELASTIC LAYER
ENDOTHELIUM

VEIN

ENDOTHELIUM

CAPILLARY

FIGURE 36-4

The structure of some important blood vessels.
A Artery.
B Vein.
C Capillary.

the cells of the body. After traversing the capillaries, the blood passes into a third kind of vessel, the **venules** or small **veins.** A network of venules and larger veins collects the circulated blood and carries it back to the heart.

Arteries

The walls of the arteries are made up of three layers of tissue (Figure 36-4, *A*). The innermost one is composed of a thin layer of **endothelial** cells. Surrounding these cells is a thick layer of smooth muscle and elastic fibers, which in turn is encased within an envelope of protective connective tissue. Because this sheath is elastic, the artery is able to expand its volume considerably in response to a pulse of fluid pressure, much as a tubular balloon might respond to air blown into it. The steady contraction of the muscle layer strengthens the wall of the vessel against overexpansion.

Arterioles

The arterioles differ from the arteries simply in being smaller in diameter. The muscle layer that surrounds an arteriole can be relaxed under the influence of hormones and metabolites to enlarge the diameter. When this happens, the blood flow can be increased, an advantage during times of high metabolic activity. Conversely, most arterioles are in contact with many nerve fibers. When stimulated, these nerves cause the muscular lining of the arteriole to contract and thus constrict the diameter of the arteriole. Such contraction limits the flow of blood to the extremities during periods of low temperature or stress. You turn pale when you are scared or need to conserve heat because contraction of this kind constricts the arterioles in your skin. You blush for just the opposite reason. When you overheat or are embarrassed, the nerve fibers connected to muscles surrounding the arterioles are inhibited, which relaxes the smooth muscle and causes the arterioles in the skin to dilate. This brings heat to the surface for escape.

Capillaries

Capillaries have the simplest structure of any element in the cardiovascular system (Figure 36-4, *C*). They are little more than tubes one cell thick and on the average about 1 millimeter long; they connect the arterioles with the venules. The internal diameter of the capillaries is, on the average, about 8 micrometers. Surprisingly, this is little more than the diameter of a red blood cell (5 to 7 micrometers). However, red blood cells squeeze through these fine tubes without difficulty (Figure 36-5). The intimate contact between the walls of the capillaries and the membranes of the red blood cells facilitates the diffusion of gases and metabolites between them. No cell of the body is more than 100 micrometers from a capillary. At any one moment, about 5% of your blood is in your capillaries. Some capillaries, called **thoroughfare channels,** connect arterioles and venules directly (Figure 36-6). From these channels, loops of true capillaries leave and return. It is through these loops that almost all exchange between the blood and the cells of the remainder of the body occurs. The entry to each loop is guarded by a ring of muscle called a **precapillary sphincter,** which, when closed, blocks flow through the capillary. Such restriction of entry to the capillaries in surface tissue is another powerful means of limiting heat loss from an animal's body during periods of cold.

The entire body is permeated with a fine mesh of these capillaries, a network that amounts to several thousand kilometers in overall length. If all of the capillaries in your body were laid end to end, they would extend across the United States. While individual capillaries have high resistance to flow because of their small diameters, the cross-sectional area of the extensive capillary network is greater than that of the arteries leading to it, so that the blood pressure is actually lower in the capillaries than in the arteries.

FIGURE 36-5

The red blood cells in this capillary are passing along in single file. Many capillaries are even smaller than that shown here from the bladder of a monkey. Red blood cells will even pass through capillaries narrower than their own diameter, pushed along by the pressure generated by a pumping heart.

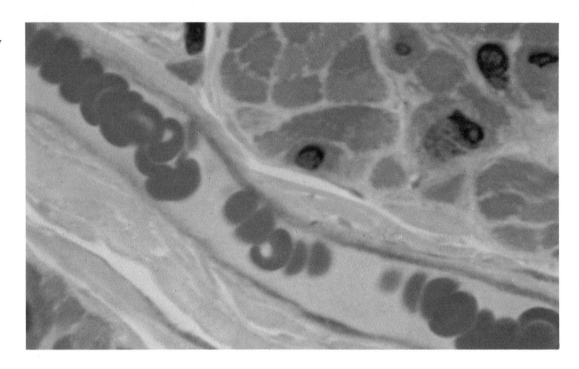

FIGURE 36-6

The capillary network connects arteries with veins. The most direct connection is via thoroughfare channels that connect arterioles directly to venules. Branching from these thoroughfare channels is a network of finer channels, the capillary network. Most of the exchange between body and red blood cells occurs while the red blood cells are in this capillary network. Entrance to the capillary network is controlled by bands of muscle, called precapillary sphincters, at the entrance to each capillary. When a sphincter is contracted, it closes off the capillary. By contracting these sphincters, the body can limit the amount of blood in the capillary network of a particular tissue, and thus control the rate of exchange in that tissue.

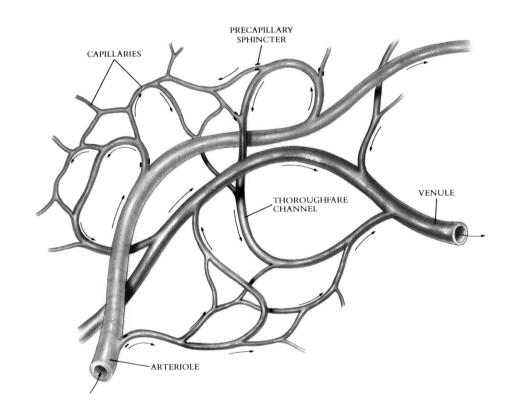

CAPILLARIES

PRECAPILLARY SPHINCTER

THOROUGHFARE CHANNEL

VENULE

ARTERIOLE

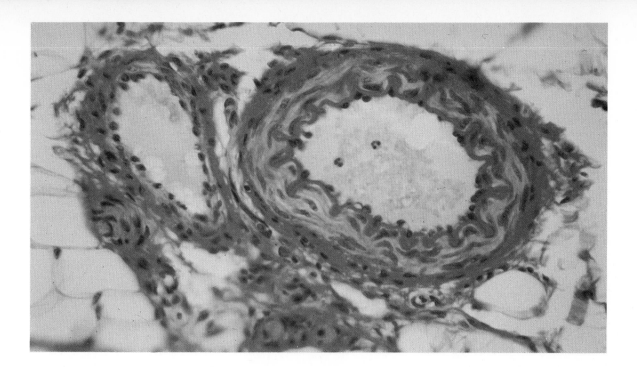

Veins and Venules

Veins do not have to accommodate the pulsing pressures that arteries do, because much of the force of the heartbeat is weakened by the high resistance and great cross-sectional area of the capillary network. The walls of veins, although similar in structure to those of the arteries, have much thinner layers of muscle and elastic fiber (Figures 36-4, *B,* and 36-7). An empty artery is still a hollow tube, like a pipe, but when a vein is empty, its walls collapse like an empty balloon.

The internal passageway of veins is often quite large. The diameter of the largest veins in the human body, the **venae cavae,** which lead into the heart, is fully 3 centimeters. The reason that veins are so much larger than arteries is related to the lower pressure of blood flowing within veins back toward the heart—it is advantageous to minimize any further resistance to its flow, and a larger tube presents much less resistance to flow than does a smaller tube. Veins are large because larger veins present less resistance to the flow of blood back to the heart.

> **The major vessels of the circulatory system are tubes of cells encased within three sheaths: (1) a layer of elastic fibers, which renders the diameter of the vessel elastic to accommodate pulses of blood pumped from the heart; (2) a layer of muscle serviced by nerves, which permits the body to control the diameter of the vessel and strengthens the wall against overexpansion; and (3) a layer of connective tissue, which protects the vessel.**

The Lymphatic System

We speak of the cardiovascular system as closed because all of its tubes are connected with one another and none are simply open-ended. In another sense, however, the system is open—it is open to diffusion through the walls of the capillaries. This diffusion is a necessary part of the functioning of the circulatory system, but it poses difficulties in maintaining the integrity of that system.

Difficulties arise because diffusion from the capillaries is accompanied by the loss of large quantities of liquid from the cardiovascular system. When blood passes through the capillaries, it loses more water to the body than it reabsorbs from it. In a human being, about 3 liters of fluid leave the cardiovascular system in this way each day, a quantity amounting to more than half the body's total supply of about 5.6 liters of blood. In counteracting the effects of this process, the body utilizes a second *open* circulatory system called the **lymphatic system.** The elements of the lymphatic system

FIGURE 36-7

The vein *(left)* has the same general structure as an artery *(right),* but much thinner layers of muscle and elastic fiber. An artery will retain its shape when empty, but a vein will collapse.

FIGURE 36-8

A The human lymphatic system. The lower portion of the body and the left side drain into a single lymphatic vessel, the thoracic duct, which empties into the left subclavian vein through a one-way valve. The upper right side of the body drains into a second collecting vessel, the right lymphatic duct, which drains into the right subclavian vein.

B Detail of blood and lymphatic capillaries, showing the nature of one-way valves.

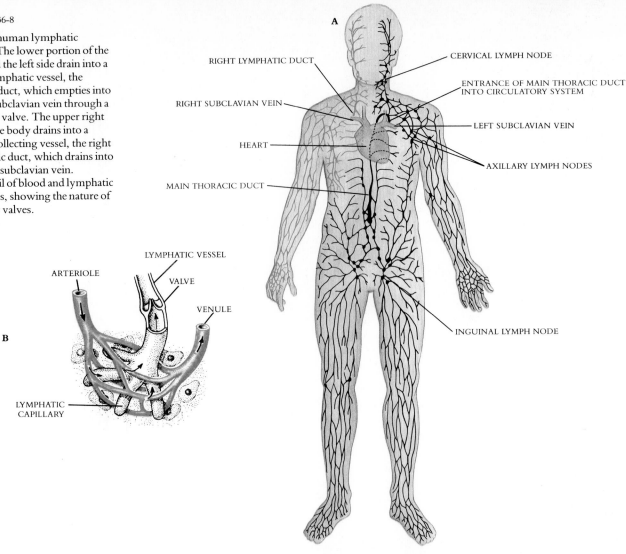

A

RIGHT LYMPHATIC DUCT

CERVICAL LYMPH NODE

ENTRANCE OF MAIN THORACIC DUCT INTO CIRCULATORY SYSTEM

RIGHT SUBCLAVIAN VEIN

LEFT SUBCLAVIAN VEIN

HEART

AXILLARY LYMPH NODES

MAIN THORACIC DUCT

B

LYMPHATIC VESSEL

ARTERIOLE

VALVE

VENULE

LYMPHATIC CAPILLARY

INGUINAL LYMPH NODE

FIGURE 36-9

A lymphatic vessel valve, magnified 25 times. Flow from left to right is not retarded since such flow tends to force open the inner cone; flow from right to left is prevented because such flow tends to force the inner cone closed.

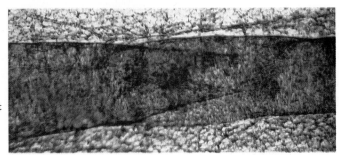

gather liquid from the body and return it to the cardiovascular circulation (Figure 36-8). Open-ended lymph capillaries gather up fluids by diffusion and carry them through a series of progressively larger vessels to two large lymphatic vessels, which resemble veins. These lymphatic vessels drain into veins in the lower part of the neck through one-way valves. No heart pumps fluid through the lymphatic system; instead, fluid is driven through it when its vessels are squeezed by the movements of the body's muscles. The lymphatic vessels contain a series of one-way valves (Figure 36-9), which permit movement only in the direction of the neck.

Much of the water within blood plasma diffuses out during passage through the capillaries. This water is collected by an open circulatory system, the lymphatic system, and is returned to the bloodstream.

THE CONTENTS OF VERTEBRATE CIRCULATORY SYSTEMS

About 8% of the body mass of most vertebrates is taken up by the blood circulating through their bodies. This blood is composed of a fluid plasma, together with several different kinds of cells that circulate within that fluid.

Blood Plasma

Blood plasma is a complex solution of water with three very different components:

1. *Metabolites and wastes.* If the circulatory system is thought of as the highway of the vertebrate body, the blood contains the traffic traveling on that highway. Dissolved within the plasma are glucose, lipids, and all of the other metabolites, vitamins, hormones, and wastes that circulate between the cells of the body.
2. *Salts and ions.* Like the water of the seas in which life arose, plasma is a dilute salt solution. The chief plasma ions are sodium, chloride, and bicarbonate. In addition, there are trace amounts of other salts, such as calcium and magnesium, as well as of metallic ions, including copper, potassium, and zinc. The composition of the plasma, therefore, is not unlike that of seawater.
3. *Proteins.* Blood plasma is 90% water. Passing by all the cells of the body, blood would soon lose most of its water to them by osmosis if it did not contain as high a concentration of proteins as the cells that it passed. Water does not move from the blood vessels into the surrounding cells, because blood plasma contains proteins. Some of these are antibody and globulin proteins that are active in the immune system, as well as a small amount of fibrinogen, a protein that plays a role in blood clotting. Taken together, however, these proteins make up less than half of the amount of protein that is necessary to balance the protein content of the other cells of the body. The rest consists of a protein called **serum albumin,** which circulates in the blood as an osmotic counterforce. Human blood contains 46 grams of serum albumin per liter.

   Blood plasma contains metabolites, wastes, and a vareity of ions and salts. It also contains high concentrations of the protein serum albumin, which functions to keep the blood plasma in osmotic equilbrium with the cells of the body.

Types of Blood Cells

Although blood is liquid, some 40% of its volume is actually occupied by cells. There are three principal types of cells in the blood (Figure 36-10): erythrocytes, leukocytes, and platelets.

Erythrocytes

Each milliliter of blood contains about 5 billion **erythrocytes,** or **red blood cells.** Each erythrocyte is a flat disk with a central depression (Figure 36-11), something like a doughnut with a hole that doesn't go all the way through. Attached to the outer membranes of the erythrocytes is a collection of polysaccharides, which determines the blood group of the individual. Almost the entire interior of each cell is packed with hemoglobin.

Mature erythrocytes in mammals contain neither a nucleus nor protein-synthesizing machinery. Because they lack a nucleus, these cells are unable to repair themselves, and they therefore have a rather short life; any one erythrocyte lives only about 4 months. New erythrocytes are constantly being synthesized and released into the blood by cells within the soft interior marrow of bones.

Leukocytes

Less than 1% of the cells in human blood are **leukocytes,** or **white blood cells;** there are about 1 or 2 leukocytes for every 1000 red blood cells. Leukocytes are larger than red blood cells; they contain no hemoglobin and are essentially colorless. There are several

FIGURE 36-10

Types of blood cells. The erythrocytes, or red blood cells, are the carriers of oxygen in the bloodstream. The erythrocytes of mammals lack nuclei, unlike those of most other vertebrates. The leukocytes, or white blood cells, are the scavengers of the bloodstream. Fragments of megakaryocytes, called platelets, play an important role in blood clotting.

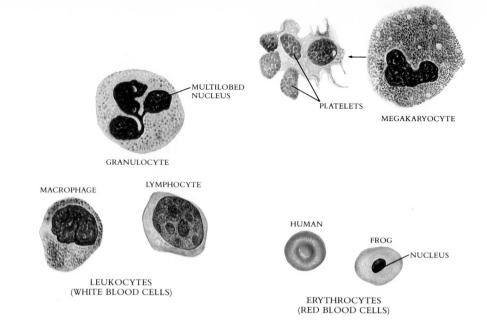

kinds of leukocytes and each has a different function. All of these functions, however, are related to the defense of the body against invading microorganisms and other foreign substances, as you will see in Chapter 37. Leukocytes are not confined to the bloodstream; they also migrate out into the interstitial fluid.

If you prick your skin, some of the injured cells release chemicals that cause the capillaries in the vicinity to expand; this **inflammatory response** makes the wound look red and feel warm. **Granulocytes,** which are circulating leukocytes, push out through the walls of the distended capillaries to the site of the injury. The granulocytes are classified into three groups by their staining properties. About 50% to 70% of them are **neutrophils,** which begin to stick to the interior walls of the blood vessels at the site of the injury. They then form projections that enable them to push their way into the infected tissues, where they engulf microorganisms and other foreign particles. **Basophils,** a second kind of leukocyte, contain granules that rupture and release chemicals that enhance the inflammatory response; they are important in causing allergic responses. The role of the third kind of leukocyte, the **eosinophils,** is not clear.

Another kind of circulating leukocyte, the **monocyte,** is also attracted to the sites of inflammation, where they are converted into **macrophages**—enlarged, amoeba-like cells that entrap microorganisms and particles of foreign matter. They usually arrive at the site of inflammation after the neutrophils, and they are important in the immune response.

Platelets

Certain large cells within the bone marrow, called **megakaryocytes,** regularly pinch off bits of their cytoplasm. These cell fragments, called **platelets,** contain no nuclei; they enter the bloodstream, where they play an important role in controlling blood clotting.

THE EVOLUTION OF THE VERTEBRATE HEART

Any closed circulatory system requires both a system of passageways through which fluid can circulate and a pump to force the fluid through them. In the circulation of blood the pump is the heart.

Early chordates, such as the lancelets, seem to have had simple tubular hearts, which amounted to little more than a specialized, muscular zone of the ventral artery, which beats in simple waves of contraction. When gills evolved in the early fishes, a

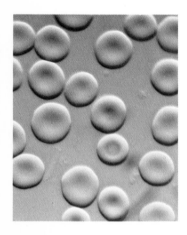

FIGURE 36-11

Human erythrocytes, magnified 1000 times. Human erythrocytes lack nuclei, which gives them a characteristic collapsed appearance, like a pillow on which someone has sat.

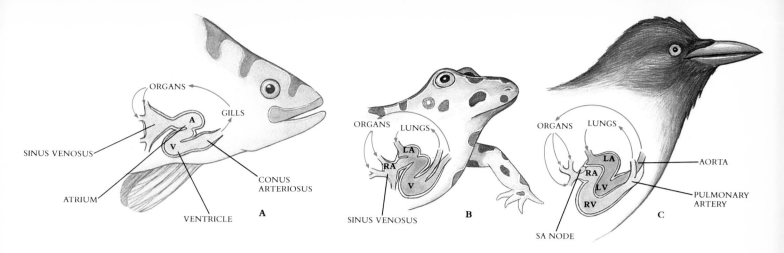

FIGURE 36-12

Evolution of the vertebrate heart.
A Fishes.
B Amphibians.
C Birds and mammals. LA =
left atrium; RA = right atrium;
LV = left ventricle; RV = right
ventricle. Blue color and blue
arrows represent unoxygenated
blood; red color and red arrows,
oxygenated blood.

more efficient pump was necessary to force blood through the fine capillary network. The fish heart can be considered a tube that has four chambers arrayed one after the other (Figure 36-12, *A*). The first two chambers (**sinus venosus** and **atrium**) are collection chambers; the second two (**ventricle** and **conus arteriosus**) are pumping chambers. When a fish heart beats, contraction starts in the rear chamber (the sinus venosus) and spreads progressively forward to the conus arteriosus. The same heartbeat sequence is characteristic of all vertebrate hearts.

After fish blood leaves the gills, it circulates through the rest of the body; in terrestrial vertebrates, however, the blood that receives oxygen and discharges carbon dioxide in the lungs is returned to the heart by two large veins, the **pulmonary veins,** and then pumped forcefully out through the body; this results in much-improved rates of circulation. To achieve this double circulation, the arrangement of the amphibian heart differs from that found in fishes in two ways (Figure 36-12, *B*). First, the atrium is divided into two chambers: a right atrium, which receives unoxygenated blood from the sinus venosus for circulation to the lungs, and a left atrium, which receives oxygenated blood from the lungs through the pulmonary vein for circulation throughout the body. This division extends partway into the ventricle. Second, the conus arteriosus is partly separated by a dividing wall, which directs oxygenated blood into the aorta and unoxygenated blood into the pulmonary arteries, which lead to the lungs. The aorta leads to the body's network of arteries. Taken together, these modifications divide the circulatory system into two separate pathways: (1) the **pulmonary circulation,** between the heart and the lungs; and (2) the **systemic circulation,** between the heart and the rest of the body. However, since the divisions in the heart ventricles and conus arteriosus are not complete, oxygenated and unoxygenated blood are mixed, reducing somewhat the oxygen level of the blood pumped to the body.

Further evolution among the vertebrates has resulted in the complete closing of the dividing wall in the ventricle, which results in a total division of the pumping chamber into two parts in birds (Figure 36-12, *C*) and mammals. In these groups, the four-chambered heart acts as a double pump, the left side pumping oxygenated blood to the general body circulation and the right side pumping unoxygenated blood from the veins to the lungs. Such efficient hearts function well in helping to maintain constant internal temperatures in birds and mammals. They also circulate blood through the lungs much more rapidly and efficiently than in other vertebrates, thus greatly increasing the efficiency with which oxygen is captured by the bloodstream.

THE HUMAN HEART

The human heart, like that of all mammals and birds, is therefore a double pump; a cross section through the heart clearly shows its organization (Figure 36-13). The left side has two connected chambers, and so does the right, but the two sides are not connected with one another.

FIGURE 36-13

The path of blood through the human heart is illustrated in this section, which shows the four chambers of the heart, completely separated from each other. Entering the right atrium by way of the superior vena cava, the blood passes into the right ventricle and then through the pulmonary valve to the pulmonary artery and the lungs. Oxygenated blood from the lungs then returns to the heart by way of the pulmonary veins, entering the left atrium and then the left ventricle, from which it enters the general circulatory system of the body by way of the aorta.

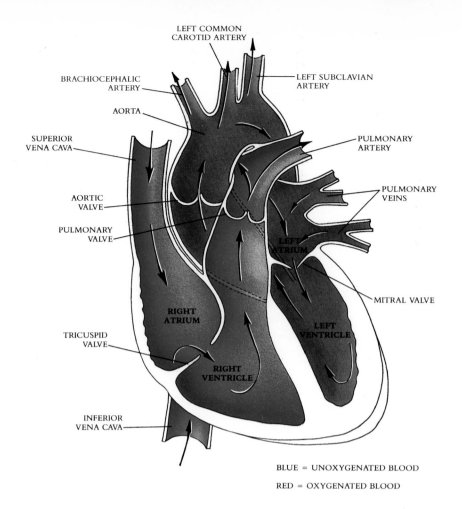

BLUE = UNOXYGENATED BLOOD

RED = OXYGENATED BLOOD

Circulation Through the Heart

Let us follow the journey of blood through the human heart, starting with the entry of oxygenated blood into the heart from the lungs. Oxygenated blood from the lungs enters the left side of the heart, emptying directly into the **left atrium** through large vessels called the **pulmonary veins.** From the atrium, blood flows through an opening into the adjoining chamber, the **left ventricle.** Most of this flow, roughly 80%, occurs while the heart is relaxed. When the heart starts to contract, the atrium contracts first, pushing the remaining 20% of its blood into the ventricle.

After a slight delay, the ventricle contracts. The walls of the ventricle are far more muscular than those of the atrium, and as a result this contraction is much stronger. It forces most of the blood out of the ventricle in a single strong pulse. The blood is prevented from going back into the atrium by a large one-way valve, the **mitral valve,** whose flaps are pushed shut as the ventricle contracts. Strong fibers that prevent the flaps from moving too far when closing are attached to their edges. If the flaps did move too far, they would project out into the atrium. The fibers that prevent this operate in much the same way as a rope might if tied from the steering wheel of a car to the driver's door handle with a bit of slack—the door can be opened only as far as the slack in the rope permits.

Prevented from reentering the atrium, the blood within the ventricle takes the only other passage out of the contracting left ventricle. It moves through a second opening that leads into a large vessel called the **aorta.** The aorta is separated from the left venticle by a one-way valve, the **aortic valve.** Unlike the mitral valve, the aortic valve is oriented to permit the flow of the blood *out* of the ventricle. Once this outward flow has occurred, the aortic valve closes, thus preventing the reentry of blood from the aorta into the heart.

The aorta and all of the other blood vessels that carry blood away from the heart are arteries. Many of these arteries branch from the aorta, carrying oxygen-rich blood to all parts of the body. The first to branch are the coronary arteries, which carry freshly oxygenated blood to the heart itself; the muscles of the heart do not obtain their supply of blood from within the heart.

The blood that flows into the arterial system eventually returns to the heart after delivering its cargo of oxygen to the cells of the body. In returning, it passes through a series of veins, eventually entering the right side of the heart. Two large veins collect blood from the systemic circulation. The **superior vena cava** drains the upper body, the **inferior vena cava** the lower body. These veins empty deoxygenated blood into the right atrium. The right side of the heart is similar in organization to the left side. Blood passes from the right atrium into the right ventricle through a one-way valve, the **tricuspid valve.** It passes out of the contracting right venticle through a second valve, the **pulmonary valve,** into the **pulmonary arteries,** which carry the deoxygenated blood to the lungs. The blood then returns from the lungs to the left side of the heart with a new cargo of oxygen, which is pumped to the rest of the body. The human circulatory system as a whole is outlined in Figure 36-14.

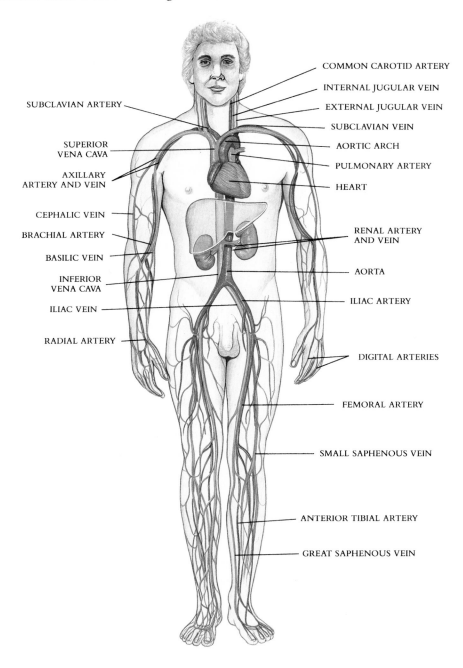

FIGURE 36-14

The human circulatory system, showing the system of veins and arteries and the organs connected to them. Although many of the individual elements are not named in the text, this figure is intended to provide a reference for the overall structure of the circulatory system.

COMMON CAROTID ARTERY
INTERNAL JUGULAR VEIN
EXTERNAL JUGULAR VEIN
SUBCLAVIAN VEIN
AORTIC ARCH
PULMONARY ARTERY
HEART
RENAL ARTERY AND VEIN
AORTA
ILIAC ARTERY
DIGITAL ARTERIES
FEMORAL ARTERY
SMALL SAPHENOUS VEIN
ANTERIOR TIBIAL ARTERY
GREAT SAPHENOUS VEIN

SUBCLAVIAN ARTERY
SUPERIOR VENA CAVA
AXILLARY ARTERY AND VEIN
CEPHALIC VEIN
BRACHIAL ARTERY
BASILIC VEIN
INFERIOR VENA CAVA
ILIAC VEIN
RADIAL ARTERY

FIGURE 36-15

Contraction of the human heart is initiated by a wave of depolarization that begins at the SA node. After passing over the right and left atria and causing their contraction, the wave of depolarization reaches the AV node. From there it passes through the bundle of His, which has a bundle branch to each ventricle. From the tip of the ventricles, the depolarization is conducted rapidly over their surfaces by the branching Purkinje fibers.

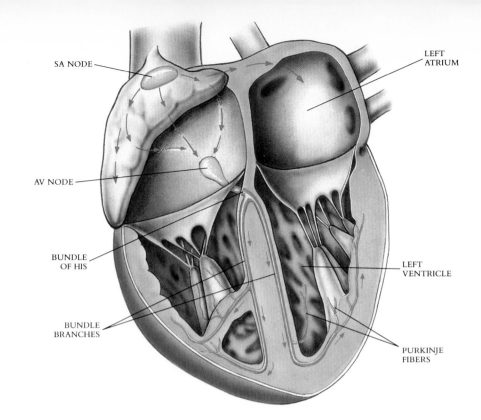

How the Heart Contracts

The overall contraction of the heart consists of a carefully orchestrated series of muscle contractions (Figure 36-15). Contraction is initiated by the **sinoatrial (SA) node,** the small cluster of excitatory cardiac muscle cells derived from the sinus venosus that is embedded in the upper wall of the right atrium. The cells of the SA node act as a pacemaker for the rest of the heart. Their membranes spontaneously depolarize with a regular rhythm that determines the rhythm of the heart's beating. Each depolarization initiated within this pacemaker region passes quickly from one cardiac muscle cell to another in a wave that envelops both the left and the right atria almost instantaneously.

> **The contraction of the heart is initiated by the periodic spontaneous depolarization of cells of the SA node; the resulting wave of depolarization passes over both the left and the right atria and causes their muscle cells to contract.**

The wave of depolarization does not immediately spread to the ventricles, however. Almost 0.1 second passes before the lower half of the heart starts to contract. The reason for the delay is that the atria of the heart are separated from the ventricles by connective tissue, and connective tissue cannot propagate a depolarization wave. The depolarization would not pass to the ventricles at all except for a slender connection of cardiac muscle cells known as the **atrioventricular (AV) node,** which connects to a strand of specialized muscle in the ventricular septum known as the **bundle of His.** Bundle branches divide to the right and left; on reaching the apex of the heart, each branch further divides into **Purkinje fibers,** which initiate contraction there. The passage of the wave triggers the almost simultaneous contraction of all the cells of the right and left venticles. The cells involved in the passage of the depolarization wave from the atria to the ventricles have small diameters; thus they propagate the depolarization slowly, causing the delay mentioned. This delay permits the atria to finish emptying their contents into the corresponding ventricles before those ventricles start to contract.

> **The passage of the wave of depolarization from the AV node via the bundle of His and bundle branches to the Purkinje fibers causes the almost simultaneous contraction of all of the cells of the right and left ventricles.**

Monitoring the Heart's Performance

As you can see, the heartbeat is not simply a squeeze-release, squeeze-release cycle, but rather a little play in which a series of events occur in a predictable order. You may watch the play in several ways, depending on the events that you are observing. The simplest way to monitor heartbeat is to listen to the heart at work. The first sound you hear, a low pitched *lub,* is the closing of the mitral and tricuspid valves at the start of ventricular contraction. A little later you hear a higher-pitched *dub,* the closing of the pulmonary and aortic valves at the end of ventricular contraction. If the valves are not closing fully, or if they open too narrowly, turbulence is created within the heart. This turbulence can be heard as a **heart murmur.** It often sounds like liquid sloshing.

A second way to examine the events of the heartbeat is to monitor the blood pressure. During the first part of the heartbeat the atria are filling and contracting. At this time the pressure in the arteries leading from the left side of the heart out to the tissues of the body decreases slightly as the blood moves out of the arteries, through the vascular system, and into the atria. This period is referred to as the **diastolic period.** During the contraction of the left ventricle, a pulse of blood is forced into the systemic arterial system, immediately raising the blood pressure within these vessels. This pushing period, which ends with the closing of the aortic valve, is referred to as the **systolic period.** Blood pressure values are measured in millimeters of mercury; they reflect the height to which a column of mercury would be raised in a tube by an equivalent pressure. By the conventional system of measurement, normal blood pressure values are 70 to 90 diastolic and 110 to 130 systolic. When the inner walls of the arteries accumulate fats, as they do in the condition known as **atherosclerosis,** the diameters of the passageways are narrowed. If this occurs, the systolic blood pressure is elevated.

A third way to monitor the progress of events during a heartbeat is to measure the waves of depolarization. Because the human body basically consists of water, it conducts electrical currents rather well. A wave of membrane depolarization passing over the surface of the heart generates an electrical current that passes in a wave throughout the body. The magnitude of this electrical pulse is tiny, but it can be detected with sensors placed on the skin. A recording made of these impulses (Figure 36-16) is called an **electrocardiogram.**

In a normal heartbeat, three successive electrical pulses are recorded. First, there is an atrial excitation, caused by the depolarization associated with atrial contraction. A tenth of a second later, there is a much stronger ventricular excitation, reflecting both the depolarization of the ventricles and the relaxation of the atria. Finally, perhaps 0.2 second later, there is a third pulse, caused by the relaxation of the ventricles.

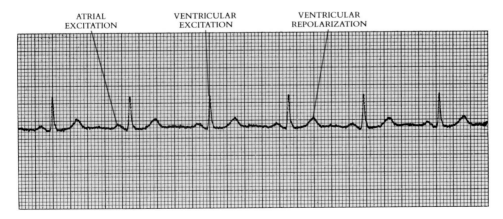

FIGURE 36-16

An electrocardiogram. In a normal heartbeat, three successive pulses are recorded, as shown here. Atrial excitation is caused by the depolarization associated with atrial contraction; ventricular excitation reflects both the depolarization of the ventricles and the relaxation of the atria; and ventricular repolarization is caused by the relaxation of the ventricles.

Cardiovascular diseases are the leading cause of death in the United States. More than 42 million people in this country have some form of cardiovascular disease—about one person in five. **Heart attacks** are the main cause of cardiovascular deaths in the United States, accounting for about a fifth of all deaths. They result from an insufficient supply of blood reaching an area of heart muscle. Heart attacks may be caused by a blood clot forming somewhere in the vessels and blocking the passage of blood through those vessels (Figure 36-A, *1*). They may also result if a vessel is blocked sufficiently by atherosclerosis. Recovery from a heart attack is possible if the segment of the heart tissue damaged was small enough that the other blood vessels in the heart can enlarge their capacity and resupply the damaged tissues. **Angina pectoris,** which literally means "chest pain," occurs for reasons similar to those which cause heart attacks, but it is not as severe. The pain may occur in the heart and often also in the left arm and shoulder.

The amount of heart damage associated with a small heart attack may be relatively slight and thus difficult to detect. It is important that such damage be detected, however, so that the overall condition of the heart can be evaluated properly. Electrocardiograms are very useful for this purpose, since they reveal abnormalities in the timing of heart contractions, abnormalities that are associated with the presence of damaged heart tissue. Damage to the AV node, for example, may delay as well as reduce the second, ventricular pulse. Unusual conduction routes may lead to continuous disorganized contractions called **fibrillations.** In many fatal heart attacks, ventricular fibrillation is the immediate cause of death.

Strokes are caused by an interference with the blood supply to the brain. They often occur when a blood vessel bursts in the brain, and they may be associated with a **thrombus,** or coagulation (clotting) of blood cells or other elements in one of the vessels. Such a thrombus may be caused by cancer or other diseases. The effects of strokes depend on how severe the damage is and where in the brain the stroke occurs.

Atherosclerosis is a buildup within the arteries (Figure 36-A, *2*). Atherosclerosis contributes both to heart attacks and to strokes. The accumulation within the arteries of fatty materials (see Figure 36-1), of abnormal amounts of smooth muscle cells, of deposits of cholesterol or fibrin, or of cellular debris of various kinds can all impair the arteries' proper functioning. When this condition is severe, the arteries can no longer expand and contract properly, and the blood moves through them with difficulty. The accumulation of cholesterol is thought to be the prime contributor to atherosclerosis, and diets low in cholesterol are now prescribed to help to prevent this condition.

Arteriosclerosis, or hardening of the arteries, occurs when calcium is deposited in arterial walls. It tends to occur when atherosclerosis is severe. Not only is flow through such arteries restricted, but they also lack the ability to expand as normal arteries do to accommodate the volume of blood pumped out by the heart. This forces the heart to work harder.

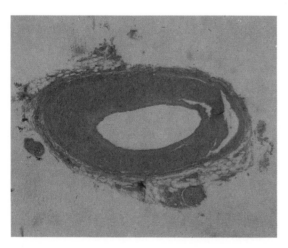
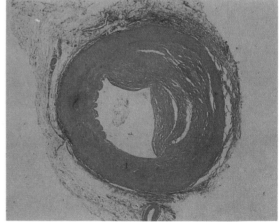

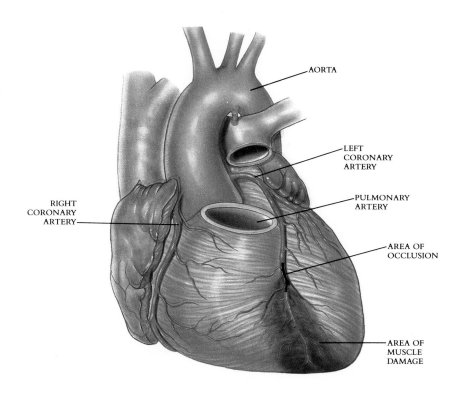

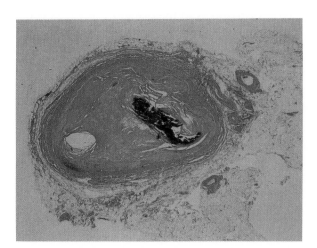

FIGURE 36-A, 1

A heart attack. This heart attack resulted from **occlusion** (blockage) of the left coronary artery. The heart muscle serviced by the portion of the coronary artery beyond the point of blockage is damaged.

AORTA

LEFT
CORONARY
ARTERY

PULMONARY
ARTERY

RIGHT
CORONARY
ARTERY

AREA OF
OCCLUSION

AREA OF
MUSCLE
DAMAGE

FIGURE 36-A, 2

Atherosclerosis is a buildup in an artery that blocks the passage of blood. The coronary artery on the left shows only minor blockage. The artery in the center exhibits severe atherosclerosis. Much of the passage is blocked by buildup on the interior walls of the artery. The coronary artery on the right is essentially completely blocked.

THE CENTRAL IMPORTANCE OF CIRCULATION

The evolution of multicellular organisms has depended critically on the ability to circulate nutrients and other materials to the various cells of the body, and to carry metabolic wastes away from them. The digestive processes described in Chapter 34, by which vertebrates obtain metabolizable foodstuffs, and the respiratory processes described in Chapter 35, by which vertebrates obtain the oxygen necessary for aerobic metabolism, both depend critically on the transport of food and oxygen to cells and the removal of the end products of their metabolism. Vertebrates carefully regulate the operation of their circulatory systems; by doing so, they are able to integrate their bodily activities. This regulation is carried out by the nervous system, the subject of chapters 38 and 39.

SUMMARY

1. The circulatory system of vertebrates is closed, permitting a more exact control of pressure and resistance in the system.

2. The plasma of the circulating blood contains the proteins and ions that are necessary to maintain the blood's osmotic equilibrium with the surrounding tissues.

3. The general flow of blood circulation through the body is a circuit starting from the heart, which pumps blood out via muscled arteries to the capillary networks that interlace the tissues of the body; the blood returns to the heart from these capillaries via the veins.

4. A second, open circulatory system, the lymphatic system, gathers liquid from the body that has been lost from the circulatory system by diffusion and returns it via a system of lymph capillaries, lymph vessels, and two large lymphatic ducts to veins in the lower part of the neck.

5. The heart of mammals and birds is a double pump, pushing both pulmonary (lung) circulation and systemic (general body) circulation. Because the two circulations are kept separate within the heart, the systemic circulation receives only fully oxygenated blood.

6. The four-chambered mammalian or bird heart evolved from two of the chambers of the four-chambered fish heart, by the creation of dividing walls within the two central chambers. The other two chambers of the fish heart were gradually lost, although the pacemaker cells of the sinus venosus have been retained in their original location.

7. The contraction of the heart is initiated at the SA node, or pacemaker, as a periodic spontaneous depolarization of these cells. The wave of depolarization spreads across the surface of the two atrial chambers, causing all of these cells to contract.

8. The passage of the wave of depolarization to the ventricles is briefly delayed by tissue that insulates the two segments of the heart from one another. Only a narrow channel of cardiac muscle cells connects the atria and the ventricles. The delay in the passage of the wave of depolarization permits the atria to empty completely into the ventricles before ventricular contraction occurs.

9. The heartbeat can be heard, or monitored, by tracking changes in blood pressure through the period of filling and contracting of the atria (the diastolic period) and the contraction of the left ventricle (the systolic period). The waves of depolarization can also be measured directly; a recording of these pulses is called an electrocardiogram.

REVIEW

1. The circulatory system of vertebrates is composed of three elements: the heart, the _____, and the blood.

2. No cell of your body is more than _____ away from a capillary.

3. The function of _____ is to keep the blood plasma in osmotic equilibrium with the cell of the body.

4. In what order does blood flow through the four chambers of your heart, starting from the lungs?

5. The sound of turbulence caused by only partial closing of heart valves is called _____.

SELF-QUIZ

1. Which of the following is *not* part of the cardiovascular system?
   (a) the heart   (d) the liver
   (b) veins   (e) capillaries
   (c) arteries

2. Which of the following is a structural component of arteries, veins, *and* capillaries?
   (a) endothelium   (d) connective tissue
   (b) an elastic layer (e) all of the above
   (c) smooth muscle

3. In general, veins are larger in diameter than arteries. Why?
   (a) to reduce resistance to flow
   (b) to increase resistance to flow
   (c) because cholesterol builds up only in arteries and not in veins
   (d) because arteries are more elastic
   (e) none of the above

4. Erythrocytes are packed with
   (a) fibrinogen
   (b) serum albumin
   (c) hemoglobin
   (d) antibody proteins
   (e) platelets

5. Blood pumped out of the left ventricle moves into the
   (a) left atrium
   (b) pulmonary veins
   (c) right ventricle
   (d) aorta
   (e) superior vena cava

THOUGHT QUESTIONS

1. Instead of evolving an entire second open circulatory system—the lymphatic system—to collect water lost from the blood plasma during passage through the capillaries, why haven't vertebrates simply increased the level of serum albumin in their blood?

2. Starving animals often exhibit swollen bodies rather than emaciated ones in early stages of their deprivation. Why?

3. The hearts of the more advanced vertebrates pump blood entirely by pushing action. Why do you suppose hearts have not evolved that act like suction pumps, drawing blood into the heart as it expands, rather than pushing it out as the heart contracts?

FOR FURTHER READING

GOLDSTEIN, G., and A.L. BELZ: "The Blood–Brain Barrier," *Scientific American,* September 1986, pages 74-83. Many chemicals will not pass from the bloodstream to the cells of the brain. This article explains that this "barrier" is the result of the structure of brain capillaries, which possess an unusual array of membrane channels.

ROBINSON, T., S. FACTOR, and E. SONNEBLINK: "The Heart As a Suction Pump," *Scientific American,* June 1986, pages 84-91. Between birth and death our hearts beat millions of times. These authors argue that the heart is aided greatly in this Herculean task by a very clever trick: contraction compresses elastic elements within the heart muscles, which then bounce back to expand the ventricles.

ZUCKER, M.B.: *"The Functioning of the Blood Platelets,"* *Scientific American,* June 1980, pages 86-103. A description of the many roles of platelets in human health, with emphasis on their role in blood clotting.

FIGURE 37-1   Edward Jenner inoculating patients with cowpox in the 1790s and thus protecting them from smallpox. The underlying principles of vaccination were not understood until more than a century later.

THE VERTEBRATE IMMUNE SYSTEM

Chapter 37

Overview

Most diseases that afflict human beings are in fact microbial infections, invasions of the body by viruses, bacteria, fungi, or protists. To defend against this onslaught, vertebrates possess a sophisticated screening system called the immune system that constantly checks the bloodstream and tissues for the presence of any foreign cells or molecules. When an infection is detected, the invading microbes are attacked and destroyed. Without such a defense against infection, no vertebrate can live for long.

For Review

Here are some important terms and concepts that you will encounter in this chapter. If you are not familiar with them, you should review them before proceeding.

AIDS (Chapters 1 and 25)

Macrophage (Chapter 32)

Lymphocyte (Chapters 32 and 36)

The lymphatic system (Chapter 36)

Few of us pass through childhood without being sick. Measles, chickenpox, mumps—these are rites of passage, childhood illnesses that most of us experience before our teens. They are diseases of childhood because most of us suffer through them as children and never catch them again. Once you have had measles, you are immune. The mechanism that provides you with this immunity to such childhood diseases is the **immune system,** your body's means of resisting infection. Your immune system is the backbone of your health, protecting you not only from measles, but also from many other far more serious diseases. It is only in the last few years, as the result of an explosion of interest and knowledge, that biologists have begun to get a clear idea of how our immune system works. What we have learned is the subject of this chapter.

DISCOVERY OF THE IMMUNE RESPONSE

In 1796 an English country doctor named Edward Jenner carried out an experiment that marks the beginning of the study of immunology (Figure 37-1). Smallpox was a common and deadly disease in those days, and only those who had previously had the disease and survived it were immune from the infection—except, Jenner observed, milkmaids. Milkmaids who had caught another, much milder form of "the pox," called cowpox (it was caught by people who associated with cows), rarely caught smallpox. It was as if they had already had the disease. Jenner set out to test the idea that cowpox conferred protection against smallpox. He deliberately infected people with material that induced cowpox, causing them to catch this mild illness, and many of them became immune to smallpox, just as he had predicted.

Jenner's work demonstrated that it is possible for the human body to protect itself against disease very effectively when it is able to make suitable preparations. We now know that smallpox is caused by a virus called *Variola* and that cowpox is caused by a different, although similar, virus. Jenner's patients injected with cowpox virus mounted a defense against the cowpox infection, a defense that was also effective against a later infection of the similar smallpox virus. Jenner's procedure of injecting a harmless microbe into a person or animal to confer resistance to a dangerous one is called **vaccination.** Modern attempts to develop resistance to malaria, herpes, and other diseases are focusing on the virus **vaccinia,** which is related to the cowpox virus used by Jenner. Such attempts are using the methods of genetic engineering (see Chapter 15) to incorporate genes encoding the protein-polysaccharide coats of the other viruses into the chromosome of vaccinia. Since vaccinia does not cause disease, it allows the body to be exposed to the protein coats of the other viruses in a harmless way, thus enabling the body to build up resistance to them.

It was a long time before people learned how one microbe can confer resistance to another, however. Further important information was added a half century after Jenner by Louis Pasteur of France. Pasteur was studying fowl cholera, a serious disease in chickens that we now know to be caused by a bacterium. From diseased chickens, Pasteur could isolate a culture of bacteria that would elicit the disease if injected into other healthy birds. One day Pasteur accidentally left his bacterial culture out on a shelf at the end of the day and went on vacation. Two weeks later he returned and injected this extract into healthy birds. The extract had been weakened; the injected birds became only slightly ill and then recovered. Surprisingly, however, the vaccinated birds could not then be infected with fowl cholera! They stayed healthy even if injected with massive doses of active fowl cholera bacteria, whereas control chickens receiving the same injections all died. Clearly something about the bacteria could elicit immunity, if only the bacteria did not kill the bird first.

FIGURE 37-2

The outer surface of a cell is not smooth, but rather a tangle of glycolipid and protein embedded within the membrane. The glycolipids are typically chains of sugar molecules attached to a lipid stalk anchored in the lipid bilayer. Glycolipid molecules often serve as highly specific cell surface markers that identify specific cell types. The two major kinds of transmembrane protein are transport channels and receptors. Transport channels import ions, sugars, and other molecules into the cell. Receptors bind hormones, growth factors, neurotransmitters, and, in the case of immune receptors, other proteins.

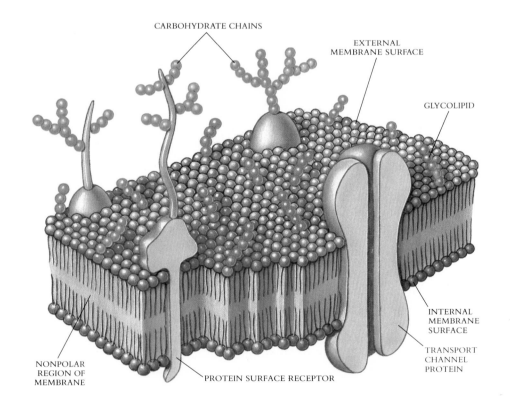

CARBOHYDRATE CHAINS
EXTERNAL MEMBRANE SURFACE
GLYCOLIPID
INTERNAL MEMBRANE SURFACE
TRANSPORT CHANNEL PROTEIN
NONPOLAR REGION OF MEMBRANE
PROTEIN SURFACE RECEPTOR

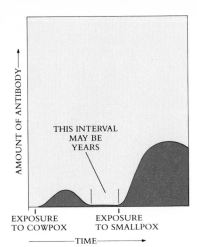

FIGURE 37-3

Immunity to smallpox in Jenner's patients occurred because their inoculation with cowpox stimulated their bodies to produce antibodies that would recognize either cowpox or smallpox. Some of these antibodies, and the cells that produce them, remain in the bloodstream for a long time. A second exposure, this time to smallpox, stimulates the body to produce very large amounts of antibody much more rapidly than before.

AMOUNT OF ANTIBODY→

THIS INTERVAL
MAY BE
YEARS

EXPOSURE
TO COWPOX

EXPOSURE
TO SMALLPOX

——TIME——→

We now know what that "something" was: molecules protruding from the surface of the bacterial cells. Every cell has on its surface a tangle of proteins, carbohydrates, and lipids (Figure 37-2), and it was the presence of foreign molecules on the surface of the cholera bacteria to which the chickens were responding. These bacterial cell surface molecules are different from any of the bird's own. "Non-self" molecules such as these are called **antigens.** Chickens injected with heat-killed fowl cholera bacteria are immune to later infection because the bacterial antigens cause the chickens to produce proteins called **antibodies.** These antibodies are able to recognize any future cholera invaders, treated or normal, and prevent them from causing disease (Figure 37-3). The production of antibodies directed against a specific antigen is called an **immune response.** The immune response is one component of a complex system of recognition and defense that we call the immune system.

> An immune response takes place when foreign proteins, called antigens, cause the production of other proteins, called antibodies, which recognize any antigens of the same sort with which they may come into contact in the future.

THE CELLS OF THE IMMUNE SYSTEM

Our immune system is not localized to one place in the body, nor is it controlled by any central organ such as the brain. Rather, it is composed of a host of individual cells, an army of defenders that rush to the site of an infection to combat invading microorganisms. These cells, the white blood cells mentioned in Chapter 36, arise in the bone marrow and circulate in blood and lymph. Of the 100 trillion cells in an adult human being, two in every hundred (2×10^{12}) are white blood cells. Although not bound together, the body's white blood cells exchange information and act in concert as a functional, integrated system. They are found not only in blood and lymph, but also in lymph nodes, spleen, liver, thymus, and bone marrow (Figure 37-4).

White blood cells are larger than the red blood cells that ferry oxygen to the body's tissues, but they are formed from the same cells in bone marrow, called **hematopoietic stem cells** (Figure 37-5). Unlike mature red blood cells, all white blood cells have a nucleus. Four kinds of white blood cells are involved in the immune system: **phagocytes, T cells, B cells,** and **natural killer cells.** (T and B cells are collectively referred to as **lymphocytes.**). Natural killer cells **lyse** bacterial cells: they destroy them by rupturing their cell membranes.

> The immune system is composed of white blood cells. Four principal classes are involved: phagocytes (including macrophages), natural killer cells, and two kinds of lymphocytes (T cells and B cells).

FIGURE 37-4

The principal organs of the
immune system are the lymph
nodes, spleen, thymus, and bone
marrow. Lymphatic vessels link
the scattered organs. Within them
a river of lymph sweeps
circulating cells through the
channels and dispatches a host of
white blood cells to battle
infection. Lymph nodes cluster
along the vessels, with major
groups in the groin, abdomen,
armpits, and neck.

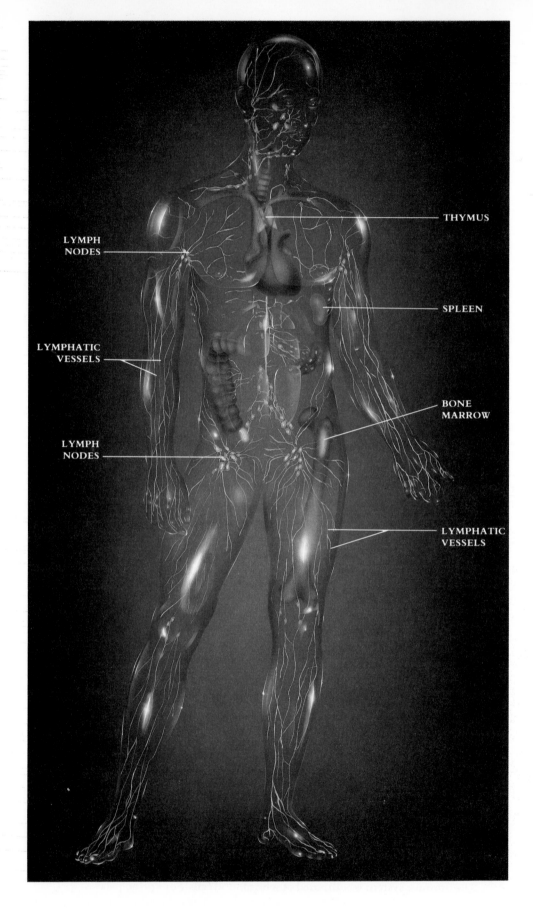

THYMUS

LYMPH
NODES

SPLEEN

LYMPHATIC
VESSELS

BONE
MARROW

LYMPH
NODES

LYMPHATIC
VESSELS

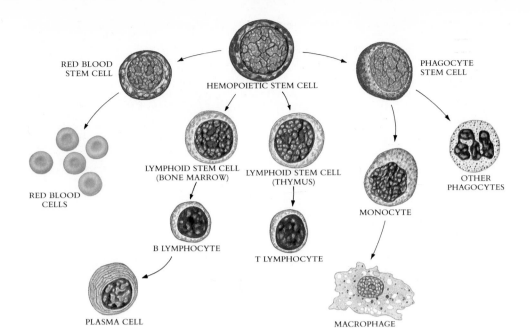

FIGURE 37-5

White blood cells (lymphocytes) and red blood cells (erythrocytes) both develop in adults from hemopoietic stem cells in red bone marrow.

THE ARCHITECTURE OF THE IMMUNE DEFENSE

The vertebrate immune system is part of a multilayered defense that uses both immediate rapid responses to infection and several levels of immune protection (Figure 37-6). The vertebrate body's first defense against infection is a patrolling army of natural killer cells and macrophages, which attack and destroy invading microorganisms and eliminate infected cells. In addition, as you will see, the immune response to infection involves many other kinds of cells.

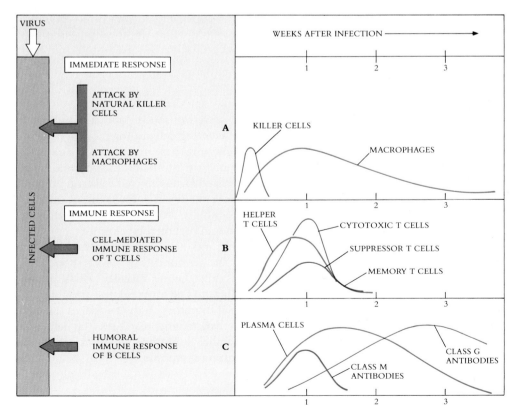

FIGURE 37-6

Time course of the immune response. This overview will serve as a guide for the discussion that follows. A virus infection is illustrated here, although other types of infection produce a similar chain of events. The immediate response of natural killer cells and macrophages (A) occurs within hours of infection and peaks in 1 or 2 days. This is followed by mobilization of helper T cells, which simultaneously initiate two parallel immune responses, cell-mediated (B) and humoral (C). These two responses both peak within a few weeks, but antibody produced in the humoral response may persist in the bloodstream far longer. The various cell classes mentioned in B and C are responsible for the persistence of the immune response.

FIGURE 37-7

How an infection causes fever. At the site of infection *(1)*, macrophages release interleukin-1, which passes through the bloodstream *(2)* to the brain. There it stimulates the hypothalamus *(3)*, the body's thermostat, triggering it to set a higher temperature. To raise the body's temperature to the new setting, the brain sends nerve impulses *(4)* to muscles, ordering them to contract; as a result, the body shivers, producing heat. Other nerve impulses *(5)* order blood vessels near the skin to constrict, minimizing heat loss. The higher temperature aids the immune response and inhibits the growth of invading microorganisms. When the immune system begins to make headway against the infection, thus reducing the numbers of invading microbes, production of interleukin-1 by macrophages stops, the fever "breaks," and body temperature soon falls to normal values.

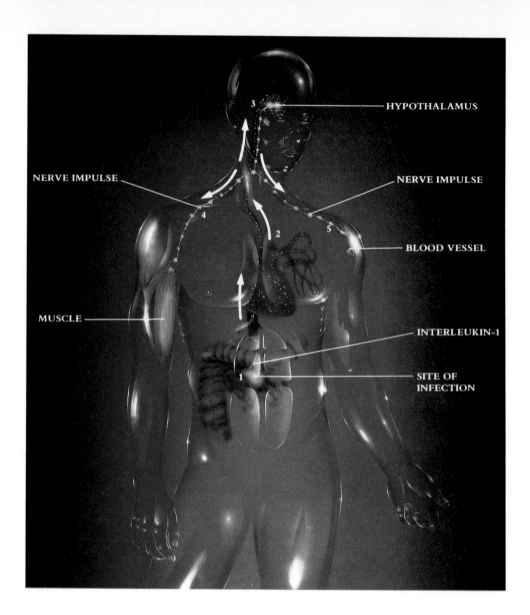

Sounding the Alarm

Macrophages respond to a virus encounter by secreting soluble proteins known as **monokines** (the name refers to the fact that the secreting cell is a form of monocyte), which signal the onset of the immune response. Among these monokines are **gamma-interferon,** which activates other monocytes to mature into macrophages, and **inter-leukin-1,** which activates T cells that have recognized virus-infected cells and prepares them to proliferate. Interleukin-1 is responsible for the onset of fever that is often associated with infection (Figure 37-7). Macrophages also attack and engulf individual viral particles and display elements of the viral protein coat on their exterior.

The immediate response of natural killer cells and macrophages is not adequate to eliminate many infections, but it buys time for the immune system to respond. The key element in this response is the class of T cells called **helper T4 cells,** which are sensitive to the alarm being broadcast by macrophages. The interleukin-1 produced by macrophages activates these helper T4 cells. When receptors on their surfaces (called T receptors) encounter viral antigens presented to them by macrophages, the helper T4 cells simultaneously initiate two different parallel immune responses, the **cell-mediated immune response** and the **humoral immune response.**

The Cell-Mediated Immune Response

The activation of helper T4 cells by interleukin-1 and the binding of viral antigen to these activated helper T4 cells unleashes a chain of events known as the **cell-mediated immune response,** in which special **cytotoxic T cells** (*cytotoxic* means "cell-poisoning") recognize and destroy infected body cells. Helper T4 cells initiate the response, activating both cytotoxic T cells and other elements of the system.

The Humoral Immune Response

When a helper T4 cell is stimulated to respond to a foreign antigen, it not only activates the T cell–mediated immune response as described above, but also simultaneously activates a second, more long-range defense, referred to as the **humoral** or antibody response. The key player in this stage of defense against infection is another kind of lymphocyte, the **B cell.**

Each B cell has on its surface about 100,000 copies of a protein called a **B receptor;** each B cell bears a different version of the receptor. B receptors are designed to bind to foreign proteins; because each cell's version is slightly different, each B cell is specialized to recognize a different foreign antigen. Because the body contains many different B cells, there is almost always at least one that will bind to the surface of *any* microorganism. For example, at the onset of a viral infection such as measles, the B receptors of one or more cells bind to virus antigens, either to free viruses or to viral antigens displayed by macrophages. These antigen-bound B cells are detected by helper T4 cells, which bind simultaneously to the attached antigen and to proteins specific to each particular individual on the B cell surface (Figure 37-8). When this happens, the helper T4 cells release factors called **lymphokines** that induce the B cell to proliferate.

After about 5 days and numerous cell divisions, a large clone of cells has been produced from each B cell that was stimulated by antigen to proliferate. All but a few of the proliferating B cells then stop reproducing and dedicate all of their resources to producing and secreting more copies of the B receptor protein that responded to the antigen. The circulating receptor proteins are called **antibodies,** and also **immunoglobulins.** The secreting B cells, called **plasma cells,** live only a few days but secrete a great deal of antibody during that time. One cell will typically secrete more than 2000 molecules per second. Antibodies constitute about 20% by weight of the total protein in blood plasma. Antibodies cover a virus or bacterium by binding to molecules on its surface, and mark it for destruction by macrophages or other mechanisms.

Some members of the clone of proliferating B cells do not go on to differentiate into plasma cells, but rather persist as circulating lymphocytes, called **memory B cells.** These provide an accelerated response to any later encounter with the stimulating antigen. This is why immune individuals are able to mount a prompt defense against infection.

> **Parallel to the cell-mediated immune response, helper T4 cells also initiate a second immune response called the humoral immune response. In this immune response B cells recognize foreign antigens and, if activated by helper T4 cells, proceed to produce large quantities of antibody molecules directed against the antigen. The antibodies bind to any antigen they encounter and mark for destruction cells or viruses bearing the antigen.**

HOW DO IMMUNE RECEPTORS RECOGNIZE ANTIGENS?

The cell surface receptors of lymphocytes (T cells and B cells) can recognize specific antigens with great precision. Even single amino acid differences between proteins can often be discriminated, with a receptor recognizing one form and not the other. This high degree of precision is a necessary property of the immune system, since without it the identification of foreign antigens would not be possible in many cases; the differences between "self" and foreign ("non-self") molecules can be very subtle.

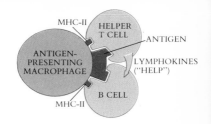

FIGURE 37-8

The humoral immune response generally involves the participation of three immune cell types: B cells, helper T cells, and antigen-presenting macrophages. B receptors on the surface of B cells recognize and bind to one portion of the antigen. The antigen is usually presented by a macrophage (although naked virus and bacterial antigens can also be effective). T receptors on the surface of helper T cells bind to a different portion of the antigen. The T cell then releases factors called lymphokines that induce the B cell to proliferate and mature into a specialized plasma cell dedicated to production of the antigen-specific antibody. Binding of the three cell types together in the humoral immune response requires mutual recognition, provided by special proteins (MHC-II) that identify cells of the humoral immune response system.

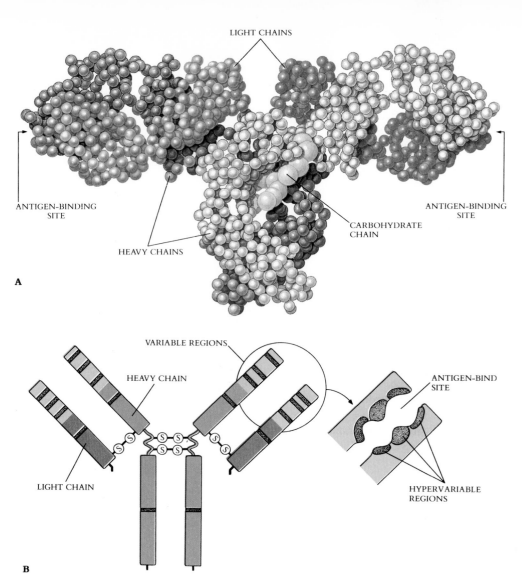

FIGURE 37-9

The structure of an antibody molecule.

A A molecular model of an antibody molecule. Each amino acid is represented by a small sphere. The heavy chains are colored blue, the light chains red. The four chains wind about one another to form a Y shape, with two identical antigen-binding sites at the arms of the Y and a tail region that serves to direct the antibody to a particular portion of the immune response.

B A schematic drawing of an antibody molecule. Each molecule is composed of two identical light *(L)* chains and two identical heavy *(H)* chains. Carbohydrate is sometimes associated with the heavy chain. While the antigen-binding sites are formed by a complex of both H and L chains, the tail region is formed by H chains alone.

LIGHT CHAINS

ANTIGEN-BINDING SITE

ANTIGEN-BINDING SITE

CARBOHYDRATE CHAIN

HEAVY CHAINS

A

VARIABLE REGIONS

HEAVY CHAIN

ANTIGEN-BIND SITE

LIGHT CHAIN

HYPERVARIABLE REGIONS

B

Antibody Structure

Antibody molecules (also called immunoglobulins) consist of four polypeptide chains (Figure 37-9). There are two identical short strands, called **light chains,** and two identical long strands, called **heavy chains.** The amino acid sequences of the two kinds of chains suggest that they evolved from a single ancestral sequence of about 110 amino acids. Modern light chains contain two of these basic 110 amino acid units or domains; and heavy chains contain three, or in some cases four, of them. The four chains are held together by disulfide (——S—S——) bonds, forming a Y-shaped molecule.

Through studies of the amino acid sequences of different antibody molecules, it has become clear that the specificity of antibodies resides in the two arms of the Y, whereas the stem determines what role the antibody plays in the immune response. The terminal half of the two arms of the Y is where most of the variation in sequence between antibodies of different specificity is found. Within this variable region, three small "hypervariable" segments come together at the end of each arm to form a cleft that acts as the binding site for the antigen. Both arms always have exactly the same cleft. The specificity of the antibody molecule for an antigen depends on the precise shape of these clefts. An antigen fits into one of the clefts like a hand into a glove; changes in the amino acid sequence of an antibody can alter the shape of its clefts and by doing so change the antigen that can bind to that antibody, just as changing the size of a glove will alter which hand can fit it.

An antibody molecule recognizes a specific antigen because it possesses two clefts or depressions into which an antigen can fit, much as a substrate fits into an enzyme's active site. Changes in the amino acid sequence at the position of these clefts alter their shape, and thus change the identity of the antigen that is able to fit into them.

Antibodies with the same variable regions have identical clefts and therefore recognize the same antigen, but they may differ from one another in the stem portion of the antibody molecule. It is the stem portion that affects whether the antibody functions in the primary or secondary immune response (Figure 37-10). The antibodies produced in the first week of an infection (the **primary immune response**) possess a class M heavy chain. These are directed at activating a group of 20 defensive proteins called the **complement system**. After the first week the class of heavy chain that the body employs switches from M to G. These new class G antibodies (the **secondary immune response**) are very efficient at eliciting macrophage attack. Three other kinds of heavy chain play minor roles in the immune response.

HOW CAN THE IMMUNE SYSTEM RESPOND TO SO MANY DIFFERENT FOREIGN ANTIGENS?

The vertebrate immune response is capable of recognizing as foreign practically any "non-self" molecule presented to it, that is, literally millions of different antigens. It is estimated that a human being is able to make between 10^6 and 10^9 different antibody molecules. It has long been a puzzle how this is done, since research has demonstrated that there are only a few hundred receptor-encoding genes on vertebrate chromosomes, not millions. What process, then, is responsible for generating the great diversity of receptor and receptor-derived antibodies? Two ideas have been proposed:

1. The **instructional theory** proposed that the antigen elicited the appropriate receptor, like a shopper ordering a custom-made suit.
2. The **clonal theory** proposed that there were indeed millions of different kinds of stem cells in the bone marrow, and that an antigen caused those few encoding an appropriate receptor to proliferate, creating a clone of descendants expressing the appropriate receptor.

We now know the clonal theory to be correct. Within our bone marrow, the stem cells destined to form B cells and T cells express an incredible diversity of receptor-encoding genes; each cell encodes only one form of B and T receptor, but every cell is different from practically every other one.

How do vertebrates generate millions of different stem cells, each producing a unique receptor, when their chromosomes encode only a few hundred copies of such

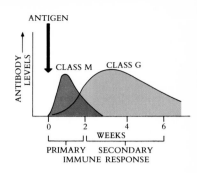

FIGURE 37-10

The first antibodies produced in the humoral immune response are class M antibodies, those with class M heavy chains. This initial wave of class M antibodies, called the primary immune response, peaks after about 1 week and is followed by a far more extended production of antibodies with a different heavy chain, class G. The production of class G antibodies is called the secondary immune response.

FIGURE 37-11

The immune receptor response library. The gene specifying the variable region of an antibody protein is assembled from segments. Each segment typically exists in several copies, and in some cases hundreds. Thus for the B receptor heavy chain there are several hundred copies of the "variable" or *V* segment, 20 copies of the "diversity" or *D* segment, and 6 copies of the "joining" or *J* segment. One specific heavy chain gene is produced in a particular stem cell by a maturation process that selects one copy of *V*, of *D*, and of *J* at random. A single gene encoding a specific light chain is similarly assembled at random from clusters of *V*, *D*, and *J* segments, and the heavy and light chains joined to form a complete B or T receptor. There are millions of different possible combinations.

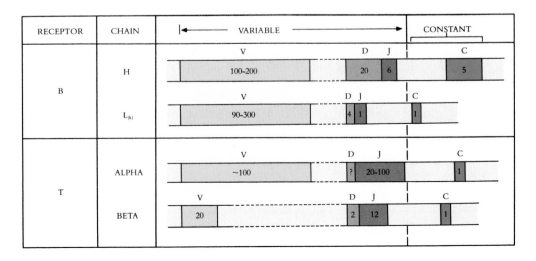

ALLERGY

Although the human immune system provides very effective protection against viruses, bacteria, parasites, and other microorganisms, sometimes it does its job too well, mounting a major defense against a harmless antigen. Such immune responses are called **allergic reactions.** Hay fever, the sensitivity that many people exhibit to proteins released from plant pollen, is a familiar example of an allergy. In response to as little as 20 pollen grains per cubic meter, a sensitive person's immune defense will swiftly mount a defense. Many other people are sensitive to proteins released from the feces of a minute house-dust mite called *Dermatophagoides* (Figure 37-A), which lives in the house dust present on mattresses and pillows and consumes the dead skin scales that all of us shed in large quantities daily. Many people sensitive to feather pillows are in reality allergic to the mites that are residents of the feathers.

What makes an allergic reaction uncomfortable, and sometimes dangerous, is the involvement of antibodies with a kind of heavy chain called "E." Antibodies with class E heavy chains are typically attached to mast cells. The binding of antigen to these antibodies initiates an inflammatory response; histamines and other powerful chemicals called **mediators** are released from the mast cells, causing dilation of blood vessels and a host of other physiological changes. Sneezing, runny noses, fever—all the symptoms of hay fever result. In some instances when the body possesses substantial amounts of class E antibody directed against an antigen, allergic reactions can be far more dangerous than hay fever, resulting in anaphylactic shock in which swelling makes breathing difficult.

Not all antigens are **allergens,** initiators of strong immune responses.

Nettle pollen, for example, is as abundant in the air as ragweed pollen, but few people are allergic to it. Nor do all people develop allergies; the sensitivity seems to run in families. It seems that allergies require both a particular kind of antigen and a high level of class E antibody: the antigen must be able to bind simultaneously to two adjacent E antibodies on the surface of the mast cell in order to trigger the mast cell's inflammatory response, and only certain antigens are able to do this; the class E antibodies must be produced in large enough amounts that many mast cells will have antibody molecules spaced closed to one another, rather than the few such cells typical of a normal immune response. Only certain people churn out these high levels of E antibody. It is this combination of appropriate antigen on the one hand and inappropriately high levels of particular class E antibodies on the other hand that produce the allergic response.

Hay fever and other allergies are often treated by injecting sufferers with extracts of the antigen, a process called desensitization. Allergy shots work best for pollen allergies and for allergy to the venom of bee and wasp stings; they are not effective against food or drug allergies. The strategy of desensitization is to produce high levels of class G antibody in the bloodstream, so that when a particular antigen is encountered, it will be mopped up by the G antibodies before encountering E antibodies on mast cells. Actually, there seems to be little correlation between levels of circulating G antibody and successful desensitization, and it is not clear why the procedure works as well as it does. A more ideal therapy would be to lower the amounts of class E antibody produced during the immune response, an approach that is being actively investigated.

FIGURE 37-A

The house-dust mite
Dermatophagoides.

genes? They accomplish this by rearranging parts of the antibody as each antibody is produced. Antibody genes do not exist as single sequences of nucleotides, as do the genes encoding all other proteins, but rather are first *assembled* by joining together three or four DNA segments. Each segment, corresponding to a region of the antibody molecule, is encoded at a different site on the chromosome. These chromosomal sites are composed of a cluster of similar sequences (Figure 37-11), each sequence varying from others in its cluster by small degrees. When an antibody is assembled, one sequence is selected at random from each cluster, and the DNA sequences selected from the various clusters are brought together by DNA recombination to form a composite gene. The process is not unlike going into a large department store and picking at random one coat, one shirt or blouse, one pair of pants, one pair of socks, and one pair of shoes—few people would come out of the store wearing the same outfit.

Because a cell may end up with any heavy-chain gene and any light-chain gene during its maturation, the total number of different antibodies possible is staggering. It is: HEAVY (16,000 combinations) × LIGHT (1200 combinations) = 19 million different possible antibodies.

Every mature stem cell divides to produce a clone of descendant lymphocytes, and each cell of the clone carries the particular rearranged gene assembled earlier, when that stem cell was undergoing maturation. As a result, all of the cells of a clone produce the specific immune receptor encoded by that stem cell, and no other. Because an adult vertebrate contains many millions of stem cells and each stem cell undergoes the maturation process independently, cells specializing in millions of different combinations of B receptor occur.

B receptors are encoded by genes that are assembled during stem cell maturation by rearrangement of the DNA. Because each component is selected at random from many possibilities, a vast array of different T receptors and B receptors are produced.

DEFEAT OF THE IMMUNE SYSTEM

All mammals and birds possess an immune system similar to the human one we have described in this chapter. During the evolutionary history of the vertebrates, however, several microbes have developed strategies, some of them quite successful, for defeating vertebrate immune defenses. As you might expect, these strategies are responsible for very serious diseases.

One of these strategies consists of a direct attack on the immune mechanism itself. If you had to design such an attack, perhaps the most sensitive target would be the T4 cells. Helper T4 cells are the key to the entire immune response (Figure 37-12), re-

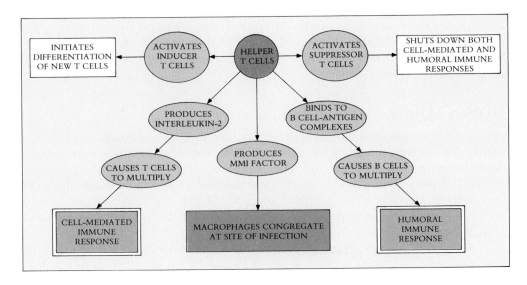

FIGURE 37-12

The many roles of the helper T cell in the immune response.

sponsible for initiating proliferation of both T cells and B cells, whereas the maturation of all T cells (including helper T4 cells) requires cooperation of inducer T4 cells. Without helper T4 cells and inducer T4 cells, the immune system is unable to mount a response to *any* foreign antigen.

AIDS is a deadly disease for just this reason. The AIDS virus mounts a direct attack on T4 lymphocytes, that is, on inducer T4 cells and helper T4 cells. Although most viruses are not specific, AIDS viruses target T4 cells, because the AIDS virus recognizes characteristic surface antigens associated with T4 cells.

AIDS-infected cells die, but only after releasing progeny viruses (see Figure 25-8) that infect other T4 cells, until the entire population of T4 cells is destroyed. In a normal individual, T4 cells make up 60% to 80% of circulating T cells; in AIDS patients T4 cells often become too rare to detect.

The combined effect of these responses to AIDS infection is to wipe out the human immune defense. An effective immune response, either a T cell–mediated cellular immune response or a B cell–mediated antibody response, is impossible without inducer T4 cells to mature the lymphocytes or helper T4 cells to initiate the response. With no defense against infection, any of a variety of otherwise commonplace infections proves fatal. It is for this reason that AIDS is a particularly devastating disease (see Figure 1-15).

> **AIDS destroys the ability of the immune system to mount a defense against any infection. The AIDS virus attacks and destroys T4 cells, without which no immune response can be initiated.**

Although the AIDS virus became a prominent cause of human disease only recently, possibly transmitted to humans from African green monkeys in Central Africa, it is already clear that AIDS is one of the most serious diseases in human history. The fatality rate of AIDS is 100%; no patient exhibiting the symptoms of AIDS has ever been known to survive more than a few years. The disease is *not* highly infectious; it is transmitted from one individual to another during the direct transfer of internal body fluids, typically in semen or vaginal fluid during sex and in blood during transfusions or by contaminated hypodermic needles. Not all individuals exposed to AIDS (as judged by antibodies in their blood directed against AIDS virus) have yet come down with the disease. There were about 1.5 million to 2 million such individuals in the United States by the mid-1980s, and it is estimated that at least 30% to 50%, and possibly all, of them will eventually contract AIDS. Most will die within 2 years of onset of the symptoms unless additional strategies for the treatment of AIDS are discovered first.

Efforts to develop a vaccine against AIDS continue, both by splicing portions of the AIDS surface protein gene into the vaccinia virus and by attempting to develop a harmless strain of AIDS. These approaches, although promising, have not yet proved successful and are limited by the fact that different strains of AIDS virus seem to possess different surface antigens. Like flu, AIDS indulges in some form of antigen shifting. Recent studies have described a stable coat protein that is a prime candidate for a vaccine. Drugs that inhibit specific genes of the AIDS virus or block the operation of enzymes critical to the synthesis of viral RNA are also being investigated. More practical advice about AIDS is presented in Chapter 42.

MONOCLONAL ANTIBODIES

Only a small portion of an antigen molecule actually fits into an antibody's recognition site. In the case of protein antigens, for example, this portion (sometimes called a **determinant**) is typically in the size range of two to six amino acids. Most antigens are much larger than this; as a result, different portions of the antigen molecule can fit into different antibody sites. A typical antigen will thus elicit many different antibodies, each fting a different portion of the antigen surface. Such an antibody response is said to be **polyclonal**.

Antibodies offer great promise in medicine and research, because they recognize biological molecules with exquisite precision. However, it is often critical that biological tools be specific in order to be useful, just as a letter must have a specific address to reach its destination—and a polyclonal response presents a whole phone book of potential addresses; to find one address, an investigator needs instead an antibody that is directed against only one determinant (a **monoclonal antibody**). In 1984 Cesar Milstein of England and George Köhler of Switzerland were awarded a Nobel Prize for learning how to engineer an antibody response that is monoclonal. They devised an easy procedure for isolating a single clone of plasma cells, all producing the same antibody molecule.

Milstein and Köhler mixed plasma cells that were producing antibody with cancer cells, malignant lymphocytes called myelomas. Neither plasma cells nor myeloma cells lived long when Milstein and Köhler mixed them together: plasma cells normally live only a few cell generations anyway, and the cancer cells they employed were mutant ones, unable to survive without a certain metabolite the cells could no longer synthesize. Some cells did live in the mixture, however, growing and dividing—these were a new kind of cell produced by the fusion of a plasma cell with a myeloma cell. These cell hybrids, called **hybridomas** (Figure 37-13), used the genes of the plasma cell to make the metabolite necessary for growth and were directed by the genes of the myeloma to ceaselessly grow and divide, as cancer cells. What made the experiment of profound importance was that the hybridoma cells continued to produce the antibody in which the plasma cell had specialized. Isolating single hybridoma cells from the mixture, Milstein and Köhler obtained rapidly growing cell lines that could be maintained in culture indefinitely, every cell of the culture producing the same antibody molecule—monoclonal antibodies.

Monoclonal antibodies have proved to be of great importance to industry, because they can be used to purify specific molecules from complex mixtures. Interferon, present in only trace amounts in tissue extracts, was first purified in this way. Monoclonal antibodies have also revolutionized many aspects of biological research. It has proved possible, for example, to generate monoclonal antibodies directed against each of the many proteins that stud a cell's surface and in this way to learn a great deal about cell surface receptors. The T receptor that plays such an important role in this discussion was first isolated in 1984 by investigators employing a monoclonal antibody. In medicine, monoclonal antibodies offer great promise as vehicles for delivering specific therapies. There is an intensive search underway, for example, for antigens that occur only, or predominantly, on cancer cells, against which radioactive monoclonal antibodies could be targeted in order to selectively kill cancer cells.

FIGURE 37-13

A hybridoma dividing.

SUMMARY

1. Although immunity was discovered almost 200 years ago, we have only recently understood that resistance to disease is achieved by populations of white blood cells, collectively called the immune system.

2. The introduction of foreign proteins, called antigens, into a vertebrate brings about the production of specific proteins, called antibodies, which recognize the original antigens from that point on. This is called the immune response.

3. There are four types of white blood cell involved in the immune system: T cells, B cells, macrophages and other phagocytes, and natural killer cells.

4. Vertebrate lymphocytes possess on their cell surfaces immune receptors that are able to bind to specific foreign molecules.

5. The immune response is initiated by those T4 cells called helper T4 cells. Helper T4 cells, when stimulated by macrophages and antigen, simultaneously activate two parallel responses: the cell-mediated immune response, in which cytotoxic T cells attack infected body cells; and the humoral immune response, a longer-range defense in which B cells secrete a free form of B receptor called antibody that binds circulating antigen and marks cells or viruses bearing it for destruction.

6. The exquisite specificity of antibodies (the circulating form of B receptors) for particular antigens reflects the three-dimensional shape of a cleft in the molecule. Slight changes in the amino acid sequence alter the shape of the cleft and thus the identity of molecules able to fit into it.

7. Vertebrates can recognize many different antigens as foreign because the bone marrow of vertebrates contains many different stem cells, and each matures independently. During maturation, a stem cell assembles the two genes encoding its particular T and B receptors by splicing together component parts, randomly selecting each part from a large library of possibilities.

8. The immune receptors present on the surface of lymphocytes are unique among proteins in that the genes encoding them are assembled by DNA rearrangement. The two principal immune receptors are T receptors (present on T cells) and B receptors (present on B cells). Antibodies are secreted forms of B receptors.

REVIEW

1. The procedure, first used by Edward Jenner, of injecting a harmless microbe that is similar to a harmful one so that the body will produce antibodies that will recognize the harmful microbe if it is encountered sometime later, is called _____.

2. The two types of lymphocytes are _____ and _____.

3. An antibody molecule recognizes a specific _____ because the antibody molecule possesses specific binding sites formed by a complex of heavy and light chains.

4. The first kind of immune system cell to respond to an invading virus is either a killer cell or a _____.

5. Allergies require both a particular type of antigen and a high level of class _____ antibody.

SELF-QUIZ

1. When Pasteur injected a 2-week-old culture of fowl cholera bacteria into chickens, the vaccinated birds did not die. What would have happened if he had heat-killed the 2-week-old culture before injecting it?
   (a) all the vaccinated chickens would have died
   (b) the heat-killing would not have changed the result
   (c) some of the vaccinated chickens would have died
   (d) none of the above

2. Which of the following is the key cell type in the immune response?
   (a) helper T4 cell
   (b) B cell
   (c) cytotoxic T cell
   (d) mast cell
   (e) natural killer cell

3. Which of the following occurs *last* in an immune response to infection?
   (a) macrophages ingest and destroy infected cells
   (b) natural killer cells attack infected cells
   (c) B cells proliferate
   (d) macrophages secrete monokines
   (e) helper T4 cells are mobilized

4. How do vertebrates generate millions of different stem cells that each produce a unique receptor, when their chromosomes encode only a few hundred copies of these genes?
   (a) by mutation
   (b) by rearrangement of segments of DNA
   (c) by antigen-elicited receptor production
   (d) by meiotic recombination
   (d) none of the above

5. The AIDS virus is remarkably effective at short-circuiting the immune response because it infects what cell type?
   (a) B cells
   (b) T4 lymphocytes
   (c) red blood cells
   (d) mast cells
   (e) stem cells

THOUGHT QUESTIONS

1. Why do you suppose the human immune system encodes only about a hundred V genes and a few D and J genes, when much more diversity could be generated by encoding a thousand copies of each?

2. AIDS is a virus that destroys the human immune system by killing helper T4 cells, which are necessary to activate the immune response. The African green monkeys from which the AIDS virus is thought to have arisen do not suffer from AIDS. How do you imagine they have escaped this?

FOR FURTHER READING

BUISSERET, P.: "Allergy," *Scientific American,* August 1982, pages 86–95. An up-to-date account of what is known about this common problem, which stresses that allergy is a disorder of the immune system.

GALLO, R.: "The AIDS Virus," *Scientific American,* January 1987, pages 47–56. A current account of what is known about the AIDS virus, from one of the two scientists who first isolated it.

JARET, P.: "Our Immune System: The Wars Within," *National Geographic,* June 1986, pages 702–736. A very readable account of current progress in the study of man's immune system, with striking photographs by Lennart Nilsson.

LAWRENCE, J.: "The Immune System in AIDS," *Scientific American,* December 1985, pages 84–93. An excellent overview of how T cells function in the immune response, and how AIDS thwarts that response.

LEDER, P.: "The Genetics of Antibody Diversity," *Scientific American,* May 1982, pages 102–116. An account of how a few hundred genes are shuffled to make millions of antibody combinations, by the man who first worked it out.

MILSTEIN, C.: "Monoclonal Antibodies," *Scientific American,* October 1980, pages 66–74. The fusion of particular antibody-producing cells with cancer cells produces clones of cells that secrete antibody directed at only one antigen.

TONEGAWA, S.: "The Molecules of the Immune System," *Scientific American,* October 1985, pages 122–131. An account of what is currently known of the structure of the B and T receptors.

FIGURE 38-1 While most vertebrate cells are small, some nerve cells are an exception. From each toe of this giraffe, a single nerve axon extends all the way up to the back of the skull—a distance of 5 meters.

HOW ANIMALS TRANSMIT INFORMATION

Chapter 38

Overview

All animals except sponges possess special cells called neurons, which are capable of transmitting information. Just as electrical pulses pass down a phone line, so waves of electrical disturbance passing over the surfaces of neurons carry information from one part of the body to another. Like the dots and dashes of Morse code, all nerve signals are the same, differing only in their frequency and their point of origin.

For Review

Here are some important terms and concepts that you will encounter in this chapter. If you are not familiar with them, you should review them before proceeding.

Depolarization (Chapters 6 and 32)

Sodium/potassium pump (Chapter 6)

Passive ion channels (Chapter 6)

Carotene (Chapter 9)

Neuron (Chapter 32)

There are several ways in which one cell of your body can communicate with another. One simple way is by direct contact, with an open channel between the two cells that permits the passage of ions and small molecules. This method of communication is similar to depending on face-to-face interactions between adjacent people in running a city—a slow and uncertain method. In the body, communication of this kind is provided by gap junctions, discussed in Chapter 6. Direct contact provides a ready means of communication between adjacent cells but is not able to provide rapid and efficient communication between distant tissues.

It would be better, in terms of distant communication, if the city manager sent a letter instructing various persons what to do. The body organizes various tissues in just this way, sending chemical instructions to these tissues. The instructions are in the form of **hormones,** small chemical molecules that act as messengers within the body. The hormone "letters" are produced by one of several different **endocrine glands,** secreted into the bloodstream and carried around through the body by the circulatory system. Like a letter, each hormone has an address, a chemical shape that only the target tissue will recognize and act on; we will discuss hormones further in Chapter 40. However, this kind of command system may be too slow, just as the mail may sometimes be when you are waiting for an important letter.

If the message to be delivered to the leg muscles of your body is "contract quickly, we are being pursued by a leopard," a quicker means of communication than hormones is desirable. In a city a person in an emergency does not mail a letter but rather uses the telephone to shout for help, dialing 911 and requesting assistance. That, in effect, is just what the vertebrate body does. All complex animals possess specialized **neurons** (Figure 38-1), which as you will recall from Chapter 32 are nerve cells specialized for signal transmission. Neurons maintain an electrical charge on their outer surface by actively pumping certain ions out across their membranes. An electrical signal can pass down the length of a neuron just as electrical impulses pass down a phone line.

The command center of higher organisms is the **brain,** a precisely ordered but complicated maze of interconnected neurons, a large biological "computer." The brain is connected by a network of neurons both to the hormone-producing glands and to the individual muscles and other tissues. This dual channel of command permits great flexibility: the signals can be slow and persistent (hormones), fast and transient (nerve signals), or any combination of the two.

> **Two forms of communication integrate body functions in vertebrates. They are neurons, which transmit rapid electrical signals that report information or initiate a quick response in a specific tissue; and hormones, slower chemical signals that initiate a widespread prolonged response, often in a variety of tissues.**

In this chapter we focus on the neuron, the chief functional element of the nervous system. In the following chapter (Chapter 39) we shall examine the senses and the ways in which they collect information and transmit it along nerves to the central nervous system. We shall then consider the functional organization of the brain and the nervous system as a whole. In Chapter 40 we shall consider how the brain uses hormones to effect long-term changes in the body's performance.

THE NEURON

The body contains many different kinds of neurons, some of them tiny with a few projections, others bushy with projections, still others with extensions meters long (Figure 38-2). Despite the fact that individual neurons often differ in appearance, all

FIGURE 38-2

Vertebrate neuron types. Vertebrate neurons differ widely in structure. Neurons within the brain often possess extensive highly-branched dendrites, whereas sensory neurons (which carry signals from sense organs to the brain) typically have dendrites only in specific receptor cells. The axons of many motor neurons (which carry commands from the brain to muscles and glands) are encased at intervals by Schwann cells, as are some of the axons of sensory neurons.

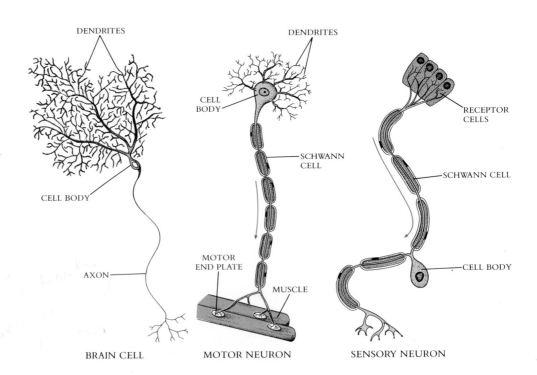

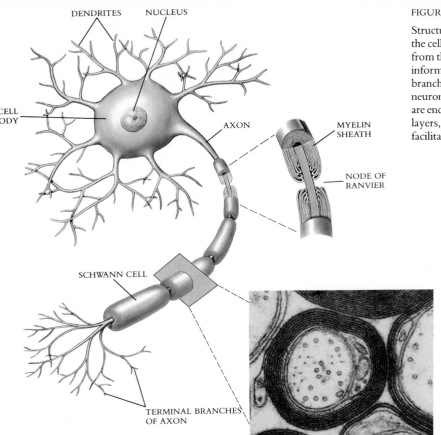

DENDRITES NUCLEUS

CELL BODY

AXON

MYELIN SHEATH

NODE OF RANVIER

SCHWANN CELL

TERMINAL BRANCHES OF AXON

FIGURE 38-3

Structure of a typical neuron. Many dendrites project out from the cell body; these serve to carry information to it. Outward from the cell body extends also a single long axon that carries information away from the cell body. This axon often ends in branches that connect to several muscles or to other nerves. In neurons that transmit information over long distances, the axons are encased at intervals by Schwann cells, whose many membrane layers, called a myelin sheath, electrically insulate the axons and so facilitate conduction of electrical impulses.

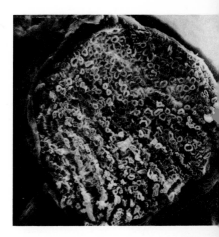

FIGURE 38-4

A nerve is a bundle of axons bound together by connective tissue. In this cross section of a bullfrog nerve magnified 1600 times, many myelinated neuron fibers are visible, each looking in cross section something like a Cheerio.

neurons have the same functional architecture. Extending from the body of all but the simplest nerve cells are one or more cytoplasmic extensions called **dendrites** (Figure 38-3). Most nerve cells possess a profusion of dendrites, many of which are branched, so that the body of the cell can receive inputs from many different sources simultaneously. Most nerve cells possess a single element called an **axon,** which may be quite long. The axons controlling motor activity in your legs are more than a meter long, and even longer ones occur in larger mammals. In a giraffe a single axon travels from the toe all the way up to the back of the skull, a distance of from 4 to 5 meters.

Most neurons are unable to survive alone for long; they require the nutritional support that is provided by companion **neuroglia cells.** More than half the volume of vertebrate nervous systems is composed of supporting neuroglia cells. In many neurons, including the motor neurons that extend from brain to muscles, the transmission of impulses along the very long axons is facilitated by neuroglial **Schwann cells,** which envelop the axon at intervals (see Figure 38-3) and act as electrical insulators. The Schwann cells form a **myelin sheath,** a flattened sheath of fatty material found in many, but not all, vertebrate neurons. Such a sheath is interrupted at frequent intervals by the **nodes of Ranvier,** where the axon is in direct contact with the surrounding intercellular fluid. An axon and its associated Schwann cells, or cells with similar properties, form a **myelinated fiber.** Bundles of nonmyelinated and myelinated neurons, which in cross section look somewhat like telephone cables (Figure 38-4), are called **nerves.**

THE NERVE IMPULSE

How does the nervous system work? The answer is not obvious from examining the amazingly complex network of nerve cells that extends throughout the vertebrate body. Looking at this network and trying to understand how the nervous system operates is a little like the experience of the American Indian who first encountered the white man's telegraph lines, the "singing wires," and tried to understand their function. The Indian understood that the wires transmitted information, but he could not see *how*—nothing seemed to move along the wires, which extended off over the horizon. Until the Indian learned of electricity, he could not appreciate that the slight humming heard from the wires was caused by the passing of information along the wires in the form of electrical currents. Understanding electricity is the key to his understanding how the telegraph works. In just this way, information passes along the nerves of the vertebrate body as electrical currents called **nerve impulses,** and the key to your understanding how the nervous system works is understanding these nerve impulses.

The Resting Potential

Electricity is generated by the separation of positive and negative charges; neurons generate electricity by setting up such a charge separation. The neurons pump positive sodium (Na^+) ions out of the cell through the active transport channels called sodium-potassium pumps (see Chapter 6). K^+ ions are pumped inward at the same time that Na^+ ions are pumped out, but the K^+ ions simply diffuse back out of the cell. Na^+ ions, however, cannot move back into the cell once pumped out, despite the low internal Na^+ ion concentration, because there are no open inward channels through the neuron membrane across which the Na^+ ions can pass. Thus, as a result of constantly pumping sodium ions out, the cell creates a concentration of positively charged sodium ions within its interior that is far less than that of the fluid surrounding it. The inside of the cell is more negatively charged than the outside; in other words, the neuron is **polarized.** This electrical potential difference across the membrane, which is called the **resting potential,** is the basis for the transmission of signals by nerves.

> Because of the activity of the Na^+ - K^+ ion transmembrane pumps and the impermeability of the cell to sodium, the surface of the neuron carries a positive charge relative to its interior. It is said to be polarized, and the electric potential difference across the membrane is defined as the resting potential.

Initiating a Nerve Impulse

The transmission of a signal by a neuron occurs in four phases, which we shall consider in order: initiation of the impulse, its transmission along a nerve fiber, its transfer to a target muscle or nerve, and its effect on the target tissue.

A neuron is stimulated when pressure, chemical activity, or some other stimulus is applied at some site on the neuron membrane. Such events alter the shape of proteins in the membrane, causing them to begin to admit Na^+ ions into the cell; these proteins are called **passive Na^+ ion channels.** The result is a temporary flood of Na^+ ions diffusing into the cell from the outside, where there are many more of them. This movement of positive sodium ions into the cell wipes out the local electrochemical gradient (polarization) and is therefore called a **depolarization** (Figure 38-5). So many ions rush

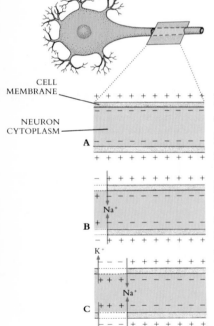

CELL MEMBRANE

NEURON CYTOPLASM

FIGURE 38-5

Transmission of a nerve impulse. Five separate sections of an axon are represented; the colored section is the interior.

A The interior of a resting neuron is more highly negatively charged than its exterior, due to the active transport of Na^+ ions outward.

B An impulse is initiated when the membrane becomes locally depolarized, Na^+ ions flooding inward across the membrane so that the neuron's interior becomes positively charged.

C The depolarization spreads, because the interior positive charge opens nearby transmembrane channels, thus permitting more Na^+ ions to enter; meanwhile, the initial site of depolarization gradually regains its original internal negative charge as a result of active transport of Na^+ ions out of the cell.

D and **E** As this process continues, a wave of depolarization moves down the neuron membrane, depolarizing in front and recovering behind. It is this self-propagating wave of depolarization that we call a nerve impulse.

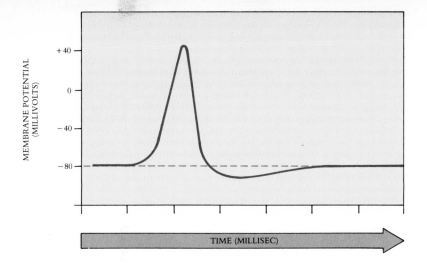

FIGURE 38-6
The changes in electrical potential associated with depolarization of a nerve membrane can be measured with minute, needle-sharp electrodes inserted into individual neurons. The change is always the same for any given neuron; each impulse alters the membrane potential by the same number of millivolts, and takes the same length of time to recover.

into the cell in response to the large concentration difference built up by the Na^+ pump, that the interior of the cell actually develops a positive charge relative to the outside.

Depolarization causes the passive Na^+ ion channels in the membrane to close, so that no more Na^+ ions enter the cell. Active transport of Na^+ ions by the sodium-potassium pumps then reestablishes the concentration differences that existed before the stimulus occurred. The whole process of stimulation and recovery is very fast—it takes only about 5 milliseconds. Fully 500 such cycles could occur, one after another, in the time it takes you to say the words "nerve impulse."

Why Transmission Occurs

During the few milliseconds when the site of stimulation is depolarized by the movement of sodium ions inward and before depolarization closes the passive Na^+ ion channels, the site of stimulation has less charge than the membrane surface surrounding it. This potential difference establishes a small, very localized electrical current in the immediate vicinity, which influences nearby closed passive Na^+ ion channels to open, permitting Na^+ ions to enter the cell and depolarizing these sites as well.

The depolarization thus spreads, depolarization at one site producing a local electrical current, which induces nearby passive Na^+ channels to open and so to depolarize the nearby site. In this way the initial depolarization passes outward over the membrane, spreading out in all directions from the site of stimulation. Like a burning fuse, the signal is usually initiated at one end and travels in one direction, but it would travel out from both directions if it were "lit" in the middle.

The time required for the Na^+–K^+ pumps to restore the original ion concentrations is called the **refractory period.** The nerve impulse moves in the outward direction only, because during the refractory period the membrane is not sensitive to depolarization. It cannot be depolarized again until the original charge difference is reestablished. By the time the recovery process has been completed, however, the signal has moved too far away for the original site to be influenced by it.

A nerve impulse arises because of a depolarization of the neuron membrane. The transient disturbance has electrical consequences to which nearby transmembrane proteins respond, spreading the disturbance.

The Action Potential

The self-propagating wave of depolarization, the nerve impulse, is an all-or-nothing affair. The amount of stimulation required to open enough passive sodium channels is called the **threshold value.** Depolarization is caused by a rapid change in membrane permeability and a corresponding shift in the balance of ions that is maintained during the resting stage. The shift of ions and consequent shift in electrical charges is called the **action potential.** (Figure 38-6). Because different neurons possess different densities of

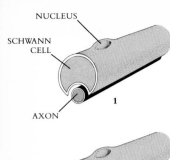

NUCLEUS

SCHWANN CELL

AXON

1

2

3

4

FIGURE 38-7

Development of the myelin sheath that surrounds many neuron axons involves the envelopment of the axon by a Schwann cell. It is the progressive growth of the Schwann cell membrane around the axon that contributes the many membrane layers characteristic of myelin sheaths. These layers act as an excellent electrical insulator.

passive Na^+ ion channels, different neurons exhibit different action potentials. For any one neuron, however, the action potential is always the same. Any stimulus that opens enough Na^+ ion channels will propagate an impulse outward as a wave of depolarization with a constant amplitude (the height or strength of the wave) corresponding to wide-open sodium ion channels. The stimulus is said to have "fired" the neuron. Every nerve impulse traversing that neuron has the same amplitude, signals differing from one another only in the frequency of these impulses.

Saltatory Conduction

As mentioned earlier, many vertebrate neurons possess axons sheathed at intervals by neuroglial Schwann cells. These cells envelop the axon (Figure 38-7), wrapping their cell membrane around it many times to produce a series of lipid-rich membrane layers, the myelin sheath. The myelin sheath acts as an electrical insulator that prevents the transport of ions across the neuron membrane beneath it, creating a region of high electrical resistance on the axon.

Schwann cells are spaced along such an axon one after the other, with nodes of Ranvier separating each Schwann cell from the next (Figure 38-8). These nodes are critical to the propagation of the nerve impulse in these cells. Within the small gap represented by each node, the surface of the axon is exposed to the intercellular fluid surrounding the nerve. The pumps and channels that move ions across the neuron membrane are concentrated in this zone. The direct fluid contact permits ion transport to occur through the pumps and an action potential to be generated. The action potential is not propagated by a wave of membrane depolarization traveling down the axon, since this is prevented by the insulating Schwann cells. Instead, the action potential jumps as an electrical current from one node to the next. When the current reaches a node, it acts as a stimulus that opens passive Na^+ ion channels; in so doing, it generates a potential difference large enough to create a current that reaches the next node. The arrival of the current at that node opens its passive sodium channels, creating another current that passes on to the next node, and so on. This very fast form of nerve impulse conduction is known as **saltatory conduction** (from the Latin *saltare*, "to jump"). An impulse conducted in this fashion moves very fast, up to 120 meters per second for large-diameter neurons. Saltatory conduction is also very cheap metabolically for the cell, since there is far less membrane depolarization for the ion pumps to deal with when only the nodes are undergoing depolarization than when the entire nerve surface is. Incidentally, the disease multiple sclerosis results in the destruction of large patches of the myelin sheath; the resulting slower transmission of signals in the nervous system results in the characteristic symptoms of the disease.

> **Not all nerve impulses propagate as a wave of depolarization spreading along the neuronal membrane. Along some vertebrate nerves, impulses travel very much faster by jumping along the membrane, leaping electrically over insulated portions.**

TRANSFERRING INFORMATION FROM NERVE TO TISSUE

An action potential passing down an axon eventually reaches the end of the axon. That end is often branched; it may be associated either with several dendrites of other nerve cells or with sites on muscle or secretory cells. These associations of nerves with other cells are called **synapses.** When the tip of a vertebrate axon is examined carefully, it becomes apparent that it does not actually make contact with the target cell it approaches. There is a narrow intercellular gap, 10 to 20 nanometers across, separating the axon tip and the target cell (Figure 38-9). This gap is called a **synaptic cleft.** Synaptic clefts are characteristic of all vertebrate nerve junctions, except for some of the synapses in the brain. When a nerve signal arrives at a synaptic cleft, it passes across the gap CHEMICALLY. The membrane on the axonal side of the synaptic cleft is called the **presynaptic membrane.** When a wave of depolarization reaches the presynaptic membrane, it

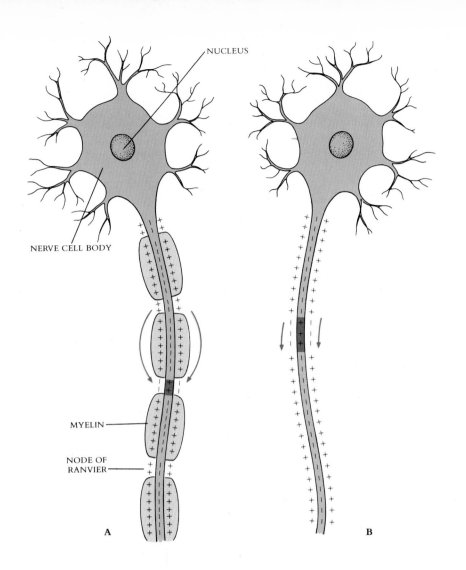

NUCLEUS

NERVE CELL BODY

MYELIN

NODE OF
RANVIER

A

B

FIGURE 38-8

The insulating properties of
myelin sheaths have a major
impact on the way in which nerve
impulses are propagated along
axons.

A In a myelinated fiber, the
wave of depolarization jumps
electrically from node to node
without ever transversing the
insulated membrane segment
between nodes. This is called
saltatory conduction. Just as
running is much faster than
walking because there is more
distance between each step, so
saltatory conduction is much
faster than direct conduction,
because there is more distance
between each current-induced
depolarization.

B In an unmyelinated fiber, the
wave of depolarization traverses
the entire length of the axon, each
portion of the membrane
becoming depolarized in turn, like
a row of falling dominos.

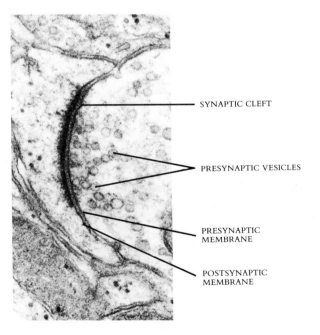

SYNAPTIC CLEFT

PRESYNAPTIC VESICLES

PRESYNAPTIC
MEMBRANE

POSTSYNAPTIC
MEMBRANE

FIGURE 38-9

A synaptic cleft between two
neurons. The right-hand neuron is
rich in presynaptic vesicles.

stimulates the release of neurotransmitter chemicals into the cleft. These chemicals rapidly pass to the other side of the gap. Once there, they combine with receptor molecules in the membrane of the target cell, which is called the **postsynaptic membrane.** By doing so, they cause ion channels to open.

> **A synapse is a junction between an axon tip and another cell, almost always including a narrow gap, which the impulse cannot bridge. Passage of the impulse across the gap is by chemical signal from the axon.**

The great advantage of a chemical junction such as this, compared with a direct electrical contact, is that the nature of the chemical transmitter can be different in different junctions, permitting different kinds of responses. More than 60 different chemicals have been identified that act as specific neurotransmitters or act to modify the activity of neurotransmitters. The events that occur within the synaptic cleft when a nerve signal arrives depend very much on the identity of the particular neurotransmitter chemical that is released into the cleft. To understand what happens, we shall first look at the junction between a nerve and a muscle cell, where the situation is a simple one, and then consider nerve-to-nerve junctions, where the situation is more complex.

Nerve-to-Muscle Connections

In synapses with muscle cells, called **neuromuscular junctions** (Figure 38-10), the neurotransmitter is **acetylcholine.** Passing across the gap, the acetylcholine molecules bind to receptors in the postsynaptic membrane, opening passive Na^+ ion channels (see Figure 33-14). During the millisecond that the channels are open, some 10^4 ions flow inward (Figure 38-11). This ion flow depolarizes the postsynaptic muscle cell membrane, which initiates a wave of depolarization that passes down the muscle cell. This wave of depolarization permits the release of calcium ions, which in turn triggers muscle contraction.

> **At a neuromuscular junction, acetylcholine released from an axon tip depolarizes the muscle cell membrane, permitting the release of calcium ions, which trigger muscle contraction.**

FIGURE 38-10

The neuromuscular junction.
A The terminal branch of the axon does not actually touch the muscle; the intervening space is the synaptic cleft. The axonal tip is typically rich in neurotransmitter-containing vesicles. In most neuromuscular junctions stimulation of the nerve occurs by fusion of these vesicles with the depolarized axonal membrane, releasing the neurotransmitter into the synaptic cleft. In other synapses, the vesicles appear to serve a storage function, the depolarization of the axonal membrane serving to open transmembrane channels specific to the neurotransmitter and so permitting cytoplasmic molecules of the neurotransmitter to stream out into the cleft.
B A neuromuscular junction, magnified 100 times. The axon has branched, different branches extending to different muscle fibers. Each junction of axon tip and muscle fiber operates an individual neuromuscular junction; such a junction is called a **motor end plate.**

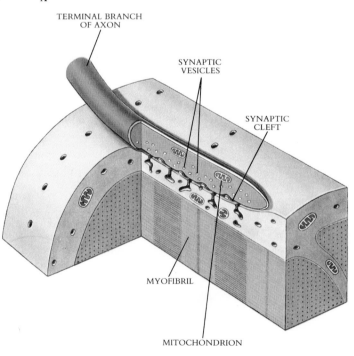

A

TERMINAL BRANCH
OF AXON

SYNAPTIC
VESICLES

SYNAPTIC
CLEFT

MYOFIBRIL

MITOCHONDRION

B

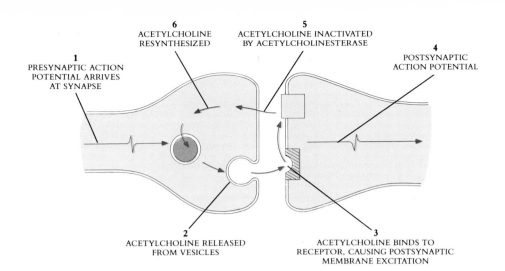

FIGURE 38-11

The sequence of events in synaptic transmission.

6 ACETYLCHOLINE RESYNTHESIZED

5 ACETYLCHOLINE INACTIVATED BY ACETYLCHOLINESTERASE

1 PRESYNAPTIC ACTION POTENTIAL ARRIVES AT SYNAPSE

4 POSTSYNAPTIC ACTION POTENTIAL

2 ACETYLCHOLINE RELEASED FROM VESICLES

3 ACETYLCHOLINE BINDS TO RECEPTOR, CAUSING POSTSYNAPTIC MEMBRANE EXCITATION

For a neuromuscular synapse to transmit more than one impulse, it is necessary to destroy the residual neurotransmitter remaining in the synaptic cleft after the last impulse. If this does not occur, the postsynaptic membrane simply remains depolarized. The removal of any leftover acetylcholine is accomplished by an enzyme, **acetylcholinesterase,** which is present in the synaptic cleft. Acetylcholinesterase is one of the fastest-acting enzymes in the vertebrate body, cleaving one acetylcholine molecule every 40 microseconds. The rapid removal of neurotransmitters by acetylcholinesterase permits as many as 1000 impulses per second to be transmitted across the neuromuscular junction. Many organic phosphate compounds such as the nerve gases tabun and sarin and the agricultural insecticide parathion are potent inhibitors of acetylcholinesterase. Because they produce continuous neuromuscular transmission, such compounds can be lethal to vertebrates. Breathing, for example, requires muscular contraction, as does blood circulation.

Nerve-to-Nerve Connections

When the axon connection is with another nerve cell rather than with a muscle, the outcome of a synaptic event is far less predictable. Vertebrate nervous systems use dozens of different kinds of neurotransmitters, as we mentioned earlier. Some of these depolarize the postsynaptic membrane, in which case the synapse is an **excitatory synapse.** Other neurotransmitters affect synapses called **inhibitory synapses** and have the reverse effect, reducing the ability of the postsynaptic membrane to depolarize.

An individual nerve cell can possess both kinds of synaptic connections to other nerve cells. When signals from both excitatory and inhibitory synapses reach the body of the neuron (Figure 38-12) as the input from the dendrites, the depolarizing effects, which cause less internal negative charge, and the stabilizing ones, which cause more internal negative charge, interact with one another. The result is a process of **integration** in which the various excitatory and inhibitory electrical effects tend to cancel or reinforce each other.

> **The integration of neural information occurs within individual neuron cell bodies. This integration is the result of the summed effect on a neuron's membrane of the electrical impulses from different synapses. Some of these synapses facilitate depolarization, whereas others inhibit it.**

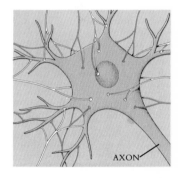

AXON

FIGURE 38-12

The cell body of a neuron serves to integrate the information arriving from many sources. In this representation, the dendrites and cell body of a neuron are contacted in many places by axonal projections from other nerves. The synapses made by some of these axons are inhibitory, tending to counteract depolarization of the cell body membrane; these are indicated in color. The synapses made by other axons are stimulatory, tending to depolarize the cell body membrane. The summed influences of all these inputs determines whether the axonal membrane will be sufficiently depolarized to intiate a propagating nerve impulse.

SUMMARY

1. A neuron is a nerve cell with an excitable membrane. It is specialized for the transmission of waves of depolarization along the membranes of processes extending out from the cell body.

2. An excitable membrane is created by the active pumping by the sodium-potassium pump of sodium ions out across the membrane and of potassium ions in. Because the membrane is not permeable to sodium ions but is permeable to potassium ions, potassium streams back out, but sodium does not flow back in. The result is a net negative charge within the cell.

3. A nerve impulse is initiated by depolarization of an excitable membrane; that is, by the opening of sodium ion channels. Such an opening equalizes the ion concentrations and thus removes the charge difference.

4. A nerve impulse propagates because the opening of some sodium ion channels facilitates the opening of other adjacent channels, causing a wave of depolarization to travel down the membrane of the nerve cell. All signals traversing a given neuron have the same amplitude, differing only in the frequency of impulses.

5. When a nerve impulse reaches the far end of a nerve cell, the axon tip, it depolarizes the membrane at the tip, causing the release of chemicals from the tip. These chemicals pass across a synaptic cleft and interact with channels in the membrane of another neuron or of muscle or gland cells. They either open these channels by depolarizing them, an "excitatory" response, or stabilize them in a closed conformation, an "inhibitory" response.

6. The integration of nerve signals occurs on the cell body membranes of individual neurons, which receive both depolarizing waves from dendrites associated with excitatory synapses and stabilizing waves from dendrites associated with inhibitory synapses. These waves tend to cancel each other out, the final amount of depolarization depending on the mix of the signals received.

REVIEW

1. The two forms of communication that integrate vertebrate body functions are _____ and _____.

2. When a nerve impulse travels down an axon, it first causes Na^+ ions to flood _____ (into/out of) the neuron cell.

3. A nerve-cell junction is called a _____.

4. In a myelinated nerve fiber the wave of depolarization jumps from one node of Ranvier to the next. This process is called _____ and is much faster than direct conduction.

5. When a wave of depolarization reaches the presynaptic membrane, how is the message transmitted to the postsynaptic membrane? _____

SELF-QUIZ

1. The specialized chemicals that carry out communication across synaptic junctions are called
   (a) synaptonemal complexes
   (b) enzymes
   (c) neuroglia
   (d) neurotransmitters
   (e) myelin

2. Acetylcholinesterase is an enzyme that is produced in neurons. Which of the following statements about acetylcholinesterase is not true?
   (a) it destroys acetylcholine
   (b) it operates only inside neuron cells
   (c) it is one of the fastest-acting enzymes in the body
   (d) it is required to maintain breathing

3. Schwann cells, which envelop the axon at intervals along its length, act as
   (a) nerves
   (b) tracts
   (c) electrical insulators
   (d) dendrites
   (e) target cells

4. A nerve impulse is
   (a) a self-propagating wave of depolarization along the neuron membrane
   (b) a result of the opening of passive Na^+ ions channels
   (c) followed by a recovery phase during which Na^+ ions are actively pumped out of the neuron
   (d) the result of a stimulus to some site on the neuron membrane
   (e) all of the above

5. Acetylcholine
   (a) passes by osmosis across the synaptic cleft
   (b) causes ATP to enter the muscle cell
   (c) directly causes K^+ ions to enter the muscle cell
   (d) is a neurotransmitter
   (e) none of the above

PROBLEM

1. If the giraffe in Figure 38-1 stubs his toe on a rock, how long until he knows it? Assume an impulse at 120 meters per second and a giraffe that is 5 meters tall.

THOUGHT QUESTIONS

1. When the brain is starved for oxygen even briefly, it dies. When the body is starved for energy, it begins to metabolize its own tissues, channeling the products preferentially to the brain. This behavior points out the importance to the brain of ongoing oxidative respiratory metabolism. Why is active oxidative respiration so very important to the continued well-being of the brain's nerve cells?

2. Why do most synapses contain gaps across which an electrical impulse cannot pass, when a direct physical connection would enable the uninhibited passage of the impulse?

FOR FURTHER READING

DUNANT, Y., and M. ISRAEL: "The Release of Acetylcholine," *Scientific American,* April 1985, pages 58-66. A challenge to the accepted theory that acetylcholine is emitted by synaptic vesicles.

MORELL, P., and T. NORTON: "Myelin," *Scientific American,* May 1980, pages 88-118. The structure and function of myelin, and how myelin deficiency leads to multiple sclerosis, are described.

STEVENS, C.F.: "The Neuron," *Scientific American,* September 1979, pages 54-65. A description of the structure and functioning of a typical nerve cell.

WAXMAN, S., and J.M. RITCHIE: "Organization of Ion Channels in the Myelinated Nerve Fiber," *Science,* vol. 228, pages 1502-1507, 1987. This article provides a superb example of how structure dictates function, describing how the distribution of ion channels within motor nerves leads to saltatory conduction of nerve impulses along these nerves.

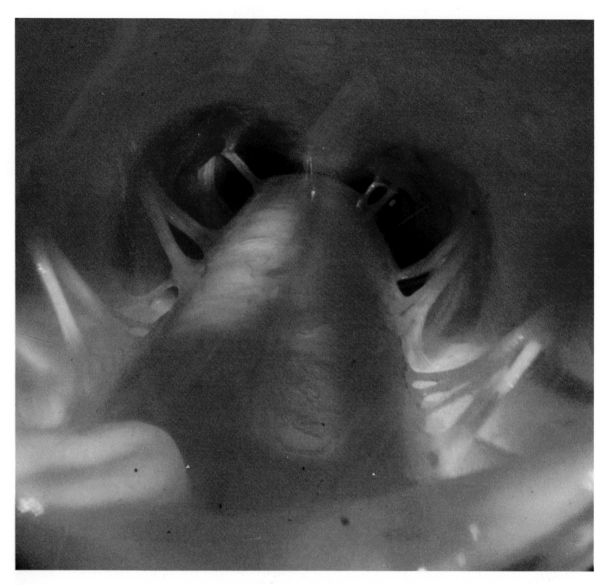

FIGURE 39-1   A view down the human spinal cord. Pairs of spinal nerves can be seen extending out from it. It is along these nerves that the central nervous system, consisting of the brain and spinal cord, communicates with the body.

THE NERVOUS SYSTEM

Chapter 39

Overview

In vertebrates, the central nervous system coordinates and regulates the diverse activities of the body, using a network of motor neurons to direct the voluntary muscles, a second network of motor neurons to control cardiac and smooth muscles, and chemical commands to effect long-term changes in physiological activities. All sensory information is acquired through the depolarization of sensory nerve endings. From a knowledge of which neurons are sending signals and how often they are doing so, the brain builds a picture of the body's internal and external environment.

For Review

Here are some important terms and concepts that you will encounter in this chapter. If you are not familiar with them, you should review them before proceeding.

Neuron (Chapters 32 and 38)

Depolarization (Chapters 32 and 38)

Muscle contraction (Chapter 33)

Vertebrates, like all other animals except sponges, use a network of neurons to gather information about the body's condition and the external environment, to process and integrate that information, and to issue commands to the body's muscles and glands. This network of neurons, the nervous system (Figure 39-1), is the subject of this chapter. The basic architecture of the nervous system is similar throughout the animal kingdom.

ORGANIZATION OF THE VERTEBRATE NERVOUS SYSTEM

All nervous systems can be said to have one underlying mechanism and three basic elements. The underlying mechanism is the nerve impulse. The three basic elements are (1) a central processing region or **brain,** (2) nerves that bring information to the brain, and (3) nerves that transmit commands from the brain.

The vertebrate nervous system consists of two contrasting functional groups, the central nervous system and the peripheral nervous system. The **central nervous system** is composed of the brain and the spinal cord (Figure 39-2), and is the site of information processing within the nervous system. The **peripheral nervous system** includes all the nerve pathways of the body outside the brain and spinal cord. These pathways are commonly divided into two groups: the **sensory** or **afferent pathways,** which transmit information to the central nervous system; and the **motor** or **efferent pathways,** which transmit commands from it (Figure 39-3). The motor pathways are in turn partitioned into the **voluntary** or **somatic nervous system,** which relays commands to skeletal muscles, and the **involuntary** or **autonomic** (Greek *auto,* "self" + *nomos,* law: self-controlling) **nervous system,** which stimulates the glands and other muscles of the body. In addition to the voluntary and involuntary systems, there is a **neuroendocrine system,** a network of **endocrine glands** whose hormone production is controlled by commands from the central nervous system. The central nervous system issues commands through these three different systems (Figure 39-4).

FIGURE 39-2

The human central nervous system is composed of the brain and spinal cord, from which extend 31 pairs of spinal nerves that direct the various actions of the body.

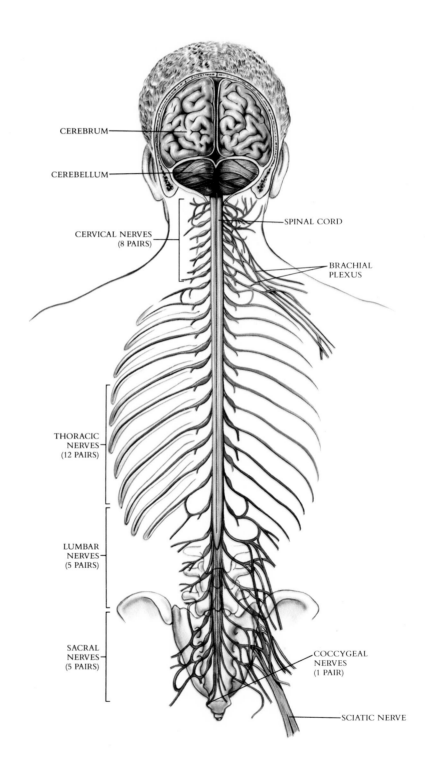

CEREBRUM

CEREBELLUM

SPINAL CORD

CERVICAL NERVES
(8 PAIRS)

BRACHIAL
PLEXUS

THORACIC
NERVES
(12 PAIRS)

LUMBAR
NERVES
(5 PAIRS)

SACRAL
NERVES
(5 PAIRS)

COCCYGEAL
NERVES
(1 PAIR)

SCIATIC NERVE

Nervous system pathways are composed of individual nerve fibers, axons, and long dendrites, all bundled together like the strands of a telephone cable. Within the central nervous system these bundles of nerve fibers are called **tracts.** In the peripheral nervous system they are called **nerves.** The cell bodies from which pathways extend are often clustered into groups, called **nuclei** if they are within the central nervous system and **ganglia** if they are in the peripheral system.

In this chapter, we begin by considering the brain and central nervous system, then consider the sensory systems that convey information to the brain, and end with a brief look at the motor pathways.

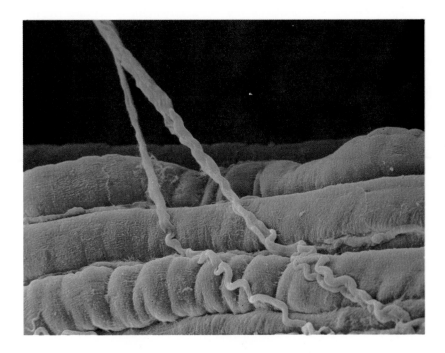

FIGURE 39-3

Motor nerves transmit commands from the central nervous system to individual muscles and glands. The slender, twisted threads are motor nerves, and the long, thick strands at the bottom are muscle fibers. Often a nerve will carry many independent fibers, which branch off to establish contact with different target muscles.

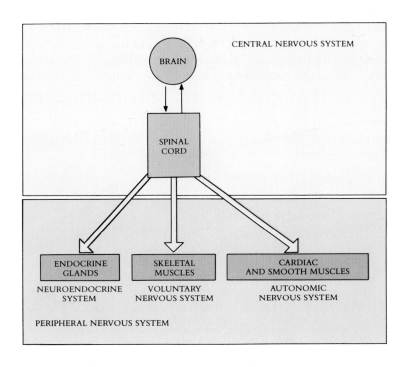

FIGURE 39-4

The central nervous system issues commands via three different systems. (1) The voluntary nervous system is a network of nerves that extend to the striated muscles. These are the nerves that carry the commands to arm muscles when you lift your arm. (2) The involuntary or autonomic nervous system is a network of nerves that extend to cardiac and smooth muscles and some glands. These are the nerves that carry the commands to the smooth muscles encasing your intestines, causing these muscles to undergo regular peristaltic contraction and to move food through your digestive tract. (3) The neuroendocrine system is a network of endocrine glands whose hormone production is controlled by commands from the central nervous system.

EVOLUTION OF THE VERTEBRATE BRAIN

The earliest vertebrates had far more complex brains than their ancestors. Casts of the interior braincases of fossil agnathans, fish that swam 500 million years ago, have revealed a lot about the early evolutionary stages of the vertebrate brain. Although they were very small, these brains already had the three principal divisions that characterize the brains of all contemporary vertebrates: (1) the hindbrain, or **rhombencephalon;** (2) the midbrain, or **mesencephalon;** and (3) the forebrain, or **prosencephalon.**

The hindbrain is the principal component of these early brains (Figure 39-5), as it still is in fishes today. Composed of the **pons** and the **medulla oblongata,** the hindbrain may be considered an extension of the spinal cord devoted primarily to coordinating motor reflexes. Tracts, cables composed of large numbers of nerve fibers, run up and down the spinal cord to the hindbrain. The hindbrain in turn integrates the afferent signals that come in from the body and determines the pattern of efferent response.

Much of this coordination is carried on in a small extension of the hindbrain called the **cerebellum** ("little cerebrum"). In the advanced vertebrates, the cerebellum plays an increasingly important role as a coordinating center and is correspondingly larger than it is in the fishes. In all vertebrates the cerebellum processes data on the current position and movement of each limb, the state of relaxation or contraction of the muscles involved, and the general position of the body and its relation to the outside world. These data are gathered in the cerebellum and synthesized, and the resulting orders are issued to efferent pathways.

In fishes, the remainder of the brain is devoted to reception and the processing of sensory information. The second major division of the fish brain, the midbrain, is composed primarily of the **optic lobes,** which receive and process visual information. The third major division, the **forebrain,** is devoted to processing olfactory (smell) information.

In early vertebrates the principal component of the brain was the hindbrain, devoted largely to coordinating motor reflexes.

The Advent of a Dominant Forebrain

Starting with the amphibians, and much more prominently in the reptiles, one can perceive a pattern that becomes a dominant evolutionary trend in the development of the vertebrate brain: the processing of sensory information becomes increasingly centered in the forebrain (Figure 39-6).

The forebrain in reptiles, amphibians, birds, and mammals is composed of two elements with distinct functions. The **diencephalon** (from the Greek word **dia,** "between") is devoted to the integration of sensory information. The **telencephalon,** or "end–brain" (Greek **telos,** "end"), is located at the front and is devoted largely to associative activity.

FIGURE 39-5

The basic organization of the vertebrate brain can be seen in the brains of primitive fishes. These brains are divided into the same regions that can be seen in different proportions in all vertebrate brains: the hindbrain, which is the largest portion of the brain in fishes; the midbrain, which in fishes is a small zone devoted to processing visual information; and the forebrain, which in fishes is devoted primarily to processing olfactory (smell) information. In the brains of terrestrial vertebrates, the forebrain plays a far more dominant role than it does in fishes.

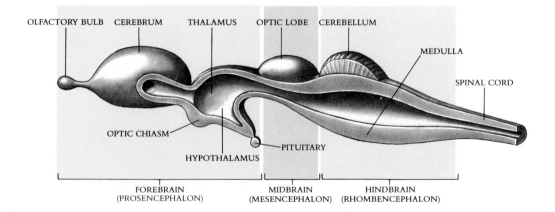

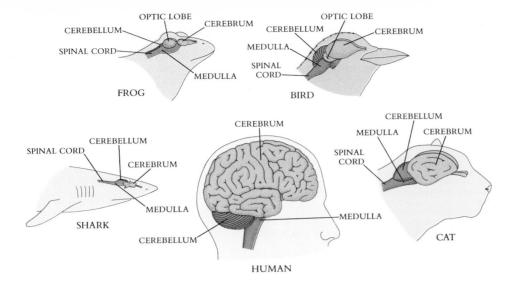

FIGURE 39-6

The evolution of the vertebrate brain has involved pronounced changes in the relative sizes of different regions of the brain. In fishes, the hindbrain (indicated in color) is predominant, the rest of the brain serving primarily to process sensory information. In amphibians and reptiles, the forebrain is far larger, and a larger cerebrum devoted to associative activity can be seen. In birds, which evolved from reptiles, the cerebrum is even more pronounced than in reptiles. In mammals, which evolved from reptiles independently of birds, the cerebrum is the largest portion of the brain. The dominance of the cerebrum is greatest in humans, where it envelops much of the rest of the brain.

In land-dwelling vertebrates the processing of information is increasingly centered in the forebrain.

The diencephalon is a "between brain" located between the telencephalon and the midbrain. It has two components, situated one atop another: the **thalamus** and the **hypothalamus.**

Sensory Integration

The **thalamus** is the primary site of sensory integration in the brain. Auditory, optical, and other information is relayed to the thalamus, passing from there to the outer surface of the brain. For example, information about posture, derived from the muscles, and information about orientation, derived from sensors within the ear, passes from the cerebellum of the hindbrain to the thalamus. The thalamus then processes the information and channels it to the appropriate motor center on the outer surface of the brain.

Integrating the Body's Responses

The **hypothalamus** integrates the visceral activities. It controls body temperature, respiration, and heartbeat; it also directs the secretions of the brain's major hormone-producing gland, the **pituitary gland.** The hypothalamus is linked by a network of neurons to the **cerebral cortex,** a layer several millimeters thick on the brain's outer surface. This network, along with the hypothalamus, is called the **limbic system.** The operations of the limbic system are responsible for many of the most deep-seated drives and emotions of vertebrates, including pain, anger, sex, hunger, thirst, and pleasure.

Stereotyped Behavior

The telencephalon of reptiles and birds consists largely of a layer of nervous tissue called the **corpus striatum,** which controls complicated stereotyped behavior. There is nothing like the corpus striatum in any other group. Much of what we think of as "behavior" in birds and reptiles is really composed of completely determined neural reactions. For example, the complex mating rituals of some birds are often choreographed by the structure of the brain; their pattern is fixed by nerve paths within the corpus striatum.

The Recent Expansion of the Cerebrum

Fishes and reptiles have brains that are small compared with the size of their bodies; mammals and birds have brains that are proportionately large. The increase in brain size in the mammals largely reflects a great enlargement of the cerebrum, the dominant part of the mammalian brain.

The **cerebrum** is the center for correlation, association, and learning in the mammalian brain. It receives sensory data from the thalamus. From the cerebrum, motor nerve fibers extend to the motor nerve columns of the spinal cord, which contain groups of neurons servicing different muscles. Efferent motor neurons extend from these motor columns directly to the muscles. The motor fibers from the cerebrum pass straight through the brain to the voluntary motor regions of the brainstem, and can be seen as a cable of fibers called the **pyramidal tract,** a structure that particularly dominates the structure of primate brains.

> **The brains of mammals and birds are unusually large relative to their body size. This reflects great enlargement of the cerebrum, the center for correlation, association, and learning.**

ANATOMY AND FUNCTION OF THE HUMAN BRAIN

The cerebrum, located at the very front of the human brain, is so large compared with the rest of the brain that it appears to envelop it (Figure 39-7). In the brains of humans and other primates the cerebrum is split into two halves, or hemispheres, which are connected only by a nerve tract called the **corpus callosum.** Each hemisphere of the brain is divided further by two deep grooves into four lobes, designated the **frontal, parietal, temporal,** and **occipital lobes** (Figure 39-8).

The Cerebral Cortex

Much of the neural activity of the cerebrum occurs within a thin, gray layer only a few millimeters thick on its outer surface, overlying a solid white region that consists of myelinated nerve fibers. This layer, the **cerebral cortex** mentioned earlier, is densely packed with neuron cell bodies. The human cerebral cortex contains more than 10 billion nerve cells, roughly 10% of all the neurons in the brain. The surface of the cerebral cortex is highly convoluted, particularly in human brains, a property that increases its surface area threefold.

By examining the effect of injuries to particular sites on the cerebrum, it has been possible to plot roughly the location of its various associative activities. There are four major regions of activity on the cerebral cortex, each of them referred to as a specialized cortex: the motor, sensory, auditory, and visual cortexes (see Figure 39-8).

The **motor cortex** straddles the rearmost portion of the frontal lobe. Each point on its surface is associated with the movement of a different part of the body. Right behind the motor cortex, on the leading edge of the parietal lobe, lies the **sensory cortex.** Each point on the surface of the sensory cortex represents sensory receptors from a different part of the body, such as the pressure sensors of the fingertips or the taste receptors of the tongue. The **auditory cortex** lies within the temporal lobe; different surface regions of this cortex correspond to different sound frequencies. The **visual cortex** lies on the occipital lobe, with different sites corresponding to different positions on the retina.

Associative Organization of the Cerebral Cortex

Only a small portion of the total surface of the cerebral cortex is occupied by the motor and sensory cortexes. The remainder of the cerebral cortex is referred to as **associative cortex.** This appears to be the site of higher mental activities, such as planning and contemplation. The associative cortex represents a far greater portion of the total cortex in primates than it does in any other mammals and reaches its greatest extent in human beings. In a mouse, for example, 95% of the surface of the cerebral cortex is occupied by motor and sensory areas. In humans, only 5% of the surface is devoted to motor and sensory functions; the remainder is associative cortex.

The two hemispheres of the cerebrum constitute two associative brains in miniature. All four of the specialized cortex regions occur on the surface of each hemisphere, with the ones on the *right hemisphere* related to the *left side* of the body and vice

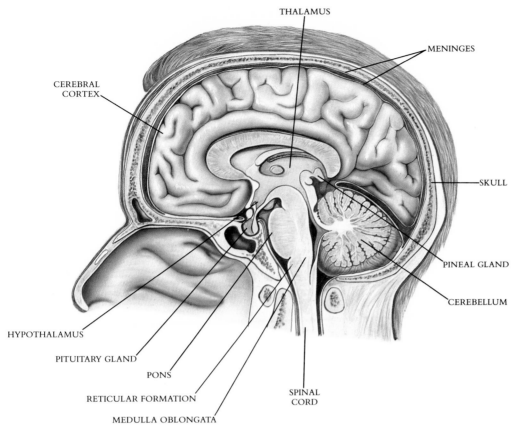

THALAMUS

MENINGES

CEREBRAL CORTEX

SKULL

PINEAL GLAND

CEREBELLUM

HYPOTHALAMUS

PITUITARY GLAND

PONS

RETICULAR FORMATION

MEDULLA OBLONGATA

SPINAL CORD

FIGURE 39-7

A section of the human brain demonstrates how the cerebrum envelops the rest of the brain. Only the cerebral cortex, part of the cerebrum, is visible in surface view. The reticular formation is the main trunk of the reticular system.

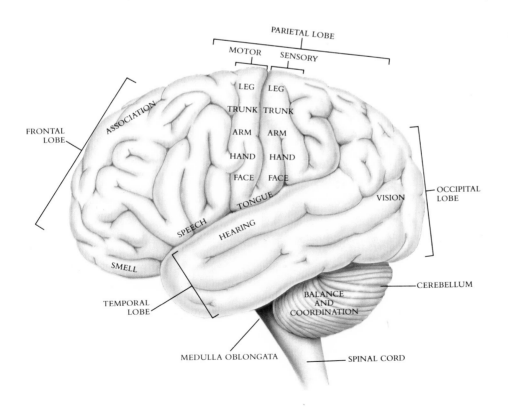

PARIETAL LOBE

MOTOR   SENSORY

LEG   LEG

TRUNK   TRUNK

ARM   ARM

HAND   HAND

FACE   FACE

TONGUE

ASSOCIATION

FRONTAL LOBE

SPEECH   HEARING

VISION

OCCIPITAL LOBE

SMELL

CEREBELLUM

BALANCE AND COORDINATION

TEMPORAL LOBE

MEDULLA OBLONGATA

SPINAL CORD

FIGURE 39-8

One hemisphere of the human brain, showing the four lobes and the major functional regions.

versa. The switch occurs because the nerve tracts descending from the hemispheres cross over in the pyramidal tract through the corpus callosum, where the two sides of the brain are connected.

Although the two hemispheres of the brain each contain the four cortexes, the hemispheres are responsible for different associative activities. Injury to the left hemisphere of the cerebrum often results in the partial or total loss of speech, but a similar injury to the right side does not. There are several **speech centers** controlling different aspects of speech. They are almost always located on the left hemisphere of right-handed people; in about one third of left-handed people they are located on the right hemisphere. An injury to one speech center produces halting but correct speech; injury to another speech center produces fluent, grammatical, but meaningless speech; injury to a third center destroys speech altogether.

Injuries to other sites on the surface of the brain's left hemisphere result in impairment of the ability to read, write, or do arithmetic. Comparable injuries to the right hemisphere have very different effects, resulting in impairment of three-dimensional vision, musical ability, or the ability to recognize patterns and solve inductive problems.

It has been a popular activity, particularly among journalists, to speculate that the human brain is composed of a "rational" left hemisphere and an "intuitive" right hemisphere, since many of the more imaginative associative activities seem to be carried out by the right hemisphere. Some people even argue that the brain actually has two consciousnesses, one dominant over the other, or that the right brain hemisphere is the more ancient, with language and reasoning having evolved much later. Actually, the significance of the clustering of associative activities in different areas of the brain is not at all clear, but it remains a subject of much interest.

Memory and Learning

One of the great mysteries of the brain is the basis of memory and learning. If portions of the brain are removed, especially the temporal lobes, memory is impaired but not lost: there is no one part of the brain in which memory appears to reside. Investigators who have tried to prove the physical mechanisms underlying memory often have felt that they were grasping a shadow. An understanding of these mechanisms continues to elude them.

However, researchers have learned some things about the physical basis of memory. The first stage, **short-term memory,** is transient, lasting only a few moments. Such memories can readily be removed from the brain by application of an electrical shock. When this is done, short-term memories are wiped from the circuits, but longer-term memories are preserved. This result suggests that short-term memories are stored electrically, in the form of short-term neural excitation. **Long-term memory,** in contrast, appears to involve structural changes in the neural connections within the brain. Very little is known about how such changes occur. Perhaps differences in electrical stimulation alter the patterns of synapse formation, with long-term memory corresponding to a pattern of nerve impulse integration.

SENSORY INFORMATION

All input from sensory neurons to the central nervous system arrives in the same form, as nerve impulses carried on afferent sensory neurons. Every arriving nerve impulse is identical to every other one. The information that the brain derives from sensory input is based on the frequency with which these impulses arrive and on the identity of the specific neuron that transmits it. A sunset, a symphony, searing pain—to the brain they are all the same, differing only in the source of the impulse and its frequency. Thus if the auditory nerve is artificially stimulated, the central nervous system perceives the stimulation as a noise. If the optic nerve is artificially stimulated in exactly the same manner and degree, the stimulation is perceived as a flash of light.

All sensory receptors are able to initiate nerve impulses by opening ion channels within sensory neuron membranes and thus depolarizing them, as you saw in Chapter

38. The receptors differ from one another with respect to the nature of the environmental input that triggers this event. Many kinds of receptors have evolved among vertebrates, with each receptor sensitive to a different aspect of the environment.

SENSING INTERNAL INFORMATION

Traditionally, the sensing of information that relates to the body itself, its internal condition and its position, is known as **interoception,** or inner perception (Figure 39-9, *A*). In contrast, the sensing of the exterior is called **exteroception** (Figure 39-9, *B*). Many of the neurons and receptors that monitor body functions are simpler than those that monitor the external environment. The simplest sensory receptors are nerve endings that depolarize in response to direct physical stimulation, to temperature, to chemicals like oxygen diffusing into the nerve cell, or to a binding or stretching of the neuron cell membrane. The vertebrate body uses a variety of such **interoceptors** to obtain information about its internal conditioning.

Temperature Change

Two kinds of nerve endings in the skin are sensitive to changes in temperature. One of these is stimulated by a lowering of temperature (cold receptors), while the other is stimulated by a raising of temperature (warm receptors).

Blood Chemistry

Special receptors called **carotid bodies** are embedded in the walls of arteries at several locations in the circulatory system. By sensing CO_2 levels, the carotid bodies provide the sensory input that the body uses to regulate its rate of respiration. When CO_2 rises above normal levels, this information is conveyed to the central nervous system, which reacts by increasing the respiration rate to get rid of the excess CO_2.

Pain

A stimulus that causes or is about to cause tissue damage is perceived as **pain.** Such a stimulus elicits an array of responses from the central nervous system, including reflexive withdrawal of a body segment from a source of stimulus and changes in heartbeat and blood pressure. The receptors that produce these effects are called **nociceptors.**

A   B

FIGURE 39-9

By tradition, the senses are grouped into two classes.
A Dancers are able to maintain their proper stance by using their internal sense of balance. Sensing of this sort is interoception.
B This tree frog *(Phyllomedusa tarsius)* learns about the world around it by using eyes to perceive patterns of light, as you are doing in reading this page. Sensing of this sort is exteroception.

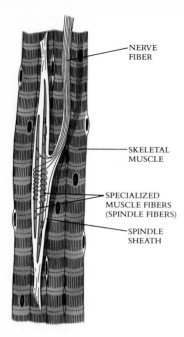

FIGURE 39-10

A stretch receptor embedded within skeletal muscle. Stretching of the muscle elongates the spindle fiber, which deforms the nerve endings, causing them to fire and send a nerve impulse out along the nerve fiber.

They consist of the ends of small, sparsely myelinated nerve fibers located within tissue, usually near the surfaces where damage is most likely to occur. Physical deformation of the ends of the nerve initiates depolarization.

Muscle Contraction

Buried deep within the muscles of all vertebrates, except the bony fishes, are specialized muscle fibers called **muscle spindles.** Wrapped around each muscle spindle is the end of an afferent neuron, called a **stretch receptor** (Figure 39-10). When a muscle is stretched, the spindle fiber elongates, stretching the spiral nerve ending of the stretch receptor and stimulating it to fire repeatedly. When the muscle contracts, the tension on the spindle lessens and the stretch receptor ceases to fire. Thus the frequency of stretch receptor discharges along the afferent nerve fiber indicates the degree of muscle contraction at any given moment. The central nervous system uses this information to control movements that involve the combined action of several muscles, such as those involved in breathing or walking.

Blood Pressure

Blood pressure is sensed by a highly branched network of nerve endings called **baroreceptors,** or **pressure receptors,** within the walls of major arteries in several locations in the circulatory system. When the blood pressure increases, the arterial wall balloons out where it is thinnest, in the region of the baroreceptor, increasing the rate of firing of the afferent neuron. A decrease in blood pressure causes the wall to move inward and lowers the rate of neuron depolarization. Thus the frequency of impulses arriving from these afferent fibers provides the central nervous system with a continuous measure of blood pressure. The greater the blood pressure, the higher the frequency of impulses.

Touch

Touch is sensed by pressure receptors below the surface of the skin. The two principal kinds of receptors, called **Meissner's corpuscles** and **Merkel cells,** are responsible for the great sensitivity of your fingertips. Meissner's corpuscles fire in response to rapid changes in pressure, and Merkel cells measure the duration and extent to which pressure is applied.

> **Mechanical stimuli of different types initiate nerve impulses by deforming the sensory neuron membrane. Such deformation opens ion channels in the membrane and initiates depolarization.**

Balance

All vertebrates possess gravity receptors known as **statocysts** or **otoliths.** In humans these are located in a series of hollow chambers within the inner ear and provide the information the brain uses to perceive balance. To illustrate how these receptors work, imagine a pencil standing in a glass. No matter which way you tip the glass, the pencil will roll along the rim, applying pressure to the lip of the glass. If you want to know the direction the glass is tipped, you need only inquire where on the rim the pressure is being applied. The body uses receptors of this sort, called **proprioceptors** (from the Greek *proprios,* meaning "self") to sense its position in space. Gravity serves as a reference point; changes in the pressure applied to a nerve ending by a heavy object within the receptor provide the stimulus.

Motion

Motion, or **angular acceleration,** is sensed by vertebrates in a way similar to that used to detect vertical position: by employing a receptor in which fluid deflects cilia in a direction opposite to that of the motion. Within the inner ear are three fluid-filled

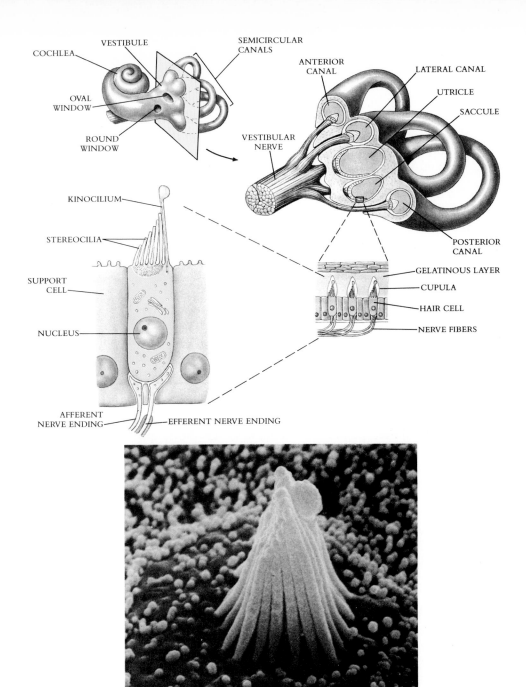

Labels in the figure:

COCHLEA
VESTIBULE
SEMICIRCULAR CANALS
ANTERIOR CANAL
LATERAL CANAL
UTRICLE
SACCULE
OVAL WINDOW
ROUND WINDOW
VESTIBULAR NERVE
POSTERIOR CANAL

KINOCILIUM
STEREOCILIA
SUPPORT CELL
NUCLEUS
AFFERENT NERVE ENDING
EFFERENT NERVE ENDING

GELATINOUS LAYER
CUPULA
HAIR CELL
NERVE FIBERS

FIGURE 39-11

Vertebrates sense movement as pressure of a long cilium, the kinocilium, against a battery of surrounding cilia, the stereocilia. As you can see in the photograph, magnified 20,000 times, these cilia form tentlike assemblies called cupolas that project out into passages within the semicircular canals of the ear's vestibular apparatus. Movement causes fluid in the canal to press against the cupola. Because there are three semicircular canals at right angles to one another, movement in any plane can be detected.

semicircular canals (Figure 39-11), each oriented in a different plane at right angles to the other two so that motion in any direction can be detected. Protruding into the canals are sensory cells, which are connected to afferent nerves. Because the three canals are oriented in all three planes, movement in any plane is sensed by at least one of them. Complex movements can be analyzed by comparing the sensory input from each canal.

> **Complex mechanical receptors can respond to pressure, to gravity, or to motion. In each case, the receptors employ mechanical devices like levers to convert the information to a mechanical stimulus, which then initiates the depolarization of a sensory membrane by deformation.**

SENSING THE EXTERNAL ENVIRONMENT

So far, most of the sensory receptors that we have examined have been directed at the internal environment of the body or at determining its position in space. Sensing the nature of other objects in the external environment presents quite a different problem. Imagine, for example, the problem of learning about an object standing some distance away. The amount of information that any sensory receptor can obtain about that object is limited both by the nature of the stimulus sensed by that receptor and by the medium, either air or water, through which the stimulus must move to reach the receptor.

The four primary senses that detect objects at a distance use different classes of receptors. Taste and smell use chemical receptors; hearing uses mechanical receptors; and vision uses electromagnetic receptors that sense photons of light. We shall consider each of these and then examine other sensory systems employed by vertebrates.

Taste and Smell

The simplest exteroceptors, like the simplest interoceptors, are chemical ones. Embedded within the membranes of afferent nerve endings or of sensory cells associated with afferent neurons are specific chemical receptors that induce depolarization when they bind particular molecules. There are two chemical sensory systems that utilize different receptors and process information at different locations in the brain: **taste,** in which the receptors are specialized sensory cells, and **smell,** in which the receptors are neurons. Despite the differences in how stimuli are received and the information processed, there is little difference in the chemical stimuli to which the two chemical senses respond. In taste, the receptors are **taste buds** located in the mouth. In smell, the receptors are neurons whose cell bodies are embedded in the epithelium of the upper portion of the nasal passage.

Hearing

Terrestrial vertebrates detect vibration in air by means of mechanical receptors located within the ear. In the ears of mammals, sound waves beat against a large membrane of the outer ear called the **tympanic membrane** (eardrum), causing corresponding vibra-

FIGURE 39-12

Structure of the human ear. Sound waves passing through the ear canal beat on the tympanic membrane, pushing a set of three small bones, or ossicles (hammer, anvil, and stirrup), against an inner membrane called the oval window. This sets up a wave motion (blue arrows) in the fluid filling the cochlea, a wave that travels through the vestibular and tympanic canals. Where the sound wave beats against the sides of the canals, the tectorial membrane is pushed against the basilar membrane, bending hair cells and firing associated neurons.

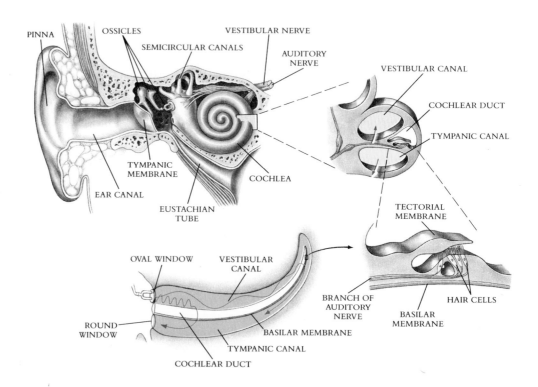

tions in three small bones: the hammer, anvil, and stirrup. These bones act together as a lever system, increasing the force of the vibration. The third in line of the levers, the stirrup, pushes against another membrane, the **oval window** (Figure 39-12). Because the oval window is smaller than the tympanic membrane, vibration against it produces more force per unit area of membrane. The chamber in which all these events occur is called the **middle ear.** It is connected to the pharynx by the eustachian tube in such a way that there is no difference in air pressure between the middle ear and the outer ear. The familiar "ear popping" associated with landing in an aircraft or with the rapid descent of an elevator in a tall building is a result of pressure equalization between the two sides of the eardrum.

The membrane of the oval window is the door to the **inner ear,** where hearing actually takes place. The chamber of the inner ear is shaped like a tightly coiled snail shell and is called the **cochlea,** from the Latin name for "snail."

The auditory receptors are hair cells located on the middle **basilar membrane,** which bisects the cochlea. The hair cells do not project into the fluid filling the cochlea. Instead, they are covered with another membrane called the **tectorial membrane.** The bending of the basilar membrane as it vibrates causes the hairs of the receptor cells pressed against the tectorial membrane to bend, depolarizing their associated afferent neurons. Sounds of different frequencies cause different portions of the basilar membrane to vibrate and thus to fire different afferent neurons. The central nervous system perceives sound intensity in terms of the frequency of discharge and sound frequency in terms of the pattern of neurons on the basilar membrane that fire afferent impulses.

Our ability to hear depends on the flexibility of the basilar membrane, a flexibility that changes as we grow older. Humans are not able to hear low-pitched sounds, below 20 vibrations or cycles per second, although some other vertebrates can. As children, human beings can hear high-pitched sounds, up to 20,000 cycles per second, but this ability decays progressively through middle age. Other vertebrates can hear sounds at far higher frequencies. Dogs, for example, readily detect sounds of 40,000 cycles per second. Thus dogs can hear a high-pitched dog whistle when it seems silent to a human observer.

Vision

Light is perhaps the most useful stimulus for learning about the external environment. Because light travels in a straight line and arrives virtually instantaneously, visual information can be used to determine both the direction and the distance of an object. No other stimulus provides as much detailed information.

The type of problem associated with perceiving light is one that we have not encountered before in our consideration of sensory systems. All receptors described to this point have been either chemical or mechanical ones. None of them are able to respond to the electromagnetic energy provided by photons of light. **Vision,** the perception of light, is carried out in vertebrates by a specialized sensory apparatus called an eye. Eyes have evolved many times independently in different groups of animals (see the Boxed Essay in Chapter 16 and Figure 16-A).

Eyes contain sensory receptors that detect photons of light. These receptors are located in the back of the eye, which is organized something like a camera. Light that falls on the eye is focused by a lens on the receptors in the rear, just as the lens of a camera focuses light on film. How does a photon-detecting sensory receptor work? Just as is the case in photosynthesis, the primary photoevent of vision is the absorption of a photon of light by a pigment. The visual pigment is called **_cis_-retinal** (Figure 39-13), a substance that we encountered as the cleavage product of carotene, a photosynthetic pigment in plants (see Figure 9-3). In vertebrates, the *cis*-retinal pigment is coupled to a transmembrane protein called **opsin** to form **rhodopsin** and **iodopsin.**

The visual pigment in vertebrate eyes is located in the tips of specialized sensory cells called **rod** and **cone cells** (Figure 39-14). Rod cells, which contain rhodopsin, are responsible for black-and-white vision. Cone cells, which contain iodopsin, function in color vision. There are three different kinds of cone cells, each with a different kind of iodopsin molecule that preferentially absorbs light of different wavelengths.

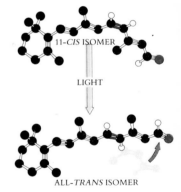

11-*CIS* ISOMER

LIGHT

ALL-*TRANS* ISOMER

FIGURE 39-13

When light is absorbed by the 11-*cis* isomer of the visual pigment retinal, the pigment undergoes a change in shape: the linear end of the molecule rotates about a double bond, indicated here in color. This new isomer is referred to as all-*trans* retinal. This change in the pigments shape causes in turn a change in the shape of the protein opsin to which the pigment is bound, initiating a chain of events that leads to the generation of a nerve impulse. The arrow in the bottom figure indicates the direction of the change in the molecule.

Vertebrate eyes are lens-focused eyes (Figure 39-15). In them, light first passes through a transparent layer, the **cornea,** which begins to focus the light onto the rear of the eye. Light then passes through the **lens,** a structure that completes the focusing. The lens is a fat disk filled with transparent jelly, somewhat resembling a flattened balloon. In mammals, the lens is attached by suspending ligaments to ciliary muscles. When these muscles contract, they change the shape of the lens (Figure 39-16) and thus the point of focus on the rear of the eye. In amphibians and fishes, the lens does not change shape. These animals instead focus their images by moving the lens in and out, thus operating the same way that a camera does. In all vertebrates, the amount of light entering the eye is controlled by a shutter, called the **iris,** between the cornea and the lens. The iris reduces the size of the transparent zone, or **pupil,** of the eye through which the light passes.

FIGURE 39-14

A The broad tubular cell diagrammed on the right is a rod; the shorter, tapered cell next to it is a cone.
B Although it is not obvious from the electron micrograph, the pigment-containing outer segments of these cells are separated from the rest of the cells by a partition through which there is only a narrow passage, the connecting cilium shown in **A**.

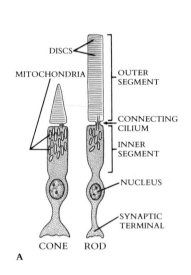

FIGURE 39-15

Structure of the human eye. Light passes through the transparent cornea and is focused by the lens on the rear surface of the eye, the retina. The retina and especially the fovea, a particular area of sharp focus on the retina, are rich in rods and cones.

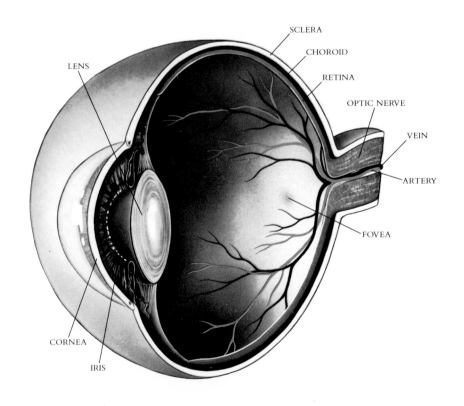

The field of receptor cells, rods and cones, that lines the back of the eye is the **retina.** The retina contains about 3 million cones, most of them located in the central region of the retina called the **fovea,** and approximately 1 billion rods. The eye forms a sharp image in the central fovea region of the retina, which is a region composed almost entirely of cone cells.

Each foveal cone cell makes a one-to-one connection with a special neuron called a **bipolar cell** (Figure 39-17). Each bipolar cell is connected in turn to an individual visual ganglion cell, whose axon is part of the **optic nerve.** The optic nerve transmits visual impulses directly to the brain. The frequency of pulses transmitted by any one receptor provides information about light intensity. The pattern of depolarization among the different foveal axons provides a point-to-point image; the different cone cells provide information about the color of the image.

FIGURE 39-16

Focusing the human eye: contraction of ciliary muscles pulls on suspensor ligaments and changes the shape of the lens, which alters its point of focus forward or backward. In many people the ciliary muscles place the point of focus in front of the fovea rather than on it. Such people are said to be nearsighted. The problem can be corrected with glasses or contact lenses, which extend the focal point back to where it should be. In others, the ciliary muscles make the opposite error, placing the point of focus behind the retina; such farsighted individuals can correct their vision with lenses that shorten the focal point.

SUSPENSOR LIGAMENT — IRIS

NORMAL DISTANT VISION   NEARSIGHTED   FARSIGHTED

RETINA — LENS

NORMAL NEAR VISION   NEARSIGHTED, CORRECTED   FARSIGHTED, CORRECTED

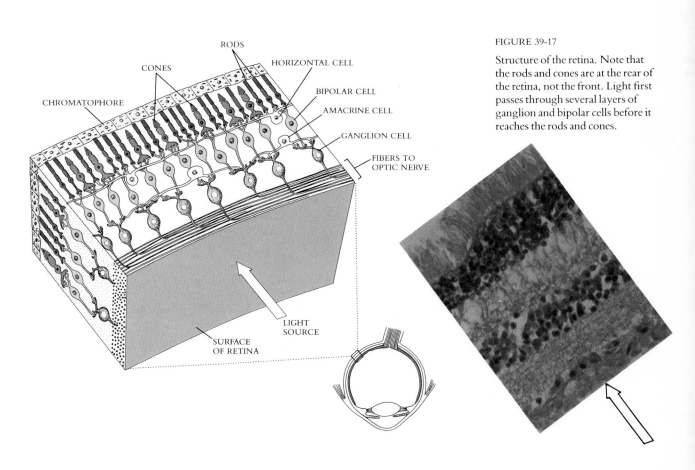

RODS
CONES
HORIZONTAL CELL
CHROMATOPHORE
BIPOLAR CELL
AMACRINE CELL
GANGLION CELL
FIBERS TO OPTIC NERVE
LIGHT SOURCE
SURFACE OF RETINA

FIGURE 39-17

Structure of the retina. Note that the rods and cones are at the rear of the retina, not the front. Light first passes through several layers of ganglion and bipolar cells before it reaches the rods and cones.

Most vertebrates have two eyes, one located on each side of the head. When both of them are trained on the same object, the image that each sees is slightly different because each eye views the object from a different angle. The bilateral arrangement of eyes, with its slight displacement of images (an effect called **parallax**) permits sensitive depth perception. By comparing the differences between the images provided by the two eyes with the physical distance to particular objects, vertebrates learn to interpret different degrees of disparity between the two images as representing different distances—stereoscopic vision. We are not born with the ability to perceive distance; we learn it. Stereoscopic vision develops in babies only over a period of months.

OTHER ENVIRONMENTAL SENSES IN VERTEBRATES

Vision is the primary sense used by all vertebrates that live in a light-filled environment, but the wavelength and intensity of visible light are by no means the only stimuli that are available to vertebrates for assessing this environment. Animals are known that sense polarized light (the octopus), ultrasound (bats), heat (snakes), electricity (catfish), and magnetism (birds). Most vertebrate sensory systems evolved early, while vertebrates were still confined to the sea. Some, such as sensors that detect electric fields, are not used by land vertebrates, presumably because air is a very poor conductor of electric currents. Others, such as sensors that respond to liquid pressure waves on fish, are used by land vertebrates to detect pressure waves in air, to hear. Because the terrestrial environment is so physically different from the sea, it has presented some new sensory opportunities. Air, for example, transmits heat much better than water does, and snakes have evolved heat sensors that provide quite detailed three-dimensional images (Figure 39-18).

No one vertebrate has a perfect repertoire of sensory systems, able to finely discriminate all potential cues. Rather, vertebrates have evolved sensory nervous systems that meet particular evolutionary challenges. Thus some hunting mammals such as dogs have evolved highly capable olfactory systems, but primates have not. It is of some interest to note, for example, the detailed sensory information in our own environment that we could potentially utilize if we possessed the proper sensory receptors. We cannot utilize thermal radiation, as snakes do, or ultrasound, as bats do, to construct a three-dimensional image.

THE PERIPHERAL NERVOUS SYSTEM

Integration of the vertebrate body's many activities is the primary function of the central nervous system. All control of bodily functions, voluntary and involuntary, is vested in this system. It directs voluntary and involuntary functions in different ways. **Voluntary** functions are movements of skeletal (striated) muscle that are directed by somatic motor pathways from the brain and spinal cord. These are the pathways that coordinate your fingers when you grasp a pencil, spin your body when you dance, and put one foot ahead of the other when you walk. The direction of these muscular movements by the central nervous system is in large measure subject to conscious control by the associative cortex.

FIGURE 39-18

The depression between the nostril and the eye of this timber rattlesnake opens into the pit organ. In the cutaway portion of the diagram you can see that the organ is composed of two chambers separated by a membrane. The organ is thought to function by constantly comparing the temperatures in the two chambers.

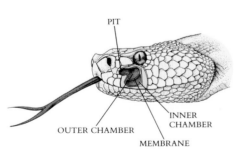

Antagonistic Control of the Autonomic Nervous System

An autonomic nerve signal crosses two synapses in traveling from the central nervous system out to its target organ, whether it travels along sympathetic or parasympathetic nerves. The first synapse is in the ganglion, between the axon of a neuron extending from the central nervous system and the dendrites of the autonomic neuron's cell body. The second synapse is between the autonomic neuron's axon and the target organ. The neurotransmitter in the ganglion is acetylcholine for both sympathetic and parasympathetic nerves. However, the neurotransmitter between the terminal autonomic neuron axon and the target organ is different in the two antagonistic elements of the autonomic nerve systems. In the parasympathetic system, the neurotransmitter at the terminal synapse is acetylcholine, just as it is in the ganglion. In the sympathetic system, the neurotransmitter at the terminal synapse is either adrenaline or noradrenaline, both of which have an effect *opposite* to that of acetylcholine. Each gland (except the adrenal gland), smooth muscle, and cardiac muscle controlled by the autonomic nervous system has synapses with *both* the sympathetic and parasympathetic systems. Thus, depending on which of the two nerve paths is selected by the central nervous system, an arriving signal will either stimulate or inhibit the organ.

Each gland, smooth muscle, and cardiac muscle constantly receives stimulatory signals through one nerve and inhibitory signals by way of the other nerve. The central nervous system controls activity in each case by varying the ratio of the two signals.

The glands and involuntary muscles of the body are innervated by two antagonistic sets of efferent nerves.

Thus an organ receiving nerves from both visceral nervous systems will be subject to the effects of two opposing neurotransmitters. If the sympathetic nerve ending excites a particular organ, the parasympathetic synapse usually inhibits it. For example, the sympathetic system speeds up the heart and slows down digestion, whereas the parasympathetic system slows down the heart and speeds up digestion. In general, the two opposing systems are organized so that the parasympathetic system stimulates the activity of normal body functions, for example, the churning of the stomach, the contractions of the intestine, and the secretions of the salivary glands. The sympathetic system, on the other hand, generally mobilizes the body for greater activity, as in increased respiration or a faster heartbeat.

In the next chapter, we shall consider neuroendocrine control, the third major way that the body's functions are controlled. Neuroendocrine control in the vertebrate body is associated with long-term changes in activity level. Such changes are achieved in most animals, including vertebrates, by chemical signals—hormones—rather than by nerve signals.

SUMMARY

1. The vertebrate nervous system is made up of the central nervous system, consisting of the brain and the spinal cord, and the peripheral nervous system. Within the peripheral nervous system, sensory pathways transmit information to the central nervous system, and motor pathways transmit commands from it.

2. The motor pathways are divided into somatic, or voluntary, pathways, which relay commands to skeletal muscles, and autonomic, or involuntary, pathways, which stimulate the glands and other muscles of the body.

3. Vertebrate brains consist of the hindbrain, or rhombenchephalon; the midbrain, or mesencephalon; and the forebrain, or prosencephalon. The hindbrain was the principal component of the brains of early vertebrates, with the forebrain becoming increasingly dominant in reptiles, amphibians, birds, and mammals.

4. In birds and mammals, the brain is much larger in proportion to the body than is the case in other vertebrates, reflecting a great increase in the size of the cerebrum.

5. In humans and other primates, the cerebrum is split into two hemispheres, connected by a nerve tract called the corpus callosum. Each hemisphere is divided further by deep grooves into four lobes, the frontal, parietal, temporal, and occipital lobes. The lobes have different functions.

6. Sensory receptors respond to three classes of stimuli: mechanical, chemical, and electromagnetic. No matter what type of stimulus, a response usually takes the same form, depolarization of a sensory neuron membrane.

7. Many of the body's internal receptors are simple receptors in which a nerve ending becomes depolarized as a response to deformation of a nerve membrane, or to a chemical-induced or temperature-induced opening of ion channels in the nerve membrane.

8. Hearing organs in terrestrial vertebrates amplify airborne sound waves and direct them at a fluid-containing chamber, the cochlea, within the ear. Mechanical receptors in the chamber are then deformed by the fluid-borne sound waves.

9. Vision, like photosynthesis, uses a pigment as a primary photoreceptor. The vertebrate eye is designed as a lens-focused camera. The fovea at the center of the retina transmits a point-to-point image to the brain.

10. Other stimuli sensed by vertebrate receptors include polarized light, ultrasound, heat, electricity, and magnetism.

11. The somatic muscles are directed by the motor neurons of the central nervous system. The activity of the central nervous system in directing the somatic muscles is modulated by stretch receptors, which are embedded in the muscles linked to afferent fibers returning to the central nervous system.

12. Smooth muscles, cardiac muscles, and glands are directed by antagonistic command nerve pairs of the autonomic nervous system, one of which stimulates while the other inhibits. In general, the parasympathetic nerves stimulate the activity of normal internal body functions and inhibit alarm responses, and the sympathetic nerves do the reverse.

REVIEW

1. The three principal divisions that characterize the brains of all living vertebrates are _____, _____, and _____.

2. What part of your forebrain is responsible for learning?

3. Short-term memory appears to be stored _____, whereas long-term memory seems to involve _____ in the neural connections within the brain.

4. The visual pigment is _____.

5. The neurotransmitter at the terminal synapse in the sympathetic nervous system is either _____ or _____.

SELF-QUIZ

1. Bundles of nerve fibers in the central nervous system are called
   (a) tracts   (d) nerves
   (b) dendrites   (e) ganglia
   (c) axons

2. The hindbrain is devoted primarily to
   (a) sensory integration
   (b) processing of olfactory information
   (c) processing of visual information
   (d) coordinating motor reflexes
   (e) integrating visceral activities

3. The hypothalamus controls
   (a) respiration
   (b) sensory integration
   (c) heartbeat
   (d) body temperature
   (e) all but b

4. The ability to hear often decreases with age
   (a) because the cilia degenerate
   (b) because the hair cells stiffen
   (c) because the flexibility of the basilar membrane changes
   (d) because the tympanic membrane breaks
   (e) because the tympanic canal straightens

5. The autonomic nervous system is composed of which two of the following elements?
   (a) the sympathetic nervous system
   (b) the sympatric nervous system
   (c) the voluntary nervous system
   (d) the peripheral nervous system
   (e) the parasympathetic nervous system

THOUGHT QUESTIONS

1. Many writers have stated that the hindbrain of the human brain is its most primitive element. How would you respond to this proposition?

2. Most of us have sensed at one time or another an oncoming storm by detecting the increase in humidity in the air. What sort of receptors detect humidity? Why do you suppose that hot days seem so much hotter when it is humid?

3. The literature of science fiction is awash with stories of extrasensory perception, and some research laboratories are actively engaged in attempts to demonstrate it. Can you describe a sensory receptor that might function in "extrasensory" perception?

4. The heat-detecting pit receptor of snakes is a very effective means of "seeing" at night. It is the same sort of sensory system employed by soldiers in "snooper-scopes" and in heat-seeking missiles. Why do you suppose other night-active vertebrates, such as bats, have not evolved this sort of sensory system?

5. In zero gravity, how would you expect a statocyst to behave? What would you expect the subjective impression of motion to be? Would the semicircular canals detect angular acceleration equally well at zero gravity?

FOR FURTHER READING

ALLPORT, S.: *Explorers of the Black Box: The Search for the Cellular Basis of Memory,* W.W. Norton & Co., New York, 1986. A vivid account of the pioneering studies of Eric Kandel and others in their efforts to demonstrate how we remember. Easy to read, this book shows scientists in action, gathering data and disputing among themselves about what the data mean.

FINKE, R.: "Mental Imagery and the Visual System," *Scientific American,* May 1986, pages 84–92. Within the human brain, the way in which visual information is perceived and carried to the brain has a strong influence on mental imagery.

GIBBONS, B.: "The Intimate Sense of Smell," *National Geographic,* September 1986, pages 324–361. A detailed account of the human sense of smell; interesting and well illustrated.

MASLAND, R.: "The Functional Architecture of the Retina," *Scientific American,* December 1986, pages 102–111. This article provides a close look at how neurobiologists are beginning to study the paths of individual neurons by selectively injecting single cells with fluorescent dyes.

WINTER, P., and J. MILLER: "Anesthesiology," *Scientific American,* April 1985, pages 124–131. How certain drugs act to block consciousness is described very clearly.

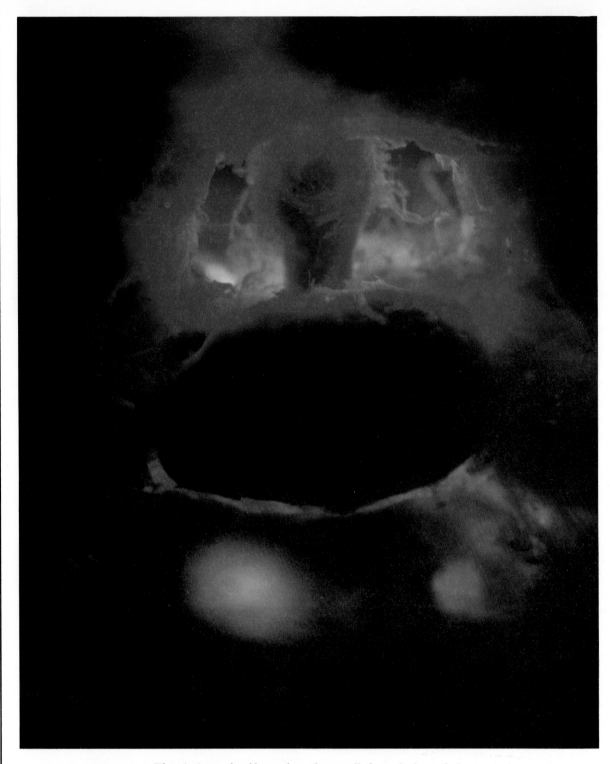

FIGURE 40-1   The pituitary gland hangs by a short stalk from the hypothalamus.
The pituitary regulates the hormone production of many of the body's endocrine glands.
Here it is enlarged 15 times.

HORMONES

Overview

In vertebrates and most other animals, the central nervous system coordinates and regulates the diverse activities of the body by using chemical signals called hormones to effect long-term changes in physiological activities. These hormones maintain physiological conditions within narrow bounds, and most of them function similarly in all vertebrates.

For Review   *Here are some important terms and concepts that you will encounter in this chapter. If you are not familiar with them, you should review them before proceeding.*

Function of the liver (Chapter 34)

Operation of neurons (Chapter 38)

Central nervous system (Chapter 39)

The tissues and organs of an adult mammal participate in a multitude of activities, all of them regulated so as to avoid conflict and to maximize interaction. This is why we think of ourselves as an *organism,* rather than as a smoothly functioning collection of organs. Integration of the many activities of the vertebrate body is the primary function of the central nervous system. All control of body functions, voluntary and involuntary, is vested in this system.

The central nervous system employs three separate motor command systems (see Figure 39-4). These systems differ in their speed of expression, duration of response, and narrowness of application. They are the voluntary nervous system and the autonomic nervous system, discussed in the preceding chapter, and the neuroendocrine system, which is discussed in this chapter (Figure 40-1). Several examples of the role of this system in maintaining the homeostasis of the body are also presented.

NEUROENDOCRINE CONTROL

The effective regulation of many of the body's functions requires not only the constant adjustment and integration of its internal activities but also the ability to bring about longer-term changes in levels of activity, such as initiating the production of milk by the mammary glands of nursing mothers or carrying out the sexual maturation that occurs when a young boy or girl goes through puberty. In the vertebrate body, these changes are achieved by chemical signals rather than by nervous signals. These signals are hormones that the circulation system delivers to the target organ (Figure 40-2). As we have noted earlier, hormones are regulating chemicals that are made at one place in

FIGURE 40-2

Chemical structure of some important hormones.

A Peptide hormones. Antidiuretic hormone (ADH) regulates water loss by the kidneys; oxytocin, which is very similar in structure to ADH, acts on the mammary glands to stimulate milk ejection. The differences between the two molecules are highlighted. Somatostatin inhibits the secretion of growth hormone.

B Steroid hormones. Many steroid hormones are similar in structure to the blood lipid cholesterol, being derived from it. Testosterone stimulates development of the genital tract; cortisone promotes the breakdown of muscle proteins in metabolism.

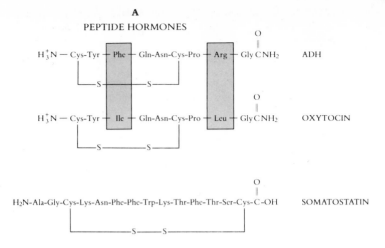

A
PEPTIDE HORMONES

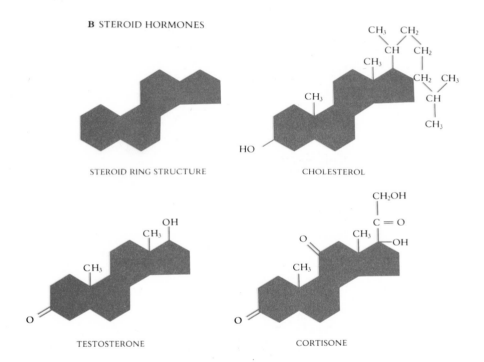

B STEROID HORMONES

the body and exert their influence at another. They are not themselves enzymes; rather they act by regulating preexisting processes. When the hormones reach their destination, they are recognized by specific receptors that only their target glands or tissues possess. Because most hormones circulating in the bloodstream are produced by endocrine glands (Figure 40-3) whose activity is under the direct control of the nervous system, *these hormones constitute a chemical extension of the nervous system.* They permit the central nervous system to issue long-term commands that may influence the level of activity of numerous organs.

The nervous system controls the activity of the endocrine glands, which in turn produce most of the hormones that circulate in the bloodstream of vertebrates.

Neuroendocrine control is vested in the hypothalamus. As you will recall from Chapter 39, the hypothalamus is part of the diencephalon region of the forebrain, which is located at the base of the brain just above the pituitary gland. Within the hy-

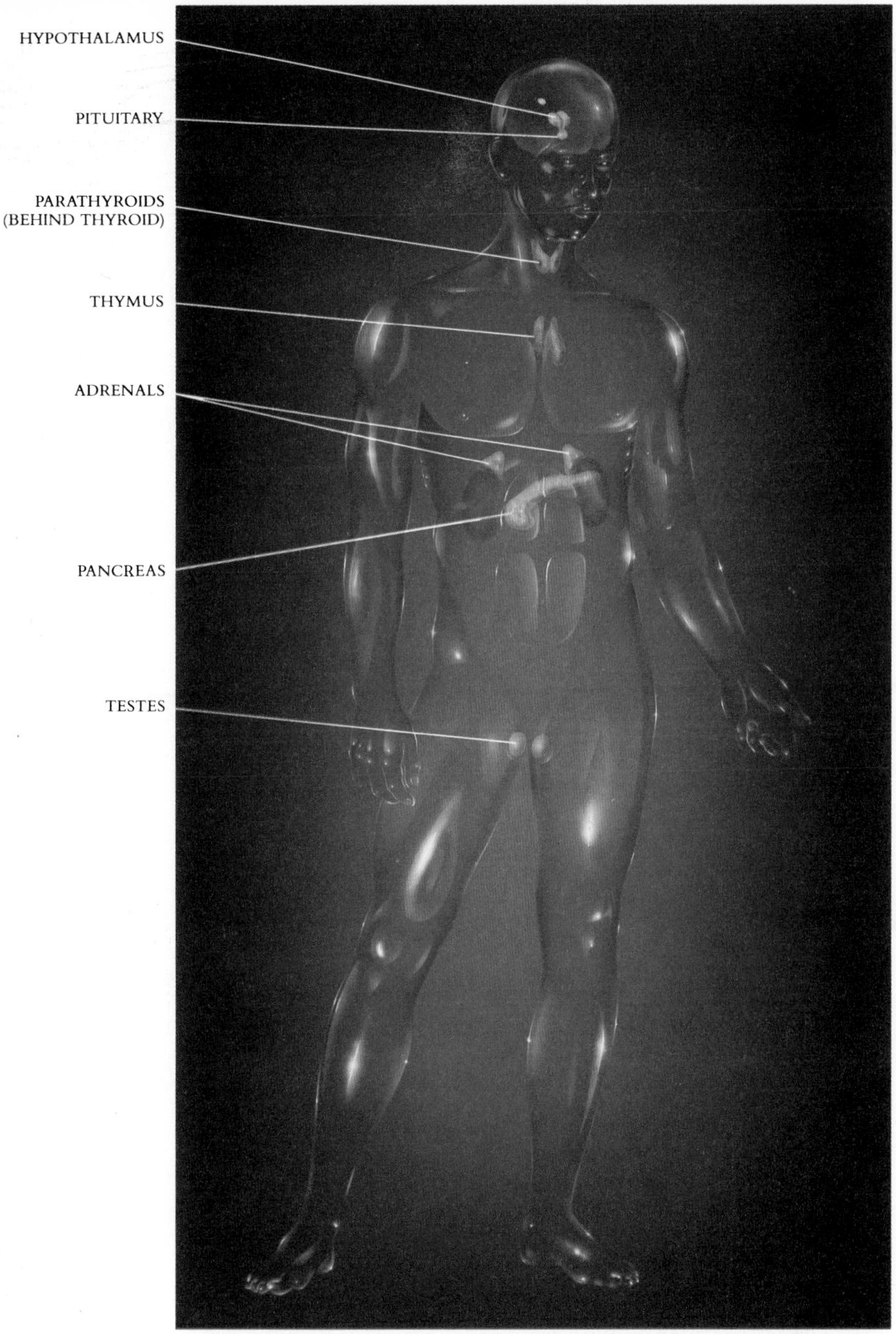

FIGURE 40-3

Location of some of the major
human endocrine glands.

HYPOTHALAMUS

PITUITARY

PARATHYROIDS
(BEHIND THYROID)

THYMUS

ADRENALS

PANCREAS

TESTES

pothalamus (see Figure 39-7), interoceptive information about the body's many internal functions is processed and regulatory commands are issued. These commands involve matters such as the regulation of body temperature, the intake of food and water,
reproductive behavior, and response to pain and emotion. The commands are issued to
the pituitary gland (see Figures 40-1 and 39-7), which in turn sends chemical signals to
the various hormone-producing glands of the body. The hypothalamus thus regulates
the body through a chain of command in the same way that a general gives orders to a
chief of staff, who relays them to lower-ranking commanders.

**The hypothalamus processes information about the body's internal functions
and issues regulatory commands that direct the pituitary gland to send further
chemical signals to the various hormone-producing glands of the body.**

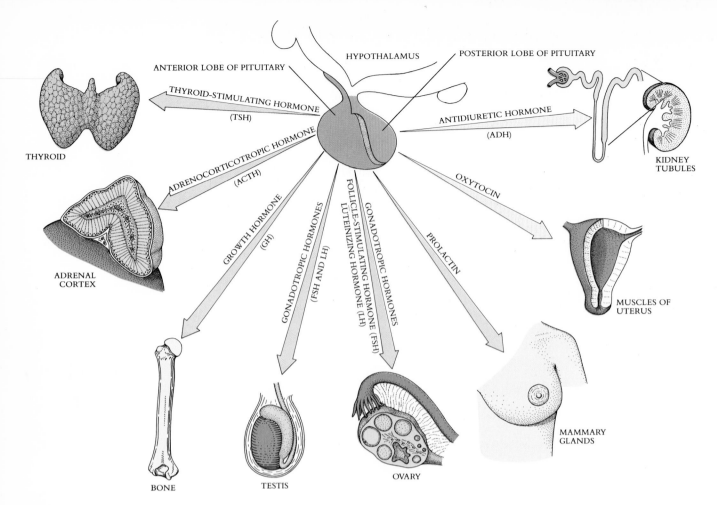

ANTERIOR LOBE OF PITUITARY

HYPOTHALAMUS

POSTERIOR LOBE OF PITUITARY

THYROID-STIMULATING HORMONE
(TSH)

THYROID

ADRENOCORTICOTROPIC HORMONE
(ACTH)

ADRENAL
CORTEX

GROWTH HORMONE
(GH)

GONADOTROPIC HORMONES
(FSH AND LH)

FOLLICLE-STIMULATING HORMONE (FSH)
LUTEINIZING HORMONE (LH)
GONADOTROPIC HORMONES

PROLACTIN

OXYTOCIN

ANTIDIURETIC HORMONE
(ADH)

KIDNEY
TUBULES

MUSCLES OF
UTERUS

MAMMARY
GLANDS

BONE

TESTIS

OVARY

FIGURE 40-4

Interactions between the hypothalamus, the anterior lobe and the posterior lobe of the pituitary (two distinct glands), and various organs of the human body.

HORMONES RELEASED BY THE POSTERIOR PITUITARY

The pituitary gland actually consists of two distinct glands, both of which secrete a number of important hormones (Figure 40-4). The first of these two glands, the **posterior lobe of the pituitary,** is linked directly to the hypothalamus by neural connections and secretes hormones into the general blood circulation. One of these hormones is **antidiuretic hormone (ADH)** (see Figure 40-2, *A*), also called vasopressin. ADH is involved in regulating the rate of water reabsorption in the kidneys and intestines. Alcohol suppresses ADH release, which is why excessive drinking leads to the production of excessive quantities of urine and eventually to dehydration.

Another hormone secreted by the posterior lobe of the pituitary is **oxytocin** (see Figure 40-2, *A*). Oxytocin aids in initiating milk release by causing the contraction of the muscles around the ducts into which mammary glands secrete milk. Oxytocin also causes the uterus to contract during the labor associated with childbirth. This is why the uterus of a nursing mother returns to normal size after its extension during pregnancy more quickly than does the uterus of a mother who does not nurse her baby.

Both vasopressin and oxytocin are actually formed within the hypothalamus. From there they are transported within nerve axons to the nerve endings of the posterior lobe of the pituitary, which releases them into the bloodstream.

The posterior lobe of the pituitary, a distinct gland, is connected with the hypothalamus by neural connections. It secretes the important hormones antidiuretic hormone and oxytocin.

HORMONES RELEASED BY THE ANTERIOR PITUITARY

The **anterior lobe of the pituitary** is connected to the hypothalamus by special blood vessels only a few millimeters long. Through these vessels pass a group of regulating hormones produced in the hypothalamus. As a group, these are called "releasing hormones." They command the anterior lobe of the pituitary to initiate the production (release) of specific hormones in distant endocrine glands (see Figure 40-4). When these distant glands have manufactured their hormones, they release them into the general circulation.

For each releasing hormone secreted by the hypothalamus there is a corresponding hormone synthesized by the anterior lobe of the pituitary. When the pituitary receives a releasing hormone from the hypothalamus, the anterior lobe responds by secreting the corresponding **pituitary hormone.** The seven principal pituitary hormones are:

Thyroid-stimulating hormone (TSH). TSH stimulates the thyroid gland to produce thyroid hormones (Figure 40-5), which in turn stimulates oxidative respiration.

Luteinizing hormone (LH). LH plays an important role in the female menstrual cycle (see Chapter 42). It also stimulates the male gonads to produce testosterone, which initiates and maintains the development of male secondary sexual characteristics, those external male sexual features not involved in reproduction.

Follicle-stimulating hormone (FSH). FSH is significant in the female menstrual cycle (see Chapter 42). In males, it stimulates certain cells in the testes to produce a hormone that regulates the development of sperm.

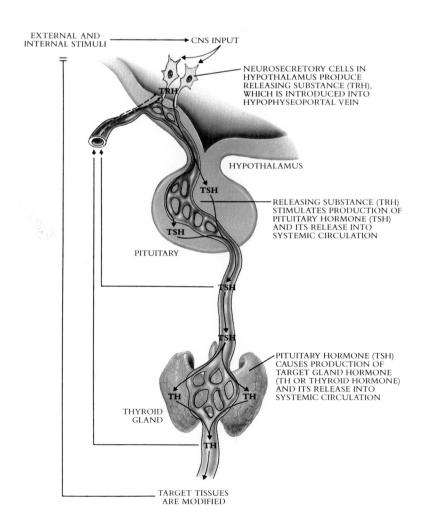

EXTERNAL AND INTERNAL STIMULI → CNS INPUT

NEUROSECRETORY CELLS IN HYPOTHALAMUS PRODUCE RELEASING SUBSTANCE (TRH), WHICH IS INTRODUCED INTO HYPOPHYSEOPORTAL VEIN

TRH

HYPOTHALAMUS

TSH

RELEASING SUBSTANCE (TRH) STIMULATES PRODUCTION OF PITUITARY HORMONE (TSH) AND ITS RELEASE INTO SYSTEMIC CIRCULATION

TSH

PITUITARY

TSH

TSH

PITUITARY HORMONE (TSH) CAUSES PRODUCTION OF TARGET GLAND HORMONE (TH OR THYROID HORMONE) AND ITS RELEASE INTO SYSTEMIC CIRCULATION

TH   TH

THYROID GLAND

TH

TARGET TISSUES ARE MODIFIED

FIGURE 40-5

The release of thyroid hormone (TH) from the thyroid gland is the end result of a long series of events mediated by the central nervous system. First, the hypothalamus produces a specific *thyroid-releasing hormone,* TRH, which circulates through special very short blood vessels to the anterior region of the pituitary. Second, the arrival of TRH stimulates the pituitary to produce the pituitary hormone *thyroid-stimulating hormone,* TSH, which it releases into the general blood circulation. This carries it to the thyroid gland. Third, the arrival of TSH stimulates the thyroid gland to produce *thyroid hormone,* TH, which is released to circulate throughout the body. Levels of both TSH and TH in the bloodstream influence the rate at which the central nervous system produces TRH.

Adrenocorticotropic hormone (ACTH). ACTH stimulates the adrenal cortex to produce corticosteroid hormones. Some of these hormones regulate the production of glucose from fat; others regulate the balance of sodium and potassium ions in the blood; and still others contribute to the development of the male secondary sexual characteristics.

Somatotropin, or *growth hormone (GH).* GH stimulates the growth of muscle and bone throughout the body.

Prolactin (PRL). PRL stimulates the breasts to produce milk.

Melatonin or *Melanocyte-stimulating hormone (MSH).* In reptiles and amphibians, MSH stimulates color changes in the epidermis. This hormone has no known function in mammals.

The seven major releasing hormones secreted by the hypothalamus each stimulate the synthesis of a corresponding hormone by the anterior lobe of the pituitary.

In addition to their endocrine functions, these and many other hormones have been shown also to be associated with particular populations of cells within the central nervous system. This is an area of very active research; biologists are attempting to uncover the as yet unknown role of these hormones in the central nervous system. Whatever their function there, it is clear that the same hormone may play different roles in different parts of the body. Hormones are signals used in different tissues for different reasons, just as raising your hand is a signal that has different meanings in different contexts: in class, it indicates you have a question; in a football game it signals a "fair catch"; and when a policeman does it on a street it means "stop." The evolving use of hormones by vertebrates has been a conservative process. Rather than produce a new hormone for every use, vertebrates have often adopted a hormone already at hand for a new use, where this new use does not cause confusion.

Control over Hormone Production

Control over the production of hormones produced by the anterior pituitary is exercised in two ways.

Antagonistic Controls

The production of the hormones GH, PRL, and MSH is controlled by both releasing and inhibitory signals produced by the hypothalamus. Consider the growth hormone somatotropin (GH) (Figure 40-6). The releasing signal for GH is the growth-hormone–releasing hormone (GHRH) produced by the hypothalamus; GHRH stimulates the anterior lobe of the pituitary to produce somatotropin. The inhibiting signal, which is also produced at the same time by the hypothalamus, is somatostatin. Somatostatin inhibits the anterior lobe of the pituitary from producing somatotropin. The hypothalamus thus regulates growth by mediating the relative rates of production of GHRH and somatostatin. In a similar way the hypothalamus regulates the production of PRL and MSH hormones. It is the brain rather than the levels of these three hormones in the blood that determines the rate of their production.

Feedback Controls

The levels of all other hormones produced by the anterior lobe of the pituitary are controlled by negative feedback from the target glands. For example, when LH stimulates the gonads to release testosterone into the bloodstream, that testosterone in turn inhibits the hypothalamus. The hypothalamus then ceases to transmit LH-releasing hormone to the pituitary.

The regulation of pituitary hormone production is achieved by (1) pairs of hypothalamic hormones with opposite effects; and (2) feedback loops, in which the hypothalamus is sensitive to hormone levels in the blood.

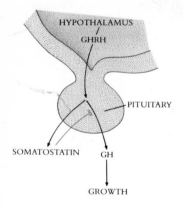

FIGURE 40-6

Levels of the growth hormone GH are regulated by both stimulatory and inhibitory signals. The production of GH by the pituitary is released by a hypothalamic releasing hormone, GHRH, and inhibited by another pituitary hormone, somatostatin. If levels of GHRH become too high, the excess somatostatin that is induced in the pituitary shuts down some of the production of GH so that effective levels of GH circulating in the blood do not fluctuate. (This effect is indicated by the colored line and double bar.)

NONPITUITARY HORMONES

The production of a small number of endocrine hormones is not controlled by the hypothalamus at all. These hormones are of two sorts.

1. *Antagonistic pairs of hormones concerned with basal metabolism.* Among them, **parathyroid hormone** increases the levels of calcium ion in blood plasma; an antagonistic hormone, **calcitonin,** lowers the concentrations of calcium ion. The final levels of calcium in the blood are determined by the balance between the two systems. The islets of Langerhans in the pancreas produce another pair of opposing hormones, **insulin** and **glucagon,** which together regulate the level of blood glucose.

2. *Noradrenaline and adrenaline.* These hormones are produced by the adrenal medulla (the inner portion of the adrenal gland located just above the kidneys in the center of the body). They are used in two ways: (1) as neurotransmitters in the sympathetic nervous system; and (2) as hormones released into the bloodstream. Circulating in the bloodstream, noradrenaline and adrenaline produce an "alarm" response throughout the body that is identical to the individual effects achieved by the sympathetic nervous system but is longer lasting. Among the effects of these hormones is an accelerated heartbeat, increased blood pressure, higher levels of blood sugar, dilated blood vessels, and increased blood flow to the heart and lungs. These adrenal hormones act as extensions of the sympathetic nervous system.

Nonpituitary hormones include antagonistic pairs of hormones associated with basal metabolism and those produced by the adrenal medulla. The latter hormones, noradrenaline and adrenaline, serve both as neurotransmitters in the sympathetic nervous system and as hormones in the bloodstream.

HOW NEUROENDOCRINE CONTROL WORKS

The body's many hormones fall into two general chemical categories. The members of each category operate by very different mechanisms:

Peptide hormones. (see Figure 40-2, *A*). Some hormones, such as adrenaline and thyroid hormone, are small molecules derived from the amino acid tyrosine; others are short polypeptide chains. Most hormones that circulate within the brain belong to this second class. They include small peptides called **enkephalins,** which are only five amino acid units long. The enkephalins appear to play a role in integrating afferent impulses from the pain receptors. A third type of peptide hormone is larger: the brain hormones called **endorphins** are polypeptides 32 amino acid units long. Endorphins appear to regulate emotional responses in the brain. Morphine has such a potent analgesic (pain-relieving) effect on the central nervous system because it mimics the effects of the endorphins. Still other hormones are proteins, which consist of even longer polypeptide chains; insulin is an example. Peptide hormones do not enter their target cells. Instead, they interact with a receptor on the cell surface and thus initiate a chain of events within the cell.

Steroid hormones (see Figure 40-2, *B*). All steroid hormones are derived from cholesterol, a complex molecule composed of three six-membered carbon rings and one five-membered carbon ring—it looks something like a fragment of chain-link fence. The hormones that promote the development of the secondary sexual characteristics are steroids. They include cortisone and testosterone, as well as the hormones estrogen and progesterone that are used in birth control pills. Steroid hormones act by altering the pattern of gene expression. They enter a target cell and penetrate its nucleus, where they initiate the transcription of some genes while repressing the transcription of others.

TABLE 40-1   PRINCIPAL ENDOCRINE GLANDS AND THEIR HORMONES

ENDOCRINE GLAND AND HORMONE	TARGET TISSUE	PRINCIPAL ACTIONS	CHEMICAL NATURE
HYPOTHALAMUS			
Oxytocin	Uterus	Stimulates contraction of uterus	Peptide (9 amino acids)
	Mammary glands	Stimulates ejection of milk	
ADH	Kidneys	Stimulates reabsorption of water; conserves water	Peptide (9 amino acids)
ANTERIOR LOBE OF PITUITARY			
Somatotropin (Growth hormone, GH)	General	Stimulates growth by promoting protein synthesis and breakdown of fatty acids	Protein
Prolactin (PRL)	Mammary glands	Stimulates milk production	Protein
Thyroid-stimulating hormone (TSH)	Thyroid gland	Stimulates secretion of thyroid hormones	Glycoprotein
Adrenocorticotropic hormone (ACTH)	Adrenal cortex	Stimulates secretion of adrenal cortical hormones	Polypeptide
Follicle-stimulating hormone (FSH)	Gonads	Stimulates ovarian follicle, spermatogenesis	Glycoprotein
Luteinizing hormone (LH)	Gonads	Stimulates ovulation and corpus luteum formation in females	Glycoprotein
OVARY			
Estrogens	General Female reproductive structures	Stimulate development of secondary sex characteristics in females and growth of sex organs at puberty; prompt monthly preparation of uterus for pregnancy	Steroid
Progesterone	Uterus	Completes preparation of uterus for pregnancy	Steroid
	Breasts	Stimulates development	
TESTIS			
Testosterone	General	Stimulates development of secondary sex characteristics in males and growth spurt at puberty	Steroid
	Male reproductive structures	Stimulates development of sex organs; stimulates spermatogenesis	

REGULATION OF GLUCOSE LEVELS IN THE BLOOD

It is important that the plasma of vertebrate blood maintain a relatively constant composition, since the different tissues of the body have evolved complex specializations that depend in important ways on the composition of the blood and particularly on its metabolites. Brain cells, for example, can store very little glucose and lack the enzymes to convert fat or amino acids into glucose. Despite these factors, brain cells are very sensitive to the level of available glucose. They are totally dependent on the blood plasma for glucose and cease to function if the level of glucose in the blood falls much below normal values.

Maintaining a constant level of metabolites in the blood plasma requires active control by the organs of the body. A moment's reflection shows why. Most vertebrates eat sporadically. In the United States, for example, most people eat three meals a day. Food enters the digestive system at intervals, and these intervals are separated by

TABLE 40-1   PRINCIPAL ENDOCRINE GLANDS AND THEIR HORMONES—cont'd

ENDOCRINE GLAND AND HORMONE	TARGET TISSUE	PRINCIPAL ACTIONS	CHEMICAL NATURE
ADRENAL MEDULLA			
Epinephrine and norepinephrine (adrenaline and noradrenaline)	Skeletal muscle, cardiac muscle, blood vessels	Initiate stress responses; increase heart rate, blood pressure, metabolic rate; dilate blood vessels; mobilize fat, raise blood sugar level	Amino acid derivatives
ADRENAL CORTEX			
Aldosterone	Kidney tubules	Maintains proper balance of sodium and phosphate ions	Steroid
Cortisol	General	Adaptation to long-term stress; raises blood glucose level; mobilizes fat	Steroid
PINEAL GLAND			
Melatonin	Gonads, pigment cells	Function not well understood: influences pigmentation in some vertebrates; may control biorhythms in some animals; may help control onset of puberty in humans	Amino acid derivative
THYROID GLAND			
Thyroxine	General	Stimulates metabolic rate; essential to normal growth and development	Iodinated amino acid
Calcitonin	Bone	Lowers blood calcium level by inhibiting loss of calcium from bone	Polypeptide (32 amino acids)
PARATHYROID GLANDS			
Parathyroid hormone	Bone, kidneys, digestive tract	Increases blood calcium level by stimulating bone breakdown; stimulates calcium reabsorption in kidneys; activates vitamin D	Polypeptide (34 amino acids)
ISLETS OF PANCREAS			
Insulin	General	Lowers blood glucose; increases storage of glycogen	Polypeptide (51 amino acids)
Glucagon	Liver, adipose tissue	Raises blood glucose level; stimulates breakdown of glycogen in liver	Polypeptide (29 amino acids)

long periods of fasting. Much of the food is digested relatively quickly, with the metabolites, such as glucose and amino acids, passing through the lining of the stomach and small intestine into the bloodstream. Without active control of the levels of these and other metabolites, they would suddenly become much more abundant in the blood right after a meal and then fall rapidly during a period of starvation, when the metabolites are being removed from the bloodstream by metabolizing cells and not being replenished.

Control over metabolite levels in the blood plasma is achieved in a very logical way: by establishing a reservoir, or metabolic bank, in the liver. Of all the liver's many functions, one of the most important is its regulation of the blood's metabolite levels. This regulation is achieved by means of a special shunt in the circulatory system. Blood returning from the stomach and small intestine flows into a special conduit called the **portal vein,** which carries it not to the vena cava and the heart but rather to the liver.

FIGURE 40-7

Hepatic-portal circulation. Blood from the stomach and small intestine, rich with the metabolites of digestion, is collected into the portal vein, which carries it to the liver. After flowing through the sinuses of the liver, the liver-processed blood is re-collected into the hepatic vein, which carries it back toward the heart.

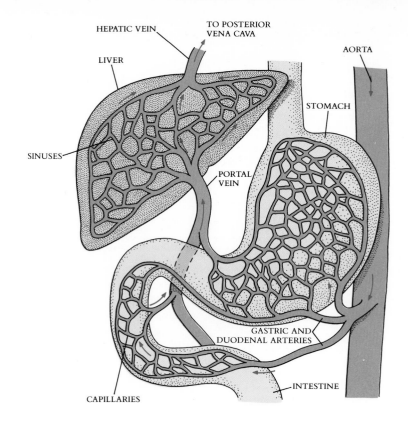

Within the liver, the blood passes through a network of fine passages called **sinuses.** Only after passing through the liver tissue is the blood collected into the **hepatic vein** and delivered to the vena cava and the heart (Figure 40-7).

When excessive amounts of glucose are present in the blood that is passing through the liver (a situation that occurs soon after a meal), the liver removes the excess glucose from the blood and converts it to a starchlike glucose polymer, glycogen (Figure 40-8) The liver then stores this glycogen. When the level of glucose in the blood falls, as it does in a period of fasting, the glucose deficit in the blood plasma is made up from the glycogen reservoir. The human liver, for example, stores enough glycogen to supply glucose to the bloodstream for about 24 hours of fasting. If fasting continues, the liver begins to convert other molecules, such as amino acids, into glucose to maintain the level of glucose in the blood. The liver, the body's metabolic reservoir, thus

FIGURE 40-8

Hormonal control of blood glucose levels.
A When blood sugar levels are low, cells within the pancreas release the hormone glucagon into the bloodstream, and other cells within the adrenal gland, situated on top of the kidneys, release the hormone adrenaline into the bloodstream. When they reach the liver, these two hormones both act to increase the liver's breakdown of glycogen to glucose.
B When blood glucose levels are high, other cells within the pancreas produce the hormone insulin, which stimulates the liver and muscles to convert blood glucose into glycogen. The glucose level in the blood determines the levels of insulin and glycogen in the blood via feedback loops (noted by the colored lines with the double bars in **A** and **B**) to the pancreas and adrenal gland.

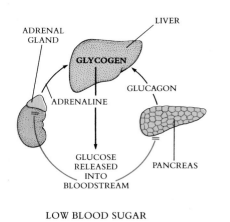

LOW BLOOD SUGAR

A

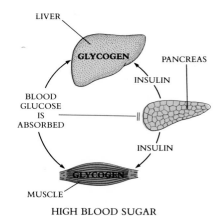

HIGH BLOOD SUGAR

B

acts much like a bank, making deposits and withdrawals in the currency of glucose molecules.

In addition to its many other roles, the liver acts to regulate the level of blood glucose, maintaining it within narrow bounds.

Also like a bank, the liver exchanges currencies, converting other molecules such as amino acids and fats to glucose for storage. The liver cannot store significant concentrations of amino acids. Instead, excess amino acids that may be present in the blood are converted to glucose by liver enzymes, and the glucose is stored as glycogen. The first step in this conversion is the removal of the amino group ($-NH_2$) from the amino acid, a process called **deamination.** Unlike plants, animals cannot reuse the nitrogen from these amino groups and must excrete it as nitrogenous waste. The product of amino acid deamination, ammonia (NH_3), chemically combined with CO_2 to form urea or further chemically modified to form uric acid, is released by the liver into the bloodstream, where the kidneys later remove the urea or uric acid for excretion from the body.

The liver has a limited storage capacity. When its glycogen reservoir is full, it continues to remove excess glucose molecules from the blood by converting them to fat, which is stored elsewhere in the body. In humans long periods of overeating and the resulting chronic oversupply of glucose frequently result in the deposition of fat around the stomach or on the hips.

When blood glucose levels fall below normal, a small group of cells in the pancreas, the islets of Langerhans (Figure 40-9), respond by releasing the hormone glucagon into the bloodstream. Simultaneously, the cells of the adrenal medulla also detect the lowered blood glucose and secrete adrenaline into the bloodstream. Both of these hormones act within the liver to increase the rate of conversion of glycogen to glucose and thereby raise the levels of glucose in the blood (see Figure 40-8).

When blood glucose levels become excessive, on the other hand, other cells within the islets of Langerhans produce a different hormone, insulin. Insulin mediates the uptake of glucose by cells from the blood and stimulates the formation of glycogen from blood glucose, causing the levels of glucose in the blood to fall. The amounts of insulin and glucagon that are produced are determined by the level of glucose in the blood—a feedback loop. The overall level of glucose is determined by the balance between the two hormones.

Diabetes is a term used to designate a diverse set of conditions, some of them hereditary, in which the affected individuals ultimately develop elevated levels of glucose in the blood. This is usually because individuals lack insulin and so are unable to take up glucose from the blood, even though the levels of glucose become very high. Such individuals lose weight and may eventually suffer brain damage and even death if their condition is untreated. Diabetes can be treated by supplying insulin to the affected individual daily. Active research on the possibility of transplanting islets of Langerhans holds much promise of a lasting treatment for diabetes.

Diabetes is a condition in which individuals are unable to obtain glucose from their blood because they lack insulin. It is a serious disease and can be fatal if untreated.

THE REGULATION OF PHYSIOLOGICAL FUNCTIONS: WATER BALANCE

The kidney is one example of an organ whose principal function is homeostasis, the maintenance of constant physiological conditions within the body. The kidney is concerned with water and ion balance. Other organs have other homeostatic functions. In almost all cases these homeostatic functions are integrated by the central nervous system, using the voluntary, autonomic, and hormonal controls we have discussed in this and the previous chapter.

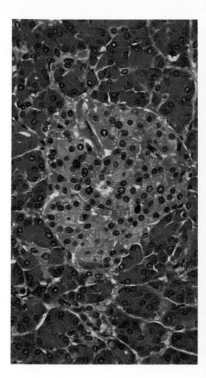

FIGURE 40-9

Glucagon and insulin are produced by clumps of cells within the pancreas called islets of Langerhans, which are stained dark in this preparation.

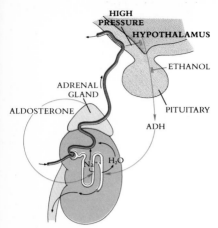

FIGURE 40-10

Control of water and salt balance within the kidney is centered in the hypothalamus. The hypothalamus produces antidiuretic hormone (ADH), which renders the collecting ducts of the kidneys freely permeable to water and so maximizes water retention. If too much water retention leads to high blood pressure, pressure-sensitive receptors in the hypothalamus detect this and cause the production of ADH to be shut down. If the level of sodium in the blood falls, the adrenal gland initiates production of the hormone aldosterone, which stimulates salt re-absorption by the renal tubules of the kidney. (Feedback loops are noted by the colored lines with the double bar.)

The kidney provides numerous examples of how integration by the central nervous system maintains homeostasis. Consider water balance (Figure 40-10). It is not always desirable for you to retain the same amount of water. If you have consumed an unusually large amount of water, the maintenance of homeostasis requires that you retain less of it than you would otherwise. When the uptake of water is excessive, the hypothalamus detects the resulting increase in blood pressure and decreases its output of the hormone ADH. ADH controls the permeability of the collecting ducts of the renal tubules in the kidney to water. When ADH is present, the collecting ducts are freely permeable to water, so that water passes out into the surrounding tissue, which has a higher osmotic concentration of solutes. A decrease in levels of ADH renders the collecting ducts of the renal tubules less permeable to water and, in that way, inhibits the reabsorption of water and increases urine volume.

Another example of integration by the central nervous system leading to homeostasis in the kidney is provided by the salt balance in your body (see Figure 40-10). The amount of salt in your diet can vary considerably, and yet it is important to many physiological processes that the salt level in your blood not vary widely. When the level of sodium ion in the blood falls, the adrenal gland increases its production of the hormone **aldosterone,** a steroid hormone that stimulates active sodium ion reabsorption across the walls of the ascending arms of your kidneys' renal tubules. In this way, it decreases the amount of sodium that is lost in the urine. In the total absence of this hormone, humans may excrete up to 25 grams of salt a day.

Chapter 41 discusses the problems of maintaining water balance and homeostasis in more detail as it is carried out by the kidneys and excretory system.

SUMMARY

1. Because most hormones circulating in the bloodstream are produced by endocrine glands whose activity is under the direct control of the nervous system, these hormones constitute a chemical extension of the nervous system.

2. The brain maintains long-term control over physiological processes by synthesizing releasing hormones in the hypothalamus. These hormones direct the synthesis of specific circulating hormones by the pituitary gland. Pituitary hormones travel out into the body and initiate the synthesis of particular hormones in target tissues.

3. The posterior lobe of the pituitary, a distinct gland, secretes hormones such as ADH into the bloodstream. This gland is linked directly to the hypothalamus by neural connections.

4. The anterior lobe of the pituitary, another distinct gland, is connected to the hypothalamus by special blood vessels a few millimeters long. It secretes seven principal kinds of hormones, each corresponding to a specific releasing hormone from the hypothalamus.

5. Most brain hormones are peptides that interact with receptors on the target cell surface, thus activating enzymes within. Many of the hormones produced by endocrine glands such as the pituitary are steroids, which enter into cells and alter their transcription patterns.

6. The liver maintains a reservoir of metabolites and regulates their level in the blood, storing excess glucose as a starchlike polymer known as glycogen.

7. The kidney is an organ concerned with water and ion balance within the body.

REVIEW

1. The neuroendocrine system is a network of _____, whose production is controlled by the central nervous system.

2. The _____ sends signals to the pituitary gland and thus controls the neuroendocrine system.

3. When the hypothalamus releases thyroid-releasing hormone to the pituitary, it causes release of _____ by the anterior pituitary.

4. Inhibitory signals for the production of growth hormone, prolactin, and melanocyte-stimulating hormone are produced by the _____.

5. If blood glucose levels are high, the islets of Langerhans in the pancreas produce the hormone _____, which stimulates the liver to convert blood glucose to glycogen.

SELF-QUIZ

1. Hormones are
   (a) enzymes
   (b) regulatory chemicals
   (c) produced at one place in the body but exert their influence at another
   (d) a and c
   (e) b and c

2. An example of a peptide hormone is
   (a) cortisone
   (b) progesterone
   (c) testosterone
   (d) oxytocin
   (e) estrogen

3. Which of the following is *not* one of the seven principal pituitary hormones?
   (a) thyroid hormone
   (b) follicle-stimulating hormone
   (c) melanocyte-stimulating hormone
   (d) somatotropin
   (e) luteinizing hormone

4. Noradrenaline and adrenaline are hormones that produce an "alarm" response. They are produced by the adrenal gland. Where is the adrenal gland?
   (a) in the brain
   (b) just above the kidneys
   (c) in the neck
   (d) in the testes or ovaries
   (e) near the thymus

5. The hypothalamus controls water and salt balance within the kidney by regulating the production of the hormone
   (a) estrogen
   (b) insulin
   (c) glucagon
   (d) antidiuretic hormone
   (e) aldosterone

THOUGHT QUESTIONS

1. If you are lost in the desert with a case of liquor and are desperately thirsty, should you drink the liquor? Explain your answer.

2. Why do you suppose the brain goes to the trouble of synthesizing releasing hormones rather than simply directing the production of the pituitary hormones immediately?

FOR FURTHER READING

BERRIDGE, M.: "The Molecular Basis of Communication Within the Cell," *Scientific American,* October 1985. An up-to-date account of what is known about second messengers in the cell.

BLOOM, F.E.: "Neuropeptides," *Scientific American,* October 1981, pages 148-168. An account of recent advances in the study of endorphins and other brain hormones.

CANTIN, M., and J. GENEST: "The Heart as an Endocrine Gland," *Scientific American,* February 1986, pages 76-81. A good example of how our body self-regulates its activities; in addition to pumping blood, the heart secretes a hormone that fine-tunes the control of blood pressure.

DAVIS, J.: *Endorphins: New Waves in Brain Chemistry,* Doubleday & Co., Inc., New York, 1984. A popular account of current research on brain hormones.

FIGURE 41-1   This African elephant is knee–deep in water, and enjoys its tromp through the mud every bit as much as you might enjoy going to the beach. For him, and us, water is both a necessity and a pleasure.

THE CONTROL OF WATER BALANCE

Chapter 41

Overview

Vertebrates live in salt water, in fresh water, and on land, and each of these environments poses different problems for balancing water retention with proper salt concentration. Vertebrates conserve or excrete water, depending on the environment in which they live, by regulating the passage of water through their excretory system.

For Review

Here are some important terms and concepts that you will encounter in this chapter. If you are not familiar with them, you should review them before proceeding.

Sodium chloride (Chapter 3)

Diffusion (Chapter 6)

Hypotonic solutions (Chapter 6)

Hormone regulation of water retention (Chapter 40)

The first vertebrates evolved in water, and the physiology of all members of the subphylum still reflects this origin (Figure 41-1). Approximately two thirds of every vertebrate's body is water. If the amount of water in the body of a vertebrate falls much lower than this, the animal will die. In this chapter we discuss the various strategies animals employ to keep from losing or gaining too much water. Although we might have chosen the evolution of the heart or another system to illustrate the increasing complexity of the adaptation of vertebrates to their varied modes of existence, we have chosen instead to consider the ways in which they control their water balance, since these strategies are so closely tied to their exploitation of the varied environments where they occur (Figure 41-2).

FIGURE 41-2

The Kangaroo rat, *Dipodomys panamintensis,* shown here has very efficient kidneys that can concentrate urine to a high degree by reabsorbing water. As a result, it avoids losing any more moisture than necessary. Since kangaroo rats live in dry or desert habitats, this feature is extremely important to them.

FIGURE 41-3

The concentration of ions is roughly similar in the bodies of different classes of vertebrates. Sharks hold the concentration of solutes in their blood at about the level in seawater, or at a slightly higher level, by adding urea to their bloodstream. In contrast, the body fluids of marine bony fishes contain a far lower ion concentration than the one characteristic of their seawater environment; they are said to be hypoosmotic with respect to seawater. Consequently, such fishes tend to lose water by osmosis and must struggle to retain as much water as possible. Freshwater fishes have the opposite problem: their body fluids contain far higher ion concentrations than the water in which they live. Such fluids are said to be hyperosmotic with respect to the surrounding water. Terrestrial vertebrates have ion concentrations not unlike those of the fishes from which they evolved.

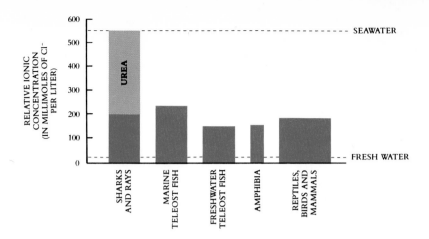

OSMOREGULATION

Plasma membranes are freely permeable to water but impermeable to salts and ions. This property of differential permeability forms the basis for many life processes, including the nerve conduction that we discussed in Chapter 38. If the concentration of salts and ions dissolved in the water surrounding a vertebrate's body was the same as that within the body, the differential permeability of the cells would present no problem; there would be no tendency for water to leave or enter the body. The osmotic pressure of the body fluids would be essentially the same as that of the surroundings. This is true of most marine invertebrates, but only of sharks and their relatives among the vertebrates. Such animals, called **osmoconformers,** maintain the osmotic concentration of their body fluids at about the same level as that of the medium in which they are living, and they change the osmotic concentrations of their body fluids when the osmotic concentration of the medium changes.

The Problems Faced by Osmoregulators

Among the aquatic vertebrates other than sharks, mechanisms have evolved by which they control the osmotic concentration of their body fluids more precisely than other aquatic animals. In other words, they are osmoregulators. **Osmoregulators** maintain an internal solute concentration that does not vary, regardless of the environment in which the vertebrate lives. The maintenance of a constant internal solute concentration has permitted the vertebrates to evolve complex patterns of internal metabolism. It does, however, require constant regulation of the animal's internal water concentration.

Freshwater vertebrates must maintain much higher concentrations of salts in their bodies than those in the water surrounding them. In other words, their body fluids are **hyperosmotic** relative to their environment, and water tends to enter their bodies. They must, therefore, exclude water to prevent the fluids within them from being diluted.

Marine vertebrates have only about one third the osmotic concentration of the surrounding seawater in their bodies. Their body fluids are therefore said to be **hypoosmotic** relative to the environment in which the animals occur—water tends to leave their body. For this reason, marine animals must retain water to prevent dehydration.

On land, surrounded by air, the bodies of vertebrates have a higher concentration of water than does the air surrounding them. They therefore tend to lose water to the air by evaporation. This situation is faced to some degree by the amphibians, which live on land only part of the time. It is also faced by all terrestrial reptiles, birds, and mammals, which must conserve water to prevent dehydration. Figure 41–3 shows the concentration of ions in the body fluids of some representative vertebrates.

How Osmoregulation Is Achieved

Animals have evolved a variety of mechanisms to cope with these problems of water balance, all of them based in one way or another on the animal's excretory system. In many animals the removal of water or salts is coupled to the removal of metabolic wastes from the body. Simple organisms, such as many protists and sponges, employ contractile vacuoles for this purpose. Many freshwater invertebrates employ **nephrid organs,** in which water and waste pass from the body across the membrane into a collecting organ, from which they are ultimately expelled to the outside through a pore. The membrane acts as a filter, retaining proteins and sugars within the body, while permitting water and dissolved waste products to leave.

Insects use a similar filtration system, with a significant improvement that helps them guard against water loss. The excretory organs in insects are the Malpighian tubules. Malpighian tubules are tubular extensions of the digestive tract that branch off before the hindgut. Potassium ion is secreted into the tubules, causing body water and organic wastes to flow into them from the body's circulatory system because of the osmotic gradient. Blood cells and protein are too large to pass across the membrane into the Malpighian tubules, so that the circulatory system is not short-circuited. Because the system of tubules empties into the hindgut, however, the water and potassium can be reabsorbed by the hindgut, and only small molecules and waste products are excreted. Malpighian tubules provide a very efficient means of water conservation.

Like insects, vertebrates use a strategy that couples water balance and salt concentration with waste excretion. Instead of relying on the secretion of salts into the excretory organ to establish an osmotic gradient, however, vertebrates rely on pressure-driven filtration. Whereas insects use an osmotic gradient to *pull* the blood through the filter, the vertebrates *push* blood through the filter, using the higher blood pressure that a closed circulatory system makes possible. Fluids are forced through a membrane that retains proteins and large molecules within the body but passes the small molecules out. Water is then reabsorbed from the filtrate as it passes through a long tube.

> **Terrestrial animals require efficient means of water conservation. In eliminating metabolic wastes, both insects and vertebrates filter the blood to retain blood cells and protein; then the water is collected from the filtrate by passing it through a long tube across whose walls it is reabsorbed.**

Like the osmotically collected waste fluid of insects, the fluid that is passed through the filter of vertebrates contains many small molecules that are of value to the organism, such as glucose, amino acids, and various salts or ions. Vertebrates have evolved a means of selectively reabsorbing these valuable small molecules without absorbing the waste molecules that are also dissolved in the filtered waste fluid or **urine.** Selective reabsorption gives the vertebrates great flexibility, since the membranes of different groups of animals can and have evolved different transport channels and thus the ability to reabsorb different molecules. This flexibility is a key factor underlying the ability of different vertebrates to function in many diverse environments. They can reabsorb small molecules that are especially valuable in their particular habitat and not absorb wastes. Among the vertebrates, the apparatus that carries out the processes of filtration, reabsorption, and secretion is the **kidney.** It can function, with modifications, in fresh water, in the sea, and on land.

THE ORGANIZATION OF THE VERTEBRATE KIDNEY

All vertebrate kidneys possess two segments that carry out different functions:

1. *Filtration,* in which blood is passed through a filter that retains blood cells and proteins but passes water and small molecules
2. *Reabsorption,* in which desirable ions and metabolites are recaptured from the filtrate, leaving metabolic wastes and water behind for later elimination
3. *Excretion,* in which the kidney excretes K^+, H^+, NH_4^+, and certain drugs and foreign organic materials

FIGURE 41-4

The basic organization of the vertebrate nephron is seen in the nephron tube of the freshwater fish, a basic design that has been retained in the kidneys of marine fishes and terrestrial vertebrates, which evolved later. Sugars, small proteins, and ions with two or more ionic bonds (bivalent ions), such as Ca^{++} and $PO_4^{\equiv}$, are recovered in the proximal arm; ions with a single ionic bond (monovalent ions), such as Na^+ and Cl^-, are recovered in the distal arm; and water is recovered in the collecting duct.

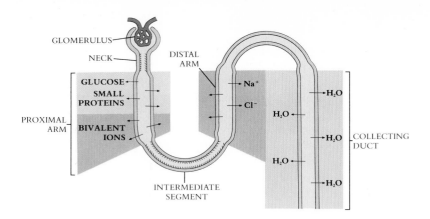

Filtration

The filtration device of the vertebrate kidney consists of a large number of individual tubular filtration–reabsorption devices called **nephrons** (Figure 41-4). At the front end of each nephron tube is a filtration apparatus called a **Malpighian corpuscle.** In each Malpighian corpuscle, an arteriole enters and splits into a fine network of capillaries called a **glomerulus** (Figure 41-5). It is the walls of these capillaries that act as a filtration device. Blood pressure forces fluid through the capillary walls, which are differentially permeable. These walls withhold the proteins and other large molecules, while passing water and small molecules such as glucose, ions, and ammonia, the primary nitrogenous waste product of metabolism.

Reabsorption

At the back end of each nephron tube is a reabsorption device that operates like the mammalian small intestine discussed in Chapter 34. The fluid that passes out of the capillaries of each glomerulus, called the **glomerular filtrate,** enters the tube portion of that nephron, within which reabsorption takes place.

FIGURE 41-5

The spherical structures in this micrograph are Malpighian corpuscles; in each, a fine network of capillaries, the glomerulus, is connected to a nephron tube by an arteriole.

Excretion

A major function of the kidney is the elimination of a variety of potentially harmful substances that animals eat, drink, or inhale. In this process, the roughly 2 million nephrons that form the bulk of the two human kidneys receive a flow of approximately 2000 liters of blood per day. The nephrons function independently, each concentrating urine and, by working together, creating large osmotic gradients within the kidney. By this means, the mammalian kidney is able to concentrate urine with a salt concentration well above that of the blood. As a part of the process, the kidney is able to build up the concentration of such materials as H^+, K^+, NH_4^+, drugs, and various foreign organic materials in the urine, and thus to excrete them from the body.

EVOLUTION OF THE VERTEBRATE KIDNEY

The same basic design has been retained in all vertebrate kidneys, although there have been some changes. In the following sections, we shall discuss the structure and function of the kidney in the different major groups of vertebrates.

Freshwater Fishes

Kidneys are thought to have evolved first among the freshwater fishes. A freshwater fish drinks little and produces large amounts of urine. Because the body fluid of a freshwater fish is hyperosmotic as compared with the water in which the fish lives, water is not reabsorbed in its nephrons. The excess water that enters its body passes instead through the nephron tubes to the bladder, from which it is eliminated as urine. Within the urine is not only the excess water but also all the small molecules that were not reabsorbed while passing through the nephron tubes. Notable among these molecules is ammonia, the principal metabolic waste product of nitrogen metabolism (present in solution as ammonium ion, NH_4^+). Ammonia in higher concentrations is toxic, but because the urine contains so much water, the concentration of ammonia is low enough not to harm the fish.

Marine Fishes

Although most groups of animals clearly seem to have evolved first in the sea, marine bony fishes probably evolved from freshwater ancestors, as was mentioned in Chapter 19. In making the transition to the sea, they faced a significant new problem of water balance, because their body fluids are hypoosmotic with respect to the water that surrounds them. For this reason, water tends to leave their bodies, in which the fluids are less concentrated osmotically than is seawater. To compensate, marine fishes drink a lot of water, excrete salts instead of reabsorbing them, and reabsorb water. This places radically different demands on their kidneys than those faced by freshwater fishes, thus turning the tables on an organ that had originally evolved to eliminate water and reabsorb salts. As a result, the kidneys of marine fishes have evolved important differences from their freshwater relatives. For example, they possess active ion transport channels that conduct ions with more than one ionic charge—molecules that are particularly abundant in seawater—*out* of the body and *into* the tube. In the sea, the water that the fish drinks is rich in ions such as Ca^{++}, Mg^{++}, $SO_4^=$, and $PO_4^=$, all of which must be excreted. Because of the functioning of these ion transport channels, the direction of movement is reversed compared with that found in the kidneys of their freshwater ancestors.

Sharks

Except for one species of shark found in Lake Nicaragua, all sharks, rays, and their relatives live in the sea (Figure 41-6). Some of these members of the class Chondrichthyes have solved the osmotic problem posed by their environment in a different way than have the bony fishes. Instead of actively pumping ions out of their bodies through their kidneys, these fishes use their kidneys to reabsorb the metabolic waste product urea, creating and maintaining urea concentrations in their blood 100 times as high as those that occur among the mammals. As a result, the sharks and their relatives become isotonic with the surrounding sea. They have evolved enzymes and tissues that tolerate these high concentrations of urea. Because they are isotonic with the water in which they swim, they avoid the problem of water loss that other marine fishes face. Since sharks do not need to drink large amounts of seawater, their kidneys do not have to remove large amounts of divalent ions from their bodies.

FIGURE 41-6

This tiger shark is swimming in the ocean in French Polynesia. Although the seawater contains a far higher concentration of ions than the shark's body does, the shark avoids losing water osmotically by maintaining such a high concentration of urea that its body fluids are osmotically similar to the sea around it or even slightly hyperosmotic.

FIGURE 41-7

The human male urinary system.

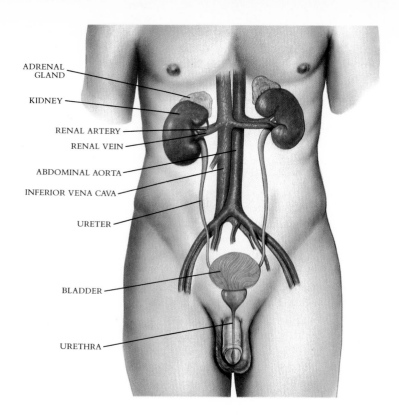

ADRENAL GLAND

KIDNEY

RENAL ARTERY

RENAL VEIN

ABDOMINAL AORTA

INFERIOR VENA CAVA

URETER

BLADDER

URETHRA

Amphibians and Reptiles

The first terrestrial vertebrates were the amphibians; the amphibian kidney is identical to that of the freshwater fishes, their ancestors. This is not surprising, since amphibians spend a significant portion of their time in fresh water, and when on land they generally stay in wet places.

Reptiles, on the other hand, live in diverse habitats, many of them very dry. The reptiles that live mainly in fresh water, like some kinds of crocodiles and alligators, occupy a habitat similar to that of the freshwater fishes and amphibians and have similar kidneys. Marine reptiles, which consist of some crocodiles, some turtles, and a few lizards and snakes, possess kidneys similar to those of their freshwater relatives. They eliminate excess salts not by kidney excretion but rather by means of salt glands located near the nose or eye.

Terrestrial reptiles, which must conserve water to survive, reabsorb much of the water in the kidney filtrate before it leaves the kidneys, and so excrete a concentrated urine. This urine cannot, however, become any more concentrated than the blood plasma; otherwise, the body water of reptiles would simply flow into the urine while it was in the kidneys. In the relatively concentrated urine of most reptiles, therefore, nitrogenous waste is no longer excreted in the form of ammonia, but either as the solid urea or as uric acid. These metabolic conversions take place in the liver.

Mammals and Birds

Your body possesses two kidneys, each about the size of a small fist, located in the lower back region (Figure 41-7). These kidneys represent a great increase in efficiency compared with those of reptiles. Mammalian, and to a lesser extent avian, kidneys can remove far more water from the glomerular filtrate than can the kidneys of reptiles and amphibians. Human urine is 4.2 times as concentrated as blood plasma. Some desert mammals achieve even greater efficiency: a camel's urine is 8 times as concentrated as its plasma, a gerbil's 14 times as concentrated, and some desert rats and mice have urine more than 20 times as concentrated as their blood plasma (see Figure 41-2). Mammals and birds achieve this remarkable degree of water conservation by using a simple but superbly designed mechanism: they greatly increase the local salt concentration in the tissue through which the nephron tube passes and then use this osmotic gradient to draw the water out of the tube.

HOW THE MAMMALIAN KIDNEY WORKS

Remarkably, mammals and birds have brought about this major improvement in efficiency by a very simple change—they bend the nephron tube. A single mammalian kidney contains about a million nephrons. Each of these is composed of a glomerulus that is connected to a nephron tube, which is called a **renal tubule** in mammals. Between its proximal and distal segments, each renal tubule is folded into a hairpin loop called the **loop of Henle** (Figure 41-8).

The kidney uses the hairpin loop of Henle to set up a countercurrent flow. Just as in the gills of a fish, as discussed in Chapter 35, but with water being absorbed instead of oxygen, countercurrent flow enables water reabsorption to occur with high efficiency. In general, the longer the hairpin loop, the more water can be reabsorbed. Animals such as desert rodents that have highly concentrated urine have exceptionally long loops of Henle. The process involves the passage of *two* solutes across the membrane of the loop: salt (NaCl) and urea, the waste product of nitrogen metabolism. It has long been known that animals fed high-protein diets, yielding large amounts of urea as waste products, can concentrate their urine better than animals excreting lower amounts of urea, a clue that urea plays a pivotal role in kidney function. Figure 41-9 illustrates the flow of materials discussed in the following paragraphs:

1. Filtrate from the glomerulus passes down the descending loop. The walls of this portion of the tubule are impermeable to either salt or urea but are freely permeable to water. Because (for reasons we will come to) the surrounding tissue has a high osmotic concentration of urea, water passes out of the descending loop by osmosis, leaving behind a more concentrated filtrate.

2. At the turn of the loop, the walls of the tubule become permeable to salt, but much less permeable to water. As the concentrated filtrate passes up the ascending arm, salt passes out into the surrounding tissue by diffusion. (The surrounding tissue, although it has lots of urea, does not contain as much salt as the concentrated filtrate.) This makes salt more concentrated at the bottom of the loop.

3. Higher in the ascending arm, the walls of the tubule contain active-transport channels that pump out even more salt. This active removal of salt from the ascending loop encourages even more water to diffuse outward from the filtrate. Left behind in the filtrate is the urea that initially passed through the glomerulus as nitrogenous waste; eventually the urea concentration becomes very high in the tubule.

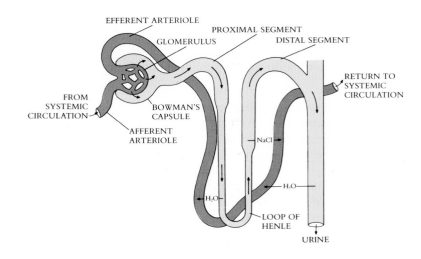

FIGURE 41-8

Organization of a mammalian renal tubule. The glomerulus is enclosed within a filtration device called Bowman's capsule. Blood pressure forces liquid through the glomerulus and into the proximal segment of the tubule, where glucose and small proteins are reabsorbed from the filtrate. The filtrate then passes through a double loop arrangement consisting of the loop of Henle and the collecting duct, which act to remove water from the filtrate. The water is then collected by blood vessels and transported out of the kidney to the systemic (body) circulation.

FIGURE 41-9

The flow of materials in the
human kidney.

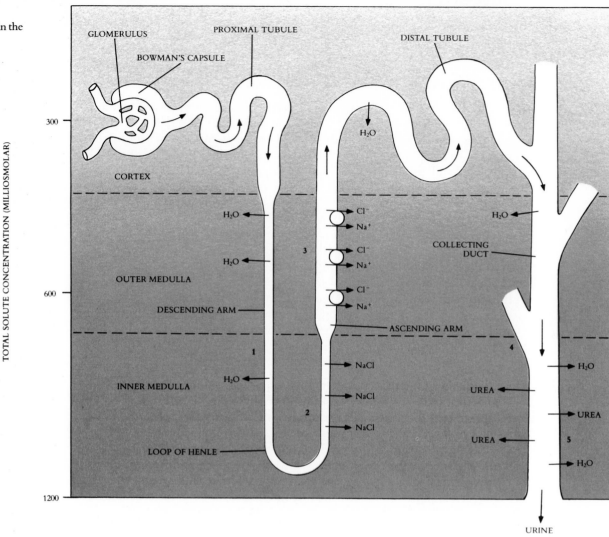

4. Finally, the tubule empties into collecting ducts that pass back through the tissue; unlike the tubule, the lower portion of the collecting ducts is permeable to urea. During this final passage, the concentrated urea in the filtrate diffuses out into the surrounding tissue, which has a lower urea concentration. A high urea concentration in the tissue results, which is what caused water to move out of the filtrate by osmosis when it first passed down the descending arm.

5. As the filtrate passes down the collecting duct, even more water passes outward by osmosis, as the osmotic concentration of the surrounding tissue reflects both the urea diffusing out from the duct *and* the salt diffusing out from the descending arm, and this sum is greater than the osmotic concentration of urea in the filtrate (the salt has already been removed).

In effect, the kidney is divided into two zones (Figure 41-10): (1) the outer portion of the kidney, called the **outer medulla,** contains the upper portion of the loop, including the upper ascending arm where reabsorption of salt from the filtrate by active transport occurs; and (2) the inner portion of the kidney, or **inner medulla,** contains the lower portion of the loop and also contains the bottom of the collecting duct, which is permeable to urea.

The active reabsorption of salt in the outer medulla of the kidney drives the process. This reabsorption of salt from the filtrate in one arm of the loop establishes a gradient of salt concentration, with the salt concentration higher in the inner medulla at the

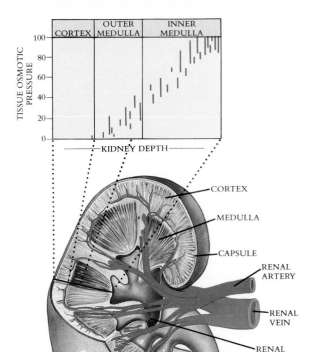

FIGURE 41-10

Structure of the human kidney. The cortex has an osmotic concentration like that of the rest of the body. The outer medulla has a somewhat higher osmotic concentration, primarily salt. The osmotic concentration within the inner medulla becomes progressively higher at greater depths; this part of the kidney maintains substantial concentrations of urea, which is a major component of its high osmotic concentration.

bottom of the loop. It is this high salt concentration that raises the total tissue osmotic concentration so high that water passes by osmosis out of the collecting duct. Just as importantly, the active reabsorption of salt in the outer medulla concentrates the renal filtrate with respect to urea. This high concentration of urea in the filtrate causes urea to diffuse outward into surrounding tissue in the only zone where it is able to do so, the lower collecting duct, creating a high urea concentration in the inner medulla. It is this high urea concentration in the inner medulla that causes water to diffuse out from the filtrate in the initial descending arm. The water is then collected by blood vessels in the kidney, which carry it into the systemic circulation.

The mammalian kidney achieves a high degree of water reabsorption by using the salts and urea in the glomerular filtrate to increase the osmotic concentration of the kidney tissue. This facilitates the movement of the water from the filtrate out into the surrounding tissue, where it is collected by blood vessels impermeable to the high urea concentration but permeable to water.

INTEGRATION AND HOMEOSTASIS

As you saw in Chapter 40 and have learned in other respects here, the kidney is an excellent example of an organ whose principal function is homeostasis, the maintenance of constant physiological conditions within the body. It is concerned both with water balance and with ion balance. Many further instances could be provided of organs and systems that function in this way, indicating that the functions of the body are highly linked and highly controlled. The interaction of all the component systems contributes to the maintenance of a functioning, healthy animal.

SUMMARY

1. Animals that live in fresh water tend to gain water from their surroundings, whereas those that live in salt water or on land tend to lose water.

2. Insects solve the problem of dehydration by conserving water. In particular, they avoid excreting water along with body wastes. They pump ions into Malpighian tubules, so that body fluids are drawn in by osmosis. These body fluids contain the wastes created by metabolism. The wall of the tubule acts as a filter, passing wastes but retaining proteins and blood cells. Insects then reabsorb the water and useful metabolites back out, leaving the wastes behind to be excreted.

3. Instead of sucking body liquid through a filter as insects do, vertebrates push it through. They are able to do this because they have a closed circulation system that operates under considerable pressure.

4. The vertebrate kidney is composed of many individual units called nephrons, each made up of two segments, the first a filter and the second a resorption tube.

5. Freshwater fishes do not reabsorb water; their bodies already gain too much by direct diffusion from the water. Marine fishes excrete the salts in the water they drink and reabsorb water. Some marine vertebrates, notably sharks, maintain high body levels of urea so that they are isotonic with the sea and do not tend to gain or lose water.

6. Amphibians and reptiles have kidneys much like those of freshwater fishes. Birds and mammals achieve much greater water conservation by bending the nephron tube, producing what is known as the loop of Henle. This creates a countercurrent flow, which greatly increases the efficiency of water reabsorption.

7. The longer the loop of Henle, the greater the osmotic concentration that can be achieved and the more water that can be reclaimed from the urine.

8. The mammalian kidney achieves a high degree of water reabsorption by using the salts and urea in the glomerular filtrate to increase the osmotic concentration of the kidney tissue. This facilitates the movement of the water from the filtrate out into the surrounding tissue, where it is collected by blood vessels impermeable to the high urea concentration but permeable to the water.

REVIEW

1. Approximately _____ of every vertebrate's body is water.

2. Among vertebrates, the organ that carries out filtration of waste and reabsorption of nonwaste molecules is the _____.

3. To help maintain their water balance, _____ fishes drink lots of water, whereas _____ fishes drink very little water.

4. Which part of the human kidney—the outer cortex, the outer medulla, or the inner medulla—has the highest concentration of urea?

5. The principal function of the kidney is to maintain homeostatic ion and _____ balance within the body.

SELF-QUIZ

1. Which of the following is an example of a vertebrate osmoconformer?
   (a) kangaroo rats
   (b) whales
   (c) freshwater fishes
   (d) sharks
   (e) amphibians

2. Which of the following is recovered in the collecting arm of the vertebrate nephron?
   (a) small proteins
   (b) glucose
   (c) bivalent ions
   (d) monovalent ions
   (e) water

3. The urine of terrestrial _____ cannot be more concentrated (hyperosmotic) than their blood.
   (a) fishes
   (b) amphibians
   (c) reptiles
   (d) birds
   (e) mammals

4. Mammalian kidneys can concentrate urine to the extent that they do because they have long loops of Henle, where two solutes pass across the membrane of the loop. These solutes are
   (a) water
   (b) salt (NaCl)
   (c) glucose
   (d) protein
   (e) urea

5. An insect conserves water by
   (a) secreting potassium ions into its Malpighian tubules
   (b) pumping sodium ions out of its nephrons
   (c) drawing urea out of its Malpighian corpuscles
   (d) filtering sodium ions through the walls of its glomerulus
   (e) reabsorbing sodium ions into its nephrons

THOUGHT QUESTIONS

1. In the mammalian kidney, water is reabsorbed from the filtrate across the walls of the collecting duct into the salty tissue near the bottom of the loops of Henle, and the water is then taken away by blood vessels. Why doesn't the blood in these vessels become very salty?

2. The mammalian kidney is, in its architecture, a high-efficiency version of a freshwater fish kidney designed to retain water. Many mammals, including whales, seals, and walruses, spend a significant portion of their lives in salt water. How well would you expect their kidneys to function there? Do you know any mammals that live for extended times in fresh water?

FOR FURTHER READING

BEEUWKES, R.: "Renal Countercurrent Mechanisms, or How to Get Something for (Almost) Nothing," *In* C.R. Taylor et al., editors: *A Companion to Animal Physiology,* Cambridge University Press, New York, 1982. A clearly presented summary of current ideas about how the human kidney works.

HEATWOLE, H.: "Adaptations of Marine Snakes," *American Scientist,* vol. 66, pages 594-604, 1978. Several groups of snakes are able to live in the sea by clever adaptations that modify salt and water balance.

SMITH, H.W.: *From Fish to Philosopher,* ed. 2, Little, Brown & Co., Boston, 1961. A broad and well-written account of the evolution of the vertebrate kidney.

FIGURE 42-1   A human fetus at 18 weeks of age is not yet halfway through the 38 weeks—about 9 months—it will spend within its mother, but already has developed many distinct behaviors, such as the sucking reflex that is so important to survival after birth.

SEX AND REPRODUCTION
Chapter 42

Overview

Almost all vertebrates reproduce sexually. Sex evolved in the sea, and its modification for organisms living on land entailed evolutionary innovations to avoid drying out. Most terrestrial vertebrates, including a few genera of primitive mammals, reproduce by means of eggs. Human beings and other placental mammals have adopted a different solution, nourishing their developing young within the mother's body. Sex in human beings can play an important role in pair bonding as well as in reproduction. Vertebrate development may be divided artificially into several stages, although in reality these stages are parts of a continuous, dynamic process. The developmental process takes longer in mammals than in other vertebrates; it lasts some 266 days from fertilization to birth in human beings and continues for some time thereafter.

For Review

Here are some important terms and concepts that you will encounter in this chapter. If you are not familiar with them, you should review them before proceeding.

Meiosis (Chapter 10)

Down syndrome (Chapter 12)

Amniotic egg (Chapter 19)

Vertebrate evolution (Chapter 19)

Mammals (Chapter 19)

Hormones (Chapter 40)

On any dark night you can hear biology happening. The cry of a cat in heat, insects chirping outside the windows, frogs croaking in swamps, wolves howling in a frozen northern scene—all of these are the sounds of evolution's essential act, reproduction. Nor are we humans immune. Few subjects pervade our everyday thinking more than sex; few urges are more insistent. They are no accident, these strong feelings—they are a natural part of being human. The frog in its swamp knows only an urgent desire to reproduce itself, a desire that has been patterned within it by a long history of evolution. It is a pattern that we share. The occasional sad perversion of these deeply ingrained feelings only serves to point out how very fundamental they are. For almost all of us, the reproduction of our families spontaneously elicits a sense of rightness and fulfillment. It is difficult not to return the smile of a new infant, not to feel warmed by it and by the look of wonder and delight to be seen on the faces of the parents. This chapter deals with sex and reproduction among the vertebrates, of which we human beings are one kind (Figure 42-1).

SEX EVOLVED IN THE SEA

Sexual reproduction first evolved among marine organisms. The eggs of most marine fishes are produced in batches by the females; when they are ripe, the eggs are simply released into the water. Fertilization is achieved by the release of sperm by the male into the water containing the eggs. Sea water itself is not a hostile environment for gametes or for young organisms.

The effective union of free gametes in the sea, an example of **external fertilization,** poses a significant problem for all marine organisms. Eggs and sperm become diluted rapidly in seawater, so their release by females and males must be almost simultaneous if successful fertilization is to occur. Because of this necessity, most fish vertebrates restrict the release of their eggs and sperm to a few brief and well-defined periods. There are few seasonal cues in the ocean that organisms can use as signals, but one that is all-pervasive is the cycle of the moon. Approximately every 28 days, the moon revolves around the earth; variations in its gravitational attraction cause the differences in ocean level that we call tides. Many different kinds of marine organisms sense the changes in water pressure that accompany the tides, and much of the reproduction that takes place in the sea is timed by the lunar cycle.

The invasion of the land by organisms from the sea meant facing for the first time in the history of life on earth the danger of drying out. This problem was all the more severe for the marine organisms' small and vulnerable reproductive cells. Obviously the gametes could not simply be released near one another on land; they would soon dry up and perish. There are many possible solutions to this problem. We have already considered the seeds of plants, the spores of fungi, and the eggs of arthropods, all successful adaptations to an existence on dry land. Vertebrates have solved the problem in yet other ways.

VERTEBRATE SEX AND REPRODUCTION: FOUR STRATEGIES

The five major groups of vertebrates have evolved reproductive strategies that are quite different, and that range from external fertilization to bearing of live young. In large measure these differences reflect different approaches to protecting gametes from drying out, since four of the five major groups of vertebrates are terrestrial.

FIGURE 42-2

When frogs mate, as these two are doing, the clasp of the male induces the female to release a large mass of mature eggs, over which the male discharges his sperm.

Fishes

Some vertebrates, of which the bony fishes are the most abundant, have remained aquatic. All fishes reproduce in the water, in essentially the same way as all other aquatic animals. Fertilization in most species is external, the eggs containing only enough yolk material to sustain the developing zygote for a short time. After the initial dowry of yolk has been exhausted, the growing individual must seek its food from the waters around it. Many thousands of eggs are fertilized in an individual mating, but few of the resulting zygotes survive their aquatic environment and grow to maturity. Some succumb to bacterial infection, many others to predation. The development of the fertilized eggs is speedy, and the young that survive achieve maturity rapidly.

Amphibians

A second group of vertebrates, the amphibians (frogs, toads, and salamanders), have invaded the land without fully adapting to the terrestrial environment. The life cycle of the amphibians is still inextricably tied to the presence of free water. Among most amphibians, fertilization is still external, just as it is among the fishes and other aquatic animals. Many female amphibians lay their eggs in a puddle or a pond of water. Among the frogs and toads, the male grasps the female and discharges fluid containing sperm onto the eggs as she releases them (Figure 42-2).

The development time of the amphibians is much longer than that of the fishes, but amphibian eggs do not include a significantly greater amount of yolk. Instead, the process of development consists of two distinct components, a larval and an adult stage, like some of the life cycles found among the insects.

FIGURE 42-3

The introduction of semen containing sperm by the male into the female's body occurs during copulation. Reptiles such as these turtles were the first terrestrial vertebrates to develop this form of reproduction, which is particularly suited to a terrestrial existence.

The development of the aquatic larval stage of the amphibians is rapid, as it utilizes yolk supplied from the egg. The larvae then function, often for a considerable period of time, as independent food-gathering machines. They scavenge nutrients from their environments and often grow rapidly. Tadpoles, which are the larvae of frogs, grow in a matter of days from creatures no bigger than the tip of a pencil into individuals as big as goldfish. When an individual larva has grown to a sufficient size, it undergoes a developmental transition, or **metamorphosis,** into the terrestrial adult form.

The invasion of land by vertebrates was tentative at first, with the amphibians keeping the external fertilization and reproduction by means of eggs that are characteristic of most of the fishes, the group from which they evolved.

Reptiles and Birds

The reptiles were the first group of vertebrates to abandon the marine habitat completely. Reptile eggs are fertilized internally within the mother before they are laid, the male introducing his **semen,** a fluid containing sperm, directly into her body (Figure 42-3). By this means, fertilization still occurs in an environment that is protected from drying out, even though the adult animals are fully terrestrial. Most vertebrates that fertilize internally utilize a tube, the **penis,** to inject semen into the female. Composed largely of tissue that can become rigid and erect, the penis penetrates into the female reproductive tract.

Many reptiles are **oviparous,** the eggs being deposited outside the body of the mother; others are **viviparous,** forming eggs that hatch within the body of the mother. The young of viviparous animals, therefore, are born alive.

Most birds lack a penis (swans are an exception) and achieve internal fertilization simply by the male slapping semen against the reproductive opening of the female before the eggs have formed their hard shells. This kind of mating occurs more quickly than that of most reptiles. All birds are oviparous, the young emerging from their eggs outside of the body of the mother. Birds encase their eggs in a harder shell than do their reptilian ancestors. They also hasten the development of the embryo within each egg by warming the eggs with their bodies. The young that hatch from the eggs of most bird species are not able to survive unaided, since their development is still incomplete. The young birds, therefore, are fed and nurtured by their parents, and they grow to maturity gradually.

The shelled eggs of reptiles and birds constitute one of their most important adaptations to life on land, since these eggs can be laid in dry places. Each egg is provided with a large amount of yolk and is encased within a membranous cover. The zygote develops within the egg, eventually achieving the form of a miniature adult before it completely uses up its supply of yolk and leaves the egg to face its fate as an adult.

The first major evolutionary change in reproductive biology among land vertebrates was that of the reptiles and their descendants, the birds. Although both groups are still oviparous like fishes, they practice internal fertilization, with the zygote encased within a watertight egg.

Mammals

The most primitive mammals, the monotremes, are oviparous like the reptiles from which they evolved. The living monotremes consist solely of the duckbilled platypus and the echidna (see Figure 19-17). No other mammals lay eggs. All other members of this class are viviparous.

The young of viviparous mammals are nourished and protected by their mother, an outstanding characteristic of the members of this class. The mammals other than monotremes have approached the problem of nourishing their young in two ways:

1. Marsupials give birth to live embryos at a very early stage of development. The tiny animals crawl to and enter pouches on the mother's body, where they continue their development. They eventually emerge when they are able to function on their own.
2. The placental mammals retain their young for a much longer period within the body of the mother. To nourish them, they have evolved a specialized, massive network of blood vessels called a placenta, through which nutrients are channeled to the embryo from the blood of the mother.

The second major evolutionary change in reproductive biology among land vertebrates was that of the marsupials and placental mammals. These groups nourish their young within a pouch or inside the body until the young have reached a fairly advanced stage of development.

SEX HORMONES

A lot of signaling goes on during sex. Unlike humans, most other viviparous females signal the male when a mature egg has been released at the beginning of an **estrous** (sexually receptive) **cycle;** it is also necessary that many different processes within the female and within the male be coordinated during the process of gametogenesis and reproduction. The delayed sexual development that is common in mammals, for example, entails a nonsexual juvenile period, after which changes occur that produce sexual maturity. These changes occur in many parts of the body, a process that requires the simultaneous coordination of further development in many different kinds of tissues. The production of gametes is another carefully orchestrated process, involving a series of carefully timed developmental events. Successful fertilization begins yet another developmental "program," in which the female body prepares itself for the many changes of pregnancy.

All of this signaling is carried out by a portion of the brain, the hypothalamus. The signals by which the hypothalamus regulates reproduction are hormones produced by the brain, which are carried by the bloodstream to the various organs of the body. Some reproductive hormones are steroids, complex carbon-ring lipids (Figure 42-4); others are peptides. On the cell surfaces of the target organs of the reproductive hormones are located specific receptors to which a particular hormone binds. In the case of steroid hormones, the receptors are located within the cytoplasm of the receptor cells. Once bound to a receptor, a steroid hormone molecule is transported to the nucleus, where it binds to specific locations on the chromosomes and alters the pattern of gene transcription, and by so doing initiates the physiological changes associated with that hormone.

The body uses hormones as signals to control various body functions in the reproductive cycle. Most reproductive hormones are steroids. Steroid hormones bind to specific receptor sites on target cells, enter the cells, and alter the pattern of gene expression in the target cells.

FIGURE 42-4

The steroid sex hormones estrogen and progesterone. Estrogen prepares and maintains the uterine lining for pregnancy; progesterone stimulates thickening of the uterine lining during pregnancy.

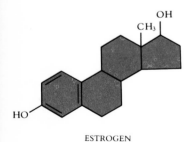

ESTROGEN

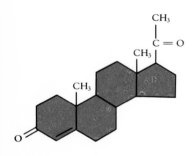

PROGESTERONE

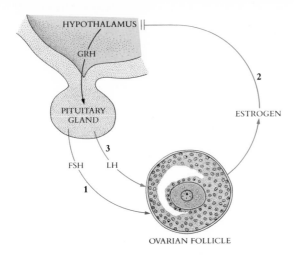

FIGURE 42-5

Mammalian egg maturation is under hormonal control.
1 Gonadotrophic releasing hormone (GRH), produced by the hypothalamus, causes the pituitary to produce the follicle-stimulating hormone (FSH) and release it into the blood circulation; when FSH reaches the ovaries, it initiates final egg development; **2** FSH also causes the ovaries to produce the hormone estrogen; rising levels of estrogen in the blood cause the pituitary to shut down its production of FSH and instead produce luteinizing hormone (LH); **3** when this LH circulates back to the ovaries, it inhibits estrogen production and initiates ovulation.

THE REPRODUCTIVE CYCLE OF MAMMALS

The reproductive cycle of female mammals, including that of human beings, is composed of two distinct phases, the follicular phase and the luteal phase. The first or **follicular phase** of the reproductive cycle is marked by the hormonally controlled development of eggs within the ovary. The anterior pituitary, after receiving a chemical signal from the hypothalamus (Figure 42-5), starts the cycle by secreting **follicle-stimulating hormone (FSH),** which binds to receptors on the surface of the follicles, initiating the final development and maturation of the egg. Normally, only a few eggs at any one time have developed far enough to respond immediately to the FSH. FSH levels are reduced before other eggs reach maturity, so that in every cycle only a few eggs ripen.

The reduction of the levels of FSH is achieved by a feedback command to the pituitary. In addition to starting final egg development, FSH also triggers the production of the female sex hormone **estrogen** by the ovary. Rising estrogen levels in the bloodstream feed back to the pituitary and cut off the further production of FSH. In this way, only the few eggs that are already developed far enough for their maturation to be initiated by the FSH are taken into the final stage of development. The rise in estrogen level and the maturation of one or more eggs completes the follicular phase of the estrous cycle.

The second or **luteal phase** of the cycle follows smoothly from the first. The pituitary responds to estrogen by secreting a second hormone, called **luteinizing hormone (LH),** which is carried in the bloodstream to the developing follicle. LH inhibits estrogen production and causes the wall of the mature follicle to burst. The egg within the follicle is released into one of the **fallopian tubes,** which extend from the ovary to the uterus. This process is called **ovulation.** Meanwhile, the ruptured follicle repairs itself, filling in and becoming yellowish. In this condition it is called the **corpus luteum** (see Figure 42-15), which is simply the Latin phrase for "yellow body." The corpus luteum soon begins to secrete a hormone, **progesterone,** which initiates the many physiological changes associated with pregnancy. The body is preparing itself for fertilization. The corpus luteum continues its production of progesterone for several weeks after fertilization. If fertilization does not occur soon after ovulation, then production of progesterone slows and eventually ceases, marking the end of the luteal phase.

> **The reproductive cycle of mammals is composed of two phases, which alternate with one another. During the follicular phase, some of the eggs within the ovary complete their development. During the following luteal phase, the mature eggs are released into the fallopian tubes, a process called ovulation. If fertilization does not occur, ovulation is followed by a new follicular phase, the start of another cycle.**

FIGURE 42-6

The human female menstrual cycle. A single cycle occurs every 28 days on the average. The growth and thickening of the uterine (endometrial) lining is governed by levels of the hormone progesterone; menstruation, the sloughing off of the blood-rich tissue, is initiated by lower levels of progesterone. Egg maturation and ovulation (egg release) are governed by the hormones FSH and LH, whose production by the pituitary is governed by the hypothalamus of the brain. When a fertilized egg implants within the endometrium, progesterone levels do not fall and pregnancy ensues.

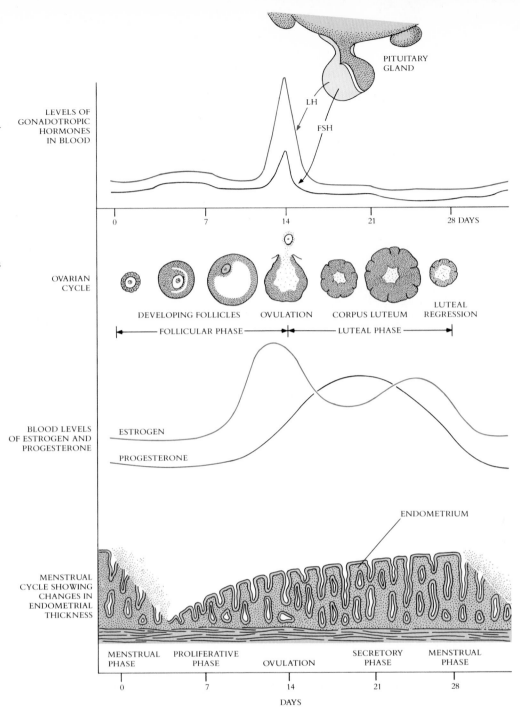

In the absence of estrogen and progesterone, the pituitary can again initiate production of FSH, thus starting another reproductive cycle. In human beings the next cycle follows immediately after the end of the preceding one. A cycle usually occurs every 28 days, or a little more frequently than once a month, although this varies in individual cases. The Latin word for month is *mens,* which is why the reproductive cycle in humans is called the **menstrual cycle,** or monthly cycle (Figure 42-6). In human beings and some other primates, the hormone progesterone has among its many effects a thickening of the walls of the uterus in preparation for the implantation of the developing embryo. When fertilization does not occur, the decreasing levels of progesterone cause this thickened layer of blood-rich tissue to be sloughed off, a process that results in the bleeding associated with **menstruation.** Menstruation, or "having a period," usually occurs about midway between successive ovulations, or roughly once a month, although its timing varies widely even for individual females.

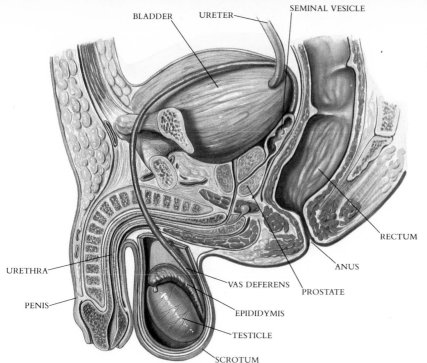

BLADDER URETER SEMINAL VESICLE

RECTUM

ANUS

URETHRA

VAS DEFERENS

PROSTATE

PENIS

EPIDIDYMIS

TESTICLE

SCROTUM

FIGURE 42-7

The human male reproductive system.

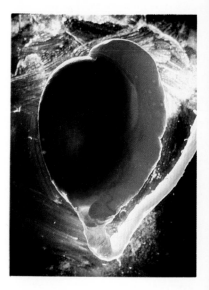

FIGURE 42-8

Human testes. The testicle is the darker sphere in the center of the photograph; within it sperm are formed. Cupped above the testicle is the epididymis, a highly coiled passageway within which sperm complete their maturation. Extending away from the epididymis is a long tube, the vas deferens, in which mature sperm are stored.

THE HUMAN REPRODUCTIVE SYSTEM

Human reproductive cells, like those of all other vertebrates, consist of **egg cells,** which are richly endowed with nutrients, and **sperm cells,** where nutrients are nearly absent. As we shall describe in more detail, the egg is fertilized within the female, and the zygote develops into a mature fetus there. Most reproductive activity, then, is centered on the female. The male contributes little physically but his sperm.

Males

The human male gamete, or **sperm,** is highly specialized for its role as a carrier of genetic information. Produced after meiosis, the sperm cells have 23 chromosomes instead of the 46 found in most cells of the human body. Unlike these other cells, sperm do not complete their development successfully at 37° C (98.6° F), the normal human body temperature. The sperm-producing organs, or **testes,** move during the course of fetal development out of the body proper and into a sac called the **scrotum** (Figure 42-7). The scrotum, which hangs between the legs of the male, maintains the testes at a temperature about 3° C cooler than that of the rest of the body.

The testes (Figure 42-8) are composed of several hundred compartments, each of which is packed with large numbers of tightly coiled tubes called **seminiferous tubules.** The tubes themselves are the sites of sperm-cell production, or **spermatogenesis** (Figure 42-9). The full process of sperm development takes about 2 months. The number

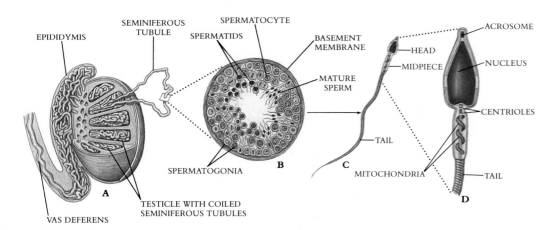

EPIDIDYMIS

SEMINIFEROUS TUBULE

SPERMATOCYTE

SPERMATIDS

BASEMENT MEMBRANE

MATURE SPERM

ACROSOME

HEAD

MIDPIECE

NUCLEUS

CENTRIOLES

TAIL

MITOCHONDRIA

TAIL

SPERMATOGONIA

B

C

D

A

TESTICLE WITH COILED SEMINIFEROUS TUBULES

VAS DEFERENS

FIGURE 42-9

The interior of the testes, site of spermatogenesis **(A).** Within the seminiferous tubules of the testes, cells called spermatogonia **(B)** develop by means of meiosis into sperm, passing through the spermatocyte and spermatid stages. Each sperm **(C)** possesses a long tail coupled to a head **(D),** which contains a haploid nucleus.

FIGURE 42-10

A penis in longitudinal section and in cross section.

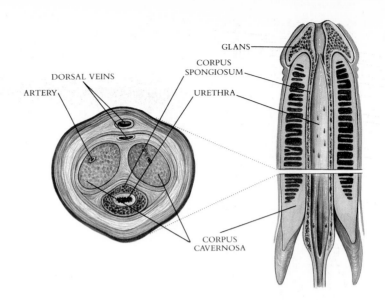

FIGURE 42-11

Human sperm. Only the heads and a portion of the long slender tails of these sperm are shown in this scanning electron micrograph.

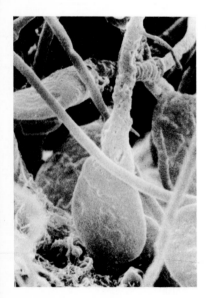

of sperm that are produced is truly incredible. A typical adult male produces several hundred million sperm each day of his life. Those that are not ejaculated from the body are resorbed, in a continual cycle of renewal.

After the sperm cells complete their differentiation within the testes, they are delivered to a long coiled tube called the **epididymis,** where they mature further. The sperm cells are not motile when they arrive in the epididymis, and they must remain there for at least 18 hours before their motility develops. From the epididymis, the sperm are delivered to another long tube, the **vas deferens,** where they are stored. When they are delivered during intercourse, the sperm travel through a tube from the vas deferens to the urethra, where the reproductive and urinary tracts join, emptying through the penis.

The penis is an external tube composed of three cylinders of spongy tissue (Figure 42-10). In cross section, the arteries and veins can be seen to run along the surface, beneath which two of the cylinders sit side by side. Below the pair of cylinders is a third cylinder, which contains in its center the **urethra,** through which both semen (during ejaculation) and urine (during liquid elimination) pass. The spongy tissue that makes up the three cylinders is riddled with small spaces between its cells. When nerve impulses from the central nervous system cause the dilation of the arterioles leading into this tissue, blood collects within the spaces. This causes the tissue to become distended and the penis to become erect and rigid. Continued stimulation by the central nervous system is required for this erection to be maintained.

Erection can be achieved without any physical stimulation of the penis. Fantasy is a common initiating factor. However, the physical stimulation of the penis usually is required for any delivery of semen to take place. Stimulation of the penis, as by repeated thrusts into the vagina of a female, leads first to the mobilization of the sperm. In this process, muscles encircling the vas deferens contract, moving the sperm into the urethra. Eventually the stimulation leads to the violent contraction of the muscles at the base of the penis. The result is ejaculation, the ejection of about 5 milliliters of semen out of the penis. Within this small volume are several hundred million sperm. Since only a single sperm cell achieves fertilization, the odds against any one individual sperm cell successfully completing the long journey to the egg and fertilizing it are extraordinarily high. Successful fertilization requires a high sperm count; males with less than 20 million sperm per milliliter are generally considered sterile (Figure 42-11).

An adult male produces sperm continuously, several hundred million each day of his life. The sperm are stored, and then are delivered during sexual intercourse.

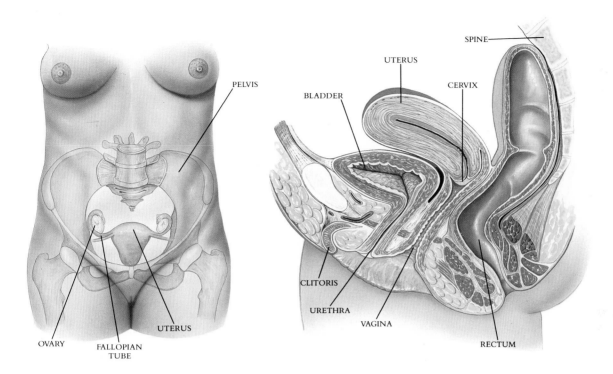

SPINE

UTERUS

CERVIX

BLADDER

PELVIS

CLITORIS

URETHRA

VAGINA

OVARY

FALLOPIAN TUBE

UTERUS

RECTUM

FIGURE 42-13

A mature egg within an ovarian follicle of a cat.

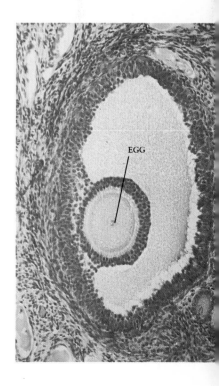

EGG

Females

Fertilization requires more than insemination. There must be a mature egg to fertilize. Eggs are produced within the ovaries of females (Figure 42-12). The ovaries are compact masses of cells, 2 to 3 centimeters long, located within the abdominal cavity. Eggs develop from cells called **oocytes,** which are located in the outer layer of the ovary. Unlike males, whose gamete-producing spermatogonia are constantly dividing, females have at birth all of the oocytes that they will ever produce. At each cycle of ovulation, one or a few of these oocytes initiate development; the others remain in a developmental holding pattern. This long maintenance period is one reason that developmental abnormalities crop up with increasing frequency in pregnancies of women who are over 35 years old. The oocytes are continually exposed to mutation throughout life, and after 35 years the odds of a harmful mutation having occurred become appreciable.

At birth a female's ovaries contain some 2 million oocytes, all of which have begun the first meiotic division. At this stage they are called **primary oocytes.** Meiosis is arrested, however, in prophase of the first meiotic division. The development of very few oocytes ever proceeds further. With the onset of puberty, the female matures sexually. At this time the release of follicle-stimulating hormone (FSH) initiates the resumption of the first meiotic division in a few oocytes, but a single oocyte soon becomes dominant, the others regressing (Figure 42-13). Approximately every 28 days after that, another oocyte matures, although the exact timing may vary from one month to another. It is rare for more than about 400 out of the approximately 2 million oocytes with which a female is born to mature during her lifetime. When they do mature, the egg cells are called **ova** (singular, **ovum**), the Latin word for "egg."

FIGURE 42-14

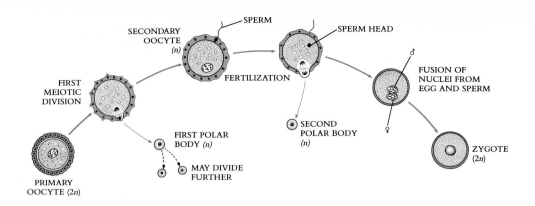

The meiotic events of oogenesis. A primary oocyte is diploid (2n, with 46 chromosomes in human beings). In its maturation the first meiotic division is completed, and one division product is eliminated as a polar body. The other product, the secondary oocyte, is released during ovulation. The second meiotic division does not occur until after fertilization and results in production of a second polar body and a single haploid (n, with 23 chromosomes in humans) egg nucleus. Fusion of the haploid egg nucleus with a haploid sperm nucleus produces a diploid zygote, from which an embryo subsequently forms.

FIGURE 42-15

The journey of an egg. Produced within a follicle and released at ovulation, an egg is swept up into a fallopian tube and carried down it by waves of contraction of the tubal walls. Fertilization occurs within the fallopian tube; the sperm journey upward to meet the egg. Several mitotic divisions occur while the fertilized egg continues its journey down the fallopian tube, so that by the time it enters the uterus, it is a hollow sphere of cells, a blastula-stage embryo. The blastula implants itself within the wall of the uterus, where it continues its development.

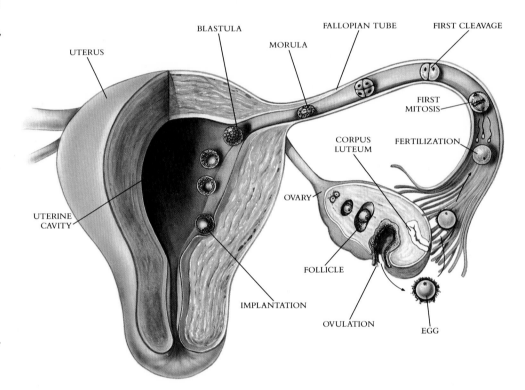

Unlike male gametogenesis, the process of meiosis in the oocytes does not result in the production of four haploid gametes. Instead, a single haploid ovum is produced, the other meiotic products being discarded as **polar bodies.** The process of meiosis is stop-and-go rather than continuous (Figure 42-14).

When the ovum is released at ovulation, it is swept by the beating of cilia into the **fallopian tubes,** which leads away from the ovary (Figure 42-15). Smooth muscles lining the fallopian tube contract rhythmically; these rhythmic, peristaltic contractions move the egg down the tube to the uterus. The journey is a slow one, taking about 3 days to complete. If the egg is unfertilized, it loses its capacity to develop. For only about 24 hours can it be successfully fertilized. For this reason the sperm cannot simply lie in wait within the uterus. Any sperm cell that is to fertilize an egg successfully must make its way up the fallopian tubes, a long passage that few survive.

Sperm are deposited within the **vagina,** a muscular tube about 7 centimeters long, which leads to the mouth of the uterus. This opening is bounded by a muscular sphincter called the **cervix.** The **uterus** is a hollow, pear-shaped organ about the size of a small fist. Its inner wall, the **endometrium,** has two layers. The outer of these layers is shed during menstruation while the one beneath it generates the next layer. Sperm entering the uterus swim and are carried upward by waves of motion in the walls of the uterus, enter the fallopian tube, and swim upward against the current generated by peristaltic contractions, which are carrying the ovum downward toward the uterus.

All of the eggs that a woman will produce during her life develop from cells that are already present at her birth. Their development is halted early in meiosis, and one or a few of these cells resume meiosis each 28 days to produce a mature egg. Upon maturation, eggs travel to the uterus. Fertilization by a sperm cell, if it occurs, happens en route.

When a successfully fertilized egg reaches the uterus, the new embryo usually attaches itself to the endometrial lining and thus starts the long developmental journey that eventually leads to the birth of a child.

THE PHYSIOLOGY OF HUMAN SEXUALITY

Few physical activities are more pleasurable to humans than sexual intercourse. It is one of the strongest drives directing human behavior, and, as such, it is circumscribed by many rules and customs. Sexual intercourse acts as a channel for the strongest of human emotions—love, tenderness and personal commitment on one hand, rage and aggression on the other. Few subjects are at the same time more private and of more general interest.

Until relatively recently, the physiology of human sexual activity was largely unknown. Perhaps because of the prevalence of strong social taboos against the open discussion of sexual matters, research on the subject was not being carried out, and detailed information was lacking. Each of us learned from anecdote, from what our parents or friends told us, and eventually from experience. Largely through the pioneering efforts of William Masters and Virginia Johnson in the last 25 years, and an army of workers who have followed them, this gap in the generally available information about the biological nature of our sexual lives has now largely been filled.

Sexual intercourse is referred to by a variety of names, including copulation and coitus, as well as a host of more informal ones. It is common to partition the physiological events that accompany intercourse into four periods, although the division is somewhat arbitrary, with no clean divisions between the four phases. The four periods are **excitement, plateau, orgasm,** and **resolution** (Figure 42-16).

Excitement

The sexual response is initiated by commands from the brain that increase the heartbeat, blood pressure, and rate of breathing. These changes are very similar to the ones that the brain induces in response to alarm. Other changes increase the diameter of blood vessels, leading to increased peripheral circulation. The nipples commonly harden and become more sensitive. In the genital area of males this increased circulation leads to the vasocongestion in the penis that produces erection. Similar swelling occurs in the **clitoris** of the female, a small knob of tissue composed of a shaft and glans much like the male penis but without the urethra running through it. The female ex-

FIGURE 42-16

The human orgasmic response. Among females, the response is highly variable. It may be typified by one of the three patterns illustrated here. **1,** One intense peak of response, with a rapid resolution phase; **2,** several peaks of intense response, with a somewhat longer resolution phase; **3,** a large number of response peaks, of less intensity, and a long resolution phase. Among males, the response does not vary as much as it does among females, with a single intense response peak being followed by a resolution phase.

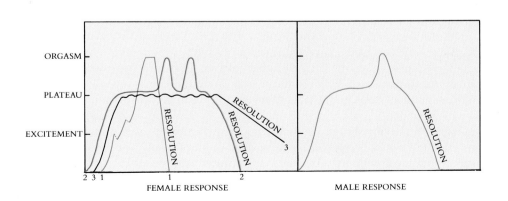

periences additional changes that prepare the vagina for sexual intercourse. The increased circulation leads to swelling and parting of the lips of tissue, or **labia,** that cover the opening to the vagina; the vaginal walls become moist; and the muscles encasing the vagina relax.

Plateau

The penetration of the vagina by the thrusting penis results in the repeated stimulation of nerve endings both in the tip of the penis and in the clitoris. The clitoris, which is now swollen, becomes very sensitive and withdraws up into a sheath or "hood." Once it has withdrawn, the stimulation of the clitoris is indirect, with the thrusting movements of the penis rubbing the clitoral hood against the clitoris. The nervous stimulation that is produced by the repeated movements of the penis within the vagina elicits a continuous sympathetic nervous system response, greatly intensifying the physiological changes that were initiated in the excitement phase. In the female, pelvic thrusts may begin, while in the male the penis maintains its rigidity.

Orgasm

The **climax** of intercourse is reached when the stimulation is sufficient to initiate a series of reflexive muscular contractions. The nerve signals producing these contractions are associated with other nervous activity within the central nervous system, activity that we experience as intense pleasure. In females, the contractions are initiated by impulses in the hypothalamus, which causes the pituitary to release large amounts of the hormone **oxytocin.** This hormone in turn causes the muscles in the uterus and around the vaginal opening to contract. Orgasmic contractions occur about one second apart. There may be one intense peak of contractions (an "orgasm") or several, or the peaks may be more numerous but less intense (see Figure 42-16).

Analogous contractions occur in the male, initiated by nerve signals from the brain. These signals first cause **emission,** in which the rhythmic peristaltic contraction of the vas deferens and of the **prostate gland** causes the sperm and seminal fluid to move to a collecting zone of the urethra. This collecting zone, which is located at the base of the penis, is called the **bulbourethra.** Shortly after the sperm move into the bulbourethra, nerve signals from the brain induce violent contractions of the muscles at the base of the penis, resulting in the ejaculation of the collected semen out through the penis. As in the female orgasm, the contractions are spaced about a second apart, although in the male they continue for a few seconds only. Unlike those of the female, the orgasmic contractions in the male do not vary in their pattern; they are restricted to the single intense wave of contractions that is associated with ejaculation.

Resolution

After ejaculation, males lose their erection and enter a **refractory period** often lasting 20 minutes or longer, in which sexual arousal is difficult to achieve and ejaculation is almost impossible. After intercourse, the bodies of both men and women return slowly, over a period of several minutes, to their normal physiological state.

CONTRACEPTION AND BIRTH CONTROL

In most vertebrates, sexual intercourse is associated solely with reproduction. In most vertebrates reflexive behavior that is deeply ingrained in the female limits sexual receptivity to those periods of the sexual cycle when she is fertile. In human beings, however, sexual behavior serves a second important function, the reinforcement of pair bonding, the emotional relationship between two individuals. The evolution of strong pair bonding is not unique to human beings, but it was probably a necessary precondition for the evolution of our increased mental capacity. The associative activities that make up human "thinking" are largely based upon learning, and learning takes time.

Human children are very vulnerable during the extended period of learning that follows their birth, and they require parental nurturing. It is perhaps for this reason that human pair bonding is a continuous process, not restricted to short periods coinciding with ovulation. Among all of the vertebrates, human females and a few species of apes are the only ones in which the characteristic of sexual receptivity throughout the reproductive cycle has evolved and has come to play a role in pair bonding.

Not all human couples want to initiate a pregnancy every time they have sexual intercourse, yet sexual intercourse may be a necessary and important part of their emotional lives together. Among some religious groups, this problem does not arise, or is not recognized, since members of these groups believe that sexual intercourse has only a reproductive function and thus should be limited to situations in which pregnancy is acceptable—among married couples wishing to have children. Most couples, however, do not limit sexual relations to procreation, and among them, unwanted pregnancy presents a real problem. The solution to this dilemma is to find a way to avoid reproduction without avoiding sexual intercourse, an approach that is commonly called **birth control** or contraception.

At least seven different approaches are commonly taken to achieve birth control. Among the commonly used methods of birth control, three are relatively *in*effective: (1) the **rhythm method,** in which the couple attempts to avoid the part of the menstrual cycle at which successful fertilization is likely to occur—this method often fails to prevent pregnancy; (2) **coitus interruptus,** in which the male withdraws his penis before ejaculation; and (3) **douches,** in which chemicals are used by the female in an effort to kill the sperm before fertilization occurs. In contrast, **condoms,** which are thin rubber bags used to cover the penis during intercourse, and **diaphragms,** which are rubber domes placed so as to cover the cervix, are effective when used correctly, but mistakes are relatively frequent. **Birth control pills,** which contain estrogen and progesterone, either together or taken sequentially, shut down the production of the pituitary hormones FSH and LH and thus prevent the ripening of the ovarian follicles. Such pills, and **intrauterine devices (IUDs),** which are coils or irregularly shaped bodies placed within the uterus to prevent embryo implantation, are very effective. IUDs, however, have a potential for serious side effects, including vaginal bleeding and tubal pregnancy; they have, for all practical purposes, been abandoned in the United States. The common operations known as vasectomies and tubal ligations (Figure 42-17) are completely effective, although reversible only with difficulty. ▪

ABORTION

Reproduction can be prevented after fertilization if the embryo is **aborted** (removed) before its development and birth. During the first trimester this can be accomplished by **vacuum suction** or by **dilation and curettage,** in which the cervix is dilated and the uterine wall is scraped with a spoon-shaped surgical knife called a curette. Chemical methods are also being developed that cause abortion early in the first trimester, apparently with complete safety; they may soon be available commercially. In the second trimester the embryo can be removed by injecting a 20% saline solution into the uterus, which induces labor and delivery of the fetus. In general, the more advanced the pregnancy, the more difficult and dangerous the abortion is to the mother.

As a method of birth control, abortion takes a great emotional toll, both on the woman undergoing the abortion and often on others who know and care for her. Abortion also presents serious moral problems for some. Many people feel that the fetus is a living person from the time of conception, and that abortion is simply murder. There are many countries in which abortions are defined as a crime. In the United States a fetus is not legally considered a person until birth, and abortions are permitted by law during the first two trimesters. They are illegal in the third trimester, however, except when the mother's life is endangered. The Supreme Court ruling that legalized abortion in the first two trimesters is a relatively recent one, and it is still the subject of intense controversy. The abortion rate in the United States (and most other developed countries) exceeds the birth rate.

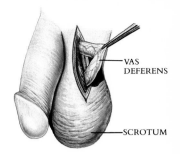

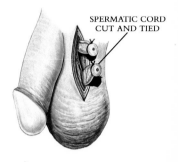

A

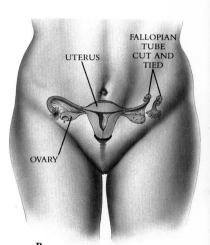

B

FIGURE 42-17

Surgical means of birth control.
A Vasectomy.
B Tubal ligation.

As most college students now know, AIDS is a serious disease that is rapidly becoming common in our country and around the world. The disease, first reported in 1981, is transmitted by a virus and is almost always fatal. Almost 38,000 individuals in the United States had contracted AIDS by the end of 1986, and nearly 20 Americans will die each day of AIDS this year. The U.S. Public Health Service estimates there will have been 270,000 cases of AIDS in the United States by 1991. Few if any of these patients will recover.

The name **AIDS** is shorthand for **acquired** (transmitted from another infected individual) **immune deficiency** (a breakdown of the body's ability to defend itself against disease) **syndrome** (a spectrum of symptoms). The disease is fatal because no one can survive for long without an immune system to defend against viral and bacterial infections and to ward off cancer. The virus also infects the central nervous system, often leading to serious mental disorders.

AIDS is not the only fatal disease to threaten humans, nor is it the most contagious. What makes AIDS an unusually serious threat is that the virus which causes the disease does not have its effect immediately upon infection. Recently infected people usually show no symptoms of the disease at all. Only much later—typically 5 years—does the virus begin to multiply and attack the immune system. During these years, however, the infected person is an unknowing carrier, able to transmit the virus to others. It is the large reservoir of undiagnosed, infected individuals that casts such a shadow over our future. Current estimates of the number of individuals in their 20s and 30s who are infected with the virus in the United States approach 1%, suggesting both a staggering load of future suffering that will be difficult to avoid and a great danger that the infection will spread further. It would be folly to assume that college campuses will escape infection. Every student should face that fact squarely.

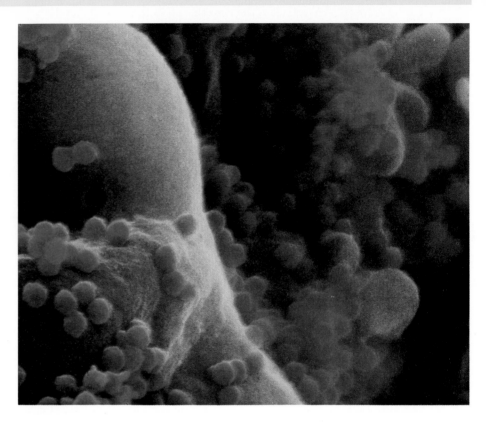

FIGURE 42-A

AIDS viruses infecting a cell of the immune system.

What causes AIDS? The virus that causes AIDS is called **Human Immunodeficiency Virus (HIV).** It is a fragile virus that does not survive outside body cells. It is present in the body fluids of infected individuals (notably in blood and in semen and vaginal fluid). The HIV virus is transmitted from one individual to another when body fluid is transferred from an infected individual. It is *not* transmitted in the air or by casual contact. You cannot catch AIDS from a bathroom seat, from a hot tub shared with an AIDS victim, from kissing an AIDS carrier, or by being bitten by a mosquito that bit an AIDS victim. In studies of 619 households of AIDS victims, not one family member contracted the AIDS virus. The *only* way you can become infected with HIV is to come into contact with the body fluids of an infected person. Among college students, there are two important routes of infection.

1. *Sexual intercourse.* Both semen and vaginal fluid of infected individuals have high levels of HIV virus. This means that vaginal, anal, and oral sex with an infected individual can all transmit the virus successfully— and to either sex. Absorption across the vaginal wall or into the penis offers a ready means for the virus to enter the body; the tiny tears produced during anal sex may facilitate entry even better. The microscopic abrasions that everybody has in their mouth from eating and chewing are a third easy means of entry, making oral sex also dangerous.

The only safe way to have sex with an infected individual is to use a condom, and use it correctly. When used correctly, condoms offer the best available protection from infection. It is important, however, that they be used properly. For some 10% of couples using condoms for birth control the woman becomes pregnant, almost always as a result of careless use. This suggests that it is not wise to mix alcohol or drugs with sexual encounters; they may cloud your judgment and lead you to do things you wouldn't do with a clearer head—such as forgetting to use a condom or using it carelessly. It is a mistake that could cost you your life.

If you are involved in a sexual relationship and don't want to use condoms, you can lessen the uncertainty (and any danger to yourself or to future partners) by sharing an AIDS antibody test with your sexual partner, well in advance of sexual intercourse. These tests, described below, detect the presence of the virus.

2. *Drug use.* A needle used more than once by an infected individual typically harbors large quantities of HIV virus, both in the fluid that remains behind in the needle and in the body of the hypodermic syringe. Anyone else who re-uses the needle will become infected. The use of intravenous drugs is itself dangerous—both illegal and life-threatening—but if you engage in such folly, do not compound the damage by employing a used needle or syringe.

Who is at risk? You are. AIDS is commonly perceived as a disease of homosexual men, because the disease first appeared in this country among the gay community. Because homosexuals tend to confine their sexual interactions to one another, the disease initially spread among homosexuals without entering the larger heterosexual community. That initial segregation appears to be ending: while only 4% of the AIDS cases diagnosed in 1986 were heterosexual non–drug users, the incidence of the HIV virus among heterosexuals is now thought to be expanding rapidly. Estimates vary widely. The Public Health Service estimates that between 1.5 million and 2 million people in the United States now harbor the virus. In Africa, where the virus first infected humans several years before it spread to this country, the epidemic has proceeded farther than in the United States, even though homosexuality is rare there. In Africa, sexual transmission of AIDS is almost exclusively heterosexual and occurs in *both* directions, female to male as well as male to female. In some central African countries, as many as 10% of the adult individuals are thought to carry the HIV virus.

The AIDS antibody test. Can a person find out if he or she has been infected with the HIV virus? Yes, easily. A simple test identifies infected individuals by detecting antibodies in their blood directed against HIV virus. Such antibodies are detectable only in the bodies of infected individuals. They are the remnants of the body's attempt to ward off the HIV infection.

How an AIDS test works. The standard test for detecting the presence of antibody directed against HIV virus is called the ELISA antibody test. In the test, about 5 ml of blood (roughly the amount that would fit in a paper drinking straw) is drawn and checked to see whether any antibodies are present that will interact with bits of HIV attached to the surface of a plastic dish (Figure 42-B). If such antibodies are present, the test is said to be positive.

If the result of the ELISA test is negative, no further tests are required, since there are almost no false-negatives unless the test is given within 6 weeks after infection, before the body has begun to produce antibodies. False-positives are also rare, but if the ELISA test result is positive, it is repeated. Two positive ELISA's are the signal for a more elaborate test, called a Western blot, which although it is far more expensive is very accurate.

How to get an AIDS test. The Red Cross offers an inexpensive walk-in AIDS antibody test at a variety of locations in most metropolitan areas. The Centers for Disease Control also maintains a national AIDS hotline that you can contact for information and advice by calling (800) 342-7154.

Confidentiality. At the Red Cross, the testing is completely confidential. The test is coded with a number that you select—the test is not associated with your name—and you telephone a few days later for the result of the test identified with that number. Maintaining the confidentiality of a positive result is very difficult, however, even though the test itself is completely confidential. When an individual does have a positive AIDS test, it is necessary to see a doctor frequently in order to monitor possibilities of disease progression. All of this has to be documented in the physician's office, along with billing information. Because these files may be accessed when you sign a medical release form or pay bills with insurance, it is impossible to maintain confidentiality over a long period.

The only way to survive AIDS is not to contract it. The only way (short of contaminated needles) that any student will contract AIDS is by having unprotected sex with someone who has the virus. But the 5-year lag prevents anyone from knowing who is infected and who is not, and the disease continues to spread. Although widespread antibody testing on college campuses has been proposed by some, in order to identify infected individuals and so dampen the spread of the virus while it is still confined to a relatively few individuals, the proposal is controversial, and the associated dangers of invasion of privacy are a real concern. In the absence of such widescale on-campus testing, any student you have sex with might be a carrier and not know it. To avoid becoming part of the epidemic, you have to accept the responsibility of protecting your health.

Continued.

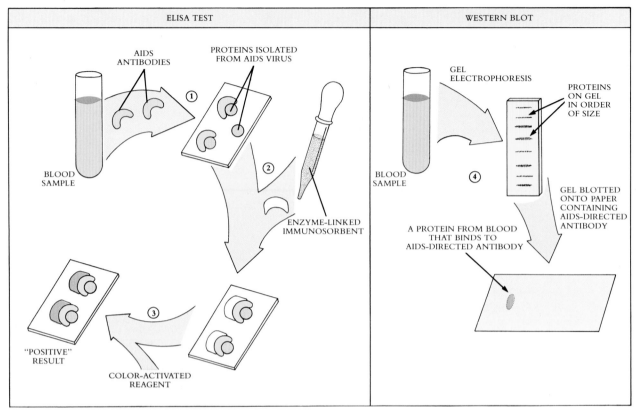

FIGURE 42-B

How an "AIDS test" works. AIDS tests are of two sorts: an inexpensive and rapid test that misses very little, but occasionally yields "false positives," and a more definitive test that is more laborious.

Step 1   The standard test for detecting the presence of antibody directed against the AIDS virus (and thus identifying infected individuals) is called ELISA. The letters stand for **E**nzyme **L**inked **I**mmuno-**S**orbent **A**ssay. In the test, about 5 ml of blood is drawn, roughly the amount that would fit in a paper drinking straw. This blood sample is exposed to a plastic surface to which AIDS virus proteins are chemically bound. Any AIDS-specific antibodies present in the blood bind to the viral proteins on the plastic surface.

Step 2   After the blood sample is washed from the plastic plate, the plate is exposed to a solution containing a special reagent, the "enzyme linked immunosorbent." Basically, it is an antibody directed against other antibody molecules, to which a color reagent is attached. This reagent will bind to antibodies on the plastic plate only if antibody molecules are already bound there—that is, only if the blood possesses antibodies directed against AIDS virus proteins.

Step 3   To detect a positive result, the plate is then treated with an enzyme that reacts with the color reagent to produce a color change. Running the full test takes about 3 hours. If the result of the ELISA test is negative, no further tests are required, since there are almost no false negatives unless the test is given within a few weeks after infection, before the body has begun to produce antibodies. False positive ELISAs are the signal for a more precise test, the Western blot.

Step 4   In the Western blot, the proteins in the blood sample are separated according to size by being induced to migrate through a gel in response to an electric field (bigger proteins migrate more slowly). The gel is then blotted onto paper containing antibodies directed against the AIDS virus. Because the size of each AIDS virus protein is known, the binding of any protein in the blood sample to the AIDS-directed antibody on the paper—in just the place on the gel to which an AIDS protein would have migrated—is considered to be a conclusively positive test.

PATHS OF EMBRYONIC DEVELOPMENT

Although some details differ from group to group, the process of development that occurs after fertilization is fundamentally the same in all vertebrates. Development occurs in six stages:

1. Fertilization   The male and female gametes form a zygote.
2. Cleavage   The zygote rapidly divides into many cells, with no overall increase in size. These divisions set the stage for development, since different cells receive different portions of the egg cytoplasm and hence different regulatory signals.
3. Gastrulation   The cells of the zygote move, forming three germ layers: ectoderm, mesoderm, and endoderm.
4. Neurulation   In all chordates the first organ to form is the notochord, followed by formation of the dorsal nerve cord.
5. Neural crest formation   The first uniquely vertebrate event is the formation of a tissue called the **neural crest.** Many of the uniquely vertebrate structures develop from it.
6. Organogenesis   The three primary cell types then proceed to combine in various ways to produce the organs of the body.

THE COURSE OF HUMAN DEVELOPMENT

Human development takes an average of 266 days from fertilization to birth, the familiar "9 months of pregnancy." What may not be so readily apparent, however, is how very early the critical stages of development occur during the course of human pregnancy. We now outline the overall process.

First Trimester
The First Month

In the first week after fertilization occurs, the fertilized egg undergoes cleavage divisions. The first of these divisions occurs about 30 hours after the fusion of the egg and the sperm, and the second 30 hours later. Cell divisions continue until an embryo forms within a ball of cells, the **trophoblast.** During this period the embryo continues the journey down the mother's oviduct, a journey that the egg initiated. On about the sixth day the trophoblast embryo reaches the uterus, releasing the protein hormone **chorionic gonadotropin.** This hormone prevents menstruation by stimulating the corpus luteum to produce estrogens and progesterone abundantly; its presence in the blood or urine is frequently used to test for pregnancy. Attaching to the uterine lining, or endometrium, the trophoblast penetrates into its tissue; this process is known as **implantation.**

Surrounded by ruptured blood vessels, the trophoblast thickens and develops fingerlike projections that penetrate the uterine lining, where they develop into membranes. One of these, the **amnion,** will enclose the developing embryo; another, the **chorion,** will interact with the uterine tissue to form the placenta that will nourish the growing embryo. The chorion becomes the chief source of chorionic gonadotropin, which prevents menstrual cycles during pregnancy.

In the second week after fertilization, **gastrulation** takes place, and the three germ layers are differentiated. Around the developing embryo the placenta starts to form.

In the third week, **neurulation** occurs. It is marked by the formation of a neural tube along the dorsal axis of the embryo, as well as by the appearance of the first somites, segmented blocks of tissue from which the muscles, vertebrae, and connective tissue develop (see Figure 28-29). By the end of the third week more than a dozen somites are evident, and the blood vessels and gut have begun to develop. At this point the embryo is about 2 millimeters long.

In the fourth week, **organogenesis**—the formation of body organs—occurs (Figure 42-18, *A*). The eyes form, and the tubular heart begins to pulsate, develops four chambers, and begins a rhythmic beating. At an average of 70 beats per minute, the little heart—beating much more rapidly at this stage—is destined to beat more than 2.5

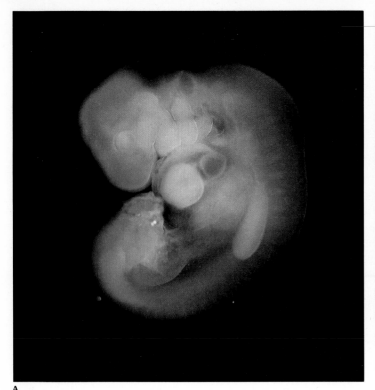

A

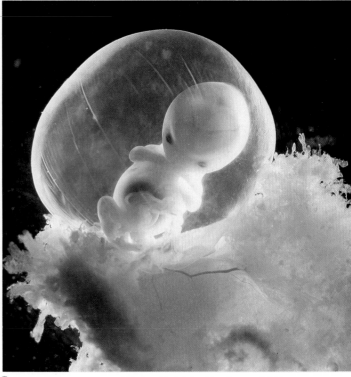

B

FIGURE 42-18

The developing human.
A 4 weeks.
B 7 weeks.
C 3 months.
D 4 months.

billion times during a human lifetime of 70 years, before it ceases. More than 30 pairs of somites are visible by the end of the fourth week, and the arm and leg buds have begun to form. The embryo more than doubles in length during this week, to about 5 millimeters.

All of the major organs of the body have begun their formation by the end of the fourth week of development. Although the developmental scenario is now far advanced, many women are not aware that they are pregnant at this stage.

Early pregnancy is a very critical time in development, since the proper course of events can be interrupted easily. In the 1960s, for example, many pregnant women took the tranquilizer **thalidomide** to minimize discomforts associated with early pregnancy. Unfortunately, this drug had not been adequately tested—it interferes with fetus limb-bud development, and its widespread use resulted in many deformed babies. Also during the first and second months of pregnancy, contraction of German measles *(Rubella)* by the mother can upset organogenesis in the developing embryo. Most **spontaneous abortions** occur in this period.

The Second Month

Morphogenesis (the formation of shape by cell migration) takes place during the second month (Figure 42-18, *B*). The miniature limbs of the embryo assume their adult shapes. The arms, legs, knees, elbows, fingers and toes can all be seen—as well as a short bony tail (see Figure 28-29)! The bones of the embryonic tail, an evolutionary reminder of our past, later fuse to form the coccyx. Within the body cavity, the major organs, including the liver, pancreas, and gall bladder, become evident. By the end of the second month the embryo has grown to about 25 millimeters in length, weighs perhaps a gram, and is beginning to look distinctly human.

The Third Month

The nervous system and sense organs begin to develop during the third month (Figure 42-18, *C*). By the end of the month, the arms and legs begin to move. The embryo begins to show facial expressions and carries out primitive reflexes such as the startle reflex and sucking. By the end of the third month, all of the major organs of the body have been established. The development of the embryo is essentially complete. From this point on, the developing human being is referred to as a **fetus** rather than an embryo. Essentially, all that remains is growth.

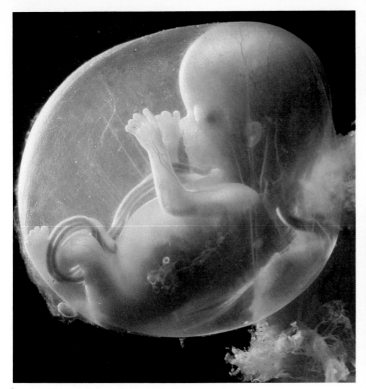

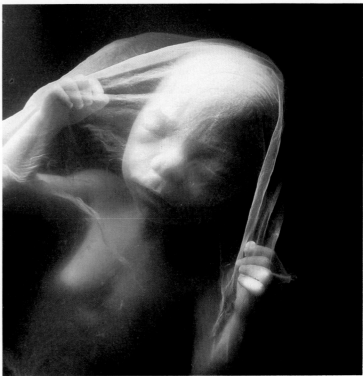

C   D

Second Trimester

In the fourth and fifth months of pregnancy the fetus grows to about 175 millimeters in length, with a body weight of about 225 grams. Bone enlargement occurs actively during the fourth month. During the fifth month the head and body become covered with fine hair. This downy body hair, called **lanugo,** is another evolutionary relict and is lost later in development. By the end of the fourth month, the mother can feel the baby kicking. By the end of the fifth month, she can hear its rapid heartbeat with a stethoscope. In the sixth month growth begins in earnest. By the end of that month, the baby weighs 0.6 kilograms—about a pound and a half—and is more than a foot long. Most of its prebirth growth is still to come, however. The fetus cannot survive outside the uterus without special medical intervention.

Third Trimester

The third trimester is predominantly a period of growth rather than one of development. In the seventh, eighth, and ninth months of pregnancy the weight of the fetus doubles several times. This increase in bulk is not the only kind of growth that occurs, however. Most of the major nerve tracts are formed within the brain during this period, as are new brain cells. All of this growth is fueled by nutrients provided by the mother's bloodstream. Within the placenta these nutrients pass into the fetal blood supply (Figure 42-19). The undernourishment of the fetus by a malnourished mother can adversely affect this growth and can result in severe retardation of the infant.

By the end of the third trimester, the neurological growth of the fetus is far from complete—in fact, it continues long after birth. By this time, however, the fetus is able to exist on its own. Why doesn't development continue within the uterus until neurological development is complete? Because physical growth would continue as well, and the fetus is probably as large as it can get and still be delivered through the pelvis (Figure 42-20) without damage to mother or child. Birth takes place as soon as the probability of survival is high. For better or worse, the fetus is on its own, a person.

The critical stages of human development take place quite early in pregnancy. All the major organs of the body have been established by the end of the third month. The following six months are essentially a period of growth.

FIGURE 42-19

Placental–fetal circulation.

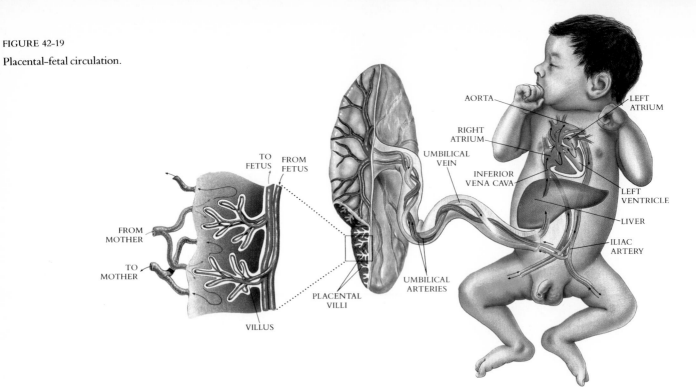

AORTA

LEFT ATRIUM

RIGHT ATRIUM

UMBILICAL VEIN

INFERIOR VENA CAVA

LEFT VENTRICLE

LIVER

ILIAC ARTERY

TO FETUS

FROM FETUS

FROM MOTHER

TO MOTHER

PLACENTAL VILLI

UMBILICAL ARTERIES

VILLUS

FIGURE 42-20

A developing fetus is a major addition to a woman's anatomy. The woman's stomach and intestines are pushed far up, and there is often considerable discomfort from pressure on the lower back. In a natural delivery the fetus exits through the vagina, which must dilate (expand) considerably to permit passage.

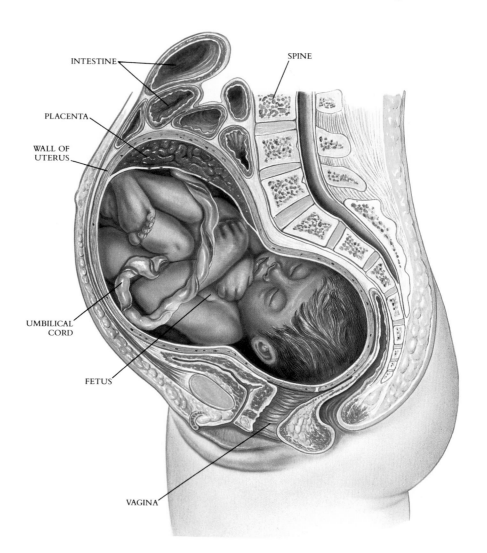

INTESTINE

SPINE

PLACENTA

WALL OF UTERUS

UMBILICAL CORD

FETUS

VAGINA

SUMMARY

1. Sexual reproduction evolved in the sea. Among most fishes and the amphibians, fertilization is external, with gametes being released into water. Amphibians and the great majority of fishes are oviparous, the young being nurtured by the egg rather than by the mother.

2. Successful invasion of the land by vertebrates involved major changes in reproductive strategy: the watertight egg and internal fertilization of the reptiles, which permit survival in dry places. The second important advance in reproductive adaptation to live on land was that of the marsupials and placental mammals. In marsupials the young are born within a few weeks of fertilization, but then they are nourished and protected during the course of their further development within a pouch; in placental mammals the young are nourished within the mother's body by means of a large, complex structure, the placenta, which exchanges material from her bloodstream with her offspring via a common circulation of blood.

3. Reproduction in mammals is regulated by hormones, which typically are produced by the pituitary on commands from the hypothalamus region of the brain.

4. The female reproductive cycle of mammals is composed of two phases: (1) a follicular phase, in which some eggs in the ovary are hormonally signaled to complete their development; and (2) a luteal phase, in which one or more mature eggs are released into the fallopian tube, a process called ovulation. A complete reproductive cycle in a human female takes about 28 days.

5. The male gametes, or sperm, of mammals are produced within the testes. In human males hundreds of millions of sperm are produced each day. Sperm mature and become motile in the epididymis, and they are stored in the vas deferens. Stimulation of the penis causes it to become distended and erect, and causes the sperm to be delivered from the vas deferens to the urethra at the base of the penis. Further stimulation causes violent muscle contractions, which ejaculate the semen from the penis.

6. At birth, female mammals contain all of the gametes, or oocytes, that they will ever have. A human female has approximately 2 million oocytes. All but a very few of these are arrested in meiotic prophase. At each ovulation the first meiotic division of one or a few eggs is completed. The second meiotic division does not occur until after fertilization.

7. Fertilization occurs within the fallopian tubes. The journey of an egg to the uterus takes 3 days, and the egg is viable for only 24 hours unless it is fertilized. Only those eggs that have been reached within 24 hours by sperm can be fertilized successfully. Fertilized eggs continue their journey down the fallopian tubes and attach to the lining of the uterus, where their development proceeds.

8. Human intercourse is marked by four physiological periods: excitement, plateau, orgasm, and resolution. The sexual response in women is highly variable. Orgasm in men is more uniform; it coincides with the ejaculation of semen.

9. A variety of birth-control procedures is practiced by humans, of which condoms used by men and birth-control pills used by women are the most common. Surgical procedures that block the delivery of gametes are becoming increasingly common. Birth control is not always effective, and some women terminate unwanted pregnancies. In the United States, as in other developed countries, the abortion rate exceeds the birth rate.

10. Most of the critical events in the development of a human occur in the first month. Cleavage occurs during the first week, gastrulation during the second week, neurulation during the third week, and organ formation during the fourth week.

11. The second and third months of the first trimester are devoted to morphogenesis and to the elaboration of the nervous system and sensory organs. By the end of this period, the development of the embryo is essentially complete.

12. The last 6 months of human pregnancy are essentially a period of growth, devoted to increase in size and to formation of nerve tracts within the brain. Most of the weight of a fetus is added in the final 3 months of pregnancy.

REVIEW

1. In both fishes and amphibians, fertilization is usually internal. True or false?

2. During which phase of the human menstrual cycle is the progesterone level in the blood highest?

3. What goes on in seminiferous tubules?

4. Where are eggs fertilized in the female human reproductive system?

5. Down syndrome is caused by a third copy of chromosome number _____, or by a critical segment of this chromosome being added to another chromosome.

SELF-QUIZ

1. Humans are
   (a) viviparous
   (b) marsupial mammals
   (c) placental mammals
   (d) a and b
   (e) a and c

2. Follicle-stimulating hormone (FSH) is produced by the _____ and binds to receptors on the _____.
   (a) hypothalamus
   (b) pituitary gland
   (c) adrenal gland
   (d) ovary
   (e) follicle

3. Human male gametogenesis produces _____ gamete(s); human female gametogenesis produces _____ gamete(s).
   (a) one
   (b) two
   (c) three
   (d) four
   (e) five

4. Human sexual response generally occurs as four phases. Put them in the order in which they take place.
   (a) resolution
   (b) orgasm
   (c) excitement
   (d) plateau

5. Which of the following does *not* occur during the first three months of human pregnancy?
   (a) organogenesis
   (b) the nervous system and sense organs develop
   (c) the placenta is formed
   (d) the major nerve tracts within the brain are formed
   (e) limb buds develop

THOUGHT QUESTIONS

1. Probably the most controversial issue to arise in our consideration of sex and reproduction is abortion. List arguments for and against abortion. Under what conditions would you permit abortions? Forbid them? Do you think the disadvantages of abortion are in any way counterbalanced by the advantages it provides to underdeveloped countries with very high birth rates and swelling populations? Should the United States promote or oppose dissemination of information about abortion in underdeveloped countries?

2. Some fishes and many reptiles are viviparous, retaining fertilized eggs within a mother's body to protect them. Birds, however, even though they evolved from reptiles, never employ this means of protecting their eggs. Can you think of a reason why?

3. Relatively few kinds of animals have both male and female sex organs, whereas most plants do. Propose an explanation for this.

4. In reptiles and birds, the fetus is basically masculine, and fetal estrogen hormones are necessary to induce the development of female characteristics. In mammals the reverse is true, the fetus being basically female, with fetal hormones acting to induce the development of male characteristics. Can you suggest a reason why the pattern that occurs in reptiles and birds would not work in mammals?

5. Female armadillos always give birth to four offspring of the same sex. Can you suggest a mechanism that would account for this?

FOR FURTHER READING

HYNES, R.: "Fibronectins," *Scientific American*, June 1986. An overview of how adhesive proteins organize the pattern of development.

LAGERCRANTZ, H., and A. SLOTKIN: "The 'Stress' of Being Born," *Scientific American*, April 1986, pages 100-107. Passage through the narrow birth canal triggers the release of hormones important to the newborn's future survival.

LEIN, A.: *The Cycling Female,* W.H. Freeman & Co., San Francisco, 1979. A short, informal description of the human menstrual cycle and its physical and emotional effects upon women.

LEISHMAN, K.: "Heterosexuals and AIDS," *The Atlantic,* February 1987, pages 39-58. A chilling account of the difficulty of modifying sexual behavior, despite the knowledge of the dangers associated with AIDS.

PATTERSON, D.: "The Causes of Down Syndrome," *Scientific American*, August 1987, pages 52-61. The genes responsible for this tragic developmental disorder have now been identified.

FIGURE 43-1   A meerkat sentinel on duty. Meerkats, *Suricata suricata*, are a species of highly social mongoose living in the semi-arid sands of the Kalahari Desert. A recent field study by Oxford University zoologist Dr. David Macdonald has revealed that theirs is an astonishingly complex and cooperative society. This meerkat is taking his turn to act as a lookout for predators; under the security of his vigilance the other group members can focus their attention on foraging.

BEHAVIOR Chapter 43

Overview

The evolution of a complex nervous system in vertebrates has led not only to a physiologically well-controlled body, capable of systematic patterns of behavior, but also to an increased ability to learn and modify new behaviors. This ability is not unique among the vertebrates. However, as the size of the associative areas of the vertebrate brain has increased, the associative "learning" abilities of the vertebrates have also increased to a level not found in any other animals. The relative contributions of learning and of innate, genetically determined traits to complex human behaviors are difficult to assess, but both doubtless play important roles.

For Review

Here are some important terms and concepts that you will encounter in this chapter. If you are not familiar with them, you should review them before proceeding.

Adaptation (Chapter 16)

Association (Chapter 39)

Memory and learning (Chapter 39)

Each of us engages continually in complex behavior. Not only do we walk and talk, we also race sailboats, sing Christmas carols, build buildings, and sometimes fight wars. More than any other animal, we are able to learn—any student who has come to the last chapter of this text has, if the authors have done their job, learned a great deal about biology. In its simplest form, behavior is conduct—the way an individual acts—and all organisms exhibit simple forms of behavior. Complex behavior, however, is limited to multicellular animals, because only they possess the mechanisms necessary to process complex information. Only with the evolution of a nervous system do the complex responses that we usually think of as **behavior** begin. A sensory system, an associative center, and a network of command fibers are the essential ingredients of an organism that exhibits complex behavior (Figure 43-1).

In this chapter we consider behavior among the vertebrates. Insects and other invertebrate animals also exhibit complex behaviors; they differ from vertebrates primarily in the relatively limited degree to which their behaviors are modified by learning, rather than by their possession of fundamentally different behavioral mechanisms. We focus first on instincts, which are genetically programmed behaviors, and on the degree to which instincts are molded by learning. We then consider the biology of social behavior, or sociobiology.

IS BEHAVIOR LEARNED, OR ARE WE RULED BY OUR GENES?

A principal concern of the social sciences (psychology, anthropology, and sociology) is the understanding of human social behavior—the ways in which human behavior acts to knit society together or disrupt it. Studying human individuals in laboratory situations, looking at broad trends in society, examining primitive societies—these have made it increasingly clear that human behavior, like animal behavior in general, possesses elements that we inherit from our parents, and other elements that we learn from experience. It is not, however, easy to know to what degree a particular behavior is influenced by heredity, the result of neural circuits "hardwired" by our genes into the brain during development. We call such programmed behaviors **instincts,** to distinguish them from behaviors that can be changed as a result of experience. We call the alteration of behavior by experience **learning** (Figure 43-2).

Learning is formally defined as the creation of long-term changes in behavior that arise as a result of experience, rather than as a result of maturation. There are two broad categories of learning. The simplest learned behaviors are **nonassociative:** they do not require the animal to form an association between two stimuli or between a stimulus and an unrelated response. Learned behaviors that require associative activity within the central nervous system are termed **associative.**

The two major forms of nonassociative learning are habituation and sensitization. **Habituation** can be regarded as learning *not* to respond to a stimulus. Learning to ignore unimportant stimuli is a critical ability to an animal confronting a barrage of stimuli in a complex environment. In many cases, the stimulus evokes a strong response when it is first encountered, but then the magnitude of response gradually declines after repeated exposure. When you sit down in a chair, you feel your own weight at first, but soon you are not conscious of it. Your brain "tunes out" this information. **Sensitization,** by contrast, is learning to be hypersensitive to a stimulus. After encountering an intense stimulus, such as an electric shock, an animal may often react vigorously to a mild stimulus that it would previously have ignored.

Associative behaviors are more complex. Thinking is associative, and so are fighting and planning and loving. Studies of nonassociative behaviors in invertebrates such as the sea slug *Aplysia* suggest that many, if not all, associative behaviors may be built up from simpler nonassociative elements.

Associative learning is the alteration of behavior by experience, leading to the formation of an association between two stimuli or between a stimulus and a response. Nonassociative learning involves no such associations.

FIGURE 43-2

Learning is one of the most important activities of young vertebrates. Much of this learning takes place during play. These lion cubs are not trying to hurt each other, but only exploring their capabilities.

FIGURE 43-3

Many complex behaviors are totally instinctive, "hard-wired" in neural circuits formed during development. The complex series of muscular contractions of tongue and mouth that this chameleon uses to capture an insect is always the same and never varies once the capture behavior has been initiated.

Ethology

The distinction between instinct and learning has had a major influence on how biologists have studied associative behavior. One approach has focused exclusively on instinct. Called **ethology,** this approach views animal behavior as governed by four basic components: (1) **sign stimuli,** which are behavioral releasers, cues from the environment that cause certain behaviors and are recognized instinctively without previous training or experience; (2) **motor programs,** which are fixed action patterns, often complex innate responses that are unchangeable, "hardwired" into the nervous system (Figure 43-3); (3) **drives,** which are motivational states; and (4) **imprinting,** which is an unmodifiable form of learning. The term *ethology* has also been used more broadly to cover all studies of animal behavior conducted in natural surroundings.

These four components can be readily seen in the behavior of geese, which were studied by Konrad Lorenz and Nikolaas Tinbergen in work for which they received a Nobel Prize, together with Karl von Frisch, another student of behavior, in 1973. Geese incubate their eggs in nests scooped out of the ground. If a goose notices an egg that has been knocked out of the nest accidentally, the goose will extend its neck towards the egg, get up, and roll the egg back into the nest with its bill. Because this behavior seems so reasonable to us, it is tempting to believe that goose saw the problem and figured out what to do. In fact, the entire behavior is entirely stereotyped and totally instinctive. Any rounded object, regardless of size or color, triggers the response. Beer bottles, for example, are very effective. In this example, the rounded nature of the object is the sign stimulus, the egg rolling response is the motor program, and the mental state of the goose during the nesting period is the drive (they do this only when laying and hatching eggs). Geese also exhibit imprinting: for a short period soon after they hatch, goslings will follow almost any receding object that murmurs "kum-kum," and will forever thereafter recognize that object as their parent (Figure 43-4). If a newborn goose is presented with a moving box containing a clock, it will adopt the box as its parent.

Behavioral Psychologists

Other students of animal behavior have focused exclusively on learning, to which they have considered instinct to be irrelevant. Called **behavioral psychologists** (or **comparative psychologists**), these behavioralists traditionally have divided complex animal behaviors into those learned by **classical conditioning** and those learned by **operant conditioning.**

FIGURE 43-4

The eager goslings following Konrad Lorenz think he is their mother. He is the first "animal" they saw when they were born, and they have become imprinted with his image; they will always recognize him as their parent.

FIGURE 43-5

The bearded man petting the harnessed dog is the Russian psychologist Ivan Pavlov. The dog is suspended in a harness while he is conditioned to salivate in response to light.

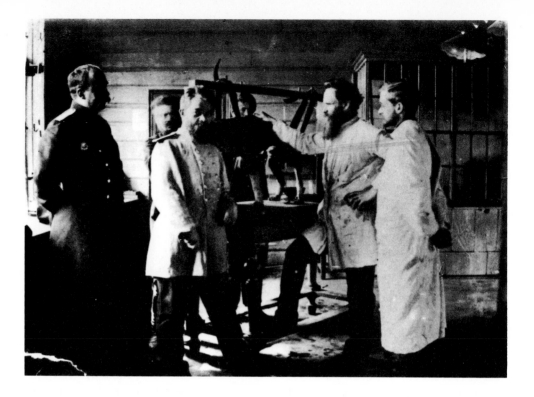

Classical Conditioning

The repeated presentation of a stimulus in association with a response can cause the brain to form an association between them, even if the stimulus and the response have never been associated before. If you present meat (stimulus) to a dog, it will salivate (response). In the classic study of this sort of **conditioning,** the Russian psychologist Ivan Pavlov also presented a second, unrelated stimulus. He shone a light on a dog at the same time that meat powder was blown into its mouth. As expected, the dog salivated. After repeated trials, the dog eventually would salivate in response to the light alone. The dog had learned to associate the unrelated light stimulus with the meat stimulus (Figure 43-5).

In this example, the meat is the sign stimulus, an innately recognized clue, and salivation is the motor program, an innately triggered behavioral act. Behavioral psychologists refer to them as the unconditioned stimulus (meat) and the unconditioned response (salivation). Both unconditioned stimulus and unconditioned response are innate, occurring without learning. Response to the light, a **conditioned stimulus,** is different in that it requires learning and will not occur without it. Early experimenters believed that *any* stimulus could be linked in this way to any unconditioned response, although as you will see below, we now know this is not true.

Operant Conditioning

In classical conditioning, successful association by the animal does not influence whether or not the animal receives the reward, the reinforcing stimulus. In **operant conditioning,** by contrast, the reward follows only after the correct behavioral response, so the animal must make the proper association before it receives the reinforcing stimulus. The American psychologist B.F. Skinner studied such conditioning in rats by placing them in a box of a sort that came to be called a "Skinner box" (Figure 43-6). Once inside, the rat would explore the box feverishly, running this way and that. Occasionally it would accidentally press a lever, and a pellet of food would appear. At first, a rat would ignore the lever and continue to move about, but soon it learned to press the lever to obtain food. When it was hungry, it would spend all of its time pushing the bar. This sort of trial-and-error learning is of major importance to most vertebrates.

FIGURE 43-6

A rat in a Skinner box. The rat rapidly learns that pressing the lever results in the appearance of a food pellet. This sort of learning—trial and error with a reward for success—is called operant conditioning.

Behavioral psychologists used to believe that animals could be conditioned by operant conditioning to perform any learnable behavior in response to any stimulus, but, as in the case of classical conditioning, we now know this is not so. In both cases, the instinct studied by ethologists also plays a major role.

How Learning and Instinct Interact to Shape Behavior

In the last 15 years, most biologists who study behavior have come to believe that the views of ethologists and of behavioral psychologists are both too extreme, and that the behavior of animals is in fact influenced importantly by both instinct and learning.

It has become clear, for example, that not all learning by association is equally easy, as classical conditioning has supposed. Particular kinds of animals have innate predispositions towards certain associations. Rats, for example, easily learn to associate smells with foods that make them ill, but they are unable to associate sounds or colors with these foods, no matter how many trials they are subjected to. Similarly, pigeons associate colors with food but cannot make associations between sounds and food; they *can* associate sounds and danger—but cannot associate colors with danger. The sorts of associations that are possible are genetically determined. That is, classical conditioning is possible only within boundaries set by instinct.

Trial-and-error learning is also restricted by inherent limits imposed by an animal's genetic makeup. Rats in a Skinner box that learn to press a lever for food cannot learn to press the same lever to avoid an electric shock; they can learn to jump in order to avoid the shock, but not in order to obtain food. Animals are innately programmed to learn some things more readily than others. Instinct determines the boundaries of learning.

It now seems clear that animals are innately programmed to respond to specific clues in particular behavioral situations. These innate programs have evolved because they represent appropriate responses. Rats, which forage at night and have a highly developed sense of smell, are better able to identify dangerous food by smell than by color. The seed that a pigeon eats may have a distinctive color that it can see, but it makes no sound that a pigeon can hear. Evolution has biased animal behavior with instincts that make suitable responses more likely.

Much of animal learning is innately guided by **instinctive learning programs.** Sparrows have an innate song that they will sing even if they are reared alone and have never heard another bird sing; they learn to perfect their song by listening as youths, but the basic song is innate (Figure 43-7). Similarly, human infants are programmed to recognize the more than two dozen consonant sounds characteristic of human speech (including those not present in the particular language they learn) while ignoring a world full of other sounds; they learn by trial and error (the "babbling" phase) how to make these sounds. Instinctive learning programs probably explain why infants find speech so much easier to learn than addition and subtraction, which are inherently much simpler tasks.

SOCIOBIOLOGY: THE BIOLOGICAL BASIS OF SOCIAL BEHAVIOR

The view of animal behavior as being the product of genes (instinct) modified by the environment (learning) in ways that are themselves regulated by heredity is a view that has become widely accepted among students of behavior only since the early 1970s. At that time, E.O. Wilson of Harvard University, Richard Alexander of the University of Michigan, and Robert Trivers of the University of California started what has become a major movement within biology, the attempt to study and understand animal behavior as a *biological* process, a process with a genetic basis that is shaped by evolution. Wilson called the study of the biological basis of social behavior "sociobiology." The application of Darwinian concepts to the study of behavior is currently a very active area of biology, and, as you might expect, studies that assess the degree to which human behavior is genetically determined are controversial. Although many aspects of human

FIGURE 43-7

The singing of a sparrow, which sounds to us so spontaneous and free, is actually a carefully orchestrated performance that the bird learns while it is immature.

NAKED MOLE RATS—A RIGIDLY ORGANIZED VERTEBRATE SOCIETY

One exception to the general rule that vertebrate societies are not highly organized is the naked mole rat, a tiny, naked rodent that lives in East Africa. Adult naked mole rats are about 8 to 13 centimeters long and weigh up to about 60 grams—about the size of a sausage (Figure 43-A). Although they are mammals, they have virtually no hair. Unlike other kinds of mole rats, which live alone or in small family groups, naked mole rats congregate in large underground colonies with a far-ranging system of tunnels and a central nesting area. It is not unusual for a colony to contain 80 or more animals.

Naked mole rats live by eating large, underground roots and tubers, which they locate by tunneling constantly. Naked mole rats tunnel in teams. Each mole rat has protruding front teeth, which make it look something like a pocket-sized walrus; it uses these teeth to chisel away the earth from the blind face at the end of a tunnel. When the leading mole rat has loosened a pile of earth, it pushes the pile between its feet and then scuttles backward through the tunnel, moving the pile of earth with its legs. When this animal finally reaches the opening, it gives the pile to another animal, which kicks the dirt out of the tunnel. Then, free of its pile of dirt, the tunneler returns to the end of the tunnel to dig again, crawling on tiptoe over the backs of a long train of

FIGURE 43-A

A naked mole rat.

other tunnelers that are moving backwards with their own dirt piles.

Naked mole rat colonies are unusual vertebrate societies not only because they are large and well-organized. They also have a breeding structure that one might normally associate with bees, termites, or other insects. All of the breeding is done by a single "queen," who has one or two male consorts. The worker group, composed of both sexes, keeps the tunnels clear and forages for food. As long as the queen is healthy, there is no fighting among the workers. This is

very unusual—few mammals surrender breeding rights without contest or challenge. If the queen is removed from the colony, however, havoc breaks out among the workers: one individual attacks another, and all of them compete to become part of the new power structure. When another female becomes dominant and starts breeding, things settle down once again and discord disappears. Biologists studying naked mole rats speculate that the dominant female secretes a chemical that prevents other individuals from maturing sexually.

behavior are genetically inherited (recent studies of identical twins in Minnesota suggest that as much as 50% of certain personality traits are inherited), in areas close to the everyday concerns of our own lives it is sometimes difficult to avoid becoming involved in value judgments.

Most social behavior is a mixture of innate and learned components. Nowhere is this relationship more obvious than when one animal interacts with other members of its own species. When one animal meets another of its species, they rarely ignore each other; each usually regards the other either as a competitor or as an ally. We call the belligerent behavior directed toward potential competitors **aggression,** and the cooperative behavior directed toward allies **sociality.** Why has evolution favored these two sorts of behavior? Students of the biology of social behavior have suggested that because animals tend to be more closely related to allies, they tend to promote the success of the genes they themselves carry when they help their ally relatives, who carry the same genes, and do not help their competitors, who carry different genes.

FIGURE 43-8

The marked queen honeybee *(bottom right center of photo with the red dot)* is laying eggs in the cells constructed by the workers, seen here surrounding her on the hive. In almost all species of the order Hymenoptera—an order that includes bees, wasps, and ants— fertilized eggs produce females, and unfertilized eggs produce males. The different castes of honeybees divide the work of the colony among them.

In insects the distinction between ally and competitor is very simple. If two individuals are part of one family, they are allies; if not, they are competitors. Many species of insects, particularly bees and ants, are organized into societies composed of highly integrated groups of individuals, each of which performs a special task. Their specialization is so extreme, and their organization so rigid, that these insect societies as a whole exhibit many of the properties of an individual organism. A human body such as yours, for example, relies on millions of individual cells that are specialized to perform many different tasks. In a similar way, a beehive or ants' nest is a coherently organized group of individuals in which certain individuals perform specialized tasks on which survival of the entire group depends (Figure 43-8). Only one component of a human body, the gonads, is responsible for reproduction. In a similar way, only one component of the beehive or ants' nest, the queen, is involved in the reproduction of that colony. All the cells of a human body are related to one another by descent from one fertilized zygote. Similarly, all of the members of a nest or hive are descended from an individual queen.

Some groups of vertebrates are also highly organized, and related individuals, or **families,** may play an important role in organizing groups of individuals. Vertebrate societies are not as rigidly constructed as are insect ones, however, and they are rarely as large. Their membership is usually not restricted to individuals that are closely related, and specialization is not as prevalent, or as pronounced in degree, as it is in bees, termites, or ants. However, vertebrate societies share with insect ones the central characteristic of any society—the ability to direct competitive behavior (aggression) outward, toward individuals that are not members of the group.

Aggression

The founder of ethology, Konrad Lorenz, proposed that aggressive behavior is innate in all vertebrates, including human beings. He considered aggression to be an inherent motivational state that is released by certain sign stimuli and whose threshold is conditioned by learning. Lorenz's idea has not been popular among social scientists, many of whom believe that human beings, at least, are born with a "clean slate." Nor has it been received well by many leaders of Western religious and political traditions. Christianity holds that human behavior is conditioned by learning and modified by "free will"; Marxism holds that human behavior is profoundly conditioned and modified by society. Both Marxists and Christians believe that, because human behavior is modifiable, people are responsible for it, either individually (according to Christians) or collectively (according to Marxists).

FIGURE 43-9

It is important not to confuse aggression with physical contact. These two male elephants *(Loxodonta africana),* although they look as if they are locked in combat, are actually greeting one another.

Despite these widespread beliefs, and although the behavioral evidence is not extensive, the weight of available evidence favors Lorenz's assertion. In many nonhuman vertebrates, aggression does appear to be an innate drive (Figure 43-9). Among certain species of cichlid fishes, for example, the males must fight before they can mate. When they have been reproductively stimulated by the sight of a female during their mating period, the male fishes will dart about furiously, searching for someone to fight. If they cannot find another male, they will frequently attack the female, often killing her. Such behavior seems to be innate; indeed, mating behavior in these fishes cannot be released except by this sort of aggressive behavior. Many other animals exhibit aggressive behavior whenever they are approached by another member of the same species.

Among vertebrates, most aggressive conflicts are not "to the death," and they usually involve little harm to the opponents (Figure 43-10). For example, deer may use their sharp hooves as very effective weapons against wolves or other predators, but they do not use them in fighting each other. Instead, they use their antlers, whose primary function is sexual display. The prongs of the antlers are usually formed in such a way that they can inflict only limited damage on an opponent. Most animals will fight back, however, if an opponent seriously attacks them.

Innate behavioral aggression is responsible for **territoriality** (Figure 43-11), a form of behavior in which individual members of a species mark off a fixed area or some other limiting resource, such as a foraging ground or a group of females, and then defend these reserved resources against invasion or use by any other member of the same species. Territoriality is very common both in birds and in mammals. You can see it when birds sit on a tree branch, spaced evenly so that no bird gets too close to its neighbor. You can see it in your own behavior, when a stranger intrudes into your "personal space"; most people back away when someone starts to talk to them from a distance of less than about 60 centimeters, although the exact distance varies among the members of different human societies.

Like any innate behavior, however, aggression in vertebrates is subject to modification by other innate behaviors, by learning, and in some animals by conscious choice. It is for this reason that a biological view of behavior is not in conflict with the central world view of Christianity (or of Marxism)—although the drive is inherent, the behavior that results is not. Our aggressive instincts are the canvas on which learning and experience paint the pattern of our behavior.

Aggression appears to be an innate drive in most if not all animals, although its expression is conditioned by many other behaviors.

FIGURE 43-10

These bighorn sheep rams are butting horns to determine which is dominant. It is no accident that their eyes are closed—the shock of contact is considerable.

FIGURE 43-11

These male hippopotami are not saying hello. They are contesting for territory, and they are in deadly earnest. They will fight until one is too injured or discouraged to continue. The winner will then dominate that particular part of the lake.

FIGURE 43-12

Baboons form cohesive social groups with a dominant male and levels of subordinate relationships. Where an individual sleeps in a tree reflects his or her position in the social group.

In large and relatively permanent social groups, aggressive interactions are often channeled into organized patterns of behavior. In conflict situations, an individual behaves according to rules that all parties understand and acknowledge, and in this way outright conflict is avoided. Thus a newly formed flock of chicken hens will quickly establish dominant-subordinate relationships among its members. These relationships constitute a dominance hierarchy, or "pecking order," which is established by the initial aggressive encounters among the hens. Such encounters are not repeated; the ground rules, which are acknowledged by all, are laid down based on the outcome of the first ones. The most dominant hen is free to peck any of the other birds without being pecked back; the next most dominant hen may peck all but the most dominant one, and so on. Conflict is avoided by a set of rules. The elimination of conflict benefits not only the dominant individual but all members of the group. The birds of a stable flock are bigger and lay more eggs than birds of flocks that are frequently disrupted by additions and removals of birds.

Dominance hierarchies are common among vertebrate societies (Figure 43-12). They are found in wolf packs, in hyena groups—and in the primates. Conflict within primate and human groups is usually lessened by such adherence to commonly recognized rules, and a sophisticated body of rules, or **laws,** is commonly held to be a hallmark of human civilization. Reduction of conflict among primates is not as complete, however, as among some other vertebrates; war is not known among any vertebrate group except the primates.

FIGURE 43-13

Not all animals are altruistic, sacrificing their own well-being for the good of the group. This ostrich, turning to run, is interested in saving its own skin more than in warning its fellows of the approaching photographer. In general, altruistic behavior is more common in societies whose members are closely related.

FIGURE 43-14

This raccoon is apparently washing its dinner before eating, a characteristic behavior for the species. To what degree these behaviors are learned—passed down from parent to offspring—and to what degree they are hereditary—reflecting the action of specific genes—is difficult to determine.

Altruistic Behavior

Some aspects of vertebrate behavior still present a puzzle to evolutionists. Particularly puzzling is the fact that vertebrates are often organized into social groups in such a way that the activities of certain individuals benefit the group at the expense of those individuals themselves. The little meerkat in Figure 43-1, for example, is maintaining a lookout on a termite mound deep in the Kalahari Desert of Southern Africa, exposed to predators and to the full glare of the sun. In doing this, the lookout draws attention to itself and thus exposes itself to greater danger than if it were not a sentry. Behaviors such as this which are not in an individual's self-interest are referred to as **altruistic.** What forces could promote the evolution of altruistic behavior? The difficulty in answering this question stems from the fact that selection acts on *individuals,* not on populations: selection favors the genes carried by those individuals that leave the most offspring. How can it be beneficial in an evolutionary sense—that is, lead to the production of more offspring bearing your genes—to sacrifice your potential offspring for the good of the group? However noble it seems to lay down one's life for a friend, the evolution of innate altruistic behaviors at first seems contrary to everything we know about how evolution proceeds.

Inclusive Fitness

A possible way out of this quandary was suggested in 1932 by a passing remark made by the great population geneticist J.B.S. Haldane. He expressed his suggestion by saying that he would willingly lay down his life for two brothers or eight first cousins. Do you see what he was getting at? Each brother and Haldane share half of their genes in common: Haldane's brothers each had a 50% chance of receiving a given allele that Haldane obtained, and so forth for all other alleles. Consequently, it is statistically true that two of his brothers would carry as many of Haldane's particular combination of alleles to the next generation as Haldane himself would. Similarly, Haldane and a first cousin would share an eighth of their alleles: their sibling parents would each share half of their alleles, and each of their children would receive half of these, of which half (on the average) would be in common—$0.5 \times 0.5 \times 0.5 = 0.125$, or one eighth. Eight first cousins would therefore pass on as many of those genes to the next generation as Haldane himself would. The point is that evolution will favor any strategy that increases the net flow of a combination of genes to the next generation. It makes "evolutionary sense" to sacrifice one's own individual fitness if doing so increases one's **inclusive fitness,** the total number of one's genes that are passed to the next generation. Thus altruistic behavior can evolve in societies whose members are related.

There has been considerable controversy among biologists about the degree to which the concept of heritable behaviors applies to animals whose behaviors are less fixed (Figure 43-14). Much of the argument has centered on altruistic behavior. The problem is that in vertebrate groups, inclusive fitness is not in itself a powerful enough genetic incentive to have fostered the evolution of altruistic behavior, since the members of vertebrate societies are not as closely related to one another as the members of insect societies are. It has been suggested that vertebrates practice a kind of "I'll scratch your back if you scratch mine" strategy known as **reciprocal altruism.** The concept of reciprocal altruism implies that individuals perform altruistic behavior in the expectation of receiving similar treatment. The difficulty with this, of course, is that the optimal strategy for transmitting *your* genes is to cheat! If someone saves you from drowning, and then you observe him drowning, you will leave the most offspring if you do not endanger your life by attempting to return the favor and rescue him. Still, the fact remains that many animals, ignorant of evolutionary niceties, practice altruistic behavior. We simply do not yet understand enough to say how such altruistic behaviors evolved. This mystery only serves to illustrate how much remains to be learned about the biological basis of behavior—and how much fun it is going to be to study it.

THE GENETIC BASIS OF BEHAVIOR

An essential element in the view that behaviors evolve as a heritable attribute of a species is the assumption that specific genes can act to determine components of specific behaviors. This point has proved to be controversial in discussions of specific human behaviors, usually because some of the behaviors—such as criminality, homosexual-

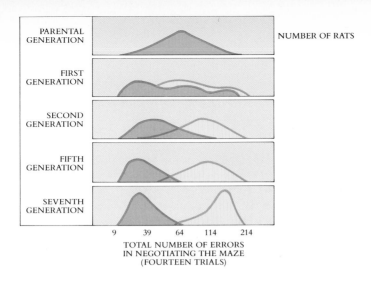

FIGURE 43-15

Robert Tryon was able to select among rats for ability to negotiate a maze, demonstrating that this ability is directly influenced by genes. He tested a large group of rats, and from them selected the few that ran the maze with the fewest errors (red) and let them breed with one another; he then tested their progenies and again selected those with the fewest maze-running errors for breeding. By seven generations he succeeded in halving the average number of errors a naive rat required to negotiate the maze. Parallel selection for more errors (yellow) was also successful—it more than doubled the average number of errors.

ity, and intelligence—are very complex ones in which the roles of genes and environment are inextricably intertwined.

Unlike complex human behaviors, simpler behaviors of other vertebrates are often clearly influenced by genes that are specific to the behavior. In a famous experiment carried out in the 1940s, Robert Tryon studied the ability of rats to find their way through a maze that had many blind alleys and only one exit, where a reward of food awaited (Figure 43-15). It took a while, as false avenues were tried and rejected, but eventually some individuals learned to zip right through the maze to the food, making few false turns. Other rats never seemed to learn the correct path. Tryon bred the "maze-bright" rats to one another, establishing a colony from the fast learners, and similarly established a second "maze-dumb" colony by breeding the slowest-learning rats to each other. He then tested the offspring in each colony to see how quickly they learned the maze. The offspring of maze-bright rats learned even more quickly than their parents had, while the offspring of maze-dumb parents were even poorer students. After repeating this artificial selection procedure over several generations, Tryon was able to produce two behaviorally distinct types of rat with very different maze-learning abilities. Clearly the ability to learn the maze was hereditary, governed by genes that were being passed from parent to offspring. And the genes are specific to this behavior, rather than being general ones that influence many behaviors—the abilities of these two groups of rats to perform other behavioral tasks did not differ.

> **Among vertebrates, some associative behaviors can be shown to have a significant genetic component, although environmental influences are also important.**

MANY VERTEBRATE BEHAVIORS ARE ADAPTIVE

A biological view of behavior as being the result of evolution would predict that the behaviors characteristic of particular animals would generally be suited to their mode of living—that is, in the term of Darwin, be adaptive. You would expect that natural selection would have favored those gene combinations that adapt animals more completely to particular habitats. This is exactly what ethologists have observed in natural populations. A number of heritable behaviors better adapt individuals to particular habitats. These include territoriality, migratory memory, dominance hierarchies, and reproductive behaviors.

Territoriality

When resources such as nesting sites or food are limited (Figure 43-16), more birds will be produced in each generation if the birds that are breeding ("breeding pairs") obtain sufficient resources, even though other birds not engaged in breeding may have to do

FIGURE 43-16

The emperor penguins in this Antarctic rookery nest in sites such as you see in the background. There are only a limited number of such sites available, which may reflect the limited amount of resources available to the breeding penguins.

FIGURE 43-17

The migratory path of California bobolinks. These birds came recently to the Far West from their more established range in the Midwest. When they migrate to South America in the winter, they do not fly directly, but rather fly over the Midwest first and then use the ancestral flyway.

without. Without a mechanism to ensure this sort of "rationing," competition for the limited resources might result in the resource being spread too thinly, with no one pair obtaining adequate resources and the entire population in danger of unsuccessful breeding. One mechanism that has evolved to deal with this problem is **territorial behavior.** Resources within a territory are reserved for a particular breeding pair or group, and other individuals are left to fend for themselves. In many mammalian societies, territoriality often plays an important role in the establishment of new populations, with the males that have been defeated by the dominant male leaving to found their own populations. Individuals surviving in the surrounding marginal habitats can repopulate any territories that become vacant, and can occupy any new habitats that they encounter.

Migratory Memory

Many birds breed in temperate or arctic habitats and spend their winters at other locations far away, nearer to the equator where winters are less severe. In many cases the birds must cross thousands of miles, sometimes over open ocean. This migratory behavior is genetically determined, although it is helped by learning, when young individuals fly for the first time with migratory flocks. When colonies of bobolinks became established in the western United States, far from their normal range in the Midwest and East, these birds did not migrate directly to their winter range in South America; instead they migrated east to their ancestral range and then south along the old flyway (Figure 43-17). The old pattern was not changed but rather added to.

Remember also the green sea turtles of Chapter 1 with which we started this text. How do the babies that hatch on Ascension Island know how to find Brazil, thousands of miles away over the open sea? The little sea turtles charge instinctively towards the sound of the sea and paddle out determinedly against the breakers in what is called a "swim frenzy" until they reach the open sea. There they spend several years, growing, as the ocean currents carry them westward. Eventually the currents bear them to Brazil. How do they find their way back as adults, when they breed, perhaps 50 years later? We don't know.

Dominance Hierarchies

In vertebrates living in groups, it is common for a few individuals—often but not always males—to dominate the group. Frequently, it is only these individuals that mate with the opposite sex (Figure 43-18). This form of dominance behavior, as you saw above, limits conflict and competition within the group, since all members of the group recognize the rules and abide by them. The patterned behaviors that identify rank within such groups are often highly exaggerated.

FIGURE 43-18
A dominant male lion with a lioness.

Reproductive Behaviors

Very early in his studies of behavior, Tinbergen noted that certain birds remove the shell fragments of broken eggs from their nests, and that this behavior is innate. Any shell fragment that Tinbergen added was quickly removed. It turned out that the odor from shell fragments attracts predators, and their prompt removal was very smart housekeeping. Many other reproductive behaviors seem to have evolved to aid in the protection and rearing of young. Male parental care, for example, is rare among vertebrates, except for birds and primates, both of whose young go through long periods of learning before they can survive on their own (Figure 43-19). Male parental care in these groups greatly increases the probability of raising the young successfully and may well have evolved for that reason.

There is little doubt that selection for specific behaviors has played an important role in vertebrate evolution, and it probably was significant in human evolution as well. The degree to which human behavior today reflects our genetic heritage is a controversial question, with no clear answer.

THE BIOLOGICAL BASIS OF HUMAN BEHAVIOR

Complex behaviors are exhibited by all animals that have nervous systems. It should come as no surprise that associative behaviors are more common among the vertebrates than among insects or other animals, since the portion of the brain devoted to associative activity is so much larger in vertebrates. The associative cortex is particularly prominent in mammals, humans being the most extreme case, and these are the animals with the most flexible behaviors, that is, those most subject to modification by learning.

There can be little doubt that much of the behavioral capacity of vertebrates is genetically determined, the neural circuits specified by genes. It is not really clear to what degree these circuits can be overridden by the associative activity of the brain, but some degree of influence is very likely. In humans, so much of whose brains are devoted to associative cortex, most biologists would say that while some simple physical behaviors are innate and invariant, few complex social behaviors are fully determined genetically. The consensus, rather, is that we carry with us as evolutionary baggage the aggression and territoriality of our ancestors, and that in any individual these behaviors are subject to change by learning.

Because the consequences of aggressive behavior among humans are so much more severe than they are among other vertebrates, the degree to which human aggressive behavior can be tempered by learning is an issue of very real importance. As you have seen in the preceding chapters, human activities are rapidly altering the complex web of interactions on which life on earth depends. Organized human conflict, common throughout recorded human history, is now able to devastate the earth, and perhaps to end the history of life there. The knowledge that enables humans to carry out conflict is itself a result of the great enlargement of the associative cortex. The evolutionary success of this modification of the vertebrate brain will depend critically on the ability of the greatly enlarged human associative cortex to modify such behaviors as aggression.

FIGURE 43-19

The young of primates and birds go through long periods of learning before they can survive on their own. This young lady, not yet 2 years old, has much yet to learn before she can wear Daddy's hat.

SUMMARY

1. Some patterns of behavior are innate (instincts), whereas others are associative responses acquired by trial and error (learned behaviors). Vertebrate behaviors combine these two kinds of behavioral elements.

2. Heredity determines the limits within which a behavior can be modified; within those limits, learning determines what that behavior will be like. These limits are narrow or nonexistent for instincts, but they may be very broad for learned behaviors.

3. Many innate behaviors consist of hierarchies of fixed action patterns; a variety of different responses are possible, depending on the environmental cues.

4. The simplest forms of learning involve sensitization and habituation. More complex associative learning may also occur in this way, by the strengthening and weakening of existing synapses, although learning may also involve the formation of entirely new synapses.

5. Behaviors directed toward other members of one's own species are usually classified as either aggressive (competitive) or social (cooperative).

7. Members of vertebrate groups often exhibit altruistic behavior, a form of social behavior in which an animal aids other members of its group, at possible risk to itself. The way in which such altruistic behaviors evolve is not clear.

8. Human behavior has both instinctive and learned components, and it is difficult if not impossible to sort out the relative contributions of these components to particular behavioral patterns.

REVIEW

1. Programmed behaviors are called _____; they are distinguished from behaviors that can change as a result of experience.

2. The study of the biological basis of social behavior is called _____.

3. The weight of available evidence favors the assertion made by Konrad Lorenz that _____ behavior is innate in all vertebrates.

4. Vertebrate societies often have _____ hierarchies, the rules of which are often laid down by aggressive interactions.

5. How do green sea turtles find their way back to Ascension Island for breeding?

SELF-QUIZ

1. Ethology views animal behavior as governed by four basic components. Which of the following is *not* one of these components?
   (a) drives
   (b) sign stimuli
   (c) habituation
   (d) motor programs
   (e) imprinting

2. Aggressive behavior in vertebrates
   (a) is responsible for territoriality
   (b) appears to be an innate drive
   (c) usually involves little harm to the opponents
   (d) is subject to modification by other innate behaviors
   (e) all of the above

3. Tryon's breeding experiments on "maze-bright" and "maze-dumb" rats resulted in the hypothesis that
   (a) some associative behavior has a significant genetic component
   (b) all associative behavior has a significant genetic component
   (c) no associative behavior has a significant genetic component
   (d) altruistic behavior has a significant genetic component
   (e) non-altruistic behavior has a significant genetic component

4. Naked mole rat colonies are unusual among vertebrate societies because
   (a) workers surrender breeding rights without challenge if a queen is present
   (b) there is a dominance heirarchy
   (c) there is cooperative behavior
   (d) there is social organization
   (e) none of the above

5. Behaviors that are not in an individual's self-interest but are of benefit to the group are called
   (a) aggressive
   (b) habituation
   (c) operant conditioning
   (d) altruistic
   (e) sensitization

THOUGHT QUESTIONS

1. War is so common among human beings that it must be considered a basic behavior of our species. It appears to be absent in all other animal groups (with the possible exception of some other primates). Do you think this behavior has a genetic basis? If so, why might its evolution have been favored by natural selection?

2. It can be argued that our current nuclear standoff is a form of "retaliator" behavior, in which each side postures but neither attacks in earnest unless it is attacked first. Does this proposition conform with what you know about the ways animals deal with aggression? Explain your answer.

3. Swallows often hunt in groups, while hawks and other predatory birds usually are solitary hunters. Can you suggest an explanation for this difference?

4. Can you suggest an evolutionary reason why many vertebrate reproductive groups are composed of one male and numerous females, rather than the reverse?

FOR FURTHER READING

BORGIA, G.: "Sexual Selection in Bower Birds," *Scientific American,* June 1986, pages 92-100. Behavior among birds is often bizarre, but none is more so than that of the Australasian bower birds, in which the females choose their mates depending on how well they adorn their future nests.

CARR, A.: "The Navigation of the Green Turtle," *Scientific American,* May 1965, pages 78-86. An interesting account of the mystery surrounding the long migration of green turtles halfway across the Atlantic Ocean. Twenty years later, we still don't know how they do it, or why.

DAWKINS, R.: *The Selfish Gene,* Oxford University Press, New York, 1976. An entertaining account of the sociobiologist's view of behavior.

GOULD, J., and P. MARLER: "Learning by Instinct," *Scientific American,* January 1987, pages 74-85. A clear and interesting account of the relative roles of instinct and learning in behavior. The authors, prominent behaviorists, argue that learning is often limited or controlled by instinct.

GRIFFIN, D.: "Animal Thinking," *American Scientist,* vol. 72, pages 456-463, 1984. An exciting discussion of the possibility that animals have conscious awareness, this article describes many examples of what appears to be conscious thinking by animals.

HUBER, F., and J. THORSON: "Cricket Auditory Communication," *Scientific American,* December 1985, pages 60-68. An unusually clear example of how nervous system activity underlies animal behavior.

JOLLY, A.: "The Evolution of Primate Behavior," *American Scientist,* vol. 73, pages 230-239, 1985. A fascinating survey of behavior among primates that indicates a progressive development of intelligence, rather than a sudden, full-blown appearance when humans evolved.

CLASSIFICATION OF ORGANISMS

The classification used in this book is explained in Chapter 24. It recognizes a separate kingdom, Monera, for the bacteria (prokaryotes), and divides the eukaryotes into four kingdoms: the diverse and predominantly unicellular Protista, and three large, characteristic multicellular groups derived from them: Fungi, Plantae, and Animalia. Viruses, which are considered nonliving, are not included in this appendix, but are treated in Chapter 25.

KINGDOM MONERA

Bacteria; the prokaryotes. Single-celled organisms, sometimes forming filaments or other forms of colonies. Bacteria lack a membrane-bound nucleus and chromosomes, sexual recombination, and internal compartmentalization of the cells; their flagella are simple, composed of a single fiber of protein. They are much more diverse metabolically than the eukaryotes. Their reproduction is predominantly asexual. About 2500 species are currently recognized, but many times that many probably exist.

KINGDOM PROTISTA

Eukaryotic organisms, including many evolutionary lines of primarily single-celled organisms. Eukaryotes have a membrane-bound nucleus and chromosomes, sexual recombination, and extensive internal compartmentalization of the cells; their flagella are complex, with $9 + 2$ internal organization. They are diverse metabolically, but much less so than bacteria; protists are heterotrophic or autotrophic, and may capture prey, absorb their food, or photosynthesize. Reproduction in protists is either sexual, involving meiosis and syngamy, or asexual.

Phylum Caryoblastea

One species of primitive amoebalike organism, *Pelomyxa palustris,* which lacks mitosis, mitochondria, and chloroplasts.

Phylum Dinoflagellata

Dinoflagellates; unicellular, photosynthetic organisms, most of which are clad in stiff, cellulose plates and have two unequal flagella that beat in grooves encircling the body at right angles. About 1000 species.

Phylum Rhizopoda

Amoebas; heterotrophic, unicellular organisms that move from place to place by cellular extensions called pseudopods and reproduce only asexually, by fission. Hundreds of species.

Phylum Sporozoa

Sporozoans; unicellular, heterotrophic, nonmotile, spore-forming parasites of animals. About 3900 species.

Phylum Acrasiomycota

Cellular slime molds; unicellular, amoebalike, heterotrophic organisms that aggregate in masses at certain stages of their life cycle and form compound sporangia. About 65 species.

Phylum Myxomycota

Plasmodial slime molds; heterotrophic organisms that move from place to place as a multicellular, gelatinous mass, which forms sporangia at times. About 450 species.

Phylum Zoomastigina

Zoomastigotes and euglenoids; a highly diverse phylum of mostly unicellular, heterotrophic or autotrophic, flagellated free-living or parasitic protists (flagella 1 to thousands). Thousands of species.

Phylum Phaeophyta

Brown algae; multicellular, photosynthetic, mostly marine protists with chlorophylls *a* and *c* and an abundant carotenoid (fucoxanthin) that colors the organisms brownish. About 1500 species.

Phylum Chrysophyta

Diatoms and related groups; mostly unicellular, photosynthetic organisms with chlorophylls *a* and *c* and fucoxanthin. About 11,500 living species.

Phylum Chlorophyta

Green algae; a large and diverse phylum of unicellular or multicellular, mostly aquatic organisms with chlorophylls *a* and *b*, carotenoids, and starch, accumulated within the plastids (as it is also in plants) as the food storage product. About 7000 species.

Phylum Ciliophora

Ciliates; diverse, mostly unicellular, heterotrophic protists, characteristically with large numbers of cilia. About 8000 species.

Phylum Oomycota

Oomycetes; water molds, white rusts, and downy mildews. Aquatic or terrestrial unicellular or multicellular parasites or saprobes that feed on dead organic matter. About 475 species.

Phylum Rhodophyta

Red algae; mostly marine, mostly multicellular protists with chloroplasts containing chlorophyll *a* and physobilins. About 4000 species.

KINGDOM FUNGI

Filamentous, multinucleate, heterotrophic eukaryotes with cell walls rich in chitin; no flagellated cells present. Mitosis in fungi takes place within the nuclei, the nuclear envelope never breaking down. The filaments of fungi grow through the substrate, secreting enzymes and digesting the products of their activity. Septa between the nuclei in the hyphae normally complete only when sexual or asexual reproductive structures are being cut off. Asexual reproduction frequent in some groups. The nuclei of fungi are haploid, with the zygote the only diploid stage in the life cycle. About 100,000 named species.

Division Zygomycota

Zygomycetes; bread molds and other microscopic fungi that occur on decaying organic matter. Hyphae aseptate except when forming sporangia or gametangia. About 600 species.

Division Ascomycota

Ascomycetes; yeasts, molds, many important plant pathogens, morels, cup fungi, and truffles. Hyphae divided by incomplete septa except when asci, the structures characteristic of sexual reproduction, are formed. Meiosis takes place within asci. About 300,000 named species.

Division Basidiomycota

Basidiomycetes; mushrooms, toadstools, bracket and shelf fungi, rusts, and smuts. Meiosis takes place within basidia. About 25,000 named species.

Fungi Inperfecti

An artificial group of about 25,000 named species; in them, the reproductive structures are not known.

Lichens

Lichens are symbiotic associations between an ascomycete (a few basidiomycetes are involved also) and either a green alga or a cyanobacterium. At least 25,000 species.

KINGDOM PLANTAE

Multicellular, photosynthetic, primarily terrestrial eukaryotes derived from the green algae (phylum Chlorophyta) and, like them, containing chlorophyll *a* and *b,* together with carotenoids, in chloroplasts and storing starch in chloroplasts. The cell walls of plants have a cellulose matrix and sometimes become lignified; cell division is by means of a cell cell plate that forms across the mitotic spindle. The vascular plants have an elaborate system of conducting cells consisting of xylem (in which water and minerals are transported) and phloem (in which carbohydrates are transported); the mosses have a reduced vascular system, whereas the liverworts and hornworts, which may not be directly related to the mosses, lack. Plants have a waxy cuticle that helps them to retain water, and most have stomata, flanked by specialized guard cells, which allow water to escape and carbon dioxide to reach the chloroplast-containing cells within their leaves and stems. All plants have an alternation of generations with reduced gametophytes and multicellular gametangia. About 270,000 species.

Division Bryophyta

Mosses, hornworts, and liverworts. Bryophytes have green, photosynthetic gametophytes and usually brownish or yellowish sporophytes with little or no chlorophyll. About 16,600 species.

Division Psilophyta

Whisk ferns, a group of vascular plants. Two genera and several species.

Division Lycophyta

Lycopods (including clubmosses and quillworts). Vascular plants. Five genera and about 1000 species.

Division Sphenophyta

Horsetails. Vascular plants. One genus *(Equisetum),* 15 species.

Division Pterophyta

Ferns. Vascular plants, often with characteristic divided, feathery leaves (fronds). About 12,000 species.

Divison Coniferophyta

Conifers. Seed-forming vascular plants; mainly trees and shrubs. About 550 species.

Division Cycadophyta

Cycads; tropical and subtropical palmlike gymnosperms. Ten genera, about 100 species.

Division Ginkgophyta

One species, the ginkgo or maidenhair tree.

Division Anthophyta

Flowering plants, or angiosperms, the dominant group of plants; characterized by a specialized reproductive system involving flowers and fruits. About 235,000 species.

KINGDOM ANIMALIA

Animals are multicellular eukaryotes that characteristically ingest their food. Their cells are usually flexible; in all of the approximately 35 phyla except sponges, these cells are organized into structural and functional units called tissues, which in turn make up organs in most animals. In animals, the cells move extensively during the development of the embryos; the blastula, a hollow ball of cells, forms early in this process, and is characteristic of the group. Most animals reproduce sexually; their nonmotile eggs are much larger than their small, flagellated sperm. The gametes fuse directly to produce a zygote and do not divide by mitosis as in plants. More than a million species of animals have described, and at least several times that many await discovery.

Phylum Porifera

Sponges. Animals that mostly lack definite symmetry, and possess neither tissues nor organs. About 10,000 marine species, mostly marine.

Phylum Cnidaria

Corals, jellyfish, hydras. Mostly marine, radially symmetrical animals that mostly have distinct tissues; two basically different body forms, polyps and medusae. About 10,100 species.

Phylum Platyhelminthes

Flatworms; bilaterally symmetrical acoelomates; the simplest animals that have organs. About 13,000 species.

Phylum Nematoda

Nematodes, eelworms, and roundworms; ubiquitous, bilaterally symmetrical, cylindrical, unsegmented, pseudocoelomate worms, including many important parasites of plants and animals. More than 12,000 described species, but the actual number is probably 500,000 or more species.

Phylum Mollusca

Mollusks; bilaterally symmetrical, protostome coelomate animals that occur in marine, fresh-water, and terrestrial habitats. Many mollusks possess a shell. At least 110,000 species.

Phylum Annelida

Annelids; segmented, bilaterally symmetrical, protostome coelomates; the segments are divided internally by septa. About 12,000 species.

Phylum Arthropoda

Arthropods; bilaterally symmetrical protostome coelomates with a segmented body, chitonous exoskeleton, complete digestive tract, dorsal brain and paired nerve cord, and jointed appendages. Arthropods are the largest phylum of animals, with nearly a million species described and many more to be found.

Phylum Echinodermata

Echinoderms; sea stars, brittle stars, sand dollars, sea cucumbers, and sea urchins. Complex deuterostome, coelomate, marine animals that are more or less radially symmetrical as adults. About 6000 living species.

Phylum Chordata

Chordates; bilaterally symmetrical, deuterostome, coelomate animals that have at some stage of their development a notochord, pharyngeal slits, a hollow nerve cord on their dorsal side, and a tail. The best-known group of animals; about 45,000 species.

MENDELIAN GENETICS PROBLEMS

1. Why did Mendel observe only two alleles of any given trait in the crosses that he carried out?

2. The illustration at right describes Mendel's cross of *wrinkled* and *round* seed characters. What is wrong with this diagram?

3. The annual plant *Haplopappus gracilis* has two pairs of chromosomes 1 and 2. In this species, the probability that two traits *a* and *b* selected at random will be on the same chromosome of *Haplopappus* is the probability that they will both be on chromosome 1 (½), times the probability that they will both be on chromosome 2 (also ½): ½ × ½ = ¼, or 25%.
This is often symbolized

$$\tfrac{1}{2}\,^{\text{(\# of pairs of chromosomes)}}$$

Human beings have 23 pairs of chromosomes. What is the probability that any two human traits selected at random will be on the same chromosome?

4. Among Hereford cattle there is a dominant allele called *polled;* the individuals that have this allele lack horns. After college, you become a cattle baron and stock your spread entirely with polled cattle. You personally make sure that each cow has no horns, and none does. Among the calves that year, however, some grow horns. Angrily you dispose of them, and make certain that no horned adult has gotten into your pasture. None has. The next year, however, more horned calves are born. What is the source of your problem? What should you do to rectify it?

5. An inherited trait among human beings in Norway causes affected individuals to have very wavy hair, not unlike that of a sheep. The trait is called *woolly*. The trait is very evident when it occurs in families; no child possesses woolly hair unless at least one parent does. Imagine you are a Norwegian judge, and that you have before you a woolly haired man suing his normal-haired wife for divorce because their first child has woolly hair but their second child has normal long, blonde hair. The husband claims this constitutes evidence of infidelity on the part of his wife. Do you accept his claim? Justify your decision.

6. In human beings, Down syndrome, a serious developmental abnormality, results from the presence of three copies of chromosome 21 rather than the usual two copies. If a female exhibiting Down syndrome mates with a normal male, what proportion of her offspring would be expected to be affected?

7. Many animals and plants bear recessive alleles for *albinism,* a condition in which homozygous individuals completely lack any pigments. An albino plant, for example, lacks chlorophyll and is white. An albino person lacks any melanin pigment. If two normally pigmented persons heterozygous for the same albinism allele marry, what proportion of their children would be expected to be albino?

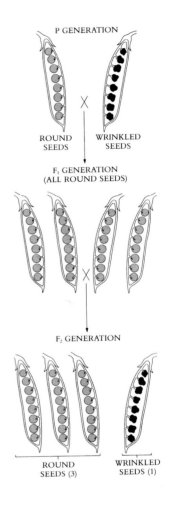

P GENERATION

ROUND SEEDS   WRINKLED SEEDS

F₁ GENERATION
(ALL ROUND SEEDS)

F₂ GENERATION

ROUND SEEDS (3)   WRINKLED SEEDS (1)

8. Your uncle dies and leaves you his race horse, Dingleberry. To obtain some money from your inheritance, you decide to put the horse out to stud. In looking over the stud book, however, you discover that Dingleberry's grandfather exhibited a rare clinical disorder that leads to brittle bones. The disorder is hereditary and results from homozygosity for a recessive allele. If Dingleberry is heterozygous for the allele, it will not be possible to use him for stud, since the genetic defect may be passed on. How would you go about determining whether Dingleberry carries this allele?

9. In the fly *Drosophila,* the allele for dumpy wings (symbolized *d*) is recessive to the normal long-wing allele (symbolized *D*). The allele for white eye (symbolized *w*) is recessive to the normal red-eye allele (symbolized *W*). In a cross of *DDWw* × *Ddww*, what proportion of the offspring are expected to be "normal" (long wing, red eye)? What proportion "dumpy, white"?

10. As a reward for being a good student, your instructor presents you with a *Drosophila* named "Oscar." Oscar has red eyes, the same color that normal flies do. You add Oscar to your fly collection, which also contains "Heidi" and "Siegfried," flies with white eyes, and "Dominique" and "Ronald," which are from a long line of red-eyed flies. Your previous work has shown that the white-eye trait exhibited by Heidi and Siegfried is caused by their being homozygous for a recessive allele. How would you determine whether or not Oscar was heterozygous for this allele?

11. In some families, children are born that exhibit recessive traits (and therefore must be homozygous for the recessive allele specifying the trait), even though one or both of the parents do not exhibit the trait. What can account for this?

12. You collect in your backyard two individuals of *Drosophila melanogaster,* one a young male and the other a young, unmated female. Both are normal in appearance, with the typical vivid red eyes of *Drosophila.* You keep the two flies in the same vial, where they mate. Two weeks later, hundreds of little offspring are flying around in the vial. They all have normal red eyes. From among them, you select 100 individuals, some male and some female. You cross each individually to a fly you know to be homozygous for a recessive allele called "sepia," which leads to black eyes when homozygous (these flies thus have the genotype *se/se*). Examining the results of your 100 crosses, you observe that in about half of them, only normal red-eyed progeny flies are produced. In the other half, however, the progeny are about 50% red eyed and 50% black eyed. What must have been the genotypes of your original backyard flies?

13. Hemophilia is a recessive sex-linked human blood disease that leads to failure of blood to clot normally. One form of hemophilia has been traced to the royal family of England, from which it spread throughout the royal families of Europe. For the purposes of this problem, assume that it originated as a mutation either in Prince Albert or in his wife, Queen Victoria (the actual explanation is in Chapter 12).

   (a) Prince Albert did not himself have hemophilia. If the disease is a sex-linked recessive abnormality, how can it have originated in Prince Albert, who is a male and therefore is expected to exhibit recessive sex-linked traits, since he did not suffer from hemophilia?

   (b) Alexis, the son of Czar Nicholas II of Russia and Empress Alexandra (a granddaughter of Victoria), had hemophilia, but their daughter Anastasia did not. Anastasia died, a victim of the Russian revolution, before she had any children. Can we assume that Anastasia would have been a carrier of the disease? How is your answer influenced if the disease originated in Nicholas II or in Alexandra?

HUMAN GENETICS PROBLEMS

1. George has Royal hemophilia and marries his mother's sister's daughter Patricia. His maternal grandfather also had hemophilia. George and Patricia have five children: two daughters are normal, and two sons and one daughter develop hemophilia. Draw the pedigree.

2. A couple with a newborn baby are troubled that the child does not appear to resemble either of them. Suspecting that a mix-up occurred at the hospital, they check the blood type of the infant. It is type O. Since the father is type A and the mother is type B, they conclude that a mistake must have been made. Are they correct?

3. Mabel's sister dies as a child from cystic fibrosis. Mabel herself is healthy, as are her parents. Mabel is pregnant with her first child. If she were to consult you as a genetic counselor, wishing to know the probability that her child will develop cystic fibrosis, what would you tell her?

4. How many chromosomes would one expect to find in the karyotype of a person with Turner's syndrome?

5. A woman is married for the second time. Her first husband was ABO blood type A, and her child by that marriage was type O. Her new husband is type B, and their child is type AB. What is the woman's ABO genotype and blood type?

6. Two bald parents have five children, three of whom eventually become bald and two who do not. Assuming that this trait is governed by a single pair of alleles, is this baldness best explained as an example of dominant or recessive inheritance?

7. In 1986, *National Geographic* magazine conducted a survey of its readers' abilities to detect odors. About 7% of whites in the United States could not smell the odor of musk. If both parents cannot smell musk, then none of their children is able to smell it. On the other hand, two parents who can smell musk generally have children who can also smell it; only a few children in each family are unable to smell it. Assuming that a single pair of alleles governs this trait, is the ability to smell musk best explained as an example of dominant or recessive inheritance?

8. Total colorblindness is a rare hereditary disorder among humans in which no color is seen, only shades of gray. It occurs in individuals homozygous for a recessive allele and is not sex linked. A man whose father is totally colorblind intends to marry a woman whose mother was totally colorblind. What are the chances that they will produce offspring who are totally colorblind?

9. A normally pigmented man marries an albino woman. They have three children, one of whom is an albino. What is the genotype of the father?

10. Four babies are born within a few minutes of each other in a large hospital, when suddenly an explosion occurs. All four babies are found alive among the rubble. None had yet been given identification bracelets. The babies prove to be of four different blood groups: A, B, AB, and O. The four pairs of parents have the following pairs of blood groups: A and B, O and O, AB and O, and B and B. Which babies belong to which parents?

11. This pedigree depicts a rare trait in which children have extra fingers and toes. Which, if any, of the following patterns of inheritance is consistent with this pedigree?
    (a) autosomal recessive
    (b) autosomal dominant
    (c) sex-linked recessive
    (d) sex-linked dominant
    (e) Y linkage

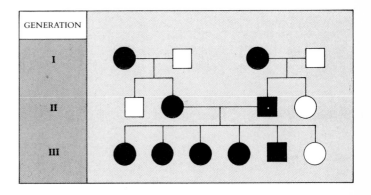

12. This pedigree shows a common form of inherited colorblindness. What pattern of inheritance best accounts for this pedigree?

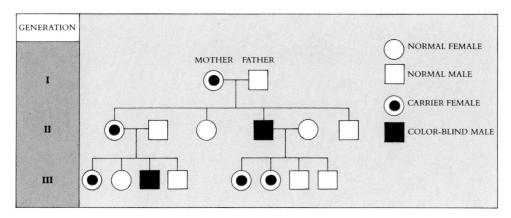

13. Of the 43 people in the five generations of this family, more than one third exhibited an inherited mental disorder. Is the trait dominant or recessive?

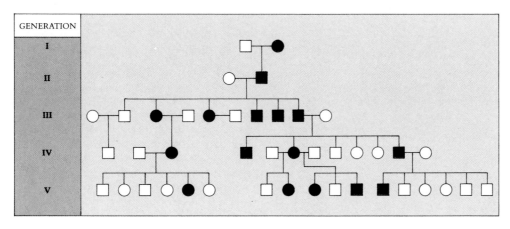

ANSWERS

Chapter 1
REVIEW
1. cellular organization, growth and metabolism, reproduction, homeostasis, and heredity
2. DNA
3. 5 billion
4. about one half
5. five sixths (83%)
6. 1 to 2 million

SELF-QUIZ
1. a, b, and d
2. c
3. c
4. a and d

Chapter 2
REVIEW
1. theory
2. scientific creationism
3. 4.5 billion
4. homologous
5. 4: Protista, Animalia, Plantae, and Fungi

SELF-QUIZ
1. b
2. b
3. a
4. d
5. a

PROBLEMS
1. 512

Chapter 3
REVIEW
1. water
2. a. neutrons, protons, and electrons
   b. neutrons and protons
3. gas
4. a. starch
   b. glycogen
5. DNA

SELF-QUIZ
1. a, c, and d
2. d
3. c
4. a
5. c, d, and e

PROBLEMS
1. a. 18
   b. 18 grams
2. 64

Chapter 4
REVIEW
1. evolution
2. 3.5
3. a. electrical sparks
   b. energetic UV light or heat
4. coacervates
5. 1.5 billion

SELF-QUIZ
1. e
2. c and e
3. c
4. b, c, and d
5. d
6. 10^{19}

Chapter 5
REVIEW
1. cell membrane
2. channels, receptors, and markers
3. a. lysosomes
   b. golgi bodies
4. mitochondria and chloroplasts
5. Nucleus, nucleolus, mitochondria (in plants, chloroplasts)

SELF-QUIZ
1. a
2. a, b, and c
3. d
4. c, d, a, and then b
5. a, b, and d

PROBLEMS
1. 125 cubic micrometers divided by 8000 cubic micrometers = 1.56%

Chapter 6
REVIEW
1. glycerol, fatty acid, alcohol
2. major histocompatibility complex proteins
3. diffusion
4. proton pump
5. hormones

SELF-QUIZ
1. e
2. a
3. d and e
4. d and e
5. e

Chapter 7
REVIEW
1. work
2. no
3. the sun
4. cellular respiration
5. a. enzymes
   b. proteins

SELF-QUIZ
1. a, b, and e
2. a or b
3. c
4. c
5. a

PROBLEMS
1. 1500 kilocalories
2. 686 calories

Chapter 8
REVIEW
1. ATP
2. glycolysis, oxidation of pyruvate, and citric acid cycle
3. pyruvate
4. mitochondrion
5. 36 molecules of ATP

SELF-QUIZ
1. a
2. b
3. e
4. a
5. b

PROBLEMS
1. 2.168×10^{25}

Chapter 9
REVIEW
1. pigments
2. fix carbon
3. $NADP^+$, NAD^+
4. photosynthesis
5. thylakoid

SELF-QUIZ
1. a
2. d
3. b and e
4. b
5. a

Chapter 10
REVIEW
1. fission
2. mitosis
3. metaphase plate
4. synapsis
5. recombinants

SELF-QUIZ
1. a
2. b
3. e
4. e and b
5. b

Chapter 11
REVIEW
1. a. genotype
   b. phenotype
2. w
3. a test cross
4. Law of Independent Assortment
5. crossing-over
6. 3:1
7. 25%
8. crossing-over
9. false
10. true
11. 25%
12. linkage

SELF-QUIZ
1. c
2. b
3. d
4. a

5. c
6. b
7. d
8. c
9. e

Chapter 12
REVIEW
1. a. 46
   b. 22
2. Barr body
3. 5%
4. Klinefelter's syndrome
5. amniocentesis
SELF-QUIZ
1. c
2. a
3. a
4. e
5. c

Chapter 13
REVIEW
1. in the nucleus
2. DNA
3. thymine, cytosine
4. enzymes
5. epistasis
SELF-QUIZ
1. a
2. d
3. a
4. d
5. e
PROBLEMS
1. tube #1

Chapter 14
REVIEW
1. genes
2. translated
3. three
4. a. mRNA
   b. tRNA
5. repressor protein
SELF-QUIZ
1. a, b, and c
2. d
3. e
4. b and d
5. c
PROBLEMS
1. DNA strand #2

Chapter 15
REVIEW
1. mutation
2. mutagen
3. RNA
4. oncogenes
5. deletions, inversions, and
   translocations
SELF-QUIZ
1. c
2. a, b, and e (or d)
3. a
4. a, d, and e
5. e

Chapter 16
REVIEW
1. microevolution
2. genetic drift, selection
3. selection
4. industrial melanism
5. parallel adaptation
SELF-QUIZ
1. b
2. d
3. e
4. e
5. a
PROBLEMS
1. a. 42%
   b. remain the same

Chapter 17
REVIEW
1. species
2. a. microevolution
   b. macroevolution or dif-
      ferent species
3. geographical isolation *now*
   or ecological isolation before 150 years ago
4. a. prezygotic
   b. postzygotic
5. macroevolution
SELF-QUIZ
1. e
2. d
3. a
4. c
5. a

Chapter 18
REVIEW
1. Pangaea
2. phyla
3. 3.5 billion
4. fungi, insects, vertebrates,
   plants
5. Permian
SELF-QUIZ
1. b, e, a, c, d
2. c
3. b
4. a(3), b(1), c(5), d(4), e(2)
5. four (a, b, d, e)

Chapter 19
REVIEW
1. amniotic
2. false
3. fishes
4. *Australopithecus*
5. three
SELF-QUIZ
1. three (b, d, e)
2. c, e, b, a, d
3. c
4. c, d, a, e, b
5. c, b, e, d, a

Chapter 20
REVIEW
1. disruptive, stabilizing,
   directional
2. true

3. more rapidly
4. no
5. *K*
SELF-QUIZ
1. d
2. c
3. b
4. b, c, a, d
5. one (c)
PROBLEMS
1. 4,684

Chapter 21
REVIEW
1. interspecific
2. theoretical
3. secondary
4. brightly
5. symbiotic
SELF-QUIZ
1. chthamalus
2. true
3. b
4. c
5. fungi and green algae
   (cyanobacteria)

Chapter 22
REVIEW
1. nitrogen
2. biomass
3. savannas or chaparral
4. physical environment
5. net primary productivity
SELF-QUIZ
1. b
2. a
3. d
4. b
5. d

Chapter 23
REVIEW
1. faster
2. 20%
3. 5 million
4. grazing, firewood gathering, lumbering
5. 1.7% annually
SELF-QUIZ
1. d
2. c
3. c
4. c
5. b

Chapter 24
REVIEW
1. Carl Linnaeus
2. a. 1.5 million
   b. 5 million
3. metabolically
4. Protista
5. viruses
SELF-QUIZ
1. b, d, g, a, e, f, c
2. A(d), B(e), C(a)
3. a
4. b, c, and d
5. a

PROBLEM
1. at least 2.9 million

Chapter 25
REVIEW
1. no
2. no
3. spirochetes
4. no
5. white blood cells (T4)
SELF-QUIZ
1. a
2. a, c, and d
3. b
4. d
5. d

Chapter 26
REVIEW
1. a. lack of chloroplasts
   b. external digestion
   c. filamentous growth
   d. complete lack of fla-
      gella
2. red algae (phylum Phaeophyta)
3. choanoflagellates (one of
   the groups of zoomastigotes)
4. hyphae
5. cyclosporine
SELF-QUIZ
1. b
2. c
3. d
4. a and c
5. d

Chapter 27
REVIEW
1. sporophyte
2. they swim
3. triploid
4. naked
5. monocots, dicots
SELF-QUIZ
1. e
2. c
3. d
4. b
5. a, b, c, d

Chapter 28
REVIEW
1. a. Arthropoda
   b. Chordata
2. endoderm
3. a. protostomes
   b. deuterostomes
4. Arthropoda
5. Pseudocoelomates
SELF-QUIZ
1. b
2. c
3. e
4. d and e
5. a, b, and c

Chapter 29

REVIEW
1. meristems
2. apical, lateral
3. cuticle
4. mesophyll
5. bark

SELF-QUIZ
1. b, c, and e
2. e
3. e
4. b
5. b

Chapter 30

REVIEW
1. mitosis
2. bees
3. imbibes water
4. gibberellins
5. outcrossing

SELF-QUIZ
1. e
2. e
3. d
4. c
5. a, b, and d

Chapter 31

REVIEW
1. hormones
2. osmosis
3. nitrogen, phosphorus, calcium, potassium, magnesium, and sulfur
4. promote
5. gravitropism

SELF-QUIZ
1. d
2. evaporative suction
3. b
4. a and c
5. b and e

Chapter 32

REVIEW
1. a. thoracic or respiratory cavity
   b. abdominal or digestive cavity
2. mesoderm
3. collagen
4. muscle
5. a. axon
   b. dendrites

SELF-QUIZ
1. d
2. b
3. c

4. e
5. d

Chapter 33

REVIEW
1. a. (chitin) exoskeleton
   b. bones
2. epidermis, dermis, subcutaneous tissue
3. spongy bone tissue or marrow
4. a. actin
   b. myosin
5. calcium ions

SELF-QUIZ
1. b
2. e
3. e
4. b
5. e

PROBLEM
1. about 9 kilograms

Chapter 34

REVIEW
1. digestion
2. pyloric sphincter
3. villi
4. fat
5. K

SELF-QUIZ
1. a
2. a and d
3. d
4. e
5. b

Chapter 35

REVIEW
1. respiration
2. true
3. pharynx
4. water loss
5. hemoglobin

SELF-QUIZ
1. e
2. c
3. d, c, e, a, b
4. d
5. a

Chapter 36

REVIEW
1. blood vessels
2. 100 micrometers
3. serum albumin
4. left atrium, left ventricle, right atrium, right ventricle

5. a heart murmur

SELF-QUIZ
1. d
2. a
3. a
4. c
5. d

Chapter 37

REVIEW
1. vaccination
2. T cells and B cells
3. antigen
4. macrophage
5. E

SELF-QUIZ
1. b
2. a
3. a
4. b
5. b

Chapter 38

REVIEW
1. neurons, hormones
2. into
3. synapse
4. saltatory conduction
5. chemically (neurotransmitters)

SELF-QUIZ
1. d
2. b
3. c
4. e
5. d

PROBLEM
1. 0.04 second

Chapter 39

REVIEW
1. a. hindbrain or rhombencephalon
   b. midbrain or mesencephalon
   c. forebrain or prosencephalon
2. cerebrum
3. a. electrically
   b. structural changes
4. cis-retinal
5. a. adrenaline
   b. noradrenaline

SELF-QUIZ
1. a
2. d
3. e
4. c
5. a and e

Chapter 40

REVIEW
1. hormones
2. hypothalamus
3. thyroid-stimulating hormone
4. hypothalamus
5. insulin

SELF-QUIZ
1. e
2. d
3. a
4. b
5. d

Chapter 41

REVIEW
1. two-thirds
2. kidney
3. a. marine
   b. freshwater
4. inner medulla
5. water

SELF-QUIZ
1. d
2. e
3. c
4. b and e
5. a

Chapter 42

REVIEW
1. false
2. secretory phase
3. sperm production
4. fallopian tubes
5. 21

SELF-QUIZ
1. e
2. b and e
3. d and a
4. c, d, b, a
5. d

Chapter 43

REVIEW
1. instinct
2. sociobiology
3. aggressive
4. dominance
5. we don't know

SELF-QUIZ
1. c
2. e
3. a
4. a
5. d

Appendix B

MENDELIAN GENETICS PROBLEMS

1. He only chose to study two pure-breeding varieties of any given trait; he could have chosen to study more. In any given cross, the maximum number of alleles that he could have observed is four, if the two parents were each heterozygous for a different pair of alleles.

2. Alleles segregate in meiosis, and the products of that segregation are contained *within* a pod. Each pea is a gamete. In this diagram, the segregation is incorrectly shown as being *between* pods, each pod shown as uniformly *wrinkled* or *round*.

3. The probability of getting two genes on the same chromosone is $\frac{1}{2}^{23}$.

4. Somewhere in your herd you have cows and bulls that are not homozygous for the dominant gene "polled." Since you have many cows and probably only one or some small number of bulls, it would make sense to concentrate on the bulls. If you have only homozygous "polled" bulls, you could never produce a horned offspring regardless of the genotype of the mother. The most expedient thing to do would be to keep track of the matings and the phenotype of the offsprings resulting from these matings and render ineffective any bull found to produce horned offspring.

5. It would not be possible on the basis of the information presented to substantiate a claim of infidelity. You do not know if the woolly trait is the result of a single gene product, or even if the trait is dominant or recessive. Assuming for the moment that it was the effect of a single dominant allele W, the man could still be a heterozygote for the gene and, when mated to a recessive homozygous female, would expect to produce woolly headed offspring only one-half the time.

6. $\frac{1}{2}$

7. Albinism, a, is a recessive gene. If heterozygotes mated you would have the following:

	A	a
A	AA	Aa
a	Aa	aa

Clearly one fourth would be expected to be albinos.

8. The best thing to do would be to mate Dingleberry to several dames homozygous for the recessive gene that causes the brittle bones. Half of the offspring would be expected to have brittle bones if Dingleberry were a heterozygous carrier of the disease gene. Although you could never be 100% certain Dingleberry was not a carrier, you could reduce the probability to a reasonable level.

9. Your mating of $DDWw$ and $Ddww$ individuals would look like the following:

	Dw	Dw	dw	dw
DW	$DDWw$	$DDWw$	$DdWw$	$DdWw$
Dw	$DDww$	$DDWw$	$Ddww$	$Ddww$
DW	$DDWw$	$DDWw$	$DdWw$	$DdWw$
Dw	$DDww$	$DDww$	$Ddww$	$Ddww$

Long-wing, red-eyed individuals would result from eight of the possible 16 combinations and dumpy, white-eyed individuals would never be produced.

10. Breed Oscar to Heidi. If half of the offspring are whited eyed, then Oscar is a heterozygote.

11. Both parents carry at least one of the recessive genes. Since it is recessive, the trait is not manifest until they produce an offspring who is homozygous.

12. To solve this problem, let's first look at the second cross, where the individuals were crossed with the homozygous recessive sepia flies se/se. In one case, all the flies were red eyed:

		Unknown genotype	
		Se	Se
Sepia	se	Se/se	Se/se
	se	Se/se	Se/se

The only way to have all red-eyed flies when bred to homozygous sepia flies is to mate the sepia fly with a homozygous red-eyed fly. In the other case, half of the offspring were black eyed and the other half red eyed:

		Unknown genotype	
		Se	se
Sepia	se	Se/se	se/se
	se	Se/se	se/se

The unknown genotype in this case must have been Se/se, since this is the only mating that will produce the proper ratio of sepia-eyed flies to red-eyed flies. Since the ratio of this unknown genotype and the one previously determined was 1:1, we must deduce the genotype of the original flies, which when mated, will produce a 1:1 ratio of Se/se to Se/Se flies.

		Unknown original 1	
		Se	Se
Original 2	Se	Se/Se	Se/Se
	se	Se/se	Se/se

You can see from this diagram that if one of the original flies was homozygous for red eyes and the other was a heterozygous individual, the proper ratio of heterozygous and homozygous offspring would be obtained.

13. a. It could have originated as a mutation in his germ cell line. b. Since their son Alex was a hemophiliac, the disease almost certainly originated with Alexandra, Nicholas II would have contributed only a silent Y chromosome to Alex's genome. There is a 50% chance that Anastasia was a carrier.

HUMAN GENETICS PROBLEMS

1.

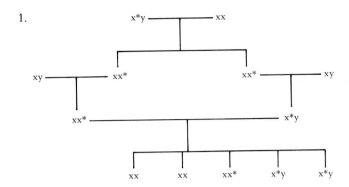

2. No. $I^A I^O \times I^B I^O \to I^O I^O$

3.

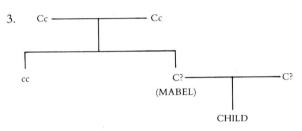

c is a rare allele, and Mabel's child can develop cystic fibrosis only if her husband also carries the allele:

$$\begin{bmatrix} \text{probability that} \\ \text{husband has} \\ \text{c allele} \end{bmatrix} = \text{frequency of c} = 1/20$$

$$\begin{bmatrix} \text{probability that} \\ \text{Mabel has} \\ \text{c allele} \end{bmatrix} = 1/4$$

In addition, the probability of transmission even if both parents carry the allele is only 25%; therefore the overall probability is $\frac{1}{20} \times \frac{1}{4} \times \frac{1}{4} = \frac{1}{320}$

4. 45 (44 autosomes + one X).

5. AO

6. Dominant.

7. Dominant.

8.

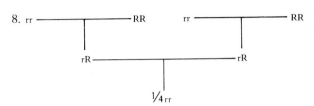

The chances are 25%.

9. Let a = albino. The genotype of the father is Aa.

10.

PARENTS	BABY
O and O	O
A and B	AB
B and B	B
AB and O	A

11. b—autosomal dominant.

12. Sex-linked recessive.

13. Dominant.

GLOSSARY

Page numbers indicate where entries may be found in the text.

abdomen The posterior portion of the body. (p. 548.)

abortion The removal or induced elimination of the embryo from the uterus before the third trimester of pregnancy. (p. 773.)

abscisic acid (ABA) (L. *ab*, away, off + *scisso*, dividing) A plant hormone with a variety of inhibitory effects; brings about dormancy in buds, maintains dormancy in seeds, effects stomatal closing, and is involved in the geotropism of roots; also known as the "stress hormone." (p. 600.)

abscission (L. *ab*, away, off + *scissio*, dividing) In vascular plants, the dropping of leaves, flowers, fruits, or stems at the end of a growing season, as the result of formation of a layer of specialized cells (the abscission zone) and the action of a hormone (ethylene). (p. 600.)

absorption (L. *absorbere*, to swallow down) The movement of water and of substances dissolved in water into a cell, tissue, or organism. (p. 593.)

absorption spectrum The characteristic array of wavelengths (colors) of light that a particular substance absorbs. (p. 167.)

acetylcholine The most important of the numerous chemical neurotransmitters responsible for the passing of nerve impulses across synaptic junctions; the neurotransmitter in neuromuscular (nerve-muscle) junctions. (p. 708.)

acid A proton donor; a substance that dissociates in water, releasing hydrogen ions (H^+) and so causing a relative increase in the concentration of H^+ ions; having a pH in solution of less than 7; the opposite of "base." (p. 47.)

acoelomate (Gr. *a*, not + *koiloma*, cavity) Without a coelom, as in flatworms and proboscis worms. (p. 537.)

acquired immune deficiency syndrome (AIDS) An infectious and usually fatal human disease caused by a retrovirus, HIV, which attacks T cells. The virus multiplies within and kills individual T cells, releasing thousands of progeny that infect and kill other T cells, until no T cells remain, leaving the affected individual helpless in the face of microbial infections because his or her immune system is now incapable of marshaling a defense against them. *See* T cell. (p. 774.)

actin (Gr. *actis*, a ray) One of the two major proteins that make up microfilaments (the other is myosin); the principal protein constituent in contractile tissues containing many microfilaments, such as muscle. (p. 96).

action potential A transient, all-or-none reversal of the electric potential across a membrane; its strength depends only on characteristics of the membrane (in nerves, the diameter) and is independent of the strength of the stimulus that triggers it; in neurons, an action potential initiates transmission of a nerve impulse. (p. 705.)

activation energy The energy that must be possessed by a molecule in order for it to undergo a specific chemical reaction. (p. 134.)

activator A regulatory molecule that is required for the efficient transcription of a gene. (p. 138.)

active site The region of an enzyme surface to which a specific set of substrates binds, lowering the activation energy required for a particular chemical reaction and so facilitating it. (p. 137.)

active transport The pumping of individual ions or other molecules across a cellular membrane from a region of lower concentration to one of higher concentration (that is, against a concentration gradient); because the ion or molecule is moved in a direction other than the one in which simple diffusion would take it, this transport process requires energy, which is typically supplied by the expenditure of ATP. (p. 117.)

adaptation (L. *adaptare*, to fit) A peculiarity of structure, physiology, or behavior that promotes the likelihood of an organism's survival and reproduction in a particular environment. (p. 7.)

adenosine triphosphate (ATP) A nucleotide consisting of adenine, ribose sugar, and three phosphate groups. ATP is the energy currency of cell metabolism in all organisms. ATP is formed from ADP + P in an enzymatic reaction that traps chemical energy released by metabolic processes or light energy captured in photosynthesis. Upon hydrolysis, ATP loses one phosphate and one hydrogen to become adenosine diphosphate (ADP), releasing energy in the process. (p. 6.)

aerobic (Gr. *aer*, air + *bios*, life) Requiring free oxygen; any biological process that can occur in the presence of gaseous oxygen (O_2). (p. 93.)

aerobic pathway A metabolic pathway at least one step of which is an oxidation/reduction reaction that depends on oxygen gas as an electron acceptor; such a pathway will not proceed in the absence of oxygen. (p. 152.)

afferent (L. *ad*, to + *ferre*, to carry) Adjective meaning leading inward or bearing toward some organ, for example, nerves conducting impulses toward the brain, or blood vessels carrying blood toward the heart; opposite of "efferent." (p. 713.)

allantois (Gr. *allas*, sausage + *eidos*, form) A membrane of the amniotic egg that functions in respiration and excretion in birds and reptiles and plays an important role in the development of the placenta in most mammals. (p. 37.)

allele (Gr. *allelon*, of one another) One of two or more alternative states of a gene. (p. 215.)

allele frequency The relative proportion of a particular allele among the chromosomes carried by individuals of a population. Not equivalent to "gene frequency," although the two are sometimes confused. (p. 309.)

allergy An inappropriately severe response by IGE antibodies to a harmless antigen, one that does not invoke a response in most people. (p. 694.)

allosteric interaction (Gr. *allos*, other + *stereos*, shape) A change in the shape of a protein resulting from the binding to the protein of a nonsubstrate molecule, called an effector. The new shape typically has different properties, becoming activated or inhibited. Cells control the activities of enzymes, transcription regulators, and other proteins by modulating the cellular concentration of particular allosteric effectors. (p. 138.)

alpha helix (Gr. *alpha*, first + L. *helix*, spiral) A right-handed coil; in DNA, the usual form of biological DNA molecules (the one originally proposed in 1961 by Watson and Crick), although left-handed forms (called "Z DNA") also occur; in proteins, a secondary structure composed of a regular right-handed coil. (p. 60.)

alternation of generations A reproductive cycle in which a haploid ($1n$) phase, the gametophyte, gives rise to gametes which, after fusion to form a zygote, germinate to produce a diploid ($2n$) phase, the sporophyte. Spores produced by meiotic division from the sporophyte give rise to new gametophytes, completing the cycle. (p. 479.)

altruism Self-sacrifice for the benefit of others; in formal terms, a behavior that increases the fitness of the recipient while reducing the fitness of the altruistic individual. (p. 794.)

alveolus, pl. **alveoli** (L. a small cavity) One of the many small, thin-walled air sacs within the lungs in which the bronchioles terminate. (p. 658.)

amino acids (Gr. *Ammon*, referring to the Egyptian sun god, near whose temple ammonium salts were first prepared from camel dung) Nitrogen-containing organic molecules. Amino acids are the units, or "building blocks," from which protein molecules are assembled; all amino acids have the same underlying structure, $H_2N-RCH-COOH$, where R stands for a side group that is different for each kind of amino acid; 20 different amino acids combine to make proteins. (p. 55.)

amniocentesis (Gr. *amnion*, membrane around the fetus + *centes*, puncture) Examination of a fetus indirectly, by the carrying out of tests on cell cultures grown from fetal cells obtained from a sample of the amniotic fluid surrounding the developing embryo; a procedure often carried out in pregnant women over the age of 35, since the examination of fetal chromosomes readily reveals Down syndrome if it is present, permitting the choice of a therapeutic abortion. (p. 238.)

amnion (Gr., membrane around the fetus) The innermost of the extraembryonic membranes; the amnion forms a fluid-filled sac around the embryo in amniotic eggs. (p. 369.)

amniotic egg An egg that is isolated and protected from the environment by a more or less impervious shell during the period of its development and that is completely self-sufficient, requiring only oxygen from the outside. (p. 369.)

anabolism (Gr. *ana*, up + *bolein*, to throw) The biosynthetic or constructive part of metabolism; those chemical reactions involved in biosynthesis; opposite of "catabolism." (p. 150.)

anaerobic (Gr. *an*, without + *aer*, air + *bios*, life) Any process that can occur without oxygen, such as anaerobic fermentation or H₂S photosynthesis; anaerobic organisms can live without free oxygen; obligate anaerobes cannot live in the presence of oxygen; facultative anaerobes can live with or without oxygen. (p. 152.)

aneuploidy (Gr. *an*, without + *eu*, good + *ploid*, multiple of) An organism whose cells have lost or gained a chromosome; cells that have a chromosome number of $2n - 1$ (one fewer than normal) or $2n + 1$ (one extra chromosome). Down syndrome, which results from an extra copy of human chromosome 21, is an example of aneuploidy in humans. (p. 291.)

anterior (L. *ante*, before) Located before or towards the front; in animals, the head end of an organism. (p. 536.)

anther (Gr. *anthos*, flower) In angiosperm flowers, the pollen-bearing portion of a stamen. (p. 527.)

antheridium, pl. **antheridia** A sperm-producing organ; antheridia occur in some plants and protists. (p. 517.)

antibody (Gr. *anti*, against) A protein called an immunoglobulin that is produced by B-lymphocytes in response to a foreign substance (antigen) and released into the bloodstream. Binding

to the antigen, antibodies mark them for distruction by other elements of the immune system. (p. 12.)

anticodon The three-nucleotide sequence at the end of a transfer RNA molecule that is complementary to, and base pairs with, an amino-acid–specifying codon in messenger RNA. (p. 268.)

antigen (Gr. *anti*, against + *genos*, origin) A foreign substance, usually a protein or polysaccharide, that stimulates one or more lymphocyte cells to begin to proliferate and secrete specific antibodies that bind to the foreign substances, labeling it as foreign and destined for destruction. (p. 687.)

aorta (Gr. *aeirein*, to lift) The major artery of vertebrate systemic blood circulation; in mammals, carries oxygenated blood away from the heart to all regions of the body except the lungs. (p. 676.)

apical meristem (L. *apex*, top + Gr. *meristos*, divided) In vascular plants, the growing point at the tip of the root or stem. (p. 563.)

archegonium, pl. **archegonia** (Gr. *archegonos*, first of a race) In bryophytes and some vascular plants, the multicellular egg-producing organ. (p. 517.)

ascus pl. **asci** (Gr. *askos*, wine-skin, bladder) A specialized cell, characteristic of the ascomycetes, in which two haploid nuclei fuse to produce a zygote that divides immediately by meiosis; at maturity, an ascus contains ascospores. (p. 507.)

asexual Without distinct sexual organs; an organism that reproduces without forming gametes—that is, without sex. Asexual reproduction, therefore, does not involve sex. (p. 470.)

atmospheric pressure (Gr. *atmos*, vapor + *sphaira*, globe) The weight of the earth's atmosphere over a unit area of the earth's surface; measured with a mercury barometer at sea level, this corresponds to the pressure required to lift a column of mercury 760 millimeters. (p. 592.)

atom (Gr. *atomos*, indivisible) The smallest particle into which a chemical element can be divided and still retain the properties characteristic of the element; consists of a central core or nucleus composed of protons and neutrons, encircled by one or more electrons that move around the nucleus in characteristic orbits whose distance from the nucleus depends on their energy. (p. 37.)

atomic mass The weight of a representative atom of an element (estimated as the average weight of all its isotopes) relative to the weight of an atom of the

most common isotope of carbon (which is by convention assigned the integral value of 12); the atomic weight of an atom is approximately equal to the number of protons plus neutrons in its nucleus. (p. 37.)

atomic number The number of protons in the nucleus of an atom; in an atom that does not bear an electric charge (i.e., one that is not an ion), the atomic number is also equal to the number of electrons. (p. 37.)

ATP *See* adenosine triphosphate (ATP).

autonomic nervous system (Gr. *autos*, self + *nomos*, law) The involuntary neurons and ganglia of the peripheral nervous system of vertebrates; regulates the heart, glands, visceral organs, and smooth muscle; subdivided into the sympathetic and parasympathetic divisions, whose effects oppose one another (one stimulating, the other inhibiting). (p. 714.)

autosome (Gr. *autos*, self + *soma*, body) Any eukaryotic chromosome that is not a sex chromosome; autosomes are present in the same number and kind in both males and females of the species. (p. 224.)

autotroph (Gr. *autos*, self + *trophos*, feeder) An organism able to build all the complex organic molecules that it requires as its own food source, using only simple inorganic compounds. Plants, algae, and some bacteria are autotrophs. Contrasts with *heterotroph*. (p. 145.)

auxin (Gr. *auxein*, to increase) A plant hormone that controls cell elongation, among other effects. (p. 598.)

axon (Gr. axle) A process extending out from a neuron that conducts impulses away from the cell body. (p. 616.)

B cell A type of lymphocyte which, when confronted with a suitable antigen, is capable of secreting a specific antibody protein; each individual lymphocyte B cell is capable of secreting only one form of antibody, but few if any B lymphocytes secrete the same form of antibody. (p. 687.)

backcross The crossing of a hybrid individual with one of its parents or with a genetically equivalent individual; the most common type of backcross is a "testcross," a cross between an individual with a dominant phenotype (which might in principle be either homozygous for a dominant allele or heterozygous for that allele and a recessive one) with an individual that is homozygous for the recessive allele. (p. 212.)

bacteriophage (Gr. *bakterion*, little rod + *phagein*, to eat) A virus that infects bacterial cells; also called a *phage*. (p. 247.)

basal body A self-reproducing cylinder-shaped cytoplasmic organelle composed of nine triplets of microtubules from which the flagella or cilia arise; identical in structure to the centriole, which is involved in mitosis and meiosis in most protists and animals and in some plants. (p. 100.)

base A substance that dissociates in water, causing a decrease in the relative number of hydrogen ions (H⁺), often by producing hydroxide ions (OH⁻), which combine with hydrogen ions to form water and thus reduce the concentration of free hydrogen ions; having a pH greater than 7; the opposite of acid. *See* alkaline. (p. 48.)

behavior A coordinated neuro-motor response to changes in external or internal conditions; a product of the integration of sensory, neural, and hormonal factors. (p. 785.)

biennial A plant that normally requires two growing seasons to complete its life cycle. Biennials flower in the second year of their lives. (p. 564.)

bilateral symmetry (L. *bi*, two + *lateris*, side; Gr. *symmetria*, symmetry) A body form in which the right and left halves of an organism are approximate mirror images of each other. (p. 536.)

bile A solution of organic salts that is secreted by the vertebrate liver and temporarily stored in the gallbladder; emulsifies fats in the small intestine. (p. 642.)

binary fission (L. *binarius*, consisting of two things or parts + *fissus*, split) Asexual reproduction by division of one cell or body into two equal, or nearly equal, parts. (p. 186.)

binomial system (L. *bi*, twice, two + Gr. *nomos*, usage, law) A system of nomenclature in which the name of a species consists of two parts, the first of which designates the genus. (p. 469.)

biomass (Gr. *bios*, life + *maza*, lump or mass) The weight of all the living organisms in a given population, area, or other unit being measured. (p. 424.)

biome One of the major terrestrial ecosystems, characterized by climatic and soil conditions; the largest ecological unit. (p. 29.)

blastopore (Gr. *blastos*, sprout + *poros*, a path or passage) In vertebrate development, the opening that connects the archenteron cavity of a gastrula stage embryo with the outside; represents the future mouth in some animals (protostomes), the future anus in others (deuterostomes). (p. 532.)

blastula (Gr. a little sprout) In vertebrates, an early embryonic stage consisting of a hollow, fluid-filled ball of cells one layer thick; a vertebrate embryo after cleavage and before gastrulation. (p. 532.)

blood type In humans, the type of cell surface antigens present on the red blood cells of an individual; genetically determined, alternative alleles yield different surface antigens. When two different blood types are mixed, the cell surfaces often interact, leading to agglutination. One genetic locus encodes the ABO blood group, another the Rh blood group, and still others encode other surface antigens. (p. 673.)

bond strength The amount of energy required to break a chemical bond; conventionally measured in kilocalories per mole. (p. 41.)

bronchus, pl. **bronchi** (Gr. *bronchos*, windpipe) One of a pair of respiratory tubes branching from the lower end of the trachea (windpipe) into either lung; the respiratory path further subdivides into progressively finer passageways, the bronchioles, culminating in the alveoli. (p. 660.)

bud An asexually produced outgrowth that develops into a new individual. In plants, an embryonic shoot, often protected by young leaves; buds may give rise to branch shoots. (p. 569.)

callus (L. *callos*, hard skin) Undifferentiated tissue; a term used in tissue culture, grafting, and wound healing. (p. 584.)

calorie (L. *calor*, heat) The amount of energy in the form of heat required to raise the temperature of 1 gram of water 1° C. (p. 130.)

Calvin cycle The dark reactions of photosynthesis; a series of enzymatically mediated photosynthetic reactions in which carbon dioxide is bound to ribulose 1,5-diphosphate and then reduced to 3-phosphoglyceraldehyde, and the carbon-dioxide–accepting ribulose 1,5-diphosphate is regenerated. For every six molecules of carbon dioxide entering the cycle, a net gain of two molecules of glyceraldehyde 3-phosphate results. Also called the Calvin-Benson cycle. (p. 176.)

calyx (Gr. *kalyx*, a husk, cup) The sepals collectively; the outermost flower whorl. (p. 526.)

cambium, pl. **cambia** In vascular plants, embryonic tissue zones (meristems) that run parallel to the sides of roots and stems; consists of the vascular cambium and the cork cambium. (p. 570.)

cancer Unrestrained invasive cell growth; a tumor or cell mass resulting from uncontrollable cell division; a result of mutational damage that destroys the normal control of cell division. A cancer cell is said to be "malignant," and in many (although by no means all) tissues, malignant cells can migrate to other body regions (metastasize), where they form secondary tumors. (p. 10.)

capillaries (L. *capillaris*, hair-like) The smallest blood vessels; the very thin walls of capillaries are permeable to many molecules, and exchanges between blood and the tissues occur across them; the vessels that connect arteries with veins. (p. 668.)

capillary action The movement of a liquid along a surface as a result of the combined effects of cohesion and adhesion. (p. 46.)

carbohydrate (L. *carbo*, charcoal + *hydro*, water) An organic compound consisting of a chain or ring of carbon atoms to which hydrogen and oxygen atoms are attached in a ratio of approximately 2:1; a compound of carbon, hydrogen, and oxygen having the generalized formula $(CH_2O)^n$; carbohydrates include sugars, starch, glycogen, and cellulose. (p. 49.)

carbon cycle The worldwide circulation and reutilization of carbon atoms. (p. 421.)

carbon fixation The conversion of CO_2 into organic compounds during photosynthesis; the first stage of the dark reactions of photosynthesis, in which carbon dioxide from the air is combined with ribulose 1,5-diphosphate. (p. 170.)

carboxyl (*carbon* + *oxygen* + *-yl*, a chemical suffix) The acid group of organic molecules, —COOH. (p. 48.)

carcinogen Any agent capable of inducing cancer; because cancer occurs as a result of mutation, most carcinogens are also potent mutagens. Programs that screen chemicals to detect potential carcinogens employ tests which measure the propensity of a chemical to cause mutations. (p. 287.)

cardiovascular system The blood circulatory system and the heart that pumps it; collectively, the blood, heart, and blood vessels. (p. 668.)

carnivore (L. *carnis*, flesh + *voro*, to devour) Any organism that eats animals; a meat eater, as opposed to a plant eater, or herbivore. (p. 406.)

carpel (Gr. *karpos*, fruit) A leaf-like organ in angiosperms that encloses one or more ovules; one of the members of the gynoecium. (p. 525.)

cartilage (L. *cartilago*, gristle) A connective tissue in skeletons of vertebrates. Cartilage forms much of the skeleton of embryos, very young vertebrates, and some adult vertebrates, such as sharks and their relatives. In most adult vertebrates much of it is converted into bone. (p. 612.)

catabolism (Gr. *katabole*, throwing down) In a cell, those metabolic reactions that result in the breakdown of complex molecules into simpler compounds, often with the release of energy. (p. 150.)

catalyst (Gr. *kata*, down + *lysis*, a loosening) A substance that accelerates the rate of a chemical reaction by lowering the activation energy, without being used up in the reaction; enzymes are the catalysts of cells. (p. 55.)

cell (L. *cella*, a chamber or small room) The structural unit of organisms; the smallest unit that can be considered living; a cell consists of cytoplasm encased within a membrane. (p. 79.)

cell plate The structure that forms at the equator of the spindle during early telophase in the dividing cells of plants and a few green algae. The cell plate becomes the middle lamella during the course of development. (p. 194.)

cell wall The rigid, outermost layer of the cells of plants, some protists, and most bacteria; the cell wall surrounds the cell (plasma) membrane. (p. 83.)

cellular respiration The metabolic harvesting of energy by oxidation; carried out by the citric acid cycle and oxidative phosphorylation, in which considerable energy is extracted from sugar molecule fragments left over from glycolysis. (p. 134.)

cellulose (L. *cellula*, a little cell) The chief constituent of the cell wall in all green plants, some algae, and a few other organisms. Cellulose is an insoluble complex carbohydrate $(C_6H_{10}O_5)^n$ formed of microfibrils of glucose molecules. (p. 49.)

central nervous system That portion of the nervous system where most association occurs; in vertebrates, it is composed of the brain and spinal cord; in invertebrates it usually consists of one or more cords of nervous tissue, together with their associated ganglia. (p. 616.)

centriole (Gr. *kentron*, center of a circle + L. *olus*, little one) A cytoplasmic organelle located outside the nuclear membrane, identical in structure to a basal body; found in animal cells and in the flagellated cells of other groups. The centriole divides and organizes spindle fibers during mitosis and meiosis. (p. 190.)

centromere (Gr. *kentron*, center + *meros*, a part) That position on a eukaryotic chromosome to which the spindle fibers are attached during cell division; composed of highly repeated DNA sequences (satellite DNA). Also called the kinetochore. (p. 188.)

cerebellum (L. little brain) The hindbrain region of the vertebrate brain, which lies above the medulla (brainstem) and behind the forebrain; it integrates information about body position and motion, coordinates muscular activities, and maintains equilibrium. (p. 716.)

cerebral cortex The thin surface layer of neurons and glial cells covering the cerebrum; well developed only in mammals, and particularly prominent in humans. The cerebral cortex is the seat of conscious sensations and voluntary muscular activity. (p. 717.)

cerebrum (L. brain) The portion of the vertebrate brain (the forebrain) that occupies the upper part of the skull, consisting of two cerebral hemispheres united by the corpus callosum. The cerebrum, which consists of paired hemispheres occupying the upper part of the skull overlying the thalamus, hypothalamus, and pituitary, is the primary association center of the brain. It coordinates and processes sensory input and coordinates motor responses. (p. 718.)

chemiosmosis The mechanism by which ATP is generated in mitochondria and chloroplasts. Energetic electrons excited by light (chloroplasts) or extracted by oxidation in the citric acid cycle (mitochondria) are used to drive proton pumps, creating a higher external proton concentration; when protons subsequently flow back in, it is through channels that couple their passage to the synthesis of ATP. (p. 147.)

chemoautotroph An autotrophic bacterium that uses chemical energy released by specific inorganic reactions to power its life processes, including the synthesis of organic molecules. (p. 485.)

chemoreceptor A sensory cell or organ that responds to the presence of a specific chemical stimulus by initiating a nerve impulse; includes taste and smell receptors. (p. 724.)

chiasma, pl. **chiasmata** (Gr. a cross) During meiosis, the region of contact between homologous chromatids where crossing-over has occurred during synapsis; a chiasma has the appearance of the letter X. (p. 198.)

chitin (Gr. *chiton*, tunic) A tough, resistant, nitrogen-containing polysaccharide that forms the cell walls of certain fungi, the exoskeleton of arthropods, and the epidermal cuticle of other surface structures of certain other invertebrates. (p. 53.)

chloroplast (Gr. *chloros*, green + *plastos*, molded) A cell-like organelle present in algae and plants that contains chlorophyll (and usually other pigments) and carries out photosynthesis. (p. 78.)

chromatid (Gr. *chroma*, color + L. *-id*, daughters of) One of the two daughter strands of a duplicated chromosome which are joined by a single centromere; separates and becomes a daughter chromosome at anaphase of mitosis or anaphase of the second meiotic division. (p. 190.)

chromatin (Gr. *chroma*, color) The complex of DNA and proteins of which eukaryotic chromosomes are composed. (p. 188.)

chromosome (Gr. *chroma*, color + *soma*, body) the vehicle by which hereditary information is physically transmitted from one generation to the next; the organelle that carries the genes. In bacteria, the chromosomes consist of a single naked circle of DNA; in eukaryotes, they consist of a single linear DNA molecule and associated proteins. (p. 84.)

cilium, pl. **cilia** (L. eyelash) A short, hairlike flagellum, especially used to refer to such structures when they are numerous. Cilia may be used in locomotion; in animals, they aid the movement of substances across surfaces. (p. 100.)

cisterna, pl. **cisternae** (L. a reservoir) In cells, a flattened or saclike space between membranes of the endoplasmic reticulum or Golgi body. (p. 91.)

citric acid cycle The cyclic series of reactions in which pyruvate, the product of glycolysis, enters the cycle to form citric acid and is then oxidized to carbon dioxide. Also called the Krebs cycle (after its discoverer) and the tricarboxylic acid (TCA) cycle (citric acid possesses three carboxyl groups). (p. 148.)

class A taxonomic category between phyla (divisions) and orders. A class contains one or more orders, and belongs to a particular phylum or division. (p. 361.)

climax community In ecology, the final stage in a successional series; the climax stage is determined primarily by the climate and soil type of the area. (p. 428.)

cloaca (L. sewer) In some animals, the common exit chamber from the digestive, reproductive, and urinary systems; in others, the cloaca may also serve as a respiratory duct. (p. 638.)

clone (Gr. *klon*, twig) A line of cells, all of which have arisen from the same single cell by mitotic division; one of a population of individuals derived by asexual reproduction from a single ancestor; one of a population of genetically identical individuals. (p. 339.)

cloning Producing a cell line or culture all of whose members contain identical copies of a particular nucleotide sequence; an essential element in genetic engineering, cloning is usually carried out by inserting the desired gene into a virus or plasmid, infecting a cell culture with the hybrid virus, and selecting for culture a cell that has taken up the gene. (p. 294.)

coacervate (L. *coacervatus*, heaped up) A spherical aggregation of lipid molecules in water, held together by hydrophobic forces. (p. 69.)

coccus, pl. **cocci** (Gr. *kokkos*, a berry) A spherical bacterium. (p. 484).

cochlea (Gr. *kochlios*, a snail) In terrestrial vertebrates, a tubular cavity of the inner ear containing the essential organs of hearing; occurs in crocodiles, birds, and mammals; spirally coiled in mammals. (p. 725.)

codominance In genetics, a situation in which the effects of both alleles at a particular locus are apparent in the phenotype of the heterozygote. (p. 229.)

codon (L. code) The basic unit ("letter") of the genetic code; a sequence of three adjacent nucleotides in DNA or mRNA that code for one amino acid, or for polypeptide chain termination. *See* anticodon. (p. 266.)

coelom (Gr. *koilos*, a hollow) A body cavity formed between layers of mesoderm and in which the digestive tract and other internal organs are suspended. (p. 537.)

coenzyme (L. *co-*, together + Gr. *en*, in + *zyme*, leaven) A nonprotein organic molecule that plays an accessory role in enzyme-catalyzed processes, often by acting as a donor or acceptor of electrons. NAD$^+$, FAD, and coenzyme A are comon cozymes. (p. 139.)

coevolution (L. *co-*, together + *e-*, out + *volvere*, to fill) The simultaneous development of adaptations in two or more populations, species, or other categories that interact so closely that each is a strong selective force on the other. (p. 407).

cofactor One or more nonprotein components required by enzymes in order to function; many cofactors are metal ions, others are coenzymes. *See* coenzyme. (p. 139.)

commensalism (L. *cum*, together with + *mensa*, table) A relationship in which one individual lives close to or on another and benefits, and the host is unaffected; a kind of symbiosis. (p. 412.)

community (L. *communitas*, community, fellowship) All of the organisms inhabiting a common environment and interacting with one another. (p. 29.)

competition Interaction between members of the same population or of two or more populations in order to obtain a mutually required resource available in limited supply; in competition, one species interferes with another enough to keep it from gaining access to the resource. (p. 406.)

competitive exclusion The hypothesis that two species with identical ecological requirements cannot exist in the same locality indefinitely, and that the more efficient of the two in utilizing the available scarce resources will exclude the other; also known as Gause's principle, after the Russian biologist G.F. Gause. (p. 406.)

complement A group of eleven proteins that act in concert to destroy foreign cells during the immune response. (p. 693.)

concentration gradient The concentration difference of a substance as a function of distance; in a cell, a greater concentration of its molecules in one region than in another. (p. 112.)

conditioning A form of learning in which a behavioral response becomes associated (by means of a reinforcing stimulus) with a new stimulus not previously capable of invoking the response. (p. 787.)

conjugation (L. *conjugare*, to yoke together) Temporary union of two unicellular organisms, during which genetic material is transferred from one cell to the other; occurs in bacteria, protists, and certain algae and fungi. (p. 501.)

connective tissues A collection of vertebrate tissues derived from the mesoderm. Some kinds of connective tissue (lymphocytes and macrophages) are mobile hunters of invading bacteria, others secrete a matrix (cartilage and bone) that provides the body with structural support, and still others provide sites of fat storage (adipose tissue) or a site for hemoglobin (red blood cells). (p. 611.)

consumer In ecology, a heterotroph that derives its energy from living or freshly killed organisms or parts thereof. Primary consumers are herbivores; higher-level consumers are carnivores. (p. 424.)

continuous variation Variation in traits to which many different genes make a contribution; in such traits, a gradation of small differences is observed; often exhibiting a "normal" or bell-shaped distribution. (p. 258.)

corolla (L. *cornea*, crown) The petals, collectively; usually the conspicuously colored flower whorl. Petals may be free or fused with one another or with the members of other floral whorls. (p. 526.)

corpus luteum (L. yellowish body) A structure that develops from a ruptured follicle in the ovary after ovulation; the corpus luteum secretes estrogens and progesterone, which maintain the uterus during pregnancy. (p. 765.)

cortex (L. bark) The outer layer of a structure; in animals, the outer, as opposed to the inner, part of an organ, as in the adrenal, kidney, and cerebral cortexes; in vascular plants, the primary ground tissue of a stem or root, bounded externally by the epidermis and internally by the central cylinder of vascular tissue. (p. 569.)

cotyledon (Gr. *kotyledon*, a cup-shaped hollow) Seed leaf; cotyledons generally store food in dicotyledons and absorb it in monocotyledons. The food is used during the course of seed germination. (p. 526.)

countercurrent exchange In organisms, the passage of heat or of molecules (such as oxygen, water, or sodium ions) from one circulation path to another moving in the opposite direction; because the flow of the two paths is in opposite directions, there is always a concentration difference between the two channels, facilitating transfer. (p. 657.)

covalent bond (L. *co-*, together + *valere*, to be strong) A chemical bond formed between atoms as a result of the sharing of one or more pairs of electrons. (p. 41.)

crista, pl. **cristae** (L. crest) In mitochondria, the enfoldings of the inner mitochondrial membrane, which form a series of "shelves" containing the electron-transport chains involved in ATP formation. (p. 93.)

crossing-over In meiosis, the exchange of corresponding chromatid segments between homologous chromosomes; responsible for genetic recombination between homologous chromosomes. (p. 197.)

cytochromes (Gr. *kytos*, hollow vessel + *chroma*, color) Any of several iron-containing protein pigments that serve as electron carriers in transport chains of photosynthesis and cellular respiration. (p. 156.)

cytokinesis (Gr. *kytos*, hollow vessel + *kinesis*, movement) Division of the cytoplasm of a cell after nuclear division. (p. 599.)

cytokinin (Gr. *kytos*, hollow vessel + *kinesis*, motion) A class of plant hormones that promote cell division, among other effects. (p. 599.)

cytoplasm (Gr. *kytos*, hollow vessel + *plasma*, anything molded) The living matter within a cell, excluding the nucleus; the protoplasm. (p. 78.)

cytoskeleton A network of protein microfilaments and microtubules within the cytoplasm of a eukaryotic cell that maintains the shape of the cell, anchors its organelles, and is involved in animal cell motility. (p. 95.)

deciduous (L. *decidere*, to fall off) In vascular plants, shedding all the leaves at a certain season. (p. 437.)

decomposers Organisms (bacteria, fungi, heterotrophic protists) that break down organic material into smaller molecules, which are then recirculated. (p. 424.)

demography (Gr. *demos*, people + *graphein*, to draw) The properties of the rate of growth and the age structure of populations. (p. 398.)

dendrite (Gr. *dendron*, tree) A process extending from the cell body of a neuron, typically branched, that conducts impulses inward toward the cell body; although they may be long, most dendrites are short, and a single neuron may possess many of them. (p. 616.)

deoxyribonucleic acid (DNA) The genetic material of all organisms; composed of two complementary chains of nucleotides wound in a double helix; local unwinding of the helix by disruption of hydrogen bonds between strands permits RNA polymerase molecules to transcribe mRNA copies of genes, and permits DNA polymerase molecules to replicate copies of the duplex molecule. *See* alpha helix. (p. 6.)

detritivores (L. *detritus*, worn down + *vorare*, to devour) Organisms that live on dead organic matter; included are large

scavengers, smaller animals such as earthworms and some insects, and decomposers (fungi and bacteria). (p. 424.)

deuterostome (Gr. *deuteros*, second + *stoma*, mouth) An animal in whose embryonic development the anus forms at or near the blastopore and the mouth forms secondarily elsewhere. Deuterostomes are also characterized by radial cleavage during the earliest stages of development and by enterocoelous formation of the coelom. *See* protostome. (p. 540.)

diaphragm (Gr. *diaphrassein*, to barricade) (1) In mammals, a sheet of muscle tissue that separates the abdominal and thoracic cavities and functions in breathing. (2) A contraceptive device used to block the entrance to the uterus temporarily and thus prevent sperm from entering during sexual intercourse. (p. 660.)

dicot Short for dicotyledon; a class of flowering plants generally characterized as having two cotyledons, net-veined leaves, and flower parts usually in fours or fives. (p. 526.)

differentially permeable membrane A membrane through which some substances can diffuse and others cannot. (p. 116.)

differentiation A developmental process by which a relatively unspecialized cell undergoes a progressive change to a more specialized form or function; differentiation in plants may be reversed under suitable conditions, but it is rarely reversible in animals. (p. 580.)

diffusion (L. *diffundere*, to pour out) The net movement of dissolved molecules or other particles from a region where they are more concentrated to a region where they are less concentrated, as a result of the random movement of individual molecules; the process tends to distribute molecules uniformly. (p. 112.)

digestion (L. *digestio*, separating out, dividing) The breakdown of complex, usually insoluble foods into molecules that can be absorbed into cells and there degraded to yield energy and the raw materials for synthetic processes. (p. 638.)

dihybrid (Gr. *dis*, twice + L. *hibrida*, mixed offspring) An individual heterozygous at two different loci; for example, A/a B/b. (p. 213.)

dioecious (Gr. *di*, two + *oikos*, house) Having the male and female elements on different individuals. (p. 585.)

diploid (Gr. *diploos*, double + *eidos*, form) Having two sets of chromosomes (2*n*); in animals,

twice the number characteristic of gametes; in plants, the chromosome number characteristic of the sporophyte generation; in contrast to haploid (1*n*). (p. 190.)

disaccharide A carbohydrate formed of two simple sugar molecules bonded covalently; sucrose is a disaccharide composed of two glucose molecules linked together. (p. 51.)

divergence Increasing separation. Species that become progressively more different from one another as the result of each accumulating a different set of DNA mutations are said to diverge. (p. 331.)

division A major taxonomic group; kingdoms are divided into divisions (or phyla, which are equivalent), and divisions are divided into classes. (p. 472.)

DNA *See* deoxyribonucleic acid.

dominant allele One allele is said to be dominant with respect to an alternative allele if a heterozygous individual bearing the two alleles is indistinguishable from an individual homozygous for the dominant allele; the alternative not detected in the heterozygote is said to be recessive with respect to the dominant allele. An allele that is dominant with respect to one alternative allele may be recessive with respect to another. (p. 215.)

dormancy (L. *dormire*, to sleep) A period during which growth ceases and is resumed only if certain requirements, as of temperature or day length, have been fulfilled. (p. 602.)

dorsal (L. *dorsum*, the back) Toward the back, or upper surface; opposite of ventral. (p. 536.)

double fertilization The fusion of the egg and sperm (resulting in a 2*n* fertilized egg, the zygote) and the simultaneous fusion of the second male gamete with the polar nuclei (resulting in a primary endosperm nucleus, which is often triploid, 3*n*); a unique characteristic of all angiosperms. (p. 525.)

ecdysis (Gr. *ekdysis*, stripping off) Shedding of outer cuticular layer; molting, as in insects or crustaceans. (p. 545.)

ecdysone (Gr. *ekdysis*, stripping off) Molting hormone of arthropods, which stimulates growth and ecdysis. (p. 545.)

ecology (Gr. *oikos*, house + *logos*, word) The study of the interactions of organisms with one another and with their physical environment. (p. 387.)

ecosystem (Gr. *oikos*, house + *systema*, that which is put together) A major interacting system that involves both or-

ganisms and their nonliving environment. (p. 419.)

ecotype (Gr. *oikos*, house + L. *typus*, image) A locally adapted variant of an organism, differing genetically from other ecotypes. (p. 339.)

ectotherm (Gr. *ectos*, outside + *therme*, heat) An organism, such as a reptile, that regulates its body temperature by taking in heat from the environment or giving it off to the environment; contrasts with endothermic; equivalent to *poikilotherm*. (p. 370.)

edaphic (Gr. *adaphos*, ground, soil) Pertaining to the soil. (p. 399.)

efferent (L. *ex*, out of + *ferre*, to bear) Leading or conveying away from some origin—for example, nerve impulses conducted away from the brain, or blood conveyed away from the heart; contrasts with afferent. (p. 714.)

electron A subatomic particle with a negative electric charge equal in magnitude to the positive charge of the proton but with a much smaller mass; electrons orbit the atom's positively charged nucleus and determine its chemical properties. *See* atom. (p. 37.)

electron transport The passage of energetic electrons through a series of membrane-associated electron-carrier molecules to proton pumps embedded within mitochondrial or chloroplast membranes; as the electrons arrive at the proton pumping channel, their energy drives the transport of protons out across the membrane, leading to the chemiosmotic synthesis of ATP. *See* chemiosmosis. (p. 156.)

element A substance composed only of atoms of the same atomic number, which cannot be decomposed by ordinary chemical means; one of more than 100 distinct natural or synthetic types of matter that, singly or in combination, compose all materials of the universe. (p. 42.)

embryo (Gr. *en*, in + *bryein*, to swell) The early developmental stage of an organism produced from a fertilized egg; in plants, a young sporophyte, before its initial period of rapid growth; in animals, a young organism before it emerges from the egg, or from the body of its mother; in humans, refers to the first 2 months of intrauterine life. (p. 580.)

embryo sac The female gametophyte of angiosperms, generally an eight-nucleate, seven-celled structure; the seven cells are the egg cell, two synergids and three antipodals (each with a single nucleus), and the central cell (with two nuclei). (p. 525.)

endergonic (Gr. *endon*, within + *ergon*, work) Describing a chemical reaction that requires energy; energy from an outside source must be added before the reaction proceeds; a thermodynamically "uphill" process; opposite of exergonic. (p. 134.)

endocrine gland (Gr. *endon*, within + *krinein*, to separate) Ductless gland that secretes hormones into the extracellular spaces, from which they diffuse into the circulatory system; in vertebrates, includes the pituitary, sex glands, adrenal, thyroid, and others. (p. 714.)

endocytosis (Gr. *endon*, within + *kytos*, hollow vessel) The uptake of material into cells by inclusion within an invagination of the plasma membrane; the material becomes trapped within a vacuole when the edges of the invagination fuse together. If solid material is included, the uptake is called phagocytosis; if dissolved material, it is called pinocytosis. (p. 114.)

endoderm (gr. *endon*, within + *derma*, skin) One of the three embryonic germ layers of early vertebrate embryos, destined to give rise to the epithelium that lines certain internal structures, such as most of the digestive tract and its outgrowths, most of the respiratory tract, and the urinary bladder, liver, pancreas, and some endocrine glands. (p. 534.)

endodermis (Gr. *endon*, within + *derma*, skin) In vascular plants, a layer of cells forming the innermost layer of the cortex in roots and some stems. The endodermis is characterized by a Casparian strip within radial and transverse walls. (p. 572.)

endometrium (Gr. *endon*, within + *metrios*, of the womb) The lining of the uterus in mammals; thickens in response to secretion of estrogens and progesterone and is sloughed off in menstruation. (p. 770.)

endoplasmic reticulum (Gr. *endon*, within + *plasma*, from cytoplasm; L *reticulum*, network) An extensive system of membranes present in most eukaryotic cells, dividing the cytoplasm into compartments and channels; those portions containing dense array of ribosomes are called "rough ER," and other portions with fewer robosomes are called "smooth". (p. 78.)

endorphin One of a group of small neuropeptides produced by the vertebrate brain; like morphine, endorphins modulate pain perception; they are also implicated in many other functions. (p. 741.)

endosperm (Gr. *endon*, within + *sperma*, seed) A storage tissue characteristic of the seeds of angiosperms, which develops from the union of a male nucleus and the polar nuclei of the embryo sac. The endosperm is digested by the growing sporophyte either before the maturation of the seed or during its germination. (p. 525.)

endotherm (Gr. *endon*, within + *therme*, heat) An organism that regulates its body temperature internally through metabolic processes, as do birds and mammals; *see also* homeotherm; contrasts with ectothermic. (p. 370.)

energy Capacity to do work. (p. 130.)

entropy (Gr. *en*, in + *tropos*, change in manner) A measure of the randomness or disorder of a system; a measure of how much energy in a system has become so dispersed (usually as evenly distributed heat) that it is no longer available to do work. (p. 131.)

enzyme (Gr. *enzymos*, leavened, from *en*, in + *zyme*, leaven) A protein that is capable of speeding up specific chemical reactions by lowering the required activation energy, but is unaltered itself in the process; a biological catalyst. (p. 55.)

epidermis (Gr. *epi*, on or over + *derma*, skin) The outermost layers of cells; in plants, the exterior primary tissue of leaves, young stems, and roots; in vertebrates, the nonvascular external layer of skin, of ectodermal origin; in invertebrates, a single layer of ectodermal epithelium. (p. 572.)

epididymis (Gr. *epi*, on + *didymos*, testicle) A sperm storage vessel; a coiled part of the sperm duct that lies near the testis. (p. 768.)

epistasis (Gr. *epi*, on + *stasis*, a standing still) Interaction between two nonallelic genes in which one of them modifies the phenotypic expression of the other; the masking or prevention of the expression of one gene by another gene at another locus. (p. 257.)

epithelium (Gr. *epi*, on + *thele*, nipple) In animals, a type of tissue that covers an exposed surface or lines a tube or cavity. (p. 608.)

equilibrium (L. *aequus*, equal + *libra*, balance) A stable condition; a system in which no further net change is occurring; the point at which a chemical reaction proceeds as rapidly in the reverse direction as it does in the forward direction, so that there is no further net change in the concentrations of products or reactants. (p. 308.)

erythrocyte (Gr. *erythros*, red + *kytos*, hollow vessel) Red blood cell, the carrier of hemoglobin. In mammals, erythrocytes lose their nuclei, whereas in other vertebrates the nuclei are retained. (p. 612.)

estrogens (Gr. *oestros*, frenzy + *genos*, origin) A group of steroid hormones, such as estradiol, that affect female secondary sex characteristics, estrus, and the human menstrual cycle. (p. 765.)

estrus (L. *oestrus*, frenzy) The period of maximum female sexual receptivity, associated with ovulation of the egg; being "in heat." (p. 764.)

ethology (Gr. *ethos*, habit or custom + *logos*, discourse) The study of patterns of animal behavior in nature. (p. 787.)

ethylene A simple hydrocarbon that is a plant hormone involved in the ripening of fruit; $H_2C{=}CH_2$. (p. 600.)

euchromatin (Gr. *eu*, good + *chroma*, color) That portion of eukaryotic chromosomes that is transcribed into mRNA; contains active genes. (p. 188.)

eukaryote (Gr. *eu*, good + *karyon*, kernel) A cell characterized by membrane-bound organelles, most notably the nucleus, and one that possesses chromosomes whose DNA is associated with proteins; an organism composed of such cells; contrasts with prokaryote. (p. 31.)

evaporation The escape of water molecules from the liquid to the gas phase at the surface of a body of water. (p. 657.)

evolution (L. *evolvere*, to unfold) Genetic change in a population of organisms; in general, evolution leads to progressive change from simple to complex. Darwin proposed that natural selection was the mechanism of evolution. (p. 7.)

exergonic (L. *ex*, out + Gr. *ergon*, work) An energy-yielding process or chemical reaction; energy is released from the reactants, so that the products contain less chemical potential energy than the reactants; a "downhill" process that will proceed spontaneously. (p. 134.)

exocytosis (Gr. *ex*, out of + *kytos*, vessel) A type of bulk transport out of cells; cytoplasmic particles are encased within membranes, forming a vacuole that is transported to the cell surface; there, the vacuole membrane fuses with the cell membrane, discharging the vacuole's contents to the outside. (p. 114.)

exon (Gr. *exo*, outside) A segment of DNA that is both transcribed into RNA and translated into protein, specifying the amino acid sequence of part of a polypeptide; contrasts with intron. Exons are characteristic of eukaryotes. (p. 273.)

exoskeleton (Gr. *exo*, outside + *skeletos*, hard) An external skeleton, as in arthropods. (p. 506.)

F₁ (first filial generation) The offspring resulting from a cross; the parents of the cross are referred to as the parental generation. (p. 208.)

F₂ (second filial generation) The offspring resulting from a cross between members of the F_1 generation; if these F_2 offspring were to mate and produce progeny, the progeny would be the F_3 generation, and so forth. (p. 208.)

facilitated diffusion Carrier-assisted diffusion; the transport of molecules across a cellular membrane through specific channels (carrier molecules embedded in the membrane) from a region of high concentration to a region of low concentration; the process is driven by the concentration difference and does not require energy. The chief difference from free diffusion is that the membrane is impermeable to the molecule except for passage through the carrier channels; contrasts with active transport, which is in the direction of higher concentration and requires energy. (p. 116.)

family A taxonomic group made up of one or more genera; one of the subdivisions of an order. The ending of family names in animals and heterotrophic protists is *-idae;* in all other organisms it is *-aceae.* (p. 361.)

fat A molecule composed of glycerol and three fatty acid molecules; the proportion of oxygen to carbon is much less in fats than it is in carbohydrates; fats in the liquid state are called oils. (p. 53.)

fatty acid A long hydrocarbon chain ending with a —COOH group; fatty acids are components of fats, oils, phospholipids, and waxes. (p. 53.)

feedback inhibition Control mechanism whereby an increase in the concentration of some molecule inhibits the synthesis of that molecule; more generally, the regulation of the level of any factor sensitive to its own magnitude; important in the regulation of enzyme and hormone levels, ion concentrations, temperature, and many other factors. (p. 740.)

fermentation (L. *fermentum*, ferment) The enzyme-catalyzed extraction of energy from organic compounds without the involvement of oxygen; the enzymatic conversion, without oxygen, of carbohydrates to alcohols, acids, and carbon dioxide; the conversion of pyruvate to ethanol or lactic acid. (p. 160.)

fertilization The fusion of two haploid gamete nuclei to form a diploid zygote nucleus. (p. 762.)

fetus (L. pregnant) An unborn or unhatched vertebrate that has passed through the earliest developmental period; in humans, a developing individual is a fetus from about the second month of gestation until birth. (p. 778.)

fission (L. a splitting) Asexual reproduction by a division of the cell or body into two or more parts of roughly equal size. *See* asexual reproduction, clone. (p. 186.)

fitness The genetic contribution of an individual to succeeding generations, relative to the contributions of other individuals in the population. (p. 313.)

flagellum, pl. **flagella** (L. *flagellum*, whip) A fine, long, threadlike organelle protruding from the surface of a cell; in bacteria, a single protein fiber, capable of rotary motion, that propels the cell through the water; in eukaryotes, an array of microtubules with a characteristic internal 9 + 2 microtubule structure, capable of vibratory but not rotary motion; used in locomotion and feeding; common in protists and motile gametes. A cilium is a small flagellum. (p. 78.)

follicle (L. *folliculus*, small ball) In a mammalian ovary, one of the spherical chambers containing an oocyte. (p. 765.)

food chain A portion of a food web. A sequence of prey species and the predators that consume them. (p. 425.)

food web The food relationships within a community. A diagram of who eats who. (p. 425.)

fovea (L. a small pit) A small depression in the center of the retina with a high concentration of cones; the area of sharpest vision. (p. 727.)

free energy change The total change in usable energy that results from a chemical reaction or other process; equal to the change in total energy (the heat content or enthalpy) minus the change in unavailable energy (the disorder or entropy times temperature). *See* entropy. (p. 135.)

fruit In angiosperms, a mature, ripened ovary (or group of ovaries), containing the seeds; also applied informally to the reproductive structures of some other kinds of organisms. (p. 581.)

gametangium, pl. **gametangia** (Gr. *gamein*, to marry + L. *tangere*, to touch) A cell or organ in

which gametes are formed. (p. 517.)

gamete (Gr. wife) A haploid reproductive cell; upon fertilization, its nucleus fuses with that of another gamete of the opposite sex; the resulting diploid cell (zygote) may develop into a new diploid individual, or, in some protists and fungi, may undergo meiosis to form haploid somatic cells. (p. 190.)

gametophyte In plants, the haploid (1*n*), gamete-producing generation, which alternates with the diploid (2*n*) sporophyte. (P. 516.)

ganglion, pl. **ganglia** (Gr. a swelling) An aggregation of nerve cell bodies; in invertebrates, ganglia are the integrative centers; in vertebrates, the term is restricted to aggregations of nerve cell bodies located outside the central nervous system. (p. 715.)

gap junction A junction between adjacent animal cells that allows the passage of materials between the cells; a system of pipes joining the cytoplasms of two adjacent cells. (p. 123.)

gastrula (Gr. little stomach) In vertebrates, the embryonic stage in which the blastula with its single layer of cells turns into a three-layered embryo made up of ectoderm, mesoderm, and endoderm, surrounding a cavity (archenteron) with one opening (blastopore). (p. 532.)

gene (Gr. *genos*, birth, race) The basic unit of heredity; a sequence of DNA nucleotides on a chromosome that encodes a protein, tRNA, or rRNA molecule, or regulates the transcription of such a sequence. (p. 210.)

gene frequency The relative occurrence of a particular allele in a population. (p. 308.)

genetic code The "language" of the genes, dictating the correspondence between nucleotide sequence in DNA and amino acid sequence in proteins; a series of 64 different three-nucleotide sequences, or triplets (codons); except for three "stop" signals, each codon corresponds to one of the 20 amino acids. (p. 269.)

genetic drift Random fluctuation in allele frequencies over time by chance. (p. 309.)

genome (Gr. *genos*, offspring + L. *oma*, abstract group) The total genetic constitution of an organism; in bacteria, all the genes in the main circular chromosome or in associated plasmids; in a eukaryote, all the genes in a haploid set of chromosomes. (p. 185.)

genotype (Gr. *genos*, offspring + *typos*, form) The total set of genes present in the cells of an

organism, as contrasted with the phenotype, which is the realized expression of these genes; often also used to refer to the genetic constitution underlying a single trait or set of traits. (p. 215.)

genus, pl. **genera** (L. race) A taxonomic group that includes species; families are divided into genera. (p. 331.)

germ layer A layer of distinctive cells in an animal embryo; each germ layer is fated to give rise to certain tissues or structures as the organism develops. The majority of multicellular animals have three embryonic layers: ectoderm, mesoderm, and endoderm. (p. 777.)

germination (L. *germinare*, to sprout) The resumption of growth and development by a spore or seed. (p. 580.)

gibberellins (*Gibberella*, a genus of fungi) A group of plant growth hormones, the best-known effect of which is on the elongation of plant stems. (p. 599.)

glomerulus (L. a little ball) In the vertebrate kidney, a cluster of capillaries enclosed by Bowman's capsule; also, a small spongy mass of tissue in the proboscis of hemichordates, presumed to have an excretory function. Also, a concentration of nerve fibers situated in the olfactory bulb. *See* Bowman's capsule, filtration. (p. 752.)

glucose A common six-carbon sugar ($C_6H_{12}O_6$); the most common monosaccharide in most organisms. (p. 50.)

glycerol Three-carbon molecule with three hydroxyl groups attached; combines with fatty acids to form fat or oil. (p. 53.)

glycogen (Gr. *glykys*, sweet + *gen*, of a kind) Animal starch; a complex branched polysaccharide that serves as a food reserve in animals, bacteria, and fungi; can be broken down readily into glucose subunits. (p. 744.)

glycolysis (Gr. *glykys*, sweet + *lyein*, to loosen) The anaerobic breakdown of glucose; the enzyme-catalyzed breakdown of glucose to two molecules of pyruvic acid, with the net liberation of two molecules of ATP. (p. 148.)

glycoxysome A small cellular organelle or microbody containing enzymes necessary for the conversion of fats into carbohydrates; glycoxysomes play an important role during seed germination in plants. (p. 91.)

Gogi body (after Camillo Golgi, Italian histologist) An organelle present in many eukaryotic cells; consisting of flat, disk-shaped sacs, tubules, and vesicles, it functions as a collecting and packaging center for substances

that the cell manufactures for export; also called dictyosome in plants. The terms "Golgi apparatus" and "Golgi complex" are used to refer collectively to all of the Golgi bodies of a given cell. (p. 90.)

grana, sing. **granum** (L. grain or seed) In chloroplasts, stacks of membrane-bound disks (thylakoids); the thylakoids contain the chlorophylls and carotenoids and are the sites of the light reactions of photosynthesis. (p. 178.)

gravitropism (L. *gravis*, heavy + *tropes*, turning) Growth response to gravity in plants; formerly called geotropism. (p. 601.)

guanine (Sp. from Quechua, *huanu*, dung) A purine base found in DNA and RNA; its name derives from the fact that it occurs in high concentration as a white crystalline base, $C_5H_5N_5O$, in guano and other animal excrements. (p. 60.)

guard cells Pairs of specialized epidermal cells that surround a stoma; when the guard cells are turgid, the stoma is open, and when they are flaccid, it is closed. (p. 565.)

gymnosperm (Gr. *gymnos*, naked + *sperma*, seed) A seed plant with seeds not enclosed in an ovary; the conifers are the most familiar group. (p. 523.)

gynoecium (Gr. *gyne*, woman + *oikos*, house) The aggregate of carpels in the flower of a seed plant. (p. 525.)

habitat (L. *habitare*, to inhabit) The environment of an organism; the place where it is usually found. (p. 432.)

habituation (L. *habitus*, condition) A form of learning; a diminishing response to a repeated stimulus; the ignoring of an often-repeated stimulus. (p. 786.)

haploid (Gr. *haploos*, single + *ploion*, vessel) Having only one set of chromosomes (*n*), in contrast to diploid (2*n*); characteristic of eukaryotic gametes, of gametophytes in plants, and of some protists and fungi. (p. 215.)

Hardy-Weinberg equilibrium A mathematical description of the fact that the relative frequencies of two or more alleles in a population do not change because of Mendelian segregation; allele and genotype frequencies remain constant in a random-mating population in the absence of inbreeding, selection, or other evolutionary forces; usually stated: if the frequency of allele *a* is *p* and the frequency of allele *b* is *q*, then the genotype frequencies after one generation of random mating will always be $p^2(a) + 2pq(ab) + q^2(b)$. (p. 308.)

hemoglobin (Gr. *haima*, blood + L. *globus*, a ball) A globular protein in vertebrate red blood cells and in the plasma of many invertebrates that carries oxygen and carbon dioxide; an essential part of each molecule is an iron-containing heme group, which both binds O_2 and CO_2 and gives blood its red color. (p. 233.)

hemophilia (Gr. *haima*, blood + *philios*, friendly) A group of hereditary diseases characterized by failure of the blood to clot and consequent excessive bleeding from even minor wounds; a mutation in a gene encoding one of the protein factors involved in blood clotting. (p. 234.)

hepatic (Gr. *hepatikos*, of the liver) Pertaining to the liver. (p. 744.)

herbivore (L. *herba*, grass + *vorare*, to devour) Any organism subsisting on plants. Adj., *herbivorous*. (p. 406.)

heredity (L. *heredis*, heir) The transmission of characteristics from parent to offspring through the gametes. (p. 7.)

heterochromatin (Gr. *heteros*, different + *chroma*, color) That portion of eukaryotic chromosomes that is not transcribed into RNA; stains intensely in histological preparations; characteristic of centromeres. (p. 188.)

heterotroph (Gr. *heteros*, other + *trophos*, feeder) An organism that cannot derive energy from photosynthesis or inorganic chemicals, and so must feed on other plants and animals, obtaining chemical energy by degrading their organic molecules; animals, fungi, and many unicellular organisms are heterotrophs; *see also* autotroph. (p. 145.)

heterozygote (Gr. *heteros*, other + *zygotos*, a pair) A diploid individual that carries two different alleles on homologous chromosomes at one or more genetic loci. Adj., *heterozygous*. Opposite of homozygote. (p. 215.)

histone (Gr. *histos*, tissue) A group of relatively small, very basic polypeptides, rich in arginine and lysine. An essential component of eukaryotic chromosomes, histones form the core of nucleosomes around which DNA is wrapped in the first stage of chromosome condensation. (p. 188.)

homeostasis (Gr. *homeos*, similar + *stasis*, standing) The maintaining of a relatively stable internal physiological environment in an organism, or steady-state equilibrium in a population or ecosystem; usually involves some form of feedback self-regulation. (p. 729.)

homeotherm (Gr. *homoios*, same

or similar + *therme*, heat) An organism, such as a bird or mammal, capable of maintaining a stable body temperature independent of the environmental temperature; at lower environmental temperatures this involves the metabolic generation of heat at considerable expense in terms of ATP utilized; "warm blooded." *See* endotherm. Contrasts with poikilotherm. (p. 370.)

hominid (L. *homo*, man) Any primate in the human family, Hominidae. *Homo sapiens* is the only living representative; *Australopithecus* is an extinct genus. (p. 376.)

hominoid (L. *homo*, man) Collectively, hominids and apes; together with the monkeys, hominoids constitute the anthropoid primates. (p. 376.)

homologous chromosome (Gr. *homologia*, agreement) In diploid cells, one chromosome of a pair that carry equivalent genes; chromosomes that associate in pairs in the first stage of meiosis; also called "homologues." (p. 190.)

homology (Gr. *homologia*, agreement) A condition in which the similarity between two structures or functions is indicative of a common evolutionary origin. Adj., *homologous*. (p. 27.)

homozygote (Gr. *homos*, same or similar + *zygotos*, a pair) A diploid individual that carries identical alleles at one or more genetic loci on its two homologous chromosomes; opposite of heterozygote. (p. 215.)

homozygous Being a homozygote; the term is usually applied to one or more specific loci, as in "homozygous with respect to the *w* locus" (that is, the genotype is *w/w*). (p. 209.)

hormone (Gr. *hormaein*, to excite) A chemical messenger; a molecule, usually a peptide or steroid, that is produced in one part of an organism and triggers a specific cellular reaction in target tissues and organs some distance away. (p. 56.)

Human Immunodeficiency Virus (HIV) The virus responsible for AIDS, a deadly disease that destroys the human immune system. HIV is a retrovirus (its genetic material is RNA) that is thought to have been introduced to humans from African green monkies. (p. 774.)

hybrid (L. *hybrida*, the offspring of a tame sow and a wild boar) Offspring of two different varieties or of two different species; alternatively, offspring of two parents that differ in one or more heritable characteristics. (p. 330.)

hybridization The mating of unlike parents. (p. 331.)

hybridoma (contraction of *hybrid* + *myeloma*) A fast-growing cell line produced by fusing a cancer cell (myeloma) to some other cell, such as an antibody-producing cell. *See* monoclonal antibody. (p. 697.)

hydrocarbon (Gr. *hydor*, water + L. *carbo*, charcoal) An organic compound consisting only of carbon and hydrogen atoms. (p. 461.)

hydrogen bond A weak and very directional molecular attraction involving hydrogen atoms; produced by the interaction of the partial positive charge of a polar hydrogen atom (typically, hydrogen atoms covalently linked to oxygen or nitrogen, which more strongly attract the shared electron and so render the hydrogen nucleus partially positive) with the partial negative charge of another polar atom (typically an oxygen or nitrogen atom without a covalently bound hydrogen). *See* polar. (p. 45.)

hydrophobic (Gr. *hydor*, water + *phobos*, hating) Repelled by water; refers to nonpolar molecules, which do not form hydrogen bonds with water and so are not soluble in water. (p. 46.)

hydrophobic interaction (L. *hydrophobos*, hated by water) The propensity for water molecules to exclude nonpolar molecules, as oil is excluded from water; water tends to form the maximum number of hydrogen bonds, and more hydrogen bonds are possible if nonpolar molecules (which do not form hydrogen bonds) are not present to interfere with the hydrogen bonds between water molecules. Hydrophobic interactions are responsible for much of the three-dimensional structure of proteins, which is why the addition of nonpolar solvents denatures (unfolds) proteins. (p. 137.)

hydroxyl group (hydrogen + oxygen + -*yl*) An OH^- group; a negatively charged ion formed by the disassociation of a water molecule. (p. 47.)

hyperosmotic (Gr. *hyper*, over + *osmos*, impulse) Refers to a hypertonic solution whose osmotic pressure is greater than that of another solution with which it is compared; contrasts with hypoosmotic. (p. 750.)

hypertonic (Gr. *hyper*, above + *tonos*, tension) Refers to a solution that contains a higher concentration of solute particles; water moves across a semipermeable membrane into hypertonic solution. (p. 113.)

hypha, pl. **hyphae** (Gr. *hyphe*, web) A filament of a fungus or

oomycete; collectively, the hyphae comprise the mycelium. (p. 505.)

hypoosmotic (Gr. *hypo*, under + *osmos*, impulse) Refers to a hypotonic solution whose osmotic pressure is less than that of another solution with which it is compared; contrasts with hyperosmotic. (p. 750.)

hypothalamus (Gr. *hypo*, under + *thalamos*, inner room) A region of the vertebrate brain just below the cerebral hemispheres, under the thalamus; a center of the autonomic nervous system, responsible for the integration and correlation of many neural and endocrine functions. (p. 717.)

hypothesis (Gr. *hypo*, under + *tithenai*, to put) A guess as to what might be; a postulated explanation of a phenomenon consistent with available information; no hypothesis is ever *proved* to be correct—all hypotheses are provisional, working ideas that are accepted for the time being but may be rejected in the future if not consistent with data generated by further experiments; a hypothesis that survives many tests and is very unlikely to be disgarded is referred to as a "theory." (p. 14.)

hypotonic (Gr. *hypo*, under + *tonos*, tension) Refers to a solution that contains a lower concentration of solute particles; water moves across a semipermeable membrane out of a hypotonic solution. (p. 113.)

immunoglobulin (L. *immunis*, free + *globus*, globe) An antibody. (p. 691.)

inbreeding The breeding of genetically related plants or animals. In plants, inbreeding results from self-pollination; in animals, inbreeding results from matings between relatives; inbreeding tends to increase homozygosity. (p. 311.)

incomplete dominance The ability of two alleles to produce a heterozygous phenotype that is different from either homozygous phenotype. (p. 259.)

independent assortment Mendel's second law; the principle that segregation of alternative alleles at one locus into gametes is independent of the segregation of alleles at other loci; only true for gene loci located on different chromosomes, or so far apart on one chromosome that crossing-over is very frequent between the loci. *See* Mendel's second law. (p. 213.)

indoleacetic acid (IAA) A naturally occurring auxin, one of the plant hormones. (p. 598.)

inflammation (L. *inflammare*, from *flamma*, flame) The mobilization of body defenses against foreign substances and infectious agents, and the repair of damage from such agents; involves phagocytosis by macrophages and is often accompanied by an increase in the local temperature. (p. 674.)

innate (L. *innatus*, inborn) Describing a characteristic based partly or wholly on inherited gene differences. (p. 787.)

instinct (L. *instinctus*, impelled) Stereotyped, predictable, genetically programmed behavior. Learning may or may not be involved. (p. 786.)

insulin A peptide hormone produced by the vertebrate pancreas which acts to promote glycogen formation and thus to lower the concentration of sugar in the blood. (p. 741.)

integration, neural The summation of the repolarizing and repolarizing effects contributed by all excitatory and inhibitory synapses acting on a neuron. *See* graded potential. (p. 709.)

integument (L. *integumentum*, covering) In plants, the outermost layer or layers of tissue enveloping the nucellus of the ovule; develops into the seed coat. (p. 534.)

interferon In vertebrates, a protein produced in virus-infected cells that inhibits viral multiplication. (p. 697.)

internode In plants, the region of a stem between two successive nodes. (p. 569.)

interphase The period between two mitotic or meiotic divisions in which a cell grows and its DNA replicates; includes G1, S, and G2 phases. (p. 190.)

intron (L. *intra*, within) Portion of mRNA as transcribed from eukaryotic DNA that is removed by enzymes before the mature mRNA is translated into protein. These untranscribed regions comprise the bulk of most eukaryotic genes; typically, the transcribed portion of the gene exists as numerous short segments called "exons" that are scattered in no particular order within a much longer stretch of nontranscribed DNA. Those segments of the background nontranscribed DNA that fall between two exons are called introns. *See* exon. (p. 272.)

inversion (L. *invertere*, to turn upside down) A reversal in order of a segment of a chromosome; also, to turn inside out, as in embryogenesis of sponges or discharge of a nematocyst. (p. 291.)

ion Any atom or molecule containing an unequal number of electrons and protons and therefore carrying a net positive or net negative charge; gain of an extra electron produces a nega-

tively charged cation, and loss of an electron produces a positively charged anion. (p. 38.)

ionic bond A chemical bond formed as a result of the mutual attraction of ions of opposite charge; ionic bonds are nondirectional and form between one ion and nearly all ions of opposite charge. (p. 40.)

isolating mechanisms Mechanisms that prevent genetic exchange between individuals of different populations or species; may be behavioral, morphological, or physiological. (p. 331.)

isomer (Gr. *isos*, equal + *meros*, part) One of a group of molecules identical in atomic composition but differing in structural arrangement, for example, glucose and fructose. (p. 49.)

isotonic (Gr. *isos*, equal + *tonos*, tension) Refers to two solutions that have equal concentrations of solute particles; if two isotonic solutions are separated by a semipermeable membrane, there will be no net flow of water across the membrane. (p. 113.)

isotope (Gr. *isos*, equal + *topos*, place) An alternative form of a chemical element; differs from other atoms of the same element in the number of neutrons in the nucleus; isotopes thus differ in atomic weight. All isotopes have the same chemical behavior, as all contain the same number of protons and electrons. Some isotopes are unstable and emit radiation. (p. 38.)

karyotype (Gr. *karyon*, kernel + *typos*, stamp or print) The morphology of the chromosomes of an organism as viewed with a light microscope. (p. 188.)

keratin (Gr. *kera*, horn + *in*, suffix used for proteins) A tough, fibrous protein formed in epidermal tissues and modified into skin, feathers, hair, and hard structures such as horns and nails. (p. 96.)

kidney In vertebrates, the organ that filters the blood to remove nitrogenous wastes and regulates the balance of water and solutes in blood plasma. *See* filtration, Bowman's capsule, glomerulus. (p. 745.)

kinetic energy Energy of motion. (p. 130.)

kinetochore (Gr. *kinetikos*, putting in motion + *choros*, chorus) Disk-shaped protein structure within the centromere to which the spindle fibers attach during mitosis or meiosis. *See* centromere. (p. 188.)

kingdom The chief taxonomic category, for example, Monera or Plantae. In this book we recognize five kingdoms. (p. 472.)

Krebs cycle Another name for

the citric acid cycle; also called the tricarboxylic acid (TCA) cycle. (p. 148.)

K-selection (from the *K* term in the logistic equation) Natural selection under conditions that favor survival when populations are controlled primarily by density-dependent factors. (p. 396.)

lamella (L. a little plate) A thin, platelike structure; in chloroplasts, a layer of chlorophyll-containing membranes; in bivalve mollusks, one of the two plates forming a gill; in vertebrates, one of the thin layers of bone laid concentrically around an osteon (Haversian) canal. (p. 93.)

larva, pl. **larvae** (L. a ghost) Immature form of an animal that is quite different from the adult and undergoes metamorphosis in reaching the adult form; examples are caterpillars and tadpoles. (p. 366.)

larynx The voice box; a cartilaginous organ that lies between the pharynx and trachea and is responsible for sound production in vertebrates. (p. 660.)

lateral meristems (L. *latus*, side + Gr. *meristos*, divided) In vascular plants, the meristems that give rise to secondary tissue; the vascular cambium and cork cambium. (p. 564.)

leaf primordium (L. *primordium*, beginning) A lateral outgrowth from the apical meristem that will eventually become a leaf. (p. 569.)

learning The modification of behavior by experience. (p. 786.)

lenticels (L. *lenticella*, a small window) Spongy areas in the cork surfaces of stem, roots, and other plant parts that allow interchange of gases between internal tissues and the atmosphere through the periderm. (p. 569.)

leukocyte (Gr. *leukos*, white + *kytos*, hollow vessel) A white blood cell; a diverse array of non-hemoglobin–containing blood cells, including phagocytic macrophages and antibody-producing lymphocytes. (p. 673.)

life cycle The sequence of phases in the growth and development of an organism, from zygote formation to gamete formation. (p. 478.)

linkage Lack of independent segregation; the tendency for two or more genes to segregate together in a cross owing to the fact that they are located on the same chromosome. (p. 219.)

lipase (Gr. *lipos*, fat + *-ase*, enzyme suffix) An enzyme that catalyzes the hydrolysis of fats. (p. 638.)

lipid (Gr. *lipos*, fat) a nonpolar organic molecules; fatlike; one of a

large variety of nonpolar hydrophobic molecules that are insoluble in water (which is polar) but that dissolve readily in nonpolar organic solvents; includes fats, oils, waxes, steroids, phospholipids, and carotenes. (p. 48.)

locus, pl. **loci** (L. place) The location of a gene on a chromosome; more precisely, the position of a particular transcription unit on a chromosome. (p. 210.)

lymph (L. *lympha*, water) In animals, a colorless fluid derived from blood by filtration through capillary walls in the tissues. (p. 688.)

lymph node In animals, a mass of spongy tissues that serve to filter the lymphatic system. Located throughout the lymphatic system, lymph nodes remove dead cells, debris, and foreign particles from the circulation. (p. 688.)

lymphocyte (L. *lympha*, water + Gr. *kytos*, hollow vessel) A type of white blood cell. Lymphocytes are responsible for the immune response; there are two principal classes—B cells (which differentiate into antibody-producing plasma cells) and T cells (which interact directly with the foreign invader and are responsible for cell-mediated immunity). (p. 611.)

lymphokines A regulatory molecule that is secreted by lymphocytes. In the immune response, lymphokines secreted by helper T cells unleash the cell-mediated immune response. (p. 691.)

lysis (Gr., a loosening) Disintegration of a cell by rupture of its cell membrane. (p. 248.)

lysosome (Gr. *lysis*, a loosening + *soma*, body) A membrane-bound cell organelle containing hydrolytic (digestive) enzymes that are released when the lysosome ruptures; important in recycling worn-out mitochondria and other cellular debris. (p. 91.)

macromolecule (Gr. *makros*, large + L. *moliculus*, a little mass) An extremely large molecule; a molecule of very high molecular weight; refers specifically to proteins, nucleic acids, polysaccharides, and complexes of these. (p. 48.)

macronutrients (Gr. *makros*, large + L. *nutrire*, to nourish) Inorganic chemical elements required in large amounts for plant growth, such as nitrogen, potassium, calcium, phosphorus, magnesium, and sulfur. (p. 596.)

Malpighian tubules (Marcello Malpighi, Italian anatomist, 1628-1694) Blind tubules opening into the hindgut of terrestrial arthropods; they function as excretory organs. (p. 752.)

mass In chemistry, the total number of protons and neutrons in the nucleus of an atom. Approximately equal to the atomic weight. (p. 38.)

mating type A strain of organisms incapable of sexual reproduction with one another but capable of such reproduction with members of other strains of the same organism. (p. 507.)

medulla (L. *marrow*) The inner portion of an organ, in contrast to the cortex or outer portion, as in the kidney or adrenal gland. Also, the most posterior region of the vertebrate brain, the hindbrain. (p. 756.)

megagametophyte (Gr. *megas,* large + *gamos,* marriage + *phyton,* plant) In heterosporous plants, the female gametophyte; located within the ovule of seed plants. (p. 519.)

megaspore (Gr. *megas,* large + *sporos,* seed) In heterosporous plants, a haploid (1*n*) spore that develops into a female gametophyte; megaspores are usually larger than microspores. (p. 519.)

meiosis (Gr. *meioun,* to make smaller) Reduction division; the two successive nuclear divisions in which a single diploid (2*n*) cell forms four haploid (1*n*) nuclei, halving the chromosome number; segregation, crossing-over, and reassortment all occur during meiosis; in animals, meiosis usually occurs in the last two divisions in the formation of the mature egg or sperm; in plants, spores—which divide by mitosis—are produced as a result of meiosis. Compare with mitosis. (p. 196.)

Mendel's first law The law of allele segregation: the factors specifying a pair of alternative characteristics (alleles) are separate, and only one may be carried in a particular gamete; gametes combine randomly in forming progeny. Chromosomes had not been observed in Mendel's time, and meiosis was unknown. Modern form: alleles segregate as chromosomes do. (p. 212.)

Mendel's second law The law of independent assortment: The inheritance of alternative characteristics (alleles) of one trait is independent of the simultaneous inheritance of other traits—different traits (genes) assort independently. Mendel never tested a pair of traits that were located close together on a chromosome (although two of the seven genes he studied in fact were), and so never discovered that only genes that are not close together assort independently. Modern form: unlinked genes assort independently. (p. 210.)

menstruation (L. *mens,* month) Periodic sloughing off of the blood-enriched lining of the uterus when pregnancy does not occur. The menstrual cycle in primates is the cycle of hormone-regulated changes in the condition of the uterine lining, which is marked by the periodic discharge of blood and disintegrated uterine lining through the vagina (menstruation). (p. 766.)

meristem (Gr. *merizein,* to divide) Undifferentiated plant tissue from which new cells arise. (p. 564.)

mesenteries (Gr. *mesos,* middle + *enteron,* gut) Double layers of mesoderm that suspend the digestive tract and other internal organs within the coelom. (p. 537.)

mesoderm (Gr. *mesos,* middle + *derma,* skin) One of the three embryonic germ layers that form in the gastrula; gives rise to muscle, bone and other connective tissue, the peritoneum, the circulatory system, and most of the excretory and reproductive systems. (p. 534.)

mesophyll (Gr. *mesos,* middle + *phyllon,* leaf) The photosynthetic parenchyma of a leaf, located within the epidermis. The vascular strands (veins) run through the mesophyll. (p. 568.)

messenger RNA (mRNA) The RNA transcribed from structural genes; RNA molecules complementary to one strand of DNA, which are translated by the ribosomes into protein. (p. 264.)

metabolism (Gr. *metabole,* change) The sum of all chemical processes occuring within a living cell or organism; includes carbon fixation (photosynthesis), digestion (catabolism), extraction of chemical energy (respiration), and synthesis of organic molecules (anabolism). (p. 6.)

metamorphosis (Gr. *meta,* after + *morphe,* form + *osis,* state of) Process in which there is a marked change in form during postembryonic development, for example, tadpole to frog or larval insect to adult. (p. 545.)

metaphase (Gr. *meta,* middle + *phasis,* form) The stage of mitosis or meiosis during which microtubules become organized into a spindle and the chromosomes come to lie in the spindle's equatorial plane. (p. 192.)

microbody A cellular organelle bounded by a single membrane and containing a variety of enzymes; generally derived from endoplasmic reticulum; includes peroxisomes and glyoxysomes. (p. 91.)

microfilament (Gr. *mikros,* small + L. *filum,* a thread) In cells, a fine protein thread composed of actin and myosin; capable of contraction when supplied with ATP. Contraction of microfilaments is the basic mechanism underlying changes in shape of eukaryotic cells, including cell motility and muscle contraction. (p. 630.)

microgametophyte (Gr. *mikros,* small + *gamos,* marriage + *phyton,* plant) In heterosporous plants, the male gametophyte. (p. 519.)

micrometer A unit of microscopic measurement convenient for describing cellular dimensions; 10⁻⁶ meter (about 1/25,000 of an inch); its symbol is μm. Replaces the now obsolete term "micron." (p. 72.)

micronutrient (Gr. *mikros,* small + L. *nutrire,* to nourish) A mineral required in only minute amounts for plant growth, such as iron, chlorine, copper, manganese, zinc, molybdenum, and boron. (p. 596.)

microspore (Gr. *mikros,* small + *sporos,* seed) In plants, a spore that develops into a male gametophyte; in seed plants, it develops into a pollen grain. (p. 519.)

microtubule (Gr. *mikros,* small + L. *tubulus,* little pipe) In eukaryotic cells, a long, hollow protein cylinder, about 25 nanometers in diameter, composed of the protein tubulin. Microtubules influence cell shape, move the chromosomes in cell division, and provide the functional internal structure of cilia and flagella. (p. 96.)

middle lamella The layer of intercellular material, rich in pectic compounds, that cements together the primary walls of adjacent plant cells. (p. 194.)

mimicry (Gr. *mimos,* mime) The resemblance in form, color, or behavior of certain organisms (mimics) to other more powerful or more protected ones (models), which results in the mimics being protected in some way. (p. 412.)

mitochondrion, pl. **mitochondria** (Gr. *mitos,* thread + *chondrion,* small grain) The site of oxidative respiration in eukaryotes; a bacterium-like organelle found within the cells of all but one species of eukaryote. Mitochondria contain the enzymes catalyzing the citric acid cycle, which generates electrons that drive proton pumps within the mitochondrial membrane and so fosters the chemiosmotic synthesis of ATP. Almost all of the ATP of nonphotosynthetic eukaryotic cells is produced in mitochondria. (p. 78.)

mitosis (Gr. *mitos,* thread) Somatic cell division; nuclear division in which the duplicated chromosomes separate to form two genetically identical daughter nuclei; usually accompanied by cytokinesis, producing two daughter cells. After mitosis, the chromosome number in each daughter nucleus is the same as it was in the original dividing cell. Mitosis is the basis of reproduction of single-celled eukaryotes, and of the physical growth of multicellular eukaryotes. (p. 189.)

mole (L. *moles,* mass) The atomic weight of a substance, expressed in grams; 1 mole is defined as the mass of 6.022 × 10²³ atoms (Avogadro's number), the number of atoms in 22.4 liters. (p. 374.)

molecule (L. *moliculus,* a small mass) A collection of two or more atoms held together by chemical bonds; the smallest unit of a compound that displays the properties of the compound. (p. 5.)

monoclonal antibody An antibody of a single type that is produced by genetically identical plasma cells (clones); a monoclonal antibody is typically produced from a cell culture derived from the fusion product of a cancer cell and an antibody-producing cell. *See* hybridoma. (p. 697.)

monocot Short for monocotyledon; a flowering plant in which the embryos have only one cotyledon, the floral parts are generally in threes, and the leaves typically are parallel-veined. Compare dicot. (p. 526.)

monocyte (Gr. *monos,* single + *kytos,* hollow vessel) A type of leukocyte that becomes a phagocytic cell (macrophage) after moving into tissues. (p. 674.)

monosaccharide (Gr. *monos,* one + *sakcharon,* sugar, from Sanskrit *sarkara,* gravel, sugar) A simple sugar that cannot be decomposed into smaller sugar molecules; the most common are five-carbon pentoses (such as ribose) and six-carbon hexoses (such as glucose). (p. 47.)

morphogenesis (Gr. *morphe,* form + *genesis,* origin) The development of form; construction of the architectural features of organisms; the formation and differentiation of tissues and organs. (p. 778.)

morphology (Gr. *morphe,* form + *logos,* discourse) The study of form and its development; includes cytology (the study of cell structure), histology (the study of tissue structure), and anatomy (the study of gross structure). (p. 340.)

motor neuron Neuron that transmits nerve impulses from the central nervous system to an effector, which is typically a muscle or a gland; an efferent neuron. (p. 714.)

mRNA *See* messenger RNA.

muscle fiber Muscle cell; a long, cylindrical, multinucleated cell containing numerous myofibrils, which is capable of contraction when stimulated. (p. 628.)

mutagen (L. *mutare*, to change + Gr. *genaio*, to produce) An agent that induces changes in DNA (mutations); includes physical agents that damage DNA and chemicals that alter one or more DNA bases, link them together, or delete one or more of them. An X ray is an example of a physical mutagen that breaks both strands of a DNA molecule; cigarette tar is an example of a chemical that produces cancer-inducing mutations in lung cells. (p. 10.)

mutalism (L. *mutuus*, lent, borrowed) The living together of two or more organisms in a symbiotic association in which both members benefit. (p. 412.)

mutant (L. *mutare*, to change) A mutated gene; alternatively, an organism carrying a gene that has undergone a mutation. (p. 11.)

mutation A permanent change in a cell's DNA; includes changes in nucleotide sequence, alteration of gene position, gene loss or duplication, and insertion of foreign sequences. (p. 10.)

mycelium (Gr. *mykes*, fungus) In fungi or oomycetes, a mass of hyphae. (p. 506.)

mycology The study of fungi. One who studies fungi is called a *mycologist*. (p. 504.)

mycorrhiza, pl. **mycorrhizae** (Gr. *mykes*, fungus + *rhiza*, root) A symbiotic association between fungi and the roots of a plant. (p. 424.)

myelin sheath (Gr. *myelinos*, full of marrow) A fatty layer surrounding the long axons of motor neurons in the peripheral nervous system of vertebrates; made up of the membranes of Schwann cells. (p. 703.)

myofibril (Gr. *myos*, muscle + L. *fibrilla*, little fiber) A contractile microfilament within muscle, composed of myosin and actin. (p. 613.)

myosin (Gr. *mys*, muscle + *in*, belonging to) One of the two protein components of microfilaments (the other is actin); a principal component of vertebrate muscle. (p. 630.)

NAD (nicotinamide adenine dinucleotide) A coenzyme that functions as an electron acceptor in many of the oxidation reactions of respiration. NAD$^+$ is the oxidized form of NAD;

NADH is the reduced form. (p. 139.)

NADP (nicotinamide adenine dinucleotide phosphate) A coenzyme that functions as an electron donor in many of the reduction reactions of biosynthesis. NADP$^+$ is the oxidized NADPH$_2$, the reduced form of NADP. (p. 173.)

natural selection The differential reproduction of genotypes; caused by factors in the environment; leads to evolutionary change. (p. 14.)

nematocyst (Gr. *nema*, thread + *kystos*, bladder) In cnidarians, a specialized cellular capsule containing a tiny barb with a poisonous, paralyzing substance, which can be discharged against predator or prey. (p. 535.)

neoteny (Gr. *neos*, new + *teinein*, to extend) The attainment of sexual maturity in the larval condition. Also, the retention of larval characters into adulthood. (p. 535.)

nephron (Gr. *nephros*, kidney) Functional unit of the vertebrate kidney; one of numerous tubules (a human kidney contains about 1 million) involved in filtration and selective reabsorption of blood; each nephron consists of a Bowman's capsule, an enclosed glomerulus, and a long attached tubule; in humans, called a renal tubule. (p. 752.)

nerve A group or bundle of nerve fibers (axons) with accompanying neuroglial cells, held together by connective tissue; located in the peripheral nervous system (a bundle of nerve fibers within the central nervous system is known as a tract). (p. 608.)

nerve fiber An axon. (p. 633.)

nerve impulse An action potential; a rapid, transient, self-propagating reversal in electric potential that travels along the membrane of a neuron. (p. 616.)

neuroglia (Gr. *neuron*, nerve + *glia*, glue) Nonconducting nerve cells that are intimately associated with neurons and appear to provide nutritional support; in vertebrates they represent at least half of the volume of the nervous system, yet the function of most is not well understood. (p. 615.)

neuron (Gr. nerve) A nerve cell specialized for signal transmission; includes cell body, dendrites, and axon. (p. 614.)

neurotransmitter (Gr. *neuron*, nerve + L. *trans*, across + *mittere*, to send) A chemical released at the axon terminal of a neuron that travels across the synaptic cleft, binds a specific receptor on the far side, and, depending on the nature of the re-

ceptor, depolarizes or hyperpolarizes a second neuron or a muscle or gland cell. (p. 633.)

neutron (L. *neuter*, neither) An uncharged subatomic particle of about the same size and mass as a proton but having no electric charge. (p. 37.)

niche The role played by a particular species in its environment. (p. 330.)

nicotinamide adenine dinucleotide *See* NAD.

nitrogen fixation The incorporation of atmospheric nitrogen into nitrogen compounds, a process that can be carried out only by certain microorganisms. (p. 422.)

nitrogenous base A nitrogen-containing molecule having basic properties; a purine or pyrimidine; one of the building blocks of nucleic acids. (p. 140.)

node (L. *nodus*, knot) The part of a plant stem where one or more leaves are attached; *see* internode. (p. 678.)

notochord (Gr. *noto*, back + L. *chorda*, cord) In chordates, a dorsal rod of cartilage that runs the length of the body and forms the primitive axial skeleton in the embryos of all chordates; in most adult chordates the notochord is replaced by a vertebral column that forms around (but not from) the notochord. (p. 362.)

nucleic acid A nucleotide polymer; a long chain of nucleotides; chief types are deoxyribonucleic acid (DNA), which is double stranded, and ribonucleic acid (RNA), which is typically single stranded. (p. 59.)

nucleolus (L. a small nucleus) In eukaryotes, the site of rRNA synthesis; a spherical body composed chiefly of rRNA in the process of being transcribed from multiple copies of rRNA genes. (p. 90.)

nucleosome (L. *nucleus*, kernel + *soma*, body) The fundamental packaging unit of eukaryotic chromosomes; a complex of DNA and histone proteins in which one-and-three-quarters turns of the double-helical DNA are wound around eight molecules of histone. Chromatin is composed of long strings of nucleosomes, like beads on a string. (p. 188.)

nucleotide A single unit of nucleic acid, composed of a phosphate, a five-carbon sugar (either ribose or deoxyribose), and a purine or a pyrimidine. (p. 48.)

nucleus In atoms, the central core, containing positively charged protons and (in all but hydrogen) electrically neutral neutrons. In eukaryotic cells, the membranous organelle

that houses the chromosomal DNA; in the central nervous system, a cluster of nerve cell bodies. (p. 37.)

obligate anaerobe An organism that is metabolically active only in the absence of oxygen. (p. 71.)

oncogene (Gr. *oncos*, cancer + *genos*, birth) A cancer-causing gene; a mutant form of a growth-regulating gene that is inappropriately "on," causing unrestrained cell growth and division. Cancer appears to result only when several such controls have been abrogated. (p. 287.)

oocyte (Gr. *oion*, egg + *kytos*, vessel) A cell that gives rise to an ovum by meiosis. (p. 769.)

operator A site of gene regulation; a sequence of nucleotides overlapping the promoter site and recognized by a repressor protein. Binding of the repressor prevents binding of the polymerase to the promoter site and so blocks transcription of the structural gene. (p. 276.)

operon (L. *operis*, work) A cluster of adjacent structural genes transcribed as a unit into a single mRNA molecule; transcription of all the genes of an operon is regulated coordinately by controlling binding of RNA polymerase to the single promoter site, typically via an adjacent and overlapping operator site. A common mode of gene organization in bacteria, but rare in eukaryotes. (p. 275.)

orbital (L. *orbis*, circle) The volume of space surrounding the atomic nucleus in which an electron will be found most of the time. (p. 38.)

order A category of classification above the level of family and below that of class; orders are composed of one or more families. (p. 361.)

organ (Gr. *organon*, tool) A body structure composed of several different tissues grouped together in a structural and functional unit. (p. 608.)

organelle (Gr. *organella*, little tool) Specialized part of a cell; literally, a small organ analogous to the organs of multicellular animals. (p. 78.)

organic Pertaining to living organisms in general, to compounds formed by living organisms, and to the chemistry of compounds containing carbon. (p. 48.)

organism Any individual living creature, either unicellular or multicellular. (p. 29.)

osmoregulation Maintenance of constant internal salt and water concentrations in an organism; the active regulation of internal osmotic pressure. (p. 750.)

osmosis (Gr. *osmos*, act of pushing, thrust) The diffusion of water across a selectively permeable membrane (a membrane that permits the free passage of water but prevents or retards the passage of a solute). In the absence of differences in pressure or volume, the net movement of water is from the side containing a lower concentration of solute to the side containing a higher concentration. (p. 112.)

osmotic pressure The potential pressure developed by a solution separated from pure water by a differentially permeable membrane. Measured as the pressure required to stop the osmotic movement of water into a solution, it is an index of the solute concentration of the solution; the higher the solute concentration, the greater the osmotic potential of the solution. *Osmotic potential* is a synonym. (p. 113.)

osteoblast (Gr. *osteon*, bone + *blastos*, bud) A bone-forming cell. (p. 627.)

ovary (L. *ovum*, egg) (1) In animals, the organ in which eggs are produced. (2) In flowering plants, the enlarged basal portion of a carpel, which contains the ovule(s); the ovary matures to become the fruit. (p. 766.)

oviparity (L. *ovum*, egg + *parere*, to bring forth) Reproduction in which unfertilized eggs are released by the female; fertilization and development of offspring occur outside the maternal body. Adj., *oviparous*. (p. 763.)

ovulation In animals, the release of an egg or eggs from the ovary. (p. 765.)

ovule (L. *ovulum*, a little egg) A structure in seed plants that contains the female gametophyte and is surrounded by the nucellus and one or two integuments; when mature, an ovule becomes a seed. (p. 523.)

ovum, pl. **ova** (L. egg) The egg cell; female gamete. (p. 769.)

oxidation (Fr. *oxider*, to oxidize) Loss of an electron by an atom or molecule. In metabolism, often associated with a gain of oxygen or loss of hydrogen. Oxidation (loss of an electron) and reduction (gain of an electron) take place simultaneously, because an electron that is lost by one atom is accepted by another. Oxidation-reduction reactions are an important means of energy transfer within living systems. (p. 66.)

pancreas (Gr. *pan*, all + *kreas*, meat, flesh) In vertebrates, the principal digestive gland; a small gland located between the stomach and the duodenum which produces digestive enzymes and

the hormones insulin and glucagon. (p. 642.)

parasite (Gr. *para*, beside + *sitos*, food) An organism that lives on or in an organism of a different species and derives nutrients from it. (p. 412.)

parasympathetic nervous system (Gr. *para*, beside + *syn*, with + *pathos*, feeling) In vertebrates, one of two subdivisions of the autonomic nervous system (the other is the sympathetic). The two subdivisions operate antagonistically, the parasympathetic system stimulating resting activities such as digestion and restoring the body to normal after emergencies by inhibiting alarm functions initiated by the sympathetic system. (p. 730.)

parenchyma (Gr. *para*, beside + *en*, in + *chein*, to pour) A plant tissue composed of *parenchyma* cells; such cells are living, thin walled, and randomly arranged, and have large vacuoles; usually photosynthetic or storage tissue. (p. 564.)

pathogen (Gr. *pathos*, suffering + *genesis*, beginning) A disease-causing organism. (p. 486.)

peptide Two or more amino acids linked by peptide bonds. (p. 57.)

peptide bond (Gr. *peptein*, to soften, digest) The type of bond that links amino acids together in proteins; formed by removing an OH from the carboxy (—COOH) group of one amino acid and an H from the amino (—NH₂) group of another to form an amide group CO═NH═. (p. 55.)

perennial (L. *per*, through + *annus*, a year) A plant that lives for more than a year and produces flowers on more than one occasion. (p. 564.)

pericycle (Gr. *peri*, around + *kykos*, circle) In vascular plants, one or more cell layers surrounding the vascular tissues of the root, bounded externally by the endodermis and internally by the phloem. (p. 572.)

periderm (Gr. *peri*, around + *derma*, skin) Outer protective tissue in vascular plants that is produced by the cork cambium and functionally replaces epidermis when it is destroyed during secondary growth; the periderm includes the cork, cork cambium, and phelloderm. (p. 570.)

peripheral nervous system (Gr. *peripherein*, to carry around) All of the neurons and nerve fibers outside the central nervous system, including motor neurons, sensory neurons, and the autonomic nervous system. (p. 616.)

peristalsis (Gr. *peristaltikos*, compressing around) Pumping by waves of contraction; in ani-

mals, a series of alternating contracting and relaxing muscle movements along the length of a tube such as the oviduct or alimentary canal that tend to force material such as an egg cell or food through the tube. (p. 640.)

permeable (L. *permeare*, to pass through) Refers to a membrane through which a specified molecule, ion, or other solute can pass freely. (p. 116.)

peroxisome A microbody that plays an important role in glycolic acid metabolism associated with photosynthesis; the site of photorespiration. (p. 91.)

petal A flower part, usually conspicuously colored; one of the units of the corolla. (p. 526.)

petiole (L. *petiolus*, a little foot) The stalk of a leaf. (p. 567.)

pH A measure of the relative concentration of hydrogen ions in a solution; equal to the negative logarithm of the hydrogen ion concentration. pH values range from 0 to 14; the lower the value, the more hydrogen ions it contains (the more acidic it is); pH 7 is neutral, less than 7 is acidic, more than 7 is alkaline. (p. 47.)

phage *See* bacteriophage.

phagocyte (Gr. *phagein*, to eat + *kytos*, hollow vessel) Any cell that engulfs and devours microorganisms or other particles. (p. 687.)

phagocytosis (Gr., cell-eating) Endocytosis of solid particles; the cell membrane folds inward around the particle (which may be another cell) and then discharges the contents into the cell interior as a vacuole; characteristic of protists, amoebas, the digestive cells of some invertebrates, and vertebrate white blood cells. (p. 114.)

pharynx (Gr., gullet) In vertebrates, a muscular tube that connects the mouth cavity and the esophagus; it serves as the gateway to the digestive tract and to the windpipe (trachea). (p. 360.)

phenotype (Gr. *phainein*, to show + *typos*, stamp or print) The realized expression of the genotype; the physical appearance or functional expression of a trait; the result of the biological activity of proteins or RNA molecules transcribed from the DNA. (p. 210.)

phloem (Gr. *phloos*, bark) In vascular plants, a food-conducting tissue basically composed of sieve elements, various kinds of parenchyma cells, fibers, and sclereids. (p. 246.)

phosphate group —PO₄; a chemical group commonly involved in high-energy bonds. (p. 48.)

phospholipid A phosphorylated lipid; similar in structure to a fat,

but only two fatty acids are attached to the glycerol backbone, with the third space linked to a phosphorylated molecule. Phospholipid molecules have a polar hydrophilic "head" end (contributed by the fatty acids); they orient spontaneously in water to form bimolecular membranes in which the nonpolar tails of the molecules are oriented inward towards one another, away from the polar water environment. Phospholipids are the foundation for all cell membranes. *See* lipid, fatty acid. (p. 106.)

photon (Gr. *photos*, light) The elementary particle of electromagnetic energy; light. (p. 166.)

photoperiodism (Gr. *photos*, light + *periodos*, a period) The tendency of biological reactions to respond to the duration and timing of day and night; a mechanism for measuring seasonal time. (p. 602.)

photophosphorylation (Gr. *photos*, light + *phosphoros*, bringing light) The formation of ATP in the chloroplast during photosynthesis. (p. 172.)

photorespiration The light-dependent production of glycolic acid in chloroplasts and its subsequent oxidation in peroxisomes; this process tends to short-circuit photosynthesis, and becomes progressively more of a drain to plants at higher temperatures. C₄ and CAM photosynthesis are evolutionary responses to this dilemma. (p. 178.)

photosynthesis (Gr. *photos*, light + *syn*, together + *tithenai*, to place) The utilization of light energy to create chemical bonds; the synthesis of organic compounds from carbon dioxide and water, using chemical energy (ATP) and reducing power (NADPH) generated by photosynthesis. (p. 6.)

phototropism (Gr. *photos*, light + *trope*, turning) In plants, a growth response to a light stimulus. (p. 600.)

phylogeny (Gr. *phylon*, race, tribe) The evolutionary relationships among any group of organisms. (p. 27.)

phylum, pl. **phyla** (Gr. *phylon*, race, tribe) A major category, between kingdom and class, of taxonomic classifications. *Division* is an equivalent term used in all groups except animals and heterotrophic protists. (p. 348.)

physiology (Gr. *physis*, nature + *logos*, a discourse) The study of life's functions; how cells, organs, or entire organisms function. (p. 340.)

pigment (L. *pigmentum*, paint) A molecule that absorbs light. (p. 167.)

pinocytosis (Gr. *pinein*, to drink + *kytos*, hollow vessel + *osis*, condition) The taking up of fluid by endocytosis; refers to cells. (p. 114.)

pistil (L. *pistillum*, pestle) Central organ of flowers, typically consisting of ovary, style, and stigma; a pistil may consist of one or more fused carpels, and is more technically and better known as the gynoecium. A flower with carpels but no functional stamens is called *pistillate*. (p. 585.)

pith The ground tissue occupying the center of the stem or root within the vascular cylinder; usually consists of parenchyma. (p. 569.)

pituitary (L. *pituita*, phlegm) Perhaps the most important of the endocrine glands in vertebrates; under the hormonal control of the hypothalamus, which directs it via releasing hormones; the anterior lobe secretes tropic hormones, growth hormone, and prolactin; the posterior lobe stores and releases oxytocin and ADH produced by the hypothalamus. (p. 717.)

placenta, pl. **placentae** (L. a flat cake) (1) In flowering plants, the part of the ovary wall to which the ovules or seeds are attached. (2) In mammals, a tissue formed in part from the inner lining of the uterus and in part from other membranes, through which the embryo (later the fetus) is nourished while in the uterus and wastes are carried away. (p. 374.)

plankton (Gr. *planktos*, wandering) Free-floating, mostly microscopic, aquatic organisms. (p. 355.)

plaque Clear area in a sheet of bacterial cells or eukaryotic cells growing in culture, resulting from the killing (lysis) of contiguous cells by viruses. (p. 487.)

plasma (Gr., form) The fluid of vertebrate blood; contains dissolved salts, metabolic wastes, hormones, and a variety of proteins, including antibodies and albumin; blood minus the blood cells. (p. 612.)

plasma cell An antibody-producing cell resulting from the multiplication and differentiation of a B lymphocyte that has interacted with an antigen; a mature plasma cell can produce from 3000 to 30,000 antibody molecules per second. (p. 691.)

plasma membrane The membrane surrounding the cytoplasm of an animal cell; the outermost membrane of a cell; consists of a single bilayer of membrane; also called cell membrane and plasmalemma. (p. 82.)

plasmid (Gr. *plasma*, a form or mold) A small fragment of extrachromosomal DNA, usually circular, that replicates independently of the main chromosome, although it may have been derived from it. Plasmids make up about 5% of the DNA of many bacteria, but are rare in eukaryotic cells. (p. 291.)

platelet (Gr. dim. of *plattus*, flat) In mammals, a fragment of a white blood cell that circulates in the blood and functions in the formation of blood clots at sites of injury. (p. 672.)

pleiotropic (Gr. *pleros*, more + *trope*, a turning) Describing a gene that produces more than one phenotypic effect. (p. 258.)

pleura (Gr. side, rib) In vertebrate animals, the membrane that lines each half of the thorax and covers the lungs. (p. 660.)

polar molecule A molecule with positively and negatively charged ends; one portion of a polar molecule attracts electrons more strongly (is more electronegative) than another portion, with the result that the electron-rich portion carries a partial negative charge contributed by the electron excess, and the electron-poor portion carries a partial positive charge because of the electron deficit. (p. 45.)

polar nuclei In flowering plants, two nuclei (usually), one derived from each end (pole) of the embryo sac, which become centrally located; they fuse with a male nucleus to form the primary ($3n$) endosperm nucleus. (p. 525.)

pollen (L. fine dust) A collective term for pollen grains. In seed plants, each pollen grain contains an immature male gametophyte enclosed within a protective outer covering; pollen grains may be two-celled or three-celled when shed. (p. 310.)

pollen tube A tube formed after germination of the pollen grain; carries the male gametes into the ovule. (p. 525.)

pollination The transfer of pollen from an anther to a stigma. (p. 578.)

polymer (Gr. *polus*, many + *meris*, part) A molecule composed of many similar or identical molecular subunits. Starch is a polymer of glucose. (p. 48.)

polypeptide (Gr. *polys*, many + *peptein*, to digest) A molecule consisting of many joined amino acids; not as complex as a protein. (p. 57.)

polyploid (Gr. *polys*, many + *ploin*, vessel) An organism, tissue, or cell with more than two complete sets of homologous chromosomes. (p. 291.)

polysaccharide (Gr. *polys*, many + *sakcharon*, sugar, from Sanskrit *sarkara*, gravel, sugar) A sugar polymer; a carbohydrate composed of many monosaccharide sugar subunits linked together in a long chain; examples are glycogen, starch, and cellulose. (p. 49.)

population (L. *populus*, the people) Any group of individuals, usually of a single species, occupying a given area at the same time. (p. 29.)

potential energy Energy that is not being used, but could be; energy in a potentially usable form; often called "energy of position." (p. 130.)

predation The eating of other organisms. The one doing the eating is called a predator, and the one being consumed is called the prey. (p. 407.)

prey (L. *prehendere*, to grasp, seize) An organism eaten by another organism. (p. 407.)

primary growth In vascular plants, growth originating in the apical meristems of shoots and roots, as contrasted with secondary growth; results in an increase in length. (p. 521.)

primary structure of a protein The amino acid sequence of a protein. (p. 57.)

primordium, pl. **primordia** (L. *primus*, first + *ordiri*, to begin to weave) A cell or organ in its earliest stage of differentiation. (p. 583.)

productivity A measure of the rate at which energy is assimilated by an organism or group of organisms. (p. 424.)

prokaryote (Gr. *pro*, before + *karyon*, kernel) A bacterium; a cell lacking a membrane-bound nucleus or membrane-bound organelles. Prokaryotic cells are more primitive than eukaryotic cells, which evolved from them. (p. 31.)

promoter A specific nucleotide sequence on a chromosome to which RNA polymerase attaches to initiate transcription of mRNA from a gene. (p. 275.)

prophase (Gr. *pro*, before + *phasis*, form) An early stage in nuclear division, characterized by the formation of a microtubule spindle along the future axis of division, the shortening and thickening of the chromosomes, and their movement toward the equator of the spindle (the "metaphase plate"). (p. 191.)

proprioceptor (L. *proprius*, one's own) In vertebrates, a sensory receptor that senses the body's position and movements; located deep within the tissues, especially muscles, tendons, and joints. (p. 722.)

prosimian (Gr. *pro*, before + L. *simia*, ape) Non-anthropoid primates, including lemurs, tarsiers, and lorises. (p. 375.)

prostate gland (Gr. *prostas*, a porch or vestibule) In male mammals, a mass of glandular tissue at the base of the urethra that secretes an alkaline fluid which has a stimulating effect on the sperm as they are released. (p. 772.)

protease (Gr. *proteios*, primary + *ase*, enzyme ending) An enzyme that digests proteins by breaking peptide bonds. Also called peptidases. (p. 638.)

protein (Gr. *proteios*, primary) A chain of amino acids joined by peptide bonds; a protein typically contains over 100 amino acids and may be composed of more than one polypeptide. (p. 48.)

protist (Gr. *protos*, first) A member of the kingdom Protista, which includes the unicellular eukaryotic organisms and some multicellular lines derived from them. (p. 33.)

proton A subatomic, or elementary, particle with a single positive charge equal in magnitude to the charge of an electron and a mass of 1, very close to that of a neutron; the nucleus of a hydrogen atom is composed of a single proton. (p. 37.)

protostome (Gr. *protos*, first + *stoma*, mouth) An animal in whose embryonic development the mouth forms at or near the blastopore. Protostomes are also characterized by spiral cleavage during the earliest stages of development and by schizocoelous formation of the coelom. (p. 540.)

pseudocoel, or **pseudocoelom** (Gr. *pseudos*, false + *koiloma*, cavity) A body cavity not lined with peritoneum and not a part of the blood or digestive systems, embryonically derived from the blastocoel. (p. 537.)

pseudopod, or **pseudopodium** (Gr. *pseudes*, false + *pous*, foot) "False foot"; a nonpermanent cytoplasmic extension of the cell body. (p. 503.)

pulmonary circulation In terrestrial vertebrates, the pathway of blood circulation leading to and from the lungs. (p. 675.)

punctuated equilibrium A model of the mechanism of evolutionary change which proposes that long periods of little or no change are punctuated by periods of rapid evolution. (p. 341.)

purine (Gr. *purinos*, fiery, sparkling) The larger of the two general kinds of nucleotide base found in DNA and RNA; a nitrogenous base with a double-ring structure, such as adenine or guanine. (p. 60.)

pyrimidine (alt. of pyridine, from Gr. *pyr*, fire) The smaller of the two general kinds of nucleotide base found in DNA and RNA; a nitrogenous base with a single-ring structure, such as cytosine, thymine, or uracil. (p. 60.)

pyruvate The three-carbon compound that is the end product of glycolysis and the starting material of the citric acid cycle. (p. 148.)

quaternary structure of a protein The level of aggregation of a globular protein molecule that consists of two or more polypeptide chains; a monomer has one subunit only, a dimer has two, a trimer three, and a tetramer four. (p. 59.)

radial symmetry (L. *radius*, a spoke of a wheel + Gr. *summetros*, symmetry) The regular arrangement of parts around a central axis such that any plane passing through the central axis divides the organism into halves that are approximate mirror images. (p. 534.)

radioisotope An unstable isotope of an element that decays or disintegrates spontaneously, emitting radiation. (p. 319.)

radula (L. scraper) Rasping tongue found in most mollusks. (p. 542.)

recessive allele (L. *recedere*, to recede) An allele whose phenotypic effect is masked in the heterozygote by that of another, dominant allele; heterozygotes, although they contain a copy of the recessive allele, are phenotypically indistinguishable from dominant homozygotes. (p. 215.)

reciprocal altruism Performance of an altruistic act with the expectation that the favor will be returned. A key and very controversial assumption of many theories dealing with the evolution of social behavior. *See* altruism. (p. 794.)

recombinant DNA Fragments of DNA from two different species, such as a bacterium and a mammal, spliced together in the laboratory into a single molecule; a key component of genetic engineering technology, in which, for example, a gene that makes a certain bacterium resistant to a chemical weed killer is transferred to a crop plant. (p. 296.)

recombination The formation of new gene combinations; in bacteria, it is accomplished by the transfer of genes into cells, often in association with viruses; in eukaryotes, it is accomplished by reassortment of chromosomes during meiosis, and by crossing-over. (p. 290.)

reduction (L. *reductio*, a bringing back; originally "bringing back"

a metal from its oxide) The gain of an electron by an atom; takes place simultaneously with oxidation (loss of an electron by an atom), because an electron that is lost by one atom is accepted by another. (p. 66.)

reflex (L. *reflectere*, to bend back) In the nervous system, an "automatic" response to a stimulus; a motor response subject to little associative modification; among the simplest neural pathways, involving only a sensory neuron, sometimes (but not always) an interneuron, and one or more motor neurons. (p. 729.)

releasing hormone A peptide hormone produced by the hypothalamus that stimulates or inhibits the secretion of specific hormones by the anterior pituitary. (p. 739.)

renal (L. *renes*, kidneys) Pertaining to the kidney. (p. 755.)

repressor (L. *reprimere*, to press back, keep back) A protein that regulates DNA transcription by preventing RNA polymerase from attaching to the promoter and transcribing the structural gene. *See* operator. (p. 264.)

respiration (L. *respirare*, to breathe) The utilization of oxygen; in terrestrial vertebrates, the inhalation of oxygen and the exhalation of carbon dioxide; in cells, the oxidation (electron removal) of food molecules, particularly pyruvate in the citric acid cycle, to obtain energy. (p. 653.)

resting membrane potential The charge difference (difference in electric potential) that exists across a neuron at rest (about 70 millivolts). (p. 704.)

restriction endonuclease An enzyme that cleaves a DNA duplex molecule at a particular base sequence. (p. 295.)

retina (L. a small net) The photosensitive layer of the vertebrate eye; contains several layers of neurons and light receptors (rods and cones); receives the image formed by the lens and transmits it to the brain via the optic nerve. (p. 727.)

retrovirus (L. *retro*, turning back) An RNA virus; a virus whose genetic material is RNA. When a retrovirus enters a cell, the cell's machinery "reads" the virus RNA, which contains a gene encoding an enzyme—reverse transcriptase—that then transcribes the virus RNA into duplex DNA, which the cell's machinery replicates as if it were its own; the DNA copies of many retroviruses appear able to enter eukaryotic chromosomes, where they act much like transposons. Retroviruses are associated with many human diseases,

including cancer and AIDS. (p. 285.)

reverse transcriptase An enzyme that transcribes RNA into DNA; found only in association with retroviruses. (p. 272.)

rhizome (Gr. *rhizoma*, mass of roots) In vascular plants, a usually more or less horizontal underground stem; may be enlarged for storage, or may function in vegetative reproduction. (p. 584.)

ribonucleic acid (RNA) A class of nucleic acids characterized by the presence of the sugar ribose (DNA contains deoxyribose instead) and the pyrimidine uracil (DNA contains thymine instead); includes mRNA, tRNA, and rRNA. (p. 59.)

ribose A five-carbon sugar. (p. 140.)

ribosomal RNA (rRNA) A class of RNA molecules found, together with characteristic proteins, in ribosomes; transcribed from the DNA of the nucleolus. (p. 89.)

ribosome The molecular machine that carries out protein synthesis; the most complicated aggregation of proteins in a cell, also containing three different rRNA molecules; may be free in the cytoplasm or in eukaryotes sometimes attached to the membranes of the endoplasmic reticulum. (p. 87.)

RNA *See* ribonucleic acid.

RNA polymerase An enzyme that catalyzes the assembly of an mRNA molecule, the sequence of which is complementary to a DNA molecule used as a temperature. *See* Transcription. (p. 59.)

rod Light-sensitive nerve cell found in the vertebrate retina; sensitive to very dim light; responsible for "night vision." (p. 725.)

root The usually descending axis of a plant, normally below ground, which anchors the plant and serves as the major point of entry for water and minerals. (p. 562.)

root hairs In vascular plants, tubular outgrowths of the epidermal cells of the root just back of its apex; most water enters through the root hairs. (p. 509.)

root pressure In vascular plants, the pressure that develops in roots as the result of osmosis, which causes guttation of water from leaves and exudation from cut stumps. (p. 595.)

r-selection (from the *r* term in the logistic equation) Natural selection under conditions that favor survival when populations are controlled primarily by density-independent factors; contrast with *K*-selection. (p. 397.)

sarcomere (Gr. *sarx*, flesh + *meris*, part of) Fundamental unit of contraction in skeletal muscle; repeating bands of actin and myosin that appear between two Z lines. (p. 632.)

sclereid (Gr. *skleros*, hard) In vascular plants, a sclerenchyma cell with a thick, lignified, secondary wall having many pits; not elongate like a fiber, the other principal kind of sclerenchyma cell. (p. 565.)

sclerenchyma cell (Gr. *skleros*, hard + *en*, in + *chymein*, to pour) A cell of variable form and size with more or less thick, often lignified, secondary walls; may or may not be living at maturity; includes fibers and sclereids. Collectively, sclerenchyma cells may make up a kind of tissue called sclerenchyma. (p. 565.)

secondary growth In vascular plants, an increase in stem and root diameter made possible by cell division of the lateral meristems. Secondary growth produces the secondary plant body. (p. 521.)

secondary sex characteristics External differences between male and female animals; not directly involved in reproduction. (p. 739.)

secondary structure of a protein The twisting or folding of a polypeptide chain; results from the formation of hydrogen bonds between different amino acid side groups of a chain; the most common structures that form are a single-stranded helix, an extended sheet, or a cable containing three (as in collagen) or more strands. (p. 58.)

seed A structure that develops from the mature ovule of a seed plant; seeds generally consist of seed coat, embryo, and a food reserve. (p. 521.)

seed coat The outer layer of a seed, developed from the integuments of the ovule. (p. 580.)

segregation of alleles *See* Mendel's first law.

selectively permeable membrane (L. *seligere*, to gather apart + *permeare*, to go through) A membrane that permits passage of water and some solutes but blocks passage of one or more solutes. Same as differentially permeable. Used to be referred to as "semipermeable." (p. 116.)

self-fertilization The union of egg and sperm produced by a single hermaphroditic organism. (p. 207.)

self-pollination The transfer of pollen from another to stigma in the same flower or to another flower of the same plant, leading to self-fertilization. (p. 207.)

semipermeable membrane *See* selectively permeable membrane.

sepal (L. *sepalum*, a covering) A member of the outermost floral whorl of a flowering plant; collectively, the sepals constitute the calyx. (p. 526.)

sessile (L. *sessilis*, of or fit for sitting, low, dwarfed) Attached; not free to move about. In vascular plants, a leaf lacking a petiole is said to be sessile. (p. 533.)

sex chromosomes Chromosomes that are different in the two sexes and that are involved in sex determination. All other chromosomes are called autosomes. *See* autosome. (p. 224.)

sex-linked characteristic A genetic characteristic, such as color blindness in humans or white eye in fruit flies, that is determined by a gene located on a sex chromosome and that therefore shows a different pattern of inheritance in males than in females. (p. 216.)

sexual reproduction The fusion of gametes followed by meiosis and recombination of some point in the life cycle. (p. 478.)

shoot In vascular plants, the aboveground portions, such as the stem and leaves. (p. 562.)

short-day plants Plants that must be exposed to light periods shorter than some critical length for flowering to occur; they flower primarily in autumn. (p. 602.)

sieve cell In the phloem (food-conducting tissue) of vascular plants, a long, slender sieve element with relatively unspecialized sieve areas and with tapering end walls that lack sieve plates; found in all vascular plants except angiosperms, which have sieve-tube members. (p. 566.)

sieve tube In the phloem of angiosperms, a series of sieve-tube members arranged end-to-end and interconnected by sieve plates. (p. 566.)

sinus (L. curve) A cavity or space in tissues or in bone. (p. 743.)

smooth muscle Nonstriated muscle; lines the walls of internal organs and arteries and is under involuntary control. (p. 627.)

solute A molecule dissolved in some solution. As a general rule, solutes dissolve only in solutions of similar polarity—for example, glucose (polar) dissolves in (forms hydrogen bonds with) water (also polar), but not in vegetable oil (nonpolar). (p. 112.)

solution A homogeneous mixture of the molecules of two or more substances; the substance present in the greatest amount (usually a liquid) is called the solvent, and the substances pres-

ent in lesser amounts are called solutes. (p. 112.)

somatic cells (Gr. *soma*, body) The differentiated cells composing body tissues of multicellular plants and animals; all body cells except those giving rise to gametes. (p. 196.)

somatic nervous system (Gr. *soma*, body) In vertebrates, the neurons of the peripheral nervous system that control skeletal muscle; the "voluntary" system, as contrasted with the "involuntary," or autonomic, nervous system. (p. 714.)

species, pl. **species** (L. kind, sort) A kind of organism; species are designated by binomial names written in italics. (p. 331.)

specificity Selectivity; as in the highly specific choice of substrate by an enzyme. (p. 692.)

spermatid (Gr. *sperma*, seed) In animals, each of four haploid (1*n*) cells that result from the meiotic divisions of a spermatocyte; each spermatid differentiates into a sperm cell. (p. 767.)

spermatocytes (Gr. *sperma*, seed + *kytos*, vessel) In animals, the diploid (2*n*) cells formed by the enlargement of the spermatogonia; they give rise by meiotic division to the spermatids. (p. 767.)

sphincter (Gr. *sphinkter*, band, from *sphingein*, to bind tight) In vertebrate animals, a ring-shaped muscle capable of closing a tubular opening by constriction (such as the one between stomach and small intestine, or between anus and exterior). (p. 642.)

spindle The motive assembly that carries out the separation of chromosomes during cell division; composed of microtubules and assembled during prophase at the equator of the dividing cell. (p. 191.)

spindle fibers A group of microtubules that together make up the spindle. (p. 722.)

spiracle (L. *spiraculum*, from *spirare*, to breathe) External opening of a trachea in arthropods. (p. 548.)

spongy parenchyma A leaf tissue composed of loosely arranged, chloroplast-bearing cells. *See* palisade parenchyma. (p. 568.)

sporangium, pl. **sporangia** (Gr. *spora*, seed + *angeion*, a vessel) A structure in which spores are produced. (p. 507.)

spore A haploid reproductive cell, usually unicellular, capable of developing into an adult without fusion with another cell. Spores result from meiosis, as do gametes, but gametes fuse immediately to produce a new diploid cell. (p. 484.)

sporophyte (Gr. *spora*, seed +

phyton, plant) The spore-producing, diploid (2*n*) phase in the life cycle of a plant having alternation of generations. (p. 516.)

stamen (L. thread) The organ of a flower that produces the pollen; usually consists of another and filament; collectively, the stamens make up the androecium. (p. 527.)

starch (Mid. Eng. *sterchen*, to stiffen) An insoluble polymer of glucose; the chief food storage substance of plants; typically composed of 1000 or more glucose units. (p. 51.)

statocyst (Gr. *statos*, standing + *kystis*, sac) A sensory receptor sensitive to gravity and motion; consists of a vesicle containing granules of sand (statoliths) or some other material that stimulates surrounding tufts of cilia when the organism moves. (p. 722.)

stem The aboveground axis of vascular plants; stems are sometimes below ground (as in rhizomes and corms). (p. 562.)

steroid (Gr. *stereos*, solid + L. *ol*, from *oleum*, oil) One of a group of lipids having a molecular skeleton of four fused carbon rings and, often, a hydrocarbon tail. Cholesterol, sex hormones, and the hormones of the adrenal cortex are steroids. (p. 54.)

stigma (Gr., mark, tattoo mark) (1) In angiosperm flowers, the region of a carpel that serves as a receptive surface for pollen grains. (2) Light-sensitive eyespot of some algae. (p. 526.)

stolon (L. *stolo*, shoot) (1) A stem that grows horizontally along the ground surface and may form adventitious roots, such as runners of the strawberry plant. (2) A functionally similar structure that occurs in some colonial cnidarians and ascidians. (p. 584.)

stoma, pl. **stomata** (Gr. mouth) In plants, a minute opening bordered by guard cells in the epidermis of leaves and stems; water passes out of a plant mainly through the stomata, and CO_2 passes in chiefly by the same pathway. (p. 515.)

striated muscle (L. from *striare*, to groove) Skeletal voluntary muscle and cardiac muscle. The name derives from its striped appearance, which reflects the arrangement of contractile elements. *See* voluntary muscle. (p. 628.)

stroma (Gr. anything spread out) (1) The ground substance of plastids. (2) The supporting connective tissue framework of an animal organ; filmy framework of red blood corpuscles and certain cells. (p. 178.)

subspecies A subdivision of a

species, often a geographically distinct race. (p. 338.)

substrate (L. *substratus*, strewn under) The foundation to which an organism is attached; a molecule upon which an enzyme acts. (p. 137.)

substrate level phosphorylation Formation of ATP that takes place via a coupled reaction during glycolysis. (p. 146.)

succession In ecology, the slow, orderly progression of changes in community composition that takes place through time. *Primary succession* occurs in nature over long periods of time; *secondary succession* occurs when a climax community has been disturbed. (p. 427.)

sucrose Cane sugar; a common disaccharide found in many plants; a molecule of glucose linked to a molecule of fructose. (p. 45.)

sugar Any monosaccharide or disaccharide. (p. 49.)

surface tension A tautness of the surface of a liquid, caused by the cohesion of the molecules of liquid. Water has an extremely high surface tension. *See* cohesion. (p. 46.)

symbiosis (Gr. *syn*, together with + *bios*, life) The living together in close association of two or more dissimilar organisms; includes parasitism (in which the association is harmful to one of the organisms), commensalism (in which it is beneficial to one, of no significance to the other), and mutualism (in which the association is advantageous to both). (p. 92.)

sympathetic nervous system A subdivision of the autonomic nervous system of vertebrates that functions as an alarm response; increases heartbeat and dilates blood vessels while putting the body's everyday functions, such as digestion, on hold; usually operates antagonistically with parasympathetic nerves; in times of stress, danger, or excitement, it mobilizes the body for rapid response. (p. 730.)

synapse (Gr. *synapsis*, a union) A junction between a neuron and another neuron or muscle cell; the two cells do not touch, the gap being bridged by neurotransmitter molecules. (p. 706.)

synapse, excitatory A synapse in which the receiving cell's receptors respond to the arrival of neurotransmitter molecules by increasing the receptor cell membrane's permeability to potassium or chloride or both; this drives its membrane potential toward threshold; excitatory synapses increase the likelihood that a receiving cell will fire an action potential. (p. 709.)

synapse, inhibitory A synapse in which the receiving cell's receptors respond to the arrival of neurotransmitter molecules by decreasing the receptor cell membrane's permeability to sodium and potassium; this drives its membrane potential away from threshold. Inhibitory synapses decrease the likelihood that a receiving cell will fire an action potential. (p. 709.)

synapsis (Gr. contact, union) The point-by-point alignment (pairing) of homologous chromosomes that occurs before the first meiotic division; crossing-over occurs during synapsis. (p. 197.)

syngamy (Gr. *syn*, together with + *gamos*, marriage) The process by which two haploid cells fuse to form a diploid zygote; fertilization. (p. 196.)

synthesis (Gr. *syntheke*, a putting together) The formation of a more complex molecule from simpler ones. (p. 277.)

systemic circulation The circulation path of blood leading to and from all body parts except the lungs. (p. 675.)

T cell A type of lymphocyte involved in cell-mediated immunity and interactions with B cells; the "T" refers to the fact that T cells are produced in the thymus; also called a T lymphocyte. (p. 687.)

taxonomy (Gr. *taxis*, arrangement + *nomos*, law) The science of the classification of organisms; equivalent to *systematics*. (p. 471.)

telophase (Gr. *telos*, end + *phasis*, form) The last stage of the nuclear division of mitosis and meiosis, during which the chromosomes become reorganized into two nuclei. (p. 193.)

tertiary structure of a protein The three-dimensional shape of a protein; primarily the result of hydrophobic interactions of amino acid side groups and, to a lesser extent, of hydrogen bonds between them; forms spontaneously. (p. 58.)

testis, pl. **testes** (L., *witness*) In mammals, the sperm-producing organ; also the source of male sex hormone. (p. 767.)

testosterone (Gr. *testis*, testicle + *steiras*, barren) The male sex hormone; a steroid hormone secreted by the testes of mammals that stimulates the development and maintenance of male sex characteristics and the production of sperm. (p. 739.)

thalamus (Gr. *thalamos*, chamber) That part of the vertebrate forebrain just posterior to the cerebrum; governs the flow of information from all other parts of the nervous system to the cerebrum. (p. 717.)

theory (Gr. *theorein*, to look at) A well-tested hypothesis, one unlikely to be rejected by future tests. (p. 14.)

thermodynamics (Gr. *therme*, heat + *dynamis*, power) The study of transformations of energy, using heat as the most convenient form of measurement of energy. The first law of thermodynamics states that the total energy of the universe remains constant. The second law of thermodynamics states that the entropy, or degree of disorder, tends to increase. (p. 130.)

thigmotropism In plants, unequal growth in some structure that comes about as a result of physical contact with an object. The unequal growth that causes a vine to curl around a fence post is an example. (p. 601.)

thorax (Gr. a breastplate) (1) In vertebrates, that portion of the trunk containing the heart and lungs. (2) In crustaceans and insects, the fused, leg-bearing segments between head and abdomen. (p. 365.)

threshold value In neurons, the minimum change in membrane potential necessary to produce an action potential. (p. 705.)

thylakoid (Gr. *thylakos*, sac + *-oides*, like) A saclike membranous structure in cyanobacteria and the chloroplasts of eukaryotic organisms; in chloroplasts, stacks of thylakoids form the grana; chlorophyll is found in the thylakoids. (p. 178.)

thymine A pyrimidine occurring in DNA but not in RNA; *see also* uracil. (p. 60.)

tight junction Region of actual fusion of cell membranes between two adjacent animal cells that prevents materials from leaking through the tissue; for example, intestinal epithelial cells are surrounded by tight junctions. (p. 129.)

tissue (L. *texere*, to weave) A group of similar cells organized into a structural and functional unit. (p. 117.)

trachea, pl. **tracheae** (L. windpipe) A tube for breathing; in terrestrial vertebrates, the windpipe that carries air between the larynx and bronchi (which leads to the lungs); in insects and some other terrestrial arthropods, a system of chitin-lined air ducts. (p. 360.)

tracheid (Gr. *tracheia*, rough) In vascular plants, an elongated, thick-walled conducting and supporting cell of xylem. Tracheids, which are dead when functional, have tapering ends and pitted walls without perfora-

tions. Compare with *vessel member*. (p. 566.)

transcription (L. *trans*, across + *scribere*, to write) The enzyme-catalyzed assembly of an RNA molecule complementary to a strand of DNA; may be either mRNA (for protein-encoding genes), tRNA, or rRNA. (p. 275.)

transcription unit The portion of a gene that is actually transcribed into mRNA; a sequence of DNA that begins with a promoter and encompasses the leader region (where the ribosome binds to mRNA) and one or more structural genes. A transcription unit does not include the regulatory portions of a gene such as operators, enhancers, or repressor-encoding sequences. (p. 275.)

transfection The transformation of eukaryotic cells in culture. (p. 287.)

transfer RNA (tRNA) (L. *trans*, across + *ferre*, to bear or carry) A class of small RNAs (about 80 nucleotides) with two functional sites; to one site an "activating enzyme" adds a specific amino acid; the other site carries the nucleotide triplet (anticodon) for that amino acid. Each type of tRNA transfers a specific amino acid to a growing polypeptide chain as specified by the nucleotide sequence of the messenger RNA being translated. *See* activating enzyme. (p. 264.)

transformation (L. *trans*, across + *formare*, to shape) The transfer of naked DNA from one organism to another; also called simply "gene transfer." First observed as uptake of DNA fragments among pneumococcal bacteria, transformation is now routinely carried out in fruitflies and other eukaryotes by microinjection of eggs, with transposons often used as vectors. (p. 247.)

translation (L. *trans*, across + *latus*, that which is carried) The assembly of a protein on the ribosomes, using mRNA to direct the order of amino acids. (p. 266.)

translocation (L. *trans*, across + *locare*, to put or place) (1) In plants, the long-distance transport of soluble food molecules (mostly sucrose), which occurs primarily in the sieve tubes of phloem tissue. (2) In genetics, the interchange of chromosome segments between nonhomologous chromosomes. (p. 596.)

transpiration (L. *trans*, across + *spirare*, to breathe) The loss of water vapor by plant parts; most transpiration occurs through the stomata. (p. 594.)

transposon (L. *transponere*, to change the position of) A DNA sequence carrying one or more genes and flanked by insertion sequences that confer the ability to move from one DNA molecule to another; an element capable of transposition, the changing of chromosomal location. (p. 288.)

tricarboxylic acid cycle or **TCA cycle** *See* Krebs cycle.

trichome (Gr. *trichos*, hair) In plants, an outgrowth of the epidermis, such as a hair, scale, or water vesicle. (p. 565.)

trophic level (Gr. *trophos*, feeder) A step in the movement of energy through an ecosystem. (p. 425.)

tropism (Gr. *trope*, a turning) A response to an external stimulus. (p. 600.)

troponin Complex of globular proteins positioned at intervals along the actin filament of skeletal muscle; thought to serve as a calcium-dependent "switch" in muscle contraction. (p. 630.)

tubulin (L. *tubulus*, small tube + *in*, belonging to) Globular protein subunit forming the hollow cylinder of microtubules. (p. 96.)

turgor (L. *turgere*, to swell) The pressure exerted on the inside of a plant cell wall by the fluid contents of the cell; the interior of the cell is hypertonic in relation to the fluids surrounding it and so gains water by osmosis. (p. 601.)

turgor pressure (L. *turgor*, a swelling) The pressure within a cell resulting from the movement of water into the cell. A cell with high turgor pressure is said to be *turgid*. *See* osmotic pressure. (p. 601.)

unicellular Composed of a single cell. (p. 473.)

uracil A pyrimidine found in RNA but not in DNA; *see also* thymine. (p. 60.)

urea (Gr. *ouron*, urine) An organic molecule formed in the vertebrate liver; the principal form of disposal of nitrogenous wastes by mammals. (p. 745.)

urethra (Gr. from *ourein*, to urinate) The tube carrying urine from the bladder to the exterior of mammals. (p. 768.)

urine (Gr. *ouron*, urine) The liquid waste filtered from the blood by the kidney and stored in the bladder pending elimination through the urethra; the principal excretory product in birds, reptiles, and insects. (p. 751.)

uterus (L. womb) In mammals, a chamber in which the developing embryo is contained and nurtured during pregnancy. (p. 770.)

vacuole (L. *vacuus*, empty) A space or cavity within the cytoplasm of a cell; typically filled with a watery fluid, the cell sap; part of the lysosomal compartment of the cell. (p. 84.)

vascular (L. *vasculum*, a small vessel) Containing or concerning vessels that conduct fluid. (p. 520.)

vascular bundle In vascular plants, a strand of tissue containing primary xylem and primary phloem (and procambium if still present) and frequently enclosed by a bundle sheath of parenchyma or fibers. (p. 521.)

vascular cambium In vascular plants, a cylindrical sheath of meristematic cells, the division of which produces secondary phloem outwardly and secondary xylem inwardly; the activity of the vascular cambium increases stem or root diameter. (pp. 564, 570.)

vas deferens (L. *vas*, a vessel + *deferre*, to carry down) In mammals, the tube carrying sperm from the testes to the urethra. (p. 768.)

vein (L. *vena*, a blood vessel) (1) In plants, a vascular bundle forming a part of the framework of the conducting and supporting tissue of a stem or leaf. (2) In animals, a blood vessel carrying blood from the tissues to the heart. (p. 669.)

ventral (L. *venter*, belly) Pertaining to the undersurface of an animal that moves on all fours, and to the front surface of an animal that holds its body erect. (p. 536.)

ventricle (L. *ventriculus*, the stomach—i.e., the belly of the heart) A muscular chamber of the heart that receives blood from an atrium and pumps blood out to either the lungs or the body tissues. (p. 574.)

vertebrate An animal having a backbone made of bony segments called vertebrae. (p. 554.)

vesicle (L. *vesicula*, a little bladder) A small, intracellular, membrane-bound sac in which various substances are transported or stored. (p. 86.)

vessel (L. *vas*, a vessel) A tube-like element in the xylem of angiosperms, composed of dead cells (vessel elements) arranged end to end. Its function is to conduct water and minerals from the soil. (p. 566.)

vessel element In vascular plants, a typically elongated cell, dead at maturity, which conducts water and solutes in the xylem; vessel elements make up vessels. *Tracheids* are less specialized conducting cells. (p. 566.)

villus, pl. **villi** (L. a tuft of hair) In vertebrates, one of the minute, fingerlike projections lining the small intestine that serve to increase the absorptive surface area of the intestine. (p. 642.)

viroids A single-stranded RNA virus with no coat. In plants, an infectious agent responsible for diseases. (p. 489.)

vitamin (L. *vita*, life + *amine*, of chemical origin) An organic substance that cannot be synthesized by a particular organism but is required in small amounts for normal metabolic function; must be supplied in the diet or synthesized by bacteria living in the intestine. (p. 646.)

viviparity (L. *vivus*, alive + *parere*, to bring forth) Reproduction in which eggs develop within the mother's body, with her nutritional aid; characteristic of mammals, many reptiles, and some fishes; offspring are born as juveniles. Adj., viviparous. (p. 763.)

water cycle Worldwide circulation of water molecules, powered by the sun. (p. 420.)

wood Accumulated secondary xylem. (p. 571.)

xylem (Gr. *xylon*, wood) In vascular plants, a specialized tissue, composed primarily of elongate, thick-walled conducting cells, which transports water and solutes through the plant body. (p. 521.)

yolk The stored food in egg cells; it nourishes the embryo. (p. 369.)

zygote (Gr. *zygotos*, paired together) The diploid (2*n*) cell resulting from the fusion of male and female gametes (fertilization). A zygote may either develop into a diploid individual by mitotic divisions or undergo meiosis to form haploid (*n*) individuals that divide mitotically to form a population of cells. (p. 196.)

ILLUSTRATION CREDITS

i George B. Schaller

ii Kjell Sandved

vii George Venable, property Smithsonian Institution

viii George Venable, property Mr. Victor Vount

ix George Venable, property Smithsonian Institution

xiii NASA
John Reader

xiv Michael Gadomski/Tom Stack & Associates
Peter Parks/Oxford Scientific Film/Animals Animals

xv T.E. Adams/Visuals Unlimited
John D. Cunningham/Visuals Unlimited

xvi National Optical Astronomy Observatories
Bob McKeever/Tom Stack & Associates

xvii John Gerard

viii Ed Reschke

xix Richard Gross
Grant Heilman Photography

xx John D. Cunningham/Visuals Unlimited

xxi Transvaal Museum, Pretoria, South Africa

xxii William L. Hamilton

xxiii NASA
William L. Hamilton

xxiv Peter Parks/Oxford Scientific Film
Walt Anderson/Tom Stack & Associates

xxv E.S. Ross
Manfred Kage/Peter Arnold

xxvi L. West
John Shaw/Tom Stack & Associates

xxvii L. West
Rod Plank/Tom Stack & Associates

xxviii Alan Briere/Tom Stack & Associates
W.J. Weber/Visuals Unlimited

xxix W.J. Weber/Visuals Unlimited
Ron Kimball

xxx Russell Mittermeier
E.S. Ross

xxxi Arthur Holzman/Animals Animals
Russell Mittermeier

xxxii John Cunningham/Visuals Unlimited

1-1 Dr. Jeanne A. Mortimer

1-2, A Mary Field
B Alan Briere/Tom Stack & Associates
C E.S. Ross
D John Cunningham/Visuals Unlimited

1-3 Y. Arthus-Bertrand/Peter Arnold, Inc.

1-4 Peter Parks/Oxford Scientific Films/Animals Animals

1-5 Runk-Schoenberger/Grant Heilman Photography

1-6 Bob Gossington/Tom Stack & Associates

1-7 Sidney Fox

1-8 E.S. Ross

1-9 E.S. Ross

1-10 Gretchen Faust

1-11 American Cancer Society

1-14 Lennart Nilsson

2-1 Christopher Ralling

2-3 Bridgeman/Art Resource

2-5 Illustration by Ron Ervin

2-6 A Nita van der Werff
B John S. Dunning

2-7 Cleveland P. Hickman

2-8 Photo Researchers

2-9 Mary Evans Picture Library/Photo Researchers

2-10 John D. Cunningham/Visuals Unlimited

2-11 William Ober

2-12 Illustration by Ron Ervin

2-13 William Ober

2-14 William Ober

2-15 Carolina Biological Supply

2-16 Cleveland P. Hickman

2-17 Cleveland P. Hickman

2-18 Marty Snyderman

2-19 William L. Hamilton

Table 2-2: 1 NASA
2 William Ober
3 William Ober
4 William Ober
5 William Ober
8 Ed Reschke
10 Ed Reschke

Table 2-3: 1 Sherman Thomson/Visuals Unlimited
2 T.E. Adams/Visuals Unlimited
3 E.S. Ross
4 John D. Cunningham/Visuals Unlimited
5 Leonard Lee Rue III

3-1 National Optical Astronomy Observatories

3-2 William Ober

3-4 Michael Godomski/Tom Stack & Associates

3-5, B National Archives/Picture Research

3-7, A E.S. Ross
B Frank Awbrey/Visuals Unlimited
C Kjell Sandved
D Gerald Corsi/Tom Stack & Associates

3-8 Michael & Patricia Fogden

3-10, A William Ober
B Clyde Smith/Peter Arnold
C E.S. Ross

3-15 E.S. Ross

3-16, B Manfred Kage/Peter Arnold

3-17, B J.D. Litvay/Visuals Unlimited

3-18 Cleveland P. Hickman

3-23, A Michael Pasdzior/Image Bank
B Peter Arnold
C Breck P. Kent/Animals Animals
D Oxford Scientific Films/Animals Animals
E Scott Blackman/Tom Stack & Associates

3-26, B Illustration by Ron Ervin

4-1 E.S. Ross

4-2 Bob McKeever/Tom Stack & Associates

4-5, A E.S. Barghoorn
B J.W. Schopf, *Journal of Paleontology*, vol. 45, pages 925-960, 1971.

4-6 S.M. Siegel

4-8 J.W. Schopf, *Journal of Paleontology*, vol. 45, pages 925-960, 1971.

4-A NASA

4-B NASA

4-C NASA

4-D California Institute of Technology

4-E National Optical Astronomy Observatories

5-1 Manfred Kage/Peter Arnold

5-2 Manfred Kage/Peter Arnold

5-3 L.L. Sims/Visuals Unlimited

5-4, A J.J. Cardamone, Jr., and B.K. Pugashetti; BPS/Tom Stack & Associates

B P.M. Phillips/Visuals Unlimited

C Ed Reschke

5-5, A Peter Arnold

5-6 Eldon H. Newcomb and T.D. Pugh

5-9 J. David Robertson, from Charles Flickinger, *Medical Cell Biology*, W.B. Saunders Co., Philadelphia, 1979.

5-10, A Richard Rodewald

B Illustration by Ron Ervin

5-11, A Illustration by Ron Ervin

B T.W. Tillack

C Charles Flickinger, *Medical Cell Biology*, W.B. Saunders Co., Philadelphia, 1979.

5-12 Ed Reschke

5-13 C.P. Morgan and R.A. Jersild, *Anatomical Record*, vol. 166, pages 575-586, 1970.

5-14, A Charles Flickinger, *Medical Cell Biology*, W.B. Saunders Co., Philadelphia, 1979.

B Illustration by Ron Ervin

5-16 K.G. Murti/Visuals Unlimited

5-17, A Illustration by Ron Ervin

B Charles Flickinger, *Medical Cell Biology*, W.B. Saunders Co., Philadelphia, 1979.

5-18, A Kenneth Miller

B William Ober

5-19 William Ober

5-20, A Klaus Weber and Mary Osborn, *Scientific American,* vol. 153, pages 110-121, 1985 and unpublished.

B Klaus Weber and Mary Osborn, *Scientific American,* vol. 153, pages 110-121, 1985, and unpublished.

C Klaus Weber and Mary Osborn, *Scientific American,* vol. 153, pages 110-121, 1985, and unpublished.

5-21, A James A. Spudich

B Susumu Ito from Charles Flickinger, *Medical Cell Biology*, W.B. Saunders Co., Philadelphia, 1979.

5-22 L.M. Pope, University of Texas at Austin; BPS/ Tom Stack & Associates

5-23, A William Dentler

B William Ober

5-A, 1 Manfred Kage/Peter Arnold, Inc.

2 Manfred Kage/Peter Arnold, Inc.

5-B John J. Skvarla, University of Oklahoma

6-1 Peter Arnold

6-4 William Ober

6-5 William Ober

6-6 William Ober

6-7 William Ober

6-12 Manfred Kage/Peter Arnold

6-13, C Gary Grimes/Taurus Photos

D Gary Grimes/Taurus Photos

E Birgit Satir

6-15 William Ober

6-16 M.M. Perry: Edinburgh Research Station

6-17 Richard Rodewald

6-18, A Camillo Peracchia, from Charles Flickinger, *Medical Cell Biology*, W.B. Saunders Co., Philadelphia, 1979.

7-1 Co Rentmeester, LIFE Magazine © 1970 TIME, Inc.

7-2 G. Ziesler/Peter Arnold

7-3 Cleveland P. Hickman

7-4 Warren Garst/Tom Stack & Associates

7-5 Z. Leszczynski/Animals Animals

7-6 William Ober

7-9 William Ober

7-10 William Ober

8-1 Grant Heilman

8-7 Grant Heilman

8-12, A William Ober

B Efraim Racker

8-17 D.M. Phillips/Visuals Unlimited

8-A, 1 E.S. Ross

2 E.S. Ross

3 E.S. Ross

4 E.S. Ross

5 E.S. Ross

9-1 David Northington

9-10, A Sherman Thomson/ Visuals Unlimited

B David Dennis/Tom Stack & Associates

9-13 Manfred Kage/Peter Arnold

9-14 Illustration by Ron Ervin

9-15 Lewis K. Shumway

10-1 Michael & Patricia Fogden

10-3 Biological Photo Service

10-4 Ulrich Laemmli

10-5 William Ober

10-6 Oscar L. Miller

10-7 William Wilson

10-9 Ed Reschke

10-10 Jan de May, Janssen Pharmaceutica, *Scientific American*, vol. 243, p age 76, August 1980.

10-12 David M. Phillips/ Visuals Unlimited

10-13, A B.A. Palevitz and E.H. Newcomb; BPS/Tom Stack & Associates

10-15 Diedrich von Wettstein; Reproduced with permission from *Annual Review of Genetics*, vol. 6 © 1972 by Annual Reviews, Inc.

10-16 James Kezer

10-17 Illustration by Ron Ervin

10-18 Walter Plaut

10-19 William Ober

11-1 Alan Carey, The Image Works

11-2 Richard Gross

11-3 National Library of Medicine/Picture Research

11-4, A Grant Heilman Photography

B Illustration by Ron Ervin

11-5 Illustration by Ron Ervin

11-7 W.F. Bedmer and L.L. Cavalli-Sforza, *Genetics, Evolution and Man*, W.H. Freeman & Co., Publishers, New York, 1979.

11-8 Illustration by M.K. Ryan

11-9 William Ober

11-10 William Ober

11-11 Illustration by M.K. Ryan

11-12 William Ober

11-13 Carolina Biological Supply

11-14 Illustration by M.K. Ryan

11-15 William Ober

11-16 William Ober

11-17 William Ober

12-1 The Bettmann Archive

12-2 Loris McGavran/Denver Children's Hospital

12-3, A Loris McGavran/ Denver Children's Hospital

B From Hughes, 1984

12-4 Richard Hutchings/ Photo Researchers, Inc.

12-6 Field Museum of Natural History

12-7 Frank Sloop

12-8 Gail Hudson

12-9 Ingalls and Salerno, *Maternal and Child Health Nursing*, ed. 6, Times Mirror/Mosby College Publishing, 1987.

12-10 Murayama 1981/ Biological Photo Service

12-12, B John S. O'Brien

12-13 David M. Phillips/ Visuals Unlimited

12-15 Michael Ochs Archives/ Venice, Calif.

12-16 William Ober

12-17 Washington University Medical School

13-1 Photographer A.C. Barrington Brown, from J.D. Watson, *The Double Helix*, page 215, Atheneum, New York, 1968 © 1968 by J.D. Watson.

13-2 Biological Photo Servie

13-4 Anne Crossway/Calgene, Inc.

13-5 William Ober

13-6 Illustration by Ron Ervin

13-7, A A.K. Kleinschmidt

B Lee D. Simon/Photo Researchers

C Lee D. Simon/Photo Researchers

13-8 William Ober

13-9 William Ober

13-10 William Ober

13-11 William Ober

22-1 John Gerlach/Tom Stack & Associates
22-6 E.S. Ross
22-7 John D. Cunningham/Visuals Unlimited
22-8 E.S. Ross
22-12 E.S. Ross
22-13 Peter Frenzen
22-14 Rod Planck/Tom Stack & Associates
22-17 William Ober
22-18 Marty Snyderman
22-19 Rick Harbo
22-20 Jim and Cathy Church
22-24, A E.S. Ross
   B Richard Gross
   C William Ober
   D William Ober
   E E.S. Ross
   F William Ober
22-25 Cleveland P. Hickman
22-26 John Gerard
22-27 John D. Cunningham/Visuals Unlimited
22-28 E.S. Ross
22-29, A Cleveland P. Hickman
   B Leonard Lee Rue III
22-30, A William J. Weber/Visuals Unlimited
   B Rick McIntyre/Tom Stack & Associates
   C Cleveland P. Hickman

23-1 Steve McCurry
23-3 E.S. Ross
23-5 Earthscan/Glen Edwards
23-6 J.B. Forbes, *St. Louis Post-Dispatch*
23-7 Stephanie Maze/Woodfin Camp, Inc
23-8, A & B E.S. Ross
   C D.O. Hall, King's College, University of London
23-9 Nigel Smith/Earth Scenes
23-10 Hugh Iltis
23-12 D.O. Hall, King's College, University of London
23-13 E.S. Ross
23-14 Harold Moore
23-15, A Jim Blair/National Geographic Society
   B C.W. Rettenmeyer
   C Inga Hedberg
23-16 Mickey Gibson/Tom Stack & Associates
23-17 John D. Cunningham/Visuals Unlimited
23-18 E.S. Ross

24-1 EXXON vol. 3, 3rd Quarter 1985.
24-2 Library of Congress/Picture Research
24-3 Illustration by M.K. Ryan
24-4, A Tim W. Clark
   B Animals/Animals
   C Brian Parker/Tom Stack & Associates
24-6, A Rod Planck/Tom Stack & Associates
   B E.S. Ross
   C Brian Parker/Tom Stack & Associates
24-7, A John D. Cunningham/Visuals Unlimited
   B Rod Planck/Tom Stack & Associates
24-8 Abraham and Beachey: BPS/Tom Stack & Associates
24-9 William Ober
24-10, A E.S. Ross
   B Kjell Sandved
   C E.S. Ross
24-11, A J. Robert Waaland; BPS/Tom Stack & Associates
   B T.E. Adams/Visuals Unlimited
   C Marty Snyderman
24-12, A Marty Snyderman
   B John D. Cunningham/Visuals Unlimited
   C Jeff Rotman
24-13 Terry Ashley/Tom Stack & Associates
24-14 William Ober
24-15 A. Friedman/Visuals Unlimited

25-1 Lee D. Simon/Photo Researchers
25-2 David White
25-3 Ed Reschke
25-4 Centers for Disease Control, Atlanta
25-5, A K.G. Murti/Visuals Unlimited
   B Illustration by Ron Ervin
25-6 William Ober
25-7 William Ober
25-8 William Ober
25-9, A Holt Studios Ltd./Animals Animals
   B Centers for Disease Control
25-A Centers for Disease Control/World Health Organization
26-1 Manfred Kage/Peter Arnold

26-2 William Ober
26-3, A E.W. Daniels
   B E.S. Ross
   C John D. Cunningham/Visuals Unlimited
26-4 Illustration by M.K. Ryan
26-5, A Ed Reschke
   B E.S. Ross
26-7 William Patterson/Tom Stack & Associates
26-8 Robert Sisson/National Geographic Society
26-9 D. Davidson/Tom Stack & Associates
26-10 Illustration by Ron Ervin
26-12, A Ed Pembleton
   B Robert Simpson/Tom Stack & Associates
   C Kjell Sandved
   D Walt Anderson/Tom Stack & Associates
   E John D. Cunningham/Visuals Unlimited
26-13 Harvey C. Hoch
26-14 Kjell Sandved
26-16 David M. Phillips/Visuals Unlimited
26-17 Terry Ashley/Tom Stack & Associates
26-18 Illustration by George Venable
26-19 E.S. Ross
26-20 Ed Reschke
26-21 R. Ronacordi/Visuals Unlimited

27-1 Hoi-Sen
27-2, A E.S. Ross
   B Rod Planck/Tom Stack & Associates
   C Ken David/Tom Stack & Associates
   D E.S. Ross
27-4 Terry Ashley/Tom Stack & Associates
27-6, A Ed Reschke
   B Richard Gross
27-8, A E.S. Ross
   B Kirtley Perkins/Visuals Unlimited
27-9 William Ober
27-11 Illustration by Ron Ervin
27-12, A Kjell Sandved
   B Runk Schoenberger/Grant Heilman Photography
   C Brian Parker/Tom Stack & Associates
27-13 William Ober
27-14, A Ed Pembleton

B John D. Cunningham/Visuals Unlimited
   C Rod Planck/Tom Stack & Associates
27-16, A & B Ed Pembleton

28-1 Larry West
28-2 William Ober
28-3 William Ober
28-4, A Jim and Cathy Church
   B William Ober
28-5, A Marty Snyderman
   B William Ober
   C Neil G. McDaniel/Tom Stack & Associates
   D Rick Harbo
28-6 William Ober
28-7 William Ober
28-8 William Ober
28-10 Jim & Cathy Church
28-11, A John D. Cunningham/Visuals Unlimited
   B Centers for Disease Control
28-12 William Ober
28-13, A Kjell Sandved
   B & C William Ober
   D Milton Rand/Tom Stack & Associates
   E Kjell Sandved
   F William Ober
   G Marty Snyderman
   H Alex Kerstitch
28-14, A & B Kjell Sandved
28-15, A David M. Dennis/Tom Stack & Associates
   B Kjell Sandved
   C Kjell Sandved
   D Gwen Fidler/Tom Stack & Associates
28-16 Illustration by George Venable
28-17 Illustration by M.K. Ryan
28-18, A Kjell Sandved
   B Cleveland P. Hickman
28-19 Illustration by M.K. Ryan
28-20, A E.S. Ross
   B Kjell Sandved
   C John D. Cunningham/Visuals Unlimited
   D Anne Mareton/Tom Stack & Associates
   E Rod Planck/Tom Stack & Associates
28-21 Illustration by George Venable

28-22, **A** Kjell Sandved

    B Kjell Sandved

    C E.S. Ross

    D T.E. Adams/Visuals Unlimited

    E & F Cleveland P. Hickman

28-23, **A** Kjell Sandved

    B & C E.S. Ross

    D Alex Kerstitch

    E E.S. Ross

    F Kjell Sandved

    G J.A. Adcock/Visuals Unlimited

28-24 Illustration by George Venable

28-25, **A** Kjell Sandved

    B Thomas Eisner

28-26, **A** John Shaw/Tom Stack & Associates

    B, C, D Gwen Fidler/Tom Stack & Associates

    E John Shaw/Tom Stack & Associates

    F & G K.A. Blanchard/Visuals Unlimited

    H & I Sal Giordano III

28-27, **A** Jeff Rotman

    B Daniel W. Gotshall

    C Alex Kerstitch

    D Carl Roessler/Tom Stack & Associates

    F William Ober

28-28, **A** Illustration by George Venable

    B Kjell Sandved

28-29 Illustration by Ron Ervin

28-30, **A** Jim and Cathy Church

    B William Ober

28-31 William Ober

28-32, **A** Heather Angel

    B William Ober

29-1 E.S. Ross

29-2, **B** Terry Ashley/Tom Stack & Associates

29-3, **A** Dan Sindelac

    B William Ober

29-4, **A & B** E.S. Ross

29-5 Randy Moore/Visuals Unlimited

29-6 T. Lawrence Mellichamp/Visuals Unlimited

29-7 John D. Cunningham/Visuals Unlimited

29-8 William Ober

29-9 Randy Moore/Visuals Unlimited

29-10 E.S. Ross

29-11, **A** Kjell Sandved

    B Biological Photo Service

29-12 E.J. Cable/Tom Stack & Associates

29-13 Ed Reschke

29-14 R.F. Evert

29-15 John Shaw/Tom Stack & Associates

29-16 William and Maggie Ober

29-17 U.S. Department of Agriculture

29-18 USDA Forest Service: Forest Products Lab

29-19 Terry Ashley/Tom Stack & Associates

29-20 Randy Moore/Visuals Unlimited

29-21 E.J. Cable/Tom Stack & Associates

30-1 Keith H. Murakami/Tom Stack & Associates

30-2 Gary Milburn/Tom Stack & Associates

30-3, **A** Steve Maslawski/Visuals Unlimited

    B E.S. Ross

30-4 Kjell Sandved

30-5 John D. Cunningham/Visuals Unlimited

30-6 Barry L. Runk/Grant Heilman Photography

30-7 D.A. Priestley

30-8, **A** Ed Pembleton

    B E.S. Ross

30-9, **A** John D. Cunningham/Visuals Unlimited

    B E.S. Ross

30-9, **A** John D. Cunningham/Visuals Unlimited

    B Ed Pembleton

    C John D. Cunningham/Visuals Unlimited

30-10 Heather Angel

30-12 Illustration by M.K. Ryan

30-13 Peter Hoch

30-14 Peter Hoch

30-A, **1 & 2** George B. Schaller

31-1 E.J. Cable/Tom Stack & Associates

31-2 William Ober

31-3 William Ober

31-4 William Ober

31-5 Illustration by M.K. Ryan

31-6 Illustration by M.K. Ryan

31-7 William Ober

31-8 St. Louis Globe-Democrat

31-9, **A** William Ober

    B Silvan H. Wittwer

31-10 John D. Cunningham/Visuals Unlimited

31-11 John D. Cunningham/Visuals Unlimited

31-12 Claudia Mills, *Nature,* February 1985, page 737.

31-13 Alan Godlewski

31-A, **1** Grant Heilman

    2 Gerritt Davidse

    3 Thomas Croat

32-1 Lennart Nilsson

32-2 William Ober

32-3 Illustration by M.K. Ryan

32-4 Ed Reschke

32-5 George Johnson

32-6, **A** Makio Murayama/BPS/Tom Stack & Associates

    B-G Trent Stephens

32-7 Emma Shelton

32-8 Jerome Gross

32-9 David M. Phillips/Visuals Unlimited

32-10 Ed Reschke

32-11 Lennart Nilsson

32-12 Illustration by M.K. Ryan

32-13 Cynthia Turner/T. Sims/T. Cockerharn

33-1 Cleveland P. Hickman

33-2 William Ober

33-3 William Ober

33-4 William Ober

33-5 Illustration by Ron Ervin

33-6 Illustration by Ron Ervin

33-7 Illustration by Ron Ervin

33-8 William Ober

33-9 Susan Lowey

33-10 William Ober

33-11 William Ober

33-12 Illustration by Ron Ervin

33-13 John D. Cunningham/Visuals Unlimited

33-14 William Ober

34-1 Robert Caputo

34-2 William Ober

34-3 Illustration by Ron Ervin

34-4 Illustration by M.K. Ryan

34-5 Illustration by Ron Ervin

34-6 Illustration by Ron Ervin

34-7 Illustration by Ron Ervin

34-8 William Ober

34-9, **A** Illustration by Ron Ervin

    B D.M. Phillips—1984/Visuals Unlimited

34-10 Susumu Ito, From Charles Flickinger, *Medical Cell Biology,* W.B. Saunders Co., Philadelphia, 1979.

34-11, **A** William Ober

    B Donald Specker/Animals Animals

34-12 William Ober

34-13 William Ober

35-1 Cleveland P. Hickman

35-2 William Ober

35-3 Illustration by M.K. Ryan

35-4 Illustration by M.K. Ryan

35-5 William Ober

35-6 Courtesy C.P. Hickman, Jr., from *Integrated Principles of Zoology,* ed. 7, Times Mirror/Mosby College Publishing, St. Louis, 1984.

35-7 Illustration by George Venable

35-8 Illustration by Ron Ervin

35-9 Illustration by Ron Ervin

35-10 Ellen Dirksen/Visuals Unlimited

35-11 William Ober

35-12 Illustration by Ron Ervin

36-1 Lennart Nilsson, *The Incredible Machine,* National Geographic Society

36-3 W. Rosenberg/BPS/Tom Stack & Associates

36-4 Illustration by Ron Ervin

36-5 Ed Reschke

36-6 Illustration by Ron Ervin

36-7 Ed Reschke

36-8, **A** Illustration by Ron Ervin

36-9 Ed Reschke

36-10 Illustration by Ron Ervin

36-12 William Ober

36-13 Illustration by Ron Ervin

36-14 Illustration by Ron Ervin

36-15 Illustration by Ron Ervin

36-16 William Ober

36-A Gernsheim Collection, Harry Ranson Humanities Research Center, University of Texas at Austin

37-1 John D. Cunningham/Visuals Unlimited

37-2 William Ober

37-4 © Kirk Moldoff

37-5 William Ober

37-7 © Kirk Moldoff

37-8 William Ober

37-9 William Ober

37-11 William Ober

37-13 © David Scharf/Peter Arnold, Inc.

37-A Acarology Laboratory, The Ohio State University

38-1 E.S. Ross

38-2 Illustration by M.K. Ryan

38-3 Illustration by William Ober; micrograph by C.S. Raines

38-4 E.R. Lewis/BPS/Tom Stack & Associates

38-5 Illustration by William Ober

38-7 Illustration by M.K. Ryan

38-8 William Ober

38-9 Courtesy of Lennart Heimer, from *The Human Brain and Spinal Cord,* Springer-Verlag, Inc., New York, 1983.

38-10, A Illustration by William Ober

B Ed Reschke

39-1 Lennart Nilsson

39-2 Illustration by Ron Ervin

39-3 Lennart Nilsson

39-4 William Ober

39-5 Illustration by Ron Ervin

39-6 Illustration by William Ober

39-7 Illustration by Ron Ervin

39-8 Illustration by Ron Ervin

39-9, A Martha Swope

B Michael Fogden

39-10 Illustration by William Ober

39-11 Illustration by William Ober; micrograph by A.J. Hudspeth and R. Jacobs

39-12 Illustration by William Ober

39-13 William Ober

39-14, A William Ober

B E.R. Lewis, F.S. Werblin, and Y.Y. Zeevi, University of California

39-15 Illustration by Ron Ervin

39-16 William Ober

39-17, A Illustration by William Ober

B Martin M. Ratker/ TAURUS Photos Inc.

39-18, A Illustration by M.K. Ryan

B Leonard Lee Rue III

39-19 Illustration by Ron Ervin

40-1 Lennart Nilsson, *The Incredible Machine,* National Geographic Society

40-3 © Kirk Moldoff

40-4 William Ober

40-5 Illustration by Ron Ervin

40-6 William Ober

40-7 Illustration by William Ober

40-8 Illustration by William Ober

40-9 Ed Reschke

40-10 William Ober

41-1 Cleveland P. Hickman

41-2 John Gerlach/Tom Stack & Associates

41-3 William Ober

41-4 William Ober

41-5 Courtesy of Richard Kessel, from Richard G. Kessel and Randy H. Kardon, *Tissues and Organs: A Test-Atlas of Scanning Electron Microscopy,* W.H. Freeman & Co., Publishers, New York, 1979.

41-6 Rob McKenzie

41-7 William Ober

41-9 William Ober

41-10 Illustration by Ron Ervin

42-1 Lennart Nilsson, *Behold Man,* Little, Brown & Co., Boston.

42-2 Hans Pfletschinger/Peter Arnold, Inc.

42-3 Cleveland P. Hickman

42-4 William Ober

42-5 William Ober

42-6 William Ober

42-7 Illustration by Ron Ervin

42-8 Lennart Nilsson

42-9 William Ober

42-10 Illustration by Ron Ervin

42-11 Lennart Nilsson

42-12 Illustration by Ron Ervin

42-13 Ed Reschke

42-14 William Ober

42-15 Illustration by Ron Ervin

42-16 William Ober

42-17 Illustration by Ron Ervin

42-18 Lennart Nilsson

42-19 Illustration by Ron Ervin

42-20 Illustration by Ron Ervin

42-A Lennart Nilsson

43-1 David W. MacDonald

43-2 E.S. Ross

43-3 Stephen Dalton/National Audubon Society Collection/Photo Researchers

43-4 Thomas McAvoy, LIFE Magazine, Time, Inc.

43-5 Osler Library, McGill University

43-6 Bridgewater College

43-7 James R. Fisher/National Audubon Society Collection/Photo Researchers

43-8 E.S. Ross

43-9 Fawcett/Tom Stack & Associates

43-10 Mark Newman/Tom Stack & Associates

43-11 Cleveland P. Hickman

43-12 E.S. Ross

43-13 E.S. Ross

43-14 John D. Cunningham/ Visuals Unlimited

43-16 Arthur Holzman/ Animals Animals

43-18 Fran Allen/Animals Animals

43-19 George Johnson

43-A Christopher Springmann

INDEX

A

Abdominal cavity in mammals, 608
ABO blood groups, 229-231
Abortion, 773
spontaneous, in human embryonic development, 778
Abscisic acid, 600
Absorption
in small intestine, 642-644
of water by roots, 593, 595
Absorption spectrum, 167
Abyssal zone in oceans, 433
Acacia tree, 404
Acacias, bull's horn, and ants, mutualism in, 414-415
Acarai, 546
Acceleration, angular, receptors for, 722-723
Accessory pigment, 176
Acer, 582
Acetabularia
cells of, 79
crenulata, 244-245
mediterranea, 244-245
Acetaldehyde, 160
Acetyl group, 152
Acetylcholine
in autonomic nervous system, 731
in neuromuscular junction, 708
as neurotransmitter, 633
Acetylcholinesterase, 709
Acetyl-CoA, 153-155
oxidation of, 155
Acid, 47
hydrochloric, in digestion, 638, 642
Acid rain, 460-461
Acoelomate, 537, 537-538
Acquired immune deficiency syndrome, 12-13, 696, 774-776
incidence of, 12, 13
Acrasiomycota, 504
ACTH; *see* Adrenocorticotropic hormone
Actias luna, 530
Actin in muscle cells, 613, 630, 631
Actin filaments in cytoskeleton, 96, 97
Action potential, 705-706
Activating enzymes in protein synthesis, 268-269
Activation in gene expression, 274
Activation energy, 134-135
Activator in enzyme activity, 138
Activator protein, 138, 274
Active sites, 137
Active transport across cell membrane, 117-119
Adaptation, 6
and evolution, 306-307
overview of, 316-317
parallel, as evidence of evolution, 322-324
Adenine
in adenosine triphosphate, 140
in DNA, 250-252
in DNA and RNA, 60
Adenine monophosphate, 139, 140

Adenosine diphosphate, 141
synthesis of adenosine triphosphate from, 146
Adenosine triphosphate, 140
as energy source, 624
in photosynthesis, 170-171
production of, by cells, 146-148
in sodium-potassium pump, 117
synthesis of, by cells, 145-163
fermentation, 152, 160-161
glucose catabolism, overview of, 158-159
glycolysis, 148, 149-152
molecule of, 6
oxidative respiration, 152-156
production of ATP by cells, 146-148
use of chemical energy to drive metabolism, 146
in transpiration in plants, 595
ADH; *see* Antidiuretic hormone
Adhering junctions on cells, 123
Adhesion, as property of water, 46
Adiantum pedatum, 514
ADP; *see* Adenosine diphosphate
Adrenal medulla, 741, 745
Adrenaline, 741, 745
in autonomic nervous system, 731
Adrenocorticotropic hormone, 740
Aedes mosquito, 493
Aegeria rutilans, 413
Aegolius acadius, 371
Aerenchyma, 594
Aerobic bacteria, 93, 486
Aerobic metabolism, 152, 154, 155
Aerobic oxidation of glucose, 159
Afferent pathways, 713-714
African blister beetle, 548
African elephant, 748
Agalychnis callidryas, 367
Age
of earth, 25-27, 28
of mother and incidence of Down syndrome, 226, 234
of population
distribution of, 397-398
human, 448
Agent Orange, 599
Aggression, 791-793
in humans, 797
Aging of skin, 610
Agnatha, 363, 556
Agriculture
future of, 451-452
Green Revolution in, 452-453
pollution related to, 461
shifting, 456
and succession, 429
Agrostis tenuis, 316
AIDS; *see* Acquired immune deficiency syndrome
AIDS tests, 776, 777
AIDS virus, infection cycle of, 491-493
Air pollution, 460
Air pressure in ear, 725
Air roots of mangroves, 594
Air sac in avian lung, 659

Aix sponsa, 371
Alanine in Miller-Urey experiment, 67-68
Albinism, 210
pedigree of, 227-228
Albino, 210
Albumen in amniotic egg, 369-370
Albumin, serum, 673
Aldosterone, 746
Alexander, Richard, 789
Algae, 31, 474-476
blue-green, 71, 485
stromatolites produced by, 347
chloroplasts of, 499
photosystems in, 173-174, 175
structure of, 485
Alkaptonuria, 253
Allele(s)
for cystic fibrosis, 309
definition of, 215
frequency of
change in, 309-313
effect of evolution on, 391
in genetic disorders, 231-234
in human variation, 307
and Mendel's studies, 210, 211-212
multiple, 229-231
rare recessive, 313
for sickle cell anemia, frequency of, 314
Allergens, 694
Allergic reaction, 694
Allergy, 694
Allodona townsendiana, 541
Allosteric change, 138
Alpha-actinin in muscle cells, 630, 631
Alternation of generations, 479
Alternative traits, segregation of, 206, 208
Altruistic behavior, 794
Alvarez, Luis W., 356
Alveoli in lungs, 658
Amanita muscaria, 505
Ambystoma tigrinum, 367
Amino acid(s), 253-255, 268
in antibodies, 692
conversion of, to glucose, 745
essential, 647
in genetic code, 269
in Miller-Urey experiment, 67-68
in protein structure, 55-59
Amino acid synthesis; *see* Protein synthesis
Ammonia
in early atmosphere of earth, 66
in urine, 753
Ammonium hydroxide in bacterial growth, 70
Amniocentesis, 238
Amnion
in amniotic egg, 369-370
in human embryonic development, 777
Amniotic egg, evolution of, 369
Amoeba, 98
Amoeba proteus, 503
AMP; *see* Adenine monophosphate

Aix sponsa, 371
Amphibia, 366-367, 556
Amphibians
evolution of, 366-367
eyes of, 726
gas exchange in, 654
heart in, 675
kidney in, 755
lung in, 658
sex and reproduction in, 762-763
Amphipoda, 547
Amphiprion perideraion, 413
Ampulla of echinoderm, 553
Amylases in digestion, 638-639
Amylopectin, 51
Amyloplasts
in gravitropism, 601
in seeds, 583
Amylose, 51, 52
Anabaena, 485
Anabolic process, 150
Anaerobic bacteria, 71, 95
Anaerobic metabolism, 152
Anal sphincters, 645
Anaphase, 193
Anaphase I, in meiosis, 199
Anatomy
comparative, as evidence of evolution, 27
and function of human brain, 718-720
Anax junius, 545
Anemia
pernicious, 648
sickle cell, 233-234, 314
molecular basis of, 255
Aneuploid chromosome, 291
Angelfish, 365
Angina pectoris, 680
Angiosperms; *see* Flowering plants
Angular acceleration, receptors for, 722-723
Animalia, 31
Animals, 31, 531-558
annelids, 543-544
arthropods, 544-545
diversity of, 545-550
bilateral symmetry in, evolution of, 536-537
cells of, and bacterial and plant cells, comparison of, 101
chemical defenses in, 410
chordates, 554-556
cnidarians, 534-536
coelomates, evolution of, 539-541
deuterostomes, 550-551
digestion of food by, 637-651
large intestine in, 644-645
mouth, in vertebrate, 639-640
nature of, 638-639
nutrition in, 645-649
small intestine in, 642-644
stomach in, 640-642
echinoderms, 551-554
ecological races in, 340
evolution of body cavity of, 538-539
general features of, 532
information transmission by, 701-711
mitosis in, 195
mollusks, 11-12

Animals—cont'd
movement in, 623-635
bone in, 626-627
mechanical problems of, 624-625
muscle in, 627-629
role of nerves in, 633, 634
skin in, 626-627
oxygen, use of, by, 653-665
evolution of respiration in, 654-655
gas transport and exchange in, 660-662
gill in, 655-656
lung in, 657-659
mechanics of human breathing in, 660-662
pollination by, 578-579
segmentation in protosomes, 543
sponges, 532-533
worms, solid, 537-538
Annelid(s), 543-544
closed circulatory system in, 668
oxygen diffusion in, 655
Annelida, 543-544
oxygen diffusion in, 655
Annual, 564
Anopheles, 503-504
Anorexia nervosa, 646
Ant, bulldog, 545
Antagonistic control
of autonomic nervous system, 731
over hormones, 740
Antagonistic efferent nerves, 731
Anterior air sac in avian lung, 659
Anterior pituitary, hormones released by, 739-740
Antheridia, 517
Anthocyanin, in corn, 256
Anthophyta, 526
Anthrophyta, 524
Anthropoids, 375-376
Antibiotics, resistance to, 315-316
Antibodies, 687
to acquired immune deficiency syndrome virus, 12
monoclonal, 696-697
polyclonal, 696
structure of, 692-693
Antibody test for AIDS, 775
Anticodon, 268, 270
Antidiuretic hormone, 738, 746
Antigen(s), 687
cell surface, 229
glycoprotein, of trypanosomes, 501
recognition of, 691-693
specificity of antibody molecule for, 692-693
Ants and bull's horn acacias, mutualism in, 414-415
Anura, 366
Aorta, 676-677
Aortic valve in human heart, 676
Apes, 377, 378
Apical meristem, 562, 563
of root, 572
Appendicitis, 322
Appendix, vermiform, in humans, 322
Aquifers, 420-421
Arachnida, 546
Arachnids, 546
Arboviruses, 493
Archaeopteryx, 370, 372
fossil of, 25
Archegonia, 517
Arches, gill, 363
Arcs, reflex, 729
Arctic ground squirrel, 441
Arithmetic progression of food supply, 23
Armadillo, 21
Arrowhead spider, 546
Arrow-poison frog, 410

Arteries, 668, 669, 677
carotid bodies in, 721
hardening of, 680
pulmonary, human, 677
Arterioles, 668, 669
Arteriosclerosis, 680
Arthropods, 544-545
colonization of land by, 350
diversity of, 545-550
tracheae of, 655
Artiodactyla, 375
Ascaris, 196
Ascocarp, 507
Ascomycetes, 507
Ascomycota, 505, 507
Ascorbic acid, 646-647
Ascus, 507
Asexual reproduction
in Fungi Imperfecti, 507
and species definition, 470
Aspen, 584, 585
bigtooth, 570
Aspergillus, 507
Associative behaviors, 786
Associative cortex of human brain, 718
Assortment, independent, in Mendel's studies, 213-215
Aster in mitosis, 191
Asteroidea, 552
Atherosclerosis, 679, 680
Atmosphere
carbon dioxide in, 421-422
circulation patterns of, 430
early, of earth, 66-67
reducing, 66-67
Atmospheric pressure, role of, in water movement in plants, 592-593
Atomic mass, 37-38
Atoms, 37-38
ATP; see Adenosine triphosphate
ATP generation in glycolysis, 150
ATP molecules, 6
Atria, contraction of, 678, 679
Atrioventricular node, 678
Atrium
in fish heart, 675
left, in human heart, 676
Attachment in AIDS virus infection cycle, 492
Auditory cortex of human brain, 718
Aurelia aurita, 534
Australia and continental movement, 355
Australopithecus, 360, 361, 378, 379
boisei, 380
robustus, 380
Autoimmunity and cyclosporine, 502
Autonomic nervous system, 714, 715, 729-731
antagonistic control of, 731
Autosomes, 224
Autotrophs, 145, 159
in ecosystems, 424
Auxin, 597-598
synthetic, 598-599
AV node; see Atrioventricular node
Avery, Oswald, 247
Aves, 370-372, 556
Avincennia nitida, 594
Axil of leaf, 569
Axon, 616
in nerve cell, 703
Azolla, 453

B

B cells, 687, 691
B receptors, 691, 695
Baboons, social groups in, 793
Bacillariophyta, 501, 502
Bacilli, 484

Bacteria, 31, 483-488, 493-494
anaerobic, 95
cell division in, 185-186
chemoautotrophic, 485
classification of, 473
earliest, 69-70
ecology and metabolic diversity of, 485-486
and eukaryotic cells, comparison of, 84-85
in fermentation, 160
fossils of, 69, 70, 347-348
genetic alteration of, 299
green sulfur, photosystem in, 172-173
heterotrophic, 485-486
light reactions in, 175
methane-producing, 71
modern, origin of, 71
nitrogen-fixing, 422, 486
as pathogens, 486-488
photosynthetic, 71, 83, 84, 485
plasmids in, 292
protein synthesis in, 267
structure of, 82-84, 484-485
symbiotic, 92
transfer of gene among, 293-294
Bacterial cells
animal and plant cells and, comparison of, 101
lysis of, 687
Bacteria-like organelles in eukaryotic cells, 92-95
Bacteriophages, 247-249, 482
Balance, receptors for, 722
Balanus balanoides, 406-407
Bamboo, flowering of, and starvation of pandas, 586
Barbules in feathers, 372
Bark, structure of, 571
Barnacles, 406-407
gooseneck, 547
and gray whale, commensalism of, 413
Baroreceptors, 722
Barr body, 226
Barrel cactus, 560
Basal body of flagellum, 100
Basal metabolism, hormones concerned with, 741
Base, 48
nitrogenous, in adenosine triphosphate, 140
Bases
in DNA, 250-252
nucleotide, of nucleic acids, 60
Basidia, 507-508
in puffballs, 505
Basidiomycetes, 507-508
Basidiomycota, 505, 507-508
Basilar membrane of ear, 725
Basophils in inflammation, 674
Bates, H. W., 412
Batesian mimicry, 412
Bateson, William, 252-253
Bats, 436
Beadle, George, 253
Beagle, voyage of, 16, 20-21
Bean, germination of, 563
Beardworm, giant, 433
Bee-eater, 371
Bees, pollination by, 578, 579, 587
Beetle
African blister, 548
scarab, 545
soldier, 579
Behavior, 785-799
altruistic, 794
genetic basis of, 795-796
human, biological basis of, 797
learned, 786-789
reproductive, 797
social, biological basis of, 789-794
stereotyped, 717

Behavior—cont'd
vertebrate, adaptive, 795-797
Behavioral isolation of species, 335-336
Behavioral psychologists, 787-789
Belt, Thomas, 414
Beltian bodies, 414
Beriberi, 648
Beta-carotene, 167, 168
Betula, pollination of, 579
Bicarbonate, in digestion, 642
Biennial, 564
Bigtooth aspen, 570
Bilateral symmetry, evolution of, 536-537
Bilayer, lipid, 107-108
Bile salts, 642
Binary fission, 186
in bacteria, 484-485
Binocular vision in anthropoid primates, 375
Binomial system of classification, 468-469
Biochemical pathways, 138-139, 146
Biogeochemical cycles, 420
Biological basis of behavior, 797
social, 789-794
Biological organization, levels of, 29-31
Biology
importance of, 7-13
molecular, as evidence of evolution, 27
as science, 17-34
evolution, history of Darwin's theory of, 20-22
evolution after Darwin, 24-27
nature of, 17-20
scientific creationism and, 27-29
theory of evolution, publication of Darwin's, 24
study of, 3-7
Bioluminescence, 433
Biomass of ecosystem, 424
Biomes, 29, 435-443
distribution of, 435
Biophysics of light, 166-167
Biosphere, future of, 445-464
agriculture in, 451-452
food in
and population, 449-451
prospects for increased supply of, 452-454
pollution in, 459-461
population explosion and, 445-448
tropics in, 454-459
Biotic factors in distribution of species, 399, 401
Biotic potential, 393
Bipolar cell of fovea, 727
Birch, pollination of, 579
Birds
digestive system of, 638
evolution of, 370-372
four-chambered heart in, 675
kidney in, 755-756
migratory memory in, 796
respiration in, 659
sex and reproduction in, 763-764
stereotyped behavior in, 717
Birth control, 772-773
Birth control pills, 773
Birth rate and death rate, relationship of, 393
Birth weight in human beings, stabilizing selection for, 390
Bishop pine, 339
Biston betularia, 307, 314-315
Bivalvia, 541
Black widow spider, 546
Black willow, 573
Blackberry, 392
Black-footed ferret, 469

Blade, in leaf, 567
Blastopore, 532
Blastula, 532
Blister beetle, 354
Blood
  ABO groups, 229-231
    genetic variation in, 307, 308
  clotting of, hemophilia and, 234-235
  flow of, in gills, 656
  level of glucose in, 645-646
  path of, through human heart, 676
  regulation of glucose in, 742-745
Blood cells
  red, 612, 673, 674, 687, 689
    in capillaries, 670
    cell surface markers on, 121
    hemoglobin in, 662-663
    plasma membrane of, 86
    in sickle cell anemia, 233, 314
    types of, 673-674
  white, 611, 612, 673-674, 687, 689, 691
Blood chemistry, receptors for, 721
Blood plasma, 673
Blood pressure, 679
  receptors for, 722
Blood vessels and heart, diseases of, 680
Blue shark, 364
Blue-green algae, 485
  stromatolites produced by, 347
Bobolinks, migratory memory in, 796
Body cavity, internal, 538-539
Bohr, Niels, 37
Bombus, 578
Bond(s)
  chemical
    nature of, 40
    oxidation-reduction reaction and, 133
  covalent, 41-43
  high-energy, in ATP, 140
  hydrogen, 137
    in water, 45-46
  ionic, 40-41
  peptide, 55
Bonding, hydrophobic, 46
Bone(s), 612
  compact, 627, 628
  of ear, 725
  homologous, in vertebrates, 27
  muscle attachment to, 625
  spongy, 627, 628
Bone marrow, 688
Bone tissue, 627, 628
Bony fishes, evolution of, 364-365
Box, Skinner, 788, 789
Boyer, Herbert, 296, 297
Bradshaw, A. D., 316
Bradypus infuscatus, 436
Brain, 616, 702
  in annelids, 544
  of cat, neuron in, 700
  of hominoid primates, 376, 377
  human, 718-720
    anatomy and function of, 718-720
    size of, 379, 381
  vertebrate, 713
    evolution of, 716-718
Branching in roots and shoots, 572
Branchiostoma lanceolatum, 556
Brassica oleracea, 599
Brassicaceae, 409
Brazil, population of, 450, 451
Breathing; see also Respiration
  in birds, 659
  human, mechanics of, 660-662
Breeding population, effective, 331
Bridge, conjugation, 293, 294

Briggs, Robert, 245
Bristleworm, 544
  shiny, 543
Brittle star, 552
5-Bromouracil, mutation caused by, 284
Bronchus, in human respiration, 660
Brown, Robert, 88
Brown algae, 476
Brown recluse spider, 546
Bryophytes, 519-520
  gametangia of, 518
Bud, 569
Buds, taste, 724
Bufo marinus, 367
Bulbourethra, 772
Bulimia, 646
Bulldog ant, 545
Bumblebee, 578
Bumpus, H.C., 390
Bundle branches, 678
Bundle of His, 678
Bundles, vascular, in plants, 521
Burgess Shale, 349
Butterfly
  cabbage, 409
  checkerspot, 332
  mimicry in, 412
  monarch, 410
  species of, 331
  in tropical rain forests, 435-437

C
Cabbage, 599
Cabbage butterfly, 409
Cactoblastis, 407-408
Cactus
  barrel, 560
  prickly pear, 407-408
Calcitonin, 741
Calcium
  in arterial walls, 680-681
  in muscle contraction, 633
Calcium ions, 41
  in neuromuscular junction, 708
Callus in plant reproduction, 584
Calorie, definition of, 130
Calvin, Melvin, 176
Calvin cycle, 176, 177
Calyx in angiosperms, 526
Cambium
  cork, 564, 570
  vascular, differentiation of, 570
Cambrian Period, 348, 349, 350
Campsis radicans, 578
Canals, semicircular, in movement detection, 722-723
Cancer
  biology of, 10-12
  and mutation, 284-288
  rates of, 286
Canis
  familiaris, 330, 469
  latrans, 469
  lupus, 469
Cap, root, 572
Capillaries, 668, 669-670
Capsid, viral, 489
Capsule, bacterial, 484
Carbohydrates
  diets high in, 646
  fermentation of, 160-161
  structure of, 48, 49-55
  transport of, in plants, 595-596
Carbon
  fixation of, in photosynthesis, 170, 176
  in metabolism, 653
  in organic molecules, 48-55
Carbon cycle, 421-422
Carbon dioxide, 600
  in atmosphere, 421-422
  carotid bodies and, 721
  in early atmosphere of earth, 66
  in metabolism, 653

Carbon dioxide—cont'd
  production of organic molecules from, 176-177
  in respiration, 663
Carbon monoxide in respiration, 663
Carbon-14 in dating of rocks, 319
Carbonate ions in respiration, 663
Carbonic anhydrase, 136
  in respiration, 663
Carboxypeptidase, 139
Carcinogens, 287
Carcinoma, 284, 285
Cardiac muscle, 629
Cardiovascular system, 668-672
Caribou, 441
Caries, dental, 487
Carnivora, 374
Carnivores, 374
  in ecosystems, 424
  teeth in, 639
Carotenoids, 167, 499
Carotid bodies, 721
Carpels, in angiosperms, 525
Carpodacus purpureus, 371
Carrying capacity, 394-395
Cartilage, 612
Cassava, 451
Castanea, 471
Cat, neuron in brain of, 700
Catabolic process, 150
Catabolism, glucose, overview of, 158-159
Catalysts, enzymes as, 55, 135-136
Catalytic cycle of enzyme, 137
Caudata, 366
Cave paintings, 381, 382
Cavity, body, internal, 538-539
Cell(s), 77-103
  bacterial
    comparison of, with animal and plant, 101
    division of, 185-186
    lysis of, 687
    structure of, 82-84
  bipolar, of fovea, 727
  blood; see Red blood cells; White blood cells
  bulk passage into, 114, 115
  cancer, 285
  collenchyma, 564
  communication of, with outside world, 124
  companion, in phloem, 566, 567
  conducting, in plants, 563, 566-567
  cork, in plant stems, 570
  earliest, 69-73
  epidermal, in plants, 565
  epithelial, 608-610
  eukaryotic
    appearance of, 72-73
    bacteria and, comparison of, 84-85
    cytoskeleton of, 95-97
    endoplasmic reticulum of, 86-88
    flagella of, 98-100
    interior of, 86
    nucleus of, 88-90
    plasma membrane of, 86
  first, formation of, 69
  general characteristics of, 77-78
  germline, mutation of, 282
  glial, 616
  Golgi complex, 90-91
  guard
    of plant stoma, 595
    in plants, 565
  hereditary information in nucleus of, 244-246
  of immune system, 611, 687-688

Cell(s)—cont'd
  interaction of, with environment, 105-126
    bulk passage into, 114, 115
    communication of cell with outer world, 124
    expression of cell identity, 121
    lipid foundation of cell membranes, 106-108
    membranes in, 111
    molecules, selective transport of, 114, 116-120
    physical connections between cells, 121-123
    plasma membrane, architecture of, 108-110
    water, passage of into and out of cells, 112-114
  mast, 611
  Merkel, 722
  microbodies, 91-92
  movement of, 629-633
  muscle, 613, 627-629
  nerve, 613, 616
  neuroglia, 703
  parenchyma, 564
    in leaf, 568
    in xylem, 566
  passage of water into and out of, 112-114
  physical connections between, 121-123
  production of adenosine triphosphate by, 146-148
  protein production by, 264-265
  reproduction of, 185-203
    bacteria, cell division in, 185-186
    cytokinesis, 189, 194-195
    eukaryotes, cell division among, 186-189
    meiosis, 196
    mitosis, 189-193
    stages of meiosis, 197-200
  Schwann, 616, 703
  sclerenchyma, 565
  sieve, in phloem, 566, 567
  simple, structure of, 82-84
  size of, 78-82
  somatic, mutation of, 282
  stem, im immune response, 695
  structure of, overview of, 101
  surface of, 686
  synthesis of adenosine triphosphate by, 145-163
    fermentation, 152, 160-161
    glucose catabolism, overview of, 158-159
    glycolysis, 148, 149-152
    oxidative respiration, 152-156
    production of adenosine triphosphate by cells, 146-148
    use of chemical energy to drive metabolism, 146
  types of, in plants, 562, 564-567
  white blood, 612
Cell biology, 31
Cell cycle, 189
Cell division
  in bacteria, 185-186
  among eukaryotes, 186-189
Cell identity, expression of, 121
Cell junction, 122, 123
Cell plate in cytokinesis, 194
Cell surface antigens, 229
Cell surface markers, 121
Cell surface receptors, 120-121
  and monoclonal antibodies, 697
Cell theory, 78-79
Cell walls
  bacterial, 83, 484
  in plants, 564
Cell-mediated immune response, 690, 691
Cellular level of biological organization, 29-30

Cellular organization as property of life, 5, 7
Cellular respiration, 134, 148
Cellulase in digestion, 644
Cellulose
  digestion of, 644
  structure of, 51-53
Cenozoic Era, 348, 356-357
  subdivisions of, 375
Centimorgan, 219
Centipede, 549
Central nervous system, 616, 713-714, 716-720
Central vacuoles in eukaryotic cells, 84, 85
Centrioles in interphase, 190-191
Centromeres
  in anaphase, 193
  of chromosomes, 188
  division of, in mitosis, 192-193
  in interphase, 190-191
  in meiosis, 197-200
Cephalochordata, 555
Cephalopoda, 541
Cephalopods, 541
Ceratocystis ulni, 507
Cereals and legumes, combination of, 453
Cerebellum, 716
Cerebral cortex, 717
  associative organization of, 718, 720
  of human brain, 718
Cerebral ganglion in annelids, 544
Cerebrum, 718-720
  size of, 717-718
Certhidea olivacea, 340
Cervix, 770
Cestoda, 538
Cetacea, 375
Chagas' disease, 501
Chameleon, 787
Chancre of syphilis, 487
Change, genetic, importance of, 299-300
Channels
  chloride-transport, in cystic fibrosis, 232
  coupled, 117
    active transport by, 118, 119
  thoroughfare, 669
Chaparral, 439, 440
Chargaff, Erwin, 250
Chase, Martha, 248
Chauliognathus pennsylvanicus, 579
Checkerspot butterfly, populations of, 332
Cheese, and fungi, 505, 507
Chelicerae of arthropods, 545
Chelicerates, 545, 546
Chelonia, 367
Chemical bonds
  nature of, 40
  oxidation-reduction reaction and, 133
Chemical defenses in animals, 410
Chemical mutagens, 284
Chemiosis in chloroplast, 179
Chemiosmosis, 119, 148, 149
Chemistry of life, 37-63
  atoms, 37-38
  carbohydrates, structure of, 49-55
  chemical bond, nature of, 40
  covalent bonds, 41-43
  ionic bonds, 40-41
  molecules, 39-40
  nucleic acids, 59-61
  orbitals of electron, 38-39
  proteins, structure of, 55-59
  water, 43-48
Chemoautotrophic bacteria, 485
Chernobyl disaster, effect of, on caribou, 441
Chest pain, 680
Chestnut blight, 507

Chewing cud, 644
Chiasmata in chromosomes, 198
Chickens
  breeding experiments in, 257
  cholera in, 686-687
  pecking order in, 793
Chimera, 296
Chimpanzees, 377, 378
Chipmunk, Townsend's, 471
Chironomus tentans, 271
Chitin, 53, 351
  in arthropods, 545
  in fungi, 506
  in insects, 548
  muscle attachment to, 625
  rigidity of, 627
Chlamydia trachomatis, 487
Chlamydial infections, 487
Chlamydomonas, 90
  life cycle of, 500
Chlamys hericia, 541
Chloride channels and cystic fibrosis, 118
Chloride-transport channels in cystic fibrosis, 232
Chlorinated hydrocarbons, 461
Chlorippe kallina, 436
Chlorophyll, 168, 169
  a, 173, 176, 499
  b, 176
  c, 499
Chlorophyta, 476, 500
Chloroplasts, 78, 92, 477
  in eukaryotic cells, 94-95
  internal structure of, 179
  in leaves, 568
  photosynthesis in, 178-179
Choanocytes, 504
  in sponges, 533
Choanoflagellates, 501, 533
Cholesterol, 54, 741
Chondrichthyes, 363-364, 556
Chordata, 362, 554-556
Chordates, 362, 554-556; see also Vertebrates
  colonization of land by, 350-351
  early, heart in, 674
Chorion in human embryonic development, 777
Chorionic gonadotropin in human embryonic development, 777
Chromatids
  in meiosis, 197
  separation of, in anaphase, 193
  sister
    in interphase, 190-191
    in meiosis, 198
    in metaphase, 193
Chromatin in chromosomes, 188
Chromosome(s), 84, 85, 473, 478
  alignment of, in metaphase, 192
  conjugation of, 294
  in Down syndrome, 225-226
  eukaryotic, structure of, 186-187
  of eukaryotic cells, 89
    structure of, 186-187, 188
  giant, of Drosophila melanogaster, 262
  homologous, 190-191
  human, 224-227
  and meiosis, 196-200
  and Mendel's studies, 210, 211-212
  in Mendelian inheritance, 215-217
  in metaphase, 193
  of rat kangaroo, 192
  sex, 224, 226-227
  transposition in, 288-290
  X, 226-227
  Y, 227
Chromosome organization, mutations altering, 290-291
Chrysanthemum, 602
Chrysotoccum, 413
Chthamalus stellatus, 406-407

Cichlid fishes, mating and aggressive behavior in, 792
Cilia, 100
  respiratory, human, 661
Ciliophora, 499, 500
CIMMYT; see International Center for the Improvement of Maize and Wheat
Circulating lymphocytes, 691
Circulation
  hepatic-portal, 744
  importance of, 682
  placental-fetal, 780
  pulmonary and systemic, 675
  through heart, 676-678
  vertebrate, 667-683
    cardiovascular system, 668-672
    circulatory systems, contents of, 673-674
    circulatory systems, evolution of, 667-668
    human heart, 675-681
    importance of, 682
    vertebrate heart, evolution of, 674-675
Circulatory system
  evolution of, 667-668
  in human, 677
  vertebrate, contents of, 673-674
Circumantarctic current, 355, 357
Cisternae in Golgi complex, 91
Citric acid cycle, 148, 154, 155, 156
Clarkia
  concinna, 331
  rubicunda, 331
  speciosa, 331
Class in hierarchical system, 361
Class G antibodies, 693
Class M antibodies, 693
Classes of vertebrates, 556
Classical conditioning, 787, 788
Classification of organisms, 468-469
  taxonomic hierarchy in, 470, 471-472
Cleavage
  egg, 540
  in glycolysis, 150
Cleavage furrow in cytokinesis, 194
Cleft, synaptic, 706-708
Clements, F. E., 428
Climate
  changes in, worldwide, 357
  in distribution of species, 399, 400
  and dormancy, 602
  and ecosystems, 429-432
  and succession, 428
Climax vegetation, 428
Cliona delitrix, 533
Clitoris, 771-772
Clonal theory of immune receptor response, 693, 695
Cloning, 294
  generalized scheme for, 299
Closed circulatory systems, 667-668
Clotting of blood, hemophilia and, 234-235
Clubmoss, 573
Cnidaria, 534-536
  oxygen diffusion in, 654-655
Cnidarians, 534-536
  oxygen diffusion in, 654-655
Cnidocytes of cnidarians, 535
CoA; see Coenzyme A
Coacervate, 7, 69
Cobalt in plant nutrition, 596
Cocci, 484
Cochlea, 725
Cockroach, trachea system of, 549
Coconuts, 582
Cocos nucifera, 582
Code, genetic, 266-270

Co-dominant alleles, 229
Codon, 266, 268
  nonsense, 269
Codosiga, 501
Coelacanth, 365
  living, 346
Coelom, 537
  advantage of, 539-540
  in mammals, 608
Coelomate, 537
  evolution of, 539-541
Coenzyme A, 153-155
Coenzymes, 139-140
Coevolution, 407
Cofactors in catalysis, 139
Cohen, Stanley, 296
Coitus interruptus, 773
Cole, Lamont, 426
Coleoptera, 545, 548
Coleus, 562
Collagen, 55
Collagen fibers, 612
  in bone, 627
Collenchyma cells, 564
Colon, 644-645
Color, changes in, from natural selection, 195
Color blindness, 219
Columnar epithelium, 610
Commensalism, 413
Communicating junctions on cells, 123
Communication of cell with outside world, 124
Communities
  definition of, 419
  and ecosystems, 419-445
Community level of biological organization, 29-30
Compact bone tissue, 627, 628
Companion cells in phloem, 566, 567
Comparative anatomy as evidence of evolution, 27
Comparative psychologists, 787
Compartments, membrane-bounded, in eukaryotic cells, 84, 85
Competition between species, 406-407
Competitive exclusion, principle of, 406
Complement system, 693
Complementarity in DNA, 250-252, 265
Complex, Golgi, 90-91
Complexity as characteristic of life, 4
Composite molecules, 48
Compound microscope, 80
Concentration
  of metabolites, by active transport, 117, 119
  molar, 47
Condensation in interphase, 190-191
Conditioned stimulus, 788
Condoms, 773
Conducting cells in plants, 563, 566-567
Conduction, saltatory, 706
Condylura cristata, 374
Cones, 725, 726, 727
Conflict, reduction of, in vertebrates, 793
Conidia, 507
Coniferophyta, 524
Coniferous trees of taiga, 439
Conifers, evolution of, 354
Conjugation, 501
  of chromosome, 294
Conjugation bridge, 293, 294
Connective tissue, vertebrate, 611-612
Connell, J. H., 406-407

Continents, movement of, 354–355
Continuous variation, 258
Contraception, 772–773
Contraction
  of heart, 678
  muscle, receptors for, 722
  of myofilaments, 630–631
  of vertebrate muscles, 630
Control over hormone production, 740
Conus arteriosus in fish heart, 675
Convolvulaceae, 415
Copepoda, 547
Copepods, 547
Copulation in reptiles, 763
Corals, 501
Coriolus versicolor, 505
Cork cambium, 564, 570
Cork cells in plant stems, 570
Corn, 454
  root tip in, 572
Cornea of eye, 726
Cornus florida, 581
Corolla in angiosperms, 526
Corpus callosum, 718, 720
Corpus luteum, 765
Corpus striatum, 717
Corpuscles
  Malpighian, 752
  Meissner's, 722
Correns, Carl, 215
Cortex
  cerebral, 717
    associative organization of, 718, 720
    of human brain, 718
  of plant stems, 569
Cotyledons in angiosperms, 526
Counseling, genetic, 236–239
Countercurrent flow
  in gills, 656, 657
  in human kidney, 755–756
Coupled channels, 117
  active transport by, 118, 119
Covalent bonds, 41–43
Cowpox, 685–687
Coyote, 469
Crab, Sally lightfoot, 547
Craniate chordates, 362
Crayfish, 547
Creationism, scientific, 27–29
Cretaceous Period, flowering plants in, 354
Crick, Francis, 242, 250–252
Crinoidea, 552
Cristae, 93
Crocodile, river, 368
Crocodilia, 367
Crocodilus acutus, 368
Cro-Magnons, 381–382
Crop plants, genetic engineering of, 298
Cross(es)
  between plants, 206
  dihybrid, 213, 214, 256
  test, 212, 213
Crossing-over, 197, 217–219
Cross-link in ultraviolet radiation damage, 283
Cruciferae, 409
Crust of earth, composition of, 39
Crustacea, 547
Crustaceans, mandibulate, 545, 547–550
Cryptic coloration, 411
Ctenophora, 534
Cuboidal epithelium, 610
Cucurbita, 567
Cud chewing, 644
Cuenot, Lucien, 258
Cup fungus, 505
Currents, ocean, 431–432
Cuscuta, 415
Cuticle in plants, 515
Cutin in plants, 515

Cyanobacteria, 71, 485
  nitrogen-fixing, in food supply, 454
  stromatolites produced by, 347
Cycadophyta, 523
Cycle(s)
  biogeochemical, 420
  citric acid, 148
Cyclic photophosphorylation, 172–173
Cyclosporine, 502
Cypress, flooding of, 594
Cyrripedium calceolus, 472
Cystic fibrosis, 232, 233
  allele for, 309
  chloride channels and, 118
Cytochrome, 156
  c, evolution of gene for, 320
Cytochrome *c* oxidase complex, 156
Cytokinesis, 189, 194–195
Cytokinins, 599
Cytoplasm, 78
  in striated muscle, 632–633
Cytoplasmic streaming in fungi, 506
Cytosine
  in DNA, 250–252
  in DNA and RNA, 60
Cytoskeleton of eukaryotic cells, 95–97
Czar Nicholas II of Russia, 222

D

2,4-D; *see* 2,4-Dichlorophenoxy-acetic acid
Daisy, pollination of, 579
Danaus plexippus, 410, 412, 550–551
Dandelion, 396
  wind dispersal of, 582
Dark reactions of photosynthesis, 170, 176–177
Darwin, 330
Darwin, Charles Robert, 20–27, 597
Darwin, Francis, 597
Darwin's finches, 22, 340–341
Dating
  of fossils, 319
  of rocks, 319
Daucus carota, 584
Day neutral plants, 602
Deamination, 745
Death as characteristic of life, 4
Death rate, 398; *see also* Mortality
  birth rate and, relationship of, 393
Decarboxylation, 152
Deciduous forests, 437
  temperate, 439
Deciduous plants, nutrient use by, 401
Decomposers in ecosystems, 424
Deductive reasoning, 17
Deer, white-tailed, 375
Deforestation, tropical, 456–459
Dehydration, 750
Deletions, chromosomal, 290–291
Demography, 398–399
Dendrite(s), 616
  in nerve cell, 703
Dendrobates lehmanni, 410
Dendrobatidae, 410
Density-dependent effects on populations, 395–396
Density-independent effects on populations, 395–396
Dental plaque, 487
Denticles, 363
Depolarization
  of cardiac muscle, 629
  of heart muscle, 678
  in nerve cells, 616, 704–706
  of nerve fiber membrane, 633
  waves of, measuring, 679

Dermatophagoides, 694
Dermis, 626
Desensitization in allergy treatment, 694
Desert, 438
Design, experimental, Mendel's, 207–208
Desmosomes, 122, 123
Desiccation in evolution, 515
Determinant, antigen, 696
Determinate growth in leaves, 567
Detoxification in liver, 87–88
Detritivores in ecosystems, 424–425
Deuteranopia, 219
Deuterostomes, 540, 550–551
  egg cleavage in, 540
Development
  as characteristic of life, 4
  embryonic
    human, 777–780
    in vertebrates, 777
  as evidence of evolution, 321
Devonian Period, 348, 351
Diabetes, 745
Diamond sting ray, 364
Diaphragm in human respiration, 660
Diaphragms in birth control, 773
Diastolic period, 679
Diatoms, 501, 502
  freshwater, 475
2,4-Dichlorophenoxyacetic acid, 599
Dicots, 526–527
  of leaves, 568
Didelphis virginiana, 355
Didermoceros sumatrensis, 392
Didinium nasutum, 115
Diencephalon, 716–717
Differentiation of flowering plants, growth and, 583
Diffusion, 112
  facilitated, across cell membrane, 116
Digestion
  extracellular, 536
  of food by animals, 637–651
    large intestine in, 644–645
    nature of digestion, 638–639
    small intestine, 642–644
  nature of, 638–639
  in small intestine, 642–644
  vertebrate mouth in, 639–640
  vertebrate nutrition and, 645–649
  vertebrate stomach in, 640–642
Digestive enzymes in lysosomes, 91–92
Digestive system(s)
  of animals, 638
  coelom in, 539–540
  organization of, 641
Dihybrid cross, 213, 214, 256
Dilation and curettage in abortion, 773
Dilobderus abderus, 545
Dimer, pyrimidine, 283
Dinoflagellata, 501, 502
Dinoflagellates, 501, 502
Dinosaurs
  diversity of, 368–369
  extinction of, 355–356
Dioxin, 599
Diploid, definition of, 215
Dipodomys panamintensis, 749
Diptera, 549
Directional selection, 389–390
Disaccharide, 51
Discontinuous synthesis of DNA, 277
Disease(s)
  affecting immune system, 695–696
  of heart and blood vessels, 680

Disease(s)—cont'd
  and viruses, 491
Disorders, genetic, 231–234
  organizations studying, 239
Dispersal of fruit, 582
Disruptive selection, 391
Distribution
  patterns of, as evidence of evolution, 324
  of species, factors limiting, 399–401
Diurnal primates, 375
Divergence of populations, 331–333
Division of cells
  in bacteria, 185–186
  among eukaryotes, 186–189
DNA, 6, 48, 252–256, 473
  in antibody genes, 695
  of bacteria, 83, 185
    and viruses, 482
  in chromosomes, 188
  complementarity of, 265
  "copy," 272
  damage to, in mutation, 282–284
  discontinuous synthesis, 277
  double-stranded, in AIDS virus, 492–493
  in eukaryotes, 89
  in evolution, 27
  in gene expression, 265–266
  in genetic engineering, 294–296
  in meiosis, 197–200
  mitochondrial, 93
  model of, 242, 250–252
  nucleotides in, 59
  plasmid, 292
  recombinant, 296–297
  replication of, 250–252
  replication fork of, 277
  and RNA, similarity of, 265
  in seed germination, 583
  in viruses, 250, 489–490
DNA cross-links in mutation, 284
DNA polymerase, 277
DNase, 247
Dodder, 415
Dog, 469
Dogwood, berries of, 581
Dominance, incomplete, 259
Dominance hierarchies, 793, 796–797
Dominant allele, definition of, 215
Dominant human traits, 210
Dominant traits in Mendel's studies, 208
Dormancy in plants, 602
Dorsal nerve cord of chordates, 554
Double bonds, 41
Double helix of DNA, 60–61
Douches, 773
Down, J. Langdon, 225
Down syndrome, 188, 225–226
Dragonfly, green darner, 545
Drosophila, 389
  clavisetae, 337
  Hawaiian species of, 335–336
  heteroneura, 336–337
  melanogaster
    giant chromosomes of, 262
    studies of eye color in, 216–217
    sex-linkage in, 216–217, 218
  silvestris, 336
  transposons on chromosomes of, 289
Drug use and acquired immune deficiency syndrome, 775
Duck, wood, 371
Duck-billed platypus, 373
Duodenum, 642
Duplex, DNA, 188
Dusky seaside sparrow, 338
Dutch elm disease, 507
Dynein, 96

E

E antibodies in allergy, 694
Ear
  human, structure of, 724, 725
  motion receptors in, 722-723
Eardrum in hearing, 724
Earth
  age of, 25-27, 28
  composition of crust of, 39
Earthworms, 543, 544
Eating habits, dangerous, 646
Ecdysis in arthropods, 545
Echidna, 373
Echinarachnius parma, 552
Echinodermata, 550, 551
Echinoderms, 551-554
  basic features of, 553
  gas exchange in, 654
Echinoidea, 552
Ecological isolation of species, 334
Ecological races, 338-340
  in animals, 340
Ecological succession, 427-429
Ecology, 31
  bacterial, 485-486
  of fungi, 506
EcoR1, 296
Ecosystem(s)
  and climate, 429-432
  cycling of nutrients in, 420
  definition of, 419
  flow of energy in, 424-427
Ecotypes in plants, 339
Ectoderm, 534, 608
Ectophylla albicollis, 436
Ectothermic animals, 370
Edaphic factors in distribution of
  species, 399, 400-401
Effective breeding population,
  331
Efferent nerves, antagonistic, 731
Efferent pathways, 715
EGF; *see* Epidermal growth factor
Egg(s)
  amniotic, evolution of, 369
  of amphibians, 366-367
  mammalian, maturation of, 765
  production of, in humans, 769-
    770
  shelled, of reptiles and birds,
    763
Egg cleavage
  in deuterostomes, 540
  in protostomes, 540
Egg-laying mammals, 373
Einstein, Albert, 166
Ejaculation, 768
El Niño, 431
Elastic fibers in blood vessels,
  669, 671
Elderberry, 565, 570
Eldonia, 349
Eldredge, Niles, 341
Electrocardiogram, 679, 680
Electromagnetic spectrum, 166,
  167
Electron(s), 37-38
  in covalent bonds, 41
  generated by citric acid, 156-158
  movement of, in light reactions,
    171-172
  in oxidation-reduction reactions,
    133
  orbitals of, 38-39
  of water, 45
Electron microscope(s), 80
Electron orbitals of water, 45
Electron transport chain, 156, 157
Electronegativity of water, 45
Electron-transport system in pho-
    tocenter, 172
Electrophoresis, 307
Elements, most common, 42-43
Elephant, African, 748

ELISA; *see* Enzyme-linked immu-
    nosorbent assay
Elongation, 270
Elongation factors, 270
Embryo
  of chordate, 362
  development of, 321
  human, 554
  in plants, 580
  in seed, 563
  of vertebrates, 322
Embryo sac in angiosperms, 525
Embryonic development
  in coelomates, 540
  in deuterostomes, 540
  human, 777-780
  in vertebrates, 777
Encephalartos kosiensis, 523
Endangered species, 392-393
Endergonic reactions, 134-135
  adenosine triphosphate in, 141
Endocrine glands, 701, 714, 715
  and hormones, 742-743
Endocytosis, 114, 115
  receptor-mediated, 119
Endoderm, 534, 608
Endodermis of plant roots, 572
Endometrium of uterus, 770
Endonuclease, restriction, 295
Endoplasmic reticulum, 78
  of eukaryotic cells, 86-88
Endorphins, 741
Endosperm, in angiosperms, 525
Endosymbiont, 92
Endothelial cells of arteries, 669
Endothia parasitica, 507
Endplate, motor, 633
End-product inhibition, 139
Energy
  activation, 134-135
  chemical, use of to drive metab-
    olism, 146
  definition of, 130
  in light, 166
    absorption of, 170
  and metabolism, 129-143
    activation energy, 134-135
    definition of energy, 130
    enzymes, 135-140
    laws of thermodynamics, 131-
      132
  sources of, in origin of life, 67
  from sun, 132
Energy extraction in citric acid
  cycle, 155
Energy level, electron, 39
Engelmann spruces, 585
Engineering, genetic, 294-296
  recent progress in, 297-299
Enkephalins, 741
Entropy, 131-132
Entry in AIDS virus infection cycle, 492
Envelope
  nuclear, of eukaryotic cells, 88-
    89
  viral, 489
Environment
  effects of, in genetics, 259
  interaction of cells with, 105-126
    bulk passage into cell, 114, 115
    communication of cell with
      outside world, 124
    expression of cell identity, 121
    lipid foundation of cell mem-
      branes, 106-108
    membranes in, 111
    molecules, selective transport
      of, 114, 116-120
    physical connections between
      cells, 121-123
    plasma membrane, architecture
      of, 108-110
    water, passage of into and out
      of cells, 112-114

Environment—cont'd
  species and, interaction of, 387-
    403
    demography, 398-399
    distribution of species, factors
      limiting, 399-401
    growth of populations, 392-
      397
    mortality and survivorship,
      397-398
    population, concept of, 391
    selection, 388-391
Enzyme(s), 55, 135-140
  activating, in protein synthesis,
    268-269
  activity of
    factors affecting, 137-138
    regulation of, 138-139
  catalytic cycle of, 137
  in digestion, 638-639
  digestive, in lysosomes, 91-92
  function of, 137
  in genetic interaction, 256
  genetic traits expressed by, 254
  genetic variation in, 307, 308
  recombination, 292
  restriction
    genetic counseling for, 238
    in genetic engineering, 295-296
  secretion of, by fungi, 505
Enzyme activity test in genetic
    counseling, 238
Enzyme-linked immunosorbent
    assay for AIDS virus antibod-
    ies, 776
Eosinophils, 674
Ephemeroptera, 548
Epidermal cells, 610
  in plants, 565
Epidermal growth factor in can-
    cer, 288
Epidermis, 626
  of plant roots, 572
Epididymis, 768
Epilobium
  angustifolium, 587
  ciliatum, 587
Epistasis, 257
Epithelial cells, 608-610
Epithelial wall in sponges, 533
Epithelium, 608
  of small intestine, 642-643
Equilibria, punctuated, 341
Equilibrium, Hardy-Weinberg,
    308-309
Equisetum, 522
ER; *see* Endoplasmic reticulum
Erythrocytes; *see* Red blood cells
Escherichia coli, 473, 482
  lac operon of, 276
  protein synthesis in, 267
  restriction endonuclease number
    1, 296
Esophageal sphincter, 640
Esophagus, 640, 641
Essay on the Principles of Population,
    23
Essential amino acids, 647
Estrogen, 765
Estrous cycle, 764
Ethanol
  conversion of pyruvate to, 152
  and fermentation, 160
Ethyl alcohol and fermentation,
    160
Ethylene as plant hormone, 600
Euchromatin, 188
Eugenics, 259
Euglenophyta, 501
Eukaryotes, 31
  cell division among, 186-189
  early fossils of, 73
  evolutionary radiation of, 498
  kingdoms of, 474
  oldest fossils of, 347-348
  origin of, and symbiosis, 499

Eukaryotes—cont'd
  and prokaryotes, features of, 473
Eukaryotic cells
  appearance of, 72-73
  and bacteria, comparison of, 84-
    85
  interior of, 86
  plasma membrane of, 86
Eukaryotic chromosome, 186-187
  structure of, 188
Eukaryotic evolution, 474-477
Eumetazoans, 534
Euphorbia, 411
  pulcherrima, 578
Euphydryas editha, 332
Eurius, 547
Eutamias townsendii, 471
Evaporation, role of, in water
    movement in plants, 592-593
Evergreen plants, nutrient use by,
    401
Evidence for evolution, 305-327
  adaptation and evolution, 306-
    307
  allele frequencies, change in,
    309-313
  development, 321
  fossil record, 319
  homology, 320-321
  molecular record, 320
  natural selection in microevolu-
    tion, 313-317
  overview of, 341-342
  parallel adaptation, 322-324
  patterns of distribution, 324
  variation in nature, 307-309
  vestigial structures, 322
Evolution, 6, 14, 31, 66, 345-359
  and adaptation, 306-307
  of bilateral symmetry, 536-537
  of body cavity, 538-539
  Cenozoic Era, 348, 356-357
  of circulatory systems, 667-668
  of coelacanth, living, 346
  of coelomates, 539-541
  after Darwin, 24
  Darwin's evidence for, 22-24
  Darwin's theory of
    history of, 20-22
    publication of, 26-28
  eukaryotic, 474-477
  evidence for, 305-327
    adaptation and evolution, 306-
      307
    allele frequencies, change in,
      309-313
    development, 321
    fossil record, 319
    homology, 320-321
    molecular record, 320
    natural selection in microevo-
      lution, 313-317
    overview of, 341-342
    parallel adaptation, 322-324
    patterns of distribution, 324
    variation in nature, 307-309
    vestigial structures, 322
  of eye, homology and, 318
  of flower color and shape, 578
  fossils in, 347
  of herbivores, 409
  of hominids, 376, 379-382
  of hominoids, 376-378
  of horse, 26
  on islands, 341
  of jaws, 363-364
  of lung, 658
  in Mesozoic Era, 348, 352-356
  of multicellularity, 478, 479
  and organisms, 480
  in Paleozoic Era, 348-352
  of primates, 374-376
  of respiration, 654-655
  of sex, 762
  of sexuality, 478-479
  vertebrate, 361-384

Evolution—cont'd
 of vertebrate brain, 716-718
 of vertebrate heart, 674-675
 of vertebrate kidney, 752-755
Excitement phase of sexual re-
  sponse, 771-772
Excitatory synapse, 709
Exclusion, competitive, principle
  of, 406
Exergonic reactions, 134-135
Exhalation in human respiration,
  660
Exocytosis, 114
Exons, 273
Exoskeleton of arthropods, 545
Experiment(s)
 Fraenkel-Conrat, 249-250
 Griffith-Avery, 246-247
 Hershey-Chase, 247-249
 Meselson-Stahl, 252
 Miller-Urey, 67-68
 nuclear transplant, 246
 reciprocal graft, Hammerling's,
  245
 in science, 18-19
Experimental design, Mendel's,
  207-208
Experimental results, Mendel's,
  208-212
Exponential notation, 47
External environment, receptors
  for, 724-728
External fertilization, 762
External parasitism, 415
Exteroception, 721, 724-728
Extinction(s), 338
 of dinosaurs, 355-356
 mass, 351-352
 in tropical rain forests, rate of,
  457-458
Extracellular digestion, 536
Eye(s)
 evolution of, and homology,
  318
 human
  focusing of, 727
  structure of, 726
 in insects, 549
 vertebrate, visual pigment in,
  725
 wavelength of light visible to,
  167

F

F plasmid, 293
F₁ generation in Mendel's studies,
  208, 211
F₂ generation in Mendel's studies,
  208, 211-212
F6P; see Fructose 6-phosphate
Fabadaeae, nitrogen fixing by, 422
Facilitated diffusion across cell
  membrane, 116
FADH₂; see Flavin adenine dinucleotide
Fagaceae, 471
Fagus, 471
Falciparum malaria and sickle cell
  anemia, 234
Fallopian tubes, 770
Family(ies)
 in hierarchical system, 361-362
 social organization of, 791
Fan worm, 543
Fargesia
 nitida, 586
 spathacea, 586
Fats
 within arteries, 681
 conversion of, to glucose, 745
 digestion of, 642
 polyunsaturated, 54
 saturated, 54
 structure of, 53-55
 unsaturated, 54
Fatty acids, 53

Feather star, 552
Feathers, 370, 372
Feces, 645
Feedback controls over hormones,
  740
Feedback loop, neuromuscular,
  729
Female reproductive system, 769-
  771
Fermentation, 152, 160-161
Fern, life cycle of, 522
Ferocactus diguetii, 560
Ferredoxin, 171, 172, 173
Ferret, black-footed, 469
Fertility and sickle cell anemia,
  314
Fertilization
 external, 762
 human, 777
Fetal-placental circulation, 780
Fetus, human, 760
 development of, 778-779
 size of, 779
 ultrasound appearance of, 239
 undernourishment of, 779
Fever in infection, 690
Fiber content of food, 647
Fibers
 collagen, 612
 cytoskeleton, 96, 97
 myelinated, in nerves, 703
 Purkinje, 678
 in sclerenchyma cells, 565
 spindle
  of muscle, 722
  in prophase, 191
 supporting, of plasma membrane, 108, 110
Fibrillation, 680
Fibroblasts, 612
Ficus carica as flower, 581
Figs as flowers, 581
Filtration in vertebrate kidney,
  751, 752
Finch(es)
 Darwin's, 22, 340-341
 purple, 371
First filial generation, 208, 211
First law of thermodynamics, 131
Fishes
 bony, evolution of, 364-365
 cichlid, mating and aggressive
  behavior in, 792
 classes of, 556
 eyes of, 726
 freshwater, kidney in, 752-753
 heart in, 675
 gas exchange in, 654
 gills of, 655-656
 jawed, appearance of, 363-364
 jawless, 363
 lobe-finned, 346, 365, 366
 marine, kidney in, 753
 and sea anemones, commensalism in, 413
 sex and reproduction in, 762
Fission, binary, 186
Fitness, 313, 794
Fittest, survival of, 24
Fixation, carbon, in photosynthesis, 170, 176
Flagellin, 99
Flagellum(a), 78
 of dinoflagellates, 503
 of eukaryotic cells, 98-100
 of sponges, 533
Flatworm, 537-538
 digestive system of, 638
 oxygen diffusion in, 654-655
Flavin adenine dinucleotide, 155,
  158
Fleas, 416
 action of muscles in, 624
Fleming, Walther, 196
Flies, 548

Flight
 in birds, 372
 in vertebrates, 344
Flooding, responses of plants to,
  594
Flower fly, 413
Flowering of bamboo and starvation of pandas, 586
Flowering plants, 524-527
 evolution of, 354
 life cycle of, 526
 reproduction of, 577-589
  differentiation in, 583
  flowers in, 527, 578-580
  formation of seeds in, 580-581
  fruits in, 581-582
  germination of seeds in, 583
  growth in, 583
  pollination in, 578-580
  strategies in, 583
Flowers
 diagram of, 527
 in peas, 206-208
 pollination and, 578-580
Fluid
 gastric, 642
 intrapleural, in human respiration, 660
 lipid bilayer as, 107-108
Flukes, 538
Fly, tsetse, 501
Flying squirrel, southern, 472
Focusing of human eye, 727
Follicle-stimulating hormone, 739,
  765, 769, 770
Follicular phase of mammalian reproductive cycle, 765
Food(s)
 digestion of, by animals, 637-
  651
  large intestine, 644-645
  mouth, vertebrate, 639-640
  nature of digestion, 638-639
  nutrition, vertebrate, 645-649
  small intestine, 642-644
  stomach, vertebrate, 640-642
 metabolism of, 148-149
 and population, 449-451
 prospects for increased supply
  of, 452-454
 protein in, 649
Food chain, 425-426
Food web, 425-426
Foolish seedling disease of rice,
  599-600
Foot of mollusks, 542
Forebrain, dominance of, 716-717
Foreign antigen, recognition of,
  691
Forest
 Carboniferous Period, reconstruction of, 353
 deciduous, 437
 temperate, 439
 rain, tropical, 435-437
  destruction of, 8, 456-459
Fossil(s), 25, 347
 of Archaeopteryx, 25
 Darwin's studies of, 21, 22
 dating of, 319
 in evidence for evolution, 22,
  319
 and evolution of horses, 26
 of hominids, 378, 379
Fossil bacteria, 69, 70
Fossil record, 25
 as evidence of evolution, 319
Founder principle, 311
Fovea, 727
Fraenkel-Conrat, Heinz, 249
Fraenkel-Conrat experiment, 249-
  250
Franklin, Rosalind, 250
Free radical in ionizing radiation,
  282-283

Frequency(ies)
 allele, change in, 309-313
 in Hardy-Weinberg equilibrium,
  308
 of sickle cell allele, 314
Fresh water, 434-435
Freshwater diatom, 475
Freshwater ecosystem, energy
  flow in, 426
Freshwater fishes, kidney in, 752-
  753
Frog(s)
 arrow-poison, 410
 mating of, 762
 species of, isolation of, 334-335
Frontal lobe, 718
Fructose, 49, 50
Fructose 6-phosphate, 176
Fruits, 581-582
 ripening of, ethylene in, 600
 stones of, 565
FSH; see Follicle-stimulating hormone
Functional groups of atoms, 48
Fundulus heteroclitus, 308
Fungi, 31, 504-506
 colonization of land by, 351
 ecology of, 506
 in ecosystem, role of, 425
 in lichens, 508
 major groups of, 507-508
 mycorrhizal, 428
 similarity of, to plants, 504
 species of, 466
 structure of, 506
Fungi Imperfecti, 507
Furrow, cleavage, in cytokinesis,
  194
Fynbos plant community, 401

G

G₁ phase of cell cycle, 189
G₂ phase of cell cycle, 189, 190
G3P; see Glyceraldehyde 3-phos-
  phate
Galapagos finch, 324
Galapagos Islands, 21, 340-341
Galapagos tortoises, 23
Gall bladder, 642
Gametangia, 517
 of bryophytes, 518
 in fungi, 507
Gametes
 diversity of, from meiosis, 200-
  201
 in plant life cycle, 516
 prevention of fusion of, in species isolation, 336
Gametic meiosis, 478-479
Gametophyte(s)
 in flowering plants, 578
 in plant life cycle, 516
 specialization of, 517-519
Gamma rays, 282
Gamma-interferon, 690
Ganglia
 cerebral, in annelids, 544
 in nervous system, 715
Gangliosides in Tay-Sachs disease,
  234
Gap junctions on cells, 122, 123
Garden pea, Mendel's studies of,
  206-207
Garrod, Archibald, 252-253
Gas exchange
 in animals, 654; see also Respiration
 and gas transport, 662-663
Gas transport
 and gas exchange, 662-663
 and hemoglobin in respiration,
  662-663
Gasohol, 451
Gastric fluid, 642
Gastric pits, 640-642
Gastropoda, 541

Gastropods, 541-542
Gastrula, 532
Gastrulation in human embryonic development, 777
Gause, G.F., 406
Gecarcinus quadratus, 425
Gene expression, 265-266
transcription in, 265
translation in, 266
Gene interactions, limit of, on phenotype alterations, 312-313
Gene mobilization, transposition in, 289
Genera, 468-469
Generations
alteration of, in plants, 516-517
alternation of, 479
in Mendel's studies, 208
Genes, 6, 256, 263-279
antibody, 695
architecture of, 275-276
cancer-causing, 10, 287
changes in, 281-302
cancer and mutation, 284-288
chromosome organization, mutations altering, 290-291
genetic engineering, 294-296
genetic engineering, recent progress in, 297-299
importance of, 299-300
mutation, 282-284
transfer of gene, 291-294
transfer of genes into plasmids, 296-297
transposition, 288-290
in chromosomal theory of inheritance, 215-217
concept of, 258-260
definition of, 215
in Down syndrome, 225-226
DNA, replication of, 276-277
expression of, 265-266
genetic code, 266-270
interaction of, to produce phenotype, 256-258
introns, discovery of, 272
known cancer-causing, 289
map of positions of, 218-219
and Mendel's studies, 210, 211-212
and protein synthesis
in eukaryotes, 273
mechanism of, 270-271
regulation of expression of, 273-275
transfer of, 291-294
among bacteria, 293-294
into plasmids, 296-297
variations in, 307
Genetic basis of behavior, 795-796
Genetic change, importance of, 299-300
Genetic code, 269
Genetic counseling, 236-239
Genetic disorders, 231-234
organizations studying, 239
Genetic drift and changes in allele frequency, 309, 310-311
Genetic engineering, 8, 294-296
to improve food supply, 453-454
recent progress in, 297-299
Genetic maps, 218-219
Genetic markers, disorders associated with, 238
Genetic self-incompatibility in flowering plants, 587
Genetic therapy, 239
Genetic traits expressed by enzymes, 254
Genetic variation
in evolution, 307-309
importance of meiosis in, 200-201
Genetics, 31, 219
human
genetic counseling, 236-239

Genetics—cont'd
genetic disorders, 231-234
human chromosomes, 224-227
multiple alleles, 229-231
patterns of inheritance, 227-228
Tay-Sachs disease, 234, 235
Mendelian, 205-221
chromosomes in inheritance, 205-217
crossing-over, 217-219
experimental design, Mendel's, 207-208
experimental results, Mendel's, 208-212
garden pea, Mendel's studies of, 206-207
heredity, early ideas about, 206
independent assortment in Mendel's studies, 213-215
test cross, 212, 213
Genome, bacteria as, 185
Genotype(s), 210, 256-257
definition of, 215
and phenotype, relationship of, 258-260
Genus, 468-469
in hierarchical system, 362
Geochemists, 66
Geographical distribution in evidence for evolution, 22
Geographical isolation, 334, 357
Geometric progression of populations, 23
Geospiza fortis, 340
Geranium, 525
Gerenuk, 409
Germination of seed, 580, 583
Germline cells
meiosis in, 196
mutation of, 282
GH; *see* Growth hormone
GHRH; *see* Growth-hormone releasing hormone
Giant chromosomes of *Drosophila melanogaster,* 262
Giant red sea urchin, 552
Giant redwood, 564
Gibberella, 599-600
Gibberellins, 599-600
in seed germination, 583
Gibbon, Mueller, 377
Gila monster, 411
Gill(s), 364, 655-656
mollusk, 542
structure of, 656
in vertebrate embryos, 321, 322
Gill arches, 363
Gill slits, 364
Ginkgo biloba, 523
Giraffe, 404
Gland(s), 610
endocrine, and hormones, 742-743
pituitary, 717, 734, 737
anterior, 739-740
posterior, 738
prostate, 772
Glaucomys volans, 472
Glial cells, 616
Global level of biological organization, 29-30
Glucagon, 741
Glucose
catabolism of, overview of, 158-159
level of, in blood, 645-646
mobilization of, in glycolysis, 150
oxidation of, 134
regulation of, in blood, 742-745
structure of, 50
Glyceraldehyde 3-phosphate, 150, 176
Glycerol, 53
Glycine in Miller-Urey experiment, 67-68

Glycogen, 744, 745
Glycogen reserve, 645-646
Glycolipids, 121
exterior, of plasma membrane, 108, 110
in Golgi complex, 90-91
Glycolysis, 148, 149-152
Glycolytic pathway, 151
Glycoproteins in Golgi complex, 90-91
Glyoxysomes, 91-92
Glyptodont, 21
Goiter, 596
Golgi, Camillo, 90
Golgi bodies, 90
Golgi complex, 90-91
Gonadotropic releasing hormone, 765
Gonorrhea, 315, 487
Gooseneck barnacles, 547
Gorilla gorilla, 377
Gorillas, 377, 378
Gould, Stephen Jay, 341
Gradualism, 341
Gram, Hans Christian, 83
Gram stain, 83
Gram-negative bacteria, 83
Gram-positive bacteria, 83
Granulocytes in inflammation, 674
Grapsus grapsus, 547
Grasses, tolerance to lead in, 316
Grasshopper
action of muscles in, 624
anatomy of, 549
Grasslands, 438
Grassquit, blue-black, 22
Gravitropism, 601
Gravity receptors, 722
Gray whale and barnacles, commensalism of, 413
Gray wolf, 469
Green algae, 476, 504
Green darner dragonfly, 545
Green Revolution, 8, 9, 452-453
Green sea turtles
egg laying by, 2, 3
migration of, 796
Green sulfur bacteria, photosystem in, 172-173
Greensnake, smooth, 368
GRH; *see* Gonadotropic releasing hormone
Griffith, Frederick, 246-247
Griffith-Avery experiments, 246-247
Ground finches, Darwin's, 340
Ground tissue in plants, 563, 564-565
Groundwater, 420-421
Growth
of flowering plants, differentiation and, 583
of populations, 392-397
as property of life, 6
of vascular plants, 521
Growth curve of human population, 446
Growth hormone, 740
Growth-hormone releasing hormone, 740
Guanine in DNA, 250-252
and RNA, 60
Guano in phosphorus cycle, 424
Guard cells of plant stoma, 595
Gulf Stream, 431
Gurdon, John, 245
Gut, 532
Gymnosperms, 523-524
Gynoecium in angiosperms, 525
Gyrals, 431
Gyrinophilus palleucus, 367

H

Habitats
freshwater, 434-435
marine, 432

Habituation, 786
Hagfishes, 363
Hair of mammals, 373
Hairs, root, 509, 572, 593
Haldane, J.B.S., 794
Hallucigenia, 349
sparsa, 349
Hammerling, Joachim, 244-245
Haploid, definition of, 215
Haploid complement, 190
Hardwoods, 571
Hardy, G.H., 308
Hardy-Weinberg principle, 308-309
and sickle cell anemia, 314
Haversian canals, 627
Hawaii, introduced populations in, 392
Hay fever, 694
Hearing, 724-725
Heart
and blood vessels, diseases of, 680
contraction of, 678
human, 675-681
path of blood through, 676
vertebrate
evolution of, 674-675
muscle of, 629
Heart attacks, 680-681
Heart murmur, 679
Heart sounds, 679
Heartwood, 571
Heat, transformation of potential energy to, 131
Heavy chains of antibodies, 692, 696
Heleoderma suspectus, 411
Helianthus annuus, 569
Helix of DNA, 60-61
Helper T cells in acquired immune deficiency syndrome, 12
Helper T4 cells, 690, 691, 695-696
Hemisphere of human brain, 718-720
Hemoglobin, 612
evolution of, 320, 321
and gas transport in respiration, 662-663
in sickle cell anemia, 233, 255, 314
subunits of, 59
Hemophilia, 222, 234-236
Hemopoietic stem cells, 687, 689
Henle, loop of, 755-756
Hepatic vein, 744
Hepatic-portal circulation, 744
Herbaceous plant, 564
Herbicides
engineering resistance to, 298
plant hormones as, 598-599
Herbivores
in ecosystems, 424
evolution of, 409
plant defenses against, 409
teeth in, 639
Hereditary information
in nucleus of cells, 244-246
Heredity; *see also* Genetics
early ideas about, 206
mechanism of, 27, 243-261
DNA in, 250-256
gene, concept of, 258-260
genes, interaction of, to produce phenotype, 256-258
hereditary information in nucleus of cells, 244-246
as property of life, 6
Hermissenda crassicornis, 541
Herpesvirus 1, 491
Herrick, James B., 314
Hershey, Alfred, 248
Hershey-Chase experiment, 247-249
Hertz, Heinrich, 166

Heterochromatin, 188
Heterosporous plants, 519
Heterotrophic bacteria, 485–486
Heterotrophs, 145, 159
  in ecosystems, 424
Heterozygosity in Mendel's studies, 212–213
Heterozygote(s)
  definition of, 215
  identification of, 238
  and selection, 313
Heterozygous, definition of, 209
Hexagenia limbata, 548
Hierarchical system, 362
Hierarchy(ies)
  dominance, 793, 796–797
  taxonomic, in classification, 470, 471–472
High-energy bonds in adenosine triphosphate, 140
High-risk pregnancy, 237–238
Hindbrain, 716
Hirudinea, 543
His, bundle of, 678
Histamine, 611
Histones in DNA duplex, 188
HIV; *see* Human immunodeficiency virus
Holmes ribgrass virus, 249
Holothuroidea, 552
Homarus americanus, 547
Homeostasis
  and autonomic nervous system, 729
  kidney in, 757
  as property of life, 6
Homeotherms, 370
Hominids, evolution of, 376, 379–382
Hominoids, evolution of, 376–378
Homo, 360, 361
  erectus, 379–380
  habilis, 379, 380
  sapiens, 381
Homogentisic acid, 253
Homologous bones in vertebrates, 27
Homologous chromosomes, 190–191
Homology
  as evidence of evolution, 320–321
  and evolution of eye, 318
  in vertebrates, 27
Homosporous plants, 518, 519
Homozygosity in Mendel's studies, 212–213
Homozygote, definition of, 215
Homozygous, definition of, 209
Hormone production, control over, 740
Hormones, 54, 701–702, 735–747
  chemical structure of, 736
  and endocrine glands, 742–743
  in human embryonic development, 777
  neuroendocrine control of, 735–737, 741
  nonpituitary, 741
  peptide, 741
  pituitary, 739–740
  plant, 597–601
  and regulation of glucose in blood, 742–745
  and regulation of water balance, 745–746
  released by anterior pituitary, 739–740
  released by posterior pituitary, 738
  sex, 764
  steroid, 741, 764
Horse, evolution of, 26
  toes of, 319
Horseshoe crabs, 349
HRV; *see* Holmes ribgrass virus

Human(s), 377, 378
  action of muscles in, 624
  aggressive behavior in, 797
  birth weight in, stabilizing selection for, 390
  circulatory system in, 677
  digestive system in, 639
Human behavior, biological basis of, 797
Human body, organization of, 608
Human brain, anatomy and function of, 718–720
Human breathing, mechanics of, 660–662
Human chromosomes, 224–227
Human ear, structure of, 724, 725
Human embryo, 554
  development of, 777–780
Human eye
  focusing of, 727
  structure of, 726
Human fetus, 760
Human genetics, 223–241
  genetic counseling in, 236–239
  genetic disorders in, 231–234
  human chromosomes in, 224–227
  multiple alleles in, 229–231
  patterns of inheritance in, 227–228
  Tay-Sachs disease and, 234, 235
Human heart, 675–681
  path of blood through, 676
Human immunodeficiency virus, 774–775
Human intercourse, physiology of, 771–772
Human kidney, structure of, 757
Human lymphatic system, 672
Human menstrual cycle, 766
Human populations, 397
Human races, 230
Human reproductive system, 767–771
Human respiratory cilia, 661
Human skeleton, 628
Human spinal cord, 712
Human urinary system, 754
Human vermiform appendix, 322
Humboldt Current, 431
Hummingbirds and flowers, 578
Humoral immune response, 690, 691
Humpbacked whale, 397
Huntington's disease, 236, 237
Hybridization, prevention of, 331, 333–337
Hybridoma, 697
Hybrids, 330–331
Hydra, structure of, 535
Hydration shell, 46
Hydrocarbons, chlorinated, 461
Hydrochloric acid, 47–48
  in digestion, 638, 642
Hydrogen in early atmosphere of earth, 66
Hydrogen atoms in photosynthesis, 133
Hydrogen bond(s), 137
  in water, 45–46
Hydrogen ion concentration and enzyme activity, 138
Hydrogen molecule, 38, 41
Hydrophobic bonding, 46
Hydrophobic interactions, 137–138
Hydrostatic pressure
  in cells, 113
  in cnidocyte function, 535
Hydroxyapatite in bone, 627
Hydroxyl ions, 47–48
Hylobates muelleri, 377
Hylobatidae, 376
Hyophorbe verschaffeltii, 457
Hyperosmotic animals, 750

Hypertonic solution, 113
Hyphae of fungi, 505–506
Hypoosmotic animals, 750
Hypothalamus, 717, 736–737
  and pituitary, interactions between, 738–740
  and sex hormones, 764
Hypothesis, testing of, in science, 18–19
Hypotonic solution, 113
Hyracotheres, 26
Hyracotherium, 26

I

I gene in ABO blood groups, 229
IAA; *see* Indoleacetic acid
IAA oxidase; *see* Indoleacetic acid oxidase
Ice formation, 46
Ichthyostega, reconstruction of, 351
Immune receptor response, 693–695
Immune response, 687, 689–691
  cell-mediated, 690, 691
  discovery of, 685–687
  humoral, 690, 691
Immune system
  cells of, 611, 687–688
  diseases affecting, 695–696
  vertebrate, 685–699
Immunoglobulins, 691, 692; *see also* Antibodies
Implantation of trophoblast in human embryonic development, 777
Imprinting, 787
Inactivation, insertional, 289
Inbreeding, 311–312
Inclusive fitness, 794
Incomplete dominance, 259
Independent assortment in Mendel's studies, 213–215
Indeterminate growth in leaves, 567
Indoleacetic acid, 598
Indoleacetic acid oxidase, 598
Inductive reasoning, 17
Industrial melanism, 307, 314–315
Infant mortality rate, 447
Infection, fever in, 690
Infection cycle of AIDS virus, 491–493
Inferior vena cava in human heart, 677
Inflammation, 674
Inflammatory response, 674
Information transmission by animals, 701–711
Ingram, Vernon, 255
Inhalation in human respiration, 660, 661, 662
Inheritance; *see also* Genetics
  chromosomal theory of, 215
  patterns of, 227–228
Inhibitor of enzyme activity, 138
Inhibitory synapse, 709
Initiation complex, 270, 271
Initiation factors, 270
Inner ear, 725
Insecta, 548
Insectivora, 374
Insects, 548–550
  gas exchange in, 654
  open circulatory system in, 668
  pollination by, 578–579
  water balance in, 751
  wing movement in, 625
Insertional inactivation, 289
Instinct and learning
  distinction between, 786–789
  interaction of, 789
Instinctive learning programs, 789
Instincts, 786, 787
Instructional theory of immune receptor response, 693, 695
Insulin, 741, 745

Integration of neural information, 709
Integuments
  animal, 534
  of plant embryo, 580
Interactions of cell with environment, membranes in, 111
Interbreeding in species definition, 470–471
Intercellular connections, types of, 123
Intercourse, human, physiology of, 771–772
Interdigitation of myofilaments, 630
Interferon, 697
  genetic engineering of, 294
Interleukin-1, 690
Intermediate fibers in cytoskeleton, 96, 97
Internal body cavity, 538–539
Internal information, sensing of, 721–723
Internal parasitism, 415
International Center for the Improvement of Maize and Wheat, 452
Internodes of plant stems, 569
Interoception, 721–723
Interoceptors, 721
Interphase in mitosis, 190–192
Interspecific competition, 406–407
Intertidal zone in oceans, 433
Intestine, 532
  large, 644–645
  small, 642–644
Intrapleural fluid in human respiration, 660
Intrauterine devices, 773
Introduced populations on islands, 392
Introns, 273
Inversion, chromosomal, 291
Invertebrates
  attachment of muscles in, 625
  water balance in, 751
Involuntary nervous system, 714, 715, 729–731
Iodine, 596, 647
Iodopsin, 725
Ionic bonds, 40–41
Ionization by water, 47–48
Ionizing radiation in mutation, 282–283
Ions, 38
  in blood plasma, 673
  calcium, 41
  concentration of, in vertebrates, 750
  hydrogen, 47–48
  movement of, in sodium-potassium pump, 117
  in plant enzyme systems, 596
  sodium, 41
  transport of, in kidneys of marine fishes, 753
Ipomoea, 415
Iridium, 356
Iris of eye, 726
Islands
  evolution on, 341
  introduced populations on, 392
  new populations on, 311
  oceanic
    distribution of organisms on, as evidence of evolution, 324
    in evidence for evolution, 22
Islets of Langerhans, 741, 745
Isolating mechanisms, reproductive, 331, 333–337
Isolation, geographic, 334, 357
Isomer, structural, of glucose, 49, 50
Isopoda, 547
Isoptera, 549
Isotonic solution, 113

Isotopes, 38
IUDs; see Intrauterine devices

J

Janze, Daniel, 414
Jawed fishes, appearance of, 363-364
Jawless fishes, 363
Jaws, evolution of, 363-364
Jellyfish, 535
Jenner, Edward, 684, 685-687
Johnson, Virginia, 771
Jumping spider, 545
Junction, neuromuscular, 633, 708

K

K (carrying capacity), 394-395
K strategists, 396-397
Kagu, 393
Kakabekia umbellata, 70
Kalstroemia, 311
Kangaroo, 373
Kangaroo rat, 749
Karyotype, 188, 189, 224
of Down syndrome, 225
Katydid, 436
Kelp, 475, 496
Kelp beds, 432
Keratin, 55
in cytoskeleton, 96, 97
Kettlewell, H.B.D., 315
Kidney, 745
human, structure of, 757
mammalian, function of, 755-757
vertebrate
evolution of, 752-755
organization of, 751-752
in water balance, 745-746
Killifish, 308
Kilocalorie, definition of, 130
Kinetic energy, 130
Kinetochore, 188, 190-193, 198-200
King, Thomas, 245
Kingdoms of life, five, 31, 467-481
classification of organisms in, 468-469
eukaryotic evolution in, 474-477
evolution of, 480
multicellularity in, evolution of, 478, 479
sexuality in, evolution of, 478-479
species in
definition of, 469-471
numbers of, 471
symbiosis and origin of eukaryotic phyla in, 477
taxonomic hierarchy in classification of, 471
Klinefelter's syndrome, 226
Knee jerk, 729
Knight, T.A., 206
Kohler, George, 697
Krebs, Sir Hans, 148
Krebs cycle, 148

L

Lac operon in gene cluster, 275, 276
Lactate dehydrogenase, 160
Lactic acid, 160, 161
Lactose in enzyme transcription, 276
Lactuca
canadensis, 334
graminifolia, 334
Ladyslipper orchid, 472
Laetisaria arvalis, 506
Lamellae, 93, 627
in gills, 656
middle
in cytokinesis, 194
in plant nutrition, 596

Lancelets, 362, 555, 556
heart in, 674
Land
colonization of, 350-351
invasion of, 365-366
Landsteiner blood groups, 229
Large intestine, 644-645
Larvae
of amphibians, 366-367, 763
of tunicates, 555
Larynx in human respiration, 660
Lateral meristems, 564
Latimeria chalumnae, 346
Latrodectus mactans, 546
Law of independent assortment, 214
Law of segregation, 212
Laws of thermodynamics, 131-132
Lead, tolerance to, in grasses, 316
Leader region, 270-271
in gene cluster, 275, 276
Leaf(ves), 562
axil of, 569
carbohydrate transport in, 595-596
of dicots, 568
evolution of, 516
primordia of, 569
silica in, as defense against herbivores, 409
structure of, 567-569
veinlets in, 590
water movement to, 592-593
Learned behavior, 786-789
Learning, 720, 786
and instinct, 786-789
Lecithin, 107
Lederberg, Joshua, 293
Leech, freshwater, 543
Left atrium in human heart, 676
Left ventricle in human heart, 676
Legionella, 486, 487
Legionellosis, 486, 487
Legionnaires' disease, 486, 487
Legumes, 413, 428
and cereals, combination of, 453
nitrogen fixing by, 422
Lens of eye, 726
Lenticels in plant stems, 570
Leopard, snow, 374
Lepas anatifera, 547
Lepidoptera, 550-551
Leptotrichia buccalis, 487
Lettuce, species of, isolation of, 334
Leucaena, 453
Leukemia and radiation, 283
Leukocytes, 611, 612, 673-674, 687, 689, 691
Levels of biological organization, 29-31
LH; see Luteinizing hormone
Lichens, 412, 508-509
Life
five kingdoms of, 467-481
classification of organisms, 468-469
eukaryotic evolution, 474-477
kingdoms of eukaryotic organisms, 474
multicellularity, evolution of, 478, 479
organisms, five kingdoms of, 473
organisms and evolution, 480
sexuality, evolution of, 478-479
species, definition of, 469
species, numbers of, 471
symbiosis and origin of eukaryotic phyla, 477
taxonomic hierarchy in classification, 470, 471-472
origin of, 65-75
cells, earliest, 69-73
cells, first, origin of, 69

Life—cont'd
origin of—cont'd
eukaryotic cells, appearance of, 72-73
organic molecules, origin of, 66-68
on other worlds, possibility of, 73-74
Life cycle(s)
of angiosperm, 526
of fern, 522
of moss, 520
of pine, 340
of plants, 516
survivorship curves for, 398
types of, 478-479
Ligase, 296
Ligation, tubal, 773
Light
biophysics of, 166-167
visible, 167
Light chains, of antibodies, 692, 695
Light energy, absorption of, 170
Light reactions
of photosynthesis, 170, 171-173
of plants, 173-176
Lilium canadense, 525
Lily, leaf of, 569
Limbic system, 717
Limenitis archippus, 412
Limnetic zone in fresh water, 434
Linnaeus, Carl, 468-469
Lion, 622
isolation of, 334
Lipases in digestion, 638-639
Lipid bilayer, 107-108
Lipid foundation of cell membranes, 106-108
Lipids, 48, 53
structure of, 54-55
Lipopolysaccharide in bacteria, 484
Liriodendron tulipifera, 568, 571
Littoral zone
in fresh water, 434
in oceans, 433
Liver, 642
blood glucose regulation by, 743-745
detoxification in, 87-88
in digestion, 644
Lobe-finned fish, 346, 365, 366
Lobes
of human brain, 718
optic, 716
of pituitary, 738, 739-740
Lobster, 547
Locus
definition of, 215
and Mendel's studies, 210, 211-212
Locusta migratoria, 395
Locusts, 395
Long-day plants, 602
Longhorn beetle, 413
Long-term memory, 720
Loop of Henle, 755-756
Lorenz, Konrad, 787, 791
Lotus, seed of, 580
Loxodonta africana, 792
Loxosceles reclusa, 546
"Lucy", 378
Ludia magnifica, 553
Luna moth, 530
Lung(s), 655, 657-659
of amphibians, 366
in birds, 659
evolution of, 658
Lung cancer, 286
incidence of, 11
and smoking, 10-12
Lungfishes, 365
Luteal phase of mammalian reproductive cycle, 765
Luteinizing hormone, 739, 765

Lycophyta, 522
Lycopodium, 573
Lymph, circulation of, 688
Lymph node, 688
Lymphatic system, 671-672
human, 672
Lymphatic vessels, 672, 688
Lymphocytes, 611, 687, 689; see also White blood cells
circulating, 691
Lymphokines, 691
Lyonization, 226
Lysis, bacterial, 248-249
Lysosomes, 91-92
Lysozyme, 136
amino acid sequence of, 57

M

M phase of cell cycle, 189-193; see also Mitosis
Macroevolution, evidence of, 306, 317-324
Macromolecules, 48, 49
Macronutrients in plants, 596
Macrophages, 611, 687, 691
in inflammation, 674
virus in, 478
Macrotermes bellicosus, 549
Major histoincompatibility complex, 121
Malaria, 503-504
and sickle cell anemia, 314
falciparum, 234
Male reproductive system, human, 767-768
Malnutrition, 8, 449
Malpighian corpuscles, 752
Malpighian tubules, 751
Malthus, Thomas, 23-24
Mammalia, 370, 372-374, 556
Mammals
characteristics of, 373
evolution of, 370, 379-381
eyes of, 726
four-chambered heart in, 675
gas exchange in, 654
grazing on savannas, 437-438
kidney in, 755-756
function of, 755-757
lung of, 658
placental, 764
evolution of, 374
reproductive cycle of, 765-766
sex and reproduction in, 764
of taiga, 439
Mandibles of arthropods, 545
Mandibulate crustaceans, 545, 547-550
Manganese, 647
Mangrove, flooding of, 594
Manibot esculenta, 451
Manioc, 451
Manta ray, 364
Mantle of mollusks, 542
Map
genetic, 218-219
recombination, 219
Maranta, 601
Marchantia, 518, 519
Marginal meristems in leaf, 567
Marine fishes, kidney in, 753
Markers
cell surface, 121
genetic, disorders associated with, 238
Marmot, yellow-bellied, 471
Marmota flaviventris, 471
Marpolia, 349
Marrow, 627, 628
Marsupials, 355, 764
evolution of, 373
and placental mammals, parallel adaptation of, 322-324
Marsupium, 373
Mass
atomic, 38

Mass—cont'd
  visceral, of mollusks, 542
Mass extinctions, 351-352
Mass flow in plants, 595-596
Mast cells, 611
  in allergy, 694
Masters, William, 771
Matrix of mitochondrion, 157
Mayfly, 548
Mayr, Ernst, 330
Mechanical isolation of species, 336
Mechanical receptors, 722-723
Mechanism of heredity, 27, 243-261
  DNA, 252-256
  gene, concept of, 258-260
  genes, interaction of, to produce phenotype, 256-258
  hereditary information in nucleus of cells, 244-246
  replication of DNA, 250-252
Mediators in allergy, 694
Medulla, renal, 756-757
Medulla oblongata, 716
Medusae of cnidarians, 535
Meerkat, 794
Megagametophytes, 519
Megakaryocytes, 674
Megaptera, 397
Megasporangia, 519
Megaspores, 519
Meiofauna, 425
Meiosis, 196, 478
  in angiosperms, 525
  and chromosomal inversion, 291
  and mitosis, comparison of, 201
  in oocytes, 769-770
  in plant life cycle, 516
  stages of, 197-200
Meiosis II, 200
Meiotic events of oogenesis, 770
Meissner's corpuscles, 722
Melanin, 626
Melanism, industrial, and peppered moths, 307, 314-315
Melanocytes, 626
Melanocyte-stimulating hormone, 740
Membrane
  basilar, of ear, 725
  cell
    lipid foundation of, 106-108
    selective permeability of, 116
    transport across, 120
  in interactions of cell with environment, 111
  of mitochondria, 93
  neuron, depolarization of, 705
  nuclear, 78
  photosynthetic, 170
  plasma, 82
    architecture of, 108-110
    differential permeability of, 750
    of eukaryotic cell, 86
    of red blood cell, 86
    pleural, in human respiration, 660
    postsynaptic, 708
    presynaptic, 706-708
    secretion of proteins across, 91
    tectorial, of ear, 725
    thylakoid, 178
    tympanic, in hearing, 724
Membrane-bounded compartments in eukaryotic cells, 84, 85
Memory, 720
  migratory, 796
Memory B cells, 691
Mendel, Gregor Johann, 206
Mendel's First Law, 212
Mendelian genetics, 205-221
  chromosomes in inheritance in, 215-217
  crossing-over in, 217-219

Mendelian genetics—cont'd
  experimental design, Mendel's, 207-208
  experimental results, Mendel's, 208-212
  garden pea, Mendel's studies of, 206-207
  heredity and, early ideas about, 206
  independent assortment in Mendel's studies, 213-215
  test cross in, 212, 213
Mendelian ratio, 209, 212
  modified, 256, 257-258
Mendel's Second Law, 214
Menstrual cycle, human, 766
Menstruation, 766
Meristem
  apical, 562, 563
    of root, 572
  marginal, in leaf, 567
  types of, 564
Merkel cells, 722
Merops, 371
Meselson, Matthew, 251
Meselson-Stahl experiment, 252
Mesencephalon, 716
Mesentery, 537
Mesoderm, 534, 608
Mesophyll, in leaf, 568
Mesozoic Era, 348, 352-356
  plants in, 353-354
Messenger RNA, 264, 265, 270
  and polypeptide synthesis, 264-265
  in seed germination, 583
Met tRNA, 270, 271
Metabolism
  aerobic, 152, 154, 155
  anaerobic, 152
  basal, hormones concerned with, 741
  catabolic, of glucose, 159
  and energy, 129-143
    activation energy, 134-135
    definition of energy, 130
    enzymes, 135-140
    laws of thermodynamics, 131-132
  of food, 148-149
  oxidative, 93, 156, 159, 653
    in mitochondria, 157
  as property of life, 6
  use of chemical energy to drive, 146
Metabolites
  concentration of, by active transport, 117, 119
  and wastes, in blood plasma, 673
Metamorphosis, 763
  in arthropods, 545
  in insects, 550
Metaphase in mitosis, 192-193
Metaphase I in meiosis, 198-199
Metaphase plate, 192
Metastases, 284, 285
Meteorite, in dinosaur extinctions, 356
Methane in early atmosphere of earth, 66
Methane-producing bacteria, 71
Method, scientific, 20
Mexico City, 450
MHC; see Major histoincompatibility complex
Micranthena, 546
Microbodies, 91-92
Microevolution, 306
  divergence in, 333
  evidence for, 341-342
  natural selection in, 313-317
Microfilaments
  in muscle cells, 630
  structure of, 630
Microgametophytes, 519

Micronutrients in plants, 596
Microscopes, 80
Microspheres, 69
Microsporangia, 519
Microspores, 519
Microtubules, 197-200
  in cytoskeleton, 96, 97
  of flagella, 98
  in meiosis, 197-200
  in mitosis, 191
Microvilli, 98
  of small intestine, 642-643
Middle ear, 725
Middle lamellae
  in cytokinesis, 194
  in plant nutrition, 596
Migration and changes in allele frequency, 310
Migratory memory, 796
Milk of mammals, 373
Milkweed, insects feeding on, 411
Miller, Stanley L., 67
Miller-Urey experiment, 67-68
Millipede, 549
Milstein, Cesar, 697
Mimicry
  Batesian, 412
  Muellerian, 412, 413
Mimosa pudica, 568
Mispairing, slipped, of chromosomes, 290
Mites, 546
  allergy to, 694
Mitochondria, 78, 92, 93-94, 477, 499
  matrix of, 157
  oxidative respiration in, 152
Mitosis, 189-193
  in animal, 195
  in fungi, 506
  and meiosis, comparison of, 201
  in plant, 195
Mitotic apparatus, formation of, 191-192
Mitral valve in human heart, 676
Modern bacteria, origin of, 71
Modified Mendelian ratios, 256, 257-258
Molar concentration, 47
Mold, slime, plasmodial, 499
Mole, definition of, 47
Molecular biology as evidence of evolution, 27
Molecular level of biological organization, 29-30
Molecular motion, random, 131
Molecular record as evidence of evolution, 320
Molecular structure of water, 45
Molecules, 39-40
  organic, 48
  production of, from carbon dioxide, 176-177
  polar, 45
  selective transport of, 114, 116-120
Moles, 374
Mollusks, 541-542
Molybdenum, 647
Monachus monachus, 393
Monarch butterfly, 410, 412
  life cycle of, 550-551
Monera, 31, 473
Monkey, 377, 378
Monoclonal antibodies, 696-697
Monocots, 526-527
Monoculture, 453
Monocyte in inflammation, 674
Monokines, 690
Monosaccharides, 49
Monosynaptic reflex arcs, 729
Monotremes, 373
  sex and reproduction in, 764
Moon, cycle of, and reproduction in sea, 762
Moose, 440

Moray eel, 365
Morchella esculenta, 505
Morel, 505
Morgan, Thomas Hunt, 216-217
Morning glory, 415
Morphine, 741
Morphogenesis in human embryonic development, 778
Mortality and survivorship, 397-398
Mosquito, 503-504
  Aedes, 493
  digestive system of, 638
Moss, 519-520
  life cycle of, 520
Moth(s), 413
  luna, 530
  peppered, and industrial melanism, 307, 314-315
Mother, age of, and incidence of Down syndrome, 226, 238
Motion, receptors for, 722-723
Motor cortex of human brain, 718
Motor endplate, 633, 708
Motor pathways, 714, 715
Mount St. Helens, succession and, 428
Mouth, vertebrate, 639-640
Movement
  animal
    bone in, 626-627
    mechanical problems of, 624-625
    muscle in, 627-629
    role of nerves in, 633, 634
    skin in, 626-627
  of cells, 629-633
  as characteristic of life, 4
  turgor, of plants, 601
mRNA; see Messenger RNA
MSH; see Melanocyte-stimulating hormone
Mucosa of stomach, 640
Mueller, Fritz, 412
Muellerian mimicry, 412, 413
Mule, 330
Multicellular protists, 500
Multicellularity
  evolution of, 478, 479
  origins of, 497-511
    in fungi, 504-506, 507-508
    in lichens, 508-509
    in mycorrhizae, 509
    in protists, 498-504
Multi-enzyme complex, 152
Multiple alleles, 229-231
Multiple drug resistance, 315-316
Murmur, heart, 679
Muscle(s)
  action of
    in flea, 624
    in grasshopper, 624
    in human being, 624
  in animal movement, 627-629
  attachment of
    in invertebrates, 625
    in vertebrates, 625
  in blood vessels, 669-671
  connections to nerve, 708
  of human heart, 678
  striated, contraction of, 632-633
  vertebrate, 613
    contraction of, 630
  in wing movement in insects, 625
Muscle contraction, receptors for, 722
Mustard oil, 409
Mustela nigripes, 469
Mutagens, 10
  chemical, 284
Mutant, 216
Mutation, 253-254, 282-284
  altering chromosome organization, 290-291
  and cancer, 284-288

Mutation—cont'd
and changes in allele frequency, 309
effects of, on organism, 253-254
of genes in cancer, 10
and transposition, 289
Mutational repair, 283
Mutualism, 414
Mycelia, 506
Mycelium, 506
Mycologists, 504
Mycorrhizae, 412, 424, 428, 509, 516
Myelin sheath, 703
development of, 706
insulating properties of, 707
Myelinated fiber in nerves, 703
Myofibrils, 613
in striated muscle, 632-633
Myofilaments
contraction of, 630-631
in muscle cells, 630, 631
Myosin, 613
in muscle cells, 630, 631
Myxomatosis virus, 408-409
Myxomycota, 499, 504

N

NAD; *see* Nicotinamide adenine dinucleotide
NADH, 140, 148, 152, 156-158, 160, 161
NADH dehydrogenase, 156
NADP$^+$; *see* Nicotine adenine dinucleotide phosphate
Naked mole rats, 790
Natural killer cells, 687
Natural selection; *see* Selection
Nautilus pompilius, 541
Neanderthals, 381
Nectar, 578
Nectaries, 414
Negative tropisms in plants, 600
Neisseria, 315
gonorrhoeae, 487
Nekton, 433
Nelumbo nucifera, 580
Nematocysts of cnidarians, 535
Nematoda, 539
Nematodes, 539
oxygen diffusion in, 655
Neotiella rutilans, 196
Nepenthes albomarginata, 512
Nephrid organs, 751
Nephron, vertebrate, organization of, 752
Neritic zone in oceans, 432
Nerve(s), 703
connections between, 709
efferent, antagonistic, 731
muscle connections to, 708
optic, 727
role of, in muscle contraction, 633, 634
transferring information from tissue to, 706-709
vertebrate, 613-616
Nerve cord, dorsal, of chordates, 362, 554
Nerve fiber membrane, depolarization of, 633
Nerve impulse, 616, 703-706
initiation of, 704-705
sensory, 720-721
Nerve tracts, 715
Nervous system, 713-733
autonomic, 714, 715
antagonistic control of, 731
neurovisceral control by, 729-731
central, 616, 713-714, 716-720
external environment, receptors for, 724-728
hormones as chemical extension of, 736

Nervous system—cont'd
human brain, anatomy and function of, 718-720
in human embryonic development, 778
internal information, sensing, 721-723
involuntary, 714, 715
organization of, 713-715
peripheral, 616, 713-714, 720-731
sensory information, 720-721
somatic, 714, 715
sympathetic, adrenal hormones as extensions of, 741
vertebrate brain, evolution of, 716-718
voluntary, 714, 715
neuromuscular control by, 729
Net primary productivity of ecosystem, 424
Net productivity
of ecosystem, 424
of tropical rain forests, 437
Net venation in leaf, 567-568
Network, capillary, 670
Neuroendocrine system, 714, 715, 735-737
Neuroglia cells, 703
Neuromuscular control by voluntary nervous system, 729
Neuromuscular junction, 633, 708
Neuron(s), 613, 702-703
in autonomic nervous system, 731
in brain of cat, 700
cell body of, 709
in cerebral cortex of human brain, 718
sensory, 720
structure of, 703
vertebrate, structure of, 616
Neurospora, mutations in, 253-254
Neurotransmitter
acetylcholine as, 633
in autonomic nervous system, 731
in neuromuscular junction, 708-709
in vertebrate nervous systems, 709
Neurovisceral control by autonomic nervous system, 729-731
Neurulation in human embryonic development, 777
Neutrons, 37-38
Neutrophils in inflammation, 674
Niche, 330, 391
ecological, of Darwin's finches, 340-341
theoretical, 407
Nicholas II of Russia, Czar, 222
Nickel in plant nutrition, 596
Nicotinamide adenine dinucleotide, 139-140
Nicotinamide monophosphate, 139
Nicotine adenine dinucleotide phosphate, 173
Nighthawk, 411
Nitric acid, 48
Nitrogen in early atmosphere of earth, 66
Nitrogen cycle, 422, 423
Nitrogen fixation, 422
Nitrogen-fixing bacteria, 486
Nitrogenous base in ATP, 140
NMP; *see* Nicotinamide monophosphate
Nociceptors, 721
Node(s)
atrioventricular, 678
lymph, 688
of plant stems, 569
of Ranvier, 703
sinoatrial, 678

Nonassociative behaviors, 786
Nondisjunction, primary, in Down syndrome, 226
Nonpituitary hormones, 741
Nonrandom mating and changes in allele frequency, 309, 311-312
"Nonself" molecules and "self" molecules, differences between, 691
Nonsense codons, 269
Noradrenaline, 741
in autonomic nervous system, 731
Notochord, 362, 554
Nuclear envelope of eukaryotic cells, 88-89
Nuclear membrane, 78
Nuclear pores, 88
Nuclear transplant experiment, 246
Nucleic acids, 59-61
Nucleoli, 90
Nucleolus of eukaryotic cells, 89-90
Nucleosome, 188
Nucleotides, 48, 256
in chromosomes, 188
in genetic code, 269
in nucleic acids, 59
Nucleus(i)
of atom, 37-38
of cell, 78
hereditary information in, 244-246
in cells, 69-70
of eukaryotic cells, 88-90
in nervous system, 715
oxygen and hydrogen, in water, 45
polar, in angiosperms, 525
primary endosperm in angiosperms, 525
re-formation of, 193
Nudibranch, 541
Nut shells, 565
Nutrients
conservation of, by plants, 401
cycling of, in ecosystems, 420
in plants, 596
Nutrition
in fungi, 505
vertebrate, 645-649
Nymphaea odorata, 525

O

Oak, 468
species of, 334, 335
Obesity, 646, 647
Occipital lobe, 718
Oceanic islands in evidence for evolution, 22
distribution of organisms on, 324
Oceans, 432-433
circulation patterns in, 431-432
overexploitation of, 454
primitive, in origin of life, 68
Ocelli, 549
Octet rule, 39
Octopus, 541
Odocoileus virginianus, 375
Odonata, 545
Oenone fulgida, 543
Oil, 54
Oligochaeta, 543, 544
Omnivores, teeth in, 639
On the Origin of Species, 20, 329
Oncogene, 287
One gene–one enzyme hypothesis, 254
Onithorhynchus anatinus, 373
Oocytes, 769
Oogenesis, meiotic events of, 770
Opabinia, 349

Open circulatory systems, 667-668
Operant conditioning, 787, 789-790
Operator in gene cluster, 275, 276
Operon in gene cluster, 275, 276
Opheodrys vernalis, 368
Ophiothrix, 552
Ophiuroidea, 552
Opossum, 355, 373
Opposable thumb of anthropoid primates, 376
Opsin, 725
Optic lobes, 716
Optic nerve, 727
Optimal yield in agricultural practice, 396
Opuntia, 407-408
Orangutan, 304, 377
Orbitals of electron, 38-39
of water, 45
Orca, 375
Orchid, ladyslipper, 472
Orcinus orca, 375
Order in hierarchical system, 361
Ordovician Period, 348, 350-351
Oreaster occidentalis, 552
Organ, animal, definition of, 532
Organ system, vertebrate, 608, 609
Organ transplants, 502
Organelles, 78
bacteria-like, in eukaryotic cells, 92-95
Organic molecules, 48
origin of, 66-68
production of, from carbon dioxide, 176-177
Organismal level of biological organization, 29-30
Organisms
classification of, 468-469
effects of mutations on, 253-254
eukaryotic, kingdoms of, 474
and evolution, 480
five kingdoms of, 473
origins of major groups of, 348-350
Organization
biological, levels of, 29-31
interior, of bacteria, 83-84
levels of, in vertebrate body, 608
Organizing junctions on cells, 123
Organogenesis in human embryonic development, 777
Organs
nephrid, 751
in vertebrate body, 608
Orgasm, 772
Origin of life, 65-75
cells
earliest, 69-73
first formation of, 69
eukaryotic cells, appearance of, 72-73
life on other worlds, possibility of, 73-74
organic molecules, origin of, 66-68
Origin of Species, 20, 329
Orthoptera, 549
Osmoconformers, 750
Osmoregulation, 750-751
Osmosis, 112-113
Osmotic pressure, 113-114
in cnidocyte function, 535
Osteichthyes, 364, 556
Osteoblasts, 627
Ostrich, 371
Outcrossing, 311
in plant reproduction, 585
in species definition, 470-471
Ova, 769, 770
Oval window of ear, 725
Oviparous reptiles, 763
Ovulation, 765

Ovules in gymnosperms, 523
Ovum, 769, 770
Owl, northern saw-whet, 371
Oxalis, leaflets of, 602
Oxidation, 66, 133
  of acetyl-coenzyme A, 155
  aerobic, of glucose, 159
  of glucose, 134
  in glycolysis, 150
  of pyruvate, 152-153, 155
Oxidation-reduction reaction, 148
  and chemical bonds, 133
  enzyme-catalyzed, 139
  photosynthesis as, 171
Oxidative metabolism, 93, 156,
    159, 653
  in mitochondria, 157
Oxidative respiration, 148, 152-
    156, 159, 160-161
Oxygen
  and flooding, in plants, 594
  formation of, in photosynthesis,
    174-175
  in metabolism, 653
  partial pressure of, in respira-
    tion, 663
  use of, by animals, 653-665
    evolution of respiration, 654-
      655
    gas transport and exchange,
      662-663
    gill, 655-656
    lung, 657-659
    mechanics of human breathing,
      660-662
Oxytocin, 738
  in orgasm, 772
Oyster, shells of, evolution of,
    319
Ozone in atmosphere, 351
Ozone layer, 166

P
P_{680}, 173, 174, 175
P_{700}, 172, 175
Pain, 721-722
  chest, 680
Paintings, cave, 381, 388
Pair bonding, human, 772-773
Paleozoic Era, 348-352
Palisade parenchyma in leaf, 568
Palmitic acid, 54
Pan troglodytus, 377
Pancreas, 642, 741, 745
Pandas, starvation of, and flower-
    ing bamboo, 586
Pangaea, 354-355
Panthera
  leo, 334
  tigris, 334
Papilio dardanus, 391
Parallax, 576
Parallel adaptation as evidence of
    evolution, 322-324
Parallel venation in leaf, 567-568
Paramecium, 79, 115, 500
  cellular organization of, 5
Parasites
  diseases caused by and cyclos-
    porine, 502
  flatworms as, 538
Parasitism, 415-416
Parasympathetic nervous system,
    730-731
Parathyroid hormone, 741
Parenchyma in leaf, 568
Parenchyma cells, 564
  in leaf, 568
  in xylem, 566
Parietal lobe, 718
Partial pressure of oxygen in res-
    piration, 663
Passerines, 371
Passiflora molissima, 392
Passion flower, 392
Pasteur, Louis, 686

Pathogens, bacteria as, 486-488
Pathway
  biochemical, 138-139, 146
  glycolytic, 151
Patterns
  of distribution as evidence of ev-
    olution, 324
  of inheritance, 227-228
Pavlov, Ivan, 788
Payne, Roger, 397
Pea, garden, Mendel's studies of,
    206-207
Pecking order in chickens, 793
Pectins, 51
Pedigree of albinism, 227-228
Pedigree analysis, 228
Pelicans, 424
Pellagra, 648
Pelomyxa palustris, 499
Pelvis in whales, 322
Penicillium, 507
Penis, 763
  human, 768
Peppered moths and industrial
    melanism, 307, 314-315
Peptide bond, 55
Peptide hormones, 741
  structure of, 736
Perennial, 564
Perforations in xylem, 566
Pericycle of plant roots, 572
Periderm in plant stems, 570
Periodic table of elements, 43
Peripheral nervous system, 616,
    713-714, 720-731
Peristalsis, 640
Permafrost, 441
Permeability
  differential, of plasma mem-
    branes, 750
  selective, of cell membrane, 116
Permian Period, 348, 351
  plants in, 353-354
Pernicious anemia, 648
Peroxisomes, 91-92
Personal space, 792
Petals in angiosperms, 526
Petiole in leaf, 567-568
pH and enzyme activity, 138
pH scale, 47-48
Phaeophyta, 475
Phagocytes, 687
Phagocytosis, 114, 115
Phanerozoic time, 348
Pharyngeal slits in chordates, 362,
    554
Pharynx in chordates, 362
Phaseolus vulgaris, germination of,
    563
Phenotype, 210
  definition of, 215
  and genotype, relationship of,
    258-260
  interaction of genes to produce,
    256-258
Phenotype alterations, limit of
    gene interactions on, 312-313
Phenotypes
  in ABO blood groups, 229
  selection acting on, 313
  selection effects on, 389-390
Phenylketonuria, 234, 239
Phloem, 246, 521, 563, 566-567
  isolation of, in carrots, 584
  movement of substances in, 595
  in plant stems, 569, 570
Phosphodiester bond in nucleic
    acids, 59
Phospholipids
  bilayer sheets of, 106
  in cell membranes, 106
Phosphorus cycle, 423-424
Phosphorylation
  of adenosine diphosphate in
    chloroplasts, 278
  substrate-level, 146, 147-148

Photocenter
  evolution of, 171-172
  two-stage, 173-174
Photoelectric effect, 166
Photons, 166
Photoperiodism, 602
Photophosphorylation, cyclic,
    172-173
Photosynthesis, 6, 132, 165-182,
    421, 560, 561-562, 568-569
  biophysics of light, 166-167
  in chloroplast, 178-179
  hydrogen atoms in, 133
  in lichens, 508
  light reactions of plants in, 173-
    176
  overview of, 170-171
  as oxidation-reduction equation,
    171
  by plankton, 433
  production of organic molecules
    from carbon dioxide in, 176-
    177
  in thylakoids, 94
Photosynthetic bacteria, 71, 83, 84
Photosynthetic membranes, 170
Photosystem I, 173
Photosystem II, 173-175
Phototropism, 597, 600-601
Phycobilins, 499
Phyla, origins of, 348-349
Phyllomedusa tarsis, 721
Phyllostachys bambusoides, 586
Phylogenetic tree, 320
Phylogenies in evolution, 27
Phymatus morbilosus, 411
Physiology of human intercourse,
    771-772
Picea
  engelmannii, 523, 585
  nigra, 514
Pieridae, 409
Pigment(s)
  accessory, chlorophyll *b* as, 176
  in photocenter, 171-172
  replenishing, in photosynthesis,
    171
  types of, 167-169
  visual, in vertebrate eyes, 725
Pilus in gene transfer, 293
Pine, life cycle of, 340
Pinocytosis, 114, 115
Pinus
  longaeva, 6
  muricata, 339
  radiata, 310
Pistillate flowers, 585
Pisum sativum, 206, 580
Pit organ of timber rattlesnake,
    728
Pith of plant stems, 569
Pits
  gastric, 640-642
  in xylem, 566
Pituitary, 717, 734, 737
  anterior, hormones released by,
    739-740
  posterior, hormones released by,
    738
Pituitary hormones, 739-740
PKU; see Phenylketonuria
Placenta, 374
Placental mammals, 764
  evolution of, 374
  and marsupials, parallel adapta-
    tion of, 322-324
Placental-fetal circulation, 780
Plankton, 355-356, 433
Plant body, primary, 564
Plant cells and bacterial and ani-
    mal cells, comparison of, 101
Plantae, 31
Plantago, 249
Plants, 31, 513-529
  alternation of generations in,
    516-517

Plants—cont'd
  bryophytes, 519-520
  carbohydrate transport in, 595-
    596
  cell types in, 562, 564-567
  colonization of land by, 350
  crop, genetic engineering of, 298
  crosses between, 206
  defenses of, against herbivores,
    409
  dormancy in, 602
  ecotypes in, 339
  flowering; see Flowering plants
  function of, 591-604
  history of, 353-354
  hormones in, 597-601
  life cycle of, 516-519
  light reactions of, 173-176
  mitosis in, 195
  mosses, 519-520
  nutrients, 596
  organization of, 561-563
  photoperiodism of, 602
  photosystems in, 173-174, 175
  reproductive strategies of, 585,
    587
  response to touch, 601
  responses of, to flooding, 594
  structure of, 561-575
  tissue types in, 563
  tropisms in, 600-601
  turgor movements of, 601
  vascular, 520-524
  water movement in, 592-595
Plaque, dental, 487
Plasma, 612
  blood, 673
  regulation of glucose levels in,
    742-745
Plasma cells, 691
Plasma membrane(s), 82
  architecture of, 108-110
  differential permeability of, 750
  of eukaryotic cell, 86
  protein of, and cancer, 288
  of red blood cell, 86
Plasmids, 291, 292-295
  in plant genetic engineering, 298
  transfer of genes into, 296-297
Plasmodesmata, 123
Plasmodium, 503-504
Plastics in pollution, 460-461
Plate
  cell, in cytokinesis, 194
  metaphase, 192
Plateau phase of sexual response,
    772
Platelets, 674
Plates, continental, movement of,
    354-355
Platyhelminthes, 537-538
  oxygen diffusion in, 654-655
Platypus, duck-billed, 373
Pleiotropy, 258-259
Pleural membrane in human res-
    piration, 660
Pneumococcus, Griffith-Avery ex-
    periments with, 246-247
Pogonophora, 433
Poinsettia, 578
Point mutations, 282
Poisonous spiders, 546
Polar bodies, 770
Polar molecules, 45
Polar nuclei in angiosperms, 525
Polarity of water, 45
Polarization in nerve cell, 704
Pollen dispersal
  by bee, 311
  by wind, 310
Pollen tube in angiosperms, 525
Pollination
  by animals, 578-579
  and flowers, 578-580
  by wind, 579
Pollution, 459-461

Polychaeta, 544
Polyclonal antibodies, 696
Polymer, 48
Polymerization in cytoskeleton, 95
Polynomial system of classification, 468
Polypeptide(s), 57-59
  genes encoding, 256
  synthesis of, 266
    and messenger RNA, 264-265
Polyploid chromosome, 291
Polyps of cnidarians, 535
Polysaccharides, 51
Polytrichum, 519
Polyunsaturated fats, 54
Pomacanthus semicircularis, 365
Pongidae, 376
Pongo pygmaeus, 377
Pons, 716
Poplar, white, 571
Popper, Karl, 20
Population explosion, 7-10, 445-448
Population growth, 8
Population level of biological organization, 29-30
Population size, 398-399
Population(s)
  concept of, 391
  divergence of, 331-333
  effective breeding, 331
  and environment, interaction of, 387-403
  and food, 449-451
  geometric progression of, 23
  growth of, 392-397
  human, 397
    age structure of, 448
  introduced, on islands, 392
Populus
  grandidentata, 570
  tremuloides, 584, 585
Porcellio scaber, 547
Pores
  nuclear, 88
  in sponges, 533
Porifera, 533
  oxygen diffusion in, 654-655
Porphyrin ring in chlorophyll, 168
Portal vein, 743
Positive tropisms in plants, 600
Posterior air sac in avian lung, 659
Posterior pituitary, hormones released by, 738
Postsynaptic membrane, 708
Postzygotic factors in reproductive isolation, 333-334, 337
Potassium
  in nerve cell, 704
  in plant nutrition, 596
Potassium ions in nerve cell, 704
Potassium sodium ion channels, 704-705
Potassium-sodium pump, 117
Potassium-40 in dating of rocks, 319
Potential
  action, 705-706
  biotic, 393-394
  resting, 704
Potential energy, 130
Pouch of marsupials, 373
Poverty, 449
Power of Movement in Plants, 597
Prairies, tall-grass, 438
Prayer plant, 601
Precapillary sphincter, 669
Precipitation, acid, 460-461
Predation, 407-409
Predator and prey, interactions of, 407-412
Pregnancy
  anatomical adaptations to, 780

Pregnancy—cont'd
  high-risk, 237-238
  human, 777-780
Preparation reaction in citric acid cycle, 155
Pressure
  air, in ear, 725
  atmospheric, role of, in water movement in plants, 592-593
  blood, 679
    receptors for, 722
  hydrostatic
    in cells, 113
    in cnidocyte function, 535
  osmotic, 113-114
    in cnidocyte function, 535
  partial, of oxygen, in respiration, 663
  turgor, in plants, 601
Pressure receptors, 722
Presynaptic membrane, 706-708
Prey and predator, interactions of, 407-412
Prezygotic factors in reproductive isolation, 333-334, 335-336
Prickly pear cactus, 407-408
Primary consumers in ecosystems, 424
Primary endosperm nucleus in angiosperms, 525
Primary growth in vascular plants, 521
Primary immune response, 693
Primary mRNA transcript, 265
Primary nondisjunction in Down syndrome, 226
Primary oocytes, 769, 770
Primary plant body, 564
Primary structure of polypeptides, 57
Primary succession, 427
Primates, evolution of, 374-376
Primitive oceans in origin of life, 68
Primordia
  leaf, 569
  of plants, 583
PRL; see Prolactin
Procambarus, 547
Prochloron, 84, 485, 499
Profundal zone in fresh water, 434
Progesterone, 765, 766
Prokaryotes, 31
  as earliest cells, 69-70
  eukaryotes and, features of, 473
Prolactin, 740
Promoters in gene cluster, 275, 276
Prophase in mitosis, 191-192
Prophase I in meiosis, 197-198
Proprioceptors, 722
Prosencephalon, 716
Prosimians, 375, 376
Prostate gland, 772
Protanopia, 219
Proteases in digestion, 638-639
Protective coloration and warning, 411
Protein(s), 48
  activator, 274
  in blood plasma, 673
  essential amino acids in, 647
  in foods, 649
  globular, shape of, 58
  of plasma membrane, and cancer, 288
  repressor, 274
  secretion of, across membranes, 91
  structure of, 55-59
  transmembrane, 109, 110
    of plasma membrane, 108, 110
  use of RNA to make, 264-265
Protein coats of viruses, 686
Protein sheath, 489

Protein synthesis
  overview of, 266
  in seed germination, 583
Protista, 31, 474-476
Protists, 31, 498-500
  major groups of, 500-502
  multicellular, 500
Proton movement in chemiosmosis, 158
Proton pump, 119
Proton-pumping channels, 156
Protons, 37-38
Protostomes, 540
  egg cleavage in, 540
  segmentation in, 543, 544-545
Protozoa, 31, 474
Protozoans, 474
pSC101, 296
Pseudemys rubriventris, 368
Pseudocoel, 537
Pseudocoelomates, 537, 539
Pseudomonas, 82
  syringiae, 299
Pseudomyrmex, 414
Pseudopods, 503
Psilophyta, 522
Psychiatric disorders affecting eating habits, 646
Psychologists, behavioral, 787-789
Ptecticus trivittatus, 548
Pterophyta, 522
Pterosaurs, 344
Ptilothrix, 311
Puccinia graminis, 508
Puffball, 505
Puffer, 365
Pulex irritans, 416
Pulmonary arteries, human, 677
Pulmonary circulation, 675
Pulmonary valve in human heart, 677
Pulmonary veins, 675
  human, 676
Pump
  proton, 119
  sodium-potassium, 117, 704-705
Punctuated equilibria, 341
Punnett, Reginald Crundall, 212
Punnett square, 212
Pupil of eye, 726
Purine in DNA, 250-252
  and RNA, 60
Purkinje fibers, 678
Purple finch, 371
Pyloric sphincter, 642
Pyramidal tract, 718
Pyramids in ecosystems, 426-427
Pyrimidine in DNA, 250-252
  and RNA, 60
Pyrimidine dimer, 283
Pyrota concinna, 354
Pyruvate, 148, 160, 161
  conversion of, to ethanol, 152
  in glycolysis, 150
  oxidation of, 152-153, 155, 156
Pyruvate dehydrogenase, 152-153

Q

Quaternary structure of polypeptides, 59
Queen Victoria, 235-236
Quercus, 471
  dumosa, 334, 335
  lobata, 334, 335
  phellos, 468
  robur, 334, 335
  rubra, 468
Quinton, Paul, 118

R

r strategists, 396-397
Rabbit, 408-409
Races
  definition of, 338
  ecological, 338-340
    in animals, 340

Races—cont'd
  human, 230
Radial cleavage, 540
Radiation
  ionizing, 282-283
  ultraviolet, 283-284
Radical, free, in ionizing radiation, 282-283
Radicle in seed germination, 583
Radioisotopes in dating of rocks, 319
Radish, 565
Radula of mollusks, 542
Rain, acid, 460-461
Rain forests, tropical, 435-437
  destruction of, 8, 456-459
Ram ventilation in gills, 655
Rana, isolation of, 334-335
Range of species, 399-401
Ranvier, nodes of, 703
Raphanus sativus, 565
Ras gene, 287
Rat kangaroo, chromosomes of, 192
Ratio
  Mendelian, 209, 212
  modified Mendelian, 256, 257-258
Rats
  conditioning of, 788, 789
  maze learning by, 795
Rattlesnake, timber, pit organ of, 728
Raup, David M., 352
Ray, John, 330
Rays, 363-364
Reabsorption in vertebrate kidney, 751, 752
Reaction, oxidation-reduction, and chemical bonds, 133
Reaction rate of endergonic reaction, 135
Rearrangement, chromosomal, 282, 291
Reasoning, deductive and inductive, 17
Receptor response, immune, 693-695
Receptor-mediated endocytosis, 119
Receptors
  cell surface, 120-121
    and monoclonal antibodies, 697
  of lymphocytes, 691
  sensory, 720-723
Recessive allele, definition of, 215
Recessive human traits, 210
Recessive traits in Mendel's studies, 208
Reciprocal altruism, 794
Reciprocal graft experiment, Hammerling's, 245
Recognition site, 292
Recombinant DNA, 296-297
Recombination, rapid, transcription in, 290
Recombination enzymes, 292
Recombination map, 219
Rectum, 645
Red algae, 476
Red blood cells, 612, 673, 674, 687, 689
  cell surface markers on, 121
  hemoglobin in, 662-663
  plasma membrane of, 86
  in sickle cell anemia, 233, 314
Red oak, 468
Red squirrel, 471
Red tides, 501
Redox reactions, 133
Reducing atmosphere, 66-67
Reduction, 66
Redwood
  California, 399-400

Redwood—cont'd
giant, 564
Reflex, 729
Reflex arcs, 729
Refractory period, 705
in sexual response cycle, 772
Regulatory sites in gene expression, 274
Reindeer, 441
Release factors, 270
Releasing hormones, 739
Renal medulla, 756-757
Renal tubule in mammals, 755-756
Repair, mutational, 283
Replication
in AIDS virus infection cycle, 492-493
of DNA, 250-252
Replication fork of DNA, 277
Replication origin, 296
in bacteria, 185
Repression in gene expression, 274
Repressor protein, 274
Reproduction
asexual
in Fungi Imperfecti, 507
and species definition, 470
of cells, 185-203
bacteria, cell division in, 185-186
cytokinesis, 189, 194-195
eukaryotes, cell division among, 186-189
meiosis, 196
mitosis, 189-193
stages of meiosis, 197-200
of flowering plant, 577-589
differentiation, 583-584
flowers, 578-580
fruits, 581-582
germination, seed, 583
growth, 583-584
pollination, 578-580
seed, formation of, 580-581
strategies in, 583
in fungi, 506
as property of life, 6
sex and, 761-783
contraception, 772-773
embryonic development in vertebrates, 777
evolution of sex, 762
human embryonic development, 777-780
human reproductive system, 767-771
physiology of human intercourse, 771-772
reproductive cycle of mammals, 765-766
sex hormones, 764
in vertebrates, 762-764
sexual, and evolutionary change, 478
Reproductive behaviors, 797
Reproductive cycle of mammals, 765-766
Reproductive isolating mechanisms, 331, 333-335
Reproductive system, human, 767-771
Reptiles
evolution of, 367-369
kidney in, 755
lung of, 658
sex and reproduction in, 763-764
stereotyped behavior in, 717
Reptilia, 367-369, 556
Residual volume in lung, 658
Resistance, multiple drug, 315-316
Resolution phase of sexual response, 772
Respiration, 653-665
in birds, 659

Respiration—cont'd
cellular, 134, 148
evolution of, 654-655
oxidative, 148, 152-156, 159, 160-161
Respiratory cilia, human, 661
Respiratory system
human, 660-662
in insects, 548
Response, stimulus and, association between, 788
Resting potential, 704
Restriction endonuclease, 295
Restriction enzymes
genetic counseling for, 238
in genetic engineering, 295-296
Reticulate venation in leaf, 567-568
Reticulum
endoplasmic, 78
sarcoplasmic, 633
Retina, structure of, 727
Retinal, 167
cis-Retinal, 725
Retrovirus, cancer caused by, 285
Reverse transcriptase, 272, 492
Rhamphorhynchus, 344
Rhinoceros, Sumatran, 392
Rhizobium, 486
Rhizomes, 584
Rhizophora, 594
Rhizopoda, 503
Rhizopus, 507
Rhodophyta, 476
Rhodopsin, 725
Rhombencephalon, 716
Rhynochetus jubatus, 393
Rhythm method of birth control, 773
Ribose in adenosine triphosphate, 140
Ribosomal RNA, 90, 264, 265, 270
Ribosome(s), 89-90
on endoplasmic reticulum, 87
subunits of, 264
Ribulose 1,5-bisphosphate, 176
Rice, 9
foolish seedling disease of, 599-600
Rickets, 648
Rings, annual, in tree trunk, 571
Ripening of fruits, ethylene in, 600
River crocodile, 368
RNA
classes of, 264-265
and DNA, similarity of, 265
as genetic material of viruses, 249
in meiosis, 197-200
messenger, 264, 265, 270
nucleotide subunits of, 59
ribosomal, 90, 264, 265, 270
structure of, 268
transcription of, 265
transfer, 264, 265, 270
use of, to make protein, 264-265
in viruses, 489-490
RNA polymerase, 265, 274, 276
RNA transcript, 265
Rocks, dating of, 319
Rods, 725, 726, 727
Root, 562
Root cap, 572
Root hairs, 509, 572
of plants, 593
Roots
absorption of water by, 593, 595
mycorrhizal, 509
structure of, 572
Rotational symmetry, twofold, 295
Rough endoplasmic reticulum, 87
Roundworms, oxygen diffusion in, 655

Rous sarcoma virus, 285
Royal hemophilia, 222, 235-237
rRNA; see Ribosomal RNA
RuBP; see Ribulose 1,5-bisphosphate
Rubus penetrans, 392
Ruby-throated hummingbird, 578
Rumen, 644

S
S phase of cell cycle, 189, 190
S-1 units, 630
SA node; see Sinoatrial node
Sabella melanostigma, 543
Salamanders, 367
Saliva, 640
Salix nigra, 573
Sally lightfoot crab, 547
Salsola, 582
Salt balance, maintenance of, 746
Salt concentration in renal function, 756-757
Salt marsh, food web in, 425
Saltatory conduction, 706
Salts in blood plasma, 673
Sambucus, 565
canadensis, 570
Sand dollar, 552
Sanger, Frederick, 255
Sapwood, 571
Sarcoma, 284, 285
Sarcomeres, 632-633
Sarcoplasm, 632-633
Sarcoplasmic reticulum, 633
Saturated fats, 54
Savannas, 437-438
Scarab beetle, 545
Schistosoma, 538
Schleiden, Matthias, 79
Schwann, Theodor, 79
Schwann cells, 616, 703
in saltatory conduction, 706
Science
biology as, 17-34
nature of, 17-20
Scientific creationism, 27-29
Scientific method, 14, 20
Sciuridae, 471
Sciurus, 471
Sclereids, 565
Sclerenchyma cells, 565
Scolopendra, 549
Scorpion, 546
Scrotum, 767
Scurvy, 646, 647, 648
Sea anemones and fishes, commensalism in, 413
Sea birds in phosphorus cycle, 424
Sea cucumber, 552
Sea horse, 365
Sea star, 552, 553
Sea turtles, green, migration of, 796
Sea urchin, giant red, 552
Seal, monk, 393
Seasonal isolation, 334-335
Second filial generation, 208, 211-212
Second law of thermodynamics, 131
Secondary consumers in ecosystems, 424
Secondary growth in vascular plants, 521
Secondary immune response, 693
Secondary structure of polypeptides, 58
Secondary succession, 427
Secretion in vertebrate kidney, 751, 752
Seed, 521, 563
formation of, 580-581
germination of, 580, 583
of lotus, 580
Seed coat of plants, 580-581
Seedless vascular plants, 522-523

Segmentation in protostomes, 543, 544-545
Segregation
of alternative traits, 206, 208
law of, 222
Selection, 14, 20, 24, 306, 388-391
and changes in allele frequency, 309, 312-313
directional, 389-390
disruptive, 391
limits to, 391
in microevolution, 313-317
stabilizing, 390
Selective permeability of cell membrane, 116
Selenium, 647
"Self" molecules and "nonself" molecules, differences between, 691
Self-feeders, 145
Self-fertilization, 311
in peas, 207, 208
Self-incompatibility, genetic, in flowering plants, 587
Self-pollination in flowering plants, 580, 587
Semen, 763
Semicircular canals in movement detection, 722-723
Seminiferous tubules, 767-768
Sense organs, 778, 779
Sensitive plant, 568
Sensitivity as characteristic of life, 4
Sensitization, 786
Sensory cortex of human brain, 718
Sensory information, 720-721
Sensory integration, 717
Sensory pathways, 713-714
Sensory receptors, 720-723
Sepals in angiosperms, 526
Sepkoski, J. John, 352
Septa in fungi, 506
Sequoia sempervirens, 399-400
Sequoiadendron giganteum, 564
Serum albumin, 673
Setae of annelids, 544
Sex, 761-783
evolution of, 762
contraception and, 772-773
evolution of, 762
human reproductive system and, 767-771
physiology of human intercourse in, 771-772
reproductive cycle of mammals and, 765-766
in vertebrates, 762-764
Sex chromosomes, 224, 226-227
Sex hormones, 764
Sex-linked traits, 216-217, 218
Sexual intercourse and AIDS, 774-775
Sexual reproduction and evolutionary change, 478
Sexuality, evolution of, 478-479
Sexually transmitted diseases, 487
Shapes of bacteria, 82
Sharks, 363-364
kidney in, 753
movement of, 364
Sheath, myelin, 703
development of, 706
insulating properties of, 707
Shell(s)
of amniotic egg, 369, 370
of reptiles and birds, 763
electron, 39
hydration, 46
of mollusks, 542
Shell fungus, 505
Shifting agriculture, 456
Shoot(s), 562
structure of, 567-572
Short-day plants, 602

Short-term memory, 720
Sickle cell anemia, 233-234, 314
  molecular basis of, 255
Sickle cell trait, 314
Side groups
  in amino acid structure, 55
  of enzyme, 137
Sieve areas, 566, 567
Sieve cells, 566, 567
Sieve elements in vascular plants, 521
Sieve plates, 566, 567
Sieve tubes, 566, 567
Sieve-tube members in phloem, 566, 567
Sigmoid growth curve of populations, 394-396
Sigmoria, 549
Silene virginica, 343
Silica
  as defense against herbivores, 409
  in plant nutrition, 596
Silurian Period, 348, 350
Simple epithelium, 610
Singing of sparrow, 789
Single bonds, 41
Single-celled organisms, gas exchange in, 654
Sinoatrial node, 678
Sinus venosus in fish heart, 675
Sinuses, hepatic, 744
Sister chromatids
  in interphase, 190-191
  in meiosis, 198
  in metaphase, 193
Size
  of cell, 78-82
  population, 398-399
Skeleton
  human, 628
  of whale, 322
Skin
  aging of, 610
  cells of, 610
  of reptiles, 367, 369
  structure of, 626
  of vertebrate, 626
Skink, 368
Skinner, B.F., 788
Skinner box, 788, 789
Skunk, 411
Slash-and-burn cultivation, 456
Sleeping sickness, 501
Slime mold
  life cycle of, 504
  plasmodial, 499
Slipped mispairing of chromosomes, 290
Slits
  gill, 364
  pharyngeal, in chordates, 362, 554
Sloth, three-toed, 436
Small intestine, 642-644
Smallpox, 685-687
  eradication of, 489
Smell, receptors for, 724
Smog, 460
Smoking and lung cancer, 10-12
Smooth endoplasmic reticulum, 87-88
Smooth muscle, 627-628
Snow leopard, 374
Social behavior, biological basis of, 789-794
Sociality, 790
Sociobiology, 789-794
Sodium hydroxide, 48
Sodium ions, 38, 41
  in nerve cells, 704-706
Sodium-potassium pump, 117, 704-705
Softwoods, 571
Soil
  ammonia-rich, bacterium from, 70

Soil—cont'd
  effect of, on distribution of species, 400-401
  in tropical rain forests, 437
Soldier beetle, 579
Soldier fly, 548
Solubility, 46
Solutes, 112-113
Solution, 112-113
Solvent, 112-113
  water as, 45-46
Somatic cells, mutation of, 282
Somatic nervous system, 714, 715
Somatotropin, 740
Sowbugs, 547
Soybeans, 509
Sparrow
  dusky seaside, 338
  singing of, 789
Specialization in insect societies, 791
Species
  clusters of, 340-341
  competition between, 406-407
  Darwin's observations of, 23
  definition of, 469-471
  distribution of, factors limiting, 399-401
  endangered, 392-393
  and environment, interaction of, 387-403
    concept of population in, 391
    demography in, 398-399
    growth of populations in, 392-397
    mortality and survivorship in, 397-398
    selection in, 388-391
  formation of, 307, 329-343
    divergence of populations in, 331-333
    evidence that divergence leads to, 338-341
    reproductive isolating mechanisms in, 331, 333-337
  in hierarchical system, 362
  interaction of, with one another, 405-417
    coevolution in, 407
    competition between, 406-407
    predator and prey, 407-412
    symbiosis in, 412-416
  nature of, 330-331
  numbers of, 471
  range of, 399-401
Specificity of antibody molecule for antigen, 692-693
Speech centers in brain, 720
Sperm, 767, 768
Spermatogenesis, 767-768
Sphenomorphus, 368
Sphenophyta, 522
Spherosomes, 91
Sphincter
  anal, 645
  esophageal, 640
  precapillary, 669
  pyloric, 642
Spicules in sponges, 533
Spider
  black widow, 546
  brown recluse, 546
  jumping, 545
  poisonous, 546
Spilogale putorius, 411
Spinal cord, 616
  human, 712
Spindle apparatus in prophase, 191
Spindle fibers
  in anaphase, 193
  of muscle, 722
  in prophase, 191
Spiny anteater, 373
Spiracles in insects, 548
Spiral cleavage, 540

Spirilla, 484
Spirillum, 82
Spirogyra, 475
Spirulina, 454, 455
Spleen, 688
Sponges, 504, 532-533
  oxygen diffusion in, 654-655
Spongin in sponges, 533
Spongy bone tissue, 627, 628
Spongy parenchyma in leaf, 568
Spontaneous abortions in human embryonic development, 778
Sporangia, 519
  in fungi, 507
Spores
  bacterial, 484
  in fungi, 506
  in plant life cycle, 516
Sporic meiosis, 479
Sporophyte in plant life cycle, 516
Sporozoa, 503
Sporozoans, 503
Spring overturn, 434
Spruces, Engelmann, 585
Spurges, 411
Squamata, 367
Squamous epithelium, 610
Square, Punnett, 212
Squash, 567
Squirrel
  arctic ground, 441
  red, 471
  southern flying, 472
Stabilizing selection, 390
Stages of meiosis, 197-200
Stahl, Frank, 251
Stain, Gram, 83
Staminate flowers, 585
Stanley, Wendell, 488-489
Staphylococcus aureus, 486
Starches, structure of, 51, 52
Statocysts, 722
Stem, 562
  structure of, 569-571
Stem cell im immune response, 695
Stereoisomer glucose, 49, 50
Stereoscopic vision, 728
Stereotyped behavior, 717
Stern, Curt, 217-218
Steroid hormones, 54, 741, 764
  structure of, 736
Steward, F.C., 246, 584
Stichopus, 552
Stigma in angiosperms, 525
Stigmatella aurantiaca, 484
Stimulus and response, association between, 788
Stipules in leaf, 567-568
Stolons, 584
Stomach, vertebrate, 640-642
Stomata in plants, 515-516, 565
  in water movement regulation, 593, 594, 595
Stone cells, 565
Strategies, reproductive, of plants, 585, 587
Stratification, thermal, in fresh water, 434
Stratified epithelium, 610
Stratum corneum, 626
Streaming, cytoplasmic, in fungi, 506
Streptococcus, 82, 486
  mitis, 487
  sanguis, 487
Stretch receptor of muscle, 722
Striated muscle, 628
  contraction of, 632-633
Stroke, 680
Stroma in chloroplasts, 178
Stromatolites, 347
Stronglyocentrotus franciscanus, 552
Structural isomer of glucose, 49, 50
Struthio camelus, 371

Subcellular level of biological organization, 29-30
Subcutaneous tissue, 626
Subspecies, definition of, 338
Substrate enzyme, 137
Substrate-level phosphorylation, 146, 147-148
Subunits
  of adenosine triphosphate, 140
  in proteins, 59
Subunits of ribosome, 264
Succession, ecological, 427-429
Sucrose, 45-46
Sugars, structure of, 49-51
Sulfide in early atmosphere of earth, 66
Sumatran rhinoceros, 392
Sun, energy from, 132
Sunflower, 569
Supercoils of DNA, 188
Superior vena cava in human heart, 677
Surface of cell, 686
Surface tension of water, 46
Surface zone in oceans, 433
Surface-to-volume ratio of cells, 82
Survival of fittest, 24
Survivorship and mortality, 397-398
Sutton, Walter, 215
Swallowing, 640
Swim bladder in fishes, 364
Symbiosis, 92, 412-416
  in bacteria, 92
  and origin of eukaryotes, 477, 499
  in succession, 428
Symmetry
  bilateral, evolution of, 536-537
  radial, 534-536
  rotational, twofold, 295
Sympathetic nervous system, 730-731
  adrenal hormones as extensions of, 741
Synapse(s), 706-709
  in autonomic nervous system, 731
  excitatory, 709
  inhibitory, 709
Synapsis, 197
Synaptic cleft, 706-708
Synaptonemal complex, 196, 197
Syngamy, 196, 478
Synthesis, discontinuous, of DNA, 277
Synthetic auxins, 598-599
Syphilis, 487
Systemic circulation, 675
Systolic period, 679

T

2,4,5-T; *see* 2,4,5-Trichlorphenoxyacetic acid
T cells, 687
  helper, 690, 691, 695-696
    in acquired immune deficiency syndrome, 12
T4 cells in AIDS virus infection, 492, 493
Tachyglossus aculeatus, 373
Taiga, 439-440
Tail in vertebrate embryos, 321, 322
Tamiasciurus budsonicus, 471
Tapeworms, 538
Tarantulas, 546
Taraxacum officinale, 396, 582
Taste, receptors for, 724
Taste buds, 724
Tatum, Edward, 253, 293
Taxodium, 594
Taxonomic hierarchy in classification, 470, 471-472
Tay-Sachs disease, 234, 235
Tectorial membrane of ear, 725

Teeth in vertebrates, 639
Telencephalon, 716-717
Telophase, 193
Telophase I in meiosis, 200
Temperate deciduous forests, 439
Temperate grasslands, 438
Temperature and enzyme activity, 137-138
Temperature change, receptors for, 721
Temperature control in land animals, 370
Temporal lobe, 718
Termite, 549
Terpenes, 55
Territoriality, 792, 795-796
Tertiary structure of polypeptides, 58
Test cross, 212, 213
Testes, 767
Testing hypotheses in science, 18-19
Tetrahymena, 76, 501
Tetramer, hemoglobin molecule as, 255
Tetrapods, ancestors of, 365-366
TH; *see* Thyroid hormone
Thalamus, 717
Thalidomide in human embryonic development, 778
Theorems, 17
Theoretical niche, 407
Theory(ies), 19
   of evolution, Darwin's, publication of, 24
Thermal stratification in fresh water, 434
Thermodynamics, 130
   laws of, 131-132
Thigmotropism, 601
Thoracic cavity
   in human respiration, 660
   in mammals, 608
Thorax in insects, 548
Thoroughfare channels, 669
Threshold value of nerve impulse, 705-706
Thrombus and stroke, 680
Thumb, opposable, of anthropoid primates, 376
Thylakoid, 178
   photosynthesis in, 94
Thymine
   absorption of ultraviolet radiation by, 283
   in DNA, 60, 250-252
Thymus, 688
Thyroid hormone, 739
Thyroid-releasing hormone, 739
Thyroid-stimulating hormone, 739
Ti plasmid, 298
Tidepools, 432
Tiger salamander, 367
Tigers, isolation of, 334
Tight junctions on cells, 123
Timber rattlesnake, pit organ of, 728
Tinbergen, Nikolaas, 787
Tissue(s), 121
   animal, definition of, 532
   bone, 627, 628
   connective, 611-612
   ground, in plants, 563, 564-565
   subcutaneous, 626
   transferring information from nerve to, 706-709
   types in plants, 563
   vascular, in plants, 515, 521, 566-567
   vertebrate, 608-619
Titanotheres, 317
TMV; *see* Tobacco mosaic virus
Toad, giant, 367
Tobacco mosaic virus, 249, 488-489

Tolerance to lead in grasses, 316
Tolypocladium inflatum, 502
Tools, use of, by hominids, 379-381
Tooth decay, 487-488
Tortoises, Galapagos, 23
Touch
   receptors for, 722
   response of plants to, 601
Townsend's chipmunk, 471
Trace elements in diet, 647
Trachea
   of arthropods, 655
   in chordates, 362
   human, 661
Trachea system of cockroach, 549
Tracheary elements in vascular plants, 521
Tracheids in xylem, 566
Tracheoles in insects, 548
Tract(s)
   nerve, 715
   pyramidal, 718
Tradescantia, 114
Traits
   alternative, segregation of, 206, 208
   genetic, expressed by enzymes, 254
Transcription
   in gene expression, 265
   of genes, blocking of, 274-275
Transcription site, 290
Transcription unit in gene cluster, 275, 276
Transfection in cancer research, 287
Transfer RNA, 264, 265, 270
   met, 270, 271
Transformation, Griffith's discovery of, 247
Translation in gene expression, 266
Translocation
   of carbohydrates in plants, 595
   chromosomal, 291
Transmembrane proteins, 109, 110
   of plasma membrane, 108, 110
Transmission electron microscope, 80
Transpiration, 593
   regulation of rate of, 595
Transplants, organ, 502
Transport
   across cell membranes, 120
   active, across cell membrane, 117-119
Transport form of glucose, 51
Transposable genetic elements, 288, 289
Transposition of genes, 282, 288-290
   impact of, 289-290
Transposons, 288, 289
Tree finches, Darwin's, 340
Tree frog, red-eyed, 367
Tree trunk, annual rings in, 571
Trees, coniferous, of taiga, 439
Trematoda, 538
Treponema pallidum, 487
TRH; *see* Thyroid-releasing hormone
Triassic Period, 352
2,4,5-Trichlorphenoxyacetic acid
Trichomes in plants, 565
Tricuspid valve in human heart, 677
Triglycerides, structure of, 53
Trillium ozarkanum
   variety *pusillum,* 527
Trilobite, 346, 349, 350
Trimesters in human pregnancy, 777-780
Triphosphate group in adenosine triphosphate, 140

Trivers, Robert, 789
tRNA; *see* Transfer RNA
Trophic levels in ecosystems, 424-427
Trophoblast
   in human embryonic development, 777
Tropical deforestation, 456-459
Tropical rain forests, 435-437
   destruction of, 8, 437, 456-459
Tropics, 454-459
Tropisms
   in plants, 600-601
Troponin
   in muscle cells, 630, 631
   in muscle contraction, 633
Tryon, Robert, 795
Trypanosoma, 501
Tsetse fly, 501
TSH; *see* Thyroid-stimulating hormone
Tubal ligation, 773
Tube foot of echinoderm, 553
Tubes, fallopian, 770
Tubule
   Malpighian, 751
   renal, in mammals, 755-756
   seminiferous, 767-768
Tubulin in cytoskeleton, 96, 97
Tulip tree, 568
Tumbleweed, 582
Tumor, 284, 285
Tundra, 441
Tunicata, 555
Tunicates, 362, 555
Turbellaria, 538
Turesson, Gote, 339
Turgor movements of plants, 601
Turgor pressure in plants, 601
Turner's syndrome, 227
Turtle, 368
   green
      egg laying by, 2, 3
      migration of, 796
Twofold rotational symmetry, 295
Tympanic membrane in hearing, 724

U

Ultrasound, prenatal, 238, 239
Ultraviolet radiation, 166
   in mutation, 283-284
   in origin of life, 67
Ulva, 475
Uncia, 374
Unicellular protists, 500-501, 503-504
Universal donor, 229
Universal recipient, 229
Unsaturated fats, 54
Uracil in RNA, 60
Uranium-238 in dating of rocks, 319
Urea, 755
   formation of, 745
Urethra, 768
Urey, Harold C., 67
Urinary system, human, 754
Urine, 751
   concentration of, 754
   control of volume of, 746
Uroctonus mordas, 546
Uterus, 770

V

Vaccination, 686
Vaccine, 12
   acquired immune deficiency syndrome, 696
Vaccinia, 686
Vacuoles, central, in eukaryotic cells, 84, 85
Vacuum suction, 773
Vagina, 770

Valve(s) in human heart
   aortic, 676
   closing of, 678, 679
   mitral, 676
   pulmonary, 677
   tricuspid, 677
van Beneden, Pierre Joseph, 196
van Leeuwenhoek, Antonie, 79, 80
Vanes in feathers, 372
Vanessa
   annabela, 331
   atlanta, 331
   virginiensis, 331
Variation
   continuous, 258
   in nature, 307-309
Variola, 686
Vas deferens, 768
Vascular bundles in plants, 521
Vascular cambium, differentiation of, 564, 570
Vascular plants, 520-524
   seedless, 522-523
Vascular tissue in plants, 515, 521, 566-567
Vasectomy, 773
Vasopressin, 738
Vaxia, 349
Vegetation, climax, 428
Vein(s)
   hepatic, 744
   portal, 743
   pulmonary, 675
      human, 676
   and venules, 669, 671
Veinlets in leaf, 590
Vena cava, 671, 677
Venation, patterns of, in leaves, 567-568
Venereal diseases, 487
Ventilation, ram, in gills, 655
Ventricle(s)
   contraction of, 678
   in fish heart, 675
   left, in human heart, 676
Venules and veins, 669, 671
Vermiform appendix in humans, 322
Vertebra, embryonic development of, 556
Vertebral column in vertebrates, 362
Vertebrata, 556
Vertebrate behaviors, adaptive, 795-797
Vertebrate body, 607-621
Vertebrate heart, evolution of, 674-675
Vertebrate kidney
   evolution of, 752-755
   organization of, 751-752
Vertebrate muscles, contraction of, 630
Vertebrates, 556
   adaptive behaviors in, 795-797
   aggressive conflicts in, 792
   brain of, 713
      evolution of, 674, 716-718
   circulation in, 667-683
      cardiovascular system, 668-672
      circulatory systems, contents of, 673-674
      human heart in, 675-681
      importance of, 681
   classes of, 556
   colonization of land by, 350-351
   concentration of ions in, 750
   digestion of food by, 637-651
      large intestine in, 644-645
      mouth in, vertebrate, 639-640
      nature of, 638-639
      nutrition and, 645-649
      small intestine in, 642-644
      stomach in, 639-640
   embryos of, 322

Vertebrates—cont'd
  evolution of, 361-384
    amphibians, 366-367
    birds, 370-372
    bony fishes, 364
    heart in, 674-675
    hominids, 376, 380-382
    hominoids, 376-378
    jawed fishes, appearance of,
      363-364
    jawless fishes, 363
    land, invasion of, 365-366
    mammals, 370, 372-374
    primates, 374-376
    reptiles, 367-369
    temperature control in land an-
      imals, 370
  flight in, 344
  general characteristics of, 362
  immune system of, 685-699
  muscle attachment in, 625
  sex and reproduction in, 762-764
  skin of, 626
  tissues in, 608-619
Vesicle
  coated, formation of, 119
  and membranes, 86
Vespula arenaria, 413
Vessel elements in xylem, 566
Vessels, lymphatic, 672
Vestigial structures as evidence of
    evolution, 322
Vibrio cholerae, 99
Viceroy butterfly, 412
Villi of small intestine, 642-643
Vimentin in cytoskeleton, 96, 97
Viroids, 489
Virus(es), 488-493, 494
  AIDS, 12, 696
    infection cycle of, 491-493
  cancers caused by, 285
  classification problems of, 480
  and disease, 491
  diversity of, 490-491

Virus(es)—cont'd
  Holmes ribgrass, 249
  myxomatosis, 408-409
  protein coats of, 686
  tobacco mosaic, 249
Virus-reconstitution experiment,
    Fraenkel-Conrat's, 249
Visceral mass of mollusks, 542
Vision, 725-728
  binocular, in anthropoid pri-
    mates, 375
Visual cortex of human brain, 718
Visual pigment in vertebrate eyes,
    725
Vitamins, 167, 646-648
Viviparous mammals, 764
Viviparous reptiles, 763
Volume, residual, in lung, 658
Voluntary nervous system, 714,
    715, 728, 729
Volvox, 478
von Frisch, Karl, 787
Vorticellia, 499, 500-501

W
Wallace, Alfred Russel, 24
Warbler finch, Darwin's, 340
Warning and protective colora-
    tion, 411
Water, 43-48
  absorption of, by roots, 593, 595
  evolution of life in, 43, 44
  ionization by, 47-48
  molecular structure of, 45
  movement of, in plant, 592-595
  passage into and out of cells,
    112-114
  as solvent, 45-46
Water balance, control of, 749-759
  evolution of vertebrate kidney
    in, 752-755
    function of mammalian kidney
      in, 755-757
    homeostasis and, 757

Water balance—cont'd
  evolution of vertebrate kidney
    in—cont'd
    organization of vertebrate kid-
      ney in, 751-752
    osmoregulation in, 750-751
Water cycle, 420
Water pollution, 461
Water vascular system of echinod-
    erm, 553
Water-soluble molecules and lipid
    bilayer, 106
Watson, James, 242, 250-252
Wavelength and energy in light,
    166-167
Weeds, hormones used to kill,
    598-599
Weevils, species of, 466
Wegener, Alfred, 354
Weinberg, G., 308
Went, Frits, 597-598
Western blot test for AIDS virus
    antibodies, 776
Whale
  gray, and barnacles, commensal-
    ism of, 413
  humpbacked, 397
  skeleton of, 322
Wheat, improved strains of, 452
Wheat rust, 508
White blood cells, 611, 612, 673-
    674, 687, 689, 691
White poplar, 571
White-tailed deer, 375
Whorls in angiosperms, 524
Willow oak, 468
Wilson, E.O., 789
Wind, pollination by, 579
Window, oval, of ear, 725
Wings, insect, 548
  movement of, 625
Winter, dormancy in, 602
Wolf, gray, 469
Women, smoking and lung cancer
    in, 11

Wood, 571
Wood ducks, 371
Wood sorrel, leaflets of, 602
Woody plant, 564
Worlds, other, possibility of life
    on, 73-74
Worms
  bristle, 544
  solid, 537-538

X
X chromosome, 226-227
X rays, 282
Xenopus laevis, 296
Xeroderma pigmentosum, 284
Xylem, 521, 563, 566
  in plant stems, 569, 570
  in water absorption, 593

Y
Y chromosome, 227
Yeast, fermentation in, 160, 161
Yellow fever virus, 493
Yellow-bellied marmot, 471
Yield, optimal, in agricultural
    practice, 396
Yolk, 762-764
  in amniotic egg, 369-370

Z
Z lines, 631, 632-633
Zea
  diploperennis, 454-455
  mays, 256, 454, 572
Zebra, 409
Zinc, 647
  in plant nutrition, 596
Zoomastigina, 501
Zygomycetes, 507
Zygomycota, 507
Zygospores in fungi, 507
Zygote, 196, 333
  in plants, 580
Zygotic meiosis, 478-479